Catalysis and Reaction Engineering Division 2016

AA009250

Core Programming Area at the 2016 AIChE Annual Meeting

San Francisco, California, USA
13-18 November 2016

Volume 1 of 2

ISBN: 978-1-5108-3429-3

Printed from e-media with permission by:

Curran Associates, Inc.
57 Morehouse Lane
Red Hook, NY 12571

Some format issues inherent in the e-media version may also appear in this print version.

Copyright© (2016) by AIChE
All rights reserved.

Printed by Curran Associates, Inc. (2017)

For permission requests, please contact AIChE
at the address below.

AIChE
120 Wall Street, FL 23
New York, NY 10005-4020

Phone: (800) 242-4363
Fax: (203) 775-5177

www.aiche.org

Additional copies of this publication are available from:

Curran Associates, Inc.
57 Morehouse Lane
Red Hook, NY 12571 USA
Phone: 845-758-0400
Fax: 845-758-2633
Email: curran@proceedings.com
Web: www.proceedings.com

TABLE OF CONTENTS

VOLUME 1

(6w) Fundamental Studies and Applications of Nano-Structured Catalysts in Domestic Fuel Production 1
Cun Wen
(6x) Catalytic Hydrotreatment for the Development of Alternative Transportation Fuels 4
Lilu Funkenbusch, Michael Mullins
(6ai) First-Principles Modeling of Catalysts: Novel Algorithms and Reaction Mechanisms 5
Bryan R Goldsmith
(6aj) Development of Catalysts for Energy and Environmental Applications 6
Marc D. Porosoff, Heather D. Willauer
(6z) Upgrading Biomass-Derived Platform Chemicals By Electrochemical and Photoelectrochemical Catalytic Oxidation 7
David Chadderdon
(6aa) Rational Design of Catalytic Sites for Energy Applications 8
Timothy Van Cleve
(6ab) Theory-Guided Understanding and Design of Heterogeneous Catalysts 10
Matthew M. Montemore
(6ac) In silico Engineering of a Future Energy Infrastructure 11
Glen R. Jenness
(6ad) Insight into Pt-Bi Bimetallic Catalysts: An Experimental and DFT Study 13
Yang Xiao, Arvind Varma
(6ae) Mechanisms of Redox Catalysts for the Greener Processing of Shale Gas Via Chemical Looping 14
Luke Neal
(6af) Nanostructured Transition Metal Dichalcogenide Catalysts for Electrochemical Energy Systems 17
Mohammad Asadi
(6y) Exploring Structure-Function Correlations of Nanomaterials in Heterogeneous Catalysis 19
Weiqing Zheng
(6ah) Mechanistic, Spectroscopic and Theoretical Assessment of Porous Catalytic Materials 20
Michele L. Sarazen
(6m) In-Situ Drifts Studies on CuNi Catalyst for Ethanol Hydrogen Production 21
Anand Kumar
(6n) Semi-Permeable Membrane Reactor for Catalysis, Hydrocarbon Processing and CO2 Reuse 22
Xiao-Yu Wu
(6o) A Practical Way to Separate Uncondensed Lignin during Biomass Pretreatment and Quantitatively Depolymerize It at a Low Temperature of 120 C with a Cheap Ni/Al2O3-SiO2 Catalyst 25
Li Shuai
(6q) Combined Quantum and Classical Computational Approaches for Investigating Complex Surface Interactions Impacting Heterogeneous Catalysis 26
Thomas P. Senftle
(6a) Electrochemical Energy Transformation Processes: An Atomistic Perspective 29
Leanne D. Chen
(6r) Nanostructured Hybrid Materials: Directing Catalytic Activity and Selectivity By Design 31
Kairat Sabyrov
(6s) Towards the Computational Design of Heterogeneous Electrocatalysts 32
Zhenhua Zeng
(6t) Engineering the Electrochemical Interface for Sustainable Energy Conversion and Storage 33
Maria Escudero-Escribano
(6u) Controlled Synthesis and in-Situ Spectroscopic Study of Highly Efficient Ptfe Bimetallic and Ptrufe Trimetallic Nanocatalysts 34
Hua Zhang, Jinbao Zheng, Nuowei Zhang, Jian-Feng Li, Binghui Chen
(6b) Developing Enhanced Catalysts for Renewable Fuels through Spectroscopic Insights 35
Linsey C. Seitz
(6c) Computational Design of Hetero-Structured Catalysts for Energy 36
Liang Zhang
(6d) Efficient Catalytic Pathways for Carbon Utilization 38
Erdem Sasmaz

(6e) New Routes and Heterogeneous Catalysts Development for Biomass Conversion 39
Homer Genuino
(6f) Transition Metal-Oxides for Sustainable Energy Conversion and Storage: The Computational Catalysis Perspective 40
Michal Bajdich
(6k) Impacts of Metal-Adsorbate Bonds on Photon Activation Mechanism and Nanoparticle Reconstruction in Heterogeneous Catalysis 41
Matthew Kale
(6p) Electrocatalysis for Sustainable Energy Technologies 42
Drew Higgins
(6g) Rational Catalyst Design for Renewable Energy Technologies 46
Samira Siahrostami
(6i) Engineering Visible-Light Organic Photocatalysis for Polymers in Biomaterials, Biosensing, and Photomedicine 47
Alan Aguirre-Soto, Hadley D. Sikes
(6ak) Catalysts for Emission Control and Energy Conversion: Computational Study Based on DFT Calculations 48
Renqin Zhang
(6al) Development of Heterogeneous Catalysis towards a Sustainable Future 50
Yuran Wang
(6ao) Understanding and Improving Electrocatalysts for Energy Conversion and Waste Remediation 51
Nirala Singh
(6ap) Development of Novel Single-Site and Isolated Bimetallic Alloy Catalysts for C-H Bond Activation 54
Junjun Shan
(6aq) Smart Materials through Molecular Networking 59
Sergey N. Semenov
(6l) Condition and Support Dependent Development of Computational Methods for the Engineering of Materials 60
Christopher Paolucci
(6am) Conversion of Waste Oil to Biofuels in a Single Catalytic Process over Bifunctional Catalysts 63
Masoudeh Ahmadi
(6an) Rational Design of Catalytic Materials for Advancing the Use of Alternative Energy Sources 64
Eric Walker
(6ar) Catalysis Reactions towards Advanced Energy Applications 66
Richa Padhye
(6as) Synthesis and Spectroscopic Characterization of Heterogeneous Catalysts for Energy Production 68
Junming Sun
(23a) Control of Fe-Ni Nanoparticle Activity for the Oxygen Evolution Reaction (OER) and Methanol Electrooxidation 70
Stephanie Candelaria, Nikki S. Rentz, Lauren F. Greenlee
(23b) Transition Metal-Nitrogen-Carbon Electrocatalysts for Oxygen Reduction Reaction 71
Plamen Atanassov, Kateryna Artyushkova, Alexey Serov, Ivana Matanovic
(23c) Electrocatalytic Generation of H2O2: Carbon Based Material Synthesis and Device Design for Portable Low Cost Water Purification 72
Zhihua Chen, Shucheng Chen, Samira Siahrostami, Zhenan Bao, Jens Nørskov, Thomas F. Jaramillo
(23f) Niobium Doped Molybdenum Disulfide Catalyst for CO2 Reduction Reaction 73
Pedram Abbasi, Mohammad Asadi, Baharak Sayahpour, Amin Salehi-Khojin
(32b) Green Chemistry with Flow Reactors and Heterogeneous Catalysts for the Synthesis of Fragrance Ingredients 74
Sunitha Tadepalli, Zhe Guo, James Stevens, Geatesh Tampy
(32c) Lewis Acid Catalyzed Pyrolysis of Polystyrene in Low-Temperature Molten Salts: Experiment and Simulation 75
Avery Brown, Michael T. Timko, Ravindra Datta, N. Aaron Deskins
(32d) Microemulsion Systems As Smart Solvent Systems for Homogeneous Catalysis: Rhodium-Catalyzed Hydroformylation of Long-Chain Alkenes in Aqueous Media 76
Tobias Pogrzeba, Marcel Schmidt, Markus Illner, Prof. Dr. Reinhard Schomäcker
(32f) Synthesis of Linear Alkylbenzenes from Biomass-Derived Furan over Solid Acid Catalysts 77
Wilbur Y. Zuo, Hsi-Wu Wong

(36a) Rate-Enhancing Roles of Water Molecules in Methyltrioxorhenium-Catalyzed Olefin Epoxidation By Hydrogen Peroxide 78
Bryan R Goldsmith, Taeho Hwang, Stefan Seritan, Baron Peters, Susannah L. Scott
(36b) Acid Activation of Ferrate(VI) for Oxidation of Organic Compounds in Aqueous Solution 79
Kyriakos Manoli, George Nakhla, Ajay K. Ray, Virender K. Sharma
(36c) Kinetic Studies of Liquid Phase Ethylene Production By Hydrogenation of Acetylene Using a Selective Solvent in a Basket Reactor 87
Humayun Shariff, Muthanna Al Dahhan
(36e) Solvent Effect in Catalytic Upgrading of Biomass-Derived Furanics 88
Zheng Zhao, Mohammed Reda Bababrik, Bin Wang, Daniel Resasco
(36f) Solvent Effect on Cyclohexene Hydrogenation over Pd 89
Zheng Zhao, Steven Crossley, Daniel E. Resasco
(36g) Numerical Study of Heavy Crude Oil Pyrolysis in Supercritical Water 90
Ping He, Ashwin Raghavan, Ahmed F. Ghoniem
(39a) Multi-Scale Multi-Phase Reduced Order Models for Dual-Layered Catalytic Monolith Reactors 91
Ram R. Ratnakar, Vemuri Balakotaiah
(39b) CFD Modeling of an Industrial Furnace Reformer and 3D Multi Scale Model for Packed Bed Reactor for Synthesis Gas Production 92
Bhanu Vardhan Reddy Kuncharam, Elaine Chen, Anthony G. Dixon, Rui Liu
(39c) DFT Calculations and Reaxff Development for MD Simulations of H-BN Growth on Nickel Surfaces 93
Song Liu, Bin Liu, James H. Edgar, Adri Van Duin
(39d) Molecular-Scale Modeling of Whole Biomass Pyrolysis – Morphological Changes 94
Abdul Salam Mohammad, Joseph Biernacki, Scott Northrup, Derek Cashman
(39e) Influence of Intraparticle Transport Processes on Optimal Reactor and Catalyst Design 95
Markus Kaiser, Hannsjörg Freund
(39f) Multiscale Simulation of Heterogeneous Catalysis: Butadiene Hydrogenation from Molecular Reaction Steps to the Fluid Flow 96
Phillippe Sautet, Ludovic Briquet, Céline Chizallet, Pascal Raybaud, Jean-Marc Schweitzer, Jan J. Verstraete, Anthony Wachs, Matthieu Rolland, Abdelkader Hammouti
(39g) Multiscale Modeling for High Temperature Heterogeneous Gas Phase Synthesis Reactions: HCN Synthesis 97
Georg S. Liesche, Kai Sundmacher
(39h) An Experimental and CFD Based Study of the Pt-Catalyzed NH3-Oxidation in Different Scales 98
Thomas Heydt, Artur Wiser, Dirk Born, Alfons Drochner, Herbert Vogel, Martin Votsmeier
(39i) How Accurate Is the Mean-Field Approximation for Catalytic Kinetics? 99
Michail Stamatakis, Bernardo Oyarzún, Richard Baochang Wang
(48a) Compositional Effect of Varying Feedstocks on Heavy Oil Upgrading By Supercritical Water 100
Ramazan Oguz Caniaz, Can Erkey, Serhat Arca
(48b) Product Formation and Kinetics of the Non-Isothermal Hydrothermal Liquefaction of Soy Protein Isolate 101
James Sheehan, Phillip E. Savage
(48c) An Investigation on Short Residence Time Macro Algae Hydro Thermal Liquefaction in a Batch Reactor 104
Arash Izadpanah, Niccolo Le Brun, Klaus Hellgardt
(48d) Hydrothermal Liquefaction (HTL) of High-Ash Algal Biomass: The Effect of Ash Contents in HTL Reactions 105
Wan-Ting Chen, Wanyi Qian, Karalyn Scheppe, Zachary Mazur, Lance Schideman, Chih-Ting Kuo, Peng Zhang, Yuanhui Zhang
(48e) Simultaneous Solid, Phosphorus Recovery and Biocrude Production through Hydrothermal Liquefaction of Algae 106
Robert Hable, Sirwan Alimoradi, Shawn Benson, Colin White, Belinda S. M. Sturm, Susan M. Stagg-Williams
(48f) Supercritical Water Treatment of Fractions of Crude Oil: Quantification of the Products 107
Soumya Gudiyella, Lawrence Lai, Isaiah Borne, Jianghuai Cai, Michael T. Timko, Geoffrey Tompsett, Alison Lui, William H. Green
(48g) Supercritical Water Process - Finding a Way to Accomodate into Petroleum Industry 108
Ki-Hyouk Choi, Joo-Hyeong Lee, Ashok K. Punetha, Emad Shafei, Abdullah Abdulhadi, Muneef Qarzouh, Bader Al-Otaibi
(64a) Why John Is Right Not to Lecture 109
H. Scott Fogler
(64b) Utilizing Classroom Management Software to Handle Large Classes 110
Janet Degrazia

(64c) Enabling Numerical Analysis Calculations within Educational Materials Delivered Via PCs and Smartphones 111
Michael B. Cutlip, Mordechai Shacham, Michael Elly
(64d) Spreadsheet Template to Aid in Creating Individualized Version of Objective Homework Problems 113
John Wagner, Amanda P. Malefyt
(64e) How Students "Learn ChemE": The Impact of Departmental Culture on Student Success 114
Lisa G. Bullard
(64f) Rapid Analysis of Written Responses to Reveal Student Misconceptions in Thermodynamics 115
Carl T. Lira, Matthew Steele, Kevin Haudek, John Merrill, Mark Urban-Lurain
(64g) No More Lecturing! What John Falconer Has Done for Student Learning 116
Milo D. Koretsky
(64h) Active Learning in Chemical Engineering 117
John Falconer, J. Will Medlin, Janet Degrazia, Katherine McDanel, Michelle Medlin
(72a) Mechanism of Furfural Hydrodeoxygenation on Metal/Metal Oxide Catalysts 118
Alexander V. Mironenko, Glen R. Jenness, Dionisios G. Vlachos
(72b) Kinetic Analysis of Hydrogenation and Hydrogenolysis of Complex Furans over Supported Noble Metals 119
Ying Lin Louie, Alexis T. Bell
(72c) ZrO2 Is Preferred over TiO2 As Support for the Ru-Catalyzed Hydrogenation of Levulinic Acid to I3-Valerolactone 120
Homer Genuino, Jamal Ftouni, Li Lu, Christopher Kiely, Pieter Bruijnincx, Bert Weckhuysen
(72d) C–O Bond Hydrogenolysis of Biomass–Derived Cyclic Ethers over Ni–Based Catalysts 121
Elmira Soghrati Khorasgani, Catherine Choong, Sibudjing Kawi, Armando Borgna
(72e) A Novel Pathway to 1,5-Pentanediol without the Use of Noble Metals 122
Zachary Brentzel, Kevin J. Barnett, Kefeng Huang, Christos T. Maravelias, George W. Huber, James A. Dumesic
(72f) Evaluation and Characterization of Carbon-Supported Nobel Metals for the Hydrodeoxygenation (HDO) of Acetic Acid 123
Jose Contreras-Mora, J. R. Monnier, Christopher Williams
(72g) Direct Conversion of Methane to Methanol and Ethanol 124
Chukwuemeka Okolie, Yasmeen Belhseine, Libor Kovarik, Eli Stavitski, Carsten Sievers
(72h) Aqueous-Phase Phenol Hydrogenation over Platinum and Rhodium: Electrocatalysis Versus Thermal Catalysis 125
Nirala Singh, Yang Song, Oliver Gutiérrez, John L. Fulton, Donald M. Camaioni, Charles T. Campbell, Johannes A. Lercher
(72i) Highly Efficient Conversion of Terpenoids to High Density Bio- Jet Fuels 126
Xiaokun Yang, Hongfei Lin
(84a) Pyrolysis of Soils Contaminated with Heavy Hydrocarbons: Reaction Mechanisms and Morphology of the Produced Char 127
Kyriacos Zygourakis, Caroline A. Masiello, Julia E. Vidonish, Pedro J. Alvarez
(84b) Spatial and Temporal Features of the Catalyzed Hydrocarbon Trap 128
Po-Yu Peng, Dan Luss, Michael P Harold
(84d) Formation of Nitrosamines in the Desorber of Tertiary Amine-Based Carbon Dioxide Capture Systems 131
Kun Yu, Ning Dai
(84e) Natural Attenuation and Biostimulation of Petroleum-Hydrocarbon Contaminated Soil in a Simulated Column 132
Mohammad-Saeed Safdari, Farhad Fazlollahi, Thomas Fletcher, Mahmood Rahmati
(84f) Study of Reaction Kinetics for Elemental Mercury Vapor Oxidation over CuCl2 for Mercury 133
Vishnu Sriram, Zhouyang Liu, Joo-Youp Lee
(84g) Advances in the Use of a Bio-Physicochemical Model to Characterize, Optimize the Chemical Absorption-Biological Reduction Integrated System for NO Removal 134
Jingkai Zhao, Meifang Li, Wei Li
(89a) Photogenerated Charge Separation in Artificial Photosynthesis Systems 135
Can Li
(89b) Oxide Nanostructures: Novel Supports, Active Sites, and Tandem Catalysts 136
Justin M. Notestein
(89c) Characterization of Tungsten and Zirconium Decorated Multi-Walled Carbon Nanotube Catalysts By XAS and NOx-Temperature Programmed Desorption 137
Gary L. Haller, Patrick Kelleher, Cyril Thomas

(89d) Mechanistic Differences Between Metal-Exchanged Zeolite SCR Catalysts, and Their Practical Implications 138
Aleksey Yezerets, Ashok Kumar, Saurabh Y. Joshi, Jinyong Luo, Yadan Tang, Hongmei An, Krishna Kamasamudram, Neal Currier
(89e) The Kinetics of Elementary Steps in Condensations Catalyzed By Solid Acids: Effects of the Size, Shape, and Charge of Intermediates, Transition States and Catalytic Structures 139
Stanley Herrmann, Enrique Iglesia
(89f) Progressing from Identifying, to Synthesizing and Designing Active Sites in Heterogeneous Catalysts 142
Harold H. Kung
(95a) Analysis of Upstream Creeping Reaction Zones in Catalytic Monolith Reactors 143
Tian Gu, Vemuri Balakotaiah
(95b) Model-Based Design of Optimal Reactors Considering Catalyst Deactivation 144
Mingquan Xie, Hannsjörg Freund
(95c) Exploiting the Nonlinear Frequency Response Method to Evaluate the Potential of Forced Periodic Operation of Chemical Reactors 145
Menka Petkovska, Daliborka Nikolic, Matthias Felischak, Andreas Seidel-Morgenstern
(95d) Modeling Catalyst Deactivation in a Packed Bed Reactor for Thermo-Catalytic Hydrogenation of CO2 146
Duo Sun, David Simakov
(95e) Optimizing Methanol Synthesis - from Lab to Industrial Plants Via High-Fidelity Modeling 147
Sebastian Werner, Wolfgang Seuffert, Willi Brandstädter, Andreas Reitzmann
(95f) Statistical Activity Variations in Diluted Catalyst Beds: Influence on Reactor Behavior 148
Gunnar Ganzer, Hannsjörg Freund
(95g) Identification of a Reaction Boundary Layer in Membrane Reactors for Hydrogen Production 149
Maria Anna Murmura, Stefano Cerbelli, Maria Cristina Annesini
(97a) Reactor Scale Modeling of Liquid Phase Ethylene Production By Hydrogenation of Acetylene in a Trickle Bed Reactor 157
Humayun Shariff, Muthanna Al-Dahhan
(97b) Influence of Solid Phase and Geometry on Liquid Velocity of a Three-Phase Bubble Column Fluid Dynamic 158
Karina K. Costa, Diana Sanchez-Forero, João L. Silva Jr, Osvaldir P. Taranto, Milton Mori
(97c) A Multiphase Semi-Flow Microreactor to Study the Influence of Water on Organometalic Catalysis 159
Bruno Pinho, Ryan L. Hartman
(97d) Original Methodology for the Screening of Catalysts Presenting High Reaction Rates at High Conversions 160
Alberto Servia
(97e) Gas Phase Mixing Behavior of Gas-Liquid Upflow in a Moving Catalytic Packed/Expanded Bed Hydrotreating Reactor Using Advanced Gas Tracer Technique 164
Vineet Alexander, Hamza Al-Bazzaz, Muthanna Al-Dahhan
(97f) Improvement of a Kinetic Model for the Hydroconversion of an Atmospheric Residue in a Slurry Semi-Batch Reactor 165
Pedro Alvarez, Barbara Browning, Jérémy Maurin, Tim Jansen, Maxime Lacroix, Mélaz Tayakout, Christophe Geantet, Isabelle Pitault
(97g) Experimental Characterization of Two Phase Interfacial Mass Transfer in Novel Milli-Scale Reactors 169
Aditi Potdar, Leen C. J. Thomassen, Simon Kuhn
(97h) Process Intensification of Aqueous Biphasic Hydroformylation of Mid Chain Olefin 1-Octene in a Jet Loop Reactor Creating Large Interfacial Areas 170
Helge Warmeling, Arno Behr, Andreas J. Vorholt
(97i) Pressure Drop and Liquid Hold up in an Open-Structure Random Packed Column with Counter Current Flow 173
Letícia Pezzi Martins Loane, J. Van Der Schaaf, Jaap C. Schouten
(343a) High-Efficiency Electrochemical Conversion of CO2 to Value Added Fuels Current Status, Challenges, and Future Directions 174
Xiao-Dong Zhou
(343b) Design and Economic Analysis of IGCC Power Plants with CO2 Capture 175
Umer Zahid, Usama Ahmed, Chonghun Han
(343c) Optimization of the CCS Network Under Harsh Geographical Conditions, Using GIS and MINLP Modeling 176
Changsoo Kim, Kyeongsu Kim, Chonghun Han

(343d) Amine-Blends Screening and Characterization for CO2 Post-Combustion Capture 177
Abdullah Al Hinai, Mohammad Abu Zohra, Dang Viet Quang
(343e) Economic and Exergy Analysis of Integrated Purification Process with Carbon Capture Process (CCP) Using Selexol in a 500 MW Integrated Gasification Combined Cycle (IGCC) Power Plant 183
Woo-Sung Lee, Hyun-Taek Oh, Chang-Ha Lee
(343g) Influence of Alloying on CO2 Electroreduction on Ag-Zn System 184
Toru Hatsukade, Kendra P. Kuhl, Etosha R. Cave, David N. Abram, Jeremy T. Feaster, Christopher Hahn, Anna L. Jongerius, Thomas F. Jaramillo
(343i) In-Situ Infrared Spectroscopic Investigation of Pyridine-Mediated CO2 Reduction on Pt Electrocatalysts 185
Marco Dunwell, Bingjun Xu, Yushan Yan
(138b) Selective Hydrogenation of Bio-Oil Model Compounds over Molybdenum Carbide Supported Catalysts 186
Sarah W. Paleg, Joshua Schaidle, Levi T. Thompson
(138c) Guaiacol Deoxygenation Using Methane over Pt-Bi Catalysts: Reaction Pathways and Kinetics 189
Yang Xiao, Arvind Varma
Fluidized Bed Catalytic Glycerol Hydrodeoxygenation to 1,3-Propanediol 190
Mahesh Edake, Jean-Luc Dubois, Mohammad Jaber Darabi Mahboub, Gregory Patience
(138e) Reaction Pathways for the Hydrodeoxygenation of Anisole and Guaiacol over Zn-Pt Bimetallic Catalysts 191
Daming Shi, John Vohs
(138f) Novel Bio-Oil Hydrodeoxygenation Catalysts Based on Strong Electrostatic Adsorption 192
Yaseen Elkasabi, Qiuli Liu, Yongsuck Choi, Akwasi A. Boateng, John R. Regalbuto
(138g) Biomass Pyrolysis Interfaced with Honeycomb Structured Upgrading Reactor for the Production of High Quality Biofuels 193
Conrad Zhang, Liaoyuan Mao, Yanxin Li
(138h) Upgrading of Biomass Pryolysis Vapors of Modified ZSM-5: Effects of Metal(s) Loading and Pretreatment 194
Matthew M. Yung, Anne Starace, Calvin Mukarakate, Kristiina Ilsa, Kim Magrini, Mark Nimlos
(138i) Cross Metathesis of Unsaturated Carboxylic Acids to Bioterephthalic Acid Intermediates 195
Erisa Saraçi, Lan Wang, Klaus Theopold, Raul F. Lobo
(145a) Investigating Synergetic Effects Between Deposited PGM Nanoparticles and Transition Metal Carbide Nanotube Support through Surface Characterization and Device Performance 196
Shibely Saha, Dongmei (Katie) Li, Shuai Tan, Brian Leonard
(145b) Bimetallic Ru-Pt and Pt-Co Fuel Cell Catalysts Prepared By Strong Electrostatic Adsorption and Electroless Deposition 197
John Meynard M. Tengco, Bahareh Alsadat Tavakoli Mehrabadi, Akkarat Wongkaew, Yunya Zhang, Weijian Diao, Taylor R. Garrick, John W. Weidner, John R. Monnier, John R. Regalbuto
(145c) Understanding the Effects of pH and Alkali Cations on H/OH Adsorption and Hydrogen Oxidation on Transition Metal Surfaces 198
Ian T. McCrum, Michael Janik
(145d) Nanoparticle Catalysts Supported on Substitutionally Doped Graphene: Effects on Activity and Stability for Hydrogen Oxidation 199
Stephen A. Giles, Stavros Caratzoulas, Dionisios G. Vlachos, Yushan Yan
(145e) Impact of Combining Metal Nanoparticle Catalysts and Semiconductor Photo-Electrodes on Photo-Catalytic Performance 200
Paul Hernley, Steven Chavez, Joseph Quinn, Suljo Linic
(145f) Trends in Hydrogen Evolution Reaction Activity Among Metal Modified Carbide Thin Films and Powders 201
Brian M. Tackett, Qian Zhang, Yannick C. Kimmel, Jingguang G. Chen
(145g) Experimental Determination of Water Orientation Effects on Kinetic Barriers to Alkaline Hydrogen Oxidation and Evolution 202
Maureen H. Tang, Joshua Snyder, Kiran Vasudevan, Jennifer Gallup
(148a) Co-Oxidation of CO and Hydrocarbons on Pd/Ceria-Zirconia/Al2O3 Three-Way Catalysts: Experiments and Modeling 203
Wendy Lang, Michael P Harold, Yisun Cheng, Carolyn Hubbard, Manish Sharma, Paul Laing
(148b) Interaction Between Highly Dispersed Ionic Pd2+ and Mn-Ce Solid Solution Support for Low Temperature CO Oxidation 206
Chao Wang, Cun Wen, Erdem Sasmaz, Jochen Lauterbach
(148c) Effect of Rare-Earth Metal Dopants on Ceria Catalysts for CO Oxidation 207
Kyung-Jong Noh, Jeong Woo Han

(148d) Carbon Monoxide Oxidation and Preferential Oxidation on Pt Alloy Nanoparticle Catalyst with Engineered Surface 208
Sang Youp Hwang, Eric Yurchekfrodl, Changlin Zhang, Zhenmeng Peng
(148j) Sensitivity of XANES and XES to the local environment: Multiple adsorption of H2O and NH3 on Cu-SSZ-13 209
Renqin Zhang, Hui Li, Christopher Paolucci, Atish A. Parekh, Janos Szanyi, Feng Gao, Trunojoyo Anggara, Hui Li, William F. Schneider, Fabio Ribeiro, Jean-Sabin McEwen
(148f) Mercury Oxidation over Cu-SSZ-13 Catalysts Under SCR Conditions for Power Plant Applications 212
Benjamin Galloway, Bihter Padak
(148g) Thiophenic Species Removal over a New Zn-Offretite Zeolite Via Selective Adsorption for FCC Processes 213
Yira Aponte, Hugo De Lasa
(148h) Wash-Coat Development for Lean-Burn Engine-Exhaust Aftertreatment: A Novel Way of Incorporation of Binder to Pd Supported on Sulfated Zirconia 214
Sreshtha Sinha-Majumdar, Gokhan Celik, Anne-Marie C. Alexander, Umit S. Ozkan
(148i) Carboxylation of Propylene Oxide to Propylene Carbonate in Slurry and Trickle Bed Reactors 215
Pallavi Bobba, Xin Jin, Raghunath V. Chaudhari, Bala Subramaniam
(157a) Supported Pt-Re Catalyst for Selective Hydrogenolysis and Hydrogenation Reactions 216
Derek D. Falcone, John H. Hack, Alexander Klyushin, Axel Knop-Gericke, Robert Schlogl, Robert J. Davis
(157b) Thermally Stable, Regenerable Single Atom Catalysts Via Atom Trapping 217
Abhaya Datye
(157c) Geometric and Electronic Effects in Pd and Pt Alloy Alkane Dehydrogenation Catalysts 218
Miller Jeffrey, Zhenwei Wu, Evan Wegener, Han-Ting Tseng, Gallagher James R., Yang Ren, Rosa Diaz, James W. Harris, Fabio Ribeiro
(157f) Catalytic Treatment of Water Contaminated with Chlorinated Hydrocarbons 219
Umit S. Ozkan, Hyuntae Sohn, Gokhan Celik, Seval Gunduz, Saurabh Ailawar, Paul Edmiston
(157e) How Pt Interacts with CeO2 Under the Reducing and Oxidizing Environments at Elevated Temperature? 220
Jaeha Lee, Youngseok Ryou, Yongwoo Kim, Xiaojun Chan, Taejin Kim, Do Heui Kim
(157d) Metal Oxide-like Clusters in MOFs As Supports for Well-Defined Rhodium Complex and Rhodium Cluster Catalysts 221
Dong Yang, Sandra Varinia Bernales Candia, Gamze Gümüslü, Jun Yu, Laura Gagliardi, Bruce C. Gates
(178a) Kinetic Modeling of Propene Hydroformylation in Propane-Expanded Liquid with Rh-Based Complexes 222
Dupeng Liu, Raghunath V. Chaudhari, Bala Subramaniam
(178b) Catalytic Cracking of Dodecane in Supercritical Water 223
Azadeh Zaker, Patricia Guerra, Geoffrey Tompsett, Michael T. Timko
(178c) Demetallation of Biocrude from Hydrothermal Liquefaction of Microalgae 224
Jimeng Jiang, Phillip E. Savage
(178d) Pilot-Scale Continuous Flow Reactor for Hydrothermal Liquefaction of Algae 225
Catherine E. Brewer, Feng Cheng, Travis Le-Doux, Juanita Miller
(178e) Production of Biodiesel Fuel Via Sub/Supercritical Transesterification Reactions with Trace Amount of Homogeneous Catalysts 226
Jiuxu Liu, Yue Nan, Lawrence L. Tavlarides
(178f) Molecular and Microkinetic Modeling of Supercritical Fluid and Liquid-Phase Fischer-Tropsch Synthesis 227
Kenneth M. Benjamin, Alireza Asiaee
(178g) Intensified Hydrolysis By Employing the Synergy of Mixed Subcritical H2o and Supercritical CO2 228
Armando Quitain, Akito Hashimoto, Saya Yamafuku, Mitsuru Sasaki, Tetsuya Kida
(205a) Pd Particle Stabilization on Carbon Supports Under Hydrothermal Conditions 229
Jiajie Huo, Robert L. Johnson, Pu Duan, Hien N. Pham, Abhaya K. Datye, Klaus Schmidt-Rohr, Brent H. Shanks
(205c) Vacancy-Mediated Hydrodeoxygenation of Furfuryl Alcohol over Metal Oxide 230
Konstantinos Goulas, Glen R. Jenness, Alexander V. Mironenko, Tobias Mazal, Dionisios G. Vlachos
(205d) Influence of Water and Solvents on the Production of Levoglucosenone and 5-(Hydroxymethyl)Furfural from Cellulose 231
Jiayue He, Fei Cao, Pranav U. Karanjkar, Siddarth H. Krishna, Kefeng Huang, Christos T. Maravelias, James Dumesic, George W. Huber
(205e) Transformation of Chitin into Value-Added Chemicals 232
Ning Yan

(205f) Direct Carbon-Carbon Coupling of Furanics with Acetic Acid over BrøNsted Zeolites ... 233
Abhishek Gumidyala, Bin Wang, Steven Crossley
(205g) Supported Cu Electrocatalysts for Electrocatalytic Hydrogenation and Hydrogenolysis of Furfural ... 234
Sungyup Jung, Elizabeth J. Biddinger
(205h) Rational Synthesis and Evaluation of Pd Bimetallic Catalysts for Furfural Rearrangement to Cyclopentanone in Aqueous Phase ... 235
Qiuli Liu, John R. Regalbuto
(205i) Polymeric Solid Acid Catalysts for Efficient Hydromethyfurfural and Levulinic Acid Production from Biomass Feedstock ... 236
Anh Vu, Ranil Wickramasinghe, Xianghong Qian
(209a) Soot Nucleation and Consumption in Oxy-Coal Systems ... 237
Alexander J Josephson
(209b) Investigation of NOx Chemistry in Oxy-Combustion Flue Gas ... 238
Nujhat Choudhury, Bihter Padak
(209c) On the Role of Sodium Content of Fine Ash Particles in Determining Deposition Rates Under Air- and Oxy-Coal Combustion Conditions ... 239
Zhonghua Zhan, Andrew Fry, Jost O. L. Wendt
(209d) NOx Formation in Syngas/Air Combustion at Elevated Pressure ... 240
Nazli Asgari, Sheikh F Ahmed, Tanvir I Farouk, Bihter Padak
(209e) Formation and Decomposition Kinetics of HONO and HNO2 ... 241
C. Franklin Goldsmith
(209f) Experimental Study on NO Reduction By Iron with Methane in Wet Simualted Flue Gas ... 242
Yaxin Su
(209g) Inferential–MODEL Based FLARE SET POINT Determination ... 243
Arokiaraj Alphones, Daniel Chen, Helen Lou, Vijaya Damodara, Xianchang Li, Christopher B. Martin, Edward Fortner, Scott Evans, Matthew Johnson
(210a) Offshore Floating Packed-Bed Reactors: Current Status, Key Challenges and Potential Solutions ... 244
Amir Motamed Dashliborun, Faical Larachi, Mohsen Hamidipour, Markus Schubert
(210i) Design Optimization of BioEnergy Combustion Systems Using Sculptor® with CFD ... 245
Joseph D. Smith, Vivek M. Rao, Mark Landon
(210d) A New Model for the Depolymerisation of Polysaccharides Originating from Biomass ... 246
Tapio Salmi, Sakari Teerikoski
(210e) Recent Trends in Multi-Phase Reactor Engineering and Design ... 247
Tobias Bauer, Stefan Haase, Markus Schubert, Ruediger Lange
(210f) Gas-Liquid and Liquid-Liquid Taylor Flow in Micro Channels: Hydrodynamics and Mass Transfer ... 248
Rufat Abiev, Stanislav Svetlov, Stefan Haase, Tobias Bauer, Ruediger Lange
(210j) 3D-Printed Monolithic Adsorbents for Gas Separation Processes ... 281
Harshul Thakkar, Stephen Eastman, Amit Hajari, Ali A. Rownaghi, Fateme Rezaei
(210h) Trickle-Bed Reactors and Their Future ... 282
Ruediger Lange
(212a) Keynote - Modeling Electrochemical Interfaces: An Approach from First Principles ... 283
Phillippe Sautet, Stephan Steinmann
(212b) Feature Engineering of Machine-Learning Chemisorption Models for Bifunctional Electrocatalyst Design ... 284
Zheng Li, Siwen Wang, Hongliang Xin
(212c) Effect of Lanthanum Doping and Chlorine Addition on Strontium Titanate Perovskites for Electrocatalytically-Assisted Oxidative Dehydrogenation of Ethane ... 285
Doruk Dogu, Anshuman Fuller, Katja E. Binkley Meyer, Seval Gunduz, Dhruba Jyoti Deka, Nathaniel Kramer, Anne Co, Umit S. Ozkan
(212d) Mechanistic and Trend Analyses of Selective Furfural Electroreduction on Transition Metals from First-Principles Methods ... 286
Nannan Shan, Xiaotong Chadderdon, David Chadderdon, Wenzhen Li, Bin Liu
(212e) NO Electrochemical Reduction on Pt(100) from First Principles ... 287
Hee-Joon Chun, Vesa Apaja, Andre Clayborne, Karoliina Honkala, Jeffrey P. Greeley
(212f) Flame Synthesis and High Pressure High Temperature Annealing of Anatase TiO2 for Increased Photocatalytic Activity ... 288
Ashley M. Pennington, Katelyn A. Dagnall, Rachel A. Yang, Fuat E. Celik
(212g) Photoelectrochemical Oxidation of Biorenewable Alcohols By Nitroxyl Radical Catalysts ... 289
David Chadderdon, Ivy Wu, Matthew G. Panthani, Wenzhen Li

(216a) so2 Interactions with Pt, Pd, and Pd-Pt/Al2O3 Catalysts ... 290
William Epling, Monique Wilburn
(216b) Insights into the Deactivation of Cu/SAPO-34 By Low- and High-Temperature Hydrothermal Treatment ... 291
Yang Zheng, Aiyong Wang, Eric D. Walter, Feng Gao, James Song, Yilin Wang, Janos Szanyi, Charles H. F. Peden
(216c) Lean-Rich Switching over a Modified Three-Way Catalyst: Experiments and Modeling ... 292
Mengmeng Li, Sam Malamis, Michael P Harold, William S. Epling
(216d) Importance of Exhaust Hydrocarbon Speciation in the Studies of Low-Temperature Emission Control Technologies for Next-Generation Fuel-Efficient Vehicles ... 294
Se H. Oh, Michelle Wiebenga, Sung Bong Kang, Sung Bang Nam, In-Sik Nam
(216e) Rapidly Pulsed Reductants in Diesel NOx Reduction with Lean NOx Traps: Spatial Distribution of Species ... 295
Amin Reihani, Brent Patterson, John Hoard, Galen B Fisher
(216f) A Rational Pathway for Generic Three-Way Catalyst Model Development in Stoichiometric Natural Gas Engines ... 297
Jian Gong, Di Wang, Junhui Li, Neal Currier
(216g) Effects of Cu Density in Cu-SSZ-13 Zeolites on Low-Temperature NOx Selective Catalytic Reduction with NH3 ... 298
Ishant Khurana, Atish A. Parekh, Jonatan Albarracin, Christopher Paolucci, Arthur Shih, John R. Di Iorio, W. Nicholas Delgass, Jeffrey T. Miller, William Schneider, Rajamani Gounder, Fabio H. Ribeiro
(216h) The Effects of Catalyst Aging on Sulfur Degradation over Fully Formulated Lean NOx Trap Catalysts ... 299
Travis Wentworth, Louise Olsson, Marie Stenfeldt
(216i) A Comparative Study of ZSM-5 and BEA-Zeolites for Hydrocarbon Trap Applications Under "Cold-Start" Conditions ... 300
Eleni A. Kyriakidou, Jae-Soon Choi, Todd J. Toops, James E. Parks
(267a) Developing a Quantitative Spatial Resolution of Deactivation Effects on FT Synthesis in a Microchannel Reactor When Operating with Ultra-Low Sulfur Levels ... 301
Soumitra Deshmukh, Henning Becker, Sara Kampfe, Kenneth Cowen, Daniele Leonarduzzi, Laura Barrio, Jay Pritchard, Heinz J. Robota
(267b) Characterization of Unconventional Reservoirs: From Lab Study to Reservoir Simulations ... 306
Lin Li
(267c) New Insight into Oxidative Dehydrogenation of n-Butenes to Butadiene ... 307
Xiankuan Zhang
(267d) Novel High Active Catalyst Innovation and Low-Energy Consumption Process Technology for Ammonia Synthesis ... 308
Huazhang Liu, Chao Huo, Wenfeng Han
(267e) Maximizing Production of Premium Distillate Fuels through Selective Hydroisomerization of Normal Paraffins ... 309
William W. Lonergan
(267f) On the Effectiveness Factor Modeling for Mass Transfer-Limited Reaction Networks in Industrial Catalysis ... 310
Bo Jin, Xiaoxiang Zhu, William Licht
(274a) Active Site Requirements and Elementary Steps in the Ketonization of Carboxylic Acids on TiO2 and ZrO2 ... 311
Enrique Iglesia, Shuai Wang
(274b) Influence of Solid Acid Structure and Composition on Hydrocarbon Cracking Activity Under Supercritical Conditions ... 314
Benjamin Huang, Robert J. Davis
(274c) Improving the Hydrothermal Stability of Heterogeneous Catalysts for Conversion of Biorenewables Via Carbon Coating and Overcoating ... 315
Abhaya Datye, Hien Pham, Haifeng Xiong, Pu Duan, Xiaoyan Cao, Klaus Schmidt-Rohr, Robert Johnson, James Dumesic
(274d) Mechanistic Insights into the Reactive Sorption of Sulfur Compounds Using Metal Oxides ... 316
Dante Simonetti, Sara Azzam, Brook Wander
(274e) Influence of Solvent on Acid and Metal Catalyzed Reations ... 319
Matthew Neurock
(281a) Effect of Alcohol Structure on the Kinetics and Mechanism of Acid Catalyzed Etherification and Unimolecular Dehydration over Tungstated Zirconia ... 320
Julie Rorrer, Alexis T. Bell, Dean Toste
(281b) Pathways to Selectively Form Individual Aromatic Products from Ethanol ... 321
David W. Flaherty, Takahiko Moteki

(281c) Combined Experimental and Theoretical Study of Ethanol Catalytic Conversion to 1,3-Butadiene on MgO ... 322
William Taifan, Jonas Baltrusaitis
(281d) Aldol Condensation of Acetaldehyde over Titania, Hydroxyapatite and Magnesia ... 323
Zachary Young, Sabra Hanspal, Robert J. Davis
(281e) Single Step Conversion of Short Carboxylic Acid Mixtures Obtained from Fermentation to Aromatics and Î±-Olefins over a Cu/ZrO2 Catalyst ... 324
Bartosz Rozmyslowicz, Jeremy S. Luterbacher
(281f) Mechanistic Insights on the Formation of Maleic Anhydride from the Oxidation of Levulinic Acid over Supported Vanadates ... 325
Anargyros Chatzidimitriou, Jesse Q. Bond
(281g) P-Toluic Acid from Biomass-Derived Coumalic Acid: A Reaction Network Analysis ... 326
Toni Pfennig, Robert L. Johnson, Brent H. Shanks
(281h) Rational Design of Nanoflower Shaped Bimetallic Ptmn Nanocatalysts for Tartronic Acid Synthesis from Biomass ... 327
Xin Jin, Meng Zhao, Chun Zeng, Wenjuan Yan, Honghong Shi, Prem Thapa, Bala Subramaniam, Raghunath V. Chaudhari
(288a) Upgrading Corn Stover Pretreatment Wastes into Biodiesel Via Oleaginous Rhodococci ... 328
Arthur J. Ragauskas
(288b) Biological Design of Lignin Conversion ... 329
Bin Yang, Art Regauskas, Joshua Yuan
(288c) Selective Conversion of Guaiacol to Substituted Alkylphenols in Supercritical Ethanol over MoO3 ... 330
Kai Cui, Yongdan Li
(288d) Enabling Microbial Utilization of Thermally Depolyermized Lignin Monomers ... 331
Kirsten Davis, Marjorie Rover, Davinia Salvachua, Gregg T. Beckham, Zhiyou Wen, Ryan Smith, Laura Jarboe, Robert Brown
(288e) Fractionated Lignin Rheology and Structure ... 332
Christopher Higgins, David C. Russ, Noppadon Sathitsuksanoh, R. Eric Berson
(288f) Selective Conversion of Biomass-Derived Lignin to Cyclic Hydrocarbons ... 333
Bin Yang, Melvin Tucker, Hongliang Wang, Hao Ruan, Xiaowen Chen
(289b) The Evolution of Multi-Functional Nanoporous Metal Composite Electrocatalysts ... 334
Joshua Snyder, Yawei Li
(289c) Effect of Doping on the Activity of Nickelate Oxides Toward Surface Oxygen Exchange and Oxygen Reduction ... 335
Xiang-Kui Gu, Anirban Das, Juliana S. A. Carneiro, Eranda Nikolla
(289d) Oxyhydroxides As a Platform for Confined Oxygen Electrochemistry ... 336
Andrew Doyle, Michal Bajdich, Aleksandra Vojvodic
(289e) Large-Size Graphene Tube Catalysts for Bifuncational Oxygen Reduction and Oxygen Evolution Electrocatalysis in Alkaline Media ... 337
Gang Wu
(289f) Atomic Layer Deposition of NiAlxOy Catalysts for Electrochemical Oxidation of Water ... 338
Jonathan G. Baker, Stacey F. Bent, A. J. M. Mackus
(289g) The Role of Superoxide in the Non-Aqueous Oxygen Reduction Reaction in Li-O2 Batteries ... 339
William C. McKee, Saurin Rawal, Ye Xu
(300a) Experimental and Simulation Study of Ethanol Steam Reforming Reaction for Hydrogen Production in Large Scale Catalytic Membrane Reactor ... 340
Rui Ma, Bernardo Castro Dominguez, Ivan Mardilovich, Anthony G. Dixon, Yi Hua Ma
(300b) Membrane Reactors for Equilibrium-Limited Alcohol and Light Hydrocarbon Synthesis ... 341
Zhongtang Li, Nazanin Entesari, Theodore Tsotsis
(300c) Dual-Phase Ceramic Membrane in a Membrane Reactor for Ethane Dehydrogenation ... 342
Libin Zhuang, Yanying Wei, Haihui Wang
(300d) Multi-Scale (Pellet-Reactor Scale) Membrane Reactor Modeling and Simulation: High Temperature, and High Pressure Water-Gas Shift Reaction ... 343
Secgin Karagoz, Vasilios Manousiouthakis, Theodore Tsotsis
(300e) Pervaporation-Assisted Furfural Production from Xylose ... 344
Alex Wang, Nitash Balsara, Alexis T. Bell
(300f) Zeolite Membrane Permselective to Water Vapor and Its Application to Dehydration Reactions ... 345
Naotsugu Itoh, Shuhei Kikuchi, Takafumi Sato, Takuya Hirose
(300g) Esterification and Extraction with Supported Liquid Membrane Permeation ... 346
Marlene Kienberger, Matthaeus Siebenhofer, Sasiradee Jantasee, Nuttakul Mungma

(300h) Investigation of the Degradation of Endocrine Disrupting Compounds in a Membrane Bioreactor 347
Mohanad Kamaz, Ees Eswaranandam, Ranil Wickramasinghe, Wen Zhang, Xianghong Qian, Michael Watts
(307a) Quantum Chemical Methods for the Simulation of Reactions of Organic Molecules in Zeolites 348
Alexis T. Bell
(307b) Identification of Possible Mechanistic Pathways during the Electrochemical Reduction of CO2: Results from Constant Potential DFT Calculations 349
Jason Goodpaster, Alexis T. Bell, Martin Head-Gordon
(307c) Theoretical Challenges in Characterizing Cu Centers in the Zeolite SSZ-13 350
Florian Göltl, Alyssa Love, Sabrina Conrad, Patrick Wolf, Phillippe Sautet, Ive Hermans
(307d) Free Energies of Adsorption Calculated through Potential of Mean Force Ab Initio Molecular Dynamics in Metal-Exchanged Zeolites 351
Hui Li, Christopher Paolucci, William F. Schneider
(307e) Nonadiabatic Effects and Creation of Electronic Excitations during Dissociation on Metal Nanoparticles 352
Matthew M. Montemore, Efthimios Kaxiras
(307f) Comparing the Effect of Counter Ions, Solvent Molecules, and Electron Correlation on Homogeneous Reaction Models 353
Mitchell Groenenboom, John A. Keith
(307g) Automated Reaction Mechanism Generation for Heterogeneous Catalisys 354
Richard H. West, C. Franklin Goldsmith
(307h) Global Optimization and Hydrogen Adsorption on Pt-Cu Clusters By Using Genetic Algorithm and Density Functional Theory 357
Ezgi Erdem, S. Bengisu Barim, Riza Kizilel, Can Erkey
(332a) Strategies for the Conversion of Biomass-Derived Carbohydrates and Fatty Acids to Fuels and Lubricants 360
Alexis T. Bell
(332b) Materials for Manipulating the Composition of Biomass Pyrolysis Vapors 361
Brent H. Shanks
(332c) Coupling Chemical and Biological Catalysis to Produce Biobased Chemicals 362
Thomas J. Schwartz, Brent H. Shanks, James Dumesic
(332d) Commodity Chemicals from Biomass: Catalytic Conversion of Biomass into a,w-Diols 363
George W. Huber, Jiayue He, Pranav U. Karanjkar, Siddarth H. Krishna, Kevin J. Barnett, Zachary Brentzel, Samuel P. Burt, Kefeng Huang, Ive Hermans, William F. Banholzer, Christos T. Maravelias, James A. Dumesic
(332e) Atomically-Thin Noble Metal Coatings on Early Transition Metal Carbide Nanoparticles: a Versatile Catalytic Platform for Electro- and Thermocatalysis 364
Yuriy Román-Leshkov, Sean T. Hunt, Maria Milina, Ana Alba-Rubio, James Dumesic, Christopher H. Hendon
(337a) Biodiesel Produced from Soybean Oil with the Two Flow Reactors in Series 365
David Baah, Nader Vahdat, Kyung Kwon
(337b) Biomass Conversion to Acrylonitrile Monomer-Precursor for Production of Carbon Fibers 366
Jadid Samad, Lindsey Chatterton, Zora Govedarica, Amit Goyal
(337c) Catalytic Conversion of Polyols to Olefins 367
Bryan E. Sharkey, Friederike C. Jentoft
(337d) Catalytic Oligomerization of Ethylene: Experimental Insight into the Effect of Supercritical Reaction Conditions 368
Oliver Jan, Fernando Resende
(337e) Mesoporous Catalysts for Conversion of 2,3-Butanediol to Butene 369
Quanxing Zheng, Jason Grossardt, Haider Almkhelfe, Jiayi Xu, Brian P. Grady, Placidus B. Amama, Keith L. Hohn
(337g) Polar Aprotic Solvent Effects on the Catalytic Conversion of Fructose to 5-(hydroxymethyl)Fulfural or Lactic Acid 370
Christian G. Rivera-Goyco, Michelle M. Marrero-Vazquez, Leida M. Vazquez-Ramos, Yomaira J. Pagan-Torres, Nelson Cardona-Martinez
(348a) Investigations of Surface Chemistry for Pyridine-Catalyzed CO2 Reduction on Gallium Phosphide 371
Coleman Kronawitter, Bruce E. Koel
(348b) Implications of Surface Reconstructions Impacting Py-Catalyzed CO2 Reduction on Semiconductor Photoelectrodes 372
Thomas P. Senftle, Emily A. Carter
(348c) Activity and Stability of Pure and Modified Coooh for Oxygen Evolution Reaction in Alkaline Medium 373
Zhu Chen, Coleman Kronawitter, Yao-Wen Yeh, Xiaofang Yang, Peng Zhao, Yao Nan, Bruce Koel

(348d) Fe-Ni Core-Shell Hydroxide Nanoparticles As an Active Oxygen Evolution Reaction (OER) Catalyst ... 374
Stephanie Candelaria, Nikki S. Rentz, Lauren F. Greenlee
(348e) Layered - Hyperthin Electrocatalysts for the Water Splitting Reactions ... 375
Kevin C. Leonard
(348f) Stabilizing Electrodeposited Nanoparticle Electrocatalysts on Si MIS Photocathodes for Solar Water Splitting ... 376
Natalie Labrador, Xinxin Li, Yukun Liu, Jeffrey T Koberstein, Daniel V. Esposito
(348g) Two Distinct Types of Surface-Bound Atomic Hydrogen and Their Role in Dictating Product Selectivity in Artificial Photosynthesis Reactions ... 377
Samiksha Poudyal, Siris Laursen
(370a) Development of Bayesian Error Estimation Density Functionals with Range-Separated Exchange ... 378
Shaama Mallikarjun Sharada, Keld T Lundgaard, Johannes Voss, Jess Wellendorff, Thomas Bligaard, Jens K. Norskov
(370b) Uncertainty Quantification of the Water-Gas Shift Reaction By Pt/CeO2 Catalyst ... 379
Eric Walker, Salai C. Ammal, Gabriel Terejanu, Andreas Heyden
(370c) A Descriptor Based Approach Towards Understanding the Effect of Functionals on Reducible Oxide Reaction Kinetics ... 380
Tej S. Choksi, Jeffrey P. Greeley
(370d) Interpreting Functional-Sensitivity in Catalysis: Exchange-Tuning Vs. DFT+U ... 381
Qing Zhao, Efthymios Ioannidis, Jon Paul Janet, Heather J. Kulik
(370e) Determining the Pd Dopant Effect on the Reducibility of Fe2O3: DFT+U or Hybrid Functionals? ... 382
Alyssa Hensley, Yongchun Hong, Yong Wang, Jean-Sabin McEwen
(370f) Neural Network and Reaxff Comparison for Au Properties ... 383
Jacob R. Boes, Mitchell Groenenboom, John A. Keith, John R. Kitchin
(370g) Machine Learning Based Interatomic Potentials for Electrocatalysts ... 384
Badri Narayanan, Fatih Sen, Alper Kinaci, Stephen Gray, Maria K. Y. Chan, Subramanian K. R. S. Sankaranarayanan
(370h) BASC: Applying Bayesian Machine Learning Techniques to the Search for Global Minima on Potential Energy Surfaces ... 385
Shane F. Carr, Cynthia S. Lo
(370i) Developing a Machine-Learning Molecular Dynamics Approach for Nonadiabatic Surface Chemistry ... 386
Jiamin Wang, Hongliang Xin
(375a) Glycerol to Hydrocarbon Fuels Via Bifunctional Catalysts ... 387
Yang Xiao, Arvind Varma
(375b) Reaction Engineering of Urea Alcoholysis ... 388
Dhananjay R. Mote, Vivek V. Ranade
(375c) Gas-Phase Catalysis for Pyrolyzing Acetic Acid ... 389
Sara Jo Taylor, Phillip R. Westmoreland, Haley Key
(375d) Hybrid Reaction Pathways to Renewable Diol Monomers ... 390
Paul J. Dauenhauer, Charles S. Spanjers, Marc A. Hillmyer, Kechun Zhang, Deborah Schneiderman
(375e) Continuous Piancatelli Rearrangement of Furfuryl Alcohol: Kinetics and Reactor Modelling ... 391
Lukas T. Vömel, Marcel A. Liauw
(375f) Catalyst Structure and C-O Activation during Fischer-Tropsch Synthesis ... 392
G. T. Kasun Kalhara Gunasooriya, Arghya Banerjee, Mark Saeys
(375g) Reaction Engineering for Modified Furans ... 393
Dae Sung Park, Paul J. Dauenhauer, Kristeen Esther Joseph, Christoph Krumm, Dion Vlachos, Raul F. Lobo
(393a) C2H6 Dehydrogenation on Pt-Based Catalysts: Fundamental Mechanistic Studies ... 394
Manos Mavrikakis, Duygu Gerceker, Guowen Peng, James Dumesic
(393b) Microkinetic Analysis of γ-Valerolactone Ring Opening and Decarboxylation over Solid Acids ... 395
Jesse Q. Bond
(393c) Solvent Effects in the Catalytic Conversion of Fructose to Lactic Acid ... 396
Nelson Cardona-Martinez, Christian G. Rivera-Goyco, Yomaira J. Pagan-Torres, Michelle M. Marrero-Vazquez
(393d) Stabilization with Formaldehyde Facilitates the High-Yield Production of Monomers from Lignin during Integrated Biomass Depolymerization ... 397
Jeremy S. Luterbacher
(393e) Solvent-Enabled Control of Reactivity for Liquid-Phase Biomass Conversion Reactions ... 398
James Dumesic, Max A. Mellmer, Ali Hussain Motagamwala, Bengi Demir, David Martin Alonso

(399a) Structure-Energy-Activity Relationships in Heterogeneous Catalysis - a View from Theory 399
Phillippe Sautet
(399b) A New DFT-Based Approach That More Accurately Predicts Adsorption Energies 400
Alyssa Hensley, Kushal Ghale, Carolin Rieg, Thanh Dang, Emily Anderst, Felix Studt, Charles T. Campbell, Jean-Sabin McEwen, Ye Xu
(399c) A DFT-Based Study of Pt-Catalyzed Hydrodechlorination of 1,2-Dichloroethane: Effect of Van Der Waals Interactions 401
Lang Xu, Eric E. Stangland, Manos Mavrikakis
(399d) Computational Design of Active Site Structures to Circumvent Scaling in Ammonia Synthesis 402
Aayush R. Singh, Joseph H. Montoya, Charlie Tsai, Aleksandra Vojvodic, Jens K. Nørskov
(399e) Mechanistic Investigation of the Catalytic Hydrogenation of Levulinic Acid over Ru Model Catalyst Surfaces 403
Osman Mamun, Jesse Q. Bond, Andreas Heyden
(399f) Reaction Pathways for Biomass Derived Furanic Compounds on Ru(001) Surfaces: a DFT Study 404
Arghya Banerjee, Samir H. Mushrif
(399g) CO2 Activation on Zr-Decorated, Cu Nanoparticles 405
Natalie Austin, Giannis Mpourmpakis
(399h) Oxygen Dissociation for Selective Oxidation on Gold: Effect of Alloying and Geometrical Structure 406
Matthew M. Montemore, Cynthia M. Friend, Robert J. Madix, Efthimios Kaxiras
(432a) Catalytic Co-Pyrolysis of Microalgae and Low Density Polyethylene Waste to Aromatic Hydrocarbons Using Activated Carbon 407
Emmanuel Ansah, Lijun Wang
(432b) Modeling the Impact of Biomass Particle Residence Time on Fast Pyrolysis Yield and Composition 415
Gavin Wiggins, C. Stuart Daw, Emilio Ramirez
(432c) Ring-Locking Enables Selective Anhydrosugar Synthesis from Carbohydrate Pyrolysis 416
Li Chen, Jin-Mo Zhao, Sivaram Pradhan, Bruce E. Brinson, Gustavo E. Scuseria, Z. Conrad Zhang, Michael S. Wong
(432d) Pyrolysis Kinetics to and from Levoglucosan 417
Vikram Seshadri, Phillip R. Westmoreland
(432e) Molecular-Level Simulation of Thermogravimetric Analysis (TGA): A Cellulose Pyrolysis Example 418
Juan Lucio-Vega, Scott R. Horton, Michael T. Klein
(432f) Impact of Carrier Gas and Temperature on Biomass Fast Pyrolysis Using a New Microsphere Microreactor Approach 419
Joseph J. Biernacki, Ali Zolghadr
(432g) Study of Transmethylation As Primary Step on the Decomposition of Methoxy-Rich Lignin Model Compound over Zeolite Catalyst 420
Zhang Jiajun, Fidalgo Beatriz, Shen Dekui
(434a) Hydroxycarbonylation Vs. Methoxycarbonylation - Palladium Catalyzed Reactions in Multiphase Systems 421
Marcel Schmidt, Tobias Pogrzeba, Ariane Weber, Prof. Dr. Reinhard Schomäcker
(434b) Heterogeneously Catalysed Benzyl Alcohol Oxidation in a Three-Phase Micro-Packed Bed Reactor: Influence of Hydrodynamics and Reactor Design on Reaction Performance and Catalyst Deactivation 422
Noor Al-Rifai, Archana Venkatesh, Enhong Cao, Stefano Cattaneo, Meenakshisundaram Sankar, Graham J. Hutchings, Asterios Gavriilidis
(434c) Highly Selective Synthesis of Primary Amines from Alcohols and Ammonia By Heterogeneous Nickel Catalysis 423
Andrew Leung, Mimi Hii, Klaus Hellgardt
(434d) Hydroformylation and Tandem Isomerization-Hydroformylation of Long-Chain Olefins: Mechanism, Kinetics and Optimal Reaction Control 424
Andreas Jörke, Andreas Seidel-Morgenstern, Christof Hamel
(434e) Troubleshooting Start-up of a Newly Built Intensified Heat Exchanger Reactor - Using Reaction Engineering and Dynamic Modeling to Solve Real-Life Problems 425
Eric Grolman
(434f) Alloy Catalysis Across Composition Space 426
A. J. Gellman, Irem Sen, Gamze Gumuslu, Petro Kondratyuk, Xiaoxiao Yu
(434g) Pressure Dependent Kinetics of Magnesium Oxide Carbothermal Reduction 427
Boris Chubukov, Scott C. Rowe, Aaron W. Palumbo, Illias Hischier, Arto J. Groehn, Alan W. Weimer

(435a) Detailed Kinetic Modeling of Thermal Decomposition of Silane Via Automatic Mechanism Generation 428
Belinda L. Slakman, Pierre L. Bhoorasingh, Richard H. West

(435b) Mechanistic Insights into the Electrochemical Reduction of CO2 over Ag Using an Integrated Transport-DFT-Microkinetic Model 429
Meenesh Singh, Jason Goodpaster, Adam Weber, Martin Head-Gordon, Alexis T. Bell

(435c) Acceleration of Kinetic Monte Carlo Simulations Combining a Hybrid Approach with Scaling of Kinetic Parameters for Free Radical Copolymerization 430
Hanyu Gao, Ivan Konstantinov, Steven G. Arturo, Linda J. Broadbelt

(435d) Kinetics and Mechanism of Ketonization of Acetic Acid on HZSM-5 Catalyst 431
Abhishek Gumidyala, Tawan Sooknoi, Bin Wang, Steven Crossley

(435e) Methane Steam Reforming Unraveled By the Microkinetic Engine, a User-Friendly Kinetic Modelling Tool 432
Brigitte R. Devocht, Kenneth Toch, Naoki Kageyama, Naghmeh Fatemi, S. Ted Oyama, Guy B. Marin, Joris W. Thybaut

(435f) Severe Pyrolysis of the C12 Hydrocarbons n-Dodecane, Cyclododecane, and 2,6-Dimethyl-Decalin, with Rice-Herzfeld and Graph-Theoretic Interpretations 438
Preetinder S. Virk

(435g) An Assessment of Solvent Effects on the Selectivity of an Alkylation Reaction: A Comparison Between Experiments and Computations 439
Aikaterini Diamanti, Eliana Grant, Zara Ganase, Anita Rea, Amparo Galindo, Claire Adjiman

(435h) Automatic Discovery of Reaction Paths and Ab Initio/Rrkm-ME Calculations Towards Pyrolysis and Combustion Mechanisms 440
Adeel Jamal, Colin Grambow, William H. Green

(435i) Computational Singular Perturbation As an Analytical Tool to Study the Fischer-Tropsch Process 441
Debanjan Chakrabarti, Gouthami Senthamaraikkannan, Vinay Prasad

(435j) Mechanism, Reaction Kinetics and Multi-Scale Modelling of CO2 Hydrogenation to Methanol over Trimetallic Heterogeneous Catalysts 442
Neja Strah, Andraž Pavlišic, Matej Huš, Venkata Dasireddy, Andrej Pohar, Blaž Likozar

(435k) Mechanistic Insights into the Catalytic Elimination of Tar on Nickel and the Promotional Effect of Boron on It: A Combined Theoretical and Experimental Study Using Toluene As a Model Compound 450
Quang Thang Trinh, Anh Vu Nguyen, Dang Chinh Huynh, Thanh Huyen Pham, Samir H. Mushrif

(461a) Discovery of Superior Cu-GaOx-HoOy Catalysts for the Reduction of Carbon Dioxide to Methanol at Atmospheric Pressure 451
Bahman Zohour, Selim Senkan

(461b) Methanation of Carbon Dioxide with a Novel Nickel/Magnesium Oxide Catalyst 452
Georg Baldauf-Sommerbauer, Susanne Lux, Matthaeus Siebenhofer

(461c) Polymer-Nanocrystal Composites As Tailorable CO2 Hydrogenation Catalysts 453
Andrew Riscoe, Joshua Willis, Matteo Cargnello

(461d) Reduction of CO2 to CO and Methanol on Ceria Based Catalyst: Mechanistic Insights 454
Neetu Kumari, Suddhasatwa Basu, M. Ali Haider, Nishant Sinha, Manish Agarwal

(461e) Catalytic Conversion of Carbon Dioxide with Solar Hydrogen Under Supercritical Conditions 455
Elisabeth Hocke, Herbert Vogel, Bjoern Kommoss, Stephan Schultheis

(461f) CO2 Hydrogenation on Silica- and Titania- Supported Single-Site Heterogeneous CoII Catalyst 456
Juan Jimenez, Cun Wen, Jason Hattrick-Simpers, Jochen Lauterbach

(461g) Development of Catalysts for CO2 Hydrogenation to Olefins 457
Marc D. Porosoff, Heather D. Willauer

(461h) K2o/WOx/Al2O3 Catalyst Interactions with the Components and Byproducts of Sour Natural Gas 458
Jih-Mirn Jehng, Israel E. Wachs, Zili Wu, Jonas Baltrusaitis

(465a) Organic Photocatalysts for Atom Transfer Radical Polymerization Driven By Visible Light 459
Charles B. Musgrave, Chern-Hooi Lim, Jordan Theriot, Garret Miyake, Haishen Yang, Matthew Ryan

(465b) CO-CO Coupling on Cu Facets: Coverage, Strain and Field Effects 460
Robert B. Sandberg, Joseph H. Montoya, Karen Chan, Jens K. Nørskov

(465d) Single-Atom Catalysts for CO2 Electroreduction with Significant Activity and Selectivity Improvements 461
Seoin Back, Yousung Jung

(465e) Solvation Effects in Surface Electrochemistry for Oxygen Evolution Reaction Catalysis on Transition Metal Oxides 462
Joseph Gauthier, Colin Dickens, Andrew Doyle, Jens K. Norskov

(465f) First Principles Quantum Chemistry Calculations to Model CO2 Electroreduction on SnO2 Particles 463
Yasemin Basdogan, Karthikeyan Saravanan, James Dean, John A. Keith
(465g) Modelling the Electrochemical Interface: Applications to CO2 Reduction 464
Karen Chan
(465h) Exploring the Remarkable Heterogeneity of Bimetallic Alloy Facet Reactivity for the Electrocatalytic Reduction of CO2 465
Zachary Ulissi, Michael Tang, Mohammadreza Karamad, Karen Chan, Jens Nørskov
(465i) Exploring Possible Active Sites for IrOx/SrIrO3, a Highly Active Oxygen Evolution Catalyst 466
Linsey C. Seitz, Colin Dickens, Kazunori Nishio, Yasuyuki Hikita, Joseph H. Montoya, Andrew Doyle, Charlotte Kirk, Aleksandra Vojvodic, Harold Y. Hwang, Jens K. Nørskov, Thomas F. Jaramillo
(496a) Characterization of TFA Regenerated Cellulose with a Two-Step Treatment to Understand the Impact of Cellulose Structure on Hydrolysis 467
Ximing Zhang
(496b) The Role of Atmosphere, Solvents and Acids on the Production of 5-Hydroxymethylfurfural from Biomass Derived Fructose 468
Glen Svenningsen, Phillip Christopher, Rajeev Kumar, Charles Wyman
(496c) Lignin Depolymerization into Aromatic Monomers over Acidic Mesoporous Silicates 469
Kakasaheb Nandiwale, Andrew Danby, Anand Ramanathan, Raghunath V. Chaudhari, Bala Subramaniam
(496d) Characteristics and Activity of Steamed Zeolites for Butanol Upgrading to Higher Olefins 470
Mond Guo, Michel Gray, Vijayakumar Murugesan, Karthikeyan K. Ramasamy
(496e) Tandem Catalysis in a New, Solid-Phase Route to Renewable Aromatic Chemicals from Biobased Furanics 471
Homer Genuino, Shanmugam Thiyagarajan, Jan Van Der Waal, Ed De Jong, Jacco Van Haveren, Dan Van Es, Bert Weckhuysen, Pieter Bruijnincx
(496g) Reaction Pathways and Microkinetic Modeling of Levulinic Acid Hydrodeoxygenation over Sulfided NiMo/Al2O3 472
Miha Grilc, Blaž Likozar
(497a) The E-Factor of the Thin Film and Its Characterization to Understand Why Water Influences a Heck Alkynylation 477
Chuntian Hu, Kevin H. Shaughnessy, Ryan L. Hartman
(497b) Triphasic Millireactors for Gas-Liquid Reactions with in Situ Catalyst Synthesis 478
Wai Kuan Wong, Swee Kun Yap, Saif A. Khan

VOLUME 2

(497c) Experimental and Model Aided Approach for Development of High Pressure Direct Asymmetric Reductive Amination Reaction 479
Shujauddin M. Changi, Tohru Yokozawa, Tetsuya Yamamoto, Hikaru Nakajima, Matthew C. Embry, Radhe K. Vaid, Sze Wing Wong, Martin D. Johnson, Eric D. Moher
(497d) Kinetics and Mechanistic Understanding of the Difluoromethyl Alkylation of Hydroxychromenone 480
Daniel D. Caspi, Anuj A. Verma, Eric M. Phillips, Shashank Shekhar
(497e) Development of a Reaction Kinetic Model for an Enzymatic Reaction Using Reaction Kinetic Progress Analysis and Dynochem 481
Kathryn Meintel
(497f) Rapid Kinetic Profiling of Copper(I)-Catalyzed Alkyne-Azide Cycloaddition(CuAAC) Click Reactions Utilizing Inline IR, HPLC and MS 482
Kosi Aroh, Klavs F. Jensen
(497g) Dynamic Resolution for Separation-Free Enantiomeric Purification in Flow 483
Andrew Teixeira, Saurabh Shahane, Nopphon Weeranoppanant, Klavs F. Jensen
(517a) Tuning the Activity and Selectivity of Transition Metal Phosphide Catalysts for CO, and CO2 Hydrogenation 484
Melis S. Duyar, Eduardo Valle, Jakob Kibsgaard, Thomas F. Jaramillo
(517b) Aerosol Processing of MOF-Based Nanocomposites for Carbon Dioxide Photoreduction 485
Wei-Ning Wang, Xiang He, Zhuoran Gan
(517c) Enhanced Electrocatalytic CO2 Reduction on Thiol-Functionalized Gold 486
Yuxin Fang, Xun Cheng, Ye Xu, John C. Flake
(517d) Investigation of the Main Factor That Drives Selectivity on Rough Cu-Based Catalysts for CO2 Electroreduction 487
Alexandros Karaiskakis, Tiffany Harden, Elizabeth J. Biddinger

(517e) S Doped TiO2 synthesized By Novel One Step Liquid Flame Spray Pyrolysis (LFSP) for Photocatalytic Oxidation of CO in Visible Region 488
Siva Nagi Reddy Inturi, Thirupathi Boningari, Makram Suidan, Panagiotis Smirniotis
(517f) Optimizing the Physicochemical Properties of ZSM-11 for Methanol-to-Hydrocarbon Reactions 489
Yufeng Shen, Thuy T. Le, Jeffrey D. Rimer
(517g) Precise Deposition of Pt Promoter Onto Silica Supported Cobalt for Fischer-Tropsch Synthesis 490
Fahad Almalki, John Monnier, John R. Regalbuto
(517h) Using Inelastic Neutron Scattering to Probe the Active Site for Cobalt-Copper Catalyst for the Conversion of Syngas to Higher Oxygenates 491
Zi Wang, Phillip T. Sprunger, Richard L. Kurtz, Luke L. Daemen, Yongqiang Cheng, James J. Spivey
(520a) Selective C-O Bond Cleavage of Diphenyl Ether over the Supported Palladium Catalyst 492
Donghai Mei, Meng Wang, Johannes A. Lercher
(520b) Metal Core/Surface Oxide Shell Catalysts for C-O Bond Activation 493
Alexander V. Mironenko, Konstantinos Goulas, Raymond J. Gorte, Christopher B. Murray, Dionisios Vlachos
(520c) A Periodic DFT Study of the Influence of Lewis Acid Site Speciation on Ethanol Dehydration in Zeolites 494
Brandon Bukowski, Jason Bates, Rajamani Gounder, Jeffrey P. Greeley
(520d) Mechanistic Study of Regio-Selective Ring-Opening Reactions of Mono-Substituted Epoxides Using Lewis Acid Catalysts 495
Ying Yu, Linda J. Broadbelt
(520e) Understanding the p-Xylene Formation Mechanism from Dimethylfuran, and Ethanol 496
Pavlo Kostetskyy, Giannis Mpourmpakis, Michail Stamatakis, Shik Chi Edman Tsang And Ivo Teixeira
(520f) Role of Enol-Keto Tautomerization in m-Cresol Selective Hydrogenation on Ni(111) and Nife(111) Surface 497
Xiaoyang Liu, Wei An
(520g) Developing Multi-Scale Models of Bimetallic Catalysts for the Hydrodeoxygenation of Bio-Oil Compounds 498
Breanna Wong, Alyssa Hensley, Pierre Gaspard, Yong Wang, Jean-Sabin McEwen
(520h) Combined Density Functional Theory and Molecular Dynamics Study of Aqueous Reforming of Glycerol on Pt (111) Via Competitive C-H and O-H Dehydrogenation 500
Tianjun Xie, Cameron Bodenschatz, Rachel Getman
(520i) Molecular Dynamics Simulations of Aldol Condensation Catalyzed By Alkylamine-Functionalized Crystalline Silica Surfaces 501
Ki Chul Kim, Eric G. Moschetta, Christopher W. Jones, Seung Soon Jang
(539a) Droplet Microfluidics with a Nanoemulsion Continuous Phase: Application in Reaction System Design for Material Synthesis 502
Tonghan Gu, Saif A. Khan, T. Alan Hatton
(539b) Activating C1 Chemistry Utilizing Microreactor Technology 503
Alexandre Yokochi, Goran Jovanovic, Annette Von Jouanne, Nick Auyeung, Scott Harpool, Justin Pommerenck, Yousef Alanazi, Yu Miao, Ian Reddick, Adam Shareghi
(539c) Gas- Liquid Flow Hydrodynamics and Mass Transfer in Micro-Packed Beds 504
Jisong Zhang, Andrew Teixeira, Klavs F. Jensen
(539d) Implementation of in Situ SAXS/Waxs Analysis in Silicon-Pyrex Microreactor 505
Samuel Marre, Cyril Aymonier, Yves Garrabos, Carole Lecoutre, Thomas Beuvier, Brice Calvignac, Alain Gibaud
(539e) Automated Oscillatory Photochemical Reactor for High Throughput Studies of Visible-Light Photoredox Catalysis 506
Milad Abolhasani, Connor Coley, Hongkun Lin, Klavs F. Jensen
(539f) Kinetic Study of Reactions of Aniline and Benzoyl Chloride in a Microstructured Chemical System 507
Peijian Wang, Kai Wang, Jisong Zhang, Guangsheng Luo
(539g) Brock Copolymer Synthesis in a Microreactor By Precise Analysis and Design of Polymerization Kinetics 508
Shusaku Asano, Ryutaro Nakayama, Ryuji Utsunomiya, Taisuke Maki, Toshiharu Kuboyama, Kazuhiro Mae
(547a) Alloy-Core@Shell Nanoparticles for Catalysis 509
Liang Zhang, Graeme Henkelman
(547b) Thermal Stability of Shape-Controlled Pd-Pt Core-Shell Nanoparticles 510
Luke T. Roling, Ahmed Elnabawy, Madeline Vara, Xue Wang, Younan Xia, Manos Mavrikakis
(547c) Accelerating Nanoparticle Synthetic Development Using in Situ Small Angle X-Ray Scattering 511
Liheng Wu, Joshua Willis, Matteo Cargnello, Christopher J. Tassone
(547d) Selective Oxidation of Methane to Methanol Using Pd/Au Bimetallic Nanoparticles 512
Cody Wrasman, Joshua Willis, Jong Suk Yoo, Jens Nørskov, Matteo Cargnello

(547e) Using Bound Organic Ligands to Direct Particle Size and Understand Catalytic Activity of Gold Nanoparticles 513
Michael M. Nigra, Nidhi Kapil, Marc-Olivier Coppens

(547h) Enhancing the Ni/YSZ Catalytic Performance in the Activation of Methane: The Influence of the Electric Field and Interface Carbon 514
Fanglin Che, Su Ha, Jean-Sabin McEwen

(547g) Metal Nanocatalysts, Their Synthesis and Size Dependent Covalent Bond Catalysis. Oxide Metal Interfaces As Active Sites for Acid-Base Catalysis 516
Gabor A. Somorjai

(555a) Automated Reaction Mechanism Generation Including Nitrogen As a Heteroatom 517
Alon Grinberg Dana, Beat Buesser, Shamel S. Merchant, William H. Green

(555b) Modeling the Combustion of Otto Fuel 518
C. Franklin Goldsmith, Eric Tierney, Richard H. West

(555c) Molecular Weight Growth Kinetics of Olefin Pyrolysis Under Low to Intermediate Temperatures 519
Kun Wang, Anthony M. Dean, Stephanie Villano

(555d) Detailed Kinetic Modeling of Gas Evolution and Energy Recovery during Chemical Quenching of Acetylene Production in Thermal Plasma 520
Yan Cheng, Tianyang Li, Yi Cheng

(555e) Predictive Modeling of Thermal Decomposition of Hydrocarbon Mixtures 521
William H. Green, Kehang Han

(555g) Nondimensional Specific Energy Index for Methane Pyrolysis in Thermal Plasma 522
Tianyang Li, Christophe Rehmet, Yan Cheng, Yi Cheng

(573a) Understanding Trends in C-H Bond Activation in Heterogeneous Catalysis 523
Allegra A. Latimer, Ambarish R. Kulkarni, Hassan Aljama, Felix Studt, Frank Abild-Pedersen, Jens K. Norskov

(573b) Novel Fe-Promoted MgAl2O4 Support Material for Control of Carbon Deposition during Reforming Reactions 524
Stavros Alexandros Theofanidis, Johannes De Jong, Vladimir Galvita, Aditya N. V. R. Dharanipragada, Hilde Poelman, Guy B. Marin

(573c) Dry Reforming of Methane over Nickel-Based Catalysts for Syngas Production 528
Bruna Rego De Vasconcelos, Doan Pham Minh, Patrick Sharrock, Ange Nzihou

(573i) Probing the Effect of Mn Dispersion on the Catalytic Performance of Mn-Na-W/SiO2 in the Oxidative Coupling of Methane 531
Christine Warwar Damouny, Naseem S. Hayek, Christine Khoury, Oz M. Gazit

(573e) Methane Steam Reforming: Using External Electric Fields to Enhance the Catalytic Performance of Ni-Based Catalysts 533
Fanglin Che, Jake T Gray, Su Ha, Jean-Sabin McEwen

(573f) Exploring Ruthenium Metal-Support Interactions for the Low-Temperature Partial Oxidation of Methane 535
Emmett Goodman, Liheng Wu, Matteo Cargnello

(573g) Methane to Methanol Conversion in Small Pore Zeolites 536
Florian Göltl, Carine Michel, Bahar Ipek, Raul F. Lobo, Phillippe Sautet, Ive Hermans

(573h) Co-Aromatization of Methane with Olefins 537
Peng He, Hua Song

(574a) Production of Terephthalate Esters Via Diels-Alder Cycloaddition-Dehydration Reactions of 1,2-Furandicarboxylate Esters and Ethylene 538
Marat Orazov, Mark E. Davis

(574c) Quantifying Lewis Acid Sites in Zeolites That Catalyze Glucose Isomerization 539
James W. Harris, Michael J. Cordon, Juan C. Vega-Vila, Fabio H. Ribeiro, Rajamani Gounder

(574d) Designing Cooperative Interactions to Tune Catalytic Activity and Selectivity for Biomass Conversion 540
Nicholas Brunelli, Nitish Deshpande, Aamena Parulkar, Frédéric Perras, Marek Pruski

(574e) Lewis Acid Zeolite Catalysts for Tandem Diels-Alder Cycloaddition and Dehydration of Dimethylfuran and Ethylene to Renewable p-Xylene 541
Hong Je Cho, Chun-Chih Chang, Jingye Yu, Raymond J. Gorte, Paul J. Dauenhauer, Wei Fan

(574f) Comparative Studies of Grafted Fe(III) and Ti(IV) Active Sites on Silaceous Supports: Crystalline Delaminated Zeolite Vs. Amorphous Silica 542
Nicolás A. Grosso Giordano, Alexander Yeh, Andrew Solovyov, Alexander Okrut, Stacey I. Zones, Alexander Katz

(574g) A Liquid Phase Spectroscopic Study of the Effect of Liquid Water on Acid Sites of NaY 543
Bingjun Xu, Nicholas Gould

(578a) Computational Study of Single-Site Molybdenum Catalysts on Amorphous Silica 544
Karl Johnson, Christopher S. Ewing, Abhishek Bagusetty, Evan Patriarca, Daniel S. Lambrecht, Götz Veser

(578b) First-Principles Kinetic Monte Carlo Simulation of the Thermal Reduction of PdO(101) ... 545
Minkyu Kim, Aravind Asthagiri
(578c) Cu Site Speciation in SSZ-13 Zeolites Resolved through UV-Visible Spectroscopy, Selective Synthesis, Time-Dependent Density Functional Theory, and Ab Initio Molecular Dynamics ... 546
Christopher Paolucci, Jonatan Albarracin, Hui Li, Florian Göltl, John R. Di Iorio, Ishant Khurana, W. Nicholas Delgass, Fabio H. Ribeiro, Rajamani Gounder, William F. Schneider
(578d) Microkinetic Modeling of the Oxygen Evolution Reaction on Oxide Surfaces ... 547
Charlotte Kirk, Colin Dickens, Karen Chan, Jens K. Norskov
(578e) Water-Gas Shift Reaction on Iron Oxide Catalysts – a DFT Study ... 548
Ozgen Yalcin, Isik Onal, Israel E. Wachs
(578f) Computational Investigation of Lewis and Brønsted Acidity in Metal-Doped Mesoporous Silicates ... 549
Amy Jystad, Alessandro Biancardi, Marco Caricato
(578g) Simulated Temperature Programmed Desorption of Acetaldehyde on CeO2(111): Evidence for the Role of Oxygen Vacancy and Hydrogen Transfer ... 550
Chuanlin Zhao, Ye Xu
(578h) Electrochemical Synthesis of Hydrogen Peroxide from Two-Electron Water Oxidation ... 551
Samira Siahrostami, Guo-Ling Li, Xinjian Shi, Xiaolin Zheng, Jens Nørskov
(617cv) Theoretical Screening of Mxenes for Catalysis Using Density Functional Theory ... 552
Kurt Fredrickson, Aleksandra Vojvodic
(600a) Nanoengineering and Application of Protected but Accessible Metal Cluster Catalysts ... 553
Alexander Katz, Alexis T. Bell, Audrey Harker, Anh T. To, Ezra Clark, Bruce C. Gates, Andrew Palermo, Louise Debefve
(600b) Electronic Band Engineering in TiO2 Particulate Layers for Photocatalysis ... 554
Qilong Huang, Edmund G. Seebauer
(600c) How Charge Screening Affects Zeolite-Catalyzed C-C Coupling Reactions ... 555
Bin Wang
(600d) Analyzing Local Composition and Structural Changes Occurring in a Nanocatalyst during Reaction ... 556
Jose L. Gomez-Ballesteros, Pin Ann Lin, Juan C. Burgos, Renu Sharma, Perla B. Balbuena
(600e) Modelling of Diffusion-Reaction Interaction inside the Co-Based Catalyst Particles for the Fischer-Tropsch Synthesis ... 557
Milos Mandic, Branislav Todic, Ljiljana Zivanic, Nikola Nikacevic, Dragomir B. Bukur
(600f) Carbon Nanotube-Supported Catalysts for Improved Fischer-Tropsch Synthesis ... 561
Xu Li, Haider Almkhelfe, Jenae Tate, Jingyi Xie, Keith L. Hohn, Placidus B. Amama
(600g) Influence of Dry-Coating of Nano Additives on Particulate Solid Flow and Its Reactions: Cement System a Case Study ... 562
Sanat Chandra Maiti, Chinmay Ghoroi
(601a) Chemical Layer Deposition of Porous Alumina Overcoats Increases Activity and Stability in Liquid Phase Catalytic Conversion of Biomass-Derived Chemicals ... 563
Florent Héroguel, Benjamin Le Monnier, Jeremy S. Luterbacher
(601b) Controlled Growth of Noble Metal Nanoparticles on Graphene Nanoplatelets Via Scalable Atmospheric Pressure Atomic Layer Deposition ... 564
J. R. Van Ommen, Fabio Grillo, Hao Bui
(601c) Computational Studies of the Dry Reforming Reaction on Al2O3-Coated Transition Metal Catalysts ... 565
Saul Perez Beltran, Perla B. Balbuena
(601d) Experimentally Probing Ligand-Strain Effect Via a Novel Catalyst Platform ... 566
Shuai Tan, Shibely Saha, Lucun Wang, Gregory S. Yablonsky, John T. Gleaves, Rebecca Fushimi, Dongmei (Katie) Li
(601e) Cobalt/Alumina Interactions in ALD Synthesized Catalysts for Fischer-Tropsch Synthesis ... 567
Jacob M. Clary, Staci A. Van Norman, John L. Falconer, Charles B. Musgrave, Alan W. Weimer
(601f) Catalytic Performance of Supported Nano-Iron Catalysts Synthesized By a Gas-Expanded Liquid Deposition Technique in Fischer-Tropsch Synthesis ... 568
Rui Xu, Pranav S. Vengsarkar, David Roe, Christopher Roberts
(601g) Photocatalysis for Z-Scheme Solar Water Splitting Using New Reactor Designs ... 569
Shane Ardo
(611a) Spatially Resolved 3D-CFD Simulation of a Packed Bed Membrane Reactor for Steam Reforming ... 570
Nico Jurtz, Thomas Eppinger, Matthias Kraume

(611b) Iron-Manganese Catalysts for the CO Hydrogenation to Alkenes: Effect of the Catalyst Synthesis on the Selectivity and Activity 571
Hafsa Karroum
(611c) Effect of Potassium Promoter in Cocu-Based Catalysts for CO and CO2 Hydrogenation 573
Jenny Voss, Lei Tang, Yizhi Xiang, Norbert Kruse
(611d) Synthesis and Development of Active Nanomaterials for Tri-Reforming 575
Erdem Sasmaz, Jochen Lauterbach
(611e) Gasoline-Selective Fischer-Tropsch Synthesis with Hierarchical ZSM-5 Coated Monolith Catalysts 576
David P. Gamliel, Chunxiang Zhu, Julia A. Valla, George M. Bollas
(611f) Comparison of Cobalt Fischer-Tropsch Catalysts on Robust Silica-Alumina Supports 578
Mahmood Rahmati, Morris D. Argyle, William C Hecker, Mark J. Lindsay, Mohammad-Saeed Safdari
(611g) Surface Stabilization and Subsequent Rupturing of Co in Co-Cu Catalysts By Formation of Adsorbed Di-, Tri-, and Tetracarbonyls 579
Greg Collinge, Norbert Kruse, Jean-Sabin McEwen
(611h) Effect of Carbon Support on Supercritical Fluid Mediated Fischer-Tropsch Synthesis with an Iron-Based Nanoscale Catalys 582
David Roe, Rui Xu, Christopher B. Roberts
(617hn) Catalytic Conversion of Carbon Dioxide to Methanol over Homogeneous and Heterogeneous Hybrid Catalyst 583
Kyung-Ho Song, Eun-Kyung Choi, Soon Kwan Jeong, Ki Tae Park, Min Hye Youn, Kwan-Young Lee, Hak Joo Kim
(617a) "Ni-MFI As a Potential Zeolitic Catalyst for Olefins Hydrogenation at Low Pressures" 584
Khaled O. Sebakhy, Gerardo Vitale, Pedro Pereira-Almao
(617cm) Predicting Surface Intermixing in Alloy Nanocatalysts from First Principles Calculations 585
Luke T. Roling, Manos Mavrikakis
(617bn) On the Influence of Mass Transport in Sulfonated Carbon Catalysts Derived from Various Carbon Sources on Liquid-Phase Esterification Reactions 586
Yukei Suzuki, Isao Ogino, Shin R. Mukai
(617ff) Catalytic Hydrogenation of 2-Ethylanthraquinone Using an in Situ Synthesized Pd Catalyst 587
Hongbao Yao, Chun Shen, Yujun Wang, Guangsheng Luo
(617fp) Copper Aluminate: A Potential Catalyst to Make Dimethyldichlorosilane in a 2-Step Process 588
Krishna Janmanchi, Aaron Coppernoll
(617bv) Centrifugal Partition Chromatograph: A Novel Continuous Multiphase Reaction System 590
Juliane Merz, Gerhard Schembecker, Jonas Krause
(617bs) Synthesizing Bromobutyl Rubber By a Microreactor System 591
Pei Xie, Kai Wang, Luo Guangsheng, Peijian Wang
(617be) Integration of CFD and Polymerization for an Industrial Scale Cis-Polybutadiene Reactor 592
Chao-Zhong Xu, Jia-Jun Wang, Xue-Ping Gu, Lian-Fang Feng
(617fj) Base-Free Knoevenagel Condensation Catalyzed By Different Heterogeneous Copper Catalysts 593
Elia Schneider, Martin Zeltner, Robert N. Grass, Wendelin J. Stark
(617ge) Reaction Route Analysis of Mo Based Shift Catalyst 597
Takashi Sasaki, Tomoko Suzuki, Masaki Takaoka
(617bz) Decrosslinking and Decomposition of Silane-Crosslinked Polyethylene in Supercritical Methanol and Water 598
Giyoung Hong, Kuan Hwan Park, Jebin Ryu, Hee Suk Woo, Won-Su Son, Youn-Woo Lee
(617fq) Regeneration of Fatty Acid-Adsorbed y-Alumina Using Supercritical Methanol 599
Hee Suk Woo, Jebin Ryu, Won-Su Son, Giyoung Hong, Youn-Woo Lee
(617o) Equilibrium Shifting of Esterification with Water-Permselective Zeolite Membrane 600
Shuhei Kikuchi, Takafumi Sato, Naotsugu Itoh
(617ft) Pd/CuO Bifunctional Catalytic Membrane for Direct Synthesis of Phenol from Benzene 601
Kento Kohara, Takagiwa Naoki, Takafumi Sato, Naotsugu Itoh
(617hg) Kinetics and Spectroscopic Studies of Methane Partial Oxidation on Ni and Rh Catalysts 602
Cun Wen, Jason Hattrick-Simpers, Jochen Lauterbach
(617ee) Developing Catalysts for the Partial Electrochemical Oxidation of Methane 603
Michael Boyd
(617dt) The Characteristics of Diesel Steam Reforming at Pressurized Environment for the Operation of Solid Oxide Fuel Cell 604
Dongyeon Kim, Jiwoo Oh, Minseok Bae, Joongmyeon Bae, Sai P. Katikaneni
(617hv) Combined Process for Hydrogen Production from Coke Oven Gas (COG) 605
Hai Woong Park, Gookhee Kim, Chang-Houn Rhee

(617ap) Glycerol Hydrogenolysis with Reduced Co2AlO4 Nanoparticles Produced By Flame Spray Pyrolysis 606
Jorma Jokiniemi, Anna Lähde, Tommi Karhunen, Luis Balam Modesto-Lopez, Ricardo José Chimenton, Jordi Llorca, Dra Mayra Garcia Alvarez, Francesc Medina
(617ai) Catalytic Selective Oxidation in Biomass Conversion 607
Yanhui Yang
(617l) Highly Dispersed Nickel within Mesochannels of SBA-15 for CO Methanation with Enhanced Activity and Excellent Thermostability 608
Miao Tao, Zhong Xin
(617cz) Fischer Tropsch Synthesis: A Study of the Selectivity to Light Hydrocarbons 609
Adolph Muleja, Yali Yao, Diane Hildebrandt, David Glasser
(617ck) Arrays of Coupled Cells Performing Chemical Computing with and without Parallelism Using Discrete Turing Patterns 610
František Muzika, Lenka Schreiberova, Igor Schreiber
(617ak) Double Bond Isomerization of 1-Butene Using Supported AlBr3 Catalysts 611
Swarom Kanitkar, Nitin Kumar, James J. Spivey
(617at) Kinetic Model on Conversion of D Fructose to 5 Hydroxymethylfurfural. 612
Tatsuya Ishikawa, Eiji Furuya, Takashi Kawakita
(617m) Oxidative Coupling of Methane over Zeolite Supported Na-Mn-W 613
Naseem S. Hayek
(617cj) The Effect of the Reaction Environment on the Surface Acidity of t-ZrO2 (101) Investigated By DFT 616
Juan Manuel Arce-Ramos, Lars C. Grabow, María-Guadalupe Cárdenas-Galindo, Brent E. Handy
(617dl) Promoter Effects of Residual K and Mn in Cobalt Supported Carbon Spheres Catalysts for Fischer Tropsch Synthesis 617
Mahluli Moyo
(617em) Enhanced Electro-Kinetics of CO2 Generation for Ethanol Oxidation Using Catalysts with a Partially Oxidized Pt and Rh Core and a SnO2 Shell 618
Guangxing Yang, Xiaowei Teng
(617hj) Development of Novel Catalysts and Reactor Configurations for Methane Dry Reforming 619
Tomás Ramirez-Reina, Harvey Arellano-Garcia
(617dm) Modification of Molybdenum Phosphide Catalyst Surface with Fischer-Tropsch and Alkali Metals for the Production of Alcohols from Syngas 620
Melis S. Duyar, Eduardo Valle, Jakob Kibsgaard, Thomas F. Jaramillo
(617gg) Formation of Rhodium Nanoclusters Fabricated from Rh(CO)2(acac) on Graphene/Cu(111) 621
Homa Khosravian
(617gy) Methanol Synthesis from CO2 in Slurry Bubble Column Reactors: A Modelling/ Experimental Approach 622
Klaus Hellgardt, Dorothy Bernasko, Fessehaye Zemichael
(617dw) Contributions of Heterogeneous Catalysis to Renewable Energy: an Electrochemical Perspective 623
Carolina Romero, Alvaro Gómez, Valentina Aristizábal, Carlos A. Cardona
(617gh) Kinetic Monte Carlo Simulation of Propylene Epoxidation with Au/TiO2/SiO2 Catalysts 624
Jingjing Ji, Zheng Lu, Yu Lei, C. Heath Turner
(617fm) Comparison of Metal-Support Interaction on Pd/a-Fe2O3 Nanorod and Pd/a-Fe2-XMnxO3 nanorod Catalysts 625
Chao Wang, Emily Freeman, Jochen Lauterbach
(617er) Titanium-Based, Hierarchical, Flow-through Catalyst for Water Electrolysis Fabricated Via a Combination of Additive Manufacturing and Anodization 626
Patricia Taboada-Serrano, Xiang Li, Costas Tsouris
(617gi) Co-Oxidation of CO and Hydrocarbons on Pd/Ceria-Zirconia/Al2O3 Three-Way Catalysts 627
Wendy Lang, Michael Harold, Yisun Cheng, Carolyn Hubbard, Manish Sharma, Paul Laing
(617ek) Understanding Gas Evolution Dynamics on Planar Photoelectrodes 630
Anna Dorfi, Daniel V. Esposito, Alan West
(617cn) Multiple H2O and NH3 Adsorbed on Cu-SSZ-13: XANES and XES Study from First Principles 631
Renqin Zhang, Hui Li, Janos Szanyi, Feng Gao, Christopher Paolucci, Trunojoyo Anggara, Hui Li, William Schneider, Jean-Sabin McEwen
(79d) Catalyst Morphology Engineering: Towards a Better Understanding of the Effects of Surface Structure and Mass Transport in Copper Electrodes for the Electrochemical CO2 Reduction Reaction 633
Stephanie Nitopi, Pongkarn Chakthranont, Jakob Kibsgaard, Christopher Hahn, Thomas F. Jaramillo

(617x) A Parametric Study of Biodiesel Production Under Ultrasounds 634
Kiran Shinde, Serge Kaliaguine
(617bi) Data Auditing: Analysis Techniques for the Compatibility of Experimental Data for Molecular-Level Kinetic Models 635
Scott R. Horton, Triveni Billa, Pratyush Agarwal, Juan Lucio-Vega, Mayuresh Sahasrabudhe, Chandra Saravanan, Michael T. Klein
(617ep) Mn@Ptn (n = 1, 2, and 3) Core-Shell Nanoparticles: Synthesis, Characterization and Electrocatalysis 636
Md. Abdul Matin, Anand Kumar
(617hu) Modeling and Simulation of the Transient Reaction Behavior in Reactive Sorption Enhanced Reforming of Methane with Nano CaO-Based Sorbents for Hydrogen Production 637
Haoliang Ping, Su-Fang Wu
(617gu) Transition Metal Modified Zirconium Phosphates for the Oxygen Evolution Reaction 638
Joel Sanchez, Ieva Narkeviciute, Mario Ramos, Jorge L. Colón, Thomas F. Jaramillo
(617cl) Methanol Oxidation on Pristine and Doped MoO3 (010): A DFT and Microkinetic Analysis 639
Tej S. Choksi, Jeffrey P. Greeley
(617bm) Simultaneous Reaction Identification and Parameter Estimation 640
Zachary Wilson, Nick Sahinidis
(617dx) Design and Synthesis of Porous Carbon-Based Catalysts for the Electrochemical Reduction of O2 Via Either 2 e- or 4 e- Route 641
Shucheng Chen, Zihua Chen, Samira Siahrostami, Taeho Roy Kim, Dennis Nordlund, Dimosthenis Sokaras, John To, Robert Sinclair, Felix Studt, Jens K. Norskov, Thomas F. Jaramillo, Zhenan Bao
(617bo) Numerical Investigation on Dispersion Phenomena in Packed Bed Reactors with Low Tube-to-Particle Diameter Ratio 642
Nico Jurtz, Gregor D, Wehinger, Thomas Eppinger, Matthias Kraume
(617da) Modeling Syngas Production for Miniaturized Gas-to-Liquids Application: Extending Operation Window By Process Improvement from Different Scales 643
Chenxi Cao, Nian Zhang, Dan Dang, Yi Cheng
(617s) Effect of ZSM-11 Crystal Size on Methanol-to-Hydrocarbon Reactions 644
Yufeng Shen, Thuy T. Le, Jeffrey D. Rimer
(617gj) Effect of Dehydration on Dmc Synthesis Via Oxidative Carbonylation over Ceria Based Catalysts 645
Jong-Ho Moon, Young Cheol Park, Hyunuk Kim
(617gk) Neural Network Predictions of Oxygen Interactions on a Dynamic Pd Surface 646
Jacob R. Boes, John R. Kitchin
(617af) Novel Catalytic Etherification Reaction of Glycerol to Short–Chain Polyglycerol 647
Maral Vahdatzaman, Sunggyu Lee
(617ef) Selective Electrocatalytic Hydrogenation of Furfural on Metal Electrodes 648
Xiaotong Chadderdon, David Chadderdon, Wenzhen Li
(617au) The Selective Production of Octene By the Metathesis of 1-Pentene 649
Heondo Jeong, Min Hye Youn, Kee Young Koo, Dong Hyun Chun
(617ej) Connected Nanoparticle Catalysts Having a Porous, Hollow Capsule Structure As Carbon-Free Electrocatalysts for Polymer Electrolyte Fuel Cells 650
Takeo Yamaguchi, Hidenori Kuroki, Takanori Tamaki
(617g) Synthesis of a Highly Active Hierarchically Ordered Micro-Meso Porous Titanosilicate Via Statistically Optimized Synthesis 651
Ayomi S. Perera, Panagiotis Trogadas, Michael M. Nigra, Kaipei Qiu, Marc-Olivier Coppens
(79b) Effects of Cations on Activity and Selectivity in Carbon Dioxide Electroreduction 652
John Lin, Christopher Hahn, Thomas F. Jaramillo
(617gc) Study of Reactivity of CO+NO Reaction on Supported Pd Catalyst 653
Yimeng Lyu, Uzoma Nwabara, Xiaoyin Chen, Johannes W. Schwank
(617bp) Study of Oxidation Reactions in Preliminary Zone of Stabilization Process of Carbon Fiber Production Line 654
Khashayar Badii, Minoo Naebe, Gelayol Golkarnarenji, Hamid Khayyam
(617cd) Understanding of Reactivity of Asphaltene in H-Donor Solvents 655
Han-Byul Park, Yong-Kul Lee
(617d) Ordered Mesoporous Metal Oxide Supports Grown Around Nanocatalysts for High Temperature Stability 656
Victor V. Miller, Joshua Willis, Matteo Cargnello
(617eo) Ionic Compounds for Carbon Dioxide Electroreduction 657
Alan Landers, Thomas R. Hellstern, Jakob Kibsgaard, Christopher Hahn, Thomas F. Jaramillo

(617aw) Catalytic Upgrading of Olefin Residues with Odd Number of Carbon Atoms 658
Min Hye Youn, Heondo Jeong, Kee Young Koo, Dong Hyun Chun
(617av) Co-Processing of Heavy Residue Oil and Biomass Using Supercritical Organic Solvents 659
Doo-Wook Kim, Anton Koriakin, Ma Fanzhong, Chang-Ha Lee
(617ax) Development of Pt/Al2O3 Catalytic System for the Production of Furanic Compounds from Biomass Derived Itaconic Acid 660
Jae-Hwan Lee, Kwan-Young Lee
(617al) Synthesis of a Bi-Functional Catalyst and Application in One-Pot Conversion of Fructose into Oxygenated C9 and C15 Hydrocarbons 661
Neha Mittal, Grace M. Nisola, Wook-Jin Chung, Chosel P. Lawagon
(617gm) Nano-Sized Tungsten Carbide Catalyst for Upgrading Extra-Heavy Oil 662
Chan Hun Kim, Young Gul Hur, Gwangsik Jeong, Kwan-Young Lee
(617es) Three-Dimensional Fluid Dynamic Analysis of a Newly Designed 100 cm2 Single Cell with Internal Flow Channels for Molten Carbonate Fuel Cells 663
Chang-Whan Lee, Seong-Cheol Chang, Hyung Chul Ham, Sung-Pil Yoon, Jonghee Han
(617db) Fischer-Tropsch Synthesis: Simulation of the Diffusion of H2, CO, and Hydrocarbons in Liquid Phase Water and Hydrocarbons 666
Xiaojun Lu, Diane Hildebrandt, David Glasser
(617gx) Resolved Particle Fixed-Bed CFD with Microkinetics for Methanol Oxidation on Silver 667
Behnam Partopour, Anthony G. Dixon
(617hc) Experimental and Theoretical Analysis of Carbon Dioxide Absorption in Aqueous Sodium Hydroxide in the Millichannels 668
Durgadevi A, S. Pushpavanam
(617bj) Performance Comparison of the 60wt% Fe2O3/Al2O3 Oxygen Carrier Under Various Reducing Atmosphere for Chemical Looping Combustion 669
Ching-Ying Huang, Yau-Pin Chyou
(617t) Dimerization of Bicyclo[2.2.1]Hepta-2,5-Diene over Zeolite Catalysts 670
Jong Ki Jeon, Kwan-Hyoung Jeong, Jong Jin Kim, Choul-Ho Lee, Jeongsik Han, Byung-Hun Jeong
(617cq) Benchmarks for Adsorption on Transition Metal Oxide Surfaces: A Comparison of DFT to Experimental Data 671
Han Chen, David F. Cox
(617gf) Gold-Supported Cerium-Doped NiOx Catalysts for Water Oxidation 672
Desmond Ng, Max García-Melchor, Michal Bajdich, Pongkarn Chakthranont, Charlotte Kirk, Aleksandra Vojvodic, Thomas F. Jaramillo
(617dc) Scale up Study of Fischer Tropsch Synthesis Reactor with Reduced CO2 emissions Using Cu-K Promoted Silica Supported Fe Catalyst 673
Satyen Kumar Das, Debasis Bhattacharyya, Eswar Prasad Dalai, Jagdev Kumar Dixit, S. S. V. Ramakumar, Biswapriya Das
(617ho) A Cost-Free Catalyst for COx-Free Hydrogen Production: Utilization of Red Mud for Ammonia Decomposition 674
Samira Fatma Kurtoglu, Alper Uzun
(617u) Crystallization of One-Dimensional Zeolites: Elucidating Nonclassical Mechanisms of Growth and the Role of Structure Direction 675
Rui Li, Jeffrey D. Rimer
(617bk) Enhancing the Value of Detailed Kinetic Models through the Development of Interrogative Software Applications 676
Triveni Billa, Mayuresh Sahasrabudhe, Chandra Saravanan, Scott R. Horton, Zhen Hou, Pratyush Agarwal, Juan Lucio-Vega, Michael T. Klein
(617fz) Tailor-Designed Oxidation Catalysts for Low Temperature Combustion Engines 677
William Epling, Melanie Hazlett, Sungbong Kang, Christos Kalamaras
(617et) Reducing Greenhouse Gasses Using Low Temperature AEM Electrochemical Devices 678
Travis J. Omasta, William A. Rigdon, Neil S. Spinner, Connor A. Lewis, William E. Mustain
(617b) Hydrogen Production By Steam Reforming of Bio-Oil Derived Acetic Acid over Ni Based Zr-SBA-15 Type Mesoporous Catalysts 679
Nurbanu Cakiryilmaz, Nuray Oktar, Huseyin Arbag, Gulsen Dogu, Timur Dogu
(617dj) Effect of Fast Cycling on the Formation of HC Intermediates and NOx Conversion Using H2, Propene or Propane As Reductants 680
Allen Wei-Lun Ting, Mengmeng Li, Michael P Harold, Vemuri Balakotaiah
(617hq) Calcium Oxide As in-Situ Catalyst for Steam Gasification of Bio-Char 681
Jun Dong, Elsa Weiss-Hortala, Ange Nzihou, Yong Chi, Mingjiang Ni, Qunxing Huang, Yuanjun Tang, Marion Ducousso

(617dd) Use of Stability Diagrams to Predict Catalyst Speciation during Fischer Tropsch Reduction and Reaction 682
Joshua Gorimbo, Xiaojun Lu, Xinying Liu, Yali Yao, David Glasser, Diane Hildebrandt
(617en) Modeling the Sensitivity of Design Parameters for Photo-Electrochemical Cells Containing a Semiconductor Absorber and Metal Photocatalysts 683
Joseph Quinn, Paul Hernley, Steven Chavez, Suljo Linic
(617ay) Enhanced Carbon Sequestration with Cascade Enzymtic Reactions 684
Shunxiang Xia, Ping Wang
(617az) Global Optimization of Initial Conditions of RAFT Polymerization Process 685
Arturo Ortiz-Arroyo, Angel Castro, Fernando Pérez
(617dn) The Role of Co-Catalysts in the Metathesis of Butenes to Propylene 686
Faisal H. Alshafei, Mohammed R. Alalouni, Munir D. Khokhar, Shaikh A. Ali, Brian S. Hanna, Maxim P. Bukhovko, Sohel K. Shaikh
(617go) Elucidation of the Mechanisms for Direct Synthesis on Pd, Aupd, and Agpt Clusters 687
Neil M. Wilson, Yung-Tin Pan, Hong Yang, David W. Flaherty
(617bq) Development of a Computationally Efficient Model of a Packed-Bed Membrane Reactor for the Oxidative Coupling of Methane 688
Alberto Penteado, Mi Jin Kim, Hamid Godini, Erik Esche, Jens-Uwe Repke, Eva Speelmanns
(617ce) Experimental and Kinetic Modeling of Lean-Rich Switching Study over a Modified Twc System 689
Mengmeng Li, Sam Malamis, Michael P Harold, William S. Epling
(617gp) Aqueous-Phase Hydrodechlorination of Trichloroethene over Pd-Supported on Swellable Organically-Modified Silica (SOMS) 692
Gokhan Celik, Hyuntae Sohn, Seval Gunduz, Saurabh Ailawar, Paul Edmiston, Umit S. Ozkan
(617eu) Understanding the Mechanisms and Key Parameters That Influence the Performance of Composite Semiconductor/Electrocatalyst Photoelectrodes 693
Steven Chavez, Paul Hernley, Joseph Quinn, Suljo Linic
(617dv) Photochemical Reactions on Plasmonic Metal Nanoparticles: Underlying Mechanism of Charge Extraction from Nanoparticles to Adsorbates 694
Calvin Boerigter, Umar Aslam, Suljo Linic
(617v) Tailoring the Physicochemical Properties of ZSM-5 with Zeolite Growth Modifiers 695
Wei Qin, Jeffrey D. Rimer
(617de) Hotspot Mitigation in an Annular Microchannel Reactor (AMR) for Hydrogen and/or Syn Gas Production 696
Holly Butcher, Benjamin Wilhite, Peter Bossard, Andy Kaldor
(617gq) Catalytic Activity and Sulfur Dependence of Co/Mo Nanoparticle Catalysts in WGS Reaction 697
Seonguk Yun, Vadim Guliants
(617hk) Optimized CeZrO2-Modification of the Metal-Foam-Coated Pd-Rh/Al2O3 Catalyst for CH4 Bi-Reforming Reaction 698
Kiseok Kim, Partho S Roy, Jinwoo Song, Ji Man Kim
(617j) Liquid-Phase Stability of ZSM-5: Role of De-Alumination and De-Silication 699
Alex Maag
(617cf) Process Intensification of Ethanol Dehydration Under Liquid Phase Conditions 700
Alex Maag
(617df) Development of Fischer-Tropsch Synthesis in Microreactor 701
Yousef Alanazi, Andrew Traverso, Justin Pommerenck, Liney Arnadottir, Alex Yokochi, Goran Jovanovic
(617gr) Au@MxOy Core-Shell Nanoparticles As Catalysts for the Oxygen Evolution Reaction 702
Alaina Strickler, Maria Escudero-Escribano, Pongkarn Chakthranont, Thomas F. Jaramillo
(617ev) Electrochemical CO2 Reduction in an Oxygen-Ion Conducting Solid Oxide Electrolyzer Cell (SOEC) 703
Seval Gunduz, Doruk Dogu, Katja E. Binkley Meyer, Dhruba Jyoti Deka, Anne Co, Umit S. Ozkan
(617hl) The Mechanism of CO2 Hydrogenation over Pd/Al2O3: A Steady State Isotopic Transient Kinetic Analysis (SSITKA)/Operando-FTIR Spectroscopy Investigation 704
Xiang Wang, Janos Szanyi
(617cr) Machine-Learning-Augmented Chemisorption Model for Metal Oxides 707
Zheng Li, Hongliang Xin
(617do) Cobalt-Copper Alloy Nanoparticle Catalysts for Higher Alcohol Synthesis from Syngas 708
Jonathan Snider, Laiza Mendes, Samuel D. Fleischman, Jakob Kibsgaard, Donato Aranda, Thomas F. Jaramillo
(617eh) A Novel Method for Synthesis of Pt/Momc Catalyst for PEM Fuel Cell Application 709
Dereje Worku, Khondker Sultana, Vishwanath Deshmane, Shamsuddin Ilias
(617ew) Lattice Strain Engineering of Tin Oxides for Energy-Efficient CO2 Electroreduction 710
Siwen Wang, Hongliang Xin

(617aq) Furfuryl Alcohol Dehydration over Metal Oxide Catalysts 711
Xiaojun Chan, Taejin Kim
(617am) Low-Cost, Potassium-Based, Catalytic Biomass Cookstove for Improved Indoor Air Quality 712
Alex D. Paulsen, Tyler Kunsa, Andrew Carpenter, Nicholas R. Schwartz, Brett Alcorn, Ted J. Amundsen, Paul E. Yelvington
(617hm) Effect of Poisoning on Low Temperature Methane Dry Reforming Pt-Ni-Mg/Ceria-Zirconia Catalysts 713
Nada Elsayed, Anthony Elwell, Babu Joseph, John N. Kuhn
(617r) Understanding Direct NOx Decomposition over Pdo/SiO2: Influence of Reaction Temperature and Pretreatment 714
Krishna Reddy Gunugunuri, Chen Ling, Torin Peck, Hongfei Jia
(617fg) Metal Distribution after Drying of Supported Catalysts with High Metal Loadings 715
Cody Bishop, Hao Chen, Xue Liu, Johannes G. Khinast, Benjamin Glasser
(617hr) Glycerol Steam Reforming for Hydrogen Production over CoxNiy-MCM-41 and CoxNiy-SBA-15 Catalysts 716
Vishwanath Deshmane, Richard Abrokwah, Sara Alsalihi, William Dade, Debasish Kuila
(617dp) Development of a Graphene Supported Iron-Oxide Nanocatalyst for Fischer Tropsch Synthesis 717
Stephen Montgomery, Rui Xu, James G. Radich
(617z) Microwave-Intensified Synthesis of Biodiesel Catalyzed with Graphene Oxide 718
Yoshifumi Sumigawa, Armando Quitain, Mitsuru Sasaki, Tetsuya Kida
(617dq) Syngas Production Via Chemical Looping 719
Amey More, Saurabh Bhavsar, Charles Hansen, Götz Veser
(617fu) Controlling Carbon Affinity with Transition Metal Ceramics and Intermetallic Alloys to Modify the Degree of Olefin and Aromatic Activation 720
Yang He, Yuanjun Song, Siris Laursen
(617ba) Tuning Catalytic Activity for Meerwein-Ponndorf-Verley (MPV) Reaction through Designing Catalytic Sites 721
Nitish Deshpande, Aamena Parulkar, Frédéric Perras, Marek Pruski, Nicholas Brunelli
(617fn) Synthesis and Characterization of Au Nanoparticles for Ethanol Oxidation: Effect of Acidity and Support Structure 722
Erfan Behravesh, Narendra Kumar, Jorma Roine, Jarno Salonen, Andrey Shchukarev, Jyri-Pekka Mikkola, Atte Aho, Kari Eränen, Dmitry Yu Murzin, Tapio Salmi
(617ex) Electrochemical Control of the Nitrogen Cycle: Ammonia Synthesis from Nitrates 725
Joshua McEnaney, Jakob Kibsgaard, Thomas F. Jaramillo
(617ct) Theoretical Study on Oxidative Coupling of Methane Using MgO 726
Hassan Aljama, Jong Suk Yoo, Ambarish R. Kulkarni, Allegra A. Latimer, Frank Abild-Pedersen, Jens K. Norskov
(617ga) A Highly Active and Stable IrOx/SrIrO3 Catalyst for the Oxygen Evolution Reaction 727
Linsey C. Seitz, Colin Dickens, Kazunori Nishio, Yasuyuki Hikita, Joseph H. Montoya, Andrew Doyle, Charlotte Kirk, Aleksandra Vojvodic, Harold Y. Hwang, Jens K. Nørskov, Thomas F. Jaramillo
(617ey) Engineering Interfaces for the Activation and Stabilization of Photovoltaic-Grade Thin Film Light Absorbers for Photoelectrochemical Hydrogen Production 728
Thomas R. Hellstern, David W. Palm, Nicolas Gaillard, Thomas F. Jaramillo
(617ez) Engineering Ultra-Small Nanochannel Electrocatalyst Towards the Improvement in Oxygen Evolution and Oxygen Reduction Reaction 729
Makoto Mitani, Pongkarn Chakthranont, Thomas F. Jaramillo
(617hs) An Integrated 'One-Box' Process for Hydrogen Production 730
Ashkan Garshasbi, Doug Parsley, Richard J. Ciora, Paul K. T. Liu, Theodore Tsotsis
(617dz) Novel Catalysts for Photocatalytic Conversion of CO2/H2O and CO2/CH4 Systems to Syngas 731
Karishma Piler, Paul Bernazzani, Cristian Bahrim, Tracy J. Benson
(617dg) Computational Modelling of the Fischer-Tropsch Synthesis within a Microreactor 732
Andrew Traverso, Justin Pommerenck, Yousef Alanazi, Liney Arnadottir, Goran Jovanovic, Alexandre Yokochi
(617ea) Hydroxyl Radical Scavenging Activity of Silver Nanoparticles: Enhancing Photocatalytic Reduction and Decreasing Photocatalytic Oxidation Rate 733
Alexzander Steiniger, Hanqing Pan, Michael D. Heagy, Sanchari Chowdhury
(617gs) Sputter Deposition of Silver-Copper Nanoparticles for the Oxygen Reduction Reaction 734
Brenna Gibbons, Melissa Wette, Bruce M. Clemens, Thomas F. Jaramillo
(617fb) Rare Earth Element Doped Titania (TiO2) for Heterogeneous Photocatalytic Applications 735
Vignesh Bhethanabotla, Debtanu Maiti, Daniel Russell, John Kuhn, Venkat R. Bhethanabotla
(617gw) Experimental and Theoretical Studies of CO2 Adsorption on Hydrotalcite 736
Jack Lowd, Vasilios Manousiouthakis, Theodore Tsotsis, Richard J. Ciora

(617fo) First-Principles Study for Controlling the Cation Segregation on the Surface of Perovskite Oxide Catalysts ... 737
Hyunguk Kwon, Wonyoung Lee, Woochul Jung, Jeong Woo Han
(617hw) Low Temperature Hydrogen Production from Kinetically Limited NH3 Decomposition in Palladium Membrane Reactor ... 738
Naotsugu Itoh, Takafumi Kato, Takafumi Sato, Misaki Takahashi
(617ha) Fast Prediction of CO Oxidation Activity on Rare-Earth Metal Doped CeO2 ... 739
Kyeounghak Kim, Jeongdo Yoo, Siwon Lee, Woochul Jung, Jeong Woo Han
(617by) Diagnostics of Field-Aged Three-Way Catalyst (TWC) on Stoichiometric Natural Gas Engines ... 740
Di Wang, Hongmei An, Jian Gong, Krishna Kamasamudram, Neal Currier, Junhui Li
(617y) The Beneficial Effect of Fructose on the Catalytic Oxidation of Glucose to Gluconic Acid ... 741
Joao G. R. Poco, Fabiana Dos Santos Lima
(617bx) Modeling of Microreactors for Chemoselective Capture of Trace Volatile Organic Compounds ... 742
Qi Li, Mingxiao Li, Yizheng Chen, Xiao-An Fu
(617fc) Trace Levels of Copper in Carbon Materials Show Significant Electrochemical CO2 Reduction Activity ... 743
Yanwei Lum, Youngkook Kwon, Peter Lobaccaro, Ezra L. Clark, Alexis T. Bell, Joel W. Ager
(617cu) Engineering the Electronic Structure of Mxenes Using Density Functional Theory: Strain and Support Effect ... 744
Liang Zhang, Kurt Fredrickson, Aleksandra Vojvodic
(617ci) Effect of Carbon Addition in the Slurry Phase Hydrocracking of Vacuum Residue ... 745
Sung-Ho Kim, Ki-Duk Kim, Yong-Kul Lee
(617dr) Dispersed Sulfide Catalysts for Hydrocracking of Vacuum Residue ... 746
Ki-Duk Kim, Sung-Ho Kim, Yong-Kul Lee
(617ds) Oxidative-Extractive Desulfurization of Model Diesel Oil By CTA-Type Dawson Type Emulsion Catalyst ... 747
Farhad Banisharif, Mohammad Reza Dehghani, Matin Parvari, Jose M. Campos-Martin
(617hf) Conversion of Carbon Dioxide to Syngas Using Transition Metal Dichalcogenide Catalysts ... 754
Mohammad Asadi, Pedram Abbasi, Amin Salehi-Khojin
(617p) On the Stability of Cu-Chabazite SCR Catalysts ... 755
Gustavo A. Fuentes, M. Eugenia Hernández-Terán, Julio C. López-Curiel, Gabriela Hernández-Salgado, Angel Martinez-Hernandez, Sergio A. Gómez
(617eb) Visible Light Photoactivity of Ion Implanted Titania Nanotubes ... 756
Mohammad Mehedi Hasan, Shammi Akter Ferdousi, King Lun Yeung
(617ed) Enhancement of Photochemical Activity of One Dimensional TiO2 By Targeted Decoration of Transition Metals ... 757
Mohammad Mehedi Hasan, King Lun Yeung, Shammi Akter Ferdousi
(617cc) New Materials Technology for Sustainable Removal of Moisture from Air ... 758
Shammi Akter Ferdousi, Mohammad Mehedi Hasan, King Lun Yeung
(617hp) Towards Industry-Scale Measurements of Reactor Gas-Liquid Mass-Transfer Phenomena ... 759
Manizheh Ansari, Damon Turney, Roman Yakobov, Dinesh V. Kalaga, Sanjoy Banerjee
(617bc) Kinetic Study for Hydrogenation of Toluene to Methylcyclohexane Over Titanium Substituted SBA-15 Supported Pt-Pd Catalysts ... 760
Shyamal Roy, Dilip Kumar Mondal, Chandona Mondal
(626a) Aqueous Phase Catalytic Conversion of Methane to Methanol on Bimetallic Catalysts Under Mild Conditions ... 775
Mengwei Li, Junjun Shan, Maria Flytzani-Stephanopoulos
(626b) Gas-Phase Catalytic Oxidation of Methane into Methanol over Copper-Exchanged Zeolites with Molecular Oxygen ... 776
Karthik Narsimhan, Kimberly Dinh, Yuriy Román-Leshkov
(626c) Bi-Reforming of Methane at Low Temperatures over Pt or Pd (Ni-Mg/ceria-zirconia) Catalysts ... 777
Nada Elsayed, Babu Joseph, John Kuhn
(626d) Understanding the Role of Fe in Bi-Metallic Ni-Fe Dry Reforming Catalysts: A Combined in-Situ XAS-XRD Study ... 778
Sung Min Kim, Paula Abdala, Wouter Van Beek, Tigran Margossian, Christophe Copéret, Christoph Mueller, Qasim Imtiaz
(626e) Selective Deposition of Ni on Mo2c That Has Been Supported on Al2O3: Enhanced Stability for Dry Reforming of Methane ... 779
Jayson Keels, Khalid Askar, John R. Monnier, John R. Regalbuto

(626f) Optimization of Metal Dopants in Pyrochlore Catalysts for the Dry Reforming of Methane: Combining Computational and Experimental Methods 780
David A. Bruce, Felipe Polo-Garzon
(626g) The Investigation of the Catalytic Partial Oxidation of Methane in a Shock Tube 781
Robyn Smith, Marco J. Castaldi, Kenneth Brezinsky
(626h) Single-Atom Active Sites for Natural Gas to Liquid Fuel Conversion 783
Thalia Quinn, Joseph Shackelford, Sierra Headrick, Pabitra Choudhury
(627a) Strategies for the Encapsulation and Stabilization of Monodisperse Au Clusters within Zeolites 784
Trenton Otto, Stacey I. Zones, Enrique Iglesia
(627c) Measuring & Understanding the Atomic-Scale Environments and Interactions of Heteroatom Sites in Zeolite Catalysts 787
Zachariah Berkson, Rahul P. Sangodkar, Subramanian Prasad, Ulrike Werner-Zwanziger, Bradley F. Chmelka
(627d) A Method to Screen Nanoporous Catalysts for Transition-State Selectivity 788
Peng Bai, Matthew Neurock
(627e) Synthetic Control of Framework Aluminum Distribution in Zeolites and Consequences for Acid and Redox Catalysis 789
John R. Di Iorio, Young Gul Hur, Rajamani Gounder
(627f) Design and Synthesis of IFR-Type Zeolites with Controlled Locations of Framework Aluminum with the Aid of Theoretical Calculation 790
Koki Muraoka, Watcharop Chaikittisilp, Tatsuya Okubo
(627g) A Mixed-Solvent Approach to Delaminate Layered Zeolitic Precursors 791
Isao Ogino, Taichi Shirobe, Shin R. Mukai
(629a) Engineering Highly Active Brookite Titania Nanorods for Sustainable Hydrogen Production 792
Matteo Cargnello, Tiziano Montini, Sergey Smolin, Jacqueline Priebe, Juan J. Delgado Jaén, Vicky Doan-Nguyen, Ian McKay, Jay Schwalbe, Marga-Martina Pohl, Thomas Gordon, Jason B. Baxter, Angelika Brückner, Paolo Fornasiero, Christopher B. Murray
(629b) Hydrogen Production Via Photocatalytic Water Splitting over Mesoporous Pt/TiO2 793
Fabricio Guayaquil Sosa, Benito Serrano, Hugo De Lasa
(629c) How Ceria Redox Materials Split H2o and CO2 with Concentrated Solar Energy into H2, CO, and O2 794
Ronald Michalsky, Aldo Steinfeld
(629d) Methane Conversion to Syngas Using Dopant Modified Metal Oxide Composites in Chemical Looping Reforming 795
Zhuo Cheng, Lang Qin, Mengqing Guo, Mingyuan Xu, Jonathan A. Fan, Liang-Shih Fan
(629e) A New Thin Layered Structural Coating on a Metal Substrate for Enhanced Hydrogen Production from Steam Methane Reforming 796
Michael Lugo, Marco Castaldi, Dean Modroukas, Andrew W. Davis, Rajinder Gill, Florent Minette, Juray De Wilde
(629f) Promotion Mechanisms of Iron Oxide-Based High Temperature-Water Gas Shift (HT-WGS) Catalysts By Chromium and Copper 797
Tulio Rocha, Minghui Zhu, Axel Knop-Gericke, Israel Wachs, Robert Schlogl
(629g) Catalytic Ammonia Decomposition to Hydrogen and Nitrogen on Supported Nanosized and Single-Site Cobalt Catalysts: Effects of Atomicity, Ligand Environment and Support 798
Konstantin V. Khivantsev, Fahad Almalki, Miao Yu
(641a) Selective Activation of Primary C-H Bonds of Propane on Pdo(101) 799
Jason F. Weaver, Aravind Asthagiri, Feng Zheng, Li Pan, Juhee Choi, Vikram Mehar, John Diulus
(641b) Methane Activation on Subsurface Boron Doped Copper: First Principles Investigation of the Structure, Activity and Electronic Properties of the Catalyst 800
Quang Thang Trinh, Samir H. Mushrif
(641c) Combined DFT and Kinetic Monte Carlo Studies of Formic Acid Decomposition on Au18 Nanoclusters 801
Benjamin W J Chen, Sha Li, Michail Stamatakis, Manos Mavrikakis
(641d) Mechanistic Study of C-O Bond Rupture on Ruthenium Phosphide Surface: Transient and Steady State Measurements 802
Siwei A. Chang, David W. Flaherty
(641e) Active Site Determination for CO Oxidation on Al2O3 Supported Pt Nanoparticle Catalysts By in-Situ Quantitative FTIR Measurements 803
Matthew Kale, Phillip Christopher
(641f) Competing Forces in Chiral Catalytic Surface Chemistry: Enantiospecificity Versus Enantiomer Disproportionation 804
Andrew J. Gellman, Yongju Yun

(641g) On the Reaction Pathways and Intermediates of Selective Ringopening of Furanics By Iridium 805
Glen Jenness, Weiming Wan, Jingguang G. Chen, Dionisios G. Vlachos

(654a) Controllable Solution Combustion Synthesis of Nanoscale a/B-Bi2O3 and Its Catalytic Application 806
Xiaohong Wang, Yang Xiao, Arvind Varma

(654b) A Combined Theoretical and Experimental Investigation on the Degradation of Organophosphorus Chemical Warfare Agents on ZnO1-X 807
John Landers, Nicole Zougheib, Fuat E. Celik, Alexander V. Neimark

(654c) Magnetic Superbasic Nano Catalysts for Rapid Catalyst Removal and Direct Product Isolation 808
Elia Schneider, Corinne Hofer, Martin Zeltner, Robert N. Grass, Wendelin J. Stark

(617ac) Selective Hydrogenolysis of Furfuryl Alcohol to 2-Methylfuran over Reduced Co-Fe-Al Mixed Metal Oxides 816
Taylor Sulmonetti, Bo Hu, Pradeep K. Agrawal, Christopher W. Jones

(617fi) Supported, Homogeneously Alloyed Bimetallic Nanoparticles By Electrostatic Adsorption 817
Andrew Wong, Qiuli Liu, John R. Regalbuto

(654f) Chemoselective Transfer Hydrogenation of Nitroarenes Catalyzed By Highly Dispersed, Supported Nickel Nanoparticles 818
Zeyu Shang, Chengjun Jiang, Xinhua Liang

(654g) Controllable Synthesis of Highly Dispersed Cu/SiO2 Catalysts in Supercritical CO2 and Their Application in the Hydrogenation of Dimethyl Oxalate to Ethylene Glycol 819
Qin-Qin Xu, Guo-Yue Qiao, Jian-Zhong Yin

(669a) Single-Site Iridium Complexes on MgO Powder: Spectroscopic, Microscopic, and Catalytic Evidence of Dependence of Properties on MgO Surface Sites 828
Adam Hoffman, Louise Debefve, Shengjie Zhang, David A. Dixon, Bruce C. Gates

(669b) Design of Palladium-Gold Catalysts for Liquid Phase Selective Alkyne Hydrogenation with Single Atom Alloy Strategy 829
Jilei Liu, Junjun Shan, Maria Flytzani-Stephanopoulos

(669c) Selective Dehydrogenation of Ethanol to Acetaldehyde and Hydrogen on a Highly Diluted Nicu Alloy 830
Junjun Shan, Nare Janvelyan, Hang Li, Jilei Liu, Tobias Egle, Jianchao Ye, Monika Biener, Juergen Biener, Cynthia M. Friend, Maria Flytzani-Stephanopoulos

(669d) Heterobimetallic Complexes As Controlled Precursors for Atomically Dispersed, Promoted Methane Oxidation Catalysts 831
Joseph Maalouf, Joshua Willis, Matteo Cargnello, Eric A. Stach, Jay Schwalbe

(669e) Site-Specific IR Spectroscopy for Analyzing Pt Structure Sensitivity from Single Atoms to Fully Coordinated Surfaces 832
Leo Derita, Phillip Christopher

(669f) Atomically Dispersed Pt Catalysts on Polyphosphomolybdic Acid: Structural Features and Catalytic Properties 833
Ning Yan

(669g) Understanding Site Isolation of Pd and Ni in Zn-Based Bulk Intermetallics during the Selective Hydrogenation of Alkynes 834
Anish Dasgupta, Robert M. Rioux

(674a) Mechanism of Oxygen Activation during Selective Catalytic Reduction of NO with NH3 in Cu-SSZ-13 Zeolites 835
Christopher Paolucci, Atish A. Parekh, Ishant Khurana, Hui Li, John R. Di Iorio, W. Nicholas Delgass, Jeffrey Miller, Fabio H. Ribeiro, William F. Schneider, Rajamani Gounder

(674b) Light Hydrocarbon Reactions on Solid Bronsted Acids: Confinement and Acid Strength Effects on Ion-Pair Transition States and Their Precursors 836
Michele L. Sarazen, Enrique Iglesia

(674c) Development of a Practical Catalyst for Terminal Olefin Epoxidation Using Organic Hydroperoxide As Oxidant 837
Alexander Okrut, Nicolás A. Grosso Giordano, Xiaoying Ouyang, Stacey I. Zones, Alexander Katz

(674d) Mesoporous Nb-Silicates As Ethylene Epoxidation Catalysts: Novel Strategies to Passivate Bronsted Acidity 838
Swarup Maiti, Anand Ramanathan, Bala Subramaniam

(674e) Investigation of the Acid Catalyzed Formaldehyde-Olefin Condensation Reaction 839
Efterpi Vasiliadou, Taha Salavati-Fard, Douglas J. Doren, Dion Vlachos, Raul F. Lobo

(674f) Experimental and Computational Interrogation of Fast SCR Mechanism and Active Sites on H-Form SSZ-13 840
Sichi Li, Yang Zheng, Feng Gao, Janos Szanyi, William Schneider

(674g) Quantification of External and Pore Mouth Acid Sites in Unit-Cell Thick Pillared MFI and Pillared MWW Zeolites 841
Dongxia Liu, Yiqing Wu
(676a) Green Application of Hydrogenation Catalyst Yn-1 on C4 Cut 842
Zhou Du
(676b) "Molybdenum Carbide As a Potential Catalyst for Olefins Hydrogenation" 843
Khaled O. Sebakhy, Gerardo Vitale, Pedro Pereira-Almao
(676c) Olefin Oligomerization on Carbon Supported Cobalt Catalyst: Identification of the Reaction Pathway 844
Zhuoran Xu, Joseph P. Chada, Dongting Zhao, Carlos Carrero, Jessica Rogers, Devon C. Rosenfeld, Ive Hermans, George W. Huber
(676e) Oxidative Dehydrogenation of Propane to Propylene over VOx/CaO-yAl2O3 Catalysts 845
Mohammad Mozahar Hossain, Afees Ayandiran, Idris Bakare, Shaikh Razzak, Housam Binous
(676f) On-Purpose Propylene Production from Light Hydrocarbons with MFI-Zeolite/Metal-Oxide Composite Catalysts 846
Shinya Hodoshima
(676g) Pt-Promoted Supported Gallium Catalyst for Dow Fluidized Catalytic Dehydrogenation Process 847
Lin Luo, Brien Stears, Andre Malek, David J. Barton, Beata A. Kilos, Joo Kang
(676h) Dehydroaromatization of Ethane over Zinc Containing Zeolites 848
Ali Mehdad, Raul F. Lobo
(676i) Direct Synthesis of Propene from Ethene Feedstock: Investigation of Catalytic Concepts 849
Matthias Felischak, Tanya Wolff, Leo Alvarado Perea, Andreas Seidel-Morgenstern, Christof Hamel
(676j) Olefin Metathesis By Supported WOx/SiO2 Catalysts 850
Soe Lwin, Israel E. Wachs
(688a) Spectroscopic Investigation of Pd Structures Formed over CeO2 and MnOx-CeO2 Supports 851
Erdem Sasmaz, Chao Wang, Jochen Lauterbach
(688b) Tailoring the Catalytic Activity of Perovskites for Oxygen Reduction and Evolution 852
Liang Zhang, Aleksandra Vojvodic
(688c) Origin of the Support Effect for Methanol Oxidation for Supported Vanadium Oxide Catalysts 853
Taejin Kim, Benjamin Moskowitz, Israel Wachs
(688d) Oxygen Transmission in Solid Phase——MnO2 promoted Phase-Pure M1 Movnbte Oxide for Ethane Oxidative Dehydrogenation 854
Xin Chen, Hang An, Bozhao Chu, Yi Cheng
(688e) On the Mechanistic Aspects of Mg6MnO8-Based Redox Catalysts for Oxidative Dehydrogenation of Ethane Via a Chemical Looping Scheme 855
Luke Neal, Seif Yusuf, Fanxing Li
(688f) Experimental and Theoretical Investigations of Propene Ammoxidation to Acrylonitrile over Bismuth Molybdate 856
Alexis T. Bell, Rachel B. Licht
(688g) Dioxygen Activation Pathways in Selective Oxidations Catalyzed By Metal Oxides 857
Stephanie Kwon, Prashant Deshlahra, Enrique Iglesia
(688h) Tuning the Dehydration Activity of Alcohols Using Phosphonic Acid Self-Assembled Monolayers 858
J. Will Medlin, Lucas Ellis, Daniel Schwartz
(688i) Single-Facet-Dominated Titania, and Zirconia As Catalysts or Supports for Bio-Alcohol Dehydration and Oxidative Dehydrogenation 859
Yuan Chen, Lu Zhang, Libor Kovarik, Huamin Wang, Feng Gao, Yong Wang, Charles H. F. Peden
(689a) Mechanistic Insights into Carbon-Oxygen Bond-Breaking on Mo2c Catalysts: Ethanol Dehydration 860
Carrie A. Farberow, Connor Nash, Joshua A. Schaidle
(689b) Elucidating Particle Size Effects on the Hydrodeoxygenation (HDO) of Phenolics By Micro-Pulse Experiments and DFT Calculations 861
Qiaohua Tan, Nhung Duong, Daniel Resasco
(689c) Adsorption of Hetero-Bifunctional Urea on Ge(100)-2x1 Surface 862
Tania E. Sandoval, Allison M. Crow, Stacey F. Bent
(689d) Suppressing Pd Catalyst Deactivation in Alcohol Oxidation Using Thiolate Self-Assembled Monolayer Modification 863
Pengxiao Hao, Daniel K. Schwartz, J. Will Medlin
(689e) Facile Formation of Methyl Acrylates By Cross Coupling of Unsaturated Alcohols-Aldehydes with Methanol on Au(110) 864
Stavros Karakalos, Fanny Hiebel, Cynthia M. Friend, Robert J. Madix

(689f) Reactivity and Selectivity Descriptors for the Activation of C-H Bonds in Hydrocarbons and Oxygenates on Metal Oxides ... 865
Prashant Deshlahra, Enrique Iglesia
(689g) Selective Oxidation of n-Butane to 1-Butanol over Transition Metal Catalysts Encapsulated By Metal-Organic Frameworks ... 866
Jiazhou Zhu, Rachel Getman
(719a) Highly Selective Fischer-Tropsch Synthesis for Production of JP-8 Jet Fuel from Coal and Coal/Biomass Mixtures ... 867
Andrew Lucero, Brittany Koob, Brandon Cline, Kevin McCabe, Santosh Gangwal
(719b) Composition of Bio-Crude Oils from Hydrothermal Liquefaction of Algae in Batch Reactor ... 868
Zheng Cui, Feng Cheng, Travis Le-Doux, Kwonit Mallick, Carolina Herrera, Graham Hoffman, Jacqueline Jarvis, Neil Paz, Tanner Schaub, Nagamany Nirmalakhandan, Catherine E. Brewer
(719c) De-Alloyed Platinum-Bismuth Nanoparticles As Highly Active Electrocatalysts for Dimethyl Ether Oxidation and Oxygen Reduction ... 869
Anastasios Angelopoulos, Punit Boolchand, Zhipeng Nan
(719d) Co-Production of Methanol and Dimethyl Ether Using a Modified Copper-Based Catalyst System ... 870
Lujie Ye
(719e) Design of a Sabatier Reactor for CO2 Conversion into Synthetic Methane ... 873
Duo Sun, David Simakov
(719g) Steam Reforming of Dimethyl Ether over a Novel Plate-Type y-Al2O3/Al Monolith Supported Cu-Based Bi-Functional Catalyst ... 874
Feiyue Fan, Qi Zhang, Zibin Zhu
(720a) Low Temperature CO Oxidation with a Single Site Pt Catalyst Supported on the "29" Cuxo/Cu(111) Surface: Elucidating the Chemical Nature of the Pt Active Sites ... 875
Alyssa Hensley, Andrew Therrien, Renqin Zhang, E. Charles H. Sykes, Jean-Sabin McEwen
(720b) Identifying the Active Site for the Water-Gas Shift Reaction over TiO2 Supported Pt Catalysts: Single Pt Atoms Versus Pt Clusters ... 876
Salai C. Ammal, Andreas Heyden
(720c) Trends in Single Transition Metal Atom/Oxide Support Interaction from Density Functional Theory ... 877
Nolan O'Connor, Thomas P. Senftle, Michael J. Janik
(720d) CO Oxidation Kinetics on Supported Ir Single Atoms ... 878
Yubing Lu, Xianfeng Ma, Hongliang Xin, Ayman M. Karim
(720e) Copper Atoms Stabilized on Reduced Titanium Oxides Catalyze Carbon Monoxide Oxidation ... 879
Guoqiang Cao, Justin T. Cournoyer, Nan Yi
(720f) the Role of Au in Au-Catalyzed, Solvent-Free, Selective Epoxidation of Cyclooctene with Molecular Oxygen ... 880
Linping Qian, Zhen Wang, Evgeny Beletskiy, Jingyue Liu, Haroldo Santos, Mayfair Kung, Harold H. Kung
(720g) Structure, Reactivity and Selective Tuning of Well-Defined Rh Complexes in HY Zeolites ... 881
Konstantin V. Khivantsev, Artem D. Vityuk, Hristiyan A. Aleksandrov, Georgi N. Vayssilov, Oleg S. Alexeev, Michael D. Amiridis
(734a) Structure-Property Relationships of Palladium Catalyzed Methane Complete Combustion Using Uniform Nanoparticles ... 884
Joshua Willis, Emmett Goodman, Matteo Cargnello
(734b) Characterizing Active Sites on Chirally-Modified Supported Noble Metal Catalysts for Asymmetric Hydrogenation through Selective Poisoning ... 885
Laura Rodriguez-Garcia, Fabian Meemken, Konrad Hungerbuehler, Alfons Baiker
(734c) Carbon Decoration in Supported Palladium Catalysts: Discrepancy in STEM-Chemisorption Particle Sizes ... 886
Ritubarna Banerjee, John R. Regalbuto
(734d) 100-Fold Rate Enhancement Caused By Dioxygen Ligands on a Supported, Single-Site Tetrairidium Cluster Catalyst for Ethylene Hydrogenation ... 887
Andrew Palermo, Alexander Okrut, Andrew Solovyov, J. D. Kistler, Louise Debefve, Ron C. Runnebaum, Shengjie Zhang, Igor Busygin, Daniel Ertler, David A. Dixon, Bruce C. Gates, Alexander Katz
(734e) Structure–Function Relationships for Graphene-Supported Pt Nanoclusters ... 888
Hongbo Shi, Scott M. Auerbach, Ashwin Ramasubramaniam
(734f) Role of the Support and Reaction Conditions on the Vapor-Phase Deoxygenation of M-Cresol over Pt/C and Pt/TiO2 Catalysts ... 889
Michael B. Griffin, Glen Allen Ferguson, Daniel A. Ruddy, Mary Biddy, Gregg T. Beckham, Joshua Schaidle

(734g) Understanding Solvent Effects over Supported Metal Catalysts: Ketone Hydrogenation over Ru/SiO2 891
Omar A. Abdelrahman, Jesse Q. Bond
(734i) Hammett Study: Combined Experimental-Computational Approach to Understanding Electronic Metal-Support Interaction 892
Gaurav Kumar, Bert D Chandler, Michael J. Janik
(735i) Quantification of Bronsted and Lewis Acid Sites on Zeolites in the Presence of Solvents 893
Bingjun Xu, Nicholas Gould
(735b) Synthesis of Itaconic Acid Ester Analogues via Self-aldol Condensation of Ethyl Pyruvate Catalyzed by Hafnium BEA Zeolites 894
Yuran Wang, Jennifer Lewis, Yuriy Román-Leshkov
(735c) Probing Reactions at the Solid-Liquid Interface By Operando Solid-State NMR Spectroscopy 895
Susannah L. Scott, Long Qi, Ali Chamas, David W. Hoyt, Eric D. Walter, Nancy M. Washton, Marcus Foston
(735d) Direct Observation of Active Sites of Ru Catalyst in Hydrodeoxygenation of Furfural 896
Weiqing Zheng, Konstantinos Goulas, Ayman M. Karim, Prashant Kumar, Michael Tsapatsis, Dionisios G. Vlachos
(735e) Single Center Reactivity of Cu(II) Sites in Copper Substituted CHA Zeolite Followed By in-Situ Electron Paramagnetic Resonance Spectroscopy 897
Susanne Mossin, Oliver Isaksen, Anita Godiksen, Søren B. Rasmussen, Peter N. R. Vennestrøm
(735f) In-Situ Characterization of Intermetallic Ni5Ga3 Catalyst for CO2 Reduction to Methanol 898
Alessandro Gallo, Dimosthenis Sokaras, Dennis Nordlund, Hirohito Ogasawara, Libor Kovarik, Jonathan Snider, Thomas F. Jaramillo
(735g) Growth of Encapsulating Carbon on Supported Pt Nanoparticles Studied By in Situ TEM 899
Jason Wu, Stig Helveg, Sven Ullmann, Zhenmeng Peng, Alexis T. Bell
(745a) Enhancing Oxygen Reduction Activity on Pt Monolayer Electrocatalysts through Selective Tuning of Ligand and Lattice Effects 900
Timothy Van Cleve, Saman Moniri, Hongliang Xin, Suljo Linic
(745b) Engineering the Active Site for Oxygen Electroreduction in Fuel Cells 901
Maria Escudero-Escribano, Ifan E. L. Stephens, Ib Chorkendorff
(745c) Engineering Core-Shell Nanoparticles of Noble Metals on Transition Metal Carbides As Next-Generation Catalysts in Energy Applications 902
Maria Milina, Sean T. Hunt, Christopher H. Hendon, Yuriy Román-Leshkov
(745d) Optimal Pattern of Bimetallic Electrocatalysts for Efficient Conversion of Water and Carbon Dioxide to Hydrocarbons and Oxygenates 903
Meenesh Singh, Yanwei Lum, Ezra L. Clark, Joel W. Ager, Alexis T. Bell
(762j) Supported Cobalt Catalysts with Preferentially Exposed Crystal Facets for CO2 Hydrogenation 904
Cun Wen, Juan Jimenez, Jason Hattrick-Simpers, Jochen Lauterbach, Maryorie C. Santos
(745f) Tunable Thermodynamic Activity of LaxSr1-XMnyAl1-YO3-δ ($0 \leq x \leq 1$, $0 \leq y \leq 1$) Perovskites for Solar-Driven Splitting of CO2 and H2O 905
Miriam Ezbiri, Michael Takacs, David Theiler, Ronald Michalsky, Aldo Steinfeld
(745g) Material Prediction for Carbon Dioxide Conversion to Carbon Monoxide Using Reverse Water Gas Shift Chemical Looping (RWGS-CL) Process 906
Debtanu Maiti, Yolanda Daza, Adela E. Ramos, Bryan J. Hare, John N. Kuhn, Venkat R. Bhethanabotla
(762a) The Influence of Water on Metal/Oxide Perimeter Sites for the Selective Conversion of Phenolic and Furanic Species over Ru/TiO2 907
Nicholas M. Briggs, Taiwo Omotoso, Tyler Vann, Valeria Herrera, Steven Crossley
(762b) Olefin Oligomerization over Cobalt Catalysts: Effect of Catalyst Support 908
Dongting Zhao, Zhuoran Xu, Joseph P. Chada, Carlos Carrero, George W. Huber, Ive Hermans, Jessica Rogers, Devon C. Rosenfeld
(762c) Interfacial Perimeter Sites of Au-TiOx in Isopropanol Decomposition 909
Yi Y. (Chloe) Wu, Harold H. Kung
(762d) Theoretical Investigation of Supported Ultra-Thin Bilayer, Double Bilayer and Trilayer Metal-Oxides: Structure, Stabilty and Catalytic Activity 910
Michal Bajdich, Max García-Melchor, Jakob Fester, Phillip Plessow, Frank Abild-Pedersen, Jeppe Lauritsen, Aleksandra Vojvodic
(762f) Controlling Selectivity of Metal Catalysts By Using Self-Assembled Monolayers on the Oxide Support 911
Jing Zhang, Patrick Coan, J. Will Medlin
(762g) In-Situ Grown Metal Nanocatalysts from Oxide Host Lattices 912
Tae-Sik Oh

(762h) Designing Catalysts for Enhanced Pairwise Selectivity of Parahydrogen Addition to Unsaturated Hydrocarbons 913
Evan W. Zhao, Helena Hagelin-Weaver, Sean McCrea, Haibin Zheng, Clifford R. Bowers, Yan Xin
(762i) TiO2 Promoted Gold-Based Catalysts for Propylene Epoxidation 914
Zheng Lu, Zili Wu, C. Heath Turner, Yu Lei
(763a) Adsorbate-Mediated Strong Metal-Support Interactions in Oxide Supported Rh Catalysts 915
Phillip Christopher, John C. Matsubu
(763b) Understanding Active Sites in Ethanol Synthesis on PtCo Catalysts Prepared Using Atomic Layer Deposition 916
Joseph Singh, Stacey F. Bent, Nuoya Yang
(763c) The Role of the Support for Pt Catalysts during the Water-Gas Shift Reaction 917
Viktor Cybulskis, Yanran Cui, Mayank Shekhar, Juan Lovón Quintana, W. Nicholas Delgass, Fabio H. Ribeiro
(763d) Operando Attenuated Total Reflection-Infrared Spectroscopic Detection of Catalytic Solid-Liquid Interfaces: Design Aspects of Spectroscopic Cells 918
Fabian Meemken, Laura Rodriguez-Garcia, Konrad Hungerbuehler, Alfons Baiker
(763e) In Situ and Operando Drifts Characterization of Catalytic Sites on Metal-Doped KIT-6 during Alcohol Conversion Reactions 919
Juan J. Bravo-Suarez, Thomas Ofosu, Maria Ramirez, Marco Caricato, Alessandro Biancardi, Amy Jystad
(763f) Operando spectroscopy during Ethylene Polymerization By Supported CrOx/SiO2: Role of Promoters 920
Anisha Chakrabarti, Israel E. Wachs
(763g) Evidence for CO Insertion in Alkane and Alkenes Production through Fischer-Tropsch Reaction 921
Motahare Athariboroujeny, Greg Collinge, Jean-Sabin McEwen, Norbert Kruse
(768a) Scaled Degree of Rate Control: Identifying Elementary Steps That Control Differences in Performance of Transition Metal Catalysts 922
Talin Avanesian, Phillip Christopher
(768b) Effective Coordination Based Descriptors for Rational Design of Metal Nanocatalysts 923
Xianfeng Ma, Siwen Wang, Hongliang Xin
(768c) Tuning Surface Carbon Affinity and Olefin Activation in Intermetallic Alloys for the Semi-Hydrogenation of Acetylene to Ethylene 924
Yuanjun Song, Siris Laursen
(768d) Scaling Relationships at Bifunctional Metal/Oxide Interfaces: A Case Study of Doped Au/MgO 925
Tej S. Choksi, Paulami Majumdar, Jeffrey P. Greeley
(768e) Rational Design of Nanostructured Metallic Surfaces Via Mathematical Optimization 926
Christopher L. Hanselman, Chrysanthos E. Gounaris
(768f) Screening of Active Site Motifs for C-H Bond Activation in Nanoporous Catalysts 927
Ambarish R. Kulkarni, Allegra A. Latimer, Felix Studt, Jens K. Norskov
(768g) Design of Functionalized Metal Organic Frameworks for CO2 Hydrogenation:the Effects of MOF Topology and Functional Group 928
Jingyun Ye, Benjamin Yeh, Ronald A Reynolds, J. Karl Johnson
(769a) The Nature of Heterogeneous Palladium Catalyzed Cross-Coupling Reactions on Novel Carbon Support Systems 929
Frank Gupton
(769h) Electrostatic Adsorption of Pt onto Carbon Nanotubes and Fibers 930
Ritubarna Banerjee, Jose Contreras-Mora, John R. Regalbuto
(769c) Impact of Mixed Metal Oxides Supports on the Durability and Low-Temperature Performance of Pd-Based Diesel Oxidation Catalysts 931
Eleni A. Kyriakidou, Todd J. Toops, Michael J. Lance, Andrew J. Binder, James E. Parks
(769e) Tuning Acidity of Mesoporous Silicates with Earth-Abundant Metals and Beneficial Effects during Catalytic Dehydration of Linear Alcohols 932
Hongda Zhu, Anand Ramanathan, Bala Subramaniam
(769g) Designed Synthesis of Co3O4 Total Oxidation Catalysts 933
Kathleen Mingle, Cun Wen, Erdem Sasmaz, Jason Hattrick-Simpers, Jochen Lauterbach
(779a) Ring-Opening of Furfuryl Alcohol Towards 1,5-Pentanediol over Reduced Cu-Co-Al Mixed Metal Oxides 934
Taylor Sulmonetti, Bo Hu, Pradeep K. Agrawal, Christopher W. Jones
(779b) Mechanistic Studies on the Direct Synthesis of H2O2 on Pd and Aupd Clusters 935
Neil M. Wilson, David W. Flaherty
(779c) Influence of Sn and Bi on Pt-Catalyzed Aqueous Phase Oxidation of 1,6-Hexanediol 936
Jiahan Xie, Derek D. Falcone, Kehua Yin, Benjamin Huang, Hien Pham, Raymond Unocic, Abhaya Datye, Robert J. Davis

(779d) CeO2 Supported Metal Oxide Catalysts for NOx Removal 937
Nusnin Akter, Jorge A. Boscoboinik, Taejin Kim
(779e) Mechanistic Insights on Catalytic Ethanol Steam Reforming 938
Stephen Crowley, Marco J. Castaldi
(779f) Effect of Reduced Adsorption Strength with Bimetallic Platinum Overlayer Catalysts 939
Chen Zhang, Joseph Holles
(779h) Tuning Nanoparticle Alloys to Enhance C-H Bond Activation for the Catalytic Dehydrogenation of Ethane 940
Viktor Cybulskis, James Gallagher, Han-Ting Tseng, Brandon Bukowski, Zhenwei Wu, Evan Wegener, A. Jeremy Kropf, Bruce Ravel, Jeffrey P. Greeley, Fabio Ribeiro, Jeffrey Miller
(779i) First Principles Light Alkane Dehydrogenation on Pt:Main Group Alloys and the Effect of Hydrogen Spectators 941
Alec Hook, Fuat E. Celik
(782a) Effect of Boron Promotion on the Performance of Platinum Catalysts for Propane Dehydrogenation 942
Esteban L. Fornero, Shiwen Li, Vladimir Galvita, Mark Saeys
(782g) Physical Properties Controlling Heterogeneous Catalyst Stability in Supercritical Water 943
Jennifer Jocz, Phillip E. Savage, Levi T. Thompson
(782c) Systematic Investigation of Earth-Abundant Transition Metal Promoters for Pd-Catalyzed Methane Complete Combustion 944
Joshua Willis, Ian Naccarella, Helen Yan, Joseph Maalouf, Kester Wade, Matteo Cargnello
(782d) Unraveling the Nature of Boundary Sites of Metal-on-Oxide Catalysts 945
Prateek Mehta, Jeffrey P. Greeley, W. Nicholas Delgass, William F. Schneider
(782e) Identification of Active Sites for NaY Catalyzed Methyl Lactate Dehydration 946
Bingjun Xu, Brian Murphy
(782f) Using Non-Covalent Interactions to Control Availability of Different Catalyst Sites 947
Michael Janik, J. Will Medlin, Chih-Heng Lien, Gaurav Kumar
(784a) Cu/Chabazite Catalyst to Meet the Chemical and Mechanical Challenges for NOx Removal 948
Nusnin Akter, Taejin Kim, Jorge A. Boscoboinik
(617aa) Effect of the Preparation Method on the Porosity of Amorphous Silica-Alumina to be Used in Hydrocracking Reactions 949
Angélica Liliana Coconubo Díaz, Alexander Guzmán Monsalve, Luz Marina Ballesteros Rueda
(784d) Insight into Supported Metal Catalyst Stability By Quantifying Thermodynamic Interactions at the Solid-Liquid Interface 952
Ahana Mukhopadhyay, Robert M. Rioux
(784e) Simulation and Experimental Studies of Dry Catalyst Impregnation for Improved Content Uniformity and Scale up 953
Yangyang Shen, William G. Borghard, M. Silvina Tomassone
(784f) Reversible Ionic Surfactants for the Preparation of Monodisperse, Supported Nanoparticles 954
Kristin Bryant, Gasim Ibrahim, Steven R. Saunders
(784g) Beneficial Influence of Surface Capping Agents on the Stability and Activity of Metal/Oxide Catalysts 955
Timothy Van Cleve, J. Will Medlin
Author Index

457913 Fundamental Studies and Applications of Nano-Structured Catalysts in Domestic Fuel Production

Abstract:

Fundamental studies and Applications of Nano-structured Catalysts in Domestic Fuel Production

Cun Wen, University of South Carolina, Columbia, SC

Liquid fuels are the largest commodity on the market in terms of total volume of sales, but the **U.S. is lagging behind the curve of novel technologies for domestic fuel production and upgrading**. For instance, U.S. is projected to be a fuel exporting country in the next few years because of the booming of shale gas. However, exporting shale gas has the lowest market value per energy content. Upgrading shale gas into gasoline, on the other side, can boost the price by more than 230%. Unfortunately, this technology is yet to appear on the horizon for the U.S. shale-gas producer, while South Africa and China have already built and operated such units for decades. Besides civilian fuel production, the situation of military fuel generation is not optimistic either. U.S. still can't fly a High-hypersonic jet with speed exceeding Mach 10, while other countries have already tested their high-hypersonic vehicles for several years. Lagging behind in powering high-hypersonic jet will potentially expose our homeland to other country's high speed jets and missiles, and therefore is a serious national-security concern. One of the main problems for powering high-hypersonic jet is to provide high energy-density fuels to the engine. Another example of domestic fuel production is the biofuels. In 2013, the US Navy and Airforce were still paying more than $15 dollars/gallon for biofuels. The high production cost was a result of energy-intensive processes that were needed at the time. Domestic production of transportation fuel is one of the United States' top priorities, as underscored by President Obama in his 2012 State of the Union

Address: *"This country needs an all-out, all-of-the-above strategy that develops every available source of American energy...... and nowhere is the promise of innovation greater than in American-made energy."*

To produce fuel domestically, catalyst development resides in the heart of the solution, because a good catalyst can lead to fuels with higher through-put, lower price, and higher energy-density. **Herein, my group will be focused on developing novel catalysts for domestic fuel production and upgrading.** Through my enriched experience in energy catalysis, I have studied reactions involved in almost every aspect of domestic fuel production, such as methane partial oxidation, water gas shift, Fischer-Tropsch, and hydrodeoxygenation for biodiesel production.[1-11] One uniqueness in my problem-solving technique is design catalyst's nanostructure based on fundamental understanding of reaction mechanism and kinetics. A recent example of how I apply such strategy to solve long-standing problems is designing the first-of-its-kind self-healing catalysts for Fischer-Tropsch synthesis. Cobalt catalysts have been considered the most attractive catalyst among all candidate metals, but a bottle-neck issue with cobalt catalysts is their tendency to deactivation through oxidation by water produced during the process. Through a comprehensive study of the reaction mechanism and kinetics, my idea is to design a cobalt catalyst that can be easily reduced with hydrogen in the feedstock so that the oxidation/deactivation process can be reversed. This idea, in the end, led to the first self-healing catalysts and attracted great amount of interests from academia and industry.[11,12] On the other hand, instead of solely relying on fundamental understanding in literature, my group will interrogate the reaction of interest with a unique approach of combining spectroscopic and kinetic studies. Often people study either reaction mechanism or kinetic one at a time. However, reaction mechanism is the reaction pathway that has the highest reaction rate among all routes. Therefore, a combination of spectroscopic and kinetic studies can lead to more comprehensive understanding of the reaction, and this approach has led to our new understanding in century-old reactions, such as methane partial-oxidation.[1,8,10] With my strength in catalyst design and fundamental understanding, breakthroughs in domestic fuel production and upgrading are expected to benefit our society and prosperity.

Research Interests: Catalysis, nanomaterial, renewable energy

Teaching Interests: thermos, kinetics

Read Me More:

1. Wen C, Liu Y, Guo Y, Wang Y, Lu G. Strategy to eliminate catalyst hot-spots in the partial oxidation of methane: enhancing its activity for direct hydrogen production by reducing the reactivity of lattice oxygen. *Chem Commun.* 2010;46(6):880-882.

2. Wen C, Dunbar D, Zhang X, Lauterbach J, Hattrick-Simpers J. Self-healing catalysts: Co3O4 nanorods for Fischer-Tropsch synthesis. *Chem Commun.* 2014;50:4575-4578.

3. Wen C, Zhu Y, Ye Y, et al. Water–Gas Shift Reaction on Metal Nanoclusters Encapsulated in Mesoporous Ceria Studied with Ambient-Pressure X-ray Photoelectron Spectroscopy. *ACS Nano.* 2012;6:9305–9313.

4. Wen C, Liu Y, Tao F. Integration of surface science, nanoscience, and catalysis. *Pure Appl Chem.* Jan 2011;83(1):243-252.

5. Liu Y, Wen C, Guo Y, Lu G, Wang Y. Modulated CO Oxidation Activity of M-Doped Ceria (M = Cu, Ti, Zr, and Tb): Role of the Pauling Electronegativity of M. *J Phys Chem C.* 2010;114(21):9889-9897.

6. Liu Y, Wen C, Guo Y, Lu G, Wang Y. Effects of surface area and oxygen vacancies on ceria in CO oxidation: Differences and relationships. *J Mol Catal a-Chem.* Feb 2010;316(1-2):59-64.

7. Liu Y, Wen C, Guo Y, et al. Mechanism of CO Disproportionation on Reduced Ceria. *ChemCatChem.* 2010;2(3):336-341.

8. Pichon A. Catalysis: Lower reactivity means more hydrogen. 2010; http://www.nature.com/nchina/2010/100120/full/nchina.2010.7.html.

9. Saxton C. Enhancing catalytic activity. 2010; http://www.rsc.org/Publishing/ChemTech/Volume/2010/02/enhancing_catalytic.asp.

10. Chinese New Year. *Chem Commun.* 2010.

11. Jacoby M. Custom Catalyst Resists Oxidation. *C&EN News.* 2014;92(13):25.

12. Catalysts C. Self-healing F-T catalysts. 2014; http://www.caicatalysts.com/news-a-links/news/69-self-healing-f-t-catalysts.

-

Copyright © American Institute of Chemical Engineers. All rights reserved.

460715 Catalytic Hydrotreatment for the Development of Alternative Transportation Fuels

Abstract:

Research Interests:

Biologically-derived feedstocks are a highly desirable source of alternative transportation fuel. They can be grown renewably and can produce fuels similar in composition to conventional fossil fuels. They are also versatile and wide-ranging. Plant oils can produce renewable diesel and wood-based pyrolysis oils can be made into renewable gasoline. Catalytic hydrotreatment can be used to reduce the oxygen content of the oils and increase their viability as a drop-in transportation fuel, since they can then easily be blended with existing petroleum-based fuels. However, product distribution depends strongly on feedstock composition and processing parameters, especially temperature, hydrogen/oil flow rates and catalyst. Current literature contains little relevant information for predicting process-level data in a way that can be used for proper life cycle or techno-economic assessment. For pyrolysis oil, the associated reaction pathways have been explored via experimental studies on model compounds in a bench scale hydrotreatment reactor. The reaction kinetics of each compound were studied as a function of temperature and catalyst. This experimental data is used to determine rate constants for a hybrid, lumped-parameter kinetic model of paradigm compounds and pyrolysis oil, which can be used to scale-up this process to simulate larger, pilot-scale reactors. For plant oils, enough appropriate data was found in the literature and adapted for a preliminary model. With a systematic collection of kinetic data, hydrotreatment models can be developed that can predict important LCA inputs, such as hydrogen consumption, energy consumption and GHG production, which are necessary for regulatory and assessment purposes.

Teaching Interests:

While my research is focused on reaction engineering, my teaching experience has been focused in thermodynamics. As a TA, I lectured, held office hours, ran review sessions, graded both undergraduate and graduate level work, and helped write problems for homework and exams. I also have experience with teaching Unit Operations, and would be confident running an experimental laboratory class as well. I received the Outstanding Teaching Award from Michigan Technological University in 2015.

450820 First-Principles Modeling of Catalysts: Novel Algorithms and Reaction Mechanisms

Abstract:

Research Interests:

My research employs first-principles (ab initio) electronic-structure theory and molecular simulation to understand heterogeneous and homogeneous catalytic phenomena under realistic conditions, and to generate a platform for rational catalyst design and use in chemical synthesis and pollution abatement. Specific areas of interest include the evaluation and design of amorphous catalytic materials; the kinetic characterization of heterogeneous catalysts, with an emphasis on the use of supported nanoclusters and metals atomically-dispersed on metal-oxides; theory and simulation of catalyst dynamics under reaction conditions; Metal-organic frameworks as catalysts and catalyst supports; and the development of efficient computational methodologies for predicting accurate thermodynamics and kinetics of homogeneous reactions (e.g., C-H activation of alkanes and chiral catalysis).

Teaching Interests:

I look forward to developing and teaching courses at both the graduate and undergraduate level. Teaching has been an interest of mine since a young age, and I see classroom instruction as a natural counterpart to the mentoring that happens with students and post-docs in a research group. My teaching philosophy is to encourage students to reflect on their own understanding of the course material and to scrutinize the possibilities that arise when example problems are modified. My ultimate goal is to equip students with the engineering tools and mindset to tackle the pressing challenges facing modern society. I am enthusiastic to develop Catalysis and Molecular Modeling electives at both the graduate and undergraduate level.

Copyright © American Institute of Chemical Engineers. All rights reserved.

452811 Development of Catalysts for Energy and Environmental Applications

Abstract:

Research Interests:

Over the next 30 years, the global energy demand is expected to increase by 60% to maintain pace with current projections of population and economic growth.[1] Fossil fuels will continue to provide a significant fraction of the energy supply; however, cleaner technologies, including solar, wind and nuclear will become prominent industries to meet the required demand. As the burning of fossil fuels continues into the foreseeable future, it is necessary to develop more efficient processes to decrease energy usage and the associated CO_2 emissions. With the threat of climate change and ocean acidification becoming more apparent each year,[2] efforts must be put forth to not only develop new forms of clean energy, but also capture and utilize CO_2 as a commodity. These large-scale CO_2 conversion processes require the discovery of active and low-cost catalysts.

Development of catalysts for CO_2 conversion can be divided into three areas: synthesis, *in-situ* characterization, and discovery, with significant research opportunities available in each area. In catalyst synthesis, the most promising methods include: atomic layer deposition, colloidal synthesis and incipient wetness impregnation. Although innovative synthesis methods, such as core-shell nanoparticles, can be extremely effective, uniform bimetallic particles can be easily and uniformly synthesized via incipient wetness impregnation. Characterization of catalysts by *in-situ* techniques is necessary to identify critical intermediates and understand the effect of reaction environments on catalyst structure. A previous study on CO_2 conversion used *in-situ* X-ray absorption near edge spectroscopy (XANES) on Mo_2C to confirm the effect of the reaction environment on catalyst oxidation states.[3]

Catalyst discovery is the most general area of catalyst development, but perhaps the most important. Previous studies demonstrate that reducible oxide supports, such as CeO_2, exhibit high activity in CO_2 conversion experiments because of their ability to exchange oxygen with CO_2.[4] Molybdenum carbide (Mo_2C), a non-precious metal catalyst, also shows promise as an active and selective catalyst for CO_2 conversion.[3] These materials are strikingly different, but have a fundamental similarity, which is they are each able to uptake and exchange oxygen with CO_2. Because of the importance of oxygen exchange in reactions with CO_2, oxygen binding energy could be a valuable descriptor when searching for CO_2 conversion catalysts. The discovery of such descriptors is an extremely powerful tool that can be applied beyond CO_2 conversion and into other applications of environmental catalysis.

Teaching Interests: Courses that I would be interested in teaching are kinetics and an introductory course in chemical engineering principles. The introductory course could be basic heat and mass transfer or an overview of the state of the field. A professor at Johns Hopkins once told me, “Everything is chemical engineering. If you name an item in your life, I can tell you how chemical engineering is involved.” I believe this principle should be imparted into young chemical engineers as early as possible. The classic view of chemical engineers as designing large reactors and processes is still relevant, but preparing students for current challenges in energy and climate change is also important. Broadening students’ views of the field has the potential to attract bright minds that might not be set on chemical engineering. I would also like to design an elective course that prepares students for solving problems in energy. These problems naturally lend themselves to applications of thermodynamics, transport, and kinetics, while exposing students to the current state of the field. This type of course and thought process clearly exemplifies my dual responsibility as an academic.

467618 Upgrading Biomass-Derived Platform Chemicals By Electrochemical and Photoelectrochemical Catalytic Oxidation

Abstract:

Research Interests:

Conversion of biomass to chemicals and fuels is a potentially viable route to minimize our dependence on unsustainable petroleum resources. However, unlike petrochemicals biomass-derived platform chemicals such as HMF, glycerol, and propylene glycol are typically highly oxygenated and require development of new catalytic processes to tune their functionality. Electrocatalysis is an extremely versatile tool where the reaction activation barriers and product selectivity can be directly regulated by tuning the electrode potential. Additionally, electrochemical methods and analysis give unique insight into complex processes at the electrode/electrolyte interface which would be otherwise unavailable without complex ***in situ***spectroscopic techniques. Electrochemical systems also give rise to fuel cells, which transform chemical energy to electricity and can also coproduce valuable chemicals. Finally, electrocatalysis can take advantage of renewable energy sources directly using photoelectrochemical (PEC) cells or indirectly by using renewable electricity to drive reactions.

My doctoral research first investigated the cogeneration of electricity and chemicals in direct alcohol anion-exchange membrane fuel cells (AEMFC) with supported noble metal anode electrocatalysts.[1] [2] It was demonstrated that a propylene glycol-fed AEMFC successfully cogenerated electricity and valuable chemicals on carbon-supported platinum or gold nanoparticles (Pt/C, Au/C). Pt/C was very selective for primary alcohol group oxidation to lactic acid under fuel cell operating conditions, while Au/C gave significant amounts of pyruvic acid, a product that has previously eluded heterogeneous catalytic studies on Au. Product selectivity on Au/C was very sensitive to anode potential. Later, I showed that HMF can be successfully oxidized to FDCA, an important bio-based polymer precursor, on supported bimetallic Pd-Au/C electrocatalysts, and the reaction benefited from a synergistic effect between Pd and Au components.[2]More recently, we have explored the use of stable nitroxyl radicals including TEMPO as electron mediators for selective oxidation of biomass-derived substrates in electrochemical and PEC reactors. This route opens opportunities to use non-noble metal electrodes and to take advantage of solar energy to drive electrochemical conversions.

Teaching Interests:

My background highlights the diverse benefits of electrochemical systems for the upgrading of biomass-derived chemicals, with unique examples in fuel cells, electrolysis flow reactors, and PECs. Electrocatalysis is a thriving research field with blooming practical applications and importance. As an independent faculty researcher, I will lead an interdisciplinary team to engineer advanced electrochemical materials and systems to meet future needs in energy conversion, production and storage, and sustainable chemical manufacturing as well as develop new curriculum for undergraduate and graduate-level electrocatalysis education.

References:

(1) Chadderdon, D. J. et al., ACS Catal. (2015), 6926–6936.

(2) Han, X. T. et al., Int J Hydrogen Energy (2014), 19767–19779

(3) Chadderdon, D. J. et al., Green Chem. (2014), 16 3778–3786.

Copyright © American Institute of Chemical Engineers. All rights reserved.

468204 Rational Design of Catalytic Sites for Energy Applications

Abstract:

Motivation:

The rapidly increasing global demand for energy coupled with large population growth necessitates the development of increasingly efficient systems for utilizing our finite energy and chemical resources. While no single approach will solve these complex issues, I strongly believe that catalysts have an important role to play in addressing these challenges. Our society is already heavily reliant on these materials to produce the chemicals that we need to sustain our lifestyles. Catalysts are incredible because of their ability to facilitate selective chemical transformations while converting between chemical, thermal, radiant, and electrical forms of energy. Coupling catalytic systems with renewable energy sources presents a unique opportunity to regenerate useful high value chemicals from low value feed stocks like CO_2 and H_2O.

In order to develop these technologies, it is vital to understand the pathway through which these chemical transformations proceed and how catalytic systems respond to different energetic and environmental stimuli. It is also vital to identify key characteristics of active, stable, and selective catalysts to enable the design of materials with optimal sites for a particular reaction. I have implemented this strategy in my academic pursuits as a PhD student and postdoctoral researcher.

Research Interests:

As faculty I plan to leverage my past research experience to improve catalytic performance for various chemical, electrochemical, and photochemical applications. My efforts will focus on the design of cheap, active, selective, and stable catalysts for carbon and nitrogen fixation, renewable H_2generation, and selective upgrading of low-value oxygenates into important fuel and chemical feedstocks. I anticipate closely collaborating with theoreticians and other experimental specialists to decipher the true nature of catalytic sites under reaction conditions. With this information, structure-property relationships can be develop to give further insight to the field.

Research Experience:

Over the course of my graduate studies under Prof. Suljo Linic, we have investigated how geometric and chemical manipulations of catalytic sites influence overall catalytic performance. Specifically, we focused on understanding how changing the size, shape, and composition of Ag and Pt alloy electrocatalysts affects their activity towards the oxygen reduction reaction (O_2+4H^+/e^-â†'2H_2O, ORR) in alkaline and acidic electrolytes. First, kinetics studies of the oxygen reduction mechanism on these surfaces allowed us to identify key chemical descriptors (like OOH or OH adsorption energies) that are correlated with catalytic activity. We then utilized quantum chemical calculations to determine how these descriptors could be selectively influenced in platinum based catalysts through perturbations such as alloying or inducing geometric strain. After identifying promising catalysts, we developed sophisticated synthesis strategies to prepare these materials. We then tested their activity, with rigorous experimental protocol allowing us to obtain valid kinetic rates from the observed rates, and found that improved activity was achieved. Finally, we performed extensive in situ and ex situ characterization to confirm both that the atomic structure of the electrocatalyst matches that of our model and that the observed activity enhancement is a direct result of the improved key chemical descriptors of this particular nanostructure.

As a postdoctoral researcher working in Prof. Will Medlin group, I have focused on controlling rates and product selectivity for various thermochemical processes (such as CO oxidation, epoxidation, alcohol dehydration, and hydrodeoxygenation of phenolics) by manipulating the relative abundance and accessibility of different catalytic sites. Furthermore, the addition of chemical promoters to metal, support, or interfacial sites has been another method for altering the material flux through particular chemical pathways.

Teaching Interests:

My preference would be to teach chemical kinetics and reactor design because it closely aligns with my research background and extensive teaching experience. At Michigan, I served as a graduate teaching assistant (TA) for a graduate-level electrochemistry elective course (~70 students) as well as graduate chemical kinetics (~25), developing and evaluating homework and exam questions in addition to administering the course website and hosting review sessions. During my final year at UM, I was a Graduate Student Teaching Fellow for the undergraduate chemical reactor design course (~170). I was responsible for preparing course materials, managing personnel, and lecturing a 1/3 of the classes. Additionally, I am open to developing a graduate level elective course describing the fundamentals of catalyst design, which would be of particular interest to undergraduate and graduate students. Lastly, I have much experience as a mentor to both undergraduate and graduate students during my doctoral and postdoctoral research.

Copyright © American Institute of Chemical Engineers. All rights reserved.

469170 Theory-Guided Understanding and Design of Heterogeneous Catalysts

Abstract:

Research Interests:

-Efficient catalyst screening and design:
Adsorption energies are good predictors of catalytic performance, but calculating adsorption energies on a large number of potential catalytic surfaces remains a challenge. I have developed methods for efficiently and accurately predicting adsorption energies of important intermediates on transition metal surfaces. These methods are quite general, applying to any transition metal alloy surface, and linking any adsorbate to electronic structure or to any other adsorbate. They are also efficient and physically transparent, allowing high-throughput screening or rational design. Future work will involve applying these methods to design novel catalysts (particularly those that break traditional scaling relations), and further improving their accuracy, efficiency and applicability.

-Nonadiabatic dynamics, excitations in metal surfaces and nanoparticles:
Most ab initio studies of catalysis and other processes on metal surfaces assume that the system remains in the ground state. However, using an efficient nonadiabatic dynamics code employing density functional theory and Ehrenfest dynamics, I have tested this assumption for nitrogen dissociation on Ru. These calculations show that even in thermal catalysis, surface processes can excite electrons in metal surfaces. Future work will involve understanding the conditions necessary for these electronic excitations, and their possible role in surface chemistry, as well as studies of plasmon-driven catalysis and photocatalysis.

Teaching Interests:
-Thermodynamics
-Materials chemistry and structure
-Mathematical methods
-Fluid dynamics and transport
-Quantum and computational chemistry

Copyright © American Institute of Chemical Engineers. All rights reserved.

469497 In silico Engineering of a Future Energy Infrastructure

Abstract:

Research Interests:

The decreasing supplies of petroleum resources coupled with an increase in demand has created a need for an alternative sources of chemical building blocks and renewable energy. While current research has identified several promising technologies, high material costs and/or low yields forces our continual dependence on non-renewable sources. As a future faculty member, I seek to utilize my prior expertise to develop and distribute computational models and software which will allow for potential catalytic materials to be identified and readily examined inexpensively. During this Session I will demonstrate how my prior experience will enable my future research into the in silico design of new electrocatalytic materials for fuel cell applications.

Fuel cells are an attractive prospect for meeting our future energy demands due to their high energy density and low emissions. Utilization of H2 generated via the photo-oxidation of water further enhances the green aspects of this technology. While prior work has focused on the application of proton exchange membranes, these devices have limited applications due the high cost of the platinum based electrocatalytic materials involved. Recent work with hydroxide exchange membrane fuel cells have demonstrated that under akaline environments, cheaper base metals (i.e., nickel) can be used; however such materials are still limited due to their low activity. The future design of electrocatalytic materials will involve the detailed understanding of the kinetics of the relevant chemical processes at the anode and cathode, as well as the influence of the solvation environment. Unfortunately, such an understanding is currently lacking. My goal is to apply new computational tools to elucidate these effects.

My graduate work at the University of Pittsburgh focused on understanding the interactions between extended -systems (i.e., graphene) and polar molecules (i.e., water),[1,2] which is critical first step in understanding how aromatic carbon supports interacts with the solvation environment. During my post-doctoral work at the University of Wisconsin examined how support materials affect the catalytic properties of metal nanoparticles,[3,4] the results of which were recently used to explore the nature of novel carbon supports for fuel cell applications.[5] My current post-doctoral work at the University of Delaware has focused on the production of value-added chemicals and fuels from biomass-derived building blocks.[6–8]

Graduate Advisor: Kenneth D. Jordan (University of Pittsburgh)

Post-doctoral Advisors: Jordan R. Schmidt (University of Wisconsin),

Dionisios G. Vlachos (University of Delaware)

Select Publications (Total citations: 269):
(1) Jenness, G. R.; Karalti, O.; Jordan, K. D. Phys. Chem. Chem. Phys. 2010, 12, 6375–81.
(2) Jenness, G. R.; Karalti, O.; Al-Saidi, W. A.; Jordan, K. D. J. Phys. Chem. A 2011, 115, 5955–64.
(3) Jenness, G. R.; Schmidt, J. R. ACS Catal. 2013, 3, 2881–2890.
(4) Hermes, E. D.; Jenness, G. R.; Schmidt, J. Mol. Simul. 2014, 41, 123–133.
(5) Zhuang, Z.; Giles, S. A.; Zheng, J.; Jenness, G. R.; Caratzoulas, S.; Vlachos, D. G.; Yan, Y. Nat. Commun. 2016, 7, 10141.
(6) Jenness, G. R.; Christiansen, M. A.; Caratzoulas, S.; Vlachos, D. G.; Gorte, R. J. J. Phys. Chem. C 2014, 118, 12899–12907.
(7) Jenness, G. R.; Vlachos, D. G. J. Phys. Chem. C 2015, 119, 5938–5945.
(8) Gilkey, M. J.; Panagiotopoulou, P.; Mironenko, A. V.; Jenness, G. R.; Vlachos, D. G.; Xu, B. ACS Catal. 2015, 5, 3988–3994.

Teaching Interests:

Based on my chemistry and research background, I am most suited to teach classes that are related to physical chemistry (quantum mechanics, thermodynamics, statistical mechanics etc.), as well as classes on surface science, molecular simulations, and catalysis. Furthermore, I have an interest in formulating electives related to the application of python programming to chemical research and in the writing, preparation, and presentation of scientific results.

My approach to teaching has always been a reductionist approach with an emphasis on hands-on learning. In this approach, the student is presented with a relevant, big picture problem that is relevant to the current state of research. As an instructor, my goal is to dissect this problem into smaller problems, and instruct the student on how to solve these smaller pieces. By engaging the student with similar problems as homework and class projects, the student then learns how to perform such a decomposition on their own.

In addition to the reductionist approach in the classroom, I will employ a homework model that will emphasize both student collaboration and public speaking. Inevitably, students assigned homework problems will seek out help and advice from their fellow students. As an educator, it is critical to encourage such collaborations. To this end, I will require my students to both list those they collaborated with and their contributions. Class projects with a group component are also key, as this will not only encourage collaboration, but also give the students experience in discussing scientific literature and results in a public setting. In fostering collaboration and promoting public speaking, the student will form connections that can persist throughout their careers, while providing opportunities for the student to enhance their communication skills.

Copyright © American Institute of Chemical Engineers. All rights reserved.

471981 Insight into Pt-Bi Bimetallic Catalysts: An Experimental and DFT Study

Abstract:

Research Interests: Catalytic Reaction Engineering

Teaching Interests: Reaction Engineering

Owing to environmental concerns and scarcity of known reserves, biomass has been shown to be an important renewable source, which can be converted into both sustainable fuels and chemicals. Heterogeneous catalysis plays an important role in such processes, while the following two issues still remain challenging and limit their industrial applications: tuning the catalyst selectivity toward target products and improving its stability against coking/carbon deposit. Both issues can be addressed by use of bimetallic catalysts, containing a primary metal (e.g. Pt, Pd, Ru, Rh, etc.) and a promoter (e.g. Mo, Sn, Bi, etc.). In the present work, we report an example of Pt-Bi bimetallic catalyst that selectively converts biodiesel waste product glycerol to 1,3-dihydroyacetone (DHA) and improves catalyst stability for reactions involving CH_4 (guaiacol deoxygenation using CH_4 in place of H_2. Based on experimental observations and theoretical analysis, Bi locates in an adatom-like configuration Pt and a cooperative effect between Pt as the primary metal and Bi as a promoter was identified. Insight into Pt-Bi bimetallic catalysts also suggests other catalytic applications, such as short-chain alcohol oxidation and oxidative coupling of methane.

Copyright © American Institute of Chemical Engineers. All rights reserved.

474682 *Mechanisms of Redox Catalysts for the Greener Processing of Shale Gas Via Chemical Looping*

Abstract:

Postdoctoral Research Scholar

Research Interests:

The North American shale gas boom has stimulated renewed interest in producing value-added products from light hydrocarbons. Unfortunately, the traditional industrial approaches to produce chemicals and liquid fuels from shale gas components are energy-intensive and have severe environmental consequences. The primary challenges include: (i) selective activation of light, saturated hydrocarbons such as methane and ethane is very difficult; (ii) the production of high-enthalpy products or intermediates, such as hydrogen, requires significant external energy inputs. The push for improved, less polluting, more efficient energy conversion technologies has spurred increased interest in catalyst development. In my career, I hope to further the use of catalyst characterization techniques to applied catalysis to expand both fundamental knowledge, as well as tools for the rational design of energy-conversion catalysts. I have significant research experience in the fields of: redox catalyst for chemical looping; selective oxidation of light hydrocarbons; heterogeneous catalysis (Synthesis, Reaction testing, Characterization); platinum group catalyst formulation; surface/catalyst characterization techniques; ultra-high vacuum (UHV) system operation and maintenance; research scale reactor design, optimization, and operation.

Teaching Interests:

I have a strong interest in teaching fundamental chemical engineering subjects in addition to graduate and undergraduate courses closely related to my field (e.g. kinetics, catalysis/surface science). I also greatly enjoy training, mentoring, and collaborating with all levels of researchers down to undergraduates. In my postdoctoral work I have:

- Mentored Ph.D. students that, to date, has led to one peer review publication, two articles under review, and two nearing submission;
- Overseen interactions with facilities and environmental health and safety, and trained graduate students in best practices for laboratory safety and hazardous waste handling;
- Assisted collaboration between graduate students in the Chemical Engineering and Mechanical and Aerospace Engineering Departments;
- Successfully collaborated with graduate researchers in the Chemistry Department, leading to two high quality publications;
- Directly trained and supervised multiple undergraduates in the lab and supervised experimental plans for three master's students.

Research and Funding Successes: I have 14 authored and co-authored refereed articles across a variety of catalytic reactions. I have supervised multiple research projects, mentoring over 12 graduate and undergraduate researchers, and have participated in writing proposals, SOPO's, and reports leading to project successes. I co-invented a novel chemical looping ODH technology, which was awarded $3.8 million overall ARPA-E funding and licensed to a commercial partner (EcoCatalytic LLC).

PhD Dissertation: ***"Nano-Particle Supported Catalyst"*** Advised by Dr. Helena Hagelin-Weaver in the Department of Chemical Engineering at the University of Florida

Research Experience:

Chemical Looping Redox Catalyst: Formulated and characterized redox catalysts for the oxidative coupling of methane; Formulated and characterized a novel core-shell redox catalyst for partial oxidation of methane to syn-gas; Co-invented a licensed catalytic/chemical-looping technology for oxidative dehydrogenation of ethane.

Direct Methanol Fuel Cells: Developed test procedures for electrode characterization, including quantification of water flux and oxygen starvation, to facilitate control in a novel membrane electrode assembly that allows passive water recovery by forcing water back into the fuel stream; Performed multiple surface and electrochemical characterizations of electrodes to identify the causes of cell degradation.

Platinum Group Metal Catalyst: Prepared and characterized Pd on nano-particle oxide supports for selective oxidation; Prepared and characterized high dispersion Pt/TiO_2 catalyst for fundamental hydrogenation studies. Prepared Cu/Zn on nano-particle alumina for catalytic reforming of methanol; Designed and built gas phase reactor for catalyst activity testing; Operated and maintained UHV (Ultra High Vacuum) XPS system.

Selected Publications:

Luke M. Neal, Seif Yusuf, Fanxing Li. "Oxidative Dehydrogenation of Ethane: A Chemical Looping Approach." ***Energy Technology, June 2016 (in press).***

Ronghui Zhou, Evan W Zhao, Wei Cheng, **Luke M. Neal**, Haibin Zheng, Ryan E. Quiñones, Helena E. Hagelin-Weaver, Clifford R. Bowers. "Parahydrogen-induced polarization by pairwise replacement catalysis on Pt and Ir nanoparticles." ***Journal of the American Chemical Society, September 2015, Volume 137, Pages 1938–1946.***

Luke M Neal, Arya Shafiefarhood, Fanxing Li. "Dynamic Methane Partial Oxidation Using a Fe_2O_3@ $La_{0.8}Sr_{0.2}FeO_{3-\delta}$ Core–Shell Redox Catalyst in the Absence of Gaseous Oxygen." ***ACS Catalysis, September 2014, 4 (10), Pages 3560–3569.***

Luke M. Neal, Michael L. Everett, Gar B. Hoflund, Helena E. Hagelin-Weaver. "Characterization of Palladium Oxide Catalysts Supported on Nanoparticle Metal Oxides for the Oxidative Coupling of 4-Methlypyridine." ***Journal of Molecular Catalysis A: Chemical*, *February 2011, Volume 335, Issues 1-2, Pages 210-22.***

Samuel D. Jones, **Luke M. Neal**, Helena E. Hagelin-Weaver. "Steam Reforming of Methanol Using Cu-ZnO Catalysts Supported on Nanoparticle Alumina." ***Applied Catalysis B: Environmental*, *December 2008, Volume 84, Issues 3-4, Pages 631-642.***

Luke M. Neal, Helena E. Hagelin-Weaver. "C–H Activation and C–C Coupling of 4-Methylpyridine Using Palladium Supported on Nanoparticle Alumina." ***Journal of Molecular Catalysis A: Chemical*, *April 2008*, *Volume 284, Issues 1-2*, *Pages 141-148.***

Copyright © American Institute of Chemical Engineers. All rights reserved.

474687 Nanostructured Transition Metal Dichalcogenide Catalysts for Electrochemical Energy Systems

Abstract:

Transition Metal Dichalcogenide Catalysts for Elecrochemical Energy Systems

Mohammad Asadi

Department of Mechanical and Industrial Engineering, University of Illinois at Chicago, Chicago, IL, 60607, USA.

Abstract:

Electrochemical energy systems have shown a tremendous potential to store energy in the form of chemical bonds using renewable energy sources. In general, energy can be converted or stored into chemical bonds through catalytic electrochemical processes such as fuel cells, oxygen reduction reaction (ORR) and carbon dioxide reduction reaction (CO_2RR). However, in practice existing catalysts are too inefficient: either weak binding interactions between the reaction intermediates and the catalyst give rise to high overpotentials, or slow electron transfer kinetics result in low exchange current densities. Both of these metrics depend not only on the intrinsic electronic properties of the catalyst, but also on the solvent and the catalyst morphology.

Here, I present the nanoarchitectures two-dimensional (2D) of molybdenum disulfide (MoS_2) and other transition metal dichalcogenides (TMDs) that manifest much higher performance for electrocatalytic ORR and CO_2RR reactions in the ionic liquid 1-ethyl-3-methylimidazolium tetrafluoroborate (EMIM-BF_4). Atomic level study of TMD nanoflakes before and after electrochemical experiments using scanning transmission electron microscopy (STEM) indicate that; (i) mainly metallic edge atoms are responsible for this remarkable activity and (ii) the edge atoms remain stable during the reactions. Moreover, I have successfully tested MoS_2 nanoflakes that work with the ionic liquid EMIM-BF_4 electrolyte as an effective co-catalyst for discharge and charge in a Li-O_2 battery.

The presented TMDs/IL co-catalyst system with remarkable electrocatalytic performance could open a new route to the advancement of energy conversion and storage systems.

Research Interests:

1- Synthesis and characterization of catalysts

2- Energy conversion systems (i.e., CO_2reduction reaction, water splitting, hydrogen evolution reaction, oxygen reduction reaction, oxygen evolution reaction)

3- Energy storage systems (i.e., metal-ion and metal-air batteries)

4- Electrochemisty and photoelectrochemistry

Teaching Interests:

1- Fluid Mechanics

2- Thermodynamics

3- Heat Transfer

4- Transport Phenomena

5- Reaction Engineering

6- Mass Transfer

store energy in into the chemical bonds in the form of electricity or fuel.

Copyright © American Institute of Chemical Engineers. All rights reserved.

461299 Exploring Structure-Function Correlations of Nanomaterials in Heterogeneous Catalysis

Abstract:

Research Interests: Catalysis is the science and technology of influencing the rates of chemical reactions, which are at the core of chemical industry. Practical success of catalysis needs the interdisciplinary intersection of chemistry, chemical engineering, and material science. My research interests lie on understanding the mechanism of chemical reactions and the role of catalysts under reaction conditions in order to transform qualitative observations and knowledge into predictive science. My research program will focus on developing novel structure-reactivity relationships combining *operando* observation, materials synthesis and characterizations guided by theoretical calculations as applied to the field of heterogeneous catalysis inculding light alkanes activation, biomass conversion and energy conversion & storage.

Teaching Interests: I am confident that my academic experience makes me well prepared to teach both core and complementary courses. Given my research background, I would particularly enjoy teaching undergraduate level courses in reaction engineering, kinetics and electron microscopy. As for the graduate level, courses in interdisciplinary methods in heterogeneous catalysis is a natural fit for my background.

Copyright © American Institute of Chemical Engineers. All rights reserved.

475338 Mechanistic, Spectroscopic and Theoretical Assessment of Porous Catalytic Materials

Abstract:

Research Interests:

Porous crystalline materials such as zeolites or porous organic polymers provide a large and diverse pool of heterogeneous catalysts and catalyst supports. Thorough mechanistic investigations that employ characterization of material properties and surface structure, kinetic and isotopic studies, and computational modeling are needed in order to elucidate how these materials work as catalysts regardless of reaction or application. Further, this approach can be used in conjunction with specialized synthetic strategies in order to produce well-defined catalysts, imperative in understanding catalytic function.

One class of these materials, Brønsted acidic zeolites, catalyze a plethora of hydrocarbon and oxygenate reactions, which are mediated by ion-pair transition states. My graduate research with Prof. Enrique Iglesia investigated the factors that affect reactivity, specifically the size, shape and charge of the organic cation (derived from reactants) and anion (formed after deprotonation of the inorganic catalyst). We combined state-of-the-art theoretical methods (density functional theory) with rigorous kinetic measurements from a range of reactants and catalysts that led to a copious array of transition states and precursors. Key properties of these systems include how the anion (i) accepts the negative charge (acid strength), which results in an electrostatic interaction with the organic cation, and (ii) provides van der Waals interactions, which are dependent on the size and shape of the organic component and the surrounding pore environment (confinement). The synergy between theory and experiment exposed unprecedented mechanistic detail, such as the flexibility of zeolites to distort locally to increase van der Waals interactions with the cation. These distortions cost energy, however, and can be expensive if the transition state is too large. Thorough studies incorporating both experiment and theory can provide clarity into complex relationships, such as this one between organic moieties and inorganic zeolite catalysts.

Future plans will utilize these results and methods, along with incorporating synthetic techniques, as a framework for studying porous catalytic materials from different catalyst classes and for other reaction pathways/applications. This framework can be extended to understand effects of important catalyst properties on reactivity and selectivity in order to advance rational catalyst design.

Teaching Interests:

Teaching, in the classroom and in the laboratory, is an outlet to share my zeal for science and a major driving force in my pursuit of an academic career. One benefit of research in catalysis is that multiple pillars of chemical engineering (kinetics, transport and thermodynamics) must be mastered, though I most enjoy teaching chemical reaction engineering courses. I was a graduate student instructor at UC Berkeley for the undergraduate and graduate level as well as designed and taught a new course titled â€œEnergy and Atom Efficient Catalytic Conversions,â€ which covered both current and green(er), alternative industrial catalytic reactions for energy, fuel and chemical production.

Copyright © American Institute of Chemical Engineers. All rights reserved.

475552 In-Situ Drifts Studies on CuNi Catalyst for Ethanol Hydrogen Production

Abstract:

In-situ DRIFTS Studies on CuNi Catalyst for Ethanol Hydrogen Production

Anand Kumar

Department of Chemical Engineering, Qatar University, P O Box 2713, Doha, Qatar

Ethanol dehydrogenation mechanism was investigated over catalysts containing Cu and Ni metals prepared by solution combustion synthesis. In situ DRIFTS studies were conducted on individual metals (Cu and Ni) as well as on CuNi alloy to understand the metal-metal interaction in bimetallic systems and their implications on reaction pathway. The catalyst synthesis parameters (such as fuel to oxidizer ratio in combustion synthesis and combustion temperature etc.) were monitored and correlated with the synthesized materials properties. The amount of fuel content in the combustion solution was found to greatly affect the phase and microstructure of the synthesized catalysts. Along with the microstructure, surface phase distribution was also studied for various fuel to oxidizer ratio using XPS and changes on surface composition were monitored after catalyst activation (reduction) and after the ethanol dehydrogenation reaction. A change in reaction pathway was clearly observed over Cu and Ni catalysts. Cu mainly favors synthesis of acetaldehyde and acetate whereas Ni and CuNi both showed more tendencies towards methane and carbon monoxide formation.

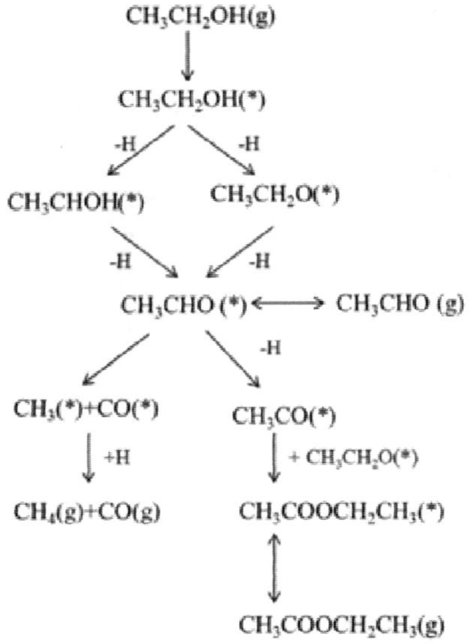

Figure 1: Postulated ethanol dehydrogenation reaction pathways over Cu and Ni based catalysts

Research Interests: Heterogeneous catalysis, combustion synthesis, microfluidic synthesis, nanomaterials, ethanol hydrogen production, methane reforming, carbon dioxide conversion, fuel cell reactions, solar fuels, electro-catalysis.

Teaching Interests: Chemical Reaction Engineering, Transport Phenomena, Chemical Process Control, Heterogeneous Catalysis, Mass Transfer, Heat Transfer, Advanced Engineering Mathematics, Numerical Methods, Natural Gas Processing, Engineering Economics

Copyright © American Institute of Chemical Engineers. All rights reserved.

475627 Semi-Permeable Membrane Reactor for Catalysis, Hydrocarbon Processing and CO2 Reuse

Abstract:

PhD expected in January 2017

PhD Dissertation: "***Syngas Production via Water and Carbon Dioxide Thermolysis on Mixed Ionic-Electronic Conductive Membranes.***"

Under supervision of Professor Ahmed F. Ghoniem, Ronald C. Crane (1972) Professor; Director, Center for Energy and Propulsion Research; Director, Reacting Gas Dynamics Laboratory, Massachusetts Institute of Technology

Research Interests:

Inorganic membrane reactor finds numerous industrial and scientific applications such as hydrogen production, hydrocarbon processing, CO2 splitting and kinetic studies. By using a semi-permeable membrane, reactants can be doped or products can be separated. Therefore, the thermodynamic limit can be overcome and the selectivity can be enhanced. For example, a mixed ionic-electronic conductive membrane facilitates H2O/CO2 splitting at elevated temperatures and increases the selectivity to syngas in partial oxidation of methane. The implementation of such an oxygen permeable membrane also reveals the kinetics of partial oxidation of methane. Experimental and numerical studies are used to understand the reaction kinetics and limiting steps and to optimize the reactor design. Ab initio studies can reveal the fundamentals of charged species transport and transition states in reactions. These information can all lead to the optimization of reactor in industrial scale.

My ultimate goal for all my potential researches is to utilize the fundamental understanding in science to guide and optimize the operation in engineering. Therefore, in the future, I foresee myself collaborating closely with scientists and industrial partners to make the findings in my own laboratory into reality.

Teaching Interests:

Thermal science and energy related subjects such as heat transfer, fluid mechanics, transport phenomenon, thermodynamics, energy conversion

Chemistry and materials related subjects, such as electronic materials, inorganic materials, chemical kinetics

Awards:

1. MIT Golden Beaver Award – Institute award for excellence in leadership (Board of MIT European Club), 2016

2. Best Presentation Award in the 2015 AIChE Annual Meeting, 2015

3. Audience Award in MIT Mechanical Engineering Research Exhibition Poster Session, 2015

4. MISTI-MITEI fellow for Internship in Germany, MIT, 2015

Research Experience:

I have been working on several fields on chemical engineering: heat/mass transfer, electrochemical reaction/diffusion, reactor analysis and chemical kinetics. My previous work in Zhejiang University was focused on experimental studies on different types of heat transfer enhancement techniques: fin tubes for falling film evaporation and nanofluids for supercritical heat transfer. Both projects were based on industry applications. Now I am working on an interdisciplinary research as my PhD thesis, which involves chemical kinetics (surface reaction) and electrochemical diffusion (ion transport membrane). Tools in various physical scales are used in my thesis: femto (Density Functional Theory), nano (Microscopy Characterization), macro (System Scale Thermodynamics). This means both experiments and simulations are used to understand the chemical process. Additionally, my work experience in Wolverine Tube, INC and Robert Bosch LLC granted me the insights in researches in industry, which will benefit my collaboration with industrial partners.

Teaching Experience:

At MIT, I took a graduate course "Instruction in Teaching Engineering " and applied the knowledge on teaching when I TA'ed two courses at MIT: Fundamental of Advanced Energy Conversion (both undergrad and grad) and Introduction to Heat Transfer (undergrad). I got evaluation from students with 6.5/7 and 6.6/7 for these two course, respectively. Besides, I have mentored three undergraduates at MIT under my thesis work: one from MIT-Egypt Exchange Fellowship Program (3 months) and two from MIT Undergraduate Research Opportunities Program (UROP) (5 and 8 months).

Selected Publications:

1. X.Y. Wu, M. Uddi, A.F. Ghoniem, "Enhancing co-production of H_2 and syngas via water splitting and POM on surface-modified oxygen permeable membranes", invited to submit to AIChE Journal as "Best paper initiative", 2016, under review

2. X.Y. Wu, L. Chang, M. Uddi, P. Kirchen, A.F. Ghoniem, "Toward enhanced hydrogen generation from water using oxygen permeating LCF membranes", ***Physical Chemistry Chemical Physics***, 17 2015 10093-10107

3. D. Huang, **X. Y. Wu**, Z. Wu, W. Li, H. T. Zhu, B. Sunden, "Experimental Study on Heat Transfer of Nanofluids in a Vertical tube at Supercritical Pressures", ***International Communications in Heat and Mass Transfer***, 63 2015 54-61

4. W. Li, **X.Y. Wu**, Z. Luo, R.L. Webb, "Falling Water Film Evaporation on Newly-Designed Enhanced Tube Bundles", ***International Journal of Heat and Mass Transfer***, 54 (13-14) 2011 2990-2997

5. W. Li, **X.Y. Wu**, Z. Luo, S.C. Yao, J.L. Xu, "Heat Transfer Characteristics of Falling Film Evaporation on Horizontal Tube Arrays", ***International Journal of Heat and Mass Transfer***, 54 (9-10) 2011 1986-1993

Personal website:

http://xywu.scripts.mit.edu/xywu/

Copyright © American Institute of Chemical Engineers. All rights reserved.

475657 A Practical Way to Separate Uncondensed Lignin during Biomass Pretreatment and Quantitatively Depolymerize It at a Low Temperature of 120°C with a Cheap Ni/Al_2O_3-SiO_2 Catalyst

Abstract:

High-yield lignin monomer production from separated lignin in practical ways could greatly increase biorefinery productivity and profitability. However, lignin condensation caused by the formation of inter-unit carbon-carbon linkages during biomass pretreatment dramatically reduces lignin monomer yields; besides, high temperatures and noble-metal catalysts used in lignin hydrogenolysis depolymerization further increase the impracticality.

Recently, we have two significant findings which can largely drive the valorization of lignin. First, adding formaldehyde during biomass pretreatment produces a soluble lignin liquor that can be converted to guaiacyl and syringyl monomers at near-theoretical yields during subsequent hydrogenolysis; second, separated lignin can be readily depolymerized at a low temperature of 120°C with a cheap Ni/Al_2O_3-SiO_2 catalyst. We achieved a yield of 45% for beech wood, and a yield of 78% for transgenic poplar wood with low levels of carbon-carbon linkages in native lignin (yields were on the basis of native lignin). These yields were 3-7 times those obtained without formaldehyde. Formaldehyde reacted with lignin side-chain alpha- and gamma- hydroxyl groups to form stable 1,3-dioxane structures and with lignin aromatic rings to block the active positions, which completely quenched the condensation of lignin and facilitate its depolymerization under mild catalytic conditions.

Research Interests: Catalytic conversion of biomass into fuels, chemicals and materials.
Teaching Interests: Biomass conversion

Copyright © American Institute of Chemical Engineers. All rights reserved.

475803 ***Combined Quantum and Classical Computational Approaches for Investigating Complex Surface Interactions Impacting Heterogeneous Catalysis***

Abstract:

2nd Year Post-Doctoral Associate

Post-Doctoral Project:Â *â€œAb initio design of semiconductor materials for the heterogeneous photo-electrocatalytic reduction of CO_2â€Â* Under the supervision of Prof. Emily A. Carter, Department of Mechanical and Aerospace Engineering, Princeton University

Ph.D. Dissertation:Â *â€œDevelopment of multi-scale computational techniques for modeling phase formation in Pd-based oxide catalystsâ€* Under the joint supervision of Prof. Michael J. Janik and Prof. Adri C. T. van Duin, Department of Chemical Engineering, Pennsylvania State University

Research Interests:

I am interested in research that employs atomistic simulation techniques at both electronic and classical levels of theory, with a particular focus on applying such methods to address issues related to energy conversion, storage, and utilization. I began academic research as an undergraduate under the supervision of Prof. William F. Schneider at the University of Notre Dame. There I was introduced toÂ *ab initioÂ* simulation techniques while completing an undergraduate thesis entitled â€œ*Ab initio*Â insights into ionic-liquid/CO_2Â binding trends for CO_2Â capture applications.â€ My training in atomistic simulation continued at the Pennsylvania State University, where I completed my Ph.D. in Chemical Engineering under the advisement of Prof. Michael J. Janik and Prof. Adri C. T. van Duin. My graduate research included the joint application of quantum and classical simulation techniques for investigating phase formation affecting the overall activity of oxide-supported metal catalysts, with a particular emphasis on identifying design principles for methane conversion catalysts. I am currently a post-doctoral associate under the advisement of Prof. Emily A. Carter at Princeton University, where I am conductingÂ *ab initio*Â investigations of heterogeneous catalytic mechanisms for CO_2Â reduction over semiconductor photo-electrodes. The aforementioned projects have led to 15 total publications (7 first-authored, 1 joint first-authored, 5 second-authored, and 2 co-authored), many of which were written with experimental collaborators.

Future Research Directions:

Through my previous research experiences, I have become well versed in both quantum and classical simulation techniques, which I will continue to apply in research aiding the development of sustainable energy technologies. I intend to integrate computational techniques I have learned from the Carter, Janik, and van Duin groups into my independent research program. I foresee that my group will collaborate with experimental researchers in the fields of catalysis, materials science, and surface characterization. In particular, my research plan consists of three principal focus areas:

(1) *Extending multi-physical simulation methodologies to the emerging field of heterogeneous catalysis over photo-electrode surfaces.* Combined quantum/classical simulation techniques will be used to assess the structural evolution of catalytically active photo-electrode surfaces that are in contact with a reactive aqueous environment. Full consideration of the reactive environment is essential, as the nature of active sites on the surface will be affected by structural changes induced by interaction with the solvent, and a reactive force field approach will be indispensable for assessing the complexities of the surface-solvent interface. Once the overall surface morphology is understood, *ab initio* methods will be used to elucidate the electronic structure of excited states that are accessible to photo-excited electrons involved in the catalytic mechanism. These studies will be relevant for H_2O splitting and CO_2 reduction applications.

(2) *Applying quantum and classical simulation techniques in tandem to assess strong metal/support interactions impacting the catalytic behavior of metal oxides.* I will continue to use combined quantum and classical simulation methods to characterize metal/support interactions leading to emergent catalytic behavior. My previous research determined the origin of highly active sites derived from unique interactions between Pd nanoclusters and a CeO_2 support. Preliminary results, obtained by an undergraduate student under my supervision, indicate that similar interaction trends occur between other metal/support pairs, thus affecting sintering rates and cluster size distributions. My group will further explore these trends, with the objective of developing novel and concrete optimization principles informing the rational design of oxide-supported metal catalysts.

(3) *Developing accelerated simulation methodologies for modeling kinetically limited processes with the ReaxFF potential.* I have developed various methods tailored to the classical ReaxFF potential for assessing the thermodynamic stability of active sites on catalytic surfaces. These methods, however, are not able to identify kinetic limitations present in the system, as processes involving significant kinetic barriers require prohibitively long simulation timescales to model. However, various well-known computational â€œtricksâ€ are available for accelerating kinetic sampling during a simulation. Some have been implemented in the literature for ReaxFF already, such as parallel replica dynamics, meta-dynamics, and bond-boost methods. My group will continue to develop such methods for ReaxFF, and in particular will implement an â€œon-the-flyâ€ kinetic Monte Carlo technique (OF-kMC) for simulating phase formation on catalyst surfaces. During OF-kMC simulations, reaction barriers are calculated directly as the state of the system evolves, thus allowing the simulation to explore phase space without the restriction of having pre-defined reaction paths. ReaxFF is uniquely suited for this, as the required barrier calculations, which typically are prohibitively expensive at the quantum level, become computationally tractable at the ReaxFF level.

Teaching Interests:

My previous teaching experiences have prepared me to serve as the instructor for both undergraduate and graduate courses traditional to the Chemical Engineering core; namely, thermodynamics, transport phenomena, and reaction engineering. While a graduate student at Penn State, I served as a Teaching Fellow forÂ *ChE 220 Introduction to Chemical Engineering Thermodynamics*. As a Teaching Fellow, I co-instructed â€" with a ChE faculty member (Prof. Scott Milner) â€" a core undergraduate thermodynamics course, where I was responsible for developing and implementing a significant portion of the class material. I served as a Teaching Assistant forÂ *ChE 544 Transport Phenomena*, which is a core graduate transport course covering momentum, heat, and mass transport. Although I have never formally taught reaction kinetics, I am confident that I can do so given my primary background in catalysis research. In addition to teaching traditional ChE courses, I assisted my advisors, Prof. Mike Janik and Prof. Adri van Duin, in teaching their co-instructed graduate elective course, titledÂ *Atomistic Simulations for Engineers*, for two semesters. I was responsible for designing and delivering lectures covering quantum mechanical simulation techniques. As a ChE faculty member, I hope to design a similar graduate course for introducing students to atomistic simulation techniques covering both quantum and classical methodologies.

Selected Publications:

Review Article:

Senftle, T. P.; Hong, S.; Islam, M. M.; Kylasa, S. B.; Zheng, Y.; Shin, Y. K.; Junkermeier, C.; Engel-Herbert, R.; Janik, M. J.; Aktulga, H. M.; Verstraelen, T.; Grama, A.; van Duin, A. C. T., â€œThe ReaxFF reactive force-field: development, applications and future directions.â€Â *npj Computational Materials*Â 2016, 2, 15011.

Research Articles:

Senftle, T. P.; van Duin, A. C. T.; Janik, M. J., â€œThe role of site stability in describing methane activation on $Pd_xCe_{1-x}O_{Î´}$.â€Â *ACS Catalysis*Â 2015, 5 (10), 6187-6199.

Strayer, M. E.;Â **Senftle, T. P.**; Winterstein, J. P.; Vargas-Barbosa, N. M.; Sharma, R.; Janik, M. J.; Mallouk, T. E., â€œStrong adsorption of late transition metal nanoparticles correlates with sintering resistance on niobium and silicon oxides.â€Â *Journal of the American Chemical Society*Â 2015, 137 (51), 16216-16224.

Addou, R.*;Â **Senftle, T. P.***; Oâ€™Connor, N.; Janik, M. J.; van Duin, A. C. T.; Batzill, M., "Influence of hydroxyls on Pd atom mobility and clustering on rutile TiO_2(011)-(2Ã—1).â€Â *ACS Nano*Â 2014, 8(6), 6321-6333. *Equal author contribution.

Senftle, T. P.; Meyer, R. J.; Janik, M. J.; van Duin, A. C. T., â€œDevelopment of a ReaxFF potential for Pd/O and application to palladium oxide formation.â€Â *Journal of Chemical PhysicsÂ* 2013,Â 139 (4), 044109-15.

Book Chapter:

Senftle, T. P.; van Duin, A. C. T.; Janik, M. J., â€œApplication of computational methods to supported metal-oxide catalysis.â€Â *Computational CatalysisÂ* 2014, Royal Society of Chemistry, 157-191.

475819 Electrochemical Energy Transformation Processes: An Atomistic Perspective

Abstract:

Research Interests:

The upward trend in worldwide energy demand, combined with an increasingly urgent need to attenuate climate change, drives the search for renewable sources of energy. To this end, electrochemical energy conversion is a powerful technique that connects chemical energy (which serves as storage) to electrical energy, currency we can spend on a desired process. As an example of energy up-conversion, electrochemical CO_2 reduction is attractive because it potentially decreases atmospheric CO_2 concentrations while providing a carbon-neutral means to generate liquid fuels and fine chemicals. On the side of energy down-conversion and to prevent further large-scale CO_2 emissions, metal-air batteries are promising for electrifying transportation.

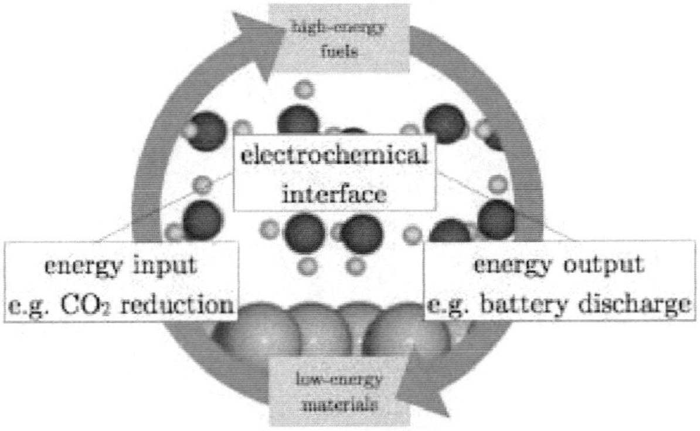

Yet even today, these energy transformation processes generally have low efficiencies. Pinpointing the source of these inefficiencies requires a detailed understanding of the stepwise mechanisms, which is where density functional theory (DFT) has been instrumental in providing insight. As illustrated by the graphic, my PhD work has three distinct yet interconnected topics with a central theme on studying these **electrochemical energy transformation processes from an atomistic perspective**:

(1) Anodic dissolution in metal-air batteries where electrical energy is the output: with DFT, I was able to elucidate the fundamental limitations on total energy output in the anodic dissolution of Al and Mg in aqueous electrolytes.

(2) Electrochemical CO_2 reduction where electrical energy is the input: I demonstrated how including the effects of electric field (in addition to potential alone) in the DFT model can significantly alter the free energy landscape of a catalytic process.

(3) The electrochemical interface, an indispensable component of these energy transformation processes: I have made progress in describing charged species at the outer Helmholtz plane and their implications on reaction energetics with DFT.

Teaching Interests:

Sharing knowledge and communicating effectively is very important to me as a scholar. In my graduate career, I have served as a Teaching Assistant (TA) for three different undergraduate courses: Organic Chemistry Laboratory, Organic Mechanisms Lecture, and Biochemistry. The most valuable undergraduate teaching experience for me came from the Organic Mechanisms Lecture course, for which I led two-hour discussions twice a week. Preparation was key for these sessions to engage my students in group problem-solving and fruitful discussions. I read up on the course material as thoroughly as possible to answer probing questions from my students.

I also served as a TA in a graduate course on Molecular Modeling. Here, the challenge came in the form of creating problem sets with the appropriate level of scope and difficulty for students who only had one fewer year of experience. This required some independent work on my part searching through textbooks and online resources for inspiration to problems that would be both in-line with the course content and interesting.

Proposal Experience:

Multidisciplinary University Research Initiatives (MURI) Program 2015

Copyright © American Institute of Chemical Engineers. All rights reserved.

475913 Nanostructured Hybrid Materials: Directing Catalytic Activity and Selectivity By Design

Abstract:

Research Interests: and Teaching Interests:

Nanostructured Hybrid Materials: Directing Catalytic Activity and Selectivity by Design

In the field of catalysis, there is a need for highly active, selective, and stable catalysts for more efficient and cleaner use of existing and alternative energy sources. In this respect, hybrid materials with multifunctional properties show great promise and offer several important advantages over traditional catalysts. A rational design of hybrid systems provides a platform (i) to combine heterogeneous and homogeneous catalysis, which have been explored rather separately, and employ the best of the both fields, (ii) to produce extraordinary or synergystic properties at the interface of different types of materials, (iii) to carry out consecutive reactions in tandem, and (iv) to exploit the flexibility and diversity of organic, inorganic, and metallic materials. In addition, recent breakthrough in the synthesis of nanostructures and discovery of exotic materials renders the investigation of hybrid systems an attractive and promising area of research.

In this poster, I will present a brief overview of my graduate and postdoctoral studies, and my research interests and future plans on the topic. My graduate research primarily focused on the synthesis of TiO_2 nanocrystals with well-defined particle size and characterization of particle growth and phase transformation mechanisms. First, I experimentally demonstrated that the particle size and particle interface produced during nanocrystal growth play a crucial role in the thermodynamic stability of nanocrystals. Then, I developed a mathematical model to describe the types of mechanisms and characterize the kinetics of phase transformation.

With the expertise and experience in the synthesis and characterization of nanostructured materials acquired during my graduate years, as a postdoctoral researcher I work towards designing multicomponent catalysts and testing them on scientifically and industrially important chemical reactions. I have synthesized highly selective bifunctional catalyst comprising platinum nanoparticles with finely tuned particle size and mesoporous silica post-synthetically modified with aluminum, and investigated its catalytic properties on n-hexane (n-C_6H_{14}) and n-hexadecane (n-$C_{16}H_{34}$) conversions, model reactions for industrial reforming of naphtha and heavy-gas oil, respectively. In addition, metal-organic framework (MOF) was demonstrated to encapsulate and disperse a large amount of phosphotungstic acid ($H_3PW_{12}O_{40}$), a solid acid with high catalytic activity and strong acidity, but with extremely low surface area. Combined with platinum nanoparticles, the MOF-based material was catalytically active towards gas-phase isomerization of n-hexane.

In the future, I aim to establish a research program on the fundamental investigation of molecular-level electronic interaction and structure-property relationship between organic and inorganic materials and the design of hybrid catalysts for environmentally and industrially important chemical transformations. I will utilize the unique structural, chemical, and compositional features of metal-organic and covalent organic frameworks, chelating ligands and capping agents, and electronically and catalytically active nanostructures to achieve the objectives mentioned in the introduction.

Copyright © American Institute of Chemical Engineers. All rights reserved.

475963 Towards the Computational Design of Heterogeneous Electrocatalysts

Abstract:

Research Interests:

Design of cost-effective electrocatalysts with enhanced activity and stability is of paramount importance for the next generation of energy conversion systems, including fuel cells and electrolyzers. Ultrathin (hydroxy)oxide films on transition metal substrates possess impressive activities for these reactions. These hybrid systems with novel properties are often prepared and characterized under very specific conditions, such as ultrahigh vacuum. When subsequently used in various applications, however, significant structural evolution, which depends strongly on the reaction conditions in-situ, is expected but remains largely unexplored. In order to identify structure-property relationships and, ultimately design/screen new materials with improved performance, the development of such understanding is essential.

The complex interfaces have long been a challenge for modeling, because of the complications introduced by strongly-correlated transition metal oxides and mysterious interface structures. As a consequence, even many ancient questions are still unanswered, for example the origin of the strong metal support interaction (SMSI) that plays a central role on the stability of the interfaces.

Recently, I have resolved the major challenge toward self-consistent and highly accurate description of strongly-correlated transition metal oxides, hydroxides and oxyhydroxides, through synergistic use of a Hubbard U correction, a state-of-the-art dispersion correction, and a water-based bulk reference state for the calculations. The strong performance is illustrated on a series of bulk transition metal (Mn, Fe, Co and Ni) hydroxides, oxyhydroxides, binary, and ternary oxides, where the corresponding thermodynamics of redox and (de)hydration are described with standard errors of 0.04 eV per (reaction) formula unit.

I have applied these methodologies to investigate the growth of monolayer Ti- and Fe-(hydroxy)oxide films on VIIB group and IB group noble transitional metal surfaces, and have obtained mechanistic understanding of the origin of strong metal support interaction. This understanding has been extended to predict the structure evolution of the other monolayer transitional metal (Mn, Co, and Ni) (hydroxy)oxide films on a variety of substrates. It has been demonstrated that the structure and oxidation state of the films can be systematically tuned by changing the applied electrode potential, the nature of substrates and/or the size of film cluster, which is further confirmed by synchrotron experiment on the monolayer Ni-(hydroxy)oxide/Pt, a hydrogen evolution (HER) electrocatalyst with improved performance in comparison with traditional Pt catalyst in alkaline conditions.

The identified interface under hydrogen evolution HER condition, NiOH/Pt, is subsequently used for a detailed structural and electrocatalytic analysis at three-phase boundaries of NiOH films, Pt substrates, and the surrounding water. The kinetic analysis shows that water can be dissociated to adsorbed OH* and H* groups in a bifunctional mechanism that significantly lowers the activation barriers for alkaline HER and provides a clearly measureable acceleration of the catalyst process as compared to mono-functional Pt catalysts. The mechanistic understanding and kinetic analysis also suggest a new catalyst, NiOH/Pt_3Ni, with improved performance in comparison with NiOH/Pt, which is further confirmed by electrochemical experiment.

The radical development on highly accurate description of strongly-correlated transition metal oxides, the mechanistic understanding of interface structure evolution and structure-properties relationships, and the successful prediction of new electrocatalyst has provided the foundation towards rational design of novel catalysts with increased activity, improved selectivity and/or under-controlled stability. In my future research, by leveraging and combining my experience, we will open up fundamentally new research thrusts, providing fundamental scientific insight into the interfacial properties of the relevant systems, and ultimately identifying new electrocatalysts of fuel cells, electrolyzers, CO oxidation and CO2 reduction, and heterogeneous catalyst of CO oxidation and water gas shift reactions.

Teaching Interests:

Albert Einstein once said "Most of the fundamental ideas of science are essentially simple, and may, as a rule, be expressed in a language comprehensible to everyone". I would like to say it is also a rule for my teaching. I find that most courses are essentially made up of concepts, not simply equations or mathematics. The concepts may be expressed in many specific ways and used in different fields, but the essence does not change, which is exactly like the second law in thermodynamics. The essential simplicity and generality make the concepts can be retained in students' memories for a long time period, and can be smoothly used to different fields later in life. Thus, good teaching should focus on the concepts; this is what I have learned from my study and research experience. For example, there are so many equations and derivations I learned but never use in my research; however, I still clearly remember conceptual mechanisms behind and can use them correctly when necessary. I have found this strategy useful when I teach "Advanced Modeling for Catalysis Studies" class and when I supervised graduate and undergraduate students. It has been highly effective when I first introduced the essential concept, then explained the role of each individual equation, and left the details for the students themselves to deepen the understanding. Thus my future teaching will mainly focus on the ideas and concepts, and using mathematics as a tool to express them more simply.

Copyright © American Institute of Chemical Engineers. All rights reserved.

475985 Engineering the Electrochemical Interface for Sustainable Energy Conversion and Storage

Abstract:

Center for Individual Nanoparticle Functionality, Technical University of Denmark, Maria.Escudero@fysik.dtu.dk

Department of Chemical Engineering, Stanford University, mariaesc@stanford.edu

Research Interests:

One of the most important challenges facing humanity is the switch from fossil fuels to sustainable energy. Electrochemical energy conversion and storage devices such as fuel cells, electrolyzers and batteries should play a key role in this switch. My research combines electrochemical methods with surface science X-ray based characterization techniques, in situ spectroscopy and scanning probe microscopy, in order to investigate problems pertaining to interfacial electrochemistry and electrocatalysis. Reactions of interest include water splitting, oxygen reduction, hydrogen oxidation, oxidation of carbon monoxide and methanol, carbon dioxide reduction and hydrogen peroxide production. My scientific interests can be summarized as follows:

(a) Studying the descriptors determining the activity, stability and selectivity of electrocatalysts for energy conversion and storage devices [1-3].

(b) Determining and designing the catalytically active site [1-5].

(c) Investigating the role of the geometric structure (atomic ensemble effects [1]) and electronic properties of the surface atoms (electronic effects [2,3]) in electrocatalysis.

(d) Engineering the electrode-electrolyte interface [1,6].

Teaching Interests:

Not only have I a strong dedication to my scientific research, but also to teaching. Since 2013 I have taught a graduate course on Materials for hydrogen production and storage at the Technical University of Denmark. As a Chemical Engineer with a PhD in Physical Chemistry, I am interested in teaching both undergraduate and graduate courses in Sustainable Chemistry, Physical Chemistry, Catalysis, Spectroscopy and Chemical Engineering.

References:

[1] D. Strmcnik, **M. Escudero-Escribano**, K. Kodama, V. Stamenkovic, A. Cuesta, *Nature Chem*. **2010**, *2*, 880.

[2] **M. Escudero-Escribano**, et al., *J. Am. Chem. Soc*. **2012**, *134*, 16476.

[3] **M. Escudero-Escribano**, et al., *Science* **2016**, *352*, 73.

[4] A.F. Pedersen, et al., *Nano Energy* **2016**, doi:10.1016/j.nanoen.2016.05.026

[5] P. Malacrida, **M. Escudero-Escribano**, A. Verdaguer-Casadevall, I.E.L. Stephens, I. Chorkendorff, *J. Mater. Chem. A* **2014**, *2*, 4234.

[6] **M. Escudero-Escribano**, et al., *ChemPhysChem* **2011**, *12*, 2230.

Copyright © American Institute of Chemical Engineers. All rights reserved.

476000 Controlled Synthesis and in-Situ Spectroscopic Study of Highly Efficient Ptfe Bimetallic and Ptrufe Trimetallic Nanocatalysts

Abstract:

Research Interests: Heterogeneous catalysis

Teaching Interests: Heterogeneous catalysis

Controlled synthesis of highly efficient nanocatalysts and studying their structure-activity relationship are of significant importance in both fundamental research and industrial applications. In this study, we developed several well defined Pt-based nanocatalysts with quasi-core-shell structure or nanoalloy structure for the preferential oxidation of CO in the presence of H_2 (PROX), which is a key reaction for the production of clean H_2. By means of various ex- or in-situ characterization method, such as HR-TEM, quasi-in-situ XPS and in-situ DRIFTS, the structure-activity relationships of these catalysts have been extensively studied. It was concluded that ferrous iron species could work as oxygen activation sites, and Ru could lower the CO adsorption strength. Furthermore, the utilization of Pt was much higher on nanoalloy catalysts as compared with quasi-core-shell catalysts. Based on these findings, highly efficient PtFe and PtRuFe nanoalloy catalysts with ultra-low Pt loading were designed, on which CO can be completely removed even under -50 ºC. We have also developed a novel strategy to in-situ monitoring the adsorbed species on the nanocatalysts by shell-isolated nanoparticles-enhanced Raman spectroscopy (SHINERS), which was then proved to be a general strategy for the in-situ study of heterogeneous nanocatalytic process. By this method, the interactions between CO/O_2 and metal can be studied, and for the first time, direct spectroscopic evidence for the activation of O_2 on ferrous center has been observed.

Copyright © American Institute of Chemical Engineers. All rights reserved.

476052 Developing Enhanced Catalysts for Renewable Fuels through Spectroscopic Insights

Abstract:

Research Interests:

My research interests focus on the pursuit of fundamental understanding of catalytic reactions towards production of renewable fuels using insights from *in situ* and *operando* spectroscopy techniques uniquely adapted to investigate these catalytic materials. Increased global energy consumption along with heightened awareness of environmental, health, and political issues with fossil fuels are driving the need for efficient and economic conversion of renewable energy sources. Wind and sun provide more than enough energy to meet the growing demand, provided that challenges with intermittency, scale, and cost effectiveness can be overcome. These obstacles can be mitigated through development of highly active catalysts using abundant and inexpensive materials to convert solar and wind energy into fuels and chemicals.

My Ph.D. dissertation with Professor Thomas Jaramillo at Stanford University in the Department of Chemical Engineering focused on development of enhanced catalysts for electrochemical water splitting. Using a combination of electrochemical and spectroscopic techniques I was able to draw correlations between the structure, oxidation state, and catalytic activity of various transition metal oxide systems to gain a better understanding of their fundamental operation. For example, using x-ray absorption spectroscopy, I was able to track the changing oxidation state of a manganese-based catalyst in the presence and absence of a small amount of gold to understand why the addition of gold causes an order of magnitude increase in catalytic activity. During my Ph.D. I also developed a model to quantify loss mechanisms in photoelectrochemical water splitting based on the current state of materials research and to calculate maximum solar-to-hydrogen conversion efficiencies. Knowledge of practical limits for this process can help researchers assess their technology and guide future directions.

Currently, I am in my first year of postdoctoral research with Professor Clemens Heske and Dr. Lothar Weinhardt at the Karlsruhe Institute for Technology in the Institute for Photon Science and Synchrotron Radiation. I was awarded a Fellowship by the Helmholtz Association Postdoc Programme to pursue my proposed project of using advanced spectroscopic techniques to investigate the chemical and electronic structure of catalysts under operating conditions. During my time here I plan to assist in the development of advanced experimental set-ups for *in situ* and *operando* soft x-ray spectroscopy measurements and expand the application of these techniques to study catalyst materials for electrochemical and thermochemical CO/CO_2 reduction as well as NO_x selective catalytic reduction. *In situ* hard and soft x-ray experiments for these reactions are just beginning to be realized and because these techniques are such a powerful probe to study electronic structure, there is great potential to expand work and knowledge in this area.

At the poster session I will present a more thorough overview of my past and present research, as well as my plan for future research activities which will continue to seek fundamental insight towards improving efficiency for the conversion of renewable energy sources to fuels and chemicals. By studying, understanding, and improving the various components of these technologies, it is my goal to help drive the field forward and provide an efficient and economical solution that can shift our global energy dependence away from fossil fuels and towards renewable energy.

Teaching Interests:

In addition to serving as a teaching assistant during my undergraduate and graduate studies for computer programming and reaction kinetics courses, I have been an invited guest lecturer for various courses at Stanford University including graduate level Fundamentals and Applications of Spectroscopy, Electrochemical Energy Conversion, undergraduate Chemical Engineering Lab, and a Solar Fuels Master Class. I have also enjoyed mentoring graduate rotation students and a visiting masterâ€™s student as well as volunteering at local schools for science demonstrations.

Copyright © American Institute of Chemical Engineers. All rights reserved.

476055 Computational Design of Hetero-Structured Catalysts for Energy

Abstract:

Research Interests:

Given the crucial role of catalysis science in our society, understanding how to design catalysts with the proper structure for desired catalytic activity and selectivity is a grand challenge. Hetero-structured materials such as core shell nanoparticles provide an extra degrees of freedom for activity tuning and control.

My expertise is in computational materials and catalyst design in close collaboration with experimental collaborators. My PhD research was dedicated to developing a fundamental and detailed understanding of the relationship between the structure of nanoscopic metallic catalysts and their function using first participial simulations. Instead of quantifying absolute catalytic activity; rather, I showed in a quantitative way how reactivity descriptors (which are known to correlate with the activity) vary with the particle structure and composition. During my post-doc at Stanford University and University of Pennsylvania, I have extended my research to a broader pool of materials with a higher complexity, such as metal oxides, metal carbides, and two-dimensional materials, and diversified my computational tool set.

My future research directions will focus, but not be limited to:

1) Development of a general picture for structure-reactivity correlation of different classes of hetero-structures including core shell nanoparticles.

2) Development of an understanding of dynamic catalytic processes on nanoparticles. The descriptor-based screening approach only provides a static and mean-field description of the catalysis on a nanoparticle. In reality, the morphology of the nanoparticle is varying during reaction, where a dynamic understanding is crucial to establish an accurate access of the catalytic function of the nanoparticle.

3) Development of a long-time scale simulation method suitable for distributed computing environments and its application to catalytic processes of complex materials.

Post-doctoral Experiences

University of Pennsylvania, Chemical and Biomolecular Engineering Supervisor: Aleksandra Vojvodic

Projects (a. Hetero-structured materials)

Stanford University, Chemical Engineering

Supervisor: Jens. K. Nørskov

Projects (a. Tailoring the Catalytic Activity of Perovskites b. CO activation under confinement)

PhD Thesis:

University of Texas at Austin, Chemistry

Supervisor: Graeme A. Henkelman

Dissertation: "Theoretical study of correlation between structure and function for nanoparticle catalysts"

Teaching Interests:

I have two years teaching experiences as a research educator of the Freshman Research Initiative (FRI) program at UT-Austin, where undergraduate students are exposed to authentic theoretical chemistry research. In addition to lecturing classes, I have TA:ed Physical Chemistry lab classes at UTAustin. In addition, I have also mentored graduate students in my group.

Selected Publications:

1. **L. Zhang** and G. Henkelman, Computational design of alloy-core@shell metal nanoparticle catalysts, *ACS Catal.* 5, 655-660 (2015).
2. **L. Zhang**, S. Chill, and G. Henkelman, Distributed Replica Dynamics *J. Chem. Phys.* 143, 174112 (2015).
3. S. García, **L. Zhang**, G. W. Piburn, G. Henkelman, and S. M. Humphrey, Microwave Synthesis of Classically Immiscible Rhodium-Silver and Rhodium-Gold Alloy Nanoparticles: Highly Active Hydrogenation Catalysts, *ACS Nano* 8, 11512-11521 (2014).
4. G. M. Mullen, **L. Zhang**, E. J. Evans Jr., T. Yan, G. Henkelman, and C. B. Mullins, Oxygen and hydroxyl species induce multiple reaction pathways for the partial oxidation of allyl alcohol over Au(111), *J. Am. Chem. Soc.* 136, 6489-6498 (2014)
5. **L. Zhang**, R. Iyyamperumal, D. F. Yancey, R. M. Crooks, and G. Henkelman, Design of Pt-shell nanoparticles with alloy cores for the oxygen reduction reaction, *ACS Nano* 7, 9168-9172 (2013).
6. **L. Zhang**, H.-Y. Kim, and G. Henkelman, CO oxidation at the A...
7. R. Iyyamperumal, **L. Zhang**, G. Henkelman, and R. M. Crooks, Efficient electrocatalytic oxidation of formic acid using Au@Pt dendrimer-encapsulated nanoparticles, *J. Am. Chem. Soc.* 135, 5521-5524 (2013).

Copyright © American Institute of Chemical Engineers. All rights reserved.

*****Formatting Issue - Best Available Paper/Graphic*****

476074 Efficient Catalytic Pathways for Carbon Utilization

Abstract:

Research Interests:

Development of emission control technologies for remediation of toxic gases from stationary and mobile combustion sources is of great interest due to the stricter regulations and changes in the emissions as a result of the improvements in burner and engine technologies. Within my previous research efforts, one of my main objectives was to understand pollutant formation mechanisms to develop better emission remediation technologies. In particular, I have worked on removal of trace metals from coal-fired power plants, and elucidated the heterogeneous reaction mechanism of mercury over activated carbon. Additionally, I have developed a low temperature hydrocarbon oxidation catalyst for diesel exhaust applications. Aside from emission control technologies, my research focused on spectroscopic investigation of CO adsorption and oxidation over single crystal surfaces at different pressures, developing a catalyst for cracking of hydrocarbons to valuable low hydrocarbons and designing analytical instruments for high-throughput experimentation. The wide range of research topics that I have studied has provided me with vast experience in catalysis, spectroscopy, and instrument design and development.

As a tenure-track faculty, I would like to continue my research on developing nano materials and novel processes in the area of low temperature exhaust remediation, CO_2 utilization, tri-reforming of methane and alternative synthesis routes for the production of C_2-C_4 hydrocarbons. Currently, in most of the states, coal and natural gas power plants still remain the largest energy source for electricity production and play a crucial role to compensate the base loads in the electricity grid. Therefore, there is an urgent need for developing breakthrough technologies for CO_2 utilization and capture, and advanced combustion systems that can be implemented to the existing fossil fuel-fired power plants. I will initially develop my research program focusing on the aforementioned areas in carbon utilization and exhaust remediation and expand my research to create alternative energy pathways. My goal is to create a collaborative research environment interacting with national laboratories and industrial partners, and ultimately bridge the gap between fundamental science and industrial applications. In my poster presentation, I will discuss different projects to demonstrate my expertise in the area of catalysis and spectroscopy.

Teaching Interests:

Teaching has been my passion throughout my education and academic career. I gave lectures in various chemical engineering courses: Chemical Engineering Thermodynamics I, Chemical Kinetics and Modeling, Chemical Engineering Design, Chemical Plant Design Project, Mass Transfer, Kinetics and Reactor Design and Molecular Modeling. I am very interested in teaching core chemical engineering courses such as thermodynamics, heat transfer and chemical kinetics. I also plan to develop a special topics course on in-situ spectroscopy for surface science applications. I will discuss my teaching vision during the poster presentation.

References

1. Sasmaz, E.; Mingle, K.; Lauterbach, J. â€œHigh-throughput screening using Fourier transform infrared imagingâ€ *Engineering,* 2015, 1, 234.
2. Wang, C.; Sasmaz, E.; Wen, C.; Lauterbach, J. â€œPd supported on SnO_2-MnO_x-CeO_2 catalysts for low temperature CO oxidationâ€ *Catalysis Today,* doi:10.1016/j.cattod.2015.02.021.
3. Kim, S.; Mayeda, K.; Sasmaz, E.; Lauterbach, J. â€œOne step process for the production of BTEX and LPG-like fuel from pentanediolâ€ *ACS Sustainable Chemical Engineering*, 2015, 3, 381.
4. Kim, S.; Sasmaz, E.; Lauterbach, J. â€œEffect of Pt and Gd on coke formation and regeneration during JP-8 cracking over ZSM-5 catalystâ€ *Applied Catalysis B: Environmental,* 2015, 168, 212.
5. Galloway, B.D.; Sasmaz, E.; Padak, B. â€œBinding of SO_3 to Fly Ash Components: CaO, MgO, Na_2O, and K_2Oâ€ *Fuel*, 2015, 145, 179.
6. Lauterbach, J.; Sasmaz, E.; Bedenbaugh,J.; Kim, S.; Hattrick-Simpers, J. â€œDiscovery and optimization of coking and sulfur resistant bi-metallic catalyst for cracking JP-8: from thin film libraries to single powdersâ€, Modern Applications of High Throughput R&D in Heterogeneous Catalysis, 2014, 84-112
7. Bedenbaugh,J.; Kim, S.; Sasmaz, E.; Lauterbach, J. â€œHigh-throughput investigation of zeolite catalysts for military aviation fuel cracking to liquefied petroleum gasâ€ *ACS Combinatorial Science*, 2013, 15, 491.
8. Sasmaz, E.; Kirchofer, A.; Saha, A.; Abram, D.; Jaramillo, T. F. Wilcox, J. â€œMercury chemistry on brominated activated carbonâ€ *Fuel,* 2012, 99,189.
9. Sasmaz, E.; Aboud, S.; Wilcox, J. â€œMercury binding on Pd alloys and overlaysâ€ *J. Phys. Chem. C,* 2009, 113, 7813.
10. Sasmaz, E.; Wilcox, J. â€œMercury species and SO_2 adsorption on CaO(100)â€ *J. Phys. Chem. C*, 2008, 112, 16484.

476116 New Routes and Heterogeneous Catalysts Development for Biomass Conversion

Abstract:

Research Interests: Heterogeneous Catalysis, Biomass Conversion, Materials Science, Platform Technology

Teaching Interests: Inorganic Chemistry and Catalysis, Applied Chemical Technology

Crude oil market volatility, together with pressing concerns about greenhouse gas effects offer opportunities for the industrial-scale production of renewable fuels and chemical building blocks from biomass. In this light, the development of active, selective and stable solid catalysts, and bio-based approaches is necessary to achieve a more sustainable and economically viable catalytic conversion of such resource. My doctoral research was about the design and development of multifunctional materials, catalysts and technologies for energy-related and environmental applications, conducted in the University of Connecticut (USA) under the supervision of Prof. Steven L. Suib and supported by the US DOE. After obtaining my PhD in Chemistry in 2013, I moved to the Netherlands and worked with Dr. Pieter C. A. Bruijnincx and Prof. Bert M. Weckhuysen as a postdoctoral research fellow in Utrecht University (Inorganic Chemistry and Catalysis). For my first project, I explored new avenues in developing strategies and heterogeneous catalysts such as bifunctional solid acids for the efficient (tandem-catalytic) Diels-Alder aromatization of biomass-derived furanics in the liquid and the solid phase. This project was in collaboration with Wageningen University and Research Center - Food and Biobased Research and our industrial partner, Avantium Chemicals B.V. in Amsterdam. For my second and current project, I have been investigating the influence of synthesis, feed impurities, and process parameters on the stability of atomically dispersed metal catalysts anchored to transition metal oxide supports for levulinic acid hydrogenation. This project is in collaboration with University of Groningen (Engineering and Technology Institute) within the framework of the CatchBio Program and supported by BASF and DSM companies. Exciting recent results from these two postdoctoral research projects will be presented in this poster.

In the near future, my research work will continue to focus on the development of new approaches and high-performing heterogeneous catalysts for the valorization of bio-based platform molecules. It will aim to gain fundamental understanding of reaction engineering and catalytic processes across length scales, including the elucidation of structure-function relationships and expert systems, the investigation of inorganic catalyst stability, deactivation and mitigation, and the application of advanced in situ and operando techniques to understand reaction mechanism. I strongly believe that biomass catalysis and technology research will aid the transition of our chemical industry to a more sustainable one, as well as promote collaboration among various fields and sectors at both national and international levels.

Copyright © American Institute of Chemical Engineers. All rights reserved.

476170 Transition Metal-Oxides for Sustainable Energy Conversion and Storage: The Computational Catalysis Perspective

Abstract:

Research Interests: The main goal of my research is to use the state-of-the-art computational methods in tandem with experiment to achieve understanding of the structure and the reaction mechanism of the most promising heterogeneous catalysts [1,2]. The largest breakthrough is needed in the catalysis of the electro-chemical water splitting, CO_2 reduction and ammonia synthesis, where the efficient, stable and earth-abundant catalysts are still to be discovered. I am especially interested in the catalytic functionality of the novel nanostructured first-row transition-metal oxides [3], which can be precisely synthesized and exhibit high level of activity and tunability.

Teaching Interests: Core level courses Chem. E, Chemistry Physics and Material Science. Ideally envision designing my own course in either Computational heterogeneous catalysis or in the Modern approaches to computational materials design.

[1] Bajdich, Sargent,el.al., Homogeneously-Dispersed Multi-Metal Oxygen-Evolving Catalysts, Science10.1126/science.aaf1525 (2016).

[2] Bajdich, Jaramillo, el.al., Gold-supported cerium-doped NiOx catalysts for water oxida- tion, Nature Energy10.1038/NENERGY.2016.53 (2016)

[3] Bajdich, Lauritsen, ACS Nano10.1021/acsnano.5b00158 (2015); Special Issue in Topics in Catalysis (2016); Nat. Materials (under review).

Copyright © American Institute of Chemical Engineers. All rights reserved.

476301 Impacts of Metal-Adsorbate Bonds on Photon Activation Mechanism and Nanoparticle Reconstruction in Heterogeneous Catalysis

Abstract:

More than 80% of modern industrial chemical processes utilize heterogeneous catalysts, including the production and processing of fuel, pollution mitigation, and commodity chemical production. The control of selectivity in complex chemical reaction networks has become the focus of heterogeneous catalysis in the 21st century, in order to decrease unwanted byproducts and to increase energy efficiency and sustainability. My work at UC-Riverside has focused on several approaches to determine how molecules interact with catalytic metal surfaces, and ways to specifically activate targeted molecules on the surface, with the goal of achieving high catalytic selectivity. These approaches include: utilizing photocatalysis in order to activate targeted strong metal-adsorbate bonds by matching photon wavelength with bond energies of hybridized metal-adsorbate bonds, modifying metal-support interfaces to efficiently remove atmospheric contaminants in confined space pollution control applications, and exploiting *in-situ* spectroscopy techniques with a correlated analysis of reactivity in order to reveal catalytic active sites. My future work will include utilizing *in-situ*probes to reveal fundamental insights of catalytic processes for sustainable useful chemical production.

Research Interests:

Heterogeneous catalysis, *In-situ* spectroscopy, selective chemical production, conversion of solar to chemical energy

Teaching Interests:

Preferences: Chemical Kinetics, Catalysis & Reaction Engineering, Physical Chemistry, Thermodynamics, Transport Phenomena, but would be open to teaching most undergraduate courses.

Selected Publications:

[1] **Kale, M. J.**; Christopher, P. Plasmons at the interface. *Science* **2015,** *349*, 587-588.

[2] **Kale, M. J.**; Avanesian, T.; Xin, H.; Yan, J.; Christopher, P. Controlling Catalytic Selectivity on Metal Nanoparticles by Direct Photoexcitation of Adsorbate-Metal Bonds. *Nano Letters* **2014,** *14*, 5405-5412.

[3] **Kale, M. J.**; Avanesian, T.; Christopher, P. Direct Photocatalysis by Plasmonic Nanostructures. *ACS Catalysis* **2014**, *4*, 116â€“128.

[4] **Kale, M. J.**; Christopher, P. Utilizing quantitative in-situ FTIR spectroscopy to identify well-coordinated Pt atoms as the active site for CO oxidation on Al_2O_3 supported Pt catalysts. *In Review.*

[5] **Kale, M. J.**; Gidcumb, D.; Christopher, P. Evaluation of Precious Metal Catalyst Stability for Confined Space Pollution Control. *In Preparation*.

Copyright © American Institute of Chemical Engineers. All rights reserved.

475756 Electrocatalysis for Sustainable Energy Technologies

Abstract:

ELECTROCATALYSIS FOR SUSTAINABLE ENERGY TECHNOLOGIES

Drew Higgins, Postdoctoral Fellow, Stanford University

RESEARCH INTERESTS

Achieving a sustainable energy economy is among the biggest challenges of the 21st century. Electrochemical energy storage and conversion devices (***i.e.***, batteries, fuel cells, electrolyzers) will be integral components of clean energy infrastructures, as they boast zero emission operation and high theoretical efficiencies. These systems, however, are not currently techno-economically viable at scale, primarily due to electrocatalyst limitations. My research interests lie in the development and device integration of advanced, high performance electrocatalysts. These materials have the potential to address the performance, durability and cost challenges facing many emerging clean electrochemical technologies. The ultimate goal is to transform the landscape of the energy sector by providing cost competitive devices that can be applied in the transportation sector, grid scale energy storage, or industrial applications.

FUTURE PLANS

My career goal is to manage a research group at an academic institution that works on the cutting edge of clean energy technologies, particularly with respect to energy transformations and electrocatalysis. In the short term, my research will be focused on two thrust areas: (1) ***Electrochemical Carbon Dioxide Reduction***, an integral process of artificial photosynthesis that converts CO_2 directly into fuels or industrial chemicals; and (2) ***Oxygen Electrocatalysis,***the most pertinent and performance limiting processes occurring in fuel cell and electrolyzer devices. I envision early work focusing on the development of new electrocatalyst materials that will be characterized and electrochemically evaluated to understand performance capabilities. The best in class electrocatalysts will be engineered into electrode structures for device performance testing and optimization.

In the long term, my vision is to lead a research program that spans: (1) ***Fundamental Electrocatlaysis/Surface Science***; (2) ***Electrocatalyst Design and Synthesis***; and (3) ***Electrochemical Device Design, Integration and Prototype Demonstration***. These three core areas will be strategically integrated to operate in a feedback loop, whereby new knowledge generated in one area will be applied to achieve advances in other areas. Over time, this research program will be expanded to encompass different thrust areas that are synergistic with those listed above, including hydrogen electrochemistry, new battery chemistries (***i.e***., metal-air or sodium-ion) and electrochemical purification techniques. The overarching goals of this research program will be two-fold. First, is to develop innovative material solutions and technologies that address the pertinent energy and environmental challenges facing the modern world. Second, is to educate and train the next generation of scientists and engineers that will be lifelong contributors in this sector of increasing global importance. Academic, industrial and government collaborations will be fostered and leveraged to establish an interdisciplinary research environment, and to provide targeted training and avenues for research personnel to enter the workforce.

RESEARCH EXPERIENCE

Banting Postdoctoral Fellow – Chemical Engineering (start date, Sep 2015)

Stanford University, Stanford, California

Supervisor: Dr. Thomas Jaramillo

- Electrochemical reduction of carbon dioxide to produce fuels and chemical products
- Physical vapour deposition of binary metallic thin films to serve as model electrocatalyst systems
- Integration of electrodes into carbon dioxide electrolysis cell for performance testing and product detection

Visiting Scholar – Materials Physics and Applications 11 (Jan 2014 – Dec 2014)

Los Alamos National Laboratory, Los Alamos, New Mexico

Supervisor: Dr. Piotr Zelenay

- Synthesis and integration of non-precious metal catalysts into polymer electrolyte fuel cell membrane electrode assemblies for performance validation

PhD and MASc – Department of Chemical Engineering(2009 –2015)

University of Waterloo, Waterloo, Ontario

Supervisor: Dr. Zhongwei Chen

- Synthesis, characterization and electrochemical evaluation of nanostructured oxygen reduction and oxygen evolution catalysts

Undergraduate Researcher – Department of Chemical and Biochemical Engineering (2007, 2008)

University of Western Ontario, London, Ontario

Supervisor: Dr. Jingxu (Jesse) Zhu

SELECTED SUCCESSFUL PROPOSAL WRITING

- Banting Postdoctoral Fellowship
- Natural Sciences and Engineering Research Council of Canada (NSERC) Postdoctoral Fellowship
- NSERC Discovery Grants Program
- 2 x NSERC Strategic Project Grants
- 3 x NSERC Collaborative Research and Development Grants
- Ontario Centres of Excellence: Collaborative Research Program
- Ontario Centres of Excellence: Voucher for Commercialization
- Ontario Research Fund: Small Infrastructure Fund
- NSERC Postgraduate Fellowship
- Ontario Graduate Scholarship

TEACHING EXPERIENCE

- **Graduate Student Mentor**, Stanford University (2015 – present), University of Waterloo (2012 - 2015)
- **Guest Lecturer** for ChE 331, ChE 102 and NE 491, University of Waterloo (2012 - 2015)
- **Certificate of University Teaching Program,** University of Waterloo (2012 – 2013)
- **Fundamentals of University Teaching Program,** University of Waterloo (2011)
- **Teaching Assistant** for ChE 331 and ChE 290, University of Waterloo (2010 – 2011)

SELECTED PUBLICATIONS

1. **D. Higgins**, P. Zamani, A. Yu, Z. Chen, "The application of graphene and its composites in oxygen reduction electrocatalysis: a perspective and review of recent progress", ***Energy & Environmental Science***, 9 (2016) 357-390.
2. **D. Higgins**, F. Hassan, M. Seo, J. Choi, A. Hoque, D. Lee, Z. Chen, "Shape-controlled octahedral cobalt disulfide nanoparticles supported on nitrogen and sulfur-doped graphene/carbon nanotube composites for oxygen reduction in acidic electrolyte", ***Journal of Materials Chemistry A***, 3 (2015) 6340-6350. Featured on the back cover of the associated issue.
3. A. Hoque, F. Hassan, **D. Higgins**, J. Choi, M. Pritzker, S. Knights, S. Ye, Z. Chen, "Multigrain platinum nanowires consisting of oriented nanoparticles anchored on sulfur-doped graphene as a highly active and durable oxygen reduction electrocatalyst", ***Advanced Materials***, 27 (2015) 1229-1234.
4. **D. Higgins**, R. Wang, A. Hoque, P. Zamani, S. Abureden, Z. Chen, "Morphology and composition controlled platinum-cobalt alloy nanowires prepared by electrospinning as oxygen reduction catalyst", ***Nano Energy***, 10 (2014) 135-143.
5. **D. Higgins**, A. Hoque, M. Seo, R. Wang, F. Hassan, J. Choi, M. Pritzker, A. Yu, J. Zhang, Z. Chen, "Development and Simulation of Sulfur-doped Graphene Supported Platinum with Exemplary Stability and Activity Towards Oxygen Reduction", ***Advanced Functional Materials***, 24 (2014) 4325-4336. Accepted as VIP article and featured on front-piece of the associated issue.
6. **D. Higgins**, A. Hoque, F. Hassan, J. Choi, B. Kim, Z. Chen, "Oxygen Reduction on Graphene-Carbon Nanotube Composites Doped Sequentially with Nitrogen and Sulfur", ***ACS Catalysis***, 4 (2014) 2734-2740.
7. V. Chabot, **D. Higgins**, A. Yu, X. Xiao, Z. Chen, J. Zhang, "A Review of Graphene and Graphene

Oxide Sponge: Material Synthesis and Applications", ***Energy & Environmental Science***, 7 (2014) 1564-1596.

8. **D. Higgins**, Z. Chen, D. Lee, Z. Chen, "Activated and nitrogen-doped exfoliated graphene as air electrodes for metal-air battery applications", ***Journal of Materials Chemistry A***, 1 (2013) 2639-2645.
9. **D. Higgins**, Z. Chen, "Recent progress in non-precious metal catalysts for PEM fuel cell applications", ***Canadian Journal of Chemical Engineering***, 91 (2013) 1881-1895.
10. J. Liao, **D. Higgins**, G. Liu, V. Chabot, X. Xiao, Z. Chen, "Multifunctional TiO2@C/MnO2 Core@Double-shell Nanowire Array as a High Performance 3D Electrode for Li-ion batteries", ***Nano Letters***, 13 (2013) 5467-5473.
11. **D. Higgins**, J. Choi, J. Wu, A. Lopez, Z. Chen, "Titanium nitride carbon nanotube core-shell composites as effective electrocatalyst supports for low temperature fuel cells", ***Journal of Materials Chemistry***, 22 (2012) 3727-3732.
12. **D. Higgins**, J. Wu, W. Li, Z. Chen, "Cyanamide derived thin film on carbon nanotubes as metal free oxygen reduction reaction electrocatalyst", ***Electrochimica Acta***, 59 (2012) 8-13.
13. **D. Higgins**, Z. Chen, S. Ye, S. Knights, "Highly Durable Platinum-Cobalt Nanowires by Microwave Irradiation as Oxygen Reduction Catalyst for PEM Fuel Cell", ***Electrochemical and Solid-State Letters,*** 6 (2012) B83-B85.
14. Z. Chen, A. Yu, **D. Higgins**, H. Wang, H. Liu, Z. Chen, "Highly Active and Durable Core-Corona Structured Bifunctional Catalyst for Rechargeable Metal-Air Battery Application", ***Nano Letters***, 69 (2012) 1946-1952.
15. **D. Higgins**, Z. Chen, Z. Chen, "Nitrogen doped carbon nanotubes synthesized from aliphatic diamines for the oxygen reduction reaction", ***Electrochimica Acta***, 56 (2011) 1570-1575.
16. ***16.*** Z. Chen, **D. Higgins**, A. Yu, L. Zhang, J. Zhang, "A Review on Non-precious Metal Electrocatalysts for PEM Fuel Cells", ***Energy & Environmental Science***, 4 (2011) 3167-3192.
17. **D. Higgins**, D. Meza, Z. Chen, "Nitrogen Doped Carbon Nanotubes as Platinum Catalyst Supports for Oxygen Reduction Reaction in PEM Fuel Cells", ***The Journal of Physical Chemistry C***, 114 (2010) 21981-21988.
18. W. Li, A. Yu, **D. Higgins**, B. Llanos, Z. Chen, "Biologically Inspired Highly Durable Iron Phthalocyanine Catalysts for Oxygen Reduction Reaction in PEM Fuel Cells", ***Journal of the American Chemical Society***, 132 (2010) 17056-17058.

Research Interests: (see above)

Teaching Interests: (see above)

Copyright © American Institute of Chemical Engineers. All rights reserved.

476174 Rational Catalyst Design for Renewable Energy Technologies

Abstract:

Research Interests:

Today, there is an extensive interest in replacing fossil fuels to supply mankind energy needs sustainably on a long-term basis. In recent years several renewable energy technologies have been developed such as fuel cells, metal-air batteries and electrolyzers. There are, however, severe shortcomings of the present technologies, which need to be overcome to make these processes more economically attractive. One of the most important problems is the lack of active, selective and inexpensive catalysts for these processes. My research focuses on designing cost-effective and efficient heterogeneous catalysts for the application in energy conversion devices such as fuel cells, batteries, and chemical synthesis processes. The design of catalysts will be guided by computational modeling in particular DFT with an emphasis on understanding the electronic structure of the catalyst and adsorbed surface species. In particular I am interested in 1) developing efficient catalysts for oxygen reduction reaction (ORR) in fuel cells and oxygen evolution reaction (OER) in metal-air batteries and water electrolyzers. 2) Additionally I am interested in synthesizing fuels using biomass, methane oxidation and electrochemical conversion of $CO_2$to hydrocarbons.

Recent theoretical understandings indicate that there are universal correlations (scaling relations) between surface bond energies of different adsorbed species that bind to the surface through similar atom for specific reaction.[1,2] These correlations pose limitations on many of the known catalysts for reactions such as ORR, OER, CO_2 reduction and N_2 reduction reactions.[3] Therefore, catalyst design requires development of new, effective and non-precious catalysts with active site motifs, which do not follow the similar correlation. ***My work utilizes theoretical tools in particular DFT to model two-dimensional materials with different modification strategies including doping and defect design to describe the surface reactions. These strategies have not been explored so far. I seek to find a breakthrough for scaling relations and my goal is to discover novel catalysts for the following applications: 1) improving the efficiency of the devices that are related to sustainable energy applications, i.e. fuel cells[4], electrolyzers[5], and metal-air batteries[6], and 2) to discover/design new, effective and non-precious catalysts for ORR, OER, and CO_2 reduction reaction, fuel synthesis from biomass, and methane oxidation.***

Another important issue facing humanity is related to the potable water. Millions of people in the developing world are lacking access to drinking water due to its contamination with urban, industrial and agricultural waste. These pollutants can in principle be removed by their total oxidation by hydrogen peroxide (H_2O_2) as a strong oxidizing agent without leaving any residues behind. Production and transportation of H_2O_2 to the developing world, however, is challenging due to high costs and safety issues, making this potentially game-changing chemical largely inaccessible to the massive number of people who need it the most. ***My research combines the expertise from fundamental understanding of the reaction and DFT with versatile synthetic methods, electrochemical evaluation and catalyst characterization to develop an electrochemical route by designing an efficient and cost effective electrocatalyst for selectively reducing O_2 to H_2O_2 to enable continuous, small-scale and decentralized production of H_2O_2.***[7,8]

References:

(1) Viswanathan, V.; Hansen, H. A.; Rossmeisl, J.; Nørskov, J. K. ***ACS Catal.*** **2012**, ***2***(8), 1654–1660.

(2) Man, I. C.; Su, H.-Y.; Calle-Vallejo, F.; Hansen, H. A.; Martínez, J. I.; Inoglu, N. G.; Kitchin, J.; Jaramillo, T. F.; Nørskov, J. K.; Rossmeisl, J. ***ChemCatChem*** **2011**, ***3***(7), 1159–1165.

(3) Vojvodic, A.; Nørskov, J. K. ***Natl. Sci. Rev.*** **2015**, ***2***(2), 140–149.

(4) Siahrostami, S.; Björketun, M. E.; Strasser, P.; Greeley, J.; Rossmeisl, J. ***Phys. Chem. Chem. Phys.*** **2013**, ***15***(23), 9326–9334.

(5) Siahrostami, S.; Vojvodic, A. ***J. Phys. Chem. C*** **2015**, ***119***(2), 1032–1037.

(6) Siahrostami, S.; Tripković, V.; Lundgaard, K. T.; Jensen, K. E.; Hansen, H. a; Hummelshøj, J. S.; Mýrdal, J. S. G.; Vegge, T.; Nørskov, J. K.; Rossmeisl, J. ***Phys. Chem. Chem. Phys.*** **2013**, ***15***(17), 6416–6421.

(7) Siahrostami, S.; Verdaguer-Casadevall, A.; Karamad, M.; Deiana, D.; Malacrida, P.; Wickman, B.; Escudero-Escribano, M.; Paoli, E. a; Frydendal, R.; Hansen, T. W.; Chorkendorff, I.; Stephens, I. E. L. S.; Stephens, I. E.; Rossmeisl, J. ***Nat. Mater.*** **2013**, ***12***(12), 1137–1143.

(8) Verdaguer-Casadevall, A.; Deiana, D.; Karamad, M.; Siahrostami, S.; Malacrida, P.; Hansen, T. W.; Rossmeisl, J.; Chorkendorff, I.; Stephens, I. E. L. ***Nano Lett.*** **2014**, ***14*** (3), 1603–1608.

Teaching Interests:

I would like to teach courses such as Physical Chemistry, Thermodynamics, Quantum Chemistry, Molecular Spectroscopy and Chemical Kinetics. Moreover, I would be interested in developing and offering new courses in accordance to my area of specialty and research experiences.

Copyright © American Institute of Chemical Engineers. All rights reserved.

476259 Engineering Visible-Light Organic Photocatalysis for Polymers in Biomaterials, Biosensing, and Photomedicine

Abstract:

Research Interests:

Visible-light organic photocatalysis is one of the most promising methods to synthesize and modify polymers of medical and biomedical relevance. These reactions permit spatiotemporal control at biologically benign irradiation levels with low-cost biocompatible organic photocatalysts. However, we have poor control of these reactions. This stems from the general lack of knowledge of the complex reaction mechanisms around proteins. Therefore, discovery of the right photocatalyst for every biomedical application has advanced slowly through trial-and-error. Molecular engineering of photocatalysts promises to accelerate the implementation and expand the utility of visible-light photocatalysis in biomedicine. My goal is to promote specific reaction pathways that yield desired outcomes, e.g. radical initiation, polymer/protein grafting, reversible crosslinking, or protein crosslinking, by engineering organic photocatalysts.

I have discovered that the photocatalysis of methylene blue mimics photosynthesis in that it involves a $2e^{-}/H^{+}$ photocatalytic cycle, as opposed to the concerted e^{-}/H^{+} transfer that is generally assumed for photoinduced electron transfer reactions. This allows synthesis of centimeter thick optically absorptive polymers for dental and orthopedic restorative medicine. Then, I investigated the photocatalysis of Eosin to initiate radical polymerization of hydrogels against a thousand times excess oxygen around proteins, useful for biosensing in medical diagnostics. The next step is elucidating the mechanism for the crosslinking of collagen proteins with Rose Bengal useful for surgical wounds closure, corneal sealing, nerve, tendon and vessel repair, and cartilage regeneration. Applying chemical engineering principles to photochemical reactions will allow me to design photocatalysts that crosslink collagen fibers with minimal tissue inflammation.

Teaching Interests:

A startled classroom of junior and senior Chemical Engineering students looked at me as I opened a lecture with a simulation of the explosion of a chemical reactor. As TA of the Reaction Engineering and Chemical Kinetics course at the University of Colorado I wanted to do something different to get students out of our classical equation-heavy class routine. I selected the best video I could find on Runaway Reactions to engage students in the topic of incorporating heat considerations into reactor design. Office Hours were well spent observing their progress and discussing specific questions in small groups. Most questions were directly related to problem sets and midterms, but some served as discussion points to chat about chemical kinetics, numerical methods, and reaction mechanisms. I saw our students making their schematics and postulating new questions as the learning process sparked their curiosity. I believe this classical learning/teaching interaction is invaluable to stimulate active learning. I enjoyed designing online assessments to assess studentsâ€™ knowledge, critical thinking, and inventiveness as TA of the Statistics for Engineering course at the University of Colorado. A combination of Backwards Design and Constructive Alignment will allow me to create an engaging environment where my students achieve the intended learning outcomes. I work to engage all students in their own learning experience. I observed this first hand in the students I mentored as the founding president of the AIChE student Chapter at my undergraduate University. All this coalesced in my Teaching Philosophy developed during the Kaufman Teaching Certificate Program at MIT.

Copyright © American Institute of Chemical Engineers. All rights reserved.

476641 Catalysts for Emission Control and Energy Conversion: Computational Study Based on DFT Calculations

Abstract:

Research Interests:

Catalysts for emission control and energy conversion

- Emission control: Metal-exchanged zeolite for NO_x reduction; "single-atom" catalyst for CO oxidation
- Energy conversion: bimetallic catalyst for biomass conversion; photocatalyst for using solar energy; "single-atom" catalyst for fuel cell

Research Experiences:

(1) Computational K-edge X-ray absorption near edge structure (XANES) and X-ray emission spectroscopy (XES) studies on properties of Cu ions in Cu-SSZ-13 and Cu-SAPO-34, collaborating with experimental results;

(2) Characterized NO chemisorption on the catalyst of Cu/SSZ-13 by comparing computational and experimental infrared (IR) spectra;

(3) Characterized benzene adsorption on Pt step surface by integrating computational and experimental X-ray photoelectron spectroscopy (XPS);

(4) Discovered the accurate model of "29"-structure Cu oxide by performing computational scanning tunneling microscope (STM), and comparing with experimental STM;

(5) Innovatively created another single-atom catalyst of Pt/TiN suitable for reactions in proton exchange membrane fuel cells;

(6) Through a method of surface modification, to functionalize silicon namowires as photocatalysts for water splitting by using solar energy;

(7) Investigated effect of electric field and strain field on properties of silicon nanowires, such as band structure and band gap.

Teaching Interests:

Theoretical surface and interface science

Reaction at solid surface

Selected Publications:

Zhang, R.; Szanyi, J.; Gao, F.; and McEwen, J.-S. "The Interaction of Reactants, Intermediates and Products with Cu Ions in Cu-SSZ-13 NH_3 SCR Catalysts: An Energetic and ab initio X-ray Absorption Modeling Study", *Catal. Sci. Technol.*, 6 (15), 5812-5829 (2016)

Therrien, A.J.; Zhang, R.; Lucci, F.R.; Marcinkowski, M.D.; Hensley, A.J.; McEwen, J.-S.; and Sykes, E.C.H., "Structurally Accurate Model for the "29"-structure of Cu_xO/Cu(111): A DFT and STM Study", *J. Phys. Chem. C*, 120 (20), 10879-10886 (2016)

Zhang, R.; McEwen, J.-S.; Kollár, M.; Gao, F.; Wang, Y.; Szanyi, J.; and Peden, C. "NO Chemisorption on Cu/SSZ-13: A Comparative Study from Infrared Spectroscopy and DFT Calculations", *ACS Catal.*, 4 (11), 4093-4105 (2014)

Hensley, A.J.; Hong, Y.; Zhang, R.; Zhang, H.; Sun, J.; Wang, Y.; and McEwen, J.-S., "Enhanced Fe_2O_3 Reducibility via Surface Modification with Pd: Characterizing the Synergy within Pd/Fe Catalysts for Hydrodeoxygenation Reactions", *ACS Catal.*, 4 (10), 3381-3392 (2014)

Zhang, R.; Hensley, A.J.; McEwen, J.-S.; Wickert, S.; Darlatt, E.; Fischer, K.; Schoppke, M.; Denecke, R.; Streber, R.; Lorenz, M.; Papp, C.; and Steinruck, H.-P., "Integrated X-ray Photoelectron Spectroscopy and DFT Characterization of Benzene Adsorption on Pt(111), Pt(355) and Pt(322) Surfaces", *Phys. Chem. Chem. Phys.*, 15 (47), 20662-20671 (2013)

Zhang, R.; Lee, T.H.; Yu, B.D.; Stampfl, C.; and Soon, A., "The Role of Non-conventional Supports for Single-atom Platinum-Based Catalysts in Fuel cell Technology", *Phys. Chem. Chem. Phys.*, 14 (48), 16552-16557 (2012)

Zhang, R.; Liu, X.M.; Wen, Z.; and Jiang, Q., "Prediction of Silicon Nanowires as Photocatalysts for Water Splitting: Band Structure Calculated Using Density Functional Theory", *J. Phys. Chem. C*, 115 (8), 3425-3428 (2011)

Ao, Z.M.; Jiang, Q.; Zhang, R.; Tan, T.T.; and Li, S., "Al Doped Graphene: A Promising Material for Hydrogen Storage at Room Temperature", *J. Appl. Phys.*, 105 (7), 074307 (1-6) (2009)

Copyright © American Institute of Chemical Engineers. All rights reserved.

476917 Development of Heterogeneous Catalysis towards a Sustainable Future

Abstract:

Research Interests:

The prosperity of a growing population has placed significant stress on the environment and society, and led to many grand challenges in climate, energy, water, and food, etc. Due to its transforming power, catalysis is definitely a key part of the solutions to these problems. My research interests are to understand and develop heterogeneous catalysis for a sustainable future.

In my Ph.D. research with Prof. Yuriy Román-Leshkov in the Department of Chemical Engineering at MIT, I have developed effective catalytic strategies to convert biomass-derived molecules into mono- and dicarboxylic acids and esters. In particular, by exploiting the dual-function heteroatom-containing zeolites, we enabled the activation of carbonyl groups for carbon-carbon coupling to upgrade biomass-derived oxygenates into renewable monomers, and we also developed stable reducible metal oxides supported gold catalysts for base-free aqueous aerobic oxidation of glucose by maximizing the anchoring and activating benefits from the oxygen vacancies of metal oxides. In my current postdoctoral research with Prof. William H. Green in the Department of Chemical Engineering at MIT, my research interest is on the prediction of reaction kinetics in liquid phase.

Motivated by those grand challenges, I would like to extend and expand the two types of chemistry, i.e. activation of carbonyl groups and aqueous aerobic oxidation of organic molecules, to contribute to the solutions. Specifically, I am interested in developing heterogeneous catalysis for and understanding the mechanism and kinetics of 1) conversion of carbon dioxide into chemicals and fuels, and 2) remediation of pollutants in water.

Teaching Interests:

My preference includes kinetics and reaction engineering courses for undergraduate and graduate students, which naturally fit my research experience. I would also like to develop electives that focus on heterogeneous catalysis and/or kinetics. During my Ph.D., I have been the teaching assistant for undergraduate kinetics courses. I have also mentored undergraduate students and master students in my Ph.D. and current postdoctoral research.

Copyright © American Institute of Chemical Engineers. All rights reserved.

478099 Understanding and Improving Electrocatalysts for Energy Conversion and Waste Remediation

Abstract:

Research Interests:

The challenge of sustainably providing the increasing demand for energy and resources relies on improving existing renewable energy processes or developing new processes. Major issues include the storage of energy, conversion of renewable energy to storable transportation fuels, and the remediation of waste streams. Catalysis and electrocatalysis can play a major role in providing sustainable energy and remediating waste streams, but improvements are needed in electrocatalyst activity, selectivity and stability. My interest is in synthesizing and testing new classes of materials (such as doped and undoped metal sulfides) that can achieve activities, selectivities and stabilities that are unmet by the existing metal electrocatalysts. I am also interesting in characterization of electrocatalysts under operating conditions through the use of spectroscopy and kinetic analysis. To evaluate potential processes, I will also use technoeconomic analysis and sensitivity analysis to identify which factors have the largest impact on the final product. As a faculty member I will focus on the following research projects:

Project 1. Understand and improve the electrocatalytic activity of flow batteries. A focus of my graduate work was on new electrocatalysts for H_2/Br_2 flow batteries, whose activity and stability have a major impact on the feasibility of flow batteries. In this project, I hope to extend my previous work to other flow batteries including the H_2/Cerium battery (where competing oxygen evolution and changing oxidation states of the catalyst are a challenge)[1] or vanadium batteries (where the competition of hydrogen evolution is an issue). *The specific aims of this project are to test new electrocatalyst materials, and characterize them under operating conditions using techniques such as X-ray absorption spectroscopy (XAFS) to understand the effects of applied potential on oxidation state or adsorption of reactant molecules.*

Project 2. New methods to convert solar energy to transportation fuels. Research on solar fuels has focused on water splitting (to H_2 and O_2) and carbon dioxide reduction, but other processes are also achievable (and possibly more viable) if new electrocatalysts are developed. A sample reaction is using electrochemical potential provided from sunlight to hydrogenate waste biomass, which enables easier deoxygenation to fuel-grade hydrocarbons. Aqueous electrochemical hydrogenation of biomass has a low overpotential compared to CO_2 electro-reduction (~0.1 V compared to 1 V), important for solar driven processes where the voltage provided by photons is limited. However, electrocatalyst activities rates must be improved, and photoelectrochemical hydrogenation of biooil has not been demonstrated using sunlight. *The specific aims of this project are to evaluate new electrocatalysts (such as metal sulfides) for the conversion of lignin-based biooils to hydrocarbons, and coupling the reaction with light absorbers powered by sunlight.* The characterization techniques such as XAFS used to evaluate flow batteries in Project 1 can also be used here to understand reactant coverages and electronic properties of the catalyst.

Project 3. Remediation of wastewater through electrochemical treatment. One of the major concerns for a sustainable future is treatment of waste streams. Nitrate is a biological poison that is present in waste as varied as fertilizer run-off, electrochemical industrial waste and low-level nuclear waste, and can potentially be treated electrochemically through nitrate reduction. Yet, the mechanism of nitrate remediation is not well-understood, and active catalysts that can maintain activity in the presence of metal ions or other poisons do not exist. *The specific aim of this project is to determine the properties of electrocatalysts that control the activity for nitrate reduction and develop catalysts with high activity and resistance to poisoning.* Similar experimental techniques from the other two proposed projects can be applied to develop waste remediation electrocatalysts.

PhD Research:

My graduate research with Professor Eric McFarland and Professor Horia Metiu at the University of California, Santa Barbara focused on understanding new types of electrocatalysts for the conversion of energy. This consisted of identifying metal sulfides and doped metal sulfides for active and stable H_2/Br_2 flow battery electrocatalysts,[2–5] and use of catalysts for the conversion of solar energy to hydrogen.[6,7] We coupled experiments with theory to understand molecular descriptors for catalyst activity (such as hydrogen binding energy on metal sulfides) and to identify active sites.[8,9] Through selective synthesis of particular phases of rhodium sulfide we were able to improve electrocatalytic activity and stability. I also completed a technoeconomic analysis of the H_2/Br_2 flow battery to identify key components in the levelized cost of electricity.[10]

Postdoctoral Research:

My postdoctoral research with Prof. Charles T. Campbell and Prof. Johannes Lercher at the University of Washington and Pacific Northwest National Laboratory consists of the interrogation of electrocatalysts for electrocatalytic hydrogenation (ECH) of phenol. This includes *in situ* characterization of the catalysts by XAFS to identify phenol and hydrogen coverages on the surface as well as electronic and structural properties of the catalyst. In addition, we identified the cause for deactivation at elevated temperatures by kinetic analysis and modified reaction conditions (e.g. hydrogen pressure) to prevent formation of poison species, and enable unprecedented rates at temperatures relevant for ECH (<100 °C). I am currently a Washington Research Foundation Innovation Fellow with the Clean Energy Institute at UW, and successfully presented for a PNNL Initiative as well as co-wrote a proposal for Laboratory Directed Research and Development (internal funding for PNNL).

Teaching Interests:

During my time as a student, the professors of my core Chemical Engineering courses helped me develop my interest in Catalysis and Reaction Engineering, and I hope to be able to do the same for future chemical engineers. I am comfortable with the core Chemical Engineering course structure, but feel best suited towards teaching Reaction Engineering and in developing elective courses in Electrochemistry or Energy. As a graduate student I was a Teaching Assistant for Undergraduate Chemical Reaction Engineering and an Energy elective. In terms of teaching in a research environment, I supervised six undergraduate students during my graduate work, including education on the experimental aspects of catalysis, electrochemistry and photoelectrochemistry.

References:

1. Tucker, M. C., Weiss, A. & Weber, A. Z. Improvement and analysis of the hydrogen-cerium redox flow cell. *J. Power Sources* **327,** 591–598 (2016).

2. Ivanovskaya, A., **Singh N.**, Liu, R., Kreutzer, H., Baltrusaitis, J., Nguyen, T. V., Metiu, H. & McFarland, E. Transition Metal Sulfide Hydrogen Evolution Catalysts for Hydrobromic Acid Electrolysis. *Langmuir* **29,** 480–492 (2013).

3. Masud, J., Nguyen, T. V., **Singh, N.**, McFarland, E., Ikenberry, M., Hohn, K., Pan, C.-J. & Hwang, B.-J. A RhxSy/C Catalyst for the Hydrogen Oxidation and Hydrogen Evolution Reactions in HBr. *J. Electrochem. Soc.* **162,** F455–F462 (2015).

4. Nguyen, T. Van, Kreutzer, H., Yarlagadda, V., McFarland, E. & **Singh, N.** HER/HOR Catalysts for the H2-Br2 Fuel Cell System. *ECS Trans.* **53,** 75–81 (2013).

5. Masud, J. Walter, J., Nguyen, T. V., Lin, G., **Singh, N.**, McFarland, E., Metiu, H., Ikenberry, M., Hohn, K., Pan, C.-J. & Hwang, B.-J. Synthesis and Characterization of RhxSy/C Catalysts for HOR/HER in HBr. *ECS Trans.* **58,** 37–43 (2014).

6. **Singh, N.**, Mubeen, S., Lee, J., Metiu, H., Moskovits, M. & McFarland, E. W. Stable electrocatalysts for autonomous photoelectrolysis of hydrobromic acid using single-junction solar cells. *Energy Environ. Sci.* 1–4 (2014). doi:10.1039/c3ee43709d

7. Mubeen, S., **Singh, N.**, Lee, J., Stucky, G. D., Moskovits, M. & McFarland, E. W. Synthesis of chemicals using solar energy with stable photoelectrochemically active heterostructures. *Nano Lett.* **13,** 2110–5 (2013).

8. **Singh, N.**, Upham, D. C., Liu, R., Burk, J., Economou, N., Buratto, S., Metiu, H. & McFarland, E. Investigation of the active sites of rhodium sulfide for hydrogen evolution/oxidation using carbon monoxide as a probe. *Langmuir* **30,** 5662–8 (2014).

9. **Singh, N.**, Upham, D. C., Metiu, H. & McFarland, E. W. Gas-Phase Chemistry to Understand Electrochemical Hydrogen Evolution and Oxidation on Doped Transition Metal Sulfides. *J. Electrochem. Soc.* **160,** A1902–A1906 (2013).

10. **Singh, N.** & McFarland, E. W. Levelized cost of energy and sensitivity analysis for the hydrogen–bromine flow battery. *J. Power Sources* **288,** 187–198 (2015).

Copyright © American Institute of Chemical Engineers. All rights reserved.

478752 Development of Novel Single-Site and Isolated Bimetallic Alloy Catalysts for C-H Bond Activation

Abstract:

Research Interests:

My research interests are primarily focused on the development of novel single-site and isolated bimetallic alloy catalysts and their application in heterogeneous catalysis, particular in the activation of C-H bonds of alkanes. In an era of cheap and abundant alkanes, there is a growing interest in the conversion of alkanes, to value-added products, in which C-H activation is the elementary first step. Despite extensive studies, C-H activation under mild conditions continues to be the "holy-grail" of catalysis. The difficulty in C-H activation arises not only from the relatively inert nature of C-H bond, but also from the poor selectivity, which leads to unattractive products, carbon dioxide and water.

One particular important application of C-H activation is the methane activation and conversion to more valuable products, such as methanol and acetic acid. Many studies over the past several decades have failed to show promise for practical applications of the direct methane to desired liquid products. Examples include the approaches using Pd and Pt homogeneous catalysts, the conversion over CuFe/ZSM-5 catalysts using expensive H_2O_2 as the oxidant in aqueous media, and the non-catalytic two-step approaches focusing on Cu/MOR catalysts, which involves first the activation and conversion of methane at high temperatures, and second an *ex-situ* extraction process at room temperature. This stoichiometric reaction produces only trace amount of methanol; therefore is economically impractical. In my previous research, I have found the first truly catalytic cycle of the direct conversion of methane to practical yields of methanol and acetic acid with using molecular oxygen as the oxidant at low temperatures. The reactions are performed in aqueous media, under mild pressures of methane, oxygen and carbon monoxide, the latter used as catalyst promoter and reactant if the desired product is acetic acid. The catalysts are Rh/ZSM-5 with tunable high selectivity to acetic acid, and Cu promoted Ir/ZSM-5 with tunable high selectivity to methanol. The active sites are likely isolated Rh and Ir cations anchored on the internal walls of zeolite. This work opens a new door for the solving of long-standing problem of the direct catalytic conversion of methane to desirable liquid products using heterogeneous catalysts under mild operating conditions. Nevertheless, further investigations that avoid using precious metals and carbon monoxide are warranted.

My near future research will continue on the developing of novel heterogeneous catalysts for the activation of C-H bonds. In particular, I would like to using my obtained knowledge and experience to design novel single-site and isolated bimetallic alloy catalysts using non-precious metal to directly convert methane to value-added products.

Research Experience

My academic career path has been a blend of two main fields in heterogeneous catalysis. The first is the UHV study of the interaction of H_2O, CH_4, H_2, CO, CO_2 with single crystal surfaces, as well as with supported nanoparticles. The second involves synthesis of heterogeneous catalysts, performing catalytic measurements under realistic conditions, and in-situ characterizations with APXPS, XANES, EXAFS, DRIFTS, and various other techniques. In the meantime, I had trained and supervised new graduate students to start their respective projects. These experiences are very helpful to maintain a productive leading-edge research group when I start my own academic career.

Teaching Interests:

In addition to doing research, I also had a strong interest in teaching undergraduate and graduate students. I spent a semester teaching high-school students basic science when I was an undergraduate student. As a graduate student, I taught courses in physical chemistry and general chemistry. Given my background and experience in chemistry and catalysis, I am most interested in teaching classes related to physical chemistry and interdisciplinary techniques used in heterogeneous catalysis.

Selected publications

25. J. Shan, M. Li, L. F. Allard, S. Lee, M. Flytzani-Stephanopoulos

Direct conversion of methane to methanol or acetic acid with molecular oxygen on Rh- or Ir-ZSM-5 catalysts

Submitted, under the revision.

24. J. Shan, N. Janvelyan, H. Li, J. Liu, T. M. Egle, J. Ye, M. M. Biener, J. Biener, C. M. Friend, M. Flytzani-Stephanopoulos

Selective non-oxidative dehydrogenation of ethanol to acetaldehyde and hydrogen on highly dilute NiCu alloys

Submitted, under the revision.

23. J. Shan, S. Zhang, T. Choksi, L. Nguyen, C. S. Bonifacio, Y. Li, W. Zhu, Y. Tang, Y. Zhang, J. C. Yang, J. Greeley, A. I. Frenkel, F. Tao

Tuning catalytic performance through a single or sequential post-synthesis reaction(s) in a gas phase

Submitted, under the revision.

22. J. Shan, F. R. Lucci, J. Liu, M. El-Soda, M. D. Marcinkowski, L. F. Allard, E. C. H. Sykes, M. Flytzani-Stephanopoulos

Water co-catalyzed selective dehydrogenation of methanol to formaldehyde and hydrogen

Surface Science, 2016, 650, 121-129.

21. S. Zhang, **J. Shan**, L. Nie, L. Nguyen, Z. Wu, F. Tao

In situ studies of surface of $NiFe_2O_4$ *catalyst during complete oxidation of methane*

Surface Science, 2016, 648, 156-162.

20. C. S. Bonifacio, **J. Shan**, F. Tao, J. C. Yang

Atomic-scale characterization of restructured PtCu nanocubes

Microscopy and Microanalysis, 2015, 21, 1067-1068.

19. F. Tao, **J. Shan,** L. Nguyen, Z. Wang, S. Zhang, L. Zhang, Z. Wu, W. Huang, S. Zeng, P. Hu

Understanding of complete oxidation of methane on spinel oxides on a molecular level

Nature Communications, 2015, 6, 7798.

18. J. Shan, L. Nguyen, S. Zhang, F. Tao

Water–gas shift on Pd/α-MnO_2 *and Pt/α-*MnO_2

Catalysis Letters, 2015, 145, 1571-1580.

17. J. Shan, W. Huang, L. Nguyen, Y. Yu, S. Zhang, Y. Li, A. I. Frenkel, F. Tao

Conversion of methane to methanol with a bent mono(μ-oxo)dinickel anchored on internal surface of micro-pores

Langmuir, 2014, 30, 8558-8569.

16. S. Zhang, **J. Shan**, Y. Zhu, L. Nguyen, W. Huang, H. Yoshida, S. Takeda, F. Tao

Restructuring transition metal oxide nanorods for 100% selectivity in reduction of nitric oxide with carbon monoxide

Nano letters, 2013, 13, 3310-3314.

15. S. Zhang, **J. Shan**, Y. Zhu, A.I. Frenkel, A. Patlolla, W. Huang, S. Yoon, L. Wang, H. Yoshida, S.

Takeda, F. Tao

WGS catalysis and in-situ studies of CoO_{1-x}, $PtCo_n/Co_3O_4$, and $Pt_mCo_{m'}/CoO_{1-x}$ nanorod catalysts

Journal of the American Chemical Society, 2013, 135, 8283-8293.

14. J. Shan, Y. Zhu, S. Zhang, T. Zhu, R. Seigei, F. Tao

Catalytic performance and in-situ surface chemistry of pure α-MnO_2 nanorods in selective reduction of NO and N_2O with CO

Journal of Physical Chemistry C, 2013, 117,8329-8335

13. J. Shan, A. Chakradhar, K. Andeson, J. Schmidt, S. Dhuey, U. Burghaus

Butane adsorption on silica supported MoO_x clusters nanofabricated by electron beam lithography

Nanotechnology for Sustainable Energy, American Chemical Society, Washington DC, USA, 2013, Chapter 12, pp 295-310.

12. J. Shan, A. Chakradhar, M. Komaneni, M. Lu, U. Burghaus

Rim effects in the adsorption of CO_2 on silica supported copper oxide clusters - utilizing electron beam lithography

Journal of Physical Chemistry C, 2012, 116, 18930-18936.

11. A. Chakradhar, **J. Shan**, M. Komaneni, M. Lu, U. Burghaus

Identifying rims along nano-sized clusters as catalytically active sites – The case of CuO_x/silica model catalysts nanofabricated by electron beam lithography

Chemical Physics Letters, 2012, 544, 70-72.

10. M. Komaneni, **J. Shan**, A. Chakradhar, E. Kadossov, S. Cabrini, U. Burghaus

Adsorption dynamics of CO on silica supported CuO_x clusters: utilizing electron beam lithography to study methanol synthesis model systems

Journal of Physical Chemistry C, 2012, 116, 5792-5801.

9. C. Hahn, **J. Shan**, Y. Liu, O. Berg, A. W. Kleyn, L. B. F. Juurlink

Employing a cylindrical single crystal in gas-surface dynamics

Journal of Chemical Physics, 2012, 136, 114201.

8. J. Shan, A. Chakadhar, Z. Yu, U. Burghaus

Adsorption of water on a hydrophobic surface – the case of antimony(111)

Chemical Physics Letters, 2011, 517, 46-50.

7. J. Shan, M. Komameni, U. Burghaus

Adsorption dynamics of CO on copper and gold clusters supported on silica –how special is nanogold?

Chemical Physics Letters, 2011, 517, 59-61.

6. M. Komaneni, **J. Shan,** U. Burghaus

Adsorption dynamics of CO on silica–supported Cu clusters: a molecular beam scattering study

Journal of Physical Chemistry C, 2011, 115, 16590-16597.

5. C. Hahn, **J. Shan,** I. M. N. Groot, A. W. Kleyn, L. B. F. Juurlink

Selective poisoning of active sites for D_2 dissociation on platium

Catalysis Today, 2010, 154, 85-91.

4. J. Shan, A. W. Kleyn, L. B. F. Juurlink

Adsorption of molecular hydrogen on an ultrathin layer of Ni(111) hydride

Chemical Physics Letters, 2009, 474, 107-111.

3. J. Shan, A. W. Kleyn, L. B. F. Juurlink

Identification of hydroxyl on Ni(111)

Chemphyschem, 2009, 10, 270-275.

2. J. Shan, J. F. M. Aarts, A. W. Kleyn, L. B. F. Juurlink

Co-adsorption of water and hydrogen on Ni(111)

Physical Chemistry Chemical Physics, 2008, 10, 4994-5003.

1. J. Shan, J. F. M. Aarts, A. W. Kleyn, L. B. F. Juurlink

The interaction of water with Ni(111) and H/Ni(111) studied by TPD and HREELS

Physical Chemistry Chemical Physics, 2008, 10, 2227-2232.

Copyright © American Institute of Chemical Engineers. All rights reserved.

479774 Smart Materials through Molecular Networking

Abstract:

Research Experience:

I have a broad experience with chemical kinetics and reaction engineering in flow systems. I developed the first synthetic enzymatic oscillator during my first postdoc at the University of Nijmegen in the Netherlands. In this work, I precisely arranged reactions in continuous stirred tank reactor (CSTR) to get oscillatory dynamics.[1] At George Whitesides group at Harvard, I applied the similar strategy to obtain bistable and oscillatory dynamics from organic reactions in CSTR.[2] I have also worked with coupled reaction-diffusion systems, microfluidics, and soft materials.[3] In my independent research, I would like to apply my experience with engineering ofreaction kinetics in open system to problems in materials and catalysis.

Research Interests:

My research program will focus on autonomous regulation in chemical systems by regulatory chemical reaction networks— that is, cascades of chemical reactions that receive a chemical signal, process this signal, and respond with a chemical output. Chemical regulation is centrally important to cellular life; it orchestrates cellular metabolism and guides the growth and movement of cells in heterogeneous environments. Although electrical and mechanical control systems are usually the choice for regulating macroscopic devices, chemical control becomes a preferable option for the regulation at the micro- and nanoscales. When applied to catalysis, self-regulation can improve efficiency of chemical reactions. My group will exploit these opportunities in designing "smart" materials and regulating catalytic transformations.

My research will have three major thrusts:

(i) Smart soft materials;

(ii) Chemical regulation at micro- and nanoscales through the surface modification.

(iii) Self-regulation in mixtures of organometallic catalysts.

Teaching Interests:

I have had a variety of teaching experiences. I voluntarily gave two full lecture courses (8 and 6 lectures) in organometallic chemistry in the Materials Science Department of Moscow State University in 2011 and 2012. I examined and graded student that attended these courses. I worked as a lab assistant for the Inorganic chemistry laboratory course 1 at the University of Zurich. I also worked as a chemistry teacher in the School for young chemists (12-14 years old pupils), which is associated with the Moscow State University. As a prospective faculty member, I am interested in teaching thermodynamics, chemical kinetics, and reaction engineering.

Selected Publications:

1. **Sergey N. Semenov,** Albert S.Y. Wong, Martijn van der Made, Sjoerd Postma, Rik van Roekel, Tom F. A. De Greef, Wilhelm T. S. Huck "Rational design of functional and tunable oscillating enzymatic networks" // *Nature Chemistry*, (2015**)**, *7*, 160 – 165.
2. **Sergey N. Semenov**, Lewis J. Kraft, Alar Ainla, Mengxia Zhao, Mostafa Baghbanzadeh, Victoria E. Campbell, Kyungtae Kang, Jerome M. Fox, George M. Whitesides "Autocatalytic, Bistable, Oscillatory Networks of Biologically Relevant Organic Reaction" // *Nature*, (2016), *537*, 656 – 660.
3. **Sergey N. Semenov**, Albert J. Markvoort, Tom F. A. de Greef, Wilhelm T. S. Huck "Threshold Sensing Through a Synthetic Enzymatic Reaction-Diffusion Network" // *Angew. Chem. Int. Ed.* (2014), 53, 8066 – 8069.

Copyright © American Institute of Chemical Engineers. All rights reserved.

476305 Condition and Support Dependent Development of Computational Methods for the Engineering of Materials

Abstract:

Research Background:

My Ph.D. work under the supervision of Dr. William F. Schneider is an example of a system where I integrated density functional theory calculations with my own Monte Carlo code to accurately predict the distribution of active sites in a metal-exchanged zeolite post-synthesis. I then used a combination of ab initio molecular dynamics and analytical statistical mechanical models to describe the dynamic response of the active sites to varying gas exposure conditions. These findings brought the insight of solvated and mobile active sites to this system, a more malleable view of these active sites than had previously been assumed in the literature and more closely resembled a homogenous catalyst under catalytic conditions than a heterogeneous one which was critical to describing the catalytic mechanism. The thermal motion of these active sites also had direct consequences to their electronic structure, which required the integration of time-dependent density functional theory and ab initio molecular dynamics to accurately model. These â€œoperandoâ€ type computational models were validated through in situ and operando spectroscopic experiments that I played a major role in while a visiting student at Purdue University for one semester under the supervision of Dr. Fabio H. Ribeiro, Dr. W. Nicholas Delgass, and Dr. Rajamani Gounder, and at Argonne National Laboratory under the supervision of Dr. Jeffrey T. Miller.

Research Interests:Â

Development of computational models that account for the dynamic response of active sites to their contacting support and fluid phase is a critical step for being able to predict new materials that are not only active, but thermally and chemically stable and also capable of being synthesized. These are concepts often neglected in current computational models where the focus has historically been activity, in large part due to the difficulties involved with taking quantum calculations at 0K performed on small systems and scaling these results up to macroscopically significant length and time scales. Computational methods integrating interactions between active sites, supports, and the contacting environment are also important for providing a direct integration between theory and the bleeding edge of experimental methodology in catalysis which is highly focused on in situ and operando characterization of catalysts.

I provide several examples below for models informed by quantum chemical calculations and ab initio molecular simulations that can be coarse-grained to macroscopic length and time scales and a wide range of catalyst exposure conditions through the use of analytical and modern numerical statistical mechanics based theories using novel approaches that either combine computational methods that work efficiently on different energy scales, or those that work akin to classical molecular simulations but integrate quantum based energetics and structures into the model which is a new approach in this field:

I) The restructuring of supported nanoparticles at high surface coverages and temperatures is important for many catalysts, including those used for hydrocarbon chemistry and electrocatalysis, and can lead to dynamical behavior such as oscillatory rates or deactivation. A major challenge is the coupling of nanoparticle reconstruction, which involves high energy barriers and longer time scales, with surface coverage, which will influence the nanoparticle structure and is tied to the chemical potential at the interface and that in the fluid phase and must be equated. Iterative numerical schemes coupling a genetic algorithm with density functional theory computed molecular binding energies allow for rapid convergence of structure and coverage, creating accurate condition-dependent based models of these materials that naturally lend themselves to parallel computing environments such as GPU computing. Accurately modeling these systems informs support engineering at the atomistic scale and allows for mediation of stability challenges through approaches such as 3D printing of supports and chemical tethering of nanoparticles.

II) In microporous media modeling of percolation informs both the birth of the catalyst in support synthesis, metal exchange, or nanoparticle encapsulation, and its death in deactivation, from coking for example. Quantum chemical calculations describe how and where blockages occur, which are then integrated into statistical mechanical models that simulate the zoning off of portions of the support at macroscopic length scales and these geometries can be easily transferred into computational fluid dynamics software. Once these models are validated, new support geometries, for instance including mesoporous and microporous domains, can then be simulated to predict materials with advantageous synthesis properties or enhanced resistance to deactivation.

III) Computational models that incorporate fluid phase and support effects have use beyond just catalysis in the design of materials where thermodynamic properties are critical. Examples of this include supports designed for storage of gaseous NOx or COx species through chemisorption, and Lithium batteries where the formation of dendrites and ultimately shortage is obviously undesirable. Nucleation phenomena like dendrite formation is propagated by the exposure conditions of the contacting fluid and solid interface and take place over long timescales. The dendrite propagation occurs in a way that can be described by analytical fractal models once initiated, but only quantum chemical calculations can describe the initial nucleation accurately, which creates a separation of time and length scales. This can be resolved by tying these analytical models to condition-dependent quantum ones that accurately describe initiation and then use the analytical models to extend to deactivation timescales that will be macroscopically observable.

All of these processes involve systems where quantum-based methods are critical to describe the chemisorption of molecules or covalent interactions between active sites and a support, but must be integrated with analytical and numerical modeling to extend to phenomena that occur on the length and time scales of catalyst synthesis, restructuring, and deactivation.

Teaching Interests:

I have a strong background in analytical and numerical mathematical methods including:

I)analytical: solution to different classes of ODEs, PDEs, asymptotic methods, integral transforms, perturbation theory, and non-linear dynamics

II)numerical: classes of numerical PDE solvers and stability, finite element and volume methods, numerical solutions to the Navier-Stokes equations and higher order expansions of them

While many of these methods are not appropriate for an undergraduate curriculum, what my exposure to a diversity of analytical and numerical methods has taught me is that typically one methodology works well when the other does not and due to this complementary nature both analytical and numerical techniques should be emphasized in an undergraduate curriculum. In particular a strong theoretical foundation is critical to any undergraduate education since an experiment or computational simulation performed without an underlying theory behind it is worthless.

My experience performing fundamental experiments both as an undergraduate at Notre Dame and as a Graduate Student visiting at Purdue also provides me with the background necessary to teach any undergraduate chemical engineering course, including those involving experiments such as unit operations, phase equilibria, process control, and catalytic systems.

A priority of mine as an educator will be on exposing undergraduates to programming languages, with emphasis on python/matlab which are very similar and have a small learning curve, as well as interfacing with unix based operating systems. Many of the best catalysis experimental laboratories currently incorporate automated reactor design, "big" data analysis that relies on codes to facilitate the interpretation of large amounts of collected data, and working with unix based systems that rely on terminal commands, macros, and skilled use of on-site software when performing experiments at government funded laboratories. Contemporary approaches to the incorporation of coding into classical chemical engineering courses will be easy (for instance, automation of steam tables or UNIFAC) to integrate naturally into the core chemical engineering curriciulum. In the modern era some computational knowledge is key not only for theoretical models and simulation, but also for the design and analysis of experiments.

Copyright © American Institute of Chemical Engineers. All rights reserved.

477292 Conversion of Waste Oil to Biofuels in a Single Catalytic Process over Bifunctional Catalysts

Abstract:

Research Interests:

My research interest relates to two patents I filed on biofuel field. It would be a great idea to convert the waste oil to biofuels in a single catalytic process. According to data I have found zeolites are the nominate catalyst for this purpose because of the high percentage of decarboxylation and high selectivity they have for branched and aromatic compounds. The other plan I have for my future research is to utilize the nanoscience and computational fluid dynamics knowledge together to cure cancer with magnetic nanoparticles. The concept of this work is cancer hyperthermia.

Teaching Interests:

I have worked in a variety of academic institutions in a wide range of teaching areas. Teaching has always been a rewarding experience for me, mainly because of the great satisfaction that comes with sharing my knowledge with others. My teaching interest includes kinetics and reaction engineering, transport phenomena and fluid dynamics courses for undergraduate and graduate students, which naturally fit my research and teaching experience. Because of the fact that my research includes kinetics and reaction engineering, heat and mass transfer and fluid mechanics concepts, so I could give students team working projects and provide them the opportunity to improve their knowledge about the fundamental principles and technical details of basics of Chemical Engineering. I also enjoy teaching hands-on laboratory classes, like unit operation laboratory.

Copyright © American Institute of Chemical Engineers. All rights reserved.

477582 Rational Design of Catalytic Materials for Advancing the Use of Alternative Energy Sources

Abstract:

Rational Design of Catalytic Materials for Advancing the Use of Alternative Energy Sources

Research Interests:

Due to the depletion of fossil fuels and the presence of greenhouse gases in the atmosphere, alternative sources of energy which do not emit greenhouse gases will become essential in the near future. Both challenges of reducing greenhouse gas emissions and utilizing alternative energy sources will be accomplished through catalysts. My research directions focus on storage of energy which potentially comes from alternative sources. Specifically, my first research direction is determining catalytic materials for using electricity to split water for the production of hydrogen fuel. Hydrogen may be used in fuel cell vehicles for example and does not emit greenhouse gases when consumed. Solid oxide fuel cell catalysts are considered for this electrolysis of water reaction. At present, the most economical approach to the generation of hydrogen is through steam reforming of hydrocarbons followed by addition of water to allow water-gas shift ($CO + H_2O \leftrightarrow CO_2 + H_2$). However, electrolysis of water[1] ($2H_2O \leftrightarrow O_2 + 4H^+ + 4e^-$) is an alternative hydrogen production reaction which has the advantage that no hydrocarbons, i.e. fossil fuels, are required. The electricity required for electrolysis may be drawn from alternative energy sources such as wind and solar.

Figure 1. Different size metal nanoparticles have different ratio of terrace, edge and corner sites. Determining which site is active for a reaction can guide the rational design of catalysts.

Figure 1. Different size metal nanoparticles have different ratio of terrace, edge and corner sites. Determining which site is active for a reaction can guide the rational design of catalysts.

My second research direction is proton exchange membrane fuel cells (PEMFC) for generating electricity from hydrogen. Current challenges for increasing PEMFC efficiency lie in improving the catalyst. In order to rationally design an improved catalyst, a knowledge of the active site of reaction at the atomistic scale is necessary. The benefit of this knowledge can be seen from Figure 1 which shows the size of a Pt nanoparticle catalyst affects the fractions of edge, corner and terrace sites present. Knowledge of which type of site is active suggests the optimal size of Pt nanoparticles. Quantum chemistry and in particular Density Functional Theory (DFT) has shown great success in predicting the activity of catalysts and is becoming an essential tool for the rational design of catalysts. However, when conducting DFT investigations it is necessary to include a quantification of uncertainty of DFT with the reason of not being misleading in making conclusions.[2] A quantification of DFT uncertainty will be conducted in order to generate reliable results.

[1] Smith, R. D. L.; Prévot, M. S.; Fagan, R. D.; Zhang, Z.; Sedach, P. A.; Siu, M. K. J.; Trudel, S.; Berlinguette, C. P. Photochemical Route for Accessing Amorphous Metal Oxide Materials for Water Oxidation Catalysis. *Science* **2013**, 340, 60-63.

[2] Deshpande, S.; Kitchin, J. R.; Viswanathan, V. Quantifying Uncertainty in Activity Volcano Relationships for Oxygen Reduction Reaction. *ACS Catal.* **2016**, 6, 5251-5259.

Teaching Interests:

In the undergraduate curriculum, I am particularly interested in and have the background for teaching courses such as: Transport Phenomena, Statistical and Numerical Methods for Chemical Engineering, and Chemical Engineering Kinetics. In the graduate curriculum, I am interested to teach courses such as Chemical Reactor Design and Chemical Process Principles. In any course I teach, I believe the inclusion of computer programming will ensure that students learn attention to detail and how to formulate engineering problems in an explicit manner. I am interested in developing a course entitled Quantum Chemistry of Materials for Catalysis to be offered as a graduate elective. This course will begin with principles of quantum chemistry starting with the Schroedinger equation, followed by Kohn-Sham density functional theory as an approximate solution to the Schroedinger equation. Topics of entropy contributions to free energy and sources of uncertainty in quantum chemistry will be discussed. A project will be conducted at the end of the semester in which students will simulate materials using quantum chemistry for catalyzing chemical reactions of industrial interest.

Copyright © American Institute of Chemical Engineers. All rights reserved.

480133 Catalysis Reactions towards Advanced Energy Applications

Abstract:

Research Experience:

Optimizing reactive efficiencies of particles that possess catalytic properties proves to be a source of major interest from the perspective of generating energy. Energetic materials can be tailored for intended applications through such enhanced surface characteristics. These enhancements can be achieved through simple techniques such as improving the hydration layer around reactant materials and thereby adding to their polarities. This change in polarity is known to boost their thermal properties adding to the energy released during the process of combustion. Simulation models for such structured crystals proves that bonds are affected by the polar nature of the environment particles are subjected to by altering their bond distances, atomic charges and stretching frequencies due to polar and hydrogen bonding forces. It would prove to be greatly beneficial if highly functionalized reactive materials could be developed for generating more effective energetic/ thermitic composites. Figures 1. a. and b. capture the idea of functionalized surface of actively polar molecules and the pronounced effect observed on their reactive output.

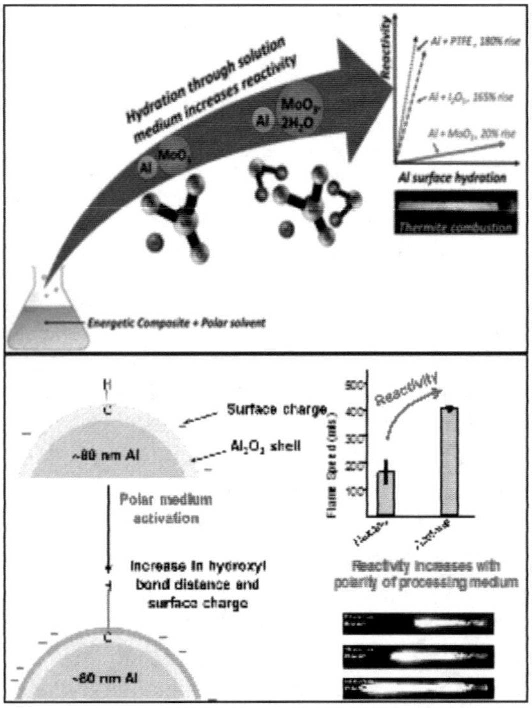

a.

THE GRAPHIC ON THIS PAGE WAS NOT AVAILABLE ON THE E-MEDIA VERSION OF THIS PAPER.

b.

Figure 1: a. Surface functionalization of alumina shell surrounding aluminum particles. b. Oxidizing agents can also be catalyzed similar to Al particles.

Research Interests:

Despite having a background in Mechanical Engineering, my research entails a lot of inputs from Chemical Engineering and Chemistry. I work with energetic compositions, with a focus on aluminum as a fuel. My dissertation has been inspired by the curious finding that the mere presence of polar media around polarized compounds like the passivating alumina shell in the fuel crystal can do wonders to improve its thermal behavior under fluorination. Through the understanding of results obtained off the Differential Scanning Calorimetry, I observed that the enthalpy of reaction could be increased significantly when water from the solution affects the reactant particle surface as against when a non-polar medium containing no trace of water was used for suspending these particles. Fourier Transform Infrared spectroscopy enabled me to detect the actually augmented hydration layer at the particle interface. I am also trained to handle a Hitachi S4300 Scanning Electron Microscope – to look at the homogeneity of dispersion matrix. Even with this hand-on experience using such highly precise instrumentation; it is extremely difficult to gauge interactions taking place at the molecular level. This brings the need for me to collaborate with

Chemists who specialize in performing ab initio calculations. Exposure to this area of research added depth to my understanding of chemical kinetics from an entirely unique perspective.

Through my knowledge of catalysis and catalysts used in various processes, I would want to focus on its applicability towards understanding basic molecular reactions and surface properties towards producing greener and cleaner fuels by maximizing conversion efficiency. Engineered materials that could be used as adsorbents and catalysts in fuel cells could effectively be used to improve their performance efficacy. Zeolites that are used in ion – exchange beds for water purification and softening can be improved for their efficiency in attracting dirt and minerals by polarizing catalysts to act as electromagnets. Understanding of catalysts through ab initio calculations helps me predict the nature of bonds affected by the hydrogen bonding forces to alter crystal structures when exposed to various environments. This information leads to the ability to tailor materials for suited applications.

In other collaborative simulation efforts, I would like to focus my attention on deriving intermediate reactions taking place which are studied in calorimetric techniques such as the DSC. From this data generated, it would be easier to predict which pathways any reaction mechanism could possibly take enabling us to better concentrate on the most desirable results from these reactions.

Teaching Experience:

I have experience teaching the undergraduate Heat Transfer and Combustion courses for Mechanical Engineering juniors and seniors respectively. I have been actively involved in mentoring masters' students in our research group where we evolve in an extremely symbiotic atmosphere.

I have found that it always important to keep students engaged through interactions so they develop more interest in what they are learning and in return take back much more than they would have. I like getting feedback from them as to how they were following what I am trying to convey in my lectures. It is really necessary for me to make classes increasingly interesting so students have the motivation to attend each class with equal zeal.

I prefer to start my classes with a short review of the previous lecture and put those in the form of points on the board so they retain what we talked about and then gradually introduce the day's topics. I think it is very important to grab the students' attention in the initial few minutes. I try to include a very relatable example for every concept so students understand its applicability. I try to involve the class in a discussion of the concept involved.

Proposal:

Submitted a grant proposal to the Army Research Office on the topic: 'Energetic Materials' (PI: Dr. Michelle L. Pantoya).

Ph.D. Dissertation:

'Altering surface properties toward enhanced aluminum reactivity'

Supervised by Dr. Michelle L. Pantoya, J. W. Wright Regents Endowed Chair Professor, Department of Mechanical Engineering, Texas Tech University, Lubbock, TX – 79409, USA

-

Publications:

Examining Hydroxyl – Alumina Bonding Toward Al Nanoparticle Reactivity; Padhye, R.; McCollum, J.; Korzeniewski, C. L.; Pantoya, M.L.; *Journal of Physical Chemistry C.* 2015, 119(47):26547-53 (DOI: 10.1021/acs.jpcc.5b08408; Publication date: November 17, 2015). Effect of Polar Environments on the Aluminum Oxide Shell Surrounding Aluminum Particles: Simulations of Surface Hydroxyl Bonding and Charge; Padhye, R.; Aquino, A. J. A.; Tunega, D.; Pantoya, M. L.; *ACS Applied Materials and Interfaces.* 2016, 8 (22), pp 13926–13933 (DOI: 10.1021/acsami.6b02665; Publication date: May 13, 2016). The Influence of Environmental Processing Conditions on the Reactivity of Aluminum with Various Oxidizing Agents; Padhye, R; Smith, D; Pantoya, M. L. (submitted to *Applied Surface Science*).

480715 Synthesis and Spectroscopic Characterization of Heterogeneous Catalysts for Energy Production

Abstract:

Research Interests:

- *In situ/operando* spectroscopic (DRIFTS, Raman, X-ray adsorption) characterization of catalysts
- Tuning surface acid-base properties of metal oxides for C-C coupling of small oxygenates to olefins
- Functional control of (bi)metallic catalysts for hydrodeoxygenation and steam reforming reactions
- Syntheses of multifunctional nano-catalysts for C1 (i.e., CO_x) activation

Successful Proposals: Washington State University new faculty seed grant

Research Experience:

My research career path has been following two major directions: materials science and catalysis. My graduate research was devoted to the structure control of nanoporous materials via self-assembly and sol-gel chemistry, as well as further synthesis of uniform metal nanoparticles catalysts confined within the pores of nanoporous materials via host-guest chemistry. During my PhD studies, I also spent three months (July, 2005-September, 2005) in the department of inorganic chemistry, Fritz Haber Institut der MPG, Germany where I received a full training on using high-resolution scanning/transmission electron microscope for materials characterizations. During my postdoc studies, I grasped the Schlenk line synthesis and *in situ* X-ray absorption spectroscope (i.e, XANES and EXAFS) characterization of structural simple supported metal clusters/mononuclear. My current research topics include the fundamental understanding and rational design of highly active and selective steam reforming catalysts, the development of bimetallic hydrogenation catalysts for removing oxygen from biomass derived phenolics, and the balance of acid-base chemistry for a cascade aldolization and self-deoxygenation of biomass derived small oxygenates. State-of-the-art *in situ/operando*characterization techniques (e.g., IR, Raman, and X-ray absorption etc.), reaction kinetics, and theoretical calculations have been used to elucidate the reaction mechanism of the catalysts. **Teaching Interests:**

During my graduate studies, I TA'd in the undergraduate physical chemistry labs taught by my advisor, and supervised a few undergraduate students to successfully finish their research projects at Dalian Institute of Chemical Physics. Since then, I have been presenting scientific results periodically in different forums, seminars and international/national conferences. As an Assistant Research Professor in the Gene and Linda Voiland School of Chemical Engineering and Bioengineering (VSCEB), I am also serving as a thesis committee member to supervise both graduate and undergraduate students independently. During the five years of working in VSCEB, I have gained valuable experiences in leading the group meetings and providing mentorship for students who are working under both government and industrial funded projects.

Selected Publications:

1. Rebecca A.L. Baylon, Junming Sun,* Kevin J. Martin, Padmesh Venkitasubramanian and Yong Wang* "Beyond ketonization: selective conversion of carboxylic acids to olefins over balanced Lewis acid–base pairs", *Chemical Communications*, 2016, 52, 4975-4978.
2. Junming Sun, Rebecca A.L. Baylon, Changjun Liu, Donghai Mei, Kevin J. Martin, Padmesh Venkitasubramanian and Yong Wang "Key roles of Lewis Acid-Base Pairs on $Zn_xZr_yO_z$ in Direct Ethanol/Acetone to Isobutene Conversion" *Journal of the American Chemical Society*, 2016, 2, 507-517.
3. He Zhang, Junming Sun,* Changjun Liu, Yong Wang* "Distinct water activation on polar/non-polar facets of ZnO nanoparticles" *Journal of Catalysis*, 2015, 331, 57.
4. Junming Sun, Ayman M. Karim, Donghai Mei, mark Engelhard, Xinhe Bao, Yong Wang "New insight into reaction mechanism of ethanol steam reforming on Co-ZrO2" *Applied Catalysis B: Environmental*, 2015, 162, 141-148.
5. Junming Sun, Ayman M. Karim, He Zhang, Libor Kovarik, Xiaohong Shari Li, Alyssa J. Hensley, Jean-Sabin McEwen, Yong Wang, "Carbon Supported Metal Catalysts for Vapor-phase hydrodeoxygenation of Guaiacol: A comparative Study" *Journal of Catalysis*, 2013, 306, 47-57.
6. Zhehao Wei, Junming Sun*, Yan Li, Abhaya K. Datye, Yong Wang* "Bimetallic catalysts for hydrogen generation" *Chemical Society Reviews*, 2012, 24, 7994-8008.
7. Kake Zhu, Junming Sun,* He Zhang, Jun Liu, Yong Wang* "Carbon as a Hard Template for Nano Materials Catalysts" *Journal of Natural Gas Chemistry*, 2012, 21, 215-232.
8. Junming Sun, Kake Zhu, Feng Gao, Chongmin Wang, Jun Liu, Charles H.F. Peden, Yong Wang "Direct Conversion of Bio-ethanol to Isobutene on Nanosized $Zn_xZr_yO_z$ Mixed Oxides with Balanced Acid-Base Sites", *Journal of the American Chemical Society*, 2011, 133, 11096.
9. Junming Sun, Miaofang Chi, Lobo L. Rodrigo, Mehraeen Shareghe, Browning Nigel, Gates C. Bruce "Tantalum Clusters Supported on Silica-Alumina: Influence of Support Composition and Chemistry on Cluster Structure", *Langmuir*, 2009, **25**, 10754-10763.
10. Junming Sun, Ding Ma, He Zhang, Xiumei Liu, Xiuwen Han, Xinhe Bao, Gisela Weinberg, Norbert Pfänder, and Dangsheng Su "Toward monodispersed silver nanoparticles with unusual thermal stability", *Journal of the American Chemical Society*, 2006, 49, 15756-15764.

Copyright © American Institute of Chemical Engineers. All rights reserved.

468388 Control of Fe-Ni Nanoparticle Activity for the Oxygen Evolution Reaction (OER) and Methanol Electrooxidation

Abstract:

The electrolysis of water for hydrogen generation and the electrochemical oxidation of alternative fuels such as methanol for alkaline fuel cells currently relies on the use of precious metal catalysts. However, the scarcity and high cost of these precious metals lead to the search for earth-abundant alternatives with similar electrocatalytic properties. Key challenges in the development of catalysts for the oxygen evolution half reaction (OER) of water electrolysis remain; these challenges include reducing the overpotential of the reaction and developing non-precious metal catalysts that are stable and active with minimal mass transport limitations. For direct electrooxidation of alternative fuels such as methanol, non-precious metal alternatives to precious metals such as platinum and palladium are often much less active. Among the non-precious metals, multimetallic nickel-based catalysts show particular promise for their high stability and high activity for the oxygen evolution reaction (OER) half-reaction of water electrolysis. In particular, it has recently been demonstrated by Trotochaud, Burke, Boettcher, and co-authors that iron incorporation into nickel hydroxide and cobalt hydroxide catalysts causes a significant increase in the OER activity of the catalyst [1-3]. It is now understood that the iron is in fact the active site of such bimetallic catalysts [2]. Additionally, nickel has exhibited high activity for methanol oxidation, particularly when in the form of a bimetallic catalyst such as NiPd [4]. However, promising active catalysts that are comprised entirely of non-precious metals such as nickel and iron have not been often reported for alternative fuel electrooxidation. Furthermore, there is a need for the development of nanostructured catalysts that can address mass transport limitations inherent in bulk or film-based catalyst materials [2]. As such, there continues to be an opportunity in alkaline electrocatalyst development to discover and understand nanostructured catalyst materials for both OER and alternative fuel electrooxidation.

In this presentation, our results on the development and optimization of an FeNi hydroxide core-shell nanoparticle

catalyst will be discussed for both OER and methanol electrooxidation. Through control of critical synthesis parameters, such as the addition rate of reducing agent, ratio of reducing agent to iron, and reaction time for nanoparticle synthesis, we have demonstrated the ability to tune nanoparticle catalyst activity for either OER or methanol electrooxidation. Results for maximum OER current as a function of the rate of addition of reducing agent sodium borohydride illustrate the sensitivity of catalyst performance to changes in synthesis parameters. Characterization of nanoparticle structure and phase will be discussed and correlated to electrochemical performance. Results suggest that tuning of the crystalline structure and order are critical to electrocatalyst performance optimization, as well as the presence of oxide and hydroxide.

References

[1] L. Trotochaud, S.L. Young, J.K. Ranney, S.W. Boettcher, Nickel-iron oxyhydroxide oxygen-evolution electrocatalysts: The role of intentional and incidental iron incorporation, J. Am. Chem. Soc., 136 (2014) 6744-6753.

[2] M.S. Burke, L.J. Enman, A.S. Batchellor, S.H. Zou, S.W. Boettcher, Oxygen evolution reaction electrocatalysis on transition metal oxides and (oxy)hydroxides: Activity trends and design principles, Chemistry of Materials, 27 (2015) 7549-7558.

[3] M.S. Burke, M.G. Kast, L. Trotochaud, A.M. Smith, S.W. Boettcher, Cobalt-iron (oxy)hydroxide oxygen evolution electrocatalysts: The role of structure and composition on activity, stability, and mechanism, J. Am. Chem. Soc., 137 (2015) 3638-3648.

[4] A. Dutta, J. Datta, Energy efficient role of Ni/NiO in PdNi nano catalyst used in alkaline DEFC, J. Mater. Chem. A, 2 (2014) 3237-3250.

Copyright © American Institute of Chemical Engineers. All rights reserved.

475202 Transition Metal-Nitrogen-Carbon Electrocatalysts for Oxygen Reduction Reaction

Abstract:

UNM has developed the Sacrificial Support Method (SSM) as a main approach for templated synthesis of hierarchically structured electrocatalysts materials. In this method the catalysts precursors are being absorbed on, impregnated within or mechanically mixed with the support (usually mono-dispersed or meso-structured structured silica), thermally processed (pyrolyzed) and then the silica support is removed by etching (in KOH or HF) to live the open frame structure of a "self-supported" material that consists of the catalysts only. Such hierarchical structures are advantageous in enhancement of the fuel cell performance since they correspond to the different levels of transport in the corrugated electrode matrixes. A wide variety of materials can be made by these methods in which not only the composition but also the microstructure can be varied. It is the combination of these attributes - control over microstructure at a number of different length scales and composition, simultaneously - that is extremely important to the performance of the electrocatalyst materials in a fuel cells.

The makeup and structure of the active site/sites of the Platinum Group Metal-free (PGM-free) ORR electro-catalysts, including geometry (coordination) and chemistry (composition and oxidation state) remain contentious even after 50 years of research. There is an emerging agreement that iron and nitrogen functionalities, displayed on the surface if the carbonaceous substrate/support, govern ORR activity. This is often combined with a broadly accepted hypothesis that micro-porous surface area plays a critical role forming the active site. Candidate structures participating in ORR include multitudes of nitrogen defect motifs in the carbon matrix of different degrees of graphitization, with metal incorporated as metal nano-particles, corresponding (native) oxides and/or as atomically dispersed, oxidized metal species, linked (coordinated) to nitrogen defects in carbonaceous matrix in a variety of possible configurations. These consist of in-plane defects such as graphitic nitrogen and Fe-coordinated to three or four nitrogen atoms and a plurality of possible edge sites such as pyridinic, pyrrolic, quaternary and $Fe-N_2$/$Fe-N_x$ sites. In this study, we have analyzed more than 45 distinct M-N-C electrocatalysts synthesized from three different families of precursors, such as (i) N-containing polymer, (ii) N-chelating macrocycles and (iii) N-containing small organic molecules (amines). Catalysts were evaluated and analyzed structurally using exactly the same protocol for deriving structure-to-property relationships. We have identified possible active sites participating in different ORR pathways: (1) metal-free electrocatalysts support partial reduction of O_2 to H_2O_2; (2) pyrrolic nitrogen acts as a site for partial O_2 reduction to H_2O_2; (3) pyridinic nitrogen displays catalytic activity in reducing H_2O_2 to H_2O; (4) Fe coordinated to N ($Fe-N_x$) serves as an active site for $4e^-$ direct reduction of O_2 to H_2O. The ratio of the amount of pyridinic and $Fe-N_x$ to the amount of pyrrolic nitrogen serves as a rational design metrics of M-N-C electrocatalytic activity in oxygen reduction reaction occurring through the preferred $4e^-$ reduction to H_2O.

There are two main hypotheses in the literature: one claiming that nitrogen functionalities on/in carbon-based support are directly responsible for their ORR activity and the second suggesting that nitrogen groups serve as the coordinating environment for metal ions, serving as reactive centers for the ORR. Theoretical calculations have confirmed that incorporation of nitrogen atoms in the carbon matrix enhances its electronic properties, but nitrogen atoms are not catalytically active towards a direct reduction of oxygen. Our mechanistic and structural studies have shown that multiple (yet similar) chemical functionalities most likely co-exists in practical M-N-C catalysts, displaying engagement in various steps of the ORR and thus obscuring the "simple model" of the "most active "***active site***".

Copyright © American Institute of Chemical Engineers. All rights reserved.

471580 Electrocatalytic Generation of H_2O_2: Carbon Based Material Synthesis and Device Design for Portable Low Cost Water Purification

Abstract:

Hydrogen peroxide, produced at a rate of 3 million tons/year, is one of the most important industrial chemicals widely used as a bleaching and oxidation agent.[1] Compared to chlorine which is the most commonly used water purification agent at the moment, hydrogen peroxide does not generate harmful byproducts after decomposition, and the higher oxidation potential makes it a more favourable reagent for treatment of water pollutants.[2] Hydrogen peroxide is mainly produced in large-scale facilities using a batch anthraquinone oxidation process, facing the challenge of high transportation and capital cost as well as explosive hazard during storage. For further applications in developing world a selective electrochemical catalyst towards H_2O_2 production is important and a prototype fuel cell/electrolyzer-solar cell device needs to be designed to enable continuous, small-scale and decentralized production of H_2O_2.

Through DFT calculation, we have predicted that different types of nitrogen-doped (N-doped) carbon or carbon with defects are able to selectively catalyze a 2-electron transfer oxygen reduction. We synthesized different types of carbon based materials and tested them along with some commercially available mesoporous carbons. Our results suggest that carbon defect sites without heteroatom doping could serve as the active center. The best catalyst by far is able to catalyse the reaction with no over-potential under alkaline condition with over 90% selectivity.

In this study, prototype devices, both fuel-cell and electrolyzer setups have also been designed and tested. Early results in a fuel-cell setup shows that H_2O_2 has higher affinity towards the aqueous phase as compared to the gas phase. Also H_2O_2 can be easily decomposed in contact with transition metals cations and reductive organic compounds. Considering the fact that electrocatalytic H_2O_2production will have the largest impact in the developing world, itâ€™s desirable to have a simple material manufacturing process and low material cost. All these considerations lead to the current electrolyzer design using plastic with no Nafion membrane and precious metal catalysts involved.

In our electrolyzer design, the produced H2O2 is up-concentrated in the electrolyte. At the current stage, the electrolyzer cell is able to operate with 50 mA overall current at pH 13 with an applied cell potential 1.7 V and faradaic efficiency close to 100%. The device can also be connected to a PV system as the energy supply. A ten-minute operation will able to accumulate 10ml solution with the concentration around 500mg/L.

1. Campos-Martin, J. M.; Blanco-Brieva, G.; Fierro, J. L. G., Hydrogen peroxide synthesis: An outlook beyond the anthraquinone process. *Angew Chem Int Edit* **2006,** *45*(42), 6962-6984.

2. Chu, S.; Majumdar, A., Opportunities and challenges for a sustainable energy future. *Nature* **2012,** *488* (7411), 294-303.

Copyright © American Institute of Chemical Engineers. All rights reserved.

473513 Niobium Doped Molybdenum Disulfide Catalyst for CO_2 Reduction Reaction

Abstract:

The world population consumes an average of 15 trillion watts of power, 85% of which comes from burning fossil resources including petroleum, natural gas, coal, and wood.[1] Despite their low cost and high quality, the use of fossil fuels is unavoidably coupled to the release of many compounds such as carbon dioxide (CO_2) that may adversely affect the environment such as acidifying oceans and climate change.[2,3] Thus, controlling CO_2 emission is essential using either renewable energy sources or remediation techniques. Among various remediation techniques, electrochemical reduction of CO_2 is known as the best approach since it not only reduces the amount of existing CO_2 in the atmosphere but also recycle the inert CO_2 molecule into energy-rich products (e.g., syngas). Layered transition metal dichalcogenides (TMDCs) have recently shown a significant potential to use as the catalyst for a wide range of electrochemical energy conversion and storage systems such as hydrogen evolution and CO_2 reduction[4,5]. However, it has been always a challenge to tailor the electronic structure of TMDCs to achieve highly promoted catalysts.

In this report, we have discovered a superior catalytic performance of Niobium (Nb) doped vertically aligned MoS_2 (VA-MoS_2) as an energy efficient and highly active catalyst for syngas production. Our results indicate that with only 5% Nb concentration at the edge atom structures, MoS_2 exhibits more than two orders of magnitude higher activity than Ag nanoparticles and one order of magnitude higher than pristine MoS_2 at low overpotentials (-0.2 V vs RHE).[6] Additionally, the product characterization results reveal that this promoted catalyst is selective for syngas (CO and H_2) formation at the potential range of -0.164 V to -0.764 V vs RHE. The effect of doping on the structure of pristine MoS_2 was studied by synthesis and testing

different concentrations of dopant for the CO_2 reduction reaction. Different characterization methods such as scanning transmission electron microscopy (STEM), X-ray photoelectron spectroscopy (XPS), electron energy loss spectra (EELS), energy-dispersive X-ray spectroscopy (EDX) and in-situ differential electrochemical mass spectrometry (DEMS) were applied to fully elucidate the effect of dopant on the electronic structure of MoS_2 and its catalytic activity and selectivity.

References:

(1) Davis, S. J.; Caldeira, K.; Matthews, H. D. Future CO2 Emissions and Climate Change from Existing Energy Infrastructure. *Science.* **2010**, *329*, 1330–1333.

(2) Haszeldine, R. S. Carbon Capture and Storage: How Green Can Black Be? *Science.* **2009**, *325*, 1647–1652.

(3) Lastoskie, C. Caging Carbon Dioxide. *Science.* **2010**, *330*, 595–596.

(4) Behranginia, A.; Asadi, M.; Liu, C.; Yasaei, P.; Kumar, B.; Phillips, P.; Foroozan, T.; Waranius, J. C.; Kim, K.; Abiade, J.; *et al.* Highly Efficient Hydrogen Evolution Reaction Using Crystalline Layered Three-Dimensional Molybdenum Disulfides Grown on Graphene Film. *Chem. Mater.* **2016**, *28*, 549–555.

(5) Asadi, M.; Kumar, B.; Liu, C.; Phillips, P.; Yasaei, P.; Behranginia, A.; Zapol, P.; Klie, R. F.; Curtiss, L. A.; Salehi-Khojin, A. Cathode Based on Molybdenum Disulfide Nanoflakes for Lithium–Oxygen Batteries. *ACS Nano,* **2016**, 10 (2), pp 2167–2175.

(6) Asadi, M.; Kumar, B.; Behranginia, A.; Rosen, B. a; Baskin, A.; Repnin, N.; Pisasale, D.; Phillips, P.; Zhu, W.; Haasch, R.; *et al.* Robust Carbon Dioxide Reduction on Molybdenum Disulphide Edges. *Nat. Com.* **2014**, *5*, 4470.

Copyright © American Institute of Chemical Engineers. All rights reserved.

461167 Green Chemistry with Flow Reactors and Heterogeneous Catalysts for the Synthesis of Fragrance Ingredients

Abstract:

International Flavors & Fragrances Inc. (IFF) is a leading global creator of flavors and fragrances for a wide variety of consumer products. Green chemistry and sustainability has long been an embedded part of molecule discovery and manufacturing processes at IFF. Example processes illustrate how green engineering technologies involving micro/flow reactors and heterogeneous catalysts are used to enhance both sustainability and economics of existing manufacturing processes for fragrance ingredients. Micro/flow reactors are significantly more energy efficient than conventional batch reactors. Their enhanced heat and mass transfer rates suppress unwanted side reactions, leading to high conversion, low waste and increased safety. These reactors combined with novel heterogenous catalysts provide greener and friendlier alternatives to many synthetic chemical processes that use conventional toxic, corrosive, hazardous reactants and catalysts.

Copyright © American Institute of Chemical Engineers. All rights reserved.

461144 Lewis Acid Catalyzed Pyrolysis of Polystyrene in Low-Temperature Molten Salts: Experiment and Simulation

Abstract:

Waste polystyrene is a potential energy and chemicals resource that is a major component of the plastics that are abundant in municipal solid waste. To valorize polystyrene waste, a major challenge is obtaining high yield and selectivity of the styrene monomer. Various reactor designs and catalysts have been reported previously. In our work, we have studied pyrolysis of polystyrene in ternary salt mixtures at or near their eutectic temperatures. Specifically, we have focused our attention on ternary mixtures of chlorides, with at least one component being $ZnCl_2$, as these mixtures have melting temperatures comparable to those used for polystyrene unzipping, i.e., 300 °C and greater. Molten salt mixtures have several important advantages for polystyrene de-polymerization, notably excellent heat transfer as well as potential for dispersion of acid and/or base co-catalysts. To improve performance further, we used an inert sweep gas that removed volatile decomposition products as they were formed.

First, we examined the benefits of polystyrene pyrolysis in a 60/20/20 $ZnCl_2$/NaCl/KCl molten salt mixture compared to thermal pyrolysis, finding that styrene yield and selectivity improved in the presence of the salt. The benefit was most pronounced at lower temperatures, with appreciable liquid yields (>60%) being achieved even at modest temperatures of 350 °C where thermal pyrolysis was much less effective. We then sought to examine the effects of polystyrene loading, $ZnCl_2$ addition, and composition of the supporting electrolytes (e.g., ratio of NaCl and KCl and/or addition of LiCl). This work revealed that polystyrene pyrolysis in a 40/30/30 $ZnCl_2$/NaCl/LiCl mixture could achieve an overall liquid yield of over 90% and a styrene yield of 75%. The molten salt pyrolysis performance compares favorably to yields reported in the literature, for example in the presence of an Fe-based catalyst or with microwave enhancement. Brønsted acids and bases were examined as catalysts. ZSM-5 and phosphotungstic acid greatly reduced the liquid yield, instead producing char. Thermal conditions converted $Ni(OH)_2$ into NiO which was removed by the inert sweep gas as green particulate material.

We then sought to understand the chemical role of the molten salt using density functional theory (DFT) simulations. In particular, we investigated the potential for the metal cations present in the molten salt acting as Lewis acids via aromatic-cationic interactions. We focused on Zn^{+2}, as this is the strongest Lewis acid in our systems. Using DFT, we calculated bonding energies of Zn^{+2} to a styrene trimer for a series of geometries. These calculations showed that the most stable arrangement was for the Zn^{+2} to intercalate between adjacent aromatic rings. The Figure is the stable geometry calculated for the intercalated configuration. However, the intercalated geometry does not result in substantial changes in carbon-carbon bonding, with only a modest changes (1-2%) in the bond lengths of several carbon-carbon bonds being observed. A second geometry, in which the Zn^{+2} positions directly on the aliphatic backbone, results in a substantial increase in the length of the carbon-carbon bond connecting a-methyl styrene to the remainder of the trimer. The backbone position is less thermodynamically stable than the intercalated configuration and we hypothesize that the intercalated configuration is first fully populated in the salt solvated polymer, followed by population of the backbone configuration. This hypothesis also may explain the benefits of the Li salt. Li^+ has a higher charge-to-surface-area ratio than either Na^+ or K^+, which allows Li^+ to participate more strongly in cation-aromatic interactions. Moreover, optimal Lewis acid loading is required to maintain high styrene yields without co-producing a-methyl styrene that reduce selectivity to the monomer. Brønsted acids, when combined with Lewis acids, catalyze carbon-carbon bond breaking and bond formation with low selectivity, leading to char production. These results form the basis for designing industrial recycling processes for obtaining high monomer yields from polystyrene pyrolysis.

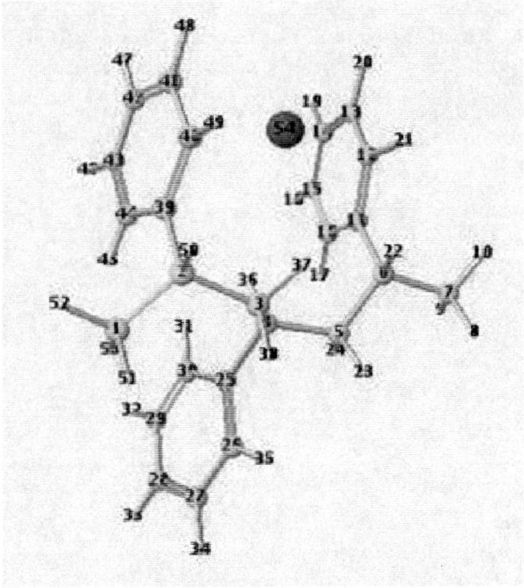

Figure 1. Optimized geometry of the styrene trimer participating in a Lewis acid/base interaction with Zn^{+2}. Carbon atoms (yellow), hydrogen (white), zinc (blue) are color labeled for clarity.

Copyright © American Institute of Chemical Engineers. All rights reserved.

455480 Microemulsion Systems As Smart Solvent Systems for Homogeneous Catalysis: Rhodium-Catalyzed Hydroformylation of Long-Chain Alkenes in Aqueous Media

Abstract:

As a part of the Collaborative Research Center InPROMPT (TRR 63) we investigate the application of microemulsion systems as smart solvent systems for homogeneous catalysis. More than 60 scientific from different universities and institutions work together in this scientific network. The long-term goal of our research is to optimize these systems in order to exploit their characteristics in multiphase chemical reaction processes. Based on the kind of catalysis, they allow for the reuse of dissolved catalysts or even for a better solubilisation of hydrophobic reactants. Examples for the application of microemulsion systems in catalysis are reactions with homogeneous catalyst complexes (e.g. rhodium-catalyzed hydroformylation [1]), with supported catalysts (e.g. Heck coupling reaction on sol-gel supported palladium catalysts [2]), or even with enzymes. One critical aspect from an economic perspective is the quantitative recycling of the catalyst complex dissolved in the multiphase system. Therefore, it is important to know the concentration of the catalyst complex in each of the single phases.

Kinetics of the rhodium-catalyzed hydroformylation of 1-dodecene in microemulsion systems have been thoroughly investigated. Furthermore, the different parameters/factors were analyzed that may have an influence on the distribution of catalyst complexes in microemulsions, e.g. temperature, type of ligand, structure of surfactant, and chain length of surfactant. Finally, the derived information was used for the design of a continuous process for the hydroformylation of long-chain alkenes in microemulsions [3].

Experimental Results

Standard experiments were carried out by using a microemulsion system consisting of water (catalyst solution), 1-dodecene, non-ionic surfactant and a sodium sulfate amount of 1 wt%. We studied the phase behaviour of the resulting microemulsion systems with optical methods. The hydroformylation of 1-dodecene was investigated in semi-batch experiments, at temperatures of 80 to 110 °C and at syngas pressures of 3 to 40 bar. The rhodium catalyst carries a bidentate, hydrophilic ligand, SulfoXantPhos. The microemulsion used for the reaction is formulated from 1-dodecene, water and the technical grade non-ionic surfactant Marlipal 24/70. Syngas is used in a ratio of 1:1 and is dispersed in the solution by a gas dispersion stirrer. Conversion and selectivity are determined by GC analysis and online measurement of syngas consumption. We investigated the influence of several parameters on the reaction, e.g. temperature, pressure, surfactant, catalyst concentration, and ligand to metal ratio. In these experiments TOFs of >1000 $h^{-1}$and selectivities of 98:2 to the linear aldehyde could be achieved. With the gained knowledge from the separation experiments, the catalyst could be successfully recycled for at least four times and rhodium loss was found to be below 1 ppm.

References

[1] T. Hamerla, A. Rost, Y. Kasaka, R. Schomäcker; ChemCatChem 5 (2013) 1854.

[2] I. Volovych, Y. Kasaka, M. Schwarze, Z. Nairoukh, J. Blum, M. Fanun, D. Avnir, R. Schomäcker; J. Mol. Catal. A Chem. 393 (2014) 210.

[3] T. Pogrzeba, D. Müller, T. Hamerla, E. Esche, N. Paul, G. Wozny, R. Schomäcker; Ind. Eng. Chem. Res. 54 (2015) 11953.

Copyright © American Institute of Chemical Engineers. All rights reserved.

469809 Synthesis of Linear Alkylbenzenes from Biomass-Derived Furan over Solid Acid Catalysts

Abstract:

Linear alkybenzenes (LABs) are the precursors of linear alkylbenzene sulfonate, the world's largest commodity biodegradable synthetic surfactant. Commercially, LABs are produced from petroleum-derived feedstocks via alkylation of benzene with C_{10}–C_{14} linear alpha olefins (i.e., 1-alkenes). Three different types of catalysts are typically employed: hydrogen fluoride (HF), aluminum chloride ($AlCl_3$), or UOP's proprietary DA-114. HF and $AlCl_3$ are both highly toxic and corrosive, while DA-114, although not toxic and corrosive, requires HF during its synthesis. There is therefore a critical need for a new LAB production process that utilizes both environmental friendly catalysts and renewable feedstocks. In this work, a new LAB synthesis pathway from biomass-derived furan and C_{12}–C_{16} linear alpha olefins over two different types of solid acid catalysts is studied. The reaction proceeds through Diel-Alder mechanism between furan and linear alpha olefins, followed by dehydration over acid catalysts. The product yields are dependent on several critical parameters, such as reaction temperature, reaction time, reactant ratio, and catalytic loading. Key reaction pathways leading to main and side reaction products are identified.

Copyright © American Institute of Chemical Engineers. All rights reserved.

450963 Rate-Enhancing Roles of Water Molecules in Methyltrioxorhenium-Catalyzed Olefin Epoxidation By Hydrogen Peroxide

Abstract:

Olefin epoxidation catalyzed by methyltrioxorhenium (MTO, CH_3ReO_3) is strongly accelerated in the presence of H_2O. The participation of H_2O in each of the elementary steps of the catalytic cycle, involving the formation of the peroxo complexes ($CH_3ReO_2(\eta^2\text{-}O_2)$, **A**, and $CH_3ReO(\eta^2\text{-}O_2)_2(H_2O)$, **B**), as well as in their subsequent epoxidation of cyclohexene, was examined in aqueous acetonitrile. Experimental measurements demonstrate that the epoxidation steps exhibit only weak $[H_2O]$-dependence, attributed by DFT calculations to hydrogen-bonding between uncoordinated H_2O and a peroxo ligand. The primary cause of the observed H_2O acceleration is the strong co-catalytic effect of water on the rates at which **A** and **B** are regenerated, and consequently on the relative abundances of the three interconverting Re-containing species at steady-state. Proton transfer from weakly coordinated H_2O_2 to the oxo ligands of MTO and **A**, resulting in peroxo complex formation, is directly mediated by solvent H_2O molecules. Computed activation parameters and kinetic isotope effects, in combination with proton-inventory experiments, suggest a proton shuttle involving one or (most favorably) two H_2O molecules in the key ligand exchange steps to form **A** and **B** from MTO and **A**, respectively.

Copyright © American Institute of Chemical Engineers. All rights reserved.

Acid Activation of Ferrate(VI) for Oxidation of Organic Compounds in Aqueous Solution

Kyriakos Manoli[1], George Nakhla[1], Ajay K. Ray[1] and Virender K. Sharma[2], (1)Chemical and Biochemical Engineering, University of Western Ontario, London, ON, Canada, (2)Environmental and Occupational Health, Texas A&M University, College Station, TX

Abstract

The unexpected enhancement of oxidation of selected organic compounds by adding simple acids (e.g. HCl and HNO_3) to the high valent iron based oxidant, ferrate(VI) ($Fe^{VI}O_4^{2-}$, Fe(VI)), is demonstrated. The activation of Fe(VI) was demonstrated by selecting organic compounds that have sluggish reactivity with Fe(VI) such as caffeine (psychostimulant, CAF), acesulfame potassium (artificial sweetener, ACE) and atenolol (β-blocker, ATL). Significantly, these pollutants exist in a variety of water systems (CAF, ATL and ACE have been detected in surface waters, groundwater and influents and effluents of wastewater treatment plants). The activated Fe(VI) resulted in increased removal of CAF, ACE, and ATL by 30 %. CAF and ACE could be removed almost completely at the highest amount of acid added. In the case of ATL, which was not removed by Fe(VI) alone, the enhancement due to H^+ ions, was also seen at the studied Fe(VI) to ATL molar ratios (4.0 and 10.0) but removal was not that high (~10 and 30 % at Fe(VI) to ATL molar ratio of 4.0 and 10.0 at 444 and 713 μM HCl respectively). The species responsible for enhancing the oxidative transformation of organic compounds were explored by conducting experiments under anoxic and oxic conditions. The aforementioned enhancement may be attributed to the increasing formation of reactive species, Fe(V)/Fe(IV) due to addition of H^+ ions to the Fe(VI)-organic mixed solution since Fe(V) is about 3 orders of magnitude more reactive than Fe(VI), probably because of its partial free radical character. The findings have wider implications in purifying water and wastewater from pollutants. The results demonstrated that the removal of pollutants can be accomplished at a shorter time scale and also at lower oxidant to organic pollutant molar ratios by adding H^+ ions than the removal of the molecules without the addition of acids. Activated Fe(VI) could decontaminate the sluggishly reactive or non-reactive pollutants. Both CAF and ACE had very low reactivity with Fe(VI) and took several hours to be oxidized. However, activated Fe(VI) was able to remove these molecules over much shorter times (e.g. 30 minutes).

Introduction

There has been increasing interest in the activation of oxidants in order to understand the potential for enhanced reactions of relevance to industrial, biological, energy, and environmental schemes.[1-6] For example, dioxygen, hydrogen peroxide, and transition metals (e.g. Cr, Mn, and Fe) participate in a wide range of oxidative reactions in metalloenzymes, which include hydrogenation, epoxidation, and N-dealklylation.[7-9] In recent years, many researchers studied the activation of iron(II)- and iron(III)-organo complexes in solutions to high-valent iron species (Fe^{IV} and Fe^{V}) with H_2O_2 in order to understand the mechanism of biological oxidation reactions and to oxidize different pollutants in the environment.[3,10-12] Researchers are interested in a simple high-valent iron(VI)-oxo species

($Fe^{VI}O_4^{2-}$, Fe(VI)) in aqueous solution because of its potential use as a green molecule in oxygen evolution, super iron batteries, organic transformations, water treatment, and disinfection of viruses and bacteria.[5,13-17]

Fe(VI) is a powerful oxidant with standard redox potential of 2.2 V and 0.72 V (vs NHE) at pH 0 and 14, respectively.[18] Under acidic conditions, Fe(VI) oxidizes water to oxygen.[19] The oxidation power of Fe(VI) strongly depends on the pH, which has been shown in the oxidation of a number of inorganic and organic compounds.[17,20] In acidic and neutral pH conditions, Fe(VI) tends to react with water rather than the target compound, thus diminishing the capability of Fe(VI) in oxidative transformations and in treating toxic chemicals. At pH 9.0-10.5, Fe(VI) does not react significantly with water (e.g. 2 % in 1h at pH 9.0). In this pH range, there is decreasing reactivity of Fe(VI)[17,21] and numerous reactions do not occur efficiently. The present paper presents for the first time the activation of Fe(VI) to enhance its oxidation capacity under slightly basic conditions.

The activation of Fe(VI) was demonstrated by selecting organic compounds that have sluggish reactivity with Fe(VI) at pH 8.0-9.0. These chosen molecules were caffeine (CAF), acesulfame potassium (ACE) and atenolol (ATL), which have different structures (Figure 1). Significantly, these compounds exist in a variety of water systems.[22] CAF is one of the two most widely used psychostimulants worldwide, with a global average consumption of 70 mg/person/day.[23] ACE is an artificial sweetener with wide application and ubiquitous occurrence in the environment due to its resistance to biodegradation and hydrolysis.[24] ATL is among the β-blockers of greatest concern due to its high toxicity to aquatic organisms.[25] CAF, ATL and ACE have been detected in surface waters, groundwater and influents and effluents of wastewater treatment plants and many researchers have explored the removal of these contaminants using various oxidation processes.[26,27] Fe(VI) is an environmentally-friendly oxidant, leaving only non-toxic Fe(III) oxides as a byproduct after its application, which is magnetically separable after the water treatment.[28] There is thus emerging interest in Fe(VI) in organic synthesis and in water treatment.[13,17]

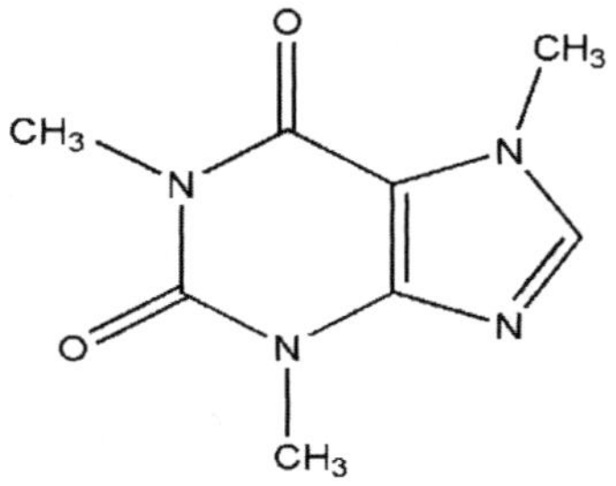

Caffeine (CAF)

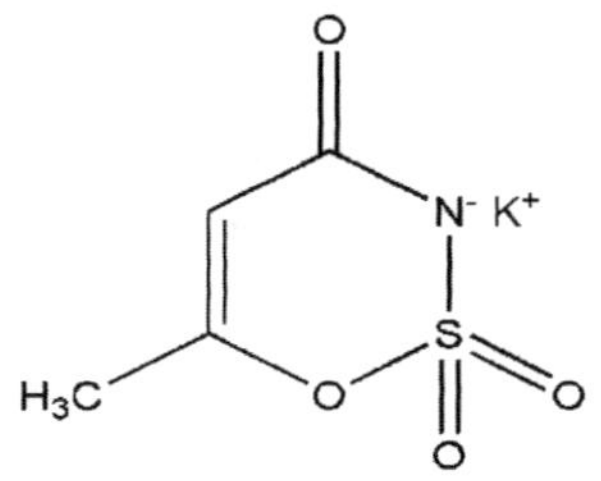

Acesulfame potassium (ACE)

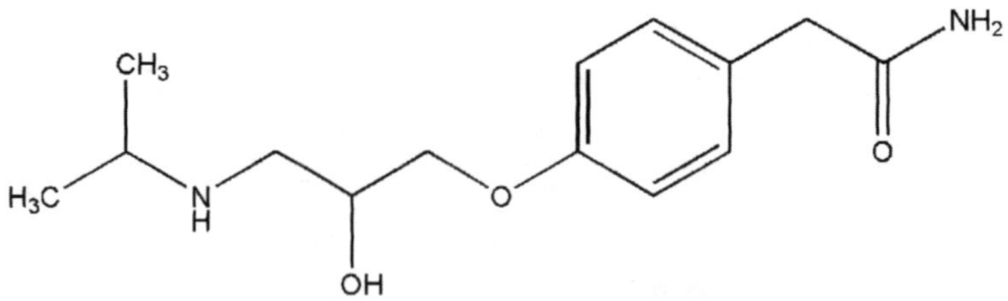

Atenolol (ATL)

Figure 1. Structures of selected organic compounds in the study.

Experimental Section

ReagentPlus grade caffeine, acesulfame potassium and atenolol powders (≥98.0 % purity), nitric acid (70 %) and acetic acid (>99.7 %) were purchased from Sigma-Aldrich, Canada. Ultrapure HPLC grade methanol (≥99.8 % purity) and 0.45 μm polypropylene filters were purchased from VWR International (Mississauga, Ontario, Canada). Hydrochloric acid (36.5-38.0 %) was purchased from Caledon Laboratory Chemicals (Georgetown, Ontario, Canada). All the chemicals mentioned above have been used without further purification. Fe(VI) was a salt of potassium (K_2FeO_4, 98% purity), which was prepared using the wet chemical method. Fe(VI) solutions were prepared in doubly distilled water that was passed through 18.2 MΩ water purification system (Thermo Scientific, Barnstead™ Easypure™ RODi), followed by centrifugation (Thermo Scietific, Sorvall Legend Centrifuge) at 3700 rpm for 2 min at 24 °C. The concentrations of Fe(VI) in filtered water were determined spectroscopically using molar absorptivity (ε_{510nm} = 1150 $M^{-1}cm^{-1}$). Fresh solutions of Fe(VI) were prepared before the experiments in order to avoid its decomposition in water. All experiments were performed at room temperature (24±1 ℃).

All experiments (Figures 3 and 4; Table 1) were conducted in 1 L beaker under dark conditions at room temperature (24±1 ℃). A 250 mL CAF solution was prepared in Milli-Q water and a certain volume of the Fe(VI) solution was added. When the pH was stabilized in up to 2 minutes, selected amount of acid was added directly to the Fe(VI)-CAF solution. After completion of the reaction, Fe(III) particulate was filtered before sample analyses.

For the control experiments, Fe(VI) in solution without CAF was first allowed to decompose to Fe(III) (with and without acid), followed by addition of CAF. After filtration of Fe(III), the concentration of CAF in filtered reaction solution was determined.

For the experiments under nitrogen-purging conditions, a 100 mL CAF solution was transferred to a gas-tight Pyrex-glass reactor that allowed us to purge gas during the reaction. The CAF solution was purged by nitrogen gas for 30 min. The Fe(VI) solution was added, followed by the addition of the acid. The mixture was allowed to react under continuous N_2 bubbling.

An Agilent Technologies (1200 series) high-performance liquid chromatography (HPLC), equipped with a ZORBAX Eclipse XDB-C18 column (4.6 × 150 mm, 5 μm) and UV detector (set up at 272 nm), was used for the quantification of caffeine. A mixture of methanol and Milli-Q water (70:30 v/v) was used as a mobile phase at a flow rate of 0.800 mL/min. The injection volume was 20 μL. UV-VIS spectrophotometer (UV-3600, Shimadzu Scientific Instruments, Maryland) was also used for the quantification of caffeine (peak absorbance at 272 nm), atenolol (peak absorbance at 223 nm) and acesulfame potassium (peak absorbance at 226 nm).

Results and Discussion

In the initial set of experiments, Fe(VI) and CAF were mixed in water and the concentration of Fe(VI) in the mixed solution was monitored (Figure 2a). The concentration of Fe(VI) without CAF was also followed (Figure 2a). At the beginning of the reactions, there was no significant difference between the Fe(VI) concentration with and without CAF. After a longer period of time, the difference in the concentration of Fe(VI) in the two mixed solutions became apparent (Figure 2a). The results showed that Fe(VI) not only reacted with water, but also reacted with CAF. However, the reaction between Fe(VI) and CAF was sluggish.

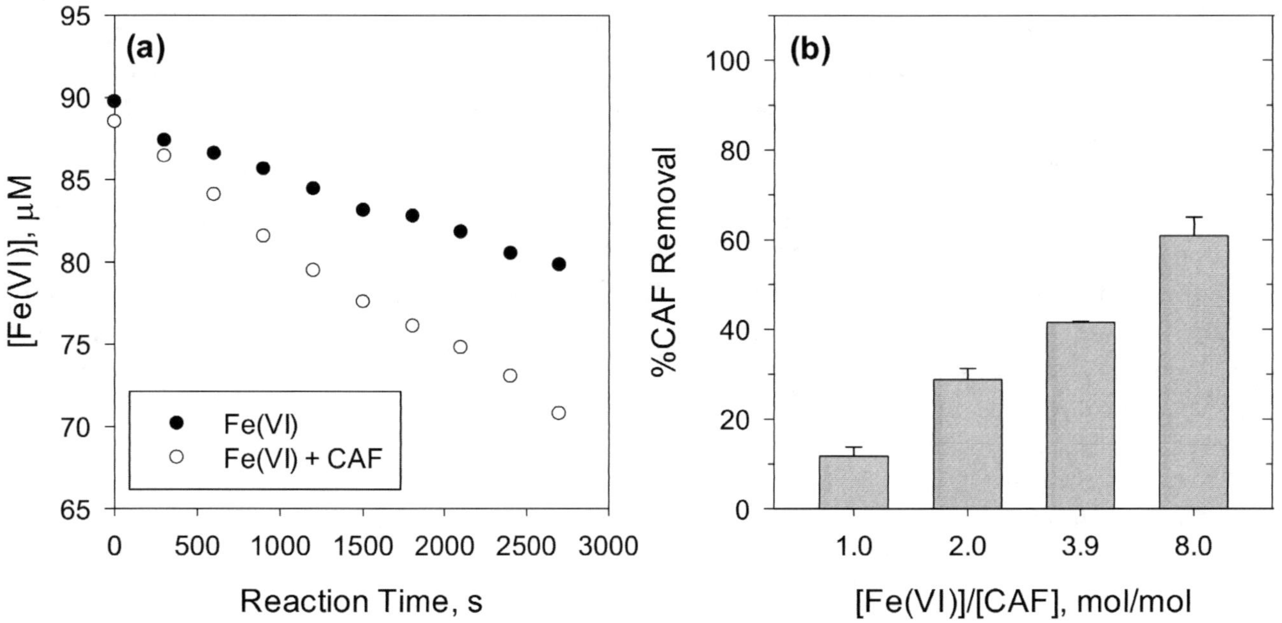

Figure 2. (a) Decay of Fe(VI) with and without CAF in solution at pH 8.5 ([CAF]=89.1 μM). (b) Oxidative transformation of CAF by Fe(VI) at different molar ratios of Fe(VI) to CAF at final pH 7.5 (Experimental conditions: [Fe(VI)]=88.5-634.5 μM; [CAF]=78.7-95.3 μM).

The oxidative transformation of CAF in mixed solutions at different molar ratios of Fe(VI) to CAF ([Fe(VI)]/[CAF] = 1.0 – 8.0) was also determined (Figure 2b). The time taken for the disappearance of the colour of Fe(VI) increased with the molar ratio of Fe(VI) to CAF. The CAF removal efficiency increased from 12% to 61% with the increase in the molar ratio of Fe(VI) to CAF (Figure 2b).

Because of the longer contact time under the conditions used and also the incomplete removal of CAF, a second set of experiments was conducted by adding a small amount of acid (0.1-0.5 mL of 0.5 M) in the Fe(VI)-CAF solution. It was expected that lowering pH would increase the redox potential, hence the oxidation of CAF (E^0(acidic)=+2.2 V versus E^0(basic)=+0.72 V).[17] The addition of acid lowered the pH by only 0.4–1.0 pH unit, depending on the concentration of the acid. Significantly, the addition of acid lowered the times for Fe(VI) color disappearance (or decay of Fe(VI)) and filtration time of Fe(III). This is shown as in Table 1.

The results of the oxidative transformation of CAF under different conditions are given in Figure 3. The results clearly demonstrated the enhancement of oxidative transformation of CAF by adding acids. Also, CAF could be removed almost completely at the highest amount of acid added. The control experiments ruled out the role of Fe(III), produced from Fe(VI), in the observed oxidation of CAF.

Table 1. Variation of pH and oxidative transformation of CAF with and without acid for a typical experiment. ([Fe(VI)]/[CAF] = 8.0).

Acid Concentration	pH(1)	pH(2)	pH(3)	pH(4)	Time (Fe(VI) Decay)	Filtration Time	%CAF Removal
No Acid	9.6	-	10.3*	7.5	4 h	8.0 h	60.9
882 μM HCl	9.5	8.8	9.1	7.5	10 s	0.5 h	96.8
888 μM HNO_3	9.4	8.7	9.3	7.5	10 s	0.5 h	95.5
879 μM CH_3COOH	9.5	8.5	9.6	7.5	10 min	1.0 h	96.8

pH(1) - Initial pH of Fe(VI)-CAF mixture; pH(2) - Immediate pH after the addition of acid
pH(3) - pH after 10 min; pH(4) - Final pH before filtration; *after 3 h;

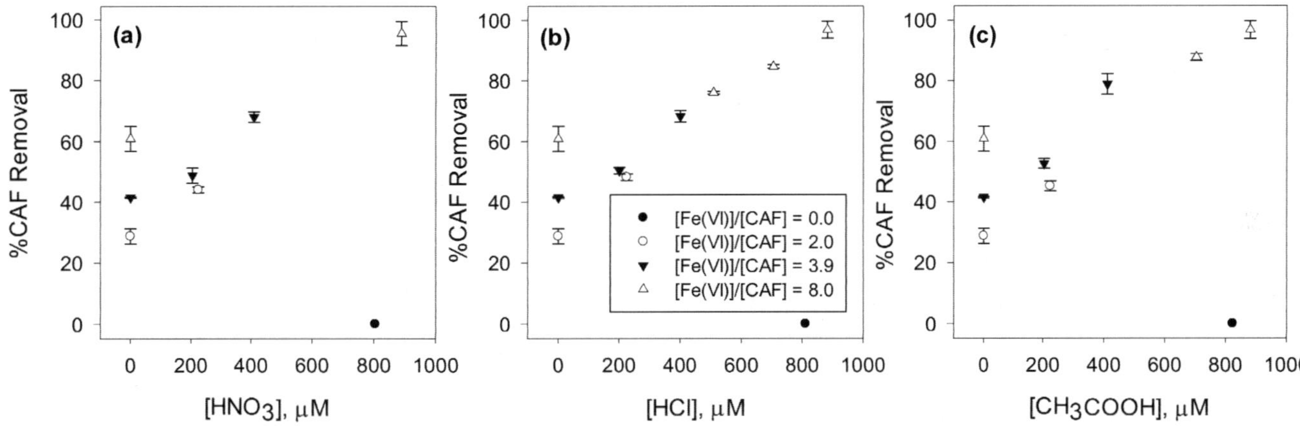

Figure 3. Effect of acids added to Fe(VI)-CAF mixed solution (a) HNO_3, (b) HCl and (c) CH_3COOH (Experimental conditions: [Fe(VI)]=177.5-634.5 μM; [CAF]=70.3-96.4 μM; final pH=7.5).

The enhancement of up to 30 % of transformation of CAF (Figure 3) was surprising because the addition of acid lowered the pH only by up to 1.0 unit. To further elucidate the enhancement, a third set of experiments was performed in which the pH of the mixed solution of Fe(VI) and CAF was varied by 1.0 pH unit (pH 8.0 – 9.0) without adding acid at a molar ratio of 2.0. The results showed no enhancement of the transformation of CAF when the pH decreased from 9.0 to 8.0 without acid addition to the reaction mixture. This suggests that the activation of Fe(VI) occurred when acid was added to the Fe(VI)-CAF mixed solution to result in the observed enhancement of the transformation of CAF. The possible species involved in activating Fe(VI) are discussed later.

Experiments on the activation of Fe(VI) by H^+ ions were also tested on other organic compounds, ACE and ATL, in order to seek a broader understanding of enhancement of oxidative transformation of the target organics. As shown in Figure 4, enhancement of the oxidation power of Fe(VI) by H^+ was observed for ACE, with more than 90 % transformation achieved. In the case of ATL, the enhancement due to H^+ ions, was also seen at a molar ratio of 10:1 ([Fe(VI)/[ATL]). Significantly, ATL could not be oxidized by Fe(VI) without acid at the studied Fe(VI) to ATL molar ratios. Interestingly, the enhancement in oxidative transformation due to H^+ ion was ~ 30 %, which is similar to the increase from 64 % to 95 % in the transformation of ACE due to acid addition (Figure 4).

Furthermore, CAF also exhibited a similar enhancement due to the activation of Fe(VI) by H^+ ions (Figures. 3a, 3b, and 3c).

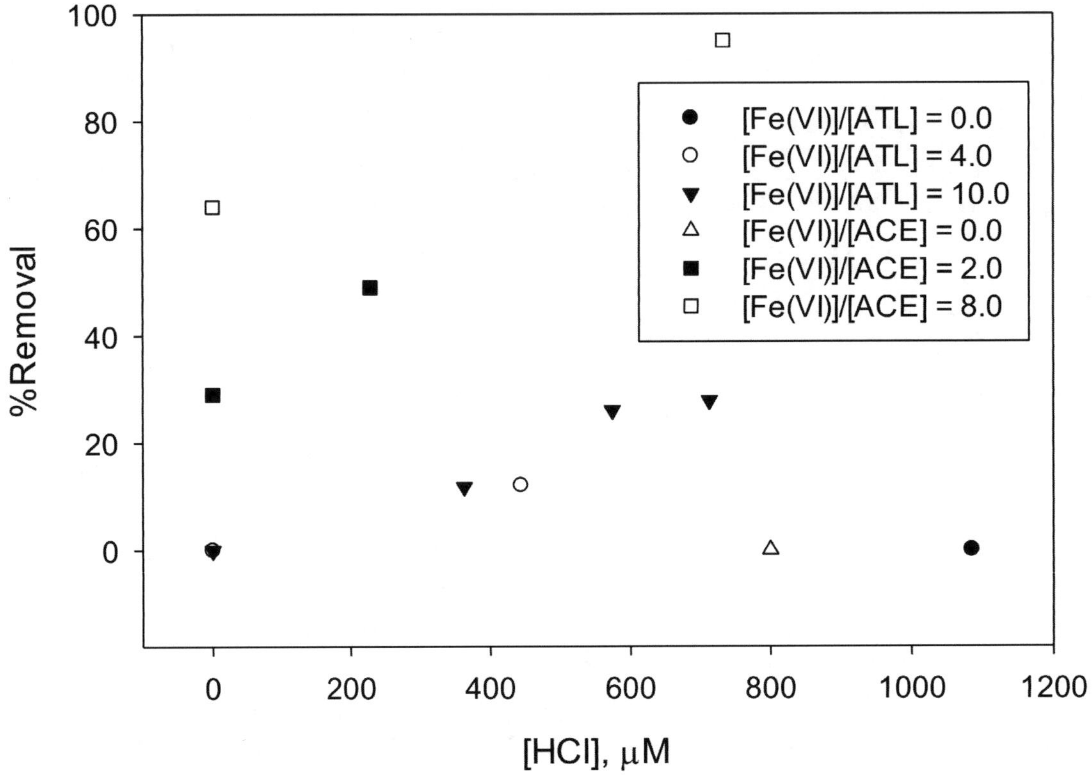

Figure 4. Effect of concentration of HCl on the transformation of acesulfame K (ACE) and atenolol (ATL) by Fe(VI). (Experimental conditions: [ACE]=49.0-81.4 μM; [ATL]=41.6-56.9 μM; [Fe(VI)]=161.9-569.2 μM; final pH=6.5-7.5).

Finally, the species responsible for enhancing the oxidative transformation of organic compounds were explored by conducting experiments under nitrogen-purging conditions. The results of the conversion of CAF with and without purging with different acid solutions are presented in Figure 5. In acidified solutions with HCl and HNO_3, the efficiency of oxidative transformation of CAF in the deoxygenated and oxygen-containing environments was similar (Figure 5). The results indicate that the dissolved oxygen played no role in the enhancement of oxidation of CAF by the activation of Fe(VI) by H^+ ions in these cases.

A possible reason for the augmentation of oxidative transformation of organics may be the increasing formation of reactive intermediate species, Fe(V)/Fe(IV), due to addition of H^+ ions to the Fe(VI)-organic mixed solution. Formation of Fe^V and Fe^{IV} species during the decomposition of Fe(VI) has been recently demonstrated.[29] Both intermediates, i.e. Fe(V) and Fe(IV) species, are known to oxidize compounds at faster rates than Fe(VI). It is well known that Fe(V)/Fe(IV) species are about 2-4 orders of magnitude more reactive than Fe(VI),[30] thus causing enhanced transformation of organics by adding acid to the Fe(VI)-organic mixtures.

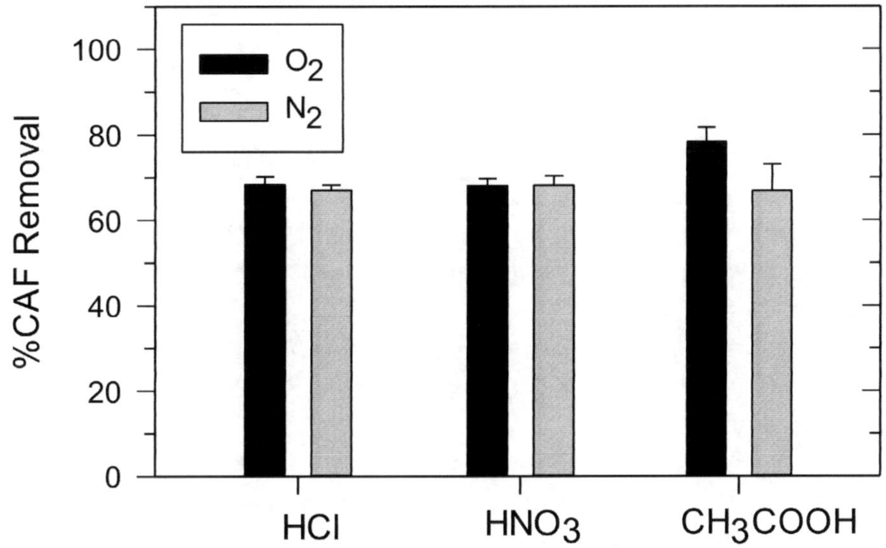

Figure 5. Oxidation of CAF by Fe(VI) under oxygenated and deoxygenated conditions using different acids. (Experimental conditions: [Fe(VI)]:[CAF]=3.9; [Acid]=393-431 μM; final pH(O_2)=7.5; final pH(N_2)=10.0).

Another possible mechanism for the formation of Fe(V)/Fe(IV) species in the reaction mixture is through the oxidation of iron species, Fe(II)/Fe(III) by possible oxidants (e.g. Fe(VI), O_2, and H_2O_2) in the system (e.g. $Fe^{VI} + Fe^{II} \rightarrow Fe^{V} + Fe^{III}$ and Fe^{II}-org/Fe^{III}-org + O_2/H_2O_2 $\rightarrow$ Fe^{IV}-org/Fe^{V}-org).[31-34] The generation of H_2O_2 is likely from the decomposition of Fe(VI) by acid.[17] Recent studies have shown that acetic acid could facilitate the generation of Fe(V) species in oxygenated system, which were able to carry out selective oxidation of organics.[34] Interestingly, our observation of 10% higher transformation of CAF by acetic acid in the oxygenated solution than that in the deoxygenated condition (under N_2 environment) is consistent with findings of other researchers (see Figure 5).

In summary, this paper reports the unexpected enhancement of the transformation of the studied organic molecules by activating the Fe(VI) in aqueous solution with an addition of a small amount of simple acid (e.g. HCl and HNO_3). Importantly, the results demonstrated that the organic transformations can be accomplished over a shorter time scale and also at lower dose of oxidant by adding H^+ ions than seen without the addition of acids. The results obtained herein have a wide range of implications in organic synthesis and in purifying water and wastewater. In conclusion, by adding a small amount of acid, Fe(VI) technology is more efficient and faster for treating water.

References

1 O. Cussó, M. Cianfanelli, X. Ribas, R. J. M. Klein Gebbink, M. Costas, *J. Am. Chem. Soc.,* 2016, *138,* 2732-2738.

2 T. -. Chen, N. Asiri, K. W. Kwong, J. Malone, R. Zhang, *Chem. Commun.,* 2015, *51,* 9949-9952.

3 O. Cussó, X. Ribas, M. Costas, *Chem. Commun.,* 2015, *51,* 14285-14298.

4 B. Sun, X. Guan, J. Fang, P. G. Tratnyek, *Environ. Sci. Technol.,* 2015, *49,* 12414-12421.
5 L. Ma, W. W. Y. Lam, P. -. Lo, K. -. Lau, T. -. Lau, *Angew. Chem. Int. Ed.,* 2016, *55,* 3012-3016.
6 Y. Zhou, J. Jiang, Y. Gao, J. Ma, S. -. Pang, J. Li, X. -. Lu, L. -. Yuan, *Environ. Sci. Technol.,* 2015, *49,* 12941-12950.
7 Z. Chen, G. Yin, *Chem. Soc. Rev.,* 2015, *44,* 1083-1100.
8 S. Fukuzumi, *Dalton Trans.* 2015, *44,* 6696-6705.
9 K. -. Cho, H. Hirao, S. Shaik, W. Nam, *Chem. Soc. Rev.,* 2016, *45,* 1197-1210.
10 W. N. Oloo, L. Que, *Acc. Chem. Res.,* 2015, *48,* 2612-2621.
11 A. R. McDonald, L. Que, *Coord. Chem. Rev.,* 2013, *257,* 414-428.
12 L. L. Tang, W. A. Gunderson, A. C. Weitz, M. P. Hendrich, A. D. Ryabov, T. J. Collins, *J. Am. Chem. Soc.,* 2015, *137,* 9704-9715.
13 V. K. Sharma, L. Chen, R. Zboril, *ACS Sustainable Chem.Eng.,* 2016, *4,* 18-34.
14 M. Farmand, D. Jiang, B. Wang, S. Ghosh, D. E. Ramaker, S. Licht, *Electrochem. Commun.,* 2011, *13,* 909-912.
15 A. Karlesa, G. A. D. De Vera, M. C. Dodd, J. Park, M. P. B. Espino, Y. Lee, *Environ. Sci. Technol.,* 2014, *48,* 10380-10389.
16 J. Q. Jiang, *J. Chem. Technol. Biotechnol.,* 2014, *89,* 165-177.
17 V. K. Sharma, R. Zboril, R. S. Varma, *Acc. Chem. Res.,* 2015, *48,* 182-191.
18 R. H. Wood, *J. Am. Chem. Soc.,* 1958, *80,* 2038-2041.
19 R. Sarma, A. M. Angeles-Boza, D. W. Brinkley, J. P. Roth, *J. Am. Chem. Soc.,* 2012, *134,* 15371-15386.
20 V. K. Sharma, *Coord. Chem. Rev.,* 2013, *257,* 495-510.
21 D. G. Lee, H. Gai, *Can. J. Chem.,* 1993, *71,* 1394-1400.
22 S. D. Richardson, S. Y. Kimura, *Anal. Chem.,* 2016, *88,* 546-582.
23 O. Gimeno, J. F. García-Araya, F. J. Beltrán, F. J. Rivas, A. Espejo, *Chem. Eng. J.,* 2016, *290,* 12-20.
24 Z. Gan, H. Sun, R. Wang, H. Hu, P. Zhang, X. Ren, *Water Res.,* 2014, *64,* 113-122.
25 A. A. Godoy, F. Kummrow, P. A. Z. Pamplin, *Chemosphere,* 2015, *138,* 281-291.
26 U. Hübner, U. von Gunten, M. Jekel, *Water Res.,* 2015, *68,* 150-170.
27 Y. Luo, W. Guo, H. H. Ngo, L. D. Nghiem, F. I. Hai, J. Zhang, S. Liang, X. C. Wang, *Sci. Total Environ.,* 2014, *473-474,* 619-641.
28 R. Prucek, J. Tucek, J. Kolarik, I. Huskova, J. Filip, R. S. Varma, V. K. Sharma, R. Zboril, *Environ. Sci. Technol.,* 2015, *49,* 2319-2327.
29 L. Machala, V. Prochazka, M. Migllenni, V.K. Sharma, Z. Marusak, H-C. Wille, R. Zboril, *Phys. Chem. Chem. Phys.* 2015, *17*, 21787-21790.
30 V. K. Sharma, R. A. Yngard, D. E. Cabelli, J. Clayton Baum, *Radiat. Phys. Chem.,* 2008, *77,* 761-767.
31 V.K. Sharma, B.H.J. Bielski, *Inorg. Chem.,* 1992, *30*, 4306-4310.
32 P. Comba, M. Kerscher, T. Krause, H.F. Scholar, *Environ. Chem.,* 2015, *12*, 381-395.
33 O. Cusso, X. Ribas, M. Costas, *Chem. Commun., 2015, 51*, 14285-14298.
34 J. Serrano-Plana, W.N. Oloo, L.A. Rueda, K.K. Meier, B. Verdejo, E.G. Espana, M.G. Basallote, E. Munck, L. Que, Jr., A. Company, M. Costas, *J. Am. Chem. Soc.*, 2015, *137*, 15833-15842.

459285 Kinetic Studies of Liquid Phase Ethylene Production By Hydrogenation of Acetylene Using a Selective Solvent in a Basket Reactor

Abstract:

Ethylene is the simplest of the olefins and a primary building block in petrochemical industries. It is produced mainly by steam cracking of naphtha with acetylene as the byproduct in the cracker effluent. Small traces of acetylene can deactivate the catalyst used in the polymerization of ethylene to manufacture polyethylene and other valuable products. Although acetylene can be absorbed from the gas mixture, it is more beneficial to selectively hydrogenate the acetylene to ethylene. Selective hydrogenation of acetylene to ethylene using supported catalysts in gas phase mostly leads to thermal runaway. Recently, liquid phase hydrogenation of acetylene has been developed as it reduces the risks of runaway conditions as the liquid phase itself acts as a heat transfer medium. The liquid phase reduces the formation of green oil due to oligomerization on the catalyst thereby improving the catalyst stability. This proposed study is novel as only the gas phase hydrogenation kinetics has been investigated as found in the open literature. The kinetics of selective hydrogenation of acetylene in the liquid phase over Pd/Al_2O_3 catalyst was investigated in a 300 mL stirred-tank basket reactor. Since acetylene is highly soluble in polar solvents, N-methyl pyrrolidone was used in our studies to selectively absorb acetylene from the gas mixture hydrogenating it to ethylene over Pd/Al_2O_3 catalyst. The reactor was operated at a temperature range of 80-120 ^{o}C under a pressure range of 200-250 psig. The liquid phase was operated in batch mode while the gas phase was continuous. A kinetic model was developed and fitted on simple power law equations and Langmuir-Hinshelwood-Hougen-Watson approach. The predictions of these approaches on the acetylene conversion was compared. The surface reaction between the adsorbed species was assumed to be the rate controlling step. The parameters studied to understand the liquid phase kinetics were temperature, H_2:C_2H_2 feed ratio, catalyst loading and operating pressure. The sensitivity of the variation of different parameters on the conversion was also studied. The results and discussions from this investigation are in progress and will be presented in the conference.

Copyright © American Institute of Chemical Engineers. All rights reserved.

472613 Solvent Effect in Catalytic Upgrading of Biomass-Derived Furanics

Abstract:

In liquid-phase heterogeneous catalysis, solvent provides an additional degree of freedom for tuning the catalytic activity and selectivity. Aqueous-phase upgrading of small oxygenates is particularly desirable for biomass conversion since pyrolysis of biomass produces large amounts of water and water-soluble small oxygenates. Fundamental understanding of the solvent effect will be valuable for further promoting the liquid-phase heterogeneous catalytic upgrading strategy. Here we will focus on furfural and discuss the solvent effect on its catalytic reactions through a synergistic combination of experiment, theory and computation. We find rate enhancement of furfural conversion and improved selectivity towards hydrogenation of the carbonyl group to form furfuryl alcohol in an aqueous phase as compared to reactions in organic solutions. We show that the water solvent has multiple, distinct effects. The aqueous phase stabilizes the intermediate and final products by forming hydrogen bonds and changes the rate-determining step. Moreover, hydrogen can also be transferred by a water molecule, which further reduces the activation barrier. We will also discuss the effect of the dielectric screening of water on stabilization of the charged transition state.

Copyright © American Institute of Chemical Engineers. All rights reserved.

472932 Solvent Effect on Cyclohexene Hydrogenation over Pd

Abstract:

We have studied the effects of different hydrogen solubility, competitive adsorption, and solvation in a model hydrogenation reaction. For this purpose, *in-situ* spectroscopy is a powerful tool to acquire high time resolution kinetic data. The ATR-IR probe is equipped on a high temperature and high pressure vessel to monitor hydrogenation rate of cyclohexene in different organic aprotic solvents. The reaction is carried out under 70°C over heterogeneous 0.5% Pd/a-Al_2O_3 and homogenous Pd(II) catalyst. The concentration of cyclohexene during reaction is estimated by the intensities of the vibrational mode at 1139cm^{-1}(wCH2 rock). The scan takes place every one minute so that the reliable kinetic analysis can be modeled. By changing different organic aprotic solvents, the hydrogenation rates are compared to study the solvent effect. The main difference in heterogeneous and homogenous catalysis is that in heterogeneous catalysis reactants are firstly adsorbed on metal surfaces and then go through surface reaction to form products. However, in homogenous catalysis reactants interact strongly with catalyst active sites and they both are solvated in solvents. The study will focus on whether solvation effect and how it affects the cyclohexene hydrogenation rate.

Copyright © American Institute of Chemical Engineers. All rights reserved.

467379 Numerical Study of Heavy Crude Oil Pyrolysis in Supercritical Water

Abstract:

Pyrolysis of heavy oil in supercritical water (SCW) occurs under elevated pressures and temperatures, when two phases often co-exist in the reacting mixture. Based on published experimental data in the literature, we propose a fractional-order lumped reaction kinetic model considering two phases (an oil-rich phase and a water-rich phase). Using our kinetic model, the pyrolysis of a heavy oil droplet (D=0.5mm) in a bulk of SCW is simulated as the coupling of kinetics, thermodynamics and transport phenomena. Results show that the phase behavior plays an important role in the pyrolysis reactions of heavy oil: (1) asphaltenes are heavily concentrated in the center of the original oil droplet, which speeds up the reaction compared to the process without water; and (2) maltenes quickly diffuse into the water-rich phase and become diluted, which slows down the coking from maltenes and thus suppress the overall coke formation. Modeling of this study can help design the heavy oil pyrolysis reactor using SCW, and indicates that phase behavior may also be important in other reactors using supercritical fluids.

Copyright © American Institute of Chemical Engineers. All rights reserved.

463308 Multi-Scale Multi-Phase Reduced Order Models for Dual-Layered Catalytic Monolith Reactors

Abstract:

We use individual phase averaging based on Lyapunov-Schmidt reduction to obtain reduced order model for describing the transient non-isothermal diffusion, convection and reaction processes in a catalytic monolith with two porous washcoat layers. We average the model over the small transverse scales and express the reduced order model in terms of physically meaningful or measurable concentration/temperature modes. We examine various limiting cases, present the solutions of the reduced order model and compare our results with the exact solution and those in the literature. Specifically, we consider the limiting case of short monolith model with flat velocity profile and compare the analytical solutions obtained from the detailed model, reduced order model and traditional multiphase phase model; and show that the reduced order model represents the detailed model more accurately while the traditional models fail to capture the effect of transverse diffusion and reactions on interfacial fluxes. Finally we compare the numerical solutions of detailed and reduced order models for general case to show that the later leads to the good results with reasonable accuracies even for the fast transient and reacting cases.

Copyright © American Institute of Chemical Engineers. All rights reserved.

467196 CFD Modeling of an Industrial Furnace Reformer and 3D Multi Scale Model for Packed Bed Reactor for Synthesis Gas Production

Abstract:

Synthesis gas (Syn gas) is used in many industrial applications such as Direct Reduction of Iron (DRI), Fischer-Tropsch synthesis, and also a source for environmental-friendly clean fuels and chemicals. In Midrex DRI process, syn gas is used to remove the chemically bound oxygen from raw iron-ore without melting. Syn gas is primarily produced from natural gas reforming in a large number of tubular packed bed reactors suspended in a furnace. Combustion of fuel takes places in the furnace supplying the requisite heat for the endothermic reforming reactions. The reactor tubes are made of special alloys and their failure due to uneven temperature or hotspots leads to the production loss or plant shutdown. Therefore, an accurate prediction of tube wall temperature is vital for preventing such tube failures. Computational Fluid Dynamics (CFD) is a powerful tool that can be used to simulate the turbulence, radiation and combustion in furnace reformer and the tube wall temperature is predicted without employing empirical heat transfer coefficients.

This talk presents a 3D CFD model of an industrial Midrex furnace reformer employing k-Îµ turbulence model, Discrete Ordinates (DO) radiation model, and finite rate/Eddy Dissipation combustion model. The reactor tubes are fully meshed in 3D and heat transferred from the furnace side to tube side is directly coupled at the tube wall. This step avoids the use of empirical heat transfer coefficients and also directly takes into account the varying tube wall heat flux due to shadowing and proximity to the burners. The flow reactor tubes are described using laminar flow porous media model with pressure drop described using Ergun equation and the reforming reaction rates are taken from the literature. The model in the tube constitutes an effective media packed bed model where a catalyst effectiveness is employed. We will present the results showing the temperature, velocity and species profiles inside the furnace as well as the performance of the packed bed reformer.

This talk also presents a 3D multiscale packed bed reformer model of a single tube developed using multiscale approach. The model takes into account dispersion in axial and radial directions as well as catalyst-fluid phase mass and energy limitations along the length of the reactor. The varying heat flux in the angular direction due to shadowing of tubes also is taken into account. Reaction, mass and heat transfer inside the catalyst are directly coupled with fluid phase equations using 1D pellet model, and thus avoiding the use of a catalyst effectiveness factor. This reactor model results will then be compared with the simpler model used to describe in the furnace reformer model. The model results will also be compared with a 2D multiscale model not accounting for varying heat flux in angular direction.

Copyright © American Institute of Chemical Engineers. All rights reserved.

460383 DFT Calculations and Reaxff Development for MD Simulations of H-BN Growth on Nickel Surfaces

Abstract:

DFT CALCULATIONS AND REAXFF DEVELOPMENT FOR MD SIMULATIONS OF H-BN GROWTH ON NICKEL SURFACES

Song Liu, Bin Liu, Adri, van Duin, and James H. Edgar

Department of Chemical Engineering, Kansas State University, Manhattan, Kansas 66506

Hexagonal boron nitride (h-BN), also known as the white graphene, is a layered, 2D material structurally analogous to the carbon-based graphene. Due to its unique structural, electronic, and optical properties, high quality h-BN single crystals have many potential applications, e.g., as ultraviolet light emitters for water purification, or as the neutron detector for sensing fissile materials. Therefore, a detailed mechanistic understanding of h-BN crystal growth is critical to improve the synthesis techniques to produce high quality crystals.

In this work, density functional theory (DFT) and reactive molecular dynamics (MD) simulations â€“ using the ReaxFF force field â€“ were employed to understand the mechanism of h-BN growth on nickel substrates. The energetics and kinetics of diffusion and B-N bond formation on flat and stepped Ni(111) and Ni(211) surfaces were calculated by DFT. Then a ReaxFF force field potential describing the B-Ni and N-Ni interactions was developed from the DFT calculations. These predict that the atomic (B, N) species and B_xN_y subunits bind much stronger at low-coordinated step-edge sites on the Ni(211) surface than the terrace Ni(111) surface, thus, suggesting that the h-BN growth primarily starts from such sites. The pathways for nucleation and growth were investigated. On the step edge sites, growth of h-BN structures with zig-zag edges are more energetically stable. On the flat terrace sites, the

calculated energetics suggests that the h-BN structures start with linear BN configurations, followed by branching and formations of 5- and 6-membered rings.

In order to validate the developed ReaxFF force field and confirm the growth pathways, NVT-MD simulations were designed and performed starting from elemental B and N species in the temperature range 1100-1500 K. The elemental B and N species are deposited to a Ni substrate stepwise. The formation of a continuous h-BN network occurs at 1500 K, and is consistent with the paths predicted by DFT calculations. Further mechanistic insights derived from the MD simulations will be presented and discussed in this work as well.

Copyright © American Institute of Chemical Engineers. All rights reserved.

468488 Molecular-Scale Modeling of Whole Biomass Pyrolysis â€“ Morphological Changes

Abstract:

As pyrolysis takes place, liquid, gas and char are formed. Volatilization is assumed to occur only from a free surface, thereby generating a gradient in liquid content, forcing solid-phase char to move in the opposite direction to maintain *volume* balance. Plant matter is clearly a complex chemically and morphologically heterogeneous materials and observed chemical and microstructure changes upon pyrolysis are likewise complex. The processes that control these effects of transport, kinetics and microstructure changes on biomass during pyrolysis are likewise complex and not yet well understood. The purpose of this study is to use molecular modeling to simulate microstructure changes during pyrolysis using modeled whole biomass structures. Biovia-Material Studio was used to create a model for whole biomass using the FORCITE module and the molecular mechanics force field pcff at 300 °K. To simulate microstructural changes, whole biomass models were optimized first and then molecular fragments were removed from the matrix. The molecular dynamics was performed on this system at 600 °K using the Molecular Operating Environment (MOE) and NAMD, and the molecular mechanics force field, AMBER. The modeled parameters, density, energy and volume at each time step were measured and compared to the initial whole biomass structures before and after from which molecular-scale morphological changes were inferred.

Copyright © American Institute of Chemical Engineers. All rights reserved.

468092 Influence of Intraparticle Transport Processes on Optimal Reactor and Catalyst Design

Abstract:

The **Multi-Level-Reactor-Design (MLRD)** methodology is a novel model-based optimization approach. It enables the development of innovative and tailored chemical reactor concepts by following a three level approach [1].The basic idea is to describe the reaction system of interest in a fluid element of arbitrary shape. Optimality with regard to a certain objective is reached by tracking this fluid element over time in state space and by manipulating its inner state accordingly. This is achieved via variation of outer fluxes (across the fluid element boundary) over time. The fluxes are either optimized directly (level one) or indirectly by optimization of suitable control variables (level two). Subsequently, an actual reactor design concept is deduced from the identified optimal control variable profiles. The resulting design parameters are then determined by optimizing the system within the specified setup (level three).

From this it follows that on level one, the system is limited by the applied reaction kinetics only. Thus the full potential of the reaction system can be revealed independent from known apparatuses without any further restrictions. On level two, the kinetics of mass and energy transport are additionally considered, thereby disclosing the positive or negative influence of these phenomena. Finally, level three represents the influence of the chosen apparatus configuration including losses due to technical feasibility.

This approach enables an early-stage detection of optimization potential and was already applied to various systems including heterogeneously catalyzed reactions, namely the oxidation of sulphur dioxide [1] and the partial oxidation of ethylene [2]. In these examples, the fluid element was described as pseudo-homogeneous system. This implicitly assumes the utilized reaction kinetics to be intrinsic, and thus neglects effects of intraparticle mass and energy transport. This, however, is a severe simplification which often is neither applicable nor justified for many industrial processes. Therefore, in this work a major extension of the optimization methodology described above is presented by incorporating the effects of intraparticle transport processes. Accordingly, the fluid element is now described as pseudo-heterogeneous system resulting in two distinct advances: Firstly, the included transport effects now affect the reactor design directly returning more realistic results. Secondly, the catalyst pellet is treated and modelled as separate subsystem, enabling its properties to be tuned and optimized as well. In our model, the catalyst pellet is regarded as non-isothermal and the diffusive fluxes inside are described using a sophisticated Stefan-Maxwell based approach. Besides common numerical methods, the pellet governing equations were also solved using the Adomian Decomposition Method (ADM). This method results in an algebraic series approximation of the differential equations reducing the computational effort during optimization.

First results originating from the application of the extended MLRD method demonstrate that the simultaneous consideration of reactor and catalyst pellet in the optimization clearly changes the resulting reactor design with regard to optimality. To conclude, this approach allows for the concurrent optimal design of reactor and catalyst and therefore represents a valuable tool to unlock the full potential of heterogeneously catalyzed reaction systems.

References

[1] Peschel, A., Freund, H., Sundmacher, K., *Ind. Eng. Chem. Res.* 49 **(2010)**, 10535-10548
[2] Peschel, A., Karst, F., Freund, H., Sundmacher, K., *Chem. Eng. J.* 66 **(2011)**, 6453-6469

Copyright © American Institute of Chemical Engineers. All rights reserved.

457943 Multiscale Simulation of Heterogeneous Catalysis: Butadiene Hydrogenation from Molecular Reaction Steps to the Fluid Flow

Abstract:

The selective hydrogenation of poly-unsaturated organic molecules is of high industrial interest in the area of fine and polymer chemistry. In order to have a good control of the reaction, it is necessary to gain a detailed understanding of all its stages and scales. On the smallest atomic scale, the reaction is governed by the interactions of the olefin with the metal catalyst surface. At a larger scale, the reaction is governed by temperature and pressure conditions and reactant and product flow in the reactor. Nevertheless the two scales are closely interlinked since local conditions at the surface of the catalyst's particle have a direct impact on the atomistic environment of the metal surface. In particular, hydrogen is known to easily cover the palladium and platinum surfaces that are used in such reactions.[1]The coverage may depend on the local temperature and pressure condition but also on the kinetics of the reaction in those conditions. In this work we have studied the reaction with a multi-scale approach, the atomic scale with electronic quantum chemical methods, and the macroscopic scale with Computational Fluid Dynamics (CFD) simulations, both scales being coupled by a mesoscopic stage using mean-field kinetic simulations.

Using periodic plane wave density functional theory calculations,[2] the hydrogenation reactivity of the simplest poly-unsaturated hydrocarbon, butadiene, on the Pd(111) surface has been investigated. The effect of the hydrogen coverage on the Pd(111) surface has been accounted for in order to model different macroscopic reaction conditions. The most important energy barriers have been exactly determined using the Nudge Elastic Band method[3] while transition states suspected to have a higher potential energy have been estimated using a Bronsted-Evan-Polanyi relationship.[4]Using transition state theory, the hydrogenation energy profile at low hydrogen coverage, developed in an earlier stage of the project, has been used to build a first implementation of a micro-kinetic model. The PeliGRIFF CFD code developed at IFP Energies Nouvelles has been upgraded in order to include for the first time a reaction term in the catalyst grain.

The obtained kinetic law has been included in PeliGRIFF and it was shown that with intra-grain diffusion coefficient one order of magnitude lower than the gas phase diffusion, the reaction mainly happens in a shell surrounding the grain.

References

1. C. Chizallet, G. Bonnard, E. Krebs, C. Thomazeau, P. Raybaud, J. Phys. Chem. B 115 (2011) 12135-12149.
2. G. Kresse, J. Furthmüller, Phys. Rev. B 54 (1996) 11169-11186.
3. D. Sheppard, P. Xiao, W. Chemelewski, D. D. Johnson, G. Henkelman, J. Chem. Phys. 136 (2012) 074103.
4. D. Loffreda, F. Delbecq, F. Vigne, P. Sautet, Angew. Chem. Int. Ed. 48 (2009) 8978-8980.

Copyright © American Institute of Chemical Engineers. All rights reserved.

466070 Multiscale Modeling for High Temperature Heterogeneous Gas Phase Synthesis Reactions: HCN Synthesis

Abstract:

The climate change challenge does not only affect energy production and mobility but also chemical production as a whole. While process design is crucial to meet long-term future challenges, process optimization of existing plants can boost resource efficiency and ecological footprint of the chemical industry today. Energy-intensive processes such as high-temperature synthesis reactions stand out in that context because small process improvements have a potentially large impact on the ecological footprint as a whole.

Steam reforming and the synthesis of hydrogen cyanide (HCN) via the BMA route (Blausäure aus Methan und Ammoniak) are classical examples for high temperature heterogeneously catalyzed gas phase reactions. HCN is an important chemical intermediate that is used in a variety of industrial processes such as the synthesis of herbicides, polymer precursors and most importantly the synthesis of food additives such as methionine. While other processes for the synthesis of HCN exist such as the Andrussow or Shawinigan process, none of those meet the high yield and product purity that is achieved in the BMA process. In the industrial process, educts methane and ammonia react in aluminum oxide tubes that are impregnated with a platinum catalyst and are suspended in stacks inside an oven that provides the energy required for the endothermic reaction. The temperature of the reaction is around 1500K to achieve optimum yield at the thermodynamic equilibrium of the involved species. HCN is the major product and stoichiometric amounts of hydrogen are produced as byproducts. The process is operated with slight ammonia excess to prevent coking of the narrow tubes [1]. Optimum heat transfer from the oven chamber to the active catalyst sites on the inside of the tube walls but also mass and heat transport inside the reactor tubes are crucial to achieve the desired yield at the minimum cost of heating fuel. For these reasons a better understanding of the tube-oven interactions is essential that can only be achieved via a coupled model of the reactor tubes and the oven chamber which constitutes an extension of the multiscale design approach [2].

In this work, a two-dimensional model of the synthesis tubes is developed based on the reaction kinetics of Hasenberg et al. [3]. The high temperature and large temperature gradients between inlet and outlet discourage the use of standard Navier-Stokes equations to describe the fluid flow inside the tubes and require the use of compressible flow and the addition of buoyancy and radiation terms in the gas phase. Application of the finite volume method for discretization ensures overall conservation of energy and mass. The tube model is used to predict the required minimum wall temperatures or respective heat flows to obtain a sufficient yield and that information serves as boundary condition of a second model in three dimensions of the oven chamber in order to account for the heating of the tubes. Additional boundary conditions are the design parameters of the oven chamber and the arrangement of the stack of tubes inside the oven. In that manner the two essential goals of the process are accounted for: a maximum synthesis yield at a minimum energy cost. At the high temperatures inside the oven, adequate description of convective and radiative heat transfer are essential and accounted for using different models for radiative heat transfer in participating media. Different reactor designs are simulated and evaluated against the industrially applied benchmark design of the reactor.
References:

[1] Diefenbach M., Brönstrup, M., Aschi, M., Schröder, D., Schwarz, H. (1999), HCN Synthesis from Methane and Ammonia: MEchanisms of Pt^{+}-Mediated C-N Coupling, Journal of the American Chemical Society

[2] Karst, F., Freund, H., Maestri, M., Sundmacher, K. (2014), Multiscale Chemical Process Design Exemplified for a PEM Fuel Cell Process, Chemie Ingenieur Technik

[3] Hasenberg, D., Schmidt, L. (1986), HCN Synthesis from CH_4 and NH_3 on Platinum, Journal of Catalysis

Copyright © American Institute of Chemical Engineers. All rights reserved.

467251 An Experimental and CFD Based Study of the Pt-Catalyzed NH_3-Oxidation in Different Scales

Abstract:

Introduction

After 100 years the Ostwald process is still the most common way for the industrial manufacturing of nitric acid, which represents the most important nitrogen source for the fertilizer and explosives production. The process is divided into three steps, whereat the first step is the catalytic combustion of ammonia with air towards nitrogen monoxide. Byproducts are N_2 and N_2O whereby the latter is a well-known greenhouse gas. 400.000 tons N_2O are emitted due to the nitric acid production yearly, which is the second largest synthetic source of N_2O after the agriculture.[1]

As catalyst mainly Pt-alloy gauzes are applied with different compositions and structures. Although the process is known for a long time and is highly optimized nowadays, the knowledge about influencing factors and kinetic details under realistic conditions is small. Reasons for that are the harsh conditions (up to 1000 °C, corrosive atmosphere) and the mass transport limitation of the reaction.[2]

The main focus of this contribution will be the investigation of the N_2O-selectivity in a lab scale reactor and the prediction of the results by CFD-simulation in consideration of the wall effects caused by heat transfer phenomena.

Methods

The investigations were carried out by using CFD-simulations and a lab scaled setup. For the CFD-simulations the Ansys® Fluent® software package was used, which allows to compute the surface reaction kinetics and the fluid-dynamics simultaneously. The use of look-up tables for the surface kinetics reduces the calculating time significantly.[3] The implemented surface kinetics are based on Krähnert.[4] The complete lab scale reactor was simulated including heat transfer phenomena.

The lab scaled setup is engineered to work under nearly realistic process conditions (tubular reactors of 1 and 3 cm diameter, woven Pt/Rh-gauzes as catalyst, 4 atm pressure, feed: 10.5 % (L L^{-1}) NH_3 in air). The temperature reaches up to 1000 K behind the catalyst. To estimate the influence of the wall effects the reaction-zone can be external heated up to 973 K. The gas phase was analysed by a special equipped Online-FTIR.

Results and discussions

The experimental results of the lab scale reactor obtain in all cases higher N_2O-selectivities than the known ones from industrial plants at similar conversions of NH_3.

In this context we introduced CFD-simulations combined with the surface kinetics. It can be shown that the N_2O-selectivity depends on the individual spatial temperature of the gauze. The formation of N_2O predominantly takes place in colder regions. An inhomogeneous radial temperature and concentration distribution along the catalyst gauze can be shown for different reactor diameters.

Accordingly in lab scale setups with small diameters (cm) the portion of cold regions is higher compared to industrial scales (m). Therefore, the area-averaged selectivity depends on the chosen scaling factor.

Further experimental results reveal a decreasing N_2O-selectivity with increasing external heating in the reaction zone. This confirms that the higher formation rate of N_2O in lab scale reactors is an effect of heat transfer phenomena.

Provided CFD-simulation and experiments pointed out that there is a limit in minimizing reactors for the investigation of the ammonia oxidation on Pt-catalysts due to the heat loss through the reactor wall. The diameter of an experimental lab scale reactor has not to be below this limit in order to avoid wall effects and provide feasible N_2O-selectivities.

[1] J. Pérez-Ramirez, F. Kaptejin, K. Schöffel, J. Moulijn, *Applied Catalysis B: Environmental*, **2003**, *44*, 117-151.

[2] R. Imbihl, A. Scheibe, Y. F. Zeng, S. Gunther, R. Kraehnert, V. A. Kondratenko, M. Baerns, W. K. Offermans, A. P. J. Janse, R. A. van Santen, *Phys. Chem. Chem. Phys.*, **2007**, *9*, 3522-3540.

[3] M. Votsmeier, *Chem. Eng. Sci.*, **2009**, *64*, 1384.

[4] R. Kraehnert, M. Baerns, *Chem. Eng. J.*, **2007**, *137*, 361-375.

Copyright © American Institute of Chemical Engineers. All rights reserved.

466241 How Accurate Is the Mean-Field Approximation for Catalytic Kinetics?

Abstract:

Modelling catalytic kinetics is indispensable for the design of reactors and chemical processes. Currently, the majority of kinetic models employ mean-field approximations and are formulated as ordinary differential equations, which leads to an approximate description of catalytic kinetics by omitting spatial correlations. On the other hand, kinetic Monte Carlo (KMC) approaches provide a discrete-space continuous-time stochastic formulation that enables a detailed treatment of spatial correlations in the adlayer, but at a significant computation expense. Such spatial correlations arise from slow diffusion in tandem with 2-site reactions, or from adsorbate-adsorbate lateral interactions, and have been shown to markedly affect the observed kinetics.[1,2] It is thus well known that mean-field models are limited, as they neglect such correlations; yet, due to their computational efficiency, such approaches are ideal in multiscale modelling frameworks (for instance as chemistry modules in computational fluid dynamics). However, it is possible to develop higher order approximations that systematically increase the accuracy of kinetic models by treating spatial correlations at a progressively higher level of detail but at the cost of higher computational effort.[3-5]

In this study, we assess the error and computational efficiency of mean-field and higher order approximations for kinetics in catalytic systems with strongly interacting adsorbates. We thus focus on a model for NO oxidation incorporating first nearest neighbor lateral interactions and construct a sequence of approximate models of progressively higher accuracy, starting from the mean-field treatment and continuing with a sequence of Bethe-Peierls models with increasing cluster sizes. By comparing the turnover frequencies of these models with those obtained from KMC simulation, we show that the mean-field predictions deviate by several orders of magnitude from the KMC simulation results. The Bethe-Peierls model, with a cluster incorporating sites up to 2nd nearest neighbors, performs well for predicting coverages; however, due to the exponential dependence of reaction rate on activation energy, the turnover frequency predictions are still inadequate. One requires Bethe-Peierls approximations with clusters of 4th or higher nearest neighbors, in order to faithfully reproduce the KMC predictions. We show that such approximations, while more computationally intense than the mean-field treatment, still enable significant computational savings compared to a KMC simulation, thereby paving the road for employing them in multiscale modelling frameworks.

References

[1] Stamatakis, M. and S. Piccinin, *Rationalising the relation between adlayer structure and observed kinetics in catalysis.* ACS Catalysis, 2016. **6**(3): p. 2105-2111.

[2] Matera, S., H. Meskine, and K. Reuter, *Adlayer inhomogeneity without lateral interactions: Rationalizing correlation effects in CO oxidation at $RuO_2(110)$ with first-principles kinetic Monte Carlo.* Journal of Chemical Physics, 2011. **134**(6): p. 064713.

[3] Are, S., M.A. Katsoulakis, P. Plechac, and L. Rey-Bellet, *Multibody Interactions in Coarse-Graining Schemes for Extended Systems.* SIAM Journal on Scientific Computing, 2008. **31**(2): p. 987-1015.

[4] Katsoulakis, M.A. and D.G. Vlachos, *Mesoscopic modeling of surface processes.* Dispersive Transport Equations and Multiscale Models, 2004. **136**: p. 179-198.

[5] Masel, R.I., *4.9.1 Bethe-Peierls Approximate Solution*, in *Principles of adsorption and reaction on solid surfaces*. 1996, Wiley: New York.

Copyright © American Institute of Chemical Engineers. All rights reserved.

466525 Compositional Effect of Varying Feedstocks on Heavy Oil Upgrading By Supercritical Water

Abstract:

Crude oil prices have shown big fluctuations around 150 and 35 $/barrel during 2008-2016. Such a volatile and unpredictable oil price tendency, which is being drastically affected by political issues, forces energy policy makers to focus on alternative energy conversion technologies for energy security. On the other hand, one third of Worldâ€™s recoverable oil reserves are in a form of heavy oil. Therefore developing alternative oil upgrading technologies with improved conversion and tunable selectivity providing low coke and improved light liquid yield are of desire. However, the variations in feedstock chemistry due to the source of the heavy oil affect economics of investments to a great extent. Supercritical fluids can be used for the development of alternative and innovative heavy oil upgrading processes and supercritical water (SCH_2O) having decreased dielectric constant and increased ionic product values is a potential candidate for the reactive extraction and thereby upgrading of heavy oils in an intensified process scheme. These altered physicochemical properties of water enable it to mimic hydrocarbons for reactive extraction purposes and also to increase [H_3O^+] concentrations for triggering acid-catalyzed cracking reactions. Domestic tar sand samples together with feedstocks of state-of-the art heavy oil upgrading units (Delayed Cokers - DCU), namely 50/70 bitumen from 2 different refineries in Turkey, 70/100 pen bitumen, and by product of Base Oil Unit, PDA bitumen, were used for the experiments. The reactions were carried out at 440 °C and 30MPa for 1 hour with water/oil ratio of 2 in a home-made 10 cm^3, SS316, bomb reactor. Characterization was done by GPC, H-NMR, DSC, TGA, FTIR, XRD, GC-Simdist, API, SEM images and SARA analysis. The coke-liquid-gas yield, coke type and boiling point distribution of liquids were shown to be influenced from the feedstock chemistry. Varying conversions with scattered selectivity values have been obtained. Modeling studies will be carried out to derive a tool for the prediction of conversion and selectivity of the reactions for a given feed under certain conditions as a part of a complementary study.

Copyright © American Institute of Chemical Engineers. All rights reserved.

467087 *Product Formation and Kinetics of the Non-Isothermal Hydrothermal Liquefaction of Soy Protein Isolate*

Abstract:

James Sheehan - Graduate Student

Dr. Phil Savage – Adviser

The Pennsylvania State University, Department Chemical Engineering

Product Formation and Kinetics of the Non-isothermal Hydrothermal Liquefaction of Soy Protein Isolate

Hydrothermal liquefaction (HTL) is a thermochemical process capable of producing biocrude oil from biomass. HTL of various biomass feedstocks including microalgae, bacteria, lignocellulose and model compounds has been studied extensively under subcritical temperatures and long holding times (>60 min).[1–7] More recently, HTL of microalgae under rapid heating conditions and short reaction times (<1 min) has achieved higher biocrude yields achieved than HTL of microalgae under subcritical conditions and longer holding times.[8,9] Protein is a major biochemical constituent of microalgae and we performed HTL of soy protein isolate to understand better the incorporation of protein derived products into biocrude.

We conducted non-isothermal, fast HTL of soy protein isolate (SPI) at temperatures ranging 140 °C to 500 °C and residence times of 10 s to 300 s. SPI solids rapidly transform into aqueous phase products and biocrude. The highest biocrude yields (38-40wt%) were obtained within 45 s to 120 s and at temperatures ranging from 375 °C to 435 °C. The largest recovery of atomic N in the aqueous phase products was observed prior to the formation of substantial yields of biocrude. Ammonia formation was significant when the hydrothermal reaction medium reached supercritical conditions. Over 50% of the atomic N was observed as ammonia under such conditions. We developed a kinetic model and determined Arrhenius parameters that describe the fast HTL of SPI. The results of the modeling are shown in Figure 1. As can be seen, the proposed reaction network and optimized Arrhenius parameters can accurately model the major product yields from fast HTL of SPI.

Figure 1: Experimental (discrete points) and model predicted yields (continuous curves) for fast HTL of SPI at set point temperatures 400°C.

References

(1) Valdez, P. J.; Nelson, M. C.; Wang, H. Y.; Lin, X. N.; Savage, P. E. *Biomass and Bioenergy* **2012**, *46*, 317–331.

(2) Valdez, P. J.; Nelson, M. C.; Faeth, J. L.; Wang, H. Y.; Lin, X. N.; Savage, P. E. *Energy & Fuels* **2014**, *28* (2013), 67–75.

(3) Biller, P.; Ross, A. B. *Bioresour. Technol.* **2011**, *102*, 215–225.

(4) Teri, G.; Luo, L.; Savage, P. E. *Energy & Fuels* **2014**, *28*, 7501–7509.

(5) Changi, S.; Zhu, M.; Savage, P. E. *ChemSusChem* **2012**, *5*, 1743–1757.

(6) Luo, L.; Sheehan, J. D.; Dai, L.; Savage, P. E. *ACS Sustain. Chem. Eng.* **2016**.

(7) Toor, S. S.; Rosendahl, L.; Rudolf, A. *Energy* **2011**, *36* (5), 2328–2342.

(8) Faeth, J. L.; Valdez, P. J.; Savage, P. E. *Energy and Fuels* **2013**, *27*, 1391–1398.

(9) Hietala, D. C.; Faeth, J. L.; Savage, P. E. *Bioresour. Technol.* **2016**, *214*, 102–111.

James Sheehan - Graduate Student

Dr. Phil Savage – Adviser The Pennsylvania State University, Department Chemical Engineering

Product Formation and Kinetics of the Non-isothermal Hydrothermal Liquefaction of Soy Protein Isolate

-

Hydrothermal liquefaction (HTL) is a thermochemical process capable of producing biocrude oil from biomass. HTL of various biomass feedstocks including microalgae, bacteria, lignocellulose and model compounds has been studied extensively under subcritical temperatures and long holding times (>60 min).[1–7] More recently, HTL of microalgae under rapid heating conditions and short reaction times (<1 min) has achieved higher biocrude yields achieved than HTL of microalgae under subcritical conditions and longer holding times.[8,9] Protein is a major biochemical constituent of microalgae and we performed HTL of soy protein isolate to understand better the incorporation of protein derived products into biocrude.

We conducted non-isothermal, fast HTL of soy protein isolate (SPI) at temperatures ranging 140 °C to 500 °C and residence times of 10 s to 300 s. SPI solids rapidly transform into aqueous phase products and biocrude. The highest biocrude yields (38-40wt%) were obtained within 45 s to 120 s and at temperatures ranging from 375 °C to 435 °C. The largest recovery of atomic N in the aqueous phase products was observed prior to the formation of substantial yields of biocrude. Ammonia formation was significant when the hydrothermal reaction medium reached supercritical conditions. Over 50% of the atomic N was observed as ammonia under such conditions. We developed a kinetic model and determined Arrhenius parameters that describe the fast HTL of SPI. The results of the modeling are shown in Figure 1. As can be seen, the proposed reaction network and optimized Arrhenius parameters can accurately model the major product yields from fast HTL of SPI.

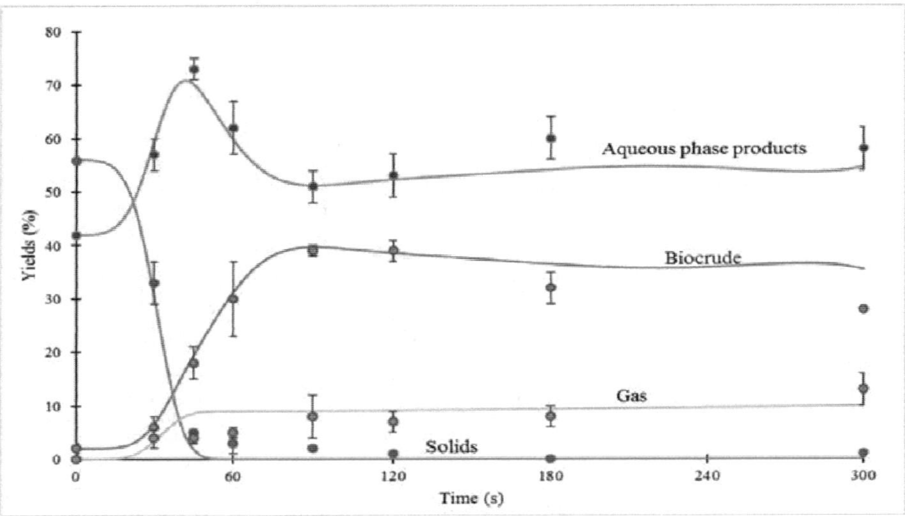

Figure 1: Experimental (discrete points) and model predicted yields (continuous curves) for fast HTL of SPI at set point temperatures 400°C.

References

(1) Valdez, P. J.; Nelson, M. C.; Wang, H. Y.; Lin, X. N.; Savage, P. E. *Biomass and Bioenergy* **2012**, *46*, 317–331.

(2) Valdez, P. J.; Nelson, M. C.; Faeth, J. L.; Wang, H. Y.; Lin, X. N.; Savage, P. E. *Energy & Fuels* **2014**, *28* (2013), 67–75.

(3) Biller, P.; Ross, A. B. *Bioresour. Technol.* **2011**, *102*, 215–225.

(4) Teri, G.; Luo, L.; Savage, P. E. *Energy & Fuels* **2014**, *28*, 7501–7509.

(5) Changi, S.; Zhu, M.; Savage, P. E. *ChemSusChem* **2012**, *5*, 1743–1757.

(6) Luo, L.; Sheehan, J. D.; Dai, L.; Savage, P. E. *ACS Sustain. Chem. Eng.* **2016**.

(7) Toor, S. S.; Rosendahl, L.; Rudolf, A. *Energy* **2011**, *36* (5), 2328–2342.

(8) Faeth, J. L.; Valdez, P. J.; Savage, P. E. *Energy and Fuels* **2013**, *27*, 1391–1398.

(9) Hietala, D. C.; Faeth, J. L.; Savage, P. E. *Bioresour. Technol.* **2016**, *214*, 102–111.

Copyright © American Institute of Chemical Engineers. All rights reserved.

467990 An Investigation on Short Residence Time Macro Algae Hydro Thermal Liquefaction in a Batch Reactor

Abstract:

AIChE annual meeting, November 2016 San Francisco

An investigation on short residence time macro algae hydro thermal liquefaction in a batch reactor

Arash Izadpanah [a], Niccol˜ Le Brun [a], Klaus Hellgardt[a]

[a] *Department of Chemical Engineering, Imperial College London, South Kensington Campus, London, SW7 2AZ, United Kingdom*

Bio crude oil production from macro algae may become a key technology to tackle tomorrowÕs renewable fuels challenge. In order to seed this technology it is important to identify and evaluate the right process pathways to convert macro algae to biofuels. In conventional algal biofuels processing, drying of the algae is a costly procedure, which could be bypassed by using Hydrothermal Liquefaction (HTL) technology with substantial advantages. For an efficient implementation of a continuous HTL process, the effect of process variables on the conversion and selectivity towards liquid biofuels needs to be understood. This is often a challenge because the timescale of the HTL process is of the order of minutes and it is difficult to be captured properly by ordinary methods. In this study we employed a molten salt bath to facilitate rapid heat transfer to the batch reactor and thus study HTL for precisely controlled residence times between 1 and 25 min. Under experimental conditions the conversion of macro algae to bio crude oil was found to be complete within 15 min reaching a maximum and then decreasing slightly (23% maximum yield after 11 min at 390 ¡C). Using different analytical techniques including CHNS-O, GC-MS, Simulated Distillation and Size Exclusion chromatography the bio crude from macro algae is characterized and compared with North Sea crude oil.

Keywords: Macro algae, Hydrothermal liquefaction, Conversion rate, Filtration, Centrifugation

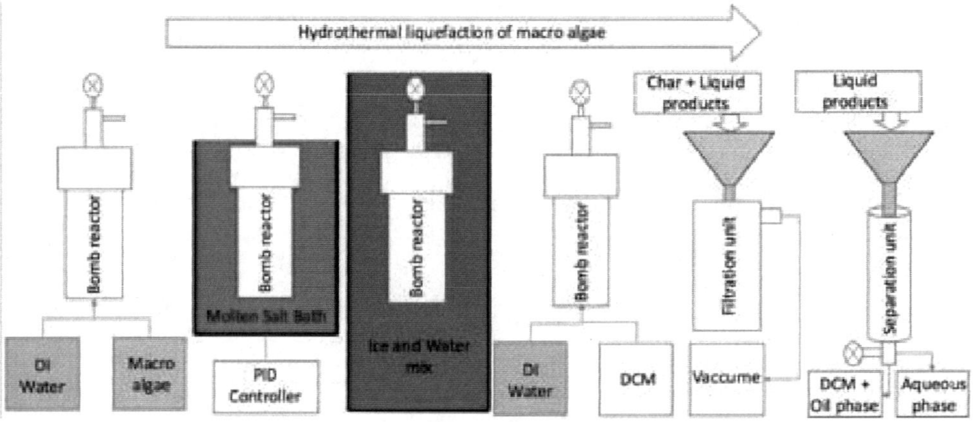

Figure 1:Graphical summery of the experiment

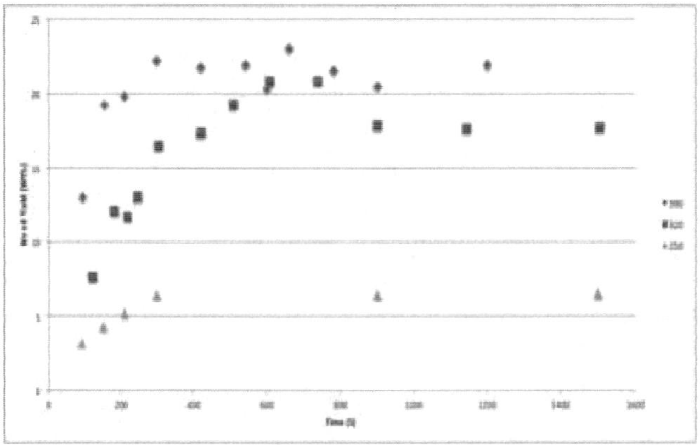

Figure 2: Bio crude oil yield as a function of residence time at 250 ℃, 320 ℃, 390 ℃.

Copyright © American Institute of Chemical Engineers. All rights reserved.

459274 Hydrothermal Liquefaction (HTL) of High-Ash Algal Biomass: The Effect of Ash Contents in HTL Reactions

Abstract:

Previous studies demonstrate that the high ash contents appeared to inhibit the formation of biocrude oil in the hydrothermal liquefaction (HTL) processes. Mixed-culture algal biomass from wastewater systems (AW) were screened into three sizes (<106 μm, 106-180 μm and >180 μm) and their ash contents were measured. AW biomass with different ash contents was converted into biocrude oil via HTL at 300 °C for a 60 minute reaction time, which is the previously determined optimum condition for producing biocrude oil from AW. Compared to AW biomass before screen pretreatments, the ash contents of AW of 106-180 μm and <106 μm were respectively decreased from 53.3 % to 38.5% and 41.8 %, while the biocrude oil yields remained almost constant. When the ash contents were reduced from 53.3 % to 38.5 % in the AW biomass, the HHV of biocrude oil was increased from 27.5 MJ/kg to 32.3 MJ/kg and the fraction of light oil (boiling point of 100-300 °C) was increased from 31% to 49%. On the other hand, GC-MS analyses of algal biocrude oil and aqueous products show that the ash contents could promote denitrogenation, catalyze the formation of hydrocarbons, as well as mitigate the recalcitrant compounds in aqueous products under the HTL processes. The above results indicate that algal biomass with certain amounts of ash contents can be converted into biocrude oil with reasonable quality and quantity. In order to quantify the limit of the ash contents under the HTL processes, further HTL conversions with model algal feedstock and representative ash contents were conducted in this study. Elemental and thermogravimetric analyses of the resulting biocrude oil both demonstrate that when the ash content in the algal feedstock was below 40%, the HHV and boiling point distribution of the algal biocrude oil could be hardly affected. This fact substantiates the feasibility of using high-ash algal biomass as HTL feedstock and diminishes the necessity of multi-step pretreatments and modifications of high-ash algal biomass for biofuel application.

Copyright © American Institute of Chemical Engineers. All rights reserved.

456792 Simultaneous Solid, Phosphorus Recovery and Biocrude Production through Hydrothermal Liquefaction of Algae

Abstract:

Subcritical water has become a highly regarded method for converting biomass, primarily algae, to an energy dense biocrude product. This conversion technique, more commonly referred to as hydrothermal liquefaction (HTL), still faces challenges of nutrient recovery, producing a practical biocrude substitute, and generating value-added coproducts before becoming a sustainable and commercially viable method. Recovering the phosphorus (P) in the solid biochar could be beneficial for secondary applications.

Previous work has shown that the inorganic content of the growth media, in particular Ca, P, and Si, and pH during harvesting dictate the fate of P in the HTL products. This paper expands on the idea of solid P-recovery via *in situ* synthesis of calcium phosphates present in the biochar, as seen previously in mixed culture, wastewater-cultivated algae. HTL was performed on diatoms, a unique group of algae that harness Si naturally within their cell walls, and mixed cultures of diatoms and *Chlorella kessleri*. Powder x-ray diffraction (XRD) characterized both the starting algal solids and resulting HTL biochar while acid digestions and inductively coupled plasma optical emission spectrometry (ICP) closed the P balance of HTL reaction.

In addition, it is hypothesized the algal carbohydrates, proteins, and lipids also play a role in the morphology and structure of the biochar. Model compounds were individually reacted via HTL, in the presence and absence of calcium phosphates, and calcium and phosphate ions. The resulting biochar, aqueous coproduct, and biocrude were characterized. Gas chromatography mass spectrometry, CHN/O elemental analysis, and simulated distillation analysis of the biocrude product were analyzed to identify potential catalytic upgrading properties of the calcium phosphate biochar.

Copyright © American Institute of Chemical Engineers. All rights reserved.

453154 Supercritical Water Treatment of Fractions of Crude Oil: Quantification of the Products

Abstract:

The Arabian Heavy (AH) crude oil was fractionated into distillate ($\leq C_{20}$) fraction and bottoms fraction. The distillate and the bottoms fraction of crude oil were treated with supercritical water (SCW) in a batch reactor setup. The gaseous and liquid phase products were analyzed using gas chromatography (GC-MS, GCxGC, FID, SCD). The change in structural composition of maltenes and asphaltenes in the bottoms fraction before and after supercritical water treatment was investigated using proton NMR, FTIR and Raman spectroscopy.

The distillate fraction primarily consisted of n-alkanes ($\leq C_{20}$) plus a minor fraction of alkyl aromatics and sulfur compounds. SCW treatment of the distillate fraction resulted in cracking of higher carbon number alkanes to lower carbon number alkanes, and production of gaseous species such as methane, ethane and propane.

SCW treatment of the bottoms fraction caused much more dramatic changes. Much of the heavy components in the crude oil cracked to form compounds in the gasoline and diesel range which are observed by GCxGC. The higher carbon number aliphatics ($\geq C_{20}$) and long chain ($\geq C_9$) alkyl aromatic compounds in the bottoms fractions underwent cracking and produced C_6-C_{18} alkanes and C_3-C_8 alkylbenzenes. Supercritical water treatment of the bottoms fraction cracked the heavy sulfur compounds, and resulted in increased recovery of lighter sulfur compounds such as alkylated benzothiophene and dibenzothiophenes. The SCW-treated bottoms also showed increased recovery of polycyclic aromatic hydrocarbons such as naphthalene. Despite being carried out at temperatures similar to those used in conventional thermal coking processes, the SCW process did not form significant amounts of coke.

The maltenes and asphaltenes were separated from the bottoms fraction and the SCW-treated bottoms fraction. The proton NMR of the SCW-treated maltenes fraction showed a decrease in the protons corresponding to aliphatic CH_2 and CH_3 groups. There was also an increase in protons in single ring and polycyclic aromatic hydrocarbons. The proton NMR results support the observations made by the GCxGC measurements.

The proton NMR, FTIR and Raman spectroscopy on untreated and SCW-treated asphaltenes showed changes in the structural composition of asphaltenes post-SCW treatment. The proton NMR experiments showed a decrease in the protons which correspond to aliphatic CH_2 and CH_3 groups. The decrease in aliphatic CH_2 and CH_3 functional groups could be due to scission of the long alkyl chains attached to the aromatic core structure of the asphaltenes. The proton NMR results also showed 49 % increase in the mono-aromatic protons and 1 % decrease in di-aromatic and poly-aromatic protons, which suggests formation of new aromatic rings and/or perhaps conversion of maltenes containing single aromatic rings into asphaltenes (e.g. by the cleavage of alkyl tails, affecting the solubility).

FT-IR spectroscopy showed similar results to proton NMR, wherein an increase in aromatic and heteroatom content (O-H/N-H) was observed in the SCW-treated asphaltenes. Raman spectroscopy also showed evidence of chemical rearrangement and higher aromatic content in the SCW-treated asphaltenes when compared to asphaltenes from untreated bottoms fraction.

Overall, the SCW treatment of the crude oil fractions resulted in cracking of heavier hydrocarbon components into more valuable lighter molecules, which could be used in transportation fuel or as chemicals.

Acknowledgement: This research was supported through a Research Agreement with Saudi Aramco, a Founding Member of the MIT Energy Initiative.

Copyright © American Institute of Chemical Engineers. All rights reserved.

458338 Supercritical Water Process - Finding a Way to Accomodate into Petroleum Industry

Abstract:

Supercritical fluid is not new to petroleum industry. Solvent extraction process utilizes supercritical fluid for enhancing selectivity and yield. However, supercritical water has not been applied in industrial scale due to its harsh condition. With progress of material and process technologies, supercritical water has been studied and evaluated in various areas to resolve technical challenges of conventional technologies. Recent research activities on heavy oil upgrading by supercritical water resulted in serious consideration of accommodating such a nonconventional way in petroleum industry. This presentation will introduce potential application areas of supercritical water technology as well as its implication in petroleum industry.

Copyright © American Institute of Chemical Engineers. All rights reserved.

465282 Why John Is Right Not to Lecture

Abstract:

The learning styles of the Millennials are quite different than the previous generations, e.g., the Xers, who relied on the lecture-inclass problem-home problem format.. This presentation will discuss some of the characteristics of the Millennials and how the materials that Professor John Falconer and his team at the University of Colorado have developed can better address the Millennials learning styles.

Copyright © American Institute of Chemical Engineers. All rights reserved.

474307 Utilizing Classroom Management Software to Handle Large Classes

Abstract:

A major problem facing instructors of large classes is handling the accompanying increased organizational details and at the same time, trying to be as accessible to the students as possible. This encompasses keeping a grade book, entering grades, and forming and grading group assignments, among others. In addition, the instructor needs to hold office hours, respond to emails, and present reviews. One method that we have adopted with great success is incorporating little-used features of a classroom or learning management system to not only save time, but also to offer our students the same atmosphere and learning environment provided in smaller classes. This is particularly important in the lower-level core courses where faculty availability and classroom resources are often essential for student success.

A classroom management system is an on-line compilation often used solely for keeping grades. There are a number of systems that are currently employed by different universities. Those used by and/or considered by the University of Colorado include Desire2Learn, Blackboard and WebCT. Currently we are working with Desire2Learn and the methods described in this presentation have been used in that system. However classroom management systems are similar enough that these methods can easily be incorporated into others. In fact, we developed many of these practices in one system and easily transferred them to our current system.

Based on both research and experience, our preferred method of teaching is the use of the "flipped classroom" approach where information delivery is outside the classroom. Faculty adopting this approach modify their role from a front-of-the-class lecturer to a collaborative and cooperative facilitator of learning. The flipped classroom results in a shift from covering curriculum to a collaborative effort to master concepts and problem-solving techniques. However, many instructors find it difficult to change their teaching styles because of a lack of time or resources. Some faculty have expressed concerns that devoting class time to active learning might not provide students with sufficient examples of problem solving or sufficient time to introduce material. A well-organized classroom management system supports the flipped classroom approach, even for those large classes.

Components of this approach include on-line quizzes that test the students' knowledge of assigned reading material; screencasts that provide supplementary guidance on particularly difficult concepts; interactive simulations that promote understanding by making abstract concepts tangible; and ConcepTests that are used in class to challenge the students' knowledge and generate discussion. Most of these materials can be found at www.learncheme.com.

The features that a classroom management system provides to insure that the instructor can retain this method of teaching for large classes without increasing the burden include:

- Modular class design with links for "information delivery"
- Group formation tools
- Dropbox, rubrics and grade items for homework assignments
- Clicker integration
- Class email
- On-line office hours capability

In this presentation, we will describe both the implementation and effectiveness of these integrated tools within an active learning environment.

Copyright © American Institute of Chemical Engineers. All rights reserved.

466087 Enabling Numerical Analysis Calculations within Educational Materials Delivered Via PCs and Smartphones

Abstract:

A major educational project called LearnChemE has been underway at the University of Colorado during the last 10 years under the direction of Prof. John L. Falconer involving Chemical Engineering and associated engineering disciplines (www.learncheme.com). The group of faculty members staff involved in this NSF supported project has recently received the 2015 HImmelblau Award of the American Institute of Chemical Engineers for new and novel contributions to computer aids for chemical engineering education. Integration of numerical problem solving has been utilized in many of the concept tests, screencasts and simulations.

User-friendly PC and Android software enabling general numerical problem solving has been under continuing development with PC software first introduced in 1984 under the name of Polymath with the efforts of the first two authors of this paper (www.polymath-software.com). This effort was initially conducted at the University of Connecticut. This overall work received the Himmelblau Award in 2010. Current Polymath related software development continues for Windows PCs and recently for the Android OS is also underway. There is increasing cooperation between the LearnChemE and PolyMath efforts.

This paper will consider current and future opportunities of interaction between current LearnChemE and PolyMath Software groups. Initial considerations are outlined here utilizing current LearnChemE educational materials and available PolyMath software. Additional future possible uses of newly developing software will also be discussed. The application of both efforts to engineering education world-wide will be highlighted. It is also hoped that this work will also be of interest to those interested in developing STEM educational materials (Science, Technology, Engineering and Math) involving PC's and Android smartphones and tablets that can provide resources such as testing, screencasts and simulations that may involve numerical problem solving.

PC Example – Use of Materials from LearnChemE with PolyMath Windows Software

The LearnChemE website has a collection of Student Resources that has a listing for POLYMATH (http://www.learncheme.com/student-resources/polymath). This site provides Introductory Videos that will run on Windows PCs and Android Smartphones and Tablets. Additionally there are currently 13 Example Problems in Kinetics/Reactor Design and four Example Problems in Thermodynamics that will work under Windows and Android devices. Most of these have been created, developed and narrated by Prof. John Falconer.

The example problem in the Kinetics/Reactor Design area titled *Multiple Reactions in a CSTR* will be onsidered. The use of a split Windows screen will be demonstrated/discussed where the video will be available and executable on the left side and the PolyMath 6.2 software will be available for use on the right side Windows screen. The ability to easily enter the program to enter and solve the material balances for the CSTR will be demonstrated and the advantages of this approach will be discussed.

An additional student challenge to the original problem is will be modify the model for the steady-state problem to accountl for the dynamics of the startup of the CSTR system to steady state. This will use the differential equation solving capabilities of PolyMath. Steady-state and the unsteady-state results will be discussed in this presentation/paper. The option to automatically generate Excel spreadsheets and MATLAB m-files for these two cases will be briefly mentioned.

Smartphone Example – Use of Materials from LearnChemE with PolyMathLite Android Software

The LearnChemE Screencast videos can be used directly on Smartphones. However, PolyMath 6.2 runs under Windows. Therefore Polymath Software has created a PolyMathLite app that has most of the capabilities of PolyMath 6.2, and executes directly on most Android smartphone and tablets. The only major changes in the problem solution code from that used in the PCs is the use of vectors to input data and some special funcions to deal with the various types of regressions.

The use of a simple smartphone with the PolyMathLite app will be demonstrated during the presentation by live use (or screen copies) of of a midsized phone display projection. The exact same set of equations as developed in the Screencast video of the *Multiple Reactions in a CSTR* problem will be ussed. The major difference between the smartphone and the PC is that one must switch between the paused video and the PolyMathLite problem entry as only one screen can be displayed at a time. The results between the PolyMathLite solution to the steady-stare CSTR and the PolyMath 6.2 solution are exactly the same. If fact, the algorithms in the Android Smartphone app and Windows PC coding are identical.

Use of the smartphone will be demonstrated in the modification of the code to the ordinary differential equations for the student challenge aspect to model the dynamic startup of the CSTR. The dynamics of the concentration within the CSTR are plotted with a separate plotting Android app or using the new Microsoft Excel Android app. All of this will be demonstrated and documented.

Summary

This presentation/paper identifies the continuing deliverables from two major educational projects coming from chemical engineering faculty and other academic departments that can combine to promote very positive enrichment activities for our students. Many additional possibilities can be considered for related subject areas. In this era of large class sizes, the educational experiences and capabilities of the individual students can continue to be enhanced. These resources from the two projects that have been discussed can also serve to encourage other STEM disciplines to provide similar educational resources, numerical calculations and other activities in support of classwork or individualized learning. E-textbooks can also be supported by PolyMath and/or PolyMathLite and other calculational software. As Android smartphones become very common among students in developing countries, the educational and problem solving capabilities and products of these two projects can be very important in international chemical engineering education.

Copyright © American Institute of Chemical Engineers. All rights reserved.

466827 Spreadsheet Template to Aid in Creating Individualized Version of Objective Homework Problems

Abstract:

Individualized homework problems ensure that each student has at least a minimal level of understanding and problem solving ability. Online programs have been developed that provide individualized problems with guided feedback. However, these services are not available for all courses, generally come at an extra cost and may tend to isolate students, losing the beneficial aspects of group collaboration. The proposed approach strikes a balance between group collaboration and individual accountability and retains the instructor's ability to create novel homework problems.

An Excel spreadsheet template has been developed to aid instructors in developing individualized versions of problem statements. The template's functions and macros make it relatively easy to develop individualized versions of problems. The spreadsheet is also flexible enough to accommodate different types of objective questions. The problem developer decides which variables to alter and the range of those variables. The program then generates 200 random sets of data for those variables. These sets are then utilized in creating the student's version of the problem in the form of a separate spreadsheet. Therefore, each student receives a homework problem that is similar in structure but different in details from the other students.

The student version contains the problem's statement and a place for the student to respond. After submitting their responses, the student receives immediate, right or wrong, feedback. Multiple submissions can be made up to a pre-set limit. Once the limit is reached, the problem is set to base-case conditions and the student is provided with detailed feedback for the base-case problem. The student can then repeat the problem with a new set of variables.

The template was used to create dozens of problems that were integrated into the homework assignments

for an "off-semester" Material Balances course with eight students in the spring semester of 2016. The students both preferred (4.5/5.0) and thought they learned more (4.5/5.0) from the spreadsheet based problems over the traditional homework problems. The spreadsheet problems will be utilized in the fall 2016 Material Balance class at Trine University and compared to similar instructor generated and traditional textbook problems.

Copyright © American Institute of Chemical Engineers. All rights reserved.

469334 How Students "Learn ChemE": The Impact of Departmental Culture on Student Success

Abstract:

Each engineering department has a unique departmental culture based on its history, faculty makeup, geography, faculty teaching styles, and a myriad of other factors. While some departments have a long history with a well-established culture and traditions, other departments that have been recently formed are faced with the challenge of creating a culture from scratch. The success of a department in creating and sustaining a desirable culture can have a significant impact on recruiting, retention, and general satisfaction of its undergraduate and graduate students. These types of activities also serve to shape students' expectations about what it means to enter the engineering profession and shapes how new students identify themselves as engineers. This presentation will discuss John Falconer's contribution to and influence on the chemical engineering culture, not only at the University of Colorado at Boulder, but also in chemical engineering departments world-wide.

Copyright © American Institute of Chemical Engineers. All rights reserved.

470670 Rapid Analysis of Written Responses to Reveal Student Misconceptions in Thermodynamics

Abstract:

Traditionally, engineering courses utilize computations to evaluate student competence. Problem solving typically involves identification of the governing principles and proper application of engineering models. Many of todayâ€™s students have challenges applying models due to misconceptions about the underlying principles. Collection and assessment of short answer written (constructed) responses to conceptual questions can provide insights into why students have difficulties solving numerical problems. Such assessments are seldom used because the evaluation process is time consuming, especially when assessing the percentages of students with certain misconceptions. Accurate automated assessment of constructed responses can provide rapid feedback to instructors about the distribution of understanding/misconceptions even in large classes, providing opportunities for the instructor to reemphasize correct application of principles. In the future, such assessments may also be valuable for self-instruction. The Automated Analysis of Constructed Response (AACR) research group (www.msu.edu/~aacr) at Michigan State University is leading a multi-institutional collaboration (TUES 3 and WIDER funding) with several connected goals: 1) create a national web portal to provide automated analysis of constructed responses (AACR) to conceptual questions in a variety of STEM areas; 2) use the resulting reports to focus community collaborations between STEM education researchers and instructors; 3) transport AACR innovations through ongoing faculty professional development; 4) expand the range of STEM disciplines in which we pursue this research from biology into chemistry, chemical engineering, physics/astronomy, and statistics; 5) engage in ongoing project evaluation for continuous quality improvement; and 6) lay the foundation for sustainability.

This presentation provides a background on the progress in automated assessment of thermodynamics concepts, initially focusing on internal energy. The presentation will provide an overview of the criteria used for question development, the methods used to refine questions, the techniques applied for automated assessment, the reports provided to the instructor and how the reports are helpful in understanding the distribution of misconceptions. The presentation will describe opportunities for participation in the question development process and the status of our current set of questions.

Copyright © American Institute of Chemical Engineers. All rights reserved.

474009 No More Lecturing! What John Falconer Has Done for Student Learning

Abstract:

This talk will provide the author's perspective of the value John Falconer has provided to enable faculty and students to engage in Active Learning of Chemical Engineering. We will discuss how such strategies and tools coincide with research from the learning sciences, as well as Prof. Falconer's contributions to our project, the AIChE Concept Warehouse. In true Falconerian spirit, it will be interactive.

Copyright © American Institute of Chemical Engineers. All rights reserved.

464047 Active Learning in Chemical Engineering

Abstract:

Freeman et al. [1] analyzed 225 studies that compared STEM courses that used active learning to those that used traditional lecturing. The studentsâ€™ performance on exams and on the concept inventory increased by 0.47 standard deviation (when active learning was used, whereas students in the traditional lecture courses were 1.5 times more likely to fail. These studies provide convincing evidence that active learning in the classroom leads to improved student performance. During a lecture, students may think they understand a concept, and only realize they donâ€™t when they try to apply it. An important objective of active learning is that essentially all students are engaged, not just a small fraction.

We have attempted to provide the resources to make it easier for new faculty to use active learning while devoting less time to preparing course materials. Complete course packages have been developed for three chemical engineering courses: material and energy balances, thermodynamics, and kinetics/reactor design. These course packages utilize active learning in class with ConcepTests , interactive simulations, and problems that students work on in groups. They also include screencasts for students to watch before class, reading quizzes, exams, and assignments. The course packages are available in Microsoft OneNote so that they can be readily modified by the user. Access to the course packages is available at www.LearnChemE.com. 1. Freeman, S, Eddy, SL, McDonough, M, Smith, MK, Okoroafor, N, Jordt, H, Wenderoth, MP, Active Learning Increases Student Performance in Science, Engineering, and Mathematics. Proc. Nat. Acad. Sci. (2014) 111, 8410-8415.

Copyright © American Institute of Chemical Engineers. All rights reserved.

456618 Mechanism of Furfural Hydrodeoxygenation on Metal/Metal Oxide Catalysts

Abstract:

Selective C-O bond activation at low temperatures is essential for oxygen removal to produce second generation biofuels and chemicals from biomass. Constraints, imposed by energy scaling relationships, make monofunctional catalysts (metals, oxides, and carbides) inefficient. At the same time, recently discovered multifunctional metal/metal oxide catalysts (e.g., Rh/MoO_x, Ir/VO_x) exhibit unusual activity and selectivity for C-O bond hydrogenolysis at temperatures < 200°C. Herein, we utilize density functional theory calculations, first principles microkinetic modeling, and electronic structure analysis to shed light on the site cooperativity in the Ru/RuO_2 catalyst, which enables up to the 76% yield of 2-methyl furan in catalytic transfer hydrogenolysis of furfural. We find the facile C-O scission to be associated with a formation of a conjugation-stabilized furfuryl radical intermediate on RuO_2 oxygen vacancies. Three types of catalytic sites required for catalytic transfer hydrogenolysis, are identified. Furthermore, we elucidate electronic factors responsible for the high catalyst activity, and establish correlations between binding energy descriptors and the catalyst activity for a variety of reducible oxides. The mechanistic picture is fully consistent with isotopic labeling experiments and kinetic studies, and the insights gained might prove useful in catalyst selection and reaction pathway design for hydrodeoxygenation of biomass.

Copyright © American Institute of Chemical Engineers. All rights reserved.

471366 Kinetic Analysis of Hydrogenation and Hydrogenolysis of Complex Furans over Supported Noble Metals

Abstract:

Upgrading furan-based compounds through mild hydrotreatment is an important step to produce biomass-derived commodity chemicals and transportation fuels. Biomass derived functionalized furans, such as furfural or 5-hydroxyfurfural, are used in a variety of carbon â€“ carbon bond coupling chemistries to produce products in the jet, diesel and lubricant range has also been established [2-4]. These products typically retain the furan functional group, which require further hydrotreatment in order to produce hydrocarbons or valued-added oxygenates.

The kinetics and mechanism of hydrogenating the furanyl moiety over carbon-supported noble metals were previously studied using 2,5-dimethylfuran (DMF) as a model substrate, in liquid phase at conditions removed of both internal and external mass transfer effects (333K - 363K, p_{H2} = 0.4-0.7MPa). Of the catalysts studied, Pt/C had the greatest activity for C-O hydrogenolysis, producing 2-hexanone as the major product, whereas Pd/C exclusively saturated the aromatic ring into its tetrahydrofuran analog. We proposed a reaction network, supported by a Langmuir Hinshelwood kinetic model, in which not only are ring opening and ring saturation of DMF competitive pathways, but the three main products observed: 2,5-dimethyltetrahydrofuran, 2-hexanone, and 2-hexanol result directly from hydrogenation of DMF over Pt/C and not from hydrogenation in series.

We apply the findings made from our model system to understand the hydrogenation behavior of more complex, substituted aromatic furans with carbon numbers in the diesel range, such as bismethylfuranylmethane (BMFM), a C_{11} compound containing two adjacent furan moieties. Similar to DMF, ring opening and ring saturation of the aromatic ring of BMFM occurred in parallel, with the former pathway preferred at elevated temperatures and low p_{H2} conditions. This behavior was determined to be due to the high activation energy and a low rate order dependence on

p_{H2} associated with the ring-opening step of DMF. Contrary to DMF, BMFM hydrogenation activity over Pt/C was sluggish when compared to activity over Pd/C, in which rate of conversion occurred 8-fold faster to produce not only ring saturated cyclic ethers, but also significant quantities of ring-opened alcohol products (~30% yield). C-O bond hydrogenolysis over Pd/C occurred selectively on the interior, aromatic C-O bond as opposed to the more accessible exterior C-O bond, accounting for 96% of the ring opened products. These differences in hydrogenation activity of BMFM suggest that geometry of the furanic substrate plays a dominant role in the reaction network.

References

[1] Y. Románn-Leshkov, J.N. Chheda, J.A. Dumesic, Phase Modifiers Promote Efficient Production of Hydroxymethylfurfural from Fructose, Science, 312 (2006) 1933-1937.

[2] A. Corma, O. de la Torre, M. Renz, Production of high quality diesel from cellulose and hemicellulose by the Sylvan process: catalysts and process variables, Energy Environ. Sci., 5 (2012) 6328.

[3] G. Li, N. Li, Z. Wang, C. Li, A. Wang, X. Wang, Y. Cong, T. Zhang, Synthesis of High-Quality Diesel with Furfural and 2-Methylfuran from Hemicellulose, ChemSusChem, 5 (2012).

[4] M. Balakrishnan, E.R. Sacia, A.T. Bell, Syntheses of Biodiesel Precursors: Sulfonic Acid Catalysts for Condensation of Biomass-Derived Platform Molecules, ChemSusChem, 7 (2014) 1078-1085.

Copyright © American Institute of Chemical Engineers. All rights reserved.

466351 ZrO_2 Is Preferred over TiO_2 As Support for the Ru-Catalyzed Hydrogenation of Levulinic Acid to Î³-Valerolactone

Abstract:

In the last few years, Î³-valerolactone (GVL), readily obtained by the hydrogenation of levulinic acid (LA), has attracted the attention of both academic researchers and industries because of its possible use in different applications related to the chemical industry (e.g., as fuel additives, food ingredients, nylon intermediate or as a solvent). Numerous approaches have been examined for LA hydrogenation, with the main focus in most cases on obtaining high GVL yields by variation of either the active phase or the catalyst support. Studies focusing on catalyst stability are much more limited in number. Indeed, while high activity/selectivity for GVL can relatively easily be achieved with most of the reported catalysts, as well as with commercial ones (e.g., Ru/C), stable performances especially under more industrially-relevant conditions is much harder to achieve. Long-term exposure to relatively harsh, polar conditions, as well as the use of non-pure, real LA feeds containing impurities from previous process steps indeed pose significant challenges to catalyst stability.

Herein, we present a study of catalyst stability and performances for the hydrogenation of LA to GVL of a Ru/ZrO_2 catalyst [1-2]. Catalyst performance has been compared with commercial Ru/C and a self-synthesized Ru/TiO_2 catalyst, which can generally considered as benchmark catalysts for this process. All catalysts show high GVL yields under standard conditions. Large differences were seen in catalyst stability; though, Ru/ZrO_2 maintanied its stability and high GVL yields were obtained even after five recycling tests. Remarkably, in the fresh Ru/ZrO_2 catalyst, Ru was found to be fully atomically dispersed on the fresh catalyst even at 1 wt.% Ru loading, with some genesis of Ru nanoparticles being observed upon recycling. Notably and in strong contrast to the benchmark catalysts (Ru/C and Ru/TiO_2), the Ru/ZrO_2 catalyst can efficiently perform the hydrogenation of LA even in the presence of impurities in

the LA feed, e.g., H_2SO_4, salts (Na_2SO_4, NaCl, and H_3PO_4), and organic compounds/contaminants derived from fractions of the biomass, which have been reported as impurities existing in real streams. To better understand these remarkable results with regards to catalytic performance, an extensive catalyst characterizatin study was performed to provide more insights into the relation between structure, performance, and long-term stability of the catalysts.

[1] J. Ftouni, A. MuÃ±oz-Murillo, A. Goryachev, J. P. Hofmann, E. J.M. Hensen, L. Lu, C. J. Kiely, P. C.A. Bruijnincx, B. M. Weckhuysen. ZrO_2 is preferred over TiO_2 as support for the Ru-catalyzed hydrogenation of levulinic acid to Î³-valerolactone,Â *submitted*.

[2] J. Ftouni, P. C.A. Bruijnincx, B. M. Weckhuysen. Method for preparing a chemical compound using a ruthenium metal catalyst on a zirconium oxide support in the presence of a contaminant, *PCT/EP2016/054031*.

Copyright © American Institute of Chemical Engineers. All rights reserved.

465502 C–O Bond Hydrogenolysis of Biomass–Derived Cyclic Ethers over Ni–Based Catalysts

Abstract:

Catalytic deoxygenation reactions play a crucial role for upgrading of biomass–derived oxygenates to high value chemicals. Among several deoxygenation processes, C–O hydrogenolysis has emerged as a viable technology to convert oxygenates such as sugar polyols and furanic derivates into target molecules. The key element to selectively cleave the C–O bond is the development of an efficient catalytic system, while minimizing the side reactions and carbon loss. Recent experimental studies suggested that rhodium and iridium catalysts modified by some oxophilic metals such as Re, Mo, W and V oxides display high activity and selectivity in the hydrogenolysis of glycerol[1], tetrahydrofurfuryl alcohol (THFA)[2] and tetrahydropyran–2–methanol (2–THPM)[2-3] to their corresponding α,ω–diols. However, due to the high cost and depleting resources of noble metals, it is necessary to develop catalytic systems based on non–noble metals. Among non–noble metals, nickel–based catalysts are widely used in hydrogenation and hydrogenolysis reactions because they exhibit good catalytic performance and are inexpensive. It was reported that C–O bond cleavage can be occurred over Ni catalysts at temperatures above 473K for both furan and tetrahydrofuran rings[4-5] . Therefore, in this contribution we focus on synthesis of Ni–based catalysts, suitable for ring opening of cyclic compounds derived from furanic conversions, particularly THFA. The effect of Ni and different support properties on their performance as THFA hydrogenolysis catalysts was examined under liquid phase reaction. The synergistic roles of the metal and acid sites of Ni/HZSM–5 and the reaction mechanism of THFA hydrogenolysis as well as their respective reaction intermediates and products are described and discussed. Ring opening of THFA into 1,5–PDO takes place over both nickel metallic sites and HZSM–5 acidic sites, followed by ring closure into tetrahydropyran (THP) product over acidic sites up to 70% yield.

[1] Y. Nakagawa, M. Tamura, K. Tomishige, *J. Mater. Chem. A.* **2014**, *2*, 6688-6702.

[2] Y. Nakagawa, M. Tamura, K. Tomishige, *ACS Catal.* **2013**, *3*, 2655-2668.

[3] T. Buntara, S. Noel, P. H. Phua, I. Melián Cabrera, J. G. de Vries, H. J. Heeres, *Angew. Chem. Int. Ed.* **2011**, *50*, 7083-7087.

[4] S. Sitthisa, D. E. Resasco, *Catal. Lett.* **2011**, *141*, 784-791.

[5] J. Lee, Y. Xu, G. W. Huber, *Appl. Catal., B.* **2013**, *140*, 98-107.

Copyright © American Institute of Chemical Engineers. All rights reserved.

469628 A Novel Pathway to 1,5-Pentanediol without the Use of Noble Metals

Abstract:

Recent focus in biomass conversion has shifted from the production of low value fuels to high value chemicals.Â Î±,Ï ‰-diols are important chemicals for their uses as polyurethane precursors, resins, and plasticizers. The petroleum industry produces 138 and 2,500 kilotons/year of 1,6-hexanediol and 1,4-butanediol, respectively. This market size and a projected market growth rate of 8% per year allow for additional production from biorenewable sources without market saturation [1,2].

We have developed a new route to synthesize 1,5-pentanediol in high yields from the hemicellulose fraction of lignocellulosic biomass. Previous work focused on direct hydrogenolysis of the fully hydrogenated form of furfural, tetrahydrofurfuryl alcohol, to 1,5-pentanediol with RhRe catalysts [3,4]. Our new multi-step process uses a combination of heterogeneous acids and supported base metal catalysts.Â The ability to use base metal catalysts stems from the improved reactivity of a key reaction intermediate.

In this presentation, we will focus on mechanistic studies used to determine why the new reaction intermediate is more active than tetrahydrofurfuryl alcohol and why base metals can be utilized for this reaction system. Experiments were performed with nuclear magnetic resonance (NMR) spectroscopy and *in situ* attenuated total reflectance Fourier transform infrared spectroscopy (*ATR-FTIR*) to help identify reaction intermediates.

[1] Lux Research, Bio-based Materials and Chemicals Intelligence Service

[2] MarketandMarkets Report on 1,6-Hexanediol

[3] Chia, M. et al. *Journal of the American Chemical Society.* **2011**, *32*, 12675-89.

[4] Koso, S. et al. Chemical Communications. **2009**, 15, 2035-2037.

Copyright © American Institute of Chemical Engineers. All rights reserved.

465036 Evaluation and Characterization of Carbon-Supported Nobel Metals for the Hydrodeoxygenation (HDO) of Acetic Acid

Abstract:

Hydrodeoxygenation (HDO) of organic acids (i.e. acetic acid, propanoic acid, fatty acids) can yield desirable products such as hydrocarbon and alcohols [1–4]. The HDO reaction pathways consist of hydrogenation, decarbonylation and decarboxylation [5]. This study has explored the catalytic chemistry of HDO of acetic acid over noble metals (Pt, Pd, Ru and Rh) on activated carbon. The reactions are evaluated in a continuous plug-flow reactor operated between 200-400ºC under atmospheric pressure with concentrations of 1% acetic acid/20%H_2/balance He. The gaseous products were analyzed by an on-line gas chromatograph. Transition metals supported on activated carbon yield primarily alkane products in the HDO of acetic acid, making them promising candidates for production of liquid fuels from organic acid feeds. The activation energies ranging from 19.7-20.3 kcal/mole and kinetics involving the reaction rate orders with respect to acetic acid (α=0 .07-0.41) and H_2 (β=0.15-.40) have been examined. The intrinsic activity based on the turn over frequency (TOF) has been evaluated for each catalyst, suggesting that Ru and Pt are the most active metals. The activity has been observed to decrease with time on stream, and post-reaction analysis by x-ray diffraction (XRD), scanning transmission electron microscopy (STEM) and temperature programmed oxidation (TPO) suggests that a combination of sintering and carbon deposition during reaction is responsible. The probable mechanism for HDO of acetic acid over these catalysts is discussed in light of these findings and those in the literature.

[1] Z. Wang, G. Li, X. Liu, Y. Huang, A. Wang, W. Chu, X. Wang, N. Li, Aqueous phase hydrogenation of acetic acid to ethanol over Ir-MoOx/SiO2 catalyst, Catal. Commun. 43 (2014) 38–41. doi:10.1016/j.catcom.2013.09.007.

[2] N. Joshi, A. Lawal, Hydrodeoxygenation of acetic acid in a microreactor, Chem. Eng. Sci. 84 (2012) 761–771. doi:10.1016/j.ces.2012.09.018.

[3] H. Olcay, L. Xu, Y. Xu, G.W. Huber, Aqueous-Phase Hydrogenation of Acetic Acid over Transition Metal Catalysts, ChemCatChem. 2 (2010) 1420–1424. doi:10.1002/cctc.201000134.

[4] Z. He, X. Wang, Required catalytic properties for alkane production from carboxylic acids: Hydrodeoxygenation of acetic acid, J. Energy Chem. 22 (2013) 883–894.

[5] Y.K. Lugo-José, J.R. Monnier, C.T. Williams, Gas-phase, catalytic hydrodeoxygenation of propanoic acid, over supported group VIII noble metals: Metal and support effects, Appl. Catal. Gen. 469 (2014) 410–418. doi:10.1016/j.apcata.2013.10.025.

Copyright © American Institute of Chemical Engineers. All rights reserved.

457028 Direct Conversion of Methane to Methanol and Ethanol

Abstract:

The enormous scale of methane reserves has motivated significant research activities focused on its conversion to fuels and chemicals. Since large amounts of natural gas are located in remote areas and transporting gases in pipelines is difficult, processes for converting methane into denser products are highly desirable. To address this challenge, we developed a bi-functional catalyst that is capable of activating the C-H bond in methane to form surface methyl groups. Some of these methyl groups undergo surface reaction that lead to the formation of higher alkyl chains. In the presence of steam, methanol and ethanol are formed by hydrolysis of the surface alkyl groups. Oxygen has to be co-fed to provide a thermodynamic driving force for this reaction. In addition to alcohols, carbon dioxide and hydrogen are formed as by-products. Importantly, formation of alcohols occurs at 450 °C under steady state conditions with a turnover frequency of at least 50 h^{-1}. This is a significant improvement from previous studies, in which a high-temperature calcination step was required for every turnover. The identification of promising catalysts and suitable reaction conditions will be discussed. Spectroscopic studies are performed to identify the active sites and the synergistic way in which they enable the reaction.

Copyright © American Institute of Chemical Engineers. All rights reserved.

470752 Aqueous-Phase Phenol Hydrogenation over Platinum and Rhodium: Electrocatalysis Versus Thermal Catalysis

Abstract:

A major future need is providing renewable energy without emitting greenhouse gases. Catalysis can play an integral role in the chemical processes required to produce transportation fuels and store energy. One method to make renewable fuels is to convert biomass to hydrocarbon fuels. Bio-oil (modeled here using phenol) is an especially promising feedstock, which can be hydrogenated and then deoxygenated to yield fuel grade hydrocarbons.[1] Although hydrogen generally comes from steam methane reforming, which emits carbon dioxide, renewable electricity can be used to drive carbon-neutral (i) electrochemical hydrogen production which is then used for thermal hydrogenation (TH), or (ii) direct electrochemical hydrogenation (ECH) of the organic substrate. Both of these routes require high rates for industrial application, and have the advantage that greenhouse gases are not emitted, and renewable energy can be stored as transportation fuels. ECH has the further advantage that there is no need to store hydrogen as a gas (other than as a side-product of ECH). TH and ECH of phenol on rhodium and platinum have similar activation barriers due to a shared rate determining step of adsorbed H adatoms reacting with adsorbed phenol.[2] The conditions (hydrogen pressure or applied potential) influence the reaction rate, but how they affect the rate and how the conditions of TH and ECH compare is not well-understood. Additionally, the high rates needed for industrial application are not currently achievable due to catalyst deactivation at high temperatures.[2] Similar behavior for gas-phase benzene TH is due to the poisoning of surface sites by dehydrogenated benzene.[3] We show here that a similar effect exists for TH of phenol, and that ECH and TH rates are controlled by the surface coverages of the reactants, which are in turn controlled by either applied potential or hydrogen pressure. We also discuss differences in ECH and TH and propose solutions to improve reaction rates.

On Pt, TH activity decreases from 60 to 80 °C, when the hydrogen pressure (P_{H2}) is below 1 bar. At 10 bar, the rate of reaction is greatly increased and the reaction takes on Arrhenius behavior from 60 to 80 °C. Based on the third-order in hydrogen, the decreased rate from 60 to 80 °C must be from poisoning species derived from dehydrogenated phenol, which are removed when the hydrogen pressure is sufficiently high (between 1.5 to 6 bar in this work) to rehydrogenate them. At pressures sufficiently high to remove the dehydrogenated phenol, the order in hydrogen is first-order, indicating a dependence on the hydrogen coverage. At even greater high pressures the hydrogen order decreases to zero, indicating further changes in the hydrogen coverage do not greatly influence the rate. The applied potential has a similar effect on the reaction rate for ECH, where initial increases in potential (from 0.05 to -0.05 V vs. RHE) greatly increase the rate of reaction, but from -0.05 to -0.45 V vs. RHE the rate increases slowly, after which it plateaus. The similarities in the maximum rate achieved for TH and ECH at a given temperature indicate the rate is controlled by the coverage and surface reaction of adsorbed phenol and an adsorbed hydrogen adatom. By comparing the rates of ECH and TH we are able to see conditions where the coverages must be similar, and thus can empirically relate applied potentials to pressure. For example, under this study it appears that 6 bar P_{H2} results in similar coverages as -0.45 V vs. RHE, and 1 bar P_{H2} is similar to -0.3 V vs. RHE.

We also present *in situ* characterization using X-Ray Absorption Spectroscopy (XAFS) on Pt nanoparticles in the presence and absence of phenol under applied potential. Through analysis of the near-edge (XANES) portion of XAFS we detect H-adsorption on Pt nanoparticles at potentials known to fully saturate the Pt surface with hydrogen. We see that the addition of phenol to the electrolyte solution lowers the H-adsorption signal, indicating phenol displaces hydrogen, lowering the hydrogen coverage. As higher applied potentials are applied, in the presence of phenol, the surface coverage of hydrogen increases. This supports our hypothesis from the reaction data that we see increased ECH rates at higher applied potentials because of an increase in hydrogen coverage, which reacts with adsorbed phenol. From the extended X-ray absorption fine structure (EXAFS) we see that the introduction of phenol also results in interaction between Pt and carbon atoms, which we believe to be adsorbed phenol. The Pt-C distance matches calculations for phenol adsorption on platinum,[4] and cannot be attributed to oxygen due to the absence of surface oxide from our XANES results. The strength of the Pt-C interaction is inversely related to the H-adsorption signal from XANES, supporting our conclusion that phenol displaces hydrogen on the surface, and that the applied potential can control the H-adsorption, and indirectly control the phenol coverage. Because the H-adsorption appears to affect phenol adsorption, we believe that phenol and hydrogen compete for sites on platinum, which is supported by electrochemical results that show phenol inhibits the H adsorption from protons at low applied potentials.[2]

References

(1) Zhao, C.; He, J.; Lemonidou, A. A.; Li, X.; Lercher, J. A. *J. Catal.* **2011**, *280* (1), 8.

(2) Song, Y.; Gutiérrez, O. Y.; Herranz, J.; Lercher, J. A. *Appl. Catal. B Environ.* **2016**, *182*, 236.

(3) Chou, P.; Vannice, M. A. *J. Catal.* **1987**, *107*, 140.

(4) Yoon, Y.; Rousseau, R.; Weber, R. S.; Mei, D.; Lercher, J. A. *J. Am. Chem. Soc.* **2014**, *136* (29), 10287.

Copyright © American Institute of Chemical Engineers. All rights reserved.

472437 Highly Efficient Conversion of Terpenoids to High Density Bio- Jet Fuels

Abstract:

The market demand for renewable aviation fuels is consistently increasing due to the growing concerns about global climate change. Bio- jet fuels produced from hydro-treating vegetable oils or Fischerâ€"Tropsch synthesis of bio-syngas are mainly straight chain alkanes with low octane numbers and volumetric densities and cannot meet the specification of petroleum based jet fuels. Herein, we develop the novel process to upgrade biomass derived terpenoids, with various cyclic hydrocarbon structures, e.g., 1,8-cineole, to produce cycloalkanes with high thermal stabilities, energy densities, and octane numbers, which can be blended to produce high-quality bio- jet fuels. We found that the effective deoxygenation of 1,8-cineole underwent three main steps: C-O ring opening, dehydration, and hydrogenation. We have designed multi-functional catalytic systems for the conversion of terpenoids to cyclic alkanes in a â€œone-potâ€ process. The conversion of monoterpenoid reached ~100% with ~100% selectivity to cycloalkanes.

Copyright © American Institute of Chemical Engineers. All rights reserved.

459925 Pyrolysis of Soils Contaminated with Heavy Hydrocarbons: Reaction Mechanisms and Morphology of the Produced Char

Abstract:

Soil contamination by petroleum and other heavy hydrocarbons is a major global environmental problem. Although offshore oil rigs and tankers are responsible for occasional large-volume spills, 98% of oil spills occur on land, with an average of 70 spills per day (more than 25,000 per year) reported to the U.S. EPA. Without adequate response, the effects of major spills could last decades. Whereas numerous remediation approaches have been proposed, most existing technologies have severe drawbacks. For example, *in situ* bioremediation of sites impacted by petroleum release can take years, particularly when recalcitrant species such as high molecular weight hydrocarbons are involved. Thermal technologies, on the other hand, can remediate sites quickly and efficiently (hours to months), often removing over 99% of a wide range of hydrocarbon fractions. However, current thermal technologies often destroy key soil constituents and severely degrade soil properties such as organic carbon content, water retention, stability and microbial activity. Thus, treated soils become unsuitable for reuse since they cannot support vegetative growth or even provide erosional stability. We have recently reported[1] that pyrolytic treatment at 400-450°C of soils contaminated with heavy petroleum crudes reduced total petroleum hydrocarbons (TPH) to levels well below regulatory standards (<0.1% by weight). A fraction of the original hydrocarbons remained in the treated soil in the form of a highly inert carbonaceous material (char). Standard EPA tests revealed that the post-pyrolysis levels of toxic polycyclic aromatic hydrocarbons (PAHs) were well below applicable standards. More importantly, plant growth studies showed significantly higher (by 80-900%) biomass production of *Arabidopsis thaliana* and *Lactuca sativa* (Simpson black-seeded lettuce) in pyrolyzed soils than in either contaminated or incinerated soils. And because it is carried out at lower temperatures, pyrolysis requires less energy than other thermal methods (like incineration) that can achieve similar TPH reduction levels. Overall, these results suggest that soil pyrolysis could be a viable thermal treatment to quickly remediate soils impacted by weathered oil while improving soil fertility, potentially enhancing revegetation.

We present here results from a follow-up study aimed at elucidating the fundamental mechanisms of the pyrolysis process and the morphology of the produced char.

The dispersion of the char in the pyrolytically treated soils will determine to a large extent its ultimate fate in the environment and modulate its interactions with water and the regenerated soil microbiome. Optical microscopy images of soils treated with pyrolysis and incineration led us to hypothesize that the produced char is distributed as a very thin film coating the soil particles. To test our "char film" hypothesis, we carried out controlled oxidation of pyrolyzed soils followed by surface analysis using XPS. We exposed treated soil to oxygen and reacted it in the TGA for one hour at temperatures between 350 and 450°C in order to progressively burn off the char film. Weight loss measurements and the XPS results confirmed our char film theory. As we burnt the film away, the percentage of C-C bonds on the surface of soil particles decreased and the amount of Si increased (our soils had high quartz content). On the other hand, the relative amounts of C and Si detected remained constant when uncontaminated "background" soil was exposed to the same treatment of controlled oxidation. Surface analysis by XPS detected the same relative amounts of C and Si when uncontaminated soil and pyrolyzed soil were oxidized at 450°C. This treatment "burnt away" the char film that covered the particles of pyrolyzed soil.

To further differentiate our process from existing thermal desorption technologies, we employed thermogravimetry with online evolved gas analysis (TG-MS) and differential scanning calorimetry (DSC) to gain a better understanding of the fundamental pyrolysis mechanisms. Our results show that as mixtures of soil and petroleum crudes are heated in an anoxic atmosphere, we first observe desorption of the lighter hydrocarbons at temperatures between 100 and 350°C. At temperatures between 350 and 450°C, however, thermal cracking and condensation reactions dominate and convert the heavier hydrocarbons (resins and asphaltenes) into char (C:H atomic ratio approximately equal to 1) and light volatiles. If the temperature rises above 550°C, however, key soil constituents (like carbonates) start to decompose and release CO_2 that becomes the dominant peak in the time-resolved MS spectra. Carbonate decomposition can explain the significant increase of the pH and the accompanying loss of fertility we observed when we incinerated contaminated soils at 650°C. MS data also show that H_2O evolves between 300 and 500°C, indicating dehydroxylation of clay constituents of soil. All these observations have significant implications for secondary reactions that may occur when pyrolytic treatment of contaminated soils is carried in commercial reactors.

Knowing the temperatures at which decomposition of soil constituents occurs will allow us to determine the least damaging thermal treatment to reach desired cleanup levels. The clearest lesson that can be gleaned from our studies is that the lowest effective treatment temperature should be used to minimize soil decomposition. While allowing for effective treatment of nearly all hydrocarbons, the high temperatures utilized by incineration and other thermal remediation technologies can cause extensive soil damage, such as decomposition of clays, carbonates, and organic content. This will physically alter the soil and change its geochemical, biological, and fertility properties.

---

[1] Vidonish, J.E., et al., "Pyrolytic treatment and fertility enhancement of soils contaminated with heavy hydrocarbons." *Environmental Science & Technology*, **50** (5), 2498–2506 (2016).

Copyright © American Institute of Chemical Engineers. All rights reserved.

468482 Spatial and Temporal Features of the Catalyzed Hydrocarbon Trap

Abstract:

Spatial and Temporal Features of the Catalyzed Hydrocarbon Trap

Po-Yu Peng, Michael P. Harold, Dan Luss***

Department of Chemical and Biomolecular Engineering, University of Houston, Houston, TX 77204

*mharold@uh.edu, **dluss@uh.edu

The diesel oxidation catalyst (DOC) system plays an important role in promoting CO and hydrocarbons (HCs) oxidation by the precious group metals Platinum (Pt) and Palladium (Pd). The DOC is designed to operate at very high HC conversion (>90%) after a sufficiently operating temperature has been reached. However, during the "cold start", most of the engine-out hydrocarbons remain unreacted and are emitted to the environment.[1, 2] A large-pore zeolite such as beta-zeolite (BEA) has been added to the DOC to trap hydrocarbons during the cold start. In particular, large molecular weight hydrocarbons are especially prone to trapping in BEA-modified DOCs. Although extensive research has been conducted on the BEA storage capacity, there is a need to gain insight on how BEA affects the performance of HC light-off (LO) and of the HC oxidation transient behaviors. [3-7]

DOC and DOC/BEA (0.5 g/in^3 BEA) monolith catalysts were used to study the transient behavior after the HCs (Dodecane, $C_{12}H_{26}$) has been pre-stored on the catalyst, The impact of BEA on the LO performance and HCs trapping were studied by flowing a mixture of 10% O_2 in Ar during temperature-programed oxidation (TPO). The transient behavior of the hydrocarbon trap was measured by both coherent optical frequency domain reflectometer (c-OFDR) and capillary inlet mass spectrometer (Spaci-MS) enabling the collection of transient temperature and concentration profiles the complete spatio-temporal data during the TPO. .

Figure 1 reports the feed HC concentration (150 ppm Dodecane, C12) and C12 effluent concentration. The figure reveals the completion of the HC pre-store and the removal of weakly-bonded HC by the lean gas purge (10% O_2 balanced by Ar). The pre-stored HC amount of 14.0 mg C12/g sample was determined by integrating the area between the feed concentration and effluent concentration. During the TPO, the effluent CO_2 concentration reveals the existence of multiple hydrocarbon LO points where the CO_2 formation occurs sequentially. Specifically, there are three major peaks that form during the TPO, indicating there are three LO stages within the monolith catalyst.

Figure 2 reports and the 3-D temperature dependence on position and time. The plot reveals the temperature rise at different locations and different times in the monolith catalyst. The three temperature hot zones occur due to the local HC combustion. Once the local HC combustion is complete the temperature drops until the next local LO. This indicates that the pre-stored HC is not release at once but occurs sequentially. We conclude by analyzing the HC LO locations and timing that the multiple HC LO phenomena is caused by a front of released HC which is trapped in the downstream and lights-off later when the local temperature exceeds the LO value.

The multiple hydrocarbon LO phenomena is strongly related to a tug-of-war between HC trapping and oxidation, thus, the steady-state HC oxidation has been investigated. Figure 3 shows the spatially-resolved C12, C_3H_6 and CO_2 concentrations at steady-state for a furnace temperature of 220 °C conditions for which most of HCs are fully oxidized. The CO_2 concentration fluctuated with a feed of 150 ppm C12 but was stationary under a feed of 0.2% C_3H_6. The effluent CO_2 concentration exhibited periodic fluctuations. These results suggest the existence of a "tug-of-war" between HC trapping and oxidation. Therefore, this phenomena provided a good opportunity to design more efficient catalyst.

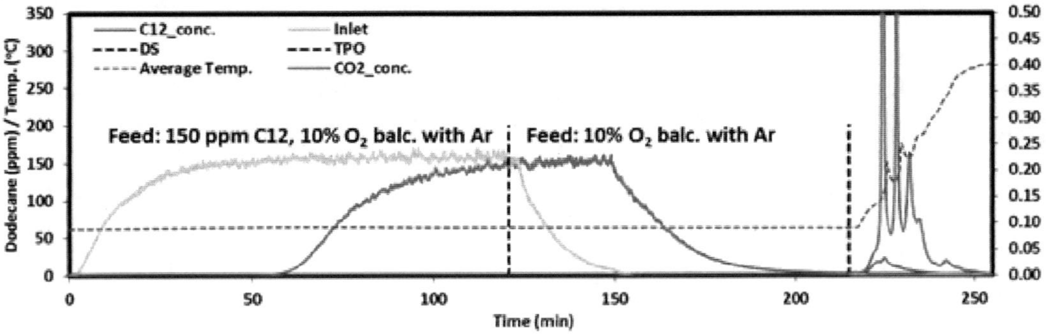

Figure 1. Dodecane and CO_2 concentration measured during adsorption and desorption followed by TPO with BEA-zeolite added Pt/Pd/Al_2O_3 monolith catalyst. Condition during saturation: 150 ppm dodecane and 10% O_2 balanced with Ar. Condition during desorption and TPO: 10% O_2 balanced with Ar.

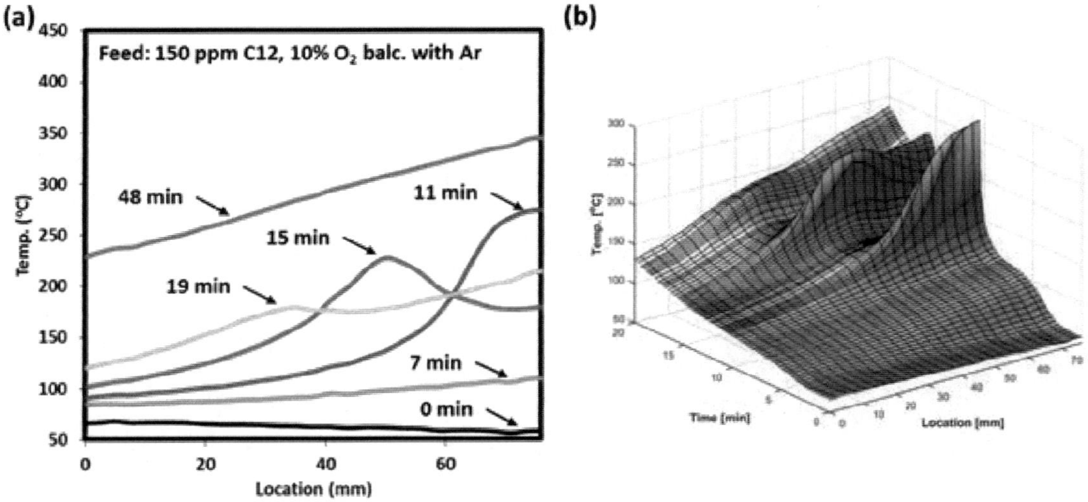

Figure 2. (a) Spatially-resolved temperature profile measured at different time in the C12 pre-stored catalyst and (b) the 3-D plot of the spatially-resolved temperature versus time in feed of 10% O_2 balanced with Ar. (Catalyst: D-05)

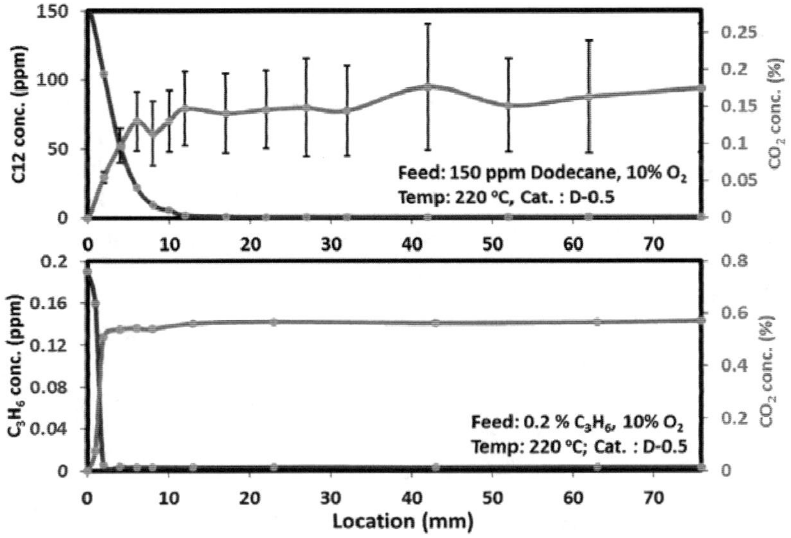

Figure 3. Spatially-resolved C12 and CO_2 concentration at steady-state at 220 °C with the feed of (a) 150 ppm C12 and 10% O_2 balanced with Ar, and (c) 0.2% C_3H_6 and O_2 balanced with Ar. Spatially-resolved C12 and CO2

conversion and selectivity at steady-state at 220 ^{o}C with the feed of (b) 150 ppm C12 and 10% O_2 balanced with Ar, and (d) 0.2% C_3H_6 and O_2 balanced with Ar.

References

1. Mukai, K., et al., *Adsorption and desorption characteristics of the adsorber to control the HC emission from a gasoline engine*. 2004, SAE Technical Paper.

2. AL Harbi, M., et al., *Competitive no, co and hydrocarbon oxidation reactions over a diesel oxidation catalyst.* The Canadian Journal of Chemical Engineering, 2012. **90**(6): p. 1527-1538.

3. Li, H.-X., et al., *Application of zeolites as hydrocarbon traps in automotive emission controls.* Studies in Surface Science and Catalysis, 2005. **158**: p. 1375-1382.

4. Chang, H., et al., *Gasoline Cold Start Concept (gCSC™) Technology for Low Temperature Emission Control.* SAE Int. J. Fuels Lubr, 2014. **7**(2): p. 480-488.

5. Murakami, K., et al., *Development of a high performance catalyzed hydrocarbon trap using Ag-zeolite*. 2004, SAE Technical Paper.

6. Elangovan, S., et al., *A comparative study of zeolites SSZ-33 and MCM-68 for hydrocarbon trap applications.* Microporous and mesoporous materials, 2006. **96**(1): p. 210-215.

7. Elangovan, S., et al., *Silicoaluminophosphate molecular sieves as a hydrocarbon trap.* Applied Catalysis B: Environmental, 2005. **57**(1): p. 31-36.

Copyright © American Institute of Chemical Engineers. All rights reserved.

463188 Formation of Nitrosamines in the Desorber of Tertiary Amine-Based Carbon Dioxide Capture Systems

Abstract:

Post-combustion CO_2 capture is one of the most developed and commonly employed technologies for power plants to reduce greenhouse gas emissions. Amine-based CO_2 absorption technology can be easily retrofitted into existing power plants and especially suitable for the flue gases containing low concentration of CO_2. Tertiary amines are considered as alternatives to the benchmark solvent monoethanolamine for CO_2 absorption, because they provide the benefits of higher absorption capacity and lower heat of reaction. However, public concerns have been raised for amine-based CO_2 capture due to the formation of nitrosamines, which are toxic to human health and ecological systems.

In this work, the influences of amine structural characteristics and desorber conditions on the formation of nitrosamines were investigated. Generally, tertiary amines were less reactive towards nitrosamines comparing with secondary amines. The formation of total nitrosamine from tertiary amine was first-order dependent on nitrite concentration and CO_2 loading, but was not dependent on amine concentration. For tertiary amines with the same number of 2-hydroxyethyl groups, less hindered amines showed faster nitrosamine formation and higher nitrosamine yields based on nitrite consumption. The distribution of specific nitrosamine formation from tertiary alkanolamines implied the preferential cleavage of 2-hydroxyethyl group over methyl group, but comparable to ethyl group, and therefore shed light on the nitrosation mechanisms.

Copyright © American Institute of Chemical Engineers. All rights reserved.

458015 Natural Attenuation and Biostimulation of Petroleum-Hydrocarbon Contaminated Soil in a Simulated Column

Abstract:

Various physiochemical and biological remediation methods can be used to remediate petroleum-hydrocarbon contaminations. Using microorganisms in biological remediation methods is considered as a cost-effective and environmental friendly solution. In the current research, two bioremediation-type processes, natural attenuation (NA) and biostimulation (BS) were applied to simulate biodegradation of petroleum-hydrocarbon contaminated soil collected from a refinery. During 60 days of operation, 88% and 65% of initial petroleum-hydrocarbon contamination biodegraded in BS and NA modes, respectively. Moreover, population of indigenous microorganisms in the soil column increased noticeably. Both bioremediation bodes showed that they are compatible with first-order reaction kinetic. In the NA mode, reaction rate constant was calculated 0.02 d^{-1} while it was 0.039 d^{-1} for the BS mode.

Copyright © American Institute of Chemical Engineers. All rights reserved.

462581 Study of Reaction Kinetics for Elemental Mercury Vapor Oxidation over $CuCl_2$ for Mercury

Abstract:

The reaction kinetics for elemental mercury (Hg(0)) vapor oxidation over $CuCl_2$/Î±-Al_2O_3 has been studied. The reaction kinetic data were obtained with respect to the reaction temperature and inlet Hg(0) concentration. The objective of this study is to determine the intrinsic reaction rate expression for the reaction of Hg(0) with $CuCl_2$. Based on the grain model, a description has been developed to represent the gas-solid reaction taking place within the $CuCl_2$ grains. During the reaction, the $CuCl_2$ grains were found to agglomerate, and thus a sintering description was added to the model in order to take into account the incomplete conversion of $CuCl_2$.Â The reaction rate constants and activation energy were determined by comparing the experimental data with the mathematical model. The detailed experimental and simulation results will be presented.

Copyright © American Institute of Chemical Engineers. All rights reserved.

453270 Advances in the Use of a Bio-Physicochemical Model to Characterize, Optimize the Chemical Absorption-Biological Reduction Integrated System for NO Removal

Abstract:

One promising technology for NO_x removal from flue gas is the chemical absorption-biological reduction (CABR) integrated process, in which NO is absorbed by aqueous solution of Fe(II)EDTA and the absorption product, Fe(II)EDTA-NO, is reduced biologically. At the same time, Fe(II)EDTA, oxidized by O_2 in flue gas, could be regenerated in a function of iron-reducing bacteria.

A kinetic model based on mass transfer coupled with chemical reaction and bioreduction was developed to simulate and predict the NO_x removal by the CABR system in a bio-trickling filter. The developed model was validated by the experimental results (the average relative deviation kept within 5.72%) under different operating conditions, e.g. NO/O_2 concentration, gas/liquid flow rate, and subsequently was used to predict the system performance. NO distribution at different inlet concentrations in the gas phase along the bio-trickling filter was also modeled and predicted.

Furthermore, the liquid flow rate and total iron concentration were optimized for >90% NO removal efficiency, which were 38.8 L h^{-1} and 10 mM under the typical operating condition (2 L min^{-1} gas stream containing 3% (v/v) oxygen and 500 ppm NO) respectively. Sensitivity analysis found that the performance of the CABR process was controlled by the bioreduction activity of Fe(III)EDTA, and the least activity has been achieved. This work could help us with the scale-up design and operation of the CABR process in the industrial application.

Graphic abstract

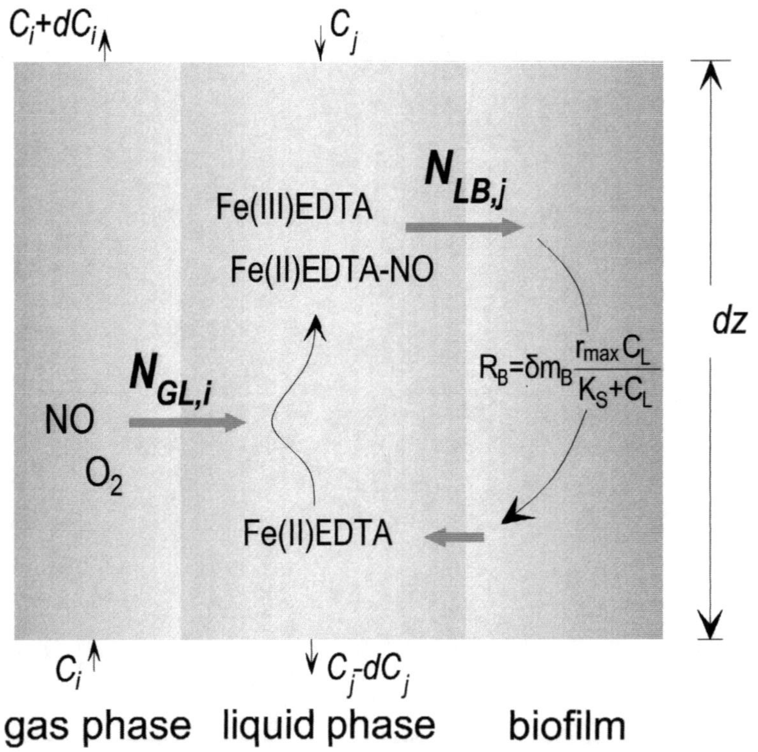

Copyright © American Institute of Chemical Engineers. All rights reserved.

474391 Photogenerated Charge Separation in Artificial Photosynthesis Systems

Abstract:

This lecture presents the research progress on solar fuel productions from artificial photosynthesis, namely photocatalysis and photoelectrocatalysis (PEC) processes with the emphasis on the photogenerated charge separation.

The great challenge of energy conversion photocatalysis lies in its complicated processes including light absorption (harvesting), charge separation and migration, and catalytic reactions. In order to gain high photon energy conversion efficiency, a photocatalyst or photocatalytic system must harmonically guarantee high efficiencies of all these three processes instead one of them. Semiconductors with appropriate phase junctions have been demonstrated to be an efficient approach for achieving efficient charge separation [1]. Cocatalysts play important roles in the assembly of efficient semiconductor photocatalyst. It has been conceptually demonstrated Pt-PdS/CdS dual cocatalyst system can achieve 93% H_2 evolution activity in the presence of sacrificial reagents under visible light irradiation (Î»>420 nm) [2,3]. Recently, it has been also demonstrated that electrons and holes can be spatially separated on the {010} and {110} facets of $BiVO_4$ crystal [4]. This finding has been demonstrated to be general for a number of catalytic systems including $SiTiO_3$, Cu_2O and TiO_2. It has been demonstrated that the intrinsic nature of charge separation between different facets of $BiVO_4$ together with the synergetic effect of dual-cocatalysts plays key role in photocatalytic activity enhancement [5]. Spatially resolved surface photovoltage spectroscopy (SRSPS) characterization on the photogenerated charge separations on different facets of a single $BiVO_4$ photocatalyst shows that the surface photovoltage signal intensity on the {011} facet can be up to 70 times stronger than that on the {010} facets[6]. This reveals that the nature of the built-in electric field in the space charge region of different facets on the anisotropic photoinduced charge transfer in a single semiconductor crystal.

References:

[1] X.Wang, H.X. Han, and C. Li, *et al. Angew. Chem. Intd. Ed.* 2012, 51, 13089.

[2] H. J. Yan, J. H. Yang and C. Li, *et al. J. Catal.* 2009, 266, 165.

[3] J. H. Yang, H. X. Han, and C. Li, *et al. Acc.Chem.Res,* 2013, 46, 1900.

[4] R. G. Li, F. X. Zhang and C. Li, *et al. Nature. Commun.* 2013, 4: 1432.

[5] R.G. Li, H. X. Han and C. Li, *et al. Energy Environ. Sci.* 2014, 7, 1369.

[6] J. Zhu, F.T. Fan, R. T. Chen and C. Li, *et al. Angew. Chem. Int. Ed.* 2015,

54, 9111.

Copyright © American Institute of Chemical Engineers. All rights reserved.

472286 Oxide Nanostructures: Novel Supports, Active Sites, and Tandem Catalysts

Abstract:

Oxide catalysts are important materials for a number of important chemical transformations. Our research seeks to better control and understand the catalytic surface of these materials through novel syntheses, probe molecules, and reaction modes. Two short vignettes reflecting current materials development will be described.

In the first area, we describe continuing efforts at creating highly dispersed metal oxide sites on silica and other supports, principally for selective oxidation (epoxidation, hydroxylation, ODH). We deomonstrate that the use of bulky and/or charged precursor complexes helps maintain small clusters. For epoxidation, isolated Nb on silica derived from a calixarene ligand are exceptional catalysts, while for ODH, we demonstrate some of the first ODH by CuOx clusters, and show that a precursor with bulky siloxane ligands outperforms traditional Cu precursors like the nitrate or EDTA complexes by forming a particular type of reduced Cu site. Going further, we are able to quantify the fraction of kinetically-relevant sites of these and other oxide materials using a newly developed phosphonic acid titration technique. Combined, synthesis and site titration allows us to organize large families of supported catalysts into quantitative, predictive relationships.

In the second area, we demonstrate a technique to easily deposit thin (<2 nm), conformal SiO_2 shells on other oxides such as TiO_2 or Al_2O_3. These core-shell materials impart a number of new properties on the material, including the creation of acidity sufficiently strong to crack alkyl benzenes (SiO_2 on Al_2O_3), and the ability to act as supports that very effectively stabilize metal nanoparticles with presenting diffusion limitations or blocking active sites. We also have shown that we can control access of molecules to the reactive surface using these layers. This trait can, for example, minimize oxidation of one speices in a mixture due to size or relative strength of adsorption.

Overall, it will be shown how advances in oxide materials synthesis can lead to improved understanding and breakout reactivity even in this very mature field of research.

Copyright © American Institute of Chemical Engineers. All rights reserved.

466737 Characterization of Tungsten and Zirconium Decorated Multi-Walled Carbon Nanotube Catalysts By XAS and NO_x-Temperature Programmed Desorption

Abstract:

The goal of this research is to develop strong solid acid catalysts supported on multi-walled carbon nanotubes (MWCNT) suitable for aqueous biomass processing. We have previously shown that zirconia supported on MWCNT (ZrO_2/MWCNT) is hydrothermally stable and can be made a solid acid catalyst by sulfating the zirconia.[1] Although this makes a relatively good acid catalyst, we have found that the sulfur species is not particularly stable in hot liquid water. On the other hand we have demonstrated that the tungsten-modified analogue (WO_x-ZrO_2/MWCNT) is hydrothermally stable and it is expected to have similar, but somewhat lower, acid catalytic activity.[2] We have studied the effects that catalyst annealing temperature and oxide loading on the oxide particle size. This was accomplished with XRD and TEM and there was good agreement in the ZrO_2 particle size of ~2 nm and the WO_x particle size of ~6 nm at high loadings, 30 wt%, (but immeasurable at low loading, < 10wt%) using these two techniques. X-Ray absorption near edge spectroscopy (XANES), has shown changes in the W L_3 edge and Zr K edge white line intensities of WO_x-ZrO_2/MWCNT upon annealing at different temperatures, indicating a change in the interaction and/or oxidation state of the tungsten and/or zirconium. To further interpret the XANES, it is essential to estimate accessible zirconia surface as a function of WO_x loading and annealing temperature at constant ZrO_2 loading and particle size of the ZrO_2/MWCNT. This was done using NO_x-TPD.[3,4]

The XAS results show that as the annealing temperature is increased (250-850 °C) for a given WO_x loading, the white line intensity of the Zr K edge decreases. Furthermore, we have found that as the WO_x loading is increased, the white line intensity of the Zr K edge again decreases. Data collected at the W L_3 edge shows similar trends to the data collected at the Zr K edge. The decrease in the white line intensity results from increased electron density at the Fermi level. This indicates an apparent reduction and/or rehybridization of Zr/W at higher temperatures and with higher WO_x loadings. Mutual reduction of the W and Zr is chemically unlikely, and moreover, it is probable that the coverage of ZrO_2 by WO_x varies with both temperature of annealing and WO_x loading. Thus, we use NO_x-TPD to estimate the accessible zirconia surface as a function of WO_x loading and annealing temperature in order to assess the WO_x-ZrO_2 interaction at constant ZrO_2 loading and constant WO_x coverage. From the NO_x-TPD data it can be deduced that: 1) As synthesized, the ZrO_2/MWCNT annealed in He at 450 °C retains amorphous C on the ZrO_2 that is removed by repetitive NO_x-TPD (presumably oxidized by NO_2), as indicated by the observed increase in the accessible ZrO_2 surface as the number of successive NO_x-TPD experiments increases, 2) the C free ZrO_2 particle size, deduced from NO_x-TPD, is 2.2 nm and consistent with that measured by XRD and TEM, and 3) this amorphous C contamination, that interferes with WO_x/ZrO_2 interaction, appears to be partially removed by interaction with WO_x (enhanced interaction by higher WO_xloading).

1. C. Liu, S. Lee, D. Su, B. Lee, S. Lee, R. E. Winans, C.Yin, S. Vajda, L. Pfefferle, G. L. Haller, Langmuir 28 (2012) 17159-17167.

2. J. Macht, R. T. Carr, E. Iglesia, J. Catal. 264 (2009) 54-66.

3. H. Y. Law, J. Blanchard, X. Carrier, C. Thomas, J. Phys. Chem. C 114 (2010) 9731-9738.

4. C. Thomas, J. Phys. Chem. C 115 (2011) 2253-2256.

5. C. Liu, S. Lee, D. Su, Z. Zhang, L. Pfefferle, G. L. Haller, J. Phys. Chem C 116 (2012) 21742-21752.

Copyright © American Institute of Chemical Engineers. All rights reserved.

471985 Mechanistic Differences Between Metal-Exchanged Zeolite SCR Catalysts, and Their Practical Implications

Abstract:

In the last decade, metal-exchanged zeolites became the technology of choice for meeting the remarkably stringent and rapidly evolving diesel emission regulations worldwide. These catalysts are used in the reaction of selective catalytic reduction of NOx with ammonia derived from urea, demonstrating outstanding efficiency across a broad range of operating conditions. Out of a large group of candidate technologies, commercial application converged to several major types of Cu- and Fe-exchanged catalysts. Due to being a part of a highly active and dynamic system, which features multiple catalysts, sensors, and actuators, and which is operated by the dedicated control and diagnostic algorithms, the successful application of these catalysts required detailed understanding of their performance and deactivation characteristics. In this presentation, we will highlight some of the significant mechanistic differences, which proved important to the performance, diagnostic characteristics, and durability of these metal-exchanged zeolites.

Copyright © American Institute of Chemical Engineers. All rights reserved.

468063 The Kinetics of Elementary Steps in Condensations Catalyzed By Solid Acids: Effects of the Size, Shape, and Charge of Intermediates, Transition States and Catalytic Structures

Abstract:

The efficient conversion of oxygen-rich feedstocks derived from biomass to valuable chemicals and fuels requires an understanding of the oxygenate chemistry that can lengthen carbon chains and decrease oxygen content. Aldol condensation of alkanones and alkanals provides a pathway to form new C-C bonds, while also removing O-atoms as H
experimentally, through spectroscopic, H/D isotopic, and kinetic measurements, in combination with density functional theory (DFT) treatments for a diverse range of solid Brønsted acid catalysts. These acids include H-FER, H-TON, H-MFI (with Al, Ga, Fe, and B at framework locations), H-Al-BEA, H-Al-FAU, H-Al-MCM-41, and supported Keggin polyoxometalate clusters ($H_3PW_{12}O_{40}/SiO_2$). Condensation rates on these solid acid catalysts decrease rapidly with time on stream as a result of secondary reactions of unstable α,β-unsaturated intermediates, leading to the formation of large, unsaturated products that block active sites. A hydrogenation metal function, present within the solid acids or as a separate component in physical mixtures, effectively scavenges these unsaturated species. In doing so, it (i) essentially eliminates deactivation; (ii) removes thermodynamic bottlenecks in C-C bond formation; (iii) avoids secondary β-scission reactions; and (iv) allows rigorous mechanistic studies and the elucidation of the consequences of acid strength and confinement within zeolite voids on the kinetically-relevant rate constants and transition states.

The selective titration of protons by non-coordinating bases (2,6-di-*tert*-butyl-pyridine) during acetone condensation, together with infrared (IR) spectra collected during catalysis, indicate that condensation reactions (473 K, 0.1-10 kPa acetone) occur on protons nearly saturated with H-bonded acetone, without detectable contributions by Lewis acid sites. These results are consistent with energies and frequencies calculated using periodic density functional theory (DFT) (T-12, Al-MFI unit cell, VASP, RPBE, D3BJ). These calculations show that acetone binds strongly (ΔG = -80 kJ mol^{-1}, 473 K) as H-bonded species and confirm the origins of the observed shift in the O-H stretch in zeolites upon acetone binding from 3600 cm^{-1} to 2440 cm^{-1}. The condensation turnover rates are proportional to acetone pressure on all samples; the resulting first-order rate constant reflects the free energy differences between a bimolecular C-C bond-formation transition state and a proton saturated with H-bonded acetone and a gaseous acetone molecule. The kinetic relevance of the C-C bond formation steps on these samples is consistent with its DFT-derived free energy barrier, which lies at the highest value along the reaction coordinate at T12 sites in H-Al-MFI structures.

These first-order condensation rate constants can be compared on solid Brønsted acids with different acid strength and confining environments to determine their consequences on the stability of the kinetically-relevant transition states and their relevant precursors. The effects of acid strength are evident from comparisons among catalysts with different acid strength and either mesopores much larger than the relevant molecular structures (Al-MCM-41 and $H_3PW_{12}O_{40}/SiO_2$) or small voids similar in structure (MFI) but with different framework heteroatoms (Al-, Ga-, Fe-, B- isomorphous substituents). In both cases, rate constants increase exponentially with decreasing deprotonation energy (DPE), a theoretically-accessible measure of catalyst acid strength [1]. These measured effects of acid strength on rate constants reflect the stability of the conjugate anion, which influences the stability of the ion-pair transition states more strongly than for the less charged H-bonded acetone precursors. The DFT-derived free energy barriers for substituted MFI systems (T-12, Al-, Fe-, Ga-, B-MFI unit cell, VASP, RPBE, D3BJ, 473 K) agree well with these experimental measurements.

Microporous aluminosilicates with different framework structures provide protons of similar acid strength [2], but very diverse confining environments. Measured condensation rate constants reach a maximum value for voids similar in size to the condensation transition states (MFI, BEA), which preferentially stabilize these transition states over smaller H-bonded acetone precursors through van der Waals interactions. The relative size (and shape) of these species and their strong effects on reactivity require DFT methods that accurately account for attractive dispersion forces in energy minimizations and that can dissect, posteriori, the quantum mechanical and dispersion components of DFT-derived Gibbs free energies (RPBE, D3BJ). This is supplemented by screening methods developed here, which use Lennard-Jones potentials to determine interaction energies between each framework oxygen atom and all atoms in the organic moieties, thus rigorously capturing the consequences of both the size and the shape of the relevant structures and the inorganic containers of these active structures. These interaction energies are then ensemble-averaged for each transition state and precursor species over all crystallographically-distinct proton locations within each microporous framework. The resulting values act as rigorous descriptors of the "fit" for all species and do so in a manner that accounts for the most favorable placement of all structures within their confining "containers". These descriptors extend those based only on the independent properties of the species (proton affinities) and the catalyst (DPE) and their ability to reorganize charge [3] to include parameters that describe how organic and inorganic components in transition state and precursors interact, which depends on the size and shape of both the organic and the inorganic moieties.

We have also addressed in this study the enduring challenge of deactivation and secondary reactions that break C-C bonds in metal-free solid acids [4]. All solid acids show significant selectivities to isobutene and acetic acid, which form via β-scission of the α,β-unsaturated products (mesityl oxide (MO) and isomesityl oxide (IMO), for acetone reactants). These secondary reactions depend strongly on the size of the void environment and are most effectively catalyzed in MFI. These reactions require the presence of protons (MO/IMO/H_2O mixtures do not react on pure-silica MFI), but β-scission selectivities at a given acetone conversion increase sharply as Al (and proton) densities decrease, even though acetone condensation turnover rates remained unchanged. These counterintuitive trends, in which lower densities of the purported active sites preferentially enhance secondary reactions, cannot reflect diffusional constraints and require the concurrent presence of protons and a vicinal confining environment that, even without active sites, can convert highly reactive intermediates into more stable β-scission products. A detailed examination of the effects of H_2O, which determines the relative concentrations of IMO/MO and their aldol precursors, shows that two β-scission routes with different site requirements are involved. One route involves the β-scission of aldol species; it occurs on all zeolites and is exclusively catalyzed by protons. A second and previously unrecognized route involves β-scission events of IMO/MO or their tautomer C_6-alkenol products. These pathways involve the formation of an initial product on protons and its subsequent reaction within vicinal confined spaces that provide stabilization for chain propagation transition states, even in the absence of a proton. In this manner, β-scission rates become proportional to both the number of initiation sites (protons) and of propagating locations (voids), a dependence that cannot be described by conventional bifunctional catalytic formalisms. Such "physical catalysis" by voids free of active sites was recently shown to account for the reactivity of pure-silica zeolites in reactions of molecules with radical character [5]; its plausible involvement here is consistent with coupled cluster (CCSD) calculations of the free energy of formation of potential chain initiators from MO/IMO and their C_6-alkenol analogs.

[1] Jones, A., Carr, R., Zones, S., and Iglesia, E., Journal of Catalysis, 312 (2014) 58.

[2] Jones, A. and Iglesia, E., ACS Catalysis, 5 (2015) 5741.

[3] Deshlahra, P. and Iglesia, E., ACS Catalysis (submitted).

[4] Kubelková, L. and Nováková, J., Journal of Molecular Catalysis, 75 (1992) 53.

[5] Artioli, N., Lobo, R., and Iglesia, E., Journal of Physical Chemistry C, 117 (2013) 20666.

Copyright © American Institute of Chemical Engineers. All rights reserved.

453427 Progressing from Identifying, to Synthesizing and Designing Active Sites in Heterogeneous Catalysts

Abstract:

Whereas being able to construct catalytic active sites by design is an ultimate goal in heterogeneous catalysis, it remains a challenge to achieve this in general. An example is presented to show the challenge in identifying the active form of the catalyst in Au-catalyzed selective epoxidation of cyclooctene with molecular oxygen. Another example illustrates success in constructing the active site of a supported Sn oxide catalyst in which the active form is known, yet the resulting catalyst exhibits unanticipated activity due to Brønsted acidity, in addition to the expected Lewis acidity.

Copyright © American Institute of Chemical Engineers. All rights reserved.

463261 Analysis of Upstream Creeping Reaction Zones in Catalytic Monolith Reactors

Abstract:

Analysis of Upstream Creeping Reaction Zones in Catalytic Monolith Reactors

Tian Gu and Vemuri Balakotaiah

Department of Chemical and Biomolecular Engineering, University of Houston, TX-77204

Abstract

Monolith reactors are widely used in catalytic after-treatment systems (e.g. TWC, DOC, LNT and SCR). One major challenge in exhaust after-treatment is reducing cold start emissions which accounts for a significant fraction of the total emissions. Recent developments in advanced combustion strategies such as low temperature combustion (LTC) make abating cold start emissions and meeting environmental regulations even more challenging.

During a cold start, ignition can occur at the front-end, in the middle or at the back-end of the reactor. After the reactor is ignited, it is desired that the reaction zone is maintained close to the inlet so that when small excursions in gas velocity, inlet temperature and concentrations do not quench the reactor. This can be achieved if front-end ignition is possible (which requires high precious metal loading and/or high inlet temperature and/or high adiabatic temperature rise). While front-end ignition is technologically possible, it may be economically impractical. In such cases, back-end ignition followed by a fast upstream creeping reaction zone provides a feasible way to reduce cold start emissions.

In this work, the upstream creeping reaction zone is investigated in detail using a modeling approach. The effects of various design parameters (e.g. solid thermal conductivity, heat capacity, etc.) and operating conditions (gas velocity, inlet temperature/concentrations) on the creep velocity are determined and summarized as a correlation. Analytical criteria for reaction zones to creep upstream are also presented. These criteria can provide guidance for both design and control of catalytic after-treatment systems.

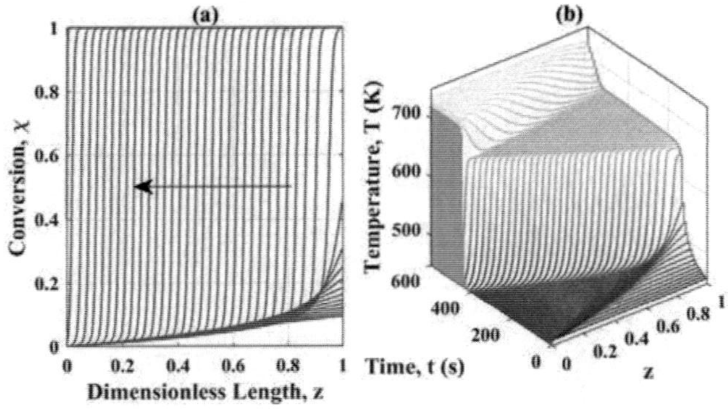

Fig.1 Snapshots of transient (a) conversion and (b) temperature profiles during a cold start showing an upstream creeping reaction zone. Snapshots are taken every 10s.

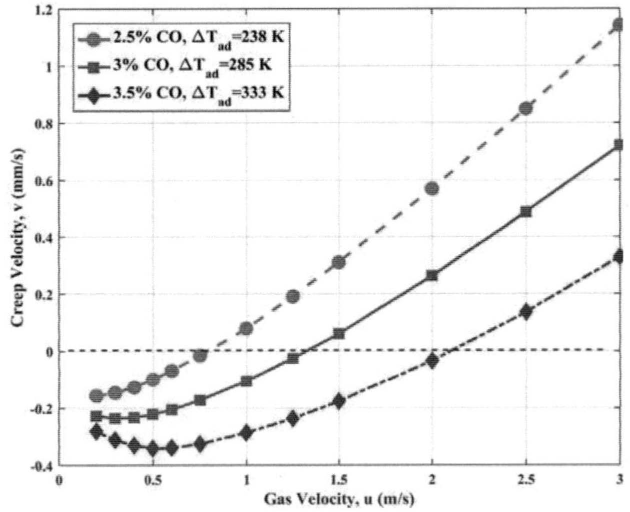

Fig.2 Creep velocity as a function of the gas velocity with different adiabatic temperature rise.

Copyright © American Institute of Chemical Engineers. All rights reserved.

468130 Model-Based Design of Optimal Reactors Considering Catalyst Deactivation

Abstract:

Catalyst deactivation causes temporal and spatial activity gradients in the reactor. This gradient changes the corresponding optimal reaction conditions, and the reaction conditions further determine the future activity profile. This interplay is commonly neglected in reactor design, however, the optimal design based on the assumption of a given (constant) reaction kinetics is usually not the optimum with regard to the overall performance for the whole catalyst lifetime. This work aims to take this interplay explicitly into consideration and provides a new optimal reactor design methodology.

In this regard, a recently proposed reactor design methodology [1] based on the concept of elementary process functions [2] serves as the basis. The fundamental idea of the methodology is to track a fluid element traveling through the reactor and to change its state within. By continuously providing the fluid element with optimal fluxes, which are adjusted according to the state of the fluid element, optimal reaction conditions in the reactor are achieved at any time. The optimal fluxes can be calculated by solving a dynamic optimization problem, which is formulated based on the apparatus-independent model equations including balance equations, reaction and transfer kinetics, thermodynamics and design bounds.

To consider deactivation of the catalyst, an additional time scale for the catalyst needs to be added to the problem. A new optimization problem with multiple time/length scales is solved and the best reactor configuration and operating conditions are determined simultaneously. Dynamic operating conditions are included in order to achieve the best performance. In addition, the potential of new catalysts can be identified by comparing the overall performance over the whole catalyst lifetime. The new method has been exemplified on different case studies for heterogeneously catalyzed reactions with different rates of catalyst deactivation for validation and illustration of its advantage. In the present contribution, special focus is on the ethylene oxide synthesis as industrially relevant example.

References

[1] A. Peschel, H. Freund, K. Sundmacher, Methodology for the design of optimal chemical reactors based on the concept of elementary process functions, Ind. Eng. Chem. Res., 49 (2010) 10535-10548.

[2] H. Freund, K. Sundmacher, Towards a methodology for the systematic analysis and design of efficient chemical processes. Part 1. From unit operations to elementary process functions, Chem. Eng. Process., 47 (2008) 2051-2060.

Copyright © American Institute of Chemical Engineers. All rights reserved.

467658 Exploiting the Nonlinear Frequency Response Method to Evaluate the Potential of Forced Periodic Operation of Chemical Reactors

Abstract:

INTRODUCTION Forced periodic operations of chemical reactors possess significant potential to improve numerous performance criteria. The corresponding analysis of reaction systems and the estimation of the magnitude of possible performance improvements have occupied many researchers since the earliest explorations by Douglas and Rippin (1966), Horn and Lin (1967) and Bailey (1973). Theoretical problems related to identification and estimation of the enhancement were typically analyzed evaluating suitable control criteria and applying four major approaches described by Watanabe et al. (1981) and Parulekar (2003). However, there are hardly industrial applications of this promising concept and there is still a need for developing generally applicable methods for identifying and analyzing promising reaction systems. NEW APPROACH We suggest applying the nonlinear frequency response (NFR) method to predict the outcome of perturbing periodically one or several input variables. Applying the concept of higher order frequency response functions (FRFs) weakly nonlinear systems can be described analytically (Petkovska and Seidel-Morgenstern, 2013). The response of a weakly nonlinear system to periodic input change(s) contains a periodic term, consisting of the basic and a number of higher harmonics, and a non-periodic (DC) component. The value and sign of the DC component of the outlet can be used as an indicator for the mean performance of the forced periodic operation, as it gives insight whether and to which extent improvements can be expected as a result of forced periodic modulation(s) of the input(s). Applying the NRF method, the non-periodic term can be efficiently estimated for modulation of single or multiple inputs. To demonstrate the strength of the theoretical concept the model of an adiabatic CSTR was analyzed considering reactions of different orders. Estimates are provided for different frequencies of sinusoidal changes of the inlet composition, the total inlet flow-rate and the feed temperature. It will be shown that it is in particular attractive to perturb simultaneously two input variables using the optimal phase shift, which can be determined using our method (Nikolic et al., 2015). EXPERIMENTAL To validate the theoretical predictions we carried out experiments in a laboratory scale CSTR using hydrolysis of acetic anhydride as a test reaction. The performance of the reaction system was evaluated with respect to different objectives under various steady state conditions, including theoretically optimal ones, and for various types of single and binary input perturbations. REFERENCES

Bailey, J.E. (1973), "Periodic Operation of Chemical Reactors: A Review", Chem. Eng. Comm., 1, 111.
Douglas, J.M. and Rippin, D.W.T. (1966), "Unsteady state process operation", Chem. Eng. Sci. 21, 305.
Horn, F. J. M. and Lin, R. C. (1967). "Periodic processes: A variation approach", Ind. Eng. Chem. Process Des. Dev., 6, 21.
Parulekar, S.J. (2003), "Systematic performance analysis of continuous processes subject to multiple input cycling", Chem. Eng. Sci., 58, 5173.
Petkovska M., Seidel-Morgenstern A. (2013). „Evaluation of Periodic Processes", in: "Periodic Operation of Chemical Reactors", Eds.: Silveston P.L., Hudgins R.R., Elsevier, Oxford, Chapter 14, 387. Watanabe, N., Onogi, K. and Matsubara, M. (1981), "Periodic control of continuous Stirred tank reactors-I. The π criterion and its application to isothermal cases", Chem. Eng. Sci., 36, 809.

Copyright © American Institute of Chemical Engineers. All rights reserved.

465019 Modeling Catalyst Deactivation in a Packed Bed Reactor for Thermo-Catalytic Hydrogenation of CO_2

Abstract:

Modeling catalyst deactivation in a packed bed reactor for thermo-catalytic hydrogenation of CO_2

Duo Sun and David Simakov

Department of Chemical Engineering, University of Waterloo, Waterloo, ON N2L 3G1, Canada

Converting CO_2 into synthetic fuels and chemicals is an attractive route to reduce CO_2 emissions and, at the same time, fossil fuels consumption. There are several pathways, including thermocatalytic conversion via Sabatier reaction ($CO_2 + 4H_2 = CH_4 + 2H_2O$) to produce synthetic natural gas [1-4]. Some technological issues are yet to be resolved in order to make this process truly viable economically, including catalyst deactivation by sintering and carbon formation. Sabatier reaction is highly exothermic which can lead to reactor overheating, catalyst sintering and accelerated methane cracking producing large amounts of coke. We developed a transient numerical model to study the effects of catalyst deactivation on the performance of the heat-exchanger type packed bed CO_2 hydrogenation reactor. Our results show that catalyst deactivation can significantly limit CO_2 conversion under certain operating conditions, changing also products distribution. On the other hand, the presence of H_2 in the reaction mixture and the generation of H_2O lead to *in situ* catalyst reactivation. Our model predicts that, with a proper design, it is possible to operate the reactor at reasonably high CO_2 conversions (e.g. 80% CO_2 conversion after 1,000 h on stream) even using Ni-based catalysts which are prone to coking.

References

[1] Rönsch, S. et al. Review on methanation – From fundamentals to current projects. *Fuel* 2016, 166, 276-296.

[2] Schaaf, T. et al. Methanation of CO_2 – storage of renewable energy in a gas distribution system. *Energy Sustain. Soc.* 2014, 4, 29.

[3] Götz, M et al. Renewable Power-to-Gas: A technological and economic review. *Renew. Energ.* 2016, 85, 1371-1390.

[4] Gahleitner, G. Hydrogen from renewable electricity: An international review of power-to-gas pilot plants for stationary applications. *Int. J. Hydrogen Energ.* 2013, 38, 2039-2061.

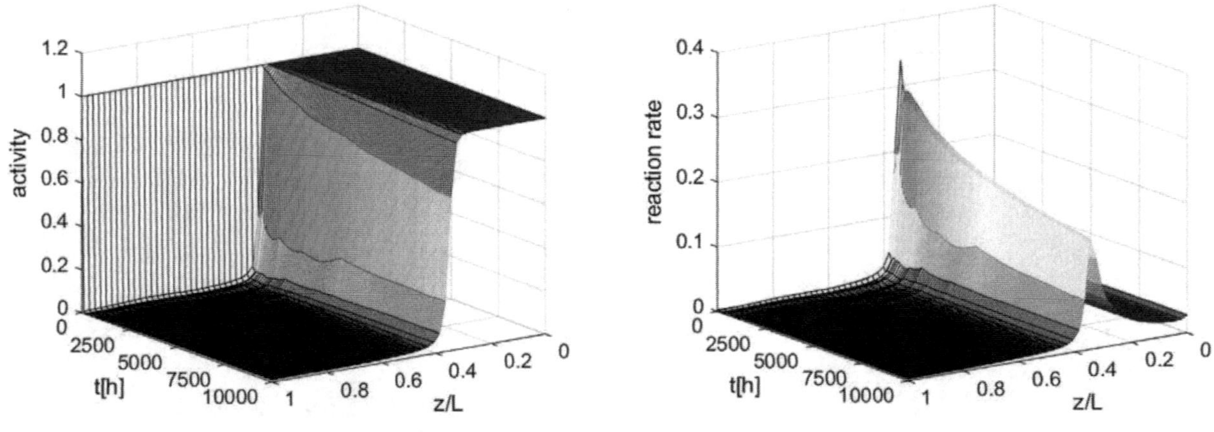

Figure 1: Simulated space-time profiles of catalyst activity and effective reaction rate (normalized) in a heat-exchanger type packed bed Sabatier reactor.

Copyright © American Institute of Chemical Engineers. All rights reserved.

463851 Optimizing Methanol Synthesis - from Lab to Industrial Plants Via High-Fidelity Modeling

Abstract:

In industry, a ternary $Cu/ZnO/Al_2O_3$ catalyst is used for the large-scale production of methanol (MeOH). These large-scale facilities with productivities of up to 5000 tons of MeOH production per day require adequate models for design and optimization. In this contribution, we want to outline the workflow from measurement of kinetic data towards parameter estimate and optimization of plants and present several case studies.

To balance the number of experiments and still preserve meaningfulness of a kinetic model [1], a design-of-experiment based approach was taken. We carried out temperature scanning experiments of catalyst pellets at varying pressures and synthesis gas compositions in a gradient-free internal recycle reactor (Berty type) to obtain quasi-stationary measurements of catalyst productivity at industrially relevant operating conditions.

In order to evaluate these data, a gRPOMS model of the experimental setup was set up and validated. Core of this approach is a micro-kinetic model combined with a representation of the mass transport in the pellet and reactor characteristics. Using a log-likelihood based optimizer 10 kinetic parameters involving rate factors of MeOH synthesis and the rWGS, adsorption constants and respective activation energies was carried out resulting in a predictive, heterogeneous catalyst model of high accuracy.

This catalyst model can be integrated in a process model of an industrial MeOH process using additional considerations on mass and heat transfer processes. The acquired plant model is subsequently used to evaluate and optimize performance of a large scale plant. Results can be used to identify catalyst and process parameters that would increase the overall process productivity under industrial conditions.

1. G.H. Graaf, E.J. Stamhuis, A.Beenackers, *Chem.Eng.Sci.* 43 (**1988**), 3185-3195.

Copyright © American Institute of Chemical Engineers. All rights reserved.

468335 Statistical Activity Variations in Diluted Catalyst Beds: Influence on Reactor Behavior

Abstract:

In multitubular fixed bed reactors for heterogeneously catalyzed highly exothermic gas phase reactions the catalyst bed is usually diluted with inert particles for improving the heat management. In order to distribute the heat generated by the reaction more evenly along the reactor length often a graded activity profile in axial direction is applied. This graded activity profile is realized by loading the tube segment by segment. Each of these segments features a different catalytic activity, which is commonly realized by different dilution levels. To help balancing the heat generation and heat removal out of the reactor with the concept of catalyst dilution, the active catalyst particles and the inert material must be perfectly mixed in the diluted zones [1, 2]. Otherwise the thermal stability of the reactor cannot be guaranteed and, e.g., catalyst deactivation may take place because of the formation of local hot spots. Since loading a mixture of active and inert particles into a reactor tube is a statistical process, non-uniform activity profiles in reactor tubes based on statistical variations cannot be avoided.

As basis for the investigation of these statistical activity fluctuations in diluted catalyst beds and their influence on reactor behavior, a numerical generation of random packings consisting of spherical particles [3, 4] is applied. The statistical process of loading a mixture into a reactor tube is mimicked by using a random number generator for the definition of active and inactive particles. The results were analyzed with regard to the statistical activity fluctuations in axial and radial direction. Additionally, the analysis of statistical activity variations in detail reveals significant non-idealities caused by accumulation of active particles in local clusters. From the geometrical point of view the extent of the non-idealities depends on the characteristics of the packing (e.g. tube-to-particle diameter ratio).

In order to study the influence of statistical activity variations on the reactor behavior, the partial oxidation of n-butane to maleic anhydride was investigated as case study. The simulation studies were carried out using 1D and 2D pseudo-homogeneous reactor models. The results illustrate significant deviations compared to the ideal case and a strong influence of operating conditions, especially for the inlet and coolant temperature. It is demonstrated that a direct link between the calculated activity profiles in axial and radial direction and the occurrence of local hot spots in the tube exists. By introducing sensitivity coefficients for reaction rates with respect to activity fluctuations based on the ideal case the extent of the deviations for the non-ideal case can be identified and described.

References

[1] E.M. Calverley, P.M. Witt, J.D. Sweeney, Reactor runaway due to statistically driven axial activity variations in graded catalyst beds, Chem. Eng. Sci., 80 (2012) 393-401.
[2] E.M. Calverley, P.M. Witt, J.D. Sweeney, Reactor runaway due to statistically driven axial activity variations in graded catalyst beds: Loading from pre-measured single tube aliquots, Chem. Eng. Sci., 90 (2013) 170-178.
[3] H. Freund, T. Zeiser, F. Huber, E. Klemm, G. Brenner, F. Durst, G. Emig, Numerical simulations of single phase reacting flows in randomly packed fixed-bed reactors and experimental validation, Chem. Eng. Sci., 58 (2003) 903-910.
[4] H. Freund, J. Bauer, T. Zeiser, G. Emig, Detailed Simulation of Transport Processes in Fixed-Beds, Ind. Eng. Chem. Res., 44 (2005) 6423-6434.

Copyright © American Institute of Chemical Engineers. All rights reserved.

Identification of a Reaction Boundary Layer in Membrane Reactors for Hydrogen Production

M.A. Murmura, S. Cerbelli, M.C. Annesini, University of Rome "Sapienza", Italy

Abstract

Integrated membrane reactors for the production of pure hydrogen at low temperatures are currently attracting much interest. In this work, we developed a model of a membrane reactor for low temperature methane steam reforming, which was solved numerically using COMSOL Multiphysics. The system studied consists of an annular reactor, packed with a catalyst. A hydrogen permeable membrane is supported on the inner wall of the reactor, allowing the selective removal of hydrogen as it is being produced through the steam reforming reaction. The behavior of the reactor may be described as the result of a competition between the steam reforming reaction, which tends to bring the system towards equilibrium, and hydrogen permeation, which moves the system away from equilibrium conditions. The presence of a region of the reactor, close to the wall on which the membrane is supported, in correspondence of which the system is not capable of reaching equilibrium, was noticed. The extent of this region, which we will refer to as a reaction boundary layer, depends on the operating conditions and on reactor geometry. The latter quantity may be defined as a function of the ratios between the length of the reactor and its inner radius, L/R_1, and the ratio between the outer and inner radii, R_2/R_1. The effects of the main operating parameters on the thickness of the reaction boundary layer has been studied, along with the influence of the ratio R_2/R_1. This procedure has allowed an optimization of reactor design when considering a fixed membrane area and feed flow rate.

Problem Description

The aim of this work is to set the ground for the optimal design of a membrane reactor for hydrogen production. The system studied consists of a double tube reactor that is fed with a gaseous mixture that is considered to be the product of a pre-reforming reactor. Its composition is therefore determined based on equilibrium conditions and evaluated by considering that the pre-reformer operated at 773 K starting from a steam /methane mixture with a 3:1 molar ratio. The reaction considered is

$$CH_4 + 2H_2O \rightleftharpoons 4H_2 + CO_2 \tag{1}$$

The mixture enters the annular gap between two co-axial cylinders. The wall of the outer cylinder (R_2) is impermeable to both components, whereas a membrane, with infinite selectivity towards hydrogen, is supported onto the outer wall of the inner cylinder (R_1). In the system considered, the aspect ratios, $L/R_1 = 10$ and $R_2/R_1 = 2$ have been adopted. The subscripts "m", "w", "h", and "c" will be used to indicate methane, water, hydrogen, and carbon dioxide, respectively.

The model of the membrane reactor has been developed under the assumptions of uniform temperature and zero partial pressure of hydrogen in the permeate. The velocity field has been evaluated by Darcy's law

$$\mathbf{v} = -\frac{\kappa}{\mu}\nabla P \tag{2}$$

where $\mathbf{v}$ if the mass average gas velocity, κ is the bed permeability, μ is the viscosity of the gas mixture, and P the pressure.

In steady-state conditions, the overall mass balance is given by:

$$\nabla \cdot (\rho \mathbf{v}) = 0 \tag{3}$$

where ρ is the mixture's mass density, which may be expressed as

$$\rho = f(\omega_i) \frac{P}{RT} \quad \text{with} \quad \frac{1}{f(\omega_i)} = \sum_i \frac{\omega_i}{W_i} \tag{4}$$

where ω_i and W_i indicate mass fraction and molecular weight, respectively, of the $i-th$ component and f is the average molecular weight. The equation of continuity in isothermal conditions becomes

$$\nabla \cdot (-fP\nabla P) = 0 \tag{5}$$

Mass balance equations in terms of mass fractions are

$$\nabla \cdot \left(-\frac{\kappa}{\mu} fP\nabla P\omega_i - fP\mathbb{D}\nabla\omega_i \right) = r_i \tag{6}$$

The rate of methane consumption has been described by the expression:

$$R_m = kP_m(1-\eta) \tag{7}$$

where η represents the distance from equilibrium, and therefore the rate of reaction for each component may be expressed as

$$r_i = \nu_i R_m \tag{8}$$

in which ν_i is the stoichiometric coefficient of the $i-th$ component, corrected for the ratio between its molecular weight and that of methane.

Re-writing the above equations in terms of dimensionless variables allows the definition of characteristic groups. The following dimensionless variables have been chosen

$$\tilde{\mathbf{v}} = \frac{\mathbf{v}}{U} \;;\; \tilde{r} = \frac{r}{R_1} \;;\; \tilde{z} = \frac{z}{R_1} \;;\; \tilde{f} = \frac{f}{W_h} \;;\; \tilde{P} = \frac{P}{P_{atm}} \;;\; \tilde{\mathbb{D}} = \frac{\mathbb{D}}{\mathcal{D}} \tag{9}$$

where P_{atm} is the atmospheric pressure and $\mathcal{D}$ is the molecular diffusivity, supposed to be equal for all components of the system.

Substituting these variables in the continuity equation gives

$$\tilde{\nabla} \cdot \left(\tilde{f}\tilde{P}\tilde{\nabla}\tilde{P} \right) = 0 \tag{10}$$

$\tilde{\nabla}$ being the dimensionless gradient operator.

The dimensionless form of the mass balance equation is instead

$$\tilde{\nabla} \cdot \left(\frac{U_c}{U} \tilde{f}\tilde{P}\omega_h \tilde{\nabla}\tilde{P} - \frac{1}{\mathrm{Pe}} \tilde{f}\tilde{P}\tilde{\mathbb{D}}\tilde{\nabla}\omega_h \right) = \nu_i Da \tilde{f}\tilde{P}\omega_m (1-\eta) \tag{11}$$

where

$$U_c = -\frac{\kappa}{\mu} \frac{P_{atm}}{R_1} \tag{12}$$

and

$$\beta = \frac{U_c}{U} \tag{13}$$

gives a measure of the pressure drops through the packed bed ($1/\beta \approx \Delta P/L$). The other dimensionless parameters that appear in the mass balance equation are the Peclet and Damkholer numbers, defined as

$$\mathrm{Pe} = \frac{UR_1}{\mathcal{D}} \quad ; \quad Da = \frac{\mathcal{R}Tk_1R_1}{U} \tag{14}$$

The continuity and mass balance equations presented above have been solved by assuming that Sieverts' law for hydrogen permeation applies, and hydrogen flux through the membrane is expressed by:

$$J_h = \mathbb{P}_m\left(\sqrt{P_h} - \sqrt{P_h^p}\right) \tag{15}$$

where $\mathbb{P}_m$ is the overall membrane permeability, P_h is the partial pressure of hydrogen in the retentate, and P_h^p is the partial pressure of hydrogen in the permeate, here set to zero. The main boundary conditions are reported

$$in \quad z = 0: \quad \left(-\frac{\kappa}{\mu}\nabla P\right)\cdot \mathbf{n_1} = U \quad ; \quad \omega_i = \omega_i^0 \tag{16}$$

$$in \quad z = L: \quad P = P_L \tag{17}$$

$$in \quad r = R_1: \quad \frac{1}{\mathcal{R}T}\left(-\frac{\kappa}{\mu}\omega_h fP\nabla P - fP\mathbb{D}\cdot\nabla\omega_h\right)\cdot \mathbf{n_2} = \mathbb{P}_m\left(\sqrt{\omega_h fP/W_h}\right) \tag{18}$$

$$\text{and} \quad \frac{1}{\mathcal{R}T}\left(-\frac{\kappa}{\mu}fP\nabla P\right)\cdot \mathbf{n_2} = \mathbb{P}_m\left(\sqrt{\omega_h fP/W_h}\right) \tag{19}$$

where $\mathbf{n_1}$ is a unit vector normal to the annular surface through which the gas flows, and directed inwards; and $\mathbf{n_2}$ is a unit vector normal to the cylindrical surface and directed towards the axis of the equipment ($r = 0$). It should be noted that, in what follows, inlet gas velocity has been maintained constant, meaning that increasing operating pressure leads to increasing inlet mass flow rate.

The dimensionless counterpart of the BCs in eqs.(16)-(18) writes

$$in \quad \tilde{z} = 0: \quad \left(-\frac{U_c}{U}\tilde{\nabla}\tilde{P}\right)\cdot \mathbf{n_1} = 1 \tag{20}$$

$$in \quad \tilde{z} = \frac{L}{R_1}: \quad \tilde{P} = \alpha \tag{21}$$

$$in \quad \tilde{r} = 1: \quad \left((U_c/U)\,\tilde{f}\tilde{P}\omega_h\tilde{\nabla}\tilde{P} + (1/Pe)\,\tilde{\mathbb{D}}\tilde{f}\tilde{P}\cdot\tilde{\nabla}\omega_h\right)\cdot \mathbf{n_2} = -\gamma\sqrt{\omega_h\tilde{f}\tilde{P}} \tag{22}$$

$$(U_c/U)\,\tilde{\nabla}\tilde{P}\cdot \mathbf{n_2} = -\gamma\sqrt{\frac{\omega_h}{\tilde{f}\tilde{P}}}$$

where the following characteristic groups have been introduced

$$\alpha = \frac{P_L}{P_{atm}} \quad ; \quad \gamma = \frac{\mathbb{P}_m\mathcal{R}TP_{atm}^{-1/2}}{W_h}\frac{1}{U} \tag{23}$$

The values of ω_i^0 are pressure-dependent, as the feed to the reactor has been considered to be the product of a pre-reforming reactor in which equilibrium conditions have been reached. As the present work has been carried out considering a uniform and constant temperature and a steam-to-carbon ratio of 3, the equilibrium conditions are uniquely determined by pressure. From this condition on the membrane reactor feed, it follows that hydrogen production can only take place as a consequence of hydrogen permeation, which causes the system to move away from equilibrium. In many cases of practical interest, the rate of the reaction is often very high compared to the rate of hydrogen transport, because of the recent development of efficient catalysts active towards methane steam reforming. In these conditions, the behavior of the reactor may be seen as a sequence of equilibrium states. A more detailed description of the problem may be found in previous works by the same authors [3, 4]. The performance of the system has been studied in terms of the hydrogen yield, defined as

$$\Psi_s = \frac{H_2\ permeate\ flow\ rate}{inlet\ H_2\ flow\ rate} \tag{24}$$

This parameter, which is not upper bound by unity, gives information regarding the efficiency of both hydrogen production and hydrogen permeation.

Results and Discussion

The observation made in previous works [4] that the behavior of the reactor may be studied as a sequence of equilibrium conditions has led to an analysis of the distance from equilibrium reached throughout the reactor's volume in different operating conditions. The first finding was that there exists a portion of the reactor, in proximity of the membrane, in which the reaction is not fast enough to re-establish equilibrium conditions. This volume is the one in correspondence of which the catalyst is clearly active. The extent of this volume, which is the result of a non-equilibrium between the rate of hydrogen removal and the rate of hydrogen production and will from here on be referred to as "reaction boundary layer", depends on the operating conditions. An example of the reaction boundary layer is shown in Figure 1, in which the value of $(1-\eta)$ is shown. When equilibrium conditions are reached, this quantity is equal to zero, since $\eta = 1$. The question that naturally arises is whether the remaining volume of catalyst is completely inactive or whether its activity is sufficiently high for the reaction rate to be virtually "infinite" compared to the rate with which hydrogen is removed. Understanding this aspect is crucial for the correct design of a membrane reactor.

In order to gain more insight on this phenomenon, the behavior of the reaction boundary layer has been studied for different operating conditions. The value $(1-\eta) = 10^{-2}$ has been chosen as the limit below which the reaction is considered to be at equilibrium equilibrium. Figure 2 illustrates how the thickness of the reaction boundary layer varies as a function of, Da, Pe, γ, and α. A comparison between Panels (a) and (b), and between (a) and (c), shows that decreasing Pe and increasing γ, both of which enhance hydrogen permeation, leads to an increase in the thickness of the boundary layer The effect of variations in Pe is more evident at low pressure values, when dispersion plays a more significant role compared to convection. Panels (c) and (d) show the effects of the reaction rate: when the Da number is decreased the thickness of the boundary layer increases as there is a larger volume of reactor in which the catalyst is unable to re-establish equilibrium after the permeation of hydrogen. The influence of pressure is more evident in the range of low values of α, when the permeate flow rate is more strongly influenced by pressure. In addition, when the rate of reaction is lower, as in Panel (d), pressure has a more significant role as it enhances the kinetics of hydrogen production.

Figure 1: Equilibrium boundary layer (red = 1, blue = 0)

The problem has also been studied for different reactor geometries, reducing the size of the annular gap in which the catalyst is present. Figure 3(a) shows the permeate flow rate as a function of pressure for different reactor configurations. The initial condition considered is: $R_2/R_1 = 2$, $\mathrm{Da} = 10$, $\gamma = 1$, and $\mathrm{Pe} = 10$. As the size of the gap is reduced the inlet velocity must increase, in order to maintain a constant volumetric inlet flow rate. The values of γ and Da decrease accordingly, whereas the value of Pe increases. The results show that the dimensionless permeate flow rate tends to increase for increasing values of R_2/R_1; however, it should be noted that this quantity is normalized with respect to the inlet velocity. For a better understanding of the behavior of the reactor it is therefore important to consider to the actual permeate flow rate. Figure 3(b) shows that this quantity remains virtually constant for different gap sizes, confirming the idea that the reactor volume outside of the reaction boundary layer does not contribute to the production of hydrogen. The result is even more evident when the permeate flow rate, and corresponding yield, per unit volume are studied (Figure 4). It appears that a lower value of R_2/R_1 improves the reactor performance per unit volume of catalyst when working with a fixed membrane area, suggesting that different aspect ratios from those commonly adopted in literature ([1, 2]), often higher than 2, should be considered.

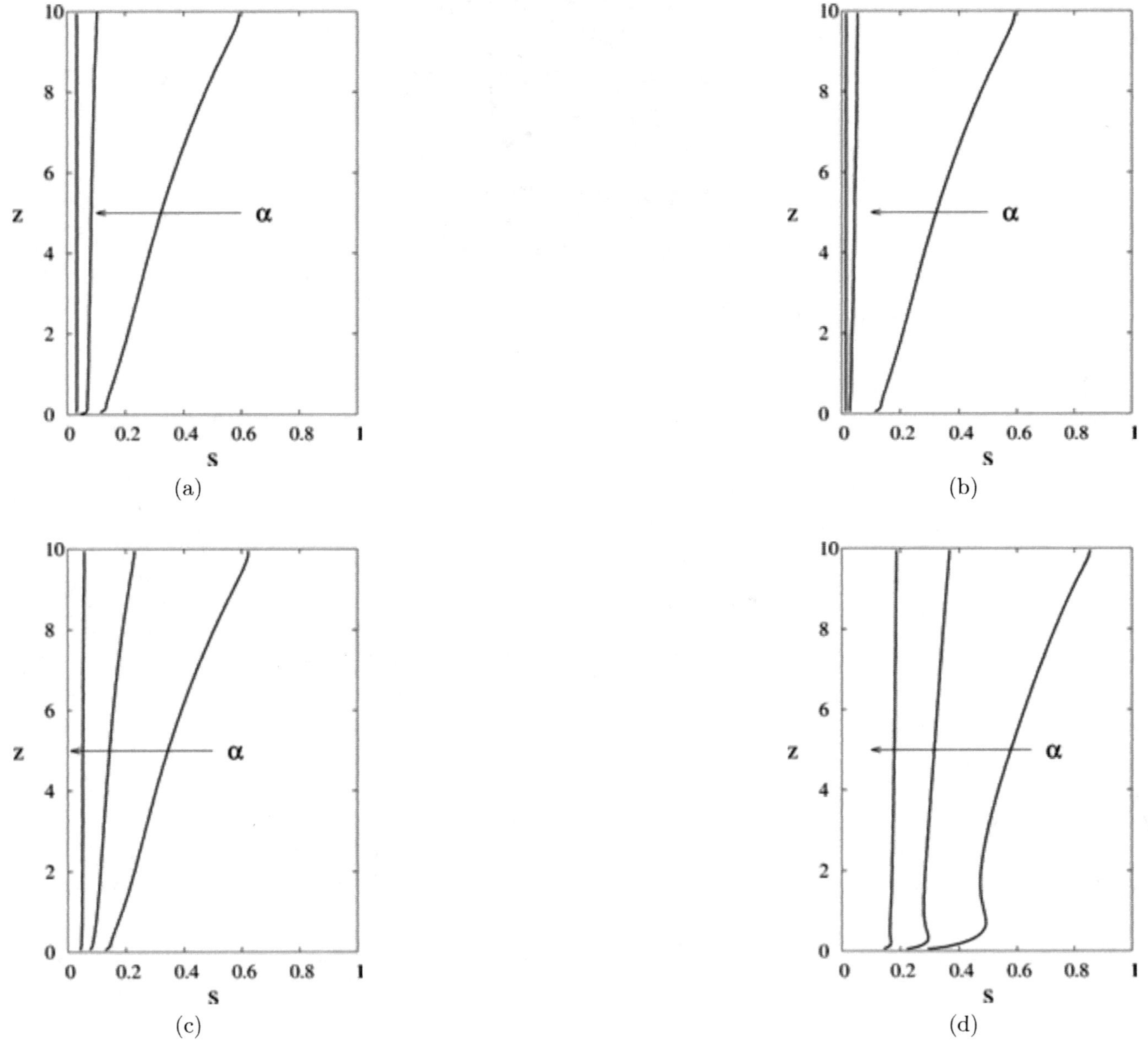

Figure 2: Level set of the function $(1-\eta)$ at 10^{-2} for $\alpha = 1;\ 10;\ 100$ and (a) $\gamma = 1$, $\mathrm{Da} = 10$, $\mathrm{Pe} = 10$; (b) $\gamma = 1$, $\mathrm{Da} = 10$, $\mathrm{Pe} = 10^2$; (c) $\gamma = 5$, $\mathrm{Da} = 10$, $\mathrm{Pe} = 10$; (d) $\gamma = 5$, $\mathrm{Da} = 1$, $\mathrm{Pe} = 10$. The axial coordinate, z, is intended in its dimensionless value and s indicates the distance from the membrane.

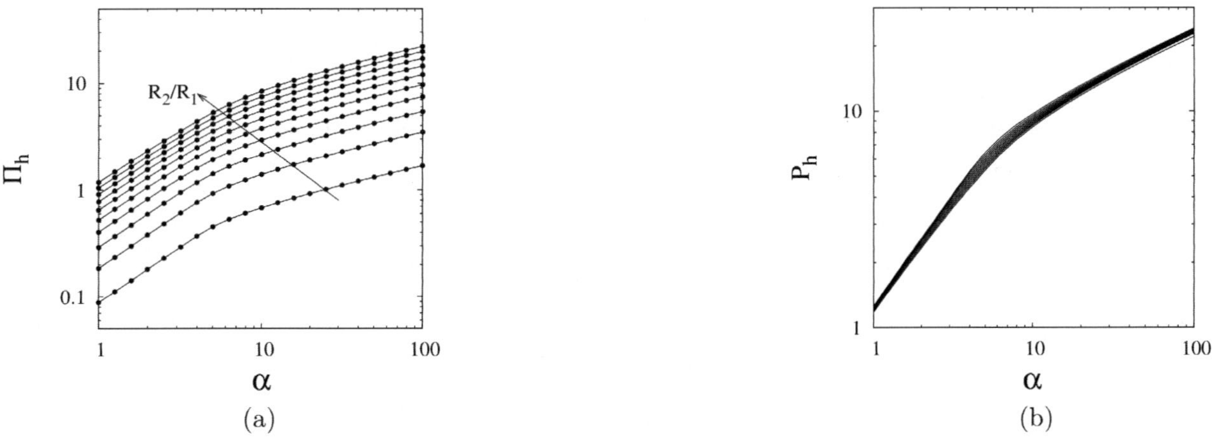

Figure 3: Dimensionless (a) and dimensional (b) permeate flow rate ($R_2/R_1 = 1.1 - 2$)

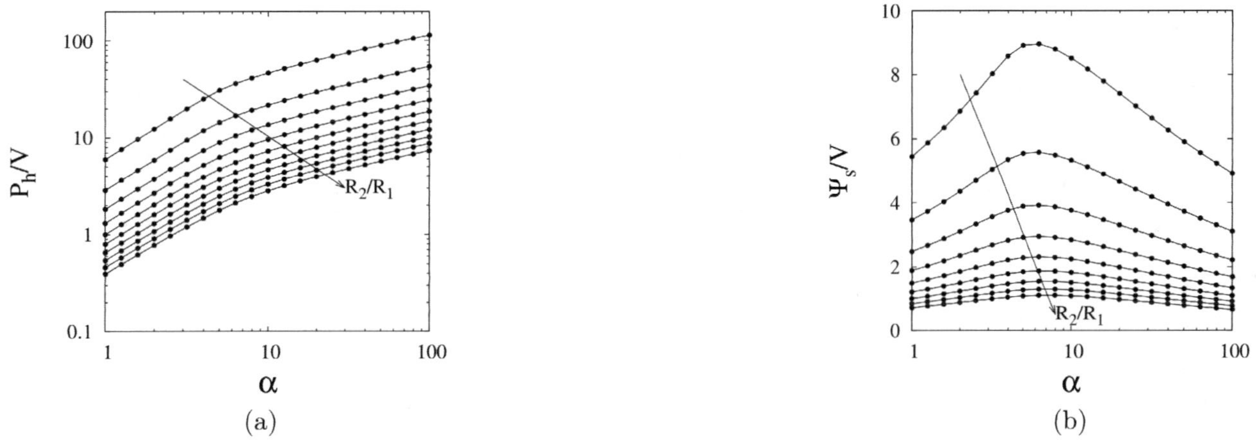

Figure 4: Permeate flow rate (a) and yield (b) per unit volume of the reactor ($R_2/R_1 = 1.1 - 2$)

Conclusions

In this work the behavior of an integrated membrane reactor for the production of pure hydrogen through methane steam reforming has been studied. The model developed considers a gaseous mixture that enters the reactor at equilibrium; as hydrogen permeates, new equilibrium conditions are restored through the methane-consuming reaction. The results show that the region identified as a reaction boundary layer gives important information regarding the reactor volume that is truly active towards hydrogen production. This is the volume in which the reaction is not sufficiently fast to compensate hydrogen permeation. Understanding its behavior is fundamental for the correct design of a membrane reactor, which, in many cases, can be be operated with a lower amount of catalyst compared to commonly adopted configurations.

Bibliography

[1] A. Iulianelli, G. Manzolini, M. De Falco, S. Campanari, T. Longo, S. Liguori, and A. Basile. H2 production by low pressure methane steam reforming in a Pd–Ag membrane reactor over a Ni-based catalyst: Experimental and modeling. *International Journal of Hydrogen Energy*, 35(20):11514–11524, oct 2010.

[2] Nobuhiko Mori, Toshiyuki Nakamura, Ken-Ichi Noda, Osamu Sakai, Akira Takahashi, Naoyuki Ogawa, Hitoshi Sakai, Yuji Iwamoto, and Tadashi Hattori. Reactor configuration and concentration polarization in methane steam reforming by a membrane reactor with a highly hydrogen-permeable membrane. *Industrial & Engineering Chemistry Research*, 46(7):1952–1958, 2007.

[3] M.A. Murmura, S. Cerbelli, and M.C. Annesini. Transport-reaction-permeation regimes in catalytic membrane reactors for hydrogen production: steam reforming of methane as a case study. *Submitted to Chemical Engineering Science*, 2016.

[4] MA Murmura, S Cerbelli, L Turchetti, and MC Annesini. An equilibrium theory for catalytic steam reforming in membrane reactors. *Submitted to Chemical Engineering Science*, 2016.

459258 Reactor Scale Modeling of Liquid Phase Ethylene Production By Hydrogenation of Acetylene in a Trickle Bed Reactor

Abstract:

Selective hydrogenation of acetylene to ethylene from the cracker effluent using Pd-based catalyst has been studied extensively due to the catalyst deactivating nature of acetylene. It acts as a catalytic poison to the polymerization catalyst used in the production of polyethylene and other valuable products from ethylene. Gas phase hydrogenation of acetylene to ethylene has been widely employed in many industries to reduce the acetylene content less than 5ppm. Liquid phase hydrogenation of acetylene is of recent interest as this process can not only avoid runaway conditions which may occur while using a gas phase but also minimize the formation of green oil. A modified reactor scale model was developed to understand and simulate the liquid phase selective hydrogenation of acetylene. A one-dimensional mathematical plug flow model was combined with a particle scale model accounting for the partial wetting of the catalyst. A sequential approach is proposed to evaluate the variables in the plug flow and particle scale model. The reactor scale model identified the local concentration along the reactor axis while the local effectiveness factor was obtained from the particle scale model for the corresponding local concentration. An algorithm was written to calculate the effluent concentration, reaction rate, and conversion in the flow direction. The model was simulated with a set of operating conditions adapted from literature. The performance of fixed-bed reactors in both trickle and upflow will be compared using appropriate correlations.

Copyright © American Institute of Chemical Engineers. All rights reserved.

468315 Influence of Solid Phase and Geometry on Liquid Velocity of a Three-Phase Bubble Column Fluid Dynamic

Abstract:

Multiphase flows have a large applicability in the industry and its studies are very important. One example of three-phase system is bubble columns operating with solid phase as catalyst in the process. In this context the hydrocracking systems are used to process heavy oil and are becoming very common at oil refineries. Its operation is done in multiple-stage each one in a different reactor. In this work, it is proposed a new geometry of bubble columns in which two different solid phases can be used as catalyst. There are two techniques used in this work: Particle Image Velocimetry (PIV) in which it is possible to measure velocity fields inside the columns without interfering with the flow; and Computational Fluid Dynamic (CFD) that allows to predict numerically the flow and compare with experimental results. In the experiments it was used Rhodamine B is a tracer for the liquid phase and the PIV is extended to the Stereo-PIV using two CCDs cameras that allows estimating the third component velocity. The liquid phase is water at 22 °C, the solid phase is FCC catalyst with diameter of 100-125 μm and air as gas phase is air injected at the bottom with 5 different flow rate between 1.5 and 10 l/min. The solid load is $4.2x10^{-4}$. The comparison of the different gas rate with solid concentration fixed is presented and discussed. Two gas rate were compared with CFD cases using the modelling of previous works and the data agreed satisfactorily.

Copyright © American Institute of Chemical Engineers. All rights reserved.

469323 A Multiphase Semi-Flow Microreactor to Study the Influence of Water on Organometalic Catalysis

Abstract:

Water is a potential solvent in homogeneous organometallic catalysis, such as Pd-catalyzed Cu-free sonogashira reaction (*a.k.a.*, heck alkynylation) [typically done in organic medium][1â€"3]. This unconventional solvent is cheap, easily accessible and non-toxic, but what makes it attractive is its ability to dissolve polar salts and products formed during the reaction. By having a device with simultaneous reaction and separation, it is expected to reduce the chemical waste produced per product formed, also known as E-factor[1]. Although water is an interesting complementary solvent, only a few efforts were placed on understanding its influence. To better understand its impact in the reaction and separation, a microfluidic semi-flow reactor has been designed and coupled with in-situ Raman and online GC.

In the present study, the role of water was analyzed using a 3D Raman spectrum chemical map of the reaction progression and species transfer. Regarding the reaction progression, the microreactor was divided in several points, and the reaction yield was acquired and treated using MatLab. For the mass transfer, both phases were trapped in a semi-flow reactor, and the species transfer were measured using image analysis tools, such as ImageJ, and Raman spectroscopy. The mass transfer measurements were validated using benchmark cases: gas-liquid and liquid-liquid.

The reaction experimental results obtained were compared with batch and flow experiments from literature (also performed in aqueous medium). In one hand, microfluidic reaction intensification experiments have shown that the reaction is often limited by mass transfer at the mili-scale. In the other hand, mass transfer experiments (indirect measurement of the molecular diffusivity) have allowed us to understand were the reaction is performed and to define the role of water in a reaction/separation process.

Bibliography

(1) Hu, C.; Shaughnessy, K. H.; Hartman, R. L. *React. Chem. Eng.* **2016**, *36*.

(2) Sabio, J. C.; Domier, R. C.; Moore, J. N.; Shaughnessy, K. H.; Hartman, R. L. *Chem. Eng. Technol.* **2015**, *38*, 1717â€"1725.

(3) Domier, R. C.; Moore, J. N.; Shaughnessy, K. H.; Hartman, R. L. *Org. Process Res. Dev.* **2013**, *17*, 1262â€"1271.

Copyright © American Institute of Chemical Engineers. All rights reserved.

456467 Original Methodology for the Screening of Catalysts Presenting High Reaction Rates at High Conversions

Abstract:

Introduction

The heterogeneous catalysts screening is the first and one of the most important steps in catalysts development. The main aim of screening is to select the catalyst that provides the best performance in terms of activity and selectivity. Thus, the experimental tests should be carried out without hydrodynamics and mass transfer limitations. The challenge appears in the case of industrial applications coupling very high reaction rates with high conversions, for which the influence of those phenomena cannot be neglected.

Objective and methodology

The aim of this work was to develop a global methodology allowing the screening of catalysts, presenting high reaction rates, at high conversion. This methodology includes both an experimental and a modeling approach, and it was validated by using the benzene hydrogenation as case study. The following figure illustrates the different steps of the methodology.

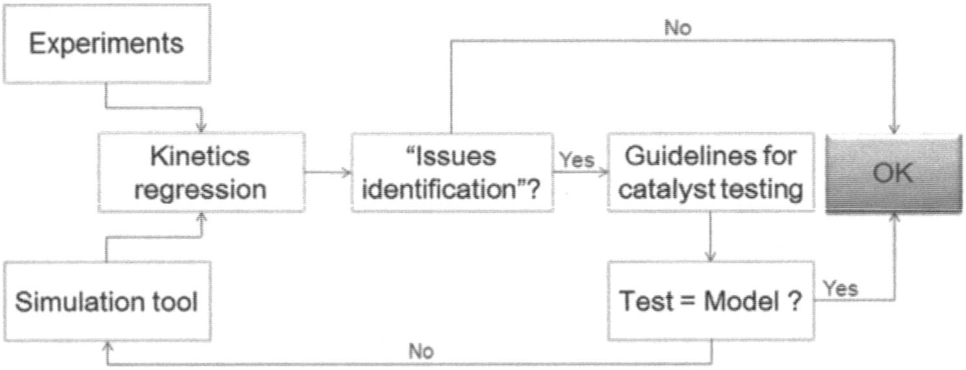

Figure 1 – Methodology description.

The first step of the methodology consist on determining the kinetics of the chemical system. Thus, experiments at different operating conditions must be carried out, and a simulation tool must be develop and coupled to an optimizer in order to regress the kinetics parameters. The simulation tool must take into account both hydrodynamics and mass transfer phenomena, and it must give access to the different species concentration along the reactor in each phase. The second step consist on analyzing the different concentration profiles in order to deduce the impact from both hydrodynamics and mass transfer phenomena on the results. If the impact is not negligible, modifications, based on numerical simulations, are made in the experimental protocol to reduce the impact from those phenomena. In case a good fit exists between model predictions and experiments, the modifications are validated and the catalyst screening can be carried out in good conditions. Otherwise, the model hypothesis must be reviewed and a new iteration must be done. This process is repeated until "convergence" achievement.

Results and discussion

The methodology described above was applied to the benzene hydrogenation reaction, which is interesting since the industrial process operates at very high conversions.

The initial experiments at different temperatures allowed the determination of an activation energy of about 6 kcal/mol for two different catalysts, which is quite low comparing to the values available in literature of about 12 kcal/mol. This may be a prove of existence of reactor limitations, internal diffusion limitations or both, even if a dilution factor of three (three volumes of inert per volume of catalyst) was used during those experiments.

A simulation tool coupling all the relevant phenomena within the reactor was developed. The kinetic law was extracted from literature and the kinetic parameters were regressed and compared with those determined in the literature. A very good agreement was found, which allows the simulation tool to be validated. The following figure presents the evolution of the different species concentration along the reactor in the gas and in the liquid phase for both catalysts.

Catalyst A Catalyst B

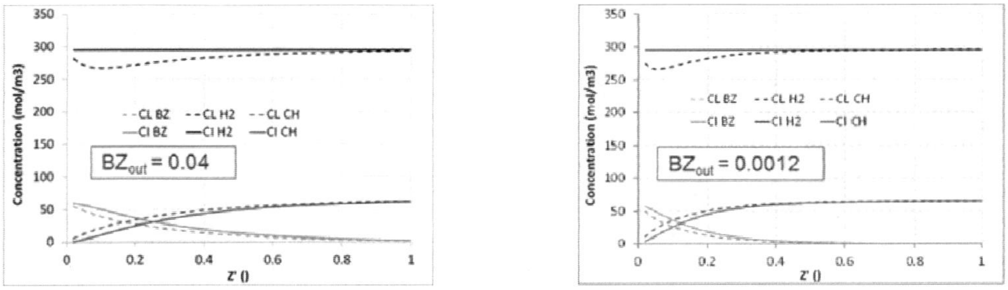

Figure 2 – Species concentration evolution along the reactor in both the gas and liquid phases at 140 °C and 30 bar for a diluted catalyst bed. BZ – benzene; CH – cyclohexane. CL – liquid concentration; CI – liquid side interface concentration

The gas-liquid mass transfer limitation is negligible with the used experimental protocol, and the liquid-solid mass transfer is unlikely to be a limiting phenomenon since other experiments carried out at different liquid rates support this hypothesis. The simulation tool considers the diffusion within the catalyst, which allows the evolution of the different species concentration within the catalyst particle to be determined at different reactor heights. This is represented in figure 3. It is possible to verify that a strong internal limitation exists at these conditions, since the value for catalyst effectiveness is between 0.3 and 0.5. This strong limitation explains the low activation energy found for both catalysts, which is around a half from the one found in literature. This result consolidates the fact that a strong limitation coming from internal mass transfer exists for those experiments. This limitation is not critical since it can be transposed to the commercial scale. In fact, this is a limitation due to the catalyst itself, and as a result, it is independent from reactor features.

Catalyst A Catalyst B

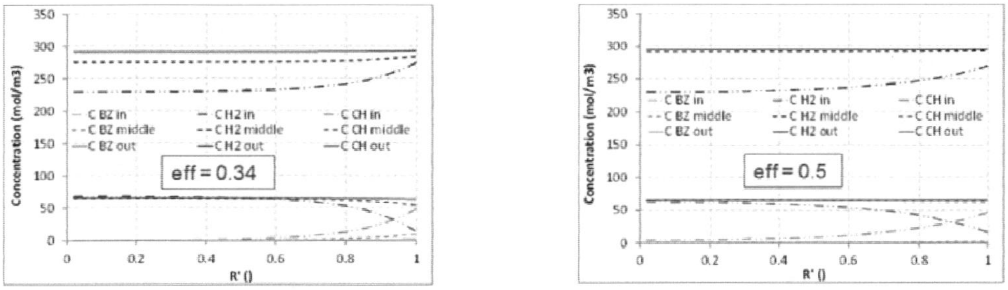

Figure 3 – Species concentration evolution within the catalyst at the reactor inlet, outlet and middle at 140 °C and 30 bar for a diluted catalyst bed. BZ – benzene; CH – cyclohexane.

The experimental protocol consist on testing those catalysts near industrial conditions in small fixed-bed reactors and by using a dilution factor of three. This allows a better performance from the thermal and from the mass transfer point of view, since the reactant converted quantity per unit of reactor length is lowered when comparing to the non-dilution case, while the transfer phenomena is kept constant. The simulation tool was used in order to generate results for the non-dilution case. The results are presented in the following figures.

Catalyst A Catalyst B

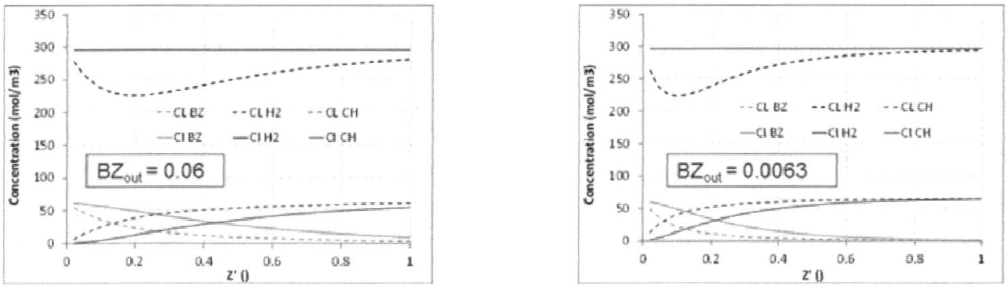

Figure 4 – Species concentration evolution along the reactor in both the gas and liquid phases at 140 °C and 30 bar for a non-diluted catalyst bed. BZ – benzene; CH – cyclohexane. CL – liquid concentration; CI – liquid side interface concentration

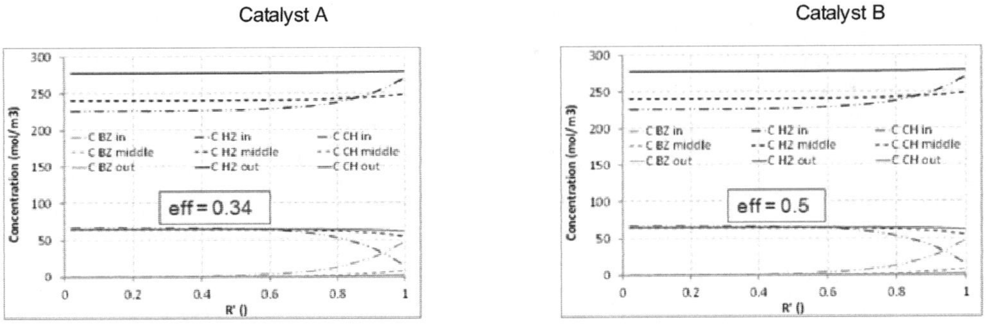

Figure 5 – Species concentration evolution within the catalyst at the reactor inlet, outlet and middle at 140 °C and 30 bar for a non-diluted catalyst bed. BZ – benzene; CH – cyclohexane.

It is possible to verify that the dilution effect is very effective mainly for catalyst B, which is actually the one presenting the highest activity. In fact, an important gas-liquid mass transfer limitation exists for benzene and hydrogen. This result is very important, since, even if both catalyst can be differentiated under undiluted conditions, the performance of the best catalyst might not be sufficiently high to justify a further development if tested under undiluted conditions. Thus, it is important to couple both the experimental and the numerical approaches in order to perform the catalyst screening in an efficient way.

The simulation tool was finally used to analyze the results at 90 °C. The results are shown in figures 6 and 7. At these conditions the gas-liquid mass transfer limitation is mainly observed for hydrogen. The results between the diluted and the non-diluted modes are quite similar, and are slightly better in the case of the undiluted bed. This result was expected since hydrogen is a reaction inhibitor under these conditions. Thus, a stronger gas-liquid mass transfer limitation on hydrogen allows the reaction to be accelerated, which was not observed at 140 °C because in this case, the benzene was affected by the gas-liquid limitation as well. The internal diffusion limitation is lower in this case, since the catalyst effectiveness is between 0.6 and 0.8. The diffusion model was validated by comparing the catalyst effectiveness of catalyst B to the experimental value of 0.85 at 80 °C. A simulated value of 0.86 was determined, which allows to consolidate the simulation tool validation.

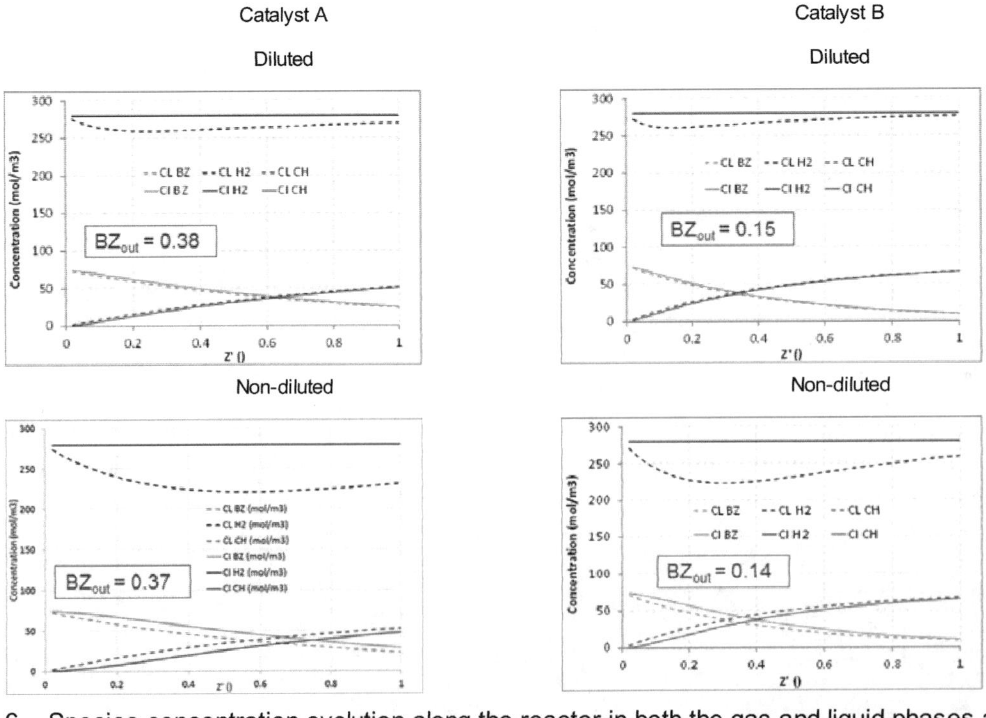

Figure 6 – Species concentration evolution along the reactor in both the gas and liquid phases at 90 °C and 30 bar for both a diluted and a non-diluted catalyst bed. BZ – benzene; CH – cyclohexane. CL – liquid concentration; CI –

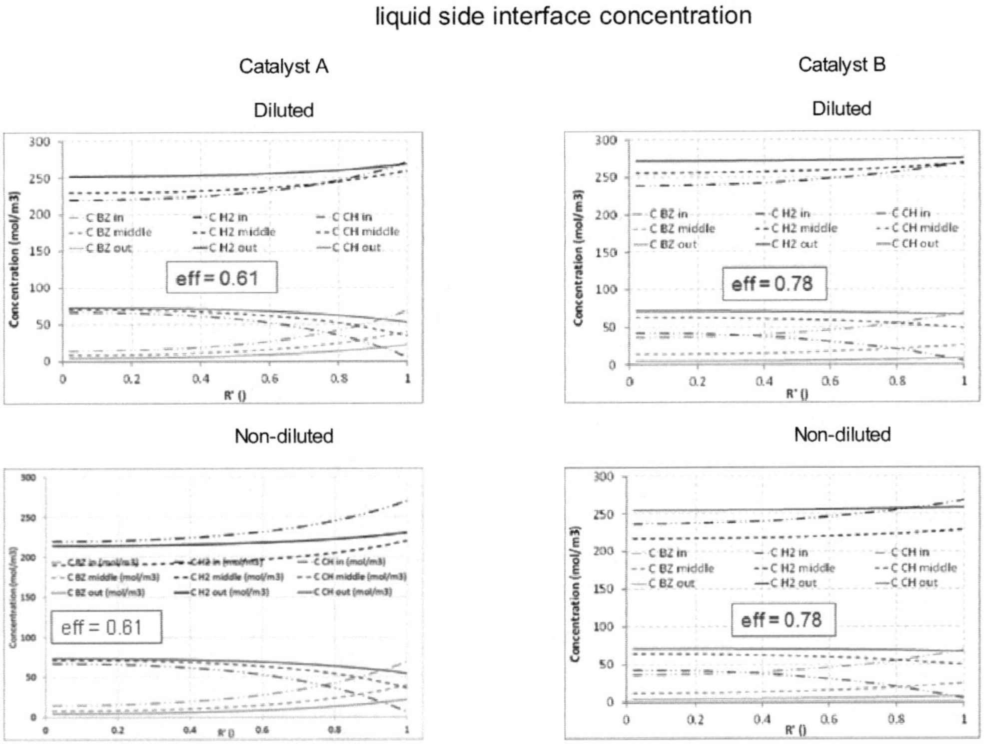

Figure 7 – Species concentration evolution within the catalyst at the reactor inlet, outlet and middle at 90 °C and 30 bar for both a diluted and a non-diluted catalyst bed. BZ – benzene; CH – cyclohexane.

Conclusion

An original methodology allowing the screening of catalysts, presenting high reaction rates, at high conversion was developed and successfully applied to benzene hydrogenation. The results show that it is quite important to perform catalyst dilution at these conditions, even if special care must be taken in order to avoid some reactor by-passing and other hydrodynamics problems. Another important conclusion is the critical character of the numerical approach, that must be used in order to avoid high experimental costs or erroneous interpretations from experiments. In fact, the numerical approach permits a better understanding of what actually happens during the experiments. A good example is what happened when two different catalysts were numerically tested by using both the dilution and non-dilution experimental protocols at the same operating conditions. Even if the results are quite similar, it does not mean that gas-liquid mass transfer limitations are negligible in the non-diluted catalytic bed.

456999 Gas Phase Mixing Behavior of Gas-Liquid Upflow in a Moving Catalytic Packed/Expanded Bed Hydrotreating Reactor Using Advanced Gas Tracer Technique

Abstract:

Gas-liquid upflow moving catalytic packed bed reactor are widely used in industries for hydrotreating of feeds with a higher level of contaminants including heavier feeds. In these reactors spent catalyst are replaced periodically by adding fresh catalyst at the top and removing spent catalyst from the conical bottom which supports catalyst bed. While the catalyst moves downwards systematically, gas and liquid phase move upwards. The catalyst is removed in small increments once a week. The other times the reactor operates in upflow packed or expanded bed condition. The problem associated with these reactors is maldistribution, which causes hotspots, sintered carbon deposition and reduces expected conversions. To address such issues, detailed studies to enhance the understanding of the hydrodynamics and mixing behavior of gas in this reactor are still required. The lab scale reactor is developed based on the combination of hydrodynamic and geometrical similarity with the industrial reactor. This reactor consists of plenums and catalyst bed section. In this work, a new methodology has been developed using advanced gas tracer technique to remove the external dispersion due to additional volume (sampling line, the top section of the reactor) and to deconvolute gas mixing characteristics of the gas phase in catalyst bed from the gas mixing characteristic of the whole reactor. The gas tracer system consists of the gas analyzer (TCD), in-house developed gas-liquid separator, water pump, air pump, sampling line and helium as a tracer.Â As part of the methodology, the helium is injected at three location (inlet, below the catalyst bed, above the catalyst bed) and sampled at the top of the reactor. These injection-sampling networks characterize residence time distribution (RTD) of various section of the reactor. Dispersion coefficient (Dg) quantifies the mixing characteristics of gas in catalyst bed section, and it is obtained using developed model for this parts. The RTDâ€™s gave the indication of ideal behavior of plenums and not much deviation from plug flow behavior in the catalyst bed. Hence, the plenums are modeled as ideal CSTR-PFR and catalyst bed as Axial Dispersion Model (ADM). Dispersion parameter estimation is done using convolution and parameter regression with the experimental data. We focus in this work on the catalyst bed section, which is a Plexiglas column of 11 inches (ID) and 30-inch height, filled with extrudate catalyst of 3mm diameter till 24-inch height. The measurements were conducted at superficial liquid (water) velocity of 0.017 cm/sec to 1.78 cm/sec and superficial gas (air) velocity of 1.27 cm/sec to 8.8 cm/sec. At low liquid superficial velocity, it is seen that the bed behaves as packed bed or slightly expanded bed and Dg increases with increasing gas superficial velocity. At higher liquid superficial velocity the bed acts as extended bed and Dg decreases with increasing gas superficial velocity. The Dg decreases with increasing liquid superficial velocity at constant gas superficial velocity. These kind of information are essential at industrial scale, to improve the performance of the real plant reactor. In this presentation, results and findings are discussed.

Copyright © American Institute of Chemical Engineers. All rights reserved.

458405 Improvement of a Kinetic Model for the Hydroconversion of an Atmospheric Residue in a Slurry Semi-Batch Reactor

Abstract:

Introduction

In the development of heavy oil upgrading technologies, such as slurry hydroconversion, the kinetic model plays an important role together with the hydrodynamic and mass transfer models, Jansen et al. (2014a). However, kinetic modeling of heavy oil hydroconversion is a complex task. Hydroconversion reactions described in the literature have various levels of detail, but a common factor: all reactions are considered non-reversible and light products are formed from heavy products, Jansen et al (2014b). Previously, using a batch reactor, Nguyen et al. (2013) proposed a kinetic model for atmospheric residue hydroconversion (see Fig. 1) wherein residue consumption rate depends on residue and dissolved hydrogen concentrations. In the present work, we improve Nguyen's kinetic model by introducing a HDS reaction and modeling the outlet gas dynamic in a semi-batch reactor. Kinetic parameters and the mass transfer coefficient (k_La) were re-estimated using a nonlinear least-squares regression on the product yields and the outlet gas dynamic. Moreover, we explain the difference between the values of kinetic parameters obtained in batch and semi-batch reactors by a free radical kinetic model taking into account coke formation and we compare also these two reactors in terms of hydrogen transfer.

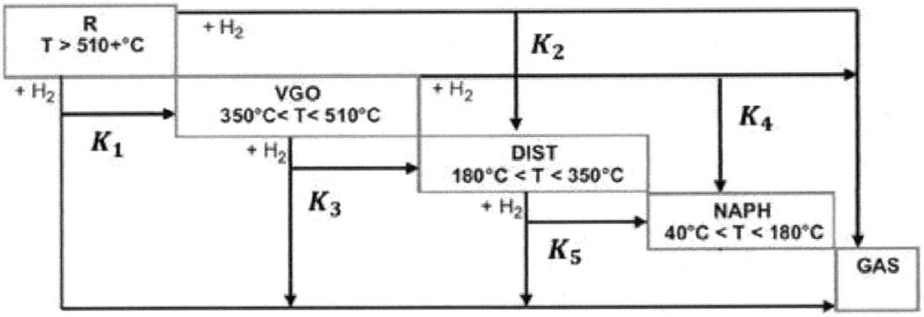

Figure 1. Kinetic scheme of Nguyen's model for atmospheric residue hydroconversion

Experimental setup

The properties of the atmospheric residue feed used for our hydroconversion experiments have already been reported by Nguyen et al (2013). Hydroconversion was performed in the presence of a Mo octoate precursor (supplied by Shepherd Chemical Co.). Experiments were conducted in a semi-batch pilot equipped with an autoclave PARR (250 cm^3), a condenser and a cold separator (see Fig. 2). For all hydroconversion experiments, the catalyst concentration (600 ppm Mo), stirring rate (900 rpm), gas/liquid ratio (60/40, about 100 g of feed), heating ramp (3°C/min), pressure (15MPa) and H_2 inlet flow (50 NL/h) were kept constant. Kinetic experiments were carried out at different temperatures (420, 430°C) and reaction times (0, 30, 60 and 120 min). The reaction time was measured from the instant the reaction temperature was reached, defined as time zero. The gas outlet flow was measured by Coriolis flow meter and analyzed by a micro-GC using two modules PoraPLOTU (8m x 0.32 mm ID) and OV1 (8m x 0.15mm ID). The total liquid products from each experiment were analyzed by simulated distillation (D-7169 ASTM). To study the hydrodynamic behavior of the gas phase, residence time distribution (RTD) experiments were carried out with N_2 as tracer under non reacting conditions (below 300°C).

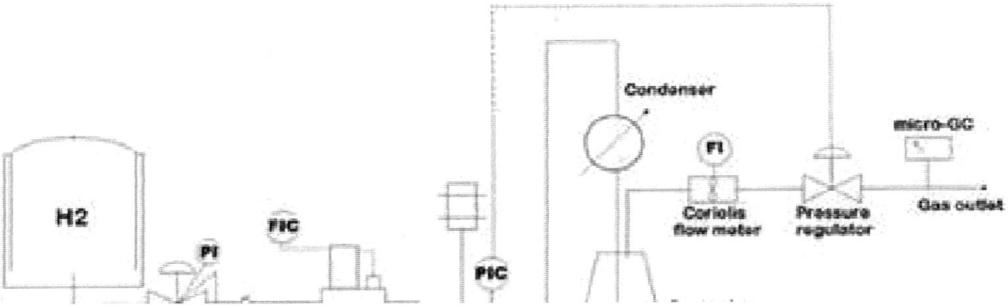

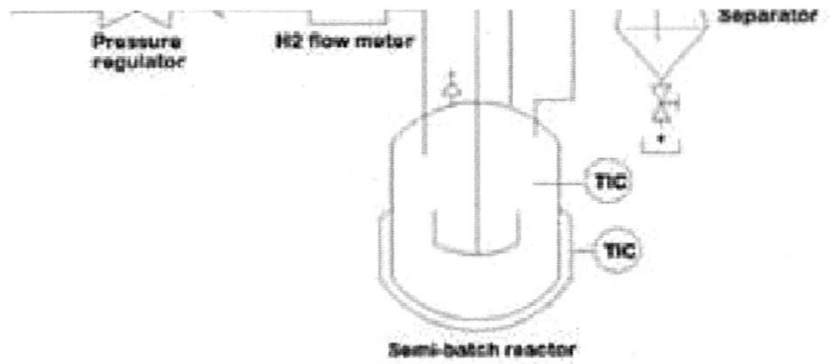

Figure 2. Experimental setup of hydroconversion of atmospheric residue

Results and discussion

The RTD experimental results show that the reactor and separator can be modeled as a Continuous Stirred-Tank Reactor and the rest of the semi-batch pilot (condenser and tubes) as a Plug Flow Reactor. The hydrodynamic model of the gas phase was coupled to the mass balance of each unit of the semi-batch pilot. For the HDS reaction, we assumed that the organic sulphur in the feed is converted into H_2S by a first order non-reversible reaction. Furthermore, the HDS reaction stoichiometry is considered assuming the HDS of 4,6-DMDBT by partial hydrogenation pathway. In Fig. 3, we can observe that the improved kinetics of Nguyen's model coupled to the semi-batch pilot model (hydrodynamic, mass transfer and thermodynamic models) predicts the product total masses and the outlet gas dynamic to a large extent.

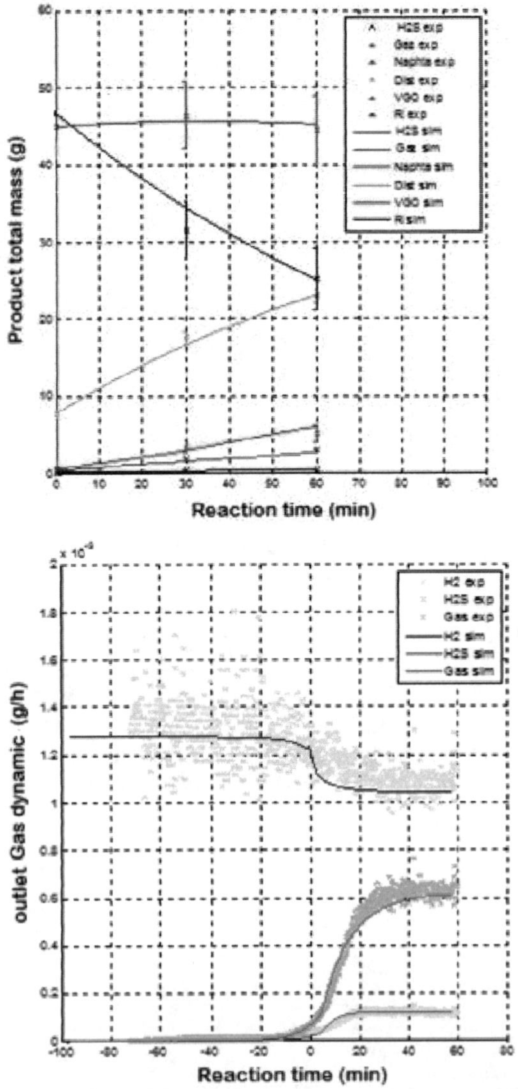

Figure 3. Comparison of experimental (Exp) and Simulated (Sim) product total mass (left) and outlet gas dynamic (right) at 430°C – 60min

The original and improved rate constants of Nguyen's model at 430°C are presented in Table 1. We can see significant differences between the original and improved rate constants of Nguyen's model. This is due to the configuration of the batch and semi-batch systems in terms of hydrogen transfer. In the case of the batch reactor, the low value of k_La suggests a strong limitation of hydrogen transfer. This causes high rate constants to be obtained with Nguyen's model. For the semi-batch reactor, the hydrogen transfer from gas to liquid is improved (high value of k_La), the reaction system is under kinetic control and the bulk hydrogen concentration is close to equilibrium hydrogen concentration. Without limitation on hydrogen transfer, the rate constants estimated from semi-batch reactor data correspond to the pseudo-intrinsic rate constants of Nguyen's model.

Table 1. Comparison of original and improved rate constants of Nguyen's model at 430°C

Reactor	Parameters	K_1	K_2	K_3	K_4	K_6	K_{HDS}	k_La
Batch	**Original**	$1.19*10^4$	$2.15*10^4$	$6.50*10^3$	$1.25*10^3$	$1.95*10^5$	-	$5.6*10^{-5}$
Semi-batch	**Improved**	$7.05*10^{-8}$	$1.59*10^{-7}$	$4.84*10^{-8}$	$6.85*10^{-9}$	$1.55*10^{-7}$	$1.87*10^{-7}$	$4.1*10^{-2}$

* Rate constants: $m^3mol^{-1}s^{-1}$, k_La: s^{-1}

The differences between the original and improved rate constants of Nguyen's model can be also explained from a purely kinetic view point. Nguyen's kinetic model can be modified to create a free radical kinetic model. In Fig. 4, reaction 1 of Nguyen's model is broken down into 3 reactions: cracking (1), hydrogenation (2) and condensation (3).

Nguyen's Kinetic Model

$$R + H_2 \xrightarrow{K_1} VGO + GAS$$

$$\frac{dC_R}{dt} = -K_1 C_{H2} C_R$$

$$\frac{dC_{VGO}}{dt} = K_1 C_{H2} C_R$$

Free Radical Kinetic Model

$$R \xrightarrow{k_{11}} R^* + GAS^* \xrightarrow[+H_2]{k_{12}} VGO + GAS$$

$$R^* \xrightarrow{k_{13}} \text{Coke Precursors}$$

$$\frac{dC_R}{dt} = -k_{11} C_R$$

$$\frac{dC_{R^\cdot}}{dt} = k_{11} C_R - k_{12} C_{H2} C_{R^\cdot} - k_{13} C_{R^\cdot} = 0$$

$$C_{R^\cdot} = \frac{k_{11}}{k_{12} C_{H2} + k_{13}} C_R$$

$$\frac{dC_{VGO}}{dt} = k_{12} C_{H2} C_{R^\cdot} = \frac{k_{11} k_{12}}{k_{12} C_{H2} + k_{13}} C_{H2} C_R$$

Figure 4. Break down of Nguyen's kinetic model into a free radical kinetic model

For the free radical kinetic model, we suppose that the radical species are produced and consumed at the same rate. Comparing the rate constants of both kinetic models, we obtain:

For Residue, R:

$$\boldsymbol{K_1} = \frac{k_{11}}{C_{H2}}$$

For Vacuum Gas Oil, VGO:

$$\boldsymbol{K_1} = \frac{k_{11} k_{12}}{k_{12} C_{H2} + k_{13}}$$

If the free radical compound hydrogenation rate (due to dissolved hydrogen concentration and/or catalyst activity) is very much higher than the coke formation rate () or the coke formation rate is intrinsically slow () then the rate constant of Nguyen's model K_1 is approximated by

$$k_{12} C_{H2} \gg k_{13}$$

$$k_3 = 0$$

$$\boldsymbol{K_1} = \frac{k_{11}}{C_{H2}}$$

The rate constants of Nguyen's model depend on the cracking rate constants and the bulk hydrogen concentration. The cracking rates are relatively similar for the batch and semi-batch systems, since they are predominately a function of temperature (see Table 2). Thus, the rate constant of Nguyen's model for a specific reaction system (K_i) depends mainly on the bulk hydrogen concentration in that system. This approach will be extended to the whole kinetic scheme and all the compounds in a further study.

Table 2. Cracking constants of Nguyen's model for the batch and semi-batch reactors at 430°C

Reactor	k_{11} (h^{-1})	k_{21} (h^{-1})	k_{31} (h^{-1})	k_{41} (h^{-1})	k_{61} (h^{-1})
Batch	0.13	0.23	0.07	0.013	0.21
Semi-batch	0.18	0.25	0.11	0.024	0.50

Conclusions

A new set of kinetic parameters for atmospheric residue hydroconversion have been estimated. The accuracy of kinetic reaction data was improved by modeling the outlet gas dynamic. A HDS reaction was introduced into Nguyen's kinetic model. In comparison to Nguyen's kinetic study performed in a batch reactor, the results of the present study do not show any limitation of the transfer of hydrogen from gas to liquid in the semi-batch reactor. A high value of k_La was obtained demonstrating the high efficiency of this kind of apparatus for estimation of the chemical kinetic constants. Using a free radical kinetic model approach, we show how Nguyen's kinetic constants depend on dissolved hydrogen concentrations that are different in the batch and semi-batch reactors.

References

Jansen, T., Dimitri, G., Gotteland, D., Bacaud, R., Lacroix, M., Ropars, M., Chantal, L., Christophe, G., Tayakout-Fayolle, M. (2014a). Characterization of a continuous micro-scale pilot unit for petroleum residue hydroconversion with dispersed catalysts: Hydrodynamics and performances in once-through and recycling mode. *Chemical Engineering Journal., 253, 493-501.*

Jansen, T., Dimitri, G., Ropars, M., Lacroix, M., Christophe, G., Tayakout-Fayolle, M. (2014b). Simulation of Petroleum Residue Hydroconversion in a Continuous Pilot Unit Using Batch Reactor Experiments and a Cold Mock-Up. *Ind. Eng. Chem. Res, 53 (41), 15852-15861.*

Nguyen, TS, Tayakout-Fayolle, M., Ropars, M., Christophe, G. (2013). Hydroconversion of an atmospheric residue with a dispersed catalyst in a batch reactor: Kinetic modeling including vapor-liquid equilibrium. *Chem. Eng. Sci., 94, 214-223.*

459139 Experimental Characterization of Two Phase Interfacial Mass Transfer in Novel Milli-Scale Reactors

Abstract:

Experimental characterization of two phase interfacial mass transfer in novel milli-scale reactors

Over the last decade, researchers are striving to make chemical process more cost efficient and safer. The ACS-GCI Pharmaceutical Round table[1] have identified process intensification as key research area for sustainable chemical industry. Moreover, demand for development of novel concept in continuous multiphase reaction system is identified which can minimize waste and energy use while maximising use of renewable raw materials. One approach to achieve this goal is based on the use of micro-scale reactors. As micro-scale reactors handle only a very small amount of substance at a time, they are capable of producing high quality goods from minimum resources with efficient management of hazardous material. However, micro-scale quantity production is not economical for the industrial scale.

The limitations of micro-structure systems can be overcome by novel milli-scale reactors based on porous micro-structured foams. The idea is to use highly porous micro-structured inserts (metal foam structures) for plug flow reactors in the millimetre range such that high throughput is provided at a comparable small pressure drop. A huge potential arises from open cell metal foams, due to their high specific surface area combined with high porosity, typically larger than 70%. The induced turbulence in the micro-structured geometry leads to an enhanced heat and mass transfer which is e.g. shown by Hutter et.al.[2] for single phase systems. However, the potential of metal foams to enhance interfacial transport processes in multiphase flow application is not yet addressed.

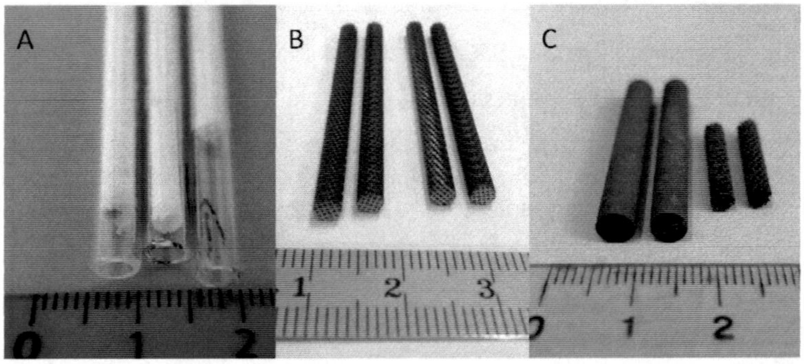

Figure 1: Representation of different micro-structured foams studied. A) Silicon scaffolds B) Metal foam by 3DFD technique C) Metal foams by SLS.

In order to optimise the performance of novel micro-structured foam reactors for multiphase systems, it is important to first gain a better understanding on the behaviour of this reactor for multiphase systems and the underlying physics of the transport processes. This can be achieved by investigating the effect of various design parameters like porosity and ligament shape, cell size etc. on interfacial mass transfer in the multiphase system. In this study, we achieve the same by evaluating a variety of porous structures (Figure 1) ranging from random silica scaffolds (manufactured by NCL, India) to custom made metal foam structures by techniques like 3-dimentional fibre deposition (3DFD, manufactured by VITO, Belgium) as well as Selective Laser Sintering (SLS, manufactured by Inspire, Switzerland). In this study, we apply a combination of PIV (Particle Image Velocimetry) and LIF (Laser Induced Fluorescence), where PIV is used to assess the fluid velocity field while LIF is used to measure the concentration distribution of a species. Consequently, the PIV data is used to characterize the turbulence intensity induced by the porous structure, while the LIF data is used to characterize the two phase flow distribution. The combination of both techniques facilitates a detailed study of interfacial mass transfer as well as its physical mechanism.

Moreover, the effect of design parameters on the mass transfer in liquid-liquid flows is quantified by physical and chemical methods. The physical method consists of two systems: i) Toluene- Water with acetone as transfer species and ii) 1-Butanol- Water with succinic acid as transfer species. On the micro-scale, the first system is represented by Taylor flow, while 1-butanol and water results in stratified flow due to the low interfacial tension. Moreover, these flow systems allow to study the influence of surface energy and resulting flow patterns in porous medium on mass transfer. As chemical method, we have investigated fast chemical reactions that occur at the interface of two phase systems; i) Hydrolysis of N-butyl formate (NBF) with sodium hydroxide and ii) Neutralization reaction of trichloroacetic acid with sodium hydroxide. The main aim of studying different two phase systems is to examine the effect of interfacial area and mass transfer coefficient independently. In addition, the performance of the novel porous structures is assessed by comparing the obtained results with the performance of milli-scale packed bed reactors under identical operating conditions.

We have observed that design parameters play vital role in determining interfacial transport processes in novel micro-structured foam reactors. We hypothesize that the porous structures lead to a promotion of merging and break up of slugs which has major impact on mass transfer through constant change in shape and size of the interfacial area as well as the renewed boundary layers. It is also observed that metal foam structures give enhanced mass transfer performance at negligible pressure drops compared to conventional milli-scale packed bed reactors.

The results of this study will not only help in the development of sustainable chemical process but will also increase the fundamental understanding of multiphase processes in a porous media.

(1) Jiménez-González, C.; Poechlauer, P.; Broxterman, Q. B.; Yang, B. S.; Am Ende, D.; Baird, J.; Bertsch, C.; Hannah, R. E.; Dell'Orco, P.; Noorman, H.; Yee, S.; Reintjens, R.; Wells, A.; Massonneau, V.; Manley, J. *Org. Process Res. Dev.* **2011**, *15*, 900–911.

(2) Hutter, C.; Allemann, C.; Kuhn, S.; von Rohr, P. R. *Chem. Eng. Sci.* **2010**, *65*, 3169–3178.

Copyright © American Institute of Chemical Engineers. All rights reserved.

455534 Process Intensification of Aqueous Biphasic Hydroformylation of Mid Chain Olefin 1-Octene in aÂ Jet Loop ReactorÂ CreatingÂ Large Interfacial Areas

Abstract:

Process Intensification of Aqueous Biphasic Hydroformylation of Mid-Chain Olefin 1-Octene in a Jet Loop Reactor Creating Large Interfacial Areas

Helge Warmeling, Arno Behr and Andreas J. Vorholt*

Chair of Technical Chemistry, Department of Biochemical and Chemical Engineering

Technical University of Dortmund, Dortmund, DE-44227, Germany

The hydroformylation is an atom-economic, homogenously catalyzed reaction where an alkene is converted under syngas pressure to form linear and branched aldehydes following figure 1. Most commonly the reaction is catalyzed by rhodium complexes modified with steric phosphine ligands to increase reactivity and selectivity. [1]

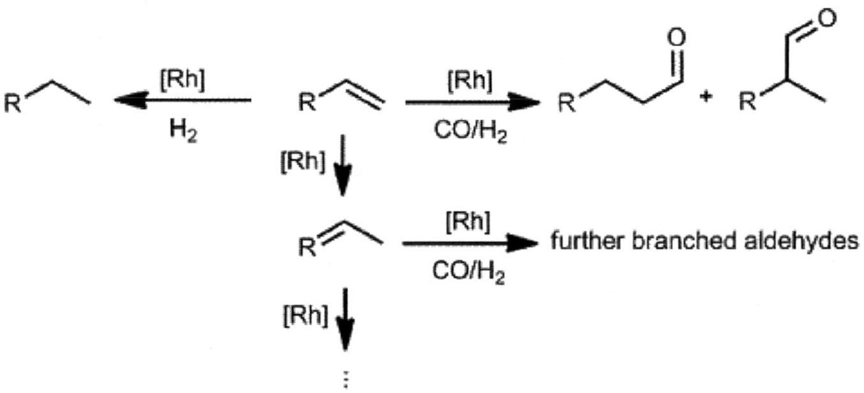

Fig. 1. Reaction scheme of the hydroformylation of an terminal alkene with prominent side reactions like isomerization and hydrogenation to alkanes.

Due to the high market price and even higher price volatility of rhodium, an efficient catalyst recycle is eminent for an economic process. A possible solution is to immobilize the rhodium in an aqueous second phase by the use of the ligand triphenylphosphine trisulfonate (TPPTS). Substrates diffuse into the aqueous catalyst phase and react at the catalyst complex. Products can subsequently be separated by a simple decantation step in an organic phase. This recycle concept works exceptional for short-chain substrates but has its limitations for alkenes with C_{6+} because the substrate solubility in the catalyst phase decreases significantly with increasing chain length. While the majority of studies working on this limitation heavily focus on the use of chemical additives like surfactants, co-solvents or phase transfer promoters to increase solubility or interfacial surface only seldom procedural approaches were investigated. [2] This work therefore focuses on an additive-free procedural approach to enhance the reaction rates of the hydroformylation of mid-chain olefin 1-octene in a phase contacting jet loop reactor.

Jet loop reactors (JLR) are a promising tool in the intensification of catalytic multiphase reactions as they provide large interfacial areas due to intense mixing and therefore excellent mass transfer characteristics. The overcoming of mass transfer limitations can be of significant economic interest for large scale industrial applications. In contrast to other phase contacting reactor concepts the jet loop reactor also offers a high energy efficiency which makes it especially interesting in process intensification projects. [3]

The principle of operation is based on the partial withdrawal and reinjection of the reaction medium through a jet nozzle as can be seen in figure 2. The recycle stream sucks gaseous media via impulse transfer into the jet stream and disperses it into the liquid phases to achieve a fine gas-liquid-liquid distribution. The small gas bubbles coagulate in an outer reactor ring called the riser and add additional circulation momentum following the principle of an airlift pump. The reactor itself features no moving parts and is hence particularly suitable for high pressure applications like the hydroformylation.

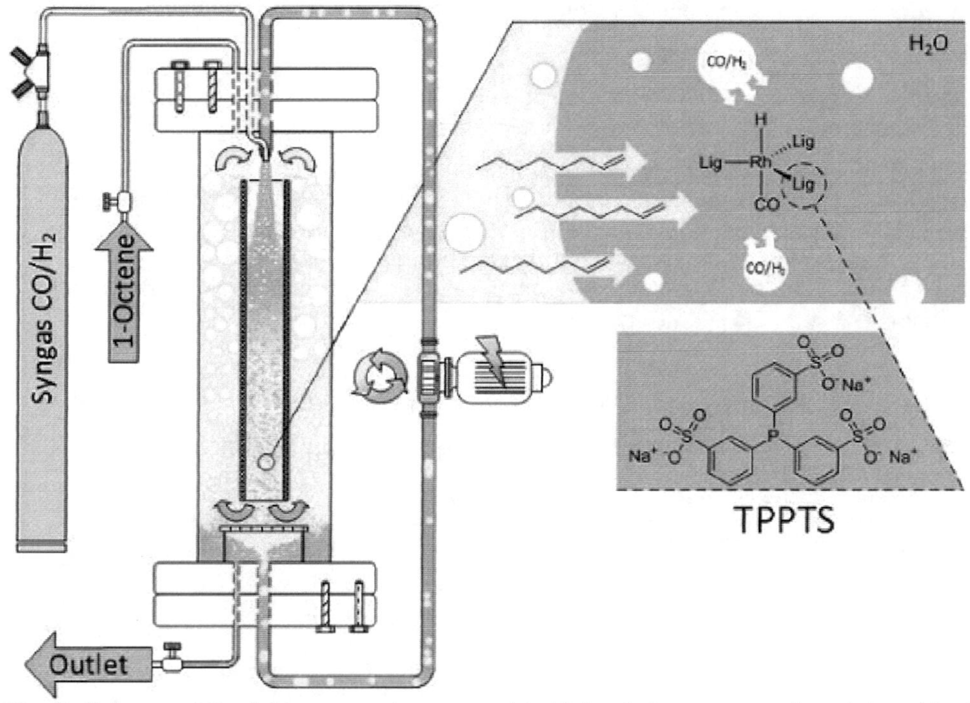

Fig. 2. Scheme of the jet loop reactor concept in biphasic homogenously catalyzed hydroformylation of terminal alkenes. Intense mixing and therefore high internal interfacial area is created through a partial external circulation of the reaction medium through a nozzle.

In this work the jet loop reactor is compared to a standard reactor type of the chemical industry, the stirred tank reactor (STR). Both reactor types were build up on miniplant scale and were successively optimized. Reaction conditions like temperature, pressure and syngas composition were kept in standard industrial range while catalyst and ligand concentration were investigated in preliminary tests. Surprisingly the reaction rates showed an inhibitory effect in regards of catalyst concentration and the highest catalytic activity and productivity were found at very low rhodium and ligand concentrations (c_{Rh}=0.00025 kmol/m$^3_{aq}$) with peak turn over frequencies (TOF) of up to 12500 in the JLR and 4500 in the STR. Due to increased leaching the investigation was continued with higher catalyst concentrations of c_{Rh}=0.0005 kmol/m$^3_{aq}$ and a ligand to rhodium ratio of 10.

While the jet loop reactor reached high turnover frequencies up to 5500 h^{-1} and space time yields (STY) up to 55.5·10^{-2} mol/m³h the stirred tank reactor only achieved lower TOFs of up to 2200 h^{-1} but with a satisfying STY of up to 44·10^{-2} mol/m³h as can be seen in Fig. 3. With a
3-pitched blade gassing stirrer the performance of the STR was improved up to a TOF of 2600 h^{-1} with a space time yield of 55.5·10^{-2} mol/m³h. The results show that in contrast to the STR the JLR is able to achieve excellent phase dispersion even with relatively small amounts of aqueous catalyst phase which results in a higher catalytic activity achieved with about a tenth of the specific energy input (0.27 kW/m³ compared to 2.2 kW/m³).

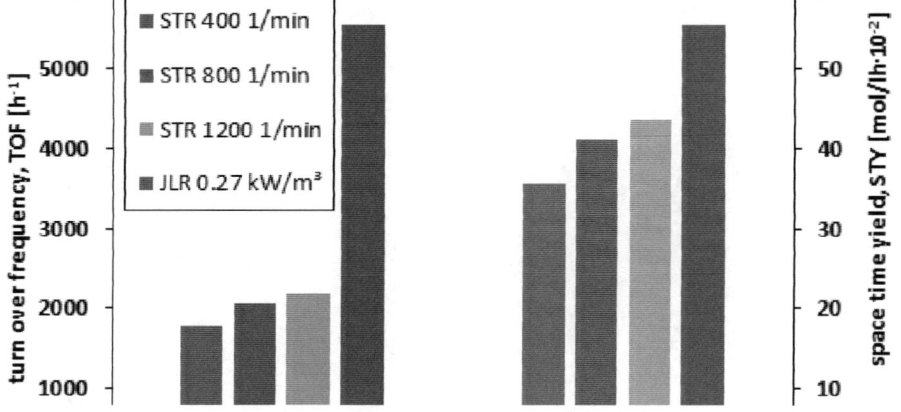

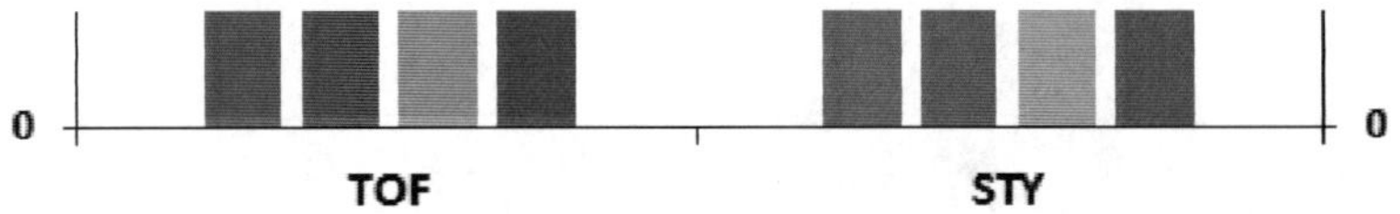

Fig. 3. Catalytic activity (TOF) and productivity (STY) of the aqueous biphasic hydroformylation of 1-octene in a jet loop reactor compared to a stirred tank reactor at different stirrer speeds.

The results point out the potential of the reactor setup compared to standard reactor types and also academic studies using additives published in the literature. The direct connection between specific energy input, created interfacial area and reaction rate is currently subject of investigation. Based on calculations of the mass transfer coefficients for all compounds, mass transfer resistance as the limiting step can be excluded due to the relatively slow reaction rates compared to substrate diffusion. The authors propose the creation of surface near film volume with elevated substrate concentration as the cause of rate acceleration.

References

[1] Franke R, Selent D, Börner A. Applied hydroformylation, *Chem. Rev.* 2012;11;5675-5732

[2] Obrecht L., Kramer P., Laan W. Alternative approaches for the aqueous–organic biphasic hydroformylation of higher alkenes, *Catal. Sci. Technol.* 2013;3;541-551

[3] H. Warmeling, A. Behr, A. Vorholt, Jet loop reactors as a versatile reactor set up - Intensifying catalytic reactions: A review, Chem. Eng. Sci., 149 (2016) 229-248

450201 Pressure Drop and Liquid Hold up in an Open-Structure Random Packed Column with Counter Current Flow

Abstract:

Pressure drop and liquid hold up in an open-structure random packed column with counter current flow

L. Pezzi Martins Loane[1], J.C. Schouten[1] and J. van der Schaaf[1,*]

[1]Eindhoven University of Technology, P.O. Box 513,

5600 MB Eindhoven, The Netherlands

*Corresponding author: J.Vanderschaaf@tue.nl

Structured foam packing has been proven to enhance mass and heat transfer and decrease pressure drop due to the intrinsic open pore nature of a foam[1]. The high accessible surface area that this type of packing offers makes it easy for fluids to access the catalyst. Furthermore, the heat generated by, or needed for the reaction is efficiently redistributed into the main stream, avoiding thus the formation of hot spots. Previous studies focus on a continuous large piece of foam of the same size as the reactor dimensions. These large blocks of foam are difficult and labor intensive to incorporate in a column which gives space to a new type of foam packing, the open-structure random packing, OSRP. OSRP are essentially small pieces of foam with a specific shape and size. Their attractiveness lies in the fact that it can be easily dumped into a column, acting as a packed bed. The liquid hold up, the flooding point, and the pressure drop of an OSRP reactor are important for assessing its performance in comparison with other packings. It is widely accepted that pressure drop in packed columns with counter current flow can be estimated using the Ergun equation[2]. In this equation the bed porosity and the superficial velocity of the gas phase are of great importance together with the kinetic loss terms, to estimate pressure drop. However for the OSRP not only the bed porosity will influence the pressure drop but also the foam porosity.

Experimental work was performed to determine pressure drop, flooding points and liquid hold up for gas superficial velocities up to 0.34 m/s and liquid superficial velocities up to 0.03 m/s for 3 different types of OSRPs.

The experiments show that the gas superficial velocity does not influence the liquid hold up, contrary to what is observed for more conventional packings. The liquid hold up is determined only by liquid superficial velocity and a simple correlation is proposed. Another remarkable difference with conventional packings is that the flooding pressure drop values remain virtually constant for one type of foam for all investigated liquid flow rates. Additionally, the dry and wet OSRP packing have different kinetic loss terms. With the experimental liquid hold up as input, the pressure drop is predicted within 15% accuracy. A revised pressure drop correlation for OSRPs is proposed.

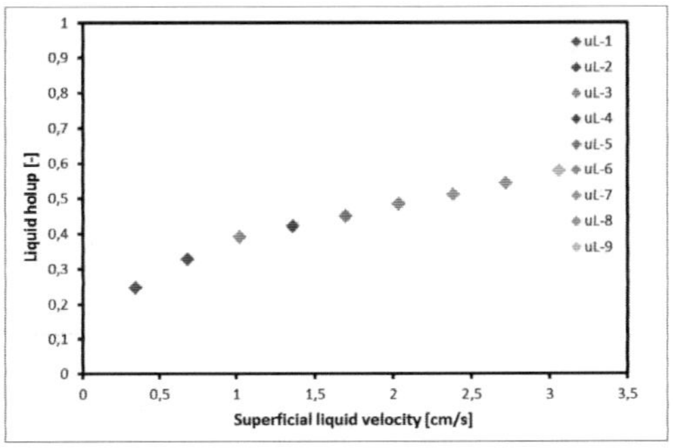

Figure SEQ Figuur * ARABIC 1: Liquid hold up versus Superficial gas velocity, for foam 'A'.

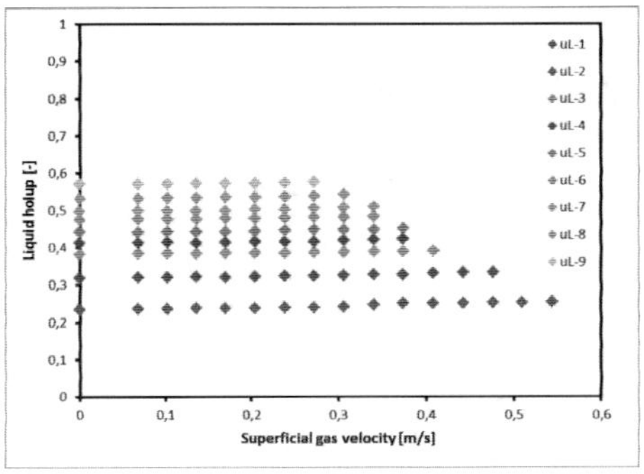

Figure SEQ Figuur * ARABIC 2: Liquid hold up versus superficial liquid velocity, for foam 'A'.

[1] C.P. Stemmet, J. van der Schaaf, B.F.M. Kuster, J.C. Schouten, Solid foam packings for multiphase reactors: Modelling of liquid holdup and mass transfer, Chem. Eng. Research and design 84 (2006) 1134-1141

[2] S. Ergun, A. Orning, Fluid Flow through randomly packed columns and fluidized beds, Industrial and Engineering Chemistry (1949), 1179-1184

Copyright © American Institute of Chemical Engineers. All rights reserved.

475724 High-Efficiency Electrochemical Conversion of CO_2 to Value Added Fuels Current Status, Challenges, and Future Directions

Abstract:

The electrochemical reduction of carbon dioxide (CO_2) into liquid fuels especially coupling with the intermittent renewable electricity offers a promising means of storing electricity in chemical form, which reduces the dependence on fossil fuels and mitigates the negative impact of anthropogenic CO_2 emissions on the planet. Although converting CO_2 to fuels is not in itself a new concept, the field has not substantially advanced in the last 30 years primarily because of the challenge of discovery of structural electrocatalysts and the development of membrane architectures for efficient collection of reactants and separation of products. In this presentation, I would summarize recent advances in electrocatalytic conversion of CO_2 and presents the challenges and future directions in producing value-added fuels.

Currently, copper, silver and tin are the most extensively scrutinized electrocatalysts, which can convert CO_2 to CO, CH_4, C_2H_4, alcohols and carbolic acid in aqueous electrolytes. A long-lasting challenge in the electrosynthesis of fuels from CO_2 is to achieve a durable electrochemical performance, an equally important characteristic of an electrode as the activity and selectivity, but it has not yet been clearly addressed in the open literature. Hypotheses for the origin of the deactivation of CO_2 reduction electrode included the formation of intermediates like oxide, hydroxide and carbon species, as well as the accumulation of a soluble CO_2 reduction product in the electrolyte causing a remarkable decrease in the selectivity. Hori et al. examined the deactivation of Cu electrode and postulated that the degradation of the cell performance was originated from the deposition of iron and zinc contaminants in the electrolyte on the active surface of the catalyst. The impurity alone, however, cannot explain the degradation since the ultrapure electrolyte coupled with pre-electrolysis was used, in which the degradation was still observed. I will discuss

approaches to simultaneously achieve high efficiency and high stability of an electrocatalyst for the conversion to CO_2 to value-added fuels.

Copyright © American Institute of Chemical Engineers. All rights reserved.

456332 Design and Economic Analysis of IGCC Power Plants with CO_2 Capture

Abstract:

Power producing industry has largely relied on coal as a fuel source and will continue to depend on it for coming decades under the current situation. However, in order to curtail CO_2 emissions while keeping high efficiency and economic feasibility of the processes is a key challenge. Post combustion coal-fired power plants usually have lower capital costs and cost of electricity without CO_2 capture. On the other hand, pre-combustion plants tend to be less expensive if the current CO_2capture systems are to be added. This analysis suggests that pre-combustion technology for the power generation can be an attractive option if carbon capture and storage (CCS) technology is implemented on a large scale. In this study IGCC process has been simulated in order to analyze the overall plant output with CO_2 capture using a physical solvent. Three design schemes have been developed to analyze the overall plant performance and economics. In order to ensure a fair evaluation of analysis, a consistent and transparent methodology has been followed for all the cases. First two cases use the water gas shift reactions scheme with sour shift catalysis process. The resulted syngas free of CO_2 can either be combusted by using air or O_2. The first case uses air as an oxidant for burning H_2 and the combustor temperature is controlled by air as well. In the second case, O_2 is used as an oxidizing agent in the H_2 combustion. However, combustion temperature control is achieved by recycling the captured CO_2. In the third model, WGS reactor has been removed. The syngas composed of CO and H_2 is sent directly for combustion which makes it similar to the oxy-fuel combustion process. In this way, captured CO_2 can be recycled back to control the combustor temperature in 2nd and 3rdcase. The results showed that the net power generation capacity and efficiencies calculated for case 1, case 2 and case 3 are 375Â MWe, 354 MWe, 388 MWe and 35.87%, 33.89%, 37.15% respectively. Finally, an economic analysis has been performed which shows thatÂ the total investment cost

(TIC) and operation and maintenance (O&M) required per unit (MW) of electricity generated for case 1, case 2 and case 3 is (2.97, 3.40, 2.64) and (0.2486, 0.2706, 0.2242) Mâ‚¬â€™s respectively. With the highest net power generation capacity and least TIC and O&M cost requirements, case 3 has been evaluated as the most feasible design. If the sole purpose of IGCC power plant is power generation with CO_2 capture, case 3 design offers a cost reduction of 8.17% and 6.59% in the total investment and O&M cost requirements respectively as compared to the base case design.

Copyright © American Institute of Chemical Engineers. All rights reserved.

468074 Optimization of the CCS Network Under Harsh Geographical Conditions, Using GIS and MINLP Modeling

Abstract:

With the evermore increasing amount of greenhouse gas emissions every year, various efforts to mitigate its amount have earned worldwide attention. Carbon capture and storage, a practical technology to achieve this task, has been studied from different viewpoints, namely capture, transportation, and utilization of CO2. While the CO2 capture technology has been studied very deeply, studies for practical implementation of the CO2 transportation infrastructure has received relatively less attention. Although the CO2 capture technology is the most critical from an economic point of view, research on the topic of economic transportation of CO2 is essential for commercialization of the CCS technology, since mass amounts of it have to be transferred across long distances. For mass amounts of CO2 transport, pipelines are deemed as the most economical option. Previous studies concerning this problem have introduced the method of pipeline network optimization regarding the geographical conditions to minimize the cost of nation-wide transport. These studies used a combination of geographical information systems and mixed-integer program (MIP) modeling to optimize the network. However, geographical conditions vary greatly according to the region of pipeline installation, and linear pipeline models are inadequate in rigorously evaluating the cost of pipelines. In this study, a pipeline network for transporting mass amounts of CO2 in harsh geographical conditions was optimized, using mixed-integer nonlinear programming (MINLP). A commercial GIS software ArcGIS was used to obtain the least cost paths from the CO2 sources to the reservoirs, then the cost factor for penalizing the pipelines built on a certain plain was segmented into more rigorous levels. This data was transferred into GAMS, an optimization software and designed into a MINLP problem, using a weight-based cost model for evaluating the pipeline costs, for more accurate calculations. A case study in South Korea was conducted to test the effect of rigorous geographical penalization and cost model usage. The results of the study showed that a more rigorous network can be obtained by implementing more segmented penalty factors corresponding to the varying elevation. Also, it showed that rather than avoiding the rough landscape entirely, it may be more economical to gather the emitted CO2 from nearby sources into local junction nodes, then transport it through mainstream pipelines to the reservoir.

Copyright © American Institute of Chemical Engineers. All rights reserved.

AMINE-BLENDS SCREENING AND CHARACTERIZATION FOR CO_2 POST-COMBUSTION CAPTURE

*Abdullah Al Hinai, Dang Viet Quang, and Mohammad R.M. Abu Zahra**
Department of Chemical and Environmental Engineering, Masdar Institute of Science and Technology, P.O. Box 54224, Masdar City, United Arab Emirates
** Tel.: +97128109181 - email: mabuzahra@masdar.ac.ae*

Abstract

This work studied the performance of novel amine blends in order to find a blend of alkanolamines, which possess all the required properties for efficient CO_2 capture. 2-(Butylamino)ethanol (2BAE), a secondary amine was selected to blend with three tertiary amines including 2-(dimethylamino)ethanol (2DMAE), N,N,N′,N′-Tetramethyl-1,3-propanediamine, (TMPDA), and N-methyldiethanolamine (MDEA). Different blends with a total amine concentration of 30 wt% were tested at 40°C using a flow micro reaction calorimeter to determine CO_2 loading and heat of absorption. Obtained results reveals that blending 2BAE with tertiary amines results in the increase of absorption heat compared to single tertiary amines. 2BAE/TMDAP blend has the highest CO_2 loading; 5, 10, and 15 wt% of 2BAE blending with 25, 20 and 15 wt% of TMDAP absorbed 1.09, 0.82 and 0.74 mole CO_2/mole amine, respectively. Results from CO_2 loading, absorption heat, and absorption kinetics revealed that blend containing 5wt% of 2BAE with TMDAP 25wt% or 2DMAE 25wt% provides the best CO_2 absorption performance.

Keywords: CO_2 capture; Absorption; CO_2 loading; Post-combustion; Amine blend;

Introduction

Concerns about global warming and climate change have urged scientists to find more effective methods to reduce anthropogenic CO_2 emission. Global CO_2 emission has increased significantly in recent years [1]. This causes great increase in atmospheric CO_2 concentration, which has reached 402 ppm in August 2016 [2]. Fossil fuel accounts for about 90% of total global CO_2 emission in which fossil fuel fired power plants are considered to be major sources of CO_2 emission [1]. Many efforts have been taken to reduce CO_2 emission in which carbon capture and storage (CCS) has been considered to be the most viable technology for CO_2 reduction. Three different CO_2 capture concepts have been introduced including pre-combustion capture, oxy-fuel combustion capture and post-combustion in which post-combustion capture is the feasible because it can be easily retrofit to the existing fleet of fossil fuel fired power plants [3]. The post combustion capture using amine solvents is the most major technology for CO_2 capture, which involves a chemical absorption of CO_2 into aqueous amine like ethanolamine [4-7]. Many works have been carried out to demonstrate the feasibility of this technology at pilot scale [8-11]. The application of this technology, however, is hindered by high energy consumption for the regeneration process, degradation, and corrosion. However, capturing CO_2 using amine solvent is still attractive due to its high absorption kinetics, which allow this technology to capture CO_2 from various sources including diluted CO_2 gas stream.

Various amine compounds including primary, secondary, and tertiary amines have been investigated for CO_2 capture application in which monoethanolamine (MEA), diethanolamine (DEA), and methyldiethanolamine (MDEA) are the most widely used amines for CO_2 absorption . The primary and secondary alkanolamines have a high reaction rate upon CO_2 absorption, which begins with the formation of a zwitterion (1) that then transfers a proton to an amine to form carbamate (2). Their maximum possible CO_2 loading is 0.5 mole of CO_2/mole of amine (two amine molecules react with one CO_2 molecule). At high pressure, the carbamate could be hydrolysed to form a free amine and bicarbonates, and then the free amine will again react with CO_2 (3).

$$R_1R_2NH + CO_2(aq) \leftrightarrow R_1R_2NH^+COO^- \quad (1)$$

$$R_1R_2NH + R_1R_2NH^+COO^- \leftrightarrow R_1R_2NH^+ + R_1R_2NCOO^- \quad (2)$$

$$R_1R_2NCOO^- + H_2O \leftrightarrow R_1R_2NH + HCO_3^- \quad (3)$$

For tertiary amines, the maximum CO_2 loading could be 1 mole of CO_2 per mole of amine (one molecule of amine reacts with one molecule of CO_2) with the formation of bicarbonates (4). Although the tertiary alkanolamines have higher CO_2 loading, they have low reactivity with CO_2 in comparison with the primary or secondary amines.

$$R_1R_2R_3 + CO_2(aq) + H_2O \leftrightarrow R_1R_2R_3NH^+ + HCO_3^- \quad (4)$$

Conventionally, aqueous MEA solution 30 wt% has been widely investigated and used for CO_2 capture [12]. However, a MEA based CO_2 capture process is very energy intensive due to the high energy requirement for MEA regeneration and this prevents this conventional post-combustion process from being used for industrial applications [13]. Recently, great effort has been spent to find alternative amines that may help increase CO_2 absorption capacity and reaction kinetics while lower regeneration energy. Amine system associated with high CO_2 absorption capacity, low regeneration energy, and high reaction kinetics results in lower cost of CO_2 capture. It is not easy to find a single amine that satisfies all above requirements. Our previous study did a survey on a large number of primary, secondary and tertiary amines bearing different functional groups and found that the primary and secondary amines have fast reaction kinetics but high absorption heat, whereas tertiary amines possess low absorption heat but slow reaction kinetics [14]. To take the advantage of both amine types, an idea to combine these amines in an amine blend system has been considered.

Following a previous experimental work which shows 2-(Butylamino)ethanol (2BAE) as a suitable amine for carbon capture as it is a secondary amine with a CO_2 loading of 0.73 mole CO_2/mole amine and a heat of absorption of -74.38 kJ/mol of CO_2. Tertiary amines including 2-(dimethylamino)ethanol (2DMAE), N,N,N′,N′-Tetramethyl-1,3-propanediamine, (TMPDA), and N-methyldiethanolamine (MDEA) have high CO_2 loading but slow reaction kinetics have been selected to blend with 2BAE. For a total of 30 wt% of aqueous amines, the 2BAE concentration is varied from 0 to 15 wt% while the respective tertiary amine contributes the remaining part to make a total of 30 wt%. CO_2 absorption performance and heat of absorption were investigated using a flow Micro Reaction Calorimeter (URC) to evaluate the effectiveness of the blends.

Experimental

Material

2-(Butylamino)ethanol, 98% (2BAE), 2-(dimethylamino)ethanol, 98% (2DMAE), N,N,N′,N′-Tetramethyl-1,3-propanediamine, 99% (TMPDA), N-methyldiethanolamine, 99% (MDEA) were purchased from Sigma Aldrich and used without further purification. CO_2 gas (99.9 %) was supplied by Gulf Industrial Gases CO. L.L.C.

Determination of CO2 loading and absorption heat

A description of the equipment can be found in the previous study [14]. A flow micro-reaction calorimeter that was supplied by Thermal Hazard Technology (UK) was used to evaluate CO_2 loading and the absorption heat for aqueous amine solutions (enthalpy of reaction/ΔH (kJ/mole of CO_2)) (Figure 1 (a)). The calorimeter is controlled by URC control software from Thermal Hazard Technology. In a typical experiment, a sample-containing cell was placed inside the calorimeter, and the temperature was set to 40 °C and atmospheric pressure. As the experiment starts, the variation in power (mW) over time (s) is recorded by the software. When the signal is stable, CO_2 gas is introduced into the cell at a rate of 0.5 ml/min (± 0.01 ml/min). The CO_2 flow is controlled by a mass flow controller, and the flow passes through a desiccant column to remove moisture before entering the sample cell. Because the CO_2 reaction with the aqueous amine solution is exothermic, the power signal initially increases, and then the signal decrease becomes constant when the reaction is complete (equilibrium) (Figure 11 (b)). The CO_2 loading (mole CO_2/mole amine) was calculated by dividing the mole of CO_2 absorbed per the mole of lean amine as described in equation 5.

$$CO_2 \text{ loading} = \frac{(m_1 - m_0)/44}{m_0 * (\frac{C1}{M1} + \frac{C2}{M2})} \quad (5)$$

Where, m_0 is the mass of lean amine and m_1 is the mass of CO_2 loaded amine that is the mass difference between the analysis cell containing amine after CO_2 adsorption and empty cell. C1 and M1 are concentration and molecular weight of amine 1 in the blend and C1 and M2 are concentration of amine 2 in the blend.

The integral heat Q (in kJ) was determined by using URC analysis software. The heat of absorption (ΔH - kJ/mole of CO_2) was calculated by dividing integral heat Q by the moles of absorbed CO_2. The absorption heat is proportional to amount of CO_2 absorption, therefore accumulative absorption heat evolved is used to monitor the CO_2 absorption progress and CO_2 absorption kinetics.

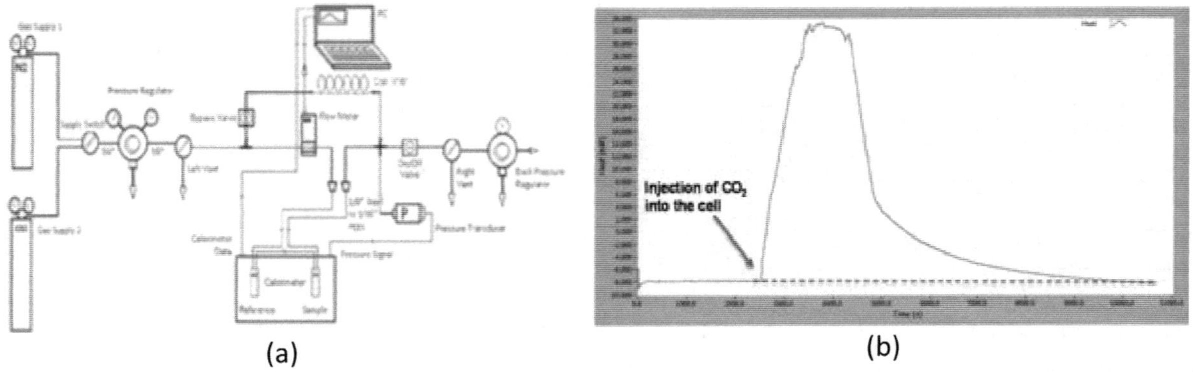

Figure 1. Schematic of the flow Micro Reaction Calorimeter (a) and the data obtained from this equipment (b) [14].

Results and discussion

Previous study screened numerous amines and indicated that 2BAE is a promising candidate for CO_2 capture [14]. Aqueous 2BAE 30 wt% solution has CO_2 loading of 0.73 mole CO_2/mole amine and absorption heat of 74.4 kJ/mole CO_2. This is better than conventional MEA solution at the same concentration, which has CO_2 loading of 0.69 mole CO_2/mole amine and absorption heat of 85 kJ/mole CO_2. Three tertiary amines with different properties have been selected for blending with 2BAE including TMDAP, 2DMAE, and MDEA. These tertiary amines have lower absorption heat compared to 2BAE, however, their CO_2 loading ranges from lower to higher than that of 2BAE. TMDAP, 2DMAE, and MDEA have CO_2 loading of 1.32, 0.77, and 0.52 mole CO_2/mole, respectively. CO_2 loading of TMDAP is much higher, while CO_2 loading of MDEA is lower than that of 2BAE, meanwhile CO_2 loading of 2DMAE is similar to that of 2BAE. Experimental CO_2 loading and absorption heat of 2BAE blended with 3 tertiary amines are presented in Table 1.

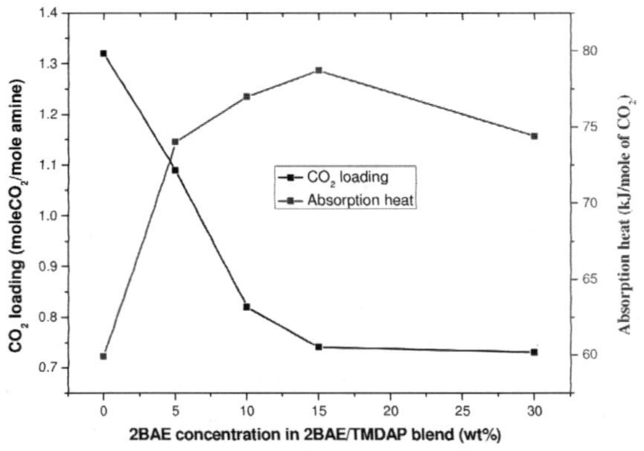

Figure 2. Variation of CO_2 loading and absorption heat as a function of 2BAE concentration in 2BAE/TMDAP blend

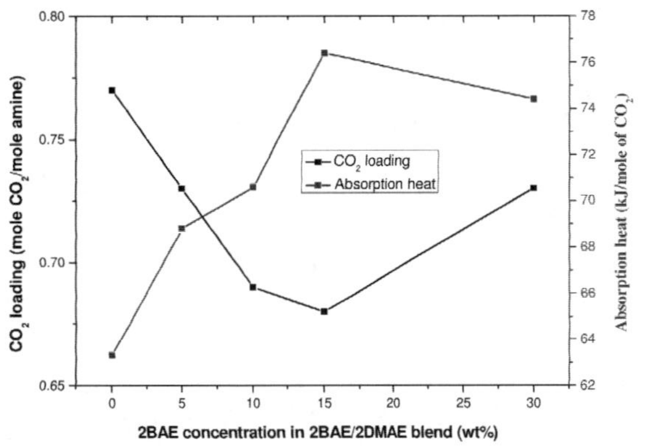

Figure 3. Variation of CO_2 loading and absorption heat as a function of 2BAE concentration in 2BAE/2DMAE blend

For 2BAE/TMDAP blend, the absorption heat of blend increased, whereas the CO_2 loading decreased compared with that of single TMDAP amine. CO_2 loading of TMDAP reduced from 1.32 to 1.09 mole CO_2/mole amine as 5 wt% of TMDAP was replaced by 2BAE and the CO_2 loading of blend is equal to that of single 2BAE as the 15 wt% of TMDAP was replaced with 2BAE (50:50 weight ratio). The result reveals that blend is efficient only when the 2BAE/TMDAP ratio is below 50%. Figure 2 shows variation of CO_2 loading and absorption heat as a function of 2BAE concentration in the blend and indicates that the best combination between 2BAE and TMDAP is 5 wt% of 2BAE and 25 wt% TMDAP in total 30 wt% amine in solution.

Table 1. CO_2 loading and absorption heat of amine blends. Tests were conducted at 40 °C using pure CO_2 at atmospheric pressure

	2BAE/TMDAP Blends				
2BAE concentration (wt%)	30	15	10	5	0
TMDAP concentration (wt%)	0	15	20	25	30
Heat of absorption (kJ/mole CO_2)	74.4	78.7	77.0	74.0	59.9
CO_2 loading (mole CO_2/mole amine)	0.73	0.74	0.82	1.09	1.32
	2BAE/2DMAE Blends				
2BAE concentration (wt%)	30	15	10	5	0
2DMAE concentration (wt%)	0	15	20	25	30
Heat of absorption (kJ/mole CO_2)	74.4	76.4	70.6	68.8	63.3
CO_2 loading (mole CO_2/mole amine)	0.73	0.68	0.69	0.73	0.77
	2BAE/MDEA Blends				
2BAE concentration (wt%)	30	15	10	5	0
MDEA concentration (wt%)	0	15	20	25	30
Heat of absorption (kJ/mole CO_2)	74.4	70.9	65.1	58.8	52.8
CO_2 loading (mole CO_2/mole amine)	0.73	0.59	0.57	0.55	0.52

Blending 2BAE with 2DMAE also results in increasing absorption heat and decreasing CO_2 loading of 2DMAE, but the change is not significant; 5 wt% of 2BAE in the blend caused reduction in CO_2 loading of 2DMAE from 0.77 to 0.73 mole CO_2/mole amine and raising absorption heat to 68.8 from 63.3 kJ/mole CO_2. Figure 3 reveals that replacing from 5 to 10 wt% of 2DMAE with 2BAE in 2DMAE 30 wt% solution will result in the best blend in term of CO_2 loading and heat of absorption. Contrast to TMDAP and 2DMAE, the addition of 2BAE in to MDEA result in increasing CO_2 loading of the blends at all 2BAE/MDEA ratios. The higher CO_2 loading of blend in comparison with single MDEA is contributed by 2BAE, whose CO_2 loading is higher than MDEA.

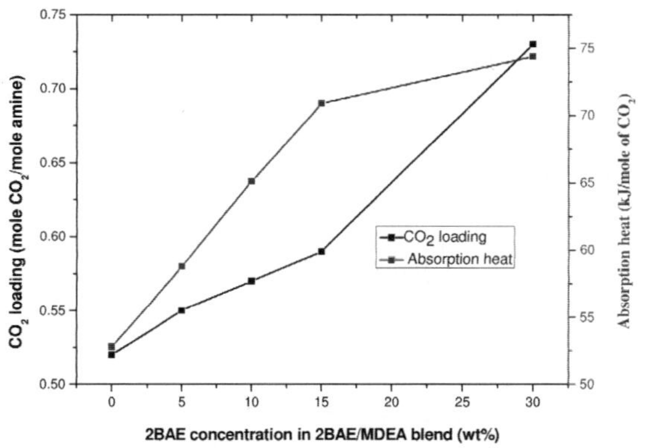

Figure 4. Variation of CO_2 loading and absorption heat as a function of 2BAE concentration in 2BAE/MDEA blend

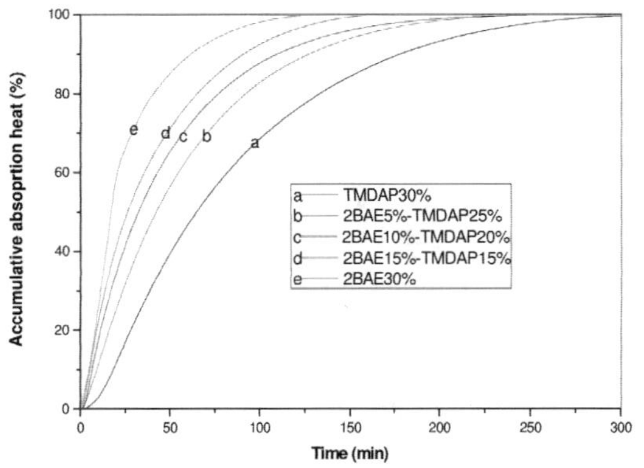

Figure 5. CO_2 absorption kinetics of 2BAE/TMDAP blends

Slow CO_2 absorption kinetics of tertiary amines is expected to be improved by blending with primary or secondary amine. In fact, a blend with low absorption heat and high CO_2 loading but slow absorption rate is still not realistic to CO_2 capture. Therefore, in addition to CO_2 loading and absorption heat, the CO_2 absorption kinetics should be investigated. In this study, heat evolved during CO_2 absorption recorded by flow micro calorimeter is proportional to amount of CO_2 absorbed. To interpret this data to kinetics, the percentage of accumulative absorption heat over absorption time was used. The CO_2 absorption kinetics of 2BAE, TMDAP, and their blends were exhibited in Figure 3. As can be seen in this figure, absorption rate of TMDAP 30 wt% solution is very slow, it took almost 178 min to complete 90% absorption, while 2BAE 30 wt% solution needed only 61 min. 2BAE5wt%/TMDAP25wt% blend required 127 min and 2BAE10wt%/TMDAP20wt% blend took 109 min to complete 90% absorption. This result indicated that replacing 5 wt% of TMDAP with 2BAE in blend helps reduced reaction time considerably. Similarly, blending with 2BAE helps improve CO_2 absorption kinetics of both 2DMAE and MDEA (Figure 6 and 7). Blending 5 wt% of 2BAE with MDEA do not significantly improve the absorption kinetics compared with single MDEA, however, the same blend concentration of 2BAE with TMDAP and 2DMAE results in considerable improvement of absorption kinetics in comparison with single counterpart tertiary amines. This further confirms that 5 wt% of 2BAE blending with TMDAP or 2DMAE was the best combination, which produced the best amine blend in terms of CO_2 loading, heat of absorption, and absorption rate.

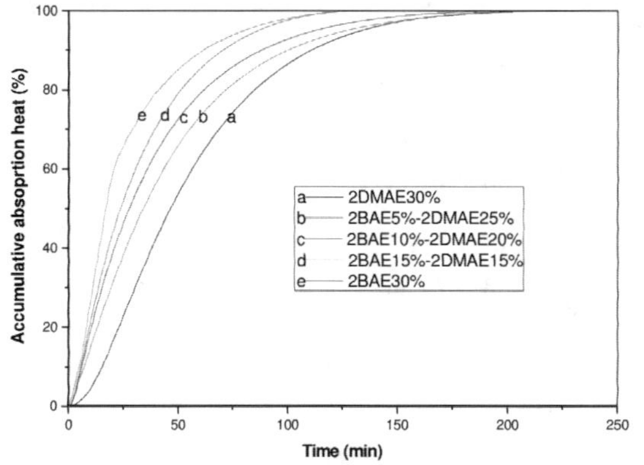

Figure 6. Absorption kinetics of 2BAE/2DMAE blend

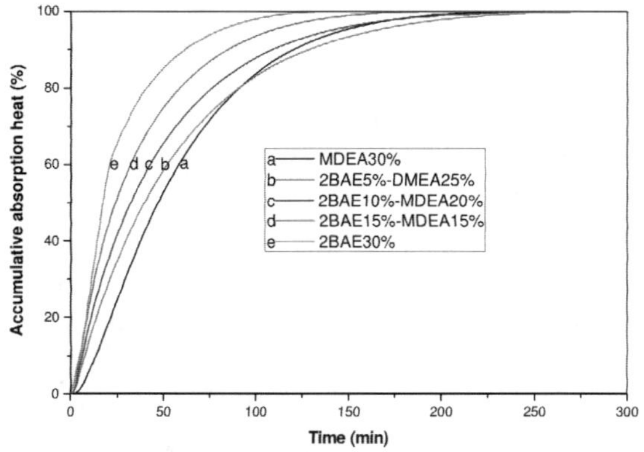

Figure 7. Absorption kinetics of 2BAE/2DMEA blend

Conclusion

Various blends of 2BAE (secondary amine) with three tertiary amines including TMDAP, 2DMAE, and MDEA were prepared and tested. Blending 2BAE with tertiary amine helps improve CO_2 absorption kinetics of tertiary amine, while remains reasonable CO_2 loading and absorption heat. Blends contain 5 wt% of 2BAE provide good CO_2 absorption performance in terms of CO_2 loading, heat of absorption, and absorption kinetics.

References

[1] J.G.J. Olivier, G. Janssens-Maenhout, J.A.H.W. Peters, Trends in global CO_2 emissions, Netherlands Environmental Assessment Agency, 2012.

[2] http://www.esrl.noaa.gov/gmd/ccgg/trends/#mlocited cited on 20/09/2016

[3] IPCC, Intergovernmental Panel on Climate Change (IPCC) Special Report on Carbon Dioxide Capture and Storage, Cambridge University press, Cambridge, UK, 2005.

[4] B. Lee, D. Kim, J. Cho, S. Park, A comparative study on the carbon dioxide capture power between 30 wt% 2-amino-2-methyl-1-propanol and 30 wt% methyldiethanol amine aqueous solutions, Korean J. Chem. Eng., 26 (2009) 818-823.

[5] M.R.M. Abu-Zahra, L.H.J. Schneiders, J.P.M. Niederer, P.H.M. Feron, G.F. Versteeg, CO_2 capture from power plants: Part I. A parametric study of the technical performance based on monoethanolamine, International journal of greenhouse gas control, 1 (2007) 37-46.

[6] M.R.M. Abu-Zahra, J.P.M. Niederer, P.H.M. Feron, G.F. Versteeg, CO_2 capture from power plants. Part II. A parametric study of the economical performance based on mono-ethanolamine, International journal of greenhouse gas control, 1 (2007) 135-142.

[7] A. Adeosun, N. El Hadri, E. Goetheer, M.R. Abu-Zahra, Absorption of CO_2 by Amine Blends Solution: An Experimental Evaluation, International Journal of Engineering and Science, 3 (2013) 12-23.

[8] P. Feron, M. Abu-Zahra, P. Alix, O. Biede, P. Broutin, H. de Jong, J. Kittel, J. Knudsen, L. Raynal, P. Vilhelmsen, Development of post-combustion capture of CO2 within the CASTOR Integrated Project: Results from the pilot plant operation using MEA, 3th International Conference on Clean Coal Technologies for our Futur, Cagliari, Italy, (2007).

[9] R. Idem, M. Wilson, P. Tontiwachwuthikul, A. Chakma, A. Veawab, A. Aroonwilas, D. Gelowitz, Pilot Plant Studies of the CO2 Capture Performance of Aqueous MEA and Mixed MEA/MDEA Solvents at the University of Regina CO2 Capture Technology Development Plant and the Boundary Dam CO2 Capture Demonstration Plant, Industrial & Engineering Chemistry Research, 45 (2005) 2414-2420.

[10] Y. Artanto, J. Jansen, P. Pearson, T. Do, A. Cottrell, E. Meuleman, P. Feron, Performance of MEA and amine-blends in the CSIRO PCC pilot plant at Loy Yang Power in Australia, Fuel, 101 (2012) 264-275.

[11] M.R. Abu-Zahra, Z. Abbas, P. Singh, P. Feron, Carbon Dioxide Post-Combustion Capture: Solvent Technologies Overview, Status and Future Directions, Materials and processes for energy: communicating current research and technological developments, Formatex Research Center2013, pp. 923-934.

[12] F.-Y. Jou, A.E. Mather, F.D. Otto, The solubility of CO2 in a 30 mass percent monoethanolamine solution, The Canadian Journal of Chemical Engineering, 73 (1995) 140-147.

[13] A. Chakma, CO2 capture processes — Opportunities for improved energy efficiencies, Energy Conversion and Management, 38, Supplement (1997) S51-S56.

[14] N. El Hadri, D.V. Quang, E.L. Goetheer, M.R.A. Zahra, Aqueous amine solution characterization for post-combustion CO 2 capture process, Applied Energy, (2016).

465184 Economic and Exergy Analysis of Integrated Purification Process with Carbon Capture Process (CCP) Using Selexol in a 500 MW Integrated Gasification Combined Cycle (IGCC) Power Plant

Abstract:

The enormous CO_2 emission from power plants raises the significant effects on global climate. In order to minimize greenhouse gas emissions from the utilization of fossils fuel, integrated coal gasification combined cycle (IGCC) with carbon capture process (CCP) takes a spotlight for high efficient and eco-friendly electricity generation. In this study, an integrated purification process with CCP for a 500MW IGCC power plant was simulated to evaluate energy efficiency and cost. First, only dual-stage Selexol processes with 80%, 85%, 90% and 95% carbon capture were preliminary simulated. According to preliminary results, energy consumption of Selexol process showed an optimum at more or less 90% carbon capture. Since each unit is integrated with the other units, the operating condition of one can affect the others. Then, as a second step, an integrated purification process with CCP simulation was carried out to evaluate energy analysis and cost estimation. It consists of contained water gas shift reactor (WGSR), dual-stage Selexol process, Claus process, scrubber unit, sour water stripper process, hydrogenation process, drying process, and compression. Finally, the exergy sankey diagram for whole integrated purification process with CCP, based on second law of thermodynamic, was performed to improve overall capture efficiency. Then, the development direction for reducing energy consumption was suggested from the exergy analysis.

Copyright © American Institute of Chemical Engineers. All rights reserved.

470573 Influence of Alloying on CO_2 Electroreduction on Ag-Zn System

Abstract:

Electrochemical reduction of CO_2 is an attractive pathway for the sustainable production of fuels and chemicals, as it allows for the effective use of renewable energy sources while mitigating the effects of CO_2 in the atmosphere. While significant decrease in overpotentials and increase in reaction rates have been reported for the two-electron reduction products, reports on improved activity and selectivity for the products that require more than two electrons (> 2 e^- products) have been limited on monometallic transition metal catalysts. Recent theoretical work indicates that this difficulty of reducing CO_2 to > 2 e^- products is due to the linear scaling of the binding energies of CO_2 reduction intermediates on transition metal catalysts, and the necessity of catalyst systems that can break such scaling has been emphasized.(1)

Alloys, which can provide bifunctional sites that selectively stabilize intermediates, are a promising class of materials, and a large amount of effort has been focused on the investigation of their effectiveness as CO_2 reduction catalysts. Recent works have reported on improved CO_2 reduction activity and selectivity on alloy systems, which were attributed to the synergy between the two metal atoms at the surface. However, most improvements reported in these works are pertaining to two-electron products, and alloy systems that demonstrate improvements for > 2 e^- products remain elusive.

Here, we report on the investigation of polycrystalline AgZn foil, which resulted in the observation of enhanced activities and selectivities for methane and methanol productions compared to Ag and Zn foils. Although CO is the main CO_2 reduction product as expected of an alloy between two metals active for CO production, the selective enhancement observed for the > 2 e^- products is noteworthy, as it indicates that alloys are promising candidates in the development of effective CO_2 electroreduction catalysts. The production of CO as major product is explained by the binding energies of intermediates, and the importance of optimal binding energies are discussed.

References

1. A. A. Peterson and J. K. Nørskov, *The Journal of Physical Chemistry Letters*, **3**, 251 (2012).

Copyright © American Institute of Chemical Engineers. All rights reserved.

470038 *In-Situ* Infrared Spectroscopic Investigation of Pyridine-Mediated CO_2 Reduction on Pt Electrocatalysts

Abstract:

The pyridine-mediated electrochemical reduction of CO_2 to methanol, formate, and formaldehyde was first demonstrated by Bocarsly et. al in 1994. Since then, many mechanistic investigations of this process have been conducted using reactivity, cyclic voltammetry, or computational methods; however, direct experimental mechanistic investigation has been lacking. We employ surface-enhanced infrared absorption spectroscopy, coupled with traditional electrochemical techniques in order to probe the electrochemical interface during the pyridine-mediated electrochemical reduction of CO_2 on Pt electrocatalysts. Previously proposed mechanisms are discussed and refined in light of new insights obtained in these studies.

Copyright © American Institute of Chemical Engineers. All rights reserved.

472304 Selective Hydrogenation of Bio-Oil Model Compounds over Molybdenum Carbide Supported Catalysts

Abstract:

The increasing complexity and cost of drilling for oil, the need to decrease global net CO2 emissions, and the desire to continue using liquid transportation fuels have motivated research to produce liquid fuels from biomass. Fast pyrolysis is a thermochemical conversion process in which biomass is reacted at high temperatures (near 500°C) at atmospheric pressure for very short residence times (on the order of seconds) in the absence of added molecular oxygen. Bio-oil from conventional fast pyrolysis is corrosive, has poor heating value, is thermally unstable, cannot be blended with conventional fuels without the addition of surfactant, and can become very viscous with aging[i]. Selective hydrogenation can improve the properties of bio-oil by increasing its thermostability and energy density, and potentially reducing acidity and therefore improving the ability to blend it with conventional fuels.

^{1}H and ^{13}C NMR analyses of pyrolysis bio-oils have shown that typical functional groups include aldehydes, ketones, hetero-aromatics, aliphatics, methoxys, alcohols, alkenes, dibenzenes, esters, benzyls, esters, and carboxylic acids[ii]. Considering the molecular character of pyrolysis bio-oil, crotonaldehyde was used as the model compound for work described in this paper as it is a simple aldehyde and can be used to probe a variety of different reaction pathways. The catalysts are based on Mo carbide (Mo_2C). Molybdenum carbide is active for a variety of hydrogenation reactions including the hydrogenation of CO to hydrocarbons (Fischer-Tropsch reaction)[iii] and naphthalene to tetralin[iv]. In addition, this material can possess, depending on preparation and treatment conditions, multiple types of surface sites (e.g. acidic, basic and metallic sites)[v],[vi],[vii] and is stable under conditions of fast pyrolysis[5]. Transition metal and alkali promotion of Mo_2C has been shown to cause shifts in the selectivity for a number of reactions, including shifting the selectivity for CO hydrogenation (Fischer-Tropsch reaction) to alcohols and heavier hydrocarbons, depending on the promoter[viii],[ix]. The goal of this work was to understand how metal promotion of Mo2C-based catalysts could be used to shift selectivity of crotonaldehyde hydrogenation.

The Mo_2C catalysts were synthesized by a temperature-programmed reaction method as described previously[x]. Metal salt was added to the native (unpassivated) Mo_2C catalyst by incipient wetness in an oxygen-free inert atmosphere. Then, the material was dried at 110°C for 2 hrs in H_2, reduced in H_2 at 450°C for 4 hrs, quenched to room temperature, and passivated in 1% O_2/He for at least 6 hrs. All of the catalysts were characterized by x-ray powder diffraction, N_2 physisorption, and temperature programmed desorption of NH_3 and CO_2. The crotonaldehyde hydrogenation experiments were performed in a flow reactor system equipped with a bubbler to introduce vapors of crotonaldehyde, and the reactor effluent was analyzed with a gas chromatograph equipped with flame ionization and thermal conductivity detectors. The catalyst was loaded into a quartz u-tube reactor and supported on quartz wool. Prior to the rate and selectivity measurements, the catalysts were reduced in 15% CH_4/H_2 for 4 h at 590 °C. All reactions were performed between 275-350°C in atmospheric pressure. The reaction mixture consisted of 0.7 mol% crotonaldehyde, 4.2 mol% H_2, and 95.1% N_2. The H_2/acetic acid molar ratio (6) was chosen to achieve approximately 2 times the stoichiometric amount of H_2required for complete saturation and deoxygenation of crotonaldehyde to butane.

The unpromoted Mo_2C favored formation of butyraldehyde and butene (ca. 40% each). Since butene is believed to be a secondary hydrogenation-dehydration product of crotyl alcohol[xi], there is not a clear preference for a specific pathway on the unpromoted catalyst. Selectivities to butyraldehyde, butene, and butadiene decreased with increasing potassium coverage, while selectivity to furan increased with potassium coverage. At coverages equivalent to 50% of a monolayer (ML), the furan selectivity was ca. 80%, while the selectivities to the other three products were each less than 10%. There was no significant further increase in the selectivity upon increasing the potassium loading, suggesting that the effect of potassium promotion was maximized at ~50% ML.

Alkali promotion is known to affect the acid-base properties of a catalyst, and acid-base properties are also known to affect selectivity[6,xii]. Acid and base site densities of the catalyst surfaces were quantified via temperature-programmed desorption of NH_3 and CO_2, respectively. While acid site densities decreased with increasing potassium loading, base site density increased with potassium promotion. The highest base site density (~$4.3x10^{14}$ molecules/cm^2) was for the K/Mo_2C catalyst with 1.1ML-equivalent coverage of potassium; the highest acid site density was on the bulk Mo_2C and was $1.5x10^{14}$ molecules/cm^2. There was a strong correlation between the furan production rate and the base site densities, implicating base sites the location of the rate-determining step. Furan appears to be produced from crotyl alcohol intermediate, indicating that potassium promotion enhanced the hydrogenation of crotonaldehyde to crotyl alcohol.

The model compound results presented in this paper demonstrate the ability to tune the product selectivity of molybdenum carbide catalysts by adding potassium. Tuning product selectivity may prove invaluable in upgrading and stabilizing bio-oils. Performing simultaneous carbon-carbon coupling and hydrogenation-dehydration reactions could ultimately lead to an upgraded bio-oil with lesser oxygen content, greater hydrogen content and a higher average carbon number.

---

[i] (a) Czernik, S.; Bridgwater, A.V.; *Energy & Fuels*, **2004**, *18*, 590-598; (b) Zhang, Q.; Chang, J.; Tiejun, W.; Yung, X. *Energy Conversion and Management*, **2007**, *48*, 87-92.

[ii] Mullen, C.A.; Strahan, G.D.; Boateng, A.A. *Energy & Fuels*, **2009**, *23*, 2707-2718.

[iii] Schaidle, J.A.; Thompson, L.T. *Journal of Catalysis*, **2015**, *329*, 325-334.

[iv] Pang, M.; Liu, C.; Xia, W.; Muhler, M.; Liang, C. *Green Chemistry*, **2012**, 14, 1272-1276.

[v] Bej, S. K.; Bennett, C. A.; Thompson, L. T., *Applied Catalysis A: General,* **2003**, *250*, 197-208.

[vi] Bej, S. K.; Thompson, L. T., *Applied Catalysis A: General,* **2004**, *264*, 141-150.

[vii] Sullivan, M. M.; Held, J. T.; Bhan, A., *Journal of Catalysis,* **2015**, *326*, 82-91.

[viii] Park, K.Y.; Seo, W.K.; Lee, J.S. *Catalysis Letters*, **1991**, *11*, 349-356.

[ix] Griboval-Constant, A.; Giraudon, J.-M.; Leclerq, G.; Leclercq, L. *Applied Catalysis A: General*, **2004**, *260*, 34-45.

[x] J.A. Schaidle, N.M. Schweitzer, O.T. Ajenifujah, L.T. Thompson, J. Catalysis 289 (2012) 210-217.

[xi] Zanella, R.; Louis, C.; Giorgio, S.; Touroude, R. *Journal of Catalysis*, **2004**, *223*, 328-339.

[xii] Kotarba, A.; Adamski, G.; Piskorz, W.; Sojka, Z.; Sayag, C.; Djega-Mariadassou, G. D. *Journal of Physical Chemistry B*, **2004**, *108*, 2885-2892.

Copyright © American Institute of Chemical Engineers. All rights reserved.

471968 Guaiacol Deoxygenation Using Methane over Pt-Bi Catalysts: Reaction Pathways and Kinetics

Abstract:

The deoxygenation of guaiacol, a typical molecule from the pyrolysis of ligno-cellulosic biomass, was investigated using methane as the reductant to overcome the high cost issue of H_2 as the reductant for the hydrodeoxygenation (HDO) process. In our prior work, a Pt-Bi/C catalyst was used to deoxygenate guaiacol and methane was found to exhibit as good deoxygenation performance as H_2. In the present work, by varying the space velocity, kinetic data was acquired. Following similar reaction pathways as in the case using H_2, rate constants and activation energies for each sub-reaction in the network were obtained. Owing to more difficult activation of C-H bond in CH_4, as compared to H-H bond in H_2, the kinetic results show higher reaction barriers when methane is used as the reductant. This work provides a new approach for bio-oil upgrading using methane as the reductant instead of hydrogen.

Copyright © American Institute of Chemical Engineers. All rights reserved.

481344 Fluidized Bed Catalytic Glycerol Hydrodeoxygenation to 1,3-Propanediol

Abstract:

1,3 propandiol (1,3-PDO) is a monomer for commercial polyesters like Corterra® (Shell) and Sorona® (DuPont). Glycerol is an attractive feedstock for this monomer because of its low cost. Here we propose a gas-phase hydrodeoxygenation process in which we atomize a dilute solution of glycerol to a fluidized bed of of Pt/WO3/Al2O3 operating above 240 °C and atmospheric pressure. Gas-solids fluidized beds have excellent heat transfer properties that vaporize the droplets within milliseconds thereby minimizing thermal degradation of the glycerol. After 2 h at 260 °C, 1,3-PDO yield reached 14%. The other products included 1,2-PDO (18%), 1-propanol (28%), and 2-propanol (15%). We postulate that glycerol first dehydrates to acrolein, then rehydrates to 3-hydroxypropanal followed by the hydrogenation to 1,3-PDO. The yield of other compounds (acrolein, ethylene glycol, propane, and acetone) increased at higher operating temperatures.

Copyright © American Institute of Chemical Engineers. All rights reserved.

454718 Reaction Pathways for the Hydrodeoxygenation of Anisole and Guaiacol over Zn-Pt Bimetallic Catalysts

Abstract:

There is an increasing interest in the use of lignin as a renewable feedstock for the production of high-value aromatic compounds; however, due to its high oxygen content, selective hydrodeoxygenation (HDO) of the aromatic oxygenates produced from lignin depolymerization is required. Metal alloys containing a group 10 metal and a more oxyphillic metal, such as Zn, have been shown to be highly selective for this class of reactions, at least under some conditions. Although, the mechanism by which alloying enhances selectivity is still poorly understood. To provide mechanistic insight, in this study we have investigated the reaction of typical lignin-derived oxygenates, anisole and guaiacol, on single-crystal Pt and Zn-Pt model catalysts by temperature programmed desorption (TPD) and high resolution electron energy loss spectroscopy (HREELS). For Zn-free Pt it was found that anisole and guaiacol interact strongly with the surface via the aromatic ring, which promotes undesired ring hydrogenation. Adding Zn to the Pt surface was found to inhibit bonding via the aromatic ring and promoted a stronger interaction with the oxygens. This in turn facilitated selective C-O bond scission without hydrogenating the aromatic ring. These results suggest that Zn-Pt alloys may be an effective catalyst for HDO of lignin-derived aromatic oxygenates with low activity for ring hydrogenation. This hypothesis was then tested and verified by investigating the reaction of anisole or guaiacol with H_2 over high surface area carbon-supported Pt and Zn-Pt catalysts.

Copyright © American Institute of Chemical Engineers. All rights reserved.

459948 Novel Bio-Oil Hydrodeoxygenation Catalysts Based on Strong Electrostatic Adsorption

Abstract:

Fast pyrolysis of biomass produces an intermediate crude oil (â€˜bio-oilâ€™) with potential to become fungible transportation fuel. Catalytic hydrodeoxygenation (HDO) has currently gained attention as a technique for bio-oil upgrading, due to its ability to convert highly oxygenated and reactive species to hydrocarbons. Most studies of HDO catalysts are process and not catalyst oriented; noble metal catalysts are typically used in conjunction with common supports, and important variables which affect catalyst performance are often not decoupled during preparation, including catalyst particle size, hydrothermal stability, and precision of bimetallic formations.

We hypothesize that a precisely synthesized combination of bimetallic base metal catalysts can possess requisite activity and selectivity for pyrolysis oil HDO. We studied the hydrotreatment of bio-oil from switchgrass (regular and catalytic fast pyrolysis) in a batch reactor, using a series of active and hydrothermally stable HDO catalysts. Strong electrostatic adsorption was used to prepare single and bimetallic combinations (Pt, Ru, PtRu, Ni, Cu, NiCu) on carbon, mesoporous alumina, and metal-doped SBA-15. Particle size was determined by XRD to fall within 3 â€“ 5 nm with narrow distribution, and in many cases, as small as 1.0 nm. Catalyst stability was determined by examining the support crystal structure before and after reaction. The selectivities of each catalyst toward deoxygenation, hydrogenation, and coke inhibition were compared, and this information was used to identify optimum HDO catalysts.

Copyright © American Institute of Chemical Engineers. All rights reserved.

460969 Biomass Pyrolysis Interfaced with Honeycomb Structured Upgrading Reactor for the Production of High Quality Biofuels

Abstract:

A reactor interfaced a pyrolysis zone with a doubly stacked honeycomb catalytic upgrading zone was designed and evaluated for biomass conversion to liquid fuels. The pyrolysis zone was operated through a bubbling bed with fine sand particles as the heat carrier. Two zeolite-coated cordierite honeycomb (ZCCH) catalyst matrices separately prepared using ZSM-5 of different Si/Al ratios were stacked in the upgrading zone. The results showed that ZCCH catalytic upgrading following the steam pyrolysis of corncob effectively reduced the oxygen content and viscosity of bio-oil. The yield, HHV and stability of the bio-oil were increased by the presence of a moderate amount of ethanol. The HHV of bio-oil was found to linearly correlate with the oxygen concentration in the oil. Models of bio-oil energy yield from corncob based on unit energy yield and yield in ethanol equivalent (YEE, in L ethanol) were developed according to the oxygen content in the produced bio-oils.

Copyright © American Institute of Chemical Engineers. All rights reserved.

471424 Upgrading of Biomass Pryolysis Vapors of Modified ZSM-5: Effects of Metal(s) Loading and Pretreatment

Abstract:

Ga/ZSM-5 and Ni/ZSM-5catalysts with varied nickel loadings were evaluated for their ability to produce aromatic hydrocarbons by upgrading of pine pyrolysis vapors. The effect of catalyst pretreatment by hydrogen reduction was also investigated. Results indicate that the addition of gallium or nickel increases the yield of aromatic hydrocarbons while simultaneously increasing the conversion of oxygenates, relative to ZSM-5, and these effects are more pronounced with increasing nickel loading. Additionally, while initial activity differences were observed between the oxidized and reduced forms of nickel on ZSM-5 (i.e., NiO/ZSM-5 vs. Ni/ZSM-5), the activity of both catalysts converge with increasing time on stream. These reaction results, coupled with characterization of pristine and spent catalysts, suggest that the catalysts reach similar active states during catalytic pyrolysis, regardless of pretreatment, as NiO under goes *in situ* reduction to Ni by biomass pyrolysis vapors. This reduction of NiO to Ni was confirmed by reaction results and characterization by NH_3 TPD, TPR, and XRD. This finding is significant in that the ability to reduce or eliminate the need for a pre-reaction H_2 reduction of Ni-modified zeolite catalysts could reduce process complexity and operating costs in a biorefinery-based vapor phase upgrading process to produce biomass-derived fuels and chemicals. The ability to monitor catalyst activity in real-time with a molecular beam mass spectrometer used to measure uncondensed, hot pyrolysis vapors, allows for improved understanding of the mechanism for improved activity with Ni-addition to ZSM-5, which is attributed to the ability to prevent deactivation by deposition of coke and capping of zeolite micropores.

Copyright © American Institute of Chemical Engineers. All rights reserved.

467798 Cross Metathesis of Unsaturated Carboxylic Acids to Bioterephthalic Acid Intermediates

Abstract:

Terephthalic acid is essential for the preparation of a variety of polyester fibers and is largely produced *via* the catalytic oxidation of petroleum-derived *p*-xylene [1]. The ongoing exchange of naphtha crackers with shale gas crackers is causing a shortage in some aromatics species, such as *p*-xylene [2]. As a consequence, alternative and sustainable routes to produce terephthalic acid from biomass derivatives, e.g. bio-sourced sorbates [3], are highly desirable. In the last decade, Ru-catalyzed olefin metathesis, and in particularly cross-metathesis, has become an important tool for the formation of carbon–carbon bonds and the formation of valuable organic molecules [4].

We have investigated the ruthenium-catalyzed cross metathesis of sorbates with acrylates to form di-ester muconates. In a later step, the reaction between alkyl muconates and ethylene affords the aromatization to dialkyl terephthalate. The effect of alkyl substituents and solvents was investigated over various organometallic ruthenium-based catalysts to increase conversion and selectivity towards the alkyl muconates. Conversions of up to 20% were achieved within the first hour of reaction with very low amounts (0.1 mol.%) of ruthenium-based catalyst containing a *N*-heterocyclic carbene (NHC) ligand. If a solvent was used, the cross metathesis products were formed in higher yields with the use of more polar solvents, *i.e.* CH_2Cl_2. With the aim to develop a greener process, the acrylate reactant was used in excess with no other solvent added. In this case the highest selectivity to the desired product was observed (~ 40%). By using different alkyl substitutents the steric effect of the substitutents was investigated. Using reactants (sorbates and acrylates) with smaller alkyl substitutents the yield of cross metathesis products was favored, possibly due to easier accessibility of the reactants to the catalyst metal site. Upon further development, this route for obtaining di-ester muconates, may provide a sustainable alternative for producing biomass derived precursors to polyethylene terephthalate (PET).

References

[2] C. Luke Williams, Chun-Chih Chang, Phuong Do, Nima Nikbin, Stavros Caratzoulas, Dionisios G. Vlachos, Raul F. Lobo, Wei Fan, and Paul J. Dauenhauer. ACS Catal. 2 (2012) 935

[4] Georgios C. Vougioukalakis, and Robert H. Grubbs. Chem. Rev. 110 (2010) 1746

Copyright © American Institute of Chemical Engineers. All rights reserved.

470919 Investigating Synergetic Effects Between Deposited PGM Nanoparticles and Transition Metal Carbide Nanotube Support through Surface Characterization and Device Performance

Abstract:

Platinum group metals (PGM) supported on transition metal carbide (TMC) surfaces (PGM/TMC) have been the focus of significant research interest, due to the similar electronic and geometric structures between TMC and Pt.[1] This presentation focuses on probing a new class of electrocatalyst fabrication using high purity Î²-molybdenum carbide nanotube (Mo_2C) as a support and atomic layer deposition (ALD) as the PGM deposition technique. Lattice spacing analysis and Pt binding energy shift in X-ray photoelectron spectroscopy (XPS) results showed strong interaction between deposited Pt particles and the TMC support. This interaction was correlated with enhanced hydrogen evolution (HER) and hydrogen oxidation reaction (HOR) activity. In addition, significantly increased durability was observed in constant potential electrolysis (CPE) and potential cycling. Finally, the Pt/Mo_2C catalyst performance was systematically evaluated in a device setting for different Pt loadings by applying it in anode and cathode of a proton exchange membrane fuel cell (PEMFC), with accelerated degradation test (ADT) being carried out between 0.6V to 1 V for 30,000 cycles at a scan rate of 500 mV s^{-1}.[2] Significantly higher power density was observed in MEAs with Pt/Mo_2C catalysts than 20% Pt/C was observed after the vigorous ADT.

References

1. R. B. Levy and M. Boudart, *Science*, 1973, 181, 547â€"549.
2. B. Han, C. E. Carlton, A. Kongkanand, R. S. Kukreja, B. R. Theobald, L. Gan, R. O'Malley, P. Strasser, F. T. Wagner and Y. Shao-Horn, *Energy Environ. Sci.*, 2014, **8**, 258â€"266.

Copyright © American Institute of Chemical Engineers. All rights reserved.

470659 Bimetallic Ru-Pt and Pt-Co Fuel Cell Catalysts Prepared By Strong Electrostatic Adsorption and Electroless Deposition

Abstract:

The methods of Strong Electrostatic Adsorption (SEA), and its incipient wetness analog, Charge Enhanced Dry Impregnation (CEDI), have been shown to yield supported metal nanoparticle catalysts with high dispersion and narrow size distribution. Catalysts prepared by SEA and CEDI therefore are desirable as seeds for addition of secondary metal using Electroless Deposition (ED), as the prepared bimetallic catalysts should be of similar dispersion as the base catalyst. This study demonstrated the use of these methods to prepare two series of carbon supported bimetallic catalysts containing Pt which were then characterized and evaluated with cyclic voltammetry. The first system used Pt as the base metal, prepared by SEA on XC72R carbon, with Ru as the secondary metal added by ED (Pt@Ru/C). The second system used Co as base metal, prepared by a modified CEDI method on Carbon Black, with Pt added by ED (Pt-Co/C)

XRD, HAADF-STEM, and XEDS characterization showed that the Pt@Ru/C catalysts did not have alloying of the component metals and that the catalysts have bimetallic surface composition. The resulting Pt-Co/C catalysts had particles with irregular morphology that were larger than the seed Co particles. These were determined to have thin alloyed Pt-Co phases and domains of pure Pt. Evaluation of these catalysts showed higher mass activities, with respect to platinum content, compared to commercially available monometallic and bimetallic catalysts. This enhancement in performance is associated with the electronic interaction between Pt and Ru on the catalyst surface and the lattice contraction for Pt-Co alloys.

Copyright © American Institute of Chemical Engineers. All rights reserved.

465251 Understanding the Effects of pH and Alkali Cations on H/OH Adsorption and Hydrogen Oxidation on Transition Metal Surfaces

Abstract:

Understanding how the complex interface between a catalytic electrode and electrolyte affects the rate and mechanism of electrocatalytic reactions is crucial to designing more active and selective electrocatalysts. As an example of this effect, it is now well known that the rate of the hydrogen oxidation reaction on a platinum electrode is orders of magnitude faster in acid than in base [1, 2] and that the rate varies continuously as a function of pH [3]. Sheng et al. have shown that the dependence of the rate on pH is correlated with what is believed to be a shift in the binding energy of hydrogen with pH, as measured by cyclic voltammetry [3]. We have recently shown using density functional theory (DFT) simulations that the low potential peaks measured by cyclic voltammetry on Pt(100) and Pt(110) are actually the co-adsorption or competitive adsorption of hydrogen and hydroxide, and that the shift of this peak with pH may not be solely related to the binding energy of hydrogen. We further show, in conjunction with our prior work on alkali cation specific adsorption [4], that potassium can specifically adsorb to Pt(111), Pt(100), and Pt(110) electrode surfaces at high pH and weaken the adsorption of hydroxide [5]. We see similar results on stepped surfaces of platinum, in particular Pt(553) and Pt(533), where both the binding strength of hydrogen on the step in the gas phase and in an electrochemical environment match only when competitive adsorption of hydroxide is considered. We extended this analysis of hydrogen and hydroxide adsorption across transition metal surfaces and find a similar effect of adsorbed alkali cation on hydroxide adsorption. Hydrogen and hydroxide competitive adsorption occurs on most surfaces which bind both species more strongly than Pt(111). Recent work by Zheng et al. has shown that the pH dependence of the hydrogen oxidation reaction and of the adsorption peaks measured by cyclic voltammetry are not unique to platinum, and occur across transition metals [6]. This has significant implications in the understanding of the hydrogen oxidation mechanism, where it has been proposed that adsorbed hydroxide may play a role in dictating the rate [7]. We will discuss our work using DFT to model hydrogen oxidation reaction mechanisms as well as other important conclusions we can draw (such as more accurately calculating electrode surface area or the surface restructuring which occurs with

platinum electrodes under electrochemical â€œconditioning/cleaningâ€) with our improved understanding of hydrogen and hydroxide adsorption on transition metal surfaces.

[1] Sheng, W.; Gasteiger, H.A.; Shao-Horn, Y. Hydrogen oxidation and evolution reaction kinetics on platinum: Acid vs Alkaline Electrolytes. *Journal of the Electrochemical Society*. **2010**, 157, B1529-B1536.

[2] Durst, J.; Siebel, A.; Simon, C., HaschÃ©, F.; Herranz, J.; Gasteiger, H.A. New insights into the electrochemical hydrogen oxidation and evolution reaction mechanism. *Energy Environ. Sci.* **2014**, 7, 2255-2260.

[3] Sheng, W.; Zhuang, Z.; Gao, M.; Zheng, J.; Chen, J. G.; Yan, Y. Correlating Hydrogen Oxidation and Evolution Activity on Platinum at Different pH with Measured Hydrogen Binding Energy. *Nat. Commun.* **2015,** *6:5848.*

[4] Mills, J.N.; McCrum, I.T.; Janik, M.J.; Alkali cation specific adsorption onto fcc(111) transition metal electrodes. *Phys. Chem. Chem. Phys.* **2014**, 16, 13699-13707.

[5] McCrum, I.T.; Janik, M. J. pH and alkali cation effects on the Pt cyclic voltammogram explained using density funtional theory. *J. Phys. Chem. C.* **2016**, 120, 457-471.

[6] Zheng, J.; Sheng, W.; Zhuang, Z.; Xu, B.; Yan, Y. Universal dependence of hydrogen oxidation and evolution reaction activity of platinum group metals on pH and hydrogen binding energy. *Sci. Adv.* **2016**, 2, e1501602.

[7] Strmcnik, D.; Uchimura, M.; Wang, C.; Subbaraman, R.; Danilovic, N.; van der, V.; Paulikas, A. P.; Stamenkovic, V. R.; Markovic, N. M. Improving the Hydrogen Oxidation Reaction Rate by Promotion of Hydroxyl Adsorption. *Nat. Chem.* **2013,** *5*, 300-306.

Copyright © American Institute of Chemical Engineers. All rights reserved.

466793 Nanoparticle Catalysts Supported on Substitutionally Doped Graphene: Effects on Activity and Stability for Hydrogen Oxidation

Abstract:

Fuel cells represent an important part of a renewable future hydrogen-based energy grid. In particular, hydroxide exchange membrane fuel cells (HEMFCs) offer the advantage of non-noble metals being more stable against dissolution than in the acidic environment. However, the rate of the hydrogen oxidation reaction (HOR) decreases by approximately two orders-of-magnitude when switching from acid to base. As a result, developing more efficient and cheaper HOR catalysts is imperative to advancing the current state-of-the-art fuel cells. In a recent publication, we have demonstrated that Ni nanoparticles supported on nitrogen-doped carbon nanotubes (N-CNT) exhibit a threefold increase in the exchange current density relative to a pristine CNT support, and a factor of 30 increase relative to an amorphous carbon support (Zhuang, Z., *et al.*, *Nature Communications* **2016**, *7*, 10141). However, a fundamental study on the synergistic interaction between the supported nanoparticle and doped CNT has yet to be done.

Herein, using graphene as a two-dimensional analog for the CNT support used experimentally, we study from first-principles the impact of graphene and doped graphene supports on the hydrogen oxidation reaction occurring on nanometer-sized catalysts. We consider nickel, copper, and silver nanoparticle compositions, and nitrogen-doped, boron-doped, and phosphorous-doped graphene supports. To understand quantitatively the effect of substitutional doping on the nanoparticle/support interaction, we study: 1) the charge transfer between the nanoparticle and support as a function of dopant and dopant location, 2) the modulation of the nanoparticle's d band center due to the presence of a dopant, 3) the interaction of metal adatoms with graphene and doped-graphene, 4) the hydrogen oxidation activity of the nanoparticle/support systems, and 5) the oxidative stability of the supported copper and nickel nanoparticles.

Copyright © American Institute of Chemical Engineers. All rights reserved.

467551 Impact of Combining Metal Nanoparticle Catalysts and Semiconductor Photo-Electrodes on Photo-Catalytic Performance

Abstract:

Semiconductors have found significant utility in the field of photo-electrochemistry as excellent optical absorbers. Metal nanoparticles are typically loaded onto these surfaces to add catalytic function to the system. The addition of metal nanoparticles also changes the optical properties of these systems since metal nanoparticles strongly interact with light through interband transitions and plasmonic phenomena. For a semiconductor substrate that is a good absorber and has long charge carrier diffusion lengths, the interaction of light with metallic nanoparticles will result in a loss of optical efficiency.

In this contribution, we will discuss our work on developing strategies for combining semiconductors and metal nanoparticle catalysts to simultaneously improve the catalytic and optical efficiencies of a photo-electrochemical device. We will present the results of our experimental and theoretical efforts based on a model system consisting of planar silicon electrodes and platinum nanoparticles as composite photo-electrodes for the hydrogen evolution reaction.

Copyright © American Institute of Chemical Engineers. All rights reserved.

462154 Trends in Hydrogen Evolution Reaction Activity Among Metal Modified Carbide Thin Films and Powders

Abstract:

Water electrolysis, coupled with renewable energy sources, is an important technology for producing CO_2-free hydrogen gas that can be used for energy storage or CO_2 conversion. One of the major barriers to widespread electrolyzer use is the high cost of platinum-group metals required to catalyze the hydrogen evolution reaction (HER) and the oxygen evolution reaction (OER) at the electrolyzer cathode and anode, respectively. In the current study, metal-modified transition metal carbides were investigated as inexpensive replacements for bulk Pt for the HER. Transition metal carbides were synthesized as thin film electrocatalysts and modified with monolayer coverage of precious and non-precious metals. The HER activities of these catalysts were measured electrochemically in both acidic and basic electrolyte, and activity trends were established based on density functional theory (DFT) calculated hydrogen binding energy (HBE). Pt-modified niobium carbide showed promise as a low-cost alternative to bulk Pt for the HER, and it also showed impurity tolerance in the presence of common tap water ions. Thin film results were then extended to industrially relevant high surface area molybdenum carbides (Mo_2C) with dispersed metal nanoparticles.
Pt-Mo_2
techniques for well-defined high surface area catalysts in order to facilitate the use of DFT to help screen for active HER catalysts.

Copyright © American Institute of Chemical Engineers. All rights reserved.

461163 Experimental Determination of Water Orientation Effects on Kinetic Barriers to Alkaline Hydrogen Oxidation and Evolution

Abstract:

It has long been recognized that the reaction rates of the hydrogen oxidation and hydrogen evolution reactions (HOR and HER) are slower in basic than acidic electrolytes, even though the surface intermediate of adsorbed hydrogen is independent of solution pH. Understanding the root of this observation is critical to designing catalysts for a multitude of electrochemical reactions with relevance to energy conversion and storage. In this work, we undertake a fundamental investigation relating interfacial water structure to the reaction barrier for hydrogen underpotential adsorption and to the HER/HOR kinetics.

The pH-dependence of HOR/HER kinetics has been attributed to a variety of reasons. The research groups of Gasteiger and Yan have both suggested that in base, hydroxide ions stabilize the Pt-H bond for stronger binding and slower catalysis [1,2]. Central to this hypothesis is an experimentally measured shift with pH in the peak potential for hydrogen underpotential deposition on the (100) and (110) steps on the Pt surface. Although it is unclear why pH would stabilize adsorbed hydrogen, stronger binding would push platinum further to the right of the volcano peak and decrease the overall activity. Markovic et al proposed instead that in base, the water recombination/dissociation step of HOR/HER requires specific adsorption of hydroxide ions and therefore optimal binding to two adsorbates [3]. One observation that has been less discussed than either hydrogen or hydroxide adsorption is the effect of pH not only on the apparent hydrogen adsorption energy, but the kinetic barrier to adsorption. Koper et al attributed this observation to a pH-dependent water orientation at the surface, and hypothesized that "H-down" in acid vs "H-up" in base resulted in slower transfer of hydrogen to and from the surface [4]. The same pH-dependence of water orientation was recently supported computationally by Rossmeisl's group[5].

In this work, we examine specifically the hypothesis that water orientation governs the rate of hydrogen adsorption and thus the overall HER/HOR kinetics. The native oxide formed on chromium metal is known from colloidal literature to have a potential of zero charge (Epzc) of 1.22V vs. RHE[6]. We hypothesize that if water orientation truly governs the alkaline HOR/HER rate, the negative oxide surface charge on chromium oxide will induce an H-down water orientation adjacent to platinum nanoparticles supported on the surface. This will result in faster hydrogen adsorption kinetics and thus, higher activity for alkaline HOR/HER than extended platinum surfaces.

To test this hypothesis, platinum nanoparticles were synthesized via pulsed electrodeposition on chromium oxide. The nanoparticle loading was controlled by the duration of deposition pulse, with longer times corresponding to higher coverage as well as larger particles. The hydrogen adsorption kinetics were measured by cyclic voltammetry in the hydrogen underpotential deposition region. Varying the sweep rate results in greater peak separation that can be used to extract rate constants for the adsorption reaction via traditional electroanalysis[7].

A typical voltammogram of Pt/CrOx in 0.1M KOH is shown in Fig. 1a. As the scan rate increases, kinetic barriers cause greater peak separation (dEp) for hydrogen adsorption on the 100 and 110 facets of platinum. For lower loadings with shorter pulse periods, both smaller currents and smaller peak separations are observed due to the lower Pt surface area. The effects of surface area can be accounted for by normalizing current to Pt surface area as measured by Hupd coulometry. The resulting surface-adsorption Tafel plot is shown in Fig 1b. The peak potential separation dEp is much greater for the larger nanoparticles with higher loading. This difference suggests that smaller particles have qualitatively more rapid adsorption kinetics due to the advantageous water orientation induced by negatively charged CrOx.

The results of this study contribute to resolving a long-standing paradox in electrocatalysis and surface science by highlighting the importance of kinetic barriers, as well as adsorption energies, and suggest the ability to modulate alkaline hydrogen activity through metal-support interactions. Future work will discuss the relation of hydrogen underpotential deposition kinetics to HER/HOR activity and the effects of different oxide supports.

(1) Durst et al. Energy Environ. Sci. 2014, 2255.

(2) Sheng et al. Nat. Commun. 2015, 5848.

(3) Strmcnik et al, Nat. Chem. 2013, 1.

(4) Van Der Niet et al. Catal. Today 2013, 105.

(5) Rossmeis et al. Phys. Chem. Chem. Phys. 2013, 10321.

(6) McCafferty, E. Electrochim. Acta 2010, 1630.

(7) Angerstein-Kozlowska et al. J. Electroanal. Chem. Interfacial Electrochem. 1979, 1.

Copyright © American Institute of Chemical Engineers. All rights reserved.

459138 Co-Oxidation of CO and Hydrocarbons on Pd/Ceria-Zirconia/Al2O3 Three-Way Catalysts: Experiments and Modeling

Abstract:

Co-oxidation of CO and Hydrocarbons on Pd/Ceria-Zirconia/Al2O3

Three-way Catalysts: Experiments and Modeling

Wendy Lang[1], Michael P. Harold[1*], Yisun Cheng[2], Carolyn Hubbard[2], Manish Sharma[2], and Paul Laing[2]

[1]*Department of Chemical and Biomolecular Engineering, University of Houston, Houston, TX*

[2]*Research and Innovation Center, Ford Motor Company, Dearborn, MI*

*mharold@uh.edu

Three-way catalyst (TWC) light-off performance is critical in cost effectively meeting future emission standards. Improvements in catalyst light-off (LO) should include catalyst modifications that minimize inhibition effects of exhaust species such as CO and hydrocarbons (HCs), lowering precious group metal (PGM) loading, or using a lower cost PGM such as Pd. Alumina (Al_2O_3) and zirconia (ZrO_2) supports provide high surface area and stability. Furthermore, ceria (CeO_2) supports for PGMs provide the key benefit of oxygen storage, compensating for deviations from stoichiometric TWC operating conditions by supplying oxygen for CO and HC oxidation under rich transient conditions. Studies have demonstrated that inclusion of ceria in PGM catalysts enhances CO and HC conversion and minimizes self-inhibition; this improved activity is mainly attributed to ceria containing more efficient oxygen activation sites at the PGM-support interface relative to other support materials such as Al_2O_3 [1-4].

Monolith catalysts having washcoats with 1 wt% Pd, 3 g/in^3 loading, 400 cpsi and formulations Pd/Al_2O_3, Pd/CZO, and Pd/CZO/Al_2O_3 (17 wt% CZO, balance Al_2O_3) were synthesized. Catalyst performance studies were conducted to elucidate the role of Pd and support (CZO, Al_2O_3) compositions and loadings on inhibition during CO and C_3H_6 LO under near-stoichiometric conditions. Transient and steady-state bench scale reactor studies with simulated exhaust gas mixtures were conducted to extract oxidation kinetics, compare LO behavior, and develop a predictive model for understanding and optimizing catalyst performance.

Catalyst activities were compared via temperature-programmed oxidation experiments. Figure 1 compares CO LO curves using feed concentrations of 0.5% CO and 1% CO ($\lambda = 1.01$ in both cases). Using all three catalysts, LO temperatures increase with feed CO concentration, showing that CO oxidation is self-inhibiting. With increasing CZO content, LO temperatures decrease, demonstrating the promotional effects of using ceria.

Figure 2 shows C_3H_6 LO using feed concentrations of 250 ppm C_3H_6 and 500 ppm C_3H_6 ($\lambda = 1.01$). Like CO, C_3H_6 is self-inhibiting. The shoulder feature in the LO curves using Pd/Al_2O_3 and Pd/CZO/Al_2O_3 is speculated to be due to a change in the oxidation state of Pd and/or the accumulation of inhibiting carbonaceous surface species. LO begins at lower temperatures and the shoulder becomes less prominent with increasing CZO content and is absent with Pd/CZO.

Comparing the individual oxidations of 1% CO and 500 ppm C_3H_6 in Figures 1 and 2 to LO behavior during their co-oxidation (1% CO + 500 ppm C_3H_6) in Figure 3, LO temperatures for both species are higher in the mixture, i.e. CO and C_3H_6 are mutually-inhibiting. During co-oxidation, CO lights off before C_3H_6 and the C_3H_6 shoulder does not appear, suggesting that CO inhibits C_3H_6 adsorption and the accumulation of carbonaceous species on the catalyst. Once CO lights off and temperatures increase, C_3H_6 oxidation proceeds. Significantly lower LO temperatures are achieved using Pd/CZO, and higher conversions of CO and C_3H_6 are reached at lower temperatures using catalysts containing CZO.

Differential kinetics studies were conducted to quantify reaction orders and activation energies. The reaction order with respect to CO was found to be approximately −1 using Pd/Al_2O_3 and Pd/CZO/Al_2O_3 and 0 to slightly negative using Pd/CZO, consistent with literature values [5-7]. The activation energies for CO oxidation was found to be 125-149 kJ/mol using Pd/Al_2O_3 and Pd/CZO/Al_2O_3, and 29-44 kJ/mol using Pd/CZO. Using the Pd/Al_2O_3 catalyst, the reaction order with respect to C_3H_6 was found to be approximately −1 and activation energy 77 kJ/mol.

Ongoing work involves the use of a mechanistic-based kinetic model incorporated into a low-dimensional monolith model [8-9]. The global reactor models are used with CO and C_3H_6 oxidation on Pd with steps involving ceria. The 1+1 dimensional model formulation comprises coupled species balances with transverse-average equations having external and internal mass transfer coefficients. As shown in Figure 4, the model verifies the experimental LO of CO on Pd/Al_2O_3. For example, it validates additional experimentally-observed oxidation behavior such as the switch in order of ignition between individual and mixture oxidations on Pd/Al_2O_3 and Pd/CZO/Al_2O_3 as well as the promotional effect of ceria.

Steady-state and transient CO and C_3H_6 oxidation experiments using fresh Pd/Al_2O_3, Pd/CZO, and Pd/CZO/Al_2O_3 monolith catalysts were conducted to compare LO behavior and kinetic parameters (i.e. reaction orders and activation energies). CO and C_3H_6 individual and mixture LO behavior was improved with increasing catalyst ceria content. A reactor model confirms experimentally-observed oxidation behavior and predicts catalyst performance. Finally, experiments involving other hydrocarbons including acetylene and toluene are being conducted to investigate inhibition during LO, extract kinetic parameters for oxidation reactions, and provide a basis for validating modeling simulations.

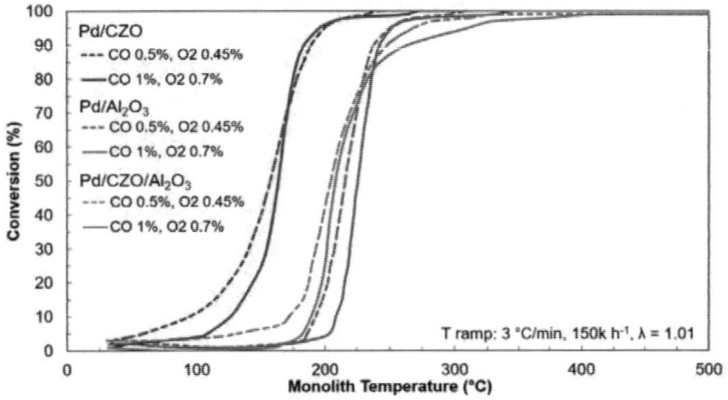

Figure 1: CO conversion (%) versus monolith temperature using fresh Pd/CZO, Pd/Al_2O_3, and Pd/CZO/Al_2O_3 catalysts for feed concentrations of 0.5% CO (λ = 1.01) and 1% CO (λ = 1.01).

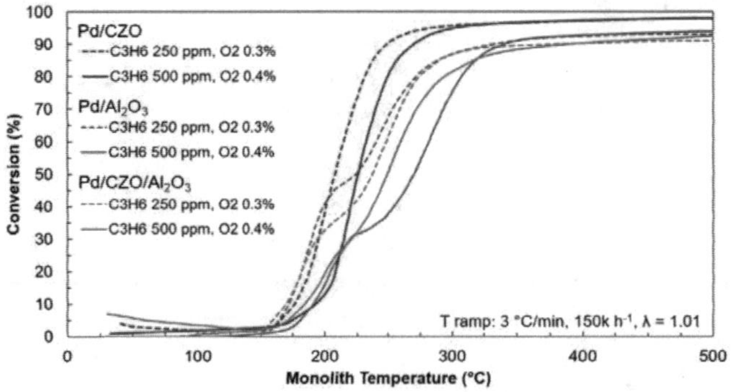

Figure 2: C_3H_6 conversion (%) versus monolith temperature using fresh Pd/CZO, Pd/Al_2O_3, and Pd/CZO/Al_2O_3 catalysts for feed concentrations of 250 ppm C_3H_6 (λ = 1.01) and 500 ppm C_3H_6 (λ = 1.01).

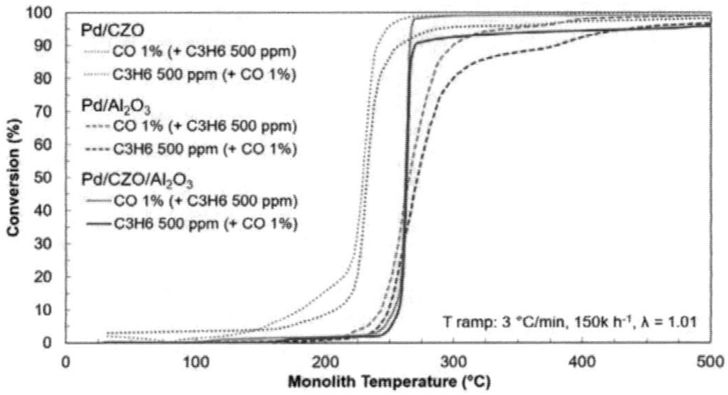

Figure 3: CO, C_3H_6 conversion (%) versus monolith temperature using fresh Pd/CZO, Pd/Al_2O_3, and Pd/CZO/Al_2O_3 catalysts for feed concentration of 500 ppm C_3H_6, 1% CO (λ = 1.01).

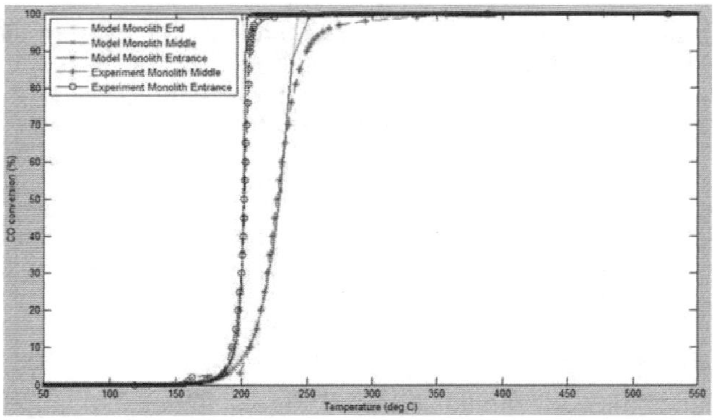

Figure 4: CO conversion (%) versus temperature using fresh Pd/Al_2O_3 catalysts for experimental and simulated feed concentration of 1% CO (λ = 1.01).

References

[1] Fernandez-Garcia, M.; Martinez-Arias, A.; Iglesias-Juez, A.; Hungria, A.B.; Anderson, J.A.; Conesa, J.C.; and Soria, J. *App. Cat. B* 2001, 3**1**, 39-50.

[2] Martinez-Arias, A., Fernandez-Garcia, M.; Iglesias-Juez, A.; Hungria, A.B.; Anderson, J.A.; Conesa, J.C.; and Soria, J.. *App. Cat. B* 2001, **31**, 51-60.

[3] Sharma, S.; Hegde, M.S.; Das, R.N.; and Pandey, M. *App. Cat. A* 2008, **337**, 130-137.

[4] Harmsen, J.M.A.; Hoebink, H.B.J.; and Shouten, J.C. *Chem. Eng. Sci.* 2001, **56**, 2019-2035.

[5] Yao, Y-.F. Y. *J. Catal.*, 1984, **87**, 152-162.

[6] Cant, N.W.; Hicks, P.C.; and Lennon, B.S. *J. Catal.* 1978, **54**, 372-383.

[7] Rainer, D.R.; Koranne, M.; Vesecky, M.; and Goodman, D.W. *J. Phys. Chem. B* 1997, **101**, 10769-10774.

[8] Joshi, S.; Harold, M.P.; Balakotaiah, V. *AIChE J.,* 2009, **55,** 1771-1783.

[9] Raj, R.; Harold, M.P.; Balakotaiah, V. *Chem. Eng. J.*, 2015, **281**, 322-333.

Copyright © American Institute of Chemical Engineers. All rights reserved.

456585 Interaction Between Highly Dispersed Ionic Pd2+ and Mn-Ce Solid Solution Support for Low Temperature CO Oxidation

Abstract:

Interaction between highly dispersed ionic Pd^{2+} and Mn-Ce solid solution support for low temperature CO oxidation

Chao Wang, Cun Wen, Erdem Sasmaz, Jochen Lauterbach*

Department of Chemical Engineering, University of South Carolina, Columbia, SC, 29208, U.S.A.

*lauteraj@cec.sc.edu

Supported Pd catalysts are widely used in automobile three-way catalysts (TWC) and diesel oxidation catalysts. While Pd catalysts perform well for CO and hydrocarbon emission control during continuous engine operation, 80â€"90% of CO and hydrocarbons are emitted into the atmosphere during the cold start phase of an automobile when the catalyst temperature is low [1]. Hence, promotion of catalytic activity at low temperatures is desired to achieve better removal of pollutants. Pd-CeO_2 catalyst is a widely studied category of supported Pd catalysts for low temperature CO oxidation. By doping Mn or both Mn and Sn into the CeO_2 lattice and forming a solid solution, we found that Pd-Ce based catalysts show excellent CO oxidation activity at even ambient temperature [2]. Metal-support interaction has been proposed as one of the most important factors that determine the Pd-CeO_2 catalyst activity at low temperature, while doping a secondary metal into the CeO_2 support may improve and change such interaction between Pd and Ce based supports. So far, the dynamic Pd structure change and interaction with the Ce based solid solution supports before and during reaction are not yet fully understood.

In this study, Pd supported on Mn doped CeO_2 solid solution (Pd-MC) and pure CeO_2 are synthesized and compared to evaluate metal-support interaction. CO oxidation tests show that Pd-MC achieves almost 100% CO conversion at 50°C, which is much higher than Pd-CeO_2 (40%). The high activity of Pd-MC can be attributed to the formation of more highly dispersed ionic Pd^{2+} on MC and strong oxygen transfer interaction between Pd species and MC support. Detailed dynamic changes of Pd sites on MC and CeO_2 before and during reaction were characterized and compared by a series of ex-situ and in-situ spectroscopic experiments, including X-ray photoelectron spectroscopy (XPS), X-ray absorption spectroscopy (XAS), and diffuse reflectance infrared Fourier transform spectroscopy (DRIFTs). H_2 temperature-programmed reduction (H_2-TPR) and CO-O_2 pulse were used to investigate the effect of metal-support interaction on the redox properties of the catalysts. In addition, for practical implementation of low temperature CO oxidation catalysts, effects of hydrothermal treatment, steam, hydrocarbon, and space velocity on the catalytic activity were evaluated on Pd-MC.

References

1. M. Shelef, R. W. McCabe, Catal. Today 62, (2000) 35-50.

2. C. Wang, E. Sasmaz, C.Wen, J. Lauterbach, Catal. Today 258 (2015) 481â€"486.

Copyright © American Institute of Chemical Engineers. All rights reserved.

472730 Effect of Rare-Earth Metal Dopants on Ceria Catalysts for CO Oxidation

Abstract:

Ceria has been intensively studied in a field of catalysis with a variety of applications, such as automotive emission control, catalytic combustion, and hydrocarbon reforming, and electrocatalytic reactions. Especially, ceria has been used as an active component in three-way catalysis (TWC) where pollutant emission from internal combustion engines is effectively reduced, due to the nature of outstanding redox behavior being amenable for adsorbing and giving off the oxygen from the atmosphere. Although most researchers have investigated transition metals-doped ceria, rare-earth metal-doped ceria nanocatalysts for CO and soot oxidation may be better support materials.

In this study, we examine CeO_2 (ceria) doped by various rare-earth metals to search for the suitable CO oxidation activity. We synthesized rare-earth metal (RE=Pr, Sm, Nd, and La) doped CeO_2 nanoparticles with two different RE concentrations, $Ce_{0.9}RE_{0.1}O_{2-\delta}$ (RE=10%) and $Ce_{0.8}RE_{0.2}O_{2-\delta}$ (RE=20%), by EDTA-citrate complexing method which is capable of producing highly pure and homogeneous nano-powders. Then, the nanoparticles were characterized by X-ray diffraction (XRD), Scanning Electron Microscopy (SEM), and Transmission Electron Microscopy (TEM). The surface area was also evaluated by Brunauer-Emmett-Teller (BET) measurement. The catalytic efficiency of RE-doped CeO_2 for CO oxidation was measured by gas chromatography (GC). Our results will be helpful for effectively exploring the RE dopant candidates of ceria and the optimal concentration for the enhanced catalytic activity.

Copyright © American Institute of Chemical Engineers. All rights reserved.

456809 Carbon Monoxide Oxidation and Preferential Oxidation on Pt Alloy Nanoparticle Catalyst with Engineered Surface

Abstract:

Carbon monoxide oxidation and preferential oxidation (PROX) are two important reactions for environmental control and for applications requiring the use of clean hydrogen, for instance hydrogen fuel cell technology and ammonia synthesis. It is because that the typical hydrogen production processes by reforming hydrocarbons and diesel fuels generate CO as a byproduct, which is notorious poisoning species and readily deactivating many catalysts. There is thus a need for selective removal of CO before hydrogen can be used for these applications. We report the study of Pt alloy nanoparticles with well-engineered surfaces as low temperature CO oxidation and PROX catalyst. In particular, the researched octahedral Pt3Ni alloy nanoparticle catalysts, which are exclusively enclosed by the (111) surface, exhibit both high PROX activity and 100% selectivity from RT all the way till around 100 °C. The much promoted CO oxidation kinetics on octahedral Pt3Ni was attributed to significantly decreased Ea comparing to spherical Pt3Ni and Pt (~40.0 kJ/mol vs. 56.9 and 66.3 kJ/mol) and the presence of both Pt and Ni active sites on the engineered (111) surface, which can work synergistically to activate and react CO and O2. The 100% PROX selectivity was associated with the unique (111) surface geometry of the octahedral Pt3Ni and the big difference in the adsorption energy of CO and H2 on the surface, which help with suppressing H2 from adsorption and oxidation.

Copyright © American Institute of Chemical Engineers. All rights reserved.

481309 Sensitivity of XANES and XES to the local environment: Multiple adsorption of H_2O and NH_3 on Cu-SSZ-13

Abstract:

Sensitivity of XANES and XES to the local environment: Multiple adsorption of H_2O and NH_3 on Cu-SSZ-13

Renqin Zhang[1], Hui Li[1], Christopher Paolucci[2], Atish A. Parekh[3], Janos Szanyi[4], Feng Gao[4], Trunojoyo Anggara[2], Hui Li[2], William F. Schneider[2], Fabio Ribeiro[3] and Jean-Sabin McEwen[1*]

1. Washington State University, Pullman, WA; 2. University of Notre Dame, Notre Dame, IN; 3. Purdue University, West Lafayette, IN; 4. PNNL, Richland, WA

*js.mcewen@wsu.edu

Nitrogen oxides (NO_x) are one of the main air pollutants present in the exhaust from diesel engines, which are popular for vehicle transportation due to their efficiency and durability [1]. However, NO_x emission control is a challenge in these "lean-burn" engines [2]. As is well known, the selective catalytic reduction (SCR) of NO_x with NH_3 is a reaction between the NO, NO_2, and O_2 oxidants and the NH_3 reductant to form N_2 and H_2O. Copper-exchanged small-pore micro-crystalline materials with the chabazite structure (Cu/CHA), such as Cu-SSZ-13, display excellent catalytic activity and hydrothermal stability in SCR of NO_x, as has been shown in a number of recent studies [3, 4]. Although Cu-SSZ-13 have been commercialized as diesel after-treatment catalysts, the fundamental chemical and physical properties need to be characterized in order to aid in the design of new and better catalysts [5].

X-ray absorption spectroscopy (XAS) is a versatile tool to determine the oxidation state and the local structure of Cu in Cu/CHA. Cu K-edge X-ray absorption near edge spectra (XANES) has been widely used to study the properties of Cu-SSZ-13 because of its high activity and selectivity in the NH_3 SCR of NO_x. It is demonstrated that XANES is extremely sensitive to the local structure of Cu ions [6, 7]. However, the exact relationship between XANES features and local structure is still unclear. In this work, we examine how the geometry of a Cu^+ ion with a linear configuration and Cu^{2+} ion with a square planar configuration in Cu-SSZ-13 correlate with the XANES features. When water and ammonia binds to a Cu^+ ion with a linear configuration, a strong intensity peak around 8983 eV in Cu K-edge XANES appears. When two ammonia molecules bind at a Cu^+ ion, it is also found that the intensity of the peak around 8983 eV decreases as one decreases the N-Cu-N bond angle from 180 to 100 degree (see Figure 1a). By analyzing the corresponding PDOS, it is found that the splitting of 4p state correlates with the N-Cu-N bond angle and the corresponding decrease in the K-edge intensity. Concerning the case of Cu^{2+} ions with a square planar configuration bounded to four ammonia molecules, it is found that there are two feature peaks around 8986 and 8993 eV in their Cu K-edge XANES. By studying this simple model, we find that changing the N-Cu-N angle between one of the bonded NH_3 and the Cu-centered ion from 180 to 100 degree results in a decrease in the intensity of both peaks around 8986 and 8993 eV. It is also found that the broken of square planar configuration results in the delocalization of 4p state.

An important limitation of XANES is that it is difficult to distinguish the nitrogen and oxygen atoms in the proximity of the metal center. X-ray emission spectroscopy (XES) could overcome this limitation. It is reported that the kbeta″ peak in XES of a Cu-SSZ-13 sample presents a blue shift with the formation of Cu-N bond [8], as shown in Figure 1b. This can be understood by the fact that the kbeta″ peak originates from the ligand 2s to a metal 1s crossover transition and the nitrogen 2s level lies higher in energy than the one for oxygen. By comparing the XES results between H_2O and NH_3 adsorbed on Cu-SSZ-13, it is found that the kbeta″ peak of Cu-O bond appears at lower energy than that of Cu-N bond. In addition, the coordination number of Cu ions significantly affects the energy difference between the position of the kbeta″ peak. By performing a density of states (DOS) analysis, these results can be understood at the electronic level. Overall, our careful elucidation of the spectroscopic properties of Cu ions in Cu-SSZ-13 via theoretical techniques can assist us in interpreting the XANES and XES experimental results so as to gain a deeper understanding of the SCR of NO_x with NH_3 throughout its catalytic cycle.

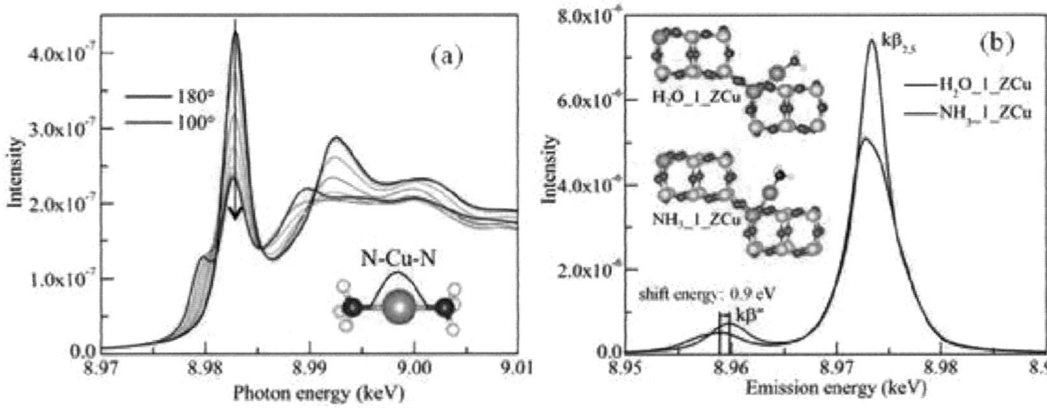

Figure 1. (a) Variation of Cu K-edge XANES for 2 NH_3 adsorbed on Cu^+-SSZ-13 while changing the N-Cu-N angle (as shown insert in the right bottom) from 180 degrees (black line) to 100 degrees (blue line). (b) Computational XES of Cu ions in H_2O and NH_3 adsorbed on Cu^+-SSZ-13. Two emission lines, kbeta" and kbeta$_{2,5}$, are presented.

References

[1] U. Deka, I. Lezcano-Gonzalez, B.M. Weckhuysen, A.M. Beale, ACS Catal., 3 (2013) 413

[2] S.i. Matsumoto, CATTECH, 4 (2000) 102

[3] J.H. Kwak, R.G. Tonkyn, D.H. Kim, J. Szanyi, C.H. Peden, J. Catal., 275 (2010) 187

[4] L. Ma, Y. Cheng, G. Cavataio, R.W. McCabe, L. Fu, J. Li, Chem. Engin. J., 225 (2013) 323

[5] F. Gao, J. Kwak, J. Szanyi, C.F. Peden, Top. Catal., 56 (2013) 1441

[6] R. Zhang, J. Szanyi, F. Gao, J.-S. McEwen, Catal. Sci. Technol., 6 (2016) 5812

[7] R. Zhang, K. Helling and J.-S. McEwen, Catal. Today, 267 (2016) 28

[8] F. Giordanino, E. Borfecchia, K.A. Lomachenko, A. Lazzarini, G. Agostini, E. Gallo, A.V. Soldatov, P. Beato, S. Bordiga and C. Lamberti, J. Phys. Chem. Lett., 5 (2014) 1552

Copyright © American Institute of Chemical Engineers. All rights reserved.

469446 Mercury Oxidation over Cu-SSZ-13 Catalysts Under SCR Conditions for Power Plant Applications

Abstract:

Many combustion-based systems, such as power plants and automobiles, utilize selective catalytic reduction (SCR) as a pollution control technology to reduce the emissions of nitrogen oxide (NO_x) species. In power plants, the reduction of the NO_x occurs via the injection of NH_3 over a V-(Mo or W)/Ti oxide catalyst. Recently, the promulgation of the Mercury and Air Toxic Standards regulates mercury, requiring a 90% reduction of Hg emissions. While activated carbon injection is one solution to limit Hg pollution, it can be costly. An alternative is the catalytic oxidation of the elemental Hg to Hg^{+2} via HCl, where the oxidized pollutant is readily water soluble and easily captured downstream in the flue gas desulfurization unit. In this way, the traditional SCR material acts as a bifunctional catalyst limiting both NO_x and Hg emissions. While previous studies have examined the oxidation activity of the commercial SCR catalyst, they often display limited activity, falling short of the required emissions limits. To address this, multiple new materials have been proposed for Hg oxidation. In this study, small pore zeolite Cu-SSZ-13 was investigated for its possible application in coal-fired power plants as a new bifunctional SCR and Hg oxidation catalyst.

In this study, SSZ-13 and Cu-exchanged SSZ-13 are compared to a commercial SCR catalyst with respect to their Hg oxidation activity under a variety of flue gas compositions and temperatures. A simulated flue gas environment is created through the combustion of methane in air in a high-temperature furnace. Other pollutants common to coal-combustion flue gas, such as HCl, SO_2, NO, and NH_3 are injected downstream of the furnace but upstream of the packed-bed reactor housing the catalyst. Concentrations of elemental and oxidized mercury were measured using a PS Analytical mercury analyzer to determine the amount of mercury oxidation. Initially, the flue gas composition was varied to examine the effect of individual flue gas components on Hg oxidation with the catalysts kept at 375°C, an intermediate SCR temperature. The compositions investigated include flue gas only (CO_2+H_2O+N_2+O_2), flue gas with HCl, with HCl+SO_2 and with HCl+NO+NH_3. Finally, the temperature effect was studied by varying the temperature between 300-450°C, representative of standard operating temperatures in SCR units for coal-fired power plants.

Copyright © American Institute of Chemical Engineers. All rights reserved.

459898 Thiophenic Species Removal over a New Zn-Offretite Zeolite Via Selective Adsorption for FCC Processes

Abstract:

This study considers the sulfur selective adsorption of a new Zn-Offretite zeolite based additive under FCC conditions. This Zn-Offretite (Zn-OFF) zeolite was prepared via direct inclusion of zinc in the OFF synthesis. The goal is to obtain a sorbent suitable for thiophenic chemical species removal via selective adsorption under FCC conditions. Physicochemical characterizations combined with reactivity studies were developed to understand the adsorption process. It was proven that the introduction of Zn in the OFF zeolite considerably increases acidity and Lewis/Brönsted acid site ratio. These two properties promote the thiophenic selective adsorption in the OFF structure. The valuable performance of the Zn-OFF additive was demonstrated in a CREC Riser Simulator unit under typical FCC conditions. Reactivity studies showed that the Zn-OFF additive is a suitable sorbent for thiophenic species adsorption with an optimum operation at 5s and 530°C. It was found that the Zn-OFF in combination with a FCC commercial catalyst can reach 51 wt% of 2-methyl-thiophene reduction with only 2wt% of sulfur left as coke. In addition, and when the FCC catalyst was blended with 5-15wt% of 3.5Zn-OFF, it yielded a 210% coke reduction. On the basis of the above, it is demonstrated that the Zn-OFF additive can contribute to both selective adsorption of sulfur containing species and lower coke. This reduced coke may become a significant Zn-OFF additive feature for mitigating SOx emissions in the FCC regenerator.

Copyright © American Institute of Chemical Engineers. All rights reserved.

468495 Wash-Coat Development for Lean-Burn Engine-Exhaust Aftertreatment: A Novel Way of Incorporation of Binder to Pd Supported on Sulfated Zirconia

Abstract:

Wash-coat development for Lean-burn Engine-exhaust Aftertreatment: A Novel Way of Incorporation of Binder to Pd Supported on Sulfated Zirconia

Sreshtha S. Majumdar, Gokhan Celik, Anne-Marie Alexander, Umit S. Ozkan*The Ohio State University, Columbus, Ohio 43210, U.S.A.

*ozkan.1@osu.edu

As the environmental regulations are getting more stringent, it is of great importance to develop a catalytic aftertreatment system for natural gas-fired lean-burn engines to reduce the emissions of nitrogen oxides (NO_x), unburned hydrocarbons (CH_4, C_2H_6, and C_3H_8) and carbon monoxide efficiently. We have developed a dual-catalyst aftertreatment system which utilizes the hydrocarbons present in the exhaust streams of lean-burn engines for reducing NO_xemissions. The dual-catalyst aftertreatment system is a physical mixture composed of a reduction catalyst, palladium supported on sulfated zirconia (Pd/SZ), and an oxidation catalyst, cobalt supported on ceria. Such a catalytic system which utilizes methane in the exhaust stream offers several advantages considering that the emissions of air-pollutant greenhouse gases are controlled in a single unit without a need of injecting and handling an external reducing agent such as ammonia [1-3]. Development of a catalytically active washcoat for monolith cores is essential for the practical use of this dual-catalyst system. However, if the washcoat adheres poorly to the monolith core, the aftertreatment unit will suffer from the irreversible loss of the catalytically active phase. In order to improve the adhesive properties, conventionally, binders are used in a catalyst slurry.

In this study, we have aimed to improve the adhesive properties at the molecular level. For this purpose, alumina, as a binder, was incorporated in situ to the sol-gel medium of Pd/SZ prior to the gelation during synthesis. Samples prepared by this novel approach have shown superior differences in terms of catalytic performance than the samples prepared by the conventional method. Addition of binder before gelation to the sol-gel medium also resulted in changes of textural and structural properties of binder-free samples, and adhesive properties of the washcoat as shown by N_2 physisorption, in-situ X-ray diffraction (XRD) calcination, ^{27}Al-NMR, laser RAMAN and infrared spectroscopy, scanning electron microscopy (SEM) and ultrasonication, Water and sulfur tolerance as well as hydrothermal stability of the binder-incorporated catalyst will be presented as well.

[1] B. Mirkelamoglu, M. Liu, U.S. Ozkan, Catalysis Today, 151 (2010) 386-394.

[2] P. Gawade, A.-M.C. Alexander, R. Silver, U.S. Ozkan, Energy and Fuels, (2012).

[3] P. Gawade, B. Bayram, A.-M.C. Alexander, U.S. Ozkan, Applied Catalysis B: Enviromental (2012).

Copyright © American Institute of Chemical Engineers. All rights reserved.

469481 Carboxylation of Propylene Oxide to Propylene Carbonate in Slurry and Trickle Bed Reactors

Abstract:

Carboxylation of Propylene oxide to Propylene carbonate in slurry and trickle bed reactorS

Pallavi Bobba, Xin Jin, Bala Subramaniam and Raghunath V. Chaudhari

Center for Environmentally Beneficial Catalysis, Department of Chemical & Petroleum Engineering, University of Kansas,
1501 Wakarusa Dr., Lawrence, KS 66047

CO_2 is naturally available carbon source and emitted from industrial processes, automobiles and petrochemical refineries. Catalytic conversion of CO_2 to value added chemicals provides an alternative green, cheap and sustainable synthesis route for industrial chemicals, otherwise produced using toxic reagents such as phosgene. An important example is the carboxylation of propylene oxide (PO) to propylene carbonate (PC), a key intermediate for dimethyl carbonate and polycarbonates. These cyclic carbonates are widely used as aprotic solvents, antifoaming agents, antifreeze, plasticizers and monomer for various commodity polymers. Several studies are known on synthesis of cyclic carbonates by carboxylation of epoxides employing homogeneous and heterogeneous catalysts. Heterogeneous catalysts consisting of metal oxides, zeolites, polymer supported quaternary onium salts, ion exchange resins, and polymer supported ionic liquids have been studied with the goal of improving catalytic activity and selectivity. However, no efforts have been made to understand the intrinsic kinetics of carboxylation and the related reaction engineering studies. Here, we report an experimental study on (a) intrinsic kinetics of carboxylation of PO to PC using ion exchange resin catalyst in a batch slurry reactor, and (b) modeling of a trickle bed reactor with experimental validation.

The experiments for kinetic studies were carried out in a stirred pressure reactor with 100 cm^3 capacity with provisions for control of agitation speed, temperature and sampling of liquids. In these experiments, the CO_2 pressure in the reactor was kept constant by continuous supply through a CO_2 reservoir using a constant pressure regulator such that the temporal reaction progress was followed from the pressure decrease in the reservoir. At the end of each experiment, liquid products were analyzed for PO and PC to assess the material balance. The effects of catalyst loading, PO concentration, pressure and temperature were studied as shown in Figures 1 and 2. Kinetic analysis of these data will be presented along with discrimination of rate models and estimation of rate parameters.

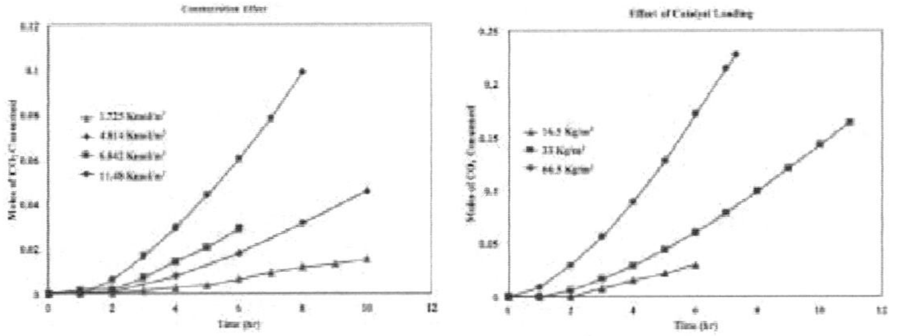

Figure 1. Effect of Concentration and Catalyst loading on CO_2 consumption as a function of time at 368.15K and 1.4 MPa

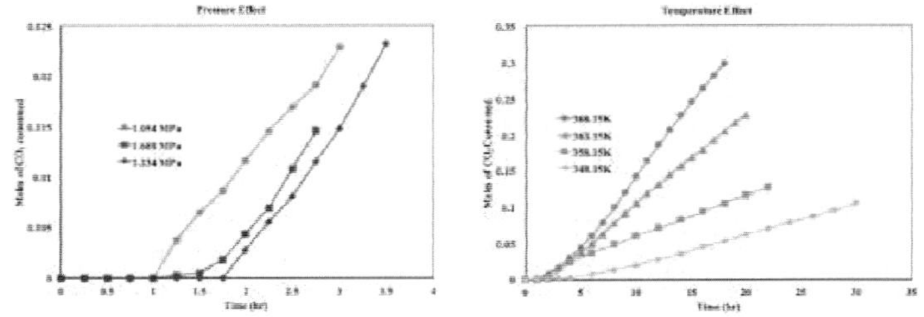

Figure 2. Effect of Pressure at 368.15K and Temperature at .14 MPa on CO_2 consumption.

For trickle bed experiments, a high-pressure trickle bed reactor of 2.5 cm diameter and 10 cm length was used with down flow of gas (pure CO_2) and liquid (PO in PC) phase. The effect of gas and liquid velocity, inlet PO concentration and temperature was studied. Steady state conversion and selectivity were observed for different reaction conditions. The analysis of reactant/products was carried out using GC. Typical results showing the effect of liquid velocity on PO conversion are shown in Figure 3. Detailed mathematical model for the trickle bed reactor based on the kinetics determined in the slurry reactor will be presented incorporating the effect of mass transfer and the wetting of catalyst particles.

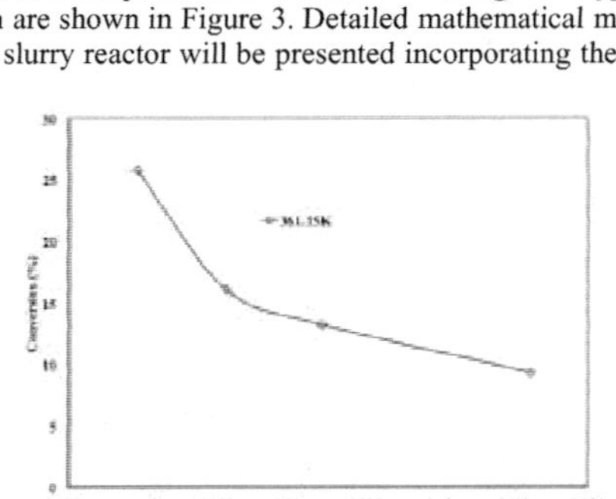

Figure 3. Conversion as function of liquid velocity at 361.15K at CO_2 velocity of 1.867 cm/s and pressure of 0.4 MPa

Copyright © American Institute of Chemical Engineers. All rights reserved.

453142 Supported Pt-Re Catalyst for Selective Hydrogenolysis and Hydrogenation Reactions

Abstract:

The addition of Re to a supported Pt catalyst promotes the rate of glycerol hydrogenolysis to propanediols in liquid water by more than an order of magnitude and shifts the initial product selectivity from predominantly terminal hydrogenolysis to internal hydrogenolysis. In this work, conventional XPS and near ambient pressure XPS were used to prove that a range of Re oxidation states was present on the Pt-Re catalysts after reduction in H_2 at 393 and 493 K, even though reduction of Re is enhanced by the Pt. The DRIFTS of adsorbed pyridine and the aqueous phase hydrolysis of propyl acetate were used to identify and quantify Brønsted acid sites that were presumably associated with the Re component of the bimetallic catalyst. The mechanism of glycerol hydrogenolysis is proposed to involve acid-catalyzed dehydration followed by Pt-catalyzed hydrogenation of the unsaturated intermediate. This proposed mechanism for glycerol hydrogenolysis is supported by kinetic results after selective neutralization of the acid sites, observation of a primary kinetic isotope effect with deuterated glycerol, and measurement of the orders of reaction with respect to glycerol and dihydrogen. A Pt-Re sample also catalyzed the hydrogenation of C=C and C=O bonds in methyl vinyl ketone, crotonaldehyde, 2-butanone, and butanal in liquid water. The relative rates of C=C and C=O bond hydrogenation over Pt-Re were significantly faster than C-O bond hydrogenolysis, supporting the mechanistic hypothesis that a hydrogenation step in the glycerol hydrogenolysis path is kinetically irrelevant. Interestingly, oxophilic Re can be used to promote the Pt-catalyzed hydrogenation of carbonyl groups in multifunctional molecules.

Copyright © American Institute of Chemical Engineers. All rights reserved.

455881 Thermally Stable, Regenerable Single Atom Catalysts Via Atom Trapping

Abstract:

Thermally Stable, Regenerable Single Atom Catalysts via Atom Trapping

Abhaya Datye

University of New Mexico, Albuquerque, NM, USA

Single atom catalysts show great promise in terms of efficient utilization of precious metals, improved reactivity and selectivity. However, single atom catalysts can deactivate due to sintering and formation of nanoparticles when the catalyst is exposed to harsh reaction conditions. Stabilization of single atom catalysts and their regeneration requires suitable trapping sites on the catalyst support. Our work has focused on alumina supports, which are extensively used in industrial catalysis.

In recent work we found that La-stabilized alumina was able to stabilize atomically dispersed Pd, leading to high activity for CO oxidation[1]. The Pd when atomically dispersed is bound to the alumina through surface oxygen atoms on the alumina surface. Lanthanum oxide which is added to alumina to stabilize the support helps to anchor the atomically dispersed Pd, making it possible to regenerate the catalyst. In the case of the Pt-Sn/alumina catalysts, we noticed a similar behavior, where the presence of Sn in atomically dispersed form on the alumina made it possible to regenerate the catalyst and restore catalyst activity and selectivity [2]. The Pt/alumina treated under similar conditions could not be regenerated to a significant extent. The major difference between these two catalysts is the presence of atomically dispersed Sn which is stable after multiple cycles of propane dehydrogenation and provides sites for trapping the Pt [2].

This lecture will provide additional examples how atom trapping can help in the generation of single atom catalysts that are thermally stable and regenerable.

References

[1] E.J. Peterson, A.T. DeLaRiva, S. Lin, R.S. Johnson, H. Guo, J.T. Miller, J. Hun Kwak, C.H.F. Peden, B. Kiefer, L.F. Allard, F.H. Ribeiro, A.K. Datye, Low-temperature carbon monoxide oxidation catalysed by regenerable atomically dispersed palladium on alumina, Nat Commun, 5 (2014).

[2] H.N. Pham, J.J.H.B. Sattler, B.M. Weckhuysen, A.K. Datye, Role of Sn in the Regeneration of Pt/γ-$A_{12}O_3$ Light Alkane Dehydrogenation Catalysts, ACS Catal., 10.1021/acscatal.5b02917 (2016).

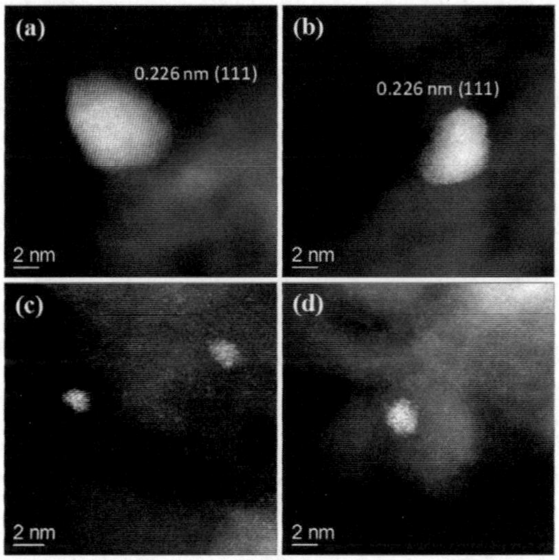

Fig. 1 Pt/γ-Al_2O_3 (a,b) and Pt-Sn/γ-Al_2O_3 after 10 cycles of propane dehydrogenation, followed by oxidative regeneration [2].

Copyright © American Institute of Chemical Engineers. All rights reserved.

462157 Geometric and Electronic Effects in Pd and Pt AlloyÂ Alkane Dehydrogenation Catalysts

Abstract:

The recent shale gas boom and corresponding decrease in light alkane prices have made on-purpose olefin production via catalytic alkane dehydrogenation economically competitive. High dehydrogenation rates while decreasing the C-C bond hydrogenolysis is critical for selectivity control to olefins. Recently, it has been found that olefin selectivity can be controlled by catalysts with specific intermetallic alloy (or intermetallic compound, IMC) structures.[1,2]

Silica supported Pd and Pd-In catalysts with different In:Pd atomic ratios and similar particle size (~2 nm) were tested for ethane dehydrogenation at 600 °C. For a monometallic Pd catalyst, the dehydrogenation selectivity was 53 %. Addition of In to Pd increased the dehydrogenation selectivity to near 100 % and also greatly increased the TOR. Combined carbon monoxide IR, *in situ* synchrotron XRD and XAS analysis were able to resolve the detailed nanoparticle geometric structure and showed that for Pd-In catalysts with increasing In loading, an increasing fraction of the Pd nanoparticle surface was transformed into PdIn IMC with a cubic CsCl structure.

A series of Pt-In catalysts with various In:Pt ratios were also prepared and compared to a monometallic Pt catalyst for ethane dehydrogenation. As with Pd, the addition of In to Pt catalysts promoted ethane dehydrogenation selectivity. However, where Pd and In formed a single intermetallic compound over a wide range of bulk In:Pd ratios, changes to the bulk In:Pt ratio led to the formation of multiple intermetallic phases. *In-situ* x-ray characterizations revealed that the Pt_3In phase formed when the bulk catalyst composition had a low In:Pt ratio whereas the Pt_2In_3 phase was formed at a high In:Pt ratio. Both phases exhibited superior catalytic performance compared to the monometallic Pt catalyst.

Formation of these PdIn and PtIn IMC structures on the catalyst surface geometrically isolates the Pd and Pt catalytic sites by non-catalytic, metallic In neighbors, which is suggested to be responsible for the high olefin selectivity. The changes in initial TOR and E_{app} seen in these IMC catalysts are also indicative of electronic changes to the catalytic atoms by the inactive In neighbors. It appears that a combination of geometric changes and electronic changes to Pt due to the formation of the IMC phases are responsible for the promotional effects of In. A fundamental understanding of these geometric and electronic changes can lead to improved catalyst design and new approaches to catalyst optimization through the formation of intermetallic compounds.

References:

1) Childers D. J., Schweitzer N. M., Shahari S. M. K., Rioux R. M., Miller J. T. and Meyer R. J., Journal of Catalysis, *2014, 318, 75-84.*

2) Gallagher J. R., Childers D. J., Zhao H., Winans R. E., Meyer R. J. and Miller J. T., Physical Chemistry Chemical Physics, *2015, 17, 28144-28153*

Copyright © American Institute of Chemical Engineers. All rights reserved.

467226 Catalytic Treatment of Water Contaminated with Chlorinated Hydrocarbons

Abstract:

Catalytic Treatment of Water Contaminated with Chlorinated Hydrocarbons

Umit. S. Ozkan[1*], Hyuntae Sohn[1], Gokhan Celik[1], Seval Gunduz[1], Saurabh A. Ailawar[1], Paul Edmiston[2]

[1]The Ohio State University, Columbus, Ohio 43210 (United States)

[2]The College of Wooster, Wooster, Ohio, 44691 (United States)

*ozkan.1@osu.edu

Contamination of groundwater by chlorinated compounds such as trichloroethylene (TCE) is an environmental concern due to their high level of toxicity and potential impact on drinking water [1]. It is estimated that TCE is present above permissible levels in 9-34% of drinking water sources in the U.S [2]. Thus, development of a remediation system to remove chlorinated compounds from groundwater has become imperative. Existing remediation techniques for treatment of contaminated water are not efficient or feasible due to low rates of remediation, high energy inputs, and media regeneration/replacement cost. [3, 4]. Although hydrodechlorination (HDC) appears to be an efficient way of groundwater remediation, it suffers kinetically due to low concentration of contaminants and catalyst deactivation due to anionic groundwater constituents [5, 6].

The focus of this study is the use of a new class of materials, namely swellable organically-modified silicates (SOMS)

as a catalyst scaffold for HDC reactions. SOMS are absorptive, swellable when contacted with organics and extremely hydrophobic [7-9]. Organometallic precursors used for catalyst synthesis facilitate deposition of the active metal inside the pores. The absorptive and hydrophobic nature of SOMS will guide the organics toward the active sites and repel water and anionic poisons. When used as catalyst scaffolds, SOMS can have potential to resolve the issues involved in HDC of TCE.

In this study, the catalytic activity and poison resistance of Pd-incorporated SOMS have been investigated for HDC of TCE. In addition to continuous flow and batch reactor experiments, a wide array of characterization techniques including N_2 physisorption, laser Raman and infrared spectroscopy, solid-state nuclear magnetic resonance (NMR), temperature programmed studies with mass spectrometer, X-ray photoelectron spectroscopy (XPS), inductively coupled plasma optical emission spectrometer (ICP-OES), X-ray absorption fine structure (XAFS) and transmission electron microscopy (TEM) has been used for a better understanding of the nature of these materials.

References

[1] R.E. Doherty, Environmental Forensics 1 (2000) 69-81.

[2] C. Wu, J. Schaum, Environmental Health Perspectives 108 (2000) 359-363.

[3] A. Elola, E. Dìlaz, S. OrdoÃ±ez, Environmental science & technology 43 (2009) 1999-2004.

[4] H.H. Russell, J.E. Matthews, G.W. Sewell, TCE removal from contaminated soil and ground water. Ground water issue, Environ. Protect. Agency, 1992, p. 12 pp.

[5] N. Munakata, M. Reinhard, Applied Catalysis B: Environmental 75 (2007) 1-10.

[6] G.V. Lowry, M. Reinhard, Environmental Science & Technology 35 (2001) 696-702.

[7] C.M. Burkett, P.L. Edmiston, Journal of Non-Crystalline Solids 351 (2005) 3174-3178.

[8] C.M. Burkett, L.A. Underwood, R.S. Volzer, J.A. Baughman, P.L. Edmiston, Chemistry of Materials 20 (2008) 1312-1321.

[9] P.L. Edmiston, L.A. Underwood, Separation and Purification Technology 66 (2009) 532-540.

Copyright © American Institute of Chemical Engineers. All rights reserved.

466611 How Pt Interacts with CeO_2 Under the Reducing and Oxidizing Environments at Elevated Temperature?

Abstract:

Platinum group metals (PGMs) are widely used as catalytically active phase in catalyst. Their unique electronic d-band structure allows them to be applied in many kinds of different reactions including CO oxidation, methane oxidation, hydrogenation, and water-gas shift reaction. PGMs should be highly dispersed on supports to increase the cost-effectiveness (catalytic activity/amounts) due to their expensiveness. When PGMs are dispersed on support, they interact electronically with each other, which brings about the physiochemical effects on their catalytic properties. In this regard, metal-support interaction (MSI) between PGM and support has been an interesting topic of many research groups for the past several decades.

It is known that the interaction between PGM and support affects the degree of sintering at high temperature, which remains a great challenge to make more durable catalysts. Especially, compared with Al_2O_3 support, PGM and ceria interact with each other more strongly, resulting in the higher PGM dispersion even after the high temperature treatment. Such strong interaction between impregnated PGM and ceria not only guarantees high metal dispersion but also can be beneficial for the catalytic reactions.

Hence, understanding the metal-ceria interaction is of practical importance to provide the rational design of ceria supported PGM catalysts. In the present research, the interaction between Pt and CeO_2 has been extensively investigated by combining various characterization methods including X-ray diffraction (XRD), N_2 adsorption/desorption (BET/BJH), H_2 temperature programmed reduction (TPR), pulsed and static CO-chemisorption, X-ray photoelectron spectroscopy (XPS), thermogravimetric analysis (TGA), and *in-situ* Raman spectroscopy. Especially, the effect of Pt-ceria interaction under the oxidizing/reducing environment on the thermal stability of Pt/CeO_2 catalyst has been focused in detail.

Copyright © American Institute of Chemical Engineers. All rights reserved.

464968 Metal Oxide-like Clusters in MOFs As Supports for Well-Defined Rhodium Complex and Rhodium Cluster Catalysts

Abstract:

The nodes of some metal-organic frameworks (MOFs) closely resemble small pieces of metal oxides, exemplified by the Zr_6 nodes of Nu-1000. Hydroxyl groups on the structural vacancies in Nu-1000 consisting of pairs of terminal OH and OH_2 groups have been evidenced by IR spectroscopy and DFT calculations; they are referred to as site 1. We developed a treatment process to convert site 1 to a new proton topology incorporating hydroxyl sites with pairs of terminal OH groups (site 2). A combination of IR spectroscopy and DFT calculations led to the identification of an intermediate structure with organic ligands coordinated to structural vacancies on the node surface which facilitated the conversion of site 1 to site 2. With $Rh(C_2H_4)_2(acac)$ (acac is acetylacetonate) as a probe of the node surfaces, reaction with the hydroxyl groups of site 1 or site 2 of Nu-1000 gave species identified by IR and EXAFS spectroscopies and DFT as mononuclear $Rh(C_2H_4)_2$ complexes. Information about the nodes as electron-donor ligands was determined in experiments in which the rhodium diethylene complexes were converted to rhodium *gem*-dicarbonyls by reaction with Co; the Î½$_{CO}$ values characterize the electron-donor properties of the support sites. The catalytic activity and selectivity of the species initially present as $Rh(C_2H_4)_2$ for ethylene hydrogenation and dimerization were investigated both experimentally and computationally and shown to be strongly influenced by the node sites. The Nu-1000-supported mononuclear rhodium complexes were treated in H_2 to form small supported rhodium clusters, characterized by changes in the IR and EXAFS spectra; the data suggest the formation of approximately Rh_2 clusters on the nodes, and these were found to be stable and 6 times more active for ethylene hydrogenation than the supported single rhodium sites.

Copyright © American Institute of Chemical Engineers. All rights reserved.

472052 Kinetic Modeling of Propene Hydroformylation in Propane-Expanded Liquid with Rh-Based Complexes

Abstract:

Conventional propylene hydroformylation to produce butyraldehyde requires polymer grade propylene as feedstock.[1] High purity propylene is obtained by energy intensive separation from propane/propylene mixtures. If propane/propylene mixtures can be used effectively as feed, process economics would be significantly enhanced. Previously, we had reported preliminary results on the beneficial effects of propane expanded liquids (with propylene/propane mixtures instead of polymer grade propylene as feed) for propylene hydroformylation with Rhodium-based catalysts. At sufficiently high pressures (>10 bar), the pressurized propane/propylene mixture forms a gas-expanded liquid phase that promotes regio-selective hydroformylation of propylene with homogenous Rh complex.[2] In this talk, we report hydroformylation of propene in propane expanded solvents in the kinetic regime in a semi batch stainless steel reactor at 70−90 °C and 6−18 bar overpressure in 2,2,4-trimethyl-1,3-pentanediol monoisobutyrate (NX-795), toluene and *n*-butanol solvent with rhodium based complexes (Rh/TPP, Rh/ CHDPP, Rh/TPPA and Rh/BIPHEPHOS). The *n/i* ratio and TOF achieved during Rh/BIPHEPHOS catalyzed hydroformylation in propane-expanded solvents is comparable with conventional processes in neat solvent. By tuning the propane/propylene partial pressure, the *n/i* ratio in propane-expanded solvents could be enhanced by 30% in Rh-catalyzed reaction with mono-phosphine ligands while maintaining a reaction rate comparable to that without propane addition. Vapor−liquid equilibria of the reaction system were also measured in a variable-volume view cell at temperatures ranging from (343.15 to 363.15) K and pressures up to 20 Bar. The measured solubilities of CO and H_2 in the liquid phases were consistent with literature values.[3] The presence of propane was found to enhance the solubilities of both CO and H_2 in the liquid phase. The enhancement factor is up to 1.56 for carbon monoxide and 1.91 for hydrogen. Based on experimentally recorded kinetic data, a kinetic model was developed to explain the effects of propane-expanded solvent on the observed rate and selectivity.

1.Krokoszinski, R., Hammon, U., & Todd, K. (2005). U.S. Patent No. 6,864,391. Washington, DC: U.S. Patent and Trademark Office.

2. Liu, D., Xie Z., Chaudhari, R.V., Subramaniam, B. Enhanced Hydroformylation of Propylene in Propane-Expanded Liquids with Rh-Based Complexes, 2015 AIChE annual meeting. Salt Lake City, UT

3. Nair, V. S., Mathew, S. P., & Chaudhari, R. V. (1999). Kinetics of hydroformylation of styrene using homogeneous rhodium complex catalyst. Journal of Molecular Catalysis A: Chemical, 143(1), 99-110.

Copyright © American Institute of Chemical Engineers. All rights reserved.

469755 Catalytic Cracking of Dodecane in Supercritical Water

Abstract:

Catalytic Cracking of Dodecane in Supercritical Water

Azadeh Zaker, Patricia Guerra, Geoffrey Tompsett, Michael T. Timko

Department of Chemical Engineering

Worcester Polytechnic Institute

Worcester MA 10609

An abstract to be submitted to the National Meeting of the AIChE (2016)

Technical Session: Reaction Engineering of Biomass and Hydrocarbons in Supercritical Water

Sponsored by: Catalysis and Reaction Engineering Division

In this work, we investigated dodecane cracking in supercritical water (SCW) using ZSM-5 as an acid catalyst. We studied the reaction in the presence and absence of water by tracking reaction progress in a batch reactor held at 400 ĪŠC and 24.1 MPa. For the same catalyst loading (5 wt. % oil-based), addition of water reduced formation rates of gas, liquid, and solid products. We identified and quantified liquid products using GC/MS and GC×GC chromatography. GC/MS was used primarily to track production of specific compounds, including toluene, xylenes, and ethylbenzene. GC×GC was used to monitor both individual compounds as well as classes of compounds: aliphatics, 1-ring aromatics, 2-ring aromatics, and 3-ring aromatics. SCW changes the observed product distribution of dodecane cracking, with increased production of aliphatics observed in the presence of water compared to increased production of aromatic compounds in its absence. Furthermore, in the absence of water, GC×GC analysis indicates increased yields of 2-ring and 3-ring products, including molecules such as naphthalene and alkyl-naphthalene. In addition to liquid phase analysis, we characterized the coke deposited on the catalyst using Temperature Programmed Oxidation (TPO) technique. Water reduces coke formation by an order of magnitude and alters its chemical characteristics. Coke deposited in the presence of water is mostly soft coke while coke formed in the absence of water is mostly graphitic hard coke. Finally, we developed a reaction network model to interpret our results. The reaction network consisted of aliphatic, aromatic, gas, and coke products. Future work will investigate catalysts other than ZSM-5 and more sophisticated molecular reaction networks.

Copyright © American Institute of Chemical Engineers. All rights reserved.

469368 Demetallation of Biocrude from Hydrothermal Liquefaction of Microalgae

Abstract:

Hydrothermal liquefaction (HTL) converts wet biomass into a crude bio-oil that could be further treated to be acceptable for processing in petroleum refineries. HTL takes advantage of the properties of hot, compressed water near its critical point to break down biomacromolecules in the feedstock into smaller molecules. Several issues presently prevent the substitution of HTL biocrude for petroleum and one is the presence of metals in the crude bio-oil. Metals could cause problems such as corrosion, fouling, catalyst deactivation during downstream processing, etc. Metal impurities in crude bio-oil are unavoidable and metals will need to be removed if biocrude is used as transportation fuels. The most abundant metals in algal biocrude after HTL include Na and Fe. Na is present in typical growth media and Fe is an element essential for microalgae cultivation.

We hydrothermally treated microalgae using different processing conditions (isothermal, rapid heating and non-isothermal conditions, different temperatures and times) and determined the metal content in the biocrude oil as well as in the other product phases. We then studied the ability of different interventions to reduce this metal content. Process options explored include treatment in supercritical water in the absence of catalyst, use of different solvents for biocrude oil recovery after HTL, and use of heterogeneous catalysts during hydrothermal treatment. The Fe and Na content was reduced over 70% by use of specific organic solvents and reduced over 98% via application of heterogeneous catalysts. Additional catalytic studies with Fe model compounds permitted post-reaction catalyst characterization via techniques such as scanning electron microscopy and Mössbauer Spectroscopy.

Copyright © American Institute of Chemical Engineers. All rights reserved.

463588 Pilot-Scale Continuous Flow Reactor for Hydrothermal Liquefaction of Algae

Abstract:

Algae-based biofuels have attracted much attention due to their advantages of not competing with land for food production, fewer terrestrial and weather limitations, higher growth rates, and strong CO_2-mitigation abilities. In recent years, research has focused on hydrothermal liquefaction (HTL) of algae since HTL does not require that the algae be dried before processing. Hydrothermal liquefaction uses high temperatures (270-350 °C) and high pressures (80-173 bar) to produce relatively high yields of bio-oil that can be upgraded to liquid transportation fuels. Most research to date has been conducted in batch reactors; shifting from batch reactors to continuous flow systems is an important next step towards commercial-scale processing.

This study presents the development and preliminary testing of a pilot-scale continuous flow reactor for hydrothermal liquefaction of fresh water microalgae at low solids loadings. A significant challenge to be overcome is the tendency of the algae feedstock and product slurries to create clogs inside filters and at bends and joints in the tubing even at very low solids loadings (<1 wt.%). To address such issues, a dual high-pressure cylinder filter system was designed to allow one filter to run while the other is being cleaned. Solid particles (char and ash), ranging from 6 to 25 μm, are captured by the cylinder filters and collected in blowdown pots. The filter elements can be cleaned in place by flushing back the product fluid. Tubing size and configuration was carefully considered to accommodate a narrow diameter desirable for the high pressure and to prevent unnecessary clogging opportunities. The goal of the reactor is to be able to perform continuous HTL on slurries with solid algae contents of 5-10 wt.%, and to produce char-free bio-oils.

Copyright © American Institute of Chemical Engineers. All rights reserved.

457442 Production of Biodiesel Fuel Via Sub/Supercritical Transesterification Reactions with Trace Amount of Homogeneous Catalysts

Abstract:

2016 AIChE Annual Meeting

November 13th to 18th, San Francisco, CA

PRODUCTION OF BIODIESEL FUEL VIA SUB/SUPERCRITICAL TRANSESTERIFICATION REACTIONS WITH TRACE AMOUNT OF HOMOGENEOUS CATALYSTS

*Jiuxu Liu, Yue Nan and Lawrence L. Tavlarides**

Department of Biomedical and Chemical Engineering, Syracuse University,

329 Link Hall, Syracuse, NY 13244, USA

* Corresponding author. Tel.: 315-443-1883; Fax: 315-443-9175

Email: lltavlar@syr.edu (Tavlarides); jliu23@syr.edu(Liu)

Abstract

Refined oils and anhydrous alcohols are required in conventional biodiesel industry since impurities as free fatty acids (FFA) and water poison the catalysts. One of the most promising advantages of the non-catalytic sub/supercritical technology (SCTE) over the conventional method is that it does not require high quality feedstocks, so low-cost waste oils which contains FFA and water, and hydrous alcohols which contains water can be used. However, it is difficult to reach a high biodiesel yield since extending reaction time at those required temperatures will thermally decompose biodiesel product, which has been proved to lower the fuel qualities such as viscosity and cold flow properties. We have known that employing catalysts in the SCTE reactions with high quality feedstocks (refined oils and anhydrous alcohol) will complete the reactions in shorter residence times to avoid decomposition, but we have not determined if those catalysts can still function under the presence of impurities such as FFA and water. Another issue related to biodiesel studies is that much fewer investigations have been done on kinetics of synthesizing FAEEs, which is a potential replacement of FAMEs, than FAMEs. Presently, the kinetic analyses of SCTE ethanolysis reactions have been based on the overall reaction and the model of first-order reaction with respect to triglycerides, and the temperature range is very limited.

In this study, we are investigating FAEEs and FAMEs biodiesel synthesis under sub and supercritical alcohol conditions catalyzed by trace amounts of homogeneous catalysts including sulfuric acid, potassium hydroxide, and potassium m/ethoxide. The objectives include: a) a kinetic study of FAEEs biodiesel synthesis at SCTE conditions with 0.1 wt% sulfuric acid, b) comparison of catalytic ability to improve yields of FAEEs and FAMEs biodiesel between sulfuric acid and other catalysts including potassium hydroxide and potassium m/ethoxide under sub and supercritical conditions, c) effect of impurities such as FFA (up to 30 wt%) and water (up to 10 v%) on these catalysts, and d) effect of reaction pressure on product yield.

At this writing we have completed the first aim which is the kinetic study of synthesizing FAEEs biodiesel under sub and supercritical conditions. The reactions were conducted at temperatures from 175 to 350 ^{o}C, residence times from 1 to 120 min, pressure of 200 bar, and 0.1 wt% sulfuric acid based on oil mass. A three-step partially reversible second-order reaction model was employed to best fit the data, and the activation energy for each reaction step was calculated. It was found that the catalytic activity of sulfuric acid decreased when switching from subcritical to supercritical conditions. Comparison of the kinetic parameters for sub and supercritical reactions will be shown, and the mechanism will be discussed. High biodiesel yields (> 90%) were reached at 175, 200, 225, 300, and 325 ^{o}C with 0.1 wt% sulfuric acid concentration.

Experiments to complete the remaining objectives are in progress and will also be reported.

Copyright © American Institute of Chemical Engineers. All rights reserved.

472237 Molecular and Microkinetic Modeling of Supercritical Fluid and Liquid-Phase Fischer-Tropsch Synthesis

Abstract:

Due to the specific and tunable properties of supercritical fluids (SCFs), they have been considered as unique media for different technical processes of extraction, separation, particle formation, and chemical reactions. In the particular case of Fischer-Tropsch synthesis (FTS), experiments have shown that utilizing supercritical solvent in this heterogeneous process would result in several distinguished improvements. Accelerating the conversion of reactants, providing more favorable product distributions, as well as preventing catalyst deactivation are some of the mentioned advantages.[1] While previous experimental studies have discussed that all these known enhancements (such as better heat distribution, improved mass transfer, selective solubility of coking products, etc.[2]) relate to solution thermodynamics and transport, not many investigations have considered the solvent effect(s) on heterogeneous kinetics of FTS reactions. In this study, we have analyzed the influence of hexane solvent on the reaction rates of FTS process over Co(0001) catalyst using different theoretical approaches of molecular dynamics and quantum mechanics. By adopting an existing microkinetic model[3] and modifying it, potential of mean forces (PMF) and density functional theory (DFT) calculation have been performed in order to estimate the equilibrium conditions and rate parameters of individual elementary reactions under different phase conditions. Ultimately, these kinetic and thermodynamic parameters determined from molecular modeling and simulation are used as input into a micro-kinetic model, to characterize the differences in catalysis in SCF and liquid hexane versus conventional gas-phase processing. Microkinetic modeling results include yield and selectivity versus reaction time and net rate and sensitivity analyses of the elementary reaction mechanisms in all the considered reaction phases. The simulation results show that the SCF and liquid phase solvent not only have significant effects on the thermodynamics of the adsorption/desorption reactions, but also they may alter the kinetics of surface reactions toward selecting certain reaction pathways.

Keywords: Fischer-Tropsch synthesis, supercritical fluid, microkinetic modeling, PMF, DFT

---

[1] D. B. Bukur, X. Lang, and L. Nowicki, Ind. Eng. Chem. Res. 44, 6038-6044 (2005).

[2] P. E. Savage, S. Gopalan, T. I. Mizan, C. J. Martino, and E. E. Brock, AIChE journal 7, 1723-1778 (1995).

[3] S, Storsæter, D. Chen, A. Holmen, Surf. Sci. 600, 2051-2063(2006).

Copyright © American Institute of Chemical Engineers. All rights reserved.

466897 Intensified Hydrolysis By Employing the Synergy of Mixed Subcritical H_2o and Supercritical CO_2

Abstract:

Hydrolysis of naturally occuring compounds into more bioactive and bioavailable materials are gaining popularity in food and pharmaceutical industries. However, conventional method employs harmful catalysts such as sulfuric acid with low yield, while separation of the target compounds from the reaction products requires rather tedious approaches. To overcome these problems, this research aims to develop, optimize and evaluate by experimental approaches the combined reaction and separation process employing the synergy of subcritical H_2O and supercritical CO_2. Application to hydrolysis of citrus bioflavonoids (hesperidin) into its aglycones (hesperitin) was elucidated. Using a batch reactor, we were successful in hydrolyzing hesperidin into hesperetin under hydrothermal conditions with supercritical CO_2. The highest yield of hesperetin was obtained at T=160 °C, P=15 MPa and t=4 h. Based on the results, a semi-batch reactor system was developed to study reaction and separation process employing the mixture of supercritical CO_2 and subcritical H_2O. After the reaction, extraction of hesperetin was carried out by passing supercritical CO_2 through the reaction products. Results indicated the possibility of separating hesperetin from the mixture, and higher yield can be obtained by simply manipulating the pressure and temperature of CO_2.

Copyright © American Institute of Chemical Engineers. All rights reserved.

463346 Pd Particle Stabilization on Carbon Supports Under Hydrothermal Conditions

Abstract:

Carbon, as a catalyst support, has been shown to provide improved hydrothermal stability relative to common metal oxide supports such as silica and alumina, making it suitable for aqueous phase biomass conversion to chemicals and fuels. However, for supported metal catalysts, not only the stability of the carbon support, but also the stability of the supported metals needs to be considered. The stability is likely dependent on the carbon surface chemistry. The current work systematically examined this premise. Three carbon coated silica supports were synthesized at different temperatures to provide a range of oxygen content and functionality followed by the deposition of Pd. Through detailed and quantitative XPS and ^{13}C NMR characterization, lower temperature synthesized carbons gave a range of oxygen functional groups, while higher temperature carbons had more aromatic groups. The hydrothermal stability was tested using treatment in water at 170°C for 24 hours. Additionally, aqueous phase hydrogenation of furfural was performed in both batch and flow reactor systems to further characterize the relative stabilities of the catalysts. The low temperature carbon with more oxygen functional groups led to better Pd dispersion and improved hydrothermal stability.

Copyright © American Institute of Chemical Engineers. All rights reserved.

469719 Vacancy-Mediated Hydrodeoxygenation of Furfuryl Alcohol over Metal Oxide

Abstract:

Hydrodeoxygenation (HDO) of biomass-derived molecules is a necessary step in their utilization as fuels or chemical intermediates. Processes in which high-pressure molecular hydrogen is used suffer from the disadvantages of cost and safety. Ru/RuOx/C materials can successfully catalyze the HDO of furfural [1] and HMF [2] in the presence of isopropanol, without the need for external hydrogen, as the alcohol acts as a hydrogen donor and solvent.

Theoretical (DFT and MKM) and experimental studies of this system show that the HDO of furfural to methylfuran (MF) proceeds via a Meerwein-Ponndorf-Verley transfer hydrogen transfer from isopropanol to furfural [3]. The furfuryl alcohol produced in this step is subsequently deoxygenated over an oxygen vacancy on the RuO_2 surface [4]. Vacancies on the RuO_2 surface form by the reaction of H_2 produced from the dehydrogenation of isopropanol on the Ru surface with surface O atoms of RuO_2. However, nucleation of the vacancies causes the complete reduction of RuO2, resulting in the suppression of the HDO reaction and poor recyclability of the catalyst [2]. As such, the discovery of a stable catalyst is of great importance for the viability of this process.

In this work, we generalize and extend our conclusions to different oxides and metals that have been reported in the literature as active for HDO reactions, such as MoO3 [5] and Pt [6]. We combine experimental measurements of the HDO rate over a range of oxides in a batch reactor with a theoretical investigation of oxygenated species on these materials. These investigations yield an excellent correlation of the rate of HDO with intrinsic properties of the materials. Based on these investigations, we are able to propose and successfully employ a rationally-designed active and stable catalyst system for the HDO of furfural to MF.

References:

1. Panagiotopoulou, Vlachos. Appl. Catal A: General, 480 (2014) 17-24

2. Jae, et al. ChemCatChem, 6 (2014) 848-856

3. Gilkey, et al. ACS Catal. 5 (2015) 3988-3994

4. Mironenko and Vlachos. J. Am. Chem. Soc. (2016) Accepted

5. Prasomsri, et al. *Energy Environ. Sci.*, 6 (2013) 1732-1738

6. Luo, et al. Catal. Lett. 146 (2016) 711-717

Copyright © American Institute of Chemical Engineers. All rights reserved.

470395 Influence of Water and Solvents on the Production of Levoglucosenone andÂ 5-(Hydroxymethyl)Furfural from Cellulose

Abstract:

Cellulose is the most abundant renewable source of carbon on the earth. 5-hydroxymethylfurfural (HMF) as a major decomposition product from cellulose hydrolysis is one of the most promising platform chemical. Anhydro-sugars could also be obtained from cellulose depolymerization via pyrolysis or dehydration under anhydrous conditions. Levoglucosenone (LGO) is an anhydro-sugars that has potential as a chiral building block, which was used for the synthesis of a wide variety of high value pharmaceuticals. Both HMF and LGO could be converted to 1,6-hexanediol, which could serve as a precusor in polymer industry. Water, which is ubiquitus in biomass feedstock and solvents both play very important role on the product selectivites during the formation of HMF and LGO from cellulose.

In the absence of water, cellulose is first dehydrated to form anhydro-sugar (Levoglucosan), which could be further dehydrated to LGO. The final product distribution could be controlled by the addition of water to the reaction system. Cellulose will be hydrolyzed to glucose, which could be further dehydrated to HMF. The selectivity to HMF will gradually increase with increasing water content in the solvent. A series of solvents have been screened for the production of HMF and LGO from cellulose, it has been shown that the use of polar, aprotic solvents is very promising.

Copyright © American Institute of Chemical Engineers. All rights reserved.

473767 Transformation of Chitin into Value-Added Chemicals

Abstract:

Chitin is the world second most abundant biopolymer widely existed in the exoskeleton of shellfish and insects containing 7 wt% of biologically fixed nitrogen. This sustainable biomass has an underestimated but remarkable potential for the production of industrially important chemicals especially nitrogen containing compounds. To date, the direct conversion of chitin into high value chemicals with high yield has yet to be demonstrated. [1] In this presentation, we will show case some of our efforts in the catalytic transformation of chitin into *N*-containing chemicals (Scheme 1). Several strategies, including dehydration, liquefaction, hydrogenolysis and hydrothermal treatment, will be described in detail. The direct, one-pot conversion of chitin into a N-containing furan derivative with 7.5 % yield was achieved in both organic and ionic liquid solvents, by using boric acid and metal chlorides/HCl additives.[2, 3] The boron-sugar complex intermediate formed during the reaction was probed by NMR analysis, which is the key intermediate in the reaction. Chitin liquefaction, which transforms solid biomass into oils, was demonstrated in ethylene glycol under the catalysis of sulphuric acid.[4] After liquefaction, the crystalline structure of chitin was efficiently damaged. The total yield of liquefied monomers was around 30 %. Moreover, chitin monomer has been converted to its corresponding amide/amino substituted sugar alcohols, over noble metal catalysts in the presence of hydrogen in water.[5] By using Ru/C catalyst, 71.9% C6 polyols (N-containing) 8.7% N-acetylmonoethanolamine (NMEA) and 6.1% C4 polyols were obtained. [6] Kinetic studies revealed four major parallel reaction pathways that lead to these various products. In an effort to enhance the product yields, a comparative, systematic investigation of chitin treatment by various methods was studied and the structure-reactivity correlation has been identified. The yield of dehydration product was increased to 4-fold as previous by ball mill treatment. This series of work demonstrated that chitin is indeed an attractive starting materials for the manufacturing of certain value added chemicals.

REFERENCES

[1] N. Yan and X. Chen, Nature, 524 (2015) 155.

[2] X. Chen, S.L. Chew, F.M. Kerton and N. Yan, Green Chem., 16 (2014) 2204.

[3] X. Chen, Y. Liu, F.M. Kerton and N. Yan, RSC Adv., 5 (2015) 20073.

[4] Y. Pierson, X. Chen, F.D. Bobbink, J. Zhang and N. Yan, ACS Sustainable Chem. Eng., 2 (2014) 2081.

[5] F.D. Bobbink, J. Zhang, Y. Pierson, X. Chen and N. Yan, Green Chem., 17 (2015) 1024.

[6] X. Chen, Y. Gao, L. Wang, H. Chen and N. Yan, ChemPlusChem, 80 (2015) 1565.

Copyright © American Institute of Chemical Engineers. All rights reserved.

472530 Direct Carbon-Carbon Coupling of Furanics with Acetic Acid over BrÃ¸Nsted Zeolites

Abstract:

Torrefaction of biomass at mild thermal conditions can selectively decompose hemicellulose, which yields a blend of furanic and carboxylic acid species that are difficult to separate. The acidity introduced by acetic acid facilitates the polymerization of furanic species at room temperature, creating obvious storage and transportation challenges[1]. One promising route to eliminate these acids was by converting them to ketones via decarboxylative ketonization over HZSM-5[2], but this results in the loss of carbon in the form of is CO_2. Here we report direct acylation of methylfuran with acetic acid as acylating agent over HZSM-5 zeolite to produce acetyl methyl furan with very high selectivity. A reaction mechanism is proposed involving the dehydration of acetic acid as rate-determining step based on a combination of temperature programmed desorption (TPD) studies, reaction kinetics, and Denisty Functional Theory (DFT) calculations. TPD experiments to generate surface acyl groups followed by pulses of methylfuran demonstrate the direct C-C coupling of the two species to form acetyl methyl furan. The apparent activation energy for acylation was estimated to be 24.5 kJ/mole over the temperature range of 220-250 °C at under steady state flow conditions while the reaction behaves as first order with respect to acetic acid. DFT calculations reveal the generation of acyl species requires a true activation energy barrier of 104 kJ/mole, with a corresponding first order an apparent barrier of 16 kJ/mole. The barrier for the formation of acetyl methyl furan is low due to its ability to stabilize the charge of the acyl group in the transition state, and thus the overall barrier is determined by the rate of acyl formation. The presence of water is inevitable in all the streams produced from biomass, and the introduction of water in the feed reavealed an inhibition on acylation reaction rate by an order of -0.7 with a positive effect on catalyst stability. This study illustrates a new path of C-C coupling to produce higher value chemicals from biomass without sacrificing carbon as CO_2in the process.

Reference

1. Stocker, M. Biofuels and Biomass-To-Liquid Fuels in the Biorefinery: Catalytic Conversion of Lignocellulosic Biomass using Porous Materials. *Angewandte Chemie-International Edition* **47**, 9200-9211, (2008).
2. Gumidyala, A., Sooknoi, T. & Crossley, S. P. Title: Selective ketonization of acetic acid over HZSM-5: The importanceof acyl species and the influence of water. *Journal of Catalysis*, (Accepted).

Copyright © American Institute of Chemical Engineers. All rights reserved.

467604 Supported Cu Electrocatalysts for Electrocatalytic Hydrogenation and Hydrogenolysis of Furfural

Abstract:

Furfural (FF) is a C_5 biomass-derived platform chemical that can be obtained by hydrolysis and dehydration of lignocellulose. By chemical hydrogenation and hydrogenolysis (CH) of FF, furfural alcohol (FA), an industrial adhesive intermediate, and 2-methyl furan (MF), a promising biofuel, can be produced[1, 2]. Traditionally, CH of FF provides significant selectivity towards to FA and MF on Cu-based catalysts, but it requires externally supplied hydrogen gas and high temperature or/and high pressure[3].

In contrast to CH, electrochemical hydrogenation and hydrogenolysis (ECH) allows for the conversion of FF to FA and MF without external hydrogen gas, high temperature or high pressure. This process becomes more viable when excess electricity from renewable sources is used for ECH of FF.

Results of previous studies showed that strongly acidic electrolytes produced both MF and FA [4], while higher pHs mainly produced FA [5] on bulk metallic electrodes. In ECH of FF, Cu foil electrodes also showed remarkable selectivity (c.a. 80%) to MF in 0.5 M H_2SO_4 [4]. Also, substantial selectivities of FA with carbonate pH 10 buffer (80%) on a Cu foil [6] and with 0.2 M NH_4Cl (79%) on a Ni foil [5] have been demonstrated.

We have experimentally shown further relationships between products and reaction conditions in acidic electrolytes on Cu foil electrodes where product yield significantly depended on the current density when equivalent charge was transferred to the system. However, product selectivity depended on pH, not anions of electrolytes.

Here, we developed high surface area supported Cu electrocatalysts to investigate the impact of

morphology on selectivity and activity in comparison with Cu foils in 0.2 M NH_4Cl and 0.5 M H_2SO_4 electrolytes. Electrocatalysts were characterized utilizing XRD, SEM, TEM and XPS to discern the differences in catalytic performance. Product analysis was performed with gas chromatography mass spectrometry (GC-MS).

References

[1] J.N. Chheda, G.W. Huber, J.A. Dumesic, Liquid-phase catalytic processing of biomass-derived oxygenated hydrocarbons to fuels and chemicals, Angew. Chem.-Int. Edit. 46 (2007) 7164-83.

[2] J.-P. Lange, E. van der Heide, J. van Buijtenen, R. Price, Furfural—A promising platform for lignocellulosic biofuels, ChemSusChem 5 (2012) 150-66.

[3] Y. Nakagawa, M. Tamura, K. Tomishige, Catalytic reduction of biomass-derived furanic compounds with hydrogen, ACS Catalysis 3 (2013) 2655-68.

[4] P. Nilges, U. Schröder, Electrochemistry for biofuel generation: production of furans by electrocatalytic hydrogenation of furfural, Energy & Environmental Science 6 (2013) 2925-31.

[5] Z. Li, S. Kelkar, C.H. Lam, K. Luczek, J.E. Jackson, D.J. Miller, C.M. Saffron, Aqueous electrocatalytic hydrogenation of furfural using a sacrificial anode, Electrochimica Acta 64 (2012) 87-93.

[6] P. Parpot, A.P. Bettencourt, G. Chamoulaud, K.B. Kokoh, E.M. Belgsir, Electrochemical investigations of the oxidation-reduction of furfural in aqueous medium: Application to electrosynthesis, Electrochimica Acta 49 (2004) 397-403.

Copyright © American Institute of Chemical Engineers. All rights reserved.

469287 Rational Synthesis and Evaluation of Pd Bimetallic Catalysts for Furfural Rearrangement to Cyclopentanone in Aqueous Phase

Abstract:

Furfural, a product of acid-catalysed hydrolysis of hemicellulose in lignocellulosic biomass, is one of most promising biochemical platforms. With multiple active functional groups, furfural (FFA) can be catalytically transformed into a series of chemicals. Cyclopentanone (CPO) is one of downstream products from furfural hydrogenation which is widely used as solvents and fine chemical starting materials for fungicides, fragrances and pharmaceuticals [1].Â Production of CPO is a good way to develop further utilization of biomass-based furfural.

Hydrogenation of furfural has been intensively reported in literature, however the rational method to prepare bimetallic catalysts for this reaction hasnâ€™t been covered much. In this work, the aqueous phase hydrogenation of FFA to CPO was investigated on supported bimetallic catalysts (Pd-Ni, Pd-Cu and Pd-Co) prepared by simultaneous Strong Electrostatic Adsorption (SEA) and incipient wetness (IW). In SEA method, hydroxyl groups populate on the surface of support are charged by adjusting pH value in the solution away from its neutral point which is termed as Point of Zero Charge (PZC), metal precursors with opposite charges will be adsorbed. Unlike incipient wetness, the strong electrostatic interaction between metal complexes and supports renders highly dispersed metal nanoparticles. X-ray diffraction, elemental mapping and H_2 chemisorption were performed to confirm the high dispersion (<1.5 nm) and intimate contact between Pd and second metal.

Furfural hydrogenation reaction was carried out in 57 mL deionized water with 3 g of FFA at 150 °C, 500 PSIG (P_{H2}= 430 PSIG) for 3 hours, products were sampled at an interval of 20 min and analyzed by GC. Reaction results showed enhanced catalytic performance with Pd bimetallics prepared by SEA compared to their IW counterparts: the yield of CPO increased by more than 5 times; nano-alloy survived and slightly sintered (~2.5 nm) in SEA catalysts while Ni and Co might have leached out from IW ones after reaction.

References

[1] M. Hronec, K. Fulajtarove, I. Vavra, T. Sotak *et al.*, App. Cataly. B : Environmental, 181 (2016) 210-219.

Copyright © American Institute of Chemical Engineers. All rights reserved.

472910 Polymeric Solid Acid Catalysts for Efficient Hydromethyfurfural and Levulinic Acid Production from Biomass Feedstock

Abstract:

Polymeric solid acid catalysts were synthesized and optimized for the production of reducing sugars, HMF and levulinic acid from biomass feed stock. The catalysts consist of a poly(sulfonic) acid chain and a poly(ionic liquid) chain grafted on a solid substrate. Crystalline cellulose as well as dilute acid, base and steam pretreated biomass samples from National Renewable Energy Laboratory were investigated with our novel tunable catalysts in ionic liquid solvent and ionic liquid/water, ionic liquid/GVL solvent mixtures. Over 95% of total reducing sugar (TRS), over 70% HMF and over 50% levulinic acid yields were obtained from pure cellulose samples at 105°C, 135°C and 155°C respectively. For the pretreated biomass samples, different yields were obtainded from three different pretreatment methods. Yields of over 80% TRS, over 60% HMF/Furfural, and over 40% levulinic acid were achieved. GVL cosolvent was found to facilitate the conversion process whereas the opposite is true for water. Our results demonstrate the high catalytic activity and versatility of these unique polymeric acid catalysts.

Copyright © American Institute of Chemical Engineers. All rights reserved.

467303 Soot Nucleation and Consumption in Oxy-Coal Systems

Abstract:

With the rising concern of carbon dioxideâ€™s effects on climate change, research involving oxy-fuel combustion has increased significantly. The environment of an oxy-fuel system can vary greatly from that of conventional air-fired systems; as a result, models to describe system physics must be adapted to account for these differences. This work focuses on some of the differences found in the oxy-coal environment and the effects this has on soot nucleation and consumption.Â The transformation of tar into soot particles along with the growth of those particles are explored and models are presented. Oxidation rates are explored and models optimized to experimental data published in the literature. Model parametersâ€™ uncertainty are also computed and presented. Gasification rates are considered and modeled in a similar way with their respective uncertainties. Effects of gasification, which is typically neglected in conventional systems, are also explored in simulation of an oxy-fuel combustor at the University of Utah; results are presented.

Copyright © American Institute of Chemical Engineers. All rights reserved.

469046 Investigation of NO_x Chemistry in Oxy-Combustion Flue Gas

Abstract:

Oxy-coal combustion technology can be a promising solution to meet the challenge of reducing carbon emissions from power plants while continuing with the fossil fuel usage. Implementation of such technology requires the removal of nitrogen (N_2) from the oxidizer and introduction of a pure oxygen (O_2) stream and a recycled flue gas stream into the boiler. In absence of N_2, the flue gas stream generated from such system will be low in volume and highly concentrated in carbon dioxide (CO_2) which in turn will facilitate the capture and sequestration process. Moreover, oxy-coal combustion process has shown promise in reducing nitric oxide (NO) emissions. Lower NO emissions are expected in the oxy-combustion process due to the absence of N_2 in the oxidizer and reduction of recycled NO in the flame zone through the reburn mechanism. In the reburn process, recycled NO interacts with fuel fragments forming different intermediates. Depending on the combustion conditions, these intermediates can form $N_2$and thus reduce the NO emission. To gain a better understanding of this reduction process under oxy-combustion conditions and to evaluate the influences of different combustion parameters, a comprehensive study focused on nitrogen chemistry in oxy-combustion is crucial.

The present study aims to provide speciation data of nitrogen oxides (NO_X) and to reveal the reaction pathways of recycled NO destruction under oxy-combustion conditions. To do so, the oxy-coal combustion environment has been simulated in a lab-scale reactor with a time-temperature history representing a plant boiler. Flue gas samples collected from different temperature points of the reactor have been analyzed by applying Fourier Transform Infrared (FTIR) spectroscopy for NO_X speciation. The influence of different combustion parameters, i.e., equivalence ratio (0.8-0.98), O_2 percentage (28-34%) and recycled NO concentration (800-2000 ppmv) has also been investigated in the course of the study. Moreover, the combustion experiments have been simulated by conducting kinetic modeling with Chemkin to evaluate the performance of reaction mechanisms reported in the literature. Sensitivity analysis has also been performed to identify the destruction routes of recycled NO. The data reported in the present study can aid in predicting NO_x emissions from power plants operating under oxy mode and to evaluate the performance of combustion mechanisms in the literature.

Copyright © American Institute of Chemical Engineers. All rights reserved.

459266 On the Role of Sodium Content of Fine Ash Particles in Determining Deposition Rates Under Air- and Oxy-Coal Combustion Conditions

Abstract:

During the combustion of coal, the sodium content in the ash is commonly thought to be the "bad actor" in determining ash deposition rates on boiler heat transfer surfaces. We have conducted 16 tests in which deposition rates of three coals, burned under air fired and oxy-fired combustion conditions, were measured. A 100kW down-fired laboratory combustor, coupled with a specially designed deposition probe was employed. Emphasis was on deposition rates of the tightly bonded "inside" deposits, rather than on the loosely bound "outside" deposits, which could not be weighed precisely. Surprisingly, the measured deposit rates correlate *neither* with the sodium content of the coal, *nor* with the sodium content of the fine ash aerosol (PM1 or PM0.6). Instead they correlate well with the *overall* flue gas concentration of PM1, the alkali composition of which varied significantly over the 16 tests. This suggests a mechanism of deposit adhesion that depends simply on the presence of submicron particles of *any* composition. Additional research is needed to confirm this, because this conclusion disagrees with nearly all existing models for ash deposition and has implications for simulations predicting boiler performance for coal or biomass fired fuels.

Copyright © American Institute of Chemical Engineers. All rights reserved.

467596 NO_x Formation in Syngas/Air Combustion at Elevated Pressure

Abstract:

Combustion of high hydrogen content (HHC) fuels is central to the implementation of integrated gasification combined cycle (IGCC) systems or to other configurations that would optimize power production while minimizing emissions. Gas turbines utilizing HHC fuels for power generation applications will be required to meet stringent pollutant emission standards, particularly with respect to nitrogen oxides (NO_x). One of the effective NO_x reduction technologies for gas turbines is lean premixed (LPM) combustion method. Accurate numerical prediction of NO_x from such combustors using multi-dimensional models has only been of limited success. This is to some extent due to the complexity of NO_x formation chemistry in LPM combustion. Experimental data is necessary for generating accurate computational models of NO_x production in combustion. Improvement of the model predictions could be accomplished by validation against data from well-defined experiments to design a very low emission process with optimum efficiency. For gas turbine conditions, reliable high-pressure data is needed to both validate and improve reaction mechanisms to simulate NO_x chemistry.

In this study, NO_x formation in post-flame gases of syngas combustion was studied at elevated pressures. A high-pressure burner facility well suited to the study of reaction mechanisms was designed and fabricated. A series of experiments were performed to investigate the NO_x formation from lean premixed syngas/air combustion with various stoichiometries (equivalence ratio between 0.5-1.0 and H_2/CO ratio between 0.25-1.0). Acquiring an accurate temperature profile throughout the system is critical for making accurate predictions via kinetic simulations. Flame temperature and two-dimensional temperature distribution were measured under various conditions. Detailed NO_x speciation measurements in the post-combustion zone were conducted by Fourier transform infrared spectrometer (FTIR). Additionally, computational fluid dynamics (CFD) simulations with detailed chemical kinetics were employed

to predict the temperature and species profiles.

Copyright © American Institute of Chemical Engineers. All rights reserved.

462880 Formation and Decomposition Kinetics of HONO and HNO2

Abstract:

The HNO_2 system â€“ cis-HONO, trans-HONO, and HNO_2 â€“ are important intermediates in the combustion of nitrogen. These isomers are typically formed either from H-abstraction or disproportionation. To date, most experimental and theoretical work on the HNO2 potential energy surface has focused on the chemically activated reaction H + NO_2â†' OH + NO. The isomerization and decomposition kinetics of the HNO2 isomers has received considerably less attention.

This talk will present recent theoretical work on the isomerization and thermal decomposition of cis-HONO, trans-HONO, and HNO_2. For the stationary points, geometry optimization and normal mode analysis was done using CCSD(T)/cc-pVQZ. Five post-coupled cluster methods were used to further improve the electronic energy. The tight transition state for isomerization between trans-HONO and HNO2 was computed using conventional transition state theory with rigid-rotor harmonic oscillator approximations. For the three H + NO_2pathways and the two OH + NO pathways, variable reaction coordinate transition state theory was used. The active space was 14 electrons in 11 orbitals.

The final part of the talk will focus on the relative importance of including all three isomers in kinetic models versus the simplicity of treating all three as a single, lumped species.

Copyright © American Institute of Chemical Engineers. All rights reserved.

447496 Experimental Study on NO Reduction By Iron with Methane in Wet Simualted Flue Gas

Abstract:

Nitrogen oxide (NO) is a serious pollutant emission from coal-fired boilers and other industrial plants. In the current SCR and SNCR technologies, the reducing agent is ammonia, which is expensive and requires special handling and storage as well as a sophisticated metering system to avoid NH_3 slip. There is a great incentive to use natural gas as the reductant in stationary SCR/SNCR units rather than NH_3. A novel and effective method to reduce NO by methane over iron and/or iron oxides was recently proposed by the authors and experimental results in dry simulated flue gas proved this method is very promising.

However, there is water vapor in the flue gas in real conditions. In the SCR-NH_3 technology, many of the catalysts will lost their activity when there is water vapor. The resistance of this method to vapor need to be demonstrated.

In this paper, the effect of water vapor on NO reduction by methane over iron was experimentally investigated at 300~1100°C in a ceramic tubular flow reactor. The iron samples after reaction were characterized by XRD, SEM and XPS. Results demonstrated that water vapor had a small effect on NO reduction by methane over iron. In N_2 atmosphere, water vapor involved in the oxidation of iron and caused decreased NO reduction efficiencies when 2.5~7% water vapor was added into the reactor, as compared to that when vapor was not added. When the water vapor increased from 2.5% to 7%, however, NO reduction efficiency increased with the increase of water vapor due to the porous surface of the iron sample caused by the oxidizing effect of water vapor on iron. Methane was involved in the reduction of the iron oxides leading to the formation of a dense layer of Fe_3O_4 and FeO, which inhabited the interaction of NO and metallic iron and caused the decreased NO reduction as compared to that without methane. In simulated flue gas atmosphere, water vapor increased the NO reduction by methane over iron. When the excess air ratios in reaction zone and burnout zone were SR_1=0.7 and SR_2=1.2 respectively, the NO reduction efficiencies were found to be 96.7% and 90.6% respectively at 1050 °C when there were 7% and 0% water vapor. SO_2 caused a slight decrease of NO reduction. Experimental data showed that more than 90% of NO was reduced by 1.14% methane over iron at 1050°C in a durable test over 50 hours in simulated flue gas atmosphere containing 7% H_2O and 0.02% SO_2.

The results proved that the present method for NO reduction by iron with methane is very effective and is not influenced by vapor.

Copyright © American Institute of Chemical Engineers. All rights reserved.

464644 Inferential–MODEL Based FLARE SET POINT Determination

Abstract:

Recent EPA rule (40 CFR parts 60 and 63) finalizes single minimum combustion zone net heating value (NHVcz) operating limit as 270 BTU/scf during any 15–minute average period for flares subject to Petroleum Refinery Maximum Achievable Control Technology standards to ensure 98% destruction efficiency (DE) or 96.5% combustion efficiency (CE) or higher at all times. In addition, refinery flares meet the flare tip velocity requirements (max. 400 ft/s), visible emission requirements (20% opacity limits) and net heating value dilution parameter (NHV_{dil}) minimum operating limit of 22 BTU/ft^2 and corrected vent gas heat content of 1212 BTU/scf for hydrogen to demonstrate the compliance. This study aims at developing robust inferential flare models to express DE, CE and opacity as a function of operating variables and to determine set point for steam and make up fuel at the incipient smoke point operation for NHVcz ≥ 270 BTU/scf and NHV_{dil} ≥ 22 BTU/ft^2. Multivariate statistical distribution and response surface models were developed for steam and air assisted flares burning propylene, propane, natural gas, methane, ethylene and typical refinery vent gas mixtures. Curve fitting tool box in MATLAB and Minitab were used in modeling. In this study, soot emissions, visible emission scale, DE, and CE test data from 1983 to 2014 (including 1983/1984/1985 EPA, 2010 TCEQ, 2009/2010 Marathon Detroit/Texas City, and 2014 Carleton University) were analyzed. The input variables include measurable variables like exit velocity, lower heating value, flare tip diameter, vent gas species (double bond, carbon to hydrogen molar ratio and molecular weight), and operating variables (NHVcz, NHV_{dil}, steam/air assists, and fuel). Response surface models developed for CE and opacity showed high coefficient of determination (R^2) values 0.91 and 0.95 respectively for steam assisted flare data.

Copyright © American Institute of Chemical Engineers. All rights reserved.

469766 Offshore Floating Packed-Bed Reactors: Current Status, Key Challenges and Potential Solutions

Abstract:

Packed beds as classical multiphase contactors and reactors are widely employed in offshore oil and gas industries to treat and refine hydrocarbons extracted from undersea reservoirs [1]. Such packed beds are embarked on ships such as floating production, storage and offloading (FPSO) and floating liquefied natural gas (FLNG) for onboard treatment in the vicinity of the extraction site [2]. Nevertheless, operation of floating reactors and separators on embarked FPSO or so-called mini-refineries undergoes remarkable technical and operational challenges stemming from the complex sea conditions.

Thanks to a new state-of-the-art experimental setup combining a hexapod robot with six-degree-of-freedom motions and an embarked column instrumented with capacitance wire-mesh sensors, the hydrodynamic behavior of floating packed beds with different gas-liquid contacting modes was examined. Ship motions and inclination initiated by marine swells were inferred to impact the performance of onboard packed-bed reactors. Translational and rotational motions of the column led to remarkable phase maldistribution and, consequently, loss of efficiency. Besides, in order to understand detailed features of the flow patterns and mixing behaviors inside a moving reactor, three-dimensional computational fluid dynamics (CFD) simulations of multiphase flows in floating packed beds were implemented (CFD). The sensitivity of the hydrodynamic characteristics to column oscillations was assessed based on the experimental results of residence time distribution (RTD) studies and CFD simulations.

Finally, to reduce those detrimental effects of ship oscillations on the performance of multiphase packed-bed reactors, two new operational techniques were proposed: (i) low-shear rotation of the reactor and (ii) modulation of liquid and/ gas flow at the reactor inlet. The superimposed low-shear rotation of packed beds allows controlling pressure drop, liquid saturation, liquid residence time and dispersion by tuning an appropriate rotational velocity depending on the type of imposed motions to the column. Liquid wave induced by the flow modulation technique showed its ability to reduce fluid maldistribution in oscillating packed beds. Hence, these two concepts can be regarded as potential methods for process intensification in offshore production units where packed-bed reactors are used.

References

[1] Kim, H.-J., Choi, D.-K., Ahn, S.-I., Kwon, H., Lim, H.-W., 2013. GTL FPSOâ€"An Alternative Solution to Offshore Stranded Gas, Offshore Technology Conference, Houston, Texas, USA, May 6-9.

[2] Leffler, W.L., Pattarozzi, R., Sterling, G., 2011. Deepwater petroleum exploration & production: A nontechnical guide. 2nd ed. PennWell Books, Tulsa, Oklahoma, USA.

Copyright © American Institute of Chemical Engineers. All rights reserved.

481204 Design Optimization of BioEnergy Combustion Systems Using Sculptor® with CFD

Abstract:

In the past, design improvements for combustion equipment including process burners, gas flares, vent incinerators, etc. has been accomplished using the "build and try" method. This tried and true approach was very expensive and took a significant time to consider three or four "new" designs before settling on the "optimal" configuration intended to reduce pressure loss and/or improve local mixing. Efficient optimization algorithms coupled with comprehensive multi-physics computational fluid dynamics (CFD) tools have been developed and applied to assist in evaluating hundreds of designs to find the "optimal" solution.

Sculptor® from Optimal Solutions Software (OSS) automates shape optimization using computer aided engineering (CAE) tools including CFD, FEA and others. This approach is particularly useful for processes where chemically reacting flows and heat transfer govern the performance of combustion equipment, chemical reactors, heat exchangers, and other process equipment.

This paper discusses design optimization of a biomass fired furnace. Various designs of a co-fired biomass/coal reactor are examined to reduce greenhouse gas emissions while improving combustion efficiency and lowering NOx formation. Results illustrate the current state-of-the-art in design optimization using coupled CAE tools to improve equipment performance. Based on this analysis, implications of various design options are discussed and recommendations are provided regarding potential improvements in burner design.

Copyright © American Institute of Chemical Engineers. All rights reserved.

461986 A New Model for the Depolymerisation of Polysaccharides Originating from Biomass

Abstract:

Controlled depolymerisation of polysaccharides is one of the key issues in the utilisation of biomass as a source for fine and specialty chemicals. The aim is to break down the glycosidic bonds of polysaccharides, such as cellulose, hemicelluloses and starch in such a way that oligomeric and monomeric sugars are obtained, but no low-molecular degradation products. This can be achieved by applying hydrolysis of polysaccharides under mild conditions in the precence of a homogeneous or heterogeneous acid catalyst. It has been demonstrated that this is possible, e.g. by using solid ion-exchangers as catalysts and interesting self-accelerating effects have been discovered. To progress further, a model for the depolymerisation kinetics is needed; a model which can predict the distribution of the oligomers and monomers as a function of the reaction time.

A new theoretical model for the depolymerisation kinetics of polysaccharides in batch and continuous reactors was developed. The main idea of the model is that the depolymerisation reactivities of the polymer bonds depend parabolically on the distance of the bonds from the central bond in the polysaccharide polymer. The model is able to describe the self-acceleration of the depolymerisation process as the polymer chains become shorter. A software was developed to be able to calculate the product distribution for an arbitrary mixture of initial polymers. The model was successfully tested on existing experimental depolymerisation data of *O*-acetylgalactoglucomannan, a hemicellulose appearing in softwood.

Copyright © American Institute of Chemical Engineers. All rights reserved.

468240 Recent Trends in Multi-Phase Reactor Engineering and Design

Abstract:

Gas-liquid-solid reactors are encountered in many production processes of the chemical, fuel, food and pharmaceutical industry. The solid catalyst is typically applied either as powder suspended in a liquid or as larger particles packed in porous beds. The reactor performance depends mainly on the solid catalyst and the interaction of the phases within the reactor vessel as well as between the phases and the reactor wall which dictate heat and mass transfer rates. This contribution reviews recent trends and the progress in gas-liquid-solid reactor engineering and design. In particular, it addresses the application of structured catalysts, micro-structured reactors, periodic operation, supercritical reaction conditions and the shift from fossil feedstocks to renewable feedstocks. The new tools and market demands foster also the transformation from traditional batch processes to continuous processes.

Copyright © American Institute of Chemical Engineers. All rights reserved.

St. Petersburg State Institute of Technology (TU),
Department of Optimization of Chemical and Biotechnological Equipment

Gas-Liquid and Liquid-Liquid Taylor Flow in Micro Channels: Hydrodynamics and Mass Transfer

Rufat Abiev, Stanislav Svetlov,

Stefan Haase, Tobias Bauer and Ruediger Lange

2016 AIChE Annual Meeting, Monday, Nov. 14, 2016: 4:55-5:15 PM
Dedicated to the 65th Birthday of Professor Ruediger Lange

Outline

1. **Introduction**
2. **Gas-Liquid flows in capillaries: theoretical approach, simulation and experiments**
3. **Liquid-Liquid flows versus Gas-Liquid flows in capillaries: similarities and peculiarities**
4. **Process intensification by means of Taylor vortices**
5. **Correlations for Taylor flow parameters**
6. **Conclusions**

Micro heat exchangers and micro reactors

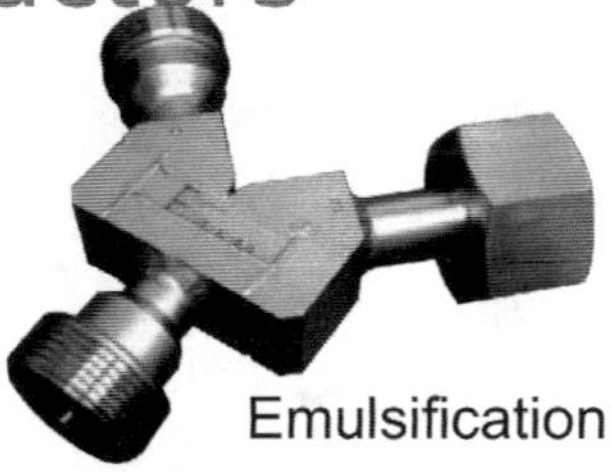

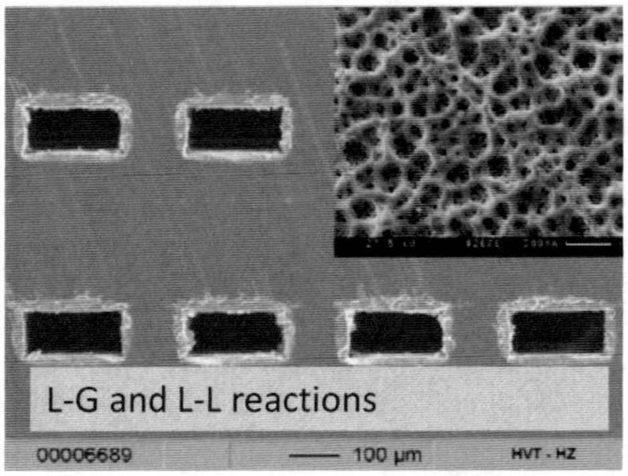

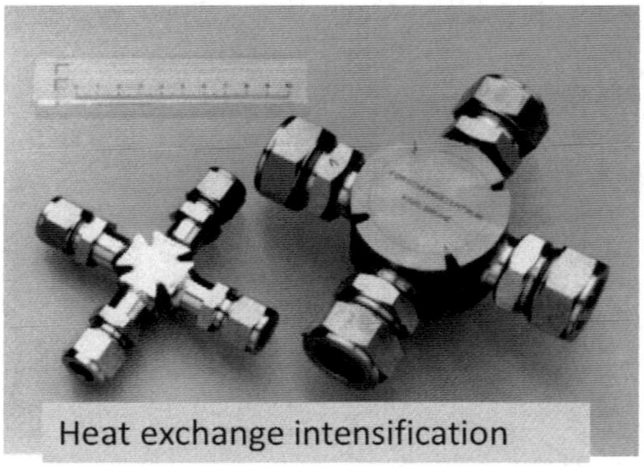

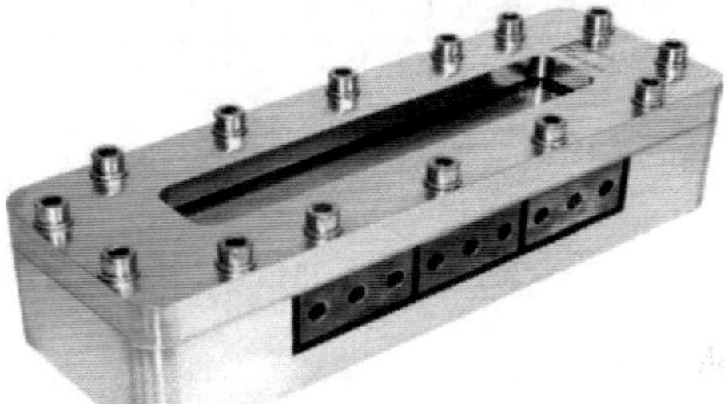

Evaporation

Gas-Liquid flow regimes in capillaries (similar for L-L flows)

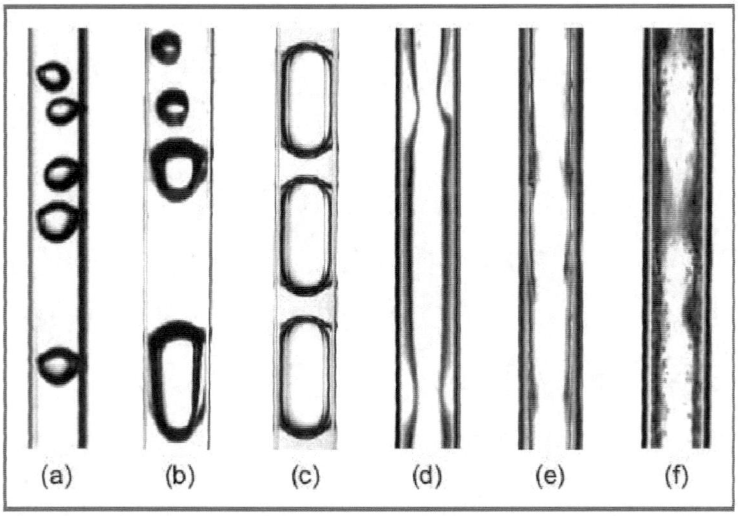

a – bubble flow; b – bubble-slug flow transition; c – slug/Taylor flow; d – slug-film flow transition; e – film flow ; f – churn flow .

The aims of the performed work

- Elaboration of the mathematical model for hydrodynamics of L-G and L-L Taylor flow in micro and milli channels.
- Development of the mathematical model for L-G (L-L), L-S mass transfer in micro channels.
- Experimental demonstration of process intensification for L-S and L-L systems by Taylor vortices.
- Discussion of opportunities to use L-G and L-L mathematical models for Research and Design of micro reactors, micro extractors and micro heat exchangers.
- Elaboration of Correlations for Taylor flow parameters

Practical applications of this work

- Better understanding of hydrodynamics of multiphase flows in micro/milli channels.
- To connect the influence of hydrodynamics on the heat and mass transfer with the overall performance of heat exchangers and reactors.
- Optimize the processes´ parameters.
- Process intensification by easy ways!

A model of Gas-Liquid slug flow in capillaries is developed

(Abiev, R. Sh., Theor. Found. of Chem. Eng., 2008, Vol. 42, No. 2)

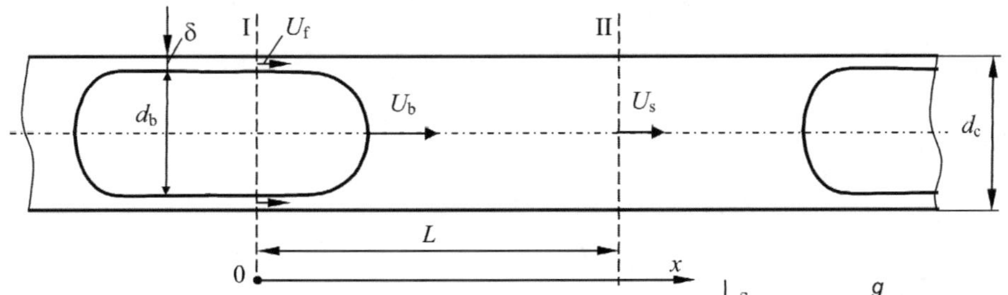

$$\frac{\partial(\rho_i \varepsilon_i)}{\partial t} + \mathbf{div}(\rho_i \alpha_i u_i) = q_{Vi} \quad (1.1)$$

$$\sum_{i=1}^{N} \varepsilon_i = 1 \quad (1.2)$$

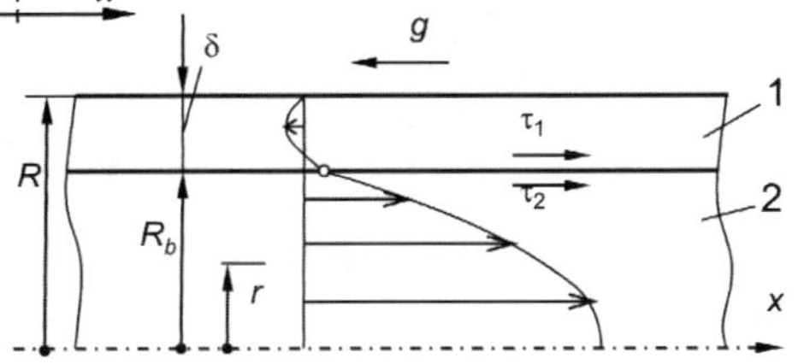

$$U_f A_f + U_b A_b = U_s A_c \quad (1.3)$$

The mathematical model of Gas-Liquid slug flow

$$\rho\left(\frac{\partial u_x}{\partial t} + u_r \frac{\partial u_x}{\partial r} + \frac{u_\theta}{r}\frac{\partial u_x}{\partial \theta} + u_x \frac{\partial u_x}{\partial x}\right) = -\frac{\partial}{\partial x}(p - \rho g_x x) + \mu \nabla^2 u_x \quad (1.4)$$

$$\frac{\partial \rho}{\partial t} + \frac{\partial(\rho u_r r)}{r \partial r} + \frac{\partial(\rho u_\theta)}{r \partial \theta} + \frac{\partial(\rho u_x)}{\partial x} = 0 \quad (1.5)$$

$$\frac{\delta}{R} = \frac{1.34 \cdot \mathrm{Ca}^{2/3}}{1 + 3.35 \cdot \mathrm{Ca}^{2/3}} \quad (1.6)$$

Boundary conditions

$$q_2(x=0) \equiv q_b = \int_{A_b} u_2 \, dA_b = 2\pi \int_0^{R_b} u_2(r) r dr \quad (1.7)$$

$$U_2(x=0) \equiv U_b = q_b / A_b \quad (1.8)$$

$$q_1(x=0) \equiv q_f = \int_{A_f} u_1 \, dA_f = 2\pi \int_{R_b}^{R} u_1(r) r dr \quad (1.9)$$

$$U_1(x=0) \equiv U_f = q_f / A_f \quad (1.10)$$

$$q_1(x=L) \equiv q_s = \int_{A_s} u_1 \, dA_s = 2\pi \int_0^{R} u_1(r) r dr \quad (1.11)$$

$$U_1(x=L) \equiv U_s = q_s / A_s \quad (1.12)$$

$\mathrm{Ca} = \mu U_b/\sigma$ – capillary number

Comparison theory with experimental data of M. Kreutzer

(exp. data: Kreutzer M.T. et al., *AIChE J.*, 2005, vol. 51)

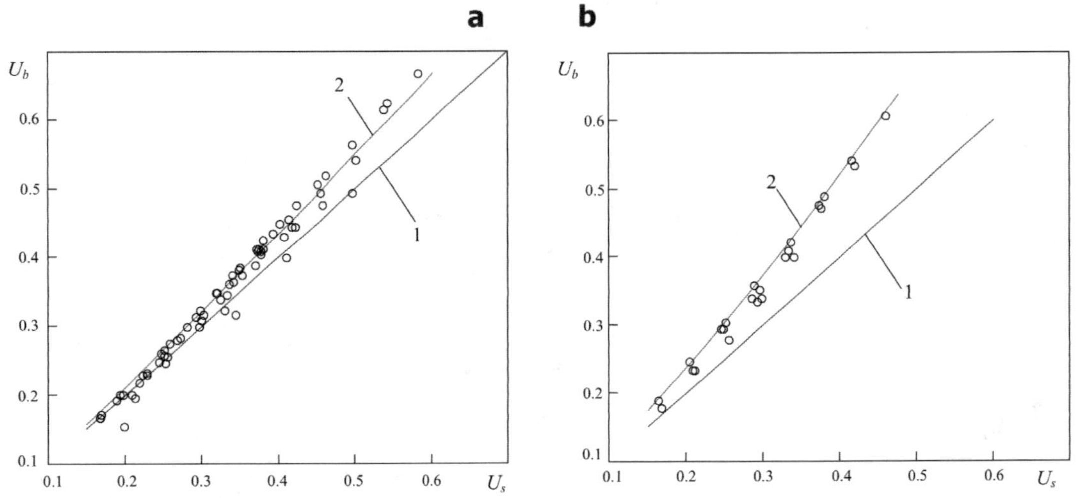

Dependence of the velocity of bubbles in a capillary of diameter 2.3 mm for the upward flow of the gas–liquid mixture on the velocity of liquid slugs for (a) water–air, (b) tetradecane–air systems: *1*, *U*s; *2*, *U*b

Comparison theory with experimental data of M. Kreutzer

(Kreutzer M.T. et al., *AIChE J.*, 2005, vol. 51)
Bubble velocity dependence on a capillary number

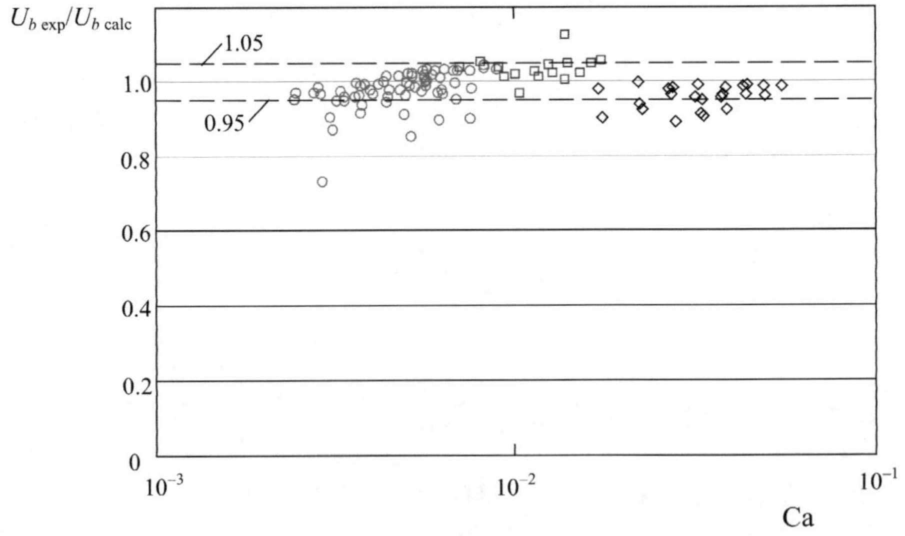

Circulation and by-pass flow modes

a

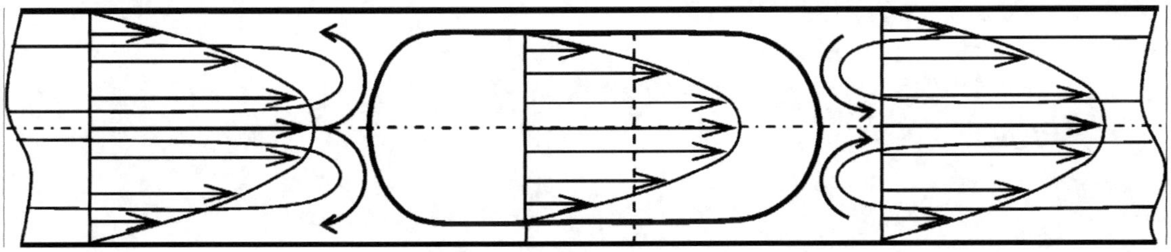

b

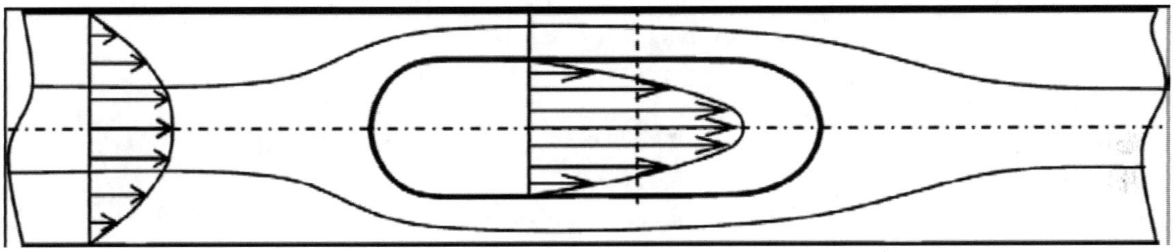

The reason of transition is:

Ratio of maximum velocity of continuous phase to the average velocity of dispersed phase (bubble/droplet): max $(u_c)/U_d$.

At max $(u_c)/U_d < 1$ there is circulation flow:

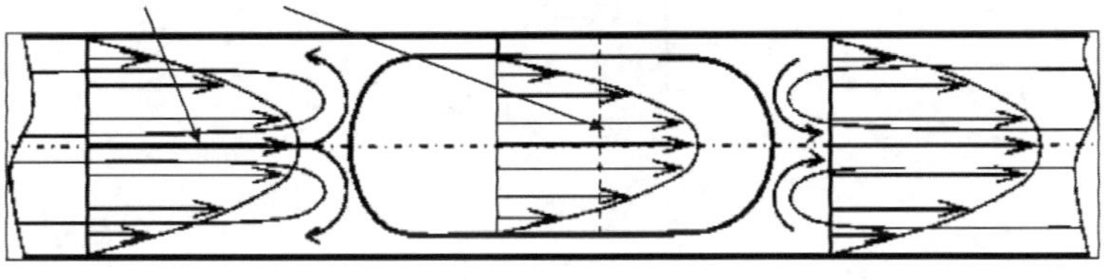

- At max $(u_c)/U_d > 1$ there is by-pass flow:

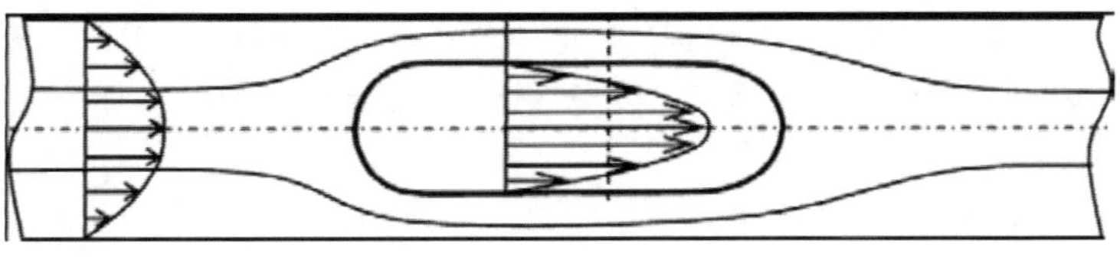

Circulation and by-pass flow modes

Trucks (bubbles) are like 'moving traffic jam'. They are slower than the most quick volumes of liquid in slugs (Ferrari)

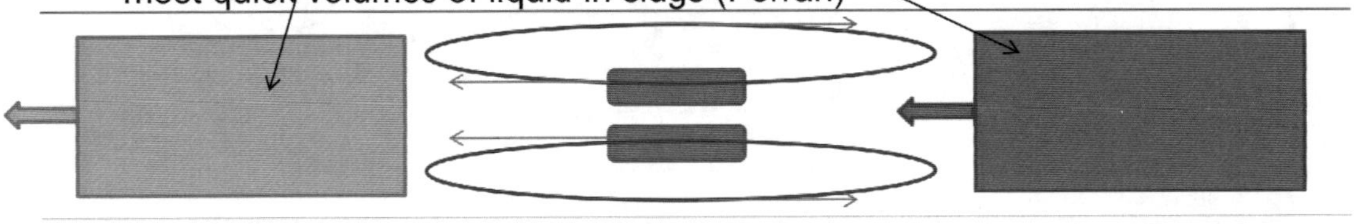

Method of calculating of critical capillary number Ca* by modified Bond number [Abiev R. Theor. Found. of Chem. Eng., 2010, V.44]

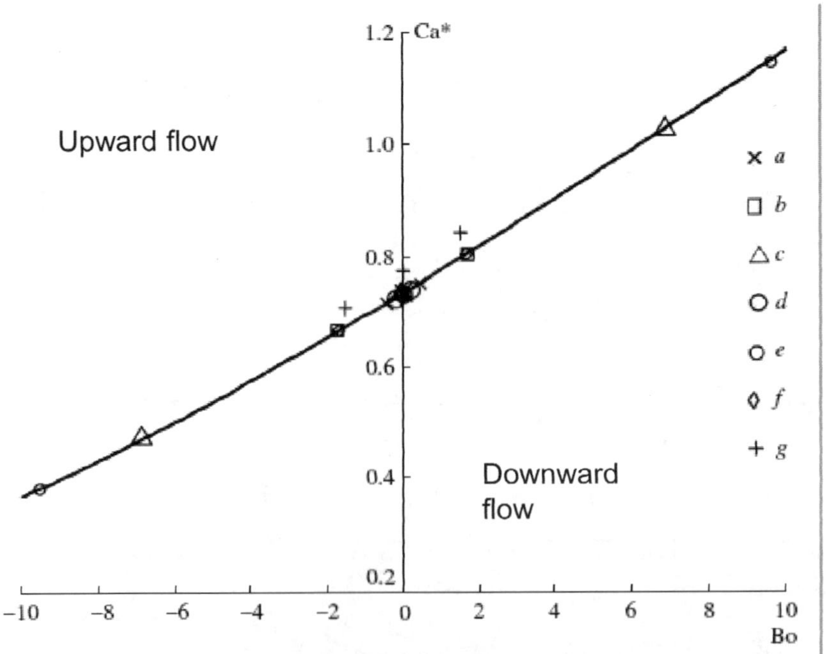

The legend is on the next slide

Abiev R.S. Circulation and Bypass Modes of the Slug Flow of a Gas–Liquid Mixture in Capillaries. Theor. Found. of Chem. Eng., 2009, Vol. 43, No. 3.

Critical capillary number Ca* is a function of modified Bond number

Influence of the Bond number Bo on the values of critical capillary numbers Ca* for upward (Bo < 0), downward (Bo > 0) and horizontal flows (Bo = 0).

$$\mathrm{Bo} = \rho_1 d_c^2 g \cos\gamma / \sigma$$

γ = (**Us**, **g**) - angle between velocity **Us** and gravity **g** vectors

Approximation for critical capillary number prediction

$$\mathrm{Ca}^* = 0.7378 + 0.0408\,\mathrm{Bo} + 3.5794\cdot 10^{-4}\,\mathrm{Bo}^2 - 1.0799\cdot 10^{-5}\,\mathrm{Bo}^3$$

The points on the graph represent calculation; the line represents approximation by polynomial: (*a*)–(*c*) silicone oil No. 3 (σ = 0.02218 N/m): (*a*) d_c = 1 mm, (*b*) d_c = 2 mm, (*c*) d_c = 4 mm, (*d*), (*e*) modifications of silicone oil No. 3: (*d*) σ = 0.2 N/m (silicone oil No. 3a), d_c = 2 mm; (*e*) σ = 0.004 N/m (silicone oil No. 3b), d_c = 2 mm; (*f*) silicone oil No. 2, d_c = 2 mm; (*g*) tetradecane, d_c = 2.3 mm.

Previous research

Usually one use simple equality of dynamic hold up and real volume hold up

$$\beta = \varepsilon_V$$

The other opportunity is the Armand-Treshev (1946) equation

$$\beta = C\varepsilon_V$$

Where C = 1.1-1.2

In fact this equation with C = const is only applicable for certain Capillary number!

Method of calculation of volume fraction of bubbles

[Abiev, R.Sh., Theor. Found. of Chem. Eng., 2010, Vol. 44, No. 1]

Dynamic gas holdup $$\beta = \frac{Q_2}{Q_1 + Q_2} \quad (3.1)$$

Exact relationship that relates β to ε_V $$\beta = \frac{U_b A_b \varepsilon_L}{U_s A_c} = \frac{U_b \varepsilon_A \varepsilon_L}{U_s} = \frac{U_b \varepsilon_V}{U_s} = \eta \varepsilon_V \quad (3.2)$$

Using experimental approximation [Liu H., Vandu C. O., Krishna R. Ind. Eng. Chem. Res. 2005. V. 44. P. 4884-4897]

$$\eta = \left(1 - 0.61\,\mathrm{Ca}_s^{0.33}\right)^{-1} \quad (3.3)$$

(or Eq. (2.2),(2.4)) on can calculate real volume gas hold-up

$$\varepsilon_V = \beta\left(1 - 0.61\,\mathrm{Ca}_s^{0.33}\right) \quad (3.4)$$

Comparison with experimental data

(exp. data from *Liu H. et al. Ind. Eng. Chem. Res. 2005. V. 44*)

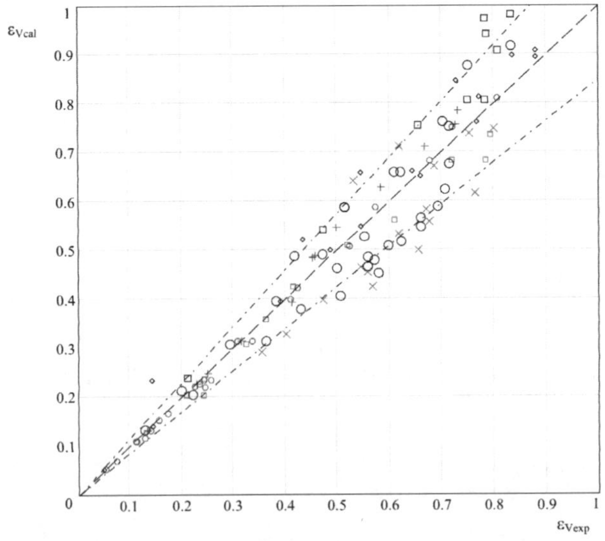

Comparison with experimental data : relative length of bubbles

(exp. data from: *Kreutzer M.T. et al. AIChE J. 2005. V. 51*).

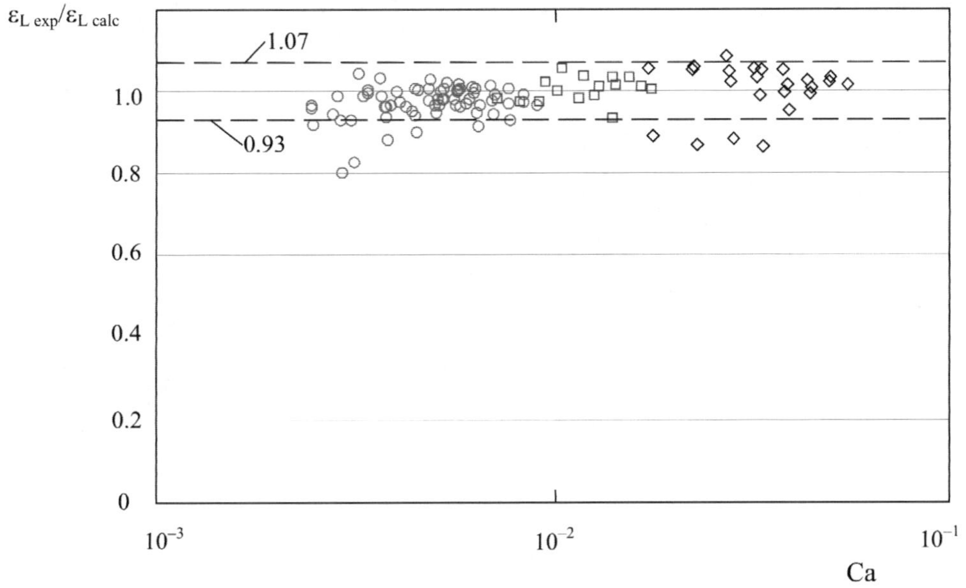

Why real volume hold up is not the same as hold up calculated by flow rate?

According to equation $\beta = \eta(Ca)\varepsilon_V$

It has been found earlier $\eta = 1..2.6$

Therefore dynamic hold up for high Ca numbers could be up to 2.5-2.6 times higher than the real volume hold up!

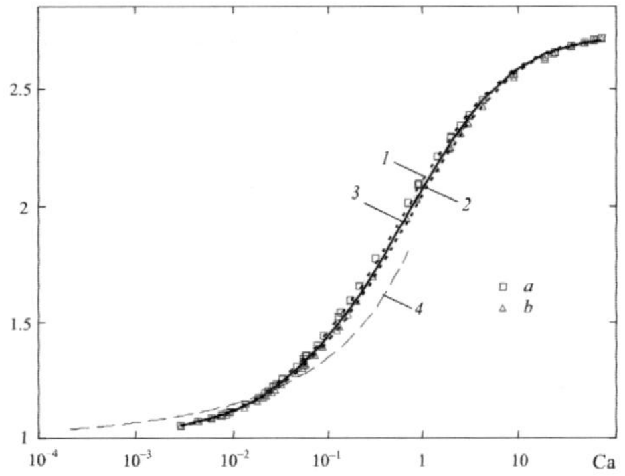

In that case the error for Armand equation is up to 160 %!

R.S. Abiev Bubbles velocity, Taylor circulation rate and mass transfer model for slug flow in milli and micro channels. Chem. Eng. J., 2013, 227, P. 66-79

Three general methods of pressure drop simulation of Taylor flow in micro channels:

Homogenous model;
Lockhart-Martinelli method;
Two-phase flow model.

Two-phase flow model is the most comprehensive method.

1) Non-parabolic velocity profile in liquid slugs

It happens due to undeveloped flow – uniform velocity distribution transforms to parabolic one. A local pressure gradient for such flow [Sliozkin, N.A., Dinamika vyazkoi neszhimaemoi zhidkosti (Dynamics of Viscous Incompressible Liquid), Moscow: Gos. Izd. Tekh.-Teor. Lit., 1955.]

$$\partial p / \partial x = -\left(8\mu U_s / R^2\right)\left[1 + \frac{1}{2}\sum_{m=1}^{\infty} \mathbf{exp}\left(-\nu\gamma_m^2 x / U_s R^2\right)\right] \quad (4.1)$$

where μ, ν – dynamic and kinematic viscosities of liquid; U_s – average liquid velocity; R – radius of channel; γ_m – roots of an 2nd order Bessel function.

Integrating (4.1) on the slug length L_S and using designations

$$a = 8\mu U_s / R^2 \qquad b = \nu\gamma_m^2 / U_s R^2$$

for the capillary with N_{UC} unit cells we have found [Abiev, R.S. Modeling of Pressure Losses for the Slug Flow of a Gas–Liquid Mixture in Mini- and Microchannels. Theor. Found. of Chem. Eng., 2011, Vol. 45, No. 2, pp. 156–163]

$$\Delta p_{trans} = -aN_{uc}L_s\left\{1 + \frac{1}{2bL_s}\left[1 - \mathbf{exp}\left(-bL_s\right)\right]\right\} \qquad (4.2)$$

2) Additional energy loss due to gas-liquid interface renewal during bubbles motion

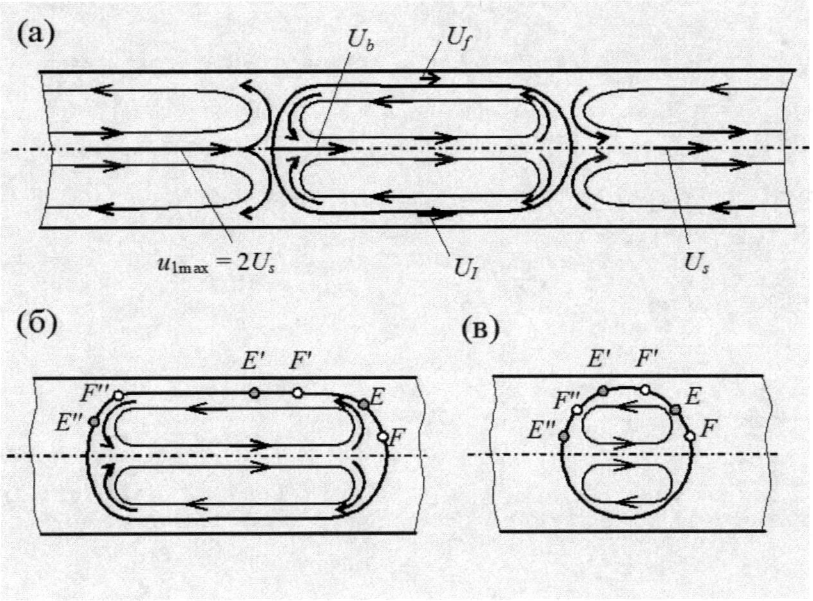

A bubble moves through a liquid slug with relative velocity $w = U_b - U_s$.

A surface dF during time increment dt is in a tale part of bubble disappearing and in the same time in a nose part of bubble the equal surface dF is appearing.

Hence, the rate of surface formation is

$$dF/dt = 2\pi R_b w \tag{4.3}$$

Energy supplied from without to capillary is

$$W_{\Delta F} = N_{uc}\sigma\, dF/dt = 2\pi R_b w \sigma L_c/L_{uc} \tag{4.4}$$

where L_c – capillary length; L_{uc} – average unit cell length, σ – surface tension.

Hence, corresponding pressure drop caused by permanent new surface formation

$$\Delta p_{\Delta F} = \frac{W_{\Delta F}}{\pi R^2 U_s} = \frac{2R_b\ w\ L_c}{R^2\ U_s\ L_{uc}}\sigma \tag{4.5}$$

Total pressure drop

$$\Delta p_{\text{cal}} = \Delta p_{trans} + \Delta p_{\Delta F} + \Delta p_b \tag{4.6}$$

3) Pressure drop along bubbles

$$\Delta p_b = \left(\frac{dp}{dx}\right)_b \sum L_b = \left(\frac{dp}{dx}\right)_b L_c \varepsilon_L \qquad (4.7)$$

The frictional losses in the gas phase are usually neglected in pressure drop models in almost all papers.

As it was revealed in our calculations, in fact their contribution more than a typical experimental error.

Pressure gradient along bubble length $dp/dx \approx 1000$ Pa/m at velocity of bubbles $U_b \approx 1$ m/s for capillary with inner diameter of 0.92 mm.

[Abiev R. Theor. Found. of Chem. Eng., 2010, V. 44, No. 1; Abiev R.S., Lavretsov I.V. Chem. Eng. Sci., 2012, V. 74, P. 59-68]

Model consistency

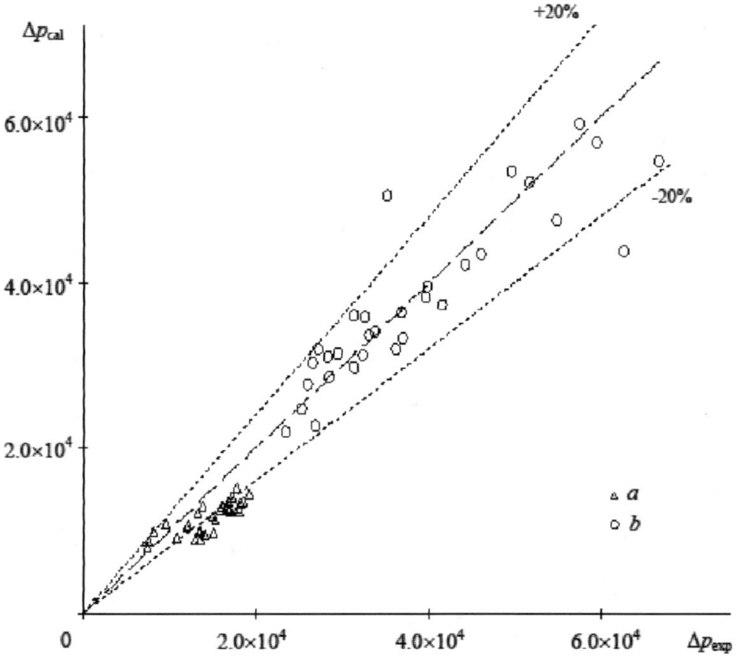

Correlation coefficient
R = 0.957

Pressure drop due to contact angle hysteresis (CAH)

[Jovanovic, J. et al. Chem. Eng. Sci. 66 (2011) 42-54]

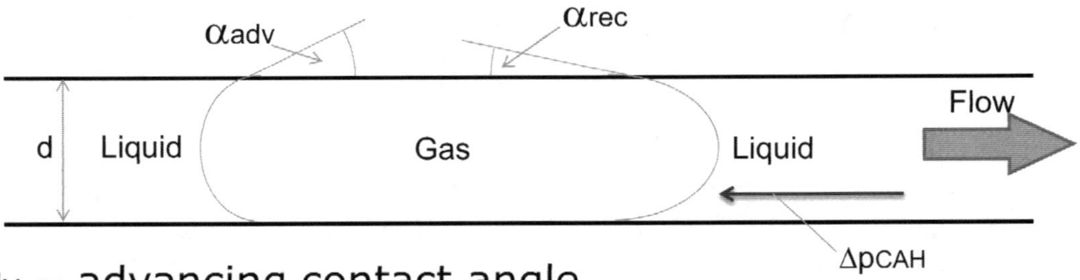

α_{adv} – advancing contact angle,
α_{rec} – receding contact angle.

For NUC number of unit cells:

$\Delta p_{CAH} = N_{UC} 4\sigma/d\ [\cos(\alpha_{ret}) - \cos(\alpha_{adv})]$.

Evidently this term applicable for Ca < 0.01 (extremely thin films)!

[Abiev R.S. Effect of Contact Angle Hysteresis on the Pressure Drop under Slug Flow Conditions in Minichannels and Microchannels, Theor. Found. of Chem. Eng., 2015, Vol. 49, No. 4]

Scheme of the experimental setup for L-L flow investigations

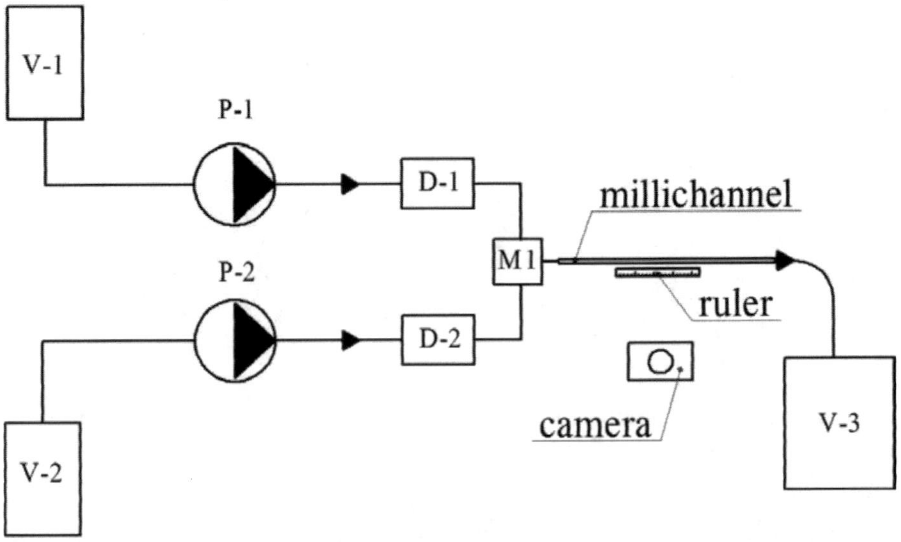

V-1 and V-2 – tanks with continuous (vegetable oil) and dispersed (water) phases; P-1 and P-2 – peristaltic pumps; D-1 and D-2 – dampers; M 1 – T-mixer; V-3 – dump tank.

Properties of liquids used in experiments

Properties	Experimental fluid		
	Sunflower oil	Water	Benzyl alcohol
Temperature, °C	20	20	20
Density, kg/m^3	888.6	998	1045.5
Dynamic viscosity, Pa s	0.054	0.001	0.0058
Coefficient of interfacial tension, N/m	0.02		
		0.0036	

[Abiev, R.S., Dymov, A.V. Modeling the Hydrodynamics of Slug Flow in a Minichannel for Liquid–Liquid Two Phase System. Theor. Found. of Chem. Eng., 2013, Vol. 47, No. 4, pp. 299–305]

Samples of L-L Taylor flows

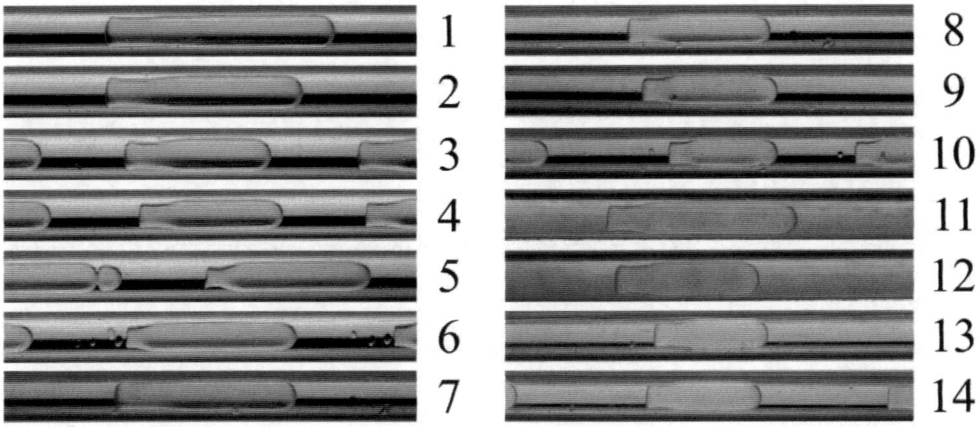

The model was corrected on the base of experimental data obtained:

I.e. an inverse problem is solved.

As a result new equation for liquid film with structure similar to Aussilous-Quere equation was found:

$$\frac{\delta}{R} = \frac{0.38Ca^{1/3}}{1+0.34Ca^{2/3}}$$

Aussilous-Quere equation:

$$\frac{\delta}{R} = \frac{1.34 \cdot \mathrm{Ca}^{2/3}}{1+3.35 \cdot \mathrm{Ca}^{2/3}}$$

Corrected mathematical model for L-L Taylor flow

$$U_f A_f + U_b A_b = U_s A_c$$

$$q_f = \frac{\pi}{8\nu_1}\left(g_x - \frac{1}{\rho_1}\frac{\partial p}{\partial x}\right)\left(R^2 - R_b^{\ 2}\right)^2 - \pi C_1\left(\frac{R^2 - R_b^{\ 2}}{2} - R_b^{\ 2}\ln\left(\frac{R}{R_b}\right)\right)$$

$$q_b = \frac{\pi}{4} g_x\left[\frac{R_b^2\left(R^2 - R_b^{\ 2}\right)}{\nu_1} + \frac{R_b^4}{2\nu_2}\right] - \frac{\pi}{4}\frac{\partial p}{\partial x}\left[\frac{R_b^2\left(R^2 - R_b^{\ 2}\right)}{\mu_1} + \frac{R_b^4}{2\mu_2}\right] - \pi C_1 R_b^2 \ln\left(\frac{R}{R_b}\right)$$

$$\frac{\partial p}{\partial x} = \frac{(G_2 - G_3)g_x - q_s}{G_1}$$

$$G_1 = \frac{\pi}{8}\left[\frac{R^4 - R_b^4}{\mu_1} + \frac{R_b^4}{\mu_2}\right] \qquad G_2 = \frac{\pi}{8}\left[\frac{R^4 - R_b^4}{\nu_1} + \frac{R_b^4}{\nu_2}\right] \qquad G_3 = \frac{\pi}{4}\frac{\rho_1 - \rho_2}{\mu_1} R_b^2\left(R^2 - R_b^{\ 2}\right)$$

New equation for liquid film with structure similar to Aussilous-Quere equation:

$$\frac{\delta}{R} = \frac{0.38Ca^{1/3}}{1+0.34Ca^{2/3}}$$

Dependence of dispersed and continuous phases' velocities ratio U_d/U_s on capillary number Ca for L-L Taylor flow *proved on several L-L systems and two materials of capillaries*

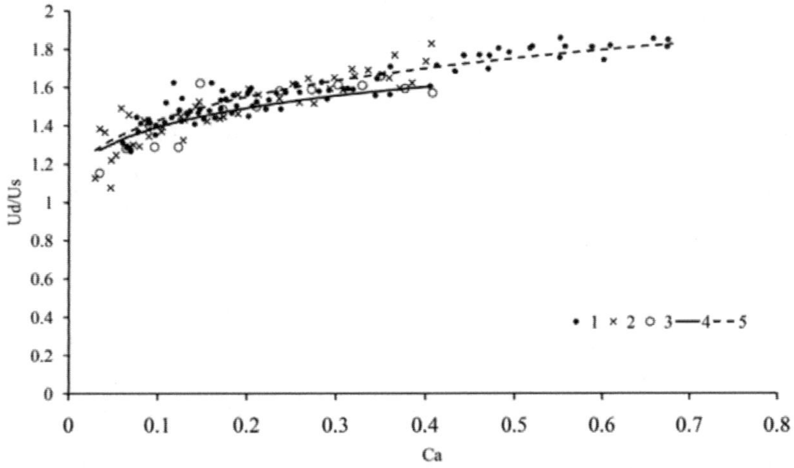

1–3 – experimental points: 1 – water – sunflower oil, PVC channel with inner diameter of 2 mm;
2 – water – sunflower oil, glass channel with inner diameter of 3.04 mm;
3 – water – benzyl alcohol, PVC channel with inner diameter of 2 mm;
4, 5 – results of calculation: 4 – water – benzyl alcohol; 5 – water – sunflower oil.

Three ways of Mass Transfer for G-L systems in capillaries

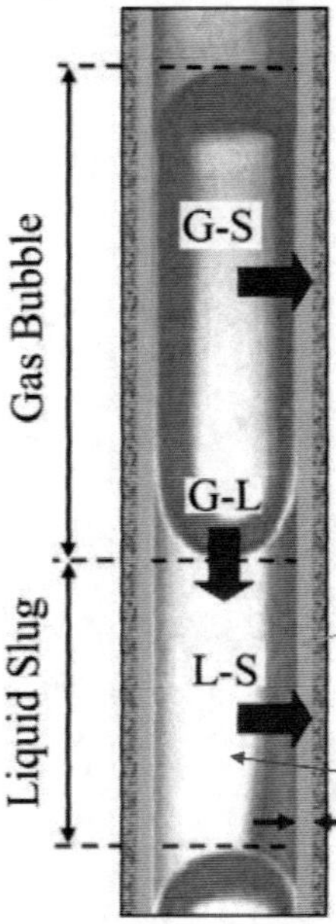

Circulation and by-pass modes (for bubbles with L > d)

Circulation mode (Ca < Ca)*

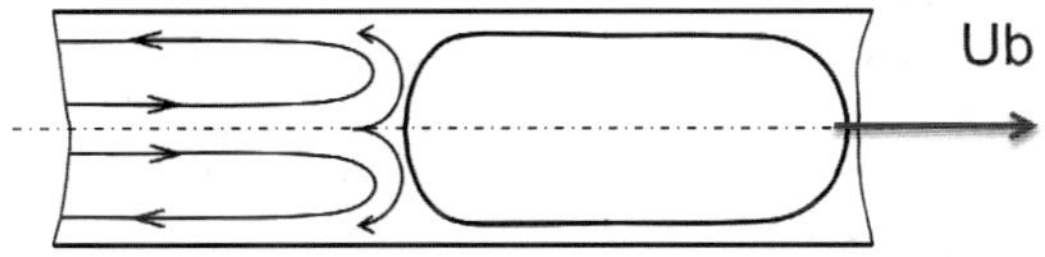

By-pass mode (Ca > Ca)*

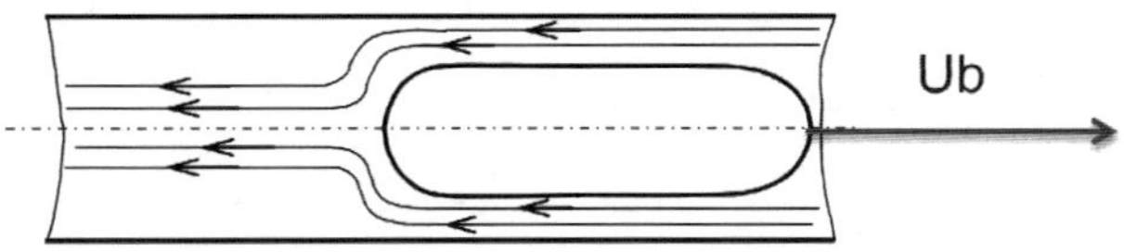

[Abiev, R.S., Lavretsov, I.V. Chem. Eng. J., 2011, Vol. 176-177, p. 57-64]

Experimental set-up

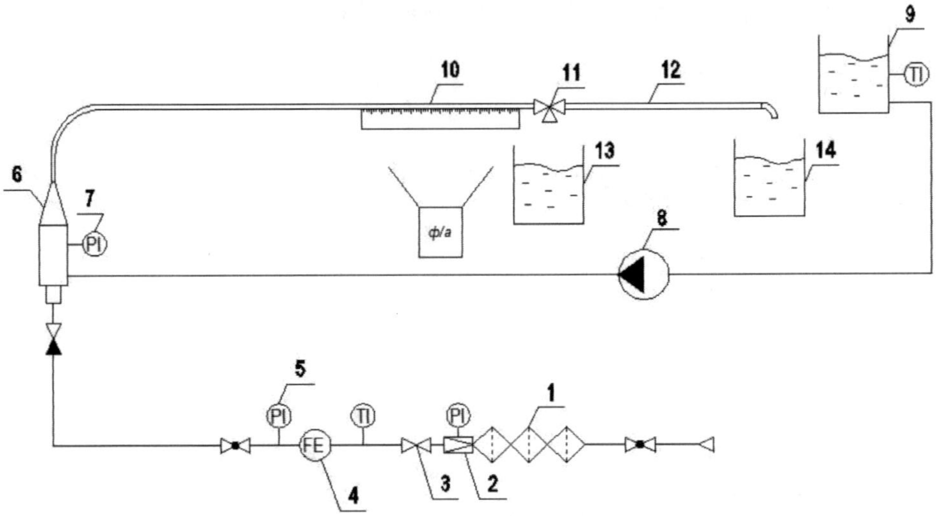

1 – filters; 2 – reducing valve; 3 – needle valve; 4 – Gas flow transducer;
5,7 – pressure gauges; 6 – mixer; 8 – peristaltic pump; 9 – **NaOH** solution;
10 glass capillary, 12 – glass capillaries coated with **benzoid acid** layer;
11 – three-way tap; 13, 14 – supporting tanks.

Mass transfer intensification by transition from by-pass to circulation mode

Circulation mode ***By-pass mode***

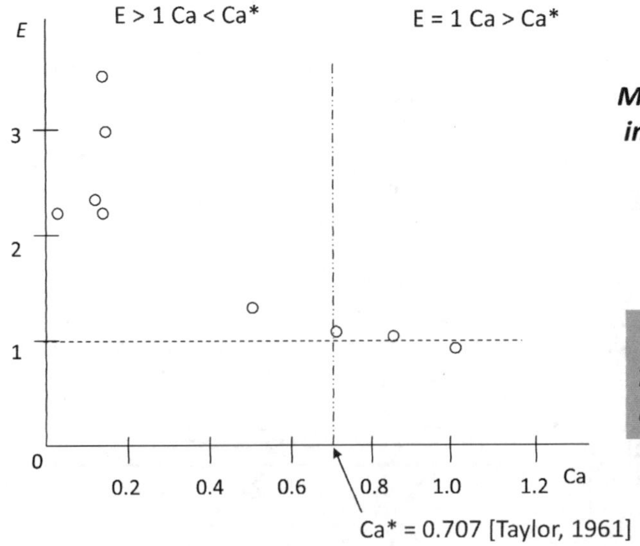

Mass transfer intensification coefficient

$$E = \frac{Sh_{TP}}{Sh_{SP}}$$

The same effect was found for heat transfer enhancement!

[Abiev, R.S., Lavretsov, I.V. Chem. Eng. Sci., 2012, V. 74, p. 59–68]

Sherwood number for L-S mass transfer by Taylor flow: A new approach

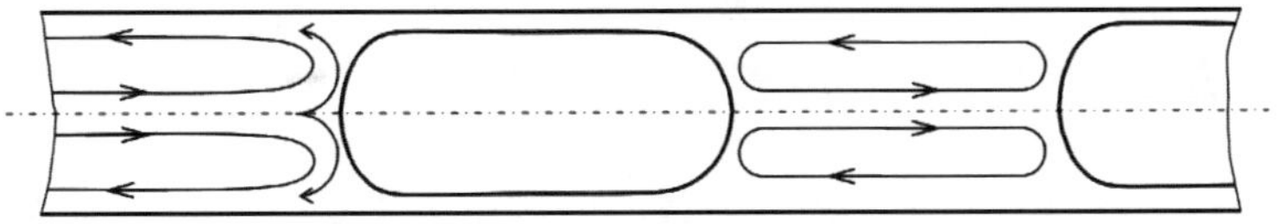

Two ways of mass transfer: in Liquid Slug and in Liquid Film

$$Sh = A_1 Sh_s + A_2 Sh_f$$

$$Sh = 3.8 Pe^{0.33} \left(\frac{d_c}{L_s} \right)^{0.05} (1 - \varepsilon_L) + 0.5 \left(\frac{d_c}{\delta} \right)^{0.3} \varepsilon_L (1 - \varepsilon_A)$$

Film Thickness [Aussillous & Quere, 2000]

$$\frac{\delta}{R} = \frac{1.34\text{Ca}^{2/3}}{1 + 3.35\text{Ca}^{2/3}}$$

Scheme of flows for mathematical model for Liquid-gas/ Liquid-liquid mass transport for Taylor flow

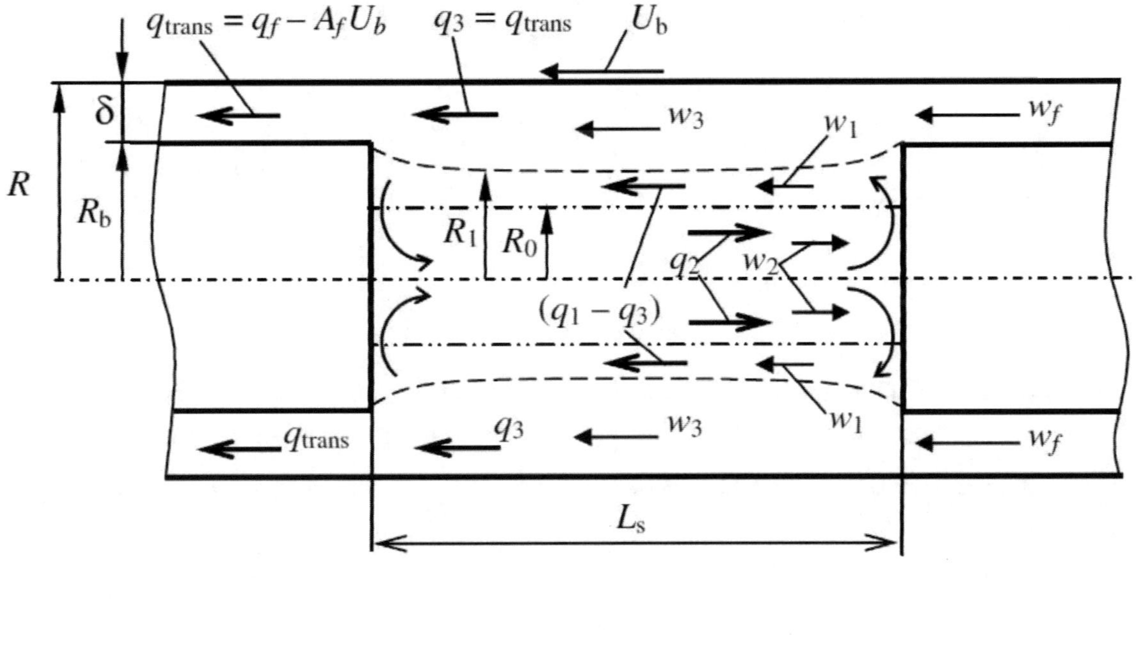

Scheme of fluxes for mathematical model for Liquid-gas/ Liquid-liquid mass transport for Taylor flow

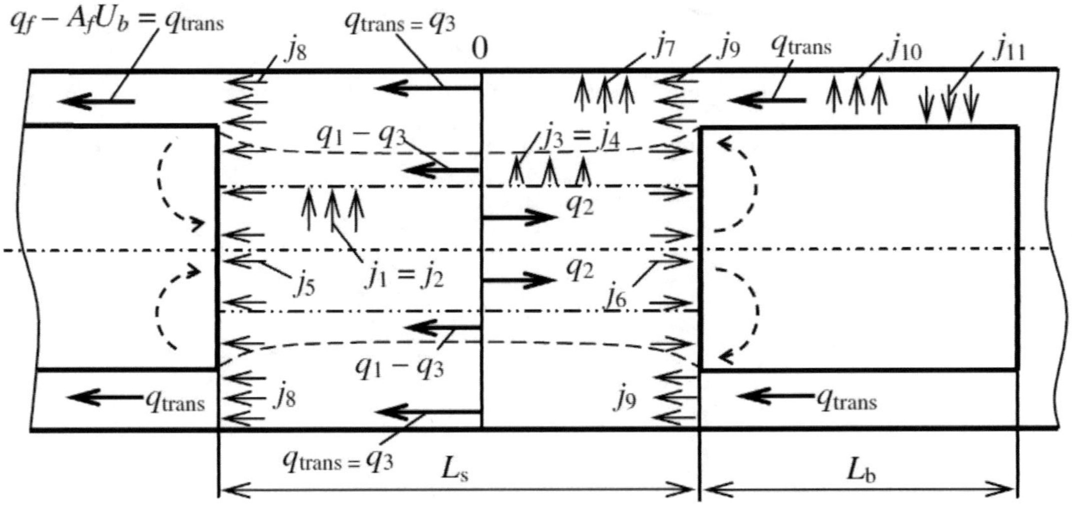

Mass transfer is described by the equation of convective diffusion in an axially symmetric form

$$\frac{\partial C}{\partial t} + V_x \frac{\partial C}{\partial x} = D\left(\frac{\partial^2 C}{\partial x^2} + \frac{\partial^2 C}{\partial r^2} + \frac{1}{r}\frac{\partial C}{\partial r}\right) \qquad 0 < x < Ls,\ 0 < r < R$$

This Equation is to be solved separately for each layer, and the conditions at the layer boundaries are described by the equations (see next page):

Velocity distribution in the channel

Mass fluxes in the layers

Summarize of transition film and outer circle flow

$$q_1 = 2\pi \int_{R_0}^{R} u(r)rdr$$

In the inner circle flow

$$q_2 = 2\pi \int_{0}^{R_0} u(r)rdr$$

In the film

$$q_3 = 2\pi \int_{R_1}^{R} u(r)rdr$$

Velocity distribution

$$u(r) = 2Us \cdot \left(1 - \frac{r^2}{R^2}\right)$$

$$w_1 = \frac{q_1 - q_3}{\pi(R_1^2 - R_0^2)}$$

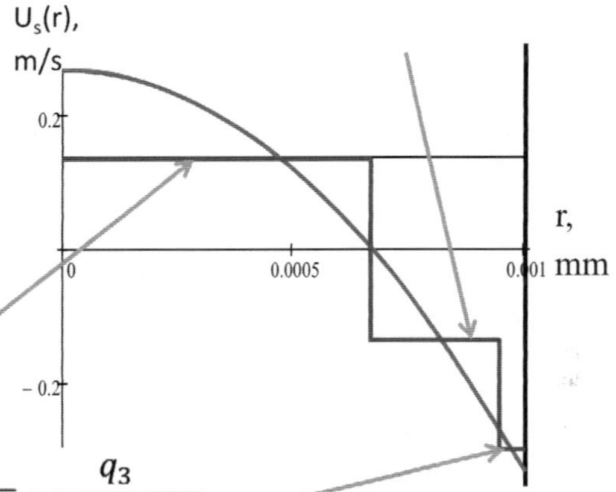

$$w_2 = \frac{q_2}{\pi R_0^2}$$

$$w_3 = \frac{q_3}{\pi(R^2 - R_1^2)}$$

Problem description

The nonstationary convection and diffusion

$$\frac{\delta C}{\delta \tau} + v_x \frac{\delta C}{\delta x} + v_y \frac{\delta C}{\delta y} + v_z \frac{\delta C}{\delta z} = D\nabla^2 C$$

$$\frac{\partial C}{\partial t} + V_x \frac{\partial C}{\partial x} = D \frac{\partial^2 C}{\partial x^2}$$

Assumptions allowing to convert axisymmetric problem in the quasi one-dimensional

- The mass transfer between gas bubbles and liquid in slugs and film can be ignored
- The radial flow makes a small contribution to the mass transfer;
- Uniform velocity distribution within a single layers is introduced
- The transport of substance in the liquid film is described by the mass conductivity equation

Problem description

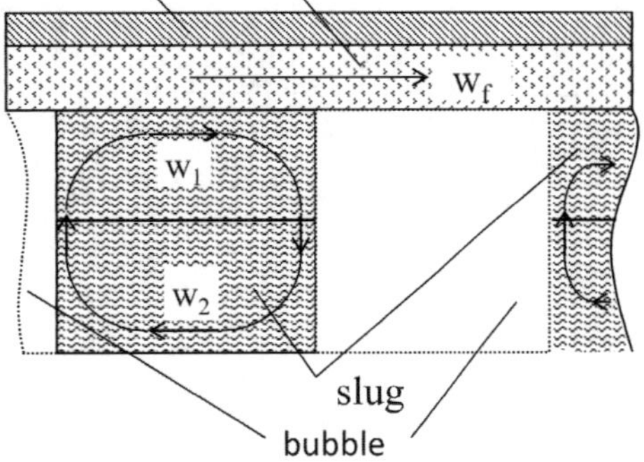

2 x 1D problem

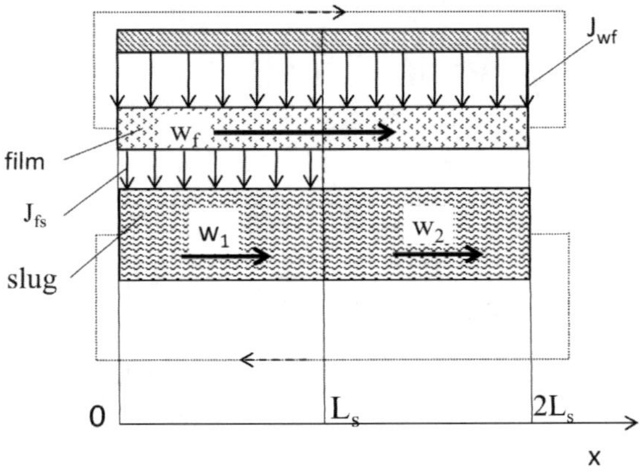

Transformation

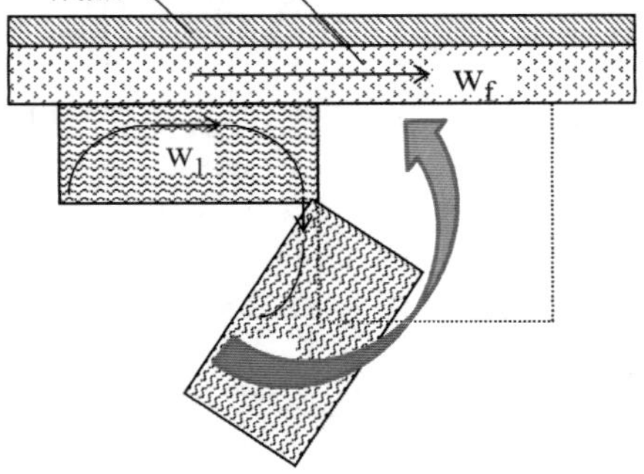

Dependence of mass transfer coefficient from capillary wall to the TP flow on TP velocity

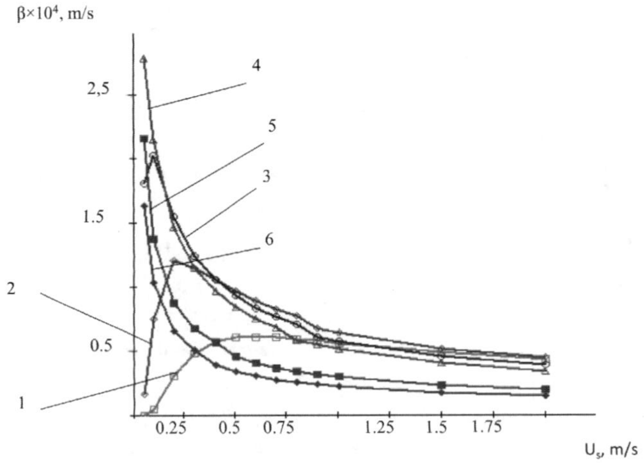

Diameter of capillary: 1 – 0.5 mm; 2 – 0.65 mm; 3 – 0.85 mm; 4 – 1 mm; 5 – 1.86 mm; 6 – 2.5 mm.

[Svetlov S.D., Abiev R.Sh. Theor. Found. of Chem. Eng. 2016 (Accepted)]

This models could be applied to the Research and Design of:

- micro reactors,
- micro extractors and
- micro heat exchangers and evaporators,
- other multiphase devices

for calculation of process parameters and process optimization and intensification.

SYNTHESIS OF 5-PHENYLTETRAZOLE AND ITS N-METHYL DERIVATIVES IN A MICROREACTOR

E.A. Popova, R.Sh. Abiev*, L.A. Lappalainen, S.D. Svetlov, T.V. Andreeva, R.E.Trifonov, V.A. Ostrovskii,
CABEQ-2014. Vol.28 No.2 May 2014. P. 241 - 246

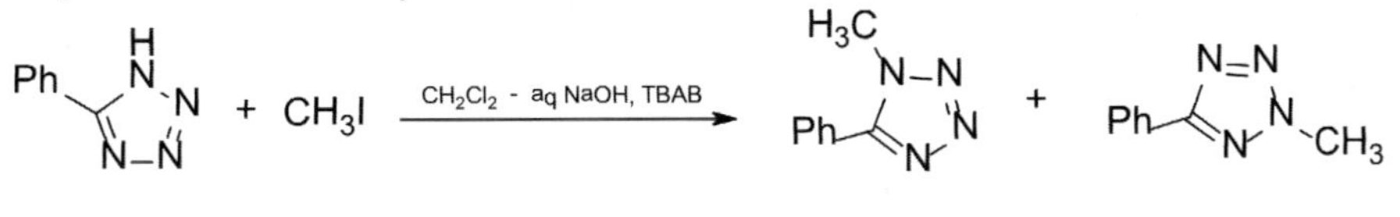

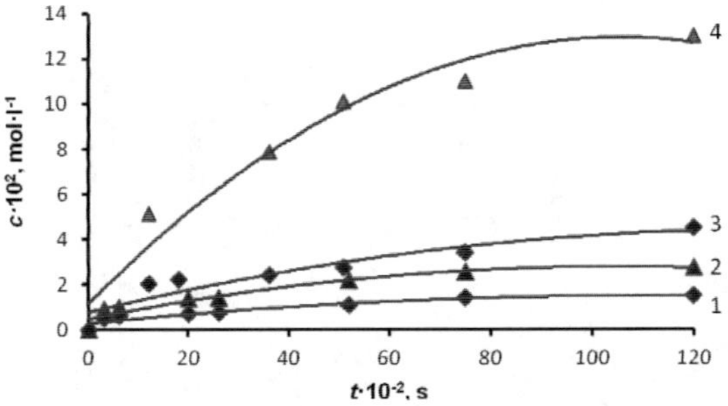

Fig. 7 – *Dependence of the concentration of 2-methyl-5-phenyl- tetrazole (1, 3) and 1-methyl-5-phenyltetrazole (2, 4) on time:*
1, 2 – in a batch reactor with TBAB;
3, 4 – in a microreactor without TBAB

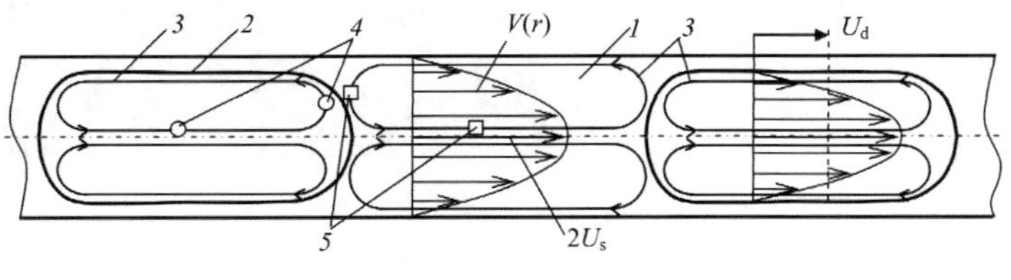

Taylor flow characteristic features

- **Gas bubble length: L_B**
- **Liquid slug length: L_S**
- Liquid film thickness
- Shape of interfacial areas
- etc.

ρ_L

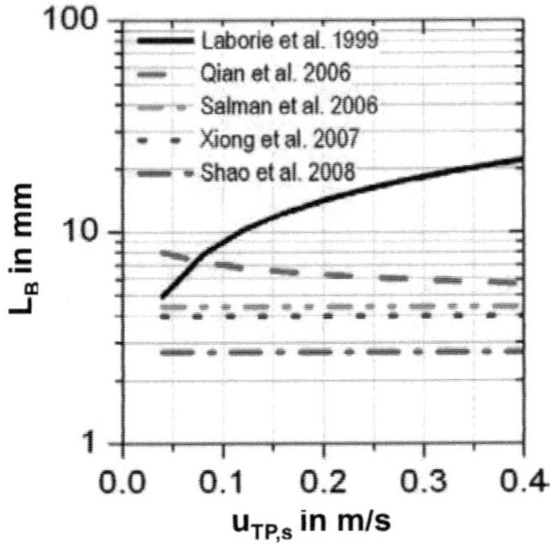

Predicted gas bubble lengths in Taylor flow mode using selected correlations

Zuber et al., 1965, J. of Heat Transfer, 87. **Laborie et al., 1999,** Chem. Eng. Sci., 54. **Triplett et al., 1999,** Int. J. Multiphase Flow, 25. **Kreutzer et al., 2001,** Chem. Eng. Sci., 56. **Yang et al., 2001,** Int. J. Multiphase Flow, 27. **Kreutzer, 2003,** PhD thesis, TU Delft. **Liu et al., 2005,** Ind. Eng. Chem. Res., 44. **Qian et al., 2006,** Chem. Eng. Sci., 61. **Salman et al., 2006,** Chem. Eng. Sci, 61. **Tsoligkas et al., 2007,** Chem. Eng. Sci, 62. **Xiong et al., 2007,** J. Micromechanics and Microengineering, 17. **Shao et al., 2008,** 1st Int. Conf. on Microfluidics. **Shao et al., 2008,** Int. J. Heat and Fluid Flow, 29. **Yue et al., 2008,** Chem. Eng. Sci., 63. **Abiev, 2010,** Theor. Found. Chem. Eng., 44. **Sobieszuk et al., 2010,** Chem. Eng. Res. Des., 88. **Keskin, et al., 2010.** AIChE Journal, 56. **Duong et al., 2013**, Computers & Fluids, 86. **Hoang et al. 2013**, J. Fluid Mechanics, 717.

Experimental

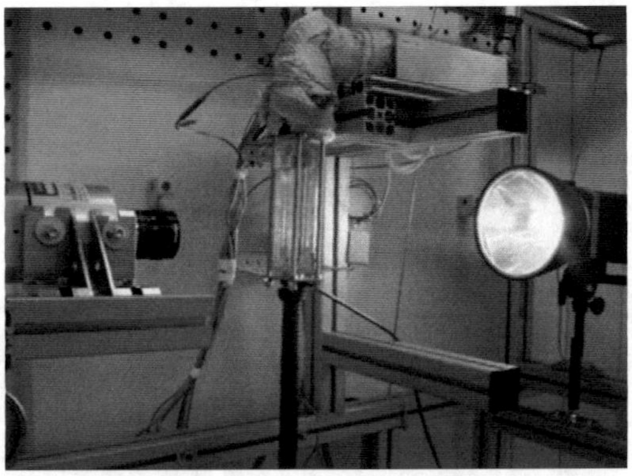

Minichannel test section

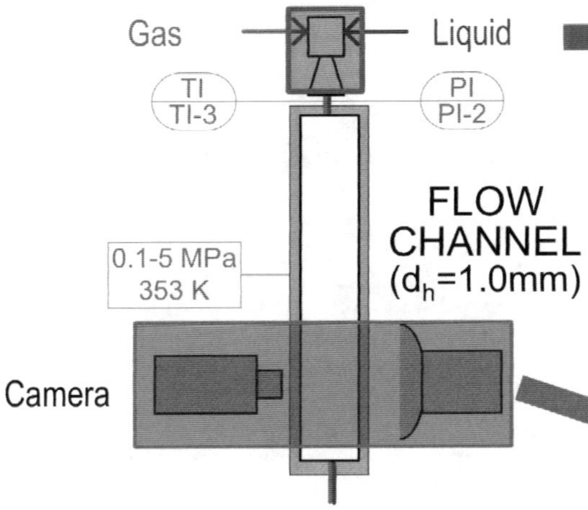

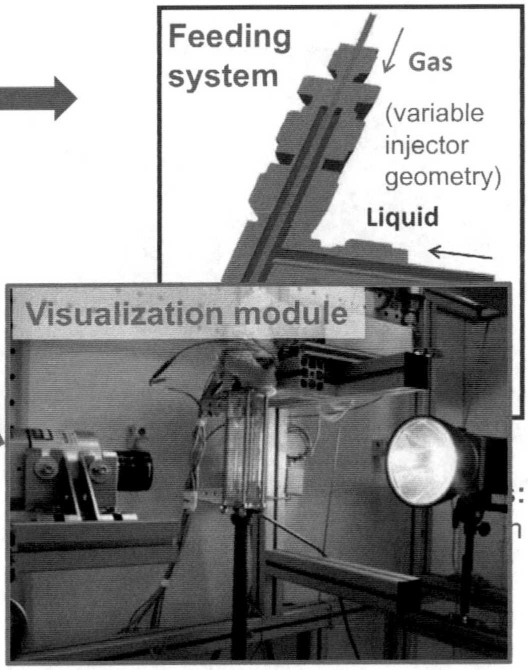

Sketch of the minichannel test section

Haase, Bauer, Lange et al., 2011. Chem. Eng. J., 176-177, 65ff.

Results and discussion

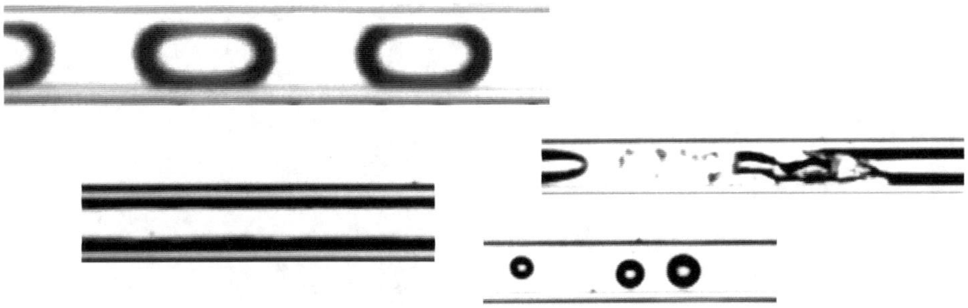

Impact of phase holdup

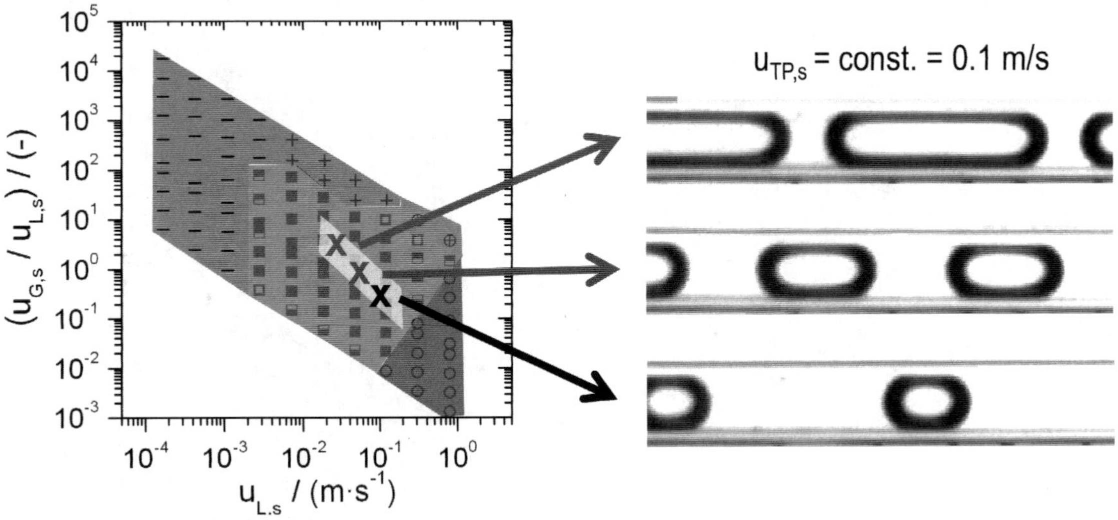

Flow map of water/hydrogen-flow
in a square minichannel (d_h=1.0 mm)

Impact of fluid velocities

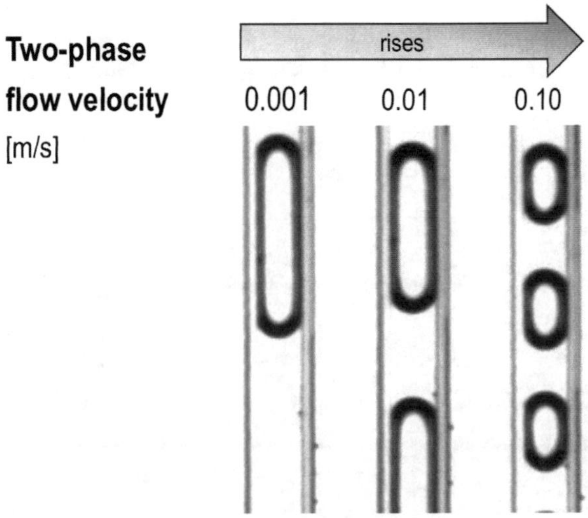

Photographs of Taylor flows
at various fluid velocities

Other parameters: gas: hydrogen, liquid: alpha-methylstyrene, d_h = 1.0 mm, $d_{In,Cl}$ = 0.32 mm, β_G = 0.50

Impact of liquid properties: viscosity

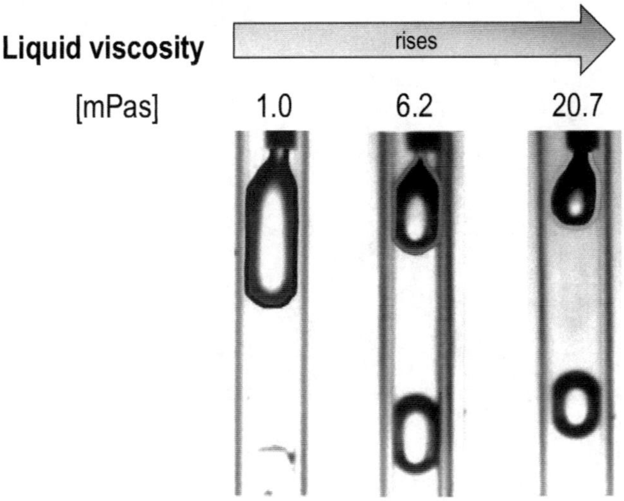

Photographs of Taylor flows for liquids with different viscosities

Other parameters: gas: hydrogen, pressure: 1 MPa, d_h = 1.0 mm, $u_{TP,s}$ = 0.01 m/s, $d_{In,Cl}$ = 0.6 mm, β_G = 0.26

Impact of liquid properties: surface tension

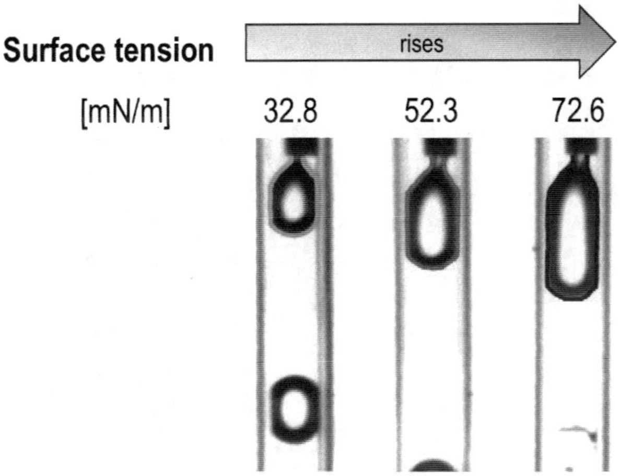

Photographs of Taylor flows for liquids with different surface tensions

Other parameters: gas: hydrogen, pressure: 1 MPa, d_h = 1.0 mm, $u_{TP,s}$ = 0.01 m/s, $d_{In,Cl}$ = 0.6 mm, β_G = 0.26

Impact of gas properties: density/viscosity

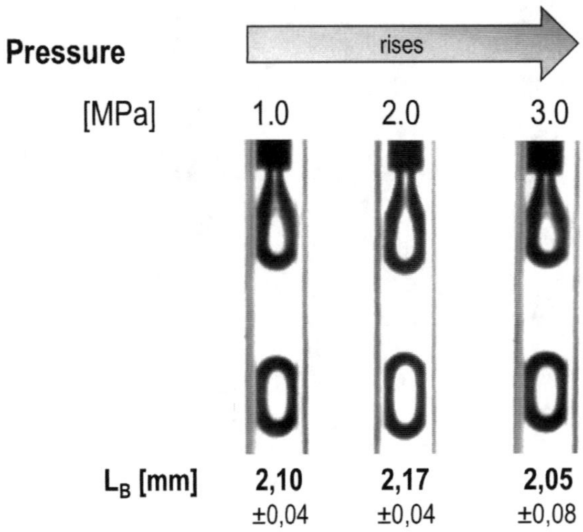

Photographs of Taylor flows
at different system pressures

Other parameters: gas: hydrogen, liquid: alpha-methylstyrene, d_h = 1.0 mm, $u_{TP,s}$ = 0.10 m/s, $d_{In,CI}$ = 0.32 mm, β_G = 0.50

Data analysis

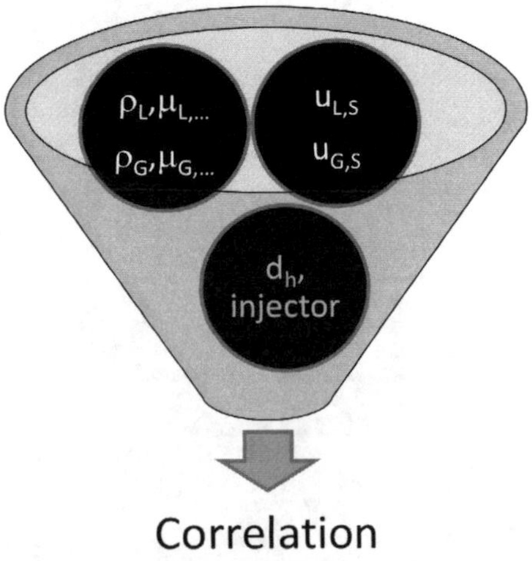

Experimental procedure

Analysis of CMOS – video files (1500 different conditions)

⇩

Taylor flow

⇩

MatLab – code
100 bubbles & slugs

⇩

Bubble and slug lengths
Mean value and deviation

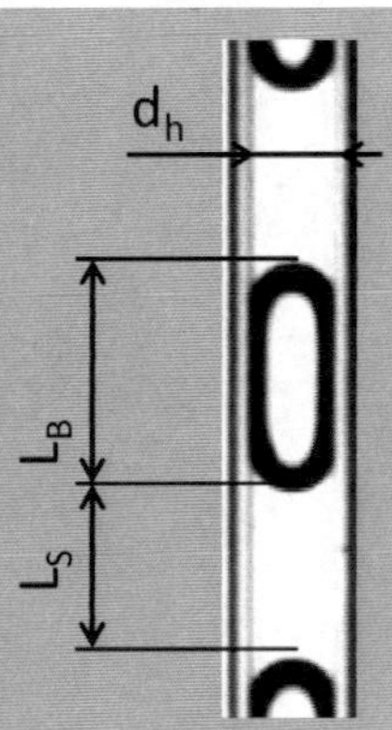

Experiments (1500)	**Relevant Parameters** (8+2)	**Dimensionless numbers** 5 (+2)
$u_{TP,s}$ = 0.001...3.5 m/s β_G = 0.01...0.99	$u_{G,s}$ $u_{L,s}$	$\frac{u_{G,s}}{u_{L,s}}$
ρ_L = 910...1180 kg/m³ μ_L = 1...20 mPas σ = 33...73 mN/m ρ_G = 0.8...11.5 kg/m³ μ_G = 8.7...17.3 mPas Θ = 0...18°	ρ_L μ_L σ θ	Re_L We_L $\Theta^* = \pi \cdot \Theta/180$
d_h = 1.0 mm $d_{In,CI}$= 0.18...0.49 mm	d_h $d_{In,CI}$	$\frac{d_{In,CI}}{d_h}$
Ethanol- water and glycerol-water mixtures	L_B L_S	$\frac{L_B}{d_h}$ $\frac{L_S}{d_h}$

Relevant process parameters → Pi-Theorem →

Correlations for Taylor flow parameters

$$\frac{L_B}{d_h} = 1 + a1\left(\frac{u_{G,s}}{u_{L,s}}\right)^{b1} Re_L^{c1}\, We_L^{d1}\left(\frac{d_{In,CI}}{d_h}\right)^{e1} \Theta^{*f1}$$

$$\frac{L_S}{d_h} = a2\left(\frac{u_{G,s}}{u_{L,s}}\right)^{b2} Re_L^{c2}\, We_L^{d2}\left(\frac{d_{In,CI}}{d_h}\right)^{e2} \Theta^{*f2}$$

→ Parameter estimation

for a1,a2,a3, b1, b2, b3, … using method of least squares

Approach according to: **S. Drobot, 1953**. Studia Mathematica, 14, 84 ff. and **L. Brand, 1957**. Archive for Rational Mechanics and Analysis, 1, 35 ff. with modification to correlate the dimensionless gas bubble length.

Empirical correlations

Superficial two-phase velocity

≤ 0.2 m/s

$$\frac{L_B}{d_h} = 1 + 0.12\left(\frac{u_{G,s}}{u_{L,s}}\right)^{0.59} Ca_L^{-0.46}\left(\frac{d_{In,CI}}{d_h}\right)^{1.10}$$

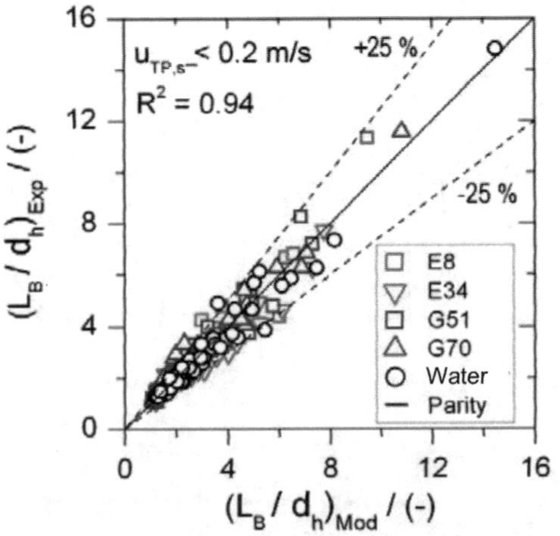

Parity plot for experimental and correlated gas bubble length in a minichannel with d_h = 1,0 mm

E8…8 wt.% ethanol in water; E34…34 wt.% ethanol in water; G51…51 wt.% glycerol in water; G70…70 wt.% glycerol in water

Empirical correlations

Superficial two-phase velocity ≤ 0.2 m/s	Superficial two-phase velocity > 0.2 m/s

$\frac{L_B}{d_h} = 1 + 0.12 \left(\frac{u_{G,s}}{u_{L,s}}\right)^{0.59} Ca_L^{-0.46} \left(\frac{d_{In,CI}}{d_h}\right)^{1.10}$ $R^2=0.94$	$\frac{L_B}{d_h} = 1 + 1.94 \left(\frac{u_{G,s}}{u_{L,s}}\right)^{0.96} Re_L^{0.12} \left(\frac{d_{In,CI}}{d_h}\right)^{1.40}$ $R^2=0.94$
$\frac{L_S}{d_h} = 0.61 \left(\frac{u_{G,s}}{u_{L,s}}\right)^{-0.85} Ca_L^{-0.18} \left(\frac{d_{In,CI}}{d_h}\right)^{0.48}$ $R^2=0.93$	$\frac{L_S}{d_h} = 1.34 \left(\frac{u_{G,s}}{u_{L,s}}\right)^{-0.58} Re_L^{0.08} We_L^{0.09} \left(\frac{d_{In,CI}}{d_h}\right)^{0.48}$ $R^2=0.92$

6. Conclusions

- Mathematical model for hydrodynamics of L-G and L-L Taylor flow in micro and milli channels is elaborated.

- Mathematical model for G-L (L-L), L-S mass transfer in micro channels is elaborated.

- Process intensification for L-S and L-L systems by Taylor vortices demonstrated experimentally and explained theoretically.

- Opportunities to use L-G and L-L mathematical models for Research and Design of micro reactors, micro extractors and micro heat exchangers are discussed.

- Correlations for Taylor flow parameters have been found

… thank you very much for your attention

… and please visit us when you are close-by

Acknowledgement:

Deutsche Forschungsgemeinschaft DFG

Max-Buchner Forschungsstiftung

481205 3D-Printed Monolithic Adsorbents for Gas Separation Processes

Abstract:

Structured adsorbents especially in the form of monolithic contactors offer an excellent gas-solid contacting strategy for development of practical and scalable CO_2 capture technologies. In this study, the fabrication of 3D-printed 13X and 5A zeolite monoliths with novel structures and their use in CO_2 removal from air are reported. The physical and structural properties of these printed monoliths are evaluated and compared with their powder counterparts. Our results indicate that 3D-printed monoliths with zeolite loading as high as 90 wt% exhibit comparable adsorption uptake to powder sorbents. The adsorption capacities of 5A and 13X monoliths were found to be 1.59 and 1.60 mmol/g, respectively using 5000 ppm (0.5%) CO_2 in nitrogen at room temperature. The dynamic CO_2/N_2 breakthrough experiments show relatively fast dynamics for monolithic structures. In addition, the printed zeolite monoliths show reasonably good mechanical stability that can eventually prevent attrition and dusting issues commonly encountered in traditional pellets and beads packing systems. Moreover, our recent work on the fabrication and evaluation of 3D-printed aminosilica adsorbents will be discussed. The 3D printing technique offers an alternative cost-effective and facile approach to fabricate structured adsorbents with tunable structural, chemical and mechanical properties for use in gas separation processes.

Copyright © American Institute of Chemical Engineers. All rights reserved.

469425 Trickle-Bed Reactors and Their Future

Abstract:

Since decades, trickle-bed reactors are successfully employed to produce various chemicals, intermediates and fuels. Nevertheless, their operability and efficiency suffers from partial wetting of the particle bed, limited mass and heat transfer, which may cause a reduced catalyst utilization or may create safety problems due to possible formation of hot spots. To intensify process efficiency, the integration of novel catalyst packings, e.g. monoliths or foams, or the periodic operation in terms of fluid throughput, component concentration or temperature is in the discussion of science and industry and will be addressed in the lecture.

This paper will present the state of the art on trickle-bed reactors and provide an outlook on their future.

Copyright © American Institute of Chemical Engineers. All rights reserved.

457329 Keynote - Modeling Electrochemical Interfaces: An Approach from First Principles

Abstract:

Electrified interfaces are central for electrocatalysis, batteries and molecular electronics. Experimental characterization of these complex interfaces with atomic resolution is highly challenging. First principles modelling could provide a link between the measurable quantities and an atomic scale understanding. However, such simulations are far from straightforward. Although approaches that include the effect of the potential and the electrolyte have been proposed, detailed validation has been scarce and "indirect" since atomically resolved experimental studies of systems that can be convincingly simulated are scarce.

We introduce in this lecture the adsorption of pyridine on Au(111) as a convenient and relevant model: the adsorption mode of pyridine switches as a function of the electrochemical potential. We demonstrate that the primitive surface-charging model gives qualitatively correct results at a low cost. For quantitative agreement, however, the model needs to include a more realistic description of the electrical double layer. Approximating the latter through the linearized Poisson-Boltzmann equation leads to a quantitative improvement, lowering the error in the transition potential from 1 V to acceptable 0.3 V. Hence, we demonstrate the qualitative usefulness of the surface charging method and the excellent agreement that can be obtained by slightly more sophisticated electrolyte models.

References

1) Stephan N. Steinmann and Philippe Sautet "Assessing a First Principles Model of an Electrochemical Interface by Comparison with Experiment" J. Phys. Chem. C 120, 5619–5623 (2016)

2) Stephan N. Steinmann, Carine Michel, Renate Schwiedernoch, Jean-Sebastien Filhol, Philippe Sautet* Modelling the HCOOH/CO2 Electrocatalytic Reaction: When Details Are Key, ChemPhysChem 16, 2307 –

2311 (2015)

Copyright © American Institute of Chemical Engineers. All rights reserved.

467825 Feature Engineering of Machine-Learning Chemisorption Models for Bifunctional Electrocatalyst Design

Abstract:

Transition metal alloys have shown great potential for catalyzing many chemical and electrochemical reactions. Attributed to the versatility of transition metals on the periodic table, the catalyst surface reactivity can be manipulated by varying its geometric characteristics, metal ligands, and extrinsic factors in the vicinity of active sites [1]. However, it is very time-consuming and costly to search for highly optimized alloy composition and geometry by high-throughput experiments and/or quantum-chemical calculations.

In this talk, we will introduce a new computational catalyst design framework, which integrates machine-learning algorithms with the descriptor-based design approach for rapid screening of transition-metal catalysts [2-3]. By engineering numerical representation of surface metal atoms using easily accessible features such as the local electronegativity and the effective coordination number that are dependent on the surroundings of an adsorption site, together with the intrinsic properties of active metal atoms including the electronegativity, ionic potential, and electron affinity, the machine-learning model optimized with ~500 *ab initio* adsorption energies on bimetallic alloys can capture complex, non-linear adsorbate/substrate interactions with the root mean squared errors (RMSE) ~0.12 eV. To validate the model, we applied the model to search for high-performance {111}-terminated bifunctional catalysts for electrocatalytic oxidation of methanol. The *CO and *OH adsorption energies represent important efficiency metrics to describe the electrocatalytic reactivity. We show that our machine-learning-augmented model exhibits outstanding prediction power in screening for the bifunctional, multimetallic surfaces with desired adsorption properties for both *CO and *OH on different adsorption sites. Compared with the traditional high-throughput computational and experimental trial-and-error approach, the machine-learning chemisorption models have great potential in accelerating the discovery of interesting catalytic materials.

[1] A. Vojvodic and J. K. Nørskov, "New design paradigm for heterogeneous catalysts," *Natl. Sci. Rev.*, p.

nwv023, Apr. 2015.

[2] X. Ma, Z. Li, L. E. K. Achenie, and H. Xin, "Machine-Learning-Augmented Chemisorption Model for CO2 Electroreduction Catalyst Screening," *J. Phys. Chem. Lett.*, vol. 6, no. 18, pp. 3528–3533, Sep. 2015.

[3] Z. Li, X. Ma, and H. Xin, "Feature Engineering of Machine-Learning Chemisorption Models for Catalyst Design," Catalysis Today, 2015 (Accepted).

Copyright © American Institute of Chemical Engineers. All rights reserved.

469263 Effect of Lanthanum Doping and Chlorine Addition on Strontium Titanate Perovskites for Electrocatalytically-Assisted Oxidative Dehydrogenation of Ethane

Abstract:

Oxidative dehydrogenation (ODH) of alkanes is an important catalytic reaction that is widely studied. It can be used to convert less valuable alkanes, such as ethane, to more valuable olefins, such as ethylene. However, the further oxidation of olefins to carbon dioxide and carbon monoxide limits the selectivity of the process [1]. Oxide ion-conducting electrolyte reactors allow for controlling oxygen availability to the reaction and thus, improve the selectivity for olefin formation [2, 3]. The reactor setup is similar to solid oxide fuel cells with an important difference: instead of generating power as in solid oxide fuel cells, an external current is applied to control the oxygen flux through the electrolyte. When the current is applied, the oxygen molecules at the cathode side are ionized to O^{-2} ions. Oxygen ions conduct through the oxygen ion-conducting electrolyte to the anode, where they react with ethane to form ethylene. Since the amount of oxygen conducted through the electrolyte can be controlled by the current applied, this can help regulate the alkane conversion and selectivity.

In this study, strontium titanate type of perovskites were tested as anode catalysts for electro ODH reaction. La-doped strontium titanates were synthesized via modified Pechini route. The effect of lanthanum doping and chlorine incorporation on these catalysts was investigated. The characteristics of the catalysts were investigated using N_2 physisorption, ambient and *in-situ* X-ray diffraction (XRD), X-ray photoelectron spectroscopy (XPS), diffuse reflectance infrared spectroscopy (DRIFTS), and electronic conductivity measurements. Besides Electro-ODH experiments to test the activities, impedance measurements were also taken using to better understand the cell characteristics and resistances. Oxygen flux measurements were also performed.

Electro-ODH tests on the cell with $La_{0.2}Sr_{0.8}TiO_{3Â±Î´}Cl_{Ïƒ}$ anode showed that ethane conversion and ethylene production rates increase with increasing current. Hydrogen production rates were also observed to be very low, indicating oxidative dehydrogenation is taking place. It was also observed that Cl addition increases the ethylene selectivity and La doping increases the conductivity of strontium titanate.

1. Kung, H.H., and Kung, M.C., Applied Catalysis A: General 157, 105 (1997).

2. Coronas, J., Menendez, M., and Santamaria, J. Industrial & Engineering Chemistry Research 34, 4229 (1995).

3. Sundmacher K., Rihko-Struckmann, L.K., Galvita, V., Catalysis Today 104, 185 (2005).

Copyright © American Institute of Chemical Engineers. All rights reserved.

461301 Mechanistic and Trend Analyses of Selective Furfural Electroreduction on Transition Metals from First-Principles Methods

Abstract:

Furfural is an important biomass-derived platform compound. Electrocatalytic hydrogenation and deep reduction, via the C-O bond cleavage, of furfural on transition metals can be developed into a sustainable strategy to obtain valuable chemicals and fuels. Further, the complex chemical nature of furfural presents challenges to understand the electroreduction pathways and the product selectivity, which are still being debated.

Density functional theory (DFT) calculations were performed on the close-packed (111) facets of Cu, Ag, Pb and Ni to investigate furfuryl alcohol (FA) and 2-methylfuran (MF) formations in acidic conditions. Two FA formation pathways via 2-methoxyfuran (mh6) and 2-hydroxymethylenefuran (mh7), and three MF formation pathways were considered in this study. The computational hydrogen electrode (CHE) model was applied to determine free energies of the reactions under an external applied potential. The energy barriers of elementary reduction steps were estimated using the BrÃ¸nstedâ€"Evansâ€"Polanyi (BEP) relationship. On Cu, Ag, Pb and Ni, it has been found that the pathway via mh6 is competitive for FA formation. On Ni, the pathway via mh7 is also competitive. For MF formation, pathways involving mh6 are more competitive on all considered metals. In order to gain insights into the product selectivity, micro-kinetic modeling was performed based on the mechanism derived from DFT calculations. The modeling confirms that FA and MF are formed via parallel pathways. The turnover frequencies of FA and MF formations were also obtained, and Cu is found to be the most active for MF production, which is supported by electrochemical measurements.

Copyright © American Institute of Chemical Engineers. All rights reserved.

464406 NO Electrochemical Reduction on Pt(100) from First Principles

Abstract:

Electrochemical denitrification is a promising technology for the removal of toxic nitrate and nitrite species from groundwater due to this processâ€™s environmental compatibility, energy efficiency, safety, and product selectivity.[1] The adsorbed NO is generally considered to be a selectivity-determining species for the reaction in acid electrolytes.[2]However, molecular-level insights into the mechanism are challenging to obtain from experiments alone, and atomic-scale studies using density functional theory (DFT) can thus provide important information about general reaction scheme of NO electrocatalysis. These basic building blocks, determined on the most common electrocatalyst for this reaction, platinum, may, in turn, serve as a starting point for future trends-based analysis on different transition metal surfaces.

While Pt(111) terraces are the lowest in energy of any single crystal Pt surface, terraces of Pt(100) have shown unusual activity and selectivity patterns for the electroreduction of NO.[3] Therefore, in this work, we use periodic, self-consistent DFT calculations to clarify the adsorption structures and energetics of NO and its reaction intermediate species on the Pt(100) surface. Solvation energies and protonation kinetics are calculated using water-incorporated systems in the DFT analysis.[4]In addition, we implement a rigorous kinetic Monte Carlo (kMC) approach that takes into account coverage-dependent interactions and activation barriers and that as that produces potential-dependent values of current from multiple reaction pathways as a function of voltage.

In our simulation, the most favorable pathway is the ammonia formation. The mechanism consists of two consecutive proton-electron transfer steps after the adsorption of NO (denoted the EE mechanism), with favorable formation of NOH* as a first intermediate. The Tafel slope, which was obtained by plotting peak potentials at different scan rates in kMC simulations, supports the EE mechanism rather than the chemical dissociation steps. Also, due to the fast

protonation, the probabilities of forming N_2 or N_2O are negligible. Finally, implications of these results for both NO and nitrate electroreduction on other transition metal surfaces will be discussed.

[1] M. Duca and M. T. M. Koper, Energ. Environ. Sci. 5 (2012) 9726.
[2] A. Cuesta, M. Escudero, Phys. Chem. Chem. Phys. 10 (2008) 3628.
[3] V. Rosca, M. T. M. Koper, J. Phys. Chem. B 109 (2005) 16750.
[4] A. Clayborne, H. J. Chun, R. B. Rankin, J. Greeley, Angew. Chem. Int. Ed. 54 (2015) 8255.

Copyright © American Institute of Chemical Engineers. All rights reserved.

472512 Flame Synthesis and High Pressure High Temperature Annealing of Anatase TiO_2 for Increased Photocatalytic Activity

Abstract:

Titanium dioxide (TiO_2) is of particular interest in photocatalysis due to its photoactivity under ultraviolet illumination and its stability under catalytic reaction conditions. As most of incident solar radiation is visible light, many researchers have sought a visible-light active photocatalyst by modifying TiO_2 in order to reduce the band gap energy (BGE). It is known that the lattice structure, particle size, dopant concentration, and supported metal nanoparticles affect the BGE of TiO_2. Through low pressure flame spray pyrolysis, carbon doping, high pressure high temperature annealing, and supporting of metals we have successfully decreased the BGE and increased the photocatalytic activity of anatase phase TiO_2.

High pressure high temperature annealing (HPHT) was conducted at 200 °C in a Parr reactor under 250 psig of hydrogen or nitrogen. We have observed that, compared to the parent sample, HPHT increases particle size, decreases the BGE, causes a slight darkening of color, and increases the absorbance throughout the UV and visible spectrum. The increase in UV and visible light absorbance indicates that HPHT samples will have a greater photocatalytic activity than un-modified anatase. This was seen via an increase in hydrogen production during methane steam reforming (MSR). Hydrogen yield increased from a negligible amount over anatase to 1.05 % and 1.12 % under UV and visible light irradiation MSR over HA-Anatase at 700 ºC.

It was found that supporting copper on anatase, 1%Cu(O)/Anatase, decreased the band gap from 3.22 eV to 3.13 eV. By combining the two modification methods, HPHT and metal supporting, we aim to decrease the BGE further and therefore increase the photocatalytic activity under UV and visible light irradiation. An order of operations variation was performed by annealing in hydrogen either before or after metals were supported and the sample calcined. All

variations of HPHT and copper supporting yielded BGE lower than unmodified anatase, with HPHT hydrogen annealed copper supported on anatase, HA-1%Cu(O)/Anatase, having the smallest BGE of 3.07 eV. EPR analysis indicates that HPHT annealing under hydrogen or nitrogen increases Ti^{3+} centers in the samples. Since it has been reported that oxygen vacancies play an important role in absorption and active sites, we hypothesize that the HPHT samples have a higher photocatalytic activity than the unmodified parent sample partially due to the effect of the Ti^{3+} centers.

A third area of interest has been the low pressure flame synthesis of carbon doped TiO_2. Various compositions and flow rates of the premixed gases are introduced to the low pressure synthesis chamber and ignited to produce a flat flame. The low pressure flame spray pyrolysis method produces various polymorphs of TiO_2 which show promise in photocatalytic conversion of methanol at moderately low temperatures. Flame temperature profiles and carbon loading are predicted via computational analysis using a SPIN code to ensure that the flame of the temperature is comparable across samples made from different pre mixed gases. Photocatalytic activity of the carbon doped samples is determined via methanol oxidation reactions under UV illumination.

Copyright © American Institute of Chemical Engineers. All rights reserved.

467273 Photoelectrochemical Oxidation of Biorenewable Alcohols By Nitroxyl Radical Catalysts

Abstract:

Transforming primary alcohols to acids is a fundamental synthetic process of interest in both the academic and industrial worlds, but it typically requires stoichiometric oxidants which are costly and generate potentially toxic waste. On the other hand, conversion of biomass-derived chemicals in a *photoelectrochemical* (PEC) cell is one of the most environmentally-friendly chemical production routes available, by taking advantage of renewable carbon sources from the earth, renewable photons from the sun, and renewable electrons from wind, geothermal, or photovoltaic sources.

Stable nitroxyl radicals such as TEMPO are a well-known class of homogeneous oxidation catalysts with widespread use in industrial and laboratory applications.[1] Furthermore, TEMPO is easily oxidized to its active form, an oxoammonium cation, at moderate electrode potentials and is very selective for C OH or C=O oxidations. It was recently reported that the overpotential required to drive TEMPO oxidation can be significantly reduced in a PEC reactor, where energy is harvested from visible light with an earth-abundant bismuth vanadate ($BiVO_4$) semiconductor electrode.[2] However, external electrical energy was still required to enhance electron-hole separation and achieve reasonable photocurrents. Additionally, the electron efficiency for TEMPO oxidation was <100% owing to competition with water oxidation. Extensive work has been done to improve the performance of $BiVO_4$ photoanodes for solar water splitting including nanostructuring, doping, and the addition of an electrocatalyst layer, but optimization to promote TEMPO oxidation has not been explored.

Previously, we have studied selective electrocatalytic oxidation of alcohols and 5-hydroxymethylfurfural (HMF) on supported noble metal catalysts.[3-4] Now, we have investigated electrochemical and PEC systems for efficient TEMPO-mediated oxidation of key biomass-derived chemicals including HMF, glycerol, and ethylene glycol. $BiVO_4$ modified with electrocatalyst layers facilitated TEMPO oxidation at overpotentials >400 mV lower than unmodified $BiVO_4$, while significantly suppressing water oxidation. The novel photoelectrocatalytic system achieved nearly 100% efficiency to TEMPO-mediated oxidations and bulk electrolysis gave high yields of organic acid products. In this talk, factors that tune the competition between TEMPO and water oxidation, including photoanode structure, applied potential, and other reaction conditions will be discussed.

References:

(1) Rafiee, M.; Miles, K. C.; Stahl, S. S., JACS (2015), 137 14751−14757.

(2) Cha, H. G.; Choi, K. S., Nature Chem. (2015), 7 328–333.

(3) Chadderdon, D. J. et al., Green Chem. (2014), 16 3778–3786.

(4) Chadderdon, D. J. et al., ACS Catal. (2015), 5 6926–6936.

Copyright © American Institute of Chemical Engineers. All rights reserved.

473457 SO_2 Interactions with Pt, Pd, and Pd-Pt/Al_2O_3 Catalysts

Abstract:

Typical natural gas engine exhaust will contain uncombusted CH_4 that will need to be oxidized. Due to their high catalytic activity for the complete oxidation of methane, Pd-based catalysts are the typical choice for such oxidation systems. These catalysts can be exposed to trace amounts of sulfur compounds present in the engine exhaust. This sulfur exposure can lead to degradation in activity. The focus of this study was how sulfur interacts with Pd-based oxidation catalysts, as well as bimetallic Pt/Pd systems since Pt has been shown to have some resistance to sulfur poisoning. Temperature programmed desorption and oxidation were used to probe these interactions, as was DRIFTS to characterize surface species. The results show that the temperature, at which sulfur species are desorbed from sulfur-treated, mono and bimetallic Pd and Pt/Al_2O_3 catalysts, is dependent on the Pd-to-Pt mole ratio of the catalyst as well as the precious metal-crystallite particle size.

Copyright © American Institute of Chemical Engineers. All rights reserved.

463008 Insights into the Deactivation of Cu/SAPO-34 By Low- and High-Temperature Hydrothermal Treatment

Abstract:

Cu/Chabazite (CHA) catalysts, Cu/SSZ-13 and Cu/SAPO-34, have been successfully commercialized for selective catalytic reduction (SCR) of NOx in lean-burn engine exhausts due to their superior activity, N_2 selectivity and durability. High-temperature (> 600 °C) hydrothermal aging treatments typically cause performance decrease for these catalysts, rationalized from partial structure degradation and loss of Cu-sites in active forms. However, recent studies have shown that the exposure of Cu/SAPO-34 to water vapor at *low temperatures* (< 100 °C) also result in an irreversible decline in NH_3-SCR activity [1, 2]. This may be attributed to structure degradation since SAPO-34 is known to hydrolyze in the presence of water vapor at such low temperatures. However, transformation of active Cu-sites to less active forms cannot be ruled out.

To gain insights into the similarity and difference between low- and high-temperature deactivation mechanisms, model Cu/SAPO-34 catalysts with varying Cu loadings were prepared using a one-pot method. These catalysts were further hydrothermally treated at either 70 or 800 °C. Effects of hydrothermal treatments on catalytic performance were examined using standard SCR and non-selective NH_3 oxidation kinetics. The zeolite framework integrity (crystallinity, porosity, coordination environments, Si-O-Al bonds, acidity) were characterized by XRD, N_2 physisorption, solid state NMR, DRIFTS and NH_3-TPD. A particular focus was given to changes in the local environments of Cu species during hydrothermal aging, examined using two-dimensional (2D) pulsed EPR. Finally, an attempt is made to elaborate the deactivation mechanisms that correlate the variation of NH_3 storage sites (acid sites) and redox sites (Cu^{2+} sites) caused by low- and high-temperature hydrothermal treatments.

[1] Wang, J., Fan, D., Tie, Y., et al., *J. Catal. 322*:84-90, 2015

[2] Leistner, K., Olsson, L., *Appl. Catal. B: Environ.* 165: 192-199, 2015

Copyright © American Institute of Chemical Engineers. All rights reserved.

464314 Lean-Rich Switching over a Modified Three-Way Catalyst:Â Experiments and Modeling

Abstract:
Lean-Rich Switching Over a Modified Three-Way Catalyst: Experiments and Modeling

Mengmeng Li, Sam Malamis, Michael P. Harold*, William Epling**

Department of Chemical and Biomolecular Engineering, University of Houston, Houston, TX 77204

* *mharold@uh.edu, ** wsepling@central.uh.edu*

The most recent CAFE standard for automobile requires an average fuel economy of 54.5 miles per gallon by 2025 [1]. At the same time, more stringent emission rules enacted by the EPA for 2017-18 will require an 80% reduction in non-methane organic gases (NMOG) plus NOx from current Tier 2 Bin 5 levels [2]. Gasoline vehicles that operate under both lean-burn and stoichiometric combustion hold promise since lean-burn combustion is more fuel efficient than conventional stoichiometric combustion. However, elimination of NOx (NO+NO_2) by the traditional three-way catalyst (TWC) is not possible under lean conditions. Advanced NOx reduction strategies must be applied to exploit the more efficient lean-burn engine combustion such as the three-way catalyst with a NOx storage function (TWNSC). The TWNSC stores NO_x storage during lean operation and reduces the trapped NOx during stoichiometric operation. To date, there have been only a few studies that report TWNSC. In this study a predictive TWNSC model is developed to predict TWNSC performance and in so doing provide operational insight and optimization.

A model of a catalytic monolith channel was developed that incorporates a TWNSC kinetic model into a low-dimensional model formulation [3–5]. The convection-diffusion-reaction equations are averaged in the transverse direction which replaces the transverse gradient term with an algebraic term that contains an internal mass transfer coefficient. Overall heat and mass transfer coefficients are used to represent the heat and mass transfer in the transverse direction between the bulk of fluid and solid phase. The TWNSC kinetics consists of a combination of elementary steps and global reactions. Following our previous experimental study of NOx reduction by propylene over a multifunctional NOx trap catalyst [6], the hybrid model includes the kinetics of the oxidation of CO, H_2, NO and hydrocarbons, of reduction of NO/NO_2 by CO, H_2 and representative hydrocarbons, and of storage of NOx and O_2. Fig. 1 shows the reaction network including 10 main paths that described the chemistry in qualitative terms.

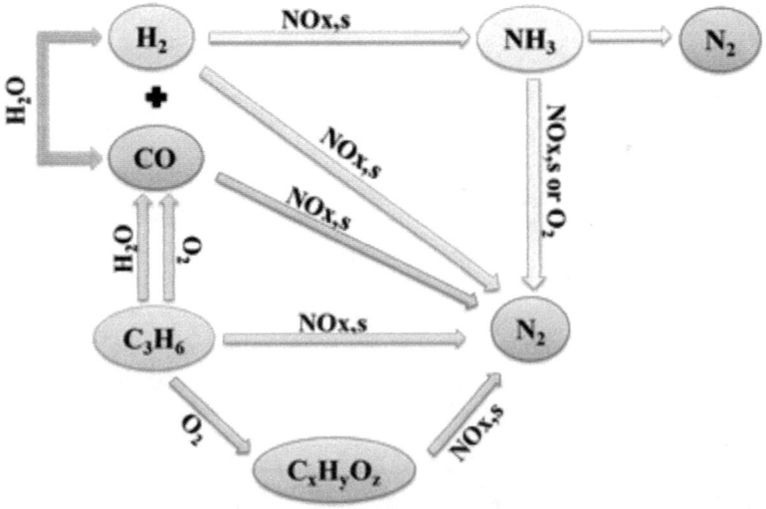

Figure 1. Simplified reaction network of NOx reduction using propylene as reductant[6]

A step-wise approach was used to develop kinetics for the oxidations of CO, H_2 and propylene. First, global kinetics are fitted tuned to transient light-off experiments following Raj et al. [5]. For example, the Langmuir-Hinshelwood expression is given by

$$R_{co} = \frac{k_1 X_{CO} X_{O_2}}{(1 + K_{CO} X_{CO})^2} \quad (1)$$

Fig. 2 shows the fitted light-off curve for CO oxidation and the comparison between experimental and model data. The model shows a reasonable fit and captures the dynamical feature of CO TPO (temperature programmed oxidation) experiment.

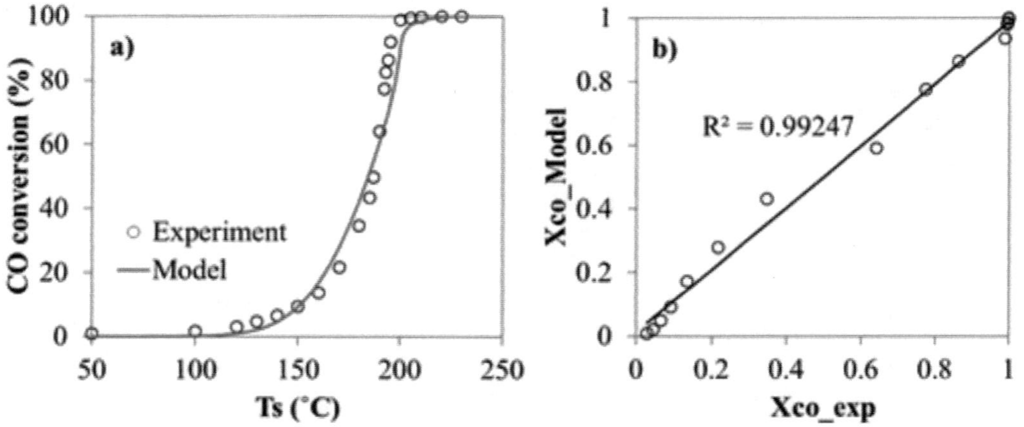

Figure 2. (a) Fitted light-off curve of CO oxidation (b) comparison between experimental and model-predicted conversions. [Conditions: 0.5% CO, 0.5% O; temperature ramp: 3°C/min]

The rate expressions for the co-oxidation of CO, H_2, and C_3H_6 include coupling effects. These are combined with a dual-site NOx storage model and incorporated into species balances.

We will show how the model is effective in capturing the main trends in the lean-rich switching data. This study advances the understanding of TWNSC catalyst and provides guidance for optimizing catalyst formulation and operation strategies.

Reference

[1] http://www.nhtsa.gov/About+NHTSA/Press+Releases/2012/Obama+Administra tion+Finalizes+Historic+54.5+mpg+Fuel+Efficiency+Standards.

[2] T. Johnson, SAE Int. J. Engines. 6 (2014) 699–715.

[3] S.Y. Joshi, M.P. Harold, V. Balakotaiah, AIChE J. 55 (2009) 1771–1783.

[4] D. Bhatia, R.D. Clayton, M.P. Harold, V. Balakotaiah, Catal. Today. 147 (2009) 250–256.

[5] R. Raj, M.P. Harold, V. Balakotaiah, Chem. Eng. J. 281 (2015) 322–333.

[6] M. Li, V.G. Easterling, M.P. Harold, Catal. Today. 267 (2016) 177–191.

453155 Importance of Exhaust Hydrocarbon Speciation in the Studies of Low-Temperature Emission Control Technologies for Next-Generation Fuel-Efficient Vehicles

Abstract:

As the fuel economy regulations being mandated all over the world require development of engines with improved fuel efficiency, the energy content in the exhaust will generally decrease and the resulting lower exhaust temperatures present significant challenges in catalytic converter performance, especially during cold-start periods. This, coupled with the more stringent future emission standards (e.g., LEV-III/Tier 3 regulations in U.S.) and recent adoption of green-house gas regulations including CH_4, makes catalytic oxidation of exhaust hydrocarbons (HCs) at low temperatures critically important.

Hydrocarbons in engine exhaust generally consist of unburned fuel components and partial combustion products. The latter includes light (C_1-C_3) saturated and unsaturated HCs such as methane, ethane, ethene and propene. Unburned fuel components typically consist of higher molecular-weight (>C_4) alkanes and aromatics. For gasoline engines, it has been commonly observed that C_5 and C_8 alkanes as well as C_7 and C_8 aromatics (toluene and xylene) are among the most abundant unburned fuel components in the exhaust. Regarding the nature of C_5 and C_8 alkanes, detailed GC-based speciation studies of gasoline exhaust HCs show that alkane isomers, such as iso-pentane and iso-octane, are the predominant species, with the concentrations of the corresponding n-alkanes (i.e., n-pentane and n-octane) being much lower.

The direct consequence of lower exhaust gas temperatures is that some of the HC species in engine exhaust are likely to remain unconverted even after the converter is fully warmed-up, thus presenting considerable challenges in HC emission control for future fuel-efficient engines. Thus, it is of practical importance to identify the types of HC species that are likely to contribute significantly to tailpipe HC emissions in cooler exhaust from next-generation gasoline engines and understand how the conversions of those HCs are affected by the type & loading of noble metals (Pd vs Rh), O_2 concentration level and catalyst aging. To that end, we conducted laboratory reactor experiments aimed at investigating the catalytic oxidation of various HCs in gasoline engine exhaust (alcohol, n-alkanes, alkane isomers, alkene and aromatics) over modern commercial three-way catalysts with variable noble metal loadings before and after aging. Another approach being considered for improved cold-start HC emission control is the passive HC adsorber technology, which is designed to store HCs during engine cold-start and oxidize (convert) the stored HCs as they are released at higher temperatures. As such, the effectiveness of this technology depends on the interplay between the adsorption, desorption and oxidation characteristics of specific HC species of interest. With this background in mind, we evaluated commercial HC adsorber catalysts for their cold-start HC emission reduction potential in a feedstream containing paraffins, olefins and aromatics.

This presentation will first discuss proper choice of "representative" HCs in gasoline engine exhaust based on HC speciation data and then examine their oxidation activities over degreened and aged commercial three-way catalysts as a function of noble metal loading. The oxidation activities were measured using a feedstream which contains not only the HC of specific interest and O_2 but the other usual exhaust constituents, such as CO, H_2, NO, CO_2 and H_2O as well. This would allow us to characterize the reactivity of each HC species under realistic conditions, namely, in the presence of the promotion effects of $H_2$and the inhibition effects of CO and NO. Also discussed in this talk are the performance characteristics of HC adsorbers in a feed containing various types of HCs including olefins, aromatics and paraffins. It was found that these types of HCs can be controlled to very different extents, due to variations in their adsorption, desorption and light-off (oxidation) properties. We also investigated the effects of hydrothermal aging as well as noble metal loading, oxygen concentration and adsorber volume on the HC conversion performance of the HC adsorber catalysts. These results will be discussed to illustrate the capabilities and limitations of the current HC adsorber technology, and to point ways toward improvement in the effectiveness of adsorber systems.

Copyright © American Institute of Chemical Engineers. All rights reserved.

471850 Rapidly Pulsed Reductants in Diesel NOx Reduction with Lean NOx Traps: Spatial Distribution of Species

Abstract:

Lean NOx Trap (LNT) is a deNOx method using a NOx storage and reduction (NSR) catalyst that can be a desirable choice for NOx reduction on light duty diesel vehicles with smaller engines (< 2.0 L). A LNT used in a conventional way with long lean (60 s) – short rich (5 s) cycles obtains poor NOx conversions at high temperatures (T>400°C) and space velocities. Rapidly Pulsed Reductants (RPR) is a method for injecting hydrocarbons in rapid pulses ahead of a LNT which has shown the potential to achieve greater than 90% NOx conversion at temperatures as high as 600°C. This approach has recently been discussed by Toyota and named Di-Air [1]; it appears to be useful for smaller diesel vehicles that do not easily accommodate SCR technologies.

Four sets of parameters have been found important in RPR operation: 1- flow field and reductant mixing uniformity [2], 2- pulsing parameters, 3- the reductant type [2], and 4- catalyst composition, PGM type/loading, NOx storage material, and oxygen storage capacity (OSC). This parameter space has been explored using a synthetic bench reactor equipped with a novel high frequency injection and mixing system capable of pulsing reductants over the catalyst at up to 100 Hz.

The main findings from our previous studies on the effect of aforementioned parameters are; (a) The NOx conversion is maximized using nearly perfect plug flow pulses with minimal axial mixing [2], (b) The pulses of injected hydrocarbon need to be large enough in magnitude to generate a rich mixture during the pulse in order to achieve high NOx conversion. However, the cycle-averaged overall A/F ratio was always net lean. (c) The NOx conversion is dependent on the type of reductant at different operating temperatures [2]. (d) There exists an optimal operating condition, in terms of pulse amplitude and frequency that maximizes NOx conversion and minimize the fuel penalty associated the pulsed reductant.

In the current study, the spatial distribution of species along the catalyst have been explored using spatially resolved capillary inlet mass spectrometry (Spaci-MS) [3]. The experiments were conducted at near optimal operating conditions found from previous studies, with maximal NOx conversion and minimal byproduct (CO, N_2O, ...) formation. Ethylene was used as the main reductant on a typical Pt/Rh/BaO/Al_2O_3 LNT sample (no Ce). Experiments were conducted under reductant pulsing conditions to find the spatial distribution of the catalyst bed temperature (T/C) and the concentration of some of the important reactants and products, such as, C_2H_6, O_2, CO, CO_2, H_2O, H_2, NO, and NH_3 in the temperature range of 150°C to 600°C. Due to the limited response time of the instruments and high frequency pulsing approach, transient measurements were not possible of products and cycle-averaged measurements were performed. As an example, Fig.1 shows the cycle-averaged NO concentration vs temperature and axial location.

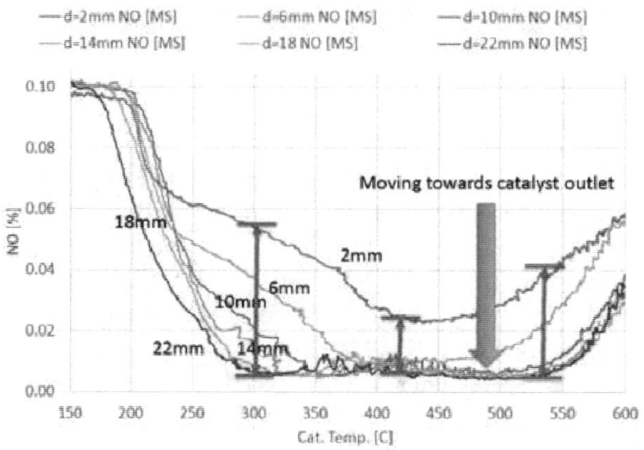

Figure 1 NO concentration vs temperature, measured at different axial locations of a LNT monolith with 25mm of length. Experimental conditions are: 30,000h^{-1} space velocity, P = 2s, 15% pulse duty cycle, and $\lambda_{pulse} = 0.73$.

As can be seen, the length of catalyst necessary to achieve high or complete NOx conversion is dependent on the operating temperature regime. At 300°C significant NOx conversion is seen 2 mm into the catalyst and complete conversion occurs 18 mm into the catalyst. At 420°C NOx conversion is high only 2 mm into the catalyst and complete at 6 mm into the catalyst, about one fourth of its length. At 540°C we find a new regime, where NOx storage capacity becomes the limiting factor over much of the length of this catalyst.

It should be mentioned that, the Spaci measurements are prone to variation due to the restriction effects of the capillary inside the channel. We have observed considerable deviation (10-20%) of these measurements from the bulk concentration at the outlet of catalyst. Thus, while these data are not perfectly quantitative, they are very useful in determining where in the catalyst product distributions change.

As described in the caption of Fig. 1, the spatial distribution provides insight into the limiting factors and different operating regimes. In addition, it enables us to locate the formation and consumption of important reaction byproducts, such as NH_3, CO and H_2. For instance, it was observed that NH_3 was formed within the first few millimeters of the catalyst, and then consumed moving downstream. Furthermore, formation of CO and H_2 was observed, around 350-400°C as an indication of steam reforming, which occurred within the first 2-6mm of the catalyst. A simple balance showed that the generated CO was then consumed moving downstream, mainly by the water gas shift reaction, forming H_2 which was observed at the catalyst outlet.

In summary, Spaci-MS is a useful technique which provides spatially resolved concentration and temperature distributions to help better understand the overall reactivity data. In the context of the RPR process, we have employed this information for determining the governing factors at different operating regimes, as well as important reaction pathways.

References

[1]Bisaiji, Yuki, et al. "Development of Di-Air-A New Diesel deNOx System by Adsorbed Intermediate Reductants."

[2] Reihani, A; Corson, B; Hoard, JW; Fisher, GB; Smirnov, E; Roemer, D; Theis, J; Lambert, C. "Rapidly Pulsed Reductants in Diesel NOx Reduction by Lean NOx Traps: Effects of Mixing Uniformity and Reductant Type."

[3] Choi, Jae-Soon, William P. Partridge, and C. Stuart Daw. "Spatially resolved in situ measurements of transient species breakthrough during cyclic, low-temperature regeneration of a monolithic Pt/K/Al 2 O 3 NO x storage-reduction catalyst."

Copyright © American Institute of Chemical Engineers. All rights reserved.

A Rational Pathway for Generic Three-way Catalyst Model Development in Stoichiometric Natural Gas Engines

Jian Gong, Di Wang, Junhui Li, Neal Currier and Aleksey Yezerets

Corporate Research and Technology, Cummins Inc., 1900 McKinley Avenue, Columbus, IN 47201, United States

Abstract

Three-way Catalysts have been widely used on stoichiometric natural gas engines for CO, hydrocarbon and NOx emission control [1]. A generic TWC model is highly desirable for design of aftertreatment systems, development of advanced control strategies and on-board diagnostics (OBD), and TWC architecture selection. However, there are many challenges in developing a generic TWC model. First, TWCs typically operate under dynamic conditions (e.g. fast lean/rich dithering). Second, TWC chemistry consists of a complex reaction network. In this work, a rational pathway in Figure 1 has been established to develop a generic TWC model that is capable of capturing the complicated chemistry as well as transient behaviors. This TWC model includes an OSC sub-model [2] and PGM related kinetic sub-model which is sensitive to catalyst state. The former is calibrated based on dynamic oxygen storage capacity (OSC) characterization performed on a bench reactor [3]. The latter is calibrated base on a complex steady state light-off performance protocol developed by the authors. The full model with calibrated kinetics is finally validated by dithering tests performed on a transient reactor and engine test data. This approach is found to be very capable and effective for generic TWC model development.

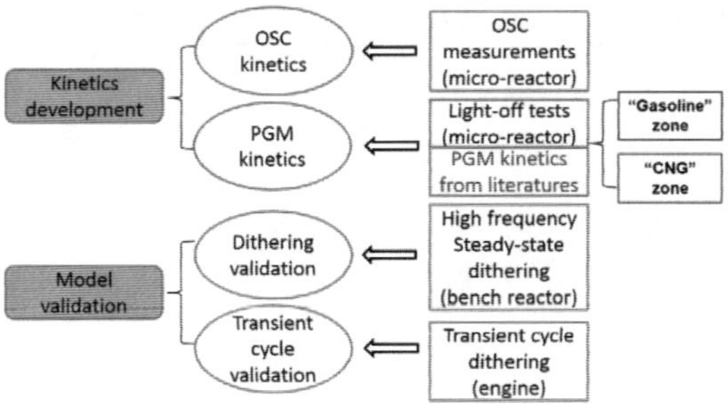

Figure 1: A schematic of the rational pathway for TWC model development

References

1. Gandhi, H.S., Graham, G.W., and McCabe, R.W., "Automotive exhaust catalysis," J. Catal. 216(1-2):433–442, 2003, doi:10.1016/S0021-9517(02) 00067-2.
2. Gong, J., Wang, D., A., Li, J., Currier, N., Yezerets, A., "Dynamic Oxygen Storage Modeling in a Three Way Catalyst for Natural Gas Engine: A Dual-site and Shrinking-core Diffusion Approach", Applied Catalysis B: Environmental, 2016 (in press), DOI: 10.1016/j.apcatb.2016.11.005
3. Wang, D., Gong, J., Junhui, L., Kamasamudram, K., Currier, N., Yezerets, A., "Controlling factors of surface and subsurface Oxygen Storage Capacity (OSC) of Three-Way Catalysts (TWCs) used in natural gas aftertreatment systems," 2015 Annual AIChE Meeting, Salt Lake City, UT, November 10, 2015.

460174 Effects of Cu Density in Cu-SSZ-13 Zeolites on Low-Temperature NOx Selective Catalytic Reduction with NH_3

Abstract:

Effects of Cu density in Cu-SSZ-13 Zeolites on Low-Temperature NOx Selective Catalytic Reduction with NH_3

Ishant Khurana[1], Atish A. Parekh[1], Jonatan D. Albarracin-Caballero[1], Christopher Paolucci[2], Arthur Shih[1], John R. Di Iorio[1], W. Nicholas Delgass[1], Jeffrey T. Miller[1], William F. Schneider[2], Rajamani Gounder[1], Fabio H. Ribeiro[1]

[1]School of Chemical Engineering, Purdue University, 480 Stadium Mall Drive, West Lafayette, IN 47907, USA

[2]Department of Chemical and Biomolecular Engineering University of Notre Dame

182 Fitzpatrick Hall, Notre Dame, IN 46556, USA

NOx selective catalytic reduction (SCR) using NH_3 as a reductant over Cu-SSZ-13 zeolites is a commercial technology used to meet emissions targets in lean-burn and diesel exhaust. Low temperature (473 K) NOx SCR on Cu-SSZ-13 occurs via a Cu(II)/Cu(I) redox cycle [1] on two types of isolated Cu sites, both divalent Cu^{2+} cations compensating paired framework Al atoms and monovalent $[CuOH]^+$ complexes at isolated framework Al atoms, which are solvated by NH_3 during reaction [2]. Both types of cationic Cu sites were identified by infrared, UV-Visible and *operando* X-ray absorption (XAS) spectroscopies and also from their exchange stoichiometry with surface protons, the residual number of which was quantified by NH_3 titration.

Under standard SCR conditions, in which O_2 is used as a four-electron oxidant in the Cu(I)/Cu(II) redox cycle, we have identified two different kinetic regimes that are characterized by different apparent orders, activation energies, and steady-state Cu(I)/Cu(II) distributions. The prevalence of standard SCR kinetic regimes depends strongly on the Cu density (determined by the Si/Al and Cu/Al) of the Cu-SSZ-13 catalysts. One regime corresponds to SCR cycles limited by Cu(I)->Cu(II) oxidation with O_2, which we probe in *operando* by varying reaction parameters (temperatures, gas pressures) while collecting steady-state kinetic data and XAS spectra, and in transient oxidation experiments of reduced Cu(I)-SSZ-13 catalysts using either O_2 and NO_2. These experiments reveal differences in the kinetics and Cu site requirements between O_2 and NO_2-catalyzed Cu(I)->Cu(II) oxidation, which respectively are four-electron and one-electron oxidants that propagate SCR cycles under "standard" and "fast" conditions. We provide evidence that O_2-oxidation requires two Cu sites and two NO molecules to be oxidized, while NO_2-oxidation occurs at single, isolated Cu sites. As a result, standard SCR rates depend on Cu cation proximity and distribution in Cu-SSZ-13 zeolites when Cu(I) oxidation steps are kinetically-relevant, but are independent of Cu density otherwise.

[1] C. Paolucci, A.A. Verma, S.A. Bates, V.F. Kispersky, J.T. Miller, R. Gounder, et al., Isolation of the Copper Redox Steps in the Standard Selective Catalytic Reduction on Cu-SSZ-13, (2014) 12022–12027.

[2] C. Paolucci, A.A. Parekh, I. Khurana, J.R. Di Iorio, J. Albarracin, A. Shih, et al., Catalysis in a Cage : Condition-Dependent Speciation and Dynamics of Exchanged Cu Cations In SSZ-13 Zeolites, J. Am. Chem. Soc. (2016) 1–63.

Copyright © American Institute of Chemical Engineers. All rights reserved.

473866 The Effects of Catalyst Aging on Sulfur Degradation over Fully Formulated Lean NO_x Trap Catalysts

Abstract:

Lean NOx trap catalysts are subjected to both hydrothermal aging and sulfur deactivation in real world environments. Studies have shown that hydrothermal aging can increase noble metal particle size and subsequently reduce the ability of the catalyst to oxidize NO. This results in reduced NOx storage, and ultimately reduced NO_x conversion. Furthermore, literature has observed that the degree of sulfur deactivation varies with changes in catalysts parameters such as barium loading[1], this is an indicator that changing catalyst morphology via another mechanism such as hydrothermal aging could also change the degree to which the catalyst is sulfated. Because, both catalyst aging and deactivation due to sulfur occur in real world environments it is important to understand the effects of these two mechanism in conjunction.

In the following study a model Pt/Ba/CeZr/Al_2O_3 catalyst was subjected to hydrothermal aging. The hydrothermally aged catalysts were subjected to 1 hour sulfation in a lean gas environment. It was found that while hydrothermal aging initially resulted in variations in, NOx uptake, total NOx conversion, as well as ammonia and N_2O production, after sulfation, the cycling trends changed, most dramatically during the rich phase. Figure 1 shows that prior to sulfation ammonia production upon rich gas introduction is greatest for the catalyst aged at 700°C and the least for the catalyst aged at 600°C. After sulfation, ammonia production for the 600°C and 700°C aged catalysts are increased substantially, while the 800°C catalyst shows only a small increase in ammonia production and as a result has the lowest total ammonia production. The effects of hydrothermal aging and sulfation have also been characterized by BET, CO-Chemisorption, XRD and TEM and used to identify the dominant factors that have resulted in the observed deviations upon lean/rich cycling.

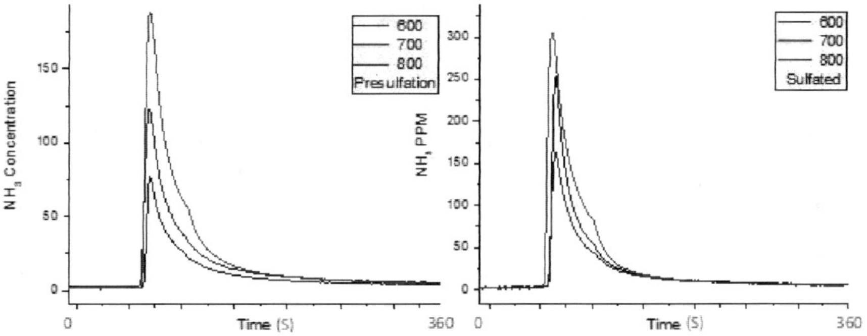

Figure 1 shows the variation in ammonia production after Lean/rich cycling at 300°C after presulfation (left) and post sulfation (right)

Acknowledgements

This project has received funding from the *European Union's Horizon 2020 research and*

innovation programme under grant agreement No 636380.

1. Kim, D. H.; Szanyi, J.; Kwak, J. H.; Szailer, T.; Hanson, J.; Wang, C. M.; Peden, C. H. F., Effect of Barium Loading on the Desulfation of Pt-BaO/Al2O3 Studied by H2 TPRX, TEM, Sulfur K-edge XANES, and in Situ TR-XRD. *The Journal of Physical Chemistry B* **2006,** *110* (21), 10441-10448.

Copyright © American Institute of Chemical Engineers. All rights reserved.

459082 A Comparative Study of ZSM-5 and BEA-Zeolites for Hydrocarbon Trap Applications Under "Cold-Start" Conditions

Abstract:

Regulations on vehicle emissions are becoming more stringent worldwide due to increasing concerns of the impact of air pollution on environment and public health. Significant attention has been paid to develop methods to treat automobile exhausts for the regulated emissions such as hydrocarbons especially during engine cold starts. An effective solution is to employ suitable porous materials which can trap and retain hydrocarbons (HCs) temporarily until emission control catalysts are lit off (i.e., exhaust gas temperature reaches high enough for catalysts to become active) [1,2]. For this purpose, zeolites have been found to be a preferred type of materials due to their affinity to HCs and relative thermal stability [3]. The goal of this work was to understand key factors controlling hydrocarbon trapping and release performance of zeolites. Gained insights can, in turn, help to design HC traps with enhanced performance suitable for addressing future emission control challenges related to low temperature exhausts expected from more efficient engines. Zeolites were evaluated using propylene (C_3H_6) as a model HC compound in simulated exhaust conditions relevant to vehicle cold starts. The specific objective was to investigate in detail trends observed between C_3H_6 trapping-release performance, zeolite framework structure (ZSM-5 vs. BEA), acidity (high vs low), exchanged cation (e.g. H^+ vs. Ag^+ vs Pd^+), synthetic procedure (incipient wetness vs. ion exchange) and gas composition. Our results show that the amount of C_3H_6 adsorbed on H-zeolites significantly decreased in the presence of water, likely due to competitive adsorption and inhibition of adsorption sites. The results demonstrated that ion-exchanged metal is necessary to achieve increased storage and release of C_3H_6 in the presence of H_2O. Increased metal loading led to increased C_3H_6 adsorption ability and decreased desorption temperature. Most of stored C_3H_6 was not released as C_3H_6, but as various products of oxidation and cracking catalyzed by zeolites. We will discuss potential strategies to design enhanced HC traps based on the gained insights.

Acknowledgements. Research sponsored by the U.S. Department of Energy, Vehicle Technologies Office.

References

[1] A. Iliyas, M.H. Zahedi-Niaki, M. Eić, S. Kaliaguine, Micropor. Mesopor. Mat. 103 (2007) 171-177.

[2] J.-H. Park, S.J. Park, I.-S. Nam, G.K. Yeo, J.K. Kil, Y.K. Youn, Micropor. Mesopor. Mat. 101 (2007) 264-270.

[3] T. Kanazawa, Catal. Today 96 (2004) 171-177.

Copyright © American Institute of Chemical Engineers. All rights reserved.

459663 Developing a Quantitative Spatial Resolution of Deactivation Effects on FT Synthesis in a Microchannel Reactor When Operating with Ultra-Low Sulfur Levels

Abstract:

Developing a quantitative spatial resolution of deactivation effects on FT synthesis in a microchannel reactor when operating with ultra-low sulfur levels

S. Deshmukh[1], H. Becker[2], S. Kampfe[1], K. Cowen[1], D. Leonarduzzi[2], L. Barrio[2], J. Pritchard[2], H.J. Robota[1]

Velocys

Plain City, OH[1] and Milton Park, UK[2]

Introduction

Fischer-Tropsch (FT) is receiving new attention for its ability to generate conventional fuel from stranded, or unconventional natural gas, and alternative renewable biomass sources. Being a catalytic process, the catalyst is subject to deactivation via multiple mechanisms [1-2]. Sulfur is the foremost among the poisons causing permanent catalyst deactivation. Despite the impact of "S", the number of published quantitative investigations related to practical conditions is quite modest. While the recommendations for industrial syngas are typically <15 ppb S, [3-5] other publications would suggest unobservable impact at much higher exposures. [6,7]

In this paper, we present results of a detailed investigation of sulfur poisoning in a fixed bed microchannel reactor, with an emphasis on quantitative verification of delivered H_2S concentrations. The poisoned catalyst was carefully extracted and analyzed to determine the spatial distribution of sulfur along the catalyst bed and within catalyst particles. This information is used to develop a quantitative model able to predict the impact in a commercially operating microchannel FT reactor.

Feeding H_2S augmented syngas

The impact of contaminants, e.g. sulfur, on active Cobalt sites does not change with catalyst operating productivity. However, high catalyst productivity implies that the catalyst is exposed to a higher flux of these contaminants. With catalyst productivities exceeding 2000 v/v/h, the Velocys reactor passes up to 10-times the number of contaminant molecules per unit time compared to a conventional fixed bed reactor.

Detection and accurate quantification of syngas contaminants at low ppbv levels is challenging, as this approaches available equipment's limit of detection. Nonetheless, it is critical to understanding the impact of poison concentration on catalyst deactivation rates. To accurately confirm the levels of sulfur delivered to our reactors, measurements were made using a Proton Transfer Reaction Mass Spectrometer (PTR-MS) at the point of delivery to the catalyst. [8] Figure 1 shows calibration of the instrument for the detection of ultra-low level H_2S in syngas feed with the PTR-MS, indicating reliable measurements at <15 ppb detection levels.

A clean syngas feed was simulated by flowing desired amounts of pure CO, H_2 and N_2 streams from cylinders using mass flow controllers and blending to the target composition. Sulfur was introduced into the clean syngas by replacing part of the N_2 feed with H_2S-doped N_2 feed. All tubing and connections used in the contaminant flow path were treated with sulfur-resistant inert coatings to avoid transmission losses to the surfaces.

Figure SEQ Figure * ARABIC 1. PTR_MS calibration and response to ultra-low ppb level H_2S

Microchannel fixed bed testing

The microchannel testing apparatus and methodology used to evaluate the catalyst and/or process performance has previously been described. [9,10] A microchannel reactor with 1 process channel and 2 adjacent coolant channels was loaded with a commercially manufactured, undiluted, particulate catalyst in the process channel as a packed bed. FT synthesis was performed by feeding clean and doped syngas as described earlier.

Figure 2 shows the impact of ultra-low levels of H_2S in the syngas feed, 13 ppb in this case, on a catalyst operating under representative commercial conditions (H_2:CO = 1.79, 28% inerts, 310 ms contact time). The deactivation rate, as measured by the decrease in CO conversion with time on stream, is higher even with such ultra-low levels of H_2S present in the reactor feed.

Figure SEQ Figure * ARABIC 2. Decline in CO conversion with time during and after low level H_2S poisoning.

In order to understand the quantitative impact of sulfur poisoning under practical extended operating conditions, an extended duration exposure of a catalyst to low level H_2S in the feed was performed. As a trade-off between readily quantifiable impact and the required operating duration, the catalyst was exposed to 100 ppb H_2S for a period of 300 hours, which is equivalent to 1 year exposure to H_2S at a level of ~4 ppb.

Catalyst deactivation commensurate with expectations from the tests using lower levels of feed H_2S was observed. Upon completion of the test, the reactor and catalyst were decommissioned to render the catalyst safe for handling and analysis.

Figure SEQ Figure * ARABIC 3. simulated long term exposure of catalyst to syngas operation in a laboratory single channel reactor. The loss of CO conversion with time is readily visible during the 300 hour H_2S exposure.

Spent catalyst analysis

Spent catalyst from the "1 year exposure" test was collected in aliquots of ~50 mg starting at the reactor inlet, with a spatial resolution of 1 cm. The samples were analyzed by ICP to quantify the sulfur accumulated in each section. Based on our prior experience with more heavily S-exposed catalyst charges, the top of the bed appears to be nearly saturated with sulfur (~5500 ppm) with a concentration of S at approximately 4500 ppm. The sulfur content then decreases linearly to ~500 ppm at about 15% of the bed length. The subsequent axial sulfur profile is relatively, flat but there is still quantifiable sulfur content at 35% of the bed.

Figure SEQ Figure * ARABIC 4. Axial distribution of sulfur along the catalyst bed.

The recovered catalyst particles were characterized on a JEOL 6480LV SEM using a back-scattering detector and a beam voltage of 15 keV at Oxford Materials Characterization Service. A sample at 10% of the bed length (1875 ppm S) was first epoxy-embedded, polished, and coated with a 15 nm carbon thickness to see the internal topology and composition by EDX analyzer. At this point in the reactor, the examined catalyst granules had sulfur concentrated near the superficial surface and not detectable in their interiors. An example is shown in Figure 5 with the S:Co ratio as quantified by EDX at the locations noted in Figure 5 summarized in Table 1.

Model development

Quantifying the impact of spatially distributed poisons in the Velocys microchannel catalyst requires the use of a model which properly accounts for the effects of both reaction kinetics and mass transfer impact. Such a model was developed using gPROMS. It accounts for the convective transport of mass and enthalpy along the catalyst bed and accounts for the diffusion of reactants into

Figure SEQ Figure * ARABIC 5. SEM/EDX analysis of catalyst particles, showing the areas of composition analysis.

Table 1. EDX analysis of sulfur on catalyst particles LINK Excel.Sheet.12 "\\\\vmplab-dc01.vmplab.lan\\Data\\FT\\Post-operative analysis\\EXP-15-DK8435 - S poisoning (1 year equiv)\\SEM\\S_SEM_quanti.xlsx" "Sheet1!R4C1:R13C9" \a \f 5 \h * MERGEFORMAT

	Molar ratio	
Spectrum #	**Co/Si**	**mmol S /mol Co**
Spectrum 1	1.1	20.9
Spectrum 2	1.4	1.5
Spectrum 3	1.2	2.0
Spectrum 4	1.0	3.4
Spectrum 5	1.3	2.6

Spectrum 6	1.2	15.0
Spectrum 7	0.5	32.4
Spectrum 8	1.8	21.0

the pores of the catalyst particle. On top of this steady state model for the main FT reaction, a dynamic model describing the effects of deactivation and poisoning is superimposed. This allows for the prediction of the time dependent integral performance of the reactor. In addition to conversion and selectivities, it also enables one to quantitatively estimate the distribution of reactants, products, temperature, and catalyst activity. The reaction kinetics were modeled using a modified form of the Yates and Satterfield equation, with a factor, "F", to account for the higher activity of the Velocys catalyst. [11-13] Deactivation was modelled as a result of the oxidation of metallic cobalt to inert cobalt oxide promoted by the presence of water as described in literature. [14] The poisoning was included by balancing the non-steady convective transport of H_2S in the gas phase and non-steady diffusive transport inside the catalyst, where the sulfur adsorbs on the active cobalt sites. This leads to a combined deactivation due to oxidation and poisoning. An initial estimation of the deactivation parameters predicts a distribution of sulfur after poisoning with 100ppb for 300h consistent with the experimental observation and can simulate the distribution at for arbitrary entry concentrations and exposure times (Figure 6).

Figure SEQ Figure * ARABIC 6. Time evolution of the effective mass fraction of sulfur along the reactor axis (left) and mass fraction profiles in the intra-particle direction at 12% of the bed length (right). Feed composition: 100 ppb H_2S, H_2:CO 1.79, 28% inert, 310 ms contact time, 205 °C.

The deactivation rate of 0.08% per day after sulfur has been switched off (Figure 2) was used to estimate the deactivation parameters assigned to cobalt oxidation. With this fitted parameter the loss in activity of 0.12% per day at 15 ppb sulfur was remarkably well predicted by the simulation with a value of 0.121% per day. This enables the predictive assessment of the effect of different levels of sulfur in the feed gas. Figure 7a shows how the CO conversion declines for different H_2S feed levels after the same overall amount of sulfur has been fed to the catalyst bed. At the lowest sulfur level, a longer exposure time allows for the oxidation contribution to have the largest impact, yielding the lowest final conversion. This can also be seen in the final axial catalyst activity distribution shown in Figure 7b. Poisoning accounts for the loss of activity close to the inlet, which is similar for all H_2S levels at equivalent exposure. Due to the longer run time at lower feed H_2S levels, the downstream part of the bed, where the highest water partial pressures appear, is affected to the greatest extent by oxidation.

Figure SEQ Figure * ARABIC 7. Conversion as a function of time for different feed sulfur levels with total delivered sulfur equivalent to 100 ppb for 100 hours (left). Final axial distribution of the relative catalyst effectiveness for each of the different exposure conditions.

Conclusions

Velocys has used a high sensitivity analytical method to accurately quantify the delivery of low ppb S levels in syngas feed to charges of FT synthesis catalyst in microchannel reactors. With this method, experiments were performed using doped and undoped syngas feeds to quantify the impact of S poisoning on the catalyst deactivation rates. Analysis of spent catalyst samples revealed gradients in concentration of S along the axial bed length and in the radial catalyst granule direction. Models were developed and validated to enable predictive assessment of the S poisoning in a commercial reactor.

References

1. Rytter, E. & Holmen, A. Deactivation and Regeneration of Commercial Type Fischer-Tropsch Co-Catalysts—A Mini-Review. Catalysts 5, 478–499 (2015).

2. Argyle, M. D. & Bartholomew, C. H. Heterogeneous Catalyst Deactivation and Regeneration: A Review. Catalysts 145–269 (2015).

3. Duvenhage, D. J. & Coville, N. J. Deactivation of a precipitated iron Fischer-Tropsch catalyst - A pilot plant study. Appl. Catal. A Gen. 298, 211–216 (2006).

4. Kritzinger, J. A. The role of sulfur in commercial iron-based Fischer-Tropsch catalysis with focus on C2-product selectivity and yield. Catal. Today 71, 307–318 (2002).

5. Ehrensperger, M. & Wintterlin, J. In situ scanning tunneling microscopy of the poisoning of a Co(0001) Fischer–Tropsch model catalyst by sulfur. J. Catal. 329, 49–56 (2015).

6. Sparks, D.E, Jacobs, G., Gnanamani, M.K., Pendyala, V.R.R., Ma, W., Kang, J., Shafer, W.D., Keogh, R.A., Graham, U.M., Gao, P. & Davis, B.H.; Poisoning of cobalt catalyst used for Fischer-Tropsch synthesis, Catalysis Today 215, 67-72 (2013)

7. Pansare, S.S. & Allison, J.D.; An investigation of the effect of ultra-low concentrations of sulfur on a Co/γ-Al_2O_3 Fischer-Tropsch synthesis catalyst, Applied Catalysis A: General 387, 224-230 (2010)

8. Herbig, J., Gutmann, R., Winkler, K., Hansel, a., & Sprachmann, G.. Real-Time Monitoring of Trace Gas Concentrations in Syngas. Oil & Gas Science and Technology – Revue d'IFP Energies Nouvelles. (2013). http://doi.org/10.2516/ogst/2012083

9. Deshmukh, S. R.; Tonkovich A. L. Y.; McDaniel, J. S.; et al., 2011 Enabling cellulosic diesel with microchannel technology. *Biofuels* 2(3): 315.

10. Deshmukh, S. R.; Tonkovich, A. L. Y.; Jarosch, K. T.; et al., 2010 Scale-up of microchannel reactors for Fischer-Tropsch synthesis. *Ind. Eng. Chem. Res.* 49*:* 10883

11. Yates, I. C. & Satterfield, C. N. Intrinsic Kinetics of the Fischer-tropsch Synthesis on a Cobalt Catalyst. Energy & Fuels 5, 168-173 (1991).

12. Kaiser, P., Pöhlmann, F., Jess, A. Intrinsic Effective Kinetics of Cobalt-Catalyzed Fischer-Tropsch Synthesis in View of a Power-to-Liquid Process Based on Renewable Energy. Chem. Eng. Technol. 37, 964-972 (2014).

13. Vervloet, D., Kapteijn, F., Nijehuis, J., van Ommen, J. R. Fischer-Tropsch reaction-diffusion in a cobalt catalyst particle: aspects of activity and selectivity for a variable chain growth probability. Catal. Sci. Technol. 2, 1221-1233 (2012).

14. Sadeqzadeh, M., Chambrey, S., Hong, J., Fongarland, P., Luck, F., Curulla-Ferre, D., Schweich, D., Bousquet, J., Khodakov A. Y. Effect of different reaction conditions on the deactivationof alumina-supported cobalt Fischer-Tropsch catalysts in a milli-fixed-bed reactor: Experiments and modeling. Ind. Eng. Chem. Res. 53, 6913-6922 (2014).

Copyright © American Institute of Chemical Engineers. All rights reserved.

459957 Characterization of Unconventional Reservoirs: From Lab Study to Reservoir Simulations

Abstract:

<span" roman"="Roman"" new="New">Unconventional resources represent an extremely important portion of future energy supply, and recent advances in horizontal well and fracking technologies have spurred the largest growth in oil and gas supply in decades.Â However, the fundamental phenomena associated with gas and oil production from unconventional reservoirs are still poorly understood, and the complexity in both composition and pore structure of unconventional rocks leads to challenges in characterizing and quantifying storage, mass transfer and fluid-rock interactions. In addition, scaling up lab data to reservoir level is more difficult than that encountered in chemical industry. This presentation will examine the key challenges for improving gas and oil production from unconventional formations, and then highlight a few areas where Reaction Engineering concepts and skills are critical. Recent results on phase behavior, mass transfer and reaction phenomena in ultra-tight porous media will be reviewed, with an emphasis on identifying key mechanisms, analyzing lab data, building adequate model and developing relevant upscaling criteria for reservoir simulations.

Copyright © American Institute of Chemical Engineers. All rights reserved.

461230 New Insight into Oxidative Dehydrogenation of n-Butenes to Butadiene

Abstract:

1,3-Butadiene is an important monomer for rubber and plastics industries. It is primarily produced as a co-product of olefins by steam cracking of naphtha. With the increasing demand on butadiene and the future impact of shale gas, the on-purpose production of butadiene has attracted great interest in both academia and industry. We have developed a high performance catalyst suitable for commercial application. In this presentation, we report our work on the synthesis and performance of the catalyst, and the studies of the reaction mechanisms with an emphasis on impact of process conditions on catalyst performance under practical conditions. We thus propose a semi-empirical kinetic model for the reactor scale-up and design.

The Cr-free, Zn-Fe-O based mixed metal oxides catalysts were synthesized by co-precipitation, spray-drying, and solid state reaction methods. The catalysts were characterized by XRF for elemental analysis, XRD for structure identification, N_2 adsorption for BET surface area, pore volume and pore size distribution, Total Carbon Analyzer for coke formation. Typically, 3.0 mL of the catalyst was tested in a fixed-bed reactor at 330-380 °C, 1 atm, and C_4H_8 GHSV of 400 h^{-1}. For catalyst screening and stability testing, the feed typically contains C_4H_8 at 30 mL/min, O_2 at 20 mL/min, H_2O at 365 mL/min, and CH_4 (used as diluent and analytical internal standard) at 30 mL/min. The feed and products were analyzed by an online GC equipped with an FID and a TCD. The kinetic studies were carried in a similar reactor by varying process conditions.

The catalyst studied has a spinel structure, ZnFe
amounts can significantly improve catalyst activity and stability. The XRD results did not show any oxide forms of Ca, Mg, and Co indicating they replaced Zn in the spinel structure. In addition to the spinel, the presence of Fe_2O_3 in a solid solution with the spinel improves performance in both catalyst activity and selectivity.

The catalyst synthesis involves many steps and conditions. Among them, the precise control of the pH value during the co-precipitation is so critical that any minor derivation could result in a significant impact on the catalyst performance. The pH value directly impacts the concentration of Zn in the catalyst. We believe the Zn concentration together the promoters incorporated into the spinel determine the catalyst performance. With the optimal compositions of the catalyst prepared under the well-controlled conditions, our catalyst showed stable performance up to 4,000 hours with a C_4H_8 conversion of over 70% and selectivity to C_4H_6 over 90% that meet the target of commercial operation.

To understand the impact of the process conditions on the catalyst performance, we conducted kinetic studies based on a reaction network of a typical selective oxidation reaction. The experiments were carried out in fixed bed reactors. Process variables studied include ratios of oxygen to butene, steam to butene, pressure, and temperature. We performed kinetic modeling of the experimental results. Key controlling parameters were identified and new reaction conditions were discovered which significantly improved the yield of butadiene up to 87%, while the literature reported a maximum yield of 75%. Based on the results, a semi-empirical kinetic model was proposed and modeled. The kinetic model provides insight into the reaction mechanisms and provides guidance on the design and operation of pilot plant for further assessment of the technology for commercial practice.

We developed a Cr-free $ZnFe_2O_4$ based catalyst promoted by multiple elements that can be scaled up for commercial use. We thoroughly studied catalyst synthesis and identified critical conditions that determine catalyst performance. Our kinetic studies evaluated the impact of the process conditions and identified new conditions that can enhance catalyst performance. Based on such studies, we proposed a semi-empirical kinetic model that can provide insight into the reaction mechanisms under practical conditions and guidance for reactor scale up and design for pilot plan studies and future commercial practice.

Copyright © American Institute of Chemical Engineers. All rights reserved.

462903 Novel High Active Catalyst Innovation and Low-Energy Consumption Process Technology for Ammonia Synthesis

Abstract:

Although catalytic ammonia synthesis technology has been matured comparatively, it still has tremendous potential for energy-saving. Further decrease of energy consumption depends mainly on high active catalyst innovation and its process for ammonia synthesis. This paper reported a ZA-5 catalyst with wüstite as precursor. It recently is a performance update of the most successful catalyst innovation for the ammonia synthesis. ZA-5 catalyst has very high activity at low temperature and pressure, so as to be competitive with costly Ru/AC catalyst. The relations between pressure, power and catalytic activity in synthesis loop were investigated. Based on the above analysis, it is proposed that new idea that the decrease in net value of ammonia resulting from the decrease in pressure can be compensated by the high activity of the novel catalyst, and a newly low-energy consumption process technology of ammonia synthesis based on high efficient ZA-5 catalyst was proposed according to optimized design. The results of optimized design indicated that, with the decrease of the pressure of synthesis loop the ammonia net value, concentration of inert gas in vent gas and ammonia condensation temperature can basically remain unchanged, the power of cycle compressor and refrigeration compressor is the same with it in high pressure, but the total power decrease mainly because of the power of the syngas compressor decreased with the decreasing pressure. And efficiency of energy-saving achieves 12.34 % and 18.31 %, saving energy achieves 1.50 $GJ.t^{-1}$ and 2.23 $GJ.t^{-1}$ or 51.34 $kgce.t^{-1}$ and 76.16 $kgce.t^{-1}$ when pressure decreasing from 30 MPa to 15 MPa and 10MPa in synthesis loop for the ammonia plant with capacity of 200 $kt.year^{-1}$, respectively.

Copyright © American Institute of Chemical Engineers. All rights reserved.

463586 Maximizing Production of Premium Distillate Fuels through Selective Hydroisomerization of Normal Paraffins

Abstract:

As the demand for clean distillate fuels grows around the world refiners are looking to optimize their refineries to produce these premium products while minimizing production of lower value products such as motor gasoline. Cetane number and cold-flow properties are critical performance specifications for distillate products, however due to the chemical nature of distillate range process streams it can be challenging to achieve both specifications without either using additive packages or blending with other product streams downstream of processing units. Both of these specifications are closely linked to paraffin concentration: cetane number benefits from high paraffin concentrations while cold-flow properties are adversely affected by high n-paraffin concentrations. Hydroisomerization is one process chemistry solution that can be employed to mitigate this dilemma through selective hydroisomerization of n-paraffins to i-paraffins which both preserves cetane number and improves cold-flow performance.

Selective hydroisomerization requires a bifunctional catalyst containing both a metal function to perform hydrogenation and dehydrogenation, and a shape-selective acid function to perform the skeletal isomerization of the olefin intermediate. These bifunctional catalysts need to be carefully designed to balance kinetics and molecular transport in order to maximize hydroisomerization selectivity. Through continuous research and development efforts over the last 25 years, ExxonMobil has developed and deployed multiple generations of MIDW^{TM} technology for production of premium distillate fuels via hydroisomerization. In addition to maintaining cetane number and improving cold-flow properties, the MIDW^{TM} technology also demonstrates high distillate selectivity through the minimization of cracking reactions that produce lower value naphtha. Recent advances in catalyst formulations and process configurations have expanded the versatility of MIDW^{TM} technology enabling refiners to process difficult feeds, including sour feedstocks.

This paper will provide the audience an industrial perspective on catalyst development; specifically, catalyst formulation considerations will be described in light of competing process chemistries, feed contaminants, and equipment constraints.

Copyright © American Institute of Chemical Engineers. All rights reserved.

469775 On the Effectiveness Factor Modeling for Mass Transfer-Limited Reaction Networks in Industrial Catalysis

Abstract:

On the Effectiveness Factor Modeling for Mass Transfer-Limited Reaction Networks in Industrial Catalysis

Bo Jin, Xiaoxiang Zhu, and Willian R. Licht

Steam-methane reforming (SMR) is an important industrial catalysis application and much work has been devoted to studying the reaction kinetics both at the molecular level and the macroscopic scales. Bridging the micro- and macroscopic activities in the catalyst is the mass transfer from the process gas into/out of the catalyst particles. Reforming reactions at high temperature are so fast that the systems are diffusion limited even at extremely fine laboratory scales which attempt to probe the intrinsic reaction kinetics. As a result, accurate incorporation of the mass transfer factors strongly affect the quantification of intrinsic reaction kinetics, the catalyst performance, and furthermore the design of reforming plants.

While a full reaction-diffusion model could be numerically solved for the coupled SMR reaction network to quantify the effectiveness factor accurately, such an approach is prohibited in practical applications due to the need of numerous iterations and the high computational cost. In this work, we present a novel approach based on perturbation analysis to analytically derive the solution of effectiveness factor for the SMR reaction network. With little computational cost, the algebraic analytical solution closely matches the numerical solution of the full diffusion-reaction model. The analytical solution is incorporated and demonstrated in a reformer tube simulation. Enabled by the analytical solution, parameter estimation in lab-conducted measurements as well as programs for plant engineering and design can efficiently and correctly take into account the mass transfer limitations in the catalyst. The approach presented here is generic and also applicable to other reaction networks.

Copyright © American Institute of Chemical Engineers. All rights reserved.

472480 Active Site Requirements and Elementary Steps in the Ketonization of Carboxylic Acids on TiO_2 and ZrO_2

Abstract:

Ketonization converts carboxylic acids into equimolar mixtures of alkanones, CO_2, and H_2O while forming a new C-C bond and removing O-atoms from the acid reactants [1], thus lengthening carbon chains and decreasing the oxygen content in biomass-derived streams [2]. Metal oxides, such as TiO_2 and ZrO_2, catalyze ketonization reactions of diverse carboxylic acids; the elementary steps involved and their kinetic relevance, as well the specific nature of the active sites involved, remain controversial [3-5]. Here, we combine kinetic, isotopic, spectroscopic, and density functional theory (DFT) methods to assess the pathways for the ketonization of C_2-C_4 carboxylic acids on anatase TiO_2 (TiO_2(a)), rutile TiO_2 (TiO_2(r)), monoclinic ZrO_2 (ZrO_2(m)), and tetragonal ZrO_2 (ZrO_2(t)). These studies show that the elementary step that forms the new C-C bond, via reactions between adsorbed 1-hydroxy-enolates and coadsorbed acids, limits ketonization rates on all these oxides. These steps require Lewis acid-base site pairs that allow the concerted stabilization of the required transition states both for the facile formation of 1-hydroxy-enolates and for their subsequent kinetically-relevant C-C coupling reactions. Consequently, both basic and acid strengths, probed here through DFT-derived binding energies of gaseous H^+ and OH^- species, respectively, influence the stability of the transition states that mediate ketonization rates and of the relevant adsorbed 1-hydroxyl-enolate precursors to such transition states.

These studies also demonstrate the intrinsic mechanistic link between ketonization reactions of carboxylic acids and aldol condensations of carbonyl compounds; such a link is evident from the common requirement that an α-C-H bond be cleaved in one reactant molecule before a new C-C bond is formed at that position via reactions with another reactant molecule [3,6]. Not unexpectedly, the metal oxides that catalyze both reactions, TiO_2 and ZrO_2 in the present case, are also similar. The demonstrated involvement of enolates in aldol condensation on TiO_2 and ZrO_2 [6] makes it plausible that similar enolate-like species (1-hydroxy-enolates) are involved in ketonization; these species are then involved in nucleophilic attack with another acid to form 3,3-dihydroxy-carboxylic acids with a new C-C bond, which decompose via dehydration and decarboxylation reactions to form alkanones, in contrast with the water elimination that forms condensation products from aldols without the loss of a C-atom. These similarities in mechanisms and the common involvement of Lewis acid-base site pairs, supported by the theoretical treatments reported here, extends the use of titration methods developed to measure the number of active sites in condensation reactions to their ketonization counterparts [6]; such site counts are used here to measure intrinsic reactivities, in the form of ketonization turnover rates and activation free energies, thus allowing rigorous reactivity comparisons among metal oxides and the rigorous benchmarking of theoretical treatments against experimental results.

Acetic acid ketonization turnover rates (1.0 kPa; 523 K) were slightly higher on ZrO_2(m) (0.51 ks^{-1}) than on ZrO_2(t) or TiO_2(a) (0.44 and 0.25 ks^{-1}), but much lower on TiO_2(r) (0.024 ks^{-1}). These reactivity trends were also reflected on larger deactivation rate constants on the more active oxides. Infrared spectra during ketonization showed that di-adsorbed acetates (*AcO*), formed via slow dissociation on acid-base site pairs and acting as unreactive spectator species, decreased the concentration of undissociated acids (AcOH*), which are the reactive intermediates in 1-hydroxy-enolate formation, during steady-state catalysis. Cu co-catalysts, as physical mixtures with these oxides, and added H_2 markedly decreased deactivation rates and led to higher steady-state ketonization rates on TiO_2 and ZrO_2; these protecting effects appear to reflect the scavenging of trace levels of ketene species, present in equilibrium with *AcO*, to form acetaldehyde and ethanol, thus providing a mechanism for balancing the depletion and formation of unreactive bound carboxylates and thus inhibiting their blocking of active sites.

Ketonization turnover rates on TiO_2(a) increased monotonically with acid pressure for all C_2-C_4 carboxylic acids; this dependence weakened with increasing pressure, ultimately leading to pressure-independent (zero-order) rates (> 1 kPa; 503-533 K); these kinetic trends were also evident on ZrO_2-based catalysts. They reflect the strong bonding of acids at acid-base site pairs, which leads to densely-covered surfaces at all practical ketonization conditions. The weak H/D kinetic isotope effects measured (acetic acid-d_4, 1.1, 523 K) and the nearly saturated monolayers of AcOH* evident in infrared spectra during catalysis, indicate that the nucleophilic attack of co-adsorbed acids by 1-hydroxy-enolates (formed via α-C-H cleavage in the acid) is the sole kinetically-relevant step in ketonization catalytic cycles. These conclusions are consistent with the measured kinetic dependences, with the observed effects of alkyl substituents on ketonization rates, and with DFT-derived activation free energy barriers using periodic slab models. The zero-order ketonization rates occur on oxide surfaces fully covered by AcOH* and represent the highest attainable rates; they reflect free energy differences ($\Delta G^{\ddagger}_{CC}$) between C-C coupling transition states (TS) and two AcOH* species.

AcOH* binds on Ti-O site pairs via concerted interactions between its carbonyl O-atom and the Ti center and the H-atom in its OH group and the basic O-atom in the site pair; *AcO*, in contrast, binds on two vicinal Ti-O pairs with its two O-atoms each interacting with one Ti center and the dissociated proton bound onto an O site. DFT-derived reaction energies and *in situ* infrared spectra during ketonization catalysis indicate that AcOH* is the prevalent species on TiO_2(a) at the nearly saturated surfaces that prevail during catalysis. The requirement for one Ti-O site pair for AcOH* and two Ti-O site pairs for *AcO* drives the strong preference to crowd surfaces with AcOH* species stabilized by lateral van der Waals and H-bonding interactions. The formation and spectroscopic detection of stranded carboxylates upon removal of gaseous acid molecules has led to the incorrect assertion that ketonization involves carboxylates as reactive intermediates [3-5].

The kinetic relevance of the C-C coupling step and the prevalence of AcOH* at coverages near saturation on TiO_2(a) surfaces were confirmed by DFT treatments of reaction and activation free energies on densely-covered anatase surfaces. These calculations give values for $\Delta G^{\ddagger}_{CC}$, for its enthalpy ($\Delta H^{\ddagger}_{CC}$) and entropy ($\Delta S^{\ddagger}_{CC}$) components, and for H/D isotope effects in quantitative agreement with measured values. The theoretical analyses reported here also illustrate how C-C coupling TS complexes become increasingly more stable relative to its relevant precursors (AcOH*) as the coverage of the latter increases. For example, the $\Delta G^{\ddagger}_{CC}$ value is 181 kJ mol^{-1} at 1/3 ML AcOH* but 160 kJ mol^{-1} at 1 ML (523 K, 1 bar AcOH), as a consequence of H-bonding between the C-C coupling TS and vicinal AcOH* species. These observations and their mechanistic interpretation provide yet another demonstration about how densely-covered surfaces provide specifically beneficial environments for chemical reactions, by decreasing differences in free energies between TS complexes and their relevant kinetic precursors [7].

These mechanistic conclusions apply also to C_3-C_4 carboxylic acids and to ZrO_2-based catalysts, as evident from their kinetic rate equations for acetic acid ketonization on TiO_2(a) and from parallel DFT treatments. Acetic acid turnover rates on ZrO_2(m) give $\Delta G^{\ddagger}_{CC}$ values smaller than on TiO_2(a) (136 ± 2 *vs.* 166 ±1 kJ mol^{-1}, 523 K), consistent with DFT-derived values on the low-index surfaces of these oxides (140 *vs.* 161 kJ mol^{-1}). These $\Delta G^{\ddagger}_{CC}$ differences may reflect the different free energies for the formation of the 1-hydroxy-enolate reactive intermediate from AcOH* on a M-O (M=Ti, Zr) site pair or of the C-C coupling transition state from the 1-hydroxy-enolate-AcOH* pair precursor bound on two vicinal M-O site pairs. These two free energies reflect, in turn, the 1-hydroxy-enolate concentrations at steady-state and its intrinsic reactivity in reactions with AcOH* to form the C-C bond, respectively. While experiments reflect only their combined contributions, DFT treatments show that the more reactive nature of ZrO_2(m) surfaces arises from higher concentrations of 1-hydroxy-enolates bound on Zr centers. DFT-derived OH^- and H^+ binding energies show that metal centers are weaker Lewis acids and O-atoms are more basic in ZrO_2(m) than TiO_2(a) surfaces, suggesting that the binding of 1-hydroxy-enolate anions is weaker on Zr than Ti centers, but that the respective proton moieties bind much more strongly with the O sites on ZrO_2(m) than TiO_2(a). These DFT-derived free energies lead us to infer that the higher 1-hydroxy-enolate concentrations on ZrO_2(m) benefit from the stronger basic O-sites that stabilize the proton moieties, which disfavor the reprotonation of 1-hydroxy-enolates to form AcOH*, consistent with the preference of ketonization reactions for acid-base site pairs containing strong basic sites [1,3].

The authors acknowledge BP for financial support through the XC^2 and ICC programs and XSEDE for access to computational facilities.

References

[1] T. N. Pham, T. Sooknoi, S. P. Crossley, D. E. Resasco, *ACS Catal.* 3 (2013) 2456–2473.

[2] D. A. Simonetti, J. A. Dumesic, *Catal. Rev.* 51 (2009) 441–484.

[3] G. Pacchioni, *ACS Catal.* 4 (2014) 2874–2888.

[4] A. Pulido, B. Oliver-Tomas, M. Renz, M. Boronat, A. Corma, *ChemSusChem*, 6 (2013) 141–151.

[5] T. N. Pham, D. Shi, D. E. Resasco, *J. Catal.* 314 (2014) 149–158.

[6] S. Wang, K. Goulas, E. Iglesia, *J. Catal.* (submitted).

[7] D. Hibbitts, E. Iglesia, *Acc. Chem. Res.* 48 (2015) 1254–1262.

Copyright © American Institute of Chemical Engineers. All rights reserved.

458924 Influence of Solid Acid Structure and Composition on Hydrocarbon Cracking Activity Under Supercritical Conditions

Abstract:

Optimization of hydrocarbon cracking catalysts under supercritical conditions is needed to develop endothermic fuel cooling systems for use on hypersonic aircraft. In this study, cracking of several different hydrocarbons over solid acid catalysts with various pore sizes and compositions was studied under supercritical conditions. The specific activity of porous aluminosilicate catalysts was substantially higher than that of non-porous tungstated zirconia (WO_x/ZrO_2) for cracking of n-dodecane, a surrogate molecule for jet fuel. Microporous H-ZSM-5 zeolite exhibited the highest turnover frequency (TOF), normalized by H+ sites titrated by n-propylamine decomposition, for cracking of linear hydrocarbons hexane and dodecane. However, the relatively small pore size of H-ZSM-5 negatively impacted the cracking of branched hydrocarbons (isooctane and isododecane) as well as military jet fuels (JP-8 and JP-10). Utilization of H-Y as an acid catalyst, which had a micropore size large enough to accommodate the kinetic diameter of JP-10 (exo-tetrahydrobicyclopentadiene), revealed excellent conversion of the hydrocarbon relative to H-ZSM-5, thus emphasizing the importance of proper catalyst/fuel pairing. The influence of temperature on the conversion of n-hexane and n-dodecane over H-ZSM-5, H-Y, and a mesoporous aluminosilicate revealed that hydrocarbon chain length and catalyst pore size significantly affected the apparent activation energy and observed product distribution. Mechanistic implications of the results will be discussed.

Copyright © American Institute of Chemical Engineers. All rights reserved.

471867 Improving the Hydrothermal Stability of Heterogeneous Catalysts for Conversion of Biorenewables Via Carbon Coating and Overcoating

Abstract:

Improving the Hydrothermal Stability of Heterogeneous Catalysts for Conversion of Biorenewables via Carbon Coating and Overcoating

Hien N. Pham[1], Haifeng Xiong[1], Pu Duan[2], Xiaoyan Cao[2], Klaus Schmidt-Rohr[2], Robert L. Johnson[3], James A. Dumesic[4] and Abhaya K. Datye[1]

(1) Department of Chemical and Biological Engineering, and Center for Microengineered Materials, University of New Mexico, Albuquerque, NM

(2) Department of Chemistry, Brandeis University, Waltham, MA

(3) Department of Chemical and Biological Engineering, Iowa State University, Ames, IA

(4) Department of Chemical and Biological Engineering, University of Wisconsin, Madison, WI

Abstract:

A major challenge for the production of biomass-derived feedstocks to value-added chemicals and fuels is the development of catalysts and supports that are hydrothermally stable during conversion of biorenewables. Conventional catalysts and supports designed for gas-phase reactions may not be suitable for aqueous-phase reactions, particularly at temperatures in excess of 200 °C, due to loss of surface area and structural integrity of the support, and sintering or leaching of the metal phase [1]. We show that we can significantly improve the hydrothermal stability of catalytic materials by modifying them with a few overlayers of carbon [2,3]. Using a CVD or liquid-phase route, we can tailor the carbon overlayers by modifying the extent of graphitization and study how the nature of the carbon affects the catalytic performance under aqueous-phase reaction conditions. The efficacy of our carbon coating approach is demonstrated by the improved stability of precious and non-precious metal catalysts for hydrogenation reactions [4].

References:

[1] Xiong, H., Pham, H.N., and Datye, A.K. *Green Chem.* 16, 4627 (2014).

[2] Pham, H.N., Anderson, A.E., Johnson, R.L., Schmidt-Rohr, K., and Datye, A.K. *Angew. Chem. Int Ed.* 51, 13163 (2012).

[3] Xiong, H., Schwartz, T.J., Andersen, N.I., Dumesic, J.A., and Datye, A.K. *Angew. Chem. Int Ed.* 54, 7939 (2015).

[4] Pham, H.N., Anderson, A.E., Johnson, R.L., Schwartz, T.J., O'Neill, B.J., Duan, P., Schmidt-Rohr, K., Dumesic, J.A., and Datye, A.K. *ACS Catal.* 5, 4546 (2015).

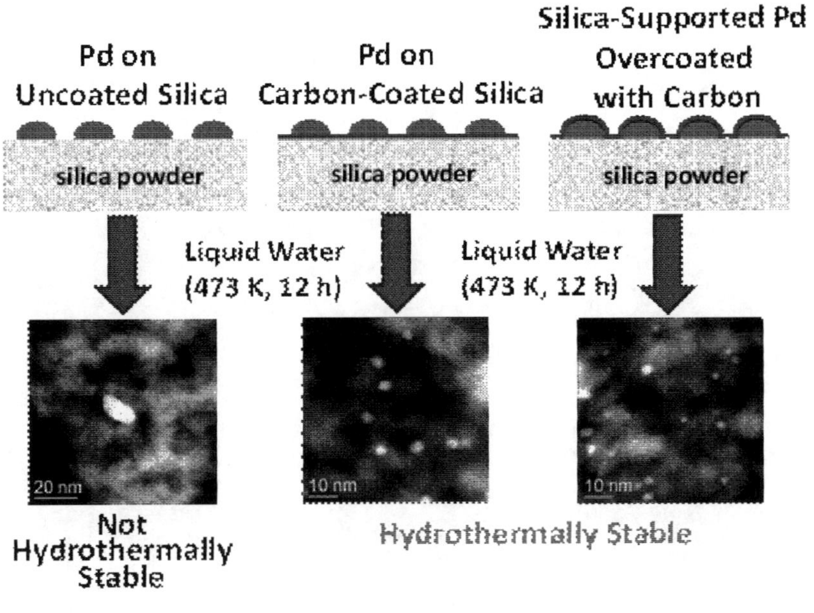

Copyright © American Institute of Chemical Engineers. All rights reserved.

470655 Mechanistic Insights into the Reactive Sorption of Sulfur Compounds Using Metal Oxides

Abstract:

Natural gas and oil typically contain a wide range of contaminants and require purification before processing because these contaminants damage equipment and catalysts and are hazardous to humans and the environment. Metal oxides are widely used to remove sulfur compounds via chemical reaction (a process known as reactive sorption), however, little research has been conducted to understand these reactions at the molecular level. In this study, we use fundamental reactor models to analyze effluent concentration profiles from fixed beds of sorbents and to derive fundamental structure-function relationships between material properties (e.g., chemical composition, metal crystallite size, and porosity) and process variables (e.g., contact time and contaminant identity and concentration). Specifically, we probe how the physico-chemical properties of CuO nanostructures produced from different precursors and using different synthesis techniques influence their performance in the reactive sorption of H_2S from gaseous streams.

Copper (II) oxide nanofibers were synthesized via electrospinning using copper nitrate trihydrate, copper acetate, or zinc acetate as precursors. Electrospinning solutions consisted of the metal salt precursor and polyvinylpyrrolidone (PVP) dissolved in ethanol/water mixtures. Electrospinning syntheses were conducted at 30-35 kV, 5 mL/hr extrusion rate, and 15-17 inch distance between the tip of the extrusion syringe needle and the collector plate. The nanofibers were subsequently treated in air at 625°C (at 0.7°C/min). CuO nanoparticles were prepared via precipitation using 0.1 M solutions of copper salts and 0.1 M NaOH solutions. The washed precipitates were dried in air at 80°C for 16 h and then treated in air at 500°C for 4 h. Breakthrough curves were collected in a tubular reactor with plug flow hydrodynamics. Adsorption models based on irreversible adsorption isotherms and linear driving force (LDF) approximations were used to analyze data from these fixed bed experiments. These models are appropriate for making quantitative assessments of reactive sorption rates because these reactions are essentially irreversible (i.e., equilibrium constants >10^{10}). The resulting equations relate the properties of the sorbent material and the experimental conditions to the breakthrough curve of the contaminant from the bed using two parameters: the sorption rate parameter (k) and the saturation concentration of the contaminant on the solid phase (q_{max}). The rate parameter can be de-convoluted into three different resistances to H_2S removal: (i) bulk phase mass transfer of H_2S, (ii) intraparticle diffusion of H_2S, and (iii) reaction/diffusion phenomena occurring at the surface of the reactive phase. These model parameters were regressed from the breakthrough curve data, while q_{max} was also determined from a mass balance of the contaminant in the vapor phase.

Fresh and spent nanostructures were analyzed via XRD and SEM to identify composition, crystallite agglomerate size, and crystallite size (using the Sherrer Equation). XRD patterns of fresh CuO nanostructures showed the characteristic peaks of tenorite (2q = 32.3°, 35.4°,38.5°, 48.5°, 53.1°, 57.9°, 61.1°, 67.5°), the monoclinic structure of CuO, for both nanofibers and nanoparticles regardless of precursor. Furthermore, nanofibers and nanoparticles consisted of CuO crystallites of similar size (16-18nm). SEM images revealed that the nanofibers exist as approximately rectangular solid shaped agglomerates of crystallites that are 250-300 nm is size and arranged in a one-dimensional stacking pattern. Nanoparticles also consist of 250-300 nm sized crystallite agglomerates, however, these agglomerates are packed into sphere like structures that are 5-10 microns in size.

The mechanism of H_2S removal was probed by observing the effects of various process parameters (i.e., pressure, H_2S concentration, contact time, and particle diameter) on H_2S breakthrough curves from fixed beds of a commercially produced Cu-Zn-Al mixed metal oxide. Breakthrough curves that exhibit symmetrical S-shaped patterns indicate sorption rates that are modeled by an LDF equation that is first order in both gas phase H_2S concentration and residual concentration of CuO in the solid phase. The rate parameters derived from this model exhibit contributions from fluid phase mass transfer and intraparticle diffusion. Sickle shaped curves result from LDF equations that are zero order in H_2S but still first order in solid phase CuO concentration. LDF equations that are zero order in gas phase H_2S concentrations indicate a sorption mechanism controlled by diffusion of S-atoms through CuS layers around unreacted CuO crystallites and agglomerates. For large granule sizes (550 micron average), the H_2S removal was described by an equation that is first order in H_2S at low linear velocities. As linear velocity was increased from 5.3 to 9.1 cm/s (holding contact time constant at 0.72 s), the removal of H_2S became governed by a zero order (in H_2S) LDF equation. This change in mechanism indicates that external and intraparticle mass transfer contribute to the overall rate of H_2S sorption for large granules, likely because mass transfer contributions are proportional to the square of the granule radius. The higher linear velocities associated with the shorter contact times reduce external and internal mass transfer effects. Thus, the reaction/diffusion phenomena at the reactive surface becomes rate controlling. Under conditions where reaction/diffusion at the crystallite-agglomerate scale dominate (i.e., 150 micron granules), rate parameters were invariant with changing total pressure (8.6×10^{-5} 1/s for P = 1-2 atm), however, rate parameters increased from 3.2×10^{-5} to 1.1×10^{-4} 1/s as the partial pressure of H_2S in the feed to the reactor increased from 620 to 2200 ppm-mol. This increase in rate parameter suggests that it is a lumped parameter that can be further de-convoluted into components from different elementary steps that are dependent upon H_2S concentration.

To further probe the reaction between H_2S and CuO at the agglomerate/crystallite level, H_2S removal using CuO nanofibers and nanoparticles was compared. These nanostructures exhibited breakthrough curves indicative of LDF equations that are zero order in H_2S. Nanofibers and nanoparticles exhibited similar maximum capacities (2.6 g H_2S per 100 g CuO), however, sorption rate parameters for nanofibers (3.0×10^{-3} 1/s) were 1.7 times larger than those for nanoparticles (1.7×10^{-3} 1/s). This faster sorption rate is likely the result of the higher intra-agglomerate void fraction in the nanofibers compared to the nanoparticles. The one-dimensional arrangement of the nanofibers provides smaller and more easily accessible reactive domains compared to the nanoparticles.

These studies have confirmed that linear driving force models can quantitatively describe the kinetics and mechanisms for the removal of H_2S from gaseous streams. Furthermore, these studies have shown that the rate parameters in these models consist of contributions from mass transfer as well as reaction/diffusion at the sub-micron scale, with the critical reactive domain at the level of the arrangement of agglomerates of crystallites. The rate parameters have been shown to be lumped parameters that are dependent upon H_2S concentration, and thus, these parameters must contain contributions from various elementary steps that are not zero order in H_2S. This knowledge can be further exploited to develop detailed microkinetic models which can be used to improve the efficiency of existing sorbents and to provide a rational framework for designing new materials for reactive sorption.

Copyright © American Institute of Chemical Engineers. All rights reserved.

473127 Influence of Solvent on Acid and Metal Catalyzed Reations

Abstract:

Many of the catalytic transformations involved in the conversion of chemicals and biomass-derived intermediates are carried out in solvents that can significantly influence the catalytic chemistry, the kinetics and the underlying mechanisms. Solvents can act to stabilize or destabilize reaction intermediates and transition states and, in some cases, they are intimately involved in co-catalyzing reactions. We will examine the role of different solvents on both metal as well as acid catalyzed reactions.

Copyright © American Institute of Chemical Engineers. All rights reserved.

458154 Effect of Alcohol Structure on the Kinetics and Mechanism of Acid Catalyzed Etherification and Unimolecular Dehydration over Tungstated Zirconia

Abstract:

Branched and linear ether molecules have attracted recent interest as diesel additives and lubricants that can be built from various structures of biomass-derived alcohols. In this study, a comprehensive investigation of the effect of alcohol structure on the mechanism and kinetics of liquid phase etherification and unimolecular dehydration reactions was performed. In this investigation, 12 wt% WO_x/ZrO_2 was identified as an effective and green solid acid catalyst for selective etherification of primary alcohols in the liquid phase, achieving ether selectivities of >96% for C_6-C_{12} linear alcohols at 120 °C, even though at this temperature unimolecular dehydration to alkenes is thermodynamically preferred. The proposed mechanisms for Brønsted acid-catalyzed unimolecular dehydration and etherification were supported by isotopic labeling and by measuring kinetic isotope effects. The catalyst was characterized with XRD, BET, and DRIFTS FTIR with pyridine adsorption. It is proposed that, while the desired etherification reaction proceeds over a small number of strong Brønsted acid sites, the surrounding Lewis acid sites help bring reactant molecules to the surface to facilitate the bimolecular reaction. The effects of alcohol structure including linear alcohol length, carbon chain branching positions, and carbon chain branching size, were probed by measuring the apparent kinetics for dehydration reactions over tungstated zirconia. Trends in apparent activation energies and selectivities were consistent with the proposed mechanism, and can elucidate feasible pathways to produce valuable ether products from biomass-derived alcohols.

Copyright © American Institute of Chemical Engineers. All rights reserved.

470138 Pathways to Selectively Form Individual Aromatic Products from Ethanol

Abstract:

Ethanol derived from the fermentation of biomass can be catalytically converted into higher molecular weight hydrocarbons and oxygenates to produce advanced biofuels and lubricants. The Guerbet reaction, which involves dehydrogenation, aldol condensation, and hydrogenation, is one pathway for this process. Cascades of subsequent self- and cross-condensation reactions between ethanol and $C_{\geq 4}$ alcohol derived intermediates provide a wide-range of possible higher carbon number (C_n) products. Successful production of these products requires high selectivities for dehydrogenation and condensation reactions to avoid the loss of carbon by forming undesirable products (e.g., C_1-, C_2-
form specific molecules which can serve as fuels and platform chemicals.

Here, we show how bifunctional catalysts, comprised of amphoteric oxides (TiO_2, hydroxyapatites, or hydrotalcites) and Cu clusters promote these networks of reactions by measuring formation rates of more than 30 distinct products as functions of reactant concentrations, temperature (523 - 623 K), and the ratios of the numbers active sites for hydrogen transfer and those for condensation reactions. The selectivities with which the C_2-units of ethanol are incorporated into growing chains is limited by the selectivity of both catalytic functions, while the isomeric structure of the products is determined by the concentrations of the unsaturated oxygenates present on the oxide surfaces and differences between the deprotonation energies of C-atoms α to the carbonyl groups of these species. Reactions among product mixtures that predominantly contain alcohol and aldehyde reactants form broad distributions of fatty alcohols whose product selectivities follow predictions from step-growth polymerization models. These diverse mixtures, however, are only suitable as fuels.

Aldol condensations directly form unsaturated aldehydes, which exist at much higher concentrations than

their saturated analogs in the absence of effective hydrogen transfer catalysts. Large differences in the deprotonation energies between α C-atoms for aldehydes and unsaturated aldehydes allow these product mixtures to be exploited to form a very small set of C_8 aromatic species (ortho- and para-tolualdehydes). The structure of the resulting products effectively halts further condensation reactions and provides a unique opportunity to make a desirable platform chemical within this complex network of reactions. The isomeric structure of the C_8 aromatics are sensitive to the presence or absence of steric hindrance about the acid-base site pairs that catalyze the reaction. Coupling hydroxyapatite, which lacks microporosity produces predominantly ortho-C_8 products with minor (< 10%) amounts of the para-isomers. These selectivities can be inverted, however, when the reactant stream include ethylene (derived from in situ dehydration of ethanol or added intentionally). The competing pathways to ortho- and para-tolualdehydes reflect the relative rates of aldol and Diels Alder additions between hexadienal and either acetaldehyde or ethylene co-reactants.

These findings provide evidence that the Guerbet reaction proceeds by dehydrogenation and aldol condensation steps and elucidates and the complex sets of surface reactions that dictate the formation rates of certain categories of products. This knowledge will help guide further research into reaction schemes to convert alcohols into larger oxygenates and hydrocarbons.

Copyright © American Institute of Chemical Engineers. All rights reserved.

452415 Combined Experimental and Theoretical Study of Ethanol Catalytic Conversion to 1,3-Butadiene on MgO

Abstract:

Biomass can serve as a sustainable and renewable carbon source. Ethanol is relatively inexpensive and widely available and can its catalytic upgrading has been proposed as a viable route of indirect biomass valorization. 1,3-butadiene, the most important monomer for synthetic rubber, has been produced via catalytic processing of ethanol during World War II by USSR and USA, using Lebedev and Ostromislensky processes, respectively. The former utilized catalytic conversion of ethanol to 1,3-butadiene in one-step over MgO/SiO_2 catalystswhile the latter utilized a two-step process with the first step of ethanol dehydrogenation to acetaldehyde over Cu/SiO_2 catalysts and transforming acetaldehyde into 1,3-butadiene over a tantalum-based catalyst during the second step. Recent abundance of shale gas resulted in a different catalytic cracker product distribution dominated by ethylene. This caused a worldwide shortage of C4 hydrocarbons such as 1,3-butadiene. Since ethanol can be produced using variety of biomass processing routes, including fermentation and gasification, it recently reemerged as the green route to catalytically form 1,3-butadiene but improvements in yields are necessary.

In this study, we identified the reaction intermediates and mechanisms of ethanol on MgO surface by combining both theoretical and experimental study by means of DFT, *in-situ* FTIR and TPSR. A complex mechanism involving acetaldehyde, crotyl species as well as unexpected pathway via ethylene are discussed.

Copyright © American Institute of Chemical Engineers. All rights reserved.

459233 Aldol Condensation of Acetaldehyde over Titania, Hydroxyapatite and Magnesia

Abstract:

Aldol condensation is a well-known reaction that results in the formation of a C-C bond and is used in the production of important chemicals such as 2-ethylhexanal, isophorone, and crotonaldehyde among others. This work aims to elucidate the mechanism of aldol condensation of acetaldehyde over three materials: anatase titania (TiO_2), hydroxyapatite (HAP), and magnesia (MgO). Reactions were carried out in a fixed bed reactor with a total system pressure of 220 kPa at temperatures between 533 and 633 K and acetaldehyde partial pressures between 0.05 and 50 kPa. Crotonaldehyde was the only product observed over all three catalysts and severe catalyst deactivation occurred at acetaldehyde partial pressures of 5 kPa or greater. The aldol condensation reaction over all three catalysts was first order at low acetaldehyde partial pressure and approached zero order at high acetaldehyde partial pressure. No kinetic isotope effect (KIE) was observed with fully deuterated acetaldehyde reacting over TiO_2 or HAP implying C-H bond activation is not kinetically relevant. These measurements are consistent with a mechanism in which adsorption and desorption steps are kinetically significant during the reaction. Characterization of the catalysts by adsorption microcalorimetry of acetaldehyde and ethanol and diffuse reflectance Fourier transformed infrared spectroscopy of adsorbed acetaldehyde, crotonaldehyde and acetic acid revealed a very high reactivity of these catalysts, even at low temperatures.

Copyright © American Institute of Chemical Engineers. All rights reserved.

467972 Single Step Conversion of Short Carboxylic Acid Mixtures Obtained from Fermentation to Aromatics and Î±-Olefins over a Cu/ZrO_2 Catalyst

Abstract:

Biomass represents an attractive feedstock for producing petrochemical substitutes. However, the high variety of functional groups found in biomass often reduces selectivity and yields during biomass conversion reactions. To overcome this problem, biological funneling can be used to drastically reduce the amount of functional groups. Using bioprocessing, biomass polysaccharides, which contain many functional groups can be converted to carboxylic acid mixtures that contain a single functional carboxylic acid group.

In this work, short carboxylic acids obtainable from a fermentation process were converted into aromatic compounds and/or alpha-olefins in a single step catalytic process with a near theoretical carbon yield. The experiments were carried out from 623 K to 673 K under hydrogen atmosphere using Cu/ZrO_2 catalyst. The combination of copper and zirconia oxide allowed full conversion of carboxylic acids via cascade C-C coupling reactions resulting in higher molecular weight (C8-C15) products, which form a water immiscible organic phase, and light alpha-olefins in the gas phase. Remaining oxygen functionalities are selectively removed by hydrogenation-dehydration reactions over copper while avoiding the hydrogenation of unsaturated C-C bonds reducing hydrogen consumption. Moreover, by tuning reaction conditions, selectivity can be controlled to favor the increased production of short alpha-olefins or heavier products.

Copyright © American Institute of Chemical Engineers. All rights reserved.

454772 Mechanistic Insights on the Formation of Maleic Anhydride from the Oxidation of Levulinic Acid over Supported Vanadates

Abstract:

With the realization that, even though biomass derived transportation fuels present a great solution to the diminishing oil reserves, they cannot compete with their petrochemical equivalents, we turn our attention to bio-based commodity chemicals. Those have the potential to compete with the existing market and present opportunities for the utilization of lignocellulosic biomass. Along those lines, we present an alternative catalytic pathway for the production of maleic anhydride (MA) through the gas phase oxidative cleavage of the methyl carbon in levulinic acid (LA). We previously showed that over a vanadia phase supported on silica, high yields of MA (71%) can be achieved. We were also intrigued to observe that the methyl carbon cleavage does not occur on monofunctional methyl ketones. Here, we try to probe the nature of this unique chemistry, by employing a number of different ketones and keto-carboxylic acids in order to decide if this unique selectivity is due to steric effects or the dual functionality of LA.

Copyright © American Institute of Chemical Engineers. All rights reserved.

458085 P-Toluic Acid from Biomass-Derived Coumalic Acid: A Reaction Network Analysis

Abstract:

The Diels-Alder/decarboxylation/dehydrogenation sequential reaction of biomass-derived coumalic acid (CMA) with propylene provides a renewable one-pot reaction pathway to *p*- toluic acid (pTA), an intermediate that can be used to produce terephthalic acid. A detailed reaction network of the formation of toluic acid (TA) was established by varying the key parameters including temperature, solvent, and reaction time, showing that high yields of TA (85 %) are possible at complete conversion of CMA and that the *para* to *meta* selectivity of TA can be controlled by adjusting reaction temperature and time. Changing the reaction media from non-polar (toluene) to polar aprotic (Î³-valerolactone) revealed a strong solvent effect with an increase in the formation of TA from 58 % to > 85 %. Performing the reaction of CMA and propylene with and without the presence of catalyst gave additional information about byproduct formation demonstrating the importance of the catalyst presence to quickly dehydrogenate TA intermediates and to prevent unwanted byproduct formation. Product yield and selectivity can be further optimized when water free reaction conditions are applied as accelerated decomposition rates of CMA were observed by adding small amounts of water (5 vol-%) to the solvent.

Copyright © American Institute of Chemical Engineers. All rights reserved.

464791 Rational Design of Nanoflower Shaped Bimetallic Ptmn Nanocatalysts for Tartronic Acid Synthesis from Biomass

Abstract:

Tartronic acid is considered as one of the important industrial building blocks for a variety of everyday chemicals such as detergents and plastics, as well as valuable pharmaceutical products. Conventionally, tartronic acid is synthesized from methanol and CO *via* multiple energy intensive and toxic steps involving cyanide agents. Significant formation of waste by-products and poor yield are major issues plaguing the traditional route. Catalytic oxidation of renewable biomass feedstocks would undoubtedly provide alternative approaches for tartronic acid production.

Several mono and bimetallic nanocatalysts have been proposed for catalytic glycerol oxidation to tartronic acid, which are often obtained from seed or polymer template-based methods and exhibit ordered core-shell and alloy structures. Most of existing nanocatalysts however show limited performances for oxidation activity and tartronic acid selectivity. In this presentation, we describe a new completely disordered morphology of PtMn nanocatalysts. Unique bud- and cauliflower-shaped bimetallic PtMn nanoclusters were synthesized using lattice mismatch induced self-assembly technique without adding polymer stabilizers. The influence of Pt/Mn ratios on the morphologies of bimetallic nanoclusters was studied systematically under various synthetic conditions. Due to the large lattice mismatch between Pt and Mn metals, anisotropic growth bimetallic crystals was obtained with high surface index numbers. The novel bimetallic nanocatalysts display synergistic an eight-fold enhancement in catalytic activity for aqueous phase oxidation of glycerol to tartronic acid compared with monometallic Pt catalysts.

Copyright © American Institute of Chemical Engineers. All rights reserved.

474591 Upgrading Corn Stover Pretreatment Wastes into Biodiesel Via Oleaginous *Rhodococci*

Abstract:

The derivation of advance biofuels and specialty oils from agricultural waste streams via oleaginous bacteria is a rapidly developing biotechnology primed for both bioenergy and biochemical sectors. However, finding viable effluents and promising microbes that produce an array of tangible oils for subsequent targeting is a current challenge. Here, we examine and compare alkaline and alkaline-peroxide pretreatments of corn stover as a feedstock for three strains of wild-type *Rhodococci*. In comparison, *R. opacus* DSM 1069 and PD 630 exhibited aggressive degradation of low molecular weight aromatics and glucose monomers compared to *R. jostii*, resulting in heightened viability and improved yields. Interestingly, all three strains yielded oils that could be implemented as a fungible diesel post-transesterification or utilized as a specialty oil for novel biochemical/bio-pharmaceutical applications. Overall, this work presents a novel method for waste stream optimization for bio-refineries.

Copyright © American Institute of Chemical Engineers. All rights reserved.

474502 Biological Design of Lignin Conversion

Abstract:

The utilization of lignin for fungible fuels and chemicals remains one of the most imminent challenges in modern biorefinery. The fundamental challenge in lignin degradation lies in the highly recalcitrant structure that requests redox (oxidation-reduction) reactions to break down the chemical linkages. First, we addressed the challenges by revealing the chemical and biological mechanisms for synergistic lignin degradation via a bacterial and enzymatic system, which significantly improved lignin consumption, cell growth and lipid yield. Chemical analysis suggested that laccase, *R. opacus* cell and Fenton reaction reagents promoted the degradation of different functional groups, elucidating the chemical basis for the synergistic effects. Second, we further improved lignin depolymerization by establishing a new enzyme-mediator system to enhance the electron transfer in enzymatic lignin depolymerization to achieve more than 35% solubilization of kraft lignin without extreme temperature, pH, and pressure. The study has manifested that an efficient enzyme-mediator system can be exploited to establish a bioprocess to solubilize lignin, cleave lignin linkages, modify the structure, and produce substrates amenable to bioconversion, all of which are vital for improving current processing and conversion methods for lignin based biofuels and biomaterials. Third, based on the aforementioned studies, consolidated lignin conversion was engineered in both *Rhodococcus opacus* and *Pseudomonas putida* using systems biology-guided biodesign. Efficient ligninase secretion system has been established through proteomics-guided optimization of promoter/rbs, secretion peptide, and others. The engineered strains have led to significantly increased cell growth and bioproduct yield on lignin substrates. Overall, the biological and chemical design of lignin conversion has opened new avenues for lignin conversion toward various bioproducts.

Copyright © American Institute of Chemical Engineers. All rights reserved.

459510 Selective Conversion of Guaiacol to Substituted Alkylphenols in Supercritical Ethanol over MoO_3

Abstract:

Recently, transition metal phosphides, nitrides, carbides and oxides have been intensively employed in biomass conversion{Ma, 2015 #41}.[1-4] The earth-abundant MoO_3 is regarded as a representative of promising alternative catalysts.[2] However, the bottleneck for the development of HDO technique over MoO_3 lies in the consumption of hydrogen and high temperature. In this work, the conversion of guaiacol to various substituted alkylphenols is investigated using MoO_3 in supercritical ethanol at a low temperature of 280 °C in inert nitrogen. Furthermore, the preliminary reaction pathways are proposed and the molybdenum active state is explored.

The conversion of guaiacol was carried out in a batch reactor at a temperature, in 260-300 °C, for a desired reaction time, 1-5 h over a MoO_3 catalyst. The product sample was injected neat into a GC-MS and a GC-FID instrument for analysis. Several possible reaction intermediates (phenol, catechol, anisole, 4-ethylphenol and 2-isopropylphenol) were employed as reactants to understand the transformation path of guaiacol. XRD, XPS and Raman spectroscopy were used to characterize the catalyst samples.

Fig. 1 exhibits that substituted phenols such as ethylphenols, iso-propylphenols, butylphenols (tert-, iso-) and tert-amylphenol were obtained as the dominating products with a high combined selectivity of 86% and at a conversion of 99% over MoO_3 under 280 °C for 4 h in supercritical ethanol.

To explore the reaction pathways of guaiacol conversion, control experiments with phenol, anisole and catechol as the starting materials under the same reaction condition were carried out. Catechol gave a complete conversion while phenol and anisole did not. Besides, a similar product distribution compared to guaiacol as the reactant is achieved with catechol. Considering phenol was never found in the guaiacol conversion process, we propose that catechol is formed as the intermediate via demethylation of guaiacol and followed by direct conversion to alkylphenols without the formation of phenol. Subsequently, we employed 4-ethylphenol and 2-isopropylphenol as the reactants and obtained higher alkylphenols as the products. There were no isopropyl, tertiary (iso) butyl or tertiary pentyl derived aliphatic compounds or n-propyl, n-butyl, n-pentyl substituted phenols detected in the liquid and gaseous products of guaiacol conversion. Thus, we speculate that the higher alkylphenols are from consecutive substitution of lower alkylphenols with methyl and ethyl groups produced by ethanol medium, instead of direct substitution with corresponding substituent groups or isomerization with interrelated alkylphenols (Fig. 2).

The Mo (3d) energy region of used MoO_3 is shown in Fig. 3. As can be seen, it contains large amounts of Mo^{5+} (56.4%) with the $3d_{5/2}$ and $3d_{3/2}$ bands located at 230.9 eV and 234.1 eV, respectively. Moreover, molybdenum trioxide (Mo^{6+} state) and molybdenum dioxide (Mo^{4+} state) have been proved to be inactive in control experiments. Therefore, Mo^{5+} was considered to be responsible for the excellent performance of guaiacol conversion.

Copyright © American Institute of Chemical Engineers. All rights reserved.

468094 Enabling Microbial Utilization of Thermally Depolyermized Lignin Monomers

Abstract:

The lignin component of biomass is an attractive source of carbon and energy for the microbial production of fuels and chemicals. This adding of value (*valorization*) to lignin by biological conversion can increase the overall economic viability of biorenewable fuel and chemical production. Thermal depolymerization of biomass through fast pyrolysis shows promise in terms of its ability to recover aromatic monomers from the lignin component of the biomass. Specialized microbes, such as *Pseudomonas putida* KT2440, can utilize many of these aromatic monomers when provided in pure culture [1,2]. However, microbes can be negatively affected by toxic impurities in biomass derived streams, and lignin rich pyrolysis fractions are insoluble in water and therefore inaccessible to microbes in liquid culture medium. Here we addressed these toxicity and insolubility issues by developing and testing an emulsion formula which granted *P. putida* KT2440 access to the phenolic monomers. The optimized emulsion formula contained 18.2 MΩ water, a 70:30 ratio of Tween20® to Span80® surfactant, and the lignin rich pyrolysis fraction. *P. putida* KT2440 was able to use this emulsion formulation as sole carbon source. Additionally, studies with a pure aromatic monomer, p-coumarate, indicated that the emulsion procedure at least partially protects *P. putida* KT2440 from the substrate toxicity of this monomer. To minimize surfactant cost, the surfactant amount was minimized to 7 wt% of the total lignin rich pyrolysis fraction weight. This emulsification method is an important step in enabling the biological valorization of biomass-derived lignin.

Refs: 1. Johnson, C. W.; Beckham, G. T. *Metabolic Engineering* **2015,** 28, 240-247. **2.** Vardon, D. R.; Franden, M. A.; Johnson, C. W.; Karp, E. M.; Guarnieri, M. T.; Linger, J. G.; Salm, M. J.; Strathmann, T. J.; Beckham, G. T. *Energy & Environmental Science* **2015,** 8, (2), 617-628.

Copyright © American Institute of Chemical Engineers. All rights reserved.

472414 Fractionated Lignin Rheology and Structure

Abstract:

Critical technological challenges remain in the development of commercially viable lignin extraction and utilization processes to yield high value commodity chemicals and novel materials such as aromatics, polymers, adhesives, and flavorings. Processing strategies need to be developed for chemical and enzymatic degradation methods, catalytic upgrading, and purification to make it feasible for commercialization. Designing strategies to process multiphase lignin slurries requires understanding the slurry characteristics, especially quantification of the rheological properties. Fractionating heterogeneous lignin structures yield relatively homogeneous sets that offer simpler processing strategies as compared to inherently heterogeneous whole lignin. The objective of this work was to characterize rheological measurements of fractionated lignin and correlate the rheology to fractionated lignin structures. Organosolv and Kraft lignins were fractionated into five fractions using solvents of increasing polarity. Rheology of lignin fractions dissolved in DMSO was characterized with a rheometer programmed to sweep a range of shear rates for non-Newtonian slurries. ^{13}C nuclear magnetic resonance (NMR) analysis was performed to determine changes in structure and functional group contents of each fractional set. Non-Newtonian behavior increased with increasing lignin concentration within any fractionated group. Although structure varied, there were no obvious rheological differences across solvents or lignin types.

Copyright © American Institute of Chemical Engineers. All rights reserved.

473182 Selective Conversion of Biomass-Derived Lignin to Cyclic Hydrocarbons

Abstract:

A potential value-added stream produced in the DMR process from NREL is the â€œnative-likeâ€ and tractable lignin residues isolated from deacetylation and enzymatic hydrolysis. Nobel metal (i.e.Ru, Rh, Pt) and earth abundant metals (i.e. Cu, Fe, Ni, Zn) immobilized on an acidic carbonaceous material act as efficient catalysts in the hydrodeoxygenation of technical lignins under batch conditions. A synergistic interaction between the metal sites and acid groups within the bifunctional and bimetallic catalyst leads to enhanced catalytic activities for the overall transformation of technical lignins to cyclic hydrocarbons as compared to the individual steps catalyzed by the separate catalytic functionalities. In addition, the selectivity of catalysts will also be addressed in our presentation.

Copyright © American Institute of Chemical Engineers. All rights reserved.

470224 The Evolution of Multi-Functional Nanoporous Metal Composite Electrocatalysts

Abstract:

The unique physicochemical properties of nanoscale materials and the continual development of synthesis and fabrication techniques has led to a renaissance in the production of complex catalytic materials with increasing degrees of multi-functionality. Nanoporous metals are an interesting subset of nanoscale electrocatalytic materials as they possess a high surface-to-volume ratio, are intrinsically conductive, are core-shell in nature, and their morphology, composition and compositional gradient are readily tuned through a manipulation of dealloying conditions. Here we present our mechanism for the formation and evolution of nanoporous metal morphology and their application to the catalysis of several elementary electrochemical reactions including the oxygen reduction reaction (ORR) and oxygen evolution reaction (OER)[1]. Nanoporous metals are inherently metastable. In addition to the standard mechanisms of catalyst deactivation and surface area loss including Ostwald ripening and coalescence, surface energy driven coarsening is a new pathway for the loss of electrochemically active surface area in nanoporous materials. With a better understanding of the interplay between nanoporous structure coarsening and transition metal corrosion, we can propose strategies to mitigate coarsening and improve operational catalyst stability. Additional utility of these porous materials lies in the ability to take advantage of the interior volume of the nanoporous metals, incorporating a material that can address the various kinetic limitations of the catalyst. Here we demonstrate the use of hydrophobic, protic ionic liquids on the ORR where the unique properties of the ionic liquid aid the optimization of the interaction between catalyst, reactive intermediates and water.

[1] Snyder, J.; Livi, K. & Erlebacher, J. *Adv. Funct. Mater.* **23,** 5494-5501 (2013).

Copyright © American Institute of Chemical Engineers. All rights reserved.

472499 Effect of Doping on the Activity of Nickelate Oxides Toward Surface Oxygen Exchange and Oxygen Reduction

Abstract:

The first-order Ruddlesden-Popper (RP) series (A_2BO_4) oxides, known as nickelate oxides (e.g. $La_2NiO_{4+Î´}$, LNO), have recently attracted increasing interest due to their good mixed ionic and electrical conductivities suitable for the electrocatalysis. For instance, they have been reported as potential cathode materials for intermediate temperature (600-800 â„ƒ) solid oxide fuel cells (IT-SOFCs)[1]. We have previously demonstrated that the morphology and surface structure of lanthanum nickelate oxide can significantly affect its activity toward the surface oxygen exchange/reduction.[2] We found that LNO nanorods highly terminated by Ni oxide (001) surfaces exhibit superior performance to the traditional nanosphere LNO catalysts. To further enhance the activity of nickelate oxides, in this contribution we explore the effect of doping on the activity toward the surface oxygen exchange/reduction process using density functional theory (DFT) calculations combined with experimental studies. We focus on the substitution of the Ni B-site with Mn, Fe, Co, and Cu transition metals, and the La A-site with Ca, Sr, and Ba to tune the activity.

The surface lattice O diffusion into the bulk to form surface O vacancy is investigated first, and two possible diffusion pathways are proposed. The apical O assisted pathway is found to be more favorable. We find that the surface chemistry of La_2NiO_4 can indeed be tuned via modification of the Ni B-site and La A-site. With the B-site metal changing from Mn, Fe, Co, to Ni, the calculated binding energy of O_2 on the surface O vacancy neighboring the transition metal becomes gradually weaker. Moreover, the barrier for O_2 dissociation becomes linearly higher as the binding energy of O_2 becomes weaker, whereas, the surface O vacancy formation energy becomes lower. A volcano-type relationship between the calculated rates and the binding energies of O_2 is found, suggesting that the binding energy of O_2 might be a good descriptor to screen for nickelate oxides with optimal oxygen exchange activity.

References

1. A. Tarancon, M. Burriel, J. Santiso, S. J. Skinner, J. A. Kilner, J. Mater. Chem. 2010, 20, 3799.

2. X. F. Ma, J. S. A. Carneiro, X. K. Gu, H. Qin, H. L. Xin, K. Sun, E. Nikolla, ACS Catal. 2015, 5, 4013.

Copyright © American Institute of Chemical Engineers. All rights reserved.

462588 Oxyhydroxides As a Platform for Confined Oxygen Electrochemistry

Abstract:

Oxyhydroxides as a Platform for Confined Oxygen Electrochemistry
Authors: Andrew D. Doyle, Michal Bajdich, Aleksandra Vojvodic

As we move towards fossil-free electrification a number of strong renewable energy technologies have emerged. However, some promising candidates (notably solar and wind power) provide only intermittent energy generation, which cannot immediately meet the total demand on a national scale. One proposed solution is to stabilize these renewable sources by coupling to external energy storage, including regenerative fuel cells which convert energy between electricity and the chemical bonds in water, oxygen, and hydrogen.

Unfortunately, researchers have shown that the catalytic efficiency of regenerative fuel cell technology is fundamentally limited by energetic scaling between reaction intermediates [1], [2]. However, our recent results have demonstrated that nanoscale channels may circumvent this scaling, and in the process allow for higher efficiency devices [3].

Additionally, experimental work has shown that doped NiOOH and similarly-structured materials have promise as possible next-generation catalysts for oxygen electrochemistry (water oxidation and oxygen reduction), possibly due to their unique crystal structure [4], [5].

In this work we will present results from density functional theory (DFT) calculations to explore NiOOH as a natural template for sub-nanometer channel formation with particular emphasis on the activity of bulk-like sites towards the oxygen evolution reaction. We will discuss the energetics of binding between hydroxide layers (relative to the surface), the impact of iron doping, and the possibility that subsurface activity

contributes to the measured activity.

[1] I. C. Man, H.-Y. Su, F. Calle-Vallejo, H. a. Hansen, J. I. Martínez, N. G. Inoglu, J. Kitchin, T. F. Jaramillo, J. K. Nørskov, and J. Rossmeisl, "Universality in Oxygen Evolution Electrocatalysis on Oxide Surfaces," *ChemCatChem*, vol. 3, no. 7, pp. 1159–1165, Jul. 2011.

[2] V. Viswanathan, H. A. Hansen, J. Rossmeisl, and J. K. Nørskov, "Universality in Oxygen Reduction Electrocatalysis on Metal Surfaces," *ACS Catal.*, vol. 2, no. 8, pp. 1654–1660, Aug. 2012.

[3] A. D. Doyle, J. H. Montoya, and A. Vojvodic, "Improving Oxygen Electrochemistry through Nanoscopic Confinement," *ChemCatChem*, vol. 94025, 2015.

[4] M. S. Burke, S. Zou, L. J. Enman, J. E. Kellon, C. a Gabor, E. Pledger, and S. W. Boettcher, "Revised Oxygen Evolution Reaction Activity Trends for First-Row Transition-Metal (Oxy)hydroxides in Alkaline Media," 2015.

[5] K. Klingan, F. Ringleb, I. Zaharieva, J. Heidkamp, P. Chernev, D. Gonzalez-Flores, M. Risch, A. Fischer, and H. Dau, "Water oxidation by amorphous cobalt-based oxides: Volume activity and proton transfer to electrolyte bases," *ChemSusChem*, vol. 7, no. 5, pp. 1301–1310, 2014.

Copyright © American Institute of Chemical Engineers. All rights reserved.

455764 Large-Size Graphene Tube Catalysts for Bifuncational Oxygen Reduction and Oxygen Evolution Electrocatalysis in Alkaline Media

Abstract:

The oxygen reduction reaction (ORR) and the oxygen evolution reaction (OER) are one pair of the most technologically important reactions for a variety of electrochemical energy storage and conversion technologies (*e.g.*, fuel cells, metal-air batteries, and water splitting). It is important to design and develop advanced nonprecious-metal catalysts capable of facilitating these sluggish reactions at sufficient rates for practical applications. In our recent efforts,[1-7] cost-effective nitrogen-doped graphene nanocomposites have been prepared *via* the graphitization of nitrogen-containing polymers or compounds, combined with transition metals (Co, Ni or Fe). Among others, a new type of N-doped carbon tubes with large diameters (up to 500 nm) and relatively thin walls (less than 10 layers), which we call graphene tubes (GTs), were prepared via a scalable graphitization process of inexpensive dicyandiamide. Due to the high ratio of diameter-to-wall thickness, much increased surface area was obtained compared to conventional multi-walled carbon nanotubes (MWNTs). An effective strategy for tuning the size of large-diameter nitrogen-doped graphene tubes (GT) from 50 to 200 nm was developed by varying the transition metals used to catalyze the graphitization. This effort demonstrates optimal manipulation of morphology and surface area, thereby providing an effective approach to further improving the performance of nonprecious metal catalysts.

Furthermore, aiming to improve the activity and stability of the state-of-the-art Pt catalysts, the ORR active GT is used as a matrix to disperse Pt nanoparticles in order to build a unique hybrid Pt cathode catalyst. This is the first demonstration of the integration of a nonprecious metal catalyst with Pt catalyst in a nanocomposite. Relative to traditional Pt/C catalyst, much improved activity and stability were achieved. This work provides a new hybrid concept to design and synthesis novel cathode catalysts for fuel cells.

Apart from dramatically enhanced ORR activity, the large-size graphene tubes demonstrated excellent OER activity comparable to state-of-the-art Ir black in alkaline media. Importantly, unlike traditional carbon blacks and 2D reduced graphene oxides, the graphene tubes can be remain stable in harsh electrochemical environments (potentials up to 1.9 V) in alkaline media. This provide a great opportunity to develop bifunctional oxygen catalysts for reversible electrochemical energy applications such as reversible alkaline media and metal-air batteries.

References

1. G. Wu, K.L. More, C.M. Johnston, P. Zelenay, *Science*, 332, 443, 2011.
2. G. Wu, P. Zelenay, *Acc. Chem. Res*., 46, 1878, 2013.
3. G. Wu, N.H. Mack, W. Gao, S. Ma, J. Han, J.K. Baldwin, P. Zelenay. *ACS Nano*. 6, 9764, 2012.
4. Q. Li, P. Xu, W. Gao, S. Ma, R. Cao, J. Cho, H.L. Wang, G Wu, *Adv. Mater*., 26, 1378,2014.
5. Q Li, H Pan, D Higgins, R Cao, G Zhang, H Lv, K Wu, J Cho, G Wu, *Small* 11, 1443, 2015.
6. Q. Li, R. Cao, J. Cho, G. Wu, *Adv. Energy Mater*., 4, 1301415,2014.
7. X Wang, Q Li, H Pan, Y Lin, Y Ke, H Sheng, MT Swihart, G Wu, *Nanoscale* 7, 20290, 2015.

Copyright © American Institute of Chemical Engineers. All rights reserved.

464805 *Atomic Layer Deposition of $NiAl_xO_y$ Catalysts for Electrochemical Oxidation of Water*

Abstract:

The electrolysis of water to form hydrogen gas and oxygen gas has been highlighted as a promising method to store electrical/solar energy in the form of chemical bonds. Of the two half-reactions involved in electrolysis, the oxygen evolution reaction (OER), 4 OH^- â†'Â 2 H_2O +O_2 + 4e^- (in alkaline conditions), suffers from slow kinetics leading to losses in efficiency. Of earth-abundant materials, nickel-iron oxide based catalysts have been found to be among the most active for OER. Recently, it has been shown that the activity of these nickel-iron oxide catalyst can be further improved through the addition of aluminum.[1]

To study the effects of aluminum on the activity of nickel oxide based catalyst, atomic layer deposition (ALD) was used to synthesize ternary $NiAl_xO_y$ films. ALD allows for the deposition of conformal thin films with angstrom-level control over thickness.Â Depositing ternary materials with ALD is shown to provide good compositional control by switching between the individual ALD chemistries. Electrochemical characterization of these synthesized thin films shows that aluminum improves the onset potential for OER and the Tafel slope when compared to nickel oxide films.

Studies of the intrinsic activity of nickel aluminum oxide catalyst are complicated by two factors: (1) nickel oxide films are prone to incidental incorporation of iron from trace impurities found in the electrolyte, (2) aluminum oxide leaches in alkaline conditions leading to changes in composition and porosity of the catalytic films. This work addresses the first issue through additional purification of the electrolyte to remove trace iron. With an iron free environment, nickel aluminum oxide catalyst is shown to outperform nickel oxide catalyst. When $NiAl_xO_y$ films are allowed to age in electrolyte containing trace iron, their activity significantly improves, performing better than as-deposited $NiFe_xO_y$ films. The second complicating factor was further explored by comparing thin and thick films to examine the effect of porosity and its effect on performance. Though the as-deposited $NiAl_xO_y$ films are initially unstable under alkaline conditions, our studies show that the films retain their high activity after aluminum leaching equilibrates with the electrolyte.

1. James B. Gerken, Sarah E. Shaner, Robert C. Massé, Nicholas J. Porubsky, and Shannon S. Stahl. *Energy Environ. Sci.,2014, 7, 2376*

Copyright © American Institute of Chemical Engineers. All rights reserved.

468156 The Role of Superoxide in the Non-Aqueous Oxygen Reduction Reaction in Li-O_2 Batteries

Abstract:

The non-aqueous Li-air (O_2) battery has attracted much research interest because it can potentially far exceed the gravimetric energy storage density than the current rechargeable battery technologies. The oxygen reduction reaction by Li (Li-ORR) is a key electrochemical process at a Li-O_2 cathode. Presently Li-O_2 batteries suffer from poor round-trip efficiency, low current density, and capacity fading,[1-3] To overcome these challenges, it is crucial to have a fundamental understanding of the reactive O_2 species involved in the ORR on different electrodes.

Recent studies have shown that the inclusion of metals in the cathode of Li-O_2 batteries can alter the cathode efficiency and possibly alter the nature of the discharge product.[1,2,4-8] We have performed a series of theoretical calculations of Li-ORR on metals and other materials. Our results show that the discharge potential is controlled by the formation of intermediates on the electrode surface. For instance, the initial O_2 reduction on Au occurs associatively with the formation of molecular lithium superoxide (LiO_2) at an equilibrium potential of ca. 2.0 V.[9] The *intrinsic* activity of Li-ORR on different metals (including Au, Ag, Pt, Pd, Ir, and Ru) forms a volcano-like trend with respect to the adsorption energy of oxygen.[9] Similar to the hydrogen-ORR, Pt and Pd are the most active of these metals, which has been borne out experimentally.10 This can be traced to the competition between the electrode on one hand and Li on the other hand for bonding with the oxygen.

The aprotic nature of typical Li-O_2 cells, however, makes the one-electron reduction of O_2 to the superoxide anion (O_2^-) possible (i.e., without concurrent Li^+ transfer), and allows O_2^- to have appreciable lifetime, particularly on a

relatively inert metal such as Au. O_2^- formation would occur at appreciably higher potential than molecular LiO_2 on many of the metals. DFT calculations, in conjunction with spectro-electrochemical studies,[11,12] therefore suggest that O_2^- formation needs to be taken into account when predicting the observed activity of ORR in Li-O_2 cells. On Au, for instance, O_2^- formation is the key process that corresponds to the discharge of a Li-O_2 cell at low overpotentials and determines how and where the solid discharge products are formed.[12] Our findings contribute to an improved understanding of the ORR in non-aqueous cells and to the further development of Li-O_2 batteries.

References

(1) Abraham, K. M.; Jiang, Z. *J. Electrochem. Soc.* **1996**, *143*, 1.

(2) Débart, A.; Bao, J.; Armstrong, G.; Bruce, P. G. *J. Power Sources* **2007**, *174*, 1177.

(3) Girishkumar, G.; McCloskey, B.; Luntz, A. C.; Swanson, S.; Wilcke, W. *J. Phys. Chem. Lett.* **2010**, *1*, 2193.

(4) Allen, C. J.; Mukerjee, S.; Plichta, E. J.; Hendrickson, M. A.; Abraham, K. M. *J. Phys. Chem. Lett.* **2011**, *2*, 2420.

(5) Lu, Y.-C.; Gasteiger, H. A.; Shao-Horn, Y. *J. Am. Chem. Soc.* **2011**, *133*, 19048.

(6) Peng, Z.; Freunberger, S. A.; Chen, Y.; Bruce, P. G. *Science* **2012**, *337*, 563.

(7) Lu, J.; Lei, Y.; Lau, K. C.; Luo, X. Y.; Du, P.; Wen, J. G.; Assary, R. S.; Das, U.; Miller, D. J.; Elam, J. W.*et al. Nat. Commun.* **2013**, *4*, 2383.

(8) Lu, J.; Jung Lee, Y.; Luo, X.; Chun Lau, K.; Asadi, M.; Wang, H.-H.; Brombosz, S.; Wen, J.; Zhai, D.; Chen, Z.*et al. Nature* **2016**, *529*, 377.

(9) Dathar, G. K. P.; Shelton, W. A.; Xu, Y. *J. Phys. Chem. Lett.* **2012**, *3*, 891.

(10) Lu, Y.-C.; Gallant, B. M.; Kwabi, D. G.; Harding, J. R.; Mitchell, R. R.; Whittingham, M. S.; Shao-Horn, Y. *Energ. Environ. Sci.* **2013**, *6*, 750.

(11) Peng, Z.; Chen, Y.; Bruce, P. G.; Xu, Y. *Angew. Chem. Int. Edit.* **2015**, *54*, 8165.

(12) Zhang, Y.; Zhang, X.; Wang, J.; McKee, W. C.; Xu, Y.; Peng, Z. *J. Phys. Chem. C* **2016**, *120*, 3690.

Copyright © American Institute of Chemical Engineers. All rights reserved.

455930 Experimental and Simulation Study of Ethanol Steam Reforming Reaction for Hydrogen Production in Large Scale Catalytic Membrane Reactor

Abstract:

Experimental and simulation study of ethanol steam reforming reaction for hydrogen production in large scale catalytic membrane reactor

Rui Ma, Bernardo Castro-Dominguez, Ivan P. Mardilovich, Anthony G. Dixon, Yi Hua Ma

Center for Inorganic Membrane Studies, Department of Chemical Engineering, Worcester Polytechnic Institute, 100 Institute Road, Worcester, MA 01609, USA.

Production of hydrogen from ethanol steam reforming (ESR) has attracted more interest due to the massive application of hydrogen and carbon neutral property of bio-ethanol. In hydrogen production, the process intensification concept has been applied to ESR by conducting the reaction in a catalytic membrane reactor (CMR) in order to generate and separate pure hydrogen in the same unit simultaneously. Since the product is being removed during the reaction, the conversion is improved.

In the present study, ESR was carried out in a large scale CMR with a 150 cm^2 membrane surface area. The process was investigated both experimentally and with Computational Fluid Dynamics (CFD) simulation. A defect-free membrane was prepared as Pd/Au/Pd/Au and placed in the center of a double screen annular cage while a nickel-based commercial catalyst (HiFUEL R110) was loaded between the double screens surrounding the membrane as shown in Figure 1, thus the catalyst particles were not in direct contact with the membrane surface. The reaction was performed under different operating conditions: liquid hourly space velocity (LHSV), operating pressure and temperature as well as steam to ethanol (S/E) ratio. Under 300 hours of operation, 100% conversion of ethanol was achieved under all conditions; H_2 with 99.9% purity was produced with the rate of 0.38 g/h at LHSV= 3.77 h^{-1}, P=5 bar, T= 500 ¡C and S/E= 5. It was shown that the process is enhanced by high pressure, high S/E ratio and high temperature. The effect of each condition is discussed in detail in this work.

Simulation models were developed in 1-D and 2-D utilizing Polymath and COMSOL Multiphysics respectively. The kinetics used in the model were validated against experimental outcomes from previous publications using the 1-D model. In the 2-D model, the hydrogen permeation process was simulated as a flux term on both shell side (retentate) and tube side (permeate) applying SievertsÕ law:

-n·Ni=PH2[*PH2shell-PH2tube*]

$$-n \cdot N_i = P_{H_2}\left[\sqrt{p_{H_2}^{shell}} - \sqrt{p_{H_2}^{tube}}\right] \quad (1)$$

in which *PH2shell* $p_{H_2}^{shell}$ and *PH2tube* $p_{H_2}^{tube}$ represent H_2 partial pressure at the retentate side and permeate side respectively.

For the 2-D simulation, the equation of motion and the species continuity equations were solved simultaneously using a finite element method. The effect of the catalytic bed was considered by implementing the Darcy-Forchheimer law to describe the extra resistance to the flow, in which Ergun equation was used to describe the pressure drop in the porous medium. The simulation results were compared with experimental data and showed an accuracy of 84% for 1-D simulation and 91% for 2-D simulation.

Applying the simulation model, the benefits of process intensification were observed by comparing the H_2 production rate from ESR in a traditional packed bed reactor (PBR) and in the CMR. While operating the reaction at higher temperatures and higher pressures, an improvement of up to 122% of H_2 generation in CMR was shown.

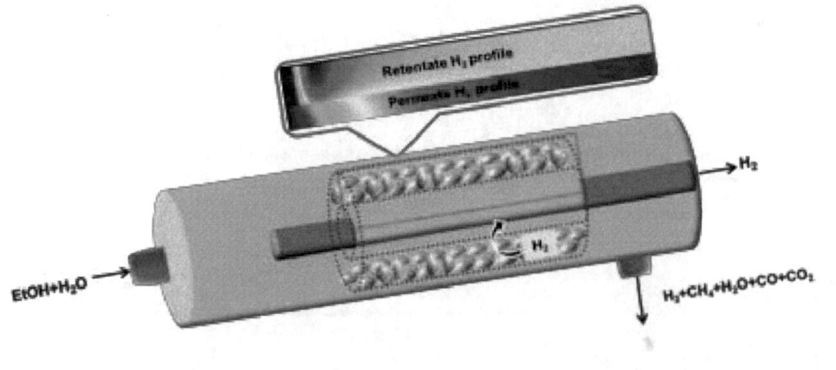

Figure 1. Representation of ESR carried out in a CMR.

Copyright © American Institute of Chemical Engineers. All rights reserved.

463670 Membrane Reactors for Equilibrium-Limited Alcohol and Light Hydrocarbon Synthesis

Abstract:

Reaction and separation are two of the most crucial steps in chemical plants. Performance of each of these two steps can strongly affect the profitability of a given process by affecting its raw materials and energy demand. To improve the overall performance of the process, reaction and separation can be combined into a multifunctional Membrane Reactor (MR). Such reactors typically consist of a catalyst bed in contact with a high-temperature membrane and increase process synergy, reduce by-product formation as well as amount of reactants consumed. Additionally, a considerable reduction in the overall energy costs is to be expected. While there are several reports on the applications of MR for both gas and liquid phase reactions, further studies are still required to move towards commercialization of these MRs.

In this talk, we will present two different examples of high-temperature MRs. The first is an extractive MR that is used for methanol (MeOH) synthesis, and the other a contactor MR that is used for carbon (CO_2) capture and utilization (CCU). Synthesis of MeOH from syngas is a severely equilibrium-limited reaction, meaning that a large fraction of the unreacted syngas feed, must be recycled to make this equilibrium-limited process economically feasible. The ability of MR to increase the per-pass yield makes them an optimal choice for process intensification, particulalry for coal-derived and biomass-derived syngas from air-blown gasifiers that contains considerable amounts of inert N_2.

Power plants are responsible for a large fraction of the man-made carbon-dioxide emissions, and are major contributors to the green house effect. The key challenge here is the dilute CO_2 concentration in the flue-gas, which make the conventional approaches such as direct sequentration technically as well as economically inefficient. Coupling of reaction and separation in a contactor MR, provides the driving force for the separation which moderates the challenge, and can also produce valuable products (e.g., synthetic natural gas, alcohols, etc.) which can be used as fuels to reduce the overall energy demand of the plant, and to improve its efficiency.

During our talk, we will present two MR configurations, and will discuss the influence of reaction conditions and separation efficiency on the final yield of the reaction. The different types of high-temperature membranes that have been investigated will be described, and the influence of membrane characteristcs on overall system performance will be described. Efforts to optimize MR performance both experimentally and numerically will be furher outlined.

Copyright © American Institute of Chemical Engineers. All rights reserved.

464110 Effect of Ni-doping on the Hydrogen Permeation Flux in La5.5WO111.25 Mixed Protonic-Electric Conductor

Abstract:

Ethylene is used as a raw material for producing various polymers. Currently, thermal steam cracking is the conventional process technology for ethylene production. However, the dehydrogenation of ethane dehydrogenation through thermal steam cracking is an expensive and energy intensive process, because the dehydrogenation of ethane is endothermic and is limited by the thermodynamic equilibrium. According to the characteristic of the reaction, the membrane reactor combines the reaction and separation process in a single step. Consequently, the limited-equilibrium reaction is able to shift to the product side. In this study, the performance of a dense ceramic hydrogen permeable membrane reactor for the ethane dehydrogenation, according to the equilibrium reaction C_2H6 $C_2H_4 + H_2$ with a Pt-Sn/Al_2O_3 catalyst was investigated. A dual-phase ceramic membrane with extremely high hydrogen permeation flux flow has been used for the in situ removal of H2 to overcome thermodynamic constraints. The conversion of ethane in membrane reactor could be increased by ~15% at 775 , in comparion with the fixed-bed reactor. Meanwhile, about 75% ethylene selectivity was obtained.

Copyright © American Institute of Chemical Engineers. All rights reserved.

472991 Multi-Scale (Pellet-Reactor Scale) Membrane Reactor Modeling and Simulation: High Temperature and High Pressure Water-Gas Shift Reaction

Abstract:

The objective of this work is to develop a mathematical model and simulation of the Membrane Reactor (MR) in order to carry out the water gas shift (WGS) reaction. MR subsystems integrate reaction and separation in a single unit, with the aim of attaining increased process efficiency, and process compactness. In the MR, a membrane that is selective to hydrogen is used to enhance the WGS reaction's kinetic rate, and to possibly overcome equilibrium conversion limitations imposed by thermodynamics.

The MR system is composed of a reaction zone packed with catalyst pellets, and a permeation zone, where the reaction products permeate. For the reaction zone (classic packed bed reactor), after completing a single-pellet isothermal/non-isothermal steady-state, stand-alone simulations, we couple our model with an isothermal/non-isothermal steady-state packed-bed reactor model to form a hybrid multi-scale reactor model. The catalyst pellet simulation is repeatedly carried out along the reactor bed length (yielding the effectiveness factor along the reactor length for the locally prevailing reaction conditions), and is coupled with a 1-D (axial) reactor model that captures species transport/reaction along the reactor length. Finally, classic packed bed reactor is coupled with a permeation zone to create full membrane reactor (MR) system.

The velocity and species' concentration profiles along the reactor length are captured by momentum/species transport models accounting for convection/reaction /diffusion mechanisms. In the derivation of the model's equations, the Reynolds Transport Theorem was applied separately to each of the domains; the pellet's domain, the reactor's domain and the permeation's domain. The rigorous Maxwell Stefan and dusty gas models are applied to describe mass diffusion fluxes. The effectiveness factors are calculated along the membrane reactor. Finally, performances of the classic packed bed reactor and the membrane reactor are

compared.

Copyright © American Institute of Chemical Engineers. All rights reserved.

469589 Pervaporation-Assisted Furfural Production from Xylose

Abstract:

Furfural produced from the biomass-derived xylose may serve as a platform molecule for the sustainable production of fuels, lubricants, and other specialty chemicals. The acid-catalyzed dehydration of xylose to furfural is burdened by side reactions that form degradation products, known as humins, which reduce the yield of furfural. The formation of humins can be minimized by removal of furfural during its production. In aqueous-phase reactions, this is conventionally achieved by steam stripping, but may also be accomplished by pervaporation, a membrane process. The goal of this study was to demonstrate improvements in pervaporation-assisted furfural production. We previously described a pervaporation-assisted reactor which improved furfural production compared to a membrane-free reactor, but it was limited by two main factors: (1) permeation-induced depletion of reaction-volume which restricted experiments to moderate xylose conversion, and (2) interstage cooling and heating required to accommodate the relatively low membrane temperature. We improved upon the system by incorporating an inlet flow to balance the permeation of material from the reactor. This permitted the study of complete-conversion batch reactions and continuous pervaporation-assisted furfural production. Furthermore, we used cross-linked polymer membranes, which allowed us to eliminate the interstage cooling and heating and to operate the pervaporation-assisted reactor isothermally. This resulted in improved membrane performance as a consequence of increased flux. We then explored the use of metal halides as homogeneous catalysts for pervaporation-assisted reactions, which have combined Lewis- and Brønsted-acid functionality and have been shown previously to produce furfural selectively. While homogeneous catalysts are traditionally thought to be cumbersome because of the need for product/catalyst separation, we show that such a separation is simplified in a pervaporation-assisted reactor because of the lack of permeation of the metal halide catalysts.

Copyright © American Institute of Chemical Engineers. All rights reserved.

472244 Zeolite Membrane Permselective to Water Vapor and Its Application to Dehydration Reactions

Abstract:

Several types of zeolite composite membranes have been developing for separation between iso- and n-isomer, olefin and paraffin, inorganic gas and hydrocarbon, and water and alcohol. Particularly, in the water removal for alcohol-water mixture it has been shown that the composite membranes composing of several kind zeolites such as LTA, ERI OFF (T), CHA, FAU can be synthesized hydrothermally on porous ceramic support. Those have higher water permselectivity, whereby being commercialized in the dehydration process for alcohol-water mixtures.

As an application of zeolite membranes to chemical reactions, various esterification reactions have already been examined mostly in liquid-phase perfectly mixed reactor to aim an equilibrium shift by selective water separation. Differently from such a batch operation, a vapor-phase one, making efficient and continuous production possible, is considered to be advantageous. Dehydration reaction of alcohols, limited thermodynamically, is also industrially important and operated at higher temperatures, producing olefins together with water. This study focuses on vapor-phase esterification and dehydration reaction of alcohol. A membrane reactor incorporated with chabazite (CHA) membrane and solid-acid catalyst is employed for removing water vapor selectively from the reacting mixture. The results in detail will be discussed.

Copyright © American Institute of Chemical Engineers. All rights reserved.

466775 Esterification and Extraction with Supported Liquid Membrane Permeation

Abstract:

Wood pulping processes often generate aqueous effluents containing carboxylic acids; particularly acetic acid. Biorefinery concepts increasingly target integrated material loops including side streams and side stream constituents, and value added products, such as esters, are desired approaches. Esterification of carboxylic acids from pulping effluents may enable production of added value products from pulping effluents. In the present project liquid supported membrane permeation is used to combine esterification reaction with subsequent extraction of the reaction product. Teerachaiyapat et al [1] showed that a hollow fiber membrane module can be used as esterification reactor in principle. The set up used in the present work, originially developed behind the background of heavy metal recycling, has been adapted for reactive membrane separations. The membrane unit offers challenging technological features in isolating value-added products from dilute effluents with a complex matrix. Since cross flow operation is applied, fouling as well as scaling is not a serious issue of process stability limiting significance [2]. The flexibility in membrane composition gives access to the transfer of non-ionogenic as well as ionogenic species. Even highly water soluble solvents, reactants and catalysts may be applied without need of raffinate treatment, since loss of solvent is limited by the physical barrier of the membrane.

The results confirm applicability of the membrane process for the combination of separation by membrane permeation with rigid membranes and esterification of acetic acid with octanol. Via combination of membrane permeation and esterification, the reaction equilibrium related to the aqueous phase may be overcome because of limited solubility of water in the membrane phase. Acetic acid is transferred from the aqueous bulk phase to the rigid interface of the membrane where it converts to the nearly water insoluble ester by releasing water to the aqueous phase while the ester diffuses into the membrane bulk phase. Esterification in the aqueous phase would need the catalyst sulfuric acid which cannot be recycled. In phase transfer catalysis diethylhexylphosphoric acid (DEHPA) or sulfonates (e.g. Dodecylbenzenesulfonic acid) may be used as catalyst in the solvent phase. DEHPA is not catalyzing this reaction sufficiently while dodecylbenzenesulphonic acid (DBSA) is catalyzing the esterification reaction to a sufficient extent. Emulsion formation of DBSA in the solvent phase is avoided by separation from the aqueous phase with the rigid membrane.

[1] T. Teerachaiyapat and P. Ramakul, "Application of hollow fiber supported liquid membrane as a chemical reactor for esterification of lactic acid and ethanol to ethyl lactate," *Korean J. Chem. Eng.*, vol. 33, no. 1, pp. 8–13, 2016.

[2] M. Siebenhofer, H. Noll, and M. Kienberger, "Liquid Membrane Permeation with Supported Liquid Membranes and their Application in Li-ion Battery Recycling," *Sep. Sci. Technol.*, vol. 50, no. 18, pp. 2937 – 2947, 2015.

Copyright © American Institute of Chemical Engineers. All rights reserved.

473103 Investigation of the Degradation of Endocrine Disrupting Compounds in a Membrane Bioreactor

Abstract:

The presence of endocrine disrupting compounds (EDCs) and pharmaceutically active compounds (PHAC) in sewage, industrial, and domestic waters is becoming a major health and environmental concern. The traditional activated sludge process is designed to eliminate solids, organic carbon and nitrogen loadings from wastewaters, but not trace contaminants such as EDCs. Membrane bioreactors (MBR) have become a preferred method for treating municipal and other industrials wastewaters. The objective of this study is to use advanced membrane technologies to develop wastewater treatment processes for removal of EDCs in order to recover and reuse wastewater to augment drinking water supplies. A set of model EDCs including acetaminophen, amoxicillin, atrazine, estrone, and triclosan were selected to study the removal by membrane bioreactor to represent different classes of compounds. Lab-scale MBR operated with real wastewater was tested under various operating conditions, such as retention time and volatile suspended solids concentrations to remove the spiked EDCs. The MBR was operated in two configurations batch and semi-continuous. In both operating modes, the module MBR can reach desired chemical oxygen demand (COD < 30 ppm), total ammonia nitrogen (TAN <5 ppm), and nitrate nitrogen (NO_3-N< 5 ppm) in different retention times. Concentrations of the targeted EDCs and PHAC in the influent, effluent, and mixed liquor suspended solids of the two reactors at different retention time were determined in the period of operation. Different modes of removal mechanisms for these EDCs were identified.

Copyright © American Institute of Chemical Engineers. All rights reserved.

456089 Quantum Chemical Methods for the Simulation of Reactions of Organic Molecules in Zeolites

Abstract:

Zeolites are used extensively to promote the cracking and reforming of hydrocarbons and the synthesis of various organic products. Given the large number of zeolite architectures and compositions, it would be highly desirable to develop a methodology for screening the effects of these variables on the zeolite activity for chosen chemical reactions. This talk will illustrate efforts devoted to this end. It will be shown that a hybrid quantum-mechanical/molecular mechanics (QM/MM) model can be used to assess the enthalpies of species adsorption and activation. Parameters for the MM portion of the model have been optimized and transportable from one zeolite to another. Using the QM/MM model, adsorption and activation enthalpies can be achieve routinely that are within a few kcal/mol of those found experimentally. More recently, a method has been developed for estimating entropies of adsorption and activation that are accurate to within a few cal/mol K of measured values. This method uses a quasi-rigid-rotor harmonic oscillator approximation and takes into account the symmetry factor for a given reaction. We will also show how these methods can be used to explain the experimentally observed effects of zeolite structure on alkane cracking and dehydrogenation. A brief discussion of future advancements will also be presented.

Copyright © American Institute of Chemical Engineers. All rights reserved.

472889 Identification of Possible Mechanistic Pathways during the Electrochemical Reduction of CO_2: Results from Constant Potential DFT Calculations

Abstract:

Identification of Possible Mechanistic Pathways during the Electrochemical Reduction of CO_2: Results from Constant Potential DFT calculations

Jason D. Goodpaster,[a,*] Martin Head-Gordon,[b] Alexis T. Bell[c]

[a]*Department of Chemistry, University of Minnesota Twin Cities, Minneapolis, Minnesota 55455, USA*[b]*Department of Chemistry, University of California at Berkeley, Berkeley, California 94720, USA*

[c]*Department of Chemical and Biomolecular Engineering, University of California at Berkeley, Berkeley, California 94720*

**Corresponding author: jgoodpas@umn.edu*

We investigated the mechanistic pathways by which CO_2 on copper surfaces is electrochemically reduced using periodic Kohn-Sham Density Functional Theory that includes the effects of the electrochemical potential, solvent, and electrolyte. Our analysis revealed that at low-applied potential, C-C bond formation occurs through a CO dimer; however, at high-applied potentials, C-C bond formation occurs through reaction of adsorbed CHO and CO. Additionally, our model allowed us to study proton transfer from water via the Heyrovsky mechanism, which we show to be an important pathway for CO_2 reduction.[1]

The electrode surface and all adsorbates were treated using DFT with the revised Perdew-Burke-Ernzerhof (RPBE) functional. The solvent was treated as a continuum dielectric and the electrolyte was described by a linearized Poisson-Boltzmann model.[2,3] The electrode potential was calculated from the Fermi energy. The Fermi energy can be varied by changing the number of electrons in the simulation cell; therefore, the electrochemical potential was set by calculating the number of electrons the simulation cell required for a specific Fermi energy. Using this model, the free energy profile for the reaction could be calculated as a function of the applied potential. The free energy of activation for each elementary step was then used to determine the rate coefficient for that step. The potential-dependent rate coefficients were then used in a microkinetic model to determine the rate of product formation.

References:

1. J. D. Goodpaster, A. T. Bell, and M. Head-Gordon J. Phys. Chem. Lett. 7, 1471 (2016).
2. K. Letchworth-Weaver and T. A. Arias Phys. Rev. B 86, 075140 (2012).
3. K. Mathew, R. Sundararaman, K. Letchworth-Weaver, T. A. Arias, and R. G. Hennig. J. Chem. Phys. 140, 084106 (2014).

Copyright © American Institute of Chemical Engineers. All rights reserved.

460323 Theoretical Challenges in Characterizing Cu Centers in the Zeolite SSZ-13

Abstract:

Understanding the true nature of active centers under realistic conditions and relate them to experimental observables is one of the major challenges in electronic structure theory today. This is especially true for characterization of heterogeneous catalysis, where a distribution of active centers is present. In this contribution we show our efforts to characterize Cu centers in the zeolite SSZ-13, a material highly efficient in cleaning car exhaust gases. It requires a combination of Density-Functional Theory (DFT), molecular dynamics simulations, thermodynamic modeling and post Hartree-Fock calculations to elucidate the true nature of the Cu-centers under different conditions.

As a first step we study vibrational spectroscopy of NO adsobed to Cu^{II} centers in zeolites using ab-initio molecular dynamics simulation. Already at room temperature the Cu cation changes its coordination regularly, which leads to a complex, multi-peak spectrum for all the studied sites, which are in excellent agreement with experimental measurements. Inspired by these results we investigated the mobilities of Cu in different oxidation states using ab-initio molecular dynamics simulations and found that Cu^{I} changes its coordination environment on a ps-timescale, while Cu^{II} remains fairly stable in its most preferred state. We furthermore tested the impact of this movement on UV-vis spectroscopy of Cu^{I} in a high Si/Al SSZ-13. Comparing theory, a combination of TDDFT and molecular dynamics simulations, and experiment for a defect-free and a normal zeolite sample reveals that upon photoexcitation the Cu-cation switches position with defect Na-atoms at finite temperature, which leads to highly different photoluminescence spectra. We combine these two ideas in studying vibrational spectra of CO adsorbed to Cu^{I} centers and find that again Cu and Na exchange positions. Additionally we develop a model to study the impact of water on the coordination environment and find that already at low temperatures Cu cations are fully solvated.

All these different aspects show the importance of including dynamical behavior of Cu cations in the computational models. Only when this mobility is accounted for correctly, it is possible to arrive at qualitatively and quantitatively correct predictions for different types of spectroscopies in these systems.

Copyright © American Institute of Chemical Engineers. All rights reserved.

472094 Free Energies of Adsorption Calculated through Potential of Mean Force *Ab Initio* Molecular Dynamics in Metal-Exchanged Zeolites

Abstract:

Dynamic adsorption and desorption of small molecules from catalytic active sites, such as exchanged metal ions in zeolites, is important in heterogeneous catalysis. Accurate calculation of free energies of adsorption (ΔG_{ads}) is critical for estimating reaction rates and condition-dependent site speciation [1]. Adsorbate free energies calculated by harmonic approximation (HA) are commonly used in addition to 0K energies calculated by density functional theory (DFT) to estimate ΔG_{ads} at finite temperatures. However this approximation becomes inaccurate when anharmonicity is present in adsorbate-site interactions [2] and when the adsorbate-site complex becomes very mobile [1,3]. In this work we show the inadequacy of HA for chemisorption of molecules on zeolite-supported exchanged metal cations. We demonstrate alternative computational methods that more reliably compute ΔG_{ads}, and develop an empirical model for quick estimation of adsorption entropy.

The reactions studied are the adsorption of small molecules (X) on the zeolite Cu-SSZ-13. We consider several small molecules, such as NH_3 and H_2O, chemically adsorbing to exchanged metal cation active sites (*) as represented by equation * + X - > *X. *Ab initio* molecular dynamics (AIMD) simulations are performed on zeolite-adsorbate complexes to explore their finite temperature behavior. Cu is mobilized within the zeolite after adsorption, resulting in a vibrational spectrum that significantly deviates from that obtained using HA.

The potential of mean force (PMF) method is thus implemented which includes anharmonic entropy in the AIMD simulations. In the PMF simulations one selected reaction coordinate (RC) is constrained. ΔG_{ads} is

then calculated by integration of the average constraint force over the range of RC from the adsorbed state to the state where no interaction exists between the adsorbate and Cu site. Metadynamics is also used to explore the free energy of the system where Gaussian potential hills are added to the free energy surface. After both the product and reactant wells are sampled the added Gaussian hills are used to reconstruct the real free energy landscape. These two methods give the same ΔG_{ads}, which are significantly more exergonic than those obtained by HA. Lastly we propose an empirical correlation between adsorbate gas phase entropies and adsorption entropies backed out from $\Delta G_{ads}=\Delta H_{ads}-T\Delta S_{ads}$. This allows a quick estimation of ΔG_{ads} using only reaction energies from static DFT calculations and gas phase entropies of the adsorbates from statistical mechanical calculations.

[1] C. Paolucci, A. A. Parekh, I. Khurana, J. R. Di Iorio, H. Li, J. D. Albarracin Caballero, A. Shih, T. Anggara, W. N. Delgass, J. T. Miller, F. H. Ribeiro, R. Gounder and W. F. Schneider, *Journal of the American Chemical Society*, (2016) DOI: 10.1021/jacs.6b02651.

[2] G. Piccini, J. Sauer, *Journal of Chemical Theory and Computation*, (2014) 10, 2479-2487.

[3] C. Paolucci, A. A. Verma, S. A. Bates, V. F. Kispersky, J. T. Miller, R. Gounder, W. N. Delgass, F. H. Ribeiro, W. F. Schneider, *Angewandte Chemie International Edition*, (2014) 53, 11828–11833.

Copyright © American Institute of Chemical Engineers. All rights reserved.

469165 Nonadiabatic Effects and Creation of Electronic Excitations during Dissociation on Metal Nanoparticles

Abstract:

The Born-Oppenheimer approximation, which assumes that the electrons are always in their ground state, is widely used in density functional theory (DFT) studies of catalytic processes. These calculations have been quite successful in explaining a large number of catalytic phenomena. However, electronic excitations are easily induced in metal surfaces, due to the lack of a band gap. Here, we show that processes on catalytic surfaces may induce electronic excitations, which are not captured by simulations using the Born-Oppenheimer approximation.

We performed non-adiabatic dynamics using real-time, time-dependent DFT, propagating the nuclei using Ehrenfest dynamics. We study several trajectories of an N_2 molecule interacting with Ru_{13} and Ru_{147} nanoparticles. These simulations show that, during the adsorption and dissociation processes, a significant amount of energy can be dissipated into electronic degrees of freedom, comparable to the amount of energy dissociated into ionic vibrations. The excitations initially take the form of electron-hole pairs, but after tens of fs they relax into hot electrons. This suggests that the electronic structure of catalytic surfaces may, transiently, be quite different from the ground state.

Copyright © American Institute of Chemical Engineers. All rights reserved.

460337 Comparing the Effect of Counter Ions, Solvent Molecules, and Electron Correlation on Homogeneous Reaction Models

Abstract:

Continuum solvation models are widely used for studying solvation in homogeneous solvents and more recently at liquid/solid interfaces. While these models are computationally efficient and straightforward to use, by virtue of being continuum models, they sometimes will not correctly model important explicit intermolecular interactions that play significant roles in solution phase reaction pathways. Using CO_2 reduction by $NaBH_4$ and $NaBH_3OH$ in aqueous solution as test cases, we prescribe a computational workflow involving nudged elastic band methods to obtain nuclear coordinates for explicitly solvated reaction pathways. Using this approach, we will discuss how explicit interactions between solutes, solvent molecules, and counter ions contribute in consort to fundamental hydride transfer processes in solution.

Copyright © American Institute of Chemical Engineers. All rights reserved.

462175 Automated Reaction Mechanism Generation for Heterogeneous Catalisys

Abstract:

Accurate prediction of the reactivity and selectivity of catalytic materials under industrially relevant conditions requires a detailed list of elementary surface reactions, or a microkinetic mechanism. The development of reliable microkinetic mechanisms is time consuming and error prone. For a given catalytic system the relevant elementary reactions must be determined, the rate coefficients must be estimated accurately yet efficiently, and the resulting mechanism must be made thermodynamically consistent. These problems are compounded because the mechanism grows exponentially with the size of the reactants, and because real-world catalysts have multiple crystalline facets with distinct kinetics.

In this talk, we present recently developed software that can generate microkinetic mechanisms for heterogeneous catalysis automatically. For a given catalyst, the user provides the initial conditions (e.g. temperature, pressure, and gas-phase composition), and the computer automatically determines which reactions are important, obtains parameterizations of the thermodynamic properties and rate coefficients, and solves the governing equations – without subsequent human intervention.

The software is based upon the open-source software Reaction Mechanism Generator (RMG) [1,2], which was originally developed for gas-phase combustion and pyrolysis. The new code is referred to as RMG-Cat. By building our software on preexisting code for automatic mechanism generation, we can take advantage of RMG's ability to solve several key problems in computer-generated chemical kinetics. RMG-Cat can automatically:

- Recognize when two or more species in the mechanism are equivalent
- Predict all the possible elementary reactions for each species and pair of species

- Determine which of the possible reactions are actually important
- Estimate all the necessary thermodynamic and kinetic parameters
- Ensure that the mechanism is thermodynamically consistent
- Include flexibility for new reactants on novel materials
- Accomplish all of the above more quickly than a human

An automatic mechanism generator like RMG-Cat requires four main components to function: a unique representation of chemical species; templates and rules to generate reactions; an algorithm to choose which reactions to include in the final mechanism; and methods to estimate all the thermodynamic and kinetic parameters in the mechanism. We describe these four aspects of the software and an example of its application to syngas production from methane on nickel.

Species representation: RMG-Cat represents adsorbates using chemical graph theory, with atoms as nodes and bonds as edges. Efficient graph theory algorithms for (sub)graph isomorphism are used throughout the code for matching reaction templates, identifying functional groups for parameter estimation, and recognizing when species are chemically equivalent.

Generation of Reactions: Chemical reactions are classified according to "reaction families". Each family contains: a template to identify when the reaction can occur, a recipe to specify the changes to be made to the molecular graphs, and rules to determine the reaction rate coefficient.

Reaction Mechanism Expansion: RMG-Cat considers all possible reactions, but to avoid the exponential explosion in mechanism size and to prevent the inclusion of reactions and intermediates that are kinetically irrelevant, it includes only those reactions that have a sufficiently high rate of reaction at the conditions at which the model will be used. The iterative, rate-based expansion algorithm is described in detail in reference [1].

Parameter Estimation: Although the final chemical kinetic mechanism may be sensitive to only a few pathways, which pathways these are is seldom known *a priori*. The number of intermediates and reactions to be considered is vast, so the parameter estimation process must be very fast. In all cases, whether seeking an enthalpy of formation, a binding energy, or a rate of reaction, RMG-Cat first checks a database of known values for an exact match. If no match is found in the database, then the properties must be estimated. For most parameters RMG uses estimation rules based on functional groups, arranged in hierarchical tree-structured databases. This allows efficient selection of the most specific rule for which data are available, falling back to broader generalizations when necessary. The rules are derived from the databases of known parameter values.

Test Case:

As a proof-of-concept, we apply RMG-Cat to the partial oxidation of methane on nickel. Three catalytic processes are considered:

steam reforming:	$CH_4 + H_2O$	$CO + 3\ H_2$
dry reforming:	$CH_4 + CO_2$	$2\ CO + 2\ H_2$
partial oxidation:	$CH_4 + \frac{1}{2}\ O_2$	$CO + 2\ H_2$

Finally, we will present future directions for RMG-Cat and how the code will be expanded to create microkinetic mechanisms for a much broader array of reactants and catalysts.

[1] C.W. Gao, J.W. Allen, W.H. Green, R.H. West. Reaction Mechanism Generator: Automatic construction of chemical kinetic mechanisms. Computer Physics Communications, 203, 212-225, (2016) http://dx.doi.org/10.1016/j.cpc.2016.02.013.

[2] RMG - Reaction Mechanism Generator, open-source software, RMG-Py version 1.0.4 http://reactionmechanismgenerator.github.io

Copyright © American Institute of Chemical Engineers. All rights reserved.

470415 Global Optimization and Hydrogen Adsorption on Pt-Cu Clusters By Using Genetic Algorithm and Density Functional Theory

Abstract:

Bimetallic transition-metal (TM) nanoparticles (NPs) can provide enhanced high activity, selectivity and good long-term stability compared to their pure counterparts in a wide range of technological applications especially for the the development of low-cost and high-performance industrial polymer electrolyte fuel-cell (PEMFC) catalysts. Platinum (Pt) has been used as a superior catalyst for hydrogen oxidation (HOR) and oxygen reduction (ORR) reactions which are the most in a fuel cell. Since the limited amount of Pt reserves and its high cost for fuel cell aplications have led to a strong demand for reducing the use of Pt. For this purpose, Pt alloyed catalysts such as Pt-Co, Pt-Ni, Pt-Cu have been studied to effectively improve both the kinetics of the catalytic reactions and utilization of Pt while reducing the expense of catalyst. In our case, copper (Cu) has been used as a second metal to improve catalytic behaviour due to the one of the most active Pt alloyed catalysts that has been developed over the past decade is the PtCu for use as electrocatalysts in PEMFC [1,2]. Previous studies showed that the particle morphology and elemental compositions of these nanoparticles plays a very important role in how a particular catalytic cycle proceeds. Moreover, the determination of the most stable structure of these nanoparticles is a crucial step for understanding their catalytic properties and designing as a catalyst for a possible industrial application. In the first part of the study, we investigated the morphology of Pt-Cu clusters with the diameter less than 1 nm (about 2-50 atoms) then we analyzed their geometries and; finally, we characterized the electronic structures by advanced computational chemistry methods with the aim to obtain atomistic understanding of the physical and chemical properties of bimetallic TM clusters. The prediction of energetically stable structures is regarded as a highly complex problem of global optimization. One of the most effective global optimization approach for finding the structure of a cluster with the minimum energy level is Density Functional Theory (DFT) which allows a close connection between theory and experiment and often leads to important clues about the geometric, electronic and spectroscopic properties of the systems being studied. Essential part of working under DFT environment is the identifying initial structure of cluster qualitatively because of feeding locally optimized structures rather than randomly arranged structures to DFT, reduces processing time and provides energetically stable clusters at global minima. For this purpose, Genetic Algorithm (GA) coupled with Gupta potential which is based on second moment approximation to tight binding theory were conducted to obtain a set of locally optimized structures for the equally weighted Pt-Cu bimetallic clusters in the range of between 2-50 atoms, all possible compositions of Pt-Cu of 10 atoms, and monometallic Pt and Cu optimized structures of 20 atoms. GA was used to solve a series of attractive and repulsive potential energy equations and find which morphology gives a minimum potential. The energy scaling parameters for Gupta potential which are A, , p and q describing heteronuclear Pt-Cu interactions, were obtained by taking weighted averages of these parameters for pure bulk Pt and Cu metals [3]. The performance of GA prediction was tested for Cu-Au TM clusters which have optimized structures and exact potential energy values are available in literature [4]. Several GA runs agreed well with previous reports and the performance of GA prediction generally was better than the previous studies for different TM systems. Very interestingly and for the first time, the optimized structures and exact potential energy values for PtCu clusters was represented by using semi-empricial Gupta potential in GA environment more quantitatively. Then, the lowest energy structures which were found empirically were reoptimised bu using quantum-mechanical calculations at the ab-initio DFT level (first principles). DFT calculations were carried out using Gaussian 09 quantum chemistry simulation package within a LANL2DZ basis, and B3PW91 exchange correlational functional (Becke Three Parameters Hybrid Function). The structural motifs calculated by Gupta potential, as a function of composition and weighted parameters, were similar with structures obtained by DFT but the distances and places of certain atoms were slightly different. To further the understanding of the structural and energetic properties of the bimetallic Pt-Cu clusters which were optimized in both GA and DFT, we calculated a large number of properties, namely, excess and binding energy, effective coordination number, average weighted bond lengths, chemical order, detachment energy and ionization potential. To obtain a better understanding of the electronic and magnetic properties of PtCu clusters radial distribution functions, energy separation between the lowest unoccupied molecular orbitals (LUMO) and the highest occupied molecular orbitals (HOMO) and their magnetic moment properties have been still studying in details. The excess energy presents negative values while binding energy is positive for all PtCu compositions and these result provide a strong evidence that the optimized structures of binary PtCu is an energetically favourable. The most stable configuration of a binary cluster was found as a Pt_3Cu_7. The structures predicted by combination of GA and DFT are randomly chemically disordered structures as minimum energy states with no apparent segregation of a particular species either to the core or to the surface. In the second part of the study, hydrogen adsorption properties in a particular 10-atom bimetallic PtCu models have been studied to investigate how hydrogen interacts with Pt–Cu clusters for improving our understanding of the factors determining the HOR activity. Firstly, initial bimetallic Pt-Cu clusters which was globally optimized was obtained from DFT results. Then, all possible adsorption sites which are bridge, top and hollow for both Pt and Cu atoms in the optimized clusters was investigated before doing any hydrogen adsorption analysis on a well defined surface. Hydrogen adsorption was carried out in Gaussian 09 program package by using the same procedure which was represented in the optimization part. We investigated the hydrogen adsorption energies on pure Pt_{10} and Cu_{10} clusters in different sites and also for a Pt_3Cu_7 cluster which was most stable composition of 10 atoms. We obtained that adsorption energies takes negative values for all optimized cluster models. More negative adsorption energy indicates the stronger the adsorption since the adsorption energy measures the magnitude of the binding energy of the species to the cluster. Our preliminary results indicates that Pt_3Cu_7 cluster have more negative adsorption energy when it was compared with pure Pt_{10} and Cu_{10} clusters. Based on our calculations, the most favorable adsorption site for Pt_{10} is located at the top site and for Cu_{10} is located at the bridge site. For the case of bimetallic Pt_3Cu_7, the hollow site are more favorable for both Pt and Cu atoms but Pt-hollow site is more energetically favorable according to its adsorption energy calculations. This result indicates that H atoms show a preference for Pt atoms in bimetallic clusters.

References

[1] Mani P, Srivastava R and Strasser P. J. Phys Chem C 2008;112:2770-8

[2] Oezaslan M and Strasser P, J. Power Sources 2011;196;5240-9

[3] L. O. Paz-Borbo´n, , A. Gupta and R. L. Johnston, J. Mater. Chem., 2008, 18, 4154–4164

[4] S. Darby, T. V. Mortimer-Jones, R. L. Johnston, and Christopher Roberts, J. Chem. Phys., Vol. 116, No. 4, 22 January 2002

Copyright © American Institute of Chemical Engineers. All rights reserved.

460747 Strategies for the Conversion of Biomass-Derived Carbohydrates and Fatty Acids to Fuels and Lubricants

Abstract:

Biomass represents a renewable source of carbon that can be used to produce fuels and lubricants in a sustainable manner. We have recently shown that C5 and C6 sugars derived from lignocellulosic biomass and fatty acids derived from a variety of plant oils can be used to produce alcohols, aldehydes, and ketones of different molecular weights. These synthons can be used in various combinations to produce oxygen-containing compounds and hydrocarbons that are suitable for addition as octane boosters for gasoline, cetane enhancers for diesel, and high-quality lubricants. We have found that a wide variety of compounds can be produced using a limited set of catalysts. This talk will illustrate how this portfolio of catalysts and synthons can be used to produce desired compounds with a minimum consumption of hydrogen. For several of the applications presented, life-cycle analyses have been carried out to show that the proposed approach is expected to result in lower GHG emissions.

Copyright © American Institute of Chemical Engineers. All rights reserved.

461828 Materials for Manipulating the Composition of Biomass Pyrolysis Vapors

Abstract:

Upgrading the chemical species generated from the fast pyrolysis of biomass either through post-condensation or *in situ* catalytic conversions has received a significant amount of attention. Recently, conversion strategies have increasingly focused on *ex situ* conversion of the pyrolysis vapors prior to their condensation. Many of these studies have concentrated on model compound studies. While this is a reasonable starting point, the compositional complexity of the pyrolysis vapors coupled with relatively low abundance of any single species creates a significant gap between model compounds and realistic vapor streams, which will likely impact what constitutes an effective catalytic material. Within the real streams important targets are: 1) carbon-efficient conversions; 2) significant reduction in acid content and 3) selective removal of oxygen. Presented will be our development and use of modified $CaCO_3$ to selectively remove organic acid species from real pyrolysis vapors. Specific surface modifications of the $CaCO_3$ were required to achieve the specificity. Also discussed will be the use of MoO_3 with low pressure hydrogen to effectively remove oxygen from real biomass pyrolysis vapor streams.

Copyright © American Institute of Chemical Engineers. All rights reserved.

470325 Coupling Chemical and Biological Catalysis to Produce Biobased Chemicals

Abstract:

The prevalence of â€œlightâ€ (C_1-C_3) hydrocarbons obtained from shale gas motivates a need to find alternative sources of higher-carbon-number molecules suitable for producing commodity and specialty chemicals. Biomass is attractive in this regard because it is composed of monomers with five or more carbons, and it natively contains the oxygenated functionality that is common to high-value chemicals. However, with a carbon-to-oxygen ratio near unity, selective de-functionalization of biomass is a key challenge for producing biobased chemicals. One attractive strategy for obtaining value-added chemicals from biomass uses a combination of chemical and biological catalysis, where biological catalysts are used to produce selectively-functionalized platform molecules that are subsequently upgraded using heterogeneous chemical catalysts. We will present an analysis that compares the moieties that can be accessed by biological catalysis with the conversions that are available using heterogeneous catalysis. Based on this analysis we suggest that biologically-derived platform molecules possessing three distinct functional groups provide the most flexibility for subsequent catalytic upgrading. However, the successful upgrading of multi-functional molecules will require the careful design of catalytically active sites. For example, using transition state theory applied to thermodynamically non-ideal systems, we have analyzed the reaction kinetics for the hydrogenation of both lactic acid and triacetic acid lactone (TAL) by catalysts intercalated with poly(vinyl alcohol). Based on this analysis, it can be shown that the activity of catalytically active sites can be tuned by solvating transition states and abundant surface intermediates using polymer-derived microenvironments. Similarly, bimetallic nanoparticles contain active sites suitable for the selective hydrogenation of unsaturated carbon-carbon bonds in aromatic-containing platform molecules. For one such reaction, the hydrogenation of 4-hydroxycoumarin, the use of a bimetallic PdAu catalyst leads to a doubling of the turnover frequency coupled with an increase in selectivity from 93% to 97% (at 82-86% conversion).

Copyright © American Institute of Chemical Engineers. All rights reserved.

470326 Commodity Chemicals from Biomass: Catalytic Conversion of Biomass into α,ω-Diols

Abstract:

In this presentation we describe a multi-step catalytic approach for conversion of cellulose into 1,6-hexanediol and hemicellulose into 1,5 pentanediol. These α,ω-diols are high-volume (130,000 tons/year), high value ($4,600/ton) commodity chemicals used in the polymer industry. Cellulose is first converted levoglucosan which is then dehydrated into levoglucosenone (LGO) in the condensed phase with dilute acid (5-20 mM acid concentration) using a polar, aprotic solvent. The product selectivity is a function of the water concentration, the solvent type and the cellulose loading. Increasing the water content leads to the production of 5-hydroxymethylfurfural. The LGO is then hydrogenated into dihydrolevoglucosenone, levoglucosanol, and tetrahydropyran-2-methanol (THPM). The THPM then undergoes selective C-O-C hydrogenolysis to produce 1,6-hexanediol using a bifunctional (reducible metal with an oxophilic promoter) catalyst with > 90% selectivity to 1,6 hexanediol. The hemicellulose is converted into furfural which then undergoes a series of three reactions to produce 1,5 pentanediol.

Copyright © American Institute of Chemical Engineers. All rights reserved.

472356 Atomically-Thin Noble Metal Coatings on Early Transition Metal Carbide Nanoparticles: a Versatile Catalytic Platform for Electro- and Thermocatalysis

Abstract:

In an increasingly carbon-constrained world, lignocellulosic biomass, natural gas, water, and carbon dioxide have emerged as attractive options to supply energy, fuels, and chemicals at scale in a cleaner and more sustainable manner. However, the unique chemical makeup of these alternative energy sources has created daunting conversion challenges, requiring the development a new generation of robust, active, and selective catalysts.

Here, new developments in the use of heterometallic early transition metal carbide (TMC) nanoparticles will be described as a novel platform to replace (or at least drastically reduce) noble metal utilization in electro- and thermo-catalytic applications. A new method to synthesize TMCs and core-shell TMC-noble metal structures with exquisite control over composition, size, crystal phase, and purity will be demonstrated. We will show how advanced synthesis techniques can be coupled with rigorous reactivity and characterization studies to uncover unique synergies in nanostructured catalysts. Specifically, we will show that Pt and PtRu monolayers supported on bimetallic titanium tungsten carbide (TiWC) cores are highly CO-tolerant electrocatalysts for hydrogen oxidation and that they achieve an order of magnitude increase in methanol electrooxidation activity over commercial catalysts even after 10,000 cycles with superb resistance to sintering. Collectively, these core-shell catalysts present a novel platform to drastically enhance utilization of noble metals while substantially enhancing their reactivity and stability

Copyright © American Institute of Chemical Engineers. All rights reserved.

469926 Biodiesel Produced from Soybean Oil with the Two Flow Reactors in Series

Abstract:

Viscous vegetable oils containing triglycerides of various fatty acids and free fatty acids are mostly converted to less viscous fatty acid methyl esters (FAME) through the methanol trans-esterification process, using homogeneous catalysts in batch reactors. FAME biofuels produced with homogeneous alkali catalysts through transesterifcation processes contain undesirable substances such as soap and homogeneous catalysts in addition to glycerol and water. Additional refining and purification processes are needed to remove these undesirable substances from FAME biofuels, generating wastewater. In our research, viscous soybean oil is converted to less viscous FAME biofuels with methanol and honeycomb monolithic catalysts in a two-flow reactor assembly in series at moderate temperatures and pressures to develop monolithic catalysts for efficient methanol transesterification processes.

Reaction mixtures from the first catalytic flow reactor consist of biodiesel (fatty methyl esters), glycerol, unreacted soybean oil, and excess methanol. Glycerol is removed from reaction mixtures from the first catalytic flow reactor to feed the second catalytic flow reactor. Reaction mixtures of biodiesel and unreacted soybean oil produced from the first catalytic flow reactor are fed into the second catalytic flow reactor to increase yields of biodiesel by shifting equilibrium conversion of soybean oil in favor of increasing conversion of soybean oil into biodiesel. Active and stable monolithic catalysts are formulated with various active metals and honeycomb monolithic catalyst supports to produce FAME biofuels from soybean oil.

Copyright © American Institute of Chemical Engineers. All rights reserved.

459906 Biomass Conversion to Acrylonitrile Monomer-Precursor for Production of Carbon Fibers

Abstract:

Biomass Conversion to Acrylonitrile Monomer-Precursor for Production of Carbon Fibers

Jadid E. Samad, Lindsey Chatterton, Govedarica Zora and Amit Goyal

Energy and Environment Division

Southern Research, Durham, NC 27712

Over the last few years, the global demand for carbon fibers has been growing at an impressive rate. With newer applications in the horizon and the major consumers (e.g., aerospace, defense, automotive, construction) already gearing towards more use, this trend is projected to continue in the foreseeable future [1]. Of the several types of carbon fibers, Polyacrylonitrile (PAN) based fibers are of particular interest for its specific mechanical properties (250 ksi tensile strength and 25 Msi Youngâ€™s modulus) required for automotive applications.

Acrylonitrile (ACN), a monomer of PAN is commercially produced from fossil sources (e.g., propylene). This process has come under scrutiny due to volatile propylene price and non-sustainable nature of petroleum based feedstock. An alternative production route using biomass derived starting materials e.g., glycerol has gained steam of late. This route enables high selectivity to ACN with an array of valuable byproducts but is limited to purified glycerol obtained from biodiesel production. As a result, although promising, this route has yet to become economically attractive [2].

Southern Research is developing a biomass to ACN (B2ACN) process under a cooperative agreement with the

Department of Energy [3]. The goal of this work is to pursue a cost effective process for ACN production at mild condition using biomass derived sugars as raw materials. From sugar to ACN, the entire B2ACN process consists of multiple catalytic reaction steps. In the first reaction step, several novel multi-functional catalysts have been used to convert sugar to. The produced oxygenates are converted to a gas phase intermediate (acrolein) which will be converted to acrylonitrile in the subsequent step. Based on preliminary results significantly reduced cost (~15-22%) as well as greenhouse gas (GHG) emissions (~37%) compared to typical ACN processes have been forecasted. The work is currently at lab scale development/testing phase. A scaled up integrated bench-scale reactor system will be designed and demonstrated after successful completion of this phase.

In the previous paper, results of catalyst development and optimization from the first reaction (sugars to oxygenates) has been discussed. This paper will primarily focus on the results obtained from the second reaction (oxygenates to acrolein) involving oxygenate conversion. Several catalyst candidates have been identified and tested for optimum catalytic performance in this part of the study.

[1] Mark Holmes, Global carbon fiber market remains on upward trend, Reinforced Plastics, Volume 58, Issue 6, Novemberâ€"December 2014, Pages 38-45

[2] M. Olga Guerrero-PÃ©rez, Miguel A. BaÃ±ares, Metrics of acrylonitrile: From biomass vs. petrochemical route, Catalysis Today, Volume 239, 1 January 2015, Pages 25-30

[3] http://energy.gov/eere/articles/energy-department-announces-11-million-advance-renewable-carbon-fiber-production

Copyright © American Institute of Chemical Engineers. All rights reserved.

470088 Catalytic Conversion of Polyols to Olefins

Abstract:

Catalytic Conversion of Polyols to Olefins

Bryan E. Sharkey and Friederike C. Jentoft*

Department of Chemical Engineering, University of Massachusetts, Amherst, MA 01003, USA

**Corresponding author: fcjentoft@umass.edu*

Hydroxyl groups are abundant in the cellulosic fraction of lignocellulosic biomass. To obtain fuels or a wider variety of chemicals from lignocellulose-derived feedstocks, the hydroxyl groups in these molecules must be removed or replaced by other, more useful functionalities. This conversion is best achieved after hydrolysis of biomass, which delivers a stream consisting of sugars or, after hydrogenation, sugar alcohols. Neighboring OH groups in sugars or sugar alcohols can be converted via deoxydehydration (DODH), that is, the simultaneous removal of one oxygen atom and a molecule of water (Eq. 1). The product is an olefin and thus a versatile chemical intermediate. The oxidation of the reductant, which is needed in stoichiometric quantities, and the elimination of a molecule of water make the reaction thermodynamically feasible.

$R\text{-}CHOH\text{-}CH_2OH + Red$ à $R\text{-}CH{=}CH_2 + RedO + H_2O$ (Eq. 1)

Original catalysts for this conversion were molecular (homogeneous) and include high-oxidation-state rhenium,[1,2] molybdenum[3], or vanadium complexes.[4,5] Heterogeneous catalysts have been reported only in recent years, and exclusively contain rhenium as the active transition metal species.[6,7] These catalysts suffer from leaching and contain expensive rhenium. In the present work, a second generation of rhenium-containing catalysts and a first generation of molybdenum-containing catalysts were developed and characterized.

Perrhenate or molybdate were deposited on various oxidic supports via incipient wetness impregnation or equilibrium adsorption of ammonium salts. The materials were optionally calcined. The conversion of diols was performed in benzene at temperatures of 135 to 200 °C, using various reductants including triphenylphosphine. The catalysts were characterized by various spectroscopic methods.

Catalysts consisting of oxide-supported ReO_x were highly selective in converting alkane diols to the respective terminal olefins. Decanediol could be converted to decenes with yields of more than 80% (at 100% diol conversion). Differences in conversion and yield could in part be explained by adsorption of both the diol and the olefin on the supports. The reaction progressed rapidly and selectively at 150 °C, whereas at higher temperature (175 °C), double bond shift occurred and internal olefins were produced in addition to 1-decene. The reductant efficiency was excellent with the ratio of decenes to triphenylphosphine oxide being close to one. Titania and silica were identified as the best oxide supports with respect to product yields, surpassing oxides of iron, zirconium, and aluminum.

Leach tests indicated some loss of rhenium, in particular for silica-supported catalysts. Consistently, the activity of recovered catalysts was generally lower than that of fresh catalysts, but there was always residual activity even after several uses. Spectroscopic characterization gave some indication that the structure of the supported oxo-rhenium species determines the proclivity for leaching.

A first generation of molybdenum-containing catalysts showed promise although they were less active than rhenium catalysts, and a higher reaction temperature was required to achieve DODH of decanediol to decene. More side products were observed, which included partially deoxygenated molecules and heavy species.

The activity of both oxo-rhenium and oxo-molybdenum species supports the hypothesis that an active catalyst must contain a metal that can form a dioxo unit on the surface and can cycle between two states by changing its oxidation state by ± 2. Deoxydehydration is confirmed as an attractive route to convert 1,2 diols into α-olefins.

References

1. G.K. Cook, M.A. Andrews, *J. Am. Chem. Soc.* **118**, 9448-9449 (1996).
2. J.E. Ziegler, M.J. Zdilla, A.J. Evans, M.M. Abu-Omar, *Inorg. Chem.* **48**, 9998-10000 (2009).
3. J.R. Dethlefsen, D. Lupp, B.-C. Oh, P. Fristrup, *ChemSusChem* **7**, 425-428 (2014).
4. G. Chapman, Jr., K.M. Nicholas, *Chem. Comm.* **49**, 8199-8201 (2013).
5. V. Tirupathi, K.M. Nicholas, *ACS Catalysis*, **6**, 1901-1904 (2016).
6. A.L. Denning, H. Dang, Z. Liu, K.M. Nicholas, F.C. Jentoft, *ChemCatChem* **5**, 3567-3570 (2013).
7. L. Sandbrink, E. Klindtworth, H.-U. Islam, A.M. Beale, *ACS Catal.* **6**, 677–680 (2016).

Copyright © American Institute of Chemical Engineers. All rights reserved.

469634 Catalytic Oligomerization of Ethylene: Experimental Insight into the Effect of Supercritical Reaction Conditions

Abstract:

Ethylene oligomerization is a valuable chemical process for the production of carbon backbones for a variety of commodity chemicals ranging from fuels to surfactants and lubricants. Traditionally, this reaction has been covered extensively in homogeneous catalysis. The development of a heterogeneous catalytic process, however, is of significant academic and industrial relevance as it abides by green chemistry principles and is more easily scalable. However, coke deposition is a major issue for most heterogeneous systems. It has been determined previously that supercritical fluids have a unique ability to solvate coke. In this talk, we will explore the effect of supercritical reaction conditions on ethylene oligomerization. Ni-Hb catalysts were synthesized in the laboratory via a dropwise impregnation technique and used in a single-pass, high pressure packed bed flow reactor. Below its critical point, ethylene was a reactant for oligomerization reactions into C4-C12 olefins. As the partial pressure of ethylene exceeded its critical pressure, ethylene gas was observed to participate both as a reactant and as a solvent. This solvation effect was most evident as the amount of carbon deposition observed on the catalysts at supercritical conditions was found to be considerably lower than that of non-supercritical conditions.

Copyright © American Institute of Chemical Engineers. All rights reserved.

466898 Mesoporous Catalysts for Conversion of 2,3-Butanediol to Butene

Abstract:

2,3-butanediol has significant potential as a platform chemical for production of renewable fuels and chemicals since it can be produced with high productivity via fermentation and provides a C4 building block for further synthesis. Previous work has shown that Cu/ZSM-5 can convert 2,3-butanediol to butene with high selecitivty (~70%) [1]. This study expands on these results by investigating the role of pore size in 2,3-butanediol conversion to butene. Three different types of mesoporous materials (Al-MCM-48, Al-SBA-15 and meso-ZSM-5) were loaded with ~20wt% CuO and tested in the conversion of 2,3-butanediol to butene. Mesoporous aluminosilicate molecular sieves Al-MCM-48 with different SiO_2/Al_2O_3 ratios of 23, 50, 100 and 200 were synthesized at room temperature by using N-Hexadecyltrimethylammonium bromide (CTAB) as the surfactant, tetraethyl orthosilicate (TEOS) and aluminum isopropoxide ($Al(^iProx)_3$) as Si and Al source [2]. Mesoporous aluminosilicate Al-SBA-15 with different SiO_2/Al_2O_3 ratios of 23, 50, 100 and 200 were prepared by the "pH-adjusting" method [3], using triblock copolymer Pluronic 123 surfactant ($EO_{20}PO_{70}EO_{20}$) as the structure-directing agent in a strong acidic HCl solution (2M), TEOS and $Al(^iProx)_3$ as Si and Al source. Meso-ZSM-5 was prepared by using a simple NaOH treatment of parent HZSM-5 zeolite [4]. Al-MCM-48 and Al-SBA-15 materials were characterized by small-angle X-ray scattering (SAXS), nitrogen adsorption and transmission electron microscopy (TEM), which clearly showed that the obtained Al-MCM-48 and Al-SBA-15 materials have highly ordered three-dimensional (3D) *Ia3d* cubic and two-dimensional (2D) *p6mm* hexagonal mesostructure, respectively. The pore (mesopore) size of Al-MCM-48, Al-SBA-15 and meso-ZSM-5 are about 2 nm, 10 nm and 20 nm, respectively, which is determined by the pore size distributions derived from the N_2adsorption-desorption isotherms using BJH method. It should be noted that, apart from the mesopores, micropores with diameter of 0.5 nm were observed on meso-ZSM-5 zeolite, which is typical for the conventional ZSM-5.

Reactions were performed over reduced catalysts at the same reaction conditions (feed rate of 2,3-butanediol of 3.0 mL/h, hydrogen to 2,3-butanediol molar ratio of 5, reaction temperature 250 ^oC). The catalytic results demonstrated that 20%CuO/Al-SBA-15(50) exhibited the highest initial selectivity of butenes, which is 76.6% at 10 min of reaction. In addition, the results showed that the existence of mesopores on the catalysts (Al-MCM-48 and Al-SBA-15 types) could decrease the selectivities of products from cracking reactions, especially $C_3^=$ and $C_5^=$- $C_7^=$, by comparison with the previous report on the catalyst 20%CuO/ZSM-5(280) [1]; meanwhile, the selectivity of $C_8^=$ was found to increase with increasing pore size of the catalyst. However, the activities of the catalysts with MCM-48 and SBA-15 types were decreased dramatically over time. With respect to CuO/meso-ZSM-5(280) catalyst, it can be seen that the catalyst has the composited performance of both CuO/ZSM-5(280) catalyst and mesoporous copper catalysts (CuO/Al-MCM-48 and CuO/Al-SBA-15). Cu/meso-ZSM-5(280) is displayed high activity on the cracking reaction ($C_3^=$, $C_5^=$~$C_7^=$), and the oligomerization ($C_8^=$) as well. Interestingly, Cu/meso-ZSM-5(280) showed an excellent catalytic stability; the selectivity of butenes dropped from 71% to 61% in 670 min of reaction, which is much better than the catalyst with Cu loaded on the conventional ZSM-5(280), the selectivity of butenes dropped from 71% to 50% in 550 min.

[1] Q. Zheng, M.D. Wales, M.G. Heidlage, M. Rezac, H. Wang, S.H. Bossmann, et al., Conversion of 2,3-butanediol to butenes over bifunctional catalysts in a single reactor, J. Catal. 330 (2015) 222–237. doi:10.1016/j.jcat.2015.07.004.

[2] K. Schumacher, C.D.F. Von Hohenesche, K.K. Unger, R. Ulrich, a. Du Chesne, U. Wiesner, et al., The Synthesis of Spherical Mesoporous Molecular Sieves MCM-48 with Heteroatoms Incorporated into the Silica Framework, Adv. Mater. 11 (1999) 1194–1198.

[3] S. Wu, Y. Han, Y.C. Zou, J.W. Song, L. Zhao, Y. Di, et al., Synthesis of Heteroatom Substituted SBA-15 by the "pH-Adjusting" Method, Chem. Mater. 16 (2004) 486–492. doi:10.1021/cm0343857.

[4] J.C. Groen, J. a. Moulijn, J. Perez-Ramirez, Desilication: on the controlled generation of mesoporosity in MFI zeolites, J. Mater. Chem. 16 (2006) 2121. doi:10.1039/b517510k.

Copyright © American Institute of Chemical Engineers. All rights reserved.

465037 Polar Aprotic Solvent Effects on the Catalytic Conversion of Fructose to 5-(hydroxymethyl)Fulfural or Lactic Acid

Abstract:

The efficient production of primary renewable building blocks from biomass-derived carbohydrates must be achieved to develop economically viable biorefineries. Lactic acid (LA) and 5-(Hydroxymethyl)furfural (HMF) are two examples of primary renewable building blocks that can be obtained from fructose as starting feedstock. LA is currently used industrially for the production of green solvents and biodegradable polymers and is extensively used in the food industry. HMF is used as a building block for the manufacturing of solvents, polymers and fuel precursors. The heterogeneous catalytic conversion of fructose in water is limited, in most cases, by low selectivities, especially towards the production of these two platform molecules. Dumesic and coworkers [1] found that polar aprotic organic solvents such as g-valerolactone (GVL) and tetrahydrofural (THF) cause significant increases in reaction rates compared to water in addition to increased product selectivity for Brønsted acid-catalyzed reactions like conversion of xylose into furfural, dehydration of 1,2-propanediol to propanal and for the hydrolysis of cellobiose to glucose. Here we report a similar effect for the Lewis and Brønsted acid-catalyzed reaction of fructose to lactic acid or HMF, respectively. Sn-Beta catalyst was employed as catalyst for the Lewis acid-catalyzed reaction and a sulfonic acid-based CMK-3 catalyst was used for the Brønsted acid-catalyzed reactions. We demonstrate that the employment of aprotic polar solvents, especially GVL, in addition of the aforementioned catalysts is an effective combination for the selective catalytic conversion of fructose to LA or HMF.

The catalysts were characterized using nitrogen adsorption, X-ray Diffraction, X-ray Photoelectron Spectroscopy, Fourier Transform Infrared Spectroscopy and Inductively Coupled Plasma Atomic Emission Spectroscopy. Characterization of the modified materials shows that the desired physical and chemical properties of the catalysts were attained. Catalytic performance studies were performed in 10 mL batch reactors. The solvent effects on the catalytic performances were assessed at 160°C for the conversion of fructose to LA and at 130°C for the conversion of fructose to HMF. The results demonstrate that using aprotic polar solvents, especially GVL, as a reaction solvent significantly increases the fructose conversion turnover frequency to both, LA and HMF.

1. Mellmer, M. A., Sener, C., Gallo, J. M. R., Luterbacher, J. S., Alonso, D. M., and Dumesic, J. A. *Angew. Chem. Int.* Ed. 53, 11872 (2014).

Copyright © American Institute of Chemical Engineers. All rights reserved.

471986 Investigations of Surface Chemistry for Pyridine-Catalyzed CO_2 Reduction on Gallium Phosphide

Abstract:

The surface chemistry of N-containing heteroaromatics, molecular co-catalysts that enable the selective electrochemical reduction of CO_2 to fuels, is discussed. The presented experimental results focus on elucidating the role of the electrode surface in CO_2 reduction reactions that are co-catalyzed by pyridine. For this catalysis, exceptionally high selectivity for reduced fuels has been reported when the reaction occurs at the surface of a GaP photocathode. For this reason, experimental emphasis is placed on assessing preferential adsorption sites and bonding interactions of adsorbates on surfaces of GaP. A surface science approach is used, whereby ultrahigh vacuum conditions facilitate the fabrication of highly characterizable electrode-adsorbate systems. The use of single crystal surfaces permits analysis of surface chemistry independent of complicating factors such as grain boundaries and morphology. Surface-sensitive core-level and vibrational spectroscopy techniques, including high-resolution X-ray photoelectron spectroscopy, synchrotron-based photoemission, and high-resolution electron energy loss spectroscopy, are used to probe adsorbate-substrate and adsorbate-adsorbate interactions for pyridine, water, hydrogen, and carbon dioxide on GaP. Scanning tunneling microscopy (STM) was used to obtain molecular orbital-resolved images of adsorbed molecules. Conclusions from experimental results on these model systems are supported by calculations using density functional theory. This work assists in generating a molecular-level understanding of the heterogeneous processes important to the reaction mechanisms involved in the efficient photoelectrocatalytic generation of carbon-containing fuels with high energy densities.

Two recent highlights from these studies will be discussed. First, we report on a combined investigation through experiment and theory on the surface-bound species on GaP(110) formed upon interaction with water. Experimentally, surface-bound species over 10 orders of magnitude of pressure were spectroscopically identified *in situ* using

synchrotron-based ambient pressure photoelectron spectroscopy (APPES). Ga3d and O1s core-level spectra indicate that the interaction of GaP(110) with H_2O induces formation of a partially dissociated adlayer, characterized by the presence of both Gaâ^'OH and molecular H_2O species. Measurements of the P2p core level indicate formation of a negatively charged hydride that irreversibly bonds to surface P in vacuum. The surface densities of the hydroxide and hydride species increase with increasing pressure (surface coverage) of water. Isobaric measurements at elevated pressures were used to probe the thermal stabilities of adsorbed species as well as the oxidation of surface Ga and P. The observation of stable surface hydride formation induced by interaction with water is especially notable given the critical role of hydride transfer to catalysts and CO_2 during chemical fuel synthesis reactions in aqueous environments. Secondly, we describe a scanning tunneling microscopy (STM) and density functional theory investigation of the orbital-resolved adsorption state defining the dative bonding interaction between a GaP(110) surface and a N-containing heteroaromatic (pyridine).Â By examining the distribution of unoccupied molecular orbitals, we show that STM images can be used to positively identify the sites on pyridine susceptible to nucleophilic attack, consistent with frontier orbital theory. This indicates that STM can be used to explore the local reaction centers of adsorbed ambidentate electrophiles and nucleophiles relevant to artificial photosynthesis, and more broadly to generate critical mechanistic information for various heterogeneous acidâ^'base reactions.

Copyright © American Institute of Chemical Engineers. All rights reserved.

461727 Implications of Surface Reconstructions Impacting Py-Catalyzed CO_2 Reduction on Semiconductor Photoelectrodes

Abstract:

The (photo)electro-catalytic reduction of CO_2 to useful fuels *via* energy from sunlight has received significant attention as a promising route for generating carbon-neutral fuels and value-added chemicals. Experimental evidence demonstrates that pyridine (Py) is an effective co-catalyst during CO_2 reduction over GaP, CdTe, and $CuInS_2$ semiconductor electrodes. Identifying the role of Py during catalytic reduction is essential for optimizing the design of such photocatalytic processes. Recent studies suggest that the semiconductor surfaces themselves can facilitate a heterogeneous reduction mechanism, and for this reason a detailed understanding of the interaction between Py and the various surfaces is required. Surface reconstructions occurring during operation alter the nature of adsorption sites available for interaction with the solution, therefore impacting the performance of the electrode. To address this issue, we employ density functional theory to assess the stability of reconstructed GaP, CdTe, and $CuInS_2$ surfaces, as well as to identify adsorption trends of Py-derived intermediates across sites created by such reconstructions. We also use many-body Greenâ€™s function theory combined with calculations of work functions to determine band edge positions of the explicitly solvated, reconstructed surfaces, which we compare to calculated reduction potentials involved in proposed elementary steps in the overall CO_2 reduction mechanism. This allows us to determine which reduction steps are thermodynamically feasible based on the energy of a photo-excited electron from the conduction band of the semiconductor. These results identify stable intermediate species along the CO_2 reaction path over reconstructed surfaces, thus lending insight into the Py-catalyzed reaction mechanism.

Copyright © American Institute of Chemical Engineers. All rights reserved.

472650 Activity and Stability of Pure and Modified Coooh for Oxygen Evolution Reaction in Alkaline Medium

Abstract:

Improving the efficiency of the oxygen evolution reaction (OER) is critical for a number emerging energy conversion devices such as metal-air batteries, fuel cells and water-splitting cells. High overpotentials are required to increase the reaction kinetics of the OER in both alkaline and acidic conditions, which leads to significant efficiency lost. As a result, significant research efforts have been devoted to the development of active OER catalysts to improve device efficiency. To date, iridium and ruthenium oxides are considered the most active catalysts for OER, however their high cost and low abundance prevent large-scale application. Thus, active, stable, and affordable OER catalysts are highly sought after.

Transition metal oxides are very promising candidates for OER due to their high abundance, high OER activities, and availability in a wide range of structures and chemical compositions that allows optimization for OER activity. Recently, theoretical calculations as well as empirical evidence from Raman and X-ray absorption experiments have identified oxyhydroxides to be the OER active phase for a number of transition metal oxides in alkaline conditions under OER-relevant potentials. As a result, understanding the activity and stability of oxyhydroxides in alkaline conditions is critical and can aid in the design and optimization of OER catalysts. In this study, we wish to understand the correlation between OER activity and the structure and chemical composition of transition metal oxyhydroxides. To achieve this goal, we synthesized cobalt oxyhydroxide (CoOOH) with a well-controlled nanowire morphology and different dopant concentrations. The pure and modified CoOOH nanowires were synthesized directly onto a conductive substrate that offers excellent electrical contact between the nanowire catalysts and improves the area accessible to the electrolyteâ€”both of which are important considerations for practical OER electrodes. Pure and modified CoOOH catalysts were characterized using electron microscopy, X-ray diffraction, Raman spectroscopy, and X-ray photoelectron spectroscopy prior to and after electrochemical testing.

These electrodes showed improved OER activity in 0.1M KOH for CoOOH with nickel incorporation, whereas Mn incorporation did not affect the OER activity of CoOOH. Ni-modified CoOOH exhibited a lower Tafel slope compared to the pure and Mn-modified CoOOH. This suggests a difference in the reaction rate-determining step for Ni-modified CoOOH. Using electrochemical impedance spectroscopy, we determined that Ni modification of CoOOH lowers the charge transfer resistance and enhances the stability of surface species, whereas Mn modification did not alter the charger transfer resistance and decreases the stability of surface species. These results provide new insights regarding the effects of incorporating transition metals in CoOOH on OER activity and illustrate the stabilization of surface species as a strategy for optimization of CoOOH catalysts for the OER in alkaline conditions.

Copyright © American Institute of Chemical Engineers. All rights reserved.

468323 Fe-Ni Core-Shell Hydroxide Nanoparticles As an Active Oxygen Evolution Reaction (OER) Catalyst

Abstract:

Hydrogen production through the electrolysis of water continues to be a key next-generation energy conversion and storage strategy with advantages including minimal greenhouse gas emissions, the lack of fossil fuels, and the potential as an alternative fuel source for fuel cells and reactant for industries such as ammonia production. Recent work by the Boettcher research group and others has demonstrated that iron incorporation into nickel hydroxide based catalysts can significantly enhance the activity of the catalyst for the oxygen evolution reaction (OER) [1-3], which is the half reaction of water electrolysis known to limit the overall process with high overpotential and slow kinetics. With these recent developments in understanding the role of iron incorporation into nickel hydroxide and the importance of iron as the catalytic site for OER within these bimetallic electrocatalysts [1], there is now an opportunity to further develop nanostructured bimetallic iron-nickel hydroxide catalysts for improved OER performance in an alkaline electrochemical environment.

In this presentation, our results on the synthesis, characterization, and electrochemical testing of an iron-nickel core-shell hydroxide nanoparticle catalyst will be presented. Bimetallic iron-nickel nanoparticles were synthesized using a multi-step procedure in water under ambient conditions. When compared to monometallic iron and nickel nanoparticles, the Fe-Ni nanoparticles show enhanced catalytic activity for OER under alkaline conditions (1 M NaOH). The bimetallic nanoparticles demonstrated an improvement in OER overpotential as well as a significant increase in maximum measured current density, as compared to the monometallic iron and nickel nanoparticles. At 1 mA/cm^2, the overpotential for the monometallic iron and nickel nanoparticles was 421 mV and 476 mV, respectively, while the bimetallic Fe-Ni nanoparticles had a greatly reduced overpotential of only 256 mV. At 10 mA/cm^2, bimetallic Fe-Ni nanoparticles had an overpotential of 311 mV. Electron microscopy and elemental analysis results

will be presented with a detailed discussion of unique aspects of the FeNi nanoparticle catalyst. Results suggest that while the nanoparticles are nominally in a core-shell morphology, there is significant migration of iron into the nickel shell as well as incorporation of some phosphorus into the nanoparticle shell, likely originating from the phosphonate-based stabilizer used during nanoparticle synthesis. X-ray photoelectron spectroscopy suggests that the primary phase of nickel is nickel hydroxide, and x-ray absorption spectroscopy characterization suggests that the primary phase of nickel is the more disordered alpha phase of nickel hydroxide. The presence of a small amount of nickel (oxy)hydroxide is also likely present, based on characterization results.

References

[1] M.S. Burke, L.J. Enman, A.S. Batchellor, S.H. Zou, S.W. Boettcher, Oxygen evolution reaction electrocatalysis on transition metal oxides and (oxy)hydroxides: Activity trends and design principles, Chemistry of Materials, 27 (2015) 7549-7558.

[2] M.S. Burke, S.H. Zou, L.J. Enman, J.E. Kellon, C.A. Gabor, E. Pledger, S.W. Boettcher, Revised oxygen evolution reaction activity trends for first-row transition-metal (oxy)hydroxides in alkaline media, Journal of Physical Chemistry Letters, 6 (2015) 3737-3742.

[3] L. Trotochaud, S.L. Young, J.K. Ranney, S.W. Boettcher, Nickel-iron oxyhydroxide oxygen-evolution electrocatalysts: The role of intentional and incidental iron incorporation, J. Am. Chem. Soc., 136 (2014) 6744-6753.

Copyright © American Institute of Chemical Engineers. All rights reserved.

473700 Layered - Hyperthin Electrocatalysts for the Water Splitting Reactions

Abstract:

We report scalable, solution-processing methods for synthesizing two different layered hyperthin nanostructure electrocatalysts for the water splitting reactions. First, we show that 2D FeS_2 disc nanostructures are an efficient and stable hydrogen evolution electrocatalyst. By changing the Fe:S ratio in the precursor solution, we were able to preferentially synthesize either 1D wire or 2D disc nanostructures. The 2D FeS_2 disc structure has the highest electrocatalytic activity for the hydrogen evolution reaction, comparable to platinum in neutral pH conditions. The ability of the FeS_2 nanostructures to generate hydrogen was confirmed by scanning electrochemical microscopy (SECM), and the 2D disc nanostructures were able to generate hydrogen for over 125 h. Second, we report a novel solution-processing synthesis route for creating nickel-iron oxyhydroxides. Nickel-iron oxyhydroxides are non-precious metal electrocatalysts for the oxygen evolution reaction (OER) that have high efficiency in alkaline media. It has been proposed that the layered structure of the oxyhydroxide allows for electron hole transport from the bulk material to the solution interface, which creates very high active site densities. We compared the intrinsic electrocatalytic activity and active site densities for two $Ni_{0.8}$:$Fe_{0.2}$ oxyhydroxides with similar electrochemical surface areas (ECSA) â€" one synthesized via electrochemical conditioning of crystalline $Ni_{0.8}$:$Fe_{0.2}$ oxide and the second synthesized via our new solution-processing route. The overpotential for oxygen evolution on $Ni_{0.8}$:$Fe_{0.2}$ oxyhydroxide decreased from 366 mV on the crystal-derived structure to 268 mV on the new solution-processing structure at 1 mA cm^{-2} when normalized to the ECSA of the catalyst. We measured the active site density of both $Ni_{0.8}$:$Fe_{0.2}$ oxyhydroxides directly using the surface interrogation mode of scanning electrochemical microscopy (SI-SECM).

Copyright © American Institute of Chemical Engineers. All rights reserved.

465065 Stabilizing Electrodeposited Nanoparticle Electrocatalysts on Si MIS Photocathodes for Solar Water Splitting

Abstract:

As the cost of solar energy continues to drop, the major hurdle limiting the widespread use of intermittent renewable solar energy is the lack of efficient and cost-effective energy storage. Photoelectrochemical cells (PECs) offer one solution through their ability to convert solar energy directly into storable solar fuels such as hydrogen. Hydrogen is a versatile energy carrier which can be used for a variety of applications (*e.g.*, chemical precursor for ammonia production, fuel for transportation, or converted back to electricity). The essential component of a PEC is the semiconducting photoelectrode that absorbs sunlight to drive non-spontaneous reactions such as water electrolysis. Commercialization of PEC technology is inhibited by challenges to identify and develop photoelectrodes that have long term stability, high efficiency, and low-cost materials.

One promising approach to achieving stable and efficient PEC water splitting is the composite photoelectrode architecture known as a metal-insulating-semiconductor (MIS) photoelectrode. [1,2,3,4] Within the MIS design, the semiconductor efficiently absorbs sunlight and a thin insulating layer (< 2 nm thick) protects the underlying semiconductor from corrosion and facilitates carrier transport between the semiconductor and metal through quantum mechanical tunneling. A metal catalyst is deposited on top of the insulator, to collect photo-generated carriers and catalyze the water splitting reaction. Through this composite design, the MIS architecture is able to decouple efficiency and stability tradeoff that typically limits conventional photoelectrodes, and has been applied to c-Si and InP photoelectrodes with great success in recent years. [1,2,3,4] In prior work on MIS photoelectrodes, the metal layer is typically deposited by physical vapor deposition in vacuum, which is an expensive and non-scalable method for deposition. If this technology is to become commercially viable, a scalable fabrication process is needed in addition to further improvements in MIS photoelectrode performance. In particular, improved photovoltage and maximized photo-current density generated by MIS photoelectrodes is essential for achieving DOE solar-to-hydrogen conversion

efficiency targets.

In this work, electrodeposition has been explored as a potentially low-cost and scalable means of depositing ultra-low loadings (1 to 20 Î¼g cm^{-2}) of discontinuous Pt nanoparticles onto SiO_2-covered p-Si photoelectrodes. Well-defined, nano-scale catalytic metal islands,[3,5] such as those achieve by electrodeposition, can facilitate high optical transmission of incident light into the semiconductor to improve photo-current density. However, as-deposited MIS photoelectrodes are found to exhibit very poor stability and performance. We report how this issue can be overcome through application of a thin transparent SiO_x overlayer. The SiO_x/Pt nanoparticle structure is similar in nature to core-shell nanoparticles that have previously been used for high temperature heterogeneous catalysis to enhance thermal stability [5], as well as for reducing undesired back-reactions in photoelectrochemical applications [6]. Surprisingly, it is found that the overlayer also results in a substantial improvement in performance in comparison to as-deposited MIS photoelectrodes of identical Pt loading. The combination of electrodeposition with the transparent overlayer thus offers an exciting opportunity for improving both the efficiency and durability of photoelectrochemical energy conversion. After describing a side-by-side comparison of MIS photoelectrodes with and without overlayer, we highlight reasons for the differences in their performance and stability.

References:

[1] Y.W. Chen, P. McIntyre, et al., Nat. Mater., 10 (2011)

[2] M.H. Lee, et al., Angew. Chem. Int. Ed., 51 (2012).

[3] J.C. Lee, A. Bard, J. Ekerdt, et al., Nature Nanotech., 10, 84-90 (2015)

[4] D.V. Esposito, I. Levin, T.P. Moffat, and A.A. Talin. Nature Materials, 12, 562-568 (2013).

[5] T. D. Gould, A. Izar, A. W. Weimer, J. L. Falconer, J. W. Medlin, ACS Catal. 2014, 4, 2714â€"2717.

[6] K. Maeda, K. Teramura, D. Lu, N. Saito, Y. Inoue, K. Domen, Angew. Chemie - Int. Ed. 2006, 45, 7806â€"7809.

Copyright © American Institute of Chemical Engineers. All rights reserved.

468580 Two Distinct Types of Surface-Bound Atomic Hydrogen and Their Role in Dictating Product Selectivity in Artificial Photosynthesis Reactions

Abstract:

Understanding the surface reaction mechanisms, the nature of surface reaction sites, and the chemical physics behind differently-promoted hydrogen transfer stepsÂ in the photocatalytic reduction of CO_2 and H_2O is critical to enable the systematic design and improvement of catalysts for the production of higher hydrocarbons and oxygenates from solar energy.Â Our recent studies indicate that the product distribution, e.g., H_2 and CO vs. CH_4, in the photocatalytic reduction of CO_2 is connected to two distinct types of hydrogen transfer steps â€" electrochemically-coupled H-transfer and excited-electron mediated H-transfer.Â Presence of protonic hydrogen (H^+) on photocatalyst surface correlated with electrochemically-coupled H-transfer steps to preferentially produce H_2. The electrochemical steps for CO production in this case likely follow similar chemical physics. On the contrary, when metallic hydrogen (H^0) is encountered, results suggest that the transfer of H^0 to surface-bound organic fragments occur via excited electron induced vibrational excitation of the uncharged H^0, resulting in production of CH_4. A clear connection was observed between the presence of H^0 on surface and CH_4 evolution, and conversely, the presence of H^+ on surface and H_2 evolution. In our photoreaction experiments, we were able to modify the catalyst selectivity from CH_4 to H_2 evolution by transforming the adsorbed metallic H^0 into protonic H^+ by incorporating more electronegative reaction sites into the photocatalyst surface. Theoretical calculation results suggested a shift from excited electron-mediated H^0 transfer phenomena to electrochemical H-transfer phenomena resulting in H_2 evolution. Understanding the electronic nature of adsorbed hydrogen on photocatalyst surface and the nature of surface reaction sites will potentially lend insights into tuning the overall photocatalytic product distribution, which may lead to the sustainable production of higher

hydrocarbons and oxygenates from CO_2 and H_2O using solar radiation.

Copyright © American Institute of Chemical Engineers. All rights reserved.

469180 Development of Bayesian Error Estimation Density Functionals with Range-Separated Exchange

Abstract:

The Kohn-Sham (KS) approach to density functional theory (DFT), owing to its simplicity and low cost, was instrumental in making computational chemistry a viable field of study. Development of density functionals within the KS ansatz involves approximations to the exchange-correlation (XC) potential between interacting electrons. The XC parameters are typically determined based on constraint satisfaction and compromise in accuracy between various chemical and material properties. Machine learning methods are very promising in this field since they provide a systematic approach to developing the optimal compromise between XC model complexity and accuracy.

Bayesian error estimation functionals (BEEF) employ ridge regression with least squares or robust loss functions to determine the XC model parameters. The Bayesian approach also allows for the estimation of uncertainties in calculated material properties arising from the functional approximation. The BEEF-vdW functional is a semilocal generalized gradient approximation (GGA) that includes vdw-DF2 type nonlocal correlation. The functional demonstrates superior accuracy relative to other GGAâ€™s, especially with respect to surface adsorption chemistry. The meta-GGA implementations, mBEEF, and mBEEF-vdW further improve the description of bulk properties and surface chemistry.

Semilocal functional approximations, however, cannot accurately describe nonlocal phenomena such as bond breaking. Hybrid functionals overcome this limitation by adding a fraction of nonlocal Fock exchange to the XC energy. We are developing a hybrid implementation of BEEF that includes short-range exact exchange. The screening distance for short-range exchange is a nonlinear parameter in the XC energy. To retain the linear model framework for functional fitting and error estimation, therefore, we first scan the two-parameter space of screening distance and fraction of Fock exchange. Parameters that provide a reasonable compromise between material properties, especially the description of metallic density of states and charge separation in diatomic dissociations, are then employed in the construction of the hybrid functional. In addition, the space of available nonlocal correlation approximations is being explored in order to improve the description of noncovalent interactions. We will discuss both the cost as well as performance of the new BEEF hybrid functional with nonlocal correlation in the prediction of barrier heights and surface reaction energies.

Copyright © American Institute of Chemical Engineers. All rights reserved.

456544 Uncertainty Quantification of the Water-Gas Shift Reaction By Pt/CeO_2 Catalyst

Abstract:

An uncertainty quantification framework[1] is applied to the water-gas shift reaction by Pt/CeO_2 catalyst. The PBE functional with Hubbard U parameters 5, 4 and 2 eV and the Heyd-Scuzeria-Ernzerhof (HSE) functional are combined in a factor analysis to obtain a prior gaussian uncertainty with correlations for intermediates and transition states. The overall thermodynamics are corrected using a uniform dirichlet distribution and the correction is spread among uncertainty in the gas molecules. A Bayesian inverse problem is conducted on an interface model[2] and a Pt(111) model using experiments from Ribeiro, et al[3]. The experiments used for the Bayesian inverse are turnover frequency, reaction orders and apparent activation barrier. The Bayesian inverse problem provides a posterior uncertainty on the energies of intermediates and transition states and an evidence value. The evidence value is used to make a model comparison between the interface model and the Pt(111) model to determine which model better explains experiments. On the interface model, multiple catalytic cycles are competitive for the highest turnover frequency. Finally, uncertainties in the degrees of rate control are presented providing insight on rate-controlling steps.

[1] Walker, E.; Terejanu, G. A.; Ammal, S. C.; Heyden, A. *J. Phys. Chem. C.* accepted-

http://pubs.acs.org/doi/abs/10.1021/acs.jpcc.6b01348 (2016).

[2] Aranifard, S.; Ammal, S. C.; Heyden, A. *J. Catal.* 309 (2014) 314-324.

[3] Phatak, A. A.; Koryabkina, N.; Rai, S.; Ratts, J. L.; Ruettinger, W.; Farrauto, R. J.; Blau, G. E.; Delgass, W. N.; Ribeiro, F. H. *Catal. Today* 123 (2007) 224-234.

Copyright © American Institute of Chemical Engineers. All rights reserved.

463599 A Descriptor Based Approach Towards Understanding the Effect of Functionals on Reducible Oxide Reaction Kinetics

Abstract:

Density Functional Theory (DFT) is the workhorse for the computational design of catalysts. The inherent errors within this theory and the effect of the functional on macroscopic observables, such as reaction rates, are well documented for transition metals [1]- [3], but there have been fewer studies focused on reducible metal oxides. Herein, using methanol oxidation on MoO_3(010) as a prototypical example, we demonstrate the effect of the functional on competition between methanol oxidation and redox steps involving oxygen vacancies, within a Mars van Krevelen mechanism. A direct correspondence between reaction kinetics and the degree of binding, a descriptor for DFT functionals, is presented. Adsorption energies of 21 covalently bound intermediates that participate in the catalytic cycle, calculated with the BEEF-vdW functional, (a functional constructed for surface science), follow linear scaling with binding energies computed using a series of GGA and GGA-vdW functionals. The linear relationships have a slope of 1±0.03, and an intercept, termed as the degree of binding (δ), representing an average difference in adsorption strengths between a given XC functional and the reference functional (BEEF-vdW). Unlike adsorption energies, however, the activation energies of elementary steps for all functionals are nearly equal. GGA-vdW functionals have a more negative degree of binding (δ values of -0.50, -0.40 and -0.23 eV for optB88-vdW, optPBE-vdW, and PBE-vdWDF2, respectively) as compared with pure GGA functionals (δ values of -0.02 and 0.39 eV for PBE and RPBE, respectively). Using scaling relationships, the kinetic and thermodynamic barriers with a given functional are expressed in terms of those calculated using BEEF-vdW and the descriptor, δ (MAE of ±0.12 eV). By inputting these relationships into a detailed microkinetic model, it is found that the apparent activation energy of the reaction exhibits a parabolic dependence with δ. Hence, higher activation energies are observed for functionals with extreme values of δ. Rate determining steps and methanol/oxygen orders remain fairly constant as δ decreases from 0.39 eV to -0.23 eV. For δ below -0.23 eV, the rate determining steps shift from those associated with methanol oxidation to steps related to oxidation of vacancies. Thus,

although rates and apparent activation energies show a strong dependence on δ, shifts in the reaction mechanism are observed for only those functionals with very negative values of δ (optB88-vdW and optPBE-vdW). From these observations, we propose that it may be possible to generate very accurate microkinetic models of kinetics on reducible oxides by fitting just one parameter, δ, to very limited series of experimental data.

1. Medford, J. A.; Wellendorff, J; Vojvodic, A.; Studt, F.; Pedersen, F. A.; Jacobsen, K. W.; Bligaard T.; Nørskov, J. K.; *Science.* **2014**, *345*, 197.

2. Liu, B.; Cheng, L.; Curtiss, L.; Greeley, J.; *Surf. Sci.* **2014**, *622*, 51-519.

3. Wellendorff, J.; Silbaugh, T.; Garcia-Pintos, D.; Nørskov, J. K.; Bilgaard, T.; Studt, F.; Campbell, C.T.; *Surf. Sci.* **2015**, *640*, 36-44.

Copyright © American Institute of Chemical Engineers. All rights reserved.

464263 Interpreting Functional-Sensitivity in Catalysis: Exchange-Tuning Vs. DFT+U

Abstract:

Density functional theory is the preferred tool for the mechanistic study of heterogeneous catalysis. First-principles simulations provide valuable insight into barriers and structures associated with short-lived intermediates and transition states. However, DFT predictions of binding energetics and activation energies are highly sensitive to the form of approximate exchange-correlation functional employed. We present recent work in examining relationships between chemical composition of a catalyst (i.e., metal, ligand environment) with corresponding sensitivity of key descriptors of reactivity (e.g., small molecule binding energetics and structure) with respect to xc functional variation. These studies have enabled our development of a descriptor-based approach to identifying the reliability of a DFT prediction obtained, e.g. at reduced cost with a generalized gradient approximation (GGA) versus one obtained with an admixture of Hartree-Fock exchange. It has also allowed us to idenify how to optimally employ the affordable DFT+U approach to correct for underlying GGA shortcomings. We demonstrate these ideas on a number of paradigmatic systems including CO binding at transition metal catalysts for CO_2 reduction and in the properties of spin-crossover solid surfaces and complexes.

Copyright © American Institute of Chemical Engineers. All rights reserved.

462293 Determining the Pd Dopant Effect on the Reducibility of Fe_2O_3: DFT+U or Hybrid Functionals?

Abstract:

The most common method used for modeling highly correlated systems, like oxides, is DFT+U which applies an energy penalty to the DFT energy proportional to a set U value. Choosing a U value has typically been done in order to accurately reproduce a bulk oxidesâ€™ physical parameters, such as band gap or bulk structure. However, there are two problems with this method. First, the application of the U proportional energy penalty can cause the system to converge to a local electronic minimum as opposed to the global electronic minimum unless you perform a computationally expensive scan from a U value of 0 (i.e. no correction) up to the wanted value.[1] Second, accurately reproducing bulk parameters by applying a certain U value does not guarantee that the energetics will also be captured correctly.[2] Kitchin, et al. have proposed a method for determining the correct U value using linear response theory;[2-4] however, this method has thus far been applied to bulk oxide reactions where the U value can be applied consistently throughout the system. The applicability of the linear response method to surface oxide reactions is currently unclear. In this contribution, we study the initial reducibility of the clean and Pd doped Fe_2O_3 (0001) surface using the DFT+U method proposed by Meredig, et al.[1] as well as the hybrid HSE06[5] functional.

The creation of oxygen vacancies in the (0001) surface of clean and Pd doped Fe_2O_3 was studied using both the DFT+U method of Meredig, et al., whereby we scan over numerous U values from low to high, and the HSE06 functional. In order to have a clear picture of the initial reduction of Fe_2O_3, we considered the formation of one, two, and three oxygen vacancies within the surface and first subsurface layer in all possible combinations. For the DFT+U studies, we found that the trend in both the vacancy formation energy and surface energy for the clean Fe_2O_3 (0001) surfaces was highly dependent on the U value, making it difficult to determine the energetic pathway for the reduction of the Fe_2O_3 (0001) surface. Furthermore, when the Fe_2O_3(0001) surface was doped with Pd, the amount of charge transferred between the dopant and oxide surface, as well as the charge distribution itself, significantly changed as U was increased.

These results highlight the difficulty in obtaining accurate models using the DFT+U method for surfaces as there are few if any reliable experimental energies of surface reactions on oxides which can be used to benchmark the U value. Therefore, we have re-examined the reduction of clean and Pd doped Fe_2O_3 (0001) surfaces using the hybrid HSE06 functional in order to obtain accurate trends for oxygen vacancy formation energies in the Fe_2O_3 system. For the clean Fe_2O_3 (0001) surface, our HSE06 results show that vacancies in the subsurface oxygen layer are slightly favored over surface vacancies; however, this process is still highly endothermic. This suggests that after the first surface oxygen vacancy is formed, a subsurface oxygen will diffuse to the surface. The most significant effect of Pd doping on the initial reduction of the Fe_2O_3 (0001) surface is in the creation of the first surface oxygen vacancy which becomes exothermic due to the Pd-O interactions. As creating the first oxygen vacancy in the Fe_2O_3 (0001) surface is the most energy intensive step examined here, the addition of Pd is likely to significantly enhance the reducilibity of the Fe_2O_3 (0001) surface. This work highlights the difficulty in applying the DFT+U method to oxide surface reactions and demonstrates that the HSE06 functional is a useful alternative.

References:

(1) Meredig, B.; Thompson, A.; Hansen, H.A.; Wolverton, C.; van de Walle, A. *Phys. Rev. B* 2010, *82*, 195128.

(2) Curnan, M.T.; Kitchin, J.R. *J. Phys. Chem. C* 2014, *118*, 28776.

(3) Xu, Z.; Joshi, Y.V.; Raman, S.; Kitchin, J.R. *J. Chem. Phys.* 2015, *142*, 144701.

(4) Xu, Z.; Rossmeisl, J.; Kitchin, J.R. *J. Phys. Chem. C* 2015, *119*, 4827.

(5) Krukau, A.V.; Vydrov, O.A.; Izmaylov, A.F.; Scuseria, G.E. *J. Chem. Phys.* 2006, *125*, 224106.

Copyright © American Institute of Chemical Engineers. All rights reserved.

464545 Neural Network and Reaxff Comparison for Au Properties

Abstract:

Atomistic potentials, such as reactive force fields[1] (ReaxFF), have been in use for decades to approximate the complexities of the potential energy surface for various chemical systems. Such potentials are critical tools for studying large atomistic systems or molecular dynamic simulations. This is because the *ab-initio* tools used to train these potentials are often too computationally intensive for such applications. However, such atomistic potentials are necessarily less accurate for certain applications, since they have historically been trained to a limited amount of *ab-initio*data.

Recently, Behler and Parrinello have developed means for training a neural network atomistic potential[2]which can operate on unit-cells without a fixed number of atoms. These machine learning based potentials have already proven effective for system with few chemical systems, but much work is left to be done improving these methods.

In this work we have compared training methods for both ReaxFF and Behlerâ€"Parrinello neural network (BPNN) atomistic potentials on various single-component Au systems. We chose Au for this study because of its diversity of known nanoscale structures. The fact that long-range electronic interactions are screened in Au makes it an appropriate system to model with atomistic potentials that are less suited for long-range interactions such as ReaxFF or a single BPNN.

Both potentials were trained using subsets of 9,972 density functional theory calculations that included 905 bulk, 1,022 surface, and 8,045 cluster configurations. 848 of the total calculations are fully relaxed systems, while the remaining were relaxation steps from the initial guess of each image. Each method was used to generate multiple potentials which were validated against a set of untrained calculations.

The best performing ReaxFF potential was trained from the 848 fully relaxed images and could reliably predict surface and bulk data; however, it was substantially less accurate for molecular clusters of 126 atoms or fewer. Training the ReaxFF potentials to more data often resulted in overfitting and lower accuracy overall.

In contrast, the BPNN performance seemed only to improve with additional data. Thus, the best performing BPNN was trained to all calculations not reserved for validation. This potential performed comparably or better than ReaxFF across all regimes. However, the BPNN potential in this implementation brings significantly higher computational cost. It was also significantly slower to implement using the open-source tools available during this work.

For additional details, see published work:
Boes, Jacob R., Groenenboom, Mitchell C., Keith, John A., Kitchin, John R. "Neural network and ReaxFF comparison for Au properties." Int. J. Quant. Chem., 2 March 2016.

1. A. C. T. van Duin, S. Dasgupta, F. Lorant, W. A. Goddard, J. Phys. Chem. A 2001, 105, 9396.
2. J. Behler, M. Parrinello, Phys. Rev. Lett. 2007, 98, 146401.

Copyright © American Institute of Chemical Engineers. All rights reserved.

474311 Machine Learning Based Interatomic Potentials for Electrocatalysts

Abstract:

The design of next-generation functional materials and the quest to gain insights into dynamical processes at the atomic scale entail efficient and accurate atomistic simulations. The success of these simulations hinges on the accuracy, robustness, and transferability of the force fields (FFs) employed to describe interatomic interactions. One aspect of developing these FFs is to determine the set of variable parameters in their functional forms that closely reproduce available experimental or quantum mechanical data. In general, accurate FFs possess complex functional forms with numerous parameters, and call for sophisticated fitting strategies. Supervised machine learning methods, e.g., genetic algorithms in combination with local minimization methods (e.g., Simplex) provide an efficient way to scan the parameter space for such an optimization problem. Here, we employ such a machine learning approach to develop accurate FF for Ir-O and Ru-O systems; these Pt-group oxides are of immediate technological relevance in electrocatalysis, pH sensors, and photocatalytic splitting of water. For these oxide systems, we employ a Tersoff-like bond order potential to treat bond-directionality, and couple it with an electronegativity-equalization scheme (EEM) to accurately capture dynamic transfer for charges between various atomic species. We parameterize these FFs by training against extensive dataset obtained from first-principles calculations by employing genetic algorithms. Our newly developed Tersoff+EEM potential for these oxides accurately captures (a) structure, energetics, and elastic properties, (b) O adsorption kinetics, and (c) reaction pathways. These representative examples are used to highlight the effectiveness of evolutionary strategies in bridging the atomistic and electronic length scales.

Copyright © American Institute of Chemical Engineers. All rights reserved.

471171 BASC: Applying Bayesian Machine Learning Techniques to the Search for Global Minima on Potential Energy Surfaces

Abstract:

Controlling molecule-surface interactions is key for chemical applications ranging from catalysis to gas sensing. We present a framework -- the Bayesian Active Site Calculator (BASC) -- for rapidly and accurately searching for the global minimum on potential energy surfaces, corresponding to stable adsorbate-surface structures. We formulate the problem as finding the minimum of a low-dimensional objective function. We present a technique using Bayesian inference that enables us to predict converged density functional theory (DFT) potential energies with fewer self-consistent field (SCF) iterations. We then discuss how we apply Bayesian Optimization to search for the global minimum of the objective function. We benchmark our method against Constrained Minima Hopping, a popular method for solving the problem at hand, and present the adsorption sites found by our global optimization method for various simple hydrocarbons on the rutile TiO_2 (110) surface.

Copyright © American Institute of Chemical Engineers. All rights reserved.

471751 Developing a Machine-Learning Molecular Dynamics Approach for Nonadiabatic Surface Chemistry

Abstract:

Transferring energy from solid surfaces to chemical bonds of adsorbed species is a fundamental process in heterogeneous catalysis. To a great extent, the dynamics of molecule-surface interactions driven by thermal energy, i.e., heat, can be well described within the framework of the Born-Oppenheimer approximation where the electronic and nuclear motions can be treated separately. New reaction channels [1] often open up in response to electronic excitations via surface-mediated energy transfer. In this process, the energetic electrons or holes can scatter into surface species and heat up vibrational modes of adsorbates. This process involves not only the electronic ground state but also the excited state, and it is nonadiabatic with respect to the nuclear motions [2].

In the weak coupling limit, the nonadiabatic energy transfer on metal surfaces can be modeled using only the ground-state potential energy surfaces and the nonadiabacity is taken into account with electronic friction contributions in Langevin dynamics by solving Newtonian equations with a thermal fluctuation term [2,3]. In this situation, the electronic friction via partial populating and depopulating of the adsorbate density of states right above the Fermi level is governing the energy transfer. The crucial parameter in this formalism is the electronic friction of an excited charge with nuclear motions of surface adsorbates. However, it is extremely time consuming to compute electron-phonon coupling strength for large systems.

We are developing a machine-learning molecular dynamics approach that uses predicted forces and electronic friction coefficients on each of atoms by 'learning from data'. Machine learning algorithms, such as the artificial neural networks [4], can use past trajectories as training datasets for fast and accurate prediction of forces and electronic friction coefficients, thus allows us to perform statistical analysis of many trajectories. We test the approach for three ultrafast laser induced surface reactions, 1) activation of oxygen from hollow to bridge, 2) CO desorption, and 3) CO2 formation. We chose those systems because of strong interests in understanding nonadiabatic chemistry of CO oxidation on metal surfaces and its simplicity for fundamental studies. We will also discuss electronic factors governing nonadiabatic energy transfer processes. We found that the electronic friction coefficients and the positions of adsorbate resonance orbital energies are two most important factors governing the efficiency and branching ratio of electron-assisted surface reactions. We believe that unraveling the underlying factors governing those properties is the first step for developing predictive models and structure-function relationships in harnessing nonadiabatic surface chemistry for efficient chemical and energy transformations.

[1] M. Bonn, S. Funk, C. Hess, D. N. Denzler, C. Stampfl, M. Scheffler, M. Wolf, and G. Ertl, Science **285**, 1042 (1999).

[2] M. Head Gordon and J. C. Tully, J. Chem. Phys. **103**, 10137 (1995).

[3] M. Brandbyge, P. Hedegård, T. F. Heinz, J. A. Misewich, and D. M. Newns, Phys. Rev. B **52**, 6042 (1995).

[4] J. Behler, Phys. Chem. Chem. Phys. **13**, 17930 (2011).

Copyright © American Institute of Chemical Engineers. All rights reserved.

471969 Glycerol to Hydrocarbon Fuels Via Bifunctional Catalysts

Abstract:

Utilization of byproduct glycerol will improve biodiesel industry in terms of both economics and sustainability. Conversion of glycerol to hydrocarbon (GTH) fuels is one of the promising options, which provides additional renewable energy besides biodiesel, and contributes further to independence from fossil fuels. In the present work, bifunctional catalysts (Pt/H-ZSM-5 and Pd/H-ZSM-5) were selected, prepared, characterized and tested for GTH conversion. Both catalysts convert glycerol to hydrocarbon fuels with a high aromatic hydrocarbon yield. A mechanism via hydrodeoxygenation and aromatization was proposed and discussed on the basis of experimental observations. By varying the space velocity, kinetic data was acquired. The rate constants and activation energies for each sub-reaction in the network were obtained based on the power-law model.

Copyright © American Institute of Chemical Engineers. All rights reserved.

465965 Reaction Engineering of Urea Alcoholysis

Abstract:

Organic carbamates and carbonates are a class of compounds having variety of applications. Dialkyl carbonates (DAC) form an important branch of organic carbonates and are used as precursor to synthesize higher organic carbonates. Synthesis of DAC's by the reaction of urea and alcohol is an attractive 'green' route[1]. In this work we have investigated the reaction engineering aspects of urea alcoholysis to Methyl carbamate.

Urea alcoholysis is a reversible reaction generating alkyl carbamate and ammonia as products and it can be performed non-catalytically or in presence of catalyst. Since ammonia is highly soluble in methanol, it needs to be stripped out from the reaction zone in order to achieve substantial conversion. The first step in Reaction engineering analysis is to finalize the reactor configuration and the decision making factor to decide the reactor configuration here is to decide the mode of stripping. If the reaction is severely equilibrium limited, then it is necessary strip ammonia continuously to drive the reaction forward whereas if the value of equilibrium constant is closer to or more than 1 then equilibrium conversion is nearly complete without the need of reaction and separation taking place simultaneously. Thus the reaction can also be performed with reaction and separation taking place in sequential manner where number of reactor can be arranged in series with separation taking place between each of them removing ammonia generated in previous reactor.

Simple mathematical models were developed and simulations were performed to check the feasibility of reaction and separation happening in simultaneous and sequential manner for different values of equilibrium constant. Our analysis shows that for this reaction sequential reactor-separation is sufficient to get substantial conversion of urea. In sequential reaction-separation the reactor is still operating in the equilibrium dominated region and we would essentially get equilibrium conversions in subsequent reactors. However among the reactor types the equilibrium conversion is achieved faster in PFR than in CSTR. Also

given the advantages of using a tubular reactor with low cost and easy operation the tubular reactor – separator in sequential manner was chosen.

An experimental setup comprising tubular reactor and separator was built. Reaction of urea and methanol yielding Methyl carbamate (MC) and ammonia was selected as a model reaction. Since the reaction is slow the residence time requirements of the reaction demand operation in laminar region. To achieve near plug flow conditions a coiled type tubular reactor was built. Experiments were performed with the tubular reactor configuration for different flow rates and temperatures. A kinetic analysis was performed where data obtained from experiments was fitted against the reactor model and kinetic parameters are reported. Although the urea alcoholysis to alkyl carbamate reaction is reversible, it is not severely equilibrium limited and therefore it is possible to get good conversion in sequential reactor-stripper configurations with appropriate alcohol to urea ratio. Given sufficient residence time, equilibrium conversion can be easily achieved in subsequent reactors. The analysis presented in this work can be readily applied to a general class of reversible reactions with volatile products and the methodology used to develop and design reactors for them.

References

1. Ranade, V.V.; Kelkar, A.A.; Rane, V.H.; Kinage, A.K.; Shingote, S.K.; Roy L.S.; Synthesis of methyl carbamate and dimethyl carbonate (dmc) in presence of stripping with inert gas or superheated vapours and a reactor for the same, US 20150315134A1.

Copyright © American Institute of Chemical Engineers. All rights reserved.

463360 Gas-Phase Catalysis for Pyrolyzing Acetic Acid

Abstract:

We compare our quantum-chemically predicted rate constants with new experimental measurements for pyrolysis of acetic acid to ketene. Previously, Padmanabhan et al. (1968) measured first-order kinetics with respect to acetic acid concentration, reporting different rate constants for overall conversion and ketene formation rate. Grottoli et al. (1991) combined experiments with reactor modeling, fitting the data to rate expressions for ketene formation, decarboxylation, and ketene decomposition.

In the present work, we measure conversion of acetic acid in a tubular flow reactor using the homogeneous catalyst triethyl phosphate that was studied by Padmanabhan et al. and by Grottoli et al. Exhaust gases are analyzed by molecular-beam mass spectrometry, allowing us to infer overall rate coefficients using plug-flow and CSTR models. We also compute gas-phase thermochemistry of species and transition states and kinetics for thermal and and phosphoric-acid-catalyzed conversion of acetic acid to ketene and water. Structures, vibrational frequencies, and energies are computed by quantum chemistry with Gaussian 09. Rate coefficients are computed with the NIST ChemRate code using the Gaussian output files. The results agree well, supporting our new process-chemistry model.

Copyright © American Institute of Chemical Engineers. All rights reserved.

464332 Hybrid Reaction Pathways to Renewable Diol Monomers

Abstract:

Transformation of lignocellulosic biomass to valuable chemicals and materials requires precise control to targeted molecule products. In this work, a hybrid fermentation/catalytic approach identifies reaction pathways to key diol monomers useful to the synthesis of polyester materials. Three fermentation diacid products were fermented from glucose, and subsequent catalytic reduction achieved sufficient yield and product purity for process efficacy. Supported metal catalysts were evaluated for various site/support combinations, with the best catalysts achieving over 80% yield to the targeted diol product. Novel diol monomers were shown to exhibit new physical properties in combination with biomass-derived diacids in new polyester polymers, thus providing unique biodegradable materials.

Copyright © American Institute of Chemical Engineers. All rights reserved.

460299 Continuous Piancatelli Rearrangement of Furfuryl Alcohol: Kinetics and Reactor Modelling

Abstract:

In the last few years the focus of research has been in the use of lignocellulosic biomass as an alternative feedstock for fuels, pharmaceuticals and platform chemicals. Although there has been much interest in this field of research, the subsequent valorization steps after the conversion of cellulose to various platform molecules have to be explored thoroughly. In most of these reactions in aqueous mixtures, humins often emerge as undesired polymeric by-products. In this study the Piancatelli Rearrangement to 4-hydroxycyclopent-2-enone (4HCP) starting from furfuryl alcohol as a commonly used platform molecule was investigated.

The reaction can be carried out continuously in a capillary microreactor, utilizing an acidic water-toluene two-phase system.[1,2] The major challenge of this reaction system is the formation of the undesired polymers (humins), or the high dilution of the substrate to avoid this, thus resulting in an expensive workup. Both of these issues were investigated by kinetic experiments in a tubular reactor and associated simulations.

The kinetic studies showed that the product formation can be assumed as first order, and the by-product formation as higher order, in the substrate. A corresponding system analysis revealed that the selectivity towards the product can be enhanced by diluting the substrate. Simulations and corresponding experiments could demonstrate that the use of a reactor system with better back-mixing, like a Continuously Ideally Stirred Tank Reactor (CSTR) or a fed-batch reactor, instead of a plug flow tubular reactor (PFTR) improves the selectivity towards the desired 4HCP significantly. Additionally, through the switch from a tubular to a better mixed reactor, the high dilution of the substrate in water can be circumvented, resulting in a significant cheaper downstream processing. This shows that although the current trend in flow chemistry is to apply tubular reactors, and since the choice of the reactor is a valuable tool to improve the performance of a reaction, it should be attended at an early stage.

Finally, using the example of the Piancatelli Rearrangement of furfuryl alcohol to 4HCP we demonstrated that it is possible to enhance the selectivity towards the desired product by a simple utilization of the reaction's kinetics. Simultaneously, the solvent demand was reduced and the work-up simplified by decreasing the amount of polymeric humin-like by-product. Literature

[1] K. Ulbrich, P. Kreitmeier, O. Reiser, *Synlett* **2010**, *13*, 2037–2040.

[2] M. Nettekoven, B. Püllmann, R. E. Martin, D. Wechsler, *Tetrahedron Lett.* **2012**, *53*, 1363–1366.

Copyright © American Institute of Chemical Engineers. All rights reserved.

469570 Catalyst Structure and C-O Activation during Fischer-Tropsch Synthesis

Abstract:

Catalyst structure and C-O activation during Fischer-Tropsch Synthesis

G. T. Kasun Kalhara Gunasooriya,† Arghya Banarjee,§ Mark Saeys†

†Laboratory for Chemical Technology, Ghent University, Ghent, Belgium

§Department of Chemical and Biomolecular Engineering, National University of Singapore, 4 Engineering Drive 4, 117576 Singapore

Fischer-Tropsch synthesis transforms synthesis gas, a mixture of CO and H_2, to long-chain hydrocarbons and water. In recent years Fischer-Tropsch synthesis has surged as an attractive route to convert natural gas, coal and biomass to clean transportation fuels. Supported cobalt catalysts are often preferred due to their high activity, selectivity towards long-chain hydrocarbons, and low CO_2 selectivity. The detailed reaction mechanism, a complex combination of C-C bond formation, C-O scission and hydrogenation steps, remains intensely debated.[1, 2] Moreover, the structure and coverage of the catalyst under reaction conditions differs dramatically from the ideal clean surface.[3,4] To begin to elucidate catalyst activity and selectivity, computational catalysis requires realistic surface models and account for coverage effects under reaction conditions.

To elucidate the structure of the catalyst, the effect of carbon and CO adsorption on the stability of various surface facets was considered. Carbon atoms adsorb strongly at B5 sites (with a stability of -20 kJ/mol and more) up to a coverage of 50%. The unique stability of square-planar carbon species can be attributed to sigma-aromaticity of the "Co_4C" unit.[5] The presence of square-planar carbon oxidizes the step atoms and enhances CO adsorption. The synergistic adsorption of carbon and CO make it thermodynamically favorable to break up terraces to create step sites and nano-islands.[6]

C-O activation is a crucial step FTS and several pathways have been proposed. In addition to direct CO dissociation, CO can be activated by insertion into growing hydrocarbon chains (CO insertion mechanism [2]), or through hydrogenation (hydrogen-assisted CO activation). Invariably, very high barriers are calculated for both direct C-O dissociation and for hydrogenation at the O-atom in CO* and RCO*, while hydrogenation at the C-atom leads to oxygenates. We show that the formation of OH* groups from O* is quasi-equilibrated, and the OH coverage is significant. OH* groups hydrogenate the O-atom with barriers below 100 kJ/mol, respectively, opening a novel pathway to activate C-O bonds.[7] To further elucidate the reaction mechanism, a micro-kinetic model accounting for coverage and reaction conditions is constructed. The role of OH groups as hydrogenating species is likely general and involved in several oxygenate transformation reactions.

References

[1] van Santen, R.A., Markvoort, A.J., Filot, I.A.W., Ghouri, M.M., Hensen, E.J.M. Phys. Chem. Chem. Phys., **2013**, 15, 17038

[2] Zhuo, M., Tan, K.F., Borgna, A., Saeys, M., *J. Phys. Chem. C* **2009,** 113, 8357

[3] Gunasooriya, G.T.K.K.; van Bavel, A.P.; Kuipers, H.P.C.E.; Saeys, M. *Surf. Sci.* 2015, 642, L6

[4] Wilson, J.; De Groot, C. *J. Phys. Chem.* 1995, 99, 7860

[5] Nandula, A.; Trinh, Q. T.; Saeys, M.; Alexandrova, A. N. Angew. Chem. Int. Ed. 2015, 54, 5312

[6] Banerjee, A.; van Bavel, A. P.; Kuipers, H. P. C. E.; Saeys, M. *ACS Catal.* 2015, 5, 4756

[7] Gunasooriya, G.T.K.K., van Bavel, A.P., Kuipers, H.P.C.E, Saeys, M., *ACS Catal.*2016, 6, 3660

Copyright © American Institute of Chemical Engineers. All rights reserved.

464352 Reaction Engineering for Modified Furans

Abstract:

Transformation of biomass to useful chemicals and biomaterials requires explicit selection of reaction pathways from biopolymer, to monomer, and eventually a sequence of steps to chemical products. In this work, we identify a new reaction sequence to convert biomass-derived glucose to furans, followed by furan functionalization to produce an array of derivatized furans. Three furan upgrading reactions are combined in a variety of sequences, and the resulting products are characterized providing a detailed understanding of furan structure-property relationships. Reaction steps are characterized for each catalytic method, and the ultimate process chemistry is evaluated for overall performance with regard to industrial applications.

Copyright © American Institute of Chemical Engineers. All rights reserved.

468398 C_2H_6 Dehydrogenation on Pt-Based Catalysts: Fundamental Mechanistic Studies

Abstract:

Hypersonic aircrafts require substantial cooling of engine and other vehicle components due to the large amount of heat produced during the flight. Using metal-catalyzed endothermic reactions of fuel, such as dehydrogenation and cracking, is a plausible strategy to remove the excess heat produced. However, coke formation on the catalyst causes rapid deactivation and clogging of the system. In this study, a combined experimental and computational chemistry based modeling approach is implemented to investigate stability, activity and selectivity of supported Pt and Pt-alloy catalysts for ethane (a small alkane) dehydrogenation with the goal to minimize coke formation. Experimental reaction kinetics data and DFT derived parameters are combined to develop a mean-field microkinetic model for understanding the reaction mechanism and for helping develop improved alloy catalysts for this reaction.

Copyright © American Institute of Chemical Engineers. All rights reserved.

471905 Microkinetic Analysis of γ-Valerolactone Ring Opening and Decarboxylation over Solid Acids

Abstract:

We recently reported substantial variation in apparent turnover frequencies and kinetic parameters during the catalytic decarboxylation of γ-valerolactone over a series of solid acids including MFI, SiO_2-Al_2O_3, γ-Al_2O_3, and supported phosphotungstic acid. In aluminosilicates, Brønsted sites associated with framework aluminum contribute the majority of decarboxylation activity. Relative to bridging hydroxyls, coordinatively unsaturated aluminum sites are substantially less active, and they do not contribute significantly to butene production rates in materials having both framework and extraframework aluminum. Through comparison of aluminosilicates with supported heteropolyacids, decarboxylation barriers were found to scale with the deprotonation energy of Brønsted acid sites; however, the apparent decarboxylation turnover frequency of a Brønsted site did not necessarily correlate with its deprotonation energy. Here, we attempt to resolve observed differences in apparent phenomena and parameters through microkinetic analysis of the elementary reactions governing valerolactone decarboxylation over Brønsted sites.

On acidic catalyts, valerolactone decarboxylation is likely preceded by ring opening, which forms surface bound alkene-acids that can either desorb as stable products or cleave to form 1-butene and carbon dioxide. Analogous to various alkene chemistries, we cast the reaction pathway as proceeding through a series of alkoxide intermediates. The resulting kinetic model is parameterized at an elementary level using insights from well-studied alkene chemistries and reconciled against experimental data. With a minimal set of variable parameters, we demonstrate good agreement between an elementary-level description of valerolactone decarboxylation and measured production rates in experimental reactors. In a clear nod to Jim and his graduate catalysis course, we employ one of his favorite toolkits to provide a useful and practical framework for interpreting macroscopic kinetic observations at an elementary level.

Copyright © American Institute of Chemical Engineers. All rights reserved.

460328 Solvent Effects in the Catalytic Conversion of Fructose to Lactic Acid

Abstract:

Lactic acid (LA) is an industrial bulk chemical that is catalogued as a platform molecule from which green solvents, biodegradable polymers and others products may be manufactured. The catalytic production of LA from fructose in water is limited by low LA selectivity and low catalyst activity. Dumesic and coworkers [1] found that polar aprotic organic solvents such as g-valerolactone (GVL) cause significant increases in reaction rates compared to water in addition to increased product selectivity for Brønsted acid-catalyzed reactions of biomass derived feedstocks. Inspired by Dumesic's work, here we report a similar effect for the Lewis acid-catalyzed conversion of fructose to lactic acid. We demonstrate that the combination of a Sn-Beta zeolite with mainly Lewis acidity prepared using a post-synthetic procedure and a solution of GVL and water as solvent is an effective catalytic process for the production of lactic acid from fructose. We also show that the incorporation of trace amounts of K_2CO_3 in the GVL and water solvent causes an additional improvement on the LA selectivity and yield. The effect of the solvent, tin content, and reaction temperature on the catalytic performance was assessed for the conversion of fructose to LA. Our results demonstrate that using GVL and water as a reaction solvent significantly increases the fructose conversion turnover frequency and LA production rate while decreasing the formation of degradation products as compared to the results obtained when only water is employed as solvent. The catalysts were characterized using nitrogen adsorption, X-ray Diffraction, X-ray Photoelectron Spectroscopy, Fourier Transform Infrared Spectroscopy and Inductively Coupled Plasma Atomic Emission Spectroscopy. Characterization of the modified materials shows that a highly crystalline Sn-Beta zeolite catalyst was synthesized with a surface area similar to that of the parent material. The Al content was as low as 236 ppm and the Sn content was varied from 0.4 to 4.1%.

1. Mellmer, M. A., Sener, C., Gallo, J. M. R., Luterbacher, J. S., Alonso, D. M., and Dumesic, J. A. *Angew. Chem. Int.* Ed. 53, 11872 (2014).

Copyright © American Institute of Chemical Engineers. All rights reserved.

467376 Stabilization with Formaldehyde Facilitates the High-Yield Production of Monomers from Lignin during Integrated Biomass Depolymerization

Abstract:

Practical, high-yield lignin depolymerization methods could greatly increase biorefinery productivity and profitability. However, development of these methods is limited by the presence of inter-unit carbon-carbon bonds within native lignin, and further by the formation of these linkages during lignin extraction. We show that adding formaldehyde during biomass pretreatment produces a soluble lignin fraction that can be converted to guaiacyl and syringyl monomers at near-theoretical yields during subsequent hydrogenolysis (48-78% of native lignin). We achieved the highest yields using a high-syringyl transgenic poplar with low levels of carbon-carbon linked units. These yields were 3-7 times those obtained without formaldehyde, which prevented lignin condensation primarily by forming 1,3-dioxane structures with lignin side-chain hydroxyl groups. By depolymerizing cellulose, hemicelluloses, and lignin separately, overall monomer yields between 76 and 90% were achieved for these three major biomass fractions.

Copyright © American Institute of Chemical Engineers. All rights reserved.

472451 Solvent-Enabled Control of Reactivity for Liquid-Phase Biomass Conversion Reactions

Abstract:

Catalytic upgrading of biomass to renewable chemicals and fuels requires the selective removal of oxygen from highly functionalized biomass components. This requirement of achieving high selectivity is illustrative of an important class of catalytic processes in the chemical and pharmaceutical industries, namely processes that take place through consecutive reaction steps, A â†' B â†' C, where species B is the desired product. The challenge for such catalytic processes is to identify catalysts and/or reaction conditions that allow for the selective conversion of A to B, but that do not facilitate the conversion of B to C, such that it is possible to produce species B with high yield. We will present results for BrÃ¸nsted acid-catalyzed conversions of a carbohydrate (fructose), an alcohol (t-butanol), and a furan (hydroxymethylfurfural, HMF) in liquid-phase organic solvents mixed with water, and we show that the rates of these reactions can be understood in terms of solvation effects for the acidic proton and the transition states for the catalytic reactions. Accordingly, we show that these solvent effects can be employed to achieve high yields for production of HMF by dehydration of fructose, by promoting the conversion of fructose to HMF, while not facilitating the subsequent conversion of HMF to levulinic acid. We will then show how solvent effects can be used to achieve high yields of HMF (e.g., 60%) from acid-catalyzed conversion of cellulose, combined with the oxidation of HMF to produce furan-dicarboxylic acid (FDCA). We will then show how this cellulose conversion strategy can be combined with the integrated conversion of ligno-cellulosic biomass to produce furfural (from hemicellulose), FDCA (from cellulose), and a stream of clean lignin.

Copyright © American Institute of Chemical Engineers. All rights reserved.

457290 Structure-Energy-Activity Relationships in Heterogeneous Catalysis - a View from Theory

Abstract:

The understanding of the catalytic properties of nanoparticle catalysts and the design of optimal composition and structures demands fast methods for the calculation of adsorption energies. By exploring the adsorption of O and OR (R=OH, OOH, OCH_3) adsorbates on a large range of surface sites with 9 transition metals, we propose new structure sensitive scaling relations between the adsorption energy of two adsorbates that are valid for all metals and for all surface sites.[1] This opens the way for a new class of activity volcano plots where the descriptor is not an energy but a structure-related descriptor. In addition, to better grasp finite size effects in the nanoparticles, a generalized coordination number is proposed as a leading descriptor for the adsorption strength.[2,3]

Simple descriptors as generalized coordination show a strong predictive potential exemplified in the design of optimal oxygen reduction reaction catalysts, with strong impact on efficient storage of energy.

References

1. F. Calle-Vallejo, D. Loffreda, M. T. M. Koper and P. Sautet, Nature Chemistry, 2015, **7**, 403-410

2. F. Calle-Vallejo, J. I. Martínez, J. M. García-Lastra, P. Sautet and D. Loffreda, Angewandte Chemie International Edition 2014, **53**, 8316-8319

3. F. Calle-Vallejo, J. Tymoczko, V. Colic, Q. H. Vu, M. D. Pohl, K. Morgenstern, D. Loffreda, P. Sautet, W. Schuhmann, A. S. Bandarenka, Science, 2015, **350**, 185-189

Copyright © American Institute of Chemical Engineers. All rights reserved.

463728 A New DFT-Based Approach That More Accurately Predicts Adsorption Energies

Abstract:

The development of density functional theory (DFT) functionals which accurately predict the energetics of heterogeneous catalytic processes, such as adsorption/desorption and bond breaking/forming, is crucial to the design and optimization of heterogeneous catalysts. For adsorption systems, the current DFT functionals give reasonable accuracy for selected classes of adsorbates, but far less than desired when looking at systems across the spectrum from strongly chemisorbed to strongly van der Waals bonded [1-7]. Improved accuracy across this whole spectrum could have tremendous impact on catalysis research and discovery. In this contribution, we present a new method for accurately predicting both molecular and dissociative adsorption energies based on DFT calculations with periodic boundary conditions. This method was benchmarked against a set of 39 experimental adsorption reactions with corresponding accurate calorimetric measurements. Our results show that our proposed method has a mean absolute error (MAE) and root mean squared error (RMSE) of 15 and 20 kJ/mol, respectively, making this method more accurate than the BEEF-vdW functional [3], which is the most accurate known DFT method for adsorbates, with a MAE and RMSE of 20 and 26 kJ/mol [3]. As such, this method accurately predicts adsorption energies for processes regardless of whether they are dominated by charge transfer or dispersion forces, and produces superior accuracy to any current standard DFT functional alone.

References:

[1] J. Wellendorff, T. L. Silbaugh, D. Garcia-Pintos, J. K. Nørskov, T. Bligaard, F. Studt, C. T. Campbell, Surf. Sci. 640 (2015) 36-44.

[2] A. J. Cohen, P. Mori-Sanchez, W. Yang, Chem. Rev. 112 (2012) 289-320.

[3] J. Wellendorff, K. T. Lundgaard, A. MÃ¸gelhÃ¸j, V. Petzold, D. D. Landis, J. K. NÃ¸rskov, T. Bligaard, K. W. Jacobsen, Phys. Rev. B 85 (2012) 235149.

[4] J. KlimeÅ¡, D. R. Bowler, A. Michaelides, Phys. Rev. B 83 (2011) 195131.

[5] J. KlimeÅ¡, D. R. Bowler, A. Michaelides, J. Phys.: Condens. Matter 22 (2010) 022201.

[6] M. Dion, H. Rydberg, E. SchrÃ¶der, D. C. Langreth, B. I. Lundqvist, Phys. Rev. Lett. 92 (2004) 246401.

[7] A. V. Krukau, O. A. Vydrov, A. F. Izmaylov, G. E. Scuseria, J. Chem. Phys. 125 (2006) 224106.

Copyright © American Institute of Chemical Engineers. All rights reserved.

452384 A DFT-Based Study of Pt-Catalyzed Hydrodechlorination of 1,2-Dichloroethane: Effect of Van Der Waals Interactions

Abstract:

A DFT-Based Study of Pt-Catalyzed Hydrodechlorination of 1,2-Dichloroethane: Effect of van der Waals Interactions

L. Xu[1], E. Stangland[2], M. Mavrikakis[1]

[1]Department of Chemical & Biological Engineering, University of Wisconsin-Madison

[2]Core Research and Development, The Dow Chemical Company, Midland MI

1,2-Dichloroethane (1,2-DCA) is an important intermediate in industrial chemical processes (e.g. production of PVC). It is also among the many chlorinated hydrocarbon compounds which are toxic and carcinogenic.[1] An effective and efficient treatment method for 1,2-DCA in the industrial waste streams is thus highly desired. Current combustion-based chlorocarbon treatment processes suffer from drawbacks such as high energy cost, generation of other harmful species and loss of hydrocarbon values.[2] The catalytic hydrodechlorination which utilizes hydrogen to convert chlorinated species into environmentally benign products (hydrogen chloride and hydrocarbons) is an attractive alternative. Pt-based catalysts have been extensively tested in experiments for the hydrodechlorination of 1,2-DCA.[3] However, the detailed reaction mechanism remains unclear due to the lack of theoretical investigations.

In this work, we aim to elucidate the hydrodechlorination mechanism of 1,2-DCA over the monometallic Pt catalyst using a combined approach of density functional theory (DFT) calculations, microkinetic modelling, and kinetic experiments. We perform DFT calculations using Pt(111) as a model for the catalytic surface. Based on the DFT-derived energetics, a comprehensive microkinetic model is constructed, and the model predictions are compared with our experimental results. Our initial DFT-based microkinetic model shows significant discrepancy with the experiments. Thereby, we introduce van der Waals (vdW) corrections into our DFT calculations. The vdW interactions stabilize multiple surface intermediates and transition states, which leads to a better agreement between theory and experiment. Our work here helps to elucidate the reaction pathway and the nature of active sites in Pt-catalyzed 1,2-DCA hydrodechlorination. It also demonstrates the importance of vdW corrections in accurately describing the surface energetics of chlorinated hydrocarbon species using DFT methods.

[1] E. D. Goldberg, *Sci. Total Environ.* **100**, 17–28 (1991).

[2] Y. Liu et al., *Appl. Catal. B* **29**, 61–67 (2001).

[3] V. I. Kovalchuk and J. L. d'Itri, *Appl. Catal. A* **271**, 13-25 (2004).

Copyright © American Institute of Chemical Engineers. All rights reserved.

458867 Computational Design of Active Site Structures to Circumvent Scaling in Ammonia Synthesis

Abstract:

The Haber-Bosch process for the reduction of atmospheric nitrogen to ammonia is one of the most optimized heterogeneous catalytic reactions, but there are aspects of the industrial process that remain less than ideal. Due to thermodynamic and kinetic limitations, the reaction proceeds appreciably only under harsh conditions (temperatures above 700K and pressures in excess of 100 bar), requiring centralized factories that are capital- and energy-intensive [1]. The discovery of catalysts enabling NH_3 synthesis in a distributed fashion under ambient conditions could be both economically and environmentally advantageous.

It has been shown that the activity of the industrially-relevant (211) transition metal stepped surfaces is limited by a Brønsted–Evans–Polanyi (BEP) scaling relationship between the N-N transition state energy (E_{N-N}) and the *N binding energy (E_N) [2]. E_{N-N} is consistently too high relative to E_N on all catalysts that satisfy this linear constraint, leading to a negligible production rate at ambient conditions and a modest rate under harsh conditions. In this work, taking inspiration from rutile oxides that bind *N onto under-coordinated metal top sites, we use DFT calculations in conjunction with mean-field microkinetic modeling to study the rate of NH_3 synthesis on model active sites that require the singly coordinated dissociative adsorption of N atoms onto transition metal atoms. Our results demonstrate that this "on-top" binding of nitrogen exhibits much-improved BEP scaling behavior, which can be rationalized in terms of transition state geometries, and leads to considerably higher predicted activity. While synthesis of these model systems is likely challenging, the stabilization of such an active site could greatly reduce the temperature and pressure requirements for thermochemical ammonia synthesis.

[1] Erisman, Sutton, Galloway, Klimont, Winiwarter, *Nat. Geosci.* **2008**, *1*, 636-639.

[2] Vojvodic, Medford, Studt, Abild-Pedersen, Khan, Bligaard, Nørskov, *Chem. Phys. Lett.* **2014**, *598*, 108-112.

Copyright © American Institute of Chemical Engineers. All rights reserved.

458974 Mechanistic Investigation of the Catalytic Hydrogenation of Levulinic Acid over Ru Model Catalyst Surfaces

Abstract:

Understanding the reaction kinetics governing the aqueous phase hydrodeoxygenation (HDO) of levulinic acid (LA) to Î³-Valerolactone (GVL) over Ru surfaces is crucial for designing better catalysts for the conversion of LA to GVL considering that Ru/C catalysts are the most often used catalysts for the HDO of LA. In this presentation, we report a computational investigation of the reaction mechanism of LA to GVL using DFT calculations and mean-field microkinetic modeling. Our gas phase and liquid phase results indicate that water (polar solvents) have a beneficial effect on the reaction kinetics of the hydrodeoxygenation of LA. In contrast, non-polar solvents (THF, methanol, & n-heptane etc.) have a detrimental effect on the reaction kinetics. Next, our results show that Ru (0001) is highly active for the hydrogenation of LA to its corresponding alcohol product 4-Hydroxypentanoic acid (HPA) at a high reaction temperature above 423 K. However, at a low temperature ($T < 373$ K), the hydrogenation rate is sluggish compared to experimental observations. Considering furthermore that the hydrogenation of various short chain ketones (acetone, butanone-2, and pentanone-2) over Ru (0001) also leads to reaction rates much smaller than predicted by experiment, we conclude that Ru (0001) is not the active site at low temperatures. To identify the active Ru site for the experimentally observed low temperature activity, we performed a constrained thermodynamics study and identified surface oxygen species as a possible active site for the hydrogenation of LA. The resulting two-site model can explain the hydrogenation mechanism of ketones and aldehyde to its corresponding alcohol over Ru catalysts.

Copyright © American Institute of Chemical Engineers. All rights reserved.

460957 Reaction Pathways for Biomass Derived Furanic Compounds on Ru(001) Surfaces:Â a DFT Study

Abstract:

The rapid depletion of fossil fuel resources globally, coupled with a growing energy demand has led to the search for alternative sustainable fuel sources. Fast pyrolysis of lignocellulosic biomass is being considered as an effective and low cost alternative for the production of liquid transportation fuels and chemicals. However, the bio-oil produced from pyrolysis is unsuitable for direct use due to its poor stability and low energy density, arising from its high oxygen content. Hydrodeoxygenation (HDO) is an effective technique for upgrading the bio-oil in a hydrogen rich atmosphere in which the oxygen groups are replaced with hydrogen atoms. Supported Ruthenium catalysts are excellent catalysts for the HDO process[1-3]. Insights into the underlying mechanism of the hydrodeoxygentation reaction are essential for controlling the activity and selectivity of the Ru catalyst. In this study, we focus on the potential reactions and products in the catalytic conversion of furfural, a model compound derived from bio-oil, with hydrogen over Ru(001) surface using Density Functional Theory (DFT). The energetics of multiple reaction pathways, including aliphatic side chain and furan ring C-O hydrogenolysis, C-C hydrogenation, C-C scission and the structural rearrangement of furfural to pentane diols and cyclic ketones were analysed. Hydrogenation of the aldehydic side chain of furfural to form furfuryl alcohol and the C_{ring}-C bond scission are not facile on Ru surfaces since these reactions are associated with high barriers (~200 kJ/mol). In contrast, hydrogenolysis of the C-O bond in the aldehyde side chain and in the furanic ring, has barriers of around 100 kJ/mol. The moderate ring opening barriers provides a low energy reaction pathway for the formation of 1,5 pentane diol, 1,2 pentane diol, cyclopentanone and cyclopentanol which have been observed experimentally on supported Ruthenium catalysts[4-5].

1. Wildschut, J.; Mahfud, F. H.; Venderbosch, R. H.; Heeres, H. J. *Ind. Eng. Chem. Res.* 2009, 48, 10324â˘10334.
2. Wildschut, J.; Melian-Cabrera, I.; Heeres, H. J. *Appl. Cat. B.* 2010, 99, 298-306.
3. Oh, S.; Hwang, H.; Choi, H. S.; Choi, J. W. *Fuel* 2015,153, 535-543
4. Elliott, D. C.; Hart, T. R.; *Energy and Fuels* 2009, 23, 631-637.
5. Nan, W.; Krishna, C. R.; Kim, T-J; Wang, L. J.; Mahajan, D.; *Energy and Fuels* 2014, 28, 4588-4595

Copyright © American Institute of Chemical Engineers. All rights reserved.

463566 CO_2 Activation on Zr-Decorated, Cu Nanoparticles

Abstract:

Cu-based catalysts are good candidates for the conversion of CO_2, a greenhouse gas, into fuels and chemicals, such as formic acid and methanol. In this work, density functional theory calculations have been applied to investigate the role of Zr decoration on the structural and electronic properties of Cu_{55} nanoparticles and its behavior towards CO_2 adsorption. The Zr decoration is thermodynamically preferred on the corner (surface) sites of the nanoparticles. This decoration results in strong CO_2 chemisorption which does not occur in the case of the monometallic Cu_{55}. By further increasing the Zr composition in $Cu_{55-x}Zr_x$ nanoparticles (with x=2-12) we observed a significant enhancement in the CO_2 adsorption starting at the composition $Cu_{47}Zr_8$. The CO_2 molecule is shown to be activated upon chemisorption on the Zr-decorated, Cu-based nanoparticles, transitioning from a linear to a bent configuration. Charge transfer from the nanoparticle to the CO_2 molecule was found to be responsible for the observed activation. Furthermore a linear relationship between the total charge transferred to CO_2 and the ionization potential (IP) of the nanoparticles was observed. Finally, a model has been developed that is able to predict Cu nanoparticle morphologies with Zr decoration over various nanoparticle sizes, which could potentially show enhanced CO_2 adsorption and activation.

Copyright © American Institute of Chemical Engineers. All rights reserved.

469064 Oxygen Dissociation for Selective Oxidation on Gold: Effect of Alloying and Geometrical Structure

Abstract:

Au is capable of selectively and efficiently catalyzing oxidation reactions, such as methanol coupling to methyl formate. However, many of these reactions require atomic oxygen to be adsorbed on the surface in order to proceed, which is a challenge since Au cannot normally dissociate O_2. Here, we use density functional theory and thermodynamic modeling to examine two routes for dissociating O_2 while maintaining high catalytic selectivity.

First, we examine AgAu alloys, motivated by the excellent catalytic performance of nanoporous Au, which is a nanostructured material with a small amount (~2%) of Ag. Nanoporous Au has proven capable of dissociating O_2, while maintaining high selectivity in oxidation reactions; however, the active site for O_2 dissociation remains unknown. Using atomistic thermodynamics, we find that, under oxidation reaction conditions, the surfaces of dilute AgAu alloys have Au terraces and AgAu bimetallic steps. These bimetallic steps are likely responsible for O_2 dissociation, and the calculated barrier on the AgAu(211) structure that we find in the thermodynamic calculations has a low activation barrier for O_2 dissociation that is in agreement with experiment.

We also show that the reconstruction of Au surfaces that occurs spontaneously plays an important role in preventing O_2 dissociation. For example, the barrier for O_2 dissociation on an unreconstructed Au(100) surface is lower than Ag(111) and Ag(110), both of which dissociate O_2 easily. Therefore, by preventing surface reconstruction, Au surfaces that can dissociate O_2 can be created. This may play a role in the activity of small Au nanoparticles.

Copyright © American Institute of Chemical Engineers. All rights reserved.

Catalytic Co-pyrolysis of Microalgae and Low Density Polyethylene Waste to Aromatic Hydrocarbons using Activated Carbon

Emmanuel Ansah[a], Lijun Wang[b,c*], John Eshun[a], Quazi Rahman[a], Abolghasem Shahbazi[bc]

[a] Department of Energy and Environmental Systems

[b] Department of Natural Resources and Environmental Design

[c] Department of Chemical, Biological and Bioengineering

North Carolina Agricultural and Technical State University, 1601 E Market Street

Greensboro, NC 27411, U.S.A.

Abstract

Catalytic and non-catalytic fast co-pyrolysis of micro algae (chlorella) and low density polyethylene (LDPE) waste were studied in multi-shot pyrolyzer using activated carbon as catalyst. Dried microalgae was co-pyrolyzed with LDPE in both the presence and absence of activated carbon to investigate the interaction between algae and LDPE during thermal decomposition and catalytic upgrade of the pyrolysis vapor. In the non-catalytic co-pyrolysis study in micro-pyrolyzer-GC/MS, the results indicated the interaction between microalgae and LDPE during co-pyrolysis had a pronounce effect on both microalgae derived oxygenated compounds and C_{10}-C_{20} aliphatic hydrocarbons. The total oxygenated compounds (mainly carbohydrate and lipid derived) increased by 18.7 % whiles olefins and alkanes saw an increase of 6.9% and 14.2% respectively. During activated carbon catalyzed co-pyrolysis of algae and LDPE, as temperature increased from 500°C to 700°C, the conversion of olefins and alkanes and the deoxygenation of oxygenated compounds from microalgae in the pores of the activated carbon produced high yield of aromatic hydrocarbon. The pyrolysis of algae produced relatively small aromatic hydrocarbons whilst pyrolysis of LDPE produced aromatic hydrocarbons with high yield at increasing temperatures. Addition of Algae to LDPE increased the yield of total aromatic hydrocarbons in Algae/LDPE/catalyst (an increase of 71.8% from 600°C to 700°C) compared to Algae/catalyst alone (an increase of 12.7% from 600°C to 700°C) and LDPE/catalyst alone (an increase of 6.1% from 600°C to 700°C) indicating high synergetic effect at rather higher temperature.

1. Introduction

A wide variety of biomass types such as grass, wood, agricultural crops and residues, animal waste, municipal solid waste and aquatic plants have been examined to determine their efficacy for the production of biofuels. Micro-algae are a hydrogen-deficient feedstock, thus leading to adverse impacts on the quality of liquid fuel especially low content of hydrocarbon content during pyrolysis. Liquid fuels

* Corresponding author, Tel: +1 336 285-3833, Fax: +1 336 3347270, Email: lwang@ncat.edu (L.J. Wang)

from microalgae were reported to have undesirable properties such as a high oxygen content and elevated acidity (i.e., high acid number). There is a growing awareness that the introduction of additional hydrogen-source is inevitable for improving the quality of bio-oil from microalgae feedstock. Waste commodity plastics, low and high-density polyethylene (LDPE and HDPE), polyethylene terephthalate (PET), polystyrene (PS), and polyvinyl chloride (PVC) have a high hydrogen to carbon ratio and molecular chain structures suitable for liquefaction and pyrolysis (Ballice, 2002). Polyethylene (PE) that is a common municipal solid waste and has a hydrogen to carbon ratio of 2 can potentially act as a cheap and available hydrogen source in a co-pyrolysis process (Zhang et al., 2015).

Pyrolysis vapors from the co-pyrolysis of microalgae and LDPE feedstocks can be converted to aromatic hydrocarbon by catalyst such as zeolites and activated carbon. Activated carbon structure is developed by imperfect aromatic sheets of carbon atoms, as well as incompletely saturated valences and unpaired electrons on the surface (Tsyntsarski et al., 2015). Few authors (Hu et al., 2014) and (Bu et al., 2013; Bu et al., 2014; Bu et al., 2015)) have reported methods to selectively produce syngas and phenolic-rich bio-oil respectively through catalytic pyrolysis of biomass using activated carbon as catalyst. Our study will seek to explore the non-catalytic and catalytic co-pyrolysis of microalgae and low density polyethylene in micro-pyrolyzer-GC/MS and determine how the thermal and catalytic interactions influence the yield of hydrocarbons.

2. Materials and methods

2.1 Materials

Microalgae (*Chlorella* spp) and Low density polyethylene (LDPE) were used in co-pyrolysis experiments. Proximate and ultimate analysis were performed on samples and shown in Table 1.

Table 1. Analysis of raw material

Description	Microalgae (chlorella spp)	Low Density Polyethylene
Ultimate analysis (dry basis)		
Carbon (wt. %)	42.60	84.90
Hydrogen (wt. %)	6.70	15.02
Nitrogen (wt. %)	6.71	0.01
Sulfur (wt. %)	1.25	0.02
Oxygen (wt. %)	42.7	-
Proximate analysis (dry basis)		
Moisture (wt. %)	1.73	-
Volatiles (wt. %)	63.63	99.98
Fixed carbon (wt. %)	32.38	-
Ash (wt. %)	2.26	.02

2.3 Micro-pyrolyzer-GC/MS (Py-GC/MS) experimental procedure

Fast pyrolysis was conducted in a multi shot pyrolyzer system (Frontier Laboratories Ltd, Japan) operated in a single-shot mode which is connected to GC/MS system (Agilent Technologies). The temperature of valve interface was set up at 300 ◦C and the temperature of the micro-pyrolyzer was set at temperatures of 400°C, 500 ◦C, 600 °C and 700 °C for the study. The GC front inlet temperature was set up at 300 ◦C to prevent the condensation of the products. Initial oven temperature was set at 40 ◦C and held for 2 min. The oven temperature was then ramped to 220°C at a rate of 5 ◦C/min and held at 200 ◦C for 15 min. Helium at a flow rate of 1 ml/min was used as carrier gas with a split ratio of 50:1.

Results and discussion

3.1 Effect of co-pyrolysis feedstock ratios and temperature on product yield

To investigate the influence of Microalgae/Plastic mass ratio on the product distribution from the co-pyrolysis of *Chlorella spp* and LDPE, experiments were performed at 400 , 500 and 600°C with Microalgae/LDPE mass ratios of 3:1, 1:1, 1:3 and the results shown in Fig 1. Fast pyrolysis of the pure forms of microalgae and LDPE at 400°C shows that LDPE only partially depolymerize at 400°C whilst microalgae almost completely devolatilize at same temperature. Microalgae showed optimum peak area at 500°C and decreased when temperature was increased to 600°C. LDPE on the other hand increased sharply when temperature was increased to 500°C and slightly increased further when increased to 600°C. During co-pyrolysis, it is seen that at 400°C, an increase in the ratio of LDPE to microalgae decreased the total peak area of volatiles due to the partial decomposition of LDPE at 400°C. However, at a co-pyrolysis feedstock ratio of equal mass and temperature of 500°C, the total peak area of volatiles lies between the total peak areas of microalgae and LDPE. Similar trend is observed when the temperature increased from 500°C to 600°C, however, the total peak area of feedstock ratio of equal mass was slightly lower than when microalgae was pyrolyzed alone. At temperature of 500°C and 600°C, an increase in the ratio of LDPE to microalgae increased the total peak area during pyrolysis. This indicates that the co-feeding of LDPE improved the total volatiles at the expense of bio-char due to the high percentage of volatile matter and low ash content in plastic (Xie et al., 2015).

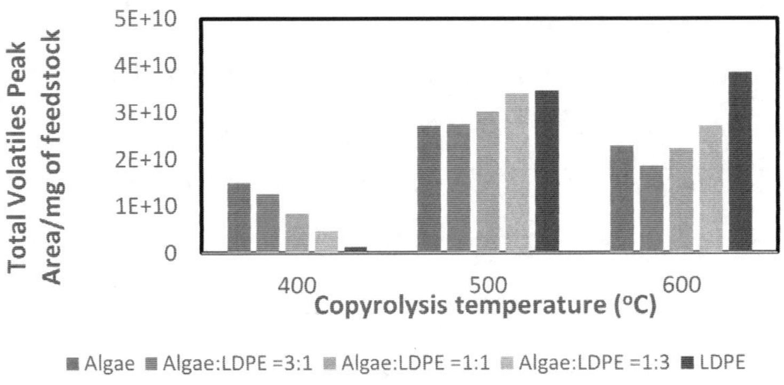

Fig. 1 Total volatiles yield at varying temperatures for co-pyrolysis of microalgae and LDPE at different feedstock ratios.

3.1.2 Product analysis

The chromatographic peaks were carefully selected and analyzed by means of the NIST MS library and literature (Kaewpengkrow et al., 2014; Lu et al., 2013; Zhang et al., 2013). For each chemical composition, a calibrated peak area% was achieved to represent its relative content. The total calibrated chromatographic peak areas could be used to reveal the changes of product yields since each peak area was normalized by the mass of feedstock during each experiment.

3.1.3 Non catalytic co-pyrolysis

Micro algae (chlorella spp) was co-pyrolyzed with LDPE to determine thermal interaction during co-pyrolysis of microalgae and PE. Typical pyrolysis products of micro algae and PE have been reported in literature (Aysu and Sanna, 2015; Chaiwong et al., 2013; Schwarzinger et al., 2015; Tan et al., 2014; Thangalazhy-Gopakumar et al., 2012; Xie et al., 2015; Xue et al.). The major pyrolysis products of microalgae (chlorella spp) obtained from NIST library in this study include CO_2, Toluene, 2-Furanmethanol, 2-hydroxy-3-methyl 2-Cyclopentene-1-one , P-cresol or 3-methyl phenol, Indole, 3-methyl 1-H-Indole, Phytol acetate, 3,7,11-trimethyl-1-Dodecanol, 3,7,11,15-Tetramethyl-2-hexadecen-1-ol, n-hexadecanoic acid. During LDPE pyrolysis, polyethylene is depolymerized to olefins and alkanes (together defined as aliphatic) with various chain lengths and therefore in the present study, we considered C_{10} to C_{20} since they were detected by GC/MS with greater certainty. The non-catalytic co-pyrolysis products used in this study were classified into seven groups namely CO_2, Toluene, Indoles, Oxygenated Compounds, Olefins, Alkanes and Aromatics. To assess synergetic thermal interaction, a ratio was calculated by dividing the GC/MS peak area of a pyrolysis product (which is proportional to the yield) during co-pyrolysis of microalgae and LDPE by the peak area of the same product when microalgae or LDPE was independently pyrolyzed. The ratios of oxygenates and aliphatic hydrocarbons for feedstock ratios of 1:1 and the optimum pyrolysis temperature at 500°C are reported and shown in fig. 2. In these graphs, a ratio higher than 1 indicates that co-pyrolysis promotes the formation of the product, where as a value lower than 1 indicates that co-pyrolysis suppresses the formation of the product. From the results, the interaction between microalgae and LDPE during co-pyrolysis had a pronounce effect on both microalgae derived oxygenated compounds and C_{10}-C_{20} aliphatic hydrocarbons from the depolymerization of LDPE. As seen from Fig 4, the co-pyrolysis resulted in an increment in yield (depicted as product ratios) of most of the $C_{10} - C_{20}$ aliphatic hydrocarbon ranging from an increment of 6.5% to 33%. Some oxygenated compounds (phytol, phytol acetate and 3, 7, 11-trimethyl-1-Dodecanol) mostly from decomposition of lipid content of microalgae showed an increase in yield ranging from 1.4% to 83%. This means that co-pyrolysis of LDPE with microalgae, enhanced the decomposition of both polyethylene and lipid content of microalgae leading to increase yield of aliphatic hydrocarbons and oxygenated compounds. From Fig 2(c), the total oxygenated compounds (carbohydrate and lipid derived) increased by 18.7 % whiles olefins and alkanes saw an increase of 6.9% and 14.2% respectively. It can be seen from same figure 2(c) that the yield of Indoles and Toluene were suppressed during co-pyrolysis as the product ratios were less than 1.

Fast pyrolysis of LDPE does not produce solid residue, however, the chain length of olefins and alkanes produced from depolymerization of LDPE can be as high as over C_{20+} during fast pyrolysis due to the short reaction time (Xue et al.). It has been reported that there is increasing probability that abstraction of hydrogen by biomass derived oxygenated compounds can facilitate the cleavage of LDPE polymer chain and its derivatives during co-pyrolysis. It is therefore probable that hydrogen transfer from LDPE to microalgae compounds likely reduces polymerization and cross-linking reactions and also suppresses decarboxylation and decarbonylation reactions (Vichaphund et al., 2014; Xue et al.).

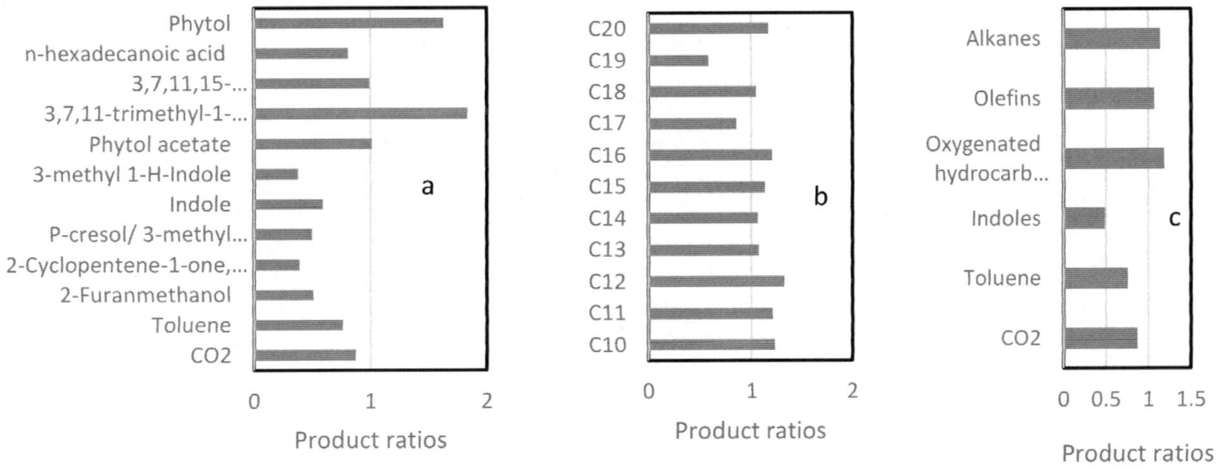

Fig 2. Ratios of product yields during co-pyrolysis of microalgae and LDPE in feedstock ratio of 1 to pyrolysis of individual feedstock determined by GC/MS peak area at pyrolysis temperature of 500°C.

3.1.4 Effect of catalytic co-pyrolysis on yield of aromatic hydrocarbons.

Pyrolysis of Algae, LDPE and Algae/LDPE mainly produced oxygenated compounds, nitrogen containing compounds and aliphatic hydrocarbons made up of olefins and alkanes. The catalytic pyrolysis of Algae/A.C, LDPE/A.C and Algae-LDPE/A.C on the other hand, mainly produced alkylated benzenes (toluene, ethylbenzene, p-xylene, propyl benzene, 1-ethyl-2-methylbenzene, 1-methyl-2-propylbenzene, butyl benzene and biphenyl), naphthalene and alkylated naphthalene (1-methylnaphthalene, 2-methylnaphthalene and 2-ethyl naphthalene) and aliphatic hydrocarbons (olefins and alkanes). These compounds from the activated carbon catalytic pyrolysis of algae, LDPE and mixture of algae-LDPE constitutes aromatic hydrocarbons and aliphatic hydrocarbons. The comparison of aromatic yield from individual algae and LDPE catalytic pyrolysis with catalytic co-pyrolysis of algae-LDPE (1:1) is presented in Figure 3. The catalytic pyrolysis of algae produced relatively small aromatic hydrocarbons whilst catalytic pyrolysis of LDPE produced aromatic hydrocarbons with high yield at increasing temperatures. Addition of Algae to LDPE increased the yield of total aromatic hydrocarbons in Algae/LDPE/catalyst (an increase of 71.8% from 600°C to 700°C) compared to Algae/catalyst alone (an increase of 12.7% from 600°C to 700°C) and LDPE/catalyst alone (an increase of 6.1% from 600°C to 700°C) indicating high synergetic effect at rather higher temperature. Similar results were reported by (Hassan et al., 2016) in pyrolysis of wood and polystyrene. During LDPE pyrolysis, free radical reactions with random chain scission would normally occur which results in the formation of olefins and alkanes. Activated carbon catalyst had a significant shape-selective effect on the products of aromatic hydrocarbons .When the primary product vapors from pyrolysis passed through the activated catalyst layer, they would diffuse into the internal pores of activated carbon catalyst and form aromatic hydrocarbons through a hydrocarbon pool mechanism in the activated carbon framework. The formation of aromatic hydrocarbons could come from two phenomena. The first one is the conversion of olefins and alkanes from LDPE pyrolysis to aromatic hydrocarbons. The second phenomenon is the deoxygenation of oxygenated organic compounds as it is processed through the catalyst layer.

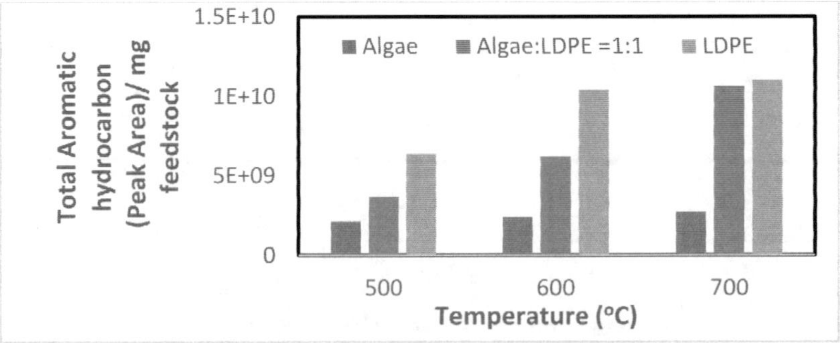

Fig 3. Yield of total aromatic hydrocarbon produced from individual catalytic pyrolysis of algae and LDPE and catalytic co-pyrolysis of algae-LDPE (1:1) at 500, 600 and 700°C.

4. Conclusion

In this study, catalytic and non-catalytic fast co-pyrolysis of microalgae and low density polyethylene were carried out using micro-pyrolyzer-GC/MS to investigate initial active pyrolysis temperature, effect of interaction of non-catalytic co-pyrolysis of microalgae and LDPE on product yields and the catalytic effect of activated carbon on the selectivity of aromatic hydrocarbons. The results show the non-catalytic interaction between microalgae and LDPE during co-pyrolysis had a pronounce effect on both microalgae derived oxygenated compounds and C_{10}-C_{20} aliphatic hydrocarbons from the depolymerization of LDPE. The pyrolysis of algae produced relatively small aromatic hydrocarbons whilst pyrolysis of LDPE produced aromatic hydrocarbons with high yield at increasing temperatures. Addition of Algae to LDPE increased the yield of total aromatic hydrocarbons in Algae/LDPE/catalyst (an increase of 71.8% from 600°C to 700°C) compared to Algae/catalyst alone (an increase of 12.7% from 600°C to 700°C) and LDPE/catalyst alone (an increase of 6.1% from 600°C to 700°C) indicating high catalytic effect at rather higher temperature.

References

Aysu, T., Sanna, A., 2015. Nannochloropsis algae pyrolysis with ceria-based catalysts for production of high-quality bio-oils. Bioresource Technology 194, 108-116.

Ballice, L., 2002. Classification of volatile products evolved during temperature-programmed co-pyrolysis of low-density polyethylene (LDPE) with polypropylene (PP). Fuel 81, 1233-1240.

Bu, Q., Lei, H., Wang, L., Wei, Y., Zhu, L., Liu, Y., Liang, J., Tang, J., 2013. Renewable phenols production by catalytic microwave pyrolysis of Douglas fir sawdust pellets with activated carbon catalysts. Bioresource Technology 142, 546-552.

Bu, Q., Lei, H., Wang, L., Wei, Y., Zhu, L., Zhang, X., Liu, Y., Yadavalli, G., Tang, J., 2014. Bio-based phenols and fuel production from catalytic microwave pyrolysis of lignin by activated carbons. Bioresource Technology 162, 142-147.

Bu, Q., Lei, H., Wang, L., Yadavalli, G., Wei, Y., Zhang, X., Zhu, L., Liu, Y., 2015. Biofuel production from catalytic microwave pyrolysis of Douglas fir pellets over ferrum-modified activated carbon catalyst. Journal of Analytical and Applied Pyrolysis 112, 74-79.

Chaiwong, K., Kiatsiriroat, T., Vorayos, N., Thararax, C., 2013. Study of bio-oil and bio-char production from algae by slow pyrolysis. Biomass and Bioenergy 56, 600-606.

Hassan, E.B., Elsayed, I., Eseyin, A., 2016. Production high yields of aromatic hydrocarbons through catalytic fast pyrolysis of torrefied wood and polystyrene. Fuel 174, 317-324.

Hu, Z., Ma, X., Li, L., Wu, J., 2014. The catalytic pyrolysis of microalgae to produce syngas. Energy Conversion and Management 85, 545-550.

Kaewpengkrow, P., Atong, D., Sricharoenchaikul, V., 2014. Catalytic upgrading of pyrolysis vapors from Jatropha wastes using alumina, zirconia and titania based catalysts. Bioresource Technology 163, 262-269.

Lu, Q., Zhang, Z.-b., Yang, X.-c., Dong, C.-q., Zhu, X.-f., 2013. Catalytic fast pyrolysis of biomass impregnated with K3PO4 to produce phenolic compounds: Analytical Py-GC/MS study. Journal of Analytical and Applied Pyrolysis 104, 139-145.

Schwarzinger, C., Hintersteiner, I., Schwarzinger, B., Buchberger, W., Moser, B., 2015. Analytical pyrolysis in the determination of the aging of polyethylene. Journal of Analytical and Applied Pyrolysis 113, 315-322.

Tan, X.-C., Zhu, C.-C., Liu, Q.-K., Ma, T.-Y., Yuan, P.-Q., Cheng, Z.-M., Yuan, W.-K., 2014. Co-pyrolysis of heavy oil and low density polyethylene in the presence of supercritical water: The suppression of coke formation. Fuel Processing Technology 118, 49-54.

Thangalazhy-Gopakumar, S., Adhikari, S., Chattanathan, S.A., Gupta, R.B., 2012. Catalytic pyrolysis of green algae for hydrocarbon production using H+ZSM-5 catalyst. Bioresource Technology 118, 150-157.

Tsyntsarski, B., Stoycheva, I., Tsoncheva, T., Genova, I., Dimitrov, M., Petrova, B., Paneva, D., Cherkezova-Zheleva, Z., Budinova, T., Kolev, H., Gomis-Berenguer, A., Ania, C.O., Mitov, I., Petrov, N., 2015. Activated carbons from waste biomass and low rank coals as catalyst supports for hydrogen production by methanol decomposition. Fuel Processing Technology 137, 139-147.

Vichaphund, S., Aht-ong, D., Sricharoenchaikul, V., Atong, D., 2014. Catalytic upgrading pyrolysis vapors of Jatropha waste using metal promoted ZSM-5 catalysts: An analytical PY-GC/MS. Renewable Energy 65, 70-77.

Xie, Q., Addy, M., Liu, S., Zhang, B., Cheng, Y., Wan, Y., Li, Y., Liu, Y., Lin, X., Chen, P., Ruan, R., 2015. Fast microwave-assisted catalytic co-pyrolysis of microalgae and scum for bio-oil production. Fuel 160, 577-582.

Xue, Y., Kelkar, A., Bai, X., Catalytic co-pyrolysis of biomass and polyethylene in a tandem micropyrolyzer. Fuel.

Zhang, B., Zhong, Z., Ding, K., Song, Z., 2015. Production of aromatic hydrocarbons from catalytic co-pyrolysis of biomass and high density polyethylene: Analytical Py–GC/MS study. Fuel 139, 622-628.

Zhang, H., Xiao, R., Jin, B., Shen, D., Chen, R., Xiao, G., 2013. Catalytic fast pyrolysis of straw biomass in an internally interconnected fluidized bed to produce aromatics and olefins: Effect of different catalysts. Bioresource Technology 137, 82-87.

458989 Modeling the Impact of Biomass Particle Residence Time on Fast Pyrolysis Yield and Composition

Abstract:

Bubbling fluidized bed and circulating fluidized bed reactors are the most common unit operations for carrying out fast pyrolysis of biomass. These reactors are designed to contact biomass particles with hot inert gas (such as nitrogen) in a bed of solids (sand) under turbulent multiphase flow conditions. In bubbling bed reactors, the gas flow is sufficient enough to cause mixing of the solids and gas without entraining most of the particles, except for char and ash generated from the biomass which can elutriate out of the sand bed then exit from the top of the reactor. In circulating bed reactors, the gas flow is much higher and the solids are typically smaller, consequently most of the solids are entrained out the top of the reactor vessel. The entrained solids are separated from the gas and most of them are recirculated back through the reactor with new gas, hence the name circulating fluidized beds.

Numerous studies have demonstrated that biomass fast pyrolysis yields depend on the residence time distribution (RTD) of the pyrolyzing particles, which is a function of the inherent feed properties as well as the reactor design and operating conditions. Experimental measurements indicate that it should be possible to represent the RTDs for biomass particles in both bubbling and circulating bed pyrolysis reactors with relatively simple functional approximations involving only two or three physically relevant parameters. We expect similar approaches can also be used to model the impact of catalyst particle residence time in vapor phase upgrading reactors.

In this presentation, we summarize literature on low-order reactor models for the particle RTDs in bubbling and circulating fluidized bed reactors and discuss the implications for the conversion of biomass particles undergoing fast pyrolysis. Based on this information, we demonstrate how particle RTDs can be combined with reaction kinetics and simplified reactor mass balances to create low-order fast pyrolysis models that can be used to summarize, interpret, and extrapolate observed pyrolysis trends in experiments and more detailed computational simulations. Our goal is to provide flexible analytical tools for consolidating information from more detailed computational reactor simulations and improving high-level process simulations of bio-oil production needed for techno-economic analyses and reactor design.

Copyright © American Institute of Chemical Engineers. All rights reserved.

472546 Ring-Locking Enables Selective Anhydrosugar Synthesis from Carbohydrate Pyrolysis

Abstract:

Anhydrosugars are common but low-yield products of carbohydrate pyrolysis. Their fixed conformation, protected stereoselectivity, and retention of most of their chemical functionality make these sugar derivatives highly desirable precursors for biologically active compounds, biodegradable surfactants, and polymers. Anhydrosugar production is very inefficient though, due to the nonselective nature of carbohydrate pyrolysis. Here we demonstrate how they form with excellent purity by ring-locking sugars prior to thermal treatment. We found that the selectivity to 1,6-anhydro-Î²-D-glucopyranose (levoglucosan, LGA) increased from 2% to greater than 90%, after alkoxy or phenoxy substituents were placed at the anomeric carbon of glucose, followed by fast pyrolysis of the resulting sugar at 600 Â°C. LGA formation becomes the dominant reaction pathway when the substituent group inhibits the pyranose ring from opening and fragmenting into non-anhydrosugar products. These findings introduce the ring-locking concept to sugar pyrolysis chemistry and offer a chemical-thermal treatment approach for upgrading simple and complex carbohydrates.

Copyright © American Institute of Chemical Engineers. All rights reserved.

471419 Pyrolysis Kinetics to and from Levoglucosan

Abstract:

Levoglucosan is a major and characteristic product of cellulose pyrolysis, formed catalytically in the absence of metal ions by R-OH molecules like water and other hydroxylated sugar-pyrolysis products [Seshadri and Westmoreland, *J. Phys. Chem. A* **116**:49 (2012) 11997-20013]. No radicals are necessary, although radical chemistry must be important at high temperatures. Other pericyclic unimolecular reactions may also occur, often involving hydroxyl groups. Here, we extend the previous work through quantum-chemistry predictions of transition states for altered glucose structures. Specifically, the OH groups on C1 and C6 that are involved in levoglucosan formation are substituted by NH_2 and CH_3. Nitrogen has one lone pair of electrons, and the methyl carbon has none; the calculations thus reveal how oxygen's two lone pairs influence the kinetics of glucose pyrolysis, and the results are then applied to cellulose. Added insights result from examining solvation by explicit water molecules, solvations by continuum models, and the effect of ionic acids. Finally, the routes by which levoglucosan may decompose are examined.

Copyright © American Institute of Chemical Engineers. All rights reserved.

464144 Molecular-Level Simulation of Thermogravimetric Analysis (TGA): A Cellulose Pyrolysis Example

Abstract:

Molecular-level Simulation of Thermogravimetric Analysis (TGA): A Cellulose Pyrolysis Example

Juan C. Lucio-Vega, Scott R. Horton, and Mike T. Klein

Department of Chemical and Biomolecular Engineering, University of Delaware

Thermogravimetric analysis (TGA) of cellulose pyrolysis was simulated using a molecular-level kinetics model. A temperature ramp between 600-1000°C was imposed in order to achieve the TGA simulation. The underlying celllose pyrolysis model accounted for polymer polydispersity whose decomposition was described using Flory-Stockmayer statistics and the degree of polymerization [1]. A recursive optimization method was utilized to calculate the degree of polymerization of the starting cellulose from literature bulk properties [1, 2]. The model's primary reaction pathways for the decomposition allowed for the clipping of the active cellulose by thermolysis in order to create levoglucasan (LGA) and glucose. The chemistry proposed by Lin et al [2] was then applied as secondary reactions where the LGA molecules reacted to produce intermediates and levoglucosenone (LGO). Further breakdown of the LGO molecules formed volatiles, furans, aldehyde, ketones, and char through a series of complex reactions. Linear free energy relationships were applied to minimize the number of model kinetic parameters. Executing the cellulose pyrolysis model in TGA mode provided a prediction of the mass loss, which in turn allowed the model parameters to be optimized to literature TGA data [2]. The optimized model showed good agreement with the experimental trends.

[1] Scott R. Horton, Rebecca J. Mohr, Yu Zhang, Francis P. Petrocelli, and Michael T. Klein. "Molecular-Level

Kinetic Modeling of Biomass Gasification." Energy & Fuels 30.3 (2016): 1647-1661.

[2] Yu-Chuan Lin, Joungmo Cho, Geoffrey A. Tompsett, Phillip R. Westmoreland, and George W. Huber, "Kinetics and Mechanism of Cellulose Pyrolysis" Journal of Physical Chemistry 113 (2009): 20097–20107.

Copyright © American Institute of Chemical Engineers. All rights reserved.

464586 Impact of Carrier Gas and Temperature on Biomass Fast Pyrolysis Using a New Microsphere Microreactor Approach

Abstract:

Biomass fast pyrolysis is emerging as a viable alternative for carbon-based resources. Yields and selectivity for fast pyrolysis is a function of numerous factors including rate of heating and quenching (residence times for heterogeneous and homogeneous events) and gas phase conditions including the composition of the carrier gas. Carrier gas composition is a parameter that has not been explored as comprehensively as other factors. In this study four gases were used, He, Ar, CO_2 and N_2, for the pyrolysis of crystal cellulose, tall fescue and switchgrass at temperatures between 500 and 900 °C. Individual manufactured biomass microspheres weighing nominally 10 micrograms each were pyrolized in a micro-reactor. The rate of hydrocarbon gas production was measured using a fast flame ionization detector (fast-FID). The effect of carrier gas on shrinkage of the biomass at different temperatures and char properties was investigated.

Copyright © American Institute of Chemical Engineers. All rights reserved.

454300 Study of Transmethylation As Primary Step on the Decomposition of Methoxy-Rich Lignin Model Compound over Zeolite Catalyst

Abstract:

In the context of the bio-based economy, lignin is a major source of aromatic compounds. Fast pyrolysis of lignin followed by bio-oil upgrading is a promising route for retrieving these aromatic chemical platforms and making them available for subsequent processing. Fast pyrolysis of lignin and upgrading of the derived bio-oil has been previously studied. However, the conversion pathway followed by each compound (or functional group) in the complex liquid product still needs to be accurately understood in order to tailor the process towards desired products.

Significant number of compounds containing methoxy functional group, such as anisole, guaiacol, syringol, and their derivatives, are found in primary liquids obtained from the thermochemical conversion of lignin. These compounds are further converted into phenolic and aromatic compounds via secondary pyrolysis and upgrading reactions. Anisole is normally used as model compound of methoxy-rich compounds present in the liquids from fast pyrolysis of lignocellulosic biomass because the only functionality in the molecule is the methoxy group. Research on anisole decomposition has been mainly focused on the deoxygenation process of the phenyl-oxygen bond. Study of the mechanism of transmethylation as reaction occurring prior to deoxygenation is less extensive, despite it is essential to understand the entire process of anisole decomposition.

In this work the transmethylation as the primary step on the catalytic decomposition of anisole is investigated. The decomposition of anisole was carried out in a fluidised bed reactor, and HZSM-5 zeolite was used as catalyst. The yield and product distribution in the liquid fraction, with particular focus on the phenolic compounds, was evaluated by gas chromatography. Non-catalytic anisole decomposition experiments were also conducted for comparison purposes.

Maximum yield of liquid products was observed at the minimum temperature at which complete conversion of anisole

is achieved (650°C for non-catalytic decomposition and 400°C for catalytic decomposition). Phenolic compounds were the primary products at these key temperatures and below. In the case of catalytic decomposition, phenol, and ortho-cresol and para-cresol were the main products. In addition, multi-methylphenols (xylenols and trimethyl-phenols) were abundantly produced over HZMS-5. Yields of aromatic compounds were larger than those of phenols at higher temperatures.

Experimental results suggest that transmethylation occurs as primary reaction when anisole conversion is not complete at low temperatures. A plausible mechanism for low-to-moderate temperatures is that, in the presence of the zeolite catalyst, the anisole is converted into phenol followed by the relocation of the methyl radical to form o-cresol and p-cresol. The addition of another methyl radical to the cresol molecule gives rise to the formation of xylenols. The rearrangement to trimethyl phenols is observed to a lesser extent. In all cases the major orientations for transmethylation are the ortho- and para-positions, and are favoured by the increase in temperature. As the temperature increases, anisole conversion is maintained but phenolic compounds yields decrease which points out to phenolics as precursors of aromatic compounds. Formation of aromatics occurs by means of deoxygenation as secondary step during anisole decomposition.

Copyright © American Institute of Chemical Engineers. All rights reserved.

461127 Hydroxycarbonylation Vs. Methoxycarbonylation - Palladium Catalyzed Reactions in Multiphase Systems

Abstract:

Introduction

The carbonylation of olefins with carbon monoxide, also known as Reppe carbonylation [1], is an important reaction for the functionalization of olefins. The palladium catalyzed, homogeneous reaction offers an easy access to esters and acids which are widely used for textile, agriculture and fine chemistry. Furthermore the reaction is extremely atom economical, all reactants are completely incorporated in the final product and no side products are formed. Besides the development of new catalyst systems to further improve the overall reaction performance [2], the recycling of the precious, homogeneously dissolved catalyst complex is in the focus of current research due to economical reasons. Many different concepts to overcome this problem are investigated like the immobilisation at a heterogeneous support or the separation by a multiphase system [3]. As a member of the collaborative research center "InPROMPT", which is mainly involved in the development of efficient production processes based on chemical reactions in liquid multiphase systems, we focus on liquid/liquid multiphase systems in order to recycle the catalyst complex and to separate the product after the reaction.

In this contribution we set our focus on the choice of an appropriate multiphase system for the hydroxycarbonylation and methoxycarbonylation of 1 dodecene in order to get high reaction rates and quantitative recycling of the catalyst complex. Different factors are analyzed to improve the reaction and separation behavior in the multiphase system for these reactions, e. g. the type of cosolvent, the polarity of the polar phase by addition of water and the kind of surfactant. Moreover, for the formation of the acid, we compare the one step hydroxycarbonylation with the two step sequence consisting of a

methoxycarbonylation with a following alkaline hydrolysis step in order to identify the better overall reaction performance.

Experimental section and results

The carbonylation experiments are performed in a 100 mL stainless steel autoclave with double jacket for adjusting the temperature and a gas dosage to perform the reactions under constant pressure. A typical reaction mixture consists of the nonpolar phase with the substrate 1 dodecene (12 g), the polar phase (methanol or water) and the catalyst system, which is formed from a palladium precursor, SulfoXantPhos as ligand (the sulfonated analogous of XantPhos) and a acid promoter. The carbonylation reaction is investigated under constant pressure at 30 bar carbon monoxide and at a temperature range between 80 °C and 110 °C.

As a result the use of a two phase system is favourable for the methoxycarbonylation of 1-dodecene. In contrast the hydroxycarbonylation is not feasible in a simple two phase system, it requires a microemulsion system to increase the interfacial area and the reaction rate. For the hydroxycarbonylation the kind and the concentration of the surfactant is crucial for the reaction performance and the separation behavior. Conversions of 30 % can be obtained after 30 h reaction time. For the methoxycarbonylation conversions over 95 % can be achieved after 20 h reaction time. An increasing amount of water in the polar phase decreases the reaction rate of the reaction, but it leads to an improvement of the separation. After the reaction 99 % of the catalyst complex can be recycled in the polar phase and 95 % of the product is in the non-polar phase. Recycling experiments were done and the catalyst is stable at least for four runs.

Acknowledgement

This work is part of the Collaborative Research Center/ Transregio 63 "Integrated Chemical Processes in Liquid Multiphase Systems" (subproject A2). Financial support by the Deutsche Forschungsgemeinschaft (DFG, German Research Foundation) is gratefully acknowledged (TRR 63).

References

[1] G. Kiss, Chem. Rev. 101 (2001) 3435–56.

[2] V. De La Fuente, M. Waugh, G.R. Eastham, J.A. Iggo, S. Castillón, C. Claver, Chem. - A Eur. J. 16 (2010) 6919–6932.

[3] A. Behr, A. J. Vorholt, N. Rentmeister, Chem. Eng. Sci. 99 (2013) 38–43.

Copyright © American Institute of Chemical Engineers. All rights reserved.

461192 Heterogeneously Catalysed Benzyl Alcohol Oxidation in a Three-Phase Micro-Packed Bed Reactor: Influence of Hydrodynamics and Reactor Design on Reaction Performance and Catalyst Deactivation

Abstract:

Micro-packed bed reactors are being increasingly used in multiphase catalysis due to their enhanced heat and mass transfer and well-controlled environments [1]. Understanding the prevailing hydrodynamics and mass transfer behaviour is pertinent to utilising the advantages of microreactors to their full potential. The aims of this study are to identify influences of hydrodynamics and reactor design on mass transfer, reaction rate and catalyst deactivation for the Au-Pd catalysed benzyl alcohol oxidation. The catalyst used is of particular interest due to its superior activity in catalysing alcohol oxidations to aldehydes [2], a class of reactions ubiquitous in the pharmaceutical and speciality chemicals industries.

The hydrodynamics had implications on the inter- and intra-phase mass transfer characteristics as well as the stability of the catalyst. Operating in a gas-continuous flow regime led to enhanced external mass transfer, resulting in high reaction rates with respect to the desired reaction pathway (TOF of 4500 hr^{-1}) [3]. Two limiting cases were found: (i) gas-limited operation that occurs at low G:L (gas:liquid flow rate) and (ii) liquid-limited operation that prevails at high G:L (**Figure 1**). The gas-liquid mass transfer coefficient under the liquid-dominated flow regime was experimentally measured to be 23.6 s^{-1}, a value that is two orders of magnitude higher than that commonly observed in conventional trickle bed reactors. The catalyst stability was drastically influenced by the hydrodynamics – possibly linked to the tendency of by-products to deposit on the catalyst surface at high G:L ratios, the extent of which was dependent on the degree of catalyst wetting.

Variations in reactor inlet design had implications on the observed reactant conversion through its influence on the hydrodynamics that developed downstream in the packed bed region. Superior performance (94% conversion in comparison to 80% conversion) was observed when the liquid was introduced directly into the catalytic bed, precluding the formation of gas-liquid slugs upstream, thus influencing the mass transfer within the packed bed.

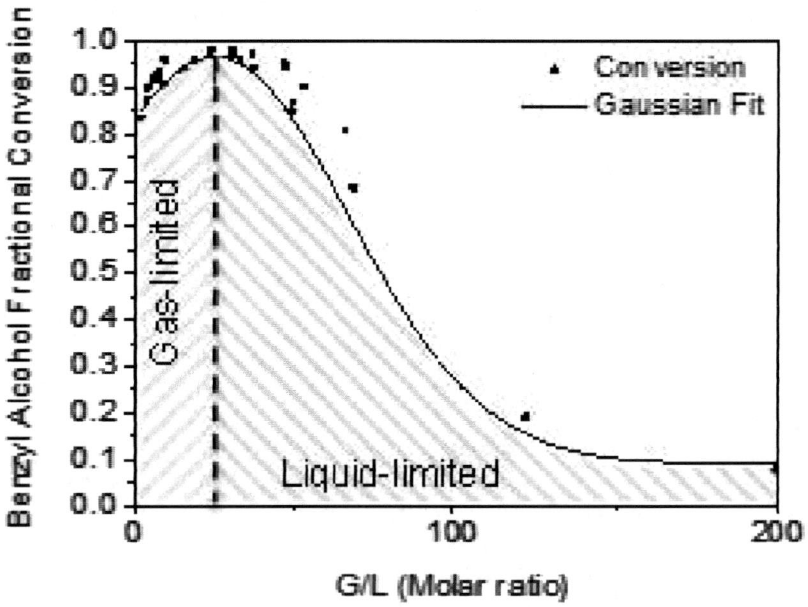

Figure 1. Benzyl alcohol conversion with increasing G/L Molar Ratio.

References

1. Al-Rifai, N., Cao, E., Dua, V., Gavriilidis, A. Microreaction technology aided catalytic process design. *Current Opinion in Chemical Engineering* **2013**, *2(3), pp. 338-345.*

2. Enache, D., Edwards, J., Landon, P., Solsona-Espriu, B., Carley, A., Herzing, A., Watanabe, M., Kiely, C., Knight, D., Hutchings, G. Solvent-free oxidation of primary alcohols to aldehydes using Au-Pd/TiO_2 catalysts. *Science* **2006**, *311, pp. 362-365.*

3. Al-Rifai, N., Galvanin, F., Morad, M., Cao, E., Cattaneo, S., Sankar, N., Dua, V., Hutchings, G., Gavriilidis, A. Hydrodynamic effects on three phase micro-packed bed reactor performance – Gold-palladium catalysed benzyl alcohol oxidation. *Chemical Engineering Science* **2016**, *149, pp. 129-142.*

Copyright © American Institute of Chemical Engineers. All rights reserved.

Imperial College London

Turning C_{7+} alcohols into primary amines: a "Hydrogen Borrowing" technique approach

Andrew Leung§, Mimi Hii*, Klaus Hellgardt§

§ACEX Building, Chemical Engineering Department, Imperial College London

*1M225, Royal College of Science, Chemistry Department, Imperial College London

Introduction

Amines, especially primary amines are important for[1]:

- Fine chemical industries
- Pharmaceutical industries
- Biochemical applications

$$ROH + NH_3 \xrightarrow[160-250\ ^{\circ}C,\ Cat.]{H_2\ 70\ bar} RNH_2, R_2NH, etc.$$

$R = C_{1-3}$

Current industrial methods start with corresponding alcohols, usually short-chained (C_{1-3}), which requires high pressures of hydrogen and ammonia.

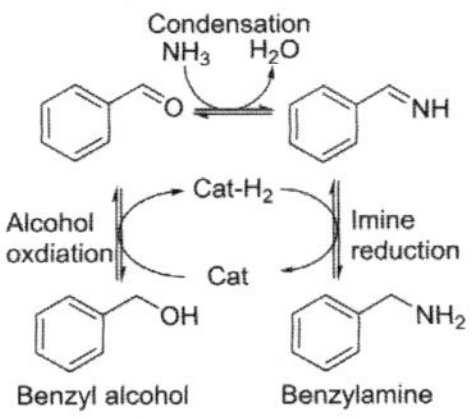

Hydrogen borrowing cycle, Cat = Au, Pd, Pt, Ni etc.

Project Aims

- Using "**Hydrogen borrowing**" technique[2]
- Using **heterogeneous catalyst**
- No **additional H_2** needed, rendering a much safer environment
- Water as the **only by-product**
- Batch reactors enable **data collection over time**
- Kinetics data can be analysed to find optimum reaction time and to aid eventual scale-up.

Previous work

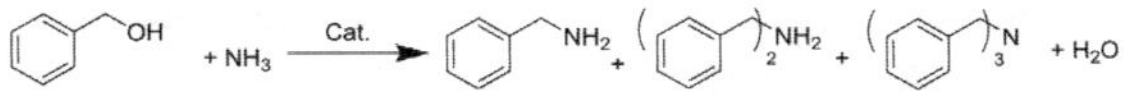

Catalyst	Time (hr)	Mol. loading	Yields to amines (%) NH$_2$	( )$_2$NH	( )$_3$N
Ir[3]	24	1%	0	0	100
Cu[4]	6	20%	n.d.	80	n.d.
Ru[5]	6	0.6%	n.d.	2	87
Ru PNP[6]	18	0.1%	95	n.d.	n.d.
Ni/θ-Al_2O_3[7]	72	5%	78	11	n.d.

Reaction Kinetics

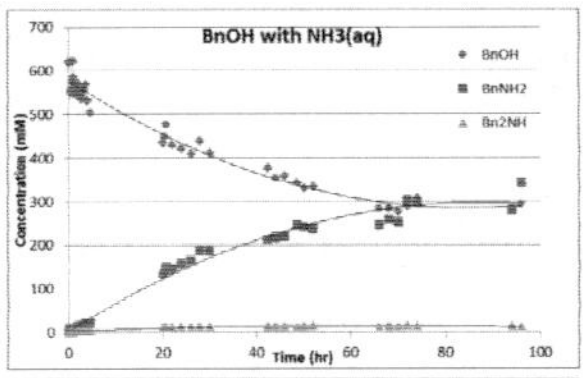

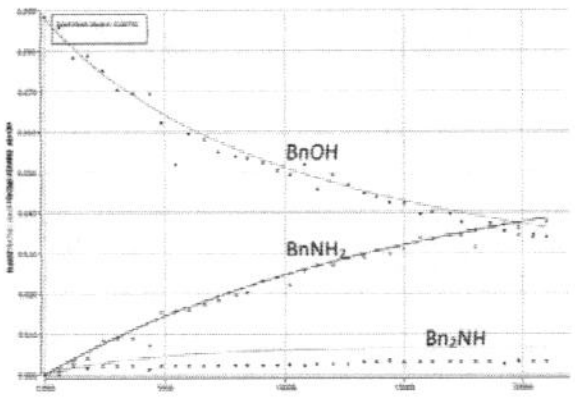

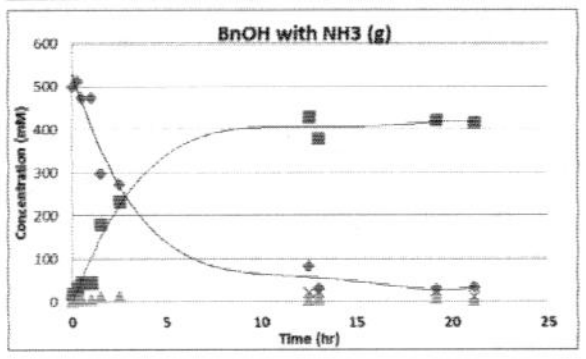

- Kinetic results collected **over time**
- Can determine the optimum point of reactions

Screening alcohols

$$R\text{-}OH \xrightarrow[160\ ^{\circ}C,\ 72\ hrs,\ o\text{-xylene}]{NH_3\ (20\ bar),\ Ni/Al_2O_3/SiO_2\ (10\ mol\%)} R\text{-}NH_2$$

0.6 M

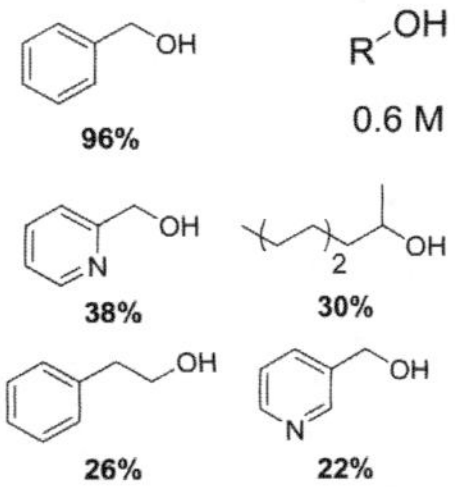

Yield determined by 1H-NMR

Alcohols were screened with 65wt%$Ni/Al_2O_3/SiO_2$

- C7+
- Primary and secondary alcohols
- Highly selective towards primary amine
- No secondary amine observed

Making new catalysts

Made with co-precipitation method, 85 and 43wt% Ni

Catalysts	Conversion	Selectivity
65%wt $Ni/Al_2O_3/SiO_2$	96%	97%
16%wt $Ni/Al_2O_3/CaO$	20%	>99%
85%wt $Ni/Al_2O_3/SiO_2$	25%	90%
43%wt $Ni/Al_2O_3/SiO_2$	4%	>99%

Reaction conditions: alcohol 0.6 M in 30 ml o-xylene, alcohol/NH_3 = 1/3, Ni 5 mol%, 160 °C, 1200 rpm, 72 hr. Yield determined by GC-FID

XRD patterns for Ni cat. (85, 65, 16 wt%)

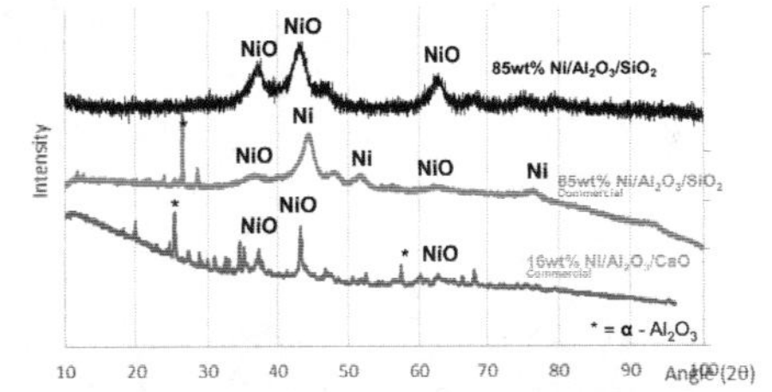

85 wt%Ni shows smaller single aggregate sizes than 65wt% catalyst

Future work

Does different Ni/Al/Si ratios show different activities based on their acid/basic sites or structure?

Does NH_3-treated, super-porous Ni materials show any catalytic activities?[8]

ACKNOWLEDGEMENTS

Many thanks to members of the **ReACT** group in chemical engineering, especially to **Dr Fessehaye Zemichael** and **Dr Sergio Lima** for their suggestions. Also many thanks to the members of **Barton Lab** for their suggestions on analysis and chemical substrates. Special thanks to **Prof, Hellgardt** and **Dr Hii** for their guidance and support throughout the project.

REFERENCES

1. K. I. Shimizu, *Catal. Sci. Technol.* 2015, **5**, 1412
2. Grigg, R.; Mitchell, T. R. B.; Sutthivaiyakit, S.; Tongpenyai, N. J. Chem. Soc., Chem. Commun. 1981, **12**, 611
3. R. Kawahara, K.-I. Fujita and R. Yamaguchi, *J. Am. Chem. Soc.*, 2010, **132**, 15108.
4. J. He, K. Yamaguchi and N. Mizuno, *Chem. Lett.*, 2010, **39**, 1182
5. K. Yamaguchi, J. He, T. Oishi and N. Mizuno, *Chem. Eur. J.*, 2010, **16**, 7199
6. C. Gunanathan and D. Milstein, *Angew. Chem., Int. Ed.*, 2008, **47**, 8661
7. K. Shimizu, K. Kon, W. Onodera, H. Yamazaki and J. N. Kondo, *ACS Catal.*, 2013, **3**, 112
8. D. J. Richardson, G. Mason, K. Hellgardt, P. A. Russell, B. A. Buffham, J. S. Bates, *J. Mater. Sci*, 2008, **43**, 665

466133 Hydroformylation and Tandem Isomerization-Hydroformylation of Long-Chain Olefins: Mechanism, Kinetics and Optimal Reaction Control

Abstract:

Introduction

The Hydroformylation of n-olefins using Rh-Bisphosphite catalysts (e.g. Rh-Biphephos) for the production of terminal aldehydes is known to suffer from side reactions, such as double-bond isomerization and hydrogenation of the substrate [1]. However, the ability to isomerize the n-olefins to their thermodynamic equilibrium composition allows such catalysts to perform the transformation of complex n-olefin mixtures with internal double-bond to valuable terminal aldehydes [2]. To exploit such processes on a technical scale, it is necessary to understand the mechanism of the reactions and tandem-effects in order to find a good kinetic model for reactor design and operation with an optimal yield of terminal aldehydes. This motivated the experimental and theoretical analysis of the hydroformylation reaction network including isomerization and hydrogenation of 1-decene and iso-decenes with terminal and internal double-bond, respectively. Based upon an extended reaction mechanism, detailed kinetic models were derived including catalyst equilibria. Fitting the model to experimental data was supported by parameter subset selection to avoid overparameterization [3]. With this model, optimal reaction control strategies were generated.

Experimental

The kinetic experiments were performed with 1-decene and an equilibrium mixture of iso-decenes in a thermomorphic multicomponent solvent system consisting of DMF, dodecane and n-decenes. A Rh-Biphephos catalyst was used in all experiments with a Rh-to-olefin ratio of 1:10000-1000. All reactions were realized in a 90 ml high pressure multiple batch reactor system (Parr Instrument Co.) under careful consideration of the catalyst pretreatment at 15-18 bar syngas and a molar Rh-to-ligand ratio of 1:3. Temperature, total Syngas pressure, partial pressures of CO and H_2 as well as initial substrate concentrations were varied systematically in 26 batch and semi-batch experiments. Gas chromatography (GC) was used for quantification.

Calculations

The parameter subset selection calculations and the parameter estimation were performed using Matlab 2012a with standard solvers. Optimal reaction control strategies were generated via maximizing the yield of undecanal subject to balance equations, kinetics and thermodynamics with time dependent state variables (temperature, partial pressures of CO/H_2) as degrees of freedom. This optimal control problem was transformed to a large NLP using orthogonal collocation and solved in AMPL.

Results

The experimental results showed the expected high isomerization activity of the catalyst as well as the highly regioselective hydroformylation. The activity towards hydrogenation was always low. The presence of CO affected the active catalyst and inhibited the isomerization reaction drastically, whereas the hydroformylation is promoted by CO and H_2. Therefore, increasing the syngas pressure lead to an increase in aldehyde yield if 1-decene was used as substrate. However, increasing syngas pressures applied to an equilibrated mixture of iso-decenes with internal double-bond counterintuitively decreased the aldehyde yield. However, the ratio of linear to branched aldehydes remained almost constant at about 97:3 in all cases. In total, 26 experiments including batch and semi-batch mode were used for parameter estimation and 14 parameters were estimated with low 95% confidence intervals (<10%). The resulting model was able to describe all reactions under all reaction conditions with low deviations (see Fig. 1).

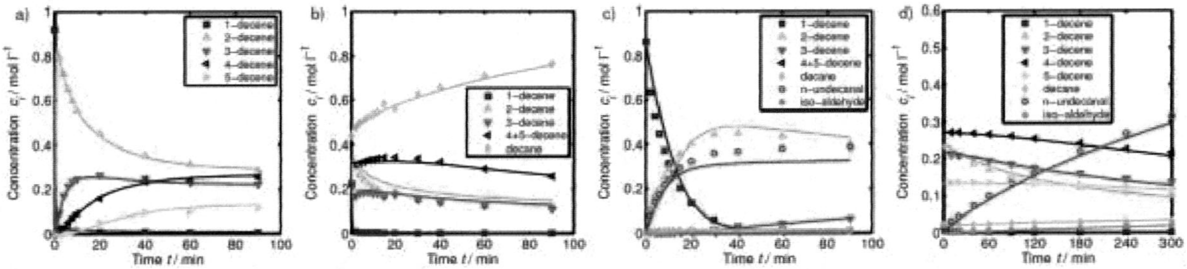

Fig. 1: Selected model calibration results: a) Isomerization of 1-decene (Rh:Olefin = 1:10000, 0 bar CO/H_2, 105 °C); b) Hydrogenation of 1/iso-decene mixture (Rh:Olefin = 1:10000, 10 bar H2, 105 °C); c) Hydroformylation of 1-decene (Rh:Olefin = 1:10000, 10 bar CO/H_2, 105 °C); d) Tandem-Isomerization-Hydroformylation of an equilibrated iso-decene mixture (Rh:Olefin = 1:1000, 5 bar CO/H_2, 105 °C).

The optimization results suggest to perform the reactions at low temperatures and high syngas pressures for maximum aldehyde yield it if 1-decene is used as substrate. The opposite is the case if an equilibrated mixture of iso-decenes with internal double-bond is used as substrate (see Fig. 2) which is a result of the coupling of isomerization and hydroformylation in the sense of a tandem-reaction.

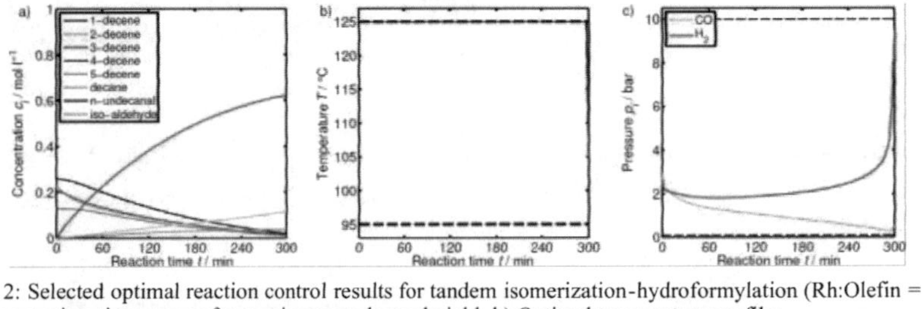

Fig. 2: Selected optimal reaction control results for tandem isomerization-hydroformylation (Rh:Olefin = 1:1000): a) Concentration-time curves for maximum undecanal yield; b) Optimal temperature profile;
c) Optimal partial pressure profiles.

Acknowledgement

This work is embedded in the collaborative research center „SFB/TRR-63 InPROMPT - Integrated Chemical Processes in Liquid Multiphase Systems" and the authors gratefully thank the German Science Foundation for financial support.

Literature

[1] R. Franke et al., Chemical Reviews 2012, 112, 5675–5732

[2] A. Behr et al., Journal of Molecular Catalysis A: Chemical 2003, 206, 179–184

[3] A. Jörke et al., Chemie Ingenieur Technik 2015, 87, 713–725

Copyright © American Institute of Chemical Engineers. All rights reserved.

466581 Troubleshooting Start-up of a Newly BuiltÂ Intensified Heat Exchanger Reactor - Using Reaction Engineering and Dynamic Modeling to Solve Real-Life Problems

Abstract:

A large existing batch process was retrofitted with an intensified heat-exchanger loop reactor, in which most of the highly exothermic reaction is performed in a small reactor volume, at well controlled conditions. The change in design was initiated to alleviated serious safety and corrosion concerns and bring the production installation up to modern day safety standards. However, after the new heat exchanger loop reactor was constructed and initial hick-ups in flow control systems were solved, startups still stranded in runaway temperatures and emergency shutdowns. After 7 months of attempted start-ups and repairs, not even 5 minutes of steady-state production was achieved and plant operations were far removed from a solution to the problem.

A new team was formed to analyze the problems and deliver a final solution and investment scenario. In a day-long meeting under significant (time) pressure, available (transient) plant data were collected, as well as reaction kinetics relations obtained from earlier pilot-plant experiments. These data were used to construct a chemical engineering, dynamic model of the loop reactor. This model was used to analyze the problem and run through various scenarios that could explain the anomalous reactor behaviour. After running about a dozen scenarios, the apparent rate of reaction after scale-up appeared to be only 1/4 of the rate obtained from the pilot plant runs, while the heat exchanger appeared to deliver only 1/2 of the expected heat duty. The model predicted that plant start-up should be possible, but at only 1/3 of the design capacity. This was demonstrated in a successful trial run a few weeks later, in which the reactor produced on-spec product for the first time. To reach full design capacity, a change in hardware was still required. After 6 months the project closed, with the reactor up and running at full capacity.

The presentation shows the hardware changes implemented, the problem solving and modeling procedures used, and compares the modeled temperature and plant temperature profiles. A plausible explanation for the strange differences in apparent reaction kinetics before and after scale-up is proposed. Finally, the presentation demonstrates that scale-up can present really nasty surprises in actual practice, but also shows how a Chemical Engineering modeling approach can be used to solve issues and find minimum investment solutions, even after an installation is built.

Copyright © American Institute of Chemical Engineers. All rights reserved.

462952 Alloy Catalysis Across Composition Space

Abstract:

Alloys are known to possess superior catalytic properties than their pure components. Finding the rational design of new alloy catalysts with optimal catalytic properties for a given application is the major challenge in multicomponent catalyst design due to the need to perform many catalyst preparation, characterization and reactivity measurements across composition space. To accelerate this search Composition Spread Alloy Films (CSAFs), thin multicomponent films that have composition gradients parallel to their surfaces, $A_xB_yC_{1-x-y}$ with x = 0 → 1 and y = 0 → 1-x, are prepared to be able to span the entire composition space. Many otherwise intractable fundamental scientific problems in alloy science and catalysis that can be effectively addressed through use of CSAFs as high throughput materials libraries.

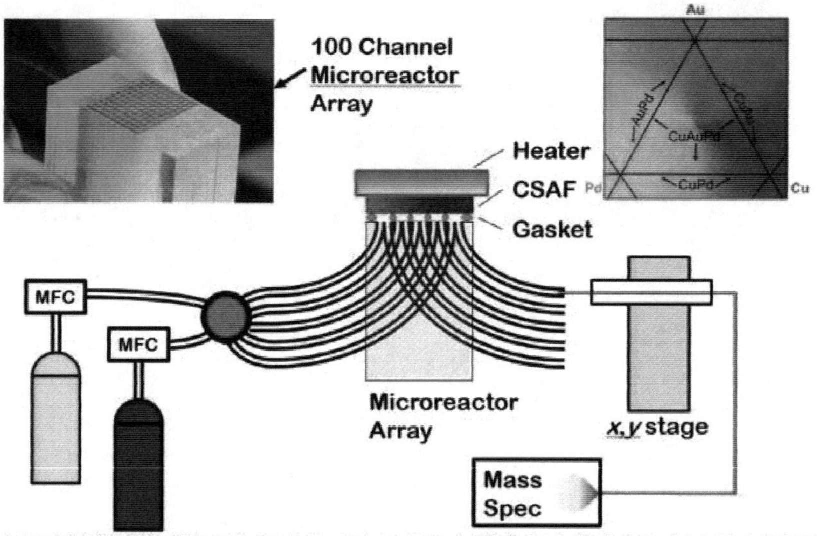

Figure 1. Illustration of the 100 channel microreactor array system. At left the gas handling system mixes and delivers gasses through 100 (5 illustrated) capillary tubes. Within the glass microreactor block the gasses are channeled to 100 different regions (5 illustrated) of the CSAF which is shown at upper right and the products from eah regions are extracted through 100 capillary channels (5 illustrated). The product gas streams are sampled sequentially by a smaller sampling capillary that transports the gases to a mass spectrometer.

Study of alloy catalysis across composition space using a CSAF requires a multichannel reactor system that can be used to run steady state catalytic reactions at many different positions or alloy compositions on the CSAF. We have developed a 100 channel microreactor array that can sample product distributions from 100 different alloy catalysts of about 10 minutes. $Cu_xAu_yPd_{1-x-y}$ CSAFs spanning all of binary and ternary composition space have been prepared using a rotating shadow mask CSAF deposition tool which is designed and developed in our group. CSAF surface composition and valence electron energy distributions are measured using X-ray Photoemission Spectroscopy (XPS).

The relationship between alloy catalyst activity and electronic structure has been investigated experimentally across a broad, continuous span of $Cu_xAu_yPd_{1-x-y}$ composition space. The CSAF was used as a catalyst library with a multichannel microreactor to measure H_2-D_2 exchange kinetics at 100 discrete compositions on the CSAFs over a temperature range of 333 – 597 K at atmospheric pressure. H_2 conversion was chosen to be the indicator of activity. It was found that H_2-D_2 exchange activity varies across the CSAF and it tends to increase with increasing Pd content.

When the activities on AuPd and CuPd binary regions are compared, it was found that more Pd is needed in CuPd than in AuPd to achieve the same activity.

A microkinetic model that has been validated using a number of single component Cu-Pd catalysts in a fixed bed reactor was used to estimate the energy barriers to dissociative adsorption ($\Delta E^{\ddagger}_{ads}$) and associative desorption ($\Delta E^{\ddagger}_{des}$) of H_2 as functions of alloy composition, x and y. On the $Cu_xAu_yPd_{1-x-y}$ CSAF, increasing Pd content from 0 to 1 was found to decrease adsorption barrier from 0.44 to 0.12 eV. Increasing Pd content from 0.25 to 1 was found to increase desorption barrier from 0.4 to 0.74 eV which suggests H_2-D_2 exchange reaction is limited by H_2 desorption step within this Pd content.

Spatially resolved X-ray photoelectron spectra were obtained from the CSAFs and used to estimate the energy of the valence-band center as a function of alloy composition. The v-band center shifted monotonically from -3.4 to 5.6 eV across the $Cu_xAu_yPd_{1-x-y}$ CSAF. The barrier to dissociative adsorption of H_2 was found to decrease as the v-band energy increases. This data provides the first experimental correlation of elementary reaction barriers with valence band energy across a continuous span of alloy composition space.

Copyright © American Institute of Chemical Engineers. All rights reserved.

467869 Pressure Dependent Kinetics of Magnesium Oxide Carbothermal Reduction

Abstract:

The rate of MgO carbothermal reduction was studied at temperatures from 1350-1650°C and pressures from 0.1-100 kPa based on product gas analysis at near isothermal conditions. For all temperatures the initial rate of carbothermal reduction increased inversely with pressure, and between conversions of 20-35% a transition occurred after which the reaction rate was maximum at 10 kPa. Analysis of reacted pellets showed that the reaction stoichiometry, the ratio of C to MgO reacted, was less than unity and decreased with pressure indicating CO_2 generation was more prevalent at elevated pressures. SEM imaging revealed the dissolution of C and MgO contact with conversion, and isoconversional analysis points to a change in the rate determining step between 1 and 10 kPa. The given experimental observations argue the importance of mass transfer and gaseous intermediates. A kinetic model is formulated based on a macroscopic species balance with CO_2 as the reaction intermediate.

Copyright © American Institute of Chemical Engineers. All rights reserved.

462002 Detailed Kinetic Modeling of Thermal Decomposition of Silane Via Automatic Mechanism Generation

Abstract:

The gas phase thermal decomposition of precursors such as silane is an important step in chemical vapor deposition for semiconductor material fabrication. Understanding the reaction pathways involved in this process can help design processes to better control the quality of the materials produced by CVD. We have extended the open-source software Reaction Mechanism Generator (RMG) [1] so that it can automatically generate detailed kinetic models for silicon hydrides and, eventually, a wide range of silicon-containing species. This work draws from and builds upon previous efforts in automatic mechanism generation for silicon hydrides, specifically in the field of silicon nanoparticle formation [2].

RMG's database was updated to include data for thermodynamics and kinetics important in silane thermal decomposition. Benson's group additivity values for thermochemistry of Si-containing molecules were added [3], and hydrogen bond increment (HBI) values for radical species were derived using quantum chemistry. Four reaction families, which RMG uses to propose reactions and estimate their kinetics, were enabled for silicon hydride chemistry:

Reaction Family	**Status**	**Example Reaction**
Hydrogen abstraction	Updated	SiH_4 + H. SiH_3. + H_2
Radical recombination	Updated	SiH_3. + H_3 SiH_4
Silylene insertion	New	SiH_2: + SiH_4 SiH_3SiH_3
Silylene-to-silene isomerization	New	SiH_3SiH: $H_2Si{=}SiH_2$

Because existing experimental and theoretical kinetic data for some of these families is sparse, additional transition state theory (TST) rates were calculated, using an automated kinetics calculator. The automated TST calculator uses a group additive scheme and distance geometry to predict transition state geometries [4], DFT to optimize and characterize them, and the software package Cantherm to apply conventional transition state theory.

Using the updated database, a detailed model was built for thermal decomposition of SiH_4. To validate the RMG-generated model, reactor simulations were compared to flow tube experimental data from Onischuk et al. [5]. Simulation conditions were matched to the experimental conditions, with base conditions of $y_0(SiH_4)$ = 1.6e-4 in argon, T = 913 K, and P = 39 kPa. The SiH_4 concentration profile variation with residence time matches reasonably well with the experimental data within a 20 K uncertainty in temperature. SiH_4 concentration dependence on temperature also matches well with experiment.

The experimental SiH_4 and Si_2H_6 concentration profiles, normalized to initial SiH_4 concentration, appear to vary with initial SiH_4 concentration suggesting a second order dependence on SiH_4 concentration. However, the simulation results approximate a first order dependence on SiH_4 concentration. Sensitivity analysis and flux analysis were employed to understand and resolve this discrepancy between the initial simulations and experiment.

The newly updated RMG and validated model allows simulation of silicon hydride thermal decomposition at different process conditions, reducing the need for future experiments. Importantly, this work provides a framework for generating detailed kinetic models of other, less well-studied CVD precursors.

References

[1] C.W. Gao, J.W. Allen, W.H. Green, R.H. West. Reaction Mechanism Generator: Automatic construction of chemical kinetic mechanisms. *Computer Physics Communications*, 203, 212-225, (2016) http://dx.doi.org/10.1016/j.cpc.2016.02.013.

[2] H.W. Wong, X. Li, M.T. Swihart, L.J. Broadbelt. Detailed kinetic modeling of silicon nanoparticle formation chemistry via automated mechanism generation. *Journal of Physical Chemistry A*, 108(46), 10122–10132, (2004) http://dx.doi.org/10.1021/jp049591w

[3] M.T. Swihart, M. T and S.L. Girshick. Thermochemistry and kinetics of silicon hydride cluster formation during thermal decomposition of silane. *Journal of Physical Chemistry B*, 103(1), 64–76, (1999) http://dx.doi.org/10.1021/jp983358e

[4] P.L. Bhoorasingh and R.H. West. Transition state geometry prediction using molecular group contributions. *Phys. Chem. Chem. Phys.*, 17, 32173-32182, (2015) http://dx.doi.org/10.1039/C5CP04706D.

[5] A.A. Onischuk, V.P. Strunin, M.A. Ushakova, V.N. Panfilov. Studying of silane thermal decomposition mechanism. *Int. J. Chem. Kinetics*, 30(2), 99-110, (1998) http://dx.doi.org/10.1002/(SICI)1097-4601(1998)30:2<99::AID-KIN1>3.0.CO;2-O

Copyright © American Institute of Chemical Engineers. All rights reserved.

462721 Mechanistic Insights into the Electrochemical Reduction of CO_2 over Ag Using an Integrated Transport-DFT-Microkinetic Model

Abstract:

Electrochemical reduction of CO_2 over Ag is affected by various physical processes such as transport of species (H^+,OH^-, CO_2, HCO_3^-, CO_3^{2-}, M^+, CO and H_2) in the electrolyte, adsorption/desorption of reactants (CO_2 and H_2O) and products (OH^-, CO and H_2), and energetics of electron/hydrogen transfer reactions involved in the formation of CO and H_2. Near cathode surface, the pH and CO_2 concentration gradients increase and the electric field decreases with decreasing potential at the cathode, which causes the coverage of adsorbed CO_2 to decrease and H to increase. Consequently, the rate of CO formation decreases and H_2 increases at higher negative applied potentials. Identification of optimal reaction conditions to obtain higher selectivity to CO requires a detailed microkinetic model based on quantum mechanical calculations coupled with the transport effects. We have used a periodic Kohn-Sham density functional theory (DFT) and a linearized Poisson-Boltzmann model to calculate coverage-dependent free energies of adsorption and energy barriers for a set of elementary reactions.[1] The rates of reactions in the microkinetic model were described using the transition-state theory. The transport of species involves diffusion, migration, convection and acid-base reactions, which was modeled according to a previously reported procedure.[2] In this talk, we will show the predictions of a fully integrated model for surface coverages and reaction rates as a function of applied potential, electrolyte properties (pH and buffer capacity) and operating conditions (temperature, mixing, CO_2 pressure and flowrate).

References:

(1) Goodpaster, J. D.; Bell, A. T.; Head-Gordon, M., Identification of Possible Pathways for Câ€"C Bond Formation during Electrochemical Reduction of CO2: New Theoretical Insights from an Improved Electrochemical Model. *The Journal of Physical Chemistry Letters* 2016, 7, (8), 1471-1477.

(2) Singh, M. R.; Clark, E. L.; Bell, A. T., Effects of electrolyte, catalyst, and membrane composition and operating conditions on the performance of solar-driven electrochemical reduction of carbon dioxide. *Physical Chemistry Chemical Physics* 2015, 17, (29), 18924-18936.

Copyright © American Institute of Chemical Engineers. All rights reserved.

467064 Acceleration of Kinetic Monte Carlo Simulations Combining a Hybrid Approach with Scaling of Kinetic Parameters for Free Radical Copolymerization

Abstract:

In recent years, there has been growing interest in kinetic Monte Carlo (KMC) methods applied to polymerization systems [1]. KMC is favored mainly for two reasons. First, KMC maintains information on the instantaneous length of each propagating radical, which enables convenient modelling of chain-length dependent rate coefficients [2]. The other reason is that polymer molecules can have different microstructures, including sequence distribution, branching and crosslinking. In KMC simulations, since reactions are explicitly simulated, it is relatively straightforward to construct and record the microstructure for each polymer chain [3].

Despite its power, KMC is computationally much more costly than deterministic modelling. Therefore, research efforts have been put forth to increase the computational efficiency of KMC algorithms. Recently there were two newly emerging methods, both of which showed strong performance in accelerating KMC simulations. Tripathi and Sundberg [4] reported a hybrid simulation method where propagation events are treated deterministically while initiation and termination events are treated stochastically, which can reduce simulation time by more than 100 times. This method was demonstrated on MMA homopolymerization, modelled with chain length dependency, and results agreed well between the hybrid algorithm and complete Monte Carlo simulations. Gao et al. [5] discovered a scaling relationship that could be applied to polymerization rate coefficients which enables the usage of a much smaller control volume than what was previously defined as the minimal volume. The method was demonstrated on copolymerization with a wide range of simulating conditions.

While the hybrid algorithm significantly accelerates KMC simulations, it treats propagation steps continuously, so in each iteration hundreds or thousands of monomers are consumed. As a result, it is only able to provide information on the chain lengths and monomer compositions on each chain, but the full sequence information is lost. Recovering sequence using information directly from a hybrid approach, however, is not a trivial task. In this work, we developed a combined model in which the hybrid approach is enhanced by the scaling method, and further extended to copolymerization with the ability to simulate the explicit polymer sequence. In a case study of the copolymerization of butyl acrylate/methyl methacrylate, we demonstrated that our model is able to combine the acceleration capability of both the hybrid approach and the scaling method, reducing the simulation time from 17.8 seconds to 0.50 seconds. Molecular weight distribution and sequence length distribution are compared between the combined model and a complete KMC simulation, and good agreement is observed for both. This method enables fast KMC simulations of free radical polymerization, which is of great help to the development of an optimization model for controlling free radical polymerization processes.

References

[1] A. L. Brandão, J. B. Soares, J. C. Pinto, and A. L. Alberton, "When Polymer Reaction Engineers Play Dice: Applications of Monte Carlo Models in PRE," *Macromolecular Reaction Engineering,* vol. 9, pp. 141-185, 2015.

[2] J. Stubbs, R. Carrier, and D. C. Sundberg, "Monte Carlo simulation of emulsion polymerization kinetics and the evolution of latex particle morphology and polymer chain architecture," *Macromolecular Theory and Simulations,* vol. 17, pp. 147-162, 2008.

[3] A. K. Tripathi, J. G. Tsavalas, and D. C. Sundberg, "Monte Carlo Simulations of Free Radical Polymerizations with Divinyl Cross-Linker: Pre-and Postgel Simulations of Reaction Kinetics and Molecular Structure," *Macromolecules,* vol. 48, pp. 184-197, 2014.

[4] A. K. Tripathi and D. C. Sundberg, "A Hybrid Algorithm for Accurate and Efficient Monte Carlo Simulations of Free-Radical Polymerization Reactions," *Macromolecular Theory and Simulations,* vol. 24, pp. 52-64, 2015.

[5] H. Gao, L. H. Oakley, I. A. Konstantinov, S. G. Arturo, and L. J. Broadbelt, "Acceleration of Kinetic Monte Carlo Method for the Simulation of Free Radical Copolymerization through Scaling," *Industrial & Engineering Chemistry Research,* vol. 54, pp. 11975-11985, 2015.

Copyright © American Institute of Chemical Engineers. All rights reserved.

472677 Kinetics and Mechanism of Ketonization of Acetic Acid on HZSM-5 Catalyst

Abstract:

Acetic acid is the most abundant and problematic compound among the light oxygenates resulting from the pyrolysis of biomass[1-2]. The acidity imparted to biomass streams from acetic acid promotes polymerization of other species that result in humin formation and subsequent catalyst deactivation in subsequent upgrading attempts. The common upgrading strategy of hydrotreating converts these light acids to low value light gases. The selective ketonization eliminates this acidity while forming acetone, which is a building block to produce longer-chain hydrocarbons via aldol condensation/hydrogenation[3-4]. In this contribution we report conditions necessary to selectively convert acetic acid to acetone over HZSM5 in the gas phase at a wide variety of temperatures ranging from 250-320°C. The reaction is investigated via a combination of temperature programmed desorption (TPD) techniques, IR of surface adsorbates, detailed reaction kinetics and Density Functional Theory (DFT) calculations. A reaction mechanism is proposed involving the formation of surface acyl species followed by coupling with an activated acid with an apparent barrier of 67 kJ/mole. TPD of acetic acid indicates an initial dehydration step in the reaction pathway, which is further confirmed by IR analysis of the catalyst surface. Isotope labeling experiments reveal the absence of a primary kinetic isotope effect involving hydrogen,[5] which when combined with the experimentally observed 2nd order Langmuir Hinshelwood rate behavior demonstrates that C-C coupling of acyl species with an activated carboxylic acid is the rate-determining step. DFT calculations reveal that acetic acid first adsorbs on HZSM-5 by forming H-bonds with Brønsted acid sites, which subsequently dehydrate to form acyl species. A second carboxylic acid is then activated via a tautomerization step followed by coupling with the adsorbed acyl species, which is the rate limiting transition state. Finally a detailed single site Langmuir-Hinshelwood model is presented in agreement with experimental and theoretical results. The influence of water on catalyst deactivation rates, competition for active sites, and surface intermediates is also discussed.

Reference

1. G.W. Huber, S. Iborra, A. Corma, Synthesis of transportation fuels from biomass: Chemistry, catalysts, and engineering, Chemical Reviews, 106 (2006) 4044-4098.
2. H.B. Goyal, D. Seal, R.C. Saxena, Bio-fuels from thermochemical conversion of renewable resources: A review, Renew Sust Energ Rev, 12 (2008) 504-517.
3. T.N. Pham, D.C. Shi, D.E. Resasco, Evaluating strategies for catalytic upgrading of pyrolysis oil in liquid phase, Appl Catal B-Environ, 145 (2014) 10-23
4. R. Pestman, R.M. Koster, J.A.Z. Pieterse, V. Ponec, Reactions of carboxylic acids on oxides .1. Selective hydrogenation of acetic acid to acetaldehyde, Journal of Catalysis, 168 (1997) 255-264.
5. A. Gumidyala, T. Sooknoi, S.P. Crossley, Selective ketonization of acetic acid over HZSM-5: The importanceof acyl species and the influence of water. Journal of Catalysis, (Accepted).

Copyright © American Institute of Chemical Engineers. All rights reserved.

Methane steam reforming unraveled by the MicroKinetic Engine, a user-friendly kinetic modeling tool

Devocht, B.R.[a], Toch, K.[a], Kageyama, N.[b], Fatemi, N.[a], Oyama, S.T.[b,c], Marin, G.B.[a], Thybaut, J.W.[a]

[a] Ghent University, Laboratory for Chemical Technology, Ghent, B-9052, Belgium

[b] The University of Tokyo, Department of Chemical System Engineering, Tokyo, Japan

[c] Environmental Catalysis and Nanomaterials Laboratory, Department of Chemical Engineering, Virginia Polytechnic Institute and State University, Blacksburg, VA 24061-0211, United States

Kinetic modeling forms the bridge between the phenomena occurring at the molecular and reactor scale. It results in a mathematical representation of the underlying reaction network and mechanism. The comparison between experimental observations and model simulations allows a critical evaluation of the assumptions made during model construction. This turns kinetic modeling into an important activity in chemical engineering. Moreover, the resulting models provide strategic information aiming at the optimization and intensification of industrial chemical reactors, and even entire processes. From an industrial point of view, global kinetic models such as power law or Langmuir-Hinshelwood-Hougen-Watson (LHHW) models often provide sufficient information for process control and optimization. However, narrowing margins and increasing computational capabilities open up perspectives for more fundamental kinetic modeling, which, up to recently, was only exploited by academia. In addition, the shift towards green production processes underlines the need for a more detailed understanding of the more complex nature of biomass conversion.

A broad spectrum of dedicated software tools with various features is available to construct such (micro)kinetic models. However, there is a significant induction period for novices in the field of kinetic modeling to get acquainted with the methodology, since a good, boundary-crossing knowledge between chemistry, mathematics, statistics and (chemo)informatics is required. In order to make fundamental kinetic modeling more accessible as well as to reduce the time spent for model construction, a user-friendly tool has been developed: the MicroKinetic Engine (μKE).

MicroKinetic Engine (μKE)

The μKE is a software package for the simulation and regression of chemical kinetics and even non-chemical applications such as solar cells characteristics simulation have been demonstrated with this tool. It has been developed during the last decade at the Laboratory for Chemical Technology, Ghent University, Belgium and was originally constructed for the detailed kinetic modeling of heterogeneously catalyzed reactions. In order to simulate different reactor types, both differential and algebraic equation solvers have been integrated in the software library. To enable model regression to experimental data, two deterministic regression routines are included, i.e., the Rosenbrock [1] and Levenberg-Marquardt algorithm [2]. A Graphical User Interface (GUI), see Figure 1 (left), is wrapped around all these routines such that no programming effort whatsoever is required from the μKE user, making it very distinct from other chemical modeling tools such as Athena Visual Studio (http://www.athenavisual.com/) or Chemkin (http://www.reactiondesign.com/products/chemkin/chemkin-2/).

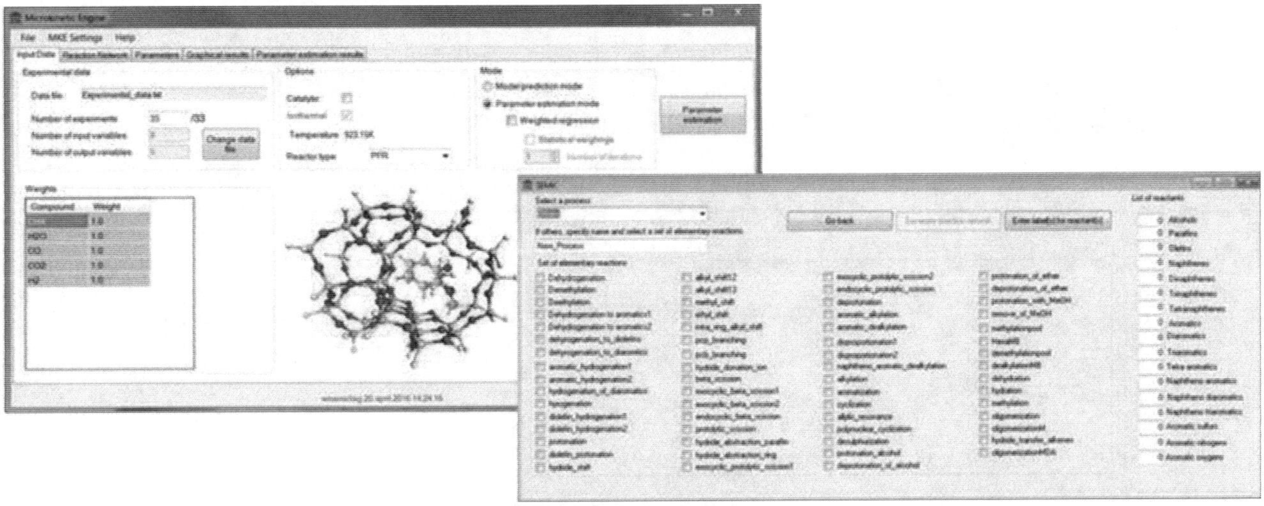

Figure 1. Graphical User Interface of the MicroKinetic Engine. Left: window for problem definitions, right: window for automatic network generation

The μKE consists of an onion structure as indicated in Figure 2. At its core, the μKE has the kinetic model which expresses the reaction rates of every (elementary) step included in the network. By default, the law of mass action is applied to calculate the corresponding rates, but also power laws or user defined functions can be included. Due to the latter, also applications other than chemical kinetics can be handled by the μKE, for example simulating a light-generated current as a function of the solar cell properties and the applied voltage. These rate equations are subsequently incorporated in a reactor model which describes the mass balance of all components over the selected reactor type. The corresponding set of algebraic and/or differential equations is solved using DASPK3.0 (https://techtransfer.universityofcalifornia.edu/NCD/10326.html) in order to calculate the individual outlet flow rates. If regression is required, an additional shell is activated in which the Rosenbrock method performs a first estimation effort, after which the Levenberg-Marquardt algorithm takes over to further optimize the parameter estimates. Both regression algorithms are based on the minimization of the (weighted) residual sum of squares of the responses in order to determine optimal parameter estimates.

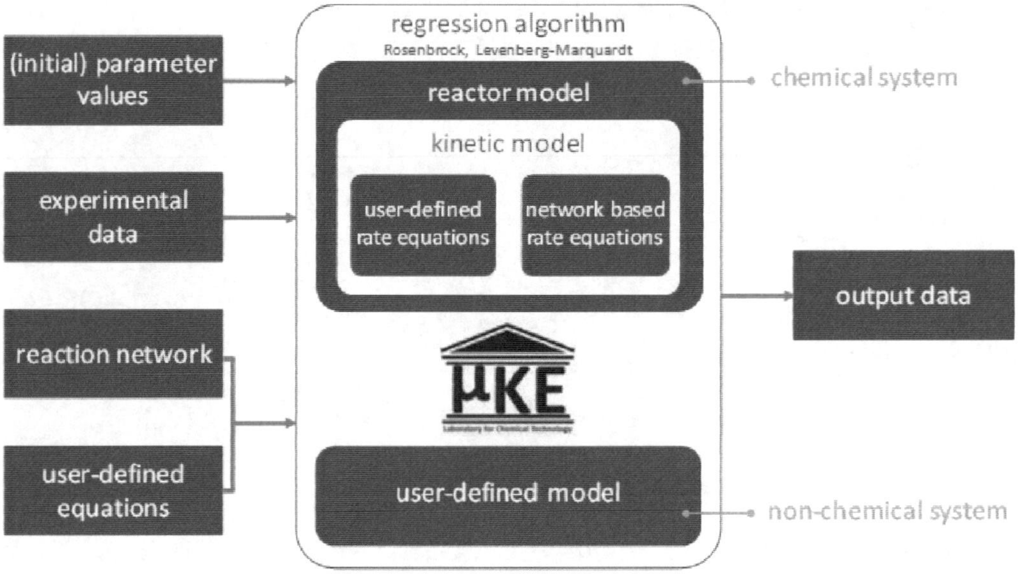

Figure 2. Structure of the MicroKinetic Engine

The input of the μKE comprises (initial) parameter values, experimental data, including independent and dependent variables (also denoted as 'operating conditions' and 'responses'), and the proposed reaction network for chemical kinetics. The reaction network can be constructed either manually via the GUI, or automatically thanks to the integration with the Reaction Network Generator (ReNGeP) [3] or, by extension, by any other network generation program, see Figure 1 (right). The μKE automatically converts the reaction network into the corresponding rate equations which are subsequently called from the reactor model. Instead of a reaction network, user-defined equations can be given which are applicable to both chemical and non-chemical systems. The output of the optimization procedure consists of the model simulations and, in case of regression, an extended statistical analysis and the corresponding optimal parameter estimates. In case of reversible steps, the parameter estimates are determined such that thermodynamic consistency is assured for each of the reaction steps. Additionally, the μKE identifies quasi-equilibrated reaction steps while no assumptions have to be made *a priori* on rate-determining step(s) or quasi-equilibria. If required, a rate of production analysis is performed by the μKE providing an additional layer of insight in the reaction network included.

Methane steam reforming

To demonstrate the features and versatility of the μKE, methane steam reforming is selected as a case study. To model the reaction mechanism of this industrially relevant process, a model is proposed with increasing complexity, i.e., going from a global power law model towards a LHHW type model.

The model is regressed to the experimental data acquired by Kageyama et al. [4]. The dataset comprises 80 experiments, performed in a tubular packed bed reactor at a total pressure of 0.4 MPa in a temperature range from 893 to 943 K with space times ranging from 0.96 to 3.36 kg s mol^{-1} and CH_4 to H_2O molar ratios between 0.125 and 0.7. In some experiments, CO, CO_2 and/or H_2 were added to the feed. The catalyst is an industrial SiO_2-MgO supported Ni catalyst.

The three reaction networks, i.e., corresponding to a power law model, a model with reactants adsorption and a Langmuir-Hinshelwood model, used for the kinetic model are shown in Figure 3. The adsorption/desorption steps in the latter two models are assumed to be in quasi-equilibrium. To reduce the number of parameters in the Langmuir-Hinshelwood model and based on literature, it is assumed that the adsorption equilibrium coefficient of methane and water are equal to $0.15K_{5C}$ and $0.05K_{5C}$ respectively, with K_{5C} the adsorption equilibrium coefficient of CO, and that CO_2 leaves the catalyst surface instantaneously ($K_{6C} = 0$), see Table 1 in bold. Moreover, the partial reaction order of hydrogen in the reforming step is set to one instead of three, as this clearly concerns a non-elementary step. The weighted regression of each model to the experimental data was found to be globally significant with statistically significant parameter estimates, see Table 1.

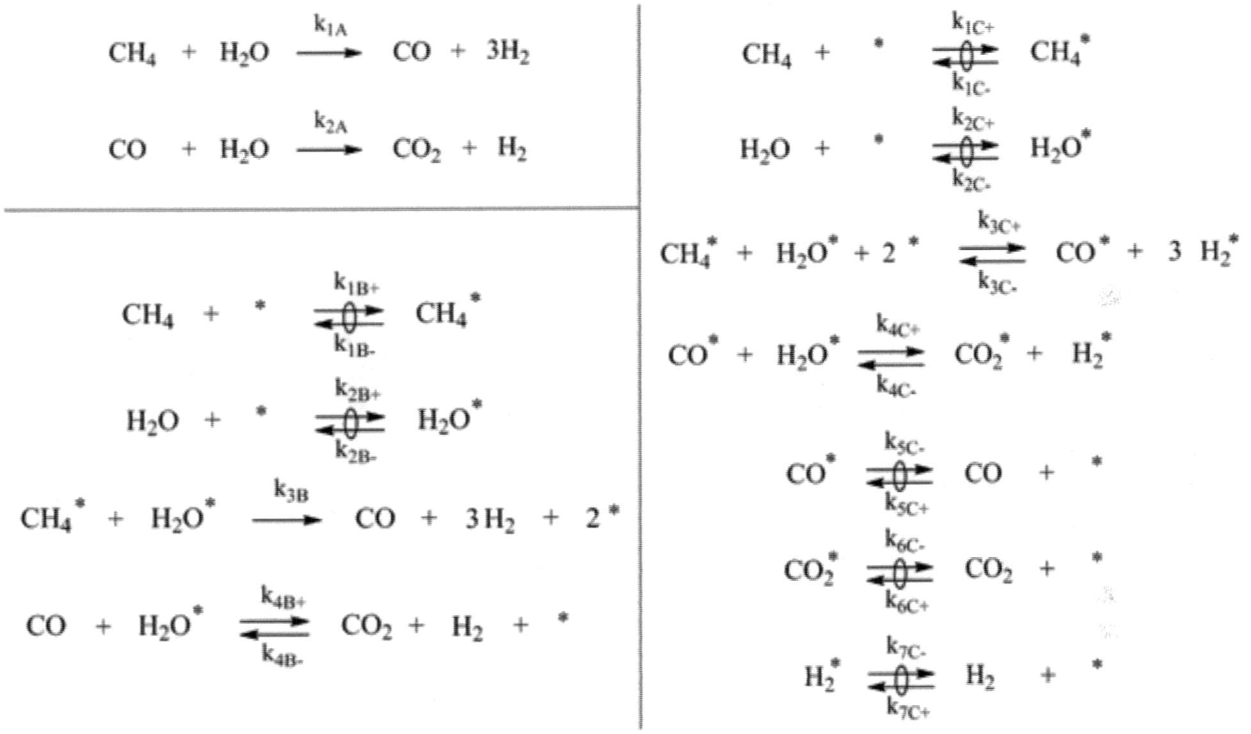

Figure 3. Reaction network for the Langmuir-Hinshelwood kinetic model for methane steam reforming. Left, up: power law model; left, down: model with reactants adsorption; right: Langmuir-Hinshelwood model

Table 1. 95% confidence interval of the model parameters of the power law model (left), the model with reactants adsorption (middle) and the Langmuir-Hinshelwood model (right).

power law model			model with reactants adsorption			Langmuir-Hinshelwood model		
	95 % confidence interval	units		95 % confidence interval	units		95% confidence interval	units
k_{1A}	1.76 ± 0.34	10^{-3} mol s^{-1} kg_{cat}^{-1} $Pa^{-0.44}$	K_{1B}	28.8 ± 4.8	MPa^{-1}	K_{1C}	**0.15 K_{5C}**	MPa^{-1}
k_{2A}	1.89 ± 1.34	10^{-5} mol s^{-1} kg_{cat}^{-1} $Pa^{-0.91}$	K_{2B}	12.5 ± 2.3	MPa^{-1}	K_{2C}	**0.05 K_{5C}**	MPa^{-1}
$p_{1,CH4}$	0.36 ± 0.05	-	k_{3B}'	1.63 ± 0.12	mol s^{-1} kg_{cat}^{-1}	k_{3C+}'	267 ± 238	mol s^{-1} kg_{cat}^{-1}
$p_{1,H2O}$	0.081 ± 0.078	-	k_{4B+}'	1.11 ± 0.03	10^{10} mol s^{-1} kg_{cat}^{-1} MPa^{-1}	k_{4C+}'	1.39 ± 0.34	10^{10} mol s^{-1} kg_{cat}^{-1}
$p_{2,CO}$	0.50 ± 0.13	-				K_{5C}	23.2 ± 19.6	MPa^{-1}
$p_{2,H2O}$	0.41 ± 0.10	-				K_{6C}	**0**	MPa^{-1}
						K_{7C}	14.7 ± 11.6	MPa^{-1}

As expected the power law model exhibited a poor performance, see Figure 4, which is attributed to the lack of reactant and product interaction with the catalyst that was accounted for. By only taking reactant adsorption and the reversibility of the water-gas shift reaction into account, the model performance improved significantly, yielding an acceptable performance combined with a clear physical meaning of the model parameter estimates. The model performance did not further significantly improve by taking product adsorption into account, indicating that the model has become overparametrized and/or insufficient information is available in the dataset.

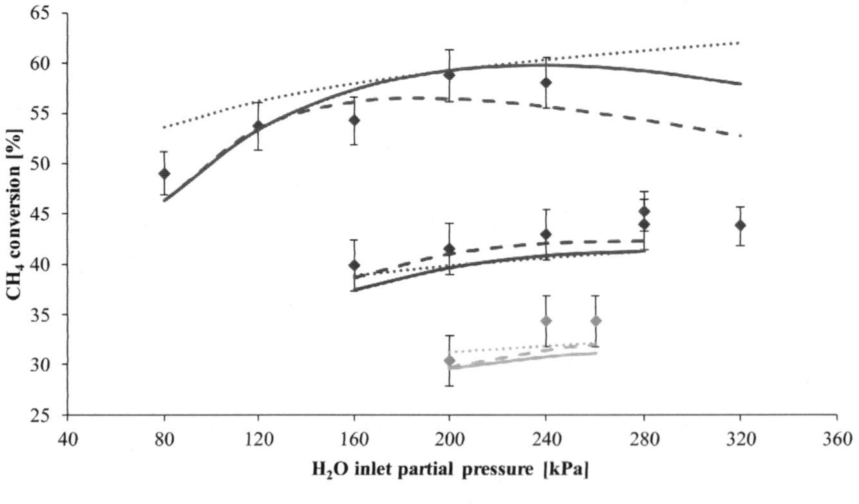

Figure 4. Methane conversion as a function of the water inlet pressure at a total pressure of 0.4 MPa and a temperature of 923 K. Symbols: experimentally observed with (blue) 40 kPa inlet partial pressure of CH_4 ($p_{CH4,in}$) and a space time of 3.36 kg_{cat} s mol_{CH4}^{-1}, (red) 80 kPa $p_{CH4,in}$ and a space time of 1.68 kg_{cat} s mol_{CH4}^{-1}, (green) 120 kPa $p_{CH4,in}$ and a space time of 1.12 kg_{cat} s mol_{CH4}^{-1}, lines: simulated via weighted regression. Dotted line: power law model, dashed line: model with reactants adsorption, full line: Langmuir-Hinshelwood model.

Conclusion

Through the case study of methane steam reforming, the μKE has proved to be a well performing and strongly user-friendly software package for the simulation and regression of chemical kinetics. With minor user intervention, regressions could be performed providing, in an automated manner, the statistical interpretation of these results, such that the user can focus on their physical interpretation. The gradual expansion of the model complexity for methane steam reforming has indicated that the most important phenomena to describe our dataset are the adsorption of the reactants and the reversibility of the water-gas shift reaction.

References

[1] H.H. Rosenbrock, AN AUTOMATIC METHOD FOR FINDING THE GREATEST OR LEAST VALUE OF A FUNCTION, Comput. J., 3 (1960) 175-184.

[2] D.W. Marquardt, AN ALGORITHM FOR LEAST-SQUARES ESTIMATION OF NONLINEAR PARAMETERS, Journal of the Society for Industrial and Applied Mathematics, 11 (1963) 431-441.

[3] J.W. Thybaut, G.B. Marin, Single-Event MicroKinetics: Catalyst design for complex reaction networks, J. Catal., 308 (2013) 352-362.

[4] N. Kageyama, B.R. Devocht, K. Toch, J.W. Thybaut, G.B. Marin, S.T. Oyama, The Interplay of Kinetics and Thermodynamics in the Catalytic Steam Methane Reforming over Ni/MgO-SiO2, Ind. Eng. Chem. Res., Submitted (2016).

465541 Severe Pyrolysis of the C12 Hydrocarbons n-Dodecane, Cyclododecane, and 2,6-Dimethyl-Decalin, with Rice-Herzfeld and Graph-Theoretic Interpretations

Abstract:

High temperature, short residence time pyrolyses of n-dodecane, cyclododecane, and 2,6-dimethyl-decalin were experimentally investigated. A coiled tube flow reactor was used, with outlet temperatures 1050-1150 K, residence times ~0.3 s, and steam/hydrocarbon ratios ~1, to simulate commercial steam cracking of naphthenic feedstocks at severities from 0.8 to 3.3. Experimental selectivities Sj = (mol j produced/100 moles of substrate decomposed) were calculated for j = methane, ethylene, propylene, butadiene, isoprene and benzene, the products of special interest in olefin plant operations. Decomposition kinetics and product selectivities were interpreted two ways. First by the classical Rice-Herzfeld types of free-radical pathways, initiated by substrate bond fissions and containing the requisite H-abstraction, beta-scission, isomerization, combination and disproportionation types of elementary steps, and second by graph-theoretic methods based upon the A(djacency) and C(ycle) matrices of the substrates. These revealed the influence of both the size and numbers of rings within each substrate upon their pyrolysis pathways and also the profound influence of either the absence or presence of methyl substituents.

Copyright © American Institute of Chemical Engineers. All rights reserved.

467113 An Assessment of Solvent Effects on the Selectivity of an Alkylation Reaction: A Comparison Between Experiments and Computations

Abstract:

An assessment of solvent effects on the selectivity of an alkylation reaction: A comparison between experiments and computations

A. Diamanti[†], E. Grant[†], Z. Ganase[†], A. M. Rea[‡], A. Galindo[†], C. S. Adjiman[†]

[†]*Centre for Process Systems Engineering, Department of Chemical Engineering Imperial College London, London, SW7 2AZ, UK*

[‡]*Process Studies Group, Technology & Engineering, Syngenta, Jealott's Hill International Research Centre, Bracknell, Berkshire RG42 6EY, United Kingdom*

The solvent in which a given reaction is conducted can influence the selectivity, rate, stability, catalyst activity and overall process performance.[1] In the production of pharmaceuticals and fine chemicals, solvents characteristically constitute between 80 and 90 % by mass of the non-aqueous materials used.[2] As a result, choosing the best overall solvent for a given process is of high importance. However, making an informed choice is challenging given that there is no universal model for predicting the solvent effects on reactions. Experimentalists can spend significant amounts of time performing reactions in various solvents to optimise a given reaction outcome in the laboratory. As a result, the suitability of computational methods to predict the solvent effects on a reaction is of great interest. Furthermore, if such methods are to enable the identification of an optimal solvent for specific reaction metrics, one must balance computational cost

and accuracy: the combination of quantum mechanical calculations with continuum solvation models is particularly appealing in this context. Indeed, the SMD continuum solvation model of Marenich et al.[3] has been found to provide a useful basis for the development of a quantum mechanical-computer aided molecular design (QM-CAMD) technique to find optimal reaction solvents. Its application to a Menschutkin reaction enabled the generation of a solvent that enhanced the rate of the reaction by 40% compared with the next best solvent that had been used previously.[4] Given that the solvent can also heavily affect the regioselectivity of a reaction, our current work aims to assess the feasibility of predicting solvent effects on selectivity so that this aspect can be incorporated in the QM-CAMD. In order to do so, we undertake a systematic study combining predictions and kinetic experiments, based on an interesting example of the solvent effects on selectivity reported in the literature for the Williamson ether synthesis of sodium beta-naphthoxide and benzyl bromide.[5] The choice of solvent significantly impacts the selectivity of reaction for alkylation at either oxygen or carbon sites leading to *O*- or *C*-alkylated products, and determines the final product ratio. Density functional theory (DFT) electronic structure calculations are used to predict the rate constants for the Williamson reaction via the two possible pathways as well as to compute the final product ratio, with the solvent represented using the SMD solvation model.[3] Twenty combinations of levels of theory and basis sets are investigated. A diverse set of solvents of varying polarity is used and detailed kinetic experiments involving in situ ^{1}HNMR data acquisition are carried out to assess the predictive results. The performance of the various models is discussed and the challenges in conducting reliable experiments are highlighted.

References

1. Reichardt, C.; Welton, T. *Solvent and Solvent Effects in Organic Chemistry*, 4th Ed.; Wiley-VCH: Weinheim, 2010.

2. Constable, D. J. C.; Jimenez-Gonzalez, C.; Henderson, R. K. *Org. Proc. Res. Dev.* **2007**, *11*, 133–137.

3. Marenich, A. V; Cramer, C. J.; Truhlar, D. G. *J. Phys. Chem. B* **2009**, *113*, 6378–6396.

4. Struebing, H.; Ganase, Z.; Karamertzanis, P. G.; Siougkrou, E.; Haycock, P.; Piccione, P. M.; Armstrong, A.; Galindo, A.; Adjiman, C. S. *Nat. Chem.* **2013**, *5*, 952–957.

5. Kornblum, N.; Seltzer, R.; Haberfield, P. *J. Am. Chem. Soc.* **1963**, *85* (8), 1148–1154.

Copyright © American Institute of Chemical Engineers. All rights reserved.

467459 Automatic Discovery of Reaction Paths and *Ab Initio*/Rrkm-ME Calculations Towards Pyrolysis and Combustion Mechanisms

Abstract:

Automatic kinetic model generators are instrumental in describing systems with many chemical species and elementary reaction steps, such as combustion, atmospheric, environmental, and astrochemical processes. Modeling software, such as Reaction Mechanism Generator (RMG), use a rate-based algorithm to keep the number of species and reactions tractable by excluding unimportant reactions based on flux, and growing the model based on known or estimated parameters for important reactions. The thermochemical and kinetic parameters can be estimated from group additivity and by using a database of known rate rules for these reaction types, respectively, allowing the inclusion of species and reaction steps as needed. Thus, unknown reaction types are omitted in the models. The discovery of unknown reaction types had to occur with human intuition of a saddle-point structure connecting reactant(s) to product(s), where some insight to atom rearrangement had to be implemented as an initial guess. Then, if successfully located, rate coefficients could be calculated and included into models to improve their reliability and predictions. Here, we introduce a strategy for the automatic discovery of unknown reaction steps (and types) without *a priori* knowledge by employing bond electron (BE) matrices combined with double-ended chain-of-state methods. This strategy allows an automatic sampling of the chemical space in converting reactant(s) to product(s), and arriving at a reasonable saddle-point structure without the need of human intuition. In BE matrices, products are generated from bond breaking and bond forming of the atoms from a given reactant. Then, a double-ended chain-of-state method attempts to connect the reactant structure to product structure by atom rearrangements passing through the saddle-point structure. This estimated saddle-point is then refined using a quasi-Newton optimization technique, followed by intrinsic reaction coordinate (IRC) calculations to confirm that the saddle-point structure indeed connects reactants to products.

The automatic discovery of reaction steps allows an unbiased exploration of the potential energy surface (PES) where novel saddle-points connecting to reactants and products can occur through unconsidered atom rearrangements or possibly through an unknown reaction type altogether. Several applications of this method to discover unexpected reaction pathways of species important in combustion and pyrolysis chemistry are reported.

Copyright © American Institute of Chemical Engineers. All rights reserved.

471269 Computational Singular Perturbation As an Analytical Tool to Study the Fischer-Tropsch Process

Abstract:

Computational Singular Perturbation as an analytical tool to study the Fischer-Tropsch process

Debanajan Chakrabarti, Gouthami Senthamaraikkannan and Vinay Prasad

The Fischer-Tropsch synthesis[1] process generates synthetic crude oil (syncrude) from carbonaceous sources such as natural gas, coal and biomass via a synthesis gas intermediate. The process involves a series of simultaneous polymerization and hydrogenation reaction steps, leading to a hydrocarbon and oxygen rich product comprising of $C_1 - C_{100+}$ hydrocarbon and oxygenate species. The product can be refined in a manner similar to crude oil to obtain naphtha, distillate and petrochemical products. However, the process is also plagued by the formation of CO_2 and CH_4 as by-products, which decrease the process efficiency significantly. While the process parameters can generally be adjusted under steady state operation to improve the system performance by increasing the desirable product yield and decreasing the CO_2 and CH_4 selectivity, there is a limit to the extent of this improvement. However, a significant improvement in the process performance can be brought about by improvement in catalyst design and by exploiting unsteady state behaviour of the system by means of periodic operations[2,3]. However, both of these strategies require an intimate understanding of the reaction behaviour.

Computational Singular Perturbation study (CSP)[4,5]is a numerical technique which can be used to study the dynamic properties of a complex reaction system. It involves discretizing a kinetic model into various reaction modes, and grouping them into active and inactive modes, based on the timescales and amplitudes of each mode. This allows the establishment of a time varying hierarchy of each reaction step and intermediate species, based on their contribution to the overall reaction behaviour, at each time

instant. This can enable the establishment of time-varying reduced kinetic models, to capture the dynamic behaviour of the reaction system.

The validity of this technique depends on the accuracy of the reaction mechanism and kinetic model, utilized to represent the reaction system. Based on observations in literature and experiments, the Fischer-Tropsch chain growth has been determined to be the result of a CO addition based mechanism, while accompanied by CO disporoption based methanation and water gas shift side reactions[1]. We have utilized the CO-insertion based kinetic model developed by Todic et al.[6]to conduct our analysis for identifying the dominant elementary reactions and species, at different time scales and operating conditions. This would enable us to determine the conditions to promote desirable reactions, as well as to design experiments for periodic operation.

References

1. Chakrabarti, D.; Prasad, V.; de Klerk, A., Mechanism of the Fischer-Tropsch process. In ***Fischer-Tropsch Synthesis, Catalysts, and Catalysis: Advances and Applications***, Davis, B. H.; Occelli, M. L., Eds. Taylor and Francis: Boca Raton, 2016; pp 183-222.

2. Adesina, A.; Hudgins, R.; Silveston, P., Feed composition modulation of hydrocarbon synthesis over a cobalt oxide catalyst. ***The Canadian Journal of Chemical Engineering* 1986,**64, (3), 447-454.

3. Feimer, J.; Silveston, P.; Hudgins, R., Influence of forced cycling on the fischer tropsch synthesis. Part I. Response to feed concentration step changes. ***The Canadian Journal of Chemical Engineering* 1984,**62, (2), 241-248.

4. Lam, S.; Goussis, D. In ***Understanding complex chemical kinetics with computational singular perturbation***, Symposium (International) on Combustion, 1989; Elsevier: 1989; pp 931-941.

5. Zagaris, A.; Kaper, H. G.; Kaper, T. J., Analysis of the computational singular perturbation reduction method for chemical kinetics. ***Journal of Nonlinear Science* 2004,**14, (1), 59-91.

6. Todic, B.; Ma, W.; Jacobs, G.; Davis, B. H.; Bukur, D. B., CO-insertion mechanism based kinetic model of the Fischer–Tropsch synthesis reaction over Re-promoted Co catalyst. ***Catalysis Today* 2014,** 228, 32-39.

Copyright © American Institute of Chemical Engineers. All rights reserved.

National Institute of Chemistry Slovenia

Mechanism, Reaction Kinetics and Multi-Scale Modelling of CO_2 Hydrogenation to Methanol over Trimetallic Heterogeneous Catalysts

Neja Strah Štefančič, Andraž Pavlišič, Matej Huš, Venkata Dasireddy, Andrej Pohar, Blaž Likozar

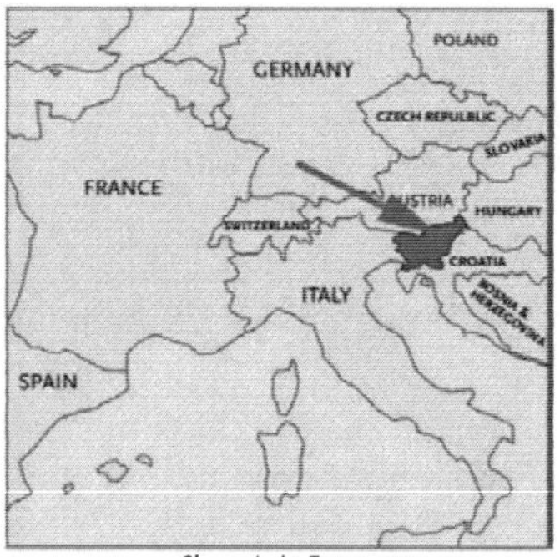

Slovenia in Europe

National Institute of Chemistry
Ljubljana
Slovenia

Ljubljana, the capital of Slovenia

San Francisco
November 15, 2016

National Institute of Chemistry Slovenia

Which catalyst to synthesize?

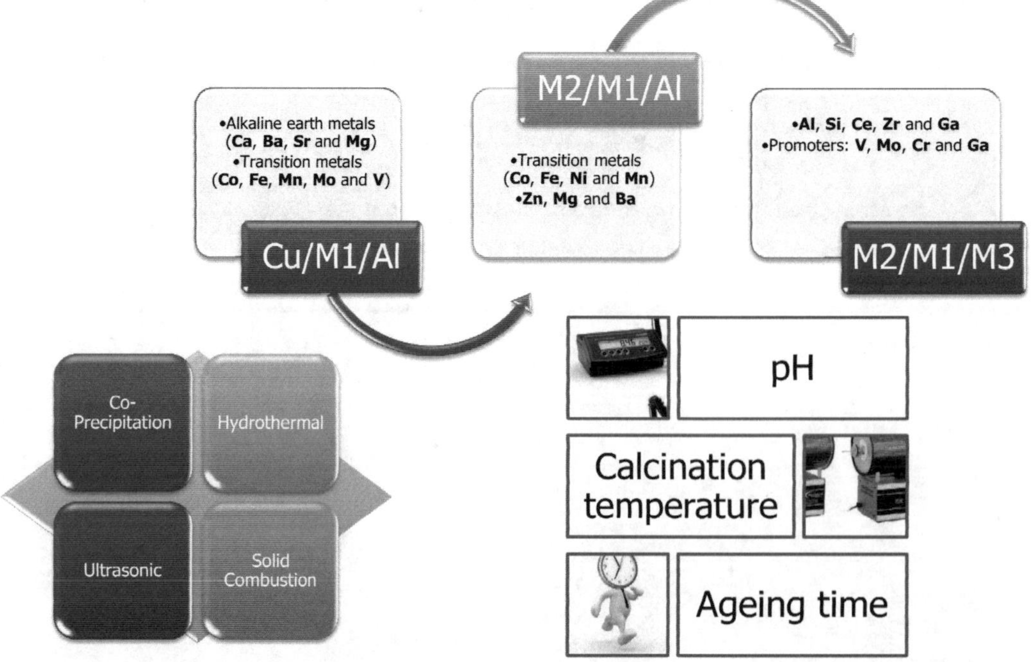

National Institute of Chemistry Slovenia

What does the Cu/Zn/Al catalyst look like?

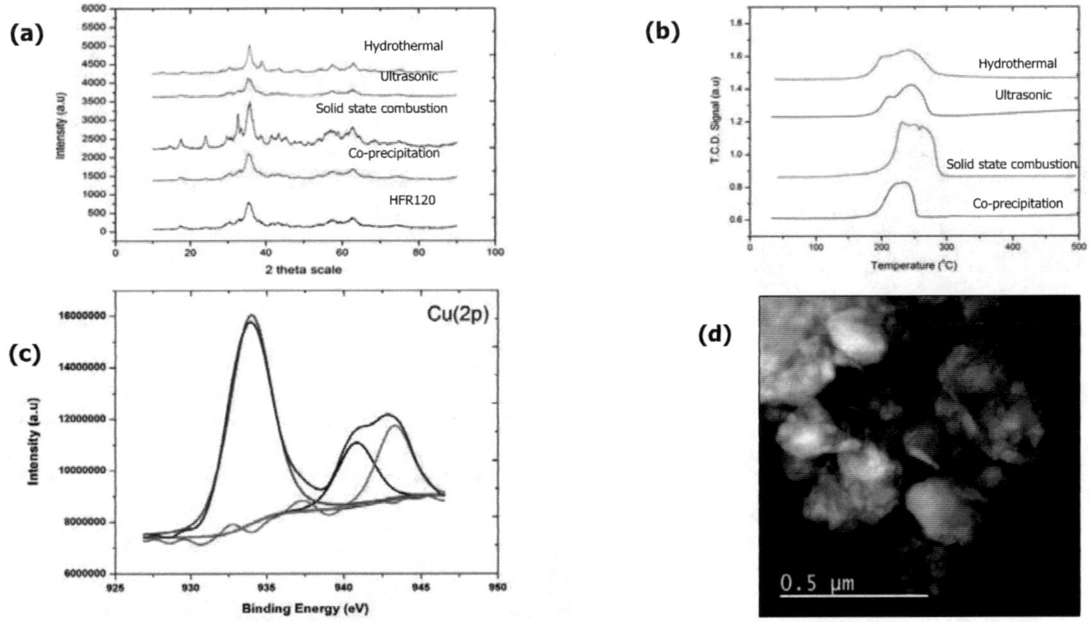

(a) and (b) TPR profiles of Cu/Zn/Al catalysts. (c) XPS spectra and (d) STEM image of Cu/Zn/Al prepared by ultrasonic method

National Institute of Chemistry Slovenia

Effect of pressure and temperature

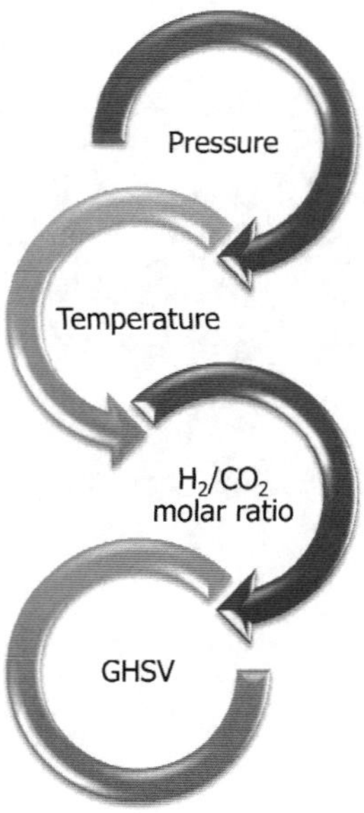

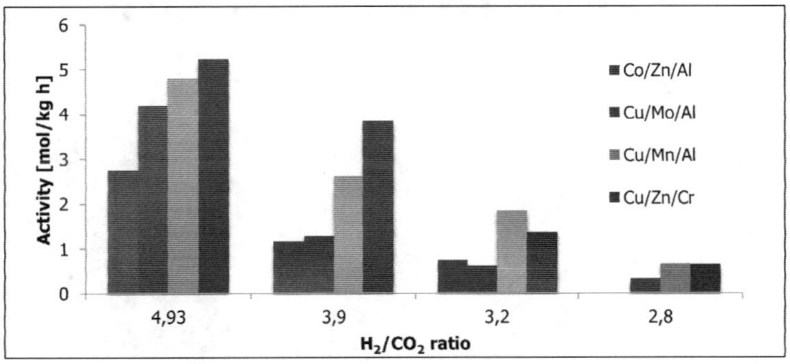

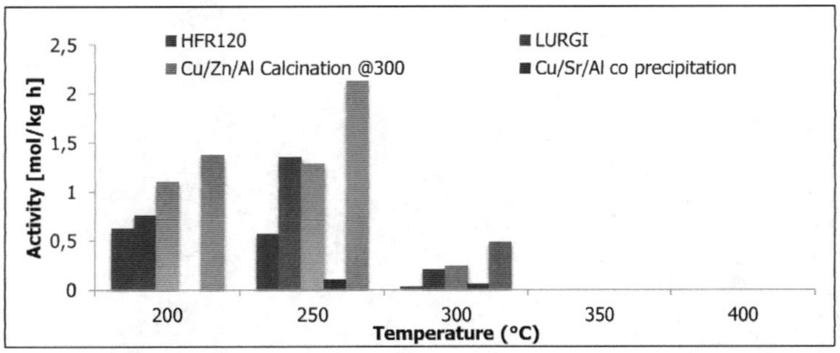

Pressure = 20 bar, GHSV of 6000 h^{-1}

National
Institute of Chemistry
Slovenia

Effect of catalyst preparation method

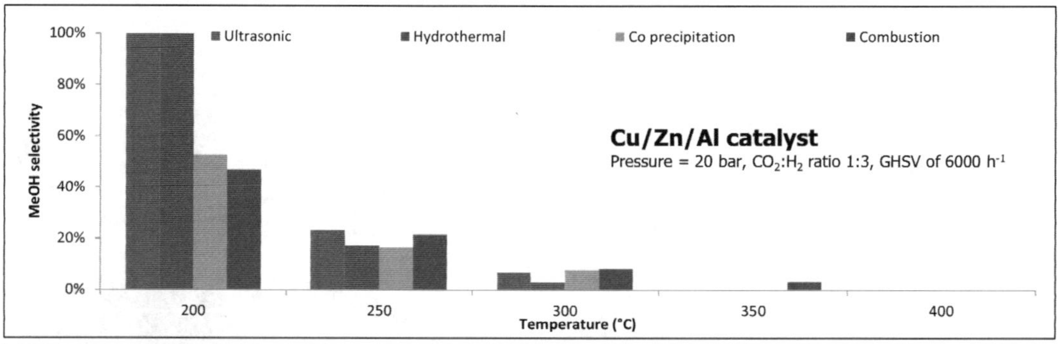

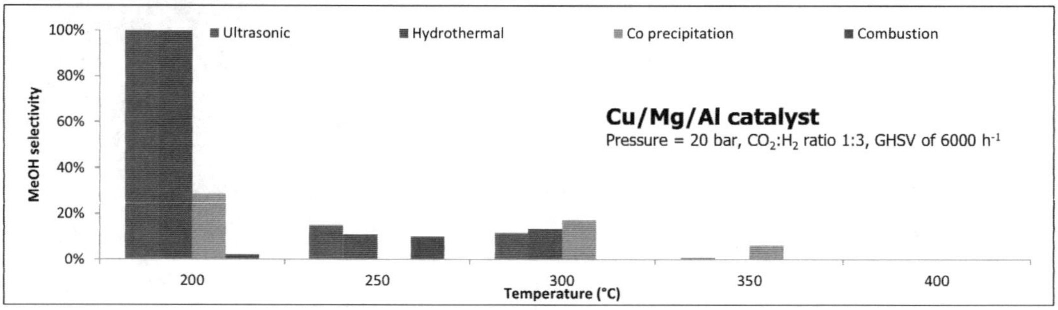

National
Institute of Chemistry
Slovenia

Atomistic model of catalyst

Challenges addressed:

- Include support effects
- Include promotor effects
- Interface effects
- Cu/Cu^+ interplay
- Large enough to avoid spurious inter-cell interactions
- Small enough to be computationally tractable

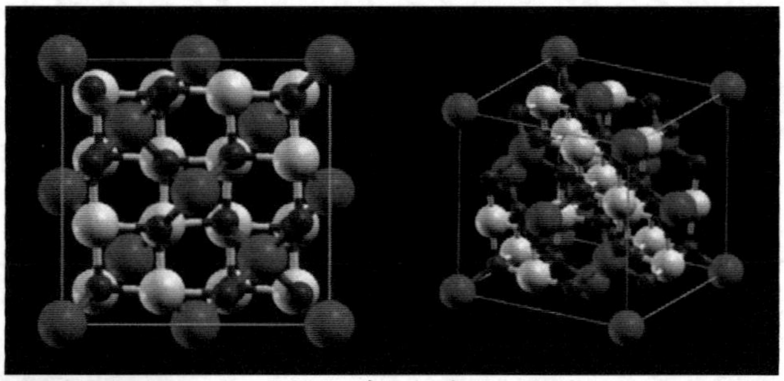

⇧ Spinel type $ZnAl_2O_4$ as a support

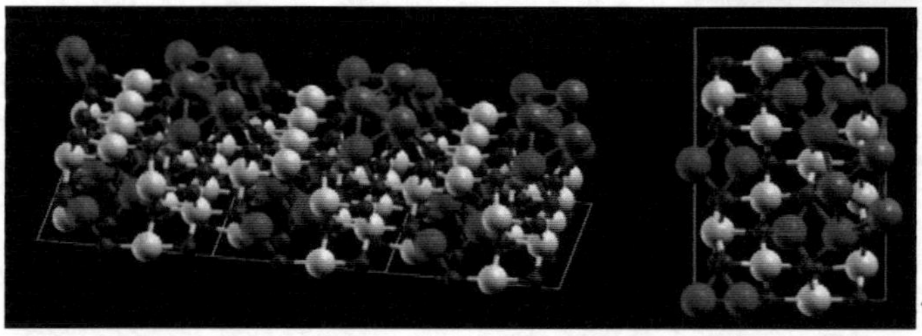

⇦ Active cluster Cu_3Zn_3 on the support

National
Institute of Chemistry
Slovenia

All possible pathways

Exact pathway not known before

Several theories

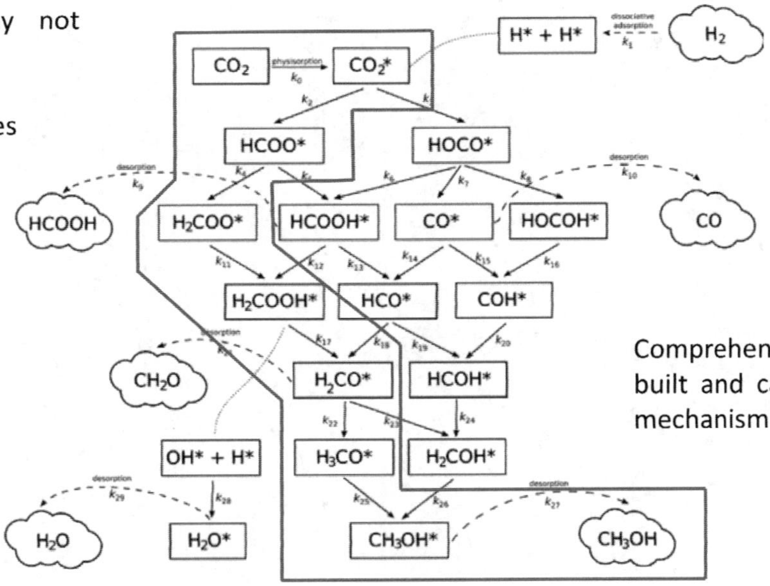

Comprehensive scheme built and calculated for mechanism elucidation

National
Institute of Chemistry
Slovenia

Formate pathway predominates

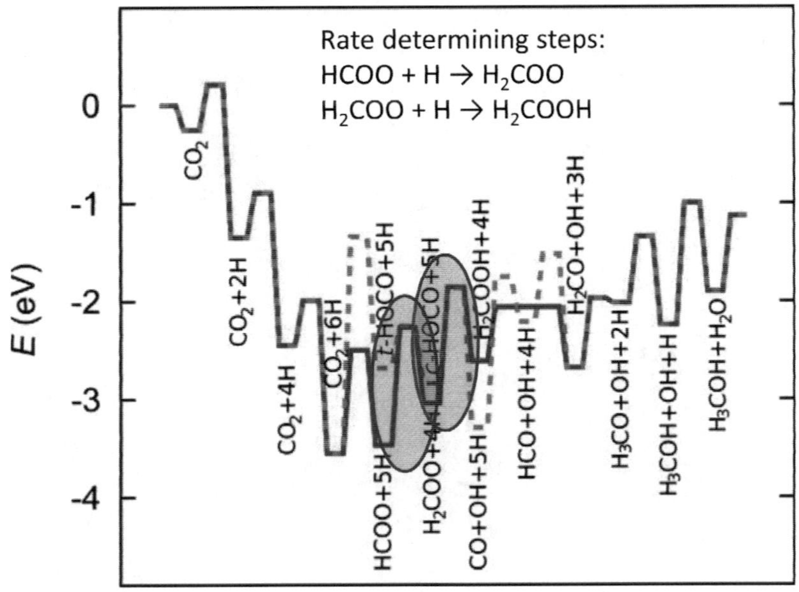

National
Institute of Chemistry
Slovenia

Simulation

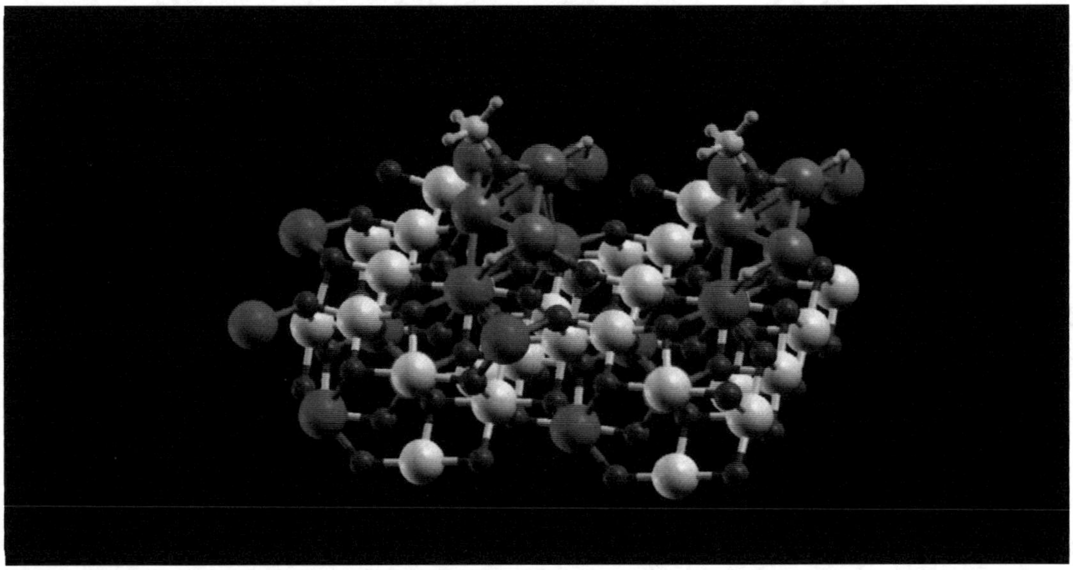

National
Institute of Chemistry
Slovenia

Kinetic Monte Carlo

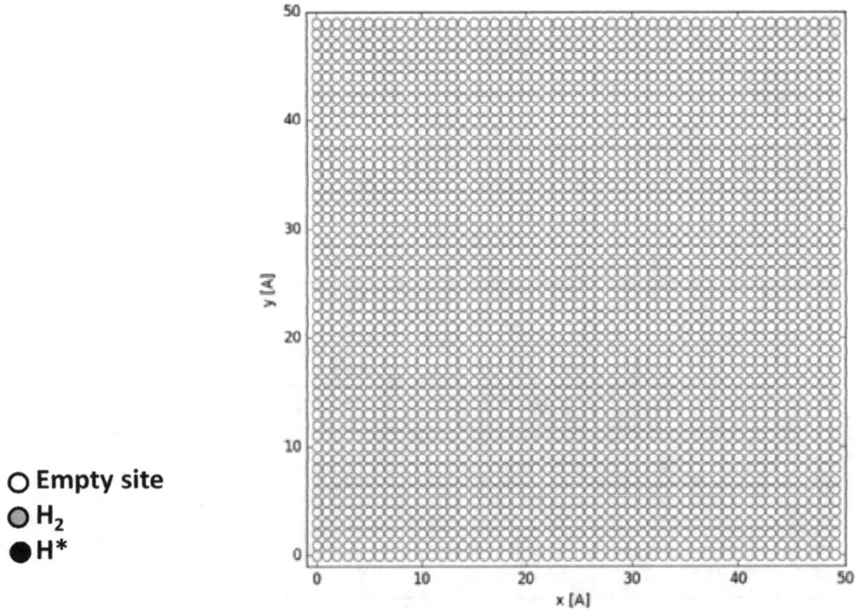

National
Institute of Chemistry
Slovenia

Microkinetic modeling

- Processes occurring on the catalyst surface
- Detailed reaction mechanism

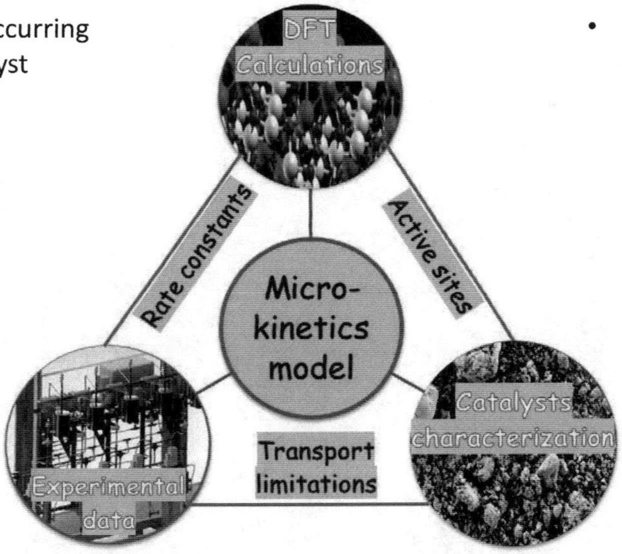

Figure xx: Scheme of microkinetic analysis.

National
Institute of Chemistry
Slovenia

Microkinetic description

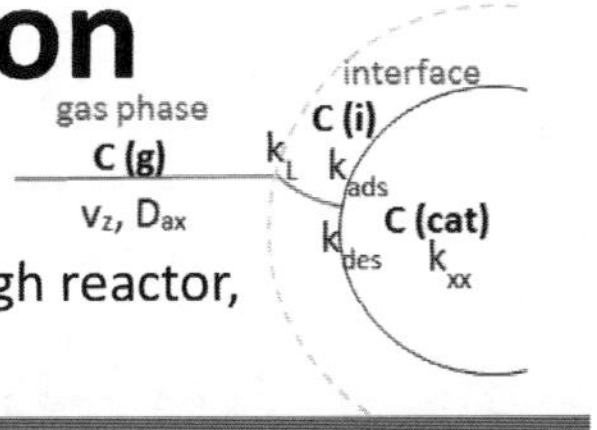

Convection-axial dispersion reactor model includes mass transfer through reactor, sorption processes and surface reactions.

Species (phase)	Balance
i, g = $CO_{2(g)}$, $H_{2(g)}$, $CO_{(g)}$, $CH_3OH_{(g)}$, $HCOOH_{(g)}$, $CH_2O_{(g)}$, $H_2O_{(g)}$ *(gas phase)*	$\frac{\partial C_{i,g}}{\partial t} = D_{i,ax}\frac{\partial^2 C_{i,g}}{\partial z^2} - v\frac{\partial C_{i,g}}{\partial z} - \frac{(1-\varepsilon)}{\varepsilon}\cdot k_L a_{cat}\cdot\left(C_{i,g} - C_{i,if}\right)$
i, if = $CO_{2(if)}$, $H_{2(if)}$, $CO_{(if)}$, $CH_3OH_{(if)}$, $HCOOH_{(if)}$, $CH_2O_{(if)}$, $H_2O_{(if)}$ *(interface)*	$\frac{dC_{i,\mathrm{if}}}{dt} = 0 = k_L a\cdot\left(C_{i,g} - C_{i,if}\right) - k_{ads}\cdot C_{i,\mathrm{if}}\cdot C^* + k_{des}\cdot C_{i,cat}$
θ_i = CO_2^*, H^*, CO^*, CH_3OH^*, $HCOOH^*$, CH_2O^*, H_2O^* *(surface coverages of stable intermediates)*	$\frac{d\theta_i}{dt} = k_{ads}\cdot p_i\cdot\theta_{free} - k_{des}\cdot\theta_i \pm R_j$
θ_i = $HCOO^*$, H_2COO^*, H_2COOH^*, H_3CO^*, H_2COH^*, $HOCO^*$, $HOCOH^*$, HCO^*, COH^*, $HCOH^*$ *(surface coverages of unstable intermediates)*	$\frac{d\theta_i}{dt} = \pm R_j$

$$\theta_i = \frac{C_{i,cat}}{C^*}$$

C^*... concentration of active sites

National
Institute of Chemistry
Slovenia

CFD packed bed - Challenges

- random packed column (Blender)

- Meshing computational domain (OpenFoam)

National
Institute of Chemistry
Slovenia

CFD packed bed – Results

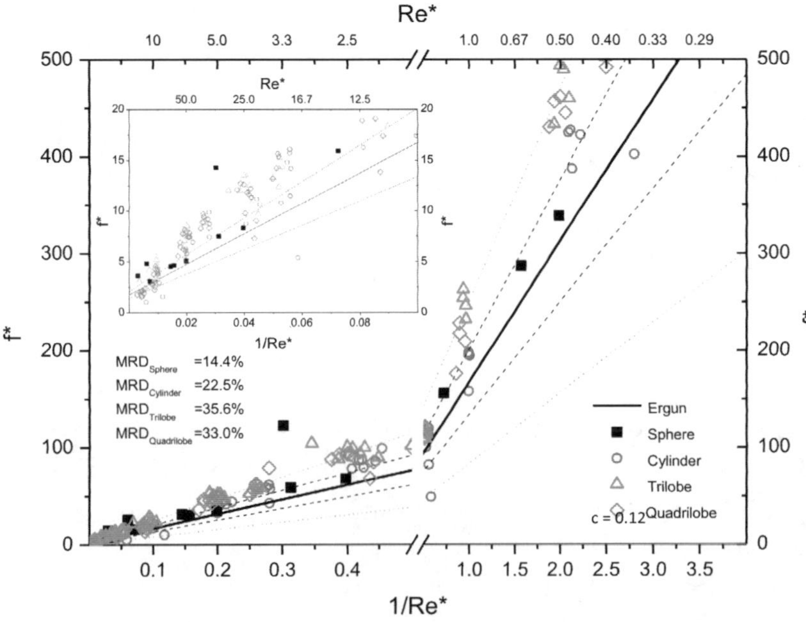

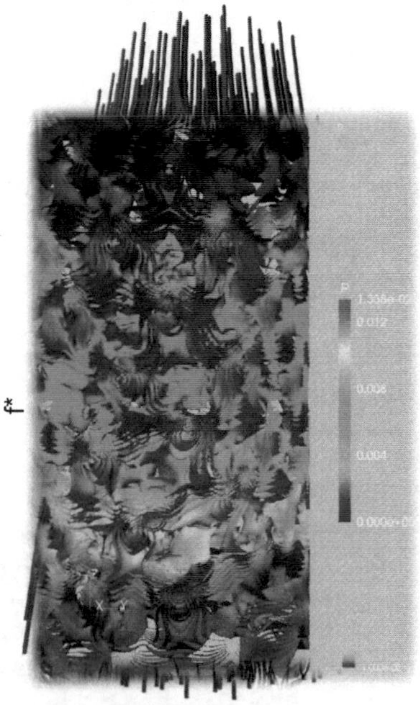

National
Institute of Chemistry
Slovenia

Conclusion

- Several different catalysts were **synthesized** with systematically **varying metals**
- **CuZnAl** and **CuMgAl**: the **best selectivity** and converisons (@ 200-250 °C)
- DFT: **kinetic parameters** were obtained – formate pathway predominates
- Microkinetic modelling and KMC: DFT kinetics was used to **simulate the reaction**
- CFD: **Pressure drop** in a packed bed reactor (PBR) as compared with correlation function.

National
Institute of Chemistry
Slovenia

Acknowledgements

Funding

- **European Union**
  (Framework Programme for Research and Innovation Horizon 2020 – Grant N 637016)
- **Republic of Slovenia**
  (Research Agency – Programme P2-0152)

458459 Mechanistic Insights into the Catalytic Elimination of Tar on Nickel and the Promotional Effect of Boron on It: A Combined Theoretical and Experimental Study Using Toluene As a Model Compound

Abstract:

Mechanistic insights into the catalytic elimination of tar on Nickel and the Promotional effect of Boron on it: A combined theoretical and experimental study using Toluene as a model compound

Quang Thang Trinh[1], Anh Vu Nguyen[2], Dang Chinh Huynh[2], Thanh Huyen Pham[2], Samir H. Mushrif [1,*]

[1]School of Chemical and Biomedical Engineering, Nanyang Technological University, 62 Nanyang Drive, Singapore 637459, Singapore.

[2]School of Chemical Engineering, Hanoi University of Science and Technology, 1 Dai Co Viet St., Hanoi, Vietnam

*Email: SHMushrif@ntu.edu.sg

Abstract: Tar is the undesired viscous black liquid produced during the gasification of biomass. Catalytic elimination of tar, via steam reforming, is an economically promising technique. Nickel (Ni) is the most widely studied catalyst for the process; however, rapid deactivation of Ni catalyst due to coking is a challenge.[1,2] Decorating metals with subsurface Boron was reported to improve the resistance of the catalyst against both, bulk carbide and graphene formation.[3,4] Motivated by that study, we performed first principles calculations to investigate the promotional effect of Boron (B) on Ni catalyst for tar decomposition.[5] Being the most abundant component in tar, Toluene is chosen as a model compound.[1] On Ni(111) surface, Toluene adsorbs strongly in a bridge configuration and the activation barrier for methyl C-H dissociation is 71.9 kJ/mol. Toluene can further decompose on Ni(111) surface via the stepwise dehydrogenation of the methyl group. The aromatic C-H bond at ortho position could only be activated after complete dehydrogenation of the methyl group and is followed by subsequent ring opening (activation barriers are 112.5 and 83.8 kJ/mol, respectively). Further C-C cleavages are also feasible to generate smaller hydrocarbon compounds. Our calculations reveal that the incorporation of subsurface B into Ni catalysts results in corrugated Ni top surface. Toluene adsorbs 10.3 kJ/mol stronger on Ni-B catalyst than on pure Ni, in the hollow configuration. The activation of Toluene is significantly promoted on Ni-B, with an activation barrier for the first methyl C-H dissociation of 51.0 kJ/mol (vs. 71.9 kJ/mol on Ni). All the subsequent C-H activations are also promoted on Ni-B catalyst, but to a smaller extent, except the aromatic C-H activation (88.7 kJ/mol vs 112.5 kJ/mol on pure Ni). The ring-opening at ortho position occurs with the activation barrier of 93.7 kJ/mol, slightly higher than that on pure Ni (83.8 kJ/mol). Additional calculations on stepped surface models of Ni show that the activation barriers of Toluene decomposition on Ni-B surface are very close to those on B5 and F4 step sites; thus suggesting that the promotional effect of B on the catalytic activity of Ni could be due to the creation of step like corrugations on Ni surface. Experimental study was also performed to validate our predictions. Ni doped with 1% wt of Boron (B-Ni) and pure Ni catalysts supported on bentonite were prepared using the method reported by Xu et al.[3] and were tested for toluene decomposition at 600°C. In excellent agreement with theoretical predictions, initial toluene conversion was higher on B-Ni (68%) than on pure Ni catalyst (62%), demonstrating the role of boron in promoting the activity of the catalyst. SEM images revealed negligible carbon formation on the surface of B-Ni, while the surface of pure Ni catalyst was completely covered with carbon filaments under the same reaction conditions. These experimental evidences confirmed that the presence of boron indeed improved both, the activity and the stability of the catalyst.

References:

1. Y. Shen, K. Yoshikawa, *Renew. Sust. Energ. Rev.* **2013**, *21*, 371–392.
2. J.R. Rostrup-Nielsen, *Catal. Today* **1993**, *18*, 305-324.
3. J. Xu, L. Chen, K. F. Tan, A. Borgna, M. Saeys, *J. Catal.* **2009**, *261*, 158–165.
4. K.F. Tan, J. Chang, A. Borgna, M. Saeys, *J. Catal.* **2011**, *280*, 50–59.
5. Q. T. Trinh, A. V. Nguyen, D. C. Huynh, T. H. Pham, S. H. Mushrif, *Catal. Sci. Tech.*, **2016**, in press, **DOI:** 10.1039/C6CY00358C.

Copyright © American Institute of Chemical Engineers. All rights reserved.

457534 Discovery of Superior Cu-GaO_x-HoO_y Catalysts for the Reduction of Carbon Dioxide to Methanol at Atmospheric Pressure

Abstract:

Catalytic conversion of carbon dioxide to liquid fuels and basic chemicals using solar-derived hydrogen at or near ambient pressure is a highly desirable goal in heterogeneous catalysis. When realized, this technology will pave the way for a sustainable society together with decentralized power generation. Here we report a novel class of superior holmium (Ho) containing multi-metal oxide Cu catalysts discovered through the application of high-throughput methods. In particular, ternary systems of Cu-GaO_x-HoO_y > Cu-CeO_x-HoO_y ~ Cu-LaO_x-HoO_y supported on Î^3-Al_2O_3 exhibited superior methanol activities (10x) with less CO production than previously reported catalysts at 1 atm pressure. Holmium was shown to form highly dispersed 1-3 atom clusters, thus exhibit a dopant effect in promoting CH_3OH formation over larger (5-7 nm) Cu and Ga clusters.

Copyright © American Institute of Chemical Engineers. All rights reserved.

463175 Methanation of Carbon Dioxide with a Novel Nickel/Magnesium Oxide Catalyst

Abstract:

Methanation of carbon dioxide with a novel nickel/magnesium oxide catalyst

Georg Baldauf-Sommerbauer, Susanne Lux, Matthaeus Siebenhofer

Institute of Chemical Engineering and Environmental Technology, Graz University of Technology

The heterogeneous nickel-catalyzed conversion of carbon monoxide (Eq. 1) and dioxide (Eq.3) into methane has first been reported more than 110 years ago by Sabatier and Senderens [1]. In the last years, reports on the performance and kinetics of different nickel-based catalysts have increased due to rising interest in carbon dioxide valorization [2, 3]. One of the reasons for this growing interest in CO_2 methanation can be found in the quest for alternative hydrogen storage systems. In this broadly discussed concept [4], renewably produced hydrogen is stored as methane, benefitting from readily available natural gas storage and distribution infrastructure.

$CO + 3\ H_2 = CH_4 + H_2O$ (1)

$CO_2 + 4\ H_2 = CH_4 + 2\ H_2O$ (2)

$CO_2 + H_2 = CO + H_2O$ (3)

Most reports in this area focus on alumina Al_2O_3 as an inert catalyst support. Alumina based nickel catalysts are well characterized but often show deactivation. Furthermore, Al_2O_3 does not intervene in the catalytic process itself.

Methanation of carbon dioxide possibly proceeds via the intermediate carbon monoxide. Thus, the reaction sequence would be CO_2 (Eq.3) --> CO (Eq.1) --> CH_4. Magnesium oxide [5] shows high activity for the reverse water-gas-shift reaction at moderate temperatures below 450 °C. Therefore, our approach is to combine the catalytic properties of MgO and nickel to facilitate the overall carbon dioxide conversion. Moreover, magnesium oxide is a cheap and abundant material that can be easily produced from mineral magnesium carbonate through calcination. Seemingly doping of catalysts with MgO increases the stability and resistance to carbon deposition of Ni/Al_2O_3 catalysts, as applied in CO methanation [6, 7].

Different types of Ni/MgO catalysts (8-20 %wt. nickel loading) were synthesized by a reactive impregnation of MgO with aqueous nickel nitrate solutions. After drying and calcination in air, nickel oxide was reduced to elemental nickel in hydrogen atmosphere. The catalytic activity was determined in screening experiments by applying a constant heating rate of 2 °C from 25 to 500 °C in a tubular reactor at process-like gas feed of $H_2:CO_2:N_2 = 4:1:5$ (15 - 60 $L_{STP}\ h^{-1}$) and at atmospheric pressure. Additional experiments were conducted at isothermal operation conditions. CO_2 conversion increases with increasing nickel loading at isothermal conditions. At high space velocity, carbon monoxide and methane is formed. Decreasing the space velocity at constant operation conditions increases the methane selectivity to above 90 %. This finding supports the suggested mechanism of intermediate carbon monoxide formation. Conversion of CO_2 of more than 80 % below operation temperature of 380 °C was achieved at 1 atm total pressure. The catalysts did not show loss of activity after more than 20 hours operation time.

[1] Sabatier P., Senderens J.-B.: Compt. Rendus Hebd. Chimie 1902 (134), 514-516

[2] Rönsch, S. et al.: Fuel 2016 (166), 276-296

[3] Wang, W.; Wang., S.; Ma, X.; Gong, J.: Chemical Society Reviews 2011 (40), 7, 3703-3027.

[4] Sterner, M.: PhD Thesis, University of Kassel 2009.

[5] Baldauf-Sommerbauer, G.; Lux, S.; Siebenhofer M.: Oral presentation, AIChE Annual Meeting 2015, Salt Lake City, Utah, USA.

[6] Hu, D. et al.: Industrial & Engineering Chemistry Research 2012 (51), 4875-4886.

[7] Fan, M. et al.: Applied Surface Science 2014 (307), 682-688

Copyright © American Institute of Chemical Engineers. All rights reserved.

464996 Polymer-Nanocrystal Composites As Tailorable CO_2 Hydrogenation Catalysts

Abstract:

The future of humankind is tied to our collective use of energy. Hydrocarbons are a proven energy vector and account for the bulk of our energy infrastructure and feedstocks to the chemical industry. Combustion of fossil based hydrocarbons is unsustainable given finite reserves and the concomitant release of greenhouse gases. Hydrogenation of waste CO_2 back in to useful fuels and feedstocks provides a pathway to use fossil resources in a sustainable way. The emerging technologies of photo and electrochemical hydrogen generation can provide the hydrogen needed for CO_2 hydrogenation renewably. However, a main issue is still in finding CO_2 hydrogenation catalysts that work at low temperature and that are selective towards desirable compounds such as methanol.

In this study, we introduce a framework with which an effective CO_2 hydrogenation catalyst may be designed. Microporous polymers provide a tunable chemical environment into which monodisperse Palladium nanocrystal catalysts are incorporated. The synergy of the confinement effect of these frameworks with well-defined metal nanoparticles affording highly active hydrogenation sites provides a platform for catalyst study and optimization under conditions outside the range of standard inorganic supports. Combining gas sorption and catalytic data provides insights into the potential that organic supports have for carbon dioxide hydrogenation.

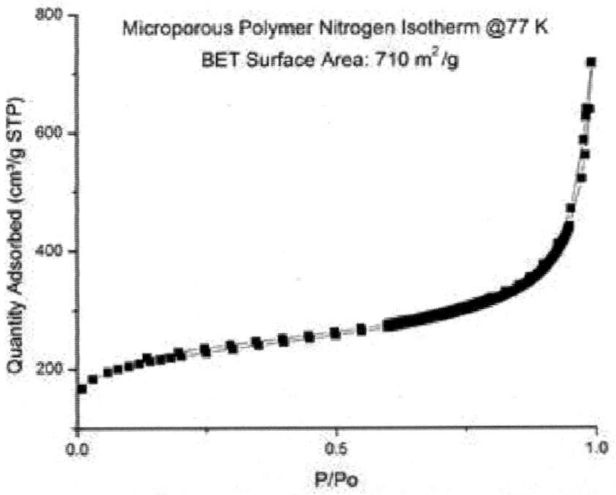

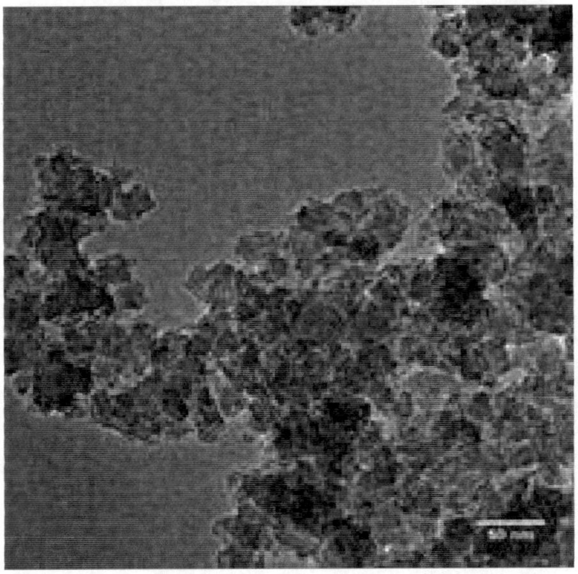

Copyright © American Institute of Chemical Engineers. All rights reserved.

466544 Reduction of CO_2 to CO and Methanol on Ceria Based Catalyst: Mechanistic Insights

Abstract:

Carbon dioxide (CO_2) utilization via reduction to commodity chemicals and fuels can be accomplished by the catalytic and electrocatalytic reduction reactions. Since CO_2 is a thermodynamically stable molecule, it requires high temperature and pressure condition to obtain favorable conversion to a desired product[1]. Ceria based materials, having high oxygen storage capacity and mixed ionic-electronic conducting property, have been suggested to play an active role in the catalytic and electrocatalytic reduction of CO_2. We have performed density function theory (DFT) calculations to study the mechanistic insights of CO_2 reduction on $CeO_2(110)$ surface. CO_2 molecule sitting in the vicinity of oxygen vacancy site on the surface, is activated to form bent carbonate CO_2^{d-} species which dissociates into CO via the incorporation of the oxygen atom into the vacancy. The calculated activation barrier and reaction energy for this redox reaction is 258.9 kJ/mole and 238.6 kJ/mole respectively. The effect of lateral interactions were studied by performing calculations for the same reaction step on two oxygen vacancy (di-vacancy) on 2x2 supercell unit. The activation barrier and reaction energy on a di-vacancy were significantly reduced to 134.3 and 127.3 kJ/mole respectively. In presence of hydrogen, CO_2 dissociation to CO is assisted by the atomic hydrogen adsorbed on the reduced ceria surface. Reaction energy diagram with intrinsic reaction and activation energy is illustrated in Figure 1A. In presence of a hydrogen atom the CO_2 dissociation reaction occurs in two exothermic steps: CO_2+H→ COOH (ΔH = -69.2 kJ/mole), COOH→CO+OH (ΔH = -80.4 kJ/mole) with activation barrier of 39.0 and 47.4 kJ/mole respectively. CO_2 or CO adsorbed on the ceria surface could hydrogenate to methanol via formate (HCOO) or carboxyl (COOH) mediated mechanisms. Formate species, produced by the hydrogenation of CO_2 (CO_2+H→ HCOO), was observed to be more stable with binding energy of -222.9 kJ/mole on the stoichiometric ceria surface as compared to the carboxyl intermediate species with binding energy of -36.0 kJ/mole[2]. Due to its highly stable structure, formate species may act as a spectator and may not participate in further hydrogenation reaction. On reduced surface COOH dissociation occur by the incorporation of carboxyl species into the surface vacancies via redox mechanism. On the other hand, carboxyl mediated mechanism involve exothermic reaction except the dissociation of COOH to CO and OH which is endothermic of -5.0 and -24.4 kJ/mole on stoichiometric and reduced ceria surface. On stoichiometric ceria surface, the endothermic dissociation step (COOH→CO+OH) has the maximum barrier of 55.6 kJ/mole among the elementary steps associated with carboxyl mediated route. On reduced ceria surface, intrinsic activation barriers of elementary reaction steps, associated with carboxyl mediated path, were calculated (Figure 1A). Hydrogen atom co-adsorbed on the surface was observed to assist CO_2 dissociation by forming a carboxyl intermediate, CO_2+H→COOH (ΔEact = 39.0 kJ/mole, ΔH = -69.2 kJ/mole) which on subsequent dissociation produces CO via the redox mechanism. On hydrogenation, CO is likely to produce methanol. The energetics of CO hydrogenation to produce methanol showed exothermic steps with activation barriers comparable to the DFT calculations reported for Cu(111) and CeO2-x/Cu(111) interface. While on the stoichiometric surface, COOH dissociation COOH→CO+OH (ΔEact = 55.6 kJ/mole, ΔH = 5.7 kJ/mole) is likely to be difficult as compared to rest of the elementary steps, whereas on the reduced surface the energetics of the same step were significantly lowered (ΔEact = 47.4 kJ/mole, ΔH = -80.4 kJ/mole). In comparison, hydrogenation of methoxy, H3CO+H→H3COH, appears to be relatively difficult (ΔEact = 58.7 kJ/mole) on the reduced surface. COOH dissociation step has the maximum barrier (126 kJ/mole) as compared to other hydrogenation elementary steps which implies that the dissociation step could be the rate determining step of this route. The activation energy of this rate determining step was calculated on reduced ceria surface which is lower (by ~50kJ/mole) than that on stoichiometric ceria surface. Ceria based materials have been suggested to possess electrocatalytic activity for CO_2 reduction which can be further improved by aliovalent metal dopants. Surface of ceria doped with aliovalent dopants such as gadolinium (Gd), praseodymium (Pr) and samarium (Sm), have been suggested to have mixed ionic-electronic conductive nature. Classical molecular dynamic simulations were utilized to determine the oxide ion diffusivity (D) in the Pr doped ceria (PDC) at different temperatures (Figure 1B). The calculated diffusivity was of 1.15×10^{-8} cm^2s^{-1} at 973 K. The activation energy of oxide ion diffusion in PDC was estimated to be of 37.0 kJ/mole. We have performed the electrocatalytic reduction to CO_2 to CO in a high temperature solid oxide electrolysis cell (SOEC). CuO-PDC composite, was used as an electrocatalyst for CO_2 reduction in the SOEC system. Physical characterization of cathode, anode and electrolyte were done by X-Ray diffraction (XRD) pattern (Figure 1C) and scanning electron microscopy (SEM). Electrochemical impedance spectroscopy (EIS) were performed in presence CO_2/CO atmosphere to determine the ohmic and polarization losses associated with electrolyte and electrode components of SOEC (Figure 1D). Combined with high catalytic activity and fast oxygen anion transport, ceria materials could be a potential candidate for catalytic or electrocatalytic reduction of CO_2.

Reference

1. Neetu Kumari, M. Ali Haider, and S. Basu. Mechanism of Catalytic and Electrocatalytic CO_2 Reduction to Fuels and Chemicals. *CRC Press* **2016**, *Chapter 6*, 267–286.

2. Kumari, N.; Sinha, N.; Haider, M. A.; Basu, S. CO_2 Reduction to Methanol on $CeO_2(110)$ Surface: A Density Functional Theory Study. *Electrochim. Acta* **2015**, *177*, 21–29.

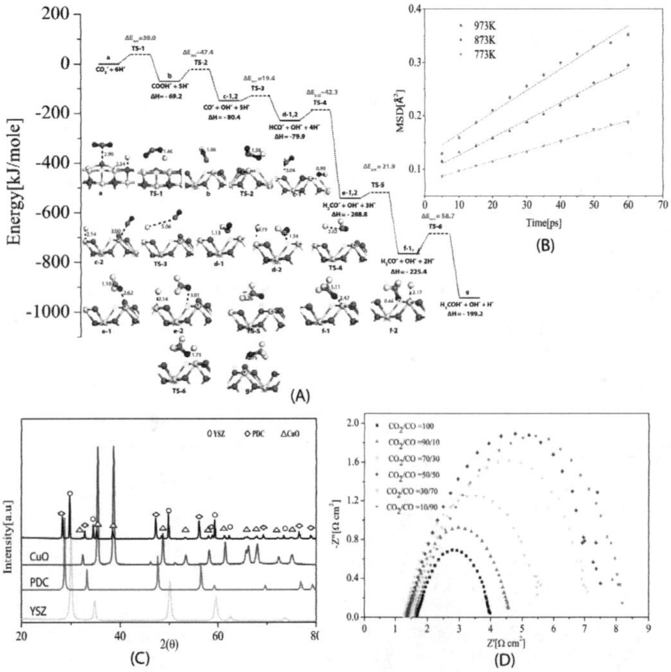

Figure 1 (A) Reaction energy of CO_2 reduction to CO and methanol on reduced ceria surface, (B) Mean square displacement plot for PDC at three different temperatures, (C) XRD pattern of PDC cathode of SOEC, and (D) EIS plot in different environment of CO_2/CO.

Copyright © American Institute of Chemical Engineers. All rights reserved.

467567 Catalytic Conversion of Carbon Dioxide with Solar Hydrogen Under Supercritical Conditions

Abstract:

Introduction

Energy storage technologies are of great interest at the present time. Excess electricity generated from wind parks and/or photovoltaic can be used to produce so called solar-hydrogen via electrolysis of water. This offers a high potential for the chemical storage of energy [1]. The solar-hydrogen can be converted with carbon dioxide to many different hydrogenation products, such as methane, methanol, dimethyl ether or hydrocarbons [2].

A favorable hydrogenation product and subject of current research is methanol. Compared with hydrogen the liquid state of methanol allows an easy storage and transportation. Methanol has excellent combustion properties making it suitable as fuel for combustion engines [3]. Furthermore, methanol can be used as commodity chemical in the chemical industry or as feedstock for intermediates such as formaldehyde, MTBE or acetic acid.

In this contribution a process for the carbon dioxide hydrogenation under supercritical conditions will be presented. Herein, the reaction equilibrium is shifted to the product side and high space-time-yields and therefore low investment costs could be achieved.

Experimental

$Cu/ZnO/Al_2O_3$-catalysts were used for the hydrogenation of carbon dioxide to methanol. The catalysts were prepared by a co-precipitation method of the corresponding metal nitrates using potassium hydroxide. The performance of these catalysts was examined in dependence of the temperature. The experiments were carried out in a tubular reactor at a pressure of 150 bar and temperatures between 200 and 380 °C. The liquid phase was analyzed via GC-analysis and the gas composition via FTIR-spectroscopy.

Results

In the hydrogenation of carbon dioxide over $Cu/ZnO/Al_2O_3$-catalysts methanol is the main product. In dependence of temperature the methanol yield and selectivity is increasing up to 300 °C. At higher temperature methanol selectivity is decreasing as expected from thermodynamics because of endothermic reverse watergas-shift-reaction. The highest methanol yield was achieved at a temperature of 300Â Â°C. The results are compared to a commercial available methanol-synthesis-catalyst.

At lower temperatures (<230Â Â°C) high selectivity to methanol and high CO_2 conversion could be achieved due to phase separation. To achieve this goal a more active catalyst is needed.

References

[1] F. Ausfelder, A. Bazzanella, Chem.-Ing.-Tech. 2009, 81 (10), 1565-1573.

[2] W. Wang, S. Wang, X. Ma, *Chem. Soc. Rev.* **2011**, 40, 3703-3727.

[3] API Report No. 4261, **1988**, *Alcohols and Ethers â€" A Technical Assessment of Their Application as Fuels and Fuel Components*

Copyright © American Institute of Chemical Engineers. All rights reserved.

456523 CO_2 Hydrogenation on Silica- and Titania- Supported Single-Site Heterogeneous Co^{II} Catalyst

Abstract:

Global climate change has been widely attributed to the rising amounts of CO_2 in the atmosphere[1,2]. Furthermore, there exist little uses for pure carbon dioxide meaning it must be chemically upgraded into a more valuable chemical. However, one of the main difficulties with reducing CO_2 via chemical conversion is the thermodynamic stability of the compound[3]. To mitigate this barrier, cobalt catalyst have been studied for the CO_2 hydrogenation reaction due to their high activity and stability[4]. Literature has shown that single-site catalysts yield higher turnover frequencies than nanoparticles catalyst for both hydrogenation and oxidation reactions[5,6]. Furthermore, a unique feature of single-site catalyst is the uniformity of the active sites, which allows for a direct correlation between catalytic activity and the structure of the sites. However, to the best of our knowledge, no work has been done on the use of single-site heterogeneous catalyst for CO_2 hydrogenation.

In this presentation, single-site Co^{II} catalyst for the CO_2 hydrogenation will be discussed. Single-site Co supported on silica are synthesized using methods adapted from literature[5]. Since the active site for this reaction is the highly stable Co^{II} center, titania was also explored as a support due to its strong metal-support interaction with Co in the oxidized state[4], leading to enhanced activity compared to silica supported Co catalyst. The stability of the Co^{II} active site comes from its tetrahedral geometry with neighboring oxygens from the oxide supports. Understanding the interactions with isolated metal ions with a uniform active site can provide insight into the mechanism of CO_2 hydrogenation, which is still widely debated. In order to obtain a detailed understanding of the catalytic activity of the single-site Co^{II} we must

verify the surface species are in an isolated geometry. To this end the techniques that will be used to characterize the surface are Raman spectroscopy, diffuse reflectance UV-vis, TPR, and STEM-HAADF. These characterization techniques coupled with reaction data can correlate the activity of single-site heterogeneous Co^{II} catalyst for CO_2 hydrogenation.

References

[1] T. F. Stocker, A. Schmittner, *Nature* **1997**, *388*, 862–865.

[2] M. Meinshausen, N. Meinshausen, W. Hare, S. C. B. Raper, K. Frieler, R. Knutti, D. J. Frame, M. R. Allen, *Nature* **2009**, *458*, 1158–62.

[3] H. Arakawa, M. Aresta, J. N. Armor, M. A. Barteau, E. J. Beckman, A. T. Bell, J. E. Bercaw, C. Creutz, E. Dinjus, D. A. Dixon, et al., *Chem. Rev.* **2001**, *101*, 953–996.

[4] G. Melaet, W. T. Ralston, C.-S. Li, S. Alayoglu, K. An, N. Musselwhite, B. Kalkan, G. a. Somorjai, *J. Am. Chem. Soc.* **2014**, *136*, 2260–2263.

[5] B. Hu, A. "Bean" Getsoian, N. M. Schweitzer, U. Das, H. Kim, J. Niklas, O. Poluektov, L. A. Curtiss, P. C. Stair, J. T. Miller, et al., *J. Catal.* **2015**, *322*, 24–37.

[6] S. F. J. Hackett, R. M. Brydson, M. H. Gass, I. Harvey, A. D. Newman, K. Wilson, A. F. Lee, *Angew. Chemie* **2007**, *119*, 8747–8750.

Copyright © American Institute of Chemical Engineers. All rights reserved.

452810 Development of Catalysts for CO_2 Hydrogenation to Olefins

Abstract:

The Naval Research Laboratory (NRL) is investigating a catalytic process to produce synthetic jet fuel from CO_2 and H_2. The current method consists of extracting CO_2 from seawater, then combining the extracted CO_2 with H_2 produced through water electrolysis. The CO_2 and H_2 are then reacted over catalysts in two steps to first produce olefins (C_2-C_4), which are oligomerized into the liquid hydrocarbon fraction (C_6-C_{17}) for synthetic jet fuel. Although there are several promising catalysts for both reaction steps, problems exist with water removal, which prevents high conversion of CO_2 and high selectivity towards liquid hydrocarbons. Future directions require carefully controlled synthesis of new catalysts that are water tolerant and improve the activity and selectivity of CO_2 hydrogenation to olefins for liquid hydrocarbon synthesis.

One of the most difficult aspects of chemistry involving CO_2 is not the catalytic activation, but the thermodynamic limitation of utilizing the highly stable molecule.[1] By coupling the endothermic reverse water-gas shift (RWGS) reaction with the slightly exothermic Fischer-Tropsch (FT) process, FT from CO_2 and H_2 (CO_2-FT) becomes more favorable as higher chain compounds are formed.[1-2] However, high conversion of CO_2 can only be achieved if the FT step is fast enough to overcome the thermodynamic limitation of RWGS, which is the main challenge for reactions involving CO_2. Therefore, to effectively synthesize jet fuel from CO_2 and H_2, new catalysts must be identified which are active for both RWGS and FT.

Because the cost of catalysts for CO_2 hydrogenation is also important, catalysts must be synthesized from low-cost materials. $Mo_2C/\gamma\text{-}Al_2O_3$ is a promising catalyst because of the ease of synthesis, low-cost, high activity, and relatively low oxygen binding energy (OBE), which allows it to freely exchange oxygen with CO_2. A previous study showed that Mo_2C was active and selective for CO_2 reduction into CO.[3] Mo_2C also has a much lower OBE than other transition metal carbides (TMCs), which contributes to its high CO_2 hydrogenation activity. By doping Mo_2C with Fe, it is possible to further enhance the activity for FT, as Fe-based catalysts are highly active for olefin production through FT and CO_2-FT.[4]

References

[1] K. Müller, L. Mokrushina, W. Arlt, *Chemie Ingenieur Technik* **2014**, *86*, 497-503.

[2] U. Rodemerck, M. Holeňa, E. Wagner, Q. Smejkal, A. Barkschat, M. Baerns, *ChemCatChem* **2013**, *5*, 1948-1955.

[3] M. D. Porosoff, X. Yang, J. A. Boscoboinik, J. G. Chen, *Angewandte Chemie International Edition* **2014**, *53*, 6705-6709.

[4] H. D. Willauer, R. Ananth, M. T. Olsen, D. M. Drab, D. R. Hardy, F. W. Williams, *Journal of CO2 Utilization* **2013**, *3–4*, 56-64.

Copyright © American Institute of Chemical Engineers. All rights reserved.

454632 $K_2o/WO_x/Al_2O_3$ Catalyst Interactions with the Components and Byproducts of Sour Natural Gas

Abstract:

Both CO_2 and H_2S acid gases are present in significant concentrations in natural â€œsourâ€ gas streams and need to be separated using energy intensive amine absorption methods. The captured CO_2 is emitted into the atmosphere acting as a major greenhouse gas while H_2S is combusted to yield elemental sulfur and water. A recent development proposed focuses on catalytic transformation of CO_2 and H_2S under reducing environments on K_2O-WO_3/Al_2O_3 catalyst to form a reactive methyl mercaptan (CH_3SH) intermediate that can be further converted to value-added hydrocarbons (alkanes, olefins and aromatics) and H_2S thus performing solventless separation and valorization of CO_2. The major problem and emphasis of this presentation is on the first reactive step. A fundamental knowledge base of acid gases - CO_2 and H_2S as well as they major reaction byproduct SO_2 - induced K_2O-WO_3/Al_2O_3 catalyst structural changes and the corresponding reactive intermediates and gas phase product distribution are being investigated. In particular, prediction and control of the CO_2 and H_2S molecular interactions, the relevant catalyst structural changes and surface chemistry are examined with *in situ* Raman, IR spectroscopy and TPSR to assess the performance of this catalyst to facilitate discovery of novel catalytic materials that can function in the presence of acid gases for CH_3SH synthesis.

Copyright © American Institute of Chemical Engineers. All rights reserved.

472823 Organic Photocatalysts for Atom Transfer Radical Polymerization Driven By Visible Light

Abstract:

We report a computational investigation of a new class of organic photocatalysts (OPCs) - Intramolecular Charge Transfer Organic Photocatalysts - for Atom Transfer Radical Polymerization (ATRP) driven by visible light. We used computational chemistry to design and understand OPCs with various desirable properties and then synthesized and characterized these OPCs for ATRP. These photocatalysts are analogous to metal-to-ligand-charge transfer complexes, but involve a photoactive organic moiety rather than a metal center, thus enabling metal-free photocatalyzed reactions and exquisite tuning of the photochemical properties of the catalysts. Organic ATRP photocatalysts have been sought to address this difficult challenge, but have not achieved the precision performance of metal catalysts. We also report experimental results that confirm the predictions of our computational designs and show that in the case of visible light photocatalyzed ATRP they achieve high initiator efficiencies to synthesize polymers with tunable molecular weights and low dispersities. Furthermore, these OPCs can act as long-lived and powerful reducing agents to catalyze a variety of reactions.

Copyright © American Institute of Chemical Engineers. All rights reserved.

462513 CO-CO Coupling on Cu Facets: Coverage, Strain and Field Effects

Abstract:

The electrochemical CO_2 reduction reaction has the potential to produce industrial chemicals and to mitigate the increasing CO_2 concentrations in the atmosphere. Polycrystalline copper, the only transition metal catalyst to produce hydrocarbons at reasonable Faradaic efficiency, produces a reasonable 1mA/cm^2 current density at ~1V overpotential.[1] Polycrystalline Cu has shown similar onset potentials for C_1 and C_2 hydrocarbons at -0.75V vs. RHE[2], suggesting both C_1 and C_2 pathways to be limited by the initial hydrogenation of *CO to form *CHO. However, the (100) facet and nanostructured Cu have been shown to have an earlier onset for C_2 products in alkaline conditions.[4-7] These results suggest a C-C coupling pathway that proceeds prior to *CO hydrogenation. Recent calculations have found CO-CO coupling to be feasible in the presence of a solvent and cation, which induces a field in the Helmholtz plane.[8] We present a DFT study on the effect of coverage, strain, and electric field on CO-CO coupling energetics on Cu (100), (111), and (211). Our calculations indicate that CO-CO coupling is facile on all three facets in the presence of a cation-induced electric field in the Helmholtz plane, with the lowest barrier on Cu(100). The CO dimerization pathway is therefore expected to play a role in C_2 formation at potentials negative of the Cu potential of zero charge, corresponding to CO_2/CO reduction conditions at high pH. Both increased *CO coverage and tensile strain further improve C-C coupling energetics on Cu (111) and (211). Since CO dimerization is facile on all 3 Cu facets, subsequent surface hydrogenation steps may also play an important role in determining the overall activity towards C_2 products. Adsorption of *CO, *H, and *OH on the 3 facets were investigated with a Pourbaix analysis. The (211) facet has the largest propensity to co-adsorb *CO and *H, which would favor surface hydrogenation following CO dimerization. These results suggest that the (100) and (211) facets should be the most active due to field and coverage effects.

REFERENCES

1. Y. Hori, Mod. Asp. Electrochem. (2008) 89–189.

2. K.P. Kuhl, E.R. Cave, D.N. Abram, T.F. Jaramillo, Energy Environ. Sci. 5 (2012) 7050.

3. J.H. Montoya, A. A. Peterson, J.K. Nørskov, ChemCatChem. 5 (2013) 737–742.

4. K.J.P. Schouten, Z. Qin, E.P. Gallent, M.T.M. Koper, J. Am. Chem. Soc. 134 (2012) 9864–9867.

5. C.W. Li, J. Ciston, M.W. Kanan, Nature. 508 (2014) 504–507.

6. R. Kas, R. Kortlever, A. Milbrat, M.T.M. Koper, G. Mul, J. Baltrusaitis, Phys. Chem. Chem. Phys. 16 (2014) 12194–201.

7. F.S. Roberts, K.P. Kuhl, A. Nilsson, Angew. Chem. 127 (2015) 5268-5271.

8. J.H. Montoya, C. Shi, K. Chan, J.K. Nørskov, J. Phys. Chem. Lett. (2015) 2032–2037.

Copyright © American Institute of Chemical Engineers. All rights reserved.

464942 Single-Atom Catalysts for CO_2 Electroreduction with Significant Activity and Selectivity Improvements

Abstract:

Single-atom catalyst (SAC) has an electronic structure that is very different from its bulk counterparts, and has shown unexpectedly high specific activity with a significant reduction of noble metal usages although physical origins of such performance enhancements are still poorly understood. Herein, by means of density functional theory calculations, we for the first time investigate the great potential of single atom catalyst for CO_2 electroreduction applications. In particular, we study a single transition metal atom anchored on the defective graphene with single or double vacancies, denoted as M@sv-Gr or M@dv-Gr, where M are various transition metals, as a CO_2 reduction catalyst. Many SACs are indeed shown to be highly selective for CO_2 reduction reaction over a competitive H_2 evolution reaction due to a favorable adsorption of carboxyl or formate over hydrogen on the catalysts. On the basis of free energies, we found that the Pt@dv-Gr catalyst shows a remarkable reduction in the limiting potential for CH_3OH production compared to any existing catalysts, synthesized or predicted (Figure 1).

We focus on the Pt@dv-Gr to investigate the origin of improvement on the SACs compared to the transition metal. As shown in Figure 1, with the Pt@dv-Gr catalyst, all the intermediates are destabilized compared to those on Pt (211), but most importantly, the destabilization of *CO (0.98 eV) is much more noticeable than that of *CHO (0.42 eV), leading to a 0.49 V reduction in the limiting potential. We also observe that with SACs, the conventional scaling relation between the *CO binding and *CHO binding established for the bulk transition metal significantly deviates from linearity (Figure 2).

We discuss features of SACs which contribute to the breakdown of scaling relation between *CO and *CHO, namely, the lack of atomic ensemble for adsorbates binding and the metal-support interactions that lead to the electronic structures conducive to the catalysis.

Atomic Ensemble:

The optimized geometries of the bare catalysts are shown for Pt@dv-Gr and Pt (211) in Figure 3. One can see that, for Pt (211), two surface Pt atoms are involved in the *CO bonding while only one Pt atom is bonding with *CHO, leading to a large destabilization in the relative free energies when going from *CO to *CHO. On the other hand, for Pt@dv-Gr, only one Pt atom is utilized for both *CO and *CHO binding, resulting in a much more moderate destabilization in relative free energies compared to Pt (211). Thus, the lack of Pt ensemble in the Pt@dv-Gr is responsible for a significantly weaker binding of *CO on the Pt@dv-Gr compared to the Pt (211) surface.

Electronic Structure:

The strong metal-support interaction affects the electronic structure of a metal atom in the SACs greatly. In Figure 4A, the Pt 5d density of states (DOS) in Pt@dv-Gr shows a significant orbital overlap with the C 2p orbitals of the graphene. The electron density isosurfaces (Figure 4B) illustrate that electron clouds of four carbon atoms surrounding the Pt atom are significantly hybridized with the Pt atom. The differential charge density map (Figure 4C) between the defective graphene and Pt@dv-Gr also suggests that the Pt atom is positively charged by electron transfer from the Pt atom to the defective graphene support.

In this paper, we investigated the single atom catalysts (SACs) as a promising CO_2 electroreduction catalyst using DFT calculations. The main findings of this work are as follows.

(i) By comparing the free energies of the initial protonation steps for the CRR and HER, we found that all SACs can selectively reduce CO_2 rather than producing H_2. In particular, the predicted limiting potential for Pt@dv-Gr (-0.27 V) for CH_3OH production is considerably less negative than the conventional transition metal catalysts.

(ii) To understand the origin of the improvements in SACs, we investigated two aspects of the Pt@dv-Gr that affect the relative stability of *CO vs. *CHO. A one-fold bonding of *CO on Pt@dv-Gr due to a lack of atomic ensemble, as compared to the two-fold *CO bonding on the Pt (211) is responsible for the significant weakening of *CO binding on the Pt@dv-Gr.

(iii) We investigated the electronic structure of Pt atom in the SAC to find the origin of the deviation of SACs from the conventional scaling relation of transition metals, which arises from the d-band center theory. We suggest that the strong electronic interaction between the d-orbital of metal atom and the p-orbital of graphene is responsible for the different behavior from the transition metal surfaces, as evidenced by the electron transfer and the overlap in the DOS.

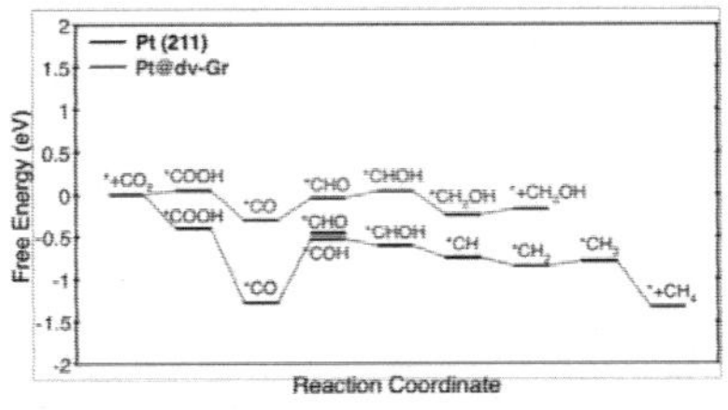

Figure 1. Free energy profiles for the CO_2 reduction reaction to CH_4 and CH_3OH on the Pt (211) and Pt@dv-Gr, respectively, at zero applied voltage (vs. RHE). Free energy changes of the PDS for Pt(211) and Pt@dv-Gr are 0.75 eV (*CO → *COH) eV and 0.27 eV (*CO → *CHO), respectively.

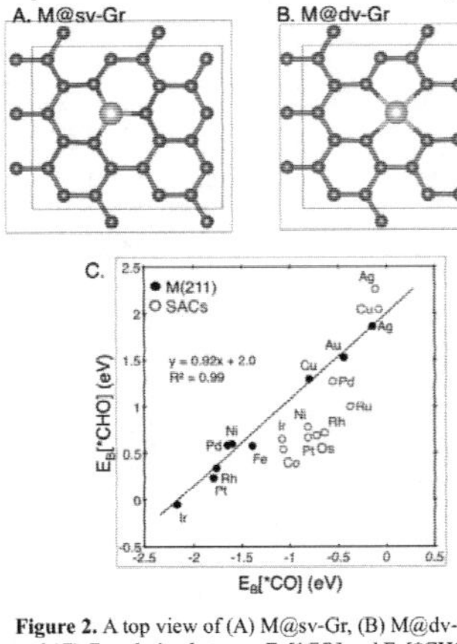

Figure 2. A top view of (A) M@sv-Gr, (B) M@dv-Gr, and (C) Correlation between E_B[*CO] and E_B[*CHO] for transition metal (211) surfaces (black) vs. SACs (blue). The conventional scaling relation for M(211) is broken in the case of SACs.

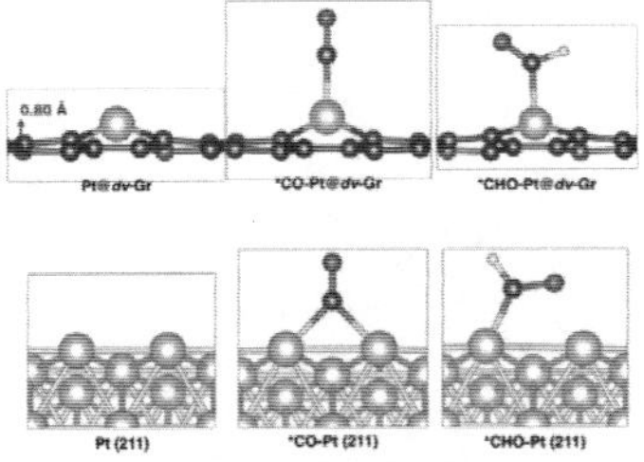

Figure 3. The optimized geometries of the Pt@dv-Gr and the Pt (211) before and after the adsorption of *CO and *CHO.

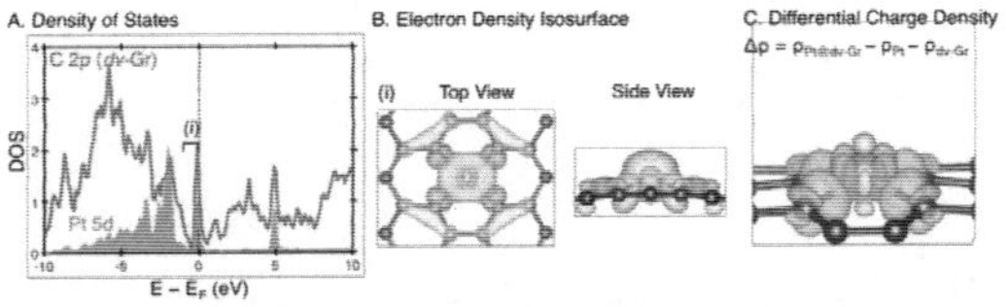

Figure 4. (A) The density of states (DOS) for Pt 5d orbital and C 2p orbital in graphene for the Pt@dv-Gr. (B) Top and side views of the electron density isosurface at the energy level as noted with (i) in A. (C) The differential charge density map upon adsorption of Pt on the dv-Gr. In the differential charge density map, yellow and blue region indicate electron accumulation and depletion, respectively.

Copyright © American Institute of Chemical Engineers. All rights reserved.

460354 Solvation Effects in Surface Electrochemistry for Oxygen Evolution Reaction Catalysis on Transition Metal Oxides

Abstract:

There is a clear need for a cost effective energy solution that can replace fossil fuels. In the past several years, electrolytic water splitting has been garnering significant attention as a potential solution to the production of hydrogen via sustainable but intermittent forms of energy such as wind or solar. This energy can be stored chemically as a bond in H_2 gas; the gas can then be used as a sustainable fuel or as a feedstock for commodity chemicals, such as the Haber-Bosch process, for which hydrogen is currently produced by steam reforming of methane.

The Oxygen Evolution Reaction (OER) takes place at the anode of electrolytic water splitting. Currently, there are no good catalysts for the OER, regardless of pH [1]. Many current ab-initio calculations for OER do not contain an explicit solvent [2], which has been shown to be significant for titanium dioxide [3,4], and for some other oxide surfaces [2]. It is necessary to understand how water impacts the stability of surface-bound OER intermediates to better design and understand electrocatalysts.

In this work we use DFT calculations and a minima hopping global optimization method to investigate how the inclusion of explicit solvent molecules affects the binding energies of OER intermediates at various intermediate coverages. Our results on IrO_2 indicate that hydrogen containing intermediates (*OH, *OOH) are significantly stabilized relative to O* by the explicit solvent. Relative stabilization of particular intermediates leads to changes in reaction energetics and ultimately in the calculation of theoretical overpotential, which is a measure of a catalysts' activity [5]. Furthermore, solvent interaction with adsorbed intermediates has potential to be important for other electrochemical reactions, such as N_2 and CO_2 reduction. As such, an understanding of these solvent interactions could greatly impact how we think about catalyst design.

1. McCrory, C. C. L., Jung, S., Peters, J. C., & Jaramillo, T. F. (2013). Benchmarking heterogeneous electrocatalysts for the oxygen evolution reaction. *Journal of the American Chemical Society*, *135*(45), 16977–16987. http://doi.org/10.1021/ja407115p
2. Siahrostami, S., & Vojvodic, A. (2015). Influence of Adsorbed Water on the Oxygen Evolution Reaction on Oxides. *The Journal of Physical Chemistry C*, *119*(2), 1032–1037. http://doi.org/10.1021/jp508932x
3. Carrasco, J., Hodgson, A., & Michaelides, A. (2012). A molecular perspective of water at metal interfaces. *Nature Materials*, *11*(8), 667–674. http://doi.org/10.1038/nmat3354
4. Sun, C., Liu, L.-M., Selloni, A., Lu, G. Q. (Max), & Smith, S. C. (2010). Titania-water interactions: a review of theoretical studies. *Journal of Materials Chemistry*, *20*(46), 10319. http://doi.org/10.1039/c0jm01491e
5. Man, I. C., Su, H. Y., Calle-Vallejo, F., Hansen, H. A., Martínez, J. I., Inoglu, N. G., Rossmeisl, J. (2011). Universality in Oxygen Evolution Electrocatalysis on Oxide Surfaces. *ChemCatChem*, *3*(7), 1159–1165. http://doi.org/10.1002/cctc.201000397

Copyright © American Institute of Chemical Engineers. All rights reserved.

466711 First Principles Quantum Chemistry Calculations to Model CO_2 Electroreduction on SnO_2 Particles

Abstract:

Electrochemical conversion of CO_2 has drawn much attention due to the urgent need to address rising CO_2 levels in the atmosphere. Electrochemical CO_2 reduction is known on Sn electrodes, but a recent study by Bocarsly and coworkers has implicated a Sn-oxide overlayer as being the active catalysis. We have investigated this system using quantum chemistry methods to understand the adsorption mechanism of CO_2 on the most stable SnO_2 surface (110). To determine the effect of oxygen concentration on the adsorption process; stoichiometric, reduced and oxidized surfaces are also investigated. We plot Pourbaix diagrams to identify relevant thermodynamic states that catalyzes the CO_2 reduction by identifying the most stable and energetically efficient electrochemical reaction conditions. We hypothesize the structures identified by using Pourbaix diagrams will be experimentally relevant structures for the CO_2 catalysis.

Copyright © American Institute of Chemical Engineers. All rights reserved.

451599 Modelling the Electrochemical Interface: Applications to CO2 Reduction

Abstract:

The electroreduction of CO2 has the potential to store energy from intermittent renewable sources and to produce carbon-neutral fuels and chemicals [1, 2]. In recent years, theoretical studies of CO2 reduction have usually applied the computational hydrogen electrode model, which allows for the determination of the energies of reaction intermediates without explicitly treating the potential and the ions in solution [3]. In this talk, I will first review the application of this approach to CO2 reduction: the determination of lowest free energy pathways and correlation of theoretical activity volcanoes with experimental onset potentials [4, 5], and the computational screening of new catalysts [6]. I will then focus on recent developments in the explicit treatment of the electrochemical interface [7], required for an understanding of charge transfer barriers, kinetics, and selectivity. I will discuss the scaling of CO2 reduction barriers, field effects, C-C coupling [8], and the integration of these effects into microkinetic models.

1. Whipple, D. T. & Kenis, P. J. a. Prospects of CO 2 Utilization via Direct Heteroge- neous Electrochemical Reduction. J. Phys. Chem. Lett. 1, 3451–3458 (Dec. 2010).
2. Graves, C., Ebbesen, S. D., Mogensen, M. & Lackner, K. S. Sustainable hydrocarbon fuels by recycling CO 2 and H 2 O with renewable or nuclear energy. Renewable and Sustainable Energy Reviews 15, 1–23 (2011).
3. Nørskov, J. K., Rossmeisl, J., Logadottir, A., Lindqvist, L., Kitchin, J. R., Bligaard, T. & Jónsson, H. Origin of the Overpotential for Oxygen Reduction at a Fuel-Cell Cathode. J. Phys. Chem. B 108, 17886–17892 (Nov. 2004).
4. Shi, C., Hansen, H. a., Lausche, A. C. & Nørskov, J. K. Trends in electrochemical CO2 reduction activity for open and close-packed metal surfaces. Phys. Chem. Chem. Phys. 16, 4720–7 (Mar. 2014).
5. Kuhl, K. P., Hatsukade, T., Cave, E. R., Abram, D. N., Kibsgaard, J. & Thomas, F. Electrocatalytic conversion of carbon dioxide to methane and methanol on transition metal surfaces (2014).
6. Chan, K., Tsai, C., Hansen, H. A. & Nørskov, J. K. Molybdenum sulfides and selenides as possible electrocatalysts for CO2 reduction. ChemCatChem 6, 1899–1905 (2014).
7. Chan, K. & Nørskov, J. K. Electrochemical Barriers Made Simple. J. Phys. Chem. Lett. 2663–2668 (2015).
8. Montoya, J. H., Shi, C., Chan, K. & Nørskov, J. K. Theoretical Insights into a CO Dimerization Mechanism in CO2 Electroreduction. The journal of physical chemistry letters 6, 2032–2037 (2015).

Copyright © American Institute of Chemical Engineers. All rights reserved.

470141 Exploring the Remarkable Heterogeneity of Bimetallic Alloy Facet Reactivity for the Electrocatalytic Reduction of CO2

Abstract:

Bimetallic alloys alloy make promising catalysts due to heterogenous active sites that allow small molecule adsorbates to be stabilized compared to single-metal active sites. The number of possible active sites for such intermetallics is much larger than for single-metal catalysts due to the compositional design space and the much larger number of possible facets resulting from a loss of some crystal symmetries. Screens that have worked for single-metal systems, such as systematically identifying activity for 111 or 211 metal facets, fail to capture the most exciting active sites for these alloys. We demonstrate this challenge for the case of nickel/gallium alloy catalysts, which have shown experimental reactivity for the electrocatalytic reduction of CO2. Using DFT, we show that the limiting potential for the rate-limiting step of CO hydrogenation can vary by approximately 1eV over alloy composition and facet, and that alloys can demonstrate a number of possible sites with interesting stabilization. We use the insights from this systematic study to develop rules and methods to make exploring future bimetallic alloys more computationally efficient in light of this large design space.

Copyright © American Institute of Chemical Engineers. All rights reserved.

468008 Exploring Possible Active Sites for $IrO_x/SrIrO_3$, a Highly Active Oxygen Evolution Catalyst

Abstract:

The Oxygen Evolution Reaction (OER) plays a key role in the renewable production of fuels and chemicals and is a source of inefficiency due to its sluggish kinetics. In addition to possessing high activity, OER catalysts must be stable for thousands of hours under operation. This is particularly challenging at the acidic conditions present in proton-exchange membrane electrolyzers. Currently, only Ir-based catalysts are practically stable for these applications,[1] and improving the intrinsic activity of these catalysts provides a pathway to decreasing Ir usage. We have discovered a novel catalyst that forms in-situ as Sr leaches from $SrIrO_3$ surface layers, leaving behind a remarkably active and acid-stable IrO_x film with multiple orders of magnitude improvement over rutile IrO_2 and RuO_2.[2]
Herein, we use Density Functional Theory calculations to propose an explanation of the exceptional observed OER activity. We find the nominally prepared $SrIrO_3$ perovskite to be thermodynamically unstable relative to aqueous Sr^{2+} and propose possible Sr-deficient overlayer structures that may form with electrochemical testing. Through a combination of lattice matching and *ab initio* molecular dynamics, we arrive at anatase IrO_2 and $IrO_{3Â}$ overlayers that possess active sites at the peak of the theoretical activity volcano.

[1]C. C. L. McCrory, S. Jung, J. C. Peters, T. F. Jaramillo (2013). Benchmarking Heterogeneous Electrocatalysts for the Oxygen Evolution Reaction. *Journal of the American Chemical Society*, *135*(45), 16977â€"16987. http://doi.org/10.1021/ja407115p

[2]L. C. Seitz, C. F. Dickens, K. Nishio, Y. Hikita, J. Montoya, A. Doyle, C. Kirk, A. Vojvodic, H. Y. Hwang, J. K.

Norskov, T. F. Jaramillo (2016). A Highly Active and Stable SrIrO3/IrOx Catalyst for the Oxygen Evolution Reaction. *In Review*.

Copyright © American Institute of Chemical Engineers. All rights reserved.

460814 Characterization of TFA Regenerated Cellulose with a Two-Step Treatment to Understand the Impact of Cellulose Structure on Hydrolysis

Abstract:

Concentrated trifluoroacetic acid (TFA) is known to depolymerize cellulose. In this paper we report the effect of temperature and treatment time on physiochemical and morphological changes to cellulose. We present x-ray diffraction, SEM, degree of polymerization, and FTIR characterizations and compare them to acid-catalyzed hydrolysis of the decrystalized cellulose and conversion of the resultant glucose to HMF and levulinic acid by maleic acid-$AlCl_3$ catalysis. Treatment with concentrated TFA for 2 hrs at 0°C resulted in a 28% reduction in crystallinity index (CrI) of the cellulose. Additional treatment for 5 hrs at 50°C resulted in a CrI of 0. Both changes of crystallinity index (CrI) resulted in about 40% improvement in overall yield of HMF and levulinic acid. Enzymatic hydrolysis study shows that TFA treatment for 2 hrs at 0°C results a 52% glucose conversion yield, which had higher conversion yield than the additional treatment for 5 hrs at 50°C with 45% glucose conversion yield. Results showed the additional reduction in CrI through heating step does not further improve yields from enzymatic hydrolysis.

Copyright © American Institute of Chemical Engineers. All rights reserved.

462550 The Role of Atmosphere, Solvents and AcidsÂ on the Production of 5-HydroxymethylfurfuralÂ from Biomass Derived Fructose

Abstract:

The Role of Atmosphere, Solvents and Acids on the Production of 5-Hydroxymethylfurfural from Biomass Derived Fructose

As global energy demands and atmospheric CO_2 levels continue to rise, sustainable green energy sources will become necessary solutions. One such alternative energy source is corn Stover, a domestically produced, cellulosic biomass which is low enough in cost and of sufficient abundance for production of liquid fuels on a large scale to be competitive with petroleum. Sugars, such as fructose, can be derived from biomass through pretreatment of biomass to glucan enriched solids, these solids can undergo saccharification and isomerization to produce fructose-enriched streams. This fructose stream can be processed to produce platform chemical feedstocks, like 5-hydroxymethylfurfural (5-HMF). 5-HMF is known as a platform chemical, as it is easily converted to biofuels or high valued commodity chemicals.

Past studies have demonstrated that it is difficult to improve the selective and/or conversion of biomass sugars to 5-HMF in aqueous solutions. The majority of these studies rely on increasing the catalyst loadings or using high-temperature processes. Furthermore, numerous studies have explored the use of organic solvents to improve the selective conversion of biomass sugars to furanic compounds. However, these studies are often times inconsistent and have not conclusively demonstrated the combined effects of acids, solvents, and atmosphere on the conversion of fructose to 5-HMF.

Presently, the highest reported yields of HMF obtained, arises from acid-catalyzed dehydration of fructose in dimethyl sulfoxide (DMSO). A combination of Brønsted acids and DMSO has been proposed to provide an effective catalytic pathway and intermediate stabilization that maximizes HMF yields. However, significant contradictions exist in literature regarding the roles of DMSO, O_2, and H_2O in the mechanism of fructose conversion. It is also unclear whether the fructose conversion mechanisms are radically different between: 1) various acids in DMSO or 2) different aprotic solvents vs. DMSO. To address these questions we utilized a 96 well plate high-throughput batch reactor system loaded and sealed under inert or atmospheric conditions. The reactor was loaded into a high pressure vessel and a Fulton electric steam boiler was utilized to rapidly heat the vessel to systematically and accurately explore the influence of pH, O_2, H_2O, solvent, time and temperature on fructose conversion and 5-HMF selectivity.

We report three primary results: 1) DMSO decomposes in O_2 at elevated temperatures to produce Brønsted acids that drive fructose conversion. (2) In O_2 free environments, increasing H_2O content does not negatively influence the HMF yield. 3) Lastly, 5-HMF production follows an almost identical, negative linear relationship between conversion and selectivity, which universally captures the behavior regardless of solvent, reaction time, reaction temperature and acid loading. These systematic studies have allowed us to identify realistic conditions where fructose conversion to 5-HMF can be executed with high yields, >80% at reasonable pH (~2), water contents, temperatures, and times. Furthermore, we utilize these findings to propose a universal mechanism for the fructose conversion pathways driven by homogeneous Brønsted acids and present insights into reactor design and conditions to optimize 5-HMF yields at minimal cost.

Copyright © American Institute of Chemical Engineers. All rights reserved.

469433 Lignin Depolymerization into Aromatic Monomers over Acidic Mesoporous Silicates

Abstract:

Lignin Depolymerization into Aromatic Monomers over Acidic Mesoporous Silicates

Kakasaheb Y. Nandiwale[a,b], Andrew M. Danby[a], Anand Ramanathan[a], Raghunath V.Chaudhari[a,b], and Bala Subramaniam[a, b]

[a]Center for Environmentally Beneficial Catalysis, The University of Kansas, Lawrence, KS 66047, USA.

[b]Department of Chemical and Petroleum Engineering, The University of Kansas, Lawrence, KS 66045, USA.

Keywords: Biomass, Lignin depolymerization, mesoporous silicates

We report on novel applications of Zr-incorporated mesoporous silicates for the one-pot conversion of lignin into value added aromatic monomers. Quantitative analysis of the depolymerized products along with mass balance based on the overall solid and liquid products are reported.

Cubic mesoporous silicates, KIT-6 and KIT-5, modified with different Zr loadings (1-6 wt%) display dominant Lewis acidity. These materials display superior lignin depolymerization activity compared to commercial zeolites that are typically Brønsted acids. Briefly, batch depolymerization of dealkaline lignin (TCI chemicals) was carried out in a 300 mL Parr reactor at 250°C, 100 psi N_2, 1000 RPM for 30 min in a methanol + water solvent mixture. The performances of 12 commercially available zeolite catalysts were evaluated. H-ZSM-5 (Si/Al = 11.5) was found to be the best catalyst among the commercial zeolites evaluated, providing 92% lignin conversion. Approximately 54 wt.% of the products are soluble in THF while the lignin substrate is insoluble in THF. The identified aromatic monomers in the THF-soluble fraction represent 4.4 wt. % of the initial lignin. These results with H-ZSM-5 are consistent with those reported in the literature. Commercial H-ZSM-5 catalyst was therefore selected for further comparative study with Zr-KIT-5 and Zr-KIT-6 catalysts. Zr-KIT-5(Si/Zr = 20) provided the best catalytic performance with 97% lignin conversion, yielding 65% of THF soluble products containing 5.5% identified monomers. Gel permeation chromatography (GPC) analysis of depolymerized product from both runs indicates a decrease in the molar mass when compared to the lignin substrate, clearly implying lignin deconstruction. Approximately 25 monomers were identified (by GC-FID, GC-MS) in the THF-soluble product mixture and 15 monomers were identified using external standards. Most of the identified products are typically expected from lignin depolymerization as reported in the literature. These results, along with fundamental insights into the correlation between catalyst activity and total acidity of catalyst are presented.

Copyright © American Institute of Chemical Engineers. All rights reserved.

472651 Characteristics and Activity of Steamed Zeolites for Butanol Upgrading to Higher Olefins

Abstract:

The increasing prevalence of renewable alcohols derived from biomass feedstocks make them an ideal platform for developing into transportation fuel. Butanol is a promising bioproduct that can be generated either from direct fermentation or ethanol condensation chemistry, and can be further upgraded to fuels and chemicals in a selective manner. Development in the methanol-to-gasoline (MTG) process has shown zeolite catalysts to be particularly effective for the production of higher olefins, capable of catalyzing both the dehydration and oligomerization reactions in a single step and directly converting butanol to jet or diesel range hydrocarbons. However, changes to the catalyst in the initial startup phase and over time are reflected in the dynamic product distribution, shrinking the optimal reaction window. Dealumination through steaming has been shown to be an effective modification method, impacting the distribution and coordination of aluminum as well as the overall zeolite structure. Tests showed that presteaming ZSM-5 zeolites prior to the reaction of butanol resulted in the immediate stabilization of the reaction product distribution. Catalyst characterization revealed significant differences in the zeolite lattice structure and acid sites from steaming and after reaction. In this presentation, the specific effects of steaming on the catalyst structure and reaction sites will be discussed along with their role in stabilizing the reaction product distribution.

Copyright © American Institute of Chemical Engineers. All rights reserved.

466460 Tandem Catalysis in a New, Solid-Phase Route to Renewable Aromatic Chemicals from Biobased Furanics

Abstract:

Efficient production routes for 'drop-in', renewable aromatics are currently bring highly sought after due to diminishing reserves and changes in the composition of fossil fuel resources. In this light, sugar-derived furanics offer attractive alternative resources for the production of aromatic chemicals. Recently, we reported on a new three-step procedure to the synthesis of (substituted) phthalic anhydrides from biobased furanics by introducing an intermediate hydrogenation step in the classical Diels-Alder (DA) aromatization route, typically run at elevated temperatures, to address the general issue of the reversible nature of the DA addition step [1]. The hydrogenated DA adduct of, e.g. 2-methylfuran and maleic anhydride (MFMA[H]), is thermally stable and can be efficiently aromatized to give 3-methylphthalic anhydride and *ortho-/ meta-* toluic acid in a tandem catalytic reaction using a physical mixture of acidic zeolites and Pd/C in toluene. Variation in acidity and textural properties of, e.g. H-Y zeolites, in the combined dehydration/dehydrogenation catalyst provides a handle on selectivity toward aromatic products [2]. A linear correlation exists between the yield of 3-methylphthalic anhydride and the product of the strong acid/total acidity ratio and mesopore volume of H-Y, indicating the need for a balanced catalyst acidity and porosity of H-Y zeolite.

Here, we show that the hydrogenated DA adducts can also be conveniently converted to renewable aromatics in a solid-phase reaction (i.e., without any solvent), using only H-Y zeolites [3]. The remarkable performance of these catalysts in the solid-phase can still be further enhanced upon the addition of a small amount of a dehydrogenation catalyst (e.g., 1 wt. % Pd/C or Pt/C) or by the use of bifunctional catalysts (e.g., 1 wt. % Pd/H-Y or Pt/H-Y). Monitoring the evolution of H_2, H_2O, and CO_2 as a function of temperature and time provided valuable insights into the mechanism of aromatization and decarboxylation. The effects of various experimental parameters, different concentrations of H-Y zeolites and dehydrogenation catalysts, on the reactivity of MFMA[H] were studied to effectively control the reaction rate and product selectivity. The use of Pt/H-Y (Si/Al=30) resulted in an excellent yield of 94% total aromatics, which is the highest reported to date. The bifunctional catalysts were also found highly selective for the aromatization of other hydrogenated DA furan derivatives. The development of new heterogeneous catalysts for the production of aromatic chemicals from biomass-derived furanics can therefore be guided by taking into account the influence of the acidic and textural properties of H-Y zeolites on selectivity.

[1] S. Thiyagarajan, H. C. Genuino, M. Śliwa, J. C. van der Waal, E. de Jong, J. van Haveren, B. M. Weckhuysen, P. C. A. Bruijnincx, D. S. van Es. Substituted phthalic anhydrides from bio-based furanics: A new approach to renewable aromatics, *ChemSusChem* **2015**, *8*, 3052-3056.

[2] H. C. Genuino, S. Thiyagarajan, J. C. van der Waal, E. de Jong, J. van Haveren, D. S. van Es, B. M. Weckhuysen, P. C. A. Bruijnincx. Selectivity control in solid acid and Pd-catalyzed tandem aromatization of bio-based furanics, *Manuscript in preparation.*

[3] S. Thiyagarajan, H. C. Genuino, J. C. van der Waal, E. de Jong, B. M. Weckhuysen, J. van Haveren, P. C. A. Bruijnincx, D. S. van Es. A facile solid-phase route to renewable aromatic chemicals from biobased furanics, *Angew. Chem. Int. Ed.* **2016**, *55*, 1368-1371.

Copyright © American Institute of Chemical Engineers. All rights reserved.

Reaction Pathway and Microkinetic Model for Levulinic Acid Hydrodeoxygenation over Sulfided $NiMo/Al_2O_3$

Miha Grilc[a], Blaž Likozar[a], Janez Levec[a]

[a] Department of Catalysis and Chemical Reaction Engineering, National Institute of Chemistry, Hajdrihova 19, 1000 Ljubljana, Slovenia, miha.grilc@ki.si

Keywords: Platform chemicals, levulinic acid, specialty chemicals;

Summary

Hydrodeoxygenation of levulinic acid was studied in a batch three-phase slurry reactor over sulphided NiMo/γ-Al_2O_3 catalyst in order to find optimal reaction conditions for high selectivity towards γ-valerolactone production and high hydrodeoxygenation rates Since two different reaction pathways are dominant, the aim of this study was to find appropriate window of parameters to catalyse the hydrodeoxygenation reactions and inhibit the thermal decarboxylation reaction. The kinetic study included the variation of temperature between 200 and 300 °C, hydrogen pressure between 25 and 75 bar, stirring speed, amount of catalyst addition, particle size etc. As expected, the temperature and pressure had the highest influence on the reaction rates of hydrodeoxygenation of decarboxylation. To obtain high selectivity of hydrodeoxygenation, temperatures must be kept below 250 °C, since above this temperature the homogeneous C-C bond cleavage occurs and results in excessive decarboxylation of carboxylic group and 2-butanone formation.

1. Introduction

In the process of acidic hydrolysis or solvolysis the cellulose, hemicellulose and sugars are converted into 5-hydroxymethylfurfural (HMF), that can be further deformylated to levulinic acid [1]. Levulinic acid (LA) is therefore the main product of lignocellulosic biomass liquefaction by acidic solvolysis or hydrolysis, where up to 83 mol% yields were reported from cellulose [2]. Levulinic acid was reported to be an ideal platform chemical, as it is a member of the gamma-keto acids and can be therefore utilized to produce various targeted bio-chemicals or just a liquid fuel.

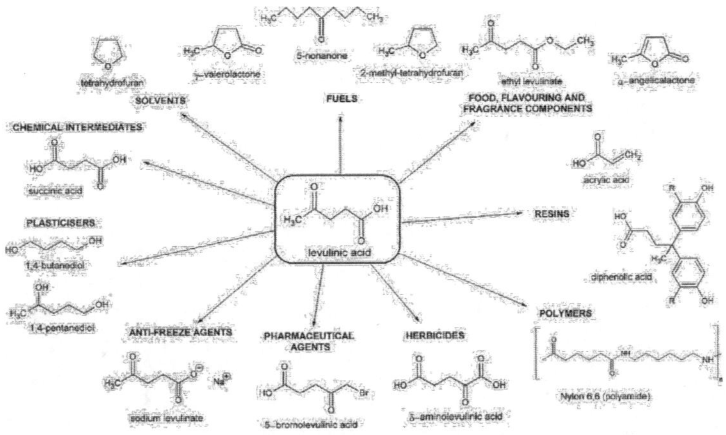

Figure 1: Levulinic acid as a platform chemical for specialty chemicals and liquid fuel [2].

2. Methods and Procedures

Catalytic hydrodeoxygenation (HDO) step is introduced to reduce the presence of chemically bonded oxygen from the feedstock. Levulinic acid can be hydrodeoxygenated by the cyclisation, resulting in formation of γ-valerolactone and water. Reaction takes place at temperatures above 200 °C in the presence of hydrogen and hydrodeoxygenation catalysts either based on transition metals (Mo, W, Ni, Co) in their sulphide, metal, oxide, nitride or carbide form or noble metals (Pt, Ru, Pd) on different supports [3]. In this study industrial sulphided NiMo/γ-Al_2O_3 catalyst was used. Figure 3 shows most important reactions for oxygen removal, where oxygen can be removed in the form of water, carbon monoxide and carbon dioxide.

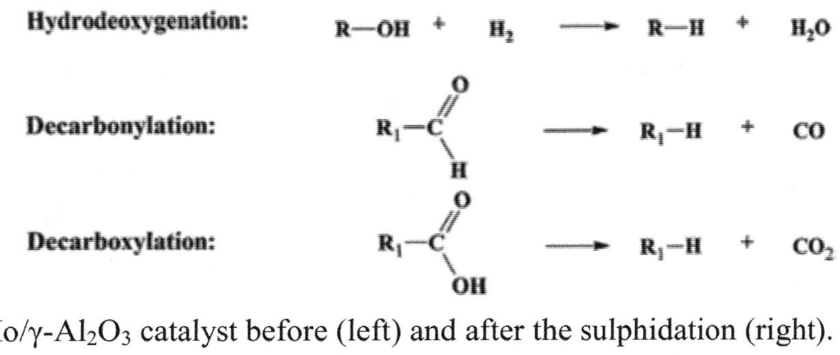

Figure 2: NiMo/γ-Al_2O_3 catalyst before (left) and after the sulphidation (right).

Decarbonylation and decarboxylation are homogeneous reactions caused by thermal decomposition, so presence of hydrogen or catalyst is not strictly necessary. HDO is most desired deoxygenation reaction, as no carbon atom is removed, which results in a higher yield of the final product.

3. Experimental

Since NiMo/γ-Al_2O_3 was obtained in its oxide form, it had to be sulphided first. The catalyst was sulphided using DMDS at 25 bar of hydrogen at temperature of 350 °C for 60 minutes in a stainless steel autoclave.

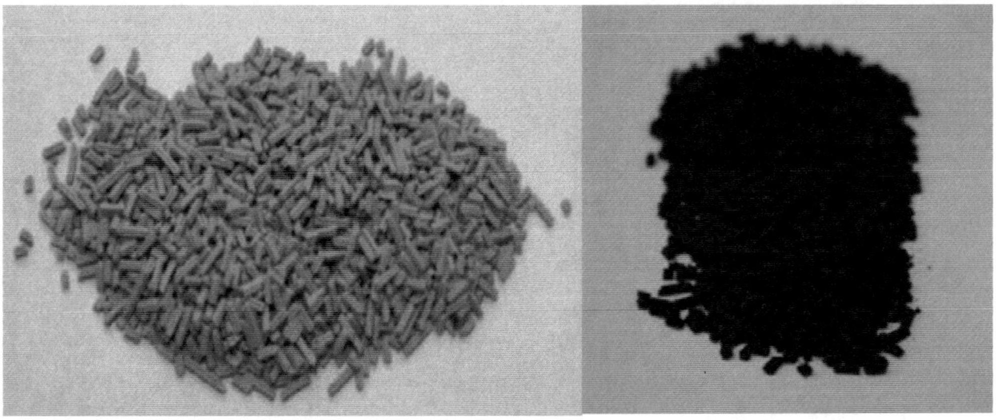

Figure 3: NiMo/γ-Al_2O_3 catalyst before (left) and after the sulphidation (right).

$$MoO_3 + 2\,H_2S + H_2 \rightarrow MoS_2 + 3\,H_2O$$

$$3\,NiO + 2\,H_2S + H_2 \rightarrow Ni_3S_2 + 3\,H_2O$$

Hydrodeoxygenation was investigated in the same 300 mL autoclave equipped with a magnetically driven turbine stirrer and electric heating jacket. Hydrodeoxygenation took place in solventless conditions, so only solid catalyst, liquid levulinic acid and gaseous hydrogen were in contact in the initial stages of the experiment.

4. Results

The analysis of gaseous (GC) and liquid phase (GC, FTIR) reveals the presence of two deoxygenation mechanisms: hydrodeoxygenation and direct decarboxylation. Hydrodeoxygenation reaction only took place in the coexistent catalyst presence and elevated hydrogen pressure. If catalyst particles were removed or nitrogen gas was used instead of hydrogen, the hydrodeoxygenation reaction was negligible. In the presence of sulphided NiMo/γ-Al_2O_3 the main product is γ-valerolactone, while other products of hydrogenation or dehydroation were rare. Few percent of valeric (pentanoic) acid was detected, while 4-hydroxypentanoic acid is obviously very reactive, since its concentration was very low during the runs. Products of decarboxylation and subsequent hydrogenation were also followed. At higher temperatures where decarboxylation takes place, 2-butanone is being formed, while its subsequent hydrogenation product (2-butanol) was never detected in the reaction mixture. Keton group is obviously resistant to hydrogenation at reaction conditions that were used in this study.

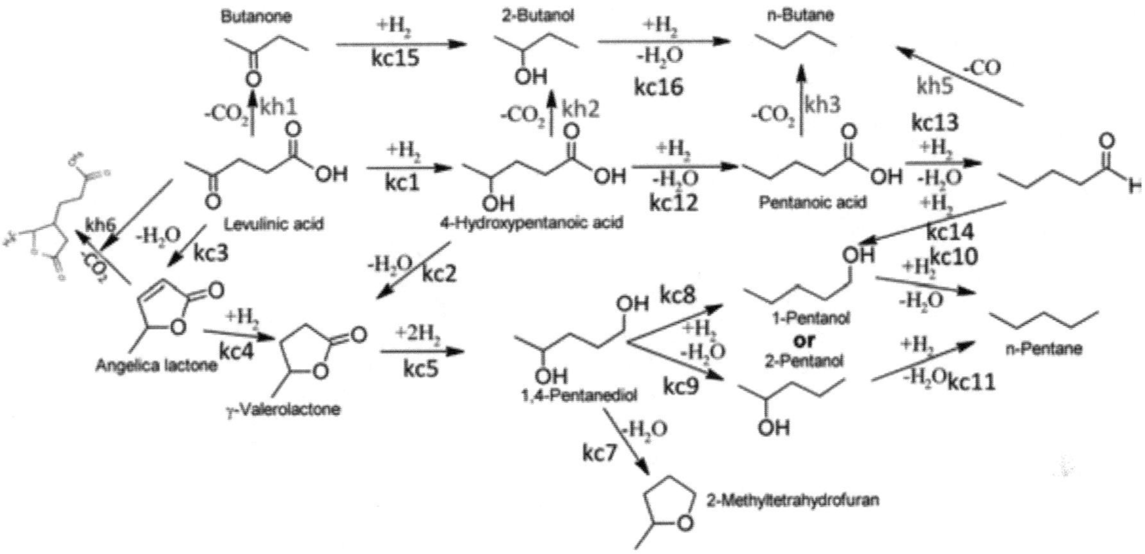

Figure 3: Proposed reaction pathway network of competitive hydrodeoxygenation and decarboxylation reaction of levulinic acid and its products.

5. Conclusions

- Levulinic acid is an important biomass-derived platform chemical
- γ-valerolactone is the main product of LA hydrodeoxygenation over NiMo/γ-Al_2O_3
- Hydrodeoxygenation and decarboxylation are competitive reactions
- Selectivity towards the γ-valerolactone can be regulated by temperature and pressure and catalyst loading
- Undesired decarboxylation is difficult to inhibit, especially at higher temperatures

Acknowledgements

The authors gratefully acknowledge the financial support of the Ministry of Education, Science and Technology of the Republic of Slovenia through the Research Program No. P2-0152.

Literature

[1] B. Girisuta, et al., Chemical Engineering Journal 217 (2013) 61-70.

[2] D.W. Rackemann, W.O.S. Doherty, Biofuels, Bioproducts and Biorefining 5 (2011) 198-214.

[3] M. Grilc, B. Likozar, J. Levec, Applied Catalysis B: Environmental 150–151 (2014) 275-287.

467198 The E-Factor of the Thin Film and Its Characterization to Understand Why Water Influences a Heck Alkynylation

Abstract:

This presentation will highlight our discoveries recently reported[1] on why water influences cationic deprotonation, anionic deprotonation, and the ionic mechanism of a Heck alkynylation. Classical chemical reaction engineering theory was applied to characterize mass transfer in the aqueous-organic system. Estimation of Hatta modulus values < 0.02 enabled analyses of Gibbs free energies for all three mechanisms. It was discovered that water potentially switches the rate-determining steps of cationic and anionic deprotonation. Oxidative addition of organohalide to form a Pd-complex, coordination of the alkyne with the oxidative addition adduct, or ligand substitution could govern either catalytic cycle. Hydrogen-bonding is one possible explanation. Molecular-level process intensification has the potential to broadly impact the continuous-flow manufacture of fine chemicals and pharmaceuticals.

Synthesis in batch and continuous-flow were also studied by defining the *E-factor* in terms of the characteristics of the aqueous-organic interfacial film. First-principle calculations, where ligand substitution is infinitely fast with respect to the diffusive flux, predicted less chemical waste in continuous-flow than batch operation. Interestingly, the concentration of hydrophilic phosphine ligand influenced mass transport limitations and the *E-factor of the thin film*. Increasing the ligand concentration beyond a critical value increased the *E-factor of the thin film* above its minimum, and it also introduced mass-transfer-limitations. Finite changes in the ligand concentration could explain ambiguous results when performing aqueous-phase catalyzed Heck alkynylations, and potentially Pd-catalyzed C-C cross-couplings in general. The reactivity of useful ligands could be masked during discovery and mass transport limitations introduced during manufacture. Our understanding of the E-*factor of the thin film* broadly impacts the sustainable discovery and manufacture of fine chemicals, materials, natural products, and pharmaceuticals.â€

[1] Hu, C., Shaughnessy, K.H., and Hartman, R.L. â€œInfluence of water on the deprotonation and the ionic

mechanisms of a Heck alkynylation and its resultant E-factorsâ€, *Reaction Chemistry & Engineering*, 1 (2016) 65-72.

Copyright © American Institute of Chemical Engineers. All rights reserved.

465814 Triphasic Millireactors for Gas-Liquid Reactions with in Situ Catalyst Synthesis

Abstract:

Metal-catalyzed gas-liquid reactions such as hydrogenation, carbonylation and oxygenation are ubiquitous in the pharmaceutical and fine chemical industries, and are conventionally carried out in batch reactors that typically face severe heat and mass transport limitations. The enhanced heat and mass transfer properties characteristic of micro/millireactors offer a potential solution to overcome these challenges. In addition to the challenges associated with heat and mass transfer, the use of cheap and abundantly available metals, such as iron, nickel and cobalt, as catalysts is of great interest. However, such metal catalysts are seldom used due to their instability, which results in rapid agglomeration and oxidation, especially when used as nanoparticles, despite their high reactivity, selectivity and ease of separation.[1] Consequently, the lifespan of these catalysts is tremendously short, of the order of minutes at times, and *in situ* synthesis followed by usage in stirred batch reactors for extended residence times is extremely challenging. Here, we present a novel method for the *in situ* synthesis of nanoparticle catalysts for hydrogenations within flow reactors. We present a triphasic millireactor that consists of aqueous solution of metallic salt, organic substrate and hydrogen gas at atmospheric pressure. The nucleation and growth of the nanoparticles is initiated upon the injection of sodium borohydride, a strong reducing agent, and the nanoparticle-catalyzed hydrogenation reaction then commences upon the formation of the nanoparticle catalyst. The rapid mixing and controlled addition of reagents into the flow reactor enables the synthesis of well-defined ultra-small nanoparticles with high catalytic activity. Using Ni-catalyzed hydrogenation of 1-hexene as a model reaction in our reactor system, we are able to achieve a substrate conversion of 87% within a total reactor residence time (nanoparticle synthesis and hydrogenation reaction) of 5 min, at ambient conditions.

References

[1] N. Yan, C. Xiao and Y. Kou; Coordination Chemistry Review, 2010, 254, pp. 1179-1218

Copyright © American Institute of Chemical Engineers. All rights reserved.

AUTHOR INDEX

Abbasi, Pedram....73, 754
Abdala, Paula....778
Abdelrahman, Omar A....891
Abdulhadi, Abdullah....108
Abiev, Rufat....248
Abild-Pedersen, Frank....523, 726, 910
Abolhasani, Milad....506
Abram, David N....184
Abrokwah, Richard....716
Adjiman, Claire....439
Agarwal, Manish....454
Agarwal, Pratyush....635, 676
Ager, Joel W....743, 903
Agrawal, Pradeep K....816, 934
Aguirre-Soto, Alan....47
Ahmadi, Masoudeh....63
Ahmed, Sheikh F....240
Ahmed, Usama....175
Aho, Atte....722
Ailawar, Saurabh....219, 692
Akter, Nusnin....937, 948
Al Dahhan, Muthanna....87
Al Hinai, Abdullah....177
Alalouni, Mohammed R....686
Alanazi, Yousef....503, 701, 732
Albarracin, Jonatan....298, 546
Alba-Rubio, Ana....364
Al-Bazzaz, Hamza....164
Alcorn, Brett....712
Al-Dahhan, Muthanna....157, 164
Aleksandrov, Hristiyan A....881
Alexander, Anne-Marie C....214
Alexander, Vineet....164
Alexeev, Oleg S....881
Ali, Shaikh A....686
Alimoradi, Sirwan....106
AlJama, Hassan....523, 726
Almalki, Fahad....490, 798
Almkhelfe, Haider....369, 561
Alonso, David Martin....398
Al-Otaibi, Bader....108
Alphones, Arokiaraj....243
Al-Rifai, Noor....422
AlSalihi, Sara....716
Alshafei, Faisal H....686
Alvarez, Dra Mayra Garcia....606
Alvarez, Pedro....165
Alvarez, Pedro J....127
Amama, Placidus B....369, 561
Amiridis, Michael D....881
Ammal, Salai C....379, 876
Amundsen, Ted J....712
An, Hang....854
An, Hongmei....138, 740
An, Wei....497
Anderst, Emily....400
Angelopoulos, Anastasios....869
Anggara, Trunojoyo....209, 631
Annesini, Maria Cristina....149
Ansah, Emmanuel....407
Ansari, Manizheh....759
Apaja, Vesa....287
Aponte, Yira....213
Aranda, Donato....708
Arbag, Huseyin....679
Arca, Serhat....100
Arce-Ramos, Juan Manuel....616
Ardo, Shane....569
Arellano-Garcia, Harvey....619
Argyle, Morris D....578
Aristizábal, Valentina....623
Arnadottir, Liney....701, 732
Aroh, Kosi....482
Arturo, Steven G....430
Artyushkova, Kateryna....71
Asadi, Mohammad....17, 73, 754
Asano, Shusaku....508
Asgari, Nazli....240
Asiaee, Alireza....227
Askar, Khalid....779
Aslam, Umar....694
Asthagiri, Aravind....545, 799
Atanassov, Plamen....71
Athariboroujeny, Motahare....921
Auerbach, Scott M....888
Austin, Natalie....405
AuYeung, Nick....503
Avanesian, Talin....922
Ayandiran, Afees....845
Aymonier, Cyril....505
Azzam, Sara....316
Baah, David....365
Bababrik, Mohammed Reda....88
Back, Seoin....461
Badii, Khashayar....654
Bae, Joongmyeon....604
Bae, Minseok....604
Bagusetty, Abhishek....544
Bahrim, Cristian....731
Bai, Peng....788
Baiker, Alfons....885, 918
Bajdich, Michal....40, 336, 672, 910
Bakare, Idris....845
Baker, Jonathan G....338
Balakotaiah, Vemuri....91, 143, 680
Balbuena, Perla B....556, 565
Baldauf-Sommerbauer, Georg....452
Balsara, Nitash....344
Baltrusaitis, Jonas....322, 458
Banerjee, Arghya....392, 404

AUTHOR INDEX

Banerjee, Ritubarna 886, 930
Banerjee, Sanjoy 759
Banholzer, William F. 363
Banisharif, Farhad 747
Bao, Zhenan 72, 641
Barim, S. Bengisu 357
Barnett, Kevin J. 122, 363
Barrio, Laura 301
Barton, David J. 847
Basdogan, Yasemin 463
Basu, Suddhasatwa 454
Bates, Jason 494
Bauer, Tobias 247, 248
Baxter, Jason B. 792
Beatriz, Fidalgo 420
Becker, Henning 301
Beckham, Gregg T. 331, 889
Behr, Arno 170
Behravesh, Erfan 722
Beletskiy, Evgeny 880
Belhseine, Yasmeen 124
Bell, Alexis T. 119, 320, 344, 348, 349, 360, 429, 553, 743, 856, 899, 903
Beltran, Saul Perez 565
Benjamin, Kenneth M. 227
Benson, Shawn 106
Benson, Tracy J. 731
Bent, Stacey F. 338, 862, 916
Berkson, Zachariah 787
Bernasko, Dorothy 622
Bernazzani, Paul 731
Berson, R. Eric 332
Beuvier, Thomas 505
Bhattacharyya, Debasis 673
Bhavsar, Saurabh 719
Bhethanabotla, Venkat R. 735, 906
Bhethanabotla, Vignesh 735
Bhoorasingh, Pierre L. 428
Biancardi, Alessandro 549, 919
Biddinger, Elizabeth J. 234, 487
Biddy, Mary 889
Biener, Juergen 830
Biener, Monika 830
Biernacki, Joseph 94
Biernacki, Joseph J. 419
Billa, Triveni 635, 676
Binder, Andrew J. 931
Binous, Housam 845
Bishop, Cody 715
Bligaard, Thomas 378
Boateng, Akwasi A. 192
Bobba, Pallavi 215
Bodenschatz, Cameron 500
Boerigter, Calvin 694
Boes, Jacob R. 383, 646
Bollas, George M. 576
Bond, Jesse Q. 325, 395, 403, 891
Boningari, Thirupathi 488
Boolchand, Punit 869
Borghard, William G. 953
Borgna, Armando 121
Born, Dirk 98
Borne, Isaiah 107
Boscoboinik, Jorge A. 937, 948
Bossard, Peter 696
Bowers, Clifford R. 913
Boyd, Michael 603
Brandstädter, Willi 147
Bravo-Suarez, Juan J. 919
Brentzel, Zachary 122, 363
Brewer, Catherine E. 225, 868
Brezinsky, Kenneth 781
Briggs, Nicholas M. 907
Brinson, Bruce E. 416
Briquet, Ludovic 96
Broadbelt, Linda J. 430, 495
Brown, Avery 75
Brown, Robert 331
Browning, Barbara 165
Bruce, David A. 780
Brückner, Angelika 792
Bruijnincx, Pieter 120, 471
Brunelli, Nicholas 540, 721
Bryant, Kristin 954
Buesser, Beat 517
Bui, Hao 564
Bukhovko, Maxim P. 686
Bukowski, Brandon 494, 940
Bukur, Dragomir B. 557
Bullard, Lisa G. 114
Burgos, Juan C. 556
Burt, Samuel P. 363
Busygin, Igor 887
Butcher, Holly 696
Cai, Jianghuai 107
Cakiryilmaz, Nurbanu 679
Calvignac, Brice 505
Camaioni, Donald M. 125
Campbell, Charles T. 125, 400
Campos-Martin, Jose M. 747
Candelaria, Stephanie 70, 374
Candia, Sandra Varinia Bernales 221
Caniaz, Ramazan Oguz 100
Cao, Chenxi 643
Cao, Enhong 422
Cao, Fei 231
Cao, Guoqiang 879
Cao, Xiaoyan 315
Caratzoulas, Stavros 199

AUTHOR INDEX

Cárdenas-Galindo, María-Guadalupe 616
Cardona, Carlos A. 623
Cardona-Martinez, Nelson 370, 396
Cargnello, Matteo 453, 511, 512, 535, 656, 792, 831, 884, 944
Caricato, Marco 549, 919
Carneiro, Juliana S. A. 335
Carpenter, Andrew 712
Carr, Shane F. 385
Carrero, Carlos 844, 908
Carter, Emily A. 372
Cashman, Derek 94
Caspi, Daniel D. 480
Castaldi, Marco 796
Castaldi, Marco J. 781, 938
Castro, Angel 685
Cattaneo, Stefano 422
Cave, Etosha R. 184
Celik, Fuat E. 288, 807, 941
Celik, Gokhan 214, 219, 692
Cerbelli, Stefano 149
Chada, Joseph P. 844, 908
Chadderdon, David 7, 286, 289, 648
Chadderdon, Xiaotong 286, 648
Chaikittisilp, Watcharop 790
Chakrabarti, Anisha 920
Chakrabarti, Debanjan 441
Chakthranont, Pongkarn 633, 672, 702, 729
Chamas, Ali 895
Chan, Karen 460, 464, 465, 547
Chan, Maria K. Y. 384
Chan, Xiaojun 220, 711
Chandler, Bert D. 892
Chang, Chun-Chih 541
Chang, Seong-Cheol 663
Chang, SiWei A. 802
Changi, Shujauddin M. 479
Chatterton, Lindsey 366
Chatzidimitriou, Anargyros 325
Chaudhari, Raghunath V. 215, 222, 327, 469
Chavez, Steven 200, 683, 693
Che, Fanglin 514, 533
Chen, Benjamin W J 801
Chen, Binghui 34
Chen, Daniel 243
Chen, Elaine 92
Chen, Han 671
Chen, Hao 715
Chen, Jingguang G. 201, 805
Chen, Leanne D. 29
Chen, Li 416
Chen, Shucheng 72, 641
Chen, Wan-Ting 105
Chen, Xiaowen 333
Chen, Xiaoyin 653
Chen, Xin 854
Chen, Yizheng 742
Chen, Yuan 859
Chen, Zhihua 72
Chen, Zhu 373
Chen, Zihua 641
Cheng, Feng 225, 868
Cheng, Xun 486
Cheng, Yan 520, 522
Cheng, Yi 520, 522, 643, 854
Cheng, Yisun 203, 627
Cheng, Yongqiang 491
Cheng, Zhuo 795
Chi, Yong 681
Chimenton, Ricardo José 606
Chizallet, Céline 96
Chmelka, Bradley F. 787
Cho, Hong Je 541
Choi, Eun-Kyung 583
Choi, Jae-Soon 300
Choi, Juhee 799
Choi, Ki-Hyouk 108
Choi, Yongsuck 192
Choksi, Tej S. 380, 639, 925
Choong, Catherine 121
Chorkendorff, Ib 901
Choudhury, Nujhat 238
Choudhury, Pabitra 783
Chowdhury, Sanchari 733
Christopher, Phillip 468, 803, 832, 915, 922
Chu, Bozhao 854
Chubukov, Boris 427
Chun, Dong Hyun 649, 658
Chun, Hee-Joon 287
Chung, Wook-Jin 661
Chyou, Yau-Pin 669
Ciora, Richard J. 730, 736
Clark, Ezra 553
Clark, Ezra L. 743, 903
Clary, Jacob M. 567
Clayborne, Andre 287
Clemens, Bruce M. 734
Cline, Brandon 867
Co, Anne 285, 703
Coan, Patrick 911
Coley, Connor 506
Collinge, Greg 579, 921
Colón, Jorge L. 638
Conrad, Sabrina 350
Contreras-Mora, Jose 123, 930
Copéret, Christophe 778
Coppens, Marc-Olivier 513, 651
Coppernoll, Aaron 588
Cordon, Michael J. 539

AUTHOR INDEX

Costa, Karina K. 158
Cournoyer, Justin T. 879
Cowen, Kenneth 301
Cox, David F. 671
Crossley, Steven 89, 233, 431, 907
Crow, Allison M. 862
Crowley, Stephen 938
Cui, Kai 330
Cui, Yanran 917
Cui, Zheng 868
Currier, Neal 138, 297, 740
Cutlip, Michael B. 111
Cybulskis, Viktor 917, 940
Dade, William 716
Daemen, Luke L. 491
Dagnall, Katelyn A. 288
Dai, Ning 131
Dalai, Eswar Prasad 673
Damodara, Vijaya 243
Damouny, Christine Warwar 531
Dana, Alon Grinberg 517
Danby, Andrew 469
Dang, Dan 643
Dang, Thanh 400
Das, Anirban 335
Das, Biswapriya 673
Das, Satyen Kumar 673
Dasgupta, Anish 834
Dashliborun, Amir Motamed 244
Dasireddy, Venkata 442
Datta, Ravindra 75
Datye, Abhaya 217, 315, 936
Datye, Abhaya K. 229
Dauenhauer, Paul J. 390, 393, 541
Davis, Andrew W. 796
Davis, Kirsten 331
Davis, Mark E. 538
Davis, Robert J. 216, 314, 323, 936
Daw, C. Stuart 415
Daza, Yolanda 906
De Jong, Ed 471
De Jong, Johannes 524
De Lasa, Hugo 213, 793
De Vasconcelos, Bruna Rego 528
De Wilde, Juray 796
Dean, Anthony M. 519
Dean, James 463
Debefve, Louise 553, 828, 887
deGrazia, Janet 110, 117
Dehghani, Mohammad Reza 747
Deka, Dhruba Jyoti 285, 703
Dekui, Shen 420
Delgass, W. Nicholas 298, 546, 835, 917, 945
Demir, Bengi 398
DeRita, Leo 832
Deshlahra, Prashant 857, 865
Deshmane, Vishwanath 709, 716
Deshmukh, Soumitra 301
Deshpande, Nitish 540, 721
Deskins, N. Aaron 75
Devocht, Brigitte R. 432
Dharanipragada, Aditya N.V.R. 524
Di Iorio, John R. 298, 546, 789, 835
Diamanti, Aikaterini 439
Diao, Weijian 197
Díaz, Angélica Liliana Coconubo 949
Diaz, Rosa 218
Dickens, Colin 462, 466, 547, 727
Dinh, Kimberly 776
Diulus, John 799
Dixit, Jagdev Kumar 673
Dixon, Anthony G. 92, 340, 667
Dixon, David A. 828, 887
Doan-Nguyen, Vicky 792
Dogu, Doruk 285, 703
Dogu, Gulsen 679
Dogu, Timur 679
Dominguez, Bernardo Castro 340
Dong, Jun 681
Doren, Douglas J. 839
Dorfi, Anna 630
Doyle, Andrew 336, 462, 466, 727
Drochner, Alfons 98
Du, Zhou 842
Duan, Pu 229, 315
Dubois, Jean-Luc 190
Ducousso, Marion 681
Dumesic, James 231, 315, 362, 364, 394, 398
Dumesic, James A. 122, 363
Dunwell, Marco 185
Duong, Nhung 861
Durgadevi, A. 668
Duyar, Melis S. 484, 620
Eastman, Stephen 281
Edake, Mahesh 190
Edgar, James H. 93
Edmiston, Paul 219, 692
Egle, Tobias 830
Elkasabi, Yaseen 192
Ellis, Lucas 858
Elly, Michael 111
Elnabawy, Ahmed 510
Elsayed, Nada 713, 777
Elwell, Anthony 713
Embry, Matthew C. 479
Entesari, Nazanin 341
Epling, William 290, 677
Epling, William S. 292, 689
Eppinger, Thomas 570, 642
Eränen, Kari 722

AUTHOR INDEX

Erdem, Ezgi 357
Erkey, Can 100, 357
Ertler, Daniel 887
Esche, Erik 688
Escudero-Escribano, Maria 33, 702, 901
Esposito, Daniel V. 376, 630
Eswaranandam, Ees 347
Evans, Scott 243
Ewing, Christopher S. 544
Ezbiri, Miriam 905
Falcone, Derek D. 216, 936
Falconer, John 117
Falconer, John L. 567
Fan, Feiyue 874
Fan, Jonathan A. 795
Fan, Liang-Shih 795
Fan, Wei 541
Fang, Yuxin 486
Fanzhong, Ma 659
Farberow, Carrie A. 860
Farouk, Tanvir I 240
Fatemi, Naghmeh 432
Fazlollahi, Farhad 132
Feaster, Jeremy T. 184
Felischak, Matthias 145, 849
Feng, Lian-Fang 592
Ferdousi, Shammi Akter 756, 757, 758
Ferguson, Glen Allen 889
Fester, Jakob 910
Fisher, Galen B. 295
Flaherty, David W. 321, 687, 802, 935
Flake, John C. 486
Fleischman, Samuel D. 708
Fletcher, Thomas 132
Flytzani-Stephanopoulos, Maria 775, 829, 830
Fogler, H. Scott 109
Fornasiero, Paolo 792
Fornero, Esteban L. 942
Fortner, Edward 243
Foston, Marcus 895
Fredrickson, Kurt 552, 744
Freeman, Emily 625
Freund, Hannsjörg 95, 144, 148
Friend, Cynthia M. 406, 830, 864
Fry, Andrew 239
Ftouni, Jamal 120
Fu, Xiao-an 742
Fuentes, Gustavo A. 755
Fuller, Anshuman 285
Fulton, John L. 125
Funkenbusch, LiLu 4
Furuya, Eiji 612
Fushimi, Rebecca 566
Gagliardi, Laura 221
Gaillard, Nicolas 728
Galindo, Amparo 439
Gallagher, James 218, 940
Gallo, Alessandro 898
Galloway, Benjamin 212
Gallup, Jennifer 202
Galvita, Vladimir 524, 942
Gamliel, David P. 576
Gan, Zhuoran 485
Ganase, Zara 439
Gangwal, Santosh 867
Ganzer, Gunnar 148
Gao, Feng 209, 291, 631, 840, 859
Gao, Hanyu 430
García-Melchor, Max 672, 910
Garrabos, Yves 505
Garrick, Taylor R. 197
Garshasbi, Ashkan 730
Gaspard, Pierre 498
Gates, Bruce C. 221, 553, 828, 887
Gauthier, Joseph 462
Gavriilidis, Asterios 422
Gazit, Oz M. 531
Geantet, Christophe 165
Gellman, A.J. 426
Gellman, Andrew J. 804
Genuino, Homer 39, 120, 471
Gerceker, Duygu 394
Getman, Rachel 500, 866
Ghale, Kushal 400
Ghoniem, Ahmed F. 90
Ghoroi, Chinmay 562
Gibaud, Alain 505
Gibbons, Brenna 734
Giles, Stephen A. 199
Gill, Rajinder 796
Giordano, Nicolás A. Grosso 542, 837
Glasser, Benjamin 715
Glasser, David 609, 666, 682
Gleaves, John T. 566
Godiksen, Anita 897
Godini, Hamid 688
Goldsmith, Bryan R 5, 78
Goldsmith, C. Franklin 241, 354, 518
Golkarnarenji, Gelayol 654
Göltl, Florian 350, 536, 546
Gómez, Alvaro 623
Gómez, Sergio A. 755
Gomez-Ballesteros, Jose L. 556
Gong, Jian 297, 740
Goodman, Emmett 535, 884
Goodpaster, Jason 349, 429
Gordon, Thomas 792
Gorimbo, Joshua 682
Gorte, Raymond J. 493, 541
Goulas, Konstantinos 230, 493, 896

AUTHOR INDEX

Gould, Nicholas 543, 893
Gounaris, Chrysanthos E. 926
Gounder, Rajamani 298, 494, 539, 546, 789, 835
Govedarica, Zora 366
Goyal, Amit 366
Grabow, Lars C. 616
Grady, Brian P. 369
Grambow, Colin 440
Grant, Eliana 439
Grass, Robert N. 593, 808
Gray, Jake T 533
Gray, Michel 470
Gray, Stephen 384
Greeley, Jeffrey P. 287, 380, 494, 639, 925, 940, 945
Green, William H. 107, 440, 517, 521
Greenlee, Lauren F. 70, 374
Gregor, D. 642
Griffin, Michael B. 889
Grilc, Miha 472
Grillo, Fabio 564
Groehn, Arto J. 427
Groenenboom, Mitchell 353, 383
Grolman, Eric 425
Grossardt, Jason 369
Gu, Tian 143
Gu, Tonghan 502
Gu, Xiang-Kui 335
Gu, Xue-Ping 592
Guangsheng, Luo 591
Gudiyella, Soumya 107
Guerra, Patricia 223
Guliants, Vadim 697
Gumidyala, Abhishek 233, 431
Gümüslü, Gamze 221, 426
Gunasooriya, G. T. Kasun Kalhara 392
Gunduz, Seval 219, 285, 692, 703
Gunugunuri, Krishna Reddy 714
Guo, Mengqing 795
Guo, Mond 470
Guo, Zhe 74
Gupton, Frank 929
Gutiérrez, Oliver 125
Ha, Su 514, 533
Haase, Stefan 247, 248
Hable, Robert 106
Hack, John H. 216
Hagelin-Weaver, Helena 913
Hahn, Christopher 184, 633, 652, 657
Haider, M. Ali 454
Hajari, Amit 281
Haller, Gary L. 137
Ham, Hyung Chul 663
Hamel, Christof 424, 849
Hamidipour, Mohsen 244
Hammouti, Abdelkader 96
Han, Chonghun 175, 176
Han, Jeong Woo 207, 737, 739
Han, Jeongsik 670
Han, Jonghee 663
Han, Kehang 521
Han, Wenfeng 308
Handy, Brent E. 616
Hanna, Brian S. 686
Hanselman, Christopher L. 926
Hansen, Charles 719
Hanspal, Sabra 323
Hao, Pengxiao 863
Harden, Tiffany 487
Hare, Bryan J. 906
Harker, Audrey 553
Harold, Michael 627
Harold, Michael P 128, 203, 292, 680, 689
Harpool, Scott 503
Harris, James W. 218, 539
Hartman, Ryan L. 159, 477
Hasan, Mohammad Mehedi 756, 757, 758
Hashimoto, Akito 228
Hatsukade, Toru 184
Hatton, T. Alan 502
Hattrick-Simpers, Jason 456, 602, 904, 933
Haudek, Kevin 115
Hayek, Naseem S. 531, 613
Hazlett, Melanie 677
He, Jiayue 231, 363
He, Peng 537
He, Ping 90
He, Xiang 485
He, Yang 720
Head-Gordon, Martin 349, 429
Headrick, Sierra 783
Heagy, Michael D. 733
Hecker, William C 578
Hellgardt, Klaus 104, 423, 622
Hellstern, Thomas R. 657, 728
Helveg, Stig 899
Hendon, Christopher H. 364, 902
Henkelman, Graeme 509
Hensley, Alyssa 382, 400, 498, 875
Hermans, Ive 350, 363, 536, 844, 908
Hernández-Salgado, Gabriela 755
Hernández-Terán, M. Eugenia 755
Hernley, Paul 200, 683, 693
Héroguel, Florent 563
Herrera, Carolina 868
Herrera, Valeria 907
Herrmann, Stanley 139
Heyden, Andreas 379, 403, 876
Heydt, Thomas 98
Hiebel, Fanny 864
Higgins, Christopher 332

AUTHOR INDEX

Higgins, Drew 42
Hii, Mimi 423
Hikita, Yasuyuki 466, 727
Hildebrandt, Diane 609, 666, 682
Hillmyer, Marc A. 390
Hirose, Takuya 345
Hischier, Illias 427
Hoard, John 295
Hocke, Elisabeth 455
Hodoshima, Shinya 846
Hofer, Corinne 808
Hoffman, Adam 828
Hoffman, Graham 868
Hohn, Keith L. 369, 561
Holles, Joseph 939
Hong, Giyoung 598, 599
Hong, Yongchun 382
Honkala, Karoliina 287
Hook, Alec 941
Horton, Scott R. 418, 635, 676
Hossain, Mohammad Mozahar 845
Hou, Zhen 676
Hoyt, David W. 895
Hu, Bo 816, 934
Hu, Chuntian 477
Huang, Benjamin 314, 936
Huang, Ching-Ying 669
Huang, Kefeng 122, 231, 363
Huang, Qilong 554
Huang, Qunxing 681
Hubbard, Carolyn 203, 627
Huber, George W. 122, 231, 363, 844, 908
Hungerbuehler, Konrad 885, 918
Hunt, Sean T. 364, 902
Huo, Chao 308
Huo, Jiajie 229
Hur, Young Gul 662, 789
Huš, Matej 442
Hutchings, Graham J. 422
Huynh, Dang Chinh 450
Hwang, Harold Y. 466, 727
Hwang, Sang Youp 208
Hwang, Taeho 78
Ibrahim, Gasim 954
Iglesia, Enrique 139, 311, 784, 836, 857, 865
Iisa, Kristiina 194
Ilias, Shamsuddin 709
Illner, Markus 76
Imtiaz, Qasim 778
Inturi, Siva Nagi Reddy 488
Ioannidis, Efthymios 381
Ipek, Bahar 536
Isaksen, Oliver 897
Ishikawa, Tatsuya 612
Itoh, Naotsugu 345, 600, 601, 738
Izadpanah, Arash 104
Jaén, Juan J. Delgado 792
Jamal, Adeel 440
Jan, Oliver 368
Janet, Jon Paul 381
Jang, Seung Soon 501
Janik, Michael 198, 947
Janik, Michael J. 877, 892
Janmanchi, Krishna 588
Jansen, Tim 165
Jantasee, Sasiradee 346
Janvelyan, Nare 830
Jaramillo, Thomas F. 72, 184, 466, 484, 620, 633, 638, 641, 652, 657, 672, 702, 708, 725, 727, 728, 729, 734, 898
Jarboe, Laura 331
Jarvis, Jacqueline 868
Jeffrey, Miller 218
Jehng, Jih-Mirn 458
Jenness, Glen 805
Jenness, Glen R. 11, 118, 230
Jensen, Klavs F. 482, 483, 504, 506
Jentoft, Friederike C. 367
Jeon, Jong Ki 670
Jeong, Byung-Hun 670
Jeong, Gwangsik 662
Jeong, Heondo 649, 658
Jeong, Kwan-Hyoung 670
Jeong, Soon Kwan 583
Ji, Jingjing 624
Jia, Hongfei 714
Jiajun, Zhang 420
Jiang, Chengjun 818
Jiang, Jimeng 224
Jimenez, Juan 456, 904
Jin, Bo 310
Jin, Xin 215, 327
Jocz, Jennifer 943
Johnson, J. Karl 928
Johnson, Karl 544
Johnson, Martin D. 479
Johnson, Matthew 243
Johnson, Robert 315
Johnson, Robert L. 229, 326
Jokiniemi, Jorma 606
Jones, Christopher W. 501, 816, 934
Jongerius, Anna L. 184
Jörke, Andreas 424
Joseph, Babu 713, 777
Joseph, Kristeen Esther 393
Josephson, Alexander J. 237
Joshi, Saurabh Y. 138
Jovanovic, Goran 503, 701, 732
Jung, Sungyup 234
Jung, WooChul 737, 739

AUTHOR INDEX

Jung, Yousung....461
Jurtz, Nico....570, 642
Jystad, Amy....549, 919
Kageyama, Naoki....432
Kaiser, Markus....95
Kalaga, Dinesh V....759
Kalamaras, Christos....677
Kaldor, Andy....696
Kale, Matthew....41, 803
Kaliaguine, Serge....634
Kamasamudram, Krishna....138, 740
Kamaz, Mohanad....347
Kampfe, Sara....301
Kang, Joo....847
Kang, Sung Bong....294
Kang, Sungbong....677
Kanitkar, Swarom....611
Kapil, Nidhi....513
Karagoz, Secgin....343
Karaiskakis, Alexandros....487
Karakalos, Stavros....864
Karamad, Mohammadreza....465
Karanjkar, Pranav U....231, 363
Karhunen, Tommi....606
Karim, Ayman M....878, 896
Karroum, Hafsa....571
Katikaneni, Sai P....604
Kato, Takafumi....738
Katz, Alexander....542, 553, 837, 887
Kawakita, Takashi....612
Kawi, Sibudjing....121
Kaxiras, Efthimios....352, 406
Keels, Jayson....779
Keith, John A....353, 383, 463
Kelleher, Patrick....137
Key, Haley....389
Khan, Saif A....478, 502
Khayyam, Hamid....654
Khinast, Johannes G....715
Khivantsev, Konstantin V....798, 881
Khokhar, Munir D....686
Khorasgani, Elmira Soghrati....121
Khosravian, Homa....621
Khoury, Christine....531
Khurana, Ishant....298, 546, 835
Kibsgaard, Jakob....484, 620, 633, 657, 708, 725
Kida, Tetsuya....228, 718
Kiely, Christopher....120
Kienberger, Marlene....346
Kikuchi, Shuhei....345, 600
Kilos, Beata A....847
Kim, Chan Hun....662
Kim, Changsoo....176
Kim, Do Heui....220
Kim, Dongyeon....604
Kim, Doo-Wook....659
Kim, Gookhee....605
Kim, Hak Joo....583
Kim, Hyunuk....645
Kim, Ji Man....698
Kim, Jong Jin....670
Kim, Ki Chul....501
Kim, Ki-Duk....745, 746
Kim, Kiseok....698
Kim, Kyeongsu....176
Kim, Kyeounghak....739
Kim, Mi Jin....688
Kim, Minkyu....545
Kim, Sung Min....778
Kim, Sung-Ho....745, 746
Kim, Taeho Roy....641
Kim, Taejin....220, 711, 853, 937, 948
Kim, YongWoo....220
Kimmel, Yannick C....201
Kinaci, Alper....384
Kirk, Charlotte....466, 547, 672, 727
Kistler, J. D....887
Kitchin, John R....383, 646
Kizilel, Riza....357
Klein, Michael T....418, 635, 676
Klyushin, Alexander....216
Knop-Gericke, Axel....216, 797
Koberstein, Jeffrey T....376
Koel, Bruce....373
Koel, Bruce E....371
Kohara, Kento....601
Kommoss, Bjoern....455
Kondratyuk, Petro....426
Konstantinov, Ivan....430
Koo, Kee Young....649, 658
Koob, Brittany....867
Koretsky, Milo D....116
Koriakin, Anton....659
Kostetskyy, Pavlo....496
Kovarik, Libor....124, 859, 898
Kramer, Nathaniel....285
Kraume, Matthias....570, 642
Krause, Jonas....590
Krishna, Siddarth H....231, 363
Kronawitter, Coleman....371, 373
Kropf, A. Jeremy....940
Krumm, Christoph....393
Kruse, Norbert....573, 579, 921
Kuboyama, Toshiharu....508
Kuhl, Kendra P....184
Kuhn, John....735, 777
Kuhn, John N....713, 906
Kuhn, Simon....169
Kuila, Debasish....716
Kulik, Heather J....381

AUTHOR INDEX

Kulkarni, Ambarish R. 523, 726, 927
Kumar, Anand 21, 636
Kumar, Ashok 138
Kumar, Gaurav 892, 947
Kumar, Narendra 722
Kumar, Nitin 611
Kumar, Prashant 896
Kumar, Rajeev 468
Kumari, Neetu 454
Kuncharam, Bhanu Vardhan Reddy 92
Kung, Harold H. 142, 880, 909
Kung, Mayfair 880
Kunsa, Tyler 712
Kuo, Chih-Ting 105
Kuroki, Hidenori 650
Kurtoglu, Samira Fatma 674
Kurtz, Richard L. 491
Kwon, Hyunguk 737
Kwon, Kyung 365
Kwon, Stephanie 857
Kwon, Youngkook 743
Kyriakidou, Eleni A. 300, 931
Labrador, Natalie 376
Lacroix, Maxime 165
Lähde, Anna 606
Lai, Lawrence 107
Laing, Paul 203, 627
Lambrecht, Daniel S. 544
Lance, Michael J. 931
Landers, Alan 657
Landers, John 807
Landon, Mark 245
Lang, Wendy 203, 627
Lange, Ruediger 247, 248, 282
Larachi, Faical 244
Latimer, Allegra A. 523, 726, 927
Lauritsen, Jeppe 910
Laursen, Siris 377, 720, 924
Lauterbach, Jochen 206, 456, 575, 602, 625, 851, 904, 933
Lawagon, Chosel P. 661
Le Brun, Niccolo 104
Le Monnier, Benjamin 563
Le, Thuy T. 489, 644
Lecoutre, Carole 505
Le-Doux, Travis 225, 868
Lee, Chang-Ha 183, 659
Lee, Chang-Whan 663
Lee, Choul-Ho 670
Lee, JaeHa 220
Lee, Jae-Hwan 660
Lee, Joo-Hyeong 108
Lee, Joo-Youp 133
Lee, Kwan-Young 583, 660, 662
Lee, Siwon 739
Lee, Sunggyu 647
Lee, Wonyoung 737
Lee, Woo-Sung 183
Lee, Yong-Kul 655, 745, 746
Lee, Youn-Woo 598, 599
Lei, Yu 624, 914
Leonard, Brian 196
Leonard, Kevin C. 375
Leonarduzzi, Daniele 301
Lercher, Johannes A. 125, 492
Leung, Andrew 423
Lewis, Connor A. 678
Lewis, Jennifer 894
Li, Can 135
Li, Dongmei (Katie) 196, 566
Li, Fanxing 855
Li, Guo-Ling 551
Li, Hang 830
Li, Hui 209, 209, 351, 546, 631, 631, 835
Li, Jian-Feng 34
Li, Junhui 297, 740
Li, Lin 306
Li, Meifang 134
Li, Mengmeng 292, 680, 689
Li, Mengwei 775
Li, Mingxiao 742
Li, Qi 742
Li, Rui 675
Li, Sha 801
Li, Shiwen 942
LI, Sichi 840
Li, Tianyang 520, 522
Li, Wei 134
Li, Wenzhen 286, 289, 648
Li, Xianchang 243
Li, Xiang 626
Li, Xinxin 376
Li, Xu 561
Li, Yanxin 193
Li, Yawei 334
Li, Yongdan 330
Li, Zheng 284, 707
Li, Zhongtang 341
Liang, Xinhua 818
Liauw, Marcel A. 391
Licht, Rachel B. 856
Licht, William 310
Lien, Chih-Heng 947
Liesche, Georg S. 97
Likozar, Blaž 442, 472
Lim, Chern-Hooi 459
Lima, Fabiana Dos Santos 741
Lin, Hongfei 126
Lin, Hongkun 506
Lin, John 652

AUTHOR INDEX

Lin, Pin Ann 556
Lindsay, Mark J. 578
Ling, Chen 714
Linic, Suljo 200, 683, 693, 694, 900
Lira, Carl T. 115
Liu, Bin 93, 286
Liu, Dongxia 841
Liu, Dupeng 222
Liu, Huazhang 308
Liu, Jilei 829, 830
Liu, Jingyue 880
Liu, Jiuxu 226
Liu, Paul K.T. 730
Liu, Qiuli 192, 235, 817
Liu, Rui 92
Liu, Song 93
Liu, Xiaoyang 497
Liu, Xinying 682
Liu, Xue 715
Liu, Yukun 376
Liu, Zhouyang 133
Llorca, Jordi 606
Lo, Cynthia S. 385
Loane, Letícia Pezzi Martins 173
Lobaccaro, Peter 743
Lobo, Raul F. 195, 393, 536, 839, 848
Lonergan, William W. 309
López-Curiel, Julio C 755
Lou, Helen 243
Louie, Ying Lin 119
Love, Alyssa 350
Lowd, Jack 736
Lu, Li 120
Lu, Xiaojun 666, 682
Lu, Yubing 878
Lu, Zheng 624, 914
Lucero, Andrew 867
Lucio-Vega, Juan 418, 635, 676
Lugo, Michael 796
Lui, Alison 107
Lum, Yanwei 743, 903
Lundgaard, Keld T 378
Luo, Guangsheng 507, 587
Luo, Jinyong 138
Luo, Lin 847
Luss, Dan 128
Luterbacher, Jeremy S. 324, 397, 563
Lux, Susanne 452
Lwin, Soe 850
Lyu, Yimeng 653
Ma, Rui 340
Ma, Xianfeng 878, 923
Ma, Yi Hua 340
Maag, Alex 699, 700
Maalouf, Joseph 831, 944
Mackus, A.J.M. 338
Madix, Robert J. 406, 864
Mae, Kazuhiro 508
Magrini, Kim 194
Mahboub, Mohammad Jaber Darabi 190
Maiti, Debtanu 735, 906
Maiti, Sanat Chandra 562
Maiti, Swarup 838
Majumdar, Paulami 925
Maki, Taisuke 508
Malamis, Sam 292, 689
Malefyt, Amanda P. 113
Malek, Andre 847
Mallick, Kwonit 868
Mamun, Osman 403
Mandic, Milos 557
Manoli, Kyriakos 79
Manousiouthakis, Vasilios 343, 736
Mao, Liaoyuan 193
Maravelias, Christos T. 122, 231, 363
Mardilovich, Ivan 340
Margossian, Tigran 778
Marin, Guy B. 432, 524
Marre, Samuel 505
Marrero-Vazquez, Michelle M. 370, 396
Martin, Christopher B. 243
Martinez-Hernandez, Angel 755
Masiello, Caroline A. 127
Matanovic, Ivana 71
Matin, Md. Abdul 636
Matsubu, John C. 915
Maurin, Jérémy 165
Mavrikakis, Manos 394, 401, 510, 585, 801
Mazal, Tobias 230
Mazur, Zachary 105
McCabe, Kevin 867
McCrea, Sean 913
McCrum, Ian T. 198
McDanel, Katherine 117
McEnaney, Joshua 725
McEwen, Jean-Sabin 209, 382, 400, 498, 514, 533, 579, 631, 875, 921
McKay, Ian 792
McKee, William C. 339
Medina, Francesc 606
Medlin, J. Will 117, 858, 863, 911, 947, 955
Medlin, Michelle 117
Meemken, Fabian 885, 918
Mehar, Vikram 799
Mehdad, Ali 848
Mehrabadi, Bahareh Alsadat Tavakoli 197
Mehta, Prateek 945
Mei, Donghai 492
Meintel, Kathryn 481
Mellmer, Max A. 398

AUTHOR INDEX

Mendes, Laiza.....708
Merchant, Shamel S.....517
Merrill, John.....115
Merz, Juliane.....590
Meyer, Katja E. Binkley.....285, 703
Miao, Yu.....503
Michalsky, Ronald.....794, 905
Michel, Carine.....536
Mikkola, Jyri-Pekka.....722
Milina, Maria.....364, 902
Miller, Jeffrey.....835, 940
Miller, Jeffrey T.....298
Miller, Juanita.....225
Miller, Victor V.....656
Minette, Florent.....796
Mingle, Kathleen.....933
Minh, Doan Pham.....528
Mironenko, Alexander V.....118, 230, 493
Mitani, Makoto.....729
Mittal, Neha.....661
Miyake, Garret.....459
Modesto-Lopez, Luis balam.....606
Modroukas, Dean.....796
Mohammad, Abdul Salam.....94
Moher, Eric D.....479
Mondal, Chandona.....760
Mondal, Dilip Kumar.....760
Moniri, Saman.....900
Monnier, J.R.....123
Monnier, John.....490
Monnier, John R.....197, 779
Monsalve, Alexander Guzmán.....949
Montemore, Matthew M.....10, 352, 406
Montgomery, Stephen.....717
Montini, Tiziano.....792
Montoya, Joseph H.....402, 460, 466, 727
Moon, Jong-Ho.....645
More, Amey.....719
Mori, Milton.....158
Moschetta, Eric G.....501
Moskowitz, Benjamin.....853
Mossin, Susanne.....897
Motagamwala, Ali Hussain.....398
Mote, Dhananjay R.....388
Moteki, Takahiko.....321
Moyo, Mahluli.....617
Mpourmpakis, Giannis.....405, 496
Mueller, Christoph.....778
Mukai, Shin R.....586, 791
Mukarakate, Calvin.....194
Mukhopadhyay, Ahana.....952
Muleja, Adolph.....609
Mullins, Michael.....4
Mungma, Nuttakul.....346
Muraoka, Koki.....790
Murmura, Maria Anna.....149
Murphy, Brian.....946
Murray, Christopher B.....493, 792
Murugesan, Vijayakumar.....470
Murzin, Dmitry Yu.....722
Musgrave, Charles B.....459, 567
Mushrif, Samir H.....404, 450, 800
Mustain, William E.....678
Muzika, František.....610
Naccarella, Ian.....944
Naebe, Minoo.....654
Nakajima, Hikaru.....479
Nakayama, Ryutaro.....508
Nakhla, George.....79
Nam, In-Sik.....294
Nam, Sung Bang.....294
Nan, Yao.....373
Nan, Yue.....226
Nan, Zhipeng.....869
Nandiwale, Kakasaheb.....469
Naoki, Takagiwa.....601
Narayanan, Badri.....384
Narkeviciute, Ieva.....638
Narsimhan, Karthik.....776
Nash, Connor.....860
Neal, Luke.....14, 855
Neimark, Alexander V.....807
Neurock, Matthew.....319, 788
Ng, Desmond.....672
Nguyen, Anh Vu.....450
Ni, Mingjiang.....681
Nigra, Michael M.....513, 651
Nikacevic, Nikola.....557
Nikolic, Daliborka.....145
Nikolla, Eranda.....335
Nimlos, Mark.....194
Nirmalakhandan, Nagamany.....868
Nishio, Kazunori.....466, 727
Nisola, Grace M.....661
Nitopi, Stephanie.....633
Noh, Kyung-Jong.....207
Nordlund, Dennis.....641, 898
Nørskov, Jens.....72, 465, 512, 551
Norskov, Jens K.....378, 462, 523, 547, 641, 726, 927
Nørskov, Jens K.....402, 460, 466, 727
Northrup, Scott.....94
Notestein, Justin M.....136
Nwabara, Uzoma.....653
Nzihou, Ange.....528, 681
O'Connor, Nolan.....877
Ofosu, Thomas.....919
Ogasawara, Hirohito.....898
Ogino, Isao.....586, 791
Oh, Hyun-Taek.....183
Oh, Jiwoo.....604

AUTHOR INDEX

Oh, Se H. 294
Oh, Tae-Sik 912
Okolie, Chukwuemeka 124
Okrut, Alexander 542, 837, 887
Oktar, Nuray 679
Okubo, Tatsuya 790
Olsson, Louise 299
Omasta, Travis J. 678
Omotoso, Taiwo 907
Onal, Isik 548
Orazov, Marat 538
Ortiz-Arroyo, Arturo 685
Otto, Trenton 784
Ouyang, Xiaoying 837
Oyama, S. Ted 432
Oyarzún, Bernardo 99
Ozkan, Umit S. 214, 219, 285, 692, 703
Padak, Bihter 212, 238, 240
Padhye, Richa 66
Pagan-Torres, Yomaira J. 370, 396
Paleg, Sarah W. 186
Palermo, Andrew 553, 887
Palm, David W. 728
Palumbo, Aaron W. 427
Pan, Hanqing 733
Pan, Li 799
Pan, Yung-Tin 687
Panthani, Matthew G. 289
Paolucci, Christopher 60, 209, 298, 351, 546, 631, 835
Parekh, Atish A. 209, 298, 835
Park, Dae Sung 393
Park, Hai Woong 605
Park, Han-Byul 655
Park, Ki Tae 583
Park, Kuan Hwan 598
Park, Young Cheol 645
Parks, James E. 300, 931
Parsley, Doug 730
Partopour, Behnam 667
Parulkar, Aamena 540, 721
Parvari, Matin 747
Patience, Gregory 190
Patriarca, Evan 544
Patterson, Brent 295
Paulsen, Alex D. 712
Pavlišic, Andraž 442
Paz, Neil 868
Peck, Torin 714
Peden, Charles H. F. 291, 859
Peng, Guowen 394
Peng, Po-Yu 128
Peng, Zhenmeng 208, 899
Pennington, Ashley M. 288
Penteado, Alberto 688
Perea, Leo Alvarado 849
Pereira-Almao, Pedro 584, 843
Perera, Ayomi S. 651
Pérez, Fernando 685
Perras, Frédéric 540, 721
Peters, Baron 78
Petkovska, Menka 145
Pfennig, Toni 326
Pham, Hien 315, 936
Pham, Hien N. 229
Pham, Thanh Huyen 450
Phillips, Eric M. 480
Piler, Karishma 731
Ping, Haoliang 637
Pinho, Bruno 159
Pitault, Isabelle 165
Plessow, Phillip 910
Poco, Joao G. R. 741
Poelman, Hilde 524
Pogrzeba, Tobias 76, 421
Pohar, Andrej 442
Pohl, Marga-Martina 792
Polo-Garzon, Felipe 780
Pommerenck, Justin 503, 701, 732
Porosoff, Marc D. 6, 457
Potdar, Aditi 169
Poudyal, Samiksha 377
Pradhan, Sivaram 416
Prasad, Subramanian 787
Prasad, Vinay 441
Priebe, Jacqueline 792
Pritchard, Jay 301
Pruski, Marek 540, 721
Punetha, Ashok K. 108
Pushpavanam, S. 668
Qarzouh, Muneef 108
Qi, Long 895
Qian, Linping 880
Qian, Wanyi 105
Qian, Xianghong 236, 347
Qiao, Guo-Yue 819
Qin, Lang 795
Qin, Wei 695
Qiu, Kaipei 651
Quang, Dang Viet 177
Quinn, Joseph 200, 683, 693
Quinn, Thalia 783
Quintana, Juan Lovón 917
Quitain, Armando 228, 718
Radich, James G. 717
Ragauskas, Arthur J. 328
Raghavan, Ashwin 90
Rahmati, Mahmood 132, 578
Ramakumar, S. S. V. 673
Ramanathan, Anand 469, 838, 932

AUTHOR INDEX

Ramasamy, Karthikeyan K. 470
Ramasubramaniam, Ashwin 888
Ramirez, Emilio 415
Ramirez, Maria 919
Ramirez-Reina, Tomás 619
Ramos, Adela E. 906
Ramos, Mario 638
Ranade, Vivek V. 388
Rao, Vivek M. 245
Rasmussen, Søren B. 897
Ratnakar, Ram R. 91
Ravel, Bruce 940
Rawal, Saurin 339
Ray, Ajay K. 79
Raybaud, Pascal 96
Razzak, Shaikh 845
Rea, Anita 439
Reddick, Ian 503
Regalbuto, John R. 192, 197, 235, 490, 779, 817, 886, 930
Regauskas, Art 329
Rehmet, Christophe 522
Reihani, Amin 295
Reitzmann, Andreas 147
Ren, Yang 218
Rentz, Nikki S. 70, 374
Repke, Jens-Uwe 688
Resasco, Daniel 88, 861
Resasco, Daniel E. 89
Resende, Fernando 368
Reynolds, Ronald A 928
Rezaei, Fateme 281
Rhee, Chang-Houn 605
Ribeiro, Fabio 209, 218, 940
Ribeiro, Fabio H. 298, 539, 546, 835, 917
Rieg, Carolin 400
Rigdon, William A. 678
Rimer, Jeffrey D. 489, 644, 675, 695
Rioux, Robert M. 834, 952
Riscoe, Andrew 453
Rivera-Goyco, Christian G. 370, 396
Roberts, Christopher 568
Roberts, Christopher B. 582
Robota, Heinz J. 301
Rocha, Tulio 797
Rodriguez-Garcia, Laura 885, 918
Roe, David 568, 582
Rogers, Jessica 844, 908
Roine, Jorma 722
Roling, Luke T. 510, 585
Rolland, Matthieu 96
Román-Leshkov, Yuriy 364, 776, 894, 902
Romero, Carolina 623
Rorrer, Julie 320
Rosenfeld, Devon C. 844, 908
Rover, Marjorie 331
Rowe, Scott C. 427
Rownaghi, Ali A. 281
Roy, Partho S 698
Roy, Shyamal 760
Rozmyslowicz, Bartosz 324
Ruan, Hao 333
Ruddy, Daniel A. 889
Rueda, Luz Marina Ballesteros 949
Runnebaum, Ron C. 887
Russ, David C. 332
Russell, Daniel 735
Ryan, Matthew 459
Ryou, Youngseok 220
Ryu, Jebin 598, 599
Sabyrov, Kairat 31
Saeys, Mark 392, 942
Safdari, Mohammad-Saeed 132, 578
Saha, Shibely 196, 566
Sahasrabudhe, Mayuresh 635, 676
Sahinidis, Nick 640
Salavati-Fard, Taha 839
Salehi-Khojin, Amin 73, 754
Salmi, Tapio 246, 722
Salonen, Jarno 722
Salvachua, Davinia 331
Samad, Jadid 366
Sanchez, Joel 638
Sanchez-Forero, Diana 158
Sandberg, Robert B. 460
Sandoval, Tania E. 862
Sangodkar, Rahul P. 787
Sankar, Meenakshisundaram 422
Sankaranarayanan, Subramanian K.R.S. 384
Santos, Haroldo 880
Santos, Maryorie C. 904
Saraçi, Erisa 195
Saravanan, Chandra 635, 676
Saravanan, Karthikeyan 463
Sarazen, Michele L. 20, 836
Sasaki, Mitsuru 228, 718
Sasaki, Takashi 597
Sasmaz, Erdem 38, 206, 575, 851, 933
Sathitsuksanoh, Noppadon 332
Sato, Takafumi 345, 600, 601, 738
Saunders, Steven R. 954
Sautet, Phillippe 96, 283, 350, 399, 536
Savage, Phillip E. 101, 224, 943
Sayahpour, Baharak 73
Schaidle, Joshua 186, 889
Schaidle, Joshua A. 860
Schaub, Tanner 868
Schembecker, Gerhard 590
Scheppe, Karalyn 105
Schideman, Lance 105

AUTHOR INDEX

Schlogl, Robert 216, 797
Schmidt, Marcel 76, 421
Schmidt-Rohr, Klaus 229, 315
Schneider, Elia 593, 808
Schneider, William 298, 631, 840
Schneider, William F. 209, 351, 546, 835, 945
Schneiderman, Deborah 390
Schomäcker, Reinhard 76, 421
Schouten, Jaap C. 173
Schreiber, Igor 610
Schreiberova, Lenka 610
Schubert, Markus 244, 247
Schultheis, Stephan 455
Schwalbe, Jay 792, 831
Schwank, Johannes W. 653
Schwartz, Daniel 858
Schwartz, Daniel K. 863
Schwartz, Nicholas R. 712
Schwartz, Thomas J. 362
Schweitzer, Jean-Marc 96
Scott, Susannah L. 78, 895
Scuseria, Gustavo E. 416
Sebakhy, Khaled O. 584, 843
Seebauer, Edmund G. 554
Seidel-Morgenstern, Andreas 145, 424, 849
Seitz, Linsey C. 35, 466, 727
Semenov, Sergey N. 59
Sen, Fatih 384
Sen, Irem 426
Senftle, Thomas P. 26, 372, 877
Senkan, Selim 451
Senthamaraikkannan, Gouthami 441
Seritan, Stefan 78
Serov, Alexey 71
Serrano, Benito 793
Servia, Alberto 160
Seshadri, Vikram 417
Seuffert, Wolfgang 147
Shacham, Mordechai 111
Shackelford, Joseph 783
Shafei, Emad 108
Shahane, Saurabh 483
Shaikh, Sohel K. 686
Shan, Junjun 54, 775, 829, 830
Shan, Nannan 286
Shang, Zeyu 818
Shanks, Brent H. 229, 326, 361, 362
Sharada, Shaama Mallikarjun 378
Shareghi, Adam 503
Shariff, Humayun 87, 157
Sharkey, Bryan E. 367
Sharma, Manish 203, 627
Sharma, Renu 556
Sharma, Virender K. 79
Sharrock, Patrick 528
Shaughnessy, Kevin H. 477
Shchukarev, Andrey 722
Sheehan, James 101
Shekhar, Mayank 917
Shekhar, Shashank 480
Shen, Chun 587
Shen, Yangyang 953
Shen, Yufeng 489, 644
Shi, Daming 191
Shi, Hongbo 888
Shi, Honghong 327
Shi, Xinjian 551
Shih, Arthur 298
Shinde, Kiran 634
Shirobe, Taichi 791
Shuai, Li 25
Siahrostami, Samira 46, 72, 551, 641
Siebenhofer, Matthaeus 346, 452
Sievers, Carsten 124
Sikes, Hadley D. 47
Silva Jr, João L. 158
Simakov, David 146, 873
Simonetti, Dante 316
Sinclair, Robert 641
Singh, Aayush R. 402
Singh, Joseph 916
Singh, Meenesh 429, 903
Singh, Nirala 51, 125
Sinha, Nishant 454
Sinha-Majumdar, Sreshtha 214
Slakman, Belinda L. 428
Smirniotis, Panagiotis 488
Smith, Joseph D. 245
Smith, Robyn 781
Smith, Ryan 331
Smolin, Sergey 792
Snider, Jonathan 708, 898
Snyder, Joshua 202, 334
Sohn, Hyuntae 219, 692
Sokaras, Dimosthenis 641, 898
Solovyov, Andrew 542, 887
Somorjai, Gabor A. 516
Son, Won-Su 598, 599
Song, Hua 537
Song, James 291
Song, Jinwoo 698
Song, Kyung-Ho 583
Song, Yang 125
Song, Yuanjun 720, 924
Sooknoi, Tawan 431
Sosa, Fabricio Guayaquil 793
Spanjers, Charles S. 390
Speelmanns, Eva 688
Spinner, Neil S. 678
Spivey, James J. 491, 611

AUTHOR INDEX

Sprunger, Phillip T. 491
Sriram, Vishnu 133
Stach, Eric A. 831
Stagg-Williams, Susan M. 106
Stamatakis, Michail 99, 496, 801
Stangland, Eric E. 401
Starace, Anne 194
Stark, Wendelin J. 593, 808
Stavitski, Eli 124
Stears, Brien 847
Steele, Matthew 115
Steinfeld, Aldo 794, 905
Steiniger, Alexzander 733
Steinmann, Stephan 283
Stenfeldt, Marie 299
Stephens, Ifan E.L. 901
Stevens, James 74
Strah, Neja 442
Strickler, Alaina 702
Studt, Felix 400, 523, 641, 927
Sturm, Belinda S.M. 106
Su, Yaxin 242
Subramaniam, Bala 215, 222, 327, 469, 838, 932
Suidan, Makram 488
Sulmonetti, Taylor 816, 934
Sultana, Khondker 709
Sumigawa, Yoshifumi 718
Sun, Duo 146, 873
Sun, Junming 68
Sundmacher, Kai 97
Suzuki, Tomoko 597
Suzuki, Yukei 586
Svenningsen, Glen 468
Svetlov, Stanislav 248
Sykes, E. Charles H. 875
Szanyi, Janos 209, 291, 631, 704, 840
Taboada-Serrano, Patricia 626
Tackett, Brian M. 201
Tadepalli, Sunitha 74
Taifan, William 322
Takacs, Michael 905
Takahashi, Misaki 738
Takaoka, Masaki 597
Tamaki, Takanori 650
Tampy, Geatesh 74
Tan, Qiaohua 861
Tan, Shuai 196, 566
Tang, Lei 573
Tang, Maureen H. 202
Tang, Michael 465
Tang, Yadan 138
Tang, Yuanjun 681
Tao, Miao 608
Taranto, Osvaldir P. 158
Tassone, Christopher J. 511
Tate, Jenae 561
Tavlarides, Lawrence L. 226
Tayakout, Mélaz 165
Taylor, Sara Jo 389
Teerikoski, Sakari 246
Teixeira, Andrew 483, 504
Teixeira, Ivo 496
Teng, Xiaowei 618
Tengco, John Meynard M. 197
Terejanu, Gabriel 379
Thakkar, Harshul 281
Thapa, Prem 327
Theiler, David 905
Theofanidis, Stavros Alexandros 524
Theopold, Klaus 195
Theriot, Jordan 459
Therrien, Andrew 875
Thiyagarajan, Shanmugam 471
Thomas, Cyril 137
Thomassen, Leen C.J. 169
Thompson, Levi T. 186, 943
Thybaut, Joris W. 432
Tierney, Eric 518
Timko, Michael T. 75, 107, 223
Ting, Allen Wei-Lun 680
To, Anh T. 553
To, John 641
Toch, Kenneth 432
Todic, Branislav 557
Tomassone, M. Silvina 953
Tompsett, Geoffrey 107, 223
Toops, Todd J. 300, 931
Toste, Dean 320
Traverso, Andrew 701, 732
Trinh, Quang Thang 450, 800
Trogadas, Panagiotis 651
Tsai, Charlie 402
Tsang, Shik Chi Edman 496
Tsapatsis, Michael 896
Tseng, Han-Ting 218, 940
Tsotsis, Theodore 341, 343, 730, 736
Tsouris, Costas 626
Tucker, Melvin 333
Turner, C. Heath 624, 914
Turney, Damon 759
Ulissi, Zachary 465
Ullmann, Sven 899
Unocic, Raymond 936
Urban-Lurain, Mark 115
Utsunomiya, Ryuji 508
Uzun, Alper 674
Vahdat, Nader 365
Vahdatzaman, Maral 647
Vaid, Radhe K. 479
Valla, Julia A. 576

AUTHOR INDEX

Valle, Eduardo 484, 620
Van Beek, Wouter 778
Van Cleve, Timothy 8, 900, 955
Van Der Schaaf, J. 173
Van Der Waal, Jan 471
Van Duin, Adri 93
Van Es, Dan 471
Van Haveren, Jacco 471
Van Norman, Staci A. 567
Van Ommen, J. R. 564
Vann, Tyler 907
Vara, Madeline 510
Varma, Arvind 13, 189, 387, 806
Vasiliadou, Efterpi 839
Vasudevan, Kiran 202
Vayssilov, Georgi N. 881
Vazquez-Ramos, Leida M. 370
Vega-Vila, Juan C. 539
Vengsarkar, Pranav S. 568
Venkatesh, Archana 422
Vennestrøm, Peter N. R. 897
Verma, Anuj A. 480
Verstraete, Jan J. 96
Veser, Götz 544, 719
Vidonish, Julia E. 127
Villano, Stephanie 519
Virk, Preetinder S. 438
Vitale, Gerardo 584, 843
Vityuk, Artem D. 881
Vlachos, Dion 393, 839
Vlachos, Dionisios 493
Vlachos, Dionisios G. 118, 199, 230, 805, 896
Vogel, Herbert 98, 455
Vohs, John 191
Vojvodic, Aleksandra 336, 402, 466, 552, 672, 727, 744, 852, 910
Vömel, Lukas T. 391
Von Jouanne, Annette 503
Vorholt, Andreas J. 170
Voss, Jenny 573
Voss, Johannes 378
Votsmeier, Martin 98
Vu, Anh 236
Wachs, Anthony 96
Wachs, Israel 797, 853
Wachs, Israel E. 458, 548, 850, 920
Wade, Kester 944
Wagner, John 113
Walker, Eric 64, 379
Walter, Eric D. 291, 895
Wan, Weiming 805
Wander, Brook 316
Wang, Aiyong 291
Wang, Alex 344
Wang, Bin 88, 233, 431, 555
Wang, Chao 206, 625, 851
Wang, Di 297, 740
Wang, Haihui 342
Wang, Hongliang 333
Wang, Huamin 859
Wang, Jia-Jun 592
Wang, Jiamin 386
Wang, Kai 507, 591
Wang, Kun 519
Wang, Lan 195
Wang, Lijun 407
Wang, Lucun 566
Wang, Meng 492
Wang, Peijian 507, 591
Wang, Ping 684
Wang, Richard Baochang 99
Wang, Shuai 311
Wang, Siwen 284, 710, 923
Wang, Wei-Ning 485
Wang, Xiang 704
Wang, Xiaohong 806
Wang, Xue 510
Wang, Yilin 291
Wang, Yong 382, 498, 859
Wang, Yujun 587
Wang, Yuran 50, 894
Wang, Zhen 880
Wang, Zi 491
Warmeling, Helge 170
Washton, Nancy M. 895
Watts, Michael 347
Weaver, Jason F. 799
Weber, Adam 429
Weber, Ariane 421
Weckhuysen, Bert 120, 471
Weeranoppanant, Nopphon 483
Wegener, Evan 218, 940
Wehinger 642
Wei, Yanying 342
Weidner, John W 197
Weimer, Alan W. 427, 567
Weiss-Hortala, Elsa 681
Wellendorff, Jess 378
Wen, Cun 1, 206, 456, 602, 904, 933
Wen, Zhiyou 331
Wendt, Jost O.L 239
Wentworth, Travis 299
Werner, Sebastian 147
Werner-Zwanziger, Ulrike 787
West, Alan 630
West, Richard H. 354, 428, 518
Westmoreland, Phillip R. 389, 417
Wette, Melissa 734
White, Colin 106
Wickramasinghe, Ranil 236, 347

AUTHOR INDEX

Wiebenga, Michelle....294
Wiggins, Gavin....415
Wilburn, Monique....290
Wilhite, Benjamin....696
Willauer, Heather D....6, 457
Williams, Christopher....123
Willis, Joshua....453, 511, 512, 656, 831, 884, 944
Wilson, Neil M....687, 935
Wilson, Zachary....640
Wiser, Artur....98
Wolf, Patrick....350
Wolff, Tanya....849
Wong, Andrew....817
Wong, Breanna....498
Wong, Hsi-Wu....77
Wong, Michael S....416
Wong, Sze Wing....479
Wong, Wai Kuan....478
Wongkaew, Akkarat....197
Woo, Hee Suk....598, 599
Worku, Dereje....709
Wrasman, Cody....512
Wu, Gang....337
Wu, Ivy....289
Wu, Jason....899
Wu, Liheng....511, 535
Wu, Su-Fang....637
Wu, Xiao-Yu....22
Wu, Yi Y. (Chloe)....909
Wu, Yiqing....841
Wu, Zhenwei....218, 940
Wu, Zili....458, 914
Wyman, Charles....468
Xia, Shunxiang....684
Xia, Younan....510
Xiang, Yizhi....573
Xiao, Yang....13, 189, 387, 806
Xie, Jiahan....936
Xie, Jingyi....561
Xie, Mingquan....144
Xie, Pei....591
Xie, Tianjun....500
Xin, Hongliang....284, 386, 707, 710, 878, 900, 923
Xin, Yan....913
Xin, Zhong....608
Xiong, Haifeng....315
Xu, Bingjun....185, 543, 893, 946
Xu, Chao-Zhong....592
Xu, Jiayi....369
Xu, Lang....401
Xu, Mingyuan....795
Xu, Qin-Qin....819
Xu, Rui....568, 582, 717
Xu, Ye....339, 400, 486, 550
Xu, Zhuoran....844, 908
Yablonsky, Gregory S....566
Yakobov, Roman....759
Yalcin, Ozgen....548
Yamafuku, Saya....228
Yamaguchi, Takeo....650
Yamamoto, Tetsuya....479
Yan, Helen....944
Yan, Ning....232, 833
Yan, Wenjuan....327
Yan, Yushan....185, 199
Yang, Bin....329, 333
Yang, Dong....221
Yang, Guangxing....618
Yang, Haishen....459
Yang, Hong....687
Yang, Nuoya....916
Yang, Rachel A....288
Yang, Xiaofang....373
Yang, Xiaokun....126
Yang, Yanhui....607
Yao, Hongbao....587
Yao, Yali....609, 682
Yap, Swee Kun....478
Ye, Jianchao....830
Ye, Jingyun....928
Ye, Lujie....870
Yeh, Alexander....542
Yeh, Benjamin....928
Yeh, Yao-wen....373
Yelvington, Paul E....712
Yeung, King Lun....756, 757, 758
Yezerets, Aleksey....138
Yi, Nan....879
Yin, Jian-Zhong....819
Yin, Kehua....936
Yokochi, Alex....701
Yokochi, Alexandre....503, 732
Yokozawa, Tohru....479
Yoo, JeongDo....739
Yoo, Jong Suk....512, 726
Yoon, Sung-Pil....663
Youn, Min Hye....583, 649, 658
Young, Zachary....323
Yu, Jingye....541
Yu, Jun....221
Yu, Kun....131
Yu, Miao....798
Yu, Xiaoxiao....426
Yu, Ying....495
Yuan, Joshua....329
Yun, SeongUk....697
Yun, Yongju....804
Yung, Matthew M....194
Yurchekfrodl, Eric....208
Yusuf, Seif....855

AUTHOR INDEX

Zahid, Umer 175
Zaker, Azadeh 223
Zeltner, Martin 593, 808
Zemichael, Fessehaye 622
Zeng, Chun 327
Zeng, Zhenhua 32
Zhan, Zhonghua 239
Zhang, Changlin 208
Zhang, Chen 939
Zhang, Conrad 193
Zhang, Hua 34
Zhang, Jing 911
Zhang, Jisong 504, 507
Zhang, Kechun 390
Zhang, Liang 36, 509, 744, 852
Zhang, Lu 859
Zhang, Nian 643
Zhang, Nuowei 34
Zhang, Peng 105
Zhang, Qi 874
Zhang, Qian 201
Zhang, Renqin 48, 209, 631, 875
Zhang, Shengjie 828, 887
Zhang, Wen 347
Zhang, Xiankuan 307
Zhang, Ximing 467
Zhang, Yuanhui 105
Zhang, Yunya 197
Zhang, Z.Conrad 416
Zhao, Chuanlin 550
Zhao, Dongting 844, 908
Zhao, Evan W 913
Zhao, Jingkai 134
Zhao, Jin-mo 416
Zhao, Meng 327
Zhao, Peng 373
Zhao, Qing 381
Zhao, Zheng 88, 89
Zheng, Feng 799
Zheng, Haibin 913
Zheng, Jinbao 34
Zheng, Quanxing 369
Zheng, Weiqing 19, 896
Zheng, Xiaolin 551
Zheng, Yang 291, 840
Zhou, Xiao-Dong 174
Zhu, Chunxiang 576
Zhu, Hongda 932
Zhu, Jiazhou 866
Zhu, Minghui 797
Zhu, Xiaoxiang 310
Zhu, Zibin 874
Zhuang, Libin 342
Zivanic, Ljiljana 557
Zohour, Bahman 451
Zohra, Mohammad Abu 177
Zolghadr, Ali 419
Zones, Stacey I. 542, 784, 837
Zougheib, Nicole 807
Zuo, Wilbur Y. 77
Zygourakis, Kyriacos 127

AIChE
120 Wall Street, FL 23
New York, NY 10005-4020

ISBN 978-1-5108-3429-3

9 781510 834293

CURRAN ASSOCIATES INC.
proceedings.com

Catalysis and Reaction Engineering Division 2016

Core Programming Area at the 2016 AIChE Annual Meeting

San Francisco, California, USA
13-18 November 2016

Volume 2 of 2

AIChE
The Global Home of Chemical Engineers

Printed from e-media with permission by:

Curran Associates, Inc.
57 Morehouse Lane
Red Hook, NY 12571

CURRAN ASSOCIATES INC.
proceedings.com

Some format issues inherent in the e-media version may also appear in this print version.

Copyright© (2016) by AIChE
All rights reserved.

Printed by Curran Associates, Inc. (2017)

For permission requests, please contact AIChE
at the address below.

AIChE
120 Wall Street, FL 23
New York, NY 10005-4020

Phone: (800) 242-4363
Fax: (203) 775-5177

www.aiche.org

Additional copies of this publication are available from:

Curran Associates, Inc.
57 Morehouse Lane
Red Hook, NY 12571 USA
Phone: 845-758-0400
Fax: 845-758-2633
Email: curran@proceedings.com
Web: www.proceedings.com

Catalysis and Reaction Engineering Division 2016

Core Programming Area at the 2016 AIChE Annual Meeting

San Francisco, California, USA
13-18 November 2016

Volume 2 of 2

AIChE
The Global Home of Chemical Engineers

Catalysis and Reaction Engineering Division 2016

Core Programming Area at the 2016 AIChE Annual Meeting

San Francisco, California, USA
13-18 November 2016

Volume 2 of 2

ISBN: 978-1-5108-3429-3

Printed from e-media with permission by:

Curran Associates, Inc.
57 Morehouse Lane
Red Hook, NY 12571

Some format issues inherent in the e-media version may also appear in this print version.

Copyright© (2016) by AIChE
All rights reserved.

Printed by Curran Associates, Inc. (2017)

For permission requests, please contact AIChE
at the address below.

AIChE
120 Wall Street, FL 23
New York, NY 10005-4020

Phone: (800) 242-4363
Fax: (203) 775-5177

www.aiche.org

Additional copies of this publication are available from:

Curran Associates, Inc.
57 Morehouse Lane
Red Hook, NY 12571 USA
Phone: 845-758-0400
Fax: 845-758-2633
Email: curran@proceedings.com
Web: www.proceedings.com

TABLE OF CONTENTS

VOLUME 1

(6w) Fundamental Studies and Applications of Nano-Structured Catalysts in Domestic Fuel Production ... 1
Cun Wen
(6x) Catalytic Hydrotreatment for the Development of Alternative Transportation Fuels ... 4
Lilu Funkenbusch, Michael Mullins
(6ai) First-Principles Modeling of Catalysts: Novel Algorithms and Reaction Mechanisms ... 5
Bryan R Goldsmith
(6aj) Development of Catalysts for Energy and Environmental Applications ... 6
Marc D. Porosoff, Heather D. Willauer
(6z) Upgrading Biomass-Derived Platform Chemicals By Electrochemical and Photoelectrochemical Catalytic Oxidation ... 7
David Chadderdon
(6aa) Rational Design of Catalytic Sites for Energy Applications ... 8
Timothy Van Cleve
(6ab) Theory-Guided Understanding and Design of Heterogeneous Catalysts ... 10
Matthew M. Montemore
(6ac) In silico Engineering of a Future Energy Infrastructure ... 11
Glen R. Jenness
(6ad) Insight into Pt-Bi Bimetallic Catalysts: An Experimental and DFT Study ... 13
Yang Xiao, Arvind Varma
(6ae) Mechanisms of Redox Catalysts for the Greener Processing of Shale Gas Via Chemical Looping ... 14
Luke Neal
(6af) Nanostructured Transition Metal Dichalcogenide Catalysts for Electrochemical Energy Systems ... 17
Mohammad Asadi
(6y) Exploring Structure-Function Correlations of Nanomaterials in Heterogeneous Catalysis ... 19
Weiqing Zheng
(6ah) Mechanistic, Spectroscopic and Theoretical Assessment of Porous Catalytic Materials ... 20
Michele L. Sarazen
(6m) In-Situ Drifts Studies on CuNi Catalyst for Ethanol Hydrogen Production ... 21
Anand Kumar
(6n) Semi-Permeable Membrane Reactor for Catalysis, Hydrocarbon Processing and CO2 Reuse ... 22
Xiao-Yu Wu
(6o) A Practical Way to Separate Uncondensed Lignin during Biomass Pretreatment and Quantitatively Depolymerize It at a Low Temperature of 120 C with a Cheap Ni/Al2O3-SiO2 Catalyst ... 25
Li Shuai
(6q) Combined Quantum and Classical Computational Approaches for Investigating Complex Surface Interactions Impacting Heterogeneous Catalysis ... 26
Thomas P. Senftle
(6a) Electrochemical Energy Transformation Processes: An Atomistic Perspective ... 29
Leanne D. Chen
(6r) Nanostructured Hybrid Materials: Directing Catalytic Activity and Selectivity By Design ... 31
Kairat Sabyrov
(6s) Towards the Computational Design of Heterogeneous Electrocatalysts ... 32
Zhenhua Zeng
(6t) Engineering the Electrochemical Interface for Sustainable Energy Conversion and Storage ... 33
Maria Escudero-Escribano
(6u) Controlled Synthesis and in-Situ Spectroscopic Study of Highly Efficient Ptfe Bimetallic and Ptrufe Trimetallic Nanocatalysts ... 34
Hua Zhang, Jinbao Zheng, Nuowei Zhang, Jian-Feng Li, Binghui Chen
(6b) Developing Enhanced Catalysts for Renewable Fuels through Spectroscopic Insights ... 35
Linsey C. Seitz
(6c) Computational Design of Hetero-Structured Catalysts for Energy ... 36
Liang Zhang
(6d) Efficient Catalytic Pathways for Carbon Utilization ... 38
Erdem Sasmaz

(6e) New Routes and Heterogeneous Catalysts Development for Biomass Conversion 39
Homer Genuino
(6f) Transition Metal-Oxides for Sustainable Energy Conversion and Storage: The Computational Catalysis Perspective 40
Michal Bajdich
(6k) Impacts of Metal-Adsorbate Bonds on Photon Activation Mechanism and Nanoparticle Reconstruction in Heterogeneous Catalysis 41
Matthew Kale
(6p) Electrocatalysis for Sustainable Energy Technologies 42
Drew Higgins
(6g) Rational Catalyst Design for Renewable Energy Technologies 46
Samira Siahrostami
(6i) Engineering Visible-Light Organic Photocatalysis for Polymers in Biomaterials, Biosensing, and Photomedicine 47
Alan Aguirre-Soto, Hadley D. Sikes
(6ak) Catalysts for Emission Control and Energy Conversion: Computational Study Based on DFT Calculations 48
Renqin Zhang
(6al) Development of Heterogeneous Catalysis towards a Sustainable Future 50
Yuran Wang
(6ao) Understanding and Improving Electrocatalysts for Energy Conversion and Waste Remediation 51
Nirala Singh
(6ap) Development of Novel Single-Site and Isolated Bimetallic Alloy Catalysts for C-H Bond Activation 54
Junjun Shan
(6aq) Smart Materials through Molecular Networking 59
Sergey N. Semenov
(6l) Condition and Support Dependent Development of Computational Methods for the Engineering of Materials 60
Christopher Paolucci
(6am) Conversion of Waste Oil to Biofuels in a Single Catalytic Process over Bifunctional Catalysts 63
Masoudeh Ahmadi
(6an) Rational Design of Catalytic Materials for Advancing the Use of Alternative Energy Sources 64
Eric Walker
(6ar) Catalysis Reactions towards Advanced Energy Applications 66
Richa Padhye
(6as) Synthesis and Spectroscopic Characterization of Heterogeneous Catalysts for Energy Production 68
Junming Sun
(23a) Control of Fe-Ni Nanoparticle Activity for the Oxygen Evolution Reaction (OER) and Methanol Electrooxidation 70
Stephanie Candelaria, Nikki S. Rentz, Lauren F. Greenlee
(23b) Transition Metal-Nitrogen-Carbon Electrocatalysts for Oxygen Reduction Reaction 71
Plamen Atanassov, Kateryna Artyushkova, Alexey Serov, Ivana Matanovic
(23c) Electrocatalytic Generation of H2O2: Carbon Based Material Synthesis and Device Design for Portable Low Cost Water Purification 72
Zhihua Chen, Shucheng Chen, Samira Siahrostami, Zhenan Bao, Jens Nørskov, Thomas F. Jaramillo
(23f) Niobium Doped Molybdenum Disulfide Catalyst for CO2 Reduction Reaction 73
Pedram Abbasi, Mohammad Asadi, Baharak Sayahpour, Amin Salehi-Khojin
(32b) Green Chemistry with Flow Reactors and Heterogeneous Catalysts for the Synthesis of Fragrance Ingredients 74
Sunitha Tadepalli, Zhe Guo, James Stevens, Geatesh Tampy
(32c) Lewis Acid Catalyzed Pyrolysis of Polystyrene in Low-Temperature Molten Salts: Experiment and Simulation 75
Avery Brown, Michael T. Timko, Ravindra Datta, N. Aaron Deskins
(32d) Microemulsion Systems As Smart Solvent Systems for Homogeneous Catalysis: Rhodium-Catalyzed Hydroformylation of Long-Chain Alkenes in Aqueous Media 76
Tobias Pogrzeba, Marcel Schmidt, Markus Illner, Prof. Dr. Reinhard Schomäcker
(32f) Synthesis of Linear Alkylbenzenes from Biomass-Derived Furan over Solid Acid Catalysts 77
Wilbur Y. Zuo, Hsi-Wu Wong

(36a) Rate-Enhancing Roles of Water Molecules in Methyltrioxorhenium-Catalyzed Olefin Epoxidation By Hydrogen Peroxide 78
Bryan R Goldsmith, Taeho Hwang, Stefan Seritan, Baron Peters, Susannah L. Scott
(36b) Acid Activation of Ferrate(VI) for Oxidation of Organic Compounds in Aqueous Solution 79
Kyriakos Manoli, George Nakhla, Ajay K. Ray, Virender K. Sharma
(36c) Kinetic Studies of Liquid Phase Ethylene Production By Hydrogenation of Acetylene Using a Selective Solvent in a Basket Reactor 87
Humayun Shariff, Muthanna Al Dahhan
(36e) Solvent Effect in Catalytic Upgrading of Biomass-Derived Furanics 88
Zheng Zhao, Mohammed Reda Bababrik, Bin Wang, Daniel Resasco
(36f) Solvent Effect on Cyclohexene Hydrogenation over Pd 89
Zheng Zhao, Steven Crossley, Daniel E. Resasco
(36g) Numerical Study of Heavy Crude Oil Pyrolysis in Supercritical Water 90
Ping He, Ashwin Raghavan, Ahmed F. Ghoniem
(39a) Multi-Scale Multi-Phase Reduced Order Models for Dual-Layered Catalytic Monolith Reactors 91
Ram R. Ratnakar, Vemuri Balakotaiah
(39b) CFD Modeling of an Industrial Furnace Reformer and 3D Multi Scale Model for Packed Bed Reactor for Synthesis Gas Production 92
Bhanu Vardhan Reddy Kuncharam, Elaine Chen, Anthony G. Dixon, Rui Liu
(39c) DFT Calculations and Reaxff Development for MD Simulations of H-BN Growth on Nickel Surfaces 93
Song Liu, Bin Liu, James H. Edgar, Adri Van Duin
(39d) Molecular-Scale Modeling of Whole Biomass Pyrolysis – Morphological Changes 94
Abdul Salam Mohammad, Joseph Biernacki, Scott Northrup, Derek Cashman
(39e) Influence of Intraparticle Transport Processes on Optimal Reactor and Catalyst Design 95
Markus Kaiser, Hannsjörg Freund
(39f) Multiscale Simulation of Heterogeneous Catalysis: Butadiene Hydrogenation from Molecular Reaction Steps to the Fluid Flow 96
Phillippe Sautet, Ludovic Briquet, Céline Chizallet, Pascal Raybaud, Jean-Marc Schweitzer, Jan J. Verstraete, Anthony Wachs, Matthieu Rolland, Abdelkader Hammouti
(39g) Multiscale Modeling for High Temperature Heterogeneous Gas Phase Synthesis Reactions: HCN Synthesis 97
Georg S. Liesche, Kai Sundmacher
(39h) An Experimental and CFD Based Study of the Pt-Catalyzed NH3-Oxidation in Different Scales 98
Thomas Heydt, Artur Wiser, Dirk Born, Alfons Drochner, Herbert Vogel, Martin Votsmeier
(39i) How Accurate Is the Mean-Field Approximation for Catalytic Kinetics? 99
Michail Stamatakis, Bernardo Oyarzún, Richard Baochang Wang
(48a) Compositional Effect of Varying Feedstocks on Heavy Oil Upgrading By Supercritical Water 100
Ramazan Oguz Caniaz, Can Erkey, Serhat Arca
(48b) Product Formation and Kinetics of the Non-Isothermal Hydrothermal Liquefaction of Soy Protein Isolate 101
James Sheehan, Phillip E. Savage
(48c) An Investigation on Short Residence Time Macro Algae Hydro Thermal Liquefaction in a Batch Reactor 104
Arash Izadpanah, Niccolo Le Brun, Klaus Hellgardt
(48d) Hydrothermal Liquefaction (HTL) of High-Ash Algal Biomass: The Effect of Ash Contents in HTL Reactions 105
Wan-Ting Chen, Wanyi Qian, Karalyn Scheppe, Zachary Mazur, Lance Schideman, Chih-Ting Kuo, Peng Zhang, Yuanhui Zhang
(48e) Simultaneous Solid, Phosphorus Recovery and Biocrude Production through Hydrothermal Liquefaction of Algae 106
Robert Hable, Sirwan Alimoradi, Shawn Benson, Colin White, Belinda S. M. Sturm, Susan M. Stagg-Williams
(48f) Supercritical Water Treatment of Fractions of Crude Oil: Quantification of the Products 107
Soumya Gudiyella, Lawrence Lai, Isaiah Borne, Jianghuai Cai, Michael T. Timko, Geoffrey Tompsett, Alison Lui, William H. Green
(48g) Supercritical Water Process - Finding a Way to Accomodate into Petroleum Industry 108
Ki-Hyouk Choi, Joo-Hyeong Lee, Ashok K. Punetha, Emad Shafei, Abdullah Abdulhadi, Muneef Qarzouh, Bader Al-Otaibi
(64a) Why John Is Right Not to Lecture 109
H. Scott Fogler
(64b) Utilizing Classroom Management Software to Handle Large Classes 110
Janet Degrazia

(64c) Enabling Numerical Analysis Calculations within Educational Materials Delivered Via PCs and Smartphones 111
Michael B. Cutlip, Mordechai Shacham, Michael Elly
(64d) Spreadsheet Template to Aid in Creating Individualized Version of Objective Homework Problems 113
John Wagner, Amanda P. Malefyt
(64e) How Students "Learn ChemE": The Impact of Departmental Culture on Student Success 114
Lisa G. Bullard
(64f) Rapid Analysis of Written Responses to Reveal Student Misconceptions in Thermodynamics 115
Carl T. Lira, Matthew Steele, Kevin Haudek, John Merrill, Mark Urban-Lurain
(64g) No More Lecturing! What John Falconer Has Done for Student Learning 116
Milo D. Koretsky
(64h) Active Learning in Chemical Engineering 117
John Falconer, J. Will Medlin, Janet Degrazia, Katherine McDanel, Michelle Medlin
(72a) Mechanism of Furfural Hydrodeoxygenation on Metal/Metal Oxide Catalysts 118
Alexander V. Mironenko, Glen R. Jenness, Dionisios G. Vlachos
(72b) Kinetic Analysis of Hydrogenation and Hydrogenolysis of Complex Furans over Supported Noble Metals 119
Ying Lin Louie, Alexis T. Bell
(72c) ZrO2 Is Preferred over TiO2 As Support for the Ru-Catalyzed Hydrogenation of Levulinic Acid to I3-Valerolactone 120
Homer Genuino, Jamal Ftouni, Li Lu, Christopher Kiely, Pieter Bruijnincx, Bert Weckhuysen
(72d) C–O Bond Hydrogenolysis of Biomass–Derived Cyclic Ethers over Ni–Based Catalysts 121
Elmira Soghrati Khorasgani, Catherine Choong, Sibudjing Kawi, Armando Borgna
(72e) A Novel Pathway to 1,5-Pentanediol without the Use of Noble Metals 122
Zachary Brentzel, Kevin J. Barnett, Kefeng Huang, Christos T. Maravelias, George W. Huber, James A. Dumesic
(72f) Evaluation and Characterization of Carbon-Supported Nobel Metals for the Hydrodeoxygenation (HDO) of Acetic Acid 123
Jose Contreras-Mora, J. R. Monnier, Christopher Williams
(72g) Direct Conversion of Methane to Methanol and Ethanol 124
Chukwuemeka Okolie, Yasmeen Belhseine, Libor Kovarik, Eli Stavitski, Carsten Sievers
(72h) Aqueous-Phase Phenol Hydrogenation over Platinum and Rhodium: Electrocatalysis Versus Thermal Catalysis 125
Nirala Singh, Yang Song, Oliver Gutiérrez, John L. Fulton, Donald M. Camaioni, Charles T. Campbell, Johannes A. Lercher
(72i) Highly Efficient Conversion of Terpenoids to High Density Bio- Jet Fuels 126
Xiaokun Yang, Hongfei Lin
(84a) Pyrolysis of Soils Contaminated with Heavy Hydrocarbons: Reaction Mechanisms and Morphology of the Produced Char 127
Kyriacos Zygourakis, Caroline A. Masiello, Julia E. Vidonish, Pedro J. Alvarez
(84b) Spatial and Temporal Features of the Catalyzed Hydrocarbon Trap 128
Po-Yu Peng, Dan Luss, Michael P Harold
(84d) Formation of Nitrosamines in the Desorber of Tertiary Amine-Based Carbon Dioxide Capture Systems 131
Kun Yu, Ning Dai
(84e) Natural Attenuation and Biostimulation of Petroleum-Hydrocarbon Contaminated Soil in a Simulated Column 132
Mohammad-Saeed Safdari, Farhad Fazlollahi, Thomas Fletcher, Mahmood Rahmati
(84f) Study of Reaction Kinetics for Elemental Mercury Vapor Oxidation over CuCl2 for Mercury 133
Vishnu Sriram, Zhouyang Liu, Joo-Youp Lee
(84g) Advances in the Use of a Bio-Physicochemical Model to Characterize, Optimize the Chemical Absorption-Biological Reduction Integrated System for NO Removal 134
Jingkai Zhao, Meifang Li, Wei Li
(89a) Photogenerated Charge Separation in Artificial Photosynthesis Systems 135
Can Li
(89b) Oxide Nanostructures: Novel Supports, Active Sites, and Tandem Catalysts 136
Justin M. Notestein
(89c) Characterization of Tungsten and Zirconium Decorated Multi-Walled Carbon Nanotube Catalysts By XAS and NOx-Temperature Programmed Desorption 137
Gary L. Haller, Patrick Kelleher, Cyril Thomas

(89d) Mechanistic Differences Between Metal-Exchanged Zeolite SCR Catalysts, and Their Practical Implications 138
Aleksey Yezerets, Ashok Kumar, Saurabh Y. Joshi, Jinyong Luo, Yadan Tang, Hongmei An, Krishna Kamasamudram, Neal Currier

(89e) The Kinetics of Elementary Steps in Condensations Catalyzed By Solid Acids: Effects of the Size, Shape, and Charge of Intermediates, Transition States and Catalytic Structures 139
Stanley Herrmann, Enrique Iglesia

(89f) Progressing from Identifying, to Synthesizing and Designing Active Sites in Heterogeneous Catalysts 142
Harold H. Kung

(95a) Analysis of Upstream Creeping Reaction Zones in Catalytic Monolith Reactors 143
Tian Gu, Vemuri Balakotaiah

(95b) Model-Based Design of Optimal Reactors Considering Catalyst Deactivation 144
Mingquan Xie, Hannsjörg Freund

(95c) Exploiting the Nonlinear Frequency Response Method to Evaluate the Potential of Forced Periodic Operation of Chemical Reactors 145
Menka Petkovska, Daliborka Nikolic, Matthias Felischak, Andreas Seidel-Morgenstern

(95d) Modeling Catalyst Deactivation in a Packed Bed Reactor for Thermo-Catalytic Hydrogenation of CO2 146
Duo Sun, David Simakov

(95e) Optimizing Methanol Synthesis - from Lab to Industrial Plants Via High-Fidelity Modeling 147
Sebastian Werner, Wolfgang Seuffert, Willi Brandstädter, Andreas Reitzmann

(95f) Statistical Activity Variations in Diluted Catalyst Beds: Influence on Reactor Behavior 148
Gunnar Ganzer, Hannsjörg Freund

(95g) Identification of a Reaction Boundary Layer in Membrane Reactors for Hydrogen Production 149
Maria Anna Murmura, Stefano Cerbelli, Maria Cristina Annesini

(97a) Reactor Scale Modeling of Liquid Phase Ethylene Production By Hydrogenation of Acetylene in a Trickle Bed Reactor 157
Humayun Shariff, Muthanna Al-Dahhan

(97b) Influence of Solid Phase and Geometry on Liquid Velocity of a Three-Phase Bubble Column Fluid Dynamic 158
Karina K. Costa, Diana Sanchez-Forero, João L. Silva Jr, Osvaldir P. Taranto, Milton Mori

(97c) A Multiphase Semi-Flow Microreactor to Study the Influence of Water on Organometalic Catalysis 159
Bruno Pinho, Ryan L. Hartman

(97d) Original Methodology for the Screening of Catalysts Presenting High Reaction Rates at High Conversions 160
Alberto Servia

(97e) Gas Phase Mixing Behavior of Gas-Liquid Upflow in a Moving Catalytic Packed/Expanded Bed Hydrotreating Reactor Using Advanced Gas Tracer Technique 164
Vineet Alexander, Hamza Al-Bazzaz, Muthanna Al-Dahhan

(97f) Improvement of a Kinetic Model for the Hydroconversion of an Atmospheric Residue in a Slurry Semi-Batch Reactor 165
Pedro Alvarez, Barbara Browning, Jérémy Maurin, Tim Jansen, Maxime Lacroix, Mélaz Tayakout, Christophe Geantet, Isabelle Pitault

(97g) Experimental Characterization of Two Phase Interfacial Mass Transfer in Novel Milli-Scale Reactors 169
Aditi Potdar, Leen C. J. Thomassen, Simon Kuhn

(97h) Process Intensification of Aqueous Biphasic Hydroformylation of Mid Chain Olefin 1-Octene in a Jet Loop Reactor Creating Large Interfacial Areas 170
Helge Warmeling, Arno Behr, Andreas J. Vorholt

(97i) Pressure Drop and Liquid Hold up in an Open-Structure Random Packed Column with Counter Current Flow 173
Letícia Pezzi Martins Loane, J. Van Der Schaaf, Jaap C. Schouten

(343a) High-Efficiency Electrochemical Conversion of CO2 to Value Added Fuels Current Status, Challenges, and Future Directions 174
Xiao-Dong Zhou

(343b) Design and Economic Analysis of IGCC Power Plants with CO2 Capture 175
Umer Zahid, Usama Ahmed, Chonghun Han

(343c) Optimization of the CCS Network Under Harsh Geographical Conditions, Using GIS and MINLP Modeling 176
Changsoo Kim, Kyeongsu Kim, Chonghun Han

(343d) Amine-Blends Screening and Characterization for CO2 Post-Combustion Capture 177
Abdullah Al Hinai, Mohammad Abu Zohra, Dang Viet Quang
(343e) Economic and Exergy Analysis of Integrated Purification Process with Carbon Capture Process (CCP) Using Selexol in a 500 MW Integrated Gasification Combined Cycle (IGCC) Power Plant 183
Woo-Sung Lee, Hyun-Taek Oh, Chang-Ha Lee
(343g) Influence of Alloying on CO2 Electroreduction on Ag-Zn System 184
Toru Hatsukade, Kendra P. Kuhl, Etosha R. Cave, David N. Abram, Jeremy T. Feaster, Christopher Hahn, Anna L. Jongerius, Thomas F. Jaramillo
(343i) In-Situ Infrared Spectroscopic Investigation of Pyridine-Mediated CO2 Reduction on Pt Electrocatalysts 185
Marco Dunwell, Bingjun Xu, Yushan Yan
(138b) Selective Hydrogenation of Bio-Oil Model Compounds over Molybdenum Carbide Supported Catalysts 186
Sarah W. Paleg, Joshua Schaidle, Levi T. Thompson
(138c) Guaiacol Deoxygenation Using Methane over Pt-Bi Catalysts: Reaction Pathways and Kinetics 189
Yang Xiao, Arvind Varma
Fluidized Bed Catalytic Glycerol Hydrodeoxygenation to 1,3-Propanediol 190
Mahesh Edake, Jean-Luc Dubois, Mohammad Jaber Darabi Mahboub, Gregory Patience
(138e) Reaction Pathways for the Hydrodeoxygenation of Anisole and Guaiacol over Zn-Pt Bimetallic Catalysts 191
Daming Shi, John Vohs
(138f) Novel Bio-Oil Hydrodeoxygenation Catalysts Based on Strong Electrostatic Adsorption 192
Yaseen Elkasabi, Qiuli Liu, Yongsuck Choi, Akwasi A. Boateng, John R. Regalbuto
(138g) Biomass Pyrolysis Interfaced with Honeycomb Structured Upgrading Reactor for the Production of High Quality Biofuels 193
Conrad Zhang, Liaoyuan Mao, Yanxin Li
(138h) Upgrading of Biomass Pryolysis Vapors of Modified ZSM-5: Effects of Metal(s) Loading and Pretreatment 194
Matthew M. Yung, Anne Starace, Calvin Mukarakate, Kristiina Ilsa, Kim Magrini, Mark Nimlos
(138i) Cross Metathesis of Unsaturated Carboxylic Acids to Bioterephthalic Acid Intermediates 195
Erisa Saraçi, Lan Wang, Klaus Theopold, Raul F. Lobo
(145a) Investigating Synergetic Effects Between Deposited PGM Nanoparticles and Transition Metal Carbide Nanotube Support through Surface Characterization and Device Performance 196
Shibely Saha, Dongmei (Katie) Li, Shuai Tan, Brian Leonard
(145b) Bimetallic Ru-Pt and Pt-Co Fuel Cell Catalysts Prepared By Strong Electrostatic Adsorption and Electroless Deposition 197
John Meynard M. Tengco, Bahareh Alsadat Tavakoli Mehrabadi, Akkarat Wongkaew, Yunya Zhang, Weijian Diao, Taylor R. Garrick, John W. Weidner, John R. Monnier, John R. Regalbuto
(145c) Understanding the Effects of pH and Alkali Cations on H/OH Adsorption and Hydrogen Oxidation on Transition Metal Surfaces 198
Ian T. McCrum, Michael Janik
(145d) Nanoparticle Catalysts Supported on Substitutionally Doped Graphene: Effects on Activity and Stability for Hydrogen Oxidation 199
Stephen A. Giles, Stavros Caratzoulas, Dionisios G. Vlachos, Yushan Yan
(145e) Impact of Combining Metal Nanoparticle Catalysts and Semiconductor Photo-Electrodes on Photo-Catalytic Performance 200
Paul Hernley, Steven Chavez, Joseph Quinn, Suljo Linic
(145f) Trends in Hydrogen Evolution Reaction Activity Among Metal Modified Carbide Thin Films and Powders 201
Brian M. Tackett, Qian Zhang, Yannick C. Kimmel, Jingguang G. Chen
(145g) Experimental Determination of Water Orientation Effects on Kinetic Barriers to Alkaline Hydrogen Oxidation and Evolution 202
Maureen H. Tang, Joshua Snyder, Kiran Vasudevan, Jennifer Gallup
(148a) Co-Oxidation of CO and Hydrocarbons on Pd/Ceria-Zirconia/Al2O3 Three-Way Catalysts: Experiments and Modeling 203
Wendy Lang, Michael P Harold, Yisun Cheng, Carolyn Hubbard, Manish Sharma, Paul Laing
(148b) Interaction Between Highly Dispersed Ionic Pd2+ and Mn-Ce Solid Solution Support for Low Temperature CO Oxidation 206
Chao Wang, Cun Wen, Erdem Sasmaz, Jochen Lauterbach
(148c) Effect of Rare-Earth Metal Dopants on Ceria Catalysts for CO Oxidation 207
Kyung-Jong Noh, Jeong Woo Han

(148d) Carbon Monoxide Oxidation and Preferential Oxidation on Pt Alloy Nanoparticle Catalyst with Engineered Surface 208
Sang Youp Hwang, Eric Yurchekfrodl, Changlin Zhang, Zhenmeng Peng
(148j) Sensitivity of XANES and XES to the local environment: Multiple adsorption of H2O and NH3 on Cu-SSZ-13 209
Renqin Zhang, Hui Li, Christopher Paolucci, Atish A. Parekh, Janos Szanyi, Feng Gao, Trunojoyo Anggara, Hui Li, William F. Schneider, Fabio Ribeiro, Jean-Sabin McEwen
(148f) Mercury Oxidation over Cu-SSZ-13 Catalysts Under SCR Conditions for Power Plant Applications 212
Benjamin Galloway, Bihter Padak
(148g) Thiophenic Species Removal over a New Zn-Offretite Zeolite Via Selective Adsorption for FCC Processes 213
Yira Aponte, Hugo De Lasa
(148h) Wash-Coat Development for Lean-Burn Engine-Exhaust Aftertreatment: A Novel Way of Incorporation of Binder to Pd Supported on Sulfated Zirconia 214
Sreshtha Sinha-Majumdar, Gokhan Celik, Anne-Marie C. Alexander, Umit S. Ozkan
(148i) Carboxylation of Propylene Oxide to Propylene Carbonate in Slurry and Trickle Bed Reactors 215
Pallavi Bobba, Xin Jin, Raghunath V. Chaudhari, Bala Subramaniam
(157a) Supported Pt-Re Catalyst for Selective Hydrogenolysis and Hydrogenation Reactions 216
Derek D. Falcone, John H. Hack, Alexander Klyushin, Axel Knop-Gericke, Robert Schlogl, Robert J. Davis
(157b) Thermally Stable, Regenerable Single Atom Catalysts Via Atom Trapping 217
Abhaya Datye
(157c) Geometric and Electronic Effects in Pd and Pt Alloy Alkane Dehydrogenation Catalysts 218
Miller Jeffrey, Zhenwei Wu, Evan Wegener, Han-Ting Tseng, Gallagher James R., Yang Ren, Rosa Diaz, James W. Harris, Fabio Ribeiro
(157f) Catalytic Treatment of Water Contaminated with Chlorinated Hydrocarbons 219
Umit S. Ozkan, Hyuntae Sohn, Gokhan Celik, Seval Gunduz, Saurabh Ailawar, Paul Edmiston
(157e) How Pt Interacts with CeO2 Under the Reducing and Oxidizing Environments at Elevated Temperature? 220
Jaeha Lee, Youngseok Ryou, Yongwoo Kim, Xiaojun Chan, Taejin Kim, Do Heui Kim
(157d) Metal Oxide-like Clusters in MOFs As Supports for Well-Defined Rhodium Complex and Rhodium Cluster Catalysts 221
Dong Yang, Sandra Varinia Bernales Candia, Gamze Gümüslü, Jun Yu, Laura Gagliardi, Bruce C. Gates
(178a) Kinetic Modeling of Propene Hydroformylation in Propane-Expanded Liquid with Rh-Based Complexes 222
Dupeng Liu, Raghunath V. Chaudhari, Bala Subramaniam
(178b) Catalytic Cracking of Dodecane in Supercritical Water 223
Azadeh Zaker, Patricia Guerra, Geoffrey Tompsett, Michael T. Timko
(178c) Demetallation of Biocrude from Hydrothermal Liquefaction of Microalgae 224
Jimeng Jiang, Phillip E. Savage
(178d) Pilot-Scale Continuous Flow Reactor for Hydrothermal Liquefaction of Algae 225
Catherine E. Brewer, Feng Cheng, Travis Le-Doux, Juanita Miller
(178e) Production of Biodiesel Fuel Via Sub/Supercritical Transesterification Reactions with Trace Amount of Homogeneous Catalysts 226
Jiuxu Liu, Yue Nan, Lawrence L. Tavlarides
(178f) Molecular and Microkinetic Modeling of Supercritical Fluid and Liquid-Phase Fischer-Tropsch Synthesis 227
Kenneth M. Benjamin, Alireza Asiaee
(178g) Intensified Hydrolysis By Employing the Synergy of Mixed Subcritical H2o and Supercritical CO2 228
Armando Quitain, Akito Hashimoto, Saya Yamafuku, Mitsuru Sasaki, Tetsuya Kida
(205a) Pd Particle Stabilization on Carbon Supports Under Hydrothermal Conditions 229
Jiajie Huo, Robert L. Johnson, Pu Duan, Hien N. Pham, Abhaya K. Datye, Klaus Schmidt-Rohr, Brent H. Shanks
(205c) Vacancy-Mediated Hydrodeoxygenation of Furfuryl Alcohol over Metal Oxide 230
Konstantinos Goulas, Glen R. Jenness, Alexander V. Mironenko, Tobias Mazal, Dionisios G. Vlachos
(205d) Influence of Water and Solvents on the Production of Levoglucosenone and 5-(Hydroxymethyl)Furfural from Cellulose 231
Jiayue He, Fei Cao, Pranav U. Karanjkar, Siddarth H. Krishna, Kefeng Huang, Christos T. Maravelias, James Dumesic, George W. Huber
(205e) Transformation of Chitin into Value-Added Chemicals 232
Ning Yan

(205f) Direct Carbon-Carbon Coupling of Furanics with Acetic Acid over BrøNsted Zeolites 233
Abhishek Gumidyala, Bin Wang, Steven Crossley
(205g) Supported Cu Electrocatalysts for Electrocatalytic Hydrogenation and Hydrogenolysis of Furfural 234
Sungyup Jung, Elizabeth J. Biddinger
(205h) Rational Synthesis and Evaluation of Pd Bimetallic Catalysts for Furfural Rearrangement to Cyclopentanone in Aqueous Phase 235
Qiuli Liu, John R. Regalbuto
(205i) Polymeric Solid Acid Catalysts for Efficient Hydromethyfurfural and Levulinic Acid Production from Biomass Feedstock 236
Anh Vu, Ranil Wickramasinghe, Xianghong Qian
(209a) Soot Nucleation and Consumption in Oxy-Coal Systems 237
Alexander J Josephson
(209b) Investigation of NOx Chemistry in Oxy-Combustion Flue Gas 238
Nujhat Choudhury, Bihter Padak
(209c) On the Role of Sodium Content of Fine Ash Particles in Determining Deposition Rates Under Air- and Oxy-Coal Combustion Conditions 239
Zhonghua Zhan, Andrew Fry, Jost O. L. Wendt
(209d) NOx Formation in Syngas/Air Combustion at Elevated Pressure 240
Nazli Asgari, Sheikh F Ahmed, Tanvir I Farouk, Bihter Padak
(209e) Formation and Decomposition Kinetics of HONO and HNO2 241
C. Franklin Goldsmith
(209f) Experimental Study on NO Reduction By Iron with Methane in Wet Simualted Flue Gas 242
Yaxin Su
(209g) Inferential–MODEL Based FLARE SET POINT Determination 243
Arokiaraj Alphones, Daniel Chen, Helen Lou, Vijaya Damodara, Xianchang Li, Christopher B. Martin, Edward Fortner, Scott Evans, Matthew Johnson
(210a) Offshore Floating Packed-Bed Reactors: Current Status, Key Challenges and Potential Solutions 244
Amir Motamed Dashliborun, Faical Larachi, Mohsen Hamidipour, Markus Schubert
(210i) Design Optimization of BioEnergy Combustion Systems Using Sculptor® with CFD 245
Joseph D. Smith, Vivek M. Rao, Mark Landon
(210d) A New Model for the Depolymerisation of Polysaccharides Originating from Biomass 246
Tapio Salmi, Sakari Teerikoski
(210e) Recent Trends in Multi-Phase Reactor Engineering and Design 247
Tobias Bauer, Stefan Haase, Markus Schubert, Ruediger Lange
(210f) Gas-Liquid and Liquid-Liquid Taylor Flow in Micro Channels: Hydrodynamics and Mass Transfer 248
Rufat Abiev, Stanislav Svetlov, Stefan Haase, Tobias Bauer, Ruediger Lange
(210j) 3D-Printed Monolithic Adsorbents for Gas Separation Processes 281
Harshul Thakkar, Stephen Eastman, Amit Hajari, Ali A. Rownaghi, Fateme Rezaei
(210h) Trickle-Bed Reactors and Their Future 282
Ruediger Lange
(212a) Keynote - Modeling Electrochemical Interfaces: An Approach from First Principles 283
Phillippe Sautet, Stephan Steinmann
(212b) Feature Engineering of Machine-Learning Chemisorption Models for Bifunctional Electrocatalyst Design 284
Zheng Li, Siwen Wang, Hongliang Xin
(212c) Effect of Lanthanum Doping and Chlorine Addition on Strontium Titanate Perovskites for Electrocatalytically-Assisted Oxidative Dehydrogenation of Ethane 285
Doruk Dogu, Anshuman Fuller, Katja E. Binkley Meyer, Seval Gunduz, Dhruba Jyoti Deka, Nathaniel Kramer, Anne Co, Umit S. Ozkan
(212d) Mechanistic and Trend Analyses of Selective Furfural Electroreduction on Transition Metals from First-Principles Methods 286
Nannan Shan, Xiaotong Chadderdon, David Chadderdon, Wenzhen Li, Bin Liu
(212e) NO Electrochemical Reduction on Pt(100) from First Principles 287
Hee-Joon Chun, Vesa Apaja, Andre Clayborne, Karoliina Honkala, Jeffrey P. Greeley
(212f) Flame Synthesis and High Pressure High Temperature Annealing of Anatase TiO2 for Increased Photocatalytic Activity 288
Ashley M. Pennington, Katelyn A. Dagnall, Rachel A. Yang, Fuat E. Celik
(212g) Photoelectrochemical Oxidation of Biorenewable Alcohols By Nitroxyl Radical Catalysts 289
David Chadderdon, Ivy Wu, Matthew G. Panthani, Wenzhen Li

(216a) so2 Interactions with Pt, Pd, and Pd-Pt/Al2O3 Catalysts 290
William Epling, Monique Wilburn
(216b) Insights into the Deactivation of Cu/SAPO-34 By Low- and High-Temperature Hydrothermal Treatment 291
Yang Zheng, Aiyong Wang, Eric D. Walter, Feng Gao, James Song, Yilin Wang, Janos Szanyi, Charles H. F. Peden
(216c) Lean-Rich Switching over a Modified Three-Way Catalyst: Experiments and Modeling 292
Mengmeng Li, Sam Malamis, Michael P Harold, William S. Epling
(216d) Importance of Exhaust Hydrocarbon Speciation in the Studies of Low-Temperature Emission Control Technologies for Next-Generation Fuel-Efficient Vehicles 294
Se H. Oh, Michelle Wiebenga, Sung Bong Kang, Sung Bang Nam, In-Sik Nam
(216e) Rapidly Pulsed Reductants in Diesel NOx Reduction with Lean NOx Traps: Spatial Distribution of Species 295
Amin Reihani, Brent Patterson, John Hoard, Galen B Fisher
(216f) A Rational Pathway for Generic Three-Way Catalyst Model Development in Stoichiometric Natural Gas Engines 297
Jian Gong, Di Wang, Junhui Li, Neal Currier
(216g) Effects of Cu Density in Cu-SSZ-13 Zeolites on Low-Temperature NOx Selective Catalytic Reduction with NH3 298
Ishant Khurana, Atish A. Parekh, Jonatan Albarracin, Christopher Paolucci, Arthur Shih, John R. Di Iorio, W. Nicholas Delgass, Jeffrey T. Miller, William Schneider, Rajamani Gounder, Fabio H. Ribeiro
(216h) The Effects of Catalyst Aging on Sulfur Degradation over Fully Formulated Lean NOx Trap Catalysts 299
Travis Wentworth, Louise Olsson, Marie Stenfeldt
(216i) A Comparative Study of ZSM-5 and BEA-Zeolites for Hydrocarbon Trap Applications Under "Cold-Start" Conditions 300
Eleni A. Kyriakidou, Jae-Soon Choi, Todd J. Toops, James E. Parks
(267a) Developing a Quantitative Spatial Resolution of Deactivation Effects on FT Synthesis in a Microchannel Reactor When Operating with Ultra-Low Sulfur Levels 301
Soumitra Deshmukh, Henning Becker, Sara Kampfe, Kenneth Cowen, Daniele Leonarduzzi, Laura Barrio, Jay Pritchard, Heinz J. Robota
(267b) Characterization of Unconventional Reservoirs: From Lab Study to Reservoir Simulations 306
Lin Li
(267c) New Insight into Oxidative Dehydrogenation of n-Butenes to Butadiene 307
Xiankuan Zhang
(267d) Novel High Active Catalyst Innovation and Low-Energy Consumption Process Technology for Ammonia Synthesis 308
Huazhang Liu, Chao Huo, Wenfeng Han
(267e) Maximizing Production of Premium Distillate Fuels through Selective Hydroisomerization of Normal Paraffins 309
William W. Lonergan
(267f) On the Effectiveness Factor Modeling for Mass Transfer-Limited Reaction Networks in Industrial Catalysis 310
Bo Jin, Xiaoxiang Zhu, William Licht
(274a) Active Site Requirements and Elementary Steps in the Ketonization of Carboxylic Acids on TiO2 and ZrO2 311
Enrique Iglesia, Shuai Wang
(274b) Influence of Solid Acid Structure and Composition on Hydrocarbon Cracking Activity Under Supercritical Conditions 314
Benjamin Huang, Robert J. Davis
(274c) Improving the Hydrothermal Stability of Heterogeneous Catalysts for Conversion of Biorenewables Via Carbon Coating and Overcoating 315
Abhaya Datye, Hien Pham, Haifeng Xiong, Pu Duan, Xiaoyan Cao, Klaus Schmidt-Rohr, Robert Johnson, James Dumesic
(274d) Mechanistic Insights into the Reactive Sorption of Sulfur Compounds Using Metal Oxides 316
Dante Simonetti, Sara Azzam, Brook Wander
(274e) Influence of Solvent on Acid and Metal Catalyzed Reations 319
Matthew Neurock
(281a) Effect of Alcohol Structure on the Kinetics and Mechanism of Acid Catalyzed Etherification and Unimolecular Dehydration over Tungstated Zirconia 320
Julie Rorrer, Alexis T. Bell, Dean Toste
(281b) Pathways to Selectively Form Individual Aromatic Products from Ethanol 321
David W. Flaherty, Takahiko Moteki

(281c) Combined Experimental and Theoretical Study of Ethanol Catalytic Conversion to 1,3-Butadiene on MgO 322
William Taifan, Jonas Baltrusaitis
(281d) Aldol Condensation of Acetaldehyde over Titania, Hydroxyapatite and Magnesia 323
Zachary Young, Sabra Hanspal, Robert J. Davis
(281e) Single Step Conversion of Short Carboxylic Acid Mixtures Obtained from Fermentation to Aromatics and Î±-Olefins over a Cu/ZrO2 Catalyst 324
Bartosz Rozmyslowicz, Jeremy S. Luterbacher
(281f) Mechanistic Insights on the Formation of Maleic Anhydride from the Oxidation of Levulinic Acid over Supported Vanadates 325
Anargyros Chatzidimitriou, Jesse Q. Bond
(281g) P-Toluic Acid from Biomass-Derived Coumalic Acid: A Reaction Network Analysis 326
Toni Pfennig, Robert L. Johnson, Brent H. Shanks
(281h) Rational Design of Nanoflower Shaped Bimetallic Ptmn Nanocatalysts for Tartronic Acid Synthesis from Biomass 327
Xin Jin, Meng Zhao, Chun Zeng, Wenjuan Yan, Honghong Shi, Prem Thapa, Bala Subramaniam, Raghunath V. Chaudhari
(288a) Upgrading Corn Stover Pretreatment Wastes into Biodiesel Via Oleaginous Rhodococci 328
Arthur J. Ragauskas
(288b) Biological Design of Lignin Conversion 329
Bin Yang, Art Regauskas, Joshua Yuan
(288c) Selective Conversion of Guaiacol to Substituted Alkylphenols in Supercritical Ethanol over MoO3 330
Kai Cui, Yongdan Li
(288d) Enabling Microbial Utilization of Thermally Depolyermized Lignin Monomers 331
Kirsten Davis, Marjorie Rover, Davinia Salvachua, Gregg T. Beckham, Zhiyou Wen, Ryan Smith, Laura Jarboe, Robert Brown
(288e) Fractionated Lignin Rheology and Structure 332
Christopher Higgins, David C. Russ, Noppadon Sathitsuksanoh, R. Eric Berson
(288f) Selective Conversion of Biomass-Derived Lignin to Cyclic Hydrocarbons 333
Bin Yang, Melvin Tucker, Hongliang Wang, Hao Ruan, Xiaowen Chen
(289b) The Evolution of Multi-Functional Nanoporous Metal Composite Electrocatalysts 334
Joshua Snyder, Yawei Li
(289c) Effect of Doping on the Activity of Nickelate Oxides Toward Surface Oxygen Exchange and Oxygen Reduction 335
Xiang-Kui Gu, Anirban Das, Juliana S. A. Carneiro, Eranda Nikolla
(289d) Oxyhydroxides As a Platform for Confined Oxygen Electrochemistry 336
Andrew Doyle, Michal Bajdich, Aleksandra Vojvodic
(289e) Large-Size Graphene Tube Catalysts for Bifuncational Oxygen Reduction and Oxygen Evolution Electrocatalysis in Alkaline Media 337
Gang Wu
(289f) Atomic Layer Deposition of NiAlxOy Catalysts for Electrochemical Oxidation of Water 338
Jonathan G. Baker, Stacey F. Bent, A. J. M. Mackus
(289g) The Role of Superoxide in the Non-Aqueous Oxygen Reduction Reaction in Li-O2 Batteries 339
William C. McKee, Saurin Rawal, Ye Xu
(300a) Experimental and Simulation Study of Ethanol Steam Reforming Reaction for Hydrogen Production in Large Scale Catalytic Membrane Reactor 340
Rui Ma, Bernardo Castro Dominguez, Ivan Mardilovich, Anthony G. Dixon, Yi Hua Ma
(300b) Membrane Reactors for Equilibrium-Limited Alcohol and Light Hydrocarbon Synthesis 341
Zhongtang Li, Nazanin Entesari, Theodore Tsotsis
(300c) Dual-Phase Ceramic Membrane in a Membrane Reactor for Ethane Dehydrogenation 342
Libin Zhuang, Yanying Wei, Haihui Wang
(300d) Multi-Scale (Pellet-Reactor Scale) Membrane Reactor Modeling and Simulation: High Temperature, and High Pressure Water-Gas Shift Reaction 343
Secgin Karagoz, Vasilios Manousiouthakis, Theodore Tsotsis
(300e) Pervaporation-Assisted Furfural Production from Xylose 344
Alex Wang, Nitash Balsara, Alexis T. Bell
(300f) Zeolite Membrane Permselective to Water Vapor and Its Application to Dehydration Reactions 345
Naotsugu Itoh, Shuhei Kikuchi, Takafumi Sato, Takuya Hirose
(300g) Esterification and Extraction with Supported Liquid Membrane Permeation 346
Marlene Kienberger, Matthaeus Siebenhofer, Sasiradee Jantasee, Nuttakul Mungma

(300h) Investigation of the Degradation of Endocrine Disrupting Compounds in a Membrane Bioreactor 347
Mohanad Kamaz, Ees Eswaranandam, Ranil Wickramasinghe, Wen Zhang, Xianghong Qian, Michael Watts
(307a) Quantum Chemical Methods for the Simulation of Reactions of Organic Molecules in Zeolites 348
Alexis T. Bell
(307b) Identification of Possible Mechanistic Pathways during the Electrochemical Reduction of CO2: Results from Constant Potential DFT Calculations 349
Jason Goodpaster, Alexis T. Bell, Martin Head-Gordon
(307c) Theoretical Challenges in Characterizing Cu Centers in the Zeolite SSZ-13 350
Florian Göltl, Alyssa Love, Sabrina Conrad, Patrick Wolf, Phillippe Sautet, Ive Hermans
(307d) Free Energies of Adsorption Calculated through Potential of Mean Force Ab Initio Molecular Dynamics in Metal-Exchanged Zeolites 351
Hui Li, Christopher Paolucci, William F. Schneider
(307e) Nonadiabatic Effects and Creation of Electronic Excitations during Dissociation on Metal Nanoparticles 352
Matthew M. Montemore, Efthimios Kaxiras
(307f) Comparing the Effect of Counter Ions, Solvent Molecules, and Electron Correlation on Homogeneous Reaction Models 353
Mitchell Groenenboom, John A. Keith
(307g) Automated Reaction Mechanism Generation for Heterogeneous Catalisys 354
Richard H. West, C. Franklin Goldsmith
(307h) Global Optimization and Hydrogen Adsorption on Pt-Cu Clusters By Using Genetic Algorithm and Density Functional Theory 357
Ezgi Erdem, S. Bengisu Barim, Riza Kizilel, Can Erkey
(332a) Strategies for the Conversion of Biomass-Derived Carbohydrates and Fatty Acids to Fuels and Lubricants 360
Alexis T. Bell
(332b) Materials for Manipulating the Composition of Biomass Pyrolysis Vapors 361
Brent H. Shanks
(332c) Coupling Chemical and Biological Catalysis to Produce Biobased Chemicals 362
Thomas J. Schwartz, Brent H. Shanks, James Dumesic
(332d) Commodity Chemicals from Biomass: Catalytic Conversion of Biomass into a,w-Diols 363
George W. Huber, Jiayue He, Pranav U. Karanjkar, Siddarth H. Krishna, Kevin J. Barnett, Zachary Brentzel, Samuel P. Burt, Kefeng Huang, Ive Hermans, William F. Banholzer, Christos T. Maravelias, James A. Dumesic
(332e) Atomically-Thin Noble Metal Coatings on Early Transition Metal Carbide Nanoparticles: a Versatile Catalytic Platform for Electro- and Thermocatalysis 364
Yuriy Román-Leshkov, Sean T. Hunt, Maria Milina, Ana Alba-Rubio, James Dumesic, Christopher H. Hendon
(337a) Biodiesel Produced from Soybean Oil with the Two Flow Reactors in Series 365
David Baah, Nader Vahdat, Kyung Kwon
(337b) Biomass Conversion to Acrylonitrile Monomer-Precursor for Production of Carbon Fibers 366
Jadid Samad, Lindsey Chatterton, Zora Govedarica, Amit Goyal
(337c) Catalytic Conversion of Polyols to Olefins 367
Bryan E. Sharkey, Friederike C. Jentoft
(337d) Catalytic Oligomerization of Ethylene: Experimental Insight into the Effect of Supercritical Reaction Conditions 368
Oliver Jan, Fernando Resende
(337e) Mesoporous Catalysts for Conversion of 2,3-Butanediol to Butene 369
Quanxing Zheng, Jason Grossardt, Haider Almkhelfe, Jiayi Xu, Brian P. Grady, Placidus B. Amama, Keith L. Hohn
(337g) Polar Aprotic Solvent Effects on the Catalytic Conversion of Fructose to 5-(hydroxymethyl)Fulfural or Lactic Acid 370
Christian G. Rivera-Goyco, Michelle M. Marrero-Vazquez, Leida M. Vazquez-Ramos, Yomaira J. Pagan-Torres, Nelson Cardona-Martinez
(348a) Investigations of Surface Chemistry for Pyridine-Catalyzed CO2 Reduction on Gallium Phosphide 371
Coleman Kronawitter, Bruce E. Koel
(348b) Implications of Surface Reconstructions Impacting Py-Catalyzed CO2 Reduction on Semiconductor Photoelectrodes 372
Thomas P. Senftle, Emily A. Carter
(348c) Activity and Stability of Pure and Modified Coooh for Oxygen Evolution Reaction in Alkaline Medium 373
Zhu Chen, Coleman Kronawitter, Yao-Wen Yeh, Xiaofang Yang, Peng Zhao, Yao Nan, Bruce Koel

(348d) Fe-Ni Core-Shell Hydroxide Nanoparticles As an Active Oxygen Evolution Reaction (OER) Catalyst 374
Stephanie Candelaria, Nikki S. Rentz, Lauren F. Greenlee
(348e) Layered - Hyperthin Electrocatalysts for the Water Splitting Reactions 375
Kevin C. Leonard
(348f) Stabilizing Electrodeposited Nanoparticle Electrocatalysts on Si MIS Photocathodes for Solar Water Splitting 376
Natalie Labrador, Xinxin Li, Yukun Liu, Jeffrey T Koberstein, Daniel V. Esposito
(348g) Two Distinct Types of Surface-Bound Atomic Hydrogen and Their Role in Dictating Product Selectivity in Artificial Photosynthesis Reactions 377
Samiksha Poudyal, Siris Laursen
(370a) Development of Bayesian Error Estimation Density Functionals with Range-Separated Exchange 378
Shaama Mallikarjun Sharada, Keld T Lundgaard, Johannes Voss, Jess Wellendorff, Thomas Bligaard, Jens K. Norskov
(370b) Uncertainty Quantification of the Water-Gas Shift Reaction By Pt/CeO2 Catalyst 379
Eric Walker, Salai C. Ammal, Gabriel Terejanu, Andreas Heyden
(370c) A Descriptor Based Approach Towards Understanding the Effect of Functionals on Reducible Oxide Reaction Kinetics 380
Tej S. Choksi, Jeffrey P. Greeley
(370d) Interpreting Functional-Sensitivity in Catalysis: Exchange-Tuning Vs. DFT+U 381
Qing Zhao, Efthymios Ioannidis, Jon Paul Janet, Heather J. Kulik
(370e) Determining the Pd Dopant Effect on the Reducibility of Fe2O3: DFT+U or Hybrid Functionals? 382
Alyssa Hensley, Yongchun Hong, Yong Wang, Jean-Sabin McEwen
(370f) Neural Network and Reaxff Comparison for Au Properties 383
Jacob R. Boes, Mitchell Groenenboom, John A. Keith, John R. Kitchin
(370g) Machine Learning Based Interatomic Potentials for Electrocatalysts 384
Badri Narayanan, Fatih Sen, Alper Kinaci, Stephen Gray, Maria K. Y. Chan, Subramanian K. R. S. Sankaranarayanan
(370h) BASC: Applying Bayesian Machine Learning Techniques to the Search for Global Minima on Potential Energy Surfaces 385
Shane F. Carr, Cynthia S. Lo
(370i) Developing a Machine-Learning Molecular Dynamics Approach for Nonadiabatic Surface Chemistry 386
Jiamin Wang, Hongliang Xin
(375a) Glycerol to Hydrocarbon Fuels Via Bifunctional Catalysts 387
Yang Xiao, Arvind Varma
(375b) Reaction Engineering of Urea Alcoholysis 388
Dhananjay R. Mote, Vivek V. Ranade
(375c) Gas-Phase Catalysis for Pyrolyzing Acetic Acid 389
Sara Jo Taylor, Phillip R. Westmoreland, Haley Key
(375d) Hybrid Reaction Pathways to Renewable Diol Monomers 390
Paul J. Dauenhauer, Charles S. Spanjers, Marc A. Hillmyer, Kechun Zhang, Deborah Schneiderman
(375e) Continuous Piancatelli Rearrangement of Furfuryl Alcohol: Kinetics and Reactor Modelling 391
Lukas T. Vömel, Marcel A. Liauw
(375f) Catalyst Structure and C-O Activation during Fischer-Tropsch Synthesis 392
G. T. Kasun Kalhara Gunasooriya, Arghya Banerjee, Mark Saeys
(375g) Reaction Engineering for Modified Furans 393
Dae Sung Park, Paul J. Dauenhauer, Kristeen Esther Joseph, Christoph Krumm, Dion Vlachos, Raul F. Lobo
(393a) C2H6 Dehydrogenation on Pt-Based Catalysts: Fundamental Mechanistic Studies 394
Manos Mavrikakis, Duygu Gerceker, Guowen Peng, James Dumesic
(393b) Microkinetic Analysis of y-Valerolactone Ring Opening and Decarboxylation over Solid Acids 395
Jesse Q. Bond
(393c) Solvent Effects in the Catalytic Conversion of Fructose to Lactic Acid 396
Nelson Cardona-Martinez, Christian G. Rivera-Goyco, Yomaira J. Pagan-Torres, Michelle M. Marrero-Vazquez
(393d) Stabilization with Formaldehyde Facilitates the High-Yield Production of Monomers from Lignin during Integrated Biomass Depolymerization 397
Jeremy S. Luterbacher
(393e) Solvent-Enabled Control of Reactivity for Liquid-Phase Biomass Conversion Reactions 398
James Dumesic, Max A. Mellmer, Ali Hussain Motagamwala, Bengi Demir, David Martin Alonso

(399a) Structure-Energy-Activity Relationships in Heterogeneous Catalysis - a View from Theory 399
Phillippe Sautet
(399b) A New DFT-Based Approach That More Accurately Predicts Adsorption Energies 400
Alyssa Hensley, Kushal Ghale, Carolin Rieg, Thanh Dang, Emily Anderst, Felix Studt, Charles T. Campbell, Jean-Sabin McEwen, Ye Xu
(399c) A DFT-Based Study of Pt-Catalyzed Hydrodechlorination of 1,2-Dichloroethane: Effect of Van Der Waals Interactions 401
Lang Xu, Eric E. Stangland, Manos Mavrikakis
(399d) Computational Design of Active Site Structures to Circumvent Scaling in Ammonia Synthesis 402
Aayush R. Singh, Joseph H. Montoya, Charlie Tsai, Aleksandra Vojvodic, Jens K. Nørskov
(399e) Mechanistic Investigation of the Catalytic Hydrogenation of Levulinic Acid over Ru Model Catalyst Surfaces 403
Osman Mamun, Jesse Q. Bond, Andreas Heyden
(399f) Reaction Pathways for Biomass Derived Furanic Compounds on Ru(001) Surfaces: a DFT Study 404
Arghya Banerjee, Samir H. Mushrif
(399g) CO2 Activation on Zr-Decorated, Cu Nanoparticles 405
Natalie Austin, Giannis Mpourmpakis
(399h) Oxygen Dissociation for Selective Oxidation on Gold: Effect of Alloying and Geometrical Structure 406
Matthew M. Montemore, Cynthia M. Friend, Robert J. Madix, Efthimios Kaxiras
(432a) Catalytic Co-Pyrolysis of Microalgae and Low Density Polyethylene Waste to Aromatic Hydrocarbons Using Activated Carbon 407
Emmanuel Ansah, Lijun Wang
(432b) Modeling the Impact of Biomass Particle Residence Time on Fast Pyrolysis Yield and Composition 415
Gavin Wiggins, C. Stuart Daw, Emilio Ramirez
(432c) Ring-Locking Enables Selective Anhydrosugar Synthesis from Carbohydrate Pyrolysis 416
Li Chen, Jin-Mo Zhao, Sivaram Pradhan, Bruce E. Brinson, Gustavo E. Scuseria, Z. Conrad Zhang, Michael S. Wong
(432d) Pyrolysis Kinetics to and from Levoglucosan 417
Vikram Seshadri, Phillip R. Westmoreland
(432e) Molecular-Level Simulation of Thermogravimetric Analysis (TGA): A Cellulose Pyrolysis Example 418
Juan Lucio-Vega, Scott R. Horton, Michael T. Klein
(432f) Impact of Carrier Gas and Temperature on Biomass Fast Pyrolysis Using a New Microsphere Microreactor Approach 419
Joseph J. Biernacki, Ali Zolghadr
(432g) Study of Transmethylation As Primary Step on the Decomposition of Methoxy-Rich Lignin Model Compound over Zeolite Catalyst 420
Zhang Jiajun, Fidalgo Beatriz, Shen Dekui
(434a) Hydroxycarbonylation Vs. Methoxycarbonylation - Palladium Catalyzed Reactions in Multiphase Systems 421
Marcel Schmidt, Tobias Pogrzeba, Ariane Weber, Prof. Dr. Reinhard Schomäcker
(434b) Heterogeneously Catalysed Benzyl Alcohol Oxidation in a Three-Phase Micro-Packed Bed Reactor: Influence of Hydrodynamics and Reactor Design on Reaction Performance and Catalyst Deactivation 422
Noor Al-Rifai, Archana Venkatesh, Enhong Cao, Stefano Cattaneo, Meenakshisundaram Sankar, Graham J. Hutchings, Asterios Gavriilidis
(434c) Highly Selective Synthesis of Primary Amines from Alcohols and Ammonia By Heterogeneous Nickel Catalysis 423
Andrew Leung, Mimi Hii, Klaus Hellgardt
(434d) Hydroformylation and Tandem Isomerization-Hydroformylation of Long-Chain Olefins: Mechanism, Kinetics and Optimal Reaction Control 424
Andreas Jörke, Andreas Seidel-Morgenstern, Christof Hamel
(434e) Troubleshooting Start-up of a Newly Built Intensified Heat Exchanger Reactor - Using Reaction Engineering and Dynamic Modeling to Solve Real-Life Problems 425
Eric Grolman
(434f) Alloy Catalysis Across Composition Space 426
A. J. Gellman, Irem Sen, Gamze Gumuslu, Petro Kondratyuk, Xiaoxiao Yu
(434g) Pressure Dependent Kinetics of Magnesium Oxide Carbothermal Reduction 427
Boris Chubukov, Scott C. Rowe, Aaron W. Palumbo, Illias Hischier, Arto J. Groehn, Alan W. Weimer

(435a) Detailed Kinetic Modeling of Thermal Decomposition of Silane Via Automatic Mechanism Generation 428
Belinda L. Slakman, Pierre L. Bhoorasingh, Richard H. West
(435b) Mechanistic Insights into the Electrochemical Reduction of CO2 over Ag Using an Integrated Transport-DFT-Microkinetic Model 429
Meenesh Singh, Jason Goodpaster, Adam Weber, Martin Head-Gordon, Alexis T. Bell
(435c) Acceleration of Kinetic Monte Carlo Simulations Combining a Hybrid Approach with Scaling of Kinetic Parameters for Free Radical Copolymerization 430
Hanyu Gao, Ivan Konstantinov, Steven G. Arturo, Linda J. Broadbelt
(435d) Kinetics and Mechanism of Ketonization of Acetic Acid on HZSM-5 Catalyst 431
Abhishek Gumidyala, Tawan Sooknoi, Bin Wang, Steven Crossley
(435e) Methane Steam Reforming Unraveled By the Microkinetic Engine, a User-Friendly Kinetic Modelling Tool 432
Brigitte R. Devocht, Kenneth Toch, Naoki Kageyama, Naghmeh Fatemi, S. Ted Oyama, Guy B. Marin, Joris W. Thybaut
(435f) Severe Pyrolysis of the C12 Hydrocarbons n-Dodecane, Cyclododecane, and 2,6-Dimethyl-Decalin, with Rice-Herzfeld and Graph-Theoretic Interpretations 438
Preetinder S. Virk
(435g) An Assessment of Solvent Effects on the Selectivity of an Alkylation Reaction: A Comparison Between Experiments and Computations 439
Aikaterini Diamanti, Eliana Grant, Zara Ganase, Anita Rea, Amparo Galindo, Claire Adjiman
(435h) Automatic Discovery of Reaction Paths and Ab Initio/Rrkm-ME Calculations Towards Pyrolysis and Combustion Mechanisms 440
Adeel Jamal, Colin Grambow, William H. Green
(435i) Computational Singular Perturbation As an Analytical Tool to Study the Fischer-Tropsch Process 441
Debanjan Chakrabarti, Gouthami Senthamaraikkannan, Vinay Prasad
(435j) Mechanism, Reaction Kinetics and Multi-Scale Modelling of CO2 Hydrogenation to Methanol over Trimetallic Heterogeneous Catalysts 442
Neja Strah, Andraž Pavlišic, Matej Huš, Venkata Dasireddy, Andrej Pohar, Blaž Likozar
(435k) Mechanistic Insights into the Catalytic Elimination of Tar on Nickel and the Promotional Effect of Boron on It: A Combined Theoretical and Experimental Study Using Toluene As a Model Compound 450
Quang Thang Trinh, Anh Vu Nguyen, Dang Chinh Huynh, Thanh Huyen Pham, Samir H. Mushrif
(461a) Discovery of Superior Cu-GaOx-HoOy Catalysts for the Reduction of Carbon Dioxide to Methanol at Atmospheric Pressure 451
Bahman Zohour, Selim Senkan
(461b) Methanation of Carbon Dioxide with a Novel Nickel/Magnesium Oxide Catalyst 452
Georg Baldauf-Sommerbauer, Susanne Lux, Matthaeus Siebenhofer
(461c) Polymer-Nanocrystal Composites As Tailorable CO2 Hydrogenation Catalysts 453
Andrew Riscoe, Joshua Willis, Matteo Cargnello
(461d) Reduction of CO2 to CO and Methanol on Ceria Based Catalyst: Mechanistic Insights 454
Neetu Kumari, Suddhasatwa Basu, M. Ali Haider, Nishant Sinha, Manish Agarwal
(461e) Catalytic Conversion of Carbon Dioxide with Solar Hydrogen Under Supercritical Conditions 455
Elisabeth Hocke, Herbert Vogel, Bjoern Kommoss, Stephan Schultheis
(461f) CO2 Hydrogenation on Silica- and Titania- Supported Single-Site Heterogeneous CoII Catalyst 456
Juan Jimenez, Cun Wen, Jason Hattrick-Simpers, Jochen Lauterbach
(461g) Development of Catalysts for CO2 Hydrogenation to Olefins 457
Marc D. Porosoff, Heather D. Willauer
(461h) K2o/WOx/Al2O3 Catalyst Interactions with the Components and Byproducts of Sour Natural Gas 458
Jih-Mirn Jehng, Israel E. Wachs, Zili Wu, Jonas Baltrusaitis
(465a) Organic Photocatalysts for Atom Transfer Radical Polymerization Driven By Visible Light 459
Charles B. Musgrave, Chern-Hooi Lim, Jordan Theriot, Garret Miyake, Haishen Yang, Matthew Ryan
(465b) CO-CO Coupling on Cu Facets: Coverage, Strain and Field Effects 460
Robert B. Sandberg, Joseph H. Montoya, Karen Chan, Jens K. Nørskov
(465d) Single-Atom Catalysts for CO2 Electroreduction with Significant Activity and Selectivity Improvements 461
Seoin Back, Yousung Jung
(465e) Solvation Effects in Surface Electrochemistry for Oxygen Evolution Reaction Catalysis on Transition Metal Oxides 462
Joseph Gauthier, Colin Dickens, Andrew Doyle, Jens K. Norskov

(465f) First Principles Quantum Chemistry Calculations to Model CO2 Electroreduction on SnO2 Particles 463
Yasemin Basdogan, Karthikeyan Saravanan, James Dean, John A. Keith
(465g) Modelling the Electrochemical Interface: Applications to CO2 Reduction 464
Karen Chan
(465h) Exploring the Remarkable Heterogeneity of Bimetallic Alloy Facet Reactivity for the Electrocatalytic Reduction of CO2 465
Zachary Ulissi, Michael Tang, Mohammadreza Karamad, Karen Chan, Jens Nørskov
(465i) Exploring Possible Active Sites for IrOx/SrIrO3, a Highly Active Oxygen Evolution Catalyst 466
Linsey C. Seitz, Colin Dickens, Kazunori Nishio, Yasuyuki Hikita, Joseph H. Montoya, Andrew Doyle, Charlotte Kirk, Aleksandra Vojvodic, Harold Y. Hwang, Jens K. Nørskov, Thomas F. Jaramillo
(496a) Characterization of TFA Regenerated Cellulose with a Two-Step Treatment to Understand the Impact of Cellulose Structure on Hydrolysis 467
Ximing Zhang
(496b) The Role of Atmosphere, Solvents and Acids on the Production of 5-Hydroxymethylfurfural from Biomass Derived Fructose 468
Glen Svenningsen, Phillip Christopher, Rajeev Kumar, Charles Wyman
(496c) Lignin Depolymerization into Aromatic Monomers over Acidic Mesoporous Silicates 469
Kakasaheb Nandiwale, Andrew Danby, Anand Ramanathan, Raghunath V. Chaudhari, Bala Subramaniam
(496d) Characteristics and Activity of Steamed Zeolites for Butanol Upgrading to Higher Olefins 470
Mond Guo, Michel Gray, Vijayakumar Murugesan, Karthikeyan K. Ramasamy
(496e) Tandem Catalysis in a New, Solid-Phase Route to Renewable Aromatic Chemicals from Biobased Furanics 471
Homer Genuino, Shanmugam Thiyagarajan, Jan Van Der Waal, Ed De Jong, Jacco Van Haveren, Dan Van Es, Bert Weckhuysen, Pieter Bruijnincx
(496g) Reaction Pathways and Microkinetic Modeling of Levulinic Acid Hydrodeoxygenation over Sulfided NiMo/Al2O3 472
Miha Grilc, Blaž Likozar
(497a) The E-Factor of the Thin Film and Its Characterization to Understand Why Water Influences a Heck Alkynylation 477
Chuntian Hu, Kevin H. Shaughnessy, Ryan L. Hartman
(497b) Triphasic Millireactors for Gas-Liquid Reactions with in Situ Catalyst Synthesis 478
Wai Kuan Wong, Swee Kun Yap, Saif A. Khan

VOLUME 2

(497c) Experimental and Model Aided Approach for Development of High Pressure Direct Asymmetric Reductive Amination Reaction 479
Shujauddin M. Changi, Tohru Yokozawa, Tetsuya Yamamoto, Hikaru Nakajima, Matthew C. Embry, Radhe K. Vaid, Sze Wing Wong, Martin D. Johnson, Eric D. Moher
(497d) Kinetics and Mechanistic Understanding of the Difluoromethyl Alkylation of Hydroxychromenone 480
Daniel D. Caspi, Anuj A. Verma, Eric M. Phillips, Shashank Shekhar
(497e) Development of a Reaction Kinetic Model for an Enzymatic Reaction Using Reaction Kinetic Progress Analysis and Dynochem 481
Kathryn Meintel
(497f) Rapid Kinetic Profiling of Copper(I)-Catalyzed Alkyne-Azide Cycloaddition(CuAAC) Click Reactions Utilizing Inline IR, HPLC and MS 482
Kosi Aroh, Klavs F. Jensen
(497g) Dynamic Resolution for Separation-Free Enantiomeric Purification in Flow 483
Andrew Teixeira, Saurabh Shahane, Nopphon Weeranoppanant, Klavs F. Jensen
(517a) Tuning the Activity and Selectivity of Transition Metal Phosphide Catalysts for CO, and CO2 Hydrogenation 484
Melis S. Duyar, Eduardo Valle, Jakob Kibsgaard, Thomas F. Jaramillo
(517b) Aerosol Processing of MOF-Based Nanocomposites for Carbon Dioxide Photoreduction 485
Wei-Ning Wang, Xiang He, Zhuoran Gan
(517c) Enhanced Electrocatalytic CO2 Reduction on Thiol-Functionalized Gold 486
Yuxin Fang, Xun Cheng, Ye Xu, John C. Flake
(517d) Investigation of the Main Factor That Drives Selectivity on Rough Cu-Based Catalysts for CO2 Electroreduction 487
Alexandros Karaiskakis, Tiffany Harden, Elizabeth J. Biddinger

(517e) S Doped TiO2 synthesized By Novel One Step Liquid Flame Spray Pyrolysis (LFSP) for Photocatalytic Oxidation of CO in Visible Region 488
Siva Nagi Reddy Inturi, Thirupathi Boningari, Makram Suidan, Panagiotis Smirniotis
(517f) Optimizing the Physicochemical Properties of ZSM-11 for Methanol-to-Hydrocarbon Reactions 489
Yufeng Shen, Thuy T. Le, Jeffrey D. Rimer
(517g) Precise Deposition of Pt Promoter Onto Silica Supported Cobalt for Fischer-Tropsch Synthesis 490
Fahad Almalki, John Monnier, John R. Regalbuto
(517h) Using Inelastic Neutron Scattering to Probe the Active Site for Cobalt-Copper Catalyst for the Conversion of Syngas to Higher Oxygenates 491
Zi Wang, Phillip T. Sprunger, Richard L. Kurtz, Luke L. Daemen, Yongqiang Cheng, James J. Spivey
(520a) Selective C-O Bond Cleavage of Diphenyl Ether over the Supported Palladium Catalyst 492
Donghai Mei, Meng Wang, Johannes A. Lercher
(520b) Metal Core/Surface Oxide Shell Catalysts for C-O Bond Activation 493
Alexander V. Mironenko, Konstantinos Goulas, Raymond J. Gorte, Christopher B. Murray, Dionisios Vlachos
(520c) A Periodic DFT Study of the Influence of Lewis Acid Site Speciation on Ethanol Dehydration in Zeolites 494
Brandon Bukowski, Jason Bates, Rajamani Gounder, Jeffrey P. Greeley
(520d) Mechanistic Study of Regio-Selective Ring-Opening Reactions of Mono-Substituted Epoxides Using Lewis Acid Catalysts 495
Ying Yu, Linda J. Broadbelt
(520e) Understanding the p-Xylene Formation Mechanism from Dimethylfuran, and Ethanol 496
Pavlo Kostetskyy, Giannis Mpourmpakis, Michail Stamatakis, Shik Chi Edman Tsang And Ivo Teixeira
(520f) Role of Enol-Keto Tautomerization in m-Cresol Selective Hydrogenation on Ni(111) and Nife(111) Surface 497
Xiaoyang Liu, Wei An
(520g) Developing Multi-Scale Models of Bimetallic Catalysts for the Hydrodeoxygenation of Bio-Oil Compounds 498
Breanna Wong, Alyssa Hensley, Pierre Gaspard, Yong Wang, Jean-Sabin McEwen
(520h) Combined Density Functional Theory and Molecular Dynamics Study of Aqueous Reforming of Glycerol on Pt (111) Via Competitive C-H and O-H Dehydrogenation 500
Tianjun Xie, Cameron Bodenschatz, Rachel Getman
(520i) Molecular Dynamics Simulations of Aldol Condensation Catalyzed By Alkylamine-Functionalized Crystalline Silica Surfaces 501
Ki Chul Kim, Eric G. Moschetta, Christopher W. Jones, Seung Soon Jang
(539a) Droplet Microfluidics with a Nanoemulsion Continuous Phase: Application in Reaction System Design for Material Synthesis 502
Tonghan Gu, Saif A. Khan, T. Alan Hatton
(539b) Activating C1 Chemistry Utilizing Microreactor Technology 503
Alexandre Yokochi, Goran Jovanovic, Annette Von Jouanne, Nick Auyeung, Scott Harpool, Justin Pommerenck, Yousef Alanazi, Yu Miao, Ian Reddick, Adam Shareghi
(539c) Gas- Liquid Flow Hydrodynamics and Mass Transfer in Micro-Packed Beds 504
Jisong Zhang, Andrew Teixeira, Klavs F. Jensen
(539d) Implementation of in Situ SAXS/Waxs Analysis in Silicon-Pyrex Microreactor 505
Samuel Marre, Cyril Aymonier, Yves Garrabos, Carole Lecoutre, Thomas Beuvier, Brice Calvignac, Alain Gibaud
(539e) Automated Oscillatory Photochemical Reactor for High Throughput Studies of Visible-Light Photoredox Catalysis 506
Milad Abolhasani, Connor Coley, Hongkun Lin, Klavs F. Jensen
(539f) Kinetic Study of Reactions of Aniline and Benzoyl Chloride in a Microstructured Chemical System 507
Peijian Wang, Kai Wang, Jisong Zhang, Guangsheng Luo
(539g) Brock Copolymer Synthesis in a Microreactor By Precise Analysis and Design of Polymerization Kinetics 508
Shusaku Asano, Ryutaro Nakayama, Ryuji Utsunomiya, Taisuke Maki, Toshiharu Kuboyama, Kazuhiro Mae
(547a) Alloy-Core@Shell Nanoparticles for Catalysis 509
Liang Zhang, Graeme Henkelman
(547b) Thermal Stability of Shape-Controlled Pd-Pt Core-Shell Nanoparticles 510
Luke T. Roling, Ahmed Elnabawy, Madeline Vara, Xue Wang, Younan Xia, Manos Mavrikakis
(547c) Accelerating Nanoparticle Synthetic Development Using in Situ Small Angle X-Ray Scattering 511
Liheng Wu, Joshua Willis, Matteo Cargnello, Christopher J. Tassone
(547d) Selective Oxidation of Methane to Methanol Using Pd/Au Bimetallic Nanoparticles 512
Cody Wrasman, Joshua Willis, Jong Suk Yoo, Jens Nørskov, Matteo Cargnello

(547e) Using Bound Organic Ligands to Direct Particle Size and Understand Catalytic Activity of Gold Nanoparticles 513
Michael M. Nigra, Nidhi Kapil, Marc-Olivier Coppens
(547h) Enhancing the Ni/YSZ Catalytic Performance in the Activation of Methane: The Influence of the Electric Field and Interface Carbon 514
Fanglin Che, Su Ha, Jean-Sabin McEwen
(547g) Metal Nanocatalysts, Their Synthesis and Size Dependent Covalent Bond Catalysis. Oxide Metal Interfaces As Active Sites for Acid-Base Catalysis 516
Gabor A. Somorjai
(555a) Automated Reaction Mechanism Generation Including Nitrogen As a Heteroatom 517
Alon Grinberg Dana, Beat Buesser, Shamel S. Merchant, William H. Green
(555b) Modeling the Combustion of Otto Fuel 518
C. Franklin Goldsmith, Eric Tierney, Richard H. West
(555c) Molecular Weight Growth Kinetics of Olefin Pyrolysis Under Low to Intermediate Temperatures 519
Kun Wang, Anthony M. Dean, Stephanie Villano
(555d) Detailed Kinetic Modeling of Gas Evolution and Energy Recovery during Chemical Quenching of Acetylene Production in Thermal Plasma 520
Yan Cheng, Tianyang Li, Yi Cheng
(555e) Predictive Modeling of Thermal Decomposition of Hydrocarbon Mixtures 521
William H. Green, Kehang Han
(555g) Nondimensional Specific Energy Index for Methane Pyrolysis in Thermal Plasma 522
Tianyang Li, Christophe Rehmet, Yan Cheng, Yi Cheng
(573a) Understanding Trends in C-H Bond Activation in Heterogeneous Catalysis 523
Allegra A. Latimer, Ambarish R. Kulkarni, Hassan Aljama, Felix Studt, Frank Abild-Pedersen, Jens K. Norskov
(573b) Novel Fe-Promoted MgAl2O4 Support Material for Control of Carbon Deposition during Reforming Reactions 524
Stavros Alexandros Theofanidis, Johannes De Jong, Vladimir Galvita, Aditya N. V. R. Dharanipragada, Hilde Poelman, Guy B. Marin
(573c) Dry Reforming of Methane over Nickel-Based Catalysts for Syngas Production 528
Bruna Rego De Vasconcelos, Doan Pham Minh, Patrick Sharrock, Ange Nzihou
(573i) Probing the Effect of Mn Dispersion on the Catalytic Performance of Mn-Na-W/SiO2 in the Oxidative Coupling of Methane 531
Christine Warwar Damouny, Naseem S. Hayek, Christine Khoury, Oz M. Gazit
(573e) Methane Steam Reforming: Using External Electric Fields to Enhance the Catalytic Performance of Ni-Based Catalysts 533
Fanglin Che, Jake T Gray, Su Ha, Jean-Sabin McEwen
(573f) Exploring Ruthenium Metal-Support Interactions for the Low-Temperature Partial Oxidation of Methane 535
Emmett Goodman, Liheng Wu, Matteo Cargnello
(573g) Methane to Methanol Conversion in Small Pore Zeolites 536
Florian Göltl, Carine Michel, Bahar Ipek, Raul F. Lobo, Phillippe Sautet, Ive Hermans
(573h) Co-Aromatization of Methane with Olefins 537
Peng He, Hua Song
(574a) Production of Terephthalate Esters Via Diels-Alder Cycloaddition-Dehydration Reactions of 1,2-Furandicarboxylate Esters and Ethylene 538
Marat Orazov, Mark E. Davis
(574c) Quantifying Lewis Acid Sites in Zeolites That Catalyze Glucose Isomerization 539
James W. Harris, Michael J. Cordon, Juan C. Vega-Vila, Fabio H. Ribeiro, Rajamani Gounder
(574d) Designing Cooperative Interactions to Tune Catalytic Activity and Selectivity for Biomass Conversion 540
Nicholas Brunelli, Nitish Deshpande, Aamena Parulkar, Frédéric Perras, Marek Pruski
(574e) Lewis Acid Zeolite Catalysts for Tandem Diels-Alder Cycloaddition and Dehydration of Dimethylfuran and Ethylene to Renewable p-Xylene 541
Hong Je Cho, Chun-Chih Chang, Jingye Yu, Raymond J. Gorte, Paul J. Dauenhauer, Wei Fan
(574f) Comparative Studies of Grafted Fe(III) and Ti(IV) Active Sites on Silaceous Supports: Crystalline Delaminated Zeolite Vs. Amorphous Silica 542
Nicolás A. Grosso Giordano, Alexander Yeh, Andrew Solovyov, Alexander Okrut, Stacey I. Zones, Alexander Katz
(574g) A Liquid Phase Spectroscopic Study of the Effect of Liquid Water on Acid Sites of NaY 543
Bingjun Xu, Nicholas Gould
(578a) Computational Study of Single-Site Molybdenum Catalysts on Amorphous Silica 544
Karl Johnson, Christopher S. Ewing, Abhishek Bagusetty, Evan Patriarca, Daniel S. Lambrecht, Götz Veser

(578b) First-Principles Kinetic Monte Carlo Simulation of the Thermal Reduction of PdO(101) 545
Minkyu Kim, Aravind Asthagiri
(578c) Cu Site Speciation in SSZ-13 Zeolites Resolved through UV-Visible Spectroscopy, Selective Synthesis, Time-Dependent Density Functional Theory, and Ab Initio Molecular Dynamics 546
Christopher Paolucci, Jonatan Albarracin, Hui Li, Florian Göltl, John R. Di Iorio, Ishant Khurana, W. Nicholas Delgass, Fabio H. Ribeiro, Rajamani Gounder, William F. Schneider
(578d) Microkinetic Modeling of the Oxygen Evolution Reaction on Oxide Surfaces 547
Charlotte Kirk, Colin Dickens, Karen Chan, Jens K. Norskov
(578e) Water-Gas Shift Reaction on Iron Oxide Catalysts – a DFT Study 548
Ozgen Yalcin, Isik Onal, Israel E. Wachs
(578f) Computational Investigation of Lewis and Brønsted Acidity in Metal-Doped Mesoporous Silicates 549
Amy Jystad, Alessandro Biancardi, Marco Caricato
(578g) Simulated Temperature Programmed Desorption of Acetaldehyde on CeO2(111): Evidence for the Role of Oxygen Vacancy and Hydrogen Transfer 550
Chuanlin Zhao, Ye Xu
(578h) Electrochemical Synthesis of Hydrogen Peroxide from Two-Electron Water Oxidation 551
Samira Siahrostami, Guo-Ling Li, Xinjian Shi, Xiaolin Zheng, Jens Nørskov
(617cv) Theoretical Screening of Mxenes for Catalysis Using Density Functional Theory 552
Kurt Fredrickson, Aleksandra Vojvodic
(600a) Nanoengineering and Application of Protected but Accessible Metal Cluster Catalysts 553
Alexander Katz, Alexis T. Bell, Audrey Harker, Anh T. To, Ezra Clark, Bruce C. Gates, Andrew Palermo, Louise Debefve
(600b) Electronic Band Engineering in TiO2 Particulate Layers for Photocatalysis 554
Qilong Huang, Edmund G. Seebauer
(600c) How Charge Screening Affects Zeolite-Catalyzed C-C Coupling Reactions 555
Bin Wang
(600d) Analyzing Local Composition and Structural Changes Occurring in a Nanocatalyst during Reaction 556
Jose L. Gomez-Ballesteros, Pin Ann Lin, Juan C. Burgos, Renu Sharma, Perla B. Balbuena
(600e) Modelling of Diffusion-Reaction Interaction inside the Co-Based Catalyst Particles for the Fischer-Tropsch Synthesis 557
Milos Mandic, Branislav Todic, Ljiljana Zivanic, Nikola Nikacevic, Dragomir B. Bukur
(600f) Carbon Nanotube-Supported Catalysts for Improved Fischer-Tropsch Synthesis 561
Xu Li, Haider Almkhelfe, Jenae Tate, Jingyi Xie, Keith L. Hohn, Placidus B. Amama
(600g) Influence of Dry-Coating of Nano Additives on Particulate Solid Flow and Its Reactions: Cement System a Case Study 562
Sanat Chandra Maiti, Chinmay Ghoroi
(601a) Chemical Layer Deposition of Porous Alumina Overcoats Increases Activity and Stability in Liquid Phase Catalytic Conversion of Biomass-Derived Chemicals 563
Florent Héroguel, Benjamin Le Monnier, Jeremy S. Luterbacher
(601b) Controlled Growth of Noble Metal Nanoparticles on Graphene Nanoplatelets Via Scalable Atmospheric Pressure Atomic Layer Deposition 564
J. R. Van Ommen, Fabio Grillo, Hao Bui
(601c) Computational Studies of the Dry Reforming Reaction on Al2O3-Coated Transition Metal Catalysts 565
Saul Perez Beltran, Perla B. Balbuena
(601d) Experimentally Probing Ligand-Strain Effect Via a Novel Catalyst Platform 566
Shuai Tan, Shibely Saha, Lucun Wang, Gregory S. Yablonsky, John T. Gleaves, Rebecca Fushimi, Dongmei (Katie) Li
(601e) Cobalt/Alumina Interactions in ALD Synthesized Catalysts for Fischer-Tropsch Synthesis 567
Jacob M. Clary, Staci A. Van Norman, John L. Falconer, Charles B. Musgrave, Alan W. Weimer
(601f) Catalytic Performance of Supported Nano-Iron Catalysts Synthesized By a Gas-Expanded Liquid Deposition Technique in Fischer-Tropsch Synthesis 568
Rui Xu, Pranav S. Vengsarkar, David Roe, Christopher Roberts
(601g) Photocatalysis for Z-Scheme Solar Water Splitting Using New Reactor Designs 569
Shane Ardo
(611a) Spatially Resolved 3D-CFD Simulation of a Packed Bed Membrane Reactor for Steam Reforming 570
Nico Jurtz, Thomas Eppinger, Matthias Kraume

(611b) Iron-Manganese Catalysts for the CO Hydrogenation to Alkenes: Effect of the Catalyst Synthesis on the Selectivity and Activity 571
Hafsa Karroum
(611c) Effect of Potassium Promoter in Cocu-Based Catalysts for CO and CO2 Hydrogenation 573
Jenny Voss, Lei Tang, Yizhi Xiang, Norbert Kruse
(611d) Synthesis and Development of Active Nanomaterials for Tri-Reforming 575
Erdem Sasmaz, Jochen Lauterbach
(611e) Gasoline-Selective Fischer-Tropsch Synthesis with Hierarchical ZSM-5 Coated Monolith Catalysts 576
David P. Gamliel, Chunxiang Zhu, Julia A. Valla, George M. Bollas
(611f) Comparison of Cobalt Fischer-Tropsch Catalysts on Robust Silica-Alumina Supports 578
Mahmood Rahmati, Morris D. Argyle, William C Hecker, Mark J. Lindsay, Mohammad-Saeed Safdari
(611g) Surface Stabilization and Subsequent Rupturing of Co in Co-Cu Catalysts By Formation of Adsorbed Di-, Tri-, and Tetracarbonyls 579
Greg Collinge, Norbert Kruse, Jean-Sabin McEwen
(611h) Effect of Carbon Support on Supercritical Fluid Mediated Fischer-Tropsch Synthesis with an Iron-Based Nanoscale Catalys 582
David Roe, Rui Xu, Christopher B. Roberts
(617hn) Catalytic Conversion of Carbon Dioxide to Methanol over Homogeneous and Heterogeneous Hybrid Catalyst 583
Kyung-Ho Song, Eun-Kyung Choi, Soon Kwan Jeong, Ki Tae Park, Min Hye Youn, Kwan-Young Lee, Hak Joo Kim
(617a) "Ni-MFI As a Potential Zeolitic Catalyst for Olefins Hydrogenation at Low Pressures" 584
Khaled O. Sebakhy, Gerardo Vitale, Pedro Pereira-Almao
(617cm) Predicting Surface Intermixing in Alloy Nanocatalysts from First Principles Calculations 585
Luke T. Roling, Manos Mavrikakis
(617bn) On the Influence of Mass Transport in Sulfonated Carbon Catalysts Derived from Various Carbon Sources on Liquid-Phase Esterification Reactions 586
Yukei Suzuki, Isao Ogino, Shin R. Mukai
(617ff) Catalytic Hydrogenation of 2-Ethylanthraquinone Using an in Situ Synthesized Pd Catalyst 587
Hongbao Yao, Chun Shen, Yujun Wang, Guangsheng Luo
(617fp) Copper Aluminate: A Potential Catalyst to Make Dimethyldichlorosilane in a 2-Step Process 588
Krishna Janmanchi, Aaron Coppernoll
(617bv) Centrifugal Partition Chromatograph: A Novel Continuous Multiphase Reaction System 590
Juliane Merz, Gerhard Schembecker, Jonas Krause
(617bs) Synthesizing Bromobutyl Rubber By a Microreactor System 591
Pei Xie, Kai Wang, Luo Guangsheng, Peijian Wang
(617be) Integration of CFD and Polymerization for an Industrial Scale Cis-Polybutadiene Reactor 592
Chao-Zhong Xu, Jia-Jun Wang, Xue-Ping Gu, Lian-Fang Feng
(617fj) Base-Free Knoevenagel Condensation Catalyzed By Different Heterogeneous Copper Catalysts 593
Elia Schneider, Martin Zeltner, Robert N. Grass, Wendelin J. Stark
(617ge) Reaction Route Analysis of Mo Based Shift Catalyst 597
Takashi Sasaki, Tomoko Suzuki, Masaki Takaoka
(617bz) Decrosslinking and Decomposition of Silane-Crosslinked Polyethylene in Supercritical Methanol and Water 598
Giyoung Hong, Kuan Hwan Park, Jebin Ryu, Hee Suk Woo, Won-Su Son, Youn-Woo Lee
(617fq) Regeneration of Fatty Acid-Adsorbed y-Alumina Using Supercritical Methanol 599
Hee Suk Woo, Jebin Ryu, Won-Su Son, Giyoung Hong, Youn-Woo Lee
(617o) Equilibrium Shifting of Esterification with Water-Permselective Zeolite Membrane 600
Shuhei Kikuchi, Takafumi Sato, Naotsugu Itoh
(617ft) Pd/CuO Bifunctional Catalytic Membrane for Direct Synthesis of Phenol from Benzene 601
Kento Kohara, Takagiwa Naoki, Takafumi Sato, Naotsugu Itoh
(617hg) Kinetics and Spectroscopic Studies of Methane Partial Oxidation on Ni and Rh Catalysts 602
Cun Wen, Jason Hattrick-Simpers, Jochen Lauterbach
(617ee) Developing Catalysts for the Partial Electrochemical Oxidation of Methane 603
Michael Boyd
(617dt) The Characteristics of Diesel Steam Reforming at Pressurized Environment for the Operation of Solid Oxide Fuel Cell 604
Dongyeon Kim, Jiwoo Oh, Minseok Bae, Joongmyeon Bae, Sai P. Katikaneni
(617hv) Combined Process for Hydrogen Production from Coke Oven Gas (COG) 605
Hai Woong Park, Gookhee Kim, Chang-Houn Rhee

(617ap) Glycerol Hydrogenolysis with Reduced Co2AlO4 Nanoparticles Produced By Flame Spray Pyrolysis 606
Jorma Jokiniemi, Anna Lähde, Tommi Karhunen, Luis Balam Modesto-Lopez, Ricardo José Chimenton, Jordi Llorca, Dra Mayra Garcia Alvarez, Francesc Medina
(617ai) Catalytic Selective Oxidation in Biomass Conversion 607
Yanhui Yang
(617l) Highly Dispersed Nickel within Mesochannels of SBA-15 for CO Methanation with Enhanced Activity and Excellent Thermostability 608
Miao Tao, Zhong Xin
(617cz) Fischer Tropsch Synthesis: A Study of the Selectivity to Light Hydrocarbons 609
Adolph Muleja, Yali Yao, Diane Hildebrandt, David Glasser
(617ck) Arrays of Coupled Cells Performing Chemical Computing with and without Parallelism Using Discrete Turing Patterns 610
František Muzika, Lenka Schreiberova, Igor Schreiber
(617ak) Double Bond Isomerization of 1-Butene Using Supported AlBr3 Catalysts 611
Swarom Kanitkar, Nitin Kumar, James J. Spivey
(617at) Kinetic Model on Conversion of D Fructose to 5 Hydroxymethylfurfural. 612
Tatsuya Ishikawa, Eiji Furuya, Takashi Kawakita
(617m) Oxidative Coupling of Methane over Zeolite Supported Na-Mn-W 613
Naseem S. Hayek
(617cj) The Effect of the Reaction Environment on the Surface Acidity of t-ZrO2 (101) Investigated By DFT 616
Juan Manuel Arce-Ramos, Lars C. Grabow, María-Guadalupe Cárdenas-Galindo, Brent E. Handy
(617dl) Promoter Effects of Residual K and Mn in Cobalt Supported Carbon Spheres Catalysts for Fischer Tropsch Synthesis 617
Mahluli Moyo
(617em) Enhanced Electro-Kinetics of CO2 Generation for Ethanol Oxidation Using Catalysts with a Partially Oxidized Pt and Rh Core and a SnO2 Shell 618
Guangxing Yang, Xiaowei Teng
(617hj) Development of Novel Catalysts and Reactor Configurations for Methane Dry Reforming 619
Tomás Ramirez-Reina, Harvey Arellano-Garcia
(617dm) Modification of Molybdenum Phosphide Catalyst Surface with Fischer-Tropsch and Alkali Metals for the Production of Alcohols from Syngas 620
Melis S. Duyar, Eduardo Valle, Jakob Kibsgaard, Thomas F. Jaramillo
(617gg) Formation of Rhodium Nanoclusters Fabricated from Rh(CO)2(acac) on Graphene/Cu(111) 621
Homa Khosravian
(617gy) Methanol Synthesis from CO2 in Slurry Bubble Column Reactors: A Modelling/ Experimental Approach 622
Klaus Hellgardt, Dorothy Bernasko, Fessehaye Zemichael
(617dw) Contributions of Heterogeneous Catalysis to Renewable Energy: an Electrochemical Perspective 623
Carolina Romero, Alvaro Gómez, Valentina Aristizábal, Carlos A. Cardona
(617gh) Kinetic Monte Carlo Simulation of Propylene Epoxidation with Au/TiO2/SiO2 Catalysts 624
Jingjing Ji, Zheng Lu, Yu Lei, C. Heath Turner
(617fm) Comparison of Metal-Support Interaction on Pd/a-Fe2O3 Nanorod and Pd/a-Fe2-XMnxO3 nanorod Catalysts 625
Chao Wang, Emily Freeman, Jochen Lauterbach
(617er) Titanium-Based, Hierarchical, Flow-through Catalyst for Water Electrolysis Fabricated Via a Combination of Additive Manufacturing and Anodization 626
Patricia Taboada-Serrano, Xiang Li, Costas Tsouris
(617gi) Co-Oxidation of CO and Hydrocarbons on Pd/Ceria-Zirconia/Al2O3 Three-Way Catalysts 627
Wendy Lang, Michael Harold, Yisun Cheng, Carolyn Hubbard, Manish Sharma, Paul Laing
(617ek) Understanding Gas Evolution Dynamics on Planar Photoelectrodes 630
Anna Dorfi, Daniel V. Esposito, Alan West
(617cn) Multiple H2O and NH3 Adsorbed on Cu-SSZ-13: XANES and XES Study from First Principles 631
Renqin Zhang, Hui Li, Janos Szanyi, Feng Gao, Christopher Paolucci, Trunojoyo Anggara, Hui Li, William Schneider, Jean-Sabin McEwen
(79d) Catalyst Morphology Engineering: Towards a Better Understanding of the Effects of Surface Structure and Mass Transport in Copper Electrodes for the Electrochemical CO2 Reduction Reaction 633
Stephanie Nitopi, Pongkarn Chakthranont, Jakob Kibsgaard, Christopher Hahn, Thomas F. Jaramillo

(617x) A Parametric Study of Biodiesel Production Under Ultrasounds 634
Kiran Shinde, Serge Kaliaguine
(617bi) Data Auditing: Analysis Techniques for the Compatibility of Experimental Data for Molecular-Level Kinetic Models 635
Scott R. Horton, Triveni Billa, Pratyush Agarwal, Juan Lucio-Vega, Mayuresh Sahasrabudhe, Chandra Saravanan, Michael T. Klein
(617ep) Mn@Ptn (n = 1, 2, and 3) Core-Shell Nanoparticles: Synthesis, Characterization and Electrocatalysis 636
Md. Abdul Matin, Anand Kumar
(617hu) Modeling and Simulation of the Transient Reaction Behavior in Reactive Sorption Enhanced Reforming of Methane with Nano CaO-Based Sorbents for Hydrogen Production 637
Haoliang Ping, Su-Fang Wu
(617gu) Transition Metal Modified Zirconium Phosphates for the Oxygen Evolution Reaction 638
Joel Sanchez, Ieva Narkeviciute, Mario Ramos, Jorge L. Colón, Thomas F. Jaramillo
(617cl) Methanol Oxidation on Pristine and Doped MoO3 (010): A DFT and Microkinetic Analysis 639
Tej S. Choksi, Jeffrey P. Greeley
(617bm) Simultaneous Reaction Identification and Parameter Estimation 640
Zachary Wilson, Nick Sahinidis
(617dx) Design and Synthesis of Porous Carbon-Based Catalysts for the Electrochemical Reduction of O2 Via Either 2 e- or 4 e- Route 641
Shucheng Chen, Zihua Chen, Samira Siahrostami, Taeho Roy Kim, Dennis Nordlund, Dimosthenis Sokaras, John To, Robert Sinclair, Felix Studt, Jens K. Norskov, Thomas F. Jaramillo, Zhenan Bao
(617bo) Numerical Investigation on Dispersion Phenomena in Packed Bed Reactors with Low Tube-to-Particle Diameter Ratio 642
Nico Jurtz, Gregor D, Wehinger, Thomas Eppinger, Matthias Kraume
(617da) Modeling Syngas Production for Miniaturized Gas-to-Liquids Application: Extending Operation Window By Process Improvement from Different Scales 643
Chenxi Cao, Nian Zhang, Dan Dang, Yi Cheng
(617s) Effect of ZSM-11 Crystal Size on Methanol-to-Hydrocarbon Reactions 644
Yufeng Shen, Thuy T. Le, Jeffrey D. Rimer
(617gj) Effect of Dehydration on Dmc Synthesis Via Oxidative Carbonylation over Ceria Based Catalysts 645
Jong-Ho Moon, Young Cheol Park, Hyunuk Kim
(617gk) Neural Network Predictions of Oxygen Interactions on a Dynamic Pd Surface 646
Jacob R. Boes, John R. Kitchin
(617af) Novel Catalytic Etherification Reaction of Glycerol to Short–Chain Polyglycerol 647
Maral Vahdatzaman, Sunggyu Lee
(617ef) Selective Electrocatalytic Hydrogenation of Furfural on Metal Electrodes 648
Xiaotong Chadderdon, David Chadderdon, Wenzhen Li
(617au) The Selective Production of Octene By the Metathesis of 1-Pentene 649
Heondo Jeong, Min Hye Youn, Kee Young Koo, Dong Hyun Chun
(617ej) Connected Nanoparticle Catalysts Having a Porous, Hollow Capsule Structure As Carbon-Free Electrocatalysts for Polymer Electrolyte Fuel Cells 650
Takeo Yamaguchi, Hidenori Kuroki, Takanori Tamaki
(617g) Synthesis of a Highly Active Hierarchically Ordered Micro-Meso Porous Titanosilicate Via Statistically Optimized Synthesis 651
Ayomi S. Perera, Panagiotis Trogadas, Michael M. Nigra, Kaipei Qiu, Marc-Olivier Coppens
(79b) Effects of Cations on Activity and Selectivity in Carbon Dioxide Electroreduction 652
John Lin, Christopher Hahn, Thomas F. Jaramillo
(617gc) Study of Reactivity of CO+NO Reaction on Supported Pd Catalyst 653
Yimeng Lyu, Uzoma Nwabara, Xiaoyin Chen, Johannes W. Schwank
(617bp) Study of Oxidation Reactions in Preliminary Zone of Stabilization Process of Carbon Fiber Production Line 654
Khashayar Badii, Minoo Naebe, Gelayol Golkarnarenji, Hamid Khayyam
(617cd) Understanding of Reactivity of Asphaltene in H-Donor Solvents 655
Han-Byul Park, Yong-Kul Lee
(617d) Ordered Mesoporous Metal Oxide Supports Grown Around Nanocatalysts for High Temperature Stability 656
Victor V. Miller, Joshua Willis, Matteo Cargnello
(617eo) Ionic Compounds for Carbon Dioxide Electroreduction 657
Alan Landers, Thomas R. Hellstern, Jakob Kibsgaard, Christopher Hahn, Thomas F. Jaramillo

(617aw) Catalytic Upgrading of Olefin Residues with Odd Number of Carbon Atoms 658
Min Hye Youn, Heondo Jeong, Kee Young Koo, Dong Hyun Chun
(617av) Co-Processing of Heavy Residue Oil and Biomass Using Supercritical Organic Solvents 659
Doo-Wook Kim, Anton Koriakin, Ma Fanzhong, Chang-Ha Lee
(617ax) Development of Pt/Al2O3 Catalytic System for the Production of Furanic Compounds from Biomass Derived Itaconic Acid 660
Jae-Hwan Lee, Kwan-Young Lee
(617al) Synthesis of a Bi-Functional Catalyst and Application in One-Pot Conversion of Fructose into Oxygenated C9 and C15 Hydrocarbons 661
Neha Mittal, Grace M. Nisola, Wook-Jin Chung, Chosel P. Lawagon
(617gm) Nano-Sized Tungsten Carbide Catalyst for Upgrading Extra-Heavy Oil 662
Chan Hun Kim, Young Gul Hur, Gwangsik Jeong, Kwan-Young Lee
(617es) Three-Dimensional Fluid Dynamic Analysis of a Newly Designed 100 cm2 Single Cell with Internal Flow Channels for Molten Carbonate Fuel Cells 663
Chang-Whan Lee, Seong-Cheol Chang, Hyung Chul Ham, Sung-Pil Yoon, Jonghee Han
(617db) Fischer-Tropsch Synthesis: Simulation of the Diffusion of H2, CO, and Hydrocarbons in Liquid Phase Water and Hydrocarbons 666
Xiaojun Lu, Diane Hildebrandt, David Glasser
(617gx) Resolved Particle Fixed-Bed CFD with Microkinetics for Methanol Oxidation on Silver 667
Behnam Partopour, Anthony G. Dixon
(617hc) Experimental and Theoretical Analysis of Carbon Dioxide Absorption in Aqueous Sodium Hydroxide in the Millichannels 668
Durgadevi A, S. Pushpavanam
(617bj) Performance Comparison of the 60wt% Fe2O3/Al2O3 Oxygen Carrier Under Various Reducing Atmosphere for Chemical Looping Combustion 669
Ching-Ying Huang, Yau-Pin Chyou
(617t) Dimerization of Bicyclo[2.2.1]Hepta-2,5-Diene over Zeolite Catalysts 670
Jong Ki Jeon, Kwan-Hyoung Jeong, Jong Jin Kim, Choul-Ho Lee, Jeongsik Han, Byung-Hun Jeong
(617cq) Benchmarks for Adsorption on Transition Metal Oxide Surfaces: A Comparison of DFT to Experimental Data 671
Han Chen, David F. Cox
(617gf) Gold-Supported Cerium-Doped NiOx Catalysts for Water Oxidation 672
Desmond Ng, Max García-Melchor, Michal Bajdich, Pongkarn Chakthranont, Charlotte Kirk, Aleksandra Vojvodic, Thomas F. Jaramillo
(617dc) Scale up Study of Fischer Tropsch Synthesis Reactor with Reduced CO2 emissions Using Cu-K Promoted Silica Supported Fe Catalyst 673
Satyen Kumar Das, Debasis Bhattacharyya, Eswar Prasad Dalai, Jagdev Kumar Dixit, S. S. V. Ramakumar, Biswapriya Das
(617ho) A Cost-Free Catalyst for COx-Free Hydrogen Production: Utilization of Red Mud for Ammonia Decomposition 674
Samira Fatma Kurtoglu, Alper Uzun
(617u) Crystallization of One-Dimensional Zeolites: Elucidating Nonclassical Mechanisms of Growth and the Role of Structure Direction 675
Rui Li, Jeffrey D. Rimer
(617bk) Enhancing the Value of Detailed Kinetic Models through the Development of Interrogative Software Applications 676
Triveni Billa, Mayuresh Sahasrabudhe, Chandra Saravanan, Scott R. Horton, Zhen Hou, Pratyush Agarwal, Juan Lucio-Vega, Michael T. Klein
(617fz) Tailor-Designed Oxidation Catalysts for Low Temperature Combustion Engines 677
William Epling, Melanie Hazlett, Sungbong Kang, Christos Kalamaras
(617et) Reducing Greenhouse Gasses Using Low Temperature AEM Electrochemical Devices 678
Travis J. Omasta, William A. Rigdon, Neil S. Spinner, Connor A. Lewis, William E. Mustain
(617b) Hydrogen Production By Steam Reforming of Bio-Oil Derived Acetic Acid over Ni Based Zr-SBA-15 Type Mesoporous Catalysts 679
Nurbanu Cakiryilmaz, Nuray Oktar, Huseyin Arbag, Gulsen Dogu, Timur Dogu
(617dj) Effect of Fast Cycling on the Formation of HC Intermediates and NOx Conversion Using H2, Propene or Propane As Reductants 680
Allen Wei-Lun Ting, Mengmeng Li, Michael P Harold, Vemuri Balakotaiah
(617hq) Calcium Oxide As in-Situ Catalyst for Steam Gasification of Bio-Char 681
Jun Dong, Elsa Weiss-Hortala, Ange Nzihou, Yong Chi, Mingjiang Ni, Qunxing Huang, Yuanjun Tang, Marion Ducousso

(617dd) Use of Stability Diagrams to Predict Catalyst Speciation during Fischer Tropsch Reduction and Reaction 682
Joshua Gorimbo, Xiaojun Lu, Xinying Liu, Yali Yao, David Glasser, Diane Hildebrandt
(617en) Modeling the Sensitivity of Design Parameters for Photo-Electrochemical Cells Containing a Semiconductor Absorber and Metal Photocatalysts 683
Joseph Quinn, Paul Hernley, Steven Chavez, Suljo Linic
(617ay) Enhanced Carbon Sequestration with Cascade Enzymtic Reactions 684
Shunxiang Xia, Ping Wang
(617az) Global Optimization of Initial Conditions of RAFT Polymerization Process 685
Arturo Ortiz-Arroyo, Angel Castro, Fernando Pérez
(617dn) The Role of Co-Catalysts in the Metathesis of Butenes to Propylene 686
Faisal H. Alshafei, Mohammed R. Alalouni, Munir D. Khokhar, Shaikh A. Ali, Brian S. Hanna, Maxim P. Bukhovko, Sohel K. Shaikh
(617go) Elucidation of the Mechanisms for Direct Synthesis on Pd, Aupd, and Agpt Clusters 687
Neil M. Wilson, Yung-Tin Pan, Hong Yang, David W. Flaherty
(617bq) Development of a Computationally Efficient Model of a Packed-Bed Membrane Reactor for the Oxidative Coupling of Methane 688
Alberto Penteado, Mi Jin Kim, Hamid Godini, Erik Esche, Jens-Uwe Repke, Eva Speelmanns
(617ce) Experimental and Kinetic Modeling of Lean-Rich Switching Study over a Modified Twc System 689
Mengmeng Li, Sam Malamis, Michael P Harold, William S. Epling
(617gp) Aqueous-Phase Hydrodechlorination of Trichloroethene over Pd-Supported on Swellable Organically-Modified Silica (SOMS) 692
Gokhan Celik, Hyuntae Sohn, Seval Gunduz, Saurabh Ailawar, Paul Edmiston, Umit S. Ozkan
(617eu) Understanding the Mechanisms and Key Parameters That Influence the Performance of Composite Semiconductor/Electrocatalyst Photoelectrodes 693
Steven Chavez, Paul Hernley, Joseph Quinn, Suljo Linic
(617dv) Photochemical Reactions on Plasmonic Metal Nanoparticles: Underlying Mechanism of Charge Extraction from Nanoparticles to Adsorbates 694
Calvin Boerigter, Umar Aslam, Suljo Linic
(617v) Tailoring the Physicochemical Properties of ZSM-5 with Zeolite Growth Modifiers 695
Wei Qin, Jeffrey D. Rimer
(617de) Hotspot Mitigation in an Annular Microchannel Reactor (AMR) for Hydrogen and/or Syn Gas Production 696
Holly Butcher, Benjamin Wilhite, Peter Bossard, Andy Kaldor
(617gq) Catalytic Activity and Sulfur Dependence of Co/Mo Nanoparticle Catalysts in WGS Reaction 697
Seonguk Yun, Vadim Guliants
(617hk) Optimized CeZrO2-Modification of the Metal-Foam-Coated Pd-Rh/Al2O3 Catalyst for CH4 Bi-Reforming Reaction 698
Kiseok Kim, Partho S Roy, Jinwoo Song, Ji Man Kim
(617j) Liquid-Phase Stability of ZSM-5: Role of De-Alumination and De-Silication 699
Alex Maag
(617cf) Process Intensification of Ethanol Dehydration Under Liquid Phase Conditions 700
Alex Maag
(617df) Development of Fischer-Tropsch Synthesis in Microreactor 701
Yousef Alanazi, Andrew Traverso, Justin Pommerenck, Liney Arnadottir, Alex Yokochi, Goran Jovanovic
(617gr) Au@MxOy Core-Shell Nanoparticles As Catalysts for the Oxygen Evolution Reaction 702
Alaina Strickler, Maria Escudero-Escribano, Pongkarn Chakthranont, Thomas F. Jaramillo
(617ev) Electrochemical CO2 Reduction in an Oxygen-Ion Conducting Solid Oxide Electrolyzer Cell (SOEC) 703
Seval Gunduz, Doruk Dogu, Katja E. Binkley Meyer, Dhruba Jyoti Deka, Anne Co, Umit S. Ozkan
(617hl) The Mechanism of CO2 Hydrogenation over Pd/Al2O3: A Steady State Isotopic Transient Kinetic Analysis (SSITKA)/Operando-FTIR Spectroscopy Investigation 704
Xiang Wang, Janos Szanyi
(617cr) Machine-Learning-Augmented Chemisorption Model for Metal Oxides 707
Zheng Li, Hongliang Xin
(617do) Cobalt-Copper Alloy Nanoparticle Catalysts for Higher Alcohol Synthesis from Syngas 708
Jonathan Snider, Laiza Mendes, Samuel D. Fleischman, Jakob Kibsgaard, Donato Aranda, Thomas F. Jaramillo
(617eh) A Novel Method for Synthesis of Pt/Momc Catalyst for PEM Fuel Cell Application 709
Dereje Worku, Khondker Sultana, Vishwanath Deshmane, Shamsuddin Ilias
(617ew) Lattice Strain Engineering of Tin Oxides for Energy-Efficient CO2 Electroreduction 710
Siwen Wang, Hongliang Xin

(617aq) Furfuryl Alcohol Dehydration over Metal Oxide Catalysts 711
Xiaojun Chan, Taejin Kim
(617am) Low-Cost, Potassium-Based, Catalytic Biomass Cookstove for Improved Indoor Air Quality 712
Alex D. Paulsen, Tyler Kunsa, Andrew Carpenter, Nicholas R. Schwartz, Brett Alcorn, Ted J. Amundsen, Paul E. Yelvington
(617hm) Effect of Poisoning on Low Temperature Methane Dry Reforming Pt-Ni-Mg/Ceria-Zirconia Catalysts 713
Nada Elsayed, Anthony Elwell, Babu Joseph, John N. Kuhn
(617r) Understanding Direct NOx Decomposition over Pdo/SiO2: Influence of Reaction Temperature and Pretreatment 714
Krishna Reddy Gunugunuri, Chen Ling, Torin Peck, Hongfei Jia
(617fg) Metal Distribution after Drying of Supported Catalysts with High Metal Loadings 715
Cody Bishop, Hao Chen, Xue Liu, Johannes G. Khinast, Benjamin Glasser
(617hr) Glycerol Steam Reforming for Hydrogen Production over CoxNiy-MCM-41 and CoxNiy-SBA-15 Catalysts 716
Vishwanath Deshmane, Richard Abrokwah, Sara Alsalihi, William Dade, Debasish Kuila
(617dp) Development of a Graphene Supported Iron-Oxide Nanocatalyst for Fischer Tropsch Synthesis 717
Stephen Montgomery, Rui Xu, James G. Radich
(617z) Microwave-Intensified Synthesis of Biodiesel Catalyzed with Graphene Oxide 718
Yoshifumi Sumigawa, Armando Quitain, Mitsuru Sasaki, Tetsuya Kida
(617dq) Syngas Production Via Chemical Looping 719
Amey More, Saurabh Bhavsar, Charles Hansen, Götz Veser
(617fu) Controlling Carbon Affinity with Transition Metal Ceramics and Intermetallic Alloys to Modify the Degree of Olefin and Aromatic Activation 720
Yang He, Yuanjun Song, Siris Laursen
(617ba) Tuning Catalytic Activity for Meerwein-Ponndorf-Verley (MPV) Reaction through Designing Catalytic Sites 721
Nitish Deshpande, Aamena Parulkar, Frédéric Perras, Marek Pruski, Nicholas Brunelli
(617fn) Synthesis and Characterization of Au Nanoparticles for Ethanol Oxidation: Effect of Acidity and Support Structure 722
Erfan Behravesh, Narendra Kumar, Jorma Roine, Jarno Salonen, Andrey Shchukarev, Jyri-Pekka Mikkola, Atte Aho, Kari Eränen, Dmitry Yu Murzin, Tapio Salmi
(617ex) Electrochemical Control of the Nitrogen Cycle: Ammonia Synthesis from Nitrates 725
Joshua McEnaney, Jakob Kibsgaard, Thomas F. Jaramillo
(617ct) Theoretical Study on Oxidative Coupling of Methane Using MgO 726
Hassan Aljama, Jong Suk Yoo, Ambarish R. Kulkarni, Allegra A. Latimer, Frank Abild-Pedersen, Jens K. Norskov
(617ga) A Highly Active and Stable IrOx/SrIrO3 Catalyst for the Oxygen Evolution Reaction 727
Linsey C. Seitz, Colin Dickens, Kazunori Nishio, Yasuyuki Hikita, Joseph H. Montoya, Andrew Doyle, Charlotte Kirk, Aleksandra Vojvodic, Harold Y. Hwang, Jens K. Nørskov, Thomas F. Jaramillo
(617ey) Engineering Interfaces for the Activation and Stabilization of Photovoltaic-Grade Thin Film Light Absorbers for Photoelectrochemical Hydrogen Production 728
Thomas R. Hellstern, David W. Palm, Nicolas Gaillard, Thomas F. Jaramillo
(617ez) Engineering Ultra-Small Nanochannel Electrocatalyst Towards the Improvement in Oxygen Evolution and Oxygen Reduction Reaction 729
Makoto Mitani, Pongkarn Chakthranont, Thomas F. Jaramillo
(617hs) An Integrated 'One-Box' Process for Hydrogen Production 730
Ashkan Garshasbi, Doug Parsley, Richard J. Ciora, Paul K. T. Liu, Theodore Tsotsis
(617dz) Novel Catalysts for Photocatalytic Conversion of CO2/H2O and CO2/CH4 Systems to Syngas 731
Karishma Piler, Paul Bernazzani, Cristian Bahrim, Tracy J. Benson
(617dg) Computational Modelling of the Fischer-Tropsch Synthesis within a Microreactor 732
Andrew Traverso, Justin Pommerenck, Yousef Alanazi, Liney Arnadottir, Goran Jovanovic, Alexandre Yokochi
(617ea) Hydroxyl Radical Scavenging Activity of Silver Nanoparticles: Enhancing Photocatalytic Reduction and Decreasing Photocatalytic Oxidation Rate 733
Alexzander Steiniger, Hanqing Pan, Michael D. Heagy, Sanchari Chowdhury
(617gs) Sputter Deposition of Silver-Copper Nanoparticles for the Oxygen Reduction Reaction 734
Brenna Gibbons, Melissa Wette, Bruce M. Clemens, Thomas F. Jaramillo
(617fb) Rare Earth Element Doped Titania (TiO2) for Heterogeneous Photocatalytic Applications 735
Vignesh Bhethanabotla, Debtanu Maiti, Daniel Russell, John Kuhn, Venkat R. Bhethanabotla
(617gw) Experimental and Theoretical Studies of CO2 Adsorption on Hydrotalcite 736
Jack Lowd, Vasilios Manousiouthakis, Theodore Tsotsis, Richard J. Ciora

(617fo) First-Principles Study for Controlling the Cation Segregation on the Surface of Perovskite Oxide Catalysts 737
Hyunguk Kwon, Wonyoung Lee, Woochul Jung, Jeong Woo Han
(617hw) Low Temperature Hydrogen Production from Kinetically Limited NH3 Decomposition in Palladium Membrane Reactor 738
Naotsugu Itoh, Takafumi Kato, Takafumi Sato, Misaki Takahashi
(617ha) Fast Prediction of CO Oxidation Activity on Rare-Earth Metal Doped CeO2 739
Kyeounghak Kim, Jeongdo Yoo, Siwon Lee, Woochul Jung, Jeong Woo Han
(617by) Diagnostics of Field-Aged Three-Way Catalyst (TWC) on Stoichiometric Natural Gas Engines 740
Di Wang, Hongmei An, Jian Gong, Krishna Kamasamudram, Neal Currier, Junhui Li
(617y) The Beneficial Effect of Fructose on the Catalytic Oxidation of Glucose to Gluconic Acid 741
Joao G. R. Poco, Fabiana Dos Santos Lima
(617bx) Modeling of Microreactors for Chemoselective Capture of Trace Volatile Organic Compounds 742
Qi Li, Mingxiao Li, Yizheng Chen, Xiao-An Fu
(617fc) Trace Levels of Copper in Carbon Materials Show Significant Electrochemical CO2 Reduction Activity 743
Yanwei Lum, Youngkook Kwon, Peter Lobaccaro, Ezra L. Clark, Alexis T. Bell, Joel W. Ager
(617cu) Engineering the Electronic Structure of Mxenes Using Density Functional Theory: Strain and Support Effect 744
Liang Zhang, Kurt Fredrickson, Aleksandra Vojvodic
(617ci) Effect of Carbon Addition in the Slurry Phase Hydrocracking of Vacuum Residue 745
Sung-Ho Kim, Ki-Duk Kim, Yong-Kul Lee
(617dr) Dispersed Sulfide Catalysts for Hydrocracking of Vacuum Residue 746
Ki-Duk Kim, Sung-Ho Kim, Yong-Kul Lee
(617ds) Oxidative-Extractive Desulfurization of Model Diesel Oil By CTA-Type Dawson Type Emulsion Catalyst 747
Farhad Banisharif, Mohammad Reza Dehghani, Matin Parvari, Jose M. Campos-Martin
(617hf) Conversion of Carbon Dioxide to Syngas Using Transition Metal Dichalcogenide Catalysts 754
Mohammad Asadi, Pedram Abbasi, Amin Salehi-Khojin
(617p) On the Stability of Cu-Chabazite SCR Catalysts 755
Gustavo A. Fuentes, M. Eugenia Hernández-Terán, Julio C. López-Curiel, Gabriela Hernández-Salgado, Angel Martinez-Hernandez, Sergio A. Gómez
(617eb) Visible Light Photoactivity of Ion Implanted Titania Nanotubes 756
Mohammad Mehedi Hasan, Shammi Akter Ferdousi, King Lun Yeung
(617ed) Enhancement of Photochemical Activity of One Dimensional TiO2 By Targeted Decoration of Transition Metals 757
Mohammad Mehedi Hasan, King Lun Yeung, Shammi Akter Ferdousi
(617cc) New Materials Technology for Sustainable Removal of Moisture from Air 758
Shammi Akter Ferdousi, Mohammad Mehedi Hasan, King Lun Yeung
(617hp) Towards Industry-Scale Measurements of Reactor Gas-Liquid Mass-Transfer Phenomena 759
Manizheh Ansari, Damon Turney, Roman Yakobov, Dinesh V. Kalaga, Sanjoy Banerjee
(617bc) Kinetic Study for Hydrogenation of Toluene to Methylcyclohexane Over Titanium Substituted SBA-15 Supported Pt-Pd Catalysts 760
Shyamal Roy, Dilip Kumar Mondal, Chandona Mondal
(626a) Aqueous Phase Catalytic Conversion of Methane to Methanol on Bimetallic Catalysts Under Mild Conditions 775
Mengwei Li, Junjun Shan, Maria Flytzani-Stephanopoulos
(626b) Gas-Phase Catalytic Oxidation of Methane into Methanol over Copper-Exchanged Zeolites with Molecular Oxygen 776
Karthik Narsimhan, Kimberly Dinh, Yuriy Román-Leshkov
(626c) Bi-Reforming of Methane at Low Temperatures over Pt or Pd (Ni-Mg/ceria-zirconia) Catalysts 777
Nada Elsayed, Babu Joseph, John Kuhn
(626d) Understanding the Role of Fe in Bi-Metallic Ni-Fe Dry Reforming Catalysts: A Combined in-Situ XAS-XRD Study 778
Sung Min Kim, Paula Abdala, Wouter Van Beek, Tigran Margossian, Christophe Copéret, Christoph Mueller, Qasim Imtiaz
(626e) Selective Deposition of Ni on Mo2c That Has Been Supported on Al2O3: Enhanced Stability for Dry Reforming of Methane 779
Jayson Keels, Khalid Askar, John R. Monnier, John R. Regalbuto

(626f) Optimization of Metal Dopants in Pyrochlore Catalysts for the Dry Reforming of Methane: Combining Computational and Experimental Methods 780
David A. Bruce, Felipe Polo-Garzon
(626g) The Investigation of the Catalytic Partial Oxidation of Methane in a Shock Tube 781
Robyn Smith, Marco J. Castaldi, Kenneth Brezinsky
(626h) Single-Atom Active Sites for Natural Gas to Liquid Fuel Conversion 783
Thalia Quinn, Joseph Shackelford, Sierra Headrick, Pabitra Choudhury
(627a) Strategies for the Encapsulation and Stabilization of Monodisperse Au Clusters within Zeolites 784
Trenton Otto, Stacey I. Zones, Enrique Iglesia
(627c) Measuring & Understanding the Atomic-Scale Environments and Interactions of Heteroatom Sites in Zeolite Catalysts 787
Zachariah Berkson, Rahul P. Sangodkar, Subramanian Prasad, Ulrike Werner-Zwanziger, Bradley F. Chmelka
(627d) A Method to Screen Nanoporous Catalysts for Transition-State Selectivity 788
Peng Bai, Matthew Neurock
(627e) Synthetic Control of Framework Aluminum Distribution in Zeolites and Consequences for Acid and Redox Catalysis 789
John R. Di Iorio, Young Gul Hur, Rajamani Gounder
(627f) Design and Synthesis of IFR-Type Zeolites with Controlled Locations of Framework Aluminum with the Aid of Theoretical Calculation 790
Koki Muraoka, Watcharop Chaikittisilp, Tatsuya Okubo
(627g) A Mixed-Solvent Approach to Delaminate Layered Zeolitic Precursors 791
Isao Ogino, Taichi Shirobe, Shin R. Mukai
(629a) Engineering Highly Active Brookite Titania Nanorods for Sustainable Hydrogen Production 792
Matteo Cargnello, Tiziano Montini, Sergey Smolin, Jacqueline Priebe, Juan J. Delgado Jaén, Vicky Doan-Nguyen, Ian McKay, Jay Schwalbe, Marga-Martina Pohl, Thomas Gordon, Jason B. Baxter, Angelika Brückner, Paolo Fornasiero, Christopher B. Murray
(629b) Hydrogen Production Via Photocatalytic Water Splitting over Mesoporous Pt/TiO2 793
Fabricio Guayaquil Sosa, Benito Serrano, Hugo De Lasa
(629c) How Ceria Redox Materials Split H2o and CO2 with Concentrated Solar Energy into H2, CO, and O2 794
Ronald Michalsky, Aldo Steinfeld
(629d) Methane Conversion to Syngas Using Dopant Modified Metal Oxide Composites in Chemical Looping Reforming 795
Zhuo Cheng, Lang Qin, Mengqing Guo, Mingyuan Xu, Jonathan A. Fan, Liang-Shih Fan
(629e) A New Thin Layered Structural Coating on a Metal Substrate for Enhanced Hydrogen Production from Steam Methane Reforming 796
Michael Lugo, Marco Castaldi, Dean Modroukas, Andrew W. Davis, Rajinder Gill, Florent Minette, Juray De Wilde
(629f) Promotion Mechanisms of Iron Oxide-Based High Temperature-Water Gas Shift (HT-WGS) Catalysts By Chromium and Copper 797
Tulio Rocha, Minghui Zhu, Axel Knop-Gericke, Israel Wachs, Robert Schlogl
(629g) Catalytic Ammonia Decomposition to Hydrogen and Nitrogen on Supported Nanosized and Single-Site Cobalt Catalysts: Effects of Atomicity, Ligand Environment and Support 798
Konstantin V. Khivantsev, Fahad Almalki, Miao Yu
(641a) Selective Activation of Primary C-H Bonds of Propane on Pdo(101) 799
Jason F. Weaver, Aravind Asthagiri, Feng Zheng, Li Pan, Juhee Choi, Vikram Mehar, John Diulus
(641b) Methane Activation on Subsurface Boron Doped Copper: First Principles Investigation of the Structure, Activity and Electronic Properties of the Catalyst 800
Quang Thang Trinh, Samir H. Mushrif
(641c) Combined DFT and Kinetic Monte Carlo Studies of Formic Acid Decomposition on Au18 Nanoclusters 801
Benjamin W J Chen, Sha Li, Michail Stamatakis, Manos Mavrikakis
(641d) Mechanistic Study of C-O Bond Rupture on Ruthenium Phosphide Surface: Transient and Steady State Measurements 802
Siwei A. Chang, David W. Flaherty
(641e) Active Site Determination for CO Oxidation on Al2O3 Supported Pt Nanoparticle Catalysts By in-Situ Quantitative FTIR Measurements 803
Matthew Kale, Phillip Christopher
(641f) Competing Forces in Chiral Catalytic Surface Chemistry: Enantiospecificity Versus Enantiomer Disproportionation 804
Andrew J. Gellman, Yongju Yun

(641g) On the Reaction Pathways and Intermediates of Selective Ringopening of Furanics By Iridium 805
Glen Jenness, Weiming Wan, Jingguang G. Chen, Dionisios G. Vlachos

(654a) Controllable Solution Combustion Synthesis of Nanoscale a/B-Bi2O3 and Its Catalytic Application 806
Xiaohong Wang, Yang Xiao, Arvind Varma

(654b) A Combined Theoretical and Experimental Investigation on the Degradation of Organophosphorus Chemical Warfare Agents on ZnO1-X 807
John Landers, Nicole Zougheib, Fuat E. Celik, Alexander V. Neimark

(654c) Magnetic Superbasic Nano Catalysts for Rapid Catalyst Removal and Direct Product Isolation 808
Elia Schneider, Corinne Hofer, Martin Zeltner, Robert N. Grass, Wendelin J. Stark

(617ac) Selective Hydrogenolysis of Furfuryl Alcohol to 2-Methylfuran over Reduced Co-Fe-Al Mixed Metal Oxides 816
Taylor Sulmonetti, Bo Hu, Pradeep K. Agrawal, Christopher W. Jones

(617fi) Supported, Homogeneously Alloyed Bimetallic Nanoparticles By Electrostatic Adsorption 817
Andrew Wong, Qiuli Liu, John R. Regalbuto

(654f) Chemoselective Transfer Hydrogenation of Nitroarenes Catalyzed By Highly Dispersed, Supported Nickel Nanoparticles 818
Zeyu Shang, Chengjun Jiang, Xinhua Liang

(654g) Controllable Synthesis of Highly Dispersed Cu/SiO2 Catalysts in Supercritical CO2 and Their Application in the Hydrogenation of Dimethyl Oxalate to Ethylene Glycol 819
Qin-Qin Xu, Guo-Yue Qiao, Jian-Zhong Yin

(669a) Single-Site Iridium Complexes on MgO Powder: Spectroscopic, Microscopic, and Catalytic Evidence of Dependence of Properties on MgO Surface Sites 828
Adam Hoffman, Louise Debefve, Shengjie Zhang, David A. Dixon, Bruce C. Gates

(669b) Design of Palladium-Gold Catalysts for Liquid Phase Selective Alkyne Hydrogenation with Single Atom Alloy Strategy 829
Jilei Liu, Junjun Shan, Maria Flytzani-Stephanopoulos

(669c) Selective Dehydrogenation of Ethanol to Acetaldehyde and Hydrogen on a Highly Diluted Nicu Alloy 830
Junjun Shan, Nare Janvelyan, Hang Li, Jilei Liu, Tobias Egle, Jianchao Ye, Monika Biener, Juergen Biener, Cynthia M. Friend, Maria Flytzani-Stephanopoulos

(669d) Heterobimetallic Complexes As Controlled Precursors for Atomically Dispersed, Promoted Methane Oxidation Catalysts 831
Joseph Maalouf, Joshua Willis, Matteo Cargnello, Eric A. Stach, Jay Schwalbe

(669e) Site-Specific IR Spectroscopy for Analyzing Pt Structure Sensitivity from Single Atoms to Fully Coordinated Surfaces 832
Leo Derita, Phillip Christopher

(669f) Atomically Dispersed Pt Catalysts on Polyphosphomolybdic Acid: Structural Features and Catalytic Properties 833
Ning Yan

(669g) Understanding Site Isolation of Pd and Ni in Zn-Based Bulk Intermetallics during the Selective Hydrogenation of Alkynes 834
Anish Dasgupta, Robert M. Rioux

(674a) Mechanism of Oxygen Activation during Selective Catalytic Reduction of NO with NH3 in Cu-SSZ-13 Zeolites 835
Christopher Paolucci, Atish A. Parekh, Ishant Khurana, Hui Li, John R. Di Iorio, W. Nicholas Delgass, Jeffrey Miller, Fabio H. Ribeiro, William F. Schneider, Rajamani Gounder

(674b) Light Hydrocarbon Reactions on Solid Bronsted Acids: Confinement and Acid Strength Effects on Ion-Pair Transition States and Their Precursors 836
Michele L. Sarazen, Enrique Iglesia

(674c) Development of a Practical Catalyst for Terminal Olefin Epoxidation Using Organic Hydroperoxide As Oxidant 837
Alexander Okrut, Nicolás A. Grosso Giordano, Xiaoying Ouyang, Stacey I. Zones, Alexander Katz

(674d) Mesoporous Nb-Silicates As Ethylene Epoxidation Catalysts: Novel Strategies to Passivate Bronsted Acidity 838
Swarup Maiti, Anand Ramanathan, Bala Subramaniam

(674e) Investigation of the Acid Catalyzed Formaldehyde-Olefin Condensation Reaction 839
Efterpi Vasiliadou, Taha Salavati-Fard, Douglas J. Doren, Dion Vlachos, Raul F. Lobo

(674f) Experimental and Computational Interrogation of Fast SCR Mechanism and Active Sites on H-Form SSZ-13 840
Sichi Li, Yang Zheng, Feng Gao, Janos Szanyi, William Schneider

(674g) Quantification of External and Pore Mouth Acid Sites in Unit-Cell Thick Pillared MFI and Pillared MWW Zeolites 841
Dongxia Liu, Yiqing Wu
(676a) Green Application of Hydrogenation Catalyst Yn-1 on C4 Cut 842
Zhou Du
(676b) "Molybdenum Carbide As a Potential Catalyst for Olefins Hydrogenation" 843
Khaled O. Sebakhy, Gerardo Vitale, Pedro Pereira-Almao
(676c) Olefin Oligomerization on Carbon Supported Cobalt Catalyst: Identification of the Reaction Pathway 844
Zhuoran Xu, Joseph P. Chada, Dongting Zhao, Carlos Carrero, Jessica Rogers, Devon C. Rosenfeld, Ive Hermans, George W. Huber
(676e) Oxidative Dehydrogenation of Propane to Propylene over VOx/CaO-yAl2O3 Catalysts 845
Mohammad Mozahar Hossain, Afees Ayandiran, Idris Bakare, Shaikh Razzak, Housam Binous
(676f) On-Purpose Propylene Production from Light Hydrocarbons with MFI-Zeolite/Metal-Oxide Composite Catalysts 846
Shinya Hodoshima
(676g) Pt-Promoted Supported Gallium Catalyst for Dow Fluidized Catalytic Dehydrogenation Process 847
Lin Luo, Brien Stears, Andre Malek, David J. Barton, Beata A. Kilos, Joo Kang
(676h) Dehydroaromatization of Ethane over Zinc Containing Zeolites 848
Ali Mehdad, Raul F. Lobo
(676i) Direct Synthesis of Propene from Ethene Feedstock: Investigation of Catalytic Concepts 849
Matthias Felischak, Tanya Wolff, Leo Alvarado Perea, Andreas Seidel-Morgenstern, Christof Hamel
(676j) Olefin Metathesis By Supported WOx/SiO2 Catalysts 850
Soe Lwin, Israel E. Wachs
(688a) Spectroscopic Investigation of Pd Structures Formed over CeO2 and MnOx-CeO2 Supports 851
Erdem Sasmaz, Chao Wang, Jochen Lauterbach
(688b) Tailoring the Catalytic Activity of Perovskites for Oxygen Reduction and Evolution 852
Liang Zhang, Aleksandra Vojvodic
(688c) Origin of the Support Effect for Methanol Oxidation for Supported Vanadium Oxide Catalysts 853
Taejin Kim, Benjamin Moskowitz, Israel Wachs
(688d) Oxygen Transmission in Solid Phase——MnO2 promoted Phase-Pure M1 Movnbte Oxide for Ethane Oxidative Dehydrogenation 854
Xin Chen, Hang An, Bozhao Chu, Yi Cheng
(688e) On the Mechanistic Aspects of Mg6MnO8-Based Redox Catalysts for Oxidative Dehydrogenation of Ethane Via a Chemical Looping Scheme 855
Luke Neal, Seif Yusuf, Fanxing Li
(688f) Experimental and Theoretical Investigations of Propene Ammoxidation to Acrylonitrile over Bismuth Molybdate 856
Alexis T. Bell, Rachel B. Licht
(688g) Dioxygen Activation Pathways in Selective Oxidations Catalyzed By Metal Oxides 857
Stephanie Kwon, Prashant Deshlahra, Enrique Iglesia
(688h) Tuning the Dehydration Activity of Alcohols Using Phosphonic Acid Self-Assembled Monolayers 858
J. Will Medlin, Lucas Ellis, Daniel Schwartz
(688i) Single-Facet-Dominated Titania, and Zirconia As Catalysts or Supports for Bio-Alcohol Dehydration and Oxidative Dehydrogenation 859
Yuan Chen, Lu Zhang, Libor Kovarik, Huamin Wang, Feng Gao, Yong Wang, Charles H. F. Peden
(689a) Mechanistic Insights into Carbon-Oxygen Bond-Breaking on Mo2c Catalysts: Ethanol Dehydration 860
Carrie A. Farberow, Connor Nash, Joshua A. Schaidle
(689b) Elucidating Particle Size Effects on the Hydrodeoxygenation (HDO) of Phenolics By Micro-Pulse Experiments and DFT Calculations 861
Qiaohua Tan, Nhung Duong, Daniel Resasco
(689c) Adsorption of Hetero-Bifunctional Urea on Ge(100)-2x1 Surface 862
Tania E. Sandoval, Allison M. Crow, Stacey F. Bent
(689d) Suppressing Pd Catalyst Deactivation in Alcohol Oxidation Using Thiolate Self-Assembled Monolayer Modification 863
Pengxiao Hao, Daniel K. Schwartz, J. Will Medlin
(689e) Facile Formation of Methyl Acrylates By Cross Coupling of Unsaturated Alcohols-Aldehydes with Methanol on Au(110) 864
Stavros Karakalos, Fanny Hiebel, Cynthia M. Friend, Robert J. Madix

(689f) Reactivity and Selectivity Descriptors for the Activation of C-H Bonds in Hydrocarbons and Oxygenates on Metal Oxides 865
Prashant Deshlahra, Enrique Iglesia
(689g) Selective Oxidation of n-Butane to 1-Butanol over Transition Metal Catalysts Encapsulated By Metal-Organic Frameworks 866
Jiazhou Zhu, Rachel Getman
(719a) Highly Selective Fischer-Tropsch Synthesis for Production of JP-8 Jet Fuel from Coal and Coal/Biomass Mixtures 867
Andrew Lucero, Brittany Koob, Brandon Cline, Kevin McCabe, Santosh Gangwal
(719b) Composition of Bio-Crude Oils from Hydrothermal Liquefaction of Algae in Batch Reactor 868
Zheng Cui, Feng Cheng, Travis Le-Doux, Kwonit Mallick, Carolina Herrera, Graham Hoffman, Jacqueline Jarvis, Neil Paz, Tanner Schaub, Nagamany Nirmalakhandan, Catherine E. Brewer
(719c) De-Alloyed Platinum-Bismuth Nanoparticles As Highly Active Electrocatalysts for Dimethyl Ether Oxidation and Oxygen Reduction 869
Anastasios Angelopoulos, Punit Boolchand, Zhipeng Nan
(719d) Co-Production of Methanol and Dimethyl Ether Using a Modified Copper-Based Catalyst System 870
Lujie Ye
(719e) Design of a Sabatier Reactor for CO2 Conversion into Synthetic Methane 873
Duo Sun, David Simakov
(719g) Steam Reforming of Dimethyl Ether over a Novel Plate-Type γ-Al2O3/Al Monolith Supported Cu-Based Bi-Functional Catalyst 874
Feiyue Fan, Qi Zhang, Zibin Zhu
(720a) Low Temperature CO Oxidation with a Single Site Pt Catalyst Supported on the "29" Cuxo/Cu(111) Surface: Elucidating the Chemical Nature of the Pt Active Sites 875
Alyssa Hensley, Andrew Therrien, Renqin Zhang, E. Charles H. Sykes, Jean-Sabin McEwen
(720b) Identifying the Active Site for the Water-Gas Shift Reaction over TiO2 Supported Pt Catalysts: Single Pt Atoms Versus Pt Clusters 876
Salai C. Ammal, Andreas Heyden
(720c) Trends in Single Transition Metal Atom/Oxide Support Interaction from Density Functional Theory 877
Nolan O'Connor, Thomas P. Senftle, Michael J. Janik
(720d) CO Oxidation Kinetics on Supported Ir Single Atoms 878
Yubing Lu, Xianfeng Ma, Hongliang Xin, Ayman M. Karim
(720e) Copper Atoms Stabilized on Reduced Titanium Oxides Catalyze Carbon Monoxide Oxidation 879
Guoqiang Cao, Justin T. Cournoyer, Nan Yi
(720f) the Role of Au in Au-Catalyzed, Solvent-Free, Selective Epoxidation of Cyclooctene with Molecular Oxygen 880
Linping Qian, Zhen Wang, Evgeny Beletskiy, Jingyue Liu, Haroldo Santos, Mayfair Kung, Harold H. Kung
(720g) Structure, Reactivity and Selective Tuning of Well-Defined Rh Complexes in HY Zeolites 881
Konstantin V. Khivantsev, Artem D. Vityuk, Hristiyan A. Aleksandrov, Georgi N. Vayssilov, Oleg S. Alexeev, Michael D. Amiridis
(734a) Structure-Property Relationships of Palladium Catalyzed Methane Complete Combustion Using Uniform Nanoparticles 884
Joshua Willis, Emmett Goodman, Matteo Cargnello
(734b) Characterizing Active Sites on Chirally-Modified Supported Noble Metal Catalysts for Asymmetric Hydrogenation through Selective Poisoning 885
Laura Rodriguez-Garcia, Fabian Meemken, Konrad Hungerbuehler, Alfons Baiker
(734c) Carbon Decoration in Supported Palladium Catalysts: Discrepancy in STEM-Chemisorption Particle Sizes 886
Ritubarna Banerjee, John R. Regalbuto
(734d) 100-Fold Rate Enhancement Caused By Dioxygen Ligands on a Supported, Single-Site Tetrairidium Cluster Catalyst for Ethylene Hydrogenation 887
Andrew Palermo, Alexander Okrut, Andrew Solovyov, J. D. Kistler, Louise Debefve, Ron C. Runnebaum, Shengjie Zhang, Igor Busygin, Daniel Ertler, David A. Dixon, Bruce C. Gates, Alexander Katz
(734e) Structure–Function Relationships for Graphene-Supported Pt Nanoclusters 888
Hongbo Shi, Scott M. Auerbach, Ashwin Ramasubramaniam
(734f) Role of the Support and Reaction Conditions on the Vapor-Phase Deoxygenation of M-Cresol over Pt/C and Pt/TiO2 Catalysts 889
Michael B. Griffin, Glen Allen Ferguson, Daniel A. Ruddy, Mary Biddy, Gregg T. Beckham, Joshua Schaidle

(734g) Understanding Solvent Effects over Supported Metal Catalysts: Ketone Hydrogenation over Ru/SiO2 891
Omar A. Abdelrahman, Jesse Q. Bond
(734i) Hammett Study: Combined Experimental-Computational Approach to Understanding Electronic Metal-Support Interaction 892
Gaurav Kumar, Bert D Chandler, Michael J. Janik
(735i) Quantification of Bronsted and Lewis Acid Sites on Zeolites in the Presence of Solvents 893
Bingjun Xu, Nicholas Gould
(735b) Synthesis of Itaconic Acid Ester Analogues via Self-aldol Condensation of Ethyl Pyruvate Catalyzed by Hafnium BEA Zeolites 894
Yuran Wang, Jennifer Lewis, Yuriy Román-Leshkov
(735c) Probing Reactions at the Solid-Liquid Interface By Operando Solid-State NMR Spectroscopy 895
Susannah L. Scott, Long Qi, Ali Chamas, David W. Hoyt, Eric D. Walter, Nancy M. Washton, Marcus Foston
(735d) Direct Observation of Active Sites of Ru Catalyst in Hydrodeoxygenation of Furfural 896
Weiqing Zheng, Konstantinos Goulas, Ayman M. Karim, Prashant Kumar, Michael Tsapatsis, Dionisios G. Vlachos
(735e) Single Center Reactivity of Cu(II) Sites in Copper Substituted CHA Zeolite Followed By in-Situ Electron Paramagnetic Resonance Spectroscopy 897
Susanne Mossin, Oliver Isaksen, Anita Godiksen, Søren B. Rasmussen, Peter N. R. Vennestrøm
(735f) In-Situ Characterization of Intermetallic Ni5Ga3 Catalyst for CO2 Reduction to Methanol 898
Alessandro Gallo, Dimosthenis Sokaras, Dennis Nordlund, Hirohito Ogasawara, Libor Kovarik, Jonathan Snider, Thomas F. Jaramillo
(735g) Growth of Encapsulating Carbon on Supported Pt Nanoparticles Studied By in Situ TEM 899
Jason Wu, Stig Helveg, Sven Ullmann, Zhenmeng Peng, Alexis T. Bell
(745a) Enhancing Oxygen Reduction Activity on Pt Monolayer Electrocatalysts through Selective Tuning of Ligand and Lattice Effects 900
Timothy Van Cleve, Saman Moniri, Hongliang Xin, Suljo Linic
(745b) Engineering the Active Site for Oxygen Electroreduction in Fuel Cells 901
Maria Escudero-Escribano, Ifan E. L. Stephens, Ib Chorkendorff
(745c) Engineering Core-Shell Nanoparticles of Noble Metals on Transition Metal Carbides As Next-Generation Catalysts in Energy Applications 902
Maria Milina, Sean T. Hunt, Christopher H. Hendon, Yuriy Román-Leshkov
(745d) Optimal Pattern of Bimetallic Electrocatalysts for Efficient Conversion of Water and Carbon Dioxide to Hydrocarbons and Oxygenates 903
Meenesh Singh, Yanwei Lum, Ezra L. Clark, Joel W. Ager, Alexis T. Bell
(762j) Supported Cobalt Catalysts with Preferentially Exposed Crystal Facets for CO2 Hydrogenation 904
Cun Wen, Juan Jimenez, Jason Hattrick-Simpers, Jochen Lauterbach, Maryorie C. Santos
(745f) Tunable Thermodynamic Activity of LaxSr1-XMnyAl1-YO3-δ ($0 \leq x \leq 1$, $0 \leq y \leq 1$) Perovskites for Solar-Driven Splitting of CO2 and H2O 905
Miriam Ezbiri, Michael Takacs, David Theiler, Ronald Michalsky, Aldo Steinfeld
(745g) Material Prediction for Carbon Dioxide Conversion to Carbon Monoxide Using Reverse Water Gas Shift Chemical Looping (RWGS-CL) Process 906
Debtanu Maiti, Yolanda Daza, Adela E. Ramos, Bryan J. Hare, John N. Kuhn, Venkat R. Bhethanabotla
(762a) The Influence of Water on Metal/Oxide Perimeter Sites for the Selective Conversion of Phenolic and Furanic Species over Ru/TiO2 907
Nicholas M. Briggs, Taiwo Omotoso, Tyler Vann, Valeria Herrera, Steven Crossley
(762b) Olefin Oligomerization over Cobalt Catalysts: Effect of Catalyst Support 908
Dongting Zhao, Zhuoran Xu, Joseph P. Chada, Carlos Carrero, George W. Huber, Ive Hermans, Jessica Rogers, Devon C. Rosenfeld
(762c) Interfacial Perimeter Sites of Au-TiOx in Isopropanol Decomposition 909
Yi Y. (Chloe) Wu, Harold H. Kung
(762d) Theoretical Investigation of Supported Ultra-Thin Bilayer, Double Bilayer and Trilayer Metal-Oxides: Structure, Stabilty and Catalytic Activity 910
Michal Bajdich, Max García-Melchor, Jakob Fester, Phillip Plessow, Frank Abild-Pedersen, Jeppe Lauritsen, Aleksandra Vojvodic
(762f) Controlling Selectivity of Metal Catalysts By Using Self-Assembled Monolayers on the Oxide Support 911
Jing Zhang, Patrick Coan, J. Will Medlin
(762g) In-Situ Grown Metal Nanocatalysts from Oxide Host Lattices 912
Tae-Sik Oh

(762h) Designing Catalysts for Enhanced Pairwise Selectivity of Parahydrogen Addition to Unsaturated Hydrocarbons 913
Evan W. Zhao, Helena Hagelin-Weaver, Sean McCrea, Haibin Zheng, Clifford R. Bowers, Yan Xin
(762i) TiO2 Promoted Gold-Based Catalysts for Propylene Epoxidation 914
Zheng Lu, Zili Wu, C. Heath Turner, Yu Lei
(763a) Adsorbate-Mediated Strong Metal-Support Interactions in Oxide Supported Rh Catalysts 915
Phillip Christopher, John C. Matsubu
(763b) Understanding Active Sites in Ethanol Synthesis on PtCo Catalysts Prepared Using Atomic Layer Deposition 916
Joseph Singh, Stacey F. Bent, Nuoya Yang
(763c) The Role of the Support for Pt Catalysts during the Water-Gas Shift Reaction 917
Viktor Cybulskis, Yanran Cui, Mayank Shekhar, Juan Lovón Quintana, W. Nicholas Delgass, Fabio H. Ribeiro
(763d) Operando Attenuated Total Reflection-Infrared Spectroscopic Detection of Catalytic Solid-Liquid Interfaces: Design Aspects of Spectroscopic Cells 918
Fabian Meemken, Laura Rodriguez-Garcia, Konrad Hungerbuehler, Alfons Baiker
(763e) In Situ and Operando Drifts Characterization of Catalytic Sites on Metal-Doped KIT-6 during Alcohol Conversion Reactions 919
Juan J. Bravo-Suarez, Thomas Ofosu, Maria Ramirez, Marco Caricato, Alessandro Biancardi, Amy Jystad
(763f) Operando spectroscopy during Ethylene Polymerization By Supported CrOx/SiO2: Role of Promoters 920
Anisha Chakrabarti, Israel E. Wachs
(763g) Evidence for CO Insertion in Alkane and Alkenes Production through Fischer-Tropsch Reaction 921
Motahare Athariboroujeny, Greg Collinge, Jean-Sabin McEwen, Norbert Kruse
(768a) Scaled Degree of Rate Control: Identifying Elementary Steps That Control Differences in Performance of Transition Metal Catalysts 922
Talin Avanesian, Phillip Christopher
(768b) Effective Coordination Based Descriptors for Rational Design of Metal Nanocatalysts 923
Xianfeng Ma, Siwen Wang, Hongliang Xin
(768c) Tuning Surface Carbon Affinity and Olefin Activation in Intermetallic Alloys for the Semi-Hydrogenation of Acetylene to Ethylene 924
Yuanjun Song, Siris Laursen
(768d) Scaling Relationships at Bifunctional Metal/Oxide Interfaces: A Case Study of Doped Au/MgO 925
Tej S. Choksi, Paulami Majumdar, Jeffrey P. Greeley
(768e) Rational Design of Nanostructured Metallic Surfaces Via Mathematical Optimization 926
Christopher L. Hanselman, Chrysanthos E. Gounaris
(768f) Screening of Active Site Motifs for C-H Bond Activation in Nanoporous Catalysts 927
Ambarish R. Kulkarni, Allegra A. Latimer, Felix Studt, Jens K. Norskov
(768g) Design of Functionalized Metal Organic Frameworks for CO2 Hydrogenation:the Effects of MOF Topology and Functional Group 928
Jingyun Ye, Benjamin Yeh, Ronald A Reynolds, J. Karl Johnson
(769a) The Nature of Heterogeneous Palladium Catalyzed Cross-Coupling Reactions on Novel Carbon Support Systems 929
Frank Gupton
(769h) Electrostatic Adsorption of Pt onto Carbon Nanotubes and Fibers 930
Ritubarna Banerjee, Jose Contreras-Mora, John R. Regalbuto
(769c) Impact of Mixed Metal Oxides Supports on the Durability and Low-Temperature Performance of Pd-Based Diesel Oxidation Catalysts 931
Eleni A. Kyriakidou, Todd J. Toops, Michael J. Lance, Andrew J. Binder, James E. Parks
(769e) Tuning Acidity of Mesoporous Silicates with Earth-Abundant Metals and Beneficial Effects during Catalytic Dehydration of Linear Alcohols 932
Hongda Zhu, Anand Ramanathan, Bala Subramaniam
(769g) Designed Synthesis of Co3O4 Total Oxidation Catalysts 933
Kathleen Mingle, Cun Wen, Erdem Sasmaz, Jason Hattrick-Simpers, Jochen Lauterbach
(779a) Ring-Opening of Furfuryl Alcohol Towards 1,5-Pentanediol over Reduced Cu-Co-Al Mixed Metal Oxides 934
Taylor Sulmonetti, Bo Hu, Pradeep K. Agrawal, Christopher W. Jones
(779b) Mechanistic Studies on the Direct Synthesis of H2O2 on Pd and Aupd Clusters 935
Neil M. Wilson, David W. Flaherty
(779c) Influence of Sn and Bi on Pt-Catalyzed Aqueous Phase Oxidation of 1,6-Hexanediol 936
Jiahan Xie, Derek D. Falcone, Kehua Yin, Benjamin Huang, Hien Pham, Raymond Unocic, Abhaya Datye, Robert J. Davis

(779d) CeO2 Supported Metal Oxide Catalysts for NOx Removal 937
Nusnin Akter, Jorge A. Boscoboinik, Taejin Kim
(779e) Mechanistic Insights on Catalytic Ethanol Steam Reforming 938
Stephen Crowley, Marco J. Castaldi
(779f) Effect of Reduced Adsorption Strength with Bimetallic Platinum Overlayer Catalysts 939
Chen Zhang, Joseph Holles
(779h) Tuning Nanoparticle Alloys to Enhance C-H Bond Activation for the Catalytic Dehydrogenation of Ethane 940
Viktor Cybulskis, James Gallagher, Han-Ting Tseng, Brandon Bukowski, Zhenwei Wu, Evan Wegener, A. Jeremy Kropf, Bruce Ravel, Jeffrey P. Greeley, Fabio Ribeiro, Jeffrey Miller
(779i) First Principles Light Alkane Dehydrogenation on Pt:Main Group Alloys and the Effect of Hydrogen Spectators 941
Alec Hook, Fuat E. Celik
(782a) Effect of Boron Promotion on the Performance of Platinum Catalysts for Propane Dehydrogenation 942
Esteban L. Fornero, Shiwen Li, Vladimir Galvita, Mark Saeys
(782g) Physical Properties Controlling Heterogeneous Catalyst Stability in Supercritical Water 943
Jennifer Jocz, Phillip E. Savage, Levi T. Thompson
(782c) Systematic Investigation of Earth-Abundant Transition Metal Promoters for Pd-Catalyzed Methane Complete Combustion 944
Joshua Willis, Ian Naccarella, Helen Yan, Joseph Maalouf, Kester Wade, Matteo Cargnello
(782d) Unraveling the Nature of Boundary Sites of Metal-on-Oxide Catalysts 945
Prateek Mehta, Jeffrey P. Greeley, W. Nicholas Delgass, William F. Schneider
(782e) Identification of Active Sites for NaY Catalyzed Methyl Lactate Dehydration 946
Bingjun Xu, Brian Murphy
(782f) Using Non-Covalent Interactions to Control Availability of Different Catalyst Sites 947
Michael Janik, J. Will Medlin, Chih-Heng Lien, Gaurav Kumar
(784a) Cu/Chabazite Catalyst to Meet the Chemical and Mechanical Challenges for NOx Removal 948
Nusnin Akter, Taejin Kim, Jorge A. Boscoboinik
(617aa) Effect of the Preparation Method on the Porosity of Amorphous Silica-Alumina to be Used in Hydrocracking Reactions 949
Angélica Liliana Coconubo Díaz, Alexander Guzmán Monsalve, Luz Marina Ballesteros Rueda
(784d) Insight into Supported Metal Catalyst Stability By Quantifying Thermodynamic Interactions at the Solid-Liquid Interface 952
Ahana Mukhopadhyay, Robert M. Rioux
(784e) Simulation and Experimental Studies of Dry Catalyst Impregnation for Improved Content Uniformity and Scale up 953
Yangyang Shen, William G. Borghard, M. Silvina Tomassone
(784f) Reversible Ionic Surfactants for the Preparation of Monodisperse, Supported Nanoparticles 954
Kristin Bryant, Gasim Ibrahim, Steven R. Saunders
(784g) Beneficial Influence of Surface Capping Agents on the Stability and Activity of Metal/Oxide Catalysts 955
Timothy Van Cleve, J. Will Medlin
Author Index

460779 Experimental and Model Aided Approach for Development of High Pressure Direct Asymmetric Reductive Amination Reaction

Abstract:

Direct Asymmetric Reductive Amination (DARA) reactions are useful in pharmaceutical routes for converting carbonyl groups into amines. Typically, these reactions involve a catalyst system to promote the hydrogenation. A high pressure DARA reaction was used to synthesize an amine as an intermediate for one of the synthetic routes for a high volume drug. Given the water sensitivity of the reaction, ketone reagent was converted to ketal prior to carrying out the DARA reaction. Effects of key reaction variables â€" temperature, catalyst type, catalyst concentration, reaction volume, and pressures were explored to understand their effect on yield and impurity formation. It was found that there were a number of undesired side reactions that took place leading to impurity levels that needed to be controlled to achieve the final product specifications downstream. A mechanistic kinetics model was developed with some empirical relationships to capture the effect of certain reaction variables on the selectivity of desired to the undesired reactions. A combination of the experiments and simulations helped provide an understanding of the complicated chemistry. Together the approach was useful for: 1) mapping out and optimizing the design space, 2) determining the critical process parameters for the process, 3) developing impurity control strategies, and 4) designing experiments for risk assessment package for regulatory agencies. Thus, fundamental understanding was gained using these tools of an otherwise challenging and complicated chemistry.

Copyright © American Institute of Chemical Engineers. All rights reserved.

455659 Kinetics and Mechanistic Understanding of the Difluoromethyl Alkylation of Hydroxychromenone

Abstract:

During a chemical reaction involving the difluoromethylation of hydroxychromenone, a problematic dimer impurity was observed. In light of the short timeline to further optimize this chemistry, a fit-for-purpose kinetics analysis was performed.

Reaction kinetics and response surface simulations were used to understand the design space required in order to maintain this dimer impurity within acceptable levels. The kinetics assessment not only led to a mechanistic understanding of this transformation, but also led to design space understanding and potential for improvements. With these insights, this process was successfully prepared on kg scale in a manufacturing environment, and a proof-of-concept using a CSTR at laboratory scale was demonstrated.

The work presented here was sponsored by AbbVie. AbbVie contributed to the design, research, and interpretation of data, writing, reviewing, and approving the presentation. All authors are employees of AbbVie.

Copyright © American Institute of Chemical Engineers. All rights reserved.

453713 Development of a Reaction Kinetic Model for an Enzymatic Reaction Using Reaction Kinetic Progress Analysis and Dynochem

Abstract:

When developing a kinetic model for a chemical reaction, it can be difficult to obtain enough data to model the full reaction mechanism; however, over-simplification of the mechanism can lead to poor model fit. Reaction Progress Kinetic Analysis can reduce the volume of data required to adequately describe a complex reaction with a mechanistic kinetic model. An enzymatic reaction was developed to produce an intermediate compound with high yield and stereospecificity while minimizing impurity formation through side reactions. Initial attempts to model the reaction with a simplified mechanism were unsuccessful, so Reaction Progress Kinetic Analysis (RPKA) was successfully utilized to better fit a just-simple-enough mechanism and the reaction kinetics were fit in Dynochem and used to optimize the process.

The final kinetic model included an improvement on power law kinetics while still reducing the complexity versus the Ping-Pong Bi-Bi mechanism as well as the thermodynamics and kinetics of dissolution of a starting material. The framework was then used to troubleshoot the process with respect to variable reaction completion times that were seen with different lots of starting material.

Copyright © American Institute of Chemical Engineers. All rights reserved.

465076 *Rapid Kinetic Profiling of Copper(I)-Catalyzed Alkyne-Azide Cycloaddition(CuAAC) Click Reactions Utilizing Inline IR, HPLC and MS*

Abstract:

Since their discovery, a great deal of interest has been placed in understanding the CuAAC reaction mechanism. Early focus was on the number of copper atoms involved in the reaction pathway, but recent studies have shown under various circumstances that only unique copper centers catalyze the reaction[1]. To develop a mechanistic understanding of the reaction pathways and quantitatively describe the kinetics, it is essential to systematically scan a wide, multidimensional parameter space as efficiently as possible.

In this work, prior techniques for data collection in continuous flow systems[2,3] are expanded upon to develop a robust method for rapid and efficient reaction monitoring. Rapid MS, HPLC and IR analytics provide inline means to generate a comprehensive quantitative understanding of the mechanism for most organic reactions. This new method is validated with the Paal-Knorr reaction before further application toward describing the click reaction. Various substrates were reacted and the kinetic contributions from several ligands were elucidated while limiting reagent consumption and overall experimental time. The results of this study demonstrates a robust, general method for systematic mechanistic studies in continuous flow systems.

1. L. Jin, D. R. Tolentino, M. Melaimi, G. Bertrand. *Isolation of bis(copper) key intermediates in Cu-catalyzed azide-alkyne "click reaction".* **Sci. Adv.** (2015)

2. J. S. Moore and K. F. Jensen. *"Batch" Kinetics in Flow: Online IR Analysis and Continuous Control.* **Angew. Chem. Int. Ed.** (2014)

3. C. J. Welch, X. Gong, W. Schafer, E. C. Pratt, T. Brkovic, Z. Pirzada, J. F. Cuff, B. Kosjek. *MISER chromatography (multiple injections in a single experimental run): the chromatogram is the graph.*

Tetrahedron Asymmetry (2010)

Copyright © American Institute of Chemical Engineers. All rights reserved.

461902 Dynamic Resolution for Separation-Free Enantiomeric Purification in Flow

Abstract:

Chiral separations of biologically active small molecules is a critical process in the synthetic pathways for pharmaceutics. Commercial processes often necessitate the use of large batch crystalizers or slow, solvent intensive chromatographic columns. In recent years, dynamic resolution of chiral compounds have demonstrated great promise in providing chemical pathways for driving racemic mixtures to single component chiral species. Typical pathways utilize a batch slurry reactor that contains two catalysts which perform the series reactions: enantioselective irreversible transesterification (enzyme) and racemization (precious metal). Such methodologies, however, are greatly limited in application by requiring the series reactions to operate homogenously and under identical reaction conditions. The heavy metals are then required to be separated downstream and the spent catalysts are discarded.

In this work, we introduce a chemical synthesis for the successful immobilization of a ruthenium metal complex for the efficient racemization of chiral molecules. The ligand-supported complex performs the series oxidation-reduction over several substrates in a model alcohol system. Additionally, the successful immobilization has enabled the introduction of a continuous flow-based technique whereby two micro-packed beds are charged in series with the immobilized enzyme and racemization catalysts. The use of a recycle system and in-line gas stripping allows for nearly quantitative enantiomeric selectivity.

Copyright © American Institute of Chemical Engineers. All rights reserved.

459411 Tuning the Activity and Selectivity of Transition Metal Phosphide Catalysts for CO and CO_2 Hydrogenation

Abstract:

CO_2 from combustion and industrial processes holds great potential as an abundant carbon feedstock for the synthesis of fuels and chemicals. CO_2 can be hydrogenated to high value products using a renewable hydrogen source, generating carbon neutral fuels and chemicals. In current industrial processes CO_2 is often present in the syngas feed or formed as a side product. Developing heterogeneous catalysts for CO_2 conversion is thus critical for improving existing processes as well as transitioning to CO_2 as a sustainable globally distributed carbon feedstock. We have investigated transition metal phosphide catalysts for the hydrogenation of CO, CO_2 and their mixtures and have developed catalysts with high stability under changing feed conditions. Transition metal phosphide catalysts were found to be active for the synthesis of a range of products including methane, hydrocarbons, methanol and higher alcohols. Promoters were found to have a large impact on product selectivity. Transition metal phosphides are novel catalysts for CO_2 activation and hold great promise for the utilization of CO_2 as a new carbon feedstock.

Copyright © American Institute of Chemical Engineers. All rights reserved.

469252 Aerosol Processing of MOF-Based Nanocomposites for Carbon Dioxide Photoreduction

Abstract:

The continuous reliance on fossil fuel-based energy is inevitable in the near future. Therefore strategies to reduce carbon dioxide (CO_2) emissions are highly demanded. Developing efficient photocatalysts that can harness solar energy appears to be a promising methodology, i.e., the so-called CO_2 photoreduction, to capture and recycle CO_2 as a fuel feedstock. The conversion efficiency of the current photocatalysts, however, is generally very low due to various limiting factors, such as fast electron-hole recombination rates, narrow light absorption range, and backward reactions. Thus, developing strategies to overcome the above limitations is an important task in this field.

Here we report a facile development of novel metal-organic framework (MOF)-based nanocomposites with hierarchical structures and multi-functionalities via a single-step aerosol process. Effects of precursor concentration, component ratio, solvent type, and synthesis temperature were investigated in detail. The as-synthesized nanocomposites were systematically characterized by SEM, TEM, XRD, UV-vis, FTIR and BET. The crystal formation mechanisms of MOFs and MOF-based nanocomposites in aerosol routes will be discussed. *In-situ* analyses by using gas chromatography (GC) and diffuse reflectance infrared Fourier transform spectroscopy (DRIFTS) will be employed to further understand the CO_2 photoreduction pathways.

Copyright © American Institute of Chemical Engineers. All rights reserved.

470019 Enhanced Electrocatalytic CO_2 Reduction on Thiol-Functionalized Gold

Abstract:

CO_2 can potentially be used as a renewable feedstock for distributed electrochemical production of fuels and chemicals, either as an energy carrier to store energy generated from intermittent sources such as wind and the sun, or as a renewable source of carbon for chemical production. The key to enabling these uses of CO_2 is maximizing the selectivity and energy efficiency of electrocatalytic CO_2 reduction. Hori and co-workers have shown that some metals (Au, Ag and Cu, etc.) can catalyze the reaction and produce CO or hydrocarbons [Hori, in *Modern Aspects of Electrochemistry*, ed. Vayenas et al., Springer (2008)]. A significant challenge is posed by the competing hydrogen evolution reaction (HER), which tends to occur with a smaller overpotential and consumes a significant fraction of the current and protons. Furthermore, Nørskov and co-workers have shown that the catalytic activity for CO_2 electroreduction is limited by the strong scaling relation that exists between the key reaction intermediate, COOH, and the reaction product, CO [Shi et al., *Phys. Chem. Chem. Phys.* 16 (2014) 4720]. In present work, we take inspiration from biological enzymes that catalyze CO_2 reduction (e.g. CO dehydrogenase) and attempt to mimic their active sites by functionalizing Au electrodes with a series of organic thiol-based ligands, to seek to potentially overcome these challenges. This has resulted in significantly altered activity for CO formation and selectivity between CO and H_2. The CO yield was increased dramatically and nearly doubled in the presence of 2-phenylethanethiol (2-PET) compared to the blank gold foil electrode. On the contrary, 2-merceptanpropanic acid (MPA) suppresses the yield of CO to negligible amounts. The 2-PET-modified gold electrode also reduced the faradaic efficiency for HER by half, whereas the MPA-modified Au electrode increased it. Density function theory based modeling suggest that ligand-induced surface reconstruction of the Au surface is a major factor in the altered catalytic activity compared to blank Au, which creates sites that favor CO production over HER. Moreover, the surface-bound thiol ligands modify the adsorption energies of surface reduction intermediates through electronic interaction and further contribute to the altered activities for CO_2 reduction and HER.

Copyright © American Institute of Chemical Engineers. All rights reserved.

465188 Investigation of the Main Factor That Drives Selectivity on Rough Cu-Based Catalysts for CO2 Electroreduction

Abstract:

Carbon dioxide (CO_2) is the residual gas of combustion, agricultural and industrial processes and is associated with global warming. There is great potential for sustainable use of this waste gaseous byproduct towards useful chemicals and low carbon fuels (methane, ethylene) with the utilization of electroreduction and renewable electricity as the energy source. The selection of the electrocatalyst is of paramount importance for CO_2 electroreduction since it is related with hindrances like selectivity, currentÂ efficiency, and stability[1]that still limit its broader application.

Recent studies[2-4] have presented the importance of catalyst surface morphology and showed that rough copper (Cu) catalysts can improve selectivity and Faradaic efficiency towards hydrocarbon production. Understanding and controlling the main factor that influences selectivity on such surfaces would lead to the improvement of catalystsâ€™ design towards the desired chemicals or fuels. This report examines the main factor that drives the selectivity on rough Cu-based catalysts surfaces. For that purpose, Cu-based catalysts were synthesized by the use of electrodeposition on copper substrates. Roughness was controlled through electrodeposition parameters of charge passed and potential applied. The evaluation of each catalyst involved the examination of surface morphology, the efficiency during the CO_2 reduction, and product selectivity. Scanning electron microscopy (SEM), X-ray diffraction (XRD), X-ray photoelectron spectroscopy (XPS) and capacitance measurements with cyclic voltammetry were used. The products analysis was conducted with a micro gas chromatograph (microGC) for gas products. Results showed a relation between roughness and product selectivity. Higher roughness was associated with the higher formation of ethylene (C_2H_4) whereas lower roughness catalysts form as main hydrocarbon product methane (CH_4).

(1) Qiao, J.; Liu, Y.; Hong, F.; Zhang, J. A review of catalysts for the electroreduction of carbon dioxide to produce low-carbon fuels. *Chemical Society Reviews* **2014,** *43*(2), 631.

(2) Tang, W.; Peterson, A. A.; Varela, A. S.; Jovanov, Z. P.; Bech, L.; Durand, W. J.; Dahl, S.; Norskov, J. K.; Chorkendorff, I. The importance of surface morphology in controlling the selectivity of polycrystalline copper for CO_2 electroreduction. *Physical Chemistry Chemical Physics* **2012,** *14*(1), 76.

(3) Qiao, J.; Jiang, P.; Liu, J.; Zhang, J. Formation of Cu nanostructured electrode surfaces by an annealingâ €“electroreduction procedure to achieve high-efficiency CO_2 electroreduction. *Electrochemistry Communications* **2014,** *38*(0), 8.

(4) Kas, R.; Kortlever, R.; Milbrat, A.; Koper, M. T. M.; Mul, G.; Baltrusaitis, J. Electrochemical CO_2 reduction on Cu_2O-derived copper nanoparticles: controlling the catalytic selectivity of hydrocarbons. *Physical Chemistry Chemical Physics* **2014,** *16* (24), 12194.

Copyright © American Institute of Chemical Engineers. All rights reserved.

470478 *S Doped TiO_2 synthesized By Novel One Step Liquid Flame Spray Pyrolysis (LFSP) for Photocatalytic Oxidation of CO in Visible Region*

Abstract:

A novel single step Liquid Spray pyrolysis (LFSP) method is used to synthesize a series of S-TiO_2 catalyst of nanosized crystals with dimensional homogeneity and high monodispersity. In order to improve the visible light response and to further narrow the band bap of TiO_2 catalyst, Sulphur is used as the dopant in this method. With the change in the amount of S incorporated into TiO_2 catalyst, we observed the generation of the successive energy bands inside the TiO_2 band gap thus lowering the bandgap energy. The effects of nanosized S doped TiO_2 catalyst on the visible light photocatalytic oxidation of gaseous CO were investigated. Our XPS results indicate that the nitrogen doping concentration could reach up to 8 wt% owing to the homogeneous doping in the mesoporous TiO_2 catalysts. The nitrogen doped TiO_2 catalyst exhibit such excellent characteristics as high specific area, relatively small particle size, pure anatase phase and excellent UV-vis absorption capacity in the range of 400-800 nm, which are all beneficial to the photocatalytic oxidation of CO under the visible light irradiation. The photo conversion of CO to CO_2 is quantitative in the presence of O_2. The photocatalytic CO oxidation rates and the amounts of N deposited on TiO_2 surface show a close correlation. The addition of water vapor dose not affects significantly the CO photooxidation kinetics in the presence of O_2. The main oxidants in CO Photocatalytic oxidation seem to be produced from adsorbed O_2. The photocatalytic oxidation kinetics was investigated in detail as a function of S loading, coated TiO_2 mass, CO concentration, O_2 concentration, light intensity, additives, and humidity. These interesting results will be discussed in the presentation.

Copyright © American Institute of Chemical Engineers. All rights reserved.

464077 Optimizing the Physicochemical Properties of ZSM-11 for Methanol-to-Hydrocarbon Reactions

Abstract:

ZSM-11 (MEL) is an emerging catalyst for a wide variety of applications. The MEL framework is very similar to ZSM-5 (MFI), which is one of the most widely used zeolite catalysts in industry. Recent studies have shown that ZSM-11 has better catalytic performance in several different reactions compared to ZSM-5 [1-3], possibly due to improved intra-crystalline diffusion within the less-tortuous straight channels of the MEL topology. However, there are relatively few systematic studies of ZSM-11 synthesis, specifically with respect to tailored physicochemical properties such as size, shape, and Si/Al ratio (SAR). Here, we will present a systematic study of ZSM-11 catalyst preparation and testing for C1 applications. Through the judicious selection of synthesis parameters, we successfully developed a facile protocol to generate a library of ZSM-11 catalysts with tunable size and SAR. More precise modification in ZSM-11 crystal morphology (e.g. aspect ratio) can be accomplished through the use of zeolite growth modifiers (ZGMs), which are molecules or macromolecules that modulate the anisotropic growth rate of crystals with concomitant alterations in bulk habit [4, 5]. These methods for tailoring zeolite crystallization will be discussed along with catalytic studies using methanol-to-hydrocarbon (MTH) reactions. Collectively, our findings are part of a broader initiative to establish quantitative structure-performance relationships as a platform for optimizing zeolite catalysts.

References

[1] Kustova, M. Y.; Rasmussen, S. B.; Kustov, A. L.; Christensen, C. H., *Applied Catalysis B-Environmental* **2006**, *67* (1-2), 60-67.

[2] Zhang, L.; Liu, H.; Li, X.; Xie, S.; Wang, Y.; Xin, W.; Liu, S.; Xu, L., *Fuel Processing Technology* **2010**, *91* (5), 449-455.

[3] Bleken, F.; Skistad, W.; Barbera, K.; Kustova, M.; Bordiga, S.; Beato, P.; Lillerud, K. P.; Svelle, S.; Olsbye, U., *Physical Chemistry Chemical Physics* **2011**, *13* (7), 2539-2549.

[4] Lupulescu, A. I.; Rimer, J. D., *Angewandte Chemie International Edition* **2012**, *51* (14), 3345-3349.

[5] Lupulescu, A. I.; Kumar, M.; Rimer, J. D., *Journal of the American Chemical Society* **2013**, *135* (17), 6608-6617.

Copyright © American Institute of Chemical Engineers. All rights reserved.

470641 Precise Deposition of Pt Promoter Onto Silica Supported Cobalt for Fischer-Tropsch Synthesis

Abstract:

Cobalt-based catalysts are commercially predominant for Fischer-Tropsch synthesis due to their stability, activity and avoidance of the water gas shift reaction [1]. This is a structure sensitive reaction exhibiting high activity, low CH_4 selectivity and high C_5^+selectivity with particles no smaller than 4-6 nm [2]. The reducibility of cobalt is another important factor which is considered as a key to improve its performance, and this might be achieved by adding a noble metal which, via hydrogen spillover, can facilitate Co oxide reduction to the metal which is more active for CO hydrogenation mechanism [3]. The promotion of Co by a noble metal will be most efficacious if the noble metal is in direct contact with Co nanoparticles, and none is wasted on the surface of the support.

In this study, we demonstrate that such a selective deposition is possible. We first employ a variation of â€œstrong electrostatic adsorptionâ€ to synthesize silica supported cobalt oxide nanoparticles of the optimum size. Second, Pt is steered onto the cobalt oxide by controlling the impregnation pH. We compare this sample to a Pt/Co catalyst prepared by the common, undirected method of co-impregnation of Pt and Co. In both methods, with fixed 10wt% of cobalt and varying 1 to 0.03 wt% of Pt, samples were characterized by XRD, TPR, and STEM.

The catalysts were tested at conventional Fischer-Tropsch conditions of 220°C and 20 bar with H_2/CO ratio 2:1. The reactivity of the catalysts clearly depends on the preparation method, which can be attributed to the intimate coordination of Pt with Co. It is suggested that hydrogen spillover effects not only Co oxide reduction but also extends to the Fischer-Tropsch reaction itself.

References

[1] H. Jahangiri, J. Bennett *et al.,* Catal. Sci. Technol., 4 (2014) 2210.

[2] K. Cheng, V. Subramanian, *et al.,*Journal of Catalysis 337 (2016) 260.

[3] D. Nabaho, J.W. Niemantsverdriet, *et al.,* Catalysis Today 261 (2016) 17.

Copyright © American Institute of Chemical Engineers. All rights reserved.

472259 Using Inelastic Neutron Scattering to Probe the Active Site for Cobalt-Copper Catalyst for the Conversion of Syngas to Higher Oxygenates

Abstract:

Cobalt-copper catalysts have been proposed for the synthesis of ethanol and higher oxygenates as a substitute of Rh and other high-cost noble metal catalysts[1,2]. Two types of sites with atomic proximity are needed to form higher oxygenates: one to dissociate CO and a second to insert CO to the intermediates to form the CH_xCO intermediate [3]. Metallic cobalt is responsible for CO dissociation, while the nature of the site for CO insertion is still under study. We have utilized inelastic neutron scattering (INS) at the VISION beamline at SNS to probe intermediate surface species of this cobalt-copper catalyst. This unique technique allows for elucidation of mechanistic details of the CO insertion and subsequent CH_xCO intermediate formation on the metal surfaces (Co^0, Co_2C and/or Cu^0). XPS results show that the amount of surface carbon increases significantly after CO hydrogenation reaction. However, in-situ XRD profiles do not show any crystallinity changes of the catalyst in the presence of syngas at 250°C. For $CuCoLa_2O_3$, freshly reduced sample and post-reaction sample showed distinct XANES profiles. EXAFS results on used $CuCoLa_2O_3$ shows a peak at lower distance for the first Co-C coordination shells. Our INS spectra show several features at the low energy region (0-150 meV), which can be assigned as the vibration of the oxygenate intermediates on the surface, including methanol, ethanol and acetaldehyde. For instance, the peak at 40 meV is identical with the bending of C-C-O in the calculated spectrum for acetaldehyde. The experimental spectra are compared with the theoretical spectra of CH_xCO on different surfaces: Co^0, Cu^0, Co_2C and dual sites of Co^0-Cu^0 and Co^0-Co_2C.

This work was sponsored through the Louisiana Consortium for Neutron Scattering supported by the U.S. Department

of Energy under EPSCoR Grant No. DE-SC0012432 with additional support from the LA BOR. This research benefited from the use of the VISION beamline at ORNLâ€™s Spallation Neutron Source, which is supported by the Scientific User Facilities Division, Office of Basic Energy Sciences, U.S. Department of Energy under Contract No. DE-AC0500OR22725.

Copyright © American Institute of Chemical Engineers. All rights reserved.

470381 Selective C-O Bond Cleavage of Diphenyl Ether over the Supported Palladium Catalyst

Abstract:

Catalytic C-O bond cleavage of model diaryl ether compounds that are derived from the depolymerization process of lignincellulosic biomass over supported transition metal catalysts is one of the most important catalytic steps in the efficient conversion of lignincellulosic biomass to transportation fuels and valuable chemicals. Previous studies have shown that the C-O bond cleavage activity and the product selectivity of diaryl ethers, which are largely determined by the dominant pathway among three parallel pathways including hydrogenolysis, hydrolysis, and hydrogenation, vary from different transition metal catalysts, solvent environments and substituents.[1-3] In the present study, the reaction mechanisms of catalytic C-O bond cleavage of diphenyl ether over supported palladium catalyst has been investigated using density functional theory (DFT) calculations. It is found that hydrolysis, which is leading to the formation of phenol and cyclohexanol, is the dominant reaction pathway for aqueous phase diphenyl ether conversion in the presence of hydrogen. The partial hydrogenation of one of aromatic rings plays the key role in the selective C-O bond cleavage. The fundamental insights from DFT calculation results not only confirm our experimental observation, but also provide a general picture of how the selective C-O bond cleavage reactivity and selectivity over transition metal catalysts in aqueous phase.

References

(1) He, J.; Zhao, C.; Lercher, J. A. Ni-catalyzed cleavage of aryl ethers in the aqueous phase. *J. Am. Chem. Soc.* **2012,** *134*, 20768-20775.

(2) He, J.; Zhao, C.; Mei, D.; Lercher, J. A. Mechanisms of selective cleavage of Câ€"O bonds in di-aryl ethers in aqueous phase. *J. Catal.* **2014,** *309*, 280-290.

(3) He, J. Y.; Lu, L.; Zhao, C.; Mei, D. H.; Lercher, J. A. Mechanisms of catalytic cleavage of benzyl phenyl ether in aqueous and apolar phases. *J. Catal.* **2014,** *311*, 41-51.

Acknowledgements

This work is supported by the US Department of Energy (DOE), Office of Basic Energy Sciences, Division of Chemical Sciences, Geosciences & Biosciences. Pacific Northwest National Laboratory (PNNL) is a multi-program national laboratory operated for DOE by Battelle. Computing time was granted by the scientific theme user project of the William R. Wiley Environmental Molecular Sciences Laboratory (EMSL) and by the National Energy Research Scientific Computing Center (NERSC). EMSL is a national scientific user facility located at PNNL and sponsored by DOE Office of Biological and Environmental Research.

Copyright © American Institute of Chemical Engineers. All rights reserved.

456629 Metal Core/Surface Oxide Shell Catalysts for C-O Bond Activation

Abstract:

C-O bond scission is a key step for a variety of industrial processes ranging from Fischer-Tropsch synthesis to valorization of biomass. Bimetallic catalysts involving a base metal and a noble metal were found to be particularly effective for C-O bond activation in a variety of O-containing molecules (diols and triols, furans, pyrans, etc.). For example, recently it was shown that monodisperse nanocrystals (NC) with the Pt_3Co_2 stoichiometry are capable of converting 5-hydroxymethylfurfural (HMF) to 2,5-dimethylfuran (DMF) in the liquid phase with up to the 99% yield at moderate temperatures and pressures (180°C, 33 bar H_2)[1].

Herein, we use a combination of the state-of-the-art density functional theory, *ab initio* molecular dynamics coupled with simulated annealing, and microkinetic modeling to elucidate the structure and properties of the NC catalytic site. In addition, we develop a geometric model that allows comparison of first principles' structure predictions with *in situ* EXAFS/XANES measurements and *ex situ* XRD data. A bimetallic alloy core/surface oxide shell structure with Co segregated to the surface is found to be consistent with the experimental data. We identify the strong interaction between the Co_xO_y surface oxide and the underneath metal to be crucial for the kinetic stability of the oxide in a highly reducing environment. Finally, we elucidate the "reverse Mars van Krevelen" mechanism of hydrogenolysis.

References:

1. Luo, Yun, Mironenko, Goulas, et al. Mechanisms for High Selectivity in Hydrodeoxygenation of 5-Hydroxymethylfurfural over PtCo Nanocrystals. Under Review.

Copyright © American Institute of Chemical Engineers. All rights reserved.

465005 A Periodic DFT Study of the Influence of Lewis Acid Site Speciation on Ethanol Dehydration in Zeolites

Abstract:

Partially-hydrolyzed â€˜openâ€™ sites in Sn-Beta have been implicated as the active site for Baeyer-Villiger oxidation (1) and glucose isomerization (2) in the liquid phase; however, few DFT studies exist which investigate the catalytic consequences of active site speciation on the kinetics and mechanism of gas phase reactions catalyzed by Lewis acid sites in zeolites. Here, the kinetics of ethanol dehydration are studied in the zeolite frameworks chabazite (CHA) and Beta. Experimental and theoretical studies of ethanol dehydration on Î³-alumina support the conclusion that Lewis acid sites are the active sites (3-4), but site heterogeneity present in the material makes identification of structure-function relations difficult. Well-defined Lewis acid sites located within zeolitic frameworks, on the other hand, can be used here as a model system to assess the effects of site speciation and the confining environment on kinetic pathways.

Periodic DFT calculations were performed with the Bayesian Error Estimation Functional (BEEF-vdw) (5) to study the kinetics and thermodynamics of ethanol dehydration in Sn-Beta and Sn-CHA. Three types of sites are considered: a tetravalent Sn, a partially hydrolyzed Sn site formed by dissociation of water, and a partially hydrolyzed Sn site formed by dissociation of ethanol. Thermodynamic and kinetic barriers were calculated for the concerted and sequential pathways which form the experimentally measured products diethyl ether, ethylene, and acetaldehyde. At 404 K, ethanol partially hydrolyzed open sites are kinetically favorable over water partially hydrolyzed open sites in both Sn-Beta and Sn-CHA. Ethylene forms through a six-membered ring transition state on both closed and partially hydrolyzed Sn sites, while DEE can be formed either from an ethanol dimer or ethanol adsorbed on a partially ethanol-hydrolyzed Sn site. Acetaldehyde formation is suppressed at 404 K due to prohibitively high kinetic barriers in cleaving C$^{Î±}$ bonds. Detailed comparisons of intrinsic barriers in Sn-CHA and Sn-Beta are presented for each

pathway. In addition to reaction free energy diagrams, a detailed discussion of zeolite entropy calculations is included. Microkinetic modeling is used to differentiate the free energy pathways in conjunction with experimental measurements to develop a probe reaction. This probe reaction, once understood for Sn-CHA and Sn-Beta, can be extended to study the catalytic influence of metal atom identity and framework type.

(1) Boronat, M.; Concepcion, P.; Corma, A.; Renz, M.; Valencia, S., *Journal of Catalysis*, **2005**, 234, 111-118.

(2) Harris, J.W.; Cordon, M.J.; Di Iorio, J.R.; Vega-Vila, J.C.; Ribeiro F.H.; Gounder, R., *Journal of Catalysis*, **2016**, 335, 141-154.

(3) Roy, S.; Mpourmpakis, G.; Hong, D.Y.; Vlachos, D.G.; Bhan, A.; Gorte R.J., *ACS Catalysis*, **2012**, 2, 1846-1853.

(4) Jenness, G.R.; Christiansen, M.A.; Caratzoulas, S.; Vlachos, D.G.; Gorte, R.J., *JPCC*, **2014**, 118, 12899-12907.

(5) Wellendorff, J.; Lundgaard, K.T.; Mogelhoj, A.; Petzold, V.; Landis, D.D.; Norskov, J.K.; Bligaard, T.; Jacobsen, K.W., *Physical Review B*, **2012**, 85, 1-23.

Copyright © American Institute of Chemical Engineers. All rights reserved.

467903 Mechanistic Study of Regio-Selective Ring-Opening Reactions of Mono-Substituted Epoxides Using Lewis Acid Catalysts

Abstract:

Polyether polyols are important intermediates in the polyurethane industry and are typically made by reacting mono-substituted epoxides with polyalcoholic initiators in the presence of a catalyst.[1] While conventional catalysts such as potassium hydroxide and double metal cyanide catalysts result in less reactive polyols terminated with secondary hydroxyl groups, a family of Lewis acid catalysts, aryl boranes, has been found to achieve high regio-selectivity towards the desired primary alcohol functionality with minimal side reactions.[2] In addition, recent experiments revealed significant additive effects on enhancing the regio-selectivity towards primary alcohol products. However, the detailed catalytic mechanism has not been fully understood. In addition, the form of the catalyst is still unclear based on existing experimental methods that are unable to monitor the *in situ* speciation of this complex reaction system. To explore the mechanism of aryl borane-catalyzed regio-selective ring-opening reactions and elucidate additive effects, an *ab initio* study was conducted, followed by comprehensive microkinetic modeling. Density functional theory (DFT) was used to investigate the reaction pathways of ring-opening reactions in a model system using tris (pentafluorophenyl) borane as catalyst. Competitive binding of ligands to the catalyst was also studied. We identified several possible catalyst forms that contribute to ring-opening reactions with different regio-selectivities. The role of different additives was identified as serving as a co-catalyst that stabilized the transition state. The contributions of different reaction pathways was decomposed into two parts: the relative reaction rate constants and the availability of different catalytic forms, which resulted from competitive binding. Our microkinetic model captured all possible competitive binding and ring-opening reactions. Model predictions of the overall regio-selectivity were in good accord with experimental results under different reaction conditions. Ultimately, this study facilitates greater understanding of catalyzed regio-selective ring-opening reactions and provides a basis for the design of new catalyst/co-catalyst systems to fine tune the overall regio-selectivity.

(1) Chinn, H. K., A.; Loechner, U. *SRI Consulting* 2006.

(2) Nakaminami, H.; Sugahara, S.; Yasuhara, S.; Murata, K. U. US20120016049A1: 2012.

Copyright © American Institute of Chemical Engineers. All rights reserved.

467716 Understanding the p-Xylene Formation Mechanism from Dimethylfuran and Ethanol

Abstract:

Understanding the p-Xylene Formation Mechanism from Dimethylfuran and Ethanol

Pavlo Kostetskyy[1], Ivo Teixeira[2], Michail Stamatakis[3], Shik Chi Edman Tsang[2] and Giannis Mpourmpakis[1]

[1]Department of Chemical Engineering, University of Pittsburgh, Pittsburgh, PA 15621, USA

[2]Wolfson Catalysis Centre, Department of Chemistry, University of Oxford, Oxford, OX1 3QR, UK

[3]Department of Chemical Engineering, University College London, Torrington Place, London WC1E 7JE, UK

Abstract:

Renewable, biomass-based chemical feedstocks as alternatives to petrol-based commodities are of high industrial relevance[1,2] and have attracted recent attention. A number of platform chemicals have been identified as industrially-important, including cyclic oxygenates such as 2,5-Dimethylfuran (DMF)[2,3], which can be catalytically converted to p-xylene, an important precursor in the production of polymers. Acidic heterogeneous catalysts can catalyze the conversion of DMF to p-xylene by a Diels-Alder (DA) reaction with ethylene. The reaction can be divided into two stages: DA cycloaddition and cycloadduct water elimination (dehydration), with the DA being the rate-controlling step in the Bronsted acid-catalyzed (BA) reactions[3]. In this work we propose an alternative route for DMF conversion to p-xylene using BA zeolite catalyst and ethanol as the dienophile source. Using electronic structure calculations, we

elucidate the detailed reaction pathways and associated energetics and demonstrate that the ethanol-based pathway is preferred to ethylene in terms of overall reaction rates. We demonstrate that the sequence of reactions in the ethanol route follows a proton affinity[4] thermodynamic preference of the reacting species, initiated by ethanol dehydration to ethylene. Improved performance is rationalized by the generation of a water molecule in this first step of the mechanism. While absent in the ethylene route, the presence of additional water in the system reduces the entropy loss in key elementary steps and facilitates proton transfer reactions. As a result, activation free energy barriers for the catalytic cycle decrease relative to the ethylene pathway. The calculated activation free energy difference (using energetic span model) between the two routes is in excellent agreement with kinetic experiments.

References:

1. Kunkes, E. L.; Simonetti, D. A.; West, R. M.; Serrano-Ruiz, J. C.; Gartner, C. A.; Dumesic, J. A., *Science* **2008,** *322*, 417-421.

2. Williams, C.L.; Chang, C.C.; Do, P.T.; Nikbin, N.; Caratzoulas, S.; Vlachos, D.G.; Lobo, R.F.; Fan, W.; Dauenhaue, P.J.; *ACS Catal.* **2012,** 2, 935-939.

3. Nikbin, N.; Do, P.T.; Caratzoulas, S.; Lobo, R.F.; Dauenhauer, P.J.; Vlachos, D.G.; *J. Catal.* **2013,** 35-43.

4. Kostetskyy, P.; Maheswari, J. P.; Mpourmpakis, G., *J. Phys. Chem. C* **2015,** *119*, 16139-16147

Copyright © American Institute of Chemical Engineers. All rights reserved.

465210 Role of Enol-Keto Tautomerization in m-Cresol Selective Hydrogenation on Ni(111) and Nife(111) Surface

Abstract:

Using density functional theory, the reaction mechanism for selective hydrogenation of m-cresol on Ni (111) and NiFe (111) has been systematically investigated. Our results show that on Ni (111), the initial hydrogenation occurs at the nearest adjacent site of phenyl-OH forming 3-methyl-1,3-cyclohexadienol followed by the formation of unsaturated intermediate 3-methyl-cyclohexadienol, whereas on NiFe(111), the first step is the enol-keto tautomerization forming 3-methyl-3,5-cyclohexadienone which is more favorable than hydrogenation of the phenyl ring. After that, two products, i.e., 3-methylcyclohexanone and 3-methylcyclohexanol, can be formed on Ni(111), where the former is through enol-keto tautomerization and the latter is through further hydrogenation. In comparison, 3-methyl-3,5-cyclohexadienone intermediate adsorbed on NiFe(111) can proceed via partial hydrogenation forming 3-methylcyclohexanone or via complete hydrogenation forming 3-methylcyclohexanol. The reaction pathway via phenoxy intermediate is less likely on either Ni (111) or NiFe(111) due to strongly-bound phenoxy intermediate which can only serve as a spectator. The calculated reaction energy scales well with activation energy for representative elementary steps, demonstrating good BrÃ¸nstedâ€"Evansâ€"Polanyi relationship. Moreover, the calculated reaction energies agree well with the experimental observations that the conversion from 3-methylcyclohexanone to 3-methylcyclohexanol or vice versa on Ni(111) is more favorable than that on NiFe(111). Overall, selective/complete hydrogenation of m-cresol on NiFe (111) towards final products of ketone/alcohol is kinetically more favorable than that on Ni(111). Our calculations confirms that alloying Fe into Ni can promote the alternative reaction pathway, i.e., via enol-keto tautomerization, rather than the conventional hydrogenation of phenyl ring of m-cresol on Ni(111).

Copyright © American Institute of Chemical Engineers. All rights reserved.

462354 Developing Multi-Scale Models of Bimetallic Catalysts for the Hydrodeoxygenation of Bio-Oil Compounds

Abstract:

With the ever-increasing need to find a sustainable source of renewable energy, bio-oil produced via the fast pyrolysis of biomass is a promising source of liquid fuel; however, the resulting bio-oil contains oxygenated products that contribute to poor fuel quality [1]. Hydrodeoxygenation (HDO) is used to refine the bio-oil by reducing the oxygen content, ideally using a minimal amount of H_2 in order to form H_2O. Bimetallic catalysts such as Pd on Fe have demonstrated synergistic behavior that contributes to a more cost-effective and longer-lasting catalyst [2-4]. Similarly, Pt/Sn catalysts have also exhibited traits favorable for the HDO of phenolic compounds, namely a propensity to promote lower temperature desorption of aromatic compounds like benzene [5]. Identifying the cause of the different synergetic effects of these systems can provide insight into the synthesis of a superior HDO bimetallic catalyst. However, the behavior of bimetallic catalysts is complex and necessitates a thorough understanding of the nanoscale behavior of these systems in order to develop a truly predictive model on relevant time scales. We present two DFT studies examining first, the interactions between a surface and its adsorbates and second, the interactions between vicinal adsorbates on a surface.

While the addition of Pd to Fe catalysts exhibits a synergistic interaction [2-4], further investigation is needed in order to conclude with greater certainty the role of Pd. We have developed a preliminary microkinetic model of the dissociation of H_2O on Fe (100) and on two surface alloys with different concentrations of Pd. We found that increasing the surface concentration of Pd from 0% to 50% caused a ~0.7 eV decrease in the adsorption strength of O, while only a ~0.1 eV decrease in the adsorption of H. This trend is consistent with other studies that investigated the influence of noble metal dopants on the adsorption of benzene and phenol [6, 7]. Moreover, after determining the activation energies for each reaction step from linear scaling relationships [8], it was found that Pd increases the barriers for both the dissociation of H_2O in OH and H and the subsequent dissociation of OH. These results show that the primary function of Pd is the disruption of the formation of an oxide, due to the aversion O demonstrates for Pd in the Pd/Fe surfaces. Consequently, the risk of catalyst deactivation due to the formation of a surface oxide decreases with the addition of a Pd dopant on an Fe surface, thereby promoting the HDO with minimal deactivation of the active Fe surface.

Quantifying the lateral interactions between adspecies is also necessary in the development of truly predictive, theoretical models for bimetallic catalysts as they will significantly affect the coverage and diffusion of adspecies and therefore the heterogeneous catalytic environment. However, such a characterization of adspecies' lateral interactions for the highly complex HDO reaction has not yet been attempted. To such end, we have characterized the benzene-benzene lateral interactions on Pt (111) and PtSn (111) surfaces. These systems have been well-characterized experimentally using temperature-programmed desorption (TPD) [5]. Here, we studied the effect of surface coverage on the adsorption energy of benzene on Pt (111) and Pt_3Sn (111) at different adsorption sites by modeling the TPD of benzene. We found that a mean field model is sufficient in describing the interactions between vicinal benzene molecules (Figure 1). And while it was previously speculated that the broad desorption peak of benzene on Pt (111) was the result of desorption from two adsorption sites of different binding strengths [5], our results indicate that the behavior is more likely due to coverage effects on the surface. Furthermore, we found that benzene's adsorption was significantly weakened on the Pt_3Sn (111) as compared to Pt (111), which is consistent with experimental TPD [5]. Our investigation of modeling the lateral interactions of an aromatic molecule on both the Pt (111) and the Pt_3Sn (111) surfaces justifies future application of our method onto systems less experimentally characterized, such as doped Fe surfaces.

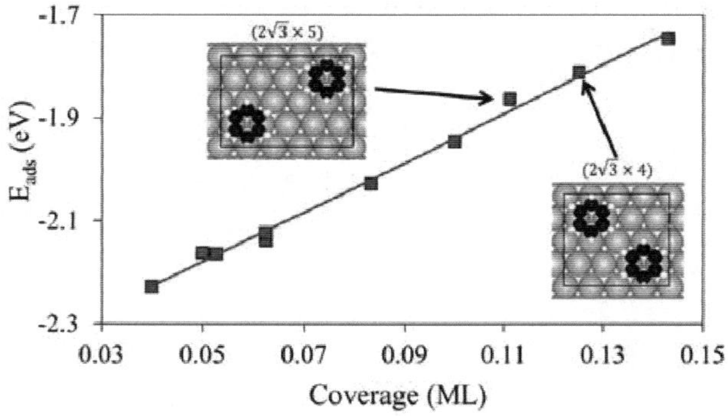

Figure 1. The adsorption energy for benzene on Pt (111) as a function of the benzene coverage.

References

[1] H. Wang, J. Male, Y. Wang, ACS Catalysis 3 (2013) 1047-1070.

[2] Y. Hong, H. Zhang, J. Sun, K. M. Ayman, A. J. R. Hensley, M. Gu, M. H. Engelhard, J.-S. McEwen, Y. Wang, ACS Catalysis 4 (2014) 3335-3345.

[3] J. Sun, A. M. Karim, H. Zhang, L. Kovarik, X. S. Li, A. J. Hensley, J.-S. McEwen, Y. Wang, Journal of Catalysis 306 (2013) 47-57.

[4] A. J. R. Hensley, Y. Hong, R. Zhang, H. Zhang, J. Sun, Y. Wang, J.-S. McEwen, ACS Catalysis 4 (2014) 3381-3392.

[5] C. Xu, Y. L. Tsai, B. E. Koel, The Journal of Physical Chemistry 98 (1994) 585-593.

[6] A. J. R. Hensley, R. Zhang, Y. Wang, J.-S. McEwen, The Journal of Physical Chemistry C 117 (2013) 24317-24328.

[7] A. J. R. Hensley, S. Schneider, Y. Wang, J.-S. McEwen, RSC Advances 5 (2015) 85705-85719.

[8] C. Tsai, K. Lee, J. S. Yoo, X. Liu, H. Aljama, L. D. Chen, C. F. Dickens, T. S. Geisler, C. J. Guido, T. M. Joseph, C. S. Kirk, A. A. Latimer, B. Loong, R. J. McCarty, J. H. Montoya, L. Power, A. R. Singh, J. J. Willis, M. M. Winterkorn, M. Yuan, Z.-J. Zhao, J. Wilcox, J. K. Nørskov, Catalysis Letters (2016) 1-7.

Copyright © American Institute of Chemical Engineers. All rights reserved.

468057 Combined Density Functional Theory and Molecular Dynamics Study of Aqueous Reforming of Glycerol on Pt (111) Via Competitive C-H and O-H Dehydrogenation

Abstract:

Glycerol is the most abundant byproduct in biodiesel production, and it can be catalyzed into H_2 and valuable liquid phase products such as 1-propanol, lactic acid, and ethylene glycol through aqueous phase reforming (APR). In this work, we use computational methods to study reactions in Pt (111)-catalyzed glycerol dehydrogenation pathways under aqueous conditions. Studying the reaction in aqueous conditions adds complexities to the analysis, since liquid systems involve large numbers of configurations that contribute to the system energies, meaning that multiple configurations must be sampled for each reaction intermediate. Given that glycerol decomposition can include dehydrogenation, dehydration, decarboxylation, hydrolysis and other processes, to study all of the reactions along all the pathways for glycerol APR would require significant amount of effort, while only a fraction of them are of importance to the overall mechanism. Because of the large reaction network and the need to study it in liquid conditions, we are seeking ways to rapidly estimate energies of reaction intermediates under aqueous conditions, in order to identify the most important intermediates and reactions. We present a two-step process aimed at rapidly estimating the energies of intermediates as well as their interaction energies with liquid water. Firstly, the binding energies of intermediates are estimated in vacuum by a linear scaling relationship obtained from density functional theory (DFT) results. Secondly, the structures are simulated in an explicit liquid water environment using molecular dynamics (MD). In this step, Lennard-Jones plus Coulombic (LJ+C) potentials are used to calculate the water/adsorbate interaction energies. Aqueous phase reaction energies are then calculated by summing the vacuum phase reaction energies and the difference in interaction energies between products and reactants. Using this approach, we evaluate intermediates and reaction steps in the aqueous phase glycerol decomposition reaction network, in an attempt to begin to understand the mechanism of glycerol APR. Our estimations are validated using a fully-DFT procedure previously reported by Bodenschatz, Sarupria, and Getman (J. Phys. Chem. C, 2015). Future work aimed at rapidly estimating the kinetics of these reactions is discussed.

Copyright © American Institute of Chemical Engineers. All rights reserved.

465601 Molecular Dynamics Simulations of Aldol Condensation Catalyzed By Alkylamine-Functionalized Crystalline Silica Surfaces

Abstract:

Molecular Dynamics Simulations of Aldol Condensation Catalyzed by Alkylamine-Functionalized Crystalline Silica Surfaces

Ki Chul Kim,[1] Eric G. Moschetta,[2] Christopher W. Jones,[2] and Seung Soon Jang[1]

[1]*School of Material Science and Engineering, Georgia Institute of Technology, 771 Ferst Drive NW, Atlanta, GA 30332-0245, USA*

[2]*School of Chemical and Biomolecular Engineering, Georgia Institute of Technology, 311 Ferst Drive NW, Atlanta, GA 30332-0100, USA*

Abstract

Molecular dynamics simulations are performed to investigate the cooperatively catalyzed aldol condensation between acetone and 4-nitrobenzaldehyde on alkylamine (or alkylenamine)-grafted silica surfaces, focusing on the mechanism of the catalytic activation of the acetone and 4-nitrobenzaldehyde by the acidic surface silanols followed by the nucleophilic attack of the basic amine functional group toward the activated reactant. From the analysis of the correlations between the catalytically active acid-base sites and reactants, it is concluded that the catalytic cooperativity of the acid-base pair can be affected by two factors: 1) the competition between the silanol and the amine (or enamine) to form a hydrogen bond with a reactant and 2) the flexibility of the alkylamine (or alkylenamine) backbone. Increasing the flexibility of the alkylamine facilitates the nucleophilic attack of the amine on the reactants. From the molecular dynamics simulations, it is found that C3 propylamine and C4 butylamine linkers exhibit the highest probability of reaction, which is consistent with the experimental observation that the activity of the aldol reaction on mesoporous silica depends on the length of alkylamine grafted on the silica surface. This simulation work serves as a pioneering study demonstrating how the molecular simulation approach can be successfully employed to investigate the cooperative catalytic activity of such bifunctional acid-base catalysts.

Copyright © American Institute of Chemical Engineers. All rights reserved.

464848 Droplet Microfluidics with a Nanoemulsion Continuous Phase: Application in Reaction System Design for Material Synthesis

Abstract:

Recently, nanoemulsions have been incorporated into the *dispersed phase* in droplet microfluidics for the synthesis of hierarchical structured materials in which the different length scales between the nano-droplets and micro-droplets are exploited. In this study, we present a novel, generalizable method that uses a nanoemulsion as the *continuous phase* to generate uniform micro-droplets. The micro-droplets act as well-mixed micro-reactors/reservoirs with walls that are 'permeable' to the surrounding nanodroplets, with versatile control of 'permeability' by chemical, thermal, and electrical forces. The droplet generation physics and mass transfer processes between the micro- and nano-droplets are studied. Finally, we demonstrate the applications of this system in material synthesis for gel micro-beads, drug crystals, and nanoparticles. Our work expands the toolbox of droplet-based microfluidics, enabling new opportunities and applications involving active colloidal continuous phases carrying chemical payloads.

Copyright © American Institute of Chemical Engineers. All rights reserved.

461406 Activating C_1 Chemistry Utilizing Microreactor Technology

Abstract:

The activation of C_1 chemistry has been the goal of engineers and scientists for over a century. The single most energy intensive step in petrochemical industry is the formation of C_2 hydrocarbons which dominates its current energy consumption [520 Petajoule] and CO_2 production [~18.5MtC, ~22% of US chemical synthesis] in the US.[i] Discovering adequate technological mitigations of this energy demand using renewable energy is one of the keys to advancing clean fuels and our economyâ€™s renewable energy production. Currently, C_1 to C_2 chemistry does not exist at an energetically competitive alternative to steam cracking for ethylene production. Microreaction technology is finding perhaps its greatest utility in pharmaceutical and fine chemical production. However, microtechnology also has tremendous potential to address energy production, storage and utilization when used in conjunction with microplasma technology, which intrinsically lends phenomenal gas processing capacity to microdevices. Microplasma microreactor technology can be applied at high efficiency for conversion of renewable electrical energy to chemical energy. Microreactor devices utilizing microcorona discharges can process sources of methane, carbon dioxide and steam into usable C_2 hydrocarbons using renewable electricity. Experiments have demonstrated high conversion of C_1 to C_2 hydrocarbons with low electrical energy consumption. Each device provides the processing capacity for several hundred standard cubic centimeters per minute. The power consumption for the engineered microcoronas is several watts. Several alternate reactions involving C_1activated chemistry, that have been explored, will be summarized and detailed in terms of design analysis, reactivity, conversion, selectivity, and gas hour space velocities as it relates to the apparent efficiency. Moving corona microreactor gas processing technology toward the forefront of industrial applications that are focused on converting renewable electricity into chemical energy will require a broader collaboration between engineers, chemists and physicists.

The fundamental length scales of microreactors reduce diffusive mass transfer resistance and allow full utilization of the corona electron cross-section which is a key factor in achieving good performance. This has the added benefit of effectively tuning the distribution of activating charges to the optimum energetic level. The conversion of renewable electrical energy to chemical energy is investigated on a variety of microreactor electrical corona discharge systems. Fields of atmospheric pressure plasmas have been engineered to convert from 5 to 70% of an inlet stream to produce higher more desirable chemical feedstocks and higher molecular weight species at 50-90% electrical efficiency. The ability to bypass activation energy barriers in microreactor microplasma systems is similar to the catalytic reduction of the activation energy for non-plasma C_1 chemistries. The microtechnology reaction platform and measurement techniques allow easy scaling of the technology. A framework for understanding C_1activation and functional microreactor design is discussed. Spectroscopic, chromatographic and electric measurements are used to establish key reaction engineering relationships between the reduced electric field, space velocity, mass transfer, kinetic rates and time scales. Condensable fuel products have been measured using gas chromatography, gas phase Raman or FTIR.

---

[i] U.S. Dept. of Energy, â€œEnergy Use and Energy Intensity of the U.S. Chemical Industry,â€ EPA/DOE- LBNL-44314 Ernest Orlando Lawrence Berkeley National Laboratory, Berkeley, CA (rev. April 2000).

Copyright © American Institute of Chemical Engineers. All rights reserved.

462081 Gas- Liquid Flow Hydrodynamics and Mass Transfer in Micro-Packed Beds

Abstract:

In this paper, the hydrodynamics and gas-liquid mass transfer characteristics of gas-liquid two-phase flow in the micro-packed bed were investigated. The hydrodynamics, including pressure drop, residence time distribution and liquid holdup were determined. The effect of gas and liquid superficial velocity, glass bead size, packed bed diameter and liquid property on the hydrodynamics has been studied. The results showed that the pressure drop and liquid holdup in micro-packed beds were higher than that in conventional packed beds due to the effect of capillary pressure. A mathematical model has been developed to describe the hydrodynamics in the micro-packed beds by taking into account the capillary force. The volumetric liquid-side mass transfer coefficients (k_La) was evaluated by the chemisorption of CO_2 in methyl diethanolamine water solution. An online measurement method was developed with in-line FTIR, which was used to measure bicarbonate ion concentration to calculate CO_2 absorption rate. The influence of gas and liquid superficial velocities, packed bed size and packing particle size on the k_La was examined. The results showed that k_La was in the range of 0.24~0.7 s^{-1}, which is about an order of magnitude than that of conventional trickle bed. An empirical correlation was developed to predict the k_La at different conditions.

Copyright © American Institute of Chemical Engineers. All rights reserved.

463959 Implementation of in Situ SAXS/Waxs Analysis in Silicon-Pyrex Microreactor

Abstract:

Optically transparent integrated microfluidic devices are widely used in biology, chemistry and material science mainly because optical in situ characterization, such as microscopic observations, dynamic light scattering, or light spectroscopy (UV-vis, Raman, photoluminescence), can be easily performed inside microchannels. Additionally, thanks to the advancement of synchrotron sources, Small-Angle and Wide-Angle X-ray scattering (SAXS/WAXS) studies utilizing microfluidic devices are more and more used, notably to characterize the morphology and the structure of proteins, nucleic acids, biomacromolecules, or inorganic nanoparticles. Although conventional transparent microfluidic systems (PDMS, PMMA, COP, PS) make excellent prototypes, they have limited operating pressure and temperature ranges, as well as poor chemical compatibility with most of the organic solvents. Alternative experimental setups can overcome some of the above difficulties by considering glass or sapphire capillaries. Their high compatibility with many organic solvents and their good resistance to aggressive chemicals extends the scope of fluids that can be used. However, glass tubings suffer from a low design flexibility and exhibit poor heat transfers. Additionally, it is extremely difficult to design fancy mixing devices in a glass capillary. The development of on-chip silicon/glass microsystems has addressed the optimization of both the experimental approaches towards the use of a large panel of fluids and solvents, as well as the extent of the operating conditions of these microsystems at high temperature and high pressure. This oral presentation will highlight the successful implementation of in situ X-ray scattering analysis of synthetized particle materials in silicon/glass microreactors is reported. Calcium carbonate ($CaCO_3$) as a model material was precipitated inside the microchannels. The synthesized calcite particles were analyzed in situ in aqueous media by combining Small Angle X-ray Scattering (SAXS) and Wide Angle X-ray Scattering (WAXS) techniques at the ESRF ID02 beam line. We will in particular focus on the benefits of such approaches for accessing continuously nano/micromaterials characteristics (shape, crystalline structures). This study

demonstrates that silicon/glass chips are potentially powerful tools for in situ SAXS/WAXS analysis and are promising for studying the structure and morphology of materials in non-conventional conditions including high pressure and high temperature.

Copyright © American Institute of Chemical Engineers. All rights reserved.

464105 Automated Oscillatory Photochemical Reactor for High Throughput Studies of Visible-Light Photoredox Catalysis

Abstract:

Over the past decade, visible-light photoredox catalysis using metal complexes (e.g., polypyridyl complexes of ruthenium and iridium) has steadily been developed as a promising strategy for sustainable and green synthesis of fine chemicals.[1] The relatively long lifetime (~ 1Î¼s) associated with the photoexcited states of metal complexes may result in a bimolecular electron transfer pathway (chemical reaction) instead of deactivation. For instance, photoredox catalysis has successfully been employed for batch scale coupling reactions, reductive dehalogenation, and oxidative hydroxylation.[2] However, the inverse correlation of the reaction vessel size and penetration depth of the irradiated light has resulted in reaction times in the order of hours. The high surface area to volume ratio offered by microscale flow chemistry technologies has addressed the aforementioned limitation of batch scale photochemical reactors by reducing the characteristic reaction vessel length scale from tens of centimeters to hundreds of micrometers.[3] Nevertheless, the direct correlation of the mixing and residence times and limited range of residence times for a pre-defined reactor length in combination with ~ mL reagent volume required per reaction condition make it challenging to employ continuous flow chemistry approaches for high-throughput screening, characterization, optimization and library development of photoredox catalysis reactions.

In this project, capitalizing on the removed residence time limitation and enhanced mixing and mass transfer advantages of oscillatory flow strategy,[4,5] a microscale photochemistry platform is developed for in-flow studies of visible-light photoredox catalysis. Position of the formed droplet (micro-reaction vessel) at the inlet and outlet of the oscillatory flow reactor is detected through a single-point optical detection, integrated within a custom-machined aluminum chuck. The optical feedback provided through the single-point position detection allows for automated switching of the flow direction of the carrier phase to ensure the droplet is always under the same irradiation intensity

over the course of the photoredox catalysis process.

The developed experimental platform allowed for the effect of irradiation light intensity on the yield (obtained using in-flow LC/MS) and selectivity of the photoredox catalysis to be precisely characterized by automatic tuning of the irradiation power of the high power LED (through LabView). In addition, utilizing gas as the carrier phase in both sides of a droplet that is pre-formed via a computer-controlled liquid handler (containing the desired photocatalyst) provided sufficient gas molecules during the photoredox catalysis using a reactive gas as an oxidant (e.g., oxygen). Through adjusting the pressure of the carrier phase, the effect of gas concentration (e.g., oxygen pressure) on the photoredox catalysis (e.g., oxidative hydroxylation of phenylboronic acids) was studied.

The proposed experimental setup enables material efficient high-throughput screening and optimization of continuous (e.g., reaction time and concentration of the photocatalyst) and discrete (e.g., different metal complexes, and reaction solvents) parameters associated with a photoredox catalysis process using only 20 Î¼L volume of the solution mixture per experimental condition. The obtained optimized parameters (e.g., photocatalyst molecule structure, concentration, solvent, irradiation power, and reaction time) will then be employed for large-scale (numbered up) continuous synthesis of the desired product under a similar characteristic length scale.

References

(1) Prier, C. K.; Rankic, D. A.; MacMillan, D. W. C. *Chemical Reviews* **2013**, *113*, 5322-5363.

(2) Zou, Y.-Q.; Chen, J.-R.; Liu, X.-P.; Lu, L.-Q.; Davis, R. L.; Jørgensen, K. A.; Xiao, W.-J. *Angewandte Chemie* **2012**, *124*, 808-812.

(3) Tucker, J. W.; Zhang, Y.; Jamison, T. F.; Stephenson, C. R. J. *Angewandte Chemie International Edition* **2012**, *51*, 4144-4147.

(4) Abolhasani, M.; Coley, C. W.; Jensen, K. F. *Analytical Chemistry* **2015**, *87*, 11130-11136.

(5) Abolhasani, M.; Bruno, N. C.; Jensen, K. F. *Chemical Communications* **2015**, *51*, 8916-8919.

Copyright © American Institute of Chemical Engineers. All rights reserved.

454255 Kinetic Study of Reactions of Aniline and Benzoyl Chloride in a Microstructured Chemical System

Abstract:

Kinetic Study of Reactions of Aniline and Benzoyl Chloride in a Microstructured Chemical System

Peijian Wang, Kai Wang, Jisong Zhang, Guangsheng Luo*

The State Key Laboratory of Chemical Engineering, Department of Chemical Engineering, Tsinghua University, Beijing 100084, China

Abstract

The reaction between aniline and benzoyl chloride is well known for producing high performance polymers and chemicals. But the fast main reaction and reversible consecutive side reaction make process rather complicated and no thorough kinetics study has ever been reported before. A microstructured chemical system, consisting of a micromixer, pre-heat exchange coiled capillaries, a delay loop and a microhydrolyzer, was designed to carry out this reaction in this study using NMP as the solvent. A kinetic model was established, and the values of kinetic constants, pre-exponential factors and activation energies of each reaction as well as their confidence intervals were acquired. With this model each substance concentration versus residence time were obtained to help understand reaction process and further optimize operating conditions. Compared to earlier reports on this reaction, this microstructured chemical system provides more accurate parameters and can develop a reliable platform for kinetic study of similar chemical reactions.

Keywords: kinetics; aniline; benzoyl chloride; NMP; microstructured chemical

---

Correspondence concerning this article should be addressed to Prof. Guangsheng Luo at gsluo@tsinghua.edu.cn

Copyright © American Institute of Chemical Engineers. All rights reserved.

452787 Brock Copolymer Synthesis in a Microreactor By Precise Analysis and Design of Polymerization Kinetics

Abstract:

Living polymerization has enabled to churn out a variation of functionalized polymers with well-defined properties. These methods of the precise polymerization are based on equilibrium approach. A kinetic aspect of the polymerization process is not well understood. Our study aimed to produce a block copolymer of norbornene by utilizing reaction kinetics. Microreactors consist of SUS tubing and tees were employed in our study due to their excellent performance for defining reaction conditions in high-resolution. The jumping operation of temperature and concentration was designed based on the rate analysis. The sequence of monomer addition, temperature and residence time of each step were optimised for maximizing the concentration of active intermediates. Gradient polymer elution chromatography results confirmed the production of block copolymers in our process.

Copyright © American Institute of Chemical Engineers. All rights reserved.

470691 Alloy-Core@Shell Nanoparticles for Catalysis

Abstract:

Alloy-core@shell nanoparticles are designed to combine the strengths of the alloy and core@shell particles, and avoid their flaws. The noble shell protects the particle core during the catalytic processes, and the alloy-core composition allows for fine tuning of the catalytic properties.[1,2] The Dendrimer-encapsulated nanopartilces (DENs) as a model catalyst is sufficiently small and well-characterized that its function can be directly predicted by theory. Specifically, our work seeks to develop a fundamental and detailed understanding of the relationship between the structure of nanoscopic electrocatalysts and their function. Two categories of stories will be given in the talk:

1) **tailoring catalytic function by tuning compositions**: Trends in reaction descriptor (binding energies of key reactants) were calculated with density functional theory (DFT) to probe the catalytic activity of two types of alloy nanoparticles structures: random alloy (X/Y)[3] and alloy-core@shell (X/Y@Z)[1,2] with various compositions. Establishing the general principal of correlation between compositions and activities provides guidelines for designing novel catalysts.[4] First-principal theory prediction of PdAu@Pt DENs for ORR were examined by experiment and leads to a great agreement with experiment results.[2]

2) **enhanced stability by alloying core:** $Pd_xAu_{140\_x}$@Pt DEN electro catalysts were tested for CO oxidation at wider alloying range.[5] Both experiments and DFT calculations suggest that this unusual behavior is caused primarily by structural changes of the DENs at high and low values of x. The alloy PdAu cores stabilize the core@shell structures by preventing Au and Pd from escaping the core.[5] These findings illustrate the importance of controlling both the stability and reactivity of the catalysts, and they provide guidance as to how core composition can be used to do that.

Overall, we demonstrate that iteration between theory and experiment can facilitate an understanding of nanoparticle catalysts and reduce the time and effort involved in the design of new catalysts.[6,7]

[1] L. Zhang, G. Henkelman, J. Phys. Chem. C **116** 20860-20865 (2012).

[2] L. Zhang, R. Iyyamperumal, D. F. Yancey, R. M. Crooks, and G. Henkelman, *ACS Nano* **7**, 9168-9172 (2013).

[3] W. Tang, L. Zhang, G. Henkelman *J. Phys. Chem. Lett.* 2011, **2,** 1328-1331.

[4] L. Zhang and G. Henkelman, *ACS Catal.* **5**, 655-660 (2015)

[5] L. Luo, L. Zhang, G. Henkelman, and R. M. Crooks, *J. Phys. Chem. Lett.* **6**, 2562-2568 (2015)

[6] L. Zhang, R. M. Anderson, R. M. Crooks, and G. Henkelman, *Surf. Sci.* **640** 65-72 (2015)

[7] R. M. Anderson, D. F. Yancey, L. Zhang, S. T. Chill, G. Henkelman, and R. M. Crooks, Acc. Chem. Res. **48** 1351-1357 (2015).

Copyright © American Institute of Chemical Engineers. All rights reserved.

451367 Thermal Stability of Shape-Controlled Pd-Pt Core-Shell Nanoparticles

Abstract:

Thermal Stability of Shape-Controlled Pd-Pt Core-Shell Nanoparticles

Luke T. Roling[a], Ahmed O. Elnabawy[a], Madeline Vara[b], Xue Wang[b], Younan Xia[b], and Manos Mavrikakis[a]

[a]Department of Chemical & Biological Engineering, University of Wisconsin-Madison, Madison, WI
[b]Department of Biomedical Engineering, Georgia Institute of Technology, Atlanta, GA

We recently reported the synthesis and electrocatalytic performance of shape-controlled Pd-Pt core-shell nanoparticles that demonstrated exceptional activity and stability toward the oxygen reduction reaction.[1,2] In this presentation, we investigate and explain the durability of these catalysts in high-temperature environments typical of a broader range of catalytic applications. We observe that the cubic and octahedral nanocatalysts are stable to at least 500 °C; interestingly, the core-shell structure and overall particle shape are each deformed at different temperatures in the two catalyst geometries.

Detailed DFT calculations rationalize the observed behaviors, attributing the differences in reconstruction to lower activation energy barriers for metal atom migration from the nanoparticle edges. Calculations also indicate that different concentrations of subsurface vacancies are seeded during nanoparticle synthesis, and that these are responsible for differences in Pd-Pt interdiffusion, which causes disruption of the core-shell structure at elevated temperatures.

1. S. Xie, S.-I. Choi, N. Lu, L. T. Roling, J. A. Herron, L. Zhang, J. Park, J. Wang, M. J. Kim, Z. Xie, M. Mavrikakis, Y. Xia, *Nano Letters* **2014**, 14, 3570
2. J. Park, L. Zhang, S. Choi, L. T. Roling, N. Lu, J. A. Herron, S. Xie, J. Wang, M. J. Kim, M. Mavrikakis, Y. Xia, *ACS Nano* **2015**, 9, 2635.

Copyright © American Institute of Chemical Engineers. All rights reserved.

462615 Accelerating Nanoparticle Synthetic Development Using in Situ Small Angle X-Ray Scattering

Abstract:

The fine control of monodisperse nanoparticles (NPs) over their size, shape, and structure is essential for understanding their properties, especially catalytic properties. In the past two decades, significant developments have been achieved in the colloidal synthesis of monodisperse NPs. However, synthesizing well-defined NPs in a predictable way is still challenging. Here, we use synchrotron-based in situ small-angle X-ray scattering (SAXS) to experimentally monitor the formation of NPs during solution phase synthesis. We choose Pd NPs as an example. Through quantitative analysis of the SAXS data, the number, size, and size distribution of Pd NPs during the synthesis are determined. By real-time probing the evolution of Pd NPs, we systematically study the effects of different surfactants (e.g., oleylamine, trioctylphosphine) on the kinetics of NP nucleation and growth. Taking advantage of this in-situ SAXS technique, we are able to better understand the growth mechanism of Pd NPs and as a result optimize the synthetic parameters to precisely control the Pd NP size from 2 to 10 nm, and these well-controlled Pd NPs serve as a model system for studying their size-dependent catalysis for methane combustion reaction.

Copyright © American Institute of Chemical Engineers. All rights reserved.

464480 Selective Oxidation of Methane to Methanol Using Pd/Au Bimetallic Nanoparticles

Abstract:

Selective Oxidation of Methane to Methanol using Pd/Au Bimetallic Nanoparticles

Cody Wrasman[1], Joshua J. Willis[1], Jong Suk Yoo[1], Jens K. Nørskov[1], Matteo Cargnello[1]

[1]*Department of Chemical Engineering and SUNCAT Center for Interface Science and Catalysis, Stanford University, Stanford, CA 94305, USA*

Methane, which makes up the largest fraction of natural gas, is an underutilized natural resource. In petroleum operations, byproduct natural gas is often flared rather than collected. The difficulty in utilizing methane fully lies in transporting gases from scattered wells to a central location for processing. Therefore, a method to convert methane to liquid products must be developed to fully utilize this resource. One promising liquid product is methanol, which is already used in the industrial chemical industry in massive quantities, can be converted into gasoline or other transportation fuels, and has potential to serve as a future transportation fuel itself. Despite significant research efforts, methods to convert methane to methanol at economically viable conditions, (T<300 °C and near ambient pressure) none have been identified. One of the largest challenges in developing processes to carry out this conversion is the fact that the oxidation of the first C-H bond in methane is the least energetically favorable step in the methane total combustion reaction that converts methane to carbon dioxide. Once one C-H bond is oxidized, the others are increasingly easy to oxidize, thus creating the undesired byproduct carbon dioxide. One possible way to address this is to synthesize a catalyst with active regions small enough to only activate and oxidize a single C-H bond in methane per catalytic cycle. Computational simulations predict that this may be accomplished by dispersing Pd onto Au nanoparticles.

In this work, controlled amounts of Pd are added to monodisperse Au nanoparticles of several sizes using colloidal synthesis methods. Pd is one of the most active methane oxidation catalysts while Au is relatively inert to the reaction. The aim of the syntheses was to add controlled coverages of Pd to Au NPs from atomic dispersion to complete Pd shells. These particles were deposited onto porous TiO_2 supports and tested for their catalytic activity for the gas phase methane to methanol reaction. These experiments allow for the elucidation of structure-activity relationships for Pd/Au systems in the methane to methanol reaction and provide a starting point for future catalyst design.

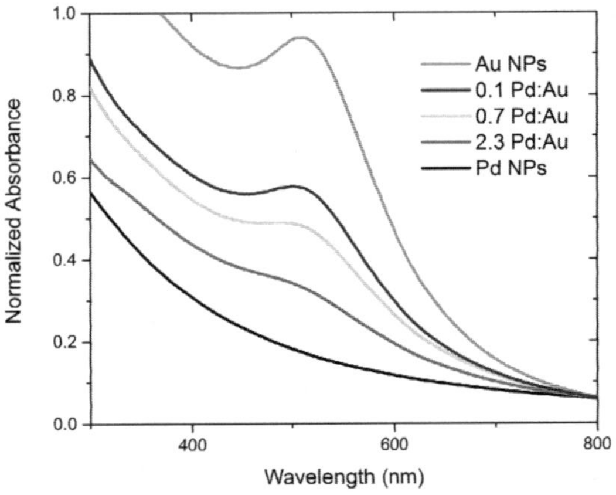

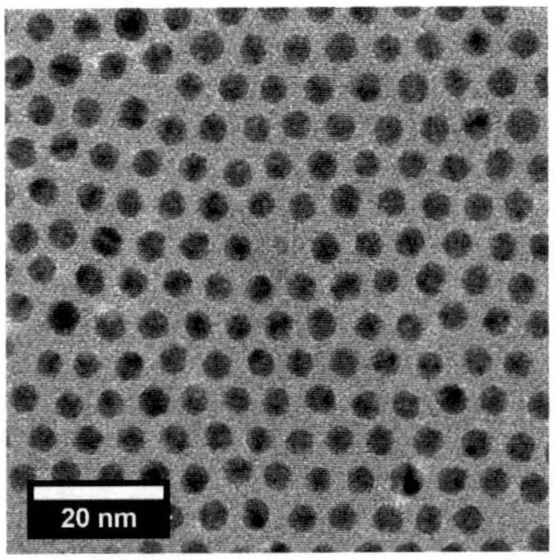

Copyright © American Institute of Chemical Engineers. All rights reserved.

457439 Using Bound Organic Ligands to Direct Particle Size and Understand Catalytic Activity of Gold Nanoparticles

Abstract:

The environment around an enzymatic active site is precisely controlled through a combination of geometric confinement as well as precise placement of organic functional groups around the active site. In a nature-inspired approach [1], organic ligands are utilized to surround a metallic active site in order to learn more about how these ligands control access to the active site as well as how they control the activity of the active site.

To investigate the effects that different organic ligand environments have around a catalytic active site, 4-nitrophenol reduction and resazurin reduction were studied using a variety of different phosphine and thiol-containing ligands bound to the gold cluster surface. 4-nitrophenol reduction is widely used to assess the catalytic properties of metallic nanoparticles, and there is debate in the literature as to where the active site is located. [2,3] This comparative study uses these organic ligands with different geometric and electronic properties to gain further insight into the factors that control catalysis in the 4-nitrophenol reduction system.

In the synthesis of metallic clusters, the steric properties of the ligands that are bound to the metal in the precursors affect the final cluster size and accessibility of the final metal clusters. This work uses different phosphine ligands to direct the synthesis of gold clusters/nanoparticles through novel synthetic procedures. [4] These different particles exhibit different amounts of accessibility to the gold cluster surface, depending on the packing of the ligand on the cluster surface. A relationship between the particle size, ligand size, and its accessibility to small probe molecules is demonstrated for these clusters.

References:

1. Trogadas, P.; Nigra, M.M.; Coppens, M.-O.; *New J. Chem.* 2016, DOI: 10.1039/C5NJ03406J
2. Nigra, M.M.; Ha, J.-M.; Katz, A.; *Catal. Sci. Technol.* **2013**, *3*, 2976.
3. Mahmoud, M.A.; Garlyyev, B.; El-Sayed, M.A.; *J. Phys. Chem. C* **2013**, *117*, 21886.
4. Nigra, M.M.; Yeh, A.J.; Okrut, A.; DiPasquale, A.G.; Yeh, S.W.; Solovyov, A.; Katz, A.; *Dalton Trans.* **2013**, *42*, 12762.

Copyright © American Institute of Chemical Engineers. All rights reserved.

481293 Enhancing the Ni/YSZ Catalytic Performance in the Activation of Methane: The Influence of the Electric Field and Interface Carbon

Abstract:

Enhancing the Ni/YSZ Catalytic Performance in the Activation of Methane: The Influence of the Electric Field and Interface Carbon

*Fanglin Che[1], Su Ha[1], and Jean-Sabin McEwen[1] **

[1]Washington State University, Pullman, WA 99163 (United States)

**js.mcewen@wsu.edu*

Understanding the oxygen vacancy formation at the triple phase boundary (TPB) of a yttrium-stabilized zirconia supported Ni nanocluster (Ni/YSZ) is of importance since such vacancies are the active sites for coke formation [1-3] and sulfur poisoning [4]. In this way, one can make a step toward designing catalysts that decrease the deactivation of the Ni/YSZ electrode via the formation of coke in a solid oxide fuel cell (SOFC) and sulfur poisoning in a solid oxide electrolysis cell (SOEC). Our previous field-dependent microkinetic model and the corresponding experimental evidence have shown that a positive electric field can enhance the methane conversion and reduce coke formation in important ways during the Ni-based methane steam reforming (MSR) process [5-10]. Therefore, considering the field effects on the oxygen vacancy formation is necessary as such fields exist in both cells [11] and can potentially be used to alter the Ni/YSZ system so as to directly modify its electrocatalytic performance.

Our computational results [12] show that a positive electric field (less than 0.6 V/Å) can significantly increase the TPB oxygen vacancy formation energy in the Ni/YSZ+O system (**Figure 1(a)**). This suggests that the presence of such a positive electric field can reduce the oxygen vacancy concentration at the TPB region, and subsequently suppress the sulfur poisoning at the TPB region of the Ni/YSZ+O electrode in a SOEC. A negative electric field in a SOFC could lead to more active TPB vacancies by reducing the vacancy formation energies in the Ni/YSZ+O model. Both charge distribution and effective dipole moments verify the qualitative findings with regard to how the field influences the formation of an oxygen vacancy in Ni/YSZ. Overall, this investigation provides guidance for designing a Ni/YSZ electrode with an improved electrocatalytic performance via a simulated positive electric field.

In addition, there has been some debate over the past decades on the role of low concentrations of interface carbon complexes in the activation of hydrocarbons over transition metal surfaces [13]. It is also a mystery as to whether or not electric fields can enhance the conversion of methane in a fuel cell or during electro-reforming at temperatures that are lower than the normal reforming conditions [14, 15]. To provide a 'bottom-up' fundamental understanding, we present the first C-H cleavage in methane over the Ni/YSZ catalysts as a case study since such reaction is the rate-limiting step during the MSR process. Our theoretical results show that the presence of carbon or carbide-like species at the interface between the Ni cluster and its metal oxide support, as well as the application of an external positive electric field, can significantly increase the local oxidation states of the Ni nanocluster. Furthermore, the first C-H cleavage in methane is thermodynamically and kinetically favored when the oxidation state of Ni is increased (**Figure 1(b)**). As a result, the presence of a low concentration of surface carbon, interfacial carbide species, or the addition of a positive electric field will facilitate the methane activation process.

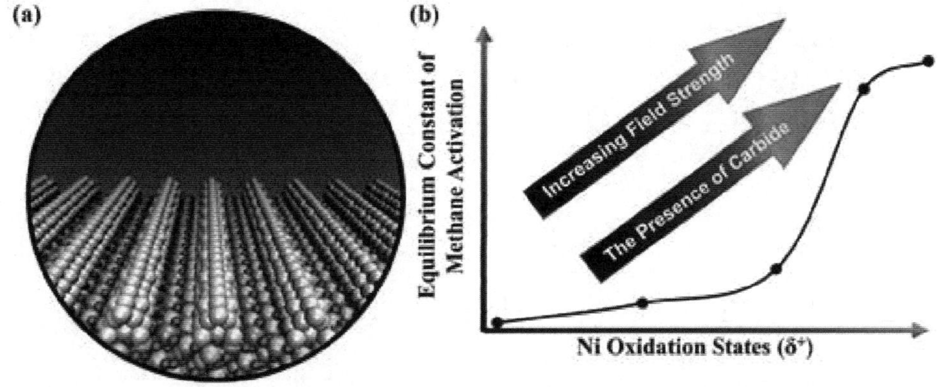

Figure 1. (a) Optimized structures used for the oxygen enriched Ni/YSZ (Ni/YSZ+O) model. Atoms with the light green, dark green, silver, and red colors represent the Zr, Y, Ni, and O species. (b) DFT results of the equilibrium constants at 873 K of the first C-H cleavage over Ni/YSZ as a function of the Ni oxidation state, which can be altered by the applied electric fields or altered by the presence of carbon species.

References

[1] Selman, J.R., *Science* **2009,** *326*, 52-53.

[2] Suzuki, T.; Hasan, Z.; Funahashi, Y.; Yamaguchi, T.; Fujishiro, Y.; Awano, M., *Science* **2009,** *325*, 852-855.

[3] Zhang, Y.; Lu, Z.; Yang, Z.; Woo, T., *J. Power Sources* **2013,** *237*, 128-131.

[4] Ebbesen, S.D.; Mogensen, M., *J. Power Sources* **2009,** *193*, 349-358.

[5] Che, F.; Ha, S.; McEwen, J.-S., *Appl. Catal. B* **2016,** *195*, 77-89.

[6] Che, F.; Gray, J.; Ha, S.; McEwen, J.-S., *J. Catal.* **2015,** *332*, 187-200.

[7] Che, F.; Zhang, R.; Hensley, A.J.; Ha, S.; McEwen, J.-S., *Phys. Chem. Chem. Phys.* **2014,** *16*, 2399-2410.

[8] Che, F.; Hensley, A.; Ha, S.; McEwen, J.-S., *Catal. Sci. Technol.* **2014,** 4020-4035.

[9] Che, F.; Gray, J.; Ha, S.; McEwen, J.-S., *In preparation* **2016**.

[10] Che, F.; Gray, J.; Ha, S.; McEwen, J.-S., *Submitted to ACS Catal.* **2016**.

[11] Stuve, E.M., *Chem. Phys. Lett.* **2012,** *519-520*, 1-17.

[12] Che, F.; Ha, S.; McEwen, J.-S., *J. Phys. Chem. C* **2016,** *120*, 14608–14620.

[13] Teschner, D.; Borsodi, J.; Wootsch, A.; Révay, Z.; Hävecker, M.; Knop-Gericke, A.; Jackson, S.D.; Schlögl, R., *Science* **2008,** *320*, 86-89.

[14] Périllat-Merceroz, C.; Gauthier, G.; Roussel, P.; Huvé, M.; Gélin, P.; Vannier, R.-N., *Chem. Mater.* **2011,** *23*, 1539-1550.

[15] Sekine, Y.; Haraguchi, M.; Tomioka, M.; Matsukata, M.; Kikuchi, E., *J. Phys. Chem. A* **2009,** *114*, 3824-3833.

Copyright © American Institute of Chemical Engineers. All rights reserved.

465433 Metal Nanocatalysts, Their Synthesis and Size Dependent Covalent Bond Catalysis. Oxide Metal Interfaces As Active Sites for Acid-Base Catalysis

Abstract:

Colloidal chemistry is used to control the size, shape and composition of metal nanoparticles usually in the 1-10 nm range. In-situ methods are used to characterize the size, structure (electronic and atomic), bonding, composition and oxidation states under reaction conditions. These methods include sum frequency generation nonlinear optical spectroscopy (SFG), ambient pressure X ray photoelectron spectroscopy (APXPS) and high pressure scanning tunneling microscope (STM). The catalytic behavior depends on the size, oxidation state, coordination number, crystallographic orientation of metal sites and bonding and orientation of surface adsorbates.

When metal nanoparticles are placed on different mezoporous or microporous oxide supports, the catalytic turnover rates and selectivities markedly change. The charge flow between the metal and the oxide ionizes the adsorbed molecules at the oxide-metal interfaces and alters the catalytic chemistry (acid-base catalysis).

The oxidation state of metal nanoparticles becomes less metallic and assumes higher oxidation states with decreasing size. The small nanoclusters behave similar to transition metal ions that are active homogeneous catalysts. Hybrid systems composed of enzymes, homogeneous and heterogeneous catalysts are constructed to study molecularly unified catalytic schemes for the future.

Copyright © American Institute of Chemical Engineers. All rights reserved.

455746 Automated Reaction Mechanism Generation Including Nitrogen As a Heteroatom

Abstract:

The open-source Reaction Mechanism Generator (RMG) software and its thermochemical and reaction kinetics databases have been extended to include nitrogen as a heteroatom. Specific changes to RMG and the mining of thermochemistry and reaction kinetics data will be discussed. This new version of RMG has been tested by generating a reaction mechanism for pyrolysis and oxidation of ethylamine ($CH_3CH_2NH_2$), and the generated model was benchmarked to recent shock tube studies. Ethylamine was selected as a model nitrogen-containing organic species since it is the simplest primary amine containing important characteristic chemical bonds (i.e., C–C, C–H, C–N, and N–H), and it is also a model molecule for nitrogen-containing species found in biofuels. Validation of the reaction network with experimental data showed that the generated model successfully reproduces measured radical levels (NH_2 and OH) as well as ignition delay times. This work presents the potential to automatically generate detailed chemical models for nitrogen-containing species, and specifically unveils the pyrolysis and oxidation mechanisms of ethylamine.

Copyright © American Institute of Chemical Engineers. All rights reserved.

465018 Modeling the Combustion of Otto Fuel

Abstract:

Otto Fuel II is a monopropellant that is used in undersea propulsion. It consists of three components: propylene glycol dinitrate, (PGDN), dibutyl sebacate (DBS), and 2-nitrodiphenylamine (NDPA). PGDN is the oxidizer, DBS is added as a plasticizer, and NDPA is serves as a stabilizer. Our modeling results suggest that there are some interesting similarities and differences between monopropellant combustion and large molecular weight transportation fuels. At high temperatures, the decomposition kinetics of the oxidizer is effectively decoupled from the oxidation of the plasticizer, and DBS effectively undergoes pyrolysis and partial oxidation in a high-pressure $CH_2O/NO_2/NO$ bath gas. The stabilizer undergoes a nitrate to nitro conversion, followed by O-NO bond fission to yield a highly stable oxy-diphenylamine radical. This radical recombines to form a large number of polycyclic aromatic amines, which are believed to be responsible for the large amount of carbon deposits that are formed during operation.

Copyright © American Institute of Chemical Engineers. All rights reserved.

455944 Molecular Weight Growth Kinetics of Olefin Pyrolysis Under Low to Intermediate Temperatures

Abstract:

In many conversion processes, such as solid oxide fuel cell operation or hot gas mixing upstream of steam reformers, hydrocarbon fuels might be exposed for sufficiently long times at temperatures between 500°C and 1000°C to allow gas-phase reactions to occur. This chemistry not only leads to a change in gas phase composition that enters the process, but it might also produce molecular weight growth species (MWG) large enough to precipitate from the gas phase on the surface to form deposits. Deposit formation can reduce the efficiency of the technical application. For example it may deactivate the reforming catalyst or fouling in the solid oxide fuel cell, or in severe cases lead to a catastrophic failure of the unit. Other examples include formation of gas-phase deposit precursors within steam crackers. These limit the time between decoking cycles. MWG species contained in exhaust gas emissions from IC engines also constitute an environmental and adverse health problem.

There have been amount of work focusing on understanding the reaction kinetics under high temperature environment, e.g., combustion. However, few studies focused on low to intermediate temperatures, in particularly the MWG kinetics for unsaturated hydrocarbons, such as alkene, is poor. In this work, propene and the three butene isomers are selected as the model compounds. Pyrolysis experiment were performed in a tabular flow reactor at pressure ~0.82atm in temperatures of ~500-800°C, with nominal residence time ~2.4s. These fuels showed substantially different reactivities, as well as product distributions. The pyrolysis data showed that the three linear alkenes (e.g., propene, 1- and 2-butene) generated similar amount of MWG species under the same fuel conversions, while the branched isobutene tended to form as much as twice amount. A detailed kinetic model was developed to understand the impact of fuel structures on the fuel conversion, product formation and especially MWG kinetics during olefin conversion. The knowledge obtained from this combined experimental and theoretical work could be applied to real-world practices including reduce deposit formation in steam cracker and understand ignition kinetics of motor engines.

Copyright © American Institute of Chemical Engineers. All rights reserved.

452102 Detailed Kinetic Modeling of Gas Evolution and Energy Recovery during Chemical Quenching of Acetylene Production in Thermal Plasma

Abstract:

Detailed Kinetic Modeling of Gas Evolution and Energy Recovery during Chemical Quenching of Acetylene Production in Thermal Plasma

Yan Cheng, Tianyang Li, Yi Cheng [*]

Department of Chemical Engineering, Beijing Key Laboratory of Green Chemical Reaction Engineering and Technology, Tsinghua University, Beijing 100084, PR China

Thermal plasma pyrolysis provides a convenient way to realize the one-step conversion from coal to acetylene. Taking advantage of the ultra-high temperature field provided by thermal plasma, the coal feedstock is heated up in milliseconds to give out volatiles. The volatiles undergo gas phase reactions afterwards to form acetylene owing to the fact that acetylene is thermodynamically stable at above 1500 K among small-molecule-hydrocarbons. However, acetylene will decompose into hydrogen and carbon black during cooling process. So efficient quenching is necessary to prevent acetylene decomposition so as to maintain the target products. It is noticed, on the other hand, that the high-temperature gas products still contain about 40% of the total energy input. So it is vital to make use of the latent energy during the quenching process.

In addition to the common physical quenching by water spray, chemical quenching method is an alternative solution to the process design. In chemical quenching, hydrocarbons are used as quenching media instead of water, which brings about co-products (e.g., ethylene) during the decomposition of hydrocarbons by the afterheat in gas products. The process profit can be accordingly improved as well.

In this work, we carried out modelling and simulation on the above-mentioned chemical quenching in thermal plasma technology. A gaseous detailed kinetics, coupled with a soot formation mechanism, was proposed to investigate the species evolutions during quenching. The model consisted of C1-C5 gas phase reactions, PAH (polycyclic aromatic hydrocarbons) growth and a simplified soot formation mechanism. The predicted acetylene decomposition and soot yield were well validated with published data. Then, numerical simulations were performed to reveal the process features, including acetylene decomposition evolution, chemical quenching behavior of different C1-C3 alkanes as well as the energy balance.

Finally, the model was used to theoretically propose an optimized chemical quenching design in a pilot-plant thermal plasma process. The results showed that under optimized operating conditions, propane quenching could recovery 35.4% of the total gas phase afterheat. At the same time, propane decomposed into smaller molecules in millisecond(s), which realized the co-production of ethylene at a ratio of 0.75 t C_2H_4/t C_2H_2. The optimal energy consumption was reduced by 43% per kilogram of C_2H_2+C_2H_4, which verified the feasibility of energy efficiency improvement via chemical quenching.

Keywords: Thermal plasma pyrolysis; Chemical quenching; Acetylene; Detailed kinetics

* Corresponding author. Address: Department of Chemical Engineering, Tsinghua University, Beijing 100084, P.R. China. Tel: +86 10 62794468, Fax: +86 10 62772051. *E-mail address*: yicheng@tsinghua.edu.cn.

Copyright © American Institute of Chemical Engineers. All rights reserved.

474760 Predictive Modeling of Thermal Decomposition of Hydrocarbon Mixtures

Abstract:

Thermal decomposition of hydrocarbon mixtures is important in a wide range of applica- tions, including geochemical conversion of oil to gas, thermal coking and cracking processes in refineries, and steam-cracking to form olefins. Most detailed studies focus on the cracking mechanism of a single molecule, to reduce the complexity. Here we present recent advances in methods for modeling mixtures, and so seeing composition effects on the chemistry. Also we present experimental data measured on thermal decomposition of various mixtures at several temperatures using the gold tube technique to maintain the system in a single phase, and discuss the development of detailed a priori kinetic models for these experiments using the open source software Reaction Mechanism Generator (RMG). Recent improvements to RMG, including a method for more rapidly and accurately estimating the thermochemistry of fused polycyclics, are presented.

Copyright © American Institute of Chemical Engineers. All rights reserved.

464972 Nondimensional Specific Energy Index for Methane Pyrolysis in Thermal Plasma

Abstract:

Acetylene is the raw material for several important chemical products including polyvinyl chloride (PVC), vinyl acetate, 1,4-Butylene glycol (BDO) and so on. Partial oxidation of natural gas is the main access to acetylene in Europe and America, yet the introduction of oxygen reduces the carbon selectivity to hydrocarbons. Methane pyrolysis in thermal plasma is an efficient way to produce acetylene. In this process, ultra-high temperature in reductive plasma allows for more preferred reactor performance such as methane conversion at above 90% and selectivity to acetylene at above 70%. A large amount of researches have been done on methane pyrolysis on various lab-scale plasma devices. However, due to the diversity of experimental conditions, it is hard to summarize the conclusions from different researches.

In this work, we introduced a nondimensional specific energy index, defined as the ratio of supplied power and enthalpy flow needed for plasma weighted with inert gas and methane content, to better evaluate the influences of supplied power and methane content in feed on the methane pyrolysis. This nondimensional value stands for the excess of energy needed for plasma and includes both the influences of energy provided for per unit of gas and methane content. It is worth noted that compared with enthalpy change of gas in plasma, reaction heat of methane to acetylene is negligible. In an ideal process, the index value should be 1, which means that the supplied power is precisely sufficient to maintain the plasma and the reaction. A series of experiments in a 2 kW DC plasma reactor were carried out to testify the applicability of the index. The experimental results showed that the conversion of methane increased from around 10% when the index was 0.5~0.6, and reached 90% when the index was 1.0~1.2, then slowly increased to a maximum around 95%~97% when the index kept increasing. This means that this nondimensional index positively correlates to the conversion and implies the energy excess state of the system as expected.

Copyright © American Institute of Chemical Engineers. All rights reserved.

463198 Understanding Trends in C-H Bond Activation in Heterogeneous Catalysis

Abstract:

Natural gas, composed primarily of methane, is a valuable energy source that currently accounts for 24% of worldwide energy needs[1]. However, due to a lack of transportability, the natural gas extracted alongside crude oil at remote drilling locations is often flared resulting in the loss of a valuable chemical feedstock and increased CO_2 emissions[1]. Discovery of an efficient catalyst for direct on-site partial methane oxidation to higher value commodity chemicals or liquids fuels could result in reduced CO_2 emissions and significant economic savings. While electronic structure calculations are becoming increasingly relevant in this search, they are often problematic due to the high cost of transition state barrier calculations. In the present work, we use density functional theory to establish a unifying framework for predicting C-H activation barriers on the wide range of catalysts that activate C-H bonds via a radical intermediate using a single universal descriptor. We demonstrate the relevance of this scaling approach by combining it with a thermodynamic analysis to generate a map of methane activation rates that successfully rationalizes available empirical data. Using this activity map as a guide, we identify several oxide materials that show favorable methane activation potential and conduct a more thorough analysis of these systems. This model will facilitate the fast and efficient screening of a wide range of potential C-H activation catalysts, bypassing the need for tedious transition state barrier calculations and allowing for greener chemistry by selectively testing promising materials.

1. Horn, R. & Schlögl, R. Methane Activation by Heterogeneous Catalysis. *Catalysis Letters* **145,** 23:39 (2015).

Copyright © American Institute of Chemical Engineers. All rights reserved.

467253 Novel Fe-Promoted $MgAl_2O_4$ Support Material for Control of Carbon Deposition during Reforming Reactions

Abstract:

Introduction

The role of the support in syngas production from methane has been widely investigated for various reforming technologies, like steam reforming (MSR), dry reforming (DRM), partial oxidation of methane (POM) and autothermal reforming (ARM) [1]. Inert supports, e.g. SiO_2, although ensuring active metal dispersion, will result in quick deactivation due to carbon deposition, since both CH_4 and CO_2 are activated on the same active site (mono-functional mechanism) [2-4]. A bi-functional mechanism exist on acidic/basic supports (Al_2O_3, La_2O_3, MgO etc.), where CH_4 is activated on active metals and CO_2 on the support, thus enhancing the oxidation of deposited carbon [5-7].

To improve the latter mechanism, the synthesis of supported reforming catalysts should be adjusted in such a way that the support inhibits carbon formation by actively participating in the reforming reaction scheme. This should include (1) high temperature resistance, (2) basicity which will enhance the adsorption of CO_2 and (3) good redox properties, in particular oxygen storage capacity [8, 9].

$MgAl_2O_4$ was widely investigated in literature as a suitable support material for reforming reactions due to its low cost. The addition of MgO to Al_2O_3 prevents interaction between active metals and alumina forming inactive spinel phases. Guo and co-workers [10] compared Ni/γ-Al_2O_3, Ni/MgO-γ-Al_2O_3 and Ni/$MgAl_2O_4$ during methane dry reforming. They found that Ni/$MgAl_2O_4$ exhibited higher activity and better stability compared to the other samples. The $MgAl_2O_4$ spinel can effectively suppress the phase transformation to form $NiAl_2O_4$ and stabilizes Ni crystallites. In order to lower significantly the carbon deposition during DRM conditions at high temperature (1023 K), the present study proposes a $MgFeAlO_x$ support, where Fe is incorporated in the magnesium aluminate lattice, as a cheaper alternative oxygen storage material.

Results

A series of Mg(a%)FeAlO_x support materials (where α =0, 2.5, 5 ,7.5 ,10 and 20 wt%) were prepared by onepot synthesis, using an aqueous solution of the corresponding nitrates and calcined at high temperature (1073 K). Supported 8wt.%Ni catalysts were prepared by incipient wetness impregnation on the Mg(α%Fe)AlO_x materials, resulting in Ni/Mg(αFe)AlO_x and were investigated during methane dry reforming at 1023 K and 121.3 kPa.

The evolution of the support and the catalyst structure during H_2-TPR and CO_2-TPO was examined using time-resolved *in-situ* XRD and *operando-XAS*. The XANES region of the support $MgFeAlO_x$ showed reduction through a shift of edge energy position towards lower values. The reduction remained partial as complete reduction to reference Fe^{+2} and Fe^0 was not observed. This could indicate spinel reduction from $MgFe^{+3}AlO_x$ to $MgFe^{+2}AlO_x$ without complete segregation of iron from the support [11], while it can be re-oxidized to $MgFe^{+3}AlO_x$ under CO_2 flow. This redox ability of the support, with iron incorporated in the spinel lattice, is beneficial for suppressing the carbon deposition during reforming reactions. The in-situ XRD during H_2-TPR of the catalyst, Ni/Mg(αFe)AlO_x, showed the formation of Fe-Ni alloy, via Fe delivery from the support material to the catalyst surface. The presence of Ni enhanced the reducibility of iron to Fe^0, indicating that part of Fe could be segregated from the support.

Figure 1 displays the space time yield, STY, of the products for the best candidate of Fe-promoted samples along with the non-promoted catalyst during a stability test (TOS of 12 h). The amount of deposited carbon was determined by O_2-TPO.

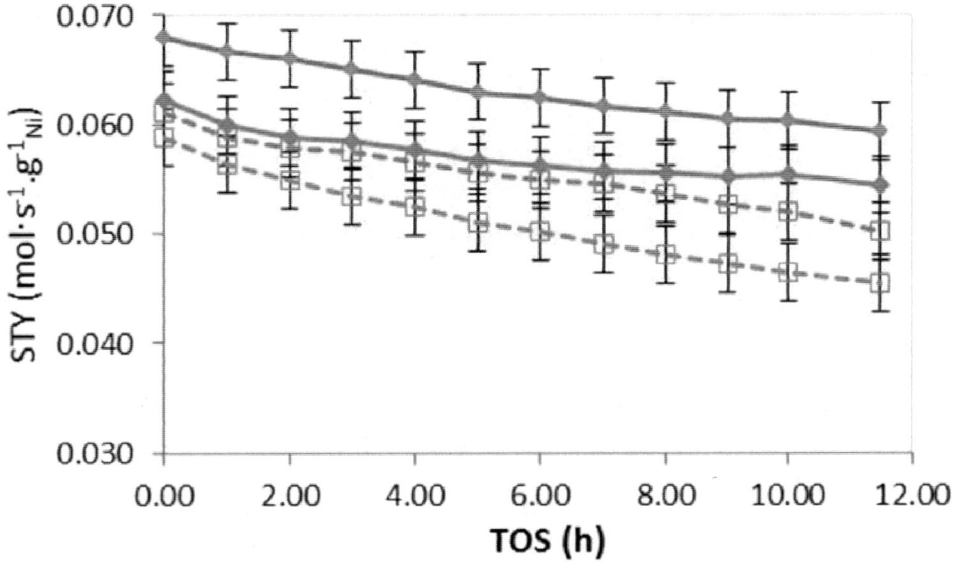

Figure 1: Stability tests during DRM after 12h time-on-stream (TOS) at 1023 K (total pressure of 121.3 kPa and

CH_4/CO_2=1/1).♦,♦: STY_{CO} (solid line); □, □: STY_{H2} (dashed line). Blue line: Ni/Mg(0Fe)AlO_x where X_{CH4} from 72% to 60%, green line: Ni/Mg(2.5Fe)AlO_x where X_{CH4} from 75% to 62%. Error bars were calculated after three independent experiments representing standard deviation (66% probability confidence interval).

Among the different supports, Ni/Mg(2.5Fe)AlO_x has an optimal composition as it results in more syngas production with a ratio of CO/H_2 equal to 1.2, due to reverse water gas shift reaction, while further increase of Fe content leads to catalyst activity decrease. The Fe addition in the support material increases the oxygen mobility through the support lattice, suppressing carbon formation during DRM. While 1.12 mol of carbon was deposited on Ni/Mg(0Fe)AlO_x after 12h, no carbon was detected on samples with Fe incorporated in the support.

The novel Fe-promoted $MgAl_2O_4$ material could be used as a support for reforming reactions, replacing conventional oxygen storage materials (i.e CeO_2) due to its low price. The difference in carbon deposition between Ni/Mg(0Fe)AlO_x and Ni/Mg(2.5Fe)AlO_x indicates the beneficial effect of Fe incorporation into the magnesium aluminate spinel lattice.

Acknowledgments

This work was supported by the FAST industrialization by Catalyst Research and Development (FASTCARD) project, which is a Large Scale Collaborative Project supported by the European Commission in the 7th Framework Programme (GA no 604277), by the "Long Term Structural Methusalem Funding by the Flemish Government", the Interuniversity Attraction Poles Programme, IAP7/5, Belgian State – Belgian Science Policy, and the Fund for Scientific Research Flanders (FWO; project 004613N). XAS experiments were performed on the ROCK beam line, at the SOLEIL Synchrotron, France (proposal number 20150256). The authors acknowledge the SOLEIL staff for smoothly running the facility, support from Prof. C. Detavernier with the in situ XRD equipment (Department of Solid State Sciences, Ghent University) and from Dr. Vitaliy Bliznuk (Department of Materials Science and Engineering, Ghent University) for the HRTEM measurements.

References

1. Littlewood, P., X. Xie, M. Bernicke, A. Thomas, and R. Schomäcker, *$Ni_{0.05}Mn_{0.95}O$ catalysts for the dry reforming of methane.* Catal. Today, 2015. **242, Part A**(0): p. 111-118.

2. Kroll, V.C.H., H.M. Swaan, S. Lacombe, and C. Mirodatos, *Methane Reforming Reaction with Carbon Dioxide over Ni/SiO2Catalyst: II. A Mechanistic Study.* Journal of Catalysis, 1996. **164**(2): p. 387-398.

3. Bitter, J.H., K. Seshan, and J.A. Lercher, *Mono and Bifunctional Pathways of CO_2/CH_4 reforming over Pt and Rh Based Catalysts.* J. Catal., 1998. **176**(1): p. 93-101.

4. Ferreira-Aparicio, P., I. Rodrguez-Ramos, J.A. Anderson, and A. Guerrero-Ruiz, *Mechanistic aspects of the dry reforming of methane over ruthenium catalysts.* Appl. Catal., A, 2000. **202**(2): p. 183-196.

5. Tsipouriari, V.A., A.M. Efstathiou, and X.E. Verykios, *Transient Kinetic Study of the Oxidation and Hydrogenation of Carbon Species Formed during CH_4/He, CO_2/He, and CH_4/CO_2 reactions over Rh/Al_2O_3 Catalyst.* J. Catal., 1996. **161**(1): p. 31-42.

6. Zhang, Z. and X.E. Verykios, *Carbon dioxide reforming of methane to synthesis gas over Ni/La_2O_3 catalysts.* Appl. Catal. A 1996. **138**(1): p. 109-133.

7. Pakhare, D. and J. Spivey, *A review of dry (CO_2) reforming of methane over noble metal catalysts.* Chem. Soc. Rev., 2014. **43**(22): p. 7813-7837.

8. Hecht, E.S., G.K. Gupta, H. Zhu, A.M. Dean, R.J. Kee, L. Maier, and O. Deutschmann, *Methane reforming kinetics within a Ni–YSZ SOFC anode support.* Appl. Catal., A, 2005. **295**(1): p. 40-51.

9. Theofanidis, S.A., V.V. Galvita, H. Poelman, and G.B. Marin, *Enhanced Carbon-Resistant Dry Reforming Fe-Ni Catalyst: Role of Fe.* ACS Catal., 2015: p. 3028-3039.

10. Guo, J., H. Lou, H. Zhao, D. Chai, and X. Zheng, *Dry reforming of methane over nickel catalysts supported on magnesium aluminate spinels.* Appl. Catal. A 2004. **273**(1–2): p. 75-82.

11. Dharanipragada, N.V.R.A., L.C. Buelens, H. Poelman, E. De Grave, V.V. Galvita, and G.B. Marin, *Mg-Fe-Al-O for advanced CO_2 to CO conversion: carbon monoxide yield vs. oxygen storage capacity.* J. Mater. Chem. A, 2015. **3**(31): p. 16251-16262.

Copyright © American Institute of Chemical Engineers. All rights reserved.

468433 Dry Reforming of Methane over Nickel-Based Catalysts for Syngas Production

Abstract:

DRY REFORMING OF METHANE OVER NICKEL-BASED CATALYSTS FOR SYNGAS PRODUCTION

B. R. DE VASCONCELOS[1], D. PHAM MINH[1], P. SHARROCK[1], A. NZIHOU[1]

[1]Mines Albi, CNRS, Centre RAPSODEE, Univ. Toulouse, Campus Jarlard, F-81013 Albi Cedex 09, France

Climate changes caused by greenhouse gases emissions, especially CO_2, have driven scientists to develop technologies for CO_2 utilization. Moreover, increases in energy demand and depletion of carbon-based resources have shown the need of substitution of fossil fuels for renewable energies. The valorization of CO_2in more valuable chemicals and fuels has become an environmentally interesting option [1][2].

Among the several chemical processes that are able to convert CO_2 into chemicals, the dry reforming of methane (Equation 1) has gained much attention. In this reaction, CO_2reacts with methane to form syngas, from which many other chemicals, such as methanol, dimethyl ether, hydrogen and liquid hydrocarbons can be produced [3]. DRM is also an interesting option for the valorization of natural gas and biogas, which are composed mainly of carbon dioxide and methane. The economic aspect is also propitious. Natural gas reserves, for example, are larger than petroleum reserves [1][4][5]. Biogas, formed by anaerobic digestion of organic biomass and wastes, is largely produced at different scales [6][7][8]. DRM reaction and its side reactions are expressed in equations (1-5):

Dry reforming of methane:

$CH_4 + CO_2 \leftrightarrow 2CO + 2H_2$ (1)

Side reactions:

Water-gas shift reaction:

$CO_2 + H_2 \leftrightarrow H_2O + CO$ (2)

Boudouard reaction:

$2CO \leftrightarrow C_{(s)} + CO_2$ (3)

Methane cracking reaction:

$CH_4 \leftrightarrow C_{(s)} + 2H_2$ (4)

Carbon gasification:

$H_2O + C_{(s)} \leftrightarrow CO + H_2$ (5)

Despite the economic and environmental benefits, no technical solution for DRM has emerged at industrial scale yet. The main reason is the catalytic deactivation by carbon deposition (Equations 3-5) and the sintering of both support and metal particles [9].

Supported nickel catalysts have been widely studied due to their low cost, high availability (compared to noble metals) and high catalytic activity. Nevertheless, they were reported to be prone to carbon deposition on the catalyst active surface. So, much effort has been spent on the design of a stable Ni catalyst for DRM. Parameters such as support chemical composition and promoters are key parameters for the development of a performing catalyst [9]. Hydroxyapatite ($Ca_{10}(PO_4)_6(OH)_2$) has gained much attention as catalyst support due to its high chemical and thermal stabilities and its ability to undergo several substitutions in its crystal lattice. So, this work focused on the development of a performing Ni supported on hydroxyapatite catalyst for DRM.

The catalyst was synthesized by incipient wetness impregnation method with a Ni loading of 5.7wt.%. Then, its performance on DRM was tested in a fixed-bed tubular reactor at 700°C for 50 h of time-on-stream (TOS) with WHSV = $15882 mLh^{-1} g_{cat}^{-1}$. The catalyst showed both methane and CO_2conversion around 80% and no deactivation was detected during the 50h of TOS.

In order to compare the performance of the Ni/hydroxyapatite catalyst synthesized in this work with the ones of the catalysts already used at industrial scale in similar reforming processes, two alumina-based catalysts (5%wt.Ni/Al_2O_3 and 5%wtNi/MgO-Al_2O_3) were also synthesized and tested on DRM in the same conditions. Ni/hydroxyapatite showed similar catalytic performance that Ni/ MgO-Al_2O_3 and better performance that Ni/ Ni/Al_2O_3. These results show that the hydroxyapatite-based catalyst could be a promising catalyst for DRM.

References

[1] M. Latifi, F. Berruti, and C. Briens, "Thermal and catalytic gasification of bio-oils in the Jiggle Bed Reactor for syngas production," *Int. J. Hydrogen Energy*, vol. 40, no. 17, pp. 5856–5868, 2015.

[2] L. Li, N. Zhao, W. Wei, and Y. Sun, "A review of research progress on CO_2 capture, storage, and utilization in Chinese Academy of Sciences," *Fuel*, vol. 108, pp. 112–130, 2013.

[3] J. Newnham, K. Mantri, M. H. Amin, J. Tardio, and S. K. Bhargava, "Highly stable and active Ni-mesoporous alumina catalysts for dry reforming of methane," *Int. J. Hydrogen Energy*, vol. 37, no. 2, pp.

1454–1464, 2012.

[4] A. D. Ballarini, S. R. De Miguel, E. L. Jablonski, O. a. Scelza, and A. a. Castro, "Reforming of CH_4 with CO_2 on Pt-supported catalysts: Effect of the support on the catalytic behaviour," *Catal. Today*, vol. 107–108, pp. 481–486, 2005.

[5] A. Ballarini, F. Basile, P. Benito, I. Bersani, G. Fornasari, S. De Miguel, S. C. P. Maina, J. Vilella, A. Vaccari, and O. a. Scelza, "Platinum supported on alkaline and alkaline earth metal-doped alumina as catalysts for dry reforming and partial oxidation of methane," *Appl. Catal. A Gen.*, vol. 433–434, pp. 1–11, 2012.

[6] J. Xu, W. Zhou, Z. Li, J. Wang, and J. Ma, "Biogas reforming for hydrogen production over nickel and cobalt bimetallic catalysts," *Int. J. Hydrogen Energy*, vol. 34, no. 16, pp. 6646–6654, 2009.

[7] P. Djinović, I. G. O. Črnivec, and A. Pintar, "Biogas to syngas conversion without carbonaceous deposits via the dry reforming reaction using transition metal catalysts," *Catal. Today*, vol. 253, pp. 155–162, 2015.

[8] M. P. Kohn, "Catalytic Reforming of Biogas for Syngas Production," Columbia University, 2012.

[9] M. M. Makri, M. a. Vasiliades, K. C. Petallidou, and A. M. Efstathiou, "Effect of support composition on the origin and reactivity of carbon formed during dry reforming of methane over 5wt% $Ni/Ce_{1-x}M_xO_{2-\delta}$ ($M=Zr^{4+}$, Pr^{3+}) catalysts," *Catal. Today*, pp. 1–15, 2015. ↔

Copyright © American Institute of Chemical Engineers. All rights reserved.

465882 Probing the Effect of Mn Dispersion on the Catalytic Performance of Mn-Na-W/SiO2 in the Oxidative Coupling of Methane

Abstract:

Probing the effect of Mn dispersion on the catalytic performance of Mn-Na-W/SiO2 in the oxidative coupling of methane

Naseem S. Hayek and Oz M. Gazit*

Chemical Engineering, Israel Institute of Technology-Technion, Haifa, Israel
ozg@technion.ac.il

Oxidative coupling of methane (OCM) has received much attention since the fundamental work of Hinsen et al.1 and Keller et al.2 (Eq. 1). Since 1981, hundreds of materials have been tested as catalysts.3 A good catalyst for this reaction has to be thermally stable and highly selective towards C2 products over the total oxidation of methane, which is by far the most thermodynamically favored reaction route under these conditions (Eq. 2).

(1) CH_4 + 1/2 O_2 -> C_2H_6 or C_2H_4 + H_2O â^†ð»°$_{298}$^{Â} = â^'140 kJ/mol

(2)Â CH_4Â + 2 O_2Â -> CO_2Â + $2H_2O$Â â^†ð»$^{0}_{298}$ = â^'890 kJ/mol

Mn-Na2WO4/SiO2 catalyst has been identified as one of the more promising catalysts having remarkable thermal and catalytic stability at the reaction conditions.[4] Moreover, this catalyst obtained performance of ~33% CH4 conversion at 80% C2 selectivity giving a total C2 yield of ~25%.6 However, the economic threshold for practical applications requires the catalyst to demonstrate enhanced properties in order to achieve above 30% yield.[5]
The performance of this catalyst was found to be dependent on a complicated synergy between the three components (Na, Mn, W) on silica, where the absence of either one of them leads to inferior catalytic performance.5 A key step in catalyst preparation is the Na induced phase transition, from amorphous silica to Î±-cristobalite. This step is deemed as crucial for the dispersion of the active metal sites which in turn are leads to the good catalytic performance.

In this study, we investigate the effect of pre-phase transition dispersion of the Mn on the catalytic performance of Mn-Na-W/SiO2 catalyst. We do this using the post-synthetic modification of Î²-zeolite (i.e. dealumination) to form vacant T- sites, in which the Mn sites are incorporated into the zeolite framework. Only after this initial dispersion, the phase transition to ð›¼-cristoballite is induced by Na impregnation and calcination. In this talk the effect of Mn precursor, either Mn acetate or Mn nitrate, is evaluated for dispersion and the formation of active phases. The various catalysts are then tested in a tubular fixed-bed reactor at 750 Â°C, ~1 bar, 4:1 CH4:O2 feed molar ratio, 20000 and 60000 ml g-1h-1 GHSV. The tested catalysts are compared to control catalyst that is prepared using conventional fumed silica as the support.

Materials are characterized using ICP-OES, N2 physisorption, DRIFT, XRD, HR-SEM and TPR. All our results point towards a significant influence of the Mn precursor on the dispersion of Mn in the dealuminated Î²-zeolite framework, which is consistent with catalytic results following the phase transition. XRD results show that Mn nitrate forms mainly bulk MnO2, while Mn acetate does not show any distinctive peaks related to MnxOy. This is supported by DRIFT measurements showing that when Mn acetate is used the available T-sites are filled while when Mn nitrate is used the T sites remain open. N2 physisorption measurements show pore contraction in the case of Mn acetate where in the case of Mn nitrate the pore volume remains largely unchanged. HR- SEM, indicates the formation of large crystals of MnO2 when Mn nitrate is used, see Figure 1a, while no apparent separate phase is seen when Mn acetate is used, see Figure 1b. These results are further corroborated by TPR experiments, in which Mn nitrate exhibits peaks characteristic of bulk MnO2 while Mn acetate material exhibit a wide peak characteristic to enhanced Mn-surface interactions. Following the impregnation with Na2WO4 and calcination, all the materials have theÂ Î±-cristobalite structure with almost complete surface area loss (450-650 to <5 m2g-1) and with Mn2O3, Na2WO4 and MnWO4 as main phases detected in XRD. The results for the OCM reaction

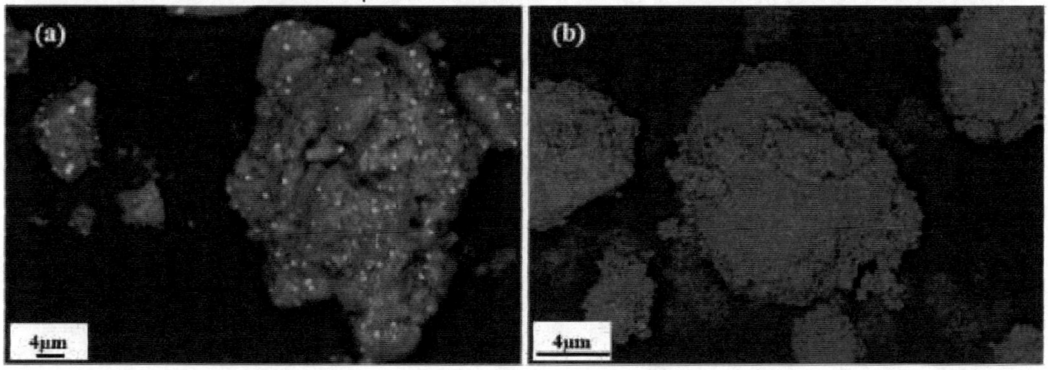

show that the incorporation of Mn into the zeolitic framework, followed by phase transition, provide superior performance in terms of CH4 conversion and C2 selectivity. Backscattering HRSEM images of Mn incorporated in the Î²-zeolite. (a) Mn Nitrate and (b) Mn Acetate. References

1 Hinsen, W. & Baerns, M. Oxidative Coupling of Methane to C2-Hydrocarbons in the Presence of Different Catalysts. *Chemiker-Zeitung* **107**, 223-226 (1983).
2 KellerG, E. & Bhasin, M. Synthesis of ethylene via oxidative coupling of methane.

JCatal **73**, 9-19 (1982).

3 Zavyalova, U., Holena, M., SchlÃ¶gl, R. & Baerns, M. Statistical Analysis of Past Catalytic Data on Oxidative Methane Coupling for New Insights into the Composition of Highâ€Performance Catalysts. *ChemCatChem* **3**, 1935-1947 (2011).
4 Liu, H. *et al.* Scale up and stability test for oxidative coupling of methane over Na 2 WO 4-Mn/SiO 2 catalyst in a 200 ml fixed-bed reactor. *Journal of Natural Gas Chemistry* **17**, 59-63 (2008).
5 Palermo, A., Vazquez, J. P. H., Lee, A. F., Tikhov, M. S. & Lambert, R. M. Critical influence of the amorphous silica-to-cristobalite phase transition on the performance of Mn/Na 2 WO 4/SiO 2 catalysts for the oxidative coupling of methane. *Journal of catalysis* **177**, 259-266 (1998).
6 San Su, Y., Ying, J. Y. & Green, W. H. Upper bound on the yield for oxidative coupling of methane. *Journal of Catalysis* **218**, 321-333 (2003).

458756 Methane Steam Reforming: Using External Electric Fields to Enhance the Catalytic Performance of Ni-Based Catalysts

Abstract:

According to the Annual Energy Outlook for 2015, natural gas production in the US is projected to continue rising through 2040. [1] To make the most of this abundant natural resource and at the same time reduce emissions of harmful greenhouse gases it is imperative that we fully understand the catalytic reactions which are used in methane processing – particularly in the case of methane steam reforming (MSR). MSR produces around 98% of the world's hydrogen gas supply. Additionally, MSR is one of main reactions that occur on the Ni-based anode for direct methane solid oxide fuel cells (SOFCs). [2] MSR is our reaction of interest because the conversion of methane to syngas greatly affects the charge-transfer chemistry and consequently influences the SOFCs' performance. There are two significant issues facing MSR: (i) Coke formation; (ii) High temperatures of above 900 K. [3] To address these issues, we are interested in the effect of an electric field on this process. A large electric field can rearrange the potential energy states of molecular orbitals, alter adsorbate-surface interactions, and directly change the overall catalytic activity of Ni-based catalysts. [4-8]

Based on the DFT results, an electric field can greatly influence the stabilities of intermediates, the activation energies and the reaction energies of the MSR-related elementary reactions. Both the established microkinetic model and the corresponding experimental evidence of the field-dependent MSR-over-Ni have shown that the presence of a positive electric field can decrease the formation of coke and reduce the MSR reaction temperature significantly while maintaining the overall methane conversion (Figure. 1).[9-14] Therefore, the importance of considering the electric field effects on the catalytic reactions in a fuel cell system is self-evident. This information can also be used to design new electrochemical systems and to enhance the catalytic performance of Ni-based steam reforming operations.

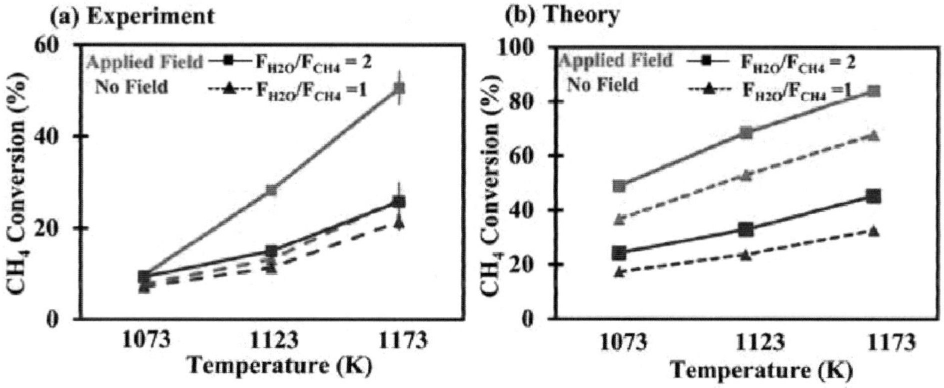

Figure 1. (a) Experimental results showing the field effects on the methane conversion at initial H_2O/CH_4 flow rates of 2 (solid line) and 1 (dotted line) in the presence (red) and the absence (blue) of an electric field. (b) Kinetic modeling results for methane conversion at steady state at the experimental conditions with a positive electric field (red) and no field (blue) and initial ratios of H_2O/CH_4 of 2 (solid line) and 1 (dotted line).

For our future work, we will focus on the field effects on the catalytic activities of Ni/YSZ catalysts since the MSR processes in SOFCs typically use Ni/YSZ as an anode material. Our preliminary results show that the oxygen vacancy formation energy of Ni/YSZ is largely influenced by the presence of an electric field [15] Moreover, the various oxidation states of the Ni atoms within the cluster will be altered by the presence of an electric field and the oxygen vacancy concentration, which can significantly alter the first C-H bond cleavage of methane (the rate-limiting step in the MSR reaction). [16] Such strong field effects increase strongly suggest that the MSR mechanism at the interface of the Ni/YSZ catalysts should have significant field effects as compared to those on a pure Ni surface. This ongoing work is important as it marks the first time that such complex and realistic electrocatalytic modeling has been undertaken.

References

[1] D. Energy, Annual energy outlook with projections 2015, Energy Dept, 2015.

[2] K. Oshima, T. Shinagawa, M. Haraguchi, Y. Sekine, Int. J. Hydrogen Energy, 38 (2013) 3003-3011.

[3] E. Nikolla, A. Holewinski, J. Schwank, S. Linic, J. Am. Chem. Soc., 128 (2006) 11354-11355.

[4] E.M. Stüve, Chem. Phys. Lett., 519-520 (2012) 1-17.

[5] H.J. Kreuzer, Surf. Sci. Anal., 36 (2004) 372-379.

[6] G. Pacchioni, J.R. Lomas, F. Illas, J. Mol. Catal. A: Chem., 119 (1997) 263-273.

[7] K.-Y. Yeh, M.J. Janik, CHAPTER 3 Density Functional Theory Methods for Electrocatalysis, Computational Catalysis, The Royal Society of Chemistry (2014), pp. 116-156.

[8] C.D. Taylor, S.A. Wasileski, J.-S. Filhol, M. Neurock, Phys. Rev. B, 73 (2006) 165402.

[9] F. Che, S. Ha, J.-S. McEwen, Appl. Catal. B, doi:10.1016/j.apcatb.2016.04.026(2016).

[10] F. Che, J. Gray, S. Ha, J.-S. McEwen, J. Catal., 332 (2015) 187-200.

[11] F. Che, R. Zhang, A.J. Hensley, S. Ha, J.-S. McEwen, Phys. Chem. Chem. Phys., 16 (2014) 2399-2410.

[12] F. Che, A. Hensley, S. Ha, J.-S. McEwen, Catal. Technol., (2014) 4020-4035.

[13] F. Che, J. T. Gray, S. Ha, J.-S. McEwen, In preparation, (2016).

[14] F. Che, J.T. Gray, S. Ha, J.-S. McEwen, In preparation, (2016).

[15] F. Che, S. Ha, J.-S. McEwen, J. Phys. Chem. C, under review (2016).

[16] F. Che, J.T. Gray, S. Ha, J.-S. McEwen, In preparation, (2016).

Copyright © American Institute of Chemical Engineers. All rights reserved.

465358 Exploring Ruthenium Metal-Support Interactions for the Low-Temperature Partial Oxidation of Methane

Abstract:

Exploring Ruthenium Metal-Support Interactions for the Low-Temperature Partial Oxidation of Methane

Emmett D. Goodman, Liheng Wu, Matteo Cargnello

Department of Chemical Engineering and SUNCAT Center for Interface Science and Catalysis, Stanford University, Stanford, CA 94305, USA

Industrially, the transformation of methane to syngas stands as an invaluable reaction, the latter products being converted downstream into valuable bulk chemicals such as methanol, ammonia, and synthetic fuels. The effective processing of methane will also limit the detrimental environmental impact of methane via industrial flaring. Steam or bireforming are common routes for methane conversion to syngas, but work at high temperatures because of their endothermic nature. Methane partial oxidation ($CH_4 + ½ O_2 = CO + 2 H_2$) is exothermic and benefits from more favorable thermodynamics. Thus, there are significant incentives to explore catalysts that can partially oxidize the pollutant methane, with high selectivity and yield, into an industrially relevant feedstock.

In order to rationally design such a catalyst, it is important to understand the underlying effect of the catalyst support. The catalyst support determines the spatial distribution, and often significantly alters the electronics of, supported nanoparticles. It therefore plays a large role in governing the overall system activity. However, in systems with differing support surface areas and nanoparticle coverages and size distributions, it is often difficult to discern the exact interaction between the nanoparticle and the support. Using monodisperse preformed Ru nanoparticles, and a system of supports with similar surface areas, here we show how to bypass such limitations and deconvolute these effects. With these methods we can back out important structure-property relations. We therefore find that selectivity and yield are strongly dependent on the support type, and mildly on Ru particle size. Using this information, we can gain fundamental mechanistic insight regarding methane partial oxidation on supported ruthenium systems, and work to design more effective catalysts.

Copyright © American Institute of Chemical Engineers. All rights reserved.

460465 Methane to Methanol Conversion in Small Pore Zeolites

Abstract:

One of the major challenges in catalysis today is the direct conversion of methane to methanol. Due to the similarities to enzymatic catalysis and the low activation barriers, most recently transition metal exchanged zeolites have drawn significant amounts of attention. However, a thorough understanding of the nature of the active sites in many of them and the impact of the confined environment is still missing. In this contribution we address two major points in this context, i.e. the nature of the active site for methane-to-methanol conversion in Cu exchanged SSZ-13 and the impact of the confining environment on enthalpy barriers in the conversion of methane to methanol over Fe-oxo centers in zeolites.

Cu_xO_y clusters embedded in different zeolite frameworks have shown significant activity in converting methane to methanol. However, depending on the zeolite structure and the exact synthesis and preparation protocols, different sites have been claimed in the literature to be most favorable. Here we develop a thermodynamic model to identify the most stable Cu_xO_y under various conditions in the zeolite SSZ-13. We find Cu_2O_2 clusters anchored at defect sites in zeolites to be most stable. We discuss the impact on several parameters on their formation and give guidelines what parameters might influence the Cu_xO_y distribution in the material. Furthermore we compare modeled UV-vis and Raman spectra to experiment and find excellent agreement for the suggested site with experimental measurements.

Additionally several studies indicate that confinement influences the activation enthalpies for this reaction in zeolites. To simplify the problem we focus on a more simple and better-understood Fe-oxo center, that also shows activity for the given reaction. We find that confinement stabilizes a methyl radical significantly more than the initially adsorbed methane molecule, which significantly lowers activation energies. We furthermore generalize these results to arrive at statements about the impact of confinement methane to methanol conversion and also other reactions.

Copyright © American Institute of Chemical Engineers. All rights reserved.

465777 Co-Aromatization of Methane with Olefins

Abstract:

The recent discoveries of large reserves of natural gas (i.e., shale gas) in North America have motivated the development of viable methods to convert this cheap energy source into higher value products. The activation and conversion of methane has been a great challenge due to its symmetric molecular structure and high C-H bond energy. The aromatization of methane has drawn increasing attention, while its industrial application is limited by the high reaction temperature. It is reported that the co-fed higher hydrocarbons would assist the conversion of methane at low temperatures (400~600 °C). In the present work, methane, the main component of natural gas, is activated with novel zeolites supported catalysts loaded with active metals, and converted to aromatics when olefin formed from heavy oil cracked distillates is present and at 400 °C. Several representative model compounds including ethylene, propylene, 1-hexene, 1-octene, 1-decene and styrene are also employed to study the co-aromatization mechanism. The participation of methane in the aromatization is proven by the increased carbon number of product molecules, as well as the isotope labelling studies. The study of the reaction mechanism is executed from multiple aspects with several analytic instruments, including the GC-MS analysis of products, which provides the composition information, and Diffuse Reflectance Infrared Fourier Transform (DRIFT) Spectroscopy, which traces the evolution of surface reaction intermediates. The co-aromatization reaction undergoes a series of steps, involving catalytic cleavage of C-H bonds in methane molecules and addition of the formed CH_x and H moieties into the carbenium ion intermediates derived from olefin feedstock. ^{13}C and ^{2}D enriched methane are also employed to track the catalytic pathways. The NMR spectra of the isotope enriched products obtained from $^{13}CH_4$ and CD_4 runs reveal how the CH_x and H species are converted to aromatics. DRIFT spectra of the catalyst, which has absorbed higher hydrocarbon molecules (i.e., olefin model compounds), acquired under $^{13}CH_4$ and CD_4 environment also disclose the evolution of methane on the catalyst surface. This work clearly demonstrates the feasibility of the conversion of methane into more valuable chemicals under mild temperature. This mechanistic study would also benefit the rational design of catalyst for direct upgrading of heavy crude oil using natural gas as a promising further work.

Copyright © American Institute of Chemical Engineers. All rights reserved.

470117 Production of Terephthalate Esters Via Diels-Alder Cycloaddition-Dehydration Reactions of 1,2-Furandicarboxylate Esters and Ethylene

Abstract:

Diels-Alder cycloaddition-dehydration (DA) reactions of dienophiles and biomass-derived furans are promising routes to renewable aromatics. Catalysis of such reactions with ethylene as the dienophile has been the topic of numerous experimental and computational studies. DA reactions of furans with reduced side groups (e.g., dimethyl furan) have been shown to be catalyzed with high selectivity by microporous materials with Brønsted or Lewis acid sites.[1] However, furans functionalized with oxygenated side groups (e.g., partially oxidized products of 5-hydroxymethyl furfural) were shown to be more prone to side-reactions.[2,3] For these reations, Lewis acidic Sn- or Zr-containing molecular sieves with the zeolite beta topology were found to be more selective catalysts than the Brønsted acidic Al-Beta. Zr-Beta yielded the highest selectivities (ca. 80%), with methyl (5-methoxymethyl) furan-2-carboxylate (MMFC) as the diene. However, the further oxidized 1,2-furandicarboxylic acid (FDCA), or esters thereof, were shown to be unreactive with such catalysts.

Recently, we have reported on the synthesis and characterization of Zn Lewis acid sites in CIT-6 (Zn-Beta).[4Â] We showed that, in addition to enabling DA reactions of MMFC, CIT-6 is able to react the dimethyl ester of FDCA (DMFDC). However, in the latter case, catalysis of decarboxylation reactions of the furan compromised selectivity towards the diester aromatic products, with the highest selectivity of 36%.

Here, we report a set of microporous catalysts for such DA reactions that do not appear to promote decarboxylation of DMFDC, and the optimization of reaction conditions that enables terephthalate diester selectivities upwards of 88%.

References:

(1) Chang, C.-C.; Je Cho, H.; Yu, J.; Gorte, R. J.; Gulbinski, J.; Dauenhauer, P.; Fan, W. *Green Chem.* **2015**.

(2) Pacheco, J. J.; Davis, M. E. *Proc. Natl. Acad. Sci. U. S. A.* **2014**, *111*(23), 8363â€"8367.

(3) Pacheco, J. J. New Catalysts for the Renewable Production of Monomers for Bioplastics, PhD Dissertation, California Institute of Technology, **2015**.

(4) Orazov, M.; Davis, M. E. *Chem. Sci.* **2016**, *7*(3), 2264â€"2274.

*Corresponding author. E-mail: mdavis@cheme.caltech.edu

Copyright © American Institute of Chemical Engineers. All rights reserved.

458119 Quantifying Lewis Acid Sites in Zeolites That Catalyze Glucose Isomerization

Abstract:

Quantifying Lewis Acid Sites in Zeolites that Catalyze Glucose Isomerization

James W. Harris, Michael J. Cordon, Juan C. Vega-Vila, Fabio H. Ribeiro, Rajamani Gounder*

School of Chemical Engineering, Purdue University, 480 Stadium Mall Drive, West Lafayette, IN 47907

*rgounder@purdue.edu

Pure-silica molecular sieves with tetravalent framework heteroatoms (M = Hf^{4+}, Sn^{4+}, Ti^{4+}, Zr^{4+}) catalyze intermolecular and intramolecular Meerwein-Ponndorf-Verley alcohol oxidation and Oppenauer aldehyde reduction (MPVO) cycles, which mediate stereoselective glucose isomerizations. Precise assessment of the reactivity of Lewis acid zeolites of different provenance, which invariably contain different densities of silanol defects and of open ((HO)-M-$(OSi)_3$) and closed (M-$(OSi)_4$) sites, require methods to accurately quantify these structures. Integrated molar extinction coefficients were determined for infrared (IR) peaks of deuterated acetonitrile (CD_3CN, 303 K) bound to open (2316 cm^{-1}) and closed (2308 cm^{-1}) Sn sites, and of pyridine (423 K) bound to Lewis acidic Sn sites (1450 cm^{-1}, 1610 cm^{-1}) in Sn-Beta zeolites [1]. Identical numbers of Lewis acidic Sn sites were titrated by CD_3CN and pyridine among Sn-Beta zeolites (>10 samples) of varying synthetic origin and composition, and by ammonia and n-propylamine only among samples with high Sn content (Si/Sn $\leq$ 150) and low-defect density. Pyridine and CD_3CN did not distinguish isolated Ti sites of different coordination in IR spectra. We will discuss our latest results using

carbon monoxide (CO, 100 K) to distinguish open and closed Sn and Ti sites in IR spectra, by extending similar reports on Zr-Beta [2].

The number of open Sn sites counted *ex situ* (CD_3CN IR) agreed quantitatively with the number of active sites in Sn-Beta zeolites that catalyze aqueous-phase glucose-fructose isomerization (373 K, 1% w/w glucose) at *initial* reaction times [1], consistent with experimental [3] and theoretical [4] evidence that open Sn sites are the dominant active sites for this reaction. Isomerization rate constants were measured on Sn-Beta and Ti-Beta in kinetic regimes that are first-order or zero-order in glucose concentration, and reflect free energy differences between the same kinetically-relevant 1,2-hydride shift isomerization transition state relative to two bound water molecules or to one bound glucose molecule at open Sn sites, respectively. Rate constants in both regimes (per active M, 373 K) were ~10-50x higher on hydrothermally-synthesized, low-defect M-Beta-F zeolites (Si/M = 100-220) than on post-synthetically prepared, high-defect M-Beta-OH zeolites (Si/M = 30-580), suggesting that isomerization rate enhancements within hydrophobic voids are caused by preferential stabilization of 1,2-hydride shift transition states. The number of silanol defects quantified from CD_3CN IR spectra (2275 cm^{-1}) correlated linearly with the molar water uptake determined from adsorption isotherms (293 K, P/P_0 = 0.2) for low-defect M-Beta-F samples, providing a quantifiable descriptor of hydrophobicity. These findings provide guidance to prepare more reactive M-Beta zeolites by maximizing the number of open sites while minimizing residual silanol defects.

[1] J. W. Harris et al. J. Catal., 335 (2016) 141-154.

[2] V. L. Sushkevich et al., J. Phys. Chem. C, 119 (2015) 17633-17639

[3] R. Bermejo-Deval et al. ACS Catal., 4 (2014) 2288-2297.

[4] N. Rai et al. ACS Catal., 3 (2013) 2294-2298; Y.P. Li et al. ACS Catal., 4 (2014) 1537-1545.

Copyright © American Institute of Chemical Engineers. All rights reserved.

471562 Designing Cooperative Interactions to Tune Catalytic Activity and Selectivity for Biomass Conversion

Abstract:

Catalytic sites incorporating multiple catalytic units can act cooperatively to achieve higher rates than isolated units. It remains an open question how to design these cooperative interactions. Specific design concepts can be gleaned from enzymes that contain multiple functional groups positioned in close proximity to provide tuned interactions with the substrates. The tuned interactions and precise spatial positioning enable the enzymes to achieve high catalytic rates and high selectivity. Understanding how to incorporate these design concepts into heterogeneous materials would improve catalyst performance.

Previously, we and others have focused on understanding the different aspects of how to tune interactions to achieve cooperative catalysis.[1â€"5]Â This work has primarily been for pharmaceutically relevant reactions such as the aldol, nitroaldol, and Knoevenagel reactions. In this presentation, we extend our synthetic approach to control catalytic activity and selectivity for directly relevant biomass production reactions that produce 5-hydroxymethylfurfural. We exploit different elements of cooperative interactions to create a novel design of the catalytic active site. These novel materials are synthesized, characterized, and tested in catalytic reactions. In particular, we utilize advanced spectroscopic techniques to demonstrate formation of the desirable catalytic site. The resulting materials achieve increased catalytic selectivity and activity for these reactions relative to current state-of-the-art materials. Overall, the increased yields and selectivities increase the potential to produce many important commodity chemicals in a sustainable manner.

References

[1] N.A. Brunelli, C.W. Jones, J. Catal. 308 (2013) 60.

[2] N.A. Brunelli, K. Venkatasubbaiah, C.W. Jones, Chem. Mater. 24 (2012) 2433.

[3] N.A. Brunelli, S.A. Didas, K. Venkatasubbaiah, C.W. Jones, J. Am. Chem. Soc. 134 (2012) 13950.

[4] R.K. Zeidan, M.E. Davis, J. Catal. 247 (2007) 379.

[5] J.D. Bass, A. Solovyov, A.J. Pascall, A. Katz, J. Am. Chem. Soc. 128 (2006) 3737.

Copyright © American Institute of Chemical Engineers. All rights reserved.

471727 Lewis Acid Zeolite Catalysts for Tandem Diels-Alder Cycloaddition and Dehydration of Dimethylfuran and Ethylene to Renewable p-Xylene

Abstract:

Recently, increasing demand for energy and commodity chemicals has led to accelerated research efforts in the conversion of renewable resource into chemicals and fuels for a sustainable economy. The processing of lignocellulosic biomass, an inexpensive, abundant and sustainable source of carbon, offers the promise of sustainable chemicals and carbon-neutral liquid transportation fuels. Lewis acid catalysts play an important role for selectively activating functional groups of the organic molecules involved in biomass conversion.

In this study,[1] Lewis acid zeolites including Zr-, Sn-, and Ti-BEA were examined for tandem [4 + 2] Dielsâ€"Alder cycloaddition of 2,5-dimethylfuran (DMF) and ethylene to form cycloadduct, which is further dehydrated to produce biorenewable *p*-xylene. *p*-Xylene is currently produced from petroleum processing, which serves as a precursor for polyethylene terephthalate (PET) polymers (e.g. PET bottle). Zr-BEA (Si/Zr = 168) offered superior performance with improved recalcitrance to deactivation, because of its low activity for the DMF oligomerization.[2] Zr-BEA also achieved the highest selectivity to *p*-xylene of 80% at 99% conversion of DMF. In the case of low catalyst loading, the reaction rate of *p*-xylene production was linearly proportional to the number of Lewis acid sites, whereas high catalyst loading was independent of Lewis acid sites. These findings were similar to the behaviors of Brønsted acid zeolite catalysts.

References

1. C.-C. Chang, H. J. Cho, J. Yu, R. J. Gorte, J. Gulbinski, P. Dauenhauer and W. Fan, *Green Chem.*, **2016**, 18,

   1368

2. J. Yu, S. Zhu, P. Dauenhauer, H. J. Cho, W. Fan and R. J. Gorte, *Catal. Sci. Technol.*, **2016**, **DOI:** 10.1039/C6CY00501B

Copyright © American Institute of Chemical Engineers. All rights reserved.

459019 Comparative Studies of Grafted Fe(III) and Ti(IV) Active Sites on Silaceous Supports: Crystalline Delaminated Zeolite Vs. Amorphous Silica

Abstract:

Silica has been widely used as a catalyst support with tunable surface area, porosity, and microscopic structure[1], making it ideal for supporting metal catalysts with high dispersion and high active site loading, while avoiding internal diffusion limitations. In contrast, zeotype materials have limited use as catalyst supports in many applications, since the majority of their surface area resides in micropores, making it inaccessible to precursors necessary for dispersing metal catalysts and to reactants and products relevant to their catalytic applications. Delamination of zeolite precursors can significantly improve accessible external surface area of zeotype materials and provides a strategy for their use as a support for dispersed metal catalysts[2]. In particular, removal of framework boron or aluminum and substitution for a different catalytically active cation has been used as a successful approach for synthesizing heterogeneous catalysts which, in contrast to amorphous silica supported metals, contain active sites embedded into well defined, crystalline, zeolitic framework positions[3].

The consequences of altering the coordination environment of a grafted cation from an amorphous support such as silica, to a crystalline framework site with well-defined structure, such as that of a zeolite, remain poorly understood. In this work, two case studies will be explored, involving Fe(III) and Ti(IV) cations grafted onto zeolite framework positions, which are compared to their grafting on amorphous silica. While grafting on amorphous silica results in cations covalently attached onto the support surface, grafting on delaminated zeolite supports results in cations that are incorporated into the crystalline framework. This incorporation is achieved by removal of framework B from delaminated zeolite supports, generating vacant tetrahedral framework positions – silanol nests. These vacancies are then repopulated with other cations, resulting in heteroatoms that are incorporated into the crystallographic position previously occupied by B, a

coordination environment different from that of amorphous silica.

During catalyst synthesis, our data demonstrate significantly higher uptake of aqueous-phase Fe(III) precursors by the delaminated zeolite framework relative to amorphous silica. DR-UV spectra demonstrate the synthesis of two different types of highly dispersed Fe(III) sites in both of these catalysts, which are both active for liquid-phase oxidation of adamantane by H_2O_2. However, after reaction, zeolitic Fe(III) active sites remain nearly unchanged, whereas Fe(III) sites on amorphous silica aggregate, as evidenced by DR-UV spectroscopy. These data support the hypothesis that zeolitic Fe(III) sites are more robust compared to their amorphous silica counterparts, presumably because grafted Fe(III) is stabilized when embedded into the zeolite framework, as opposed to simply grafted onto the silica surface.

We also investigated isolated Ti(IV) cations grafted on the same two supports, as relevant models of industrial catalysts for the epoxidation of olefins with organic hydroperoxides. In order to further facilitate a direct comparison between the two supports, we defined the coordination environment from the top of the Ti(IV) cation with a calix[4]arene macrocycle and from the bottom with either a crystalline zeolitic or amorphous silica support. We investigate both catalysts for the epoxidation of demanding terminal olefins with *tert*-btuyl hydroperoxide as oxidant, and compare with calcined variants. Regardless of whether there is a calix[4]arene ligand present or not, our results demonstrate significant catalytic consequences that favor the zeolitic support over the amorphous silica support in terms of rate. This in turn suggests that the degree of connectivity to the support is less important in determining rate than the actual support type.

Altogether, these two examples provide insights into the importance of the support crystallinity, which dictates the molecular coordination environment of a grafted metal cations, in establishing the catalytic properties of atomically dispersed supported catalyst.

(1) Soled, S. *Science* **2015**, *350* (6265), 1171–1172.

(2) Corma, A. *J. Catal.* **2003**, *216* (1-2), 298–312.

(3) Ouyang, X.; Hwang, S.-J.; Xie, D.; Rea, T.; Zones, S. I.; Katz, A. *ACS Catal.* **2015**, *5* (5), 3108–3119.

Copyright © American Institute of Chemical Engineers. All rights reserved.

456728 A Liquid Phase Spectroscopic Study of theÂ Effect of Liquid Water on Acid Sites of NaY

Abstract:

Liquid phase heterogeneously catalyzed reactions are ubiquitous, especially in biomass upgrade processes. However, establishing structure-activity relations for liquid phase reactions is challenging because the effect of solvent is poorly understood. Spectroscopy of adsorbed probe molecules like pyridine have been widely used to characterize catalyst acidity, and these experiments have almost exclusively been conducted in vacuum. This motivates a need for liquid phase probe molecule characterization, both to more accurately reflect reaction conditions, and to study the effect of solvent. Attenuated Total Reflection Fourier Transform Infrared spectroscopy (ATR-FTIR) was used to detect pyridine adsorbed on sodium faujasite (NaY) in water. The slow accumulation of pyridinium on NaY in water was observed in the ATR cell, suggesting water promotes Brønsted acidity on NaY. The accumulation of protonated pyridine in NaY pores led to the question of whether NaY can support Brønsted acid catalyzed pathways in protic solvent. The Brønsted catalyzed ring opening of 2,5-dimethylfuran (DMF) to 2,5-hexanedione is not active on NaY in water, which is attributed to the lack of basicity of DMF. A weaker base than pyridine, deuterated acetonitrile, adsorbed on NaY in water demonstrated no Brønsted character.

Copyright © American Institute of Chemical Engineers. All rights reserved.

471232 Computational Study of Single-Site Molybdenum Catalysts on Amorphous Silica

Abstract:

Amorphous silica is an extremely commonly used catalyst support yet there are very few experimental or computational studies on catalyst-support interactions for this material. This is largely due to the inherent difficulty in modeling and experimental characterization of amorphous silica. We have used a recently developed slab model for amorphous silica surfaces to study the support effects on single-site molybdenum catalysts. We have found that the local structure of the silica support in the vicinity of the Mo site has a profound effect on the energetics and kinetics of ethene metathesis. We have compared site energies, reaction energies, and reaction barriers computed from simple cluster models with results from the slab models. The cluster models show a clear relationship between Si-Si distances and the site energies and reaction energies. In contrast, the slab model shows no correlation between Si-Si distances and energetics. Likewise, the reaction barriers increase with increasing Si-Si distances in the cluster model but show no trends in the slab model. We have found that simple cluster models of the surface are not able to capture the rich heterogeneity of an extended surface model containing hundreds of atoms. As a result, trends that appear in the cluster models do not appear in the larger extended models.

Copyright © American Institute of Chemical Engineers. All rights reserved.

457446 First-Principles Kinetic Monte Carlo Simulation of the Thermal Reduction of PdO(101)

Abstract:

The impacts of oxygen vacancies on CO or alkane oxidation over transition metal oxide surfaces have been recently reported in catalysis and surface science studies. Oxygen vacancies may enhance CO and methane oxidation rates under reaction conditions, including potential autocatalytic behavior during CO oxidation over the transition metal oxide surfaces. Despite this rising attention, oxygen vacancy effects have not been closely studied. As a first step in our study, we investigate the thermal reduction of PdO(101) to elucidate oxygen vacancy effects in the absence of any reactants. First we employ density functional theory (DFT) to map out the kinetics of elementary surface processes related to the thermal reduction of PdO(101). Using the DFT calculations and harmonic transition state theory, we perform kinetic Monte Carlo (kMC) to examine the evolution of the PdO(101) surface as it begins to reduce at high temperatures. Our kMC simulations capture the experimentally observed autocatalytic behavior in which the decomposition rate increases as the surface decomposes. The simulation results show that an oxygen vacancy triggers the autocatalytic behavior by facilitating the formation of oxygen vacancies at adjacent positions. The presence of oxygen vacancies also promotes formation of O_2, which is the rate-determining step in the removal of oxygen from the surface. Moreover, we also elucidate the role of healing surface oxygen vacancies through diffusion of subsurface oxygens by conducting kMC simulations with a sublayer. The role of subsurface oxygen is found to delay and moderate the autocatalytic behavior. Future work will examine the role of surface processes identified in the thermal reduction (e.g. autocatalytic formation of oxygen vacancies, subsurface oxygen healing) on CO oxidation on PdO(101) surfaces at lower reaction temperatures.

Copyright © American Institute of Chemical Engineers. All rights reserved.

463122 Cu Site Speciation in SSZ-13 Zeolites Resolved through UV-Visible Spectroscopy, Selective Synthesis, Time-Dependent Density Functional Theory, and Ab Initio Molecular Dynamics

Abstract:

UV-visible spectroscopy on Cu-Zeolites, an industrially used catalyst for NOx selective catalytic reduction, result in complex many feature spectra, not easily assigned to specific molecular Cu species[1]. Furthermore, these spectra are highly sensitive to the contacting gaseous environment and temperature of exposure due to exchanged Cu ion mobility within the zeolite[2,3]. Here, we present a precisely coordinated study where Cu-SSZ-13 samples are synthesized and characterized through titrimetric and spectroscopic methods to contain only certain classes of Cu sites[4], and then assign these Cu species to specific configurations at the molecular level through a novel multi-step computational approach. The computational procedure employed here combines ab initio molecular dynamics, to estimate the macroscopic ensemble of Cu configurations, and time-dependent density functional theory, to compute quantitatively accurate UV-visible spectra, for these species. We show this methodology converges on spectra consistent with those observed experimentally and provides definitive molecular assignments for Cu exchanged cations in real synthesized catalysts. These calculations also provide evidence for what the dominant and minority Cu species are on other zeolite topologies, including MFI, BEA and MOR.

(1) Giordanino, F.; Vennestrøm, P. N. R.; Lundegaard, L. F., Stappen, F. N.; Mossin, S.; Beato, P.; Bordiga, S.; Lamberti, C. *Dalton Transactions* 2013, 42, 12741–61.

(2) Göltl, F.; Sautet, P.; Hermans, I. *Catal. Today* 2015, 267, 41–46.

(3) C. Paolucci, A. A. Parekh, I. Khurana, J. R. Di Iorio, H. Li, J. D. Albarracin Caballero, A. Shih, T. Anggara, W. N. Delgass, J. T. Miller, F. H. Ribeiro, R. Gounder and W. F. Schneider *Journal of the American Chemical Society*, (**2016**) DOI: 10.1021/jacs.6b02651

(4) Di Iorio, J. R.; Gounder, R. *Chem. Mater.* 2016, 28, 2236–2247.

Copyright © American Institute of Chemical Engineers. All rights reserved.

460420 Microkinetic Modeling of the Oxygen Evolution Reaction on Oxide Surfaces

Abstract:

The oxygen evolution reaction (OER) is an important anodic reaction that enables the sustainable production of H_2 from water at ambient conditions via the hydrogen evolution reaction or provides protons for cathodic CO_x reduction reactions. Extensive density functional theory (DFT) calculations have been previously carried out to determine the theoretical limiting potential (where all steps are exergonic) of the OER as a function of binding energy descriptors [1]. This work extends those analyses by feeding DFT calculated energies into a mean field, steady state microkinetic model for the OER. This model demonstrates that a simple mechanism consisting of four coupled proton electron transfers can be used to predict activity trends in relevant catalyst surfaces such as rutile and perovskite oxides.

A potential dependent barrier is assumed for each coupled proton electron transfer step [2]. The location of the peak of the volcano at OER relevant potentials agrees with thermodynamic predictions. A degree of rate control analysis and a sensitivity analysis of rate with assumed barrier are carried out. Changes in Tafel slope for a given surface have been observed experimentally for the OER. The model presented here predicts these changes in Tafel slope and demonstrates that changes in Tafel slope can be due to switches in reaction mechanism, coverage, and rate controlling step [3]. Additionally, this research outlines a method to determine the non-equilibrium, kinetic coverage of a surface under OER conditions. Generally, we find the steady state coverage corresponds to the precursor of the thermodynamically limiting step for potentials above the limiting potential.

This analysis demonstrates that qualitative trends can be predicted by a simple model and assumed proton transfer barriers. Future work in determining electrochemical barriers and their potential dependence is necessary to quantitatively predict rates and compare to experimentally measured rates. An analysis of electrochemical barriers for the OER on IrO_2 (110) is carried out to explore the assumption that all OER steps can be modeled as simple proton electron transfers. A charge extrapolation scheme is used to determine electrochemical barriers as a function of potential [4]. This analysis indicates that the barriers are low and scale with reaction energy with a slope near 0.5 as assumed in the microkinetic model. However, further work is necessary to fully quantify the electrochemical barriers for this system.

1. Man, I.C., Su, H., Calle-Vallejo, F., Hansen, H.A., Martinez, J.I., Inoglu, N.G., Kitchin, J., Jaramillo, T.F., Nørskov, J.K., Rossmeisl, J. Universality in Oxygen Evolution Electrolysis on Oxide Surfaces. *ChemCatChem* 3, 1159-1165 (Mar. 2011).
2. Tripkovic, V., Skulason, E., Siahrostami, S., Nørskov, J.K., Rossmeisl, J. The Oxygen Reduction Reaction Mechanism on Pt(111) from Density Functional Theory Calculations. *Electrochimica Acta* 55, 7975-7981 (Nov. 2010).
3. Holewinski, A., Linic, S. Elementary Mechanisms in Electrocatalysis: Revisiting the ORR Tafel Slope. *J. Electrochem. Soc.* 159, H864-H870 (Sept. 2012).
4. Chan, K., Nørskov, J.K. Electrochemical Barriers Made Simple. *J. Phys. Chem. Lett.* 6, 2663-2668 (June 2015).

Copyright © American Institute of Chemical Engineers. All rights reserved.

467187 Water-Gas Shift Reaction on Iron Oxide Catalysts â€“ a DFT Study

Abstract:

The water-gas shift (WGS) reaction is used for hydrogen enrichment and decreasing the carbon monoxide concentration in synthesis gas (H_2/CO). The copper supported chromium-iron oxide catalyst is used as the primary industrial catalyst for the high temperature (320-450 °C) WGS (HT-WGS) reaction since the Fe/Cr-based oxide catalysts can reduce the amount of CO from 10-15% to 2-3% [1]. Chromium is used as textural promoter to prevent thermal sintering and copper provides higher catalytic activity. [2]

The Feoct2 termination of the Fe_3O_4 (111) plane and Cr atoms substituted Fe_3O_4 (111) slab surfaces as the HT-WGS catalysts are investigated by means of spin-polarized periodic DFT approach using the Vienna Ab-initio Simulation Package (VASP) with the projector-augmented wave (PAW) method. Ferrimagnetic Fe_3O_4 is successfully optimized with nearly the same magnetic moment for octahedral and tetrahedral sites and the results are in line with the experimental values. The Feoct2 termination reveals two Fe atom layers located at the octahedral site (topmost) and tetrahedral site over four oxygen atom layer is studied among the six Fe_3O_4 (111) terminations. The DFT calculations indicate that the Cr atoms prefer being below the FeO_6 sites, which agrees with the LEIS surface analysis of Cr-Fe oxide catalyst activated under HT-WGS reaction conditions. The oxygen atom located at the topmost layer is energetically favorable for oxygen vacancy formation and the presence of Cr in the structure increases the energy for oxygen vacancy formation, which is in line with experimental kinetic measurements. Water adsorption on both Feoct2 and Cr-Feoct2 surfaces is dissociative with the adsorption energy values of -159 and -154 kJ/mol, respectively. There is only a minor effect of Cr on adsorption of water. The adsorption energy for CO on both surfaces are nearly same (the activation barrier for CO_2 formation on the Feoct2 site is 48 kJ/mol). Copper promotion of iron oxide based catalysts has also been investigated.

Copyright © American Institute of Chemical Engineers. All rights reserved.

468179 Computational Investigation of Lewis and BrÃ¸nsted Acidity in Metal-Doped Mesoporous Silicates

Abstract:

Mesoporous silicates are attractive materials for catalysis since they can host a variety of metals, and their adjustable pore size allows to control selectivity. Catalytic activity is enhanced by acidity around the metal centers, and the choice of metal influences both the type, i.e Lewis vs. Brønsted, and degree of acidity. However, understanding the correlation between metal and acidity is difficult due to the complexity of the amorphous structure of these materials. In this contribution, we focus on three metal dopants: Zr, Nb. and W. The first was reported to present only Lewis acidity, while the other two present both Lewis and Brønsted acidity[1-4]. Furthermore, studies of NH_3 adsorption showed that Zr-doped silicates present many more acid sites than Nb and W for the same metal loading (in mole fraction)[1-4]. We study these differences using multiscale quantum mechanical simulations based on density functional theory with the goal to understand the structure-property relations in doped mesoporous silicates. We consider multiple doping sites with varying metal coordination number and number of OH groups directly attached to the metal. Our results reproduce experimental trends, and allow us to correlate the Lewis/Brønsted acidity of these metals with their electronic structure and change in electronic density distribution in the presence of a base. Therefore, we can provide a molecular level understanding of these materials, which in turn may help improve experimental design towards increased catalytic activity and lifetime of these materials.

1. Ramanathan, *et al.* J Porous Mater 19, 961 (2012)
2. Pan, *et al.* Ind. Eng. Chem. Res. 52, 15481 (2013)
3. Ramanathan, *et al*. Microporous Mesoporous Mater. 167, 207 (2013)
4. Ramanathan, *et al.* Microporous Mesoporous Mater. 190, 240 (2014)

Copyright © American Institute of Chemical Engineers. All rights reserved.

473015 Simulated Temperature Programmed Desorption of Acetaldehyde on $CeO_2(111)$: Evidence for the Role of Oxygen Vacancy and Hydrogen Transfer

Abstract:

Typical biomass processing routes produce mixtures containing large fractions of small organic oxygenates including aldehydes, carboxylic acids, ketones and alcohols. However, their low molecular weights and high oxygen contents are undesirable in the production of fuels and higher value chemicals. Recently reducible oxides have been suggested as potential catalysts for the conversion of these organic oxygenates, via selective C-C coupling and deoxygenation reactions. We have studied the temperature programmed desorption (TPD) of acetaldehyde adsorbed on partially reduced $CeO_2(111)$ in detail using micokinetic modeling based on self-consistent, periodic density functional theory calculations at the GGA+U-PW91 level. The outcomes of our microkinetic model are in close agreement with the observed desorption activities in TPD and infrared evidence for surface species. According to our proposed mechanism, oxygen vacancies play the critical role of activating the carbonyl bond by stabilizing the negatively charged O atom with exposed Ce^{4+}, which facilitates the transfer of hydrogen atoms among the organic moieties and the formation of enolate that enables C-C coupling reactions under appropriate conditions.

Copyright © American Institute of Chemical Engineers. All rights reserved.

462264 Electrochemical Synthesis of Hydrogen Peroxide from Two-Electron Water Oxidation

Abstract:

Electrochemical Synthesis of Hydrogen peroxide from Two-electron Water Oxidation

Samira Siahrostami[1], Guo-Ling Li,[2],[3] Xinjian Shi,[4] Xiaolin Zheng,[4] Jens K. Nørskov[1],[2]

samiras@stanford.edu

[1] SUNCAT Center for Interface Science and Catalysis, Department of Chemical Engineering, Stanford University, 443 Via Ortega, Stanford, California 94305, United States

[2] SUNCAT Center for Interface Science and Catalysis, SLAC National Accelerator Laboratory, 2575 Sand Hill Road, Menlo Park, California 94025, United States

[3] School of Physics and Engineering, Henan University of Science and Technology, Luoyang 471023, China

[4] Department of Mechanical Engineering, Stanford University, 443 Via Ortega, Stanford, California 94305, United States

Hydrogen peroxide (H_2O_2), known as a green and sustainable solvent, is one of top 100 most important chemicals in the world. Hydrogen peroxide has wide applications such as in paper industry, textile industry and detoxification and color removal for wastewater.[1] The wastewater treatment application is of particular interest because only less than 1% of water resources on earth are accessible for human usages. Frequently, pollutants from urban, industrial, and agricultural human activities contaminate these available

water resources. Majority of these pollutants are only be oxidized into harmless products without generating waste in water by hydrogen peroxide. In contrast to adding H_2O_2 to water, it is very attractive to directly convert water to H_2O_2 on a photoanode (eq. 1) using sunlight, and to evolve hydrogen gas (H_2) (eq. 2) on the photocathode. Such system provides a clean process to produce two valuable chemicals: hydrogen gas and hydrogen peroxide (eq. 3) while purifying the water.

$2H_2O \text{ --> } H_2O_2 + 2(H^+ + e^-)\ E^0 = -1.76\ V$ (1)

$2(H^+ + e^-) \text{ --> } H_2\ \ E^0 = 0.0\ V$ (2)

$2H_2O \text{ --> } H_2O_2 + H_2$ (3)

In this work, we use combined density functional theory (DFT) and experiments to identify suitable photoanodes that have high selectivity towards two-electron path (H_2O_2 evolution), instead of four-electron path (O_2 evolution). Our DFT calculations suggest $BiVO_4$ as one of the best and most active catalyst for H_2O_2 production (Eq. 1). This is confirmed by our experimental results that show $BiVO_4$ has high faraday efficiency about 60% at 2.0 V vs. RHE towards hydrogen peroxide evolution.

References

(1) Campos-Martin, J. M.; Blanco-Brieva, G.; Fierro, J. L. G. *Angew. Chem. Int. Ed. Engl.* **2006**, *45*(42), 6962–6984.

(2) Martínez-Huitle, C. A.; Ferro, S. *Chem. Soc. Rev.* **2006**, *35*(12), 1324–1340.

(3) Izgorodin, A.; Izgorodina, E.; MacFarlane, D. R. *Energy Environ. Sci.* **2012**, *5*(11), 9496.

(4) McDonnell-Worth, C.; MacFarlane, D. R. *RSC Adv.* **2014**, *4*(58), 30551.

(5) Fuku, K.; Sayama, K. *Chem. Commun. (Camb).* **2016**, *52* (31), 5406–5409.

Copyright © American Institute of Chemical Engineers. All rights reserved.

459102 Theoretical Screening of Mxenes for Catalysis Using Density Functional Theory

Abstract:

MXenes (M = Transition Metal, X = C or N) are a promising family of materials that have been recently manufactured from MAX phases. MXenes have already been shown as promising candidates for use in lithium ion batteries, supercapacitors, and fuel cells (1). MXenes consist of $M_{(x)}X_{(x-1)}$, two-dimensional sheets weakly bound by van der Waals forces. However, due to the selective removal of the A ion to manufacture MXenes, they are highly reactive, with a wide variety of possible functional groups. We discuss the functionalization dependence of MXenes with respect to temperature, applied potential, and pH. We use computational screening of MXenes to predict their performance for different electrochemical reactions. This work is performed in close collaboration with several experimental groups and a comparison to their findings will be discussed.

(1) M. Naguib, V.D. Mochalin, M.W. Barsoum and Y. Gogotsi, *Adv. Mater.* **26**, 992 (2014).

Copyright © American Institute of Chemical Engineers. All rights reserved.

470976 Nanoengineering and Application of Protected but Accessible Metal Cluster Catalysts

Abstract:

The synthesis of metal clusters that are protected against aggregation while possessing exposed metal-surface sites for catalysis is crucial in both conventional catalysis as well as electrocatalysis, where the trigger for aggregation is high temperature of reducing environment and high reducing potential, respectively. We have relied on calixarene-bound metal clusters as highly accessible yet protected metal colloids in solution and when supported. Here, we demonstrate that they also serve as such under high reducing potentials, as required for CO2 reduction. For example, 0.9 nm gold clusters stabilized with five calixarene diphosphine ligands are stable at -1.4 V, conditions under which intrinsically more stable (larger) 4 nm gold nanoparticles aggregate readily. We demonstrate that for electrocatalysis, the ability to control the organic ligand sphere as well as the size and composition of the metal cluster are all key in the molecular engineering of the active site. In addition, relying on calixarene-bound iridium clusters as precursors to supported active sites, we demonstrate the synthesis of smaller nanoparticles, compared with conventional approaches. These have direct catalytic benefit as a result of their higher surface to mass ratio, when dealing with noble metals. Altogether, our results demonstrate the benefit of creating high dispersions of aggregation-resistant supported metal-cluster sites, with calixarene-ligated clusters.

Copyright © American Institute of Chemical Engineers. All rights reserved.

459193 Electronic Band Engineering in TiO_2 Particulate Layers for Photocatalysis

Abstract:

Electronic Band Engineering in TiO_2 Particulate Layers for Photocatalysis

Qilong Huang and Edmund G. Seebauer

University of Illinois at Urbana-Champaign, Urbana, IL 61801

Many large-scale photocatalysis applications employ TiO_2 deposited as a thin film by liquid-based methods from a particulate-based starting material. With solar illumination, photogenerated charge carriers form throughout the film and diffuse to nearby interior surfaces within the pore structure. For very thin films with wide pores, catalyzing gas-phase reactions at modest light intensity, reactants have little trouble diffusing to meet the charge carriers, and the photocatalyst effectiveness factor is near unity. For thicker films with narrow pores, catalyzing liquid-phase reactions of low ionic strength at high light intensity, catalyst effectiveness would improve if electric fields within the structure could be adjusted to pull minority photocarriers toward the nominal free surface, thereby circumventing the fluid transport problem. Such a strategy requires electronic band engineering of a sort not normally practiced in particulate-based films. The present work demonstrates experimentally how such band engineering may be implemented in the case of methylene blue photo-oxidation. Variations in film thickness, pore structure, light intensity, penetration depth and fluid viscosity highlight the interplay of reactant transport within the fluid and photocarrier transport within the solid. The concept of changing average electric field normal to the nominal surface via variations in the surface potential and majority carrier concentration are illustrated.

Copyright © American Institute of Chemical Engineers. All rights reserved.

469697 How Charge Screening Affects Zeolite-Catalyzed C-C Coupling Reactions

Abstract:

Zeolites are remarkable materials with adjustable confinement and Brønsted acid strength that catalyze reactions of vital importance. An issue that has attracted the attention of researchers for a long time is the role that the zeolite characteristics, such as intrinsic acid strength, structural confinement, as well as van der Waals and Coulomb interactions between the molecules and the zeolite framework, may have on the catalytic phenomena. In this study, we focus on the important influence of the charge screening effect in zeolite catalysis using acylation of biomass-derived furanic compounds as an example. Insight into the transition state obtained from first-principles density functional theory and hybrid functional calculations demonstrates the importance of the ability of aromatics in the proximity to stabilize the acylium ion by charge delocalization, which lowers the activation barrier in the C-C coupling step. This charge screening effect may be generally true for other zeolite-catalyzed C-C coupling reactions that involves a charged species in the reaction pathway.

Copyright © American Institute of Chemical Engineers. All rights reserved.

465530 Analyzing Local Composition and Structural Changes Occurring in a Nanocatalyst during Reaction

Abstract:

Understanding the steps involved in a catalytic process is crucial for materials design. The ability to discern the phenomena occurring at the catalyst surface, subsurface and interface with the support in heterogeneous catalysts provides key information to control the structure and properties of materials for the synthesis. Single-walled carbon nanotubes (SWCNTs) are materials typically synthesized by catalytic chemical vapor deposition (CCVD) and exhibit desirable properties for a variety of applications including electronics with smaller and faster transistors, sensors, biomedical, composite materials, energy-related applications and others. SWCNTsâ€™ structural features such as chirality, diameter and defects determine their properties and their suitability for a particular application. Therefore, current research efforts are focused on finding a way to grow SWCNTs with specific structures tailored by using the catalyst structure as a template. The current work combines computational techniques such as reactive molecular dynamics, density functional theory, and ab initio molecular dynamics with *in situ* real-time atomic-resolution transmission electron microscopy observations to discern the interactions of Co nanocatalyst, supported on MgO, with the growing nanotube throughout the nucleation and growth process. Evidence of the coexistence of a Co metal phase and carbide-like phases in the nanoparticle is presented along with a description of C dissolution, surface and bulk diffusion in the catalyst, and evolution of the C gradient in the nanoparticle. A mathematical model of the growth process considering the contributions of each of these phenomena from simulated data suggests that fluctuations in the growth rate can be attributed to fluctuations in the C composition of the nanoparticle. A similar correlation is found between changes in nanotube length and fluctuations in the size of the carbide-like regions in the nanoparticle. These results provide insight about the nature of the catalyst structure during growth and the role that metastable carbide phases may have in the process, which are both important factors in catalyst design.

Copyright © American Institute of Chemical Engineers. All rights reserved.

457344 Modelling of Diffusion-Reaction Interaction inside the Co-Based Catalyst Particles for the Fischer-Tropsch Synthesis

Abstract:

1. Introduction

Fischer-Tropsch synthesis (FTS) is a heterogeneous reaction used to convert synthesis gas into a range of hydrocarbon products. This reaction is a pivotal part of the Gas-to-Liquid (GTL) process in which natural gas is converted into liquid fuels and value-added chemicals. As such, FTS is often conducted inside of multi-tubular fixed bed reactors filled with cobalt-based catalyst particles. In order to avoid high pressure drops inside of the catalyst bed, relatively large particles (1-3 mm) are typically applied [1,2]. This may lead to intraparticle diffusional limitations for highly active catalysts. This means that the particle size, shape and operating process conditions may have a profound influence on the catalyst effectiveness and FTS selectivity. Several literature studies modelled diffusion with chemical reaction with Co based FTS catalysts [3,4], but focused mainly on spherical geometry and utilized rate equations based on experiments with a low activity catalyst [5].

In recent years Velocys [6] developed micro-channel reactors for FTS. These reactors utilize "super-active" Co catalysts compared to fixed-bed reactors in order to improve reactor volumetric productivity. It is expected that transport resistances could play a significant role with these reactors as well. Models of diffusion-reaction interaction are helpful in determining an optimal layer thickness that maximizes catalyst effectiveness and improves FTS product selectivity.

Here we present results from simulations utilizing one-dimensional particle models for Co-based FTS catalysts. Several critical parameters, such as the effects of shape (sphere, hollow cylinder and slab), diffusion length, temperature distribution inside the particle, diffusivity coefficients (gas- and wax-filled) and catalyst activity, and their influence on catalyst effectiveness and selectivity have been examined.

2. Diffusion-reaction model

FTS reaction can be simplified using the following stoichiometric relation:

$$nCO + (2n+1)H_2 \rightarrow C_nH_{2n+2} + nH_2O \qquad (1)$$

Our model assumes that the average hydrocarbon product is n-hexane (i.e. n = 6). Therefore, the components considered in the model are: carbon-monoxide, hydrogen, water and n-hexane.

The equation describing diffusion with chemical reaction inside FTS catalyst particle can be expressed as:

$$\frac{1}{x^g}\cdot\frac{d}{dx}\left(x^g\cdot\frac{dy_i}{dx}\right)+\alpha_i\cdot\varphi_i^2\cdot\psi=0 \qquad (2)$$

where x is the dimensionless distance, superscript g stands for geometry (g = 2, 1 and 0 for sphere, cylinder and slab, respectively), y_i is dimensionless concentration (i = CO, H_2, H_2O and n-hexane), stoichiometric coefficient (α_i = 1, 13/6, 1 and 1/6 for CO, H_2, H_2O and n-hexane, respectively), Ψ is the dimensionless CO consumption rate and φ_i is the dimensionless parameter ( ). This parameter includes catalyst density (ρ_p), characteristic length (L), reactant surface concentrations ( ), and effective diffusivity coefficients ($D_{e,i}$). Set of second order ordinary differential equations (Eq. 2) was solved in Matlab using the bvp4c solver.

$$\varphi_i = L\sqrt{\frac{\rho_p\cdot(-R_{CO}^s)}{D_{e,i}\cdot C_{CO}^s}}$$

C_{CO}^s

Reaction rate was described using the Yates-Satterfield expression [5]:

$$(-R_{CO})=\frac{k\cdot P_{CO}\cdot P_{H_2}}{(1+a\cdot P_{CO})^2}$$

(3)

The parameters *k* and *a* in Eq. (3) were estimated from experimental data with 0.48%Re–25%Co/Al_2O_3 catalyst [7]. Rates of H_2 disappearance and H_2O and n-hexane formation are calculated based on Eq. (3) and stoichiometry of Eq. (1). It is worth noting that this catalyst is approximately an order of magnitude more active under typical FTS operating conditions compared to the original Yates-Satterfield catalyst [8], and as such is more representative of modern Co catalysts used in industrial applications.

Two approaches to effective diffusion calculations were used, one assuming wax-filled and the other gas-filled catalyst pores. The former used correlations for diffusivity in FTS wax developed by Akgerman et al. [9], which is a representative of the porous catalyst transport environment over a significant portion of the reactor bed. For gas-filled particle we used multi-component diffusion approach, adjusted with Knudsen diffusion, and this is an extreme representation of segments of the reactor bed in which catalyst pores are not entirely filled with wax.

The key aspect of FTS catalyst selectivity is methane selectivity. Methane formation rate was calculated using Ma et al. [10] expression:

$$R_{CH_4} = \frac{k_M \cdot P_{CO}^{\,a_M} \cdot P_{H_2}^{\,b_M}}{(1 + m_M \cdot \frac{P_{H_2O}}{P_{H_2}})}$$

(4)

where numerical values of kinetic parameters were estimated from the same set of data as those in Eq. (3), i.e. with 0.48%Re–25%Co/Al_2O_3 catalyst [7]. Solution of Eq. (2) provides partial pressures needed to evaluate methane formation rate and calculate methane selectivity as:

$$S_{CH_4} = \frac{R_{CH_4}}{(-R_{CO})} \cdot 100\%$$

(5)

which is valid for Co-catalysts due to their low WGS activity, i.e. low CO_2 formation.

3. Results and discussion

Our simulations show that with very small spherical particles (< 150 μm) diffusional resistance is small, even with highly active Co catalyst used in our study. This is consistent with literature results [2-4]. However, with larger particle sizes (e.g. 2 mm) diffusion resistances are present both with the highly active Co catalyst and low activity catalyst used in previous studies [3,4]. Figure 1 shows model results for a 2 mm diameter spherical pellet, filled with wax, with kinetic parameters obtained with 0.48%Re–25%Co/Al_2O_3 catalyst. Reaction rate results show that FTS follows "abnormal" kinetics (increase in rate with conversion followed by decrease at high CO conversions). Ratio of H_2 to CO diffusivity is higher than the corresponding consumption ratio. This means that the ratio of H_2 and CO concentrations is increasing inside the particle. Because of this, as well as the positive reaction order in H_2 and negative order in CO, we get that the reaction rates inside the particle initially increase starting from the particle surface. However, for large particles, reactants are consumed relatively fast and reaction rate drops to zero leaving a large fraction of the particle unutilized (Figure 1). The calculated effectiveness factor for such a particle is only 0.52. Even more interesting is the behaviour of selectivity inside the spherical particle (Figure 1). Because methane selectivity follows the trend of H_2/CO ratio, S(CH_4) increases rapidly inside the particle. As concentration of CO goes to zero, H_2/CO ratio becomes so high that methane selectivity at a certain point inside the particle becomes 100%. Additional insights were obtained from simulations with a lower activity catalyst, different size particles and assumption of gas-filled pores. Results strongly suggest that the eggshell catalyst distribution is the only viable option for large catalyst particles utilized in multi-tubular fixed bed reactors.

Similar simulations were performed for a slab geometry present in wall-coated micro-channel reactors with various slab thickness and reaction rates. Figure 2 shows results obtained with 300 μm thickness layer. Catalyst effectiveness for this thickness is 0.65, suggesting that lower thickness should be used in order to fully utilize the coated catalyst. Also, methane selectivity is extremely high. This is contrary to the goal of increasing the reactor productivity by adding more catalyst (i.e. using thicker layers). Therefore, an optimal solution between high catalyst effectiveness and reactor productivity has to be found. Selectivity has to be taken into account also. Our simulation show that, similar to spherical particles, methane selectivity rises almost exponentially inside of 300 μm catalyst slab and reaches 100% well before FTS rate drops to zero. Thus, selectivity has to be considered in selection of the optimal slab thickness.

4. Conclusions

A one-dimensional model of diffusion-reaction interaction inside the FTS cobalt catalyst was developed. A number of simulations for sphere and slab geometry were performed. Results show that intraparticle diffusional limitations are severe with a highly active cobalt-based catalyst. Results suggest that large 2 mm spherical particles should utilize eggshell design in order to minimize the negative impact of pore diffusion on activity and selectivity. Also, pore diffusional limitations were found to be significant for a thin layer of catalyst (300 mm). Good understanding of diffusion-reaction interplay inside FTS catalyst is required in order to optimize reactor productivity, catalyst effectiveness and product selectivity.

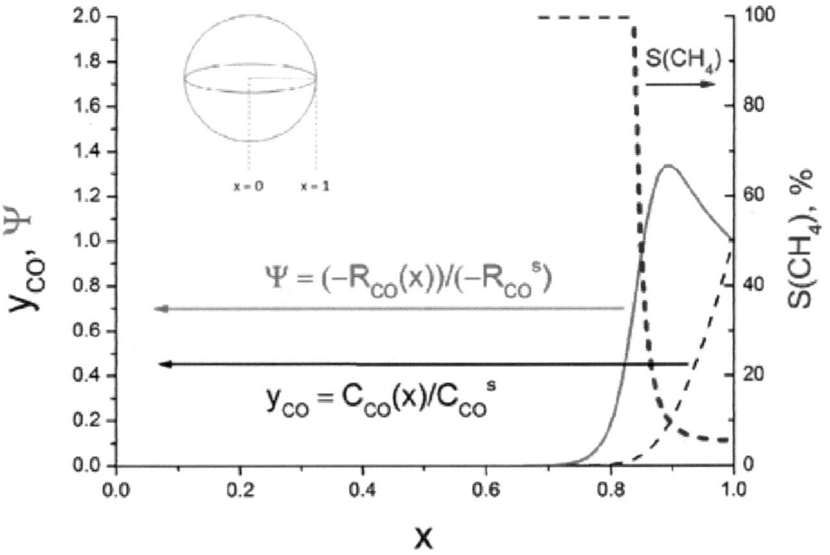

Figure SEQ Figure * ARABIC 1 – Distribution of dimensionless CO concentration (y_{CO}), dimensionless CO consumption rate (Ψ) and methane selectivity ($S(CH_4)$) within a spherical particle (d_p = 2 mm, T = 473 K, P = 25 bar, bulk H_2/CO = 2).

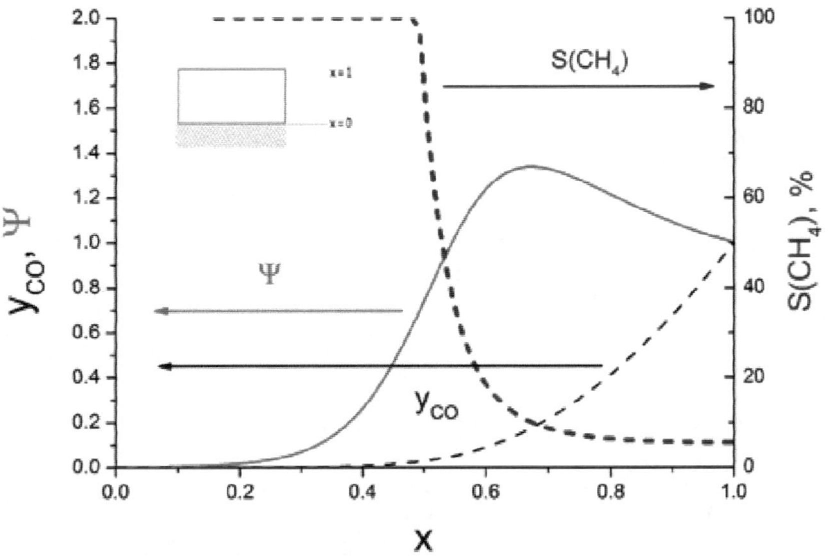

Figure 2 – Distribution of dimensionless CO concentration (y_{CO}), dimensionless CO consumption rate (Ψ) and methane selectivity ($S(CH_4)$) within a slab geometry (slab thickness = 300 mm, T = 473 K, P = 25 bar, bulk H_2/CO = 2).

Acknowledgements

This publication was made possible by NPRP grant 7-559-2-211 from the Qatar National Research Fund (a member of Qatar Foundation). The statements made herein are solely the responsibility of the authors.

References

[1] A.P. Steynberg, M.E. Dry, B.H. Davis and B.B. Breman, Stud. Surf. Sci. Catal., 152 (2004) 64.

[2] E. Rytter, N.E. Tsakoumis, A. Holmen, Catal. Today, 261 (2016) 3.

[3] D. Vervloet, F. Kapteijn, J. Nijenhuis, J.R. van Ommen, Catal. Sci. Technol., 2 (2012) 1221.

[4] S.A. Gardezi, B. Joseph, Ind. Eng. Chem. Res., 54 (2015) 8080.

[5] I.C. Yates, C.N. Satterfield, Energy Fuels, 5 (1991) 168.

[6] H.J. Robota, L.A. Richard, S. Deshmukh, S. LeViness, D. Leonarduzzi, D. Roberts, Catal. Surv. Asia, 18 (2014) 177.

[7] B. Todic, T. Bhatelia, G.F. Froment, W. Ma, G. Jacobs, B.H. Davis, D.B. Bukur, Ind. Eng. Chem. Res., 52 (2013) 669.

[8] C. Maretto, R. Krishna, Catal. Today, 52 (1999) 279.

[9] Akgerman, Final Report DOE AC22-84PC70032, (1984).

[10] W. Ma, G. Jacobs, T.K. Das, C.M. Masuku, J. Kang, V.R.R. Pendyala, B.H. Davis, J.L.S. Klettlinger, C.H. Yen, Ind. Eng. Chem. Res., 53 (2014) 2157.

464208 Carbon Nanotube-Supported Catalysts for Improved Fischer-Tropsch Synthesis

Abstract:

To address the lower reducibility and resulting lower activity of catalysts on oxidic supports during Fischer-Tropsch synthesis, there is increasing interest in utilizing carbon nanomaterials as catalyst supports. The production of clean fuel via FTS using Co and Fe catalysts supported on carbon nanotubes (CNTs) was studied and compared with similar catalysts supported on conventional substrates (alumina and silica). The synthesis of the catalysts was conducted using a variety of Co and Fe precursors and their physicochemical properties and FTS activities were investigated. The use of CNTs as catalyst supports enhanced the reducibility of the catalyst; both CNT-supported Fe and Co catalysts were fully reducible at temperatures below 550°C. Catalytic tests revealed a strong dependence of the CO conversion and C_{5+} selectivity on the type of catalyst precursor used for the various supports.

Copyright © American Institute of Chemical Engineers. All rights reserved.

469043 Influence of Dry-Coating of Nano Additives on Particulate Solid Flow and Its Reactions: Cement System a Case Study

Abstract:

The influence of dry coating of nano-additives on flow and decomposition of $CaCO_3$, reactivity of CaO towards di-calcium silicate (C_2S) phase formation and stabilization of C_2S in cement system were investigated. Influence of dry coating on flowability of $CaCO_3$ was investigated by FT4 powder rheometer. The results indicate that flowability in presence of nano-additives improves with respect to pure $CaCO_3$. The activation energy of the calcination step was calculated by Kissinger-Akahira-Sunose method using themo-kinetic analysis. The results show that 1% w/w nano-additives (nano-TiO_2 and SiO_2) reduce the activation energy by 15 to 30% compared to original $CaCO_3$. Further, the reaction model for decomposition reaction was predicted by Master plot and Malekâ€™s method and kinetic parameters were evaluated. The solid-solid reaction between CaO (originated from pure and nano-coated $CaCO_3$) and SiO_2 was then studied for formation kinetics of di-calcium silicate. The analysis of different phase formation was studied with the help of Rietveld refinement of XRD data. The results show that the reactivity of the CaO-SiO_2 system improves in presence of nano-additives which is explained by strain inside the crystal. Finally the stabilization of Î²-dicalcium silicate phase (desired cement phase) was studied. The results indicate that nano-TiO_2 effectively stabilize the Î²-phase which is explained by Hume-Rothery rules

Copyright © American Institute of Chemical Engineers. All rights reserved.

467625 Chemical Layer Deposition of Porous Alumina Overcoats Increases Activity and Stability in Liquid Phase Catalytic Conversion of Biomass-Derived Chemicals

Abstract:

Biomass-derived chemicals represent an attractive sustainable alternative to petrochemicals. However, the competitiveness of these processes has been limited by the low stability of metal catalysts under the severe conditions used during biomass upgrading. In order to prevent metal leaching and sintering, we developed a chemical layer deposition (CLD) overcoating technique based on the control over aluminum alkoxide surface reactions kinetics and stoichiometry. After deposition of 30 cycles of an alumina overcoat onto alumina supported copper nanoparticles, imaging using high angle annular dark field scanning transmission electron microscopy (HAADF STEM) revealed a conformal porous overcoat. This porosity was was characterized and quantified by nitrogen physisorption. The overcoat suppressed irreversible catalyst deactivation during the liquid-phase hydrogenation of furfural. While a significant loss of activity due to mass transfer limitations and metal coverage by the overcoat was observed by O'Neill *et al.* when using atomic layer deposition (ALD),[2] most of the surface metal sites remained accessible with CLD as confirmed by chemisorption, illustrating the benefits of overcoats porosity that can be targeted by tuning synthesis parameters. Current work is focused on tuning this porosity for the synthesis of shape selective catalyst overcoats.

(1) Héroguel, F.; Rozmysłowicz, B.; Luterbacher, J. S. *Chim. Int. J. Chem.* **2015**, *69* (10), 582–591.

(2) O'Neill, B. J.; Jackson, D. H. K.; Crisci, A. J.; Farberow, C. A.; Shi, F.; Alba-Rubio, A. C.; Lu, J.; Dietrich, P. J.; Gu, X.; Marshall, C. L.; Stair, P. C.; Elam, J. W.; Miller, J. T.; Ribeiro, F. H.; Voyles, P. M.; Greeley, J.; Mavrikakis, M.; Scott, S. L.; Kuech, T. F.; Dumesic, J. A. *Angew. Chem.* **2013**, *125* (51), 14053–14057.

Copyright © American Institute of Chemical Engineers. All rights reserved.

468951 Controlled Growth of Noble Metal Nanoparticles on Graphene Nanoplatelets Via Scalable Atmospheric Pressure Atomic Layer Deposition

Abstract:

Graphene-supported nanoparticles (NPs) are emerging as a novel class of high-performance catalysts [1,2]. For this reason, a great deal of research has been devoted to the fabrication of graphene-supported NPs of several noble and base metals for various applications such as electrocatalysis, photocatalysis, water treatment and splitting, and energy generation and storage. A number of methods have already been developed for the synthesis of metal/graphene. However, most of these methods rely on wet chemistry, in which often the metal dispersion is hard to control, and a high level of impurities arising from residual solvent and reaction byproducts. Therefore, a solvent-free and scalable method capable of depositing NPs on graphene in a controlled fashion, is highly desirable.

We demonstrate the deposition of crystalline palladium and platinum NPs on graphene nanoplatelets via atmospheric pressure atomic layer deposition (ALD). We carry this out in a reactor in which the graphene is dispersed in the gas phase, enabling the production of large amount of material [3]. The nucleation and growth of noble metal nanoparticles on the inert graphene surface was enabled by applying an ozone pretreatment step, without significantly affecting the graphene crystalline structure. Uniform nucleation on both basal planes and edges of the graphene was obtained. By analysing the evolution of the particle size distribution and spatial density, we obtained insights into the nucleation and growth of metal ALD on graphene. Such insights enabled us to tailor the particle size distribution and loading of the noble metal clusters. Finally, our approach provides a 100% solvent-free, controllable and scalable process for producing the bulk quantities of functionalized graphene. We will demonstrate the high catalytic activity of the obtained materials.

[1] Zhang, N.; Zhang, Y.; Xu, Y.-J., Recent Progress on Graphene-Based Photocatalysts: Current Status and Future Perspectives, Nanoscale 4 (2012) 5792– 5813

[2] Huang, C.; Li, C.; Shi, G., Graphene Based Catalysts, Energy Environ. Sci. 5 (2012) 8848– 8868

[3] Van Bui, H.; Grillo, F.; Helmer, R.; Goulas, A.; van Ommen, J. R., Controlled Growth of Palladium Nanoparticles on Graphene Nanoplatelets via Scalable Atmospheric Pressure Atomic Layer Deposition, J. Phys. Chem. C 120 (2016) 8832–8840

Copyright © American Institute of Chemical Engineers. All rights reserved.

464043 Computational Studies of the Dry Reforming Reaction on Al_2O_3-Coated Transition Metal Catalysts

Abstract:

More than 50 percent of man-made greenhouse gases are composed by carbon dioxide (CO_2), which makes urgent the development of effective strategies for remediation of CO_2 emissions. A promising route for chemical conversion of CO_2 is the dry reforming reaction (DRR) between CO_2 and methane (CH_4) to produce Syngas. Syngas is a mixture of H_2 and CO used as intermediate production of synthetic natural gas and ultra-clean fuels, or as precursor for ammonia, methanol, and other valued-added chemicals. Thus, the DRR reaction opens a window for remediation processes of CO_2 with high promise of producing value-added chemicals.

So far the screening of catalysts has shown that metallic catalysts from Group VIII are active for the DRR reaction. Among these catalysts the Rh-based seem to be the most active but the Ni-based catalysts have the best compromise between performance and cost. However, all of these catalysts suffer of sintering and coke formation due to the high temperature required for the DRR reaction (~800 °C), which still hinders the implementation at commercial scale of this route for chemical conversion of CO_2. In case of Ni-based catalysts two main causes are associated to sintering and coke formation. The first cause is the inability to precisely control the synthesis parameters and the interactions between the active site and the surface acting as support. The second cause is the exposure of the active catalytic sites to the extreme temperature at which the DRR reaction takes place. While the exposure of the catalysts to the high temperatures at which the DRR is performed is necessary, the minimization of sintering and coke formation can be performed through development of structures able to control the continuous structural reconstruction of the active phase.

Sintering and coke formation can be minimized by synthesis of Ni nanoparticles (NPs) surrounded by atomically precise assemblies protecting the active site. Recent experiments have reported increased resistance to sintering and coke formation for Pd-based catalysts overcoated with an Alumina layer using atomic layer deposition (ALD) methods. It was suggested that this Alumina layer may offer better protection to the active phase thanks to stabilization of the Ni nanoparticles. Those Pd catalysts covered by the Alumina layer resulted in only 5% of carbon accumulation in comparison with bared Pd catalysts, indicating increased resistance to catalyst coking as the ALD overcoating is applied. However, there is still debate regarding the effect of overcoating thickness on catalysts activity, as thinner coatings have been also associated with better overall catalytic performance.

So far the above mentioned experiments have been performed using structures composed of monometallic Pd NPs deposited on non-modified Al_2O_3 substrates without additives to stabilize the overcoating ALD AL_2O_3 structure. However, additional improvements on these structures can potentially enhance the catalyst stability and selectivity. In this work we examined a new class of Ni catalysts coated with Alumina using ALD methods. The reasons for the enhanced catalytic efficiency were investigated through state-of-the-art DFT/AIMD analysis. The active phases of Ni nanoparticles were represented by terrace and stepped slabs in order to evaluate the earlier steps of the Alumina layer formation. It is shown that the Alumina overcoating of Ni nanoparticles is viable but also Alumina deposition is preferred on stepped regions, leaving available the catalytically active terrace regions. The effects of Alumina overcoating on the mechanistic pathways for CO_2 activation were also studied, unveiling the effect of shape changes and nanoparticles reconstruction under support-nanoparticle and nanoparticle-overcoat interactions, and the possible appearance of side reactions and coke formation.

Copyright © American Institute of Chemical Engineers. All rights reserved.

470177 Experimentally Probing Ligand-Strain Effect Via a Novel Catalyst Platform

Abstract:

In this presentation, a novel catalyst fabrication platform was designed and investigated. Using a phase-pure transition metal carbide (TMC) nanotube as support [1], noncontiguous platinum group metal (PGM) particles were deposited from atomic level to 2-3 nanometers (nm) onto the TMC support by atomic layer deposition [2]. This platform enables experimental investigation on ligand-strain effect of resultant PGM/TMC catalysts. Characterization of metal-adsorbates bond strength, via diffuse reflectance infrared Fourier transform spectroscopy (DRIFTS) [3] and temperature-programmed oxidation (TPO), informs catalyst surface properties and model reaction mechanism [4]. Temporal analysis of products (TAP) technique was first applied to this catalyst platform revealing the existence of different active sites for Pt particle sizes (<3 nm). Additionally, the correlation between catalyst reactivity and change of deposited particle size was further investigated using water-gas-shift (WGS) reaction. The rarely investigated PGM particle size range and the ease of surface characterization by high resolution transmission electron microscopy (HRTEM), X-ray photoelectron spectroscopy (XPS) and other techniques, provides a unique PGM/TMC platform, which can be utilized closely with theoretical predictions to further our understanding on the interaction between PGM and non-PGM metals.

References

[1] Wan, C.; Regmi, Y. N.; Leonard, B. M. *Angew. Chemie Int. Ed.* **2014**, *53*(25), 6407â€"6410.

[2] McCormick, J. A.; Cloutier, B. L.; Weimer, A. W.; George, S. M. *J. Vac. Sci. Technol. A Vacuum, Surfaces, Film.* **2007**, *25*(1), 67.

[3] Ding, K.; Gulec, A.; Johnson, A. M.; Schweitzer, N. M.; Stucky, G. D.; Marks, L. D.; Stair, P. C. *Science*. **2015**, 350 (6257), 1688â€"1690.

[4] Schweitzer, N. M.; Schaidle, J. a; Ezekoye, O. K.; Pan, X.; Linic, S.; Thompson, L. T. *J. Am. Chem. Soc.* **2011**, *133*, 2378â€"2381.

Copyright © American Institute of Chemical Engineers. All rights reserved.

464481 Cobalt/Alumina Interactions in ALD Synthesized Catalysts for Fischer-Tropsch Synthesis

Abstract:

Cobalt metal was deposited onto an alumina (Al_2O_3) support using atomic layer deposition (ALD) to create dispersed catalysts for Fischer-Tropsch Synthesis (FTS). These catalysts are highly active for the FTS reaction, exhibiting triple the cobalt mass activity of standard incipient wetness catalysts at similar reaction conditions. HRTEM imaging shows that cobalt has simultaneously deposited during ALD in both crystalline plane and nanoparticle regimes, regardless of the number of ALD cycles. However, it was found that the number of ALD cycles greatly alters the activity of these catalysts, with 4 and 8 cycle catalysts being much more active than the 1 cycle catalyst. The known structure sensitivity of FTS indicates that the crystalline cobalt regions are the active regions for FTS. The ALD process allows deposition of metallic cobalt onto Al_2O_3 with protective organic capping ligands, preventing cobalt oxidation and reconstruction in air prior to reaction. Additional ALD experiments on different phases of Al_2O_3 followed by FTS testing proves the influence of support structure on the cobalt deposition regime and consequently FTS activity. Finally, density functional theory (DFT) calculations were done to understand the metal/support properties responsible for planar vs. nanoparticle growth during ALD and explain the high activity shown by these catalysts.

Copyright © American Institute of Chemical Engineers. All rights reserved.

467730 *Catalytic Performance of Supported Nano-Iron Catalysts Synthesized By a Gas-Expanded Liquid Deposition Technique in Fischer-Tropsch Synthesis*

Abstract:

Supported iron catalysts for low temperature Fischer Tropsch Synthesis (FTS) as prepared by impregnation of iron salt precursors on support materials have met with limited success. While the support materials provide high surface areas, iron species tend to interact strongly with traditional support materials such as SiO_2 and Al_2O_3. As such, the activity and selectivity towards long chain hydrocarbons for these catalysts were found to be inferior to those prepared by a precipitation method. However, supported iron catalysts provide a few potential advantages over precipitated catalysts, such as 1) better mechanical strength in order to withstand degradation, 2) enhanced dispersion of the iron, and 3) better control over the size and distribution of the iron species on the surface, etc. Therefore, there is a need to develop new synthesis methodologies to create improved supported iron catalysts in order to further enhance catalyst activity and selectivity towards hydrocarbons.

In this study, a new catalyst was prepared using pre-synthesized iron oxide nanoparticles that were controllably deposited onto a SiO_2 support using a gas-expanded liquid precipitation technique (GXL).Â In general, gas expanded liquids are solvent systemsâ€"typically a mixture of an organic solvent and a compressible gas, such as CO_2â€"that enable controlled deposition of pre-synthesized nanoparticles, based on their size, onto a support surface.Â By controlling the applied CO_2 pressure, and hence the solvent strength, one can systematically deposit/precipitate nanoparticles of desired sizes onto the support materials. The iron oxide nanoparticle dispersion used in this study contains oleic acid stabilized iron oxide nanoparticles dispersed in an *n*-hexane solvent. We induced the precipitation of larger sized particles from the initial nanoparticle dispersion, thereby depositing these larger particles onto the surface of the support while enabling the removal of the smaller sized particles by removal of the residual hexane solution. The ability to exclude the smaller sized particles from those that are deposited onto the catalyst support is important since supported iron FT catalysts with nanoparticles of small size have lower activity due to their stronger interaction with the support surface. As an added benefit, the lower viscosity and higher mass transfer coefficient in the GXL can enhance the diffusivity of the iron particles into the catalyst pores during deposition, resulting in a more uniform distribution of the nanoparticles compared to traditional methods.

The objective of this work is to explore the benefits of using the GXL technique in preparing SiO_2 supported catalysts with moderate iron loadings for Fischer-Tropsch synthesis applications. The catalytic performance of GXL prepared Fe/SiO_2 catalysts with different iron loadings were evaluated in terms of CO conversion, CH_4 selectivity, C5+ selectivity and C5+ productivity. In addition to the effect of iron loading, we have also investigated the effect of reaction conditions, such as reaction temperature, on the catalytic performance of our GXL catalysts in order to understand the influence of operating conditions on the selectivity and productivity of desired hydrocarbon products.

Copyright © American Institute of Chemical Engineers. All rights reserved.

474539 Photocatalysis for Z-Scheme Solar Water Splitting Using New Reactor Designs

Abstract:

Particle suspension reactors for solar water splitting can be an economical alternative to photovoltaic-driven electrolysis. One design resembles Nature's Z-scheme where together two photocatalyst nanoparticle reactor beds that are adjoined side-by-side and on the meters length scale, drive overall solar water splitting.[1,2] Electronic charge is mediated between the beds by a dissolved redox shuttle that undergoes oxidation or reduction at the particles and transports between the beds through a nanoporous material. While this design facilitates product separation (i.e. separation of H_2 and O_2) and therefore circumvents formation of an explosive mixture of gases, active transport of the redox shuttle over these distances has been projected to account for about half of the capital cost of the reactor.[1,2]

Our team is evaluating the feasibility of new reactor designs where the beds are stacked. This generates a true tandem light-absorbing reactor where the theoretical maximum solar-to-hydrogen conversion efficiency is over 50% larger than a side-by-side or single light-absorber design. Each bed is projected to be < 10 cm tall and therefore, this design greatly decreases the distance required for redox shuttle transport therefore reducing or even eliminating the need for forced convection.

In my presentation I will report on our team's progress on this design. We used finite-element numerical methods to model and simulate in two dimensions the transient mass transport processes, light absorption, and electrochemical kinetics in the proposed reactor. The developed model provided insights into the influence of the reactor geometry and operating conditions on the overall performance. The Beer–Bouger–Lambert law was applied to obtain the spatial light-intensity field and volumetric reaction rates were obtained by coupling solid-state photodiode expressions with Butler–Volmer kinetics on the surface of the particles. Model results suggested that a reactor operating at a ~1% solar-to-hydrogen conversion efficiency can operate for greater than half a year without complete loss of redox shuttle at any location in the reactor.

Experimentally, we investigated nanomaterials over many size scales, from single nanoparticles (~10 nm to ~1 µm in diameter) to mesoporous thin films (~10 µm thick) to laboratory-scale prototype particle-suspension reactors (on the scale of feet). On the single particle level we used bipolar electrodeposition to create Janus-type particles consisting of model carbon particles with metal and metal-oxide electrocatalysts for H_2 evolution and O_2 evolution at the poles. We also jammed and covalently bound TiO_2 nanoparticles into a single nanopore in a plastic sheet, wetted the particles with liquid electrolyte on both sides, and measured photovoltages that resulted from excitation of few particles. This was significant because it allowed, for the first time, in situ photoelectrochemical characterization of particle(s). We also synthesized, characterized, and evaluated the photo(electro)chemical performance of $BiVO_4$ and Rh-modified $SrTiO_3$ nanocrystallites as mesoporous thin films and particles in model reactors, and evaluated the transport properties of several redox shuttles. The $BiVO_4$ and Rh-modified $SrTiO_3$ were grown by sol gel methods and characterized by XRD, XPS, transmission-mode or diffuse-reflectance ultraviolet–visible electronic absorption spectroscopy, He II UPS, DLS (as particles), Raman, and photoelectrochemical performance. These nanocrystallites exhibited some of the best performance reported to date.

Collectively, our efforts represent strides toward achieving a high-level of techno-economic viability in solar water splitting reactors.

Acknowledgments: This work was supported by the U.S. Department of Energy, Office of Energy Efficiency and Renewable Energy, Fuel Cell Technologies Incubator Program under Award No. DE-EE0006963 and Lawrence Berkeley National Laboratory under Contract No. DE-AC02-05CH11231.

References:

(1) B. D. James, G. N. Baum, J. Perez and K. N. Baum, *Technoeconomic Analysis of Photoelectrochemical (PEC) Hydrogen Production*, Directed Technologies Inc., (US DOE Contract no. GS-10F-009J), Arlington, VA, **2009**.

(2) B. A. Pinaud, J. D. Benck, L. C. Seitz, A. J. Forman, Z. Chen, T. G. Deutsch, B. D. James, K. N. Baum, G. N. Baum, S. Ardo, H. Wang, E. Miller, and T. F. Jaramillo, *Energy & Environmental Science*, **2013**, *6*, 1983–2002.

Copyright © American Institute of Chemical Engineers. All rights reserved.

463882 Spatially Resolved 3D-CFD Simulation of a Packed Bed Membrane Reactor for Steam Reforming

Abstract:

Spatially resolved 3D-CFD simulation of a packed bed membrane reactor for steam reforming

Nico Jurtz[1], Thomas Eppinger[2], Matthias Kraume[1]

[1]Chair Chemical and Process Engineering, TU Berlin, Berlin, Germany;

[2]CD-adapco, Nuremberg, Germany;

Abstract

Due to the progressing climate change and crude oil depletion a shift to renewable resources is needed. Hydrogen and syngas are important feedstocks in the chemical and process industry and are used in a variety of processes like methanol synthesis, the Haber-Bosch process or the Fischer-Tropsch synthesis.

One route for syngas generation is the catalytic steam reforming of methane or biogas in packed bed reactors (PBR). Recent studies show the potential of process intensification by hydrogen removal with semi-permeable membranes. Packed bed catalytic membrane reactors show an increased methane conversion compared to PBR's (Chibane, 2011).

For a reliable but coast efficient design of that reactor type Computational Fluid Dynamics (CFD) is a valuable tool.

Although due to the low tube-to-particle diameter ratio local wall effects play an important role recent numerical studies only used pseudo-homogenous approaches often combined with the assumption of a constant transmembrane species flux.

The current work will overcome that simplifications and show how spatially resolved 3D-CFD simulations help to improve the reactor design by including all relevant phenomena like fluid flow through the complex bed topology, heat transfer, catalytic chemical surface reaction and a locally resolved transmembrane species flux.

For the packed bed the DEM-CFD coupled method by (Eppinger, 2011) is used. The transmembrane flux is modelled using an approach presented at the AIChE Annual Meeting 2015 (Jurtz, 2015) that uses a conformal interface to calculate the local membrane flux.

References

Chibane, L., Djellouli, B., 2011. Methane Steam Reforming Reaction Behavior in a Packed Bed Membrane Reactor. IJCEA, Vol. 2, No. 3, 147-155.

Eppinger, T., Seidler, K., Kraume, M., 2011. DEM-CFD simulations of fixed bed reac-tors with small tube to particle diameter ratios. Chem. Eng. J. 166, 324-331.

Jurtz, N., Eppinger, T., Aglave, R., 2015. A CFD Study on the Impact of Convective Flow Characteristics and Species Distribution on the Performance of Full-Scale Membrane Reactors. AIChE Annual Meeting, Salt Lake City

Copyright © American Institute of Chemical Engineers. All rights reserved.

468857 Iron-Manganese Catalysts for the CO Hydrogenation to Alkenes: Effect of the Catalyst Synthesis on the Selectivity and Activity

Abstract:

@font-face { "Arial"; }@font-face { "　　明朝"; }@font-face { "Cambria Math"; }@font-face { "Calibri"; }@font-face { "Cambria"; }@font-face { "ヒラギノ角ゴ Pro W3"; }@font-face { "MS Mincho"; }p.MsoNormal, li.MsoNormal, div.MsoNormal { margin: 0cm 0cm 0.0001pt; font-size: 12pt; ; }span.MsoFootnoteReference { vertical-align: super; }p.FreeForm, li.FreeForm, div.FreeForm { margin: 0cm 0cm 0.0001pt; font-size: 12pt; ; color: black; }p.Authors, li.Authors, div.Authors { margin: 6pt 0cm; line-height: 16pt; font-size: 11pt; ; }.MsoChpDefault { ; }div.WordSection1 { }

Iron-manganese catalysts for the CO hydrogenation to alkenes: effect of the catalyst synthesis on the selectivity and activity.

H. Karroum[1,2], V.Iablokov[1], Y. Xiang[1], V. Dubois[2], N. Kruse[1]

[1] Voiland School of Chemical Engineering, Washington State University, Pullman, WA 99164

[2] Institut Meurice, Brussels, Belgium

Fischer-Tropsch Synthesis (FTS) consists of the catalytic conversion of syngas into a large range of hydrocarbons and/or alcohols. Research into FTS has spurred in recent times due to the possibility of producing chemical raw materials such as oxygenates and olefins. These materials are being considered as important building blocks for industrial organic and polymer chemistry and provide the feedstock for lubricants, detergents, etc. The main challenge is to tune the selectivity of the Fischer-Tropsch reaction to favor one product class over others. Here we are interested in favoring short chain olefins and investigate the prospects of innovative catalyst preparation to boost olefins versus paraffins production.

It has been previously demonstrated that the size and morphology as well as the composition of metal particles have a major influence on catalyst performance (activity, selectivity and stability) [1]. The use of colloidal recipes seems most promising in controlling the size and morphology of metallic nanoparticles. Iron-manganese catalysts have been prepared in our study using this colloidal approach. Furthermore, and quite differently, the same catalysts have been prepared via oxalate co-precipitation as developed in our laboratory in recent years. The catalytic properties of the various catalysts prepared by either method will be compared in the high-pressure FTS synthesis.

The oxalate route consists of the co-precipitation of Fe^{2+} and Mn^{2+} with oxalic acid. The precipitate is known to form an extended polymeric oxalate structure incorporating both metal cations in the same structure. Subsequent hydrogen-assisted thermal decomposition (at 390°C) of the mixed-metal oxalate leads to an intimately mixed iron-manganese catalyst ('cat 1'). On the other hand, the colloidal route consists of the thermal decomposition of $Fe(acac)_3$ and $Mn(acac)_2$ using oleylamine simultaneously as solvent and reductant. Typically, the suspension is first heated under vacuum at 100°C in order to dissolve the salts, followed by increasing the temperature up to 290°C under nitrogen atmosphere. Finally, iron-manganese nanoparticles are precipitated with methanol and after centrifugation the precipitate is redisolved in chloroform. The particles have been impregnated into silica (Davisil S = $260 m^2/g$) support ('cat 2').

Prior to catalytic tests, both catalysts have been activated in situ under hydrogen using temperature programmed decomposition. XRD analyses have been performed on both activated catalysts. According to the XRD patterns, the bulk of 'cat 1' is mainly composed of spinel-type Jacobstite ($MnFe_2O_4$) while the bulk of 'cat 2' indicates broad diffraction lines tentatively assigned to Fe-oxide structures (Fe_2O_3 and Fe_3O_4). The absence of diffraction patterns associated with Mn-related structures is indicative of high Mn dispersion providing XRD-amorphous states in "cat 2'. TEM micrographs show that the particle sizes of 'cat 1' are ranged between 7-15 nm while 'cat 2' has a narrow size distribution ranged between 6-6.5nm. For both catalyst preparations metal particles appear in spherical morphology.

Catalytic tests in the CO hydrogenation have been performed in a fixed-bed plug-flow reactor after activating both catalysts under hydrogen. The catalytic performances of both catalysts have been tested at 300°C and 20 bar. 'Cat 1' shows an alkane selectivity of 27% and an alkene selectivity of 69% at 11 % CO conversion while 'cat 2' shows an alkane selectivity of 61% and an alkene selectivity of 39% at 28% CO conversion. Obviously these data demonstrate a tradeoff between conversion and olefins selectivity: while high CO conversion is associated with relatively low olefins production (39% is, however, not low!), low conversion increases the selectivities dramatically. The higher activity of 'cat 2' becomes obvious when comparing the reaction rates per gram of iron. In fact, the reaction rate of 'cat 1' is 14.7 umol g_{Fe}^{-1} s^{-1} while the reaction rate of 'cat 2' is 56.6 umol g_{Fe}^{-1} s^{-1}. The TOFs (turnover frequencies) of 'cat 1' and 'cat 2' can then be calculated to obtain 0.009 s^{-1} and 0.02 s^{-1}, respectively. Note that the reaction rate in terms of TOF values takes into account a surface iron density of 0.0613 nm^{-2} assuming Fe sites provide the active sites.

We have demonstrated here that the catalyst preparation procedures have a dramatic influence on the selectivity and activity toward the formation of the various product classes in FTS. Superior activity of silica-supported colloidal particles has been observed which is promising with regard to a further optimization of the process. Future efforts will therefore address the preparation of Fe-Mn bimetallic nanosized catalysts in different ratios to tune the production of short chain olefins.

[1] V. Iablokov, Y.Xiang, A. Meffre, P-F Fazzini, B. Chaudret, and N. Kruse, 'Size-dependent Activity and selectivity of Fe(0) Nanoparticles in the catalytic Hydrogenation of Carbon Monoxide', ACS Catal., 2016, 6 (4), pp 2496–2500

Copyright © American Institute of Chemical Engineers. All rights reserved.

469315 Effect of Potassium Promoter in Cocu-Based Catalysts for CO and CO_2 Hydrogenation

Abstract:

Effect of Potassium Promoter in CoCu-based Catalysts for CO and CO_2 Hydrogenation

By Jenny Voss, Lei Tang, Yizhi Xiang and Norbert Kruse

Gene and Linda Voiland School of Chemical Engineering and Bioengineering, Washington State University, Pullman, WA

Key Words: CO_2 Hydrogenation, reverse water gas shift, potassium-promoted CoCu-based catalysts

Conversion of CO_2 and CO to higher alcohols and olefins is driven by incentives to produce transportation fuels or additives on one hand and building blocks for the polymer industry on the other. Traditionally, alcohols are made from aldehydes which are obtained via hydroformylation of alkenes with CO and H_2 using high-cost homogenous catalysts. Previously, our group demonstrated ternary CoCuMn catalysts to favor C_8-C_{14} alcohols while, quite differently, CoCuNb catalysts turned out to favor shorter alcohols in CO hydrogenation [1-2]. Furthermore, all CoCu-based catalysts produced considerable amounts of olefins as well.

In the present contribution we explore the perspectives of turning CO_2 into CO via the reverse water gas shift reaction and using the known CO hydrogenation patterns for CoCu-based catalysts to produce oxygenates and olefins. Since the above CoCuMn and CoCuNb catalysts showed limited reverse water gas shift activity, we concentrated on alkali-promoted CoCu catalysts here.

We synthesized such catalysts using the oxalate co-precipitation route. To do so, metal nitrates were co-precipitated with oxalic acid in acetone to form a polymeric oxalate structure containing double-chelating oxalate ligands with Co or Cu within the same structure. Alkaline oxalates were precipitated at the same time using their limited solubility in acetone. To activate the precursors, we performed hydrogen-assisted temperature-programmed decomposition which usually resulted in the formation of gaseous CO_2 with insignificant amounts of concomitant CO. This observation allowed us to conclude that our CoCu catalysts are metallic, with no phase-separated oxidic structures appearing. Following previous microscopic investigations with Atom-Probe Tomography (APT) and Transmission Electron Microscopy (TEM), these catalysts are Co@Cu core-shell structured [1]. Recent results by TEM and X-ray Diffraction (XRD) seem to indicate that the presence of potassium promotes the formation of Co-carbide while running the catalyst into steady-state CO_2 hydrogenation conditions, see Table 1 below.

All high-pressure experiments were conducted in a fixed bed flow reactor operating at 40 bar, with a feed ratio of H_2/CO_2= 3 and temperatures of 200-260°C. Without a promoter, CoCu catalysts mainly produced methane. These non-promoted catalysts also showed unstable performance with extended times-on-stream. CoCuLi, CoCuNa and CoCuCs catalysts (all containing a 1:1:0.1 ratio of metals) also produced significant amounts of methane. Most importantly, however, these catalysts (except for CoCuCs) had little reverse water gas shift activity. On the contrary, potassium-promoted CoCu catalysts favored formation of CO and Fischer Tropsch products such as alkanes (RH), olefins (R=) and alcohols (ROH) as displayed in Table 1. The importance of the reverse water gas shift reaction is therefore clearly demonstrated here since appreciable amounts of alkanes, olefins and alcohols are obtained when CO formation is high. However, there resides an overarching tradeoff between high CO formation and high olefins/alcohols selectivity with low CO_2 conversion at low temperatures and improved CO_2 conversion but lower selectivity at high temperatures.

When determining the bulk composition of CoCu-based catalysts before and after CO_2 hydrogenation reaction via XRD studies, all catalysts contained a metallic cobalt phase before the reaction yet only CoCuK catalysts contained cobalt carbide (Co_2C) post reaction. Therefore, we suggest potassium acts as a structural promoter. Yet, we advocate the view that potassium helps establish the water gas shift equilibrium at sufficiently low temperatures. The combined olefin /alcohol selectivities between 220°C and 260°C are close to 40%; however, the CO conversion is quite low.

We are currently investigating the electronic and structural effects of potassium promotion in CoCu-based catalysts. We are also exploring other synthesis routes to improve catalyst activity at low temperatures where CO selectivity and valuable product formation are favored due to the reverse water gas shift activity.

Table 1: Results of a $Co_2Cu_1K_{0.1}$ catalyst under CO_2 hydrogenation conditions

Catalyst	Temperature (°C)	CO_2 (wt %)	CO (wt %)	Selectivity (wt %) without CO included			
		Conversion	Selectivity	CH_4	RH	$R_=$	ROH
$Co_2Cu_1K_{0.1}$	220	5	87	31	65	27	9
	240	8	82	35	63	26	12
	260	15	74	42	64	24	13
	280	19	42	61	78	15	8

Citations:

[1] Xiang, Y., et al., "Long-Chain Terminal Alcohols through Catalytic CO Hydrogenation," *Journal of the American Chemical Society*, 13 (19), pp. 7114-7117 (Apr. 2013)

[2] Xiang, Y., et al., "Ternary Cobalt-Copper-Niobium Catalysts for the Selective CO Hydrogenation to Higher Alcohols," *ACS Catalysis,* 5, pp. 2929-2934 (Apr. 2015)

Copyright © American Institute of Chemical Engineers. All rights reserved.

468630 Synthesis and Development of Active Nanomaterials for Tri-Reforming

Abstract:

The rising concentration of greenhouse gases in the atmosphere requires new technologies to control CO_2 emissions. The catalytic tri-reforming process has been proposed to directly utilize CO_2 emissions from a flue gas without the need of CO_2 pre-separation step while producing syngas with a suitable H_2:CO ratio desired for the Fischer-Tropsch process and methanol synthesis. Tri-reforming process requires a novel catalyst with a high surface area, good redox properties and high O_2 storage capacity. It should also be thermally stable between 700°C - 900°C and provide resistance to coke formation.

Nickel-based catalysts are typically used in tri-reforming reactions; however stabilizing Ni particles poses a substantial challenge at high reaction temperatures and under reducing reaction conditions. Ni particles covered with metal oxide shells can stabilize Ni, preventing its deactivation at high reaction temperatures. Little is known about how morphology, composition, and dispersion of Ni particles can influence the activity of the catalyst under tri-reforming conditions.

In this work, highly active and stable Ni@CeO_2 and Ni@SiO_2 core-shell structures are developed by carefully adjusting their structure and composition. The structures are dispersed over Al_2O_3 and SiO_2 supports and their activity for tri-reforming is tested at various temperatures and feed gas compositions. The catalysts are characterized via electron microscopy, x-ray and infrared spectroscopies. Coupling reaction and characterization experiments performed on the well-designed core-shell nanoparticles provide information about the structure-activity relationship.

Copyright © American Institute of Chemical Engineers. All rights reserved.

473261 Gasoline-Selective Fischer-Tropsch Synthesis with Hierarchical ZSM-5 Coated Monolith Catalysts

Abstract:

Energy independence is important for economic stability and security. Despite significant incentives and developments for the decrease of crude oil imports, the U.S. still imports about 24% of its oil (2015 EIA data) [1]. Moreover, the infrastructure of the U.S. transportation sector imposes requirements for high-octane, gasoline-range hydrocarbons. Gasoline consumption accounts for about 50% of the total U.S. oil consumption [2]. Fischer-Tropsch Synthesis (FTS) is a well-known process for the conversion of (U.S. abundant) natural and shale gas to hydrocarbons [3,4]. However, FTS is historically conceived as a diesel-producing process, with long-chain hydrocarbon products favored due to the polymerization nature of the reaction network [3,5,6]. Also, FTS produces mainly linear olefins and paraffins [7]; and therefore, low-octane gasoline. These limitations can be overcome by the addition of an acidic co-catalyst to enhance cracking, oligomerization, isomerization and aromatization. *In-situ* upgrading of primary FTS products is possible by use of bifunctional or hybrid catalysts comprising the FTS active phase and an acid catalyst. Acid catalysts enhance oligomerization, cyclization, aromatization, cracking, and isomerization reactions and provide shape selectivity towards gasoline-range hydrocarbons. In this work, multilayered structured catalysts were synthesized, composed of a supported FTS active phase (Co catalyst), and coated with a layer of ZSM-5 catalyst. Both layers were formed on the internal surface of a monolith support. The structure of these multilayered catalysts is conceptually presented in Figure 1. The synthesized multilayered catalysts enable: a) relaxation of heat and mass transfer limitations; b) high activity with low water-gas shift potential; c) control of product size through manipulation of the catalyst outer film thickness; and d) enhanced selectivity to gasoline-range hydrocarbons. Early work has shown excellent selectivity and performance of these catalysts. The bi-functional, multi-layered catalysts with Co dispersed on an Al_2O_3 layer over the monolith and coated with ZSM-5 doubled the selectivity to gasoline range compounds, without affecting the overall conversion of the process.

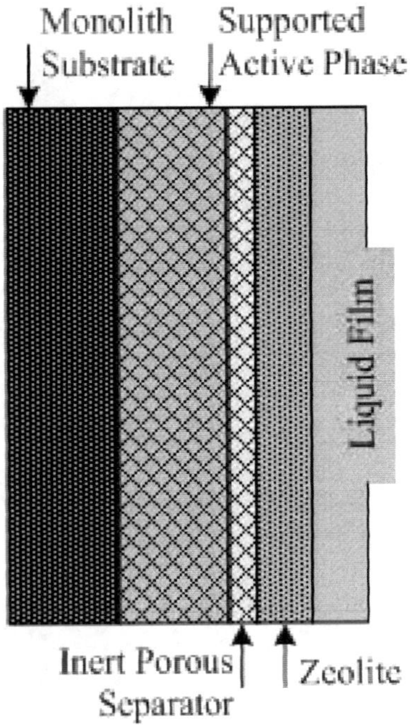

Figure 1: *Bi-functional, multi-layered, structured catalyst.*

The production of gasoline-range hydrocarbons depends on the extent of cracking and isomerization reactions, which in turn depend on the ZSM-5 to intermediate hydrocarbons ratio in the local neighborhoods of the catalyst acid sites. Increasing the ZSM-5 loading (layer thickness) can possibly improve the gasoline selectivity of the process. However, increase of the ZSM-5 membrane thickness also contributes to diffusion limitations. In essence, a thick layer of ZSM-5 has the potential to enhance selectivity, but negatively impacts conversion because the process becomes diffusion-limited. For that reason, we explored mesoporous ZSM-5 as an acidic catalyst layer capable of relaxing diffusion limitations. Hierarchical zeolitic materials have demonstrated improved mass transfer properties at high catalytic activity for a number of applications [8,9]. Inclusion of mesopores in the ZSM-5 structure eases transport, specifically of molecules in the C5-C12 range [10]. Thus, it can enable higher loadings of ZSM-5 on the FTS monolith catalysts without affecting conversion and diffusional characteristics; i.e. the transport of syngas to the catalyst and hydrocarbons out of it. We will present the synthesis methods, process configuration and results of the effect of hierarchical ZSM-5 catalytic films, as selective and reactive membranes for FTS. The hierarchical ZSM-5 was prepared using a finely controlled desilication technique, with mesopore diameters between 100 and 300 Å. They were then coated over FTS monoliths, using wash-coating methods and were tested for their reactivity, selectivity and stability as gasoline-producing FTS catalysts. The effect of pressure, temperature and gas residence time will be presented. The effect of thickness and the improvement of diffusion tolerances of the mesoporous catalysts will be presented and explained on the basis of materials characterization results. FTS with these catalysts maintains CO conversion, but may also shift selectivity from diesel-range compounds to high-octane gasoline.

Acknowledgment

Support from the ACS PRF 53648-DNI5 is gratefully acknowledged.

References

[1] U.S. Energy Information Administration, U.S. Imports by Country of Origin. (2016). http://www.eia.gov (accessed 09 May 2016).

[2] U.S. Energy Information Administration, Short-Term Energy and Summer Fuels Outlook, 2016.

[3] M.E. Dry, J. Chem. Technol. Biotechnol. 77 (2002) 43–50.

[4] A.Y. Khodakov, W. Chu, P. Fongarland, ChemInform 107 (2007) 1692–1744.

[5] M.E. Dry, Catal. Today 71 (2002) 227–241.

[6] S. A. Kondrat, R.P. Marin, J.R. Gallagher, D.I. Enache, T.E. Davies, J.K. Bartley, S.H. Taylor, M.J. Rosseinsky, G.J. Hutchings, J. Am. Chem. Soc. 3 (2013) 764–772.

[7] M.E. Dry, J.R. Boudart, M. Anderson (Eds.), Catalsis Science and Technology, vol. I, Springer, New York, 1981.

[8] K. Li, J. Valla, J. Garcia-Martinez, ChemCatChem 6 (2014) 46–66.

[9] D.P. Gamliel, H.J. Cho, W. Fan, J.A. Valla, Appl. Catal. A Gen. (2016).

[10] J.C. Groen, W. Zhu, S. Brouwer, S.J. Huynink, F. Kapteijn, J.A. Moulijn, J. Pérez-Ramírez, J. Am. Chem. Soc. 129 (2007) 355–60.

Copyright © American Institute of Chemical Engineers. All rights reserved.

464437 Comparison of Cobalt Fischer-Tropsch Catalysts on Robust Silica-Alumina Supports

Abstract:

Fischer-Tropsch Synthesis (FTS) is a key step in gas-to-liquid processes to produce clean and environmentally friendly fuels. The Fischer-Tropsch (FT) reaction is catalyzed, typically by supported cobalt catalysts. The support's physical properties, including pore size and thermal stability, and chemical characteristics, like surface acidity, can greatly affect the catalyst's performance [1, 2]. Recently, we have developed a new support which is simple to prepare and has unique characteristics, including high surface area and pore volume, very high thermal stability, and low surface acidity [3]. This bicomponent support, AlSi, containing 95 wt% alumina and 5 wt% silica, has previously been incorporated into successful supported iron FT catalysts [4]. The goal of this research is to compare the performance of cobalt FT catalysts using AlSi versus other commonly used alumina supports, including those available commercially from Sasol and St. Gobein. To focus only on the support effects, all of the catalysts were synthesized using the same incipient wetness impregnation method, with 20 wt% cobalt and 0.15 wt% ruthenium as a reduction promoter. Each catalyst was characterized by TPO, TPR, XRD, TEM, BET, EOR, and CO and H_2 uptake experiments to compare the nanostructure and other physical and chemical features of the catalysts. Catalytic activity, selectivity, and stability were investigated in a packed bed reactor operated at industrial FTS conditions (220°C and 20 atm at a 2.1:1 H_2/CO ratio) during 240 h runs. XRD data show that with calcination to temperatures even higher than 1100°C, the AlSi support still consists of predominantly the gamma phase, which allows the support to maintain high surface area (>95 m^2/g). Further, this thermal stability allows high temperature oxidative treatments to remove surface acid sites, which appears to produce supports that favor increased FT reaction rates and improved C_2+ selectivity. Nitrogen physisorption data also indicate that the AlSi surface area and pore volume are relatively high, about 3 times more than commercially available Sasol alumina-silica supports, which provide better selectivity and activity.

1. Khodakov, A.Y., *Fischer-Tropsch synthesis: Relations between structure of cobalt catalysts and their catalytic performance.* Catalysis Today, 2009. **144**(3–4): 251-257.

2. Soled, S.L., et al., *Control of Metal Dispersion and Structure by Changes in the Solid-State Chemistry of Supported Cobalt Fischer–Tropsch Catalysts.* Topics in Catalysis. **26**(1): 101-109.

3. Mardkhe, M.K., et al., *Acid site properties of thermally stable, silica-doped alumina as a function of silica/alumina ratio and calcination temperature.* Applied Catalysis A: General, 2014. **482**: 16-23.

4. Keyvanloo, K., et al., *Supported Iron Fischer–Tropsch Catalyst: Superior Activity and Stability Using a Thermally Stable Silica-Doped Alumina Support.* ACS Catalysis, 2014. **4**(4): 1071-1077.

Copyright © American Institute of Chemical Engineers. All rights reserved.

465794 Surface Stabilization and Subsequent Rupturing of Co in Co-Cu Catalysts By Formation of Adsorbed Di-, Tri-, and Tetracarbonyls

Abstract:

Surface Stabilization and Subsequent Rupturing of Co in CoCu Catalysts by Formation of Adsorbed Di-, Tri-, and Tetracarbonyls

Greg Collinge, Norbert Kruse, Jean-Sabin McEwen

The Gene and Linda Voiland School of Chemical Engineering and Bioengineering

Washington State University, Pullman, WA

Long-chain hydrocarbons with terminal oxygen functionalization (oxygenates) are very desirable chemical products, evident from the over 6 million tons/year produced in the mega-process of hydroformylation. In the 1970s, the Institut Francais du Petrole patented a heterogeneous CoCu catalyst for the production of short-chain alcohols (i.e. oxygenates), following CO hydrogenation [1]. Recent developments in our group suggest that both short- and long-chain oxygenates can be produced at both high selectivity and yield using CoCu-based (either CoCuNb or CoCuMn) catalysts in Fischer-Tropsch (FT) synthesis [2-4]. To make this technology truly robust and to direct future research into highly selective long-chain oxygenate synthesis, the fundamental role each metal plays during catalysis needs to be elucidated. However, little is theoretically known about the interaction of Co and Cu beyond the single-atom impurities in high and low index crystal facets investigated by Ruban et al. and later by Nilekar et al., respectively [5, 6]. To this end, our present work uses density functional theory (DFT) calculations to investigate the segregation tendency of a Cu monolayer on both Co(0001) and Co(10-12) model catalysts (henceforth Cu/Co(0001) and Cu/Co(10-12)) as a function of CO pressure. Reconstruction is further investigated by determining the thermodynamic feasibility of CO-induced Co rupturing.

Figure 1. Co rupturing from step sites of Cu/Co(755) as result of geminal tetracarbonyl formation and stabilization. Here, Co has been ruptured (right side) from the step site and exists as a subcarbonyl on the terrace surface. This allows the adsorbed CO on the Cu terrace sites (left side) to move toward the step and pump up new step Co, which can then also be subsequently ruptured (center). Orange spheres are terrace Cu; brown, step Cu; dark blue, subsurface Co; light blue, surface Co.

We have shown that CO adsorption causes a reversal in segregation on flat Cu/Co(0001) wherein Co is effectively pumped to the surface by the presence of the adsorbed CO [7]. In this work, we have expanded this research to include the stepped Cu/Co(10-12) surface and connected both results to experimentally relevant conditions through the construction of phase diagrams. These phase diagrams show that surface Co enrichment is limited at the most relevant reaction conditions, but is nonetheless highly favorable in the presence of even a small CO partial pressure. We also address another phenomenon, which adds another layer of complexity: elucidating how CO induces not just reconfiguration of the surface, but also the reconstruction of the surface. This latter investigation is performed using the vicinal Cu/Co(755) surface, and our results suggest that geminal tri- and tetracarbonyls can favorably rupture Co from the step sites (see Figure 1), producing mobile subcarbonyls on the surface and providing a mechanistic explanation for catalyst restructuring.

The next challenge facing the design of FT catalysts for the production of long-chain oxygenates is the question of CO dissociation on CoCu as evidenced by X-ray photoelectron spectroscopy in our group [3]. The results presented suggest that CO bond cleavage is prohibitively endothermic on both the flat and stepped surfaces, regardless of coverage. We present preliminary evidence for CO dissociation driven by carbide formation and discuss implications in relation to mounting experimental and theoretical evidence [8-13] for carbide formation in certain catalyst formulations for higher alcohol synthesis.

REFERENCES

1. Sugier, A. and E. Freund, Process for manufacturing alcohols, particularly linear saturated primary alcohols, from synthesis gas. 1978: US.

2. Xiang, Y., et al., Long-Chain Terminal Alcohols through Catalytic CO Hydrogenation. Journal of the American Chemical Society, 2013. 135(19): p. 7114-7117.

3. Xiang, Y., R. Barbosa, and N. Kruse, Higher Alcohols through CO Hydrogenation over CoCu Catalysts: Influence of Precursor Activation. ACS Catalysis, 2014. 4(8): p. 2792-2800.

4. Xiang, Y., et al., Ternary Cobalt–Copper–Niobium Catalysts for the Selective CO Hydrogenation to Higher Alcohols. ACS Catalysis, 2015. 5(5): p. 2929-2934.

5. Ruban, A.V., H.L. Skriver, and J.K. Norskov, Surface segregation energies in transition-metal alloys. Physical Review B, 1999. 59(24): p. 15990-16000.

6. Nilekar, A.U., A.V. Ruban, and M. Mavrikakis, Surface segregation energies in low-index open surfaces of bimetallic transition metal alloys. Surface Science, 2009. 603(1): p. 91-96.

7. Collinge, G., et al., CO-induced inversion of the layer sequence of a model CoCu catalyst. Surface Science, 2016. 648: p. 74-83.

8. Wang, Z., N. Kumar, and J.J. Spivey, Preparation and characterization of lanthanum-promoted cobalt–copper catalysts for the conversion of syngas to higher oxygenates: Formation of cobalt carbide. Journal of Catalysis, 2016. 339: p. 1-8.

9. G.G. Volkova, T.M.Y., L.M. Plyasova, M.I. Naumova, V.I. Zaikovskii, Role of the Cu–Co alloy and cobalt carbide in higher alcohol synthesis. Journal of Molecular Catalysis A: Chemical, 2000. 158(1): p. 389-393.

10. Lebarbier, V.M., et al., Effects of La2O3on the Mixed Higher Alcohols Synthesis from Syngas over Co Catalysts: A Combined Theoretical and Experimental Study. The Journal of Physical Chemistry C, 2011. 115(35): p. 17440-17451.

11. Pei, Y., et al., Study on the effect of alkali promoters on the formation of cobalt carbide (Co_2C) and on the performance of Co_2C via CO hydrogenation reaction. Reaction Kinetics, Mechanisms and Catalysis, 2013. 111(2): p. 505-520.

12. Pei, Y.-P., et al., High Alcohols Synthesis via Fischer–Tropsch Reaction at Cobalt Metal/Carbide Interface. ACS Catalysis, 2015. 5(6): p. 3620-3624.

13. Banerjee, A., et al., Origin of the Formation of Nanoislands on Cobalt Catalysts during Fischer–Tropsch Synthesis. ACS Catalysis, 2015. 5(8): p. 4756-4760.

Copyright © American Institute of Chemical Engineers. All rights reserved.

467552 Effect of Carbon Support on Supercritical Fluid Mediated Fischer-Tropsch Synthesis with an Iron-Based Nanoscale Catalys

Abstract:

Fischer-Tropsch synthesis (FTS) is a well-established catalytic process to produce hydrocarbon fuels and chemicals from syngas. Traditionally, FTS has been conducted in either the gas phase or slurry phase, but in each case, the heterogeneous nature of the process can lead to heat and mass transfer limitations. These limitations in turn can cause poor selectivity, shorten catalyst life, or necessitate a more complex reactor system. As an alternative, conducting the reaction in a supercritical fluid reaction medium allows for single phase, fixed-bed operation and has been shown to extend catalyst life, reduce CH_4 and CO_2 selectivity, and aid in the extraction of reaction intermediates as well as heavy products.

Because FTS is a surface-catalyzed reaction, it stands to benefit from the significant surface area per mass offered by nanoparticle catalysts. These nanoparticles, however, must be supported in order to reduce attrition, maintain dispersion, and provide mechanical strength. Traditional supports such as SiO_2 and Al_2O_3 can be used, but can have overly strong interactions with small particles, inhibiting reduction and reducing activity. Carbon-based supports such as nanotubes, nanofibers, and graphene offer a promising alternative, because they are stable and inert under FTS reaction conditions and do not adversely affect catalyst reduction. This paper presents the systematic characterization of multiple carbon-supported, iron-based catalysts coupled with a comparison of their FTS performance in both gas phase and supercritical phase operation.

Copyright © American Institute of Chemical Engineers. All rights reserved.

448577 Catalytic Conversion of Carbon Dioxide to Methanol over Homogeneous and Heterogeneous Hybrid Catalyst

Abstract:

Carbon dioxide is the final and inevitable product of the utilization of fossil fuels. Since it is thermo-dynamically stable, there seems to be difficult to convert it back to useful chemical products. Hydrogenation of carbon dioxide to fuels such as methanol, higher alcohols and hydrocarbons can be considered as a promising route to use carbon dioxide as a feedstock to produce valuable chemicals rather than as a waste with a cost of disposal. In this work, modified copper-based catalysts were prepared by conventional co-precipitation method with the addition of small amount of transition metal for the hydrogenation of carbon dioxide to alcohols via reverse water gas shift reaction. The catalysts were characterized by TPR, TEM, ICP, XRD, BET, PSA, FT-IR, XPS methods. The prepared heterogeneous catalysts were tested in a CSTR with the addition of alcohols used as homogeneous catalyst at low temperature of 160oC. The effect of the type of alcohols as homogeneous catalyst on the activity toward hydrogenation was observed profoundly. In-situ DRIFTS(Diffuse Reflectance Infrared Fourier Transform Spectroscopy) analysis was conducted to trace the reaction mechanism for CO2 gydrogenation to methanol.

Copyright © American Institute of Chemical Engineers. All rights reserved.

450871 "Ni-MFI As a Potential Zeolitic Catalyst for Olefins Hydrogenation at Low Pressures"

Abstract:

The Elimination of Olefins from Processed Light Oil Fractions Is a Necessity. These Deleterious Molecules Can Form Gum Deposits on Equipment and Polymerize in Pipelines over Time during These Streams Transportation, Resulting in the Fouling of Heat Exchangers and Possible Formation of Coke. Most of the Catalysts Used so Far in Olefins Hydrogenation, Are Expensive Noble Metals, Such As Pt, Pd and Ru. Ni Catalysts Supported on Zeolite Carriers Have Been Previously Used in Hydrogenation Reactions.[1] However, in Order to Obtain a Good Hydrogenation Activity and Selectivity, Metallic Ni Species Have to be Well Dispersed on the Surface of the Zeolite Carrier. Hence, the Method of Ni Incorporation on the Support Surface Plays a Key Role in Obtaining an Active and Selective Catalyst. the Aim of This Work Is to Develop Novel Catalysts for Reducing Olefins Content without Affecting Octane Numbers of Light Distillates (no hydrogenation of aromatics). Ni-Doped Siliceous MFI Materials Have Thus Far Shown Interesting Performances As It Is Here Presented. the Ni Was Added during the Zeolite Formation Step and Anchored during the Crystallization Process.[2] Ni-MFI Catalysts Were Prepared with Various Metal Loadings. All Catalysts Were Characterized By XRD, TPR, ICP, FTIR and BET. Activity Tests for Toluene and 1-Octene Hydrogenation Were Conducted at Low Pressures (< 100 psi) in a Bench Scale Packed Bed Reactor. the Hydrogenated Products Were then Analyzed By a Gas Chromatograph (GC) Equipped with a Flame-Ionization Detector (FID). the Spent Catalysts Were Characterized By XRD, BET, DSC, TGA and DTA. When Using 1 Wt% Ni-MFI the Hydrogenation of Toluene Was Zero below 200 °C but Negligible at 200 ^{o} C (1 wt %). However, Increasing the Ni Content to 2 % Resulted in a Moderate Activity of the Catalyst with 100 % Selectivity Towards Methyl-Cyclohexane (MCH). Increasing the Ni to 3 % Resulted in the Complete Conversion of Toluene to MCH at Very Low Pressures. This Directed Our Attention to the Possibility of Hydrogenating Olefins at Milder Conditions. 1-Octene Was Selected As a Model Molecule for the Hydrogenation Tests. the Conversion of 1-Octene to n-Octane Was Observed after a Very Short Time (~ 20 min) with

All the Ni-MFI Catalysts Tested. When Using 1 Wt. % Ni-MFI Isomerization of 1-Octene Began to be Observed after 13 h on Stream. Increasing the Ni to 2 Wt. % Increased the Time at Which Isomerization Occurred, to 40 h. DSC Data in air Showed a Small Exothermic Peak of Coke at ~ 300 ^{o} C. Isomerization of 1-Octene on Ni-MFI Might be Attributed to the Formation of Some New Acid-Sites Promoted By the Formed C-Layer. the Catalyst with 1 % Ni-MFI Can be Used for Reducing Olefins Content in Upgraded Feedstocks without Hydrogenation of Aromatics and Retaining High Octane Numbers of the Processed Feedstock.

References

(1) A. Masalska, *J. Appl. Catal. A* 2005, 294, 260-272.

(2) G. Vitale, H. Molero, E. Hernandez, S. Aquino, V. Birss, P. Pereira-Almao *J. Appl. Catal. A: Gen* 2013, 452, 75-87

Copyright © American Institute of Chemical Engineers. All rights reserved.

451366 Predicting Surface Intermixing in Alloy Nanocatalysts from First Principles Calculations

Abstract:

Predicting Surface Intermixing in Alloy Nanocatalysts from First Principles Calculations

Luke T. Roling and Manos Mavrikakis

Department of Chemical & Biological Engineering, University of Wisconsin-Madison, Madison, WI

Controlled catalyst design and synthesis are enhanced by a detailed fundamental understanding of phenomena occurring at the catalyst surface. One such class of phenomena that has long attracted attention is metal adatom diffusion across metal substrates, and the impact these phenomena have on catalyst growth.[1-2] We recently described the synthesis and electrochemical activity of hollow Pt-Pd nanocages with exceptional performance for the oxygen reduction reaction.[3] Detailed DFT calculations predicted that the formation of these nanostructures, which occurred through etching Pd from a Pd-Pt core-shell structure, requires intermixing of Pt and Pd during the core-shell synthesis. This intermixing was accommodated by the ease of substitution of Pt into the underlying Pd substrate relative to the diffusion of Pt across the surface.

In this presentation, we extend these calculations to bimetallic combinations of ten transition metals to predict similar intermixing phenomena in the synthesis of other catalytically-relevant alloy nanocatalysts. In particular, we present results for metal adatom surface diffusion and substitution for all combinations of eight *fcc* and two *hcp* metals, considering both close-packed and open crystal facets. We also present results at two different surface coverages of metal adatoms. Our results can guide future simulation and inorganic synthesis efforts toward the atomic-scale design of nanocatalysts with controlled structures and compositions.

1. P. J. Feibelman, *Physical Review Letters* **1990**, 65, 729.
2. T. Ala-Nissila, R. Ferrando, S. Ying, *Advances in Physics* **2002**, 51, 949.
3. L. Zhang, L. T. Roling, X. Wang, M. Vara, M. Chi, J. Liu, S. Choi, J. Park, J. A. Herron, Z. Xie, M. Mavrikakis, Y. Xia, *Science* **2015**, 349, 412.

Copyright © American Institute of Chemical Engineers. All rights reserved.

451573 On the Influence of Mass Transport in Sulfonated Carbon Catalysts Derived from Various Carbon Sources on Liquid-Phase Esterification Reactions

Abstract:

Sulfonated carbon catalysts have been widely investigated for their potential applications such as biomass conversion. Various carbons and carbon precursors have been used to synthesize such catalysts, leading to a wide range of properties and performance of catalysts. Our goal is to gain some insight into the structure and performance relationship of these catalysts and use them to design a new catalyst. In this presentation, we will report results from synthesis and characterization of sulfonated carbon catalysts derived from various sources: pyrolyzed phenolic resin (sulfonated carbon gel, SCG) [1], pyrolyzed cellulose (SCel) and glucose (SGlu), graphite oxide (SGO), activated carbon (SAC), carbon derived by hydrothermal treatment of glucose (SHTGlu), and multi-walled carbon nanotube (SCNT). Catalytic performance of these catalysts was tested in three liquid-phase esterification reactions at 333 K with substantially different size of reactants (acetic acid with ethanol, oleic acid with methanol, and levulinic acid with ethanol), and the results are correlated with catalyst properties characterized by Boehm titration, elemental analysis, Raman and IR spectroscopies, and N_2 and water vapor adsorption conducted at 77 and 298 K, respectively. A combined analysis of N_2 and water vapor adsorption data indicates that the catalysts derived from carbohydrates (SCel, SGlu, and SHTGlu) as well as graphite oxide (SGO) swell significantly in the presence of water vapor. Thus, although they show low surface areas in dry form, this property appears to allow easy access of small and polar reactants to their $-SO_3H$ sites, and the esterification reaction of acetic acid with ethanol proceeded at essentially the same initial turnover frequency per $-SO_3H$ group ($\approx 0.02\ s^{-1}$) for all catalysts regardless of significantly different properties of the catalysts. However, the reaction of oleic acid with methanol proceeded efficiently only on catalysts that have a large mesopore volume in dry form presumably because the catalysts derived from carbohydrates and graphite oxide do not swell efficiently in the reaction mixture, which limits the access of reactants to their $-SO_3H$ sites. In contrast, the esterification reaction of levulinic acid with ethanol proceeded slightly faster on catalysts that can swell or have high concentrations of surface oxygen-containing functional groups than other catalysts. These results indicate that both mass transfer and surface properties influence their catalytic property for this reaction. Collectively, our results show the performance of active sites ($-SO_3H$ groups) in sulfonated carbon catalysts, a role of surface oxygen-containing functional groups, and the influence of porous and swelling properties of the catalysts on mass transport of reactants for the liquid-phase esterification reactions.

[1] Ogino, I., Suzuki, Y., Mukai, S.R. ***ACS Catal***. 5, 4951–4958 (2015)

Copyright © American Institute of Chemical Engineers. All rights reserved.

450877 Catalytic Hydrogenation of 2-Ethylanthraquinone Using an in Situ Synthesized Pd Catalyst

Abstract:

An in situ synthesized Pd catalyst with egg-shell structure was employed to catalyze the hydrogenation of 2-ethylanthraquinone in a fixed bed. The effects of residence time, gas to liquid phase ratio, reaction temperature and Pd loading amounts on the yield and hydrogenation efficiency were investigated specifically. The yield and hydrogenation efficiency attained values of 35.5% and 6.6 g/L using the optimum Pd loading amount of 0.32%, respectively, with a residence time of less than 4 seconds exhibiting excellent catalytic activity. Meanwhile, within the temperature range of 50 to 70 , the Thiele modulus and effective internal diffusion factors were calculated to be about 0.008 to 0.033 and 0.9996 to 0.9999, respectively, which indicates that the prepared catalyst with egg-shell structure weaken the internal diffusion resistance effectively and is suitable for fast reactions. The maximum space time yield value reached up to 567.5 g_{H2O2} g_{Pd}^{-1} h^{-1} and was much higher than reported results from other literature.

Copyright © American Institute of Chemical Engineers. All rights reserved.

451830 Copper Aluminate: A Potential Catalyst to Make Dimethyldichlorosilane in a 2-Step Process

Abstract:

The present study relates to the development of a new generation of copper aluminate type spinel catalysts for the production of dimethyldichlorosilane (Me_2SiCl_2) from $SiCl_4$ in a 2-step process. Copper aluminates were found to exhibit superior activity over other conventional catalysts and hold better particle integrity in lab scale fixed bed and fluid bed reactors. In the present study, copper aluminates were synthesized by incipient wetness impregnation of $Cu(NO_3)_2$ over Al_2O_3 support followed by air calcination as well as by physical mixing of CuO and Al_2O_3 followed by high temperature calcination. Both γ-alumina and copper aluminate possess the same structure (cubic-close packed), which results automatically in a dispersed CuO on alumina single phase, which was confirmed by XRD analysis. In the reduction of spinel compound, $CuAl_2O_4$ and part of CuO closely associated with $CuAl_2O_4$ were reduced much slower than bulk CuO or supported CuO catalysts. However, once reduced, a phase change is forced and most of the alumina is converted to alpha phase. Calcination at as high as 1000°C has prevented sintering of Cu and minimized $AlCl_3$ formation in step-1 reaction with H_2 and $SiCl_4$. High rate of methylchlorosilanes production on spinel catalysts in step-2 was attributed to the smaller particle sizes and "meta-stable" copper silicide (Cu_3Si) that is formed during the step-1 reaction. The catalysts were characterized by XRD, SEM/EDS, N_2O pulse chemisorptions and particle size distribution analysis and correlated with activity results in the 2-step process to make Me_2SiCl_2 from $SiCl_4$. In-situ XRD analysis confirmed the high reactivity of copper in spinel compared to other conventional catalysts. SEM analysis has shown the Cu association with alumina in fresh catalysts and the dissociation of Cu from support was less in the spent catalysts compared to conventional silica supported copper catalysts. Spinel catalyst was found to be active even at as low as 240°C step-2 reaction temperature with MeCl, producing high selectivity towards Me_2SiCl_2.

INTRODUCTION

Dimethyldichlorosilane (Me_2SiCl_2) is typically produced by Rochow direct process [1], which comprises the reaction of methyl chloride with elemental silicon in the presence of promoted copper catalyst. However, the feedstock for direct process to produce Me_2SiCl_2, metallurgical grade silicon, is produced by carbothermic reduction of silica. This is a very energy intensive process and requires 12 kWh of electrical energy and 12 kWh of carbon feedstock based energy, thus leading to relatively high production cost, per kg of Si produced.

An alternate process to produce silicones was investigated by Dow Corning [2-5] and the process is energy conservative in polydimethylsiloxanes (PDMS) production. The process involves the production of chlorosilanes from silica followed by reaction with methyl chloride to generate the PDMS. This new process capability showed a range of potential impacts from flexibility of monomers produced, improved selectivity of $Me_2SiCl_2$and more efficient raw material utilization.

The following reactions are carried out on a metal or metal supported catalyst (CAT).

$H_2 + SiCl_4 + CAT \text{ --> } Si\text{-}CAT + HCl + HSiCl_3$ (>600°C) ------ step 1

$CH_3Cl + Si\text{-}CAT \text{ --> } (CH_3)_xSiCl_{4-x} + CAT$ (300°C), x=1-4 ---- step 2

The catalyst selection for the 2-step process was challenging as the process involves severe reaction conditions which include reacting with H_2 and $SiCl_4$ feed gases at above 650°C in a fluid bed reactor. Copper based catalysts were evaluated in the 2-step process due to potential similarities of step-2 reaction with the direct process. High surface area, large pore volume and homogeneous dispersion are all important characteristics of the catalyst for high activity and cycle stability.

g-Al_2O_3 is the most widely used as a support material for several hydrotreating commercial processes. Notable features of alumina supports are their ability to disperse high loading of active metal phase, good mechanical properties, and high thermal stability and can interact with active phase resulting in less sintering of active phase in hydrogen atmosphere [6-9]. However a considerable drawback with g-Al_2O_3 is that it is not resistant to $SiCl_4$ during the course of reaction and will be transformed to aluminum chloride. Thermodynamic calculations show the reaction between Al_2O_3 and $SiCl_4$ can take place around 600°C to make SiO_2 and $AlCl_3$. It has been reported that silicon tetrachloride can de-aluminate aluminum from the Al_2O_3 support causing $AlCl_3$formation [10]. However, one would not expect this reaction to take place in the presence of hydrogen.

Alumina has several advantages as a support material and if all these advantages associated with this support are incorporated in a catalyst system, it will be an ideal catalyst for the 2-step process. If g-alumina support is transformed in to alpha-alumina (or a more stable phase) and associate with copper, the resulting catalyst may inherit the favorable physical properties and retaining hydrogenation activity of $SiCl_4$in step-1.

In this study, various compositions of copper aluminate type spinel materials were synthesized by wet impregnation and physical mixing methods and investigated in the 2-step process. The catalytic performance of these copper aluminates was superior to those of conventional copper catalysts. More uniform distribution of $CuAl_2O_4$ and complete formation of $CuAl_2O_4$ phase in the catalyst/ supports due to high calcination temperatures thus higher Cu mobility/diffusion are believed to help strengthen the particle integrity and could minimize $AlCl_3$formation in the process.

REFERENCES

1. E.G. Rochow, J Amer. Chem. Soc., 67, 1945, 963-965
2. D. Katsoulis, R Larsen, US patent 8772525
3. A. Coppernoll, C. Horner, K. Janmanchi, D. Katsoulis, R. Larsen, US patent application 20150158890
4. K. Janmanchi, C. Horner, A. Coppernoll, WIPO patent application WO/2015/073213
5. A. Coppernoll, C. Horner, K. Janmanchi, WIPO patent application WO/2014/099125
6. B. Dhandapani, S.T. Oyama, Catal Lett., 35 (1995) 353
7. P. Bodnariuk, B. Coq, G. Ferrat, F. Figueras, J. Catal., 116 (1989) 459
8. B. Coq, J.M. Cognion, F. Figueras, D. Tournigant, J. Catal., 141 (1993) 21
9. B. Coq, F. Figueras, S. Hub, D. Tournigant, J. Phys. Chem., 99 (1995) 1115
10. D. Gencev, K.S. nee Mogyorosi, S. Riederauer, J. Szepovolgyi, US patent 4,416,862

Copyright © American Institute of Chemical Engineers. All rights reserved.

453100 Centrifugal Partition Chromatograph: A Novel Continuous Multiphase Reaction System

Abstract:

Enzymatic reactions are of main interest due to their substrate specificity and gentle reaction conditions. However, substrates, which are poorly soluble in aqueous media, can be hardly catalyzed by enzymes due to the fact that most enzymes are active and soluble in aqueous media only. With the use of aqueous-organic two-phase reaction systems it is possible to continuously feed hydrophobic substrate, solved in the organic phase, and simultaneously remove product from the aqueous, catalytically active phase. Hence, product inhibition can be avoided and the reaction equilibrium shifted towards the product side.

To enhance the catalytic efficiency and long term stability as well as to enable continuous processing the biocatalysts are commonly immobilized. Unfortunately, most of the immobilization techniques reduce the initial catalytic activity of the system as steric and/or mass transport hindrances are present or the conformation of the enzymes changed after immobilization. Besides, multiphase reaction systems lead to different design aspects compared to monophasic aqueous systems. Efficient mixing to ensure the mass transport of substrate to the biocatalyst (reaction) and the product to the organic phase (product removal) is needed. For continuous processing a clear and continuous separation of the phases is mandatory to remove the product from the system without losing biocatalyst or product.

In recent years the Centrifugal Partition Chromatograph (CPC) as an alternative multiphase reaction device is investigated. Commonly the CPC is used for liquid-liquid chromatography, where one phase of a two-phase-system is immobilized in a chamber system that is arranged around a rotary axis. By rotating the system a centrifugal field is generated which keeps one phase of any two-phase system in the chamber cascade. The second phase is pumped through the stationary one and separation is commonly achieved via the partitioning of the components in the phase system. Due to the CPC set-up, the immobilization of biocatalysts in their natural environment, namely an aqueous

solution, is easily feasible. The intense mixing of the two phases and the efficient phase separation for *in situ* product removal is ensured simultaneously. Hence, the CPC combines all design aspects for an efficient continuous multiphase reactor, namely immobilization of the biocatalyst, intense mixing, complete phase separation and *in situ* product removal in one compact device. additionally, as no solid phase is used to immobilize the biocatalyst structure changes due to covalent binding, loss of the hydrate shell or steric hindrances may not occur.

The present study will address the process design procedure to systematically select the phase system to achieve stable and efficient reaction rates and the CPC operation conditions to enable efficient mixing and separation of the phases. The procedure is applied to several biocatalysts ranging from enzyme to whole cell systems to produce pharmaceutical building blocks. The applicability of the CPC as a multiphase reactor is evaluated by comparing and discussing classical stirred tank reactor experiments.

Copyright © American Institute of Chemical Engineers. All rights reserved.

451651 Synthesizing Bromobutyl Rubber By a Microreactor System

Abstract:

Abstract

Bromobutyl rubber (BIIR), an important synthetic rubber, not only maintains many desirable characteristics of traditional butyl rubber (IIR), but also has faster vulcanization rate. However, there exists two critical problems in the process of brominating IIR in n-hexane, that are strong corrosion caused by Br_2 and HBr and high reactant viscosity. A PTFE-lined microreactor platform was developed to solve the corrosion problem and obtain high quality BIIR based on excellent mixing ability. The results showed that low reaction temperature and quickly eliminating HBr from the reacting solution were crucial to obtain high selectivity of demanded molecule structure and prevent polymer from decomposition. To remove HBr from the reaction system in time, a water assistant technology was successfully implemented in the microreactor system, which produced high quality BIIR with almost 100% selectivity and less reduced molecule weight.

Copyright © American Institute of Chemical Engineers. All rights reserved.

454217 Integration of CFD and Polymerization for an Industrial Scale Cis-Polybutadiene Reactor

Abstract:

Integration of CFD and Polymerization for an Industrial Scale Cis-polybutadiene Reactor

Chao-zhong Xu, Jia-jun Wang, Xue-ping Gu, Lian-fang Feng

State Key Laboratory of Chemical Engineering, College of Chemical & Biological Engineering, Zhejiang University, Hangzhou P.R. China 310027

jiajunwang@zju.edu.cn

Abstract A major objective of polymerization reaction engineering is to understand how the reaction mechanism, the physical transport processes, reactor configuration and operating conditions affect the polymer properties that assessed product quality. In particular, cis-polybutadiene polymerization in industrial scale reactors becomes quite complicated due to complex mechanism and metastable behaviors. However, computational fluid dynamics (CFD) is barely exploited to forecast the polymer product quality in the general modeling approach though some efforts have been done toward a better understanding of polymerization process. Hence, modeling polymerization by CFD, which takes polymer quality indices into consideration, still remains very challenging and significant.

A CFD model coupled with polymerization kinetics was developed to investigate butadiene solution polymerization using three-component system (Ni, Al, B) as catalyst and raffinate oil as solvent not previously available for a 12 m^3

industrial scale reactor equipped with a double helical ribbon agitator. The integration of kinetic model with two catalytic active sites and CFD model was accomplished by solving continuity, momentum, energy and species equations simultaneously, in which the reactive source terms related to polymerization were implemented by a user-defined function with the method of moments. Model validation was successfully conducted by comparing CFD results with plant data and the developed model was further extended to discuss the effects of operation conditions including inlet temperature, impeller as well as inlet flowrate of solvent, and finally shed light on the differences of reactor behaviors from ideal CSTR model.

The simulation indicated that the reactor behaviors with double helical ribbon agitator were very similar to traditional CSTR model in the middle and the top of the reactor, while at the inlet near the reactor bottom, there apparently existed a deteriorated area for polymerization in the present production process. The distinction between double helical ribbon agitator and frame impeller mainly embodied in axial circulation capability and it seemed to be easier to keep the uniformity of polymerization process when helical ribbon agitator was used. It was also found when compared with inlet temperature, the fluctuation of reactor performance was more sensitive to inlet flowrate of solvent and therefore it was more advisable to regulate the inlet temperature for normal operations in industrial production in most cases. The results showed the proposed method could be served as a guidance to achieve a better control on cis-polybutadiene production, thereby maintaining the stability of polymerization process and optimizing the quality of products.

[1]

This work was supported financially by the National Natural Science Foundation of China (21276222), National Basic Research Program of China (2011CB606001) and State Key Laboratory of Chemical Engineering (SKL-ChE-13D01).

Copyright © American Institute of Chemical Engineers. All rights reserved.

ChemComm

COMMUNICATION

View Article Online
View Journal | View Issue
CrossMark click for updates

Cite this: *Chem. Commun.*, 2015, 51, 10695

Received 27th March 2015, Accepted 16th May 2015

DOI: 10.1039/c5cc02541a

www.rsc.org/chemcomm

Base-free Knoevenagel condensation catalyzed by copper metal surfaces†

E. M. Schneider,[a] M. Zeltner,[a] N. Kränzlin,[b] R. N. Grass[a] and W. J. Stark*[a]

For the first time Knoevenagel condensation has been catalyzed by elemental copper with unexpected activity and excellent isolated yields. Inexpensive, widely available copper powder was used to catalyze the condensation of cyanoacetate and benzaldehyde under mild conditions. To ensure general applicability, a wide variety of different substrates was successfully reacted.

Copper has been a fascinating element in catalysis for decades.[1] In the early days of copper catalysis, famous name reactions such as the Ullmann coupling,[2] the Sandmeyer reaction and the Chan–Evans–Lam coupling have emerged.[3] Around the millennium, Sharpless and coworkers developed the extremely important "click chemistry", in which the copper catalyzed azide–alkyne Huisgen cycloaddition (CuAAc) plays the most essential role.[4] However, much more expensive noble metals such as platinum, palladium and gold have received most attention in the field of catalysis, while the inexpensive yet semi-noble copper has been more or less neglected.[5] Knoevenagel condensation (Scheme 1) is a widely used reaction in research and industry and has been of importance for several pharmaceutical products.[6] Generally this reaction is catalyzed by organo-bases, such as pyridine or piperidine. But using these homogeneous base catalysts often leads to time consuming work-up procedures. Additionally, undesired side-reactions such as oligomerizations can occur, high temperatures are necessary, and catalyst recovery is difficult.[7] Thus, numerous accounts on heterogeneous Knoevenagel catalysts, for example modified zeolites, ionic liquids or magnetic base analogues, have been reported.[8] This leads to cleaner products while complex neutralization procedures can be avoided.[4b] Moreover, these catalysts can be recovered and regenerated. Within one of these studies it was shown that carbon-coated

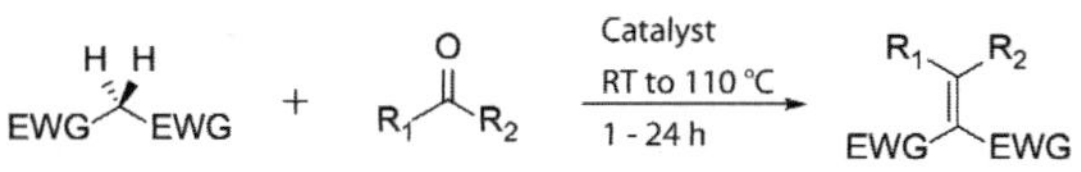

Scheme 1 Knoevenagel condensation of an active methylene compound with an aldehyde or ketone.

cobalt nanoparticles (C/Co)[9] showed some activity in the Knoevenagel condensation if compared to the uncatalyzed reaction (entries 1 and 2, Table 1).[8e] As a consequence of these results, several other carbon coated nanoparticles were tested; amongst which carbon-coated copper nanoparticles (C/Cu)[10] showed the highest activity. As reference, we compared the nanoparticles with pure metal powders. These experiments afforded the insight that commercially available, dendritic copper powder Cu(D) catalyses the Knoevenagel condensation of benzaldehyde **1**

Table 1 Activity of different catalysts for the Knoevenagel reaction of benzaldehyde and ethyl cyanoacetate[a]

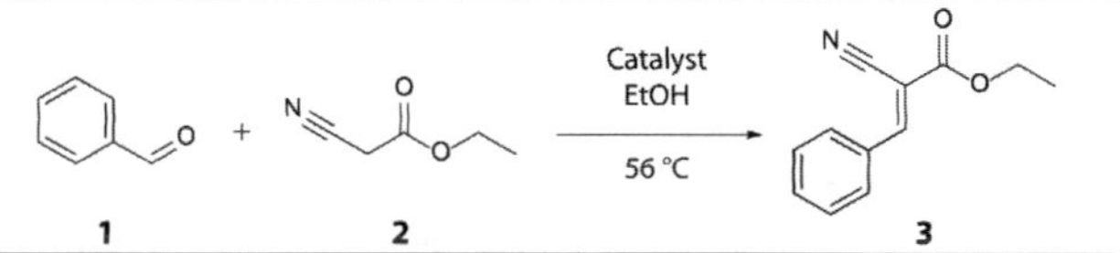

1 **2** **3**

Entry	Catalyst	Time (h)	Yield[b] (%)
1	None	2	3
2	C/Co	2	8
3	Co powder	2	9
4	C/Fe	4	4
5	$FeCl_3$	4	4
6	Fe powder	4	6
7	C/Cu	4	12
8	Cu powder	2	32
9	Brass alloy 260	2	5
10	Ag powder	4	7
11	Au powder	4	6
12	$ZnCl_2$	6	4

[a] Reaction conditions: catalyst (4 mg), benzaldehyde (0.12 mmol) and ethyl cyanoacetate (0.1 mmol) in 1 mL of EtOH for 2 h. [b] Only the *E*-isomer was detected, yield determined *via* HPLC-UV (MS) using commercial product **3** as the reference standard.

[a] *Institute for Chemical and Bioengineering, ETH Zurich, Vladimir-Prelog-Weg 1, 8093 Zurich, Switzerland. E-mail: wendelin.stark@chem.ethz.ch*
[b] *Laboratory for Multifunctional Materials, Department of Materials, ETH Zurich, Vladimir-Prelog-Weg 5, 8093 Zurich, Switzerland*

† Electronic supplementary information (ESI) available: Experimental procedures, supplementary kinetic figures, SEM micrographs, X-ray crystallographic analysis and ^{1}H and ^{13}C NMR spectra. See DOI: 10.1039/c5cc02541a

This journal is © The Royal Society of Chemistry 2015

Table 2 Optimization of the reaction conditions[a]

Entry	Solvent	Loading (mg)	t (h)	Cu[b] (ppm)	Yield[c] (%)
1	EtOH	0.4	2	3	6
2	EtOH	2	2	38	14
3	EtOH	4	2	44	32
4	EtOH	10	2	216	96 (95)[d]
5	EtOH	40	2	114	99 (85)[d]
6	EtOH	4	6	44	99 (95)[d]
7	Acetone	4	20	18	1
8	Cyclohexane	4	20	32	3
9	Toluene	4	20	30	3
10	DMF	4	6	17	99 (98)[d]
11	DMSO	4	2	241	99
12	MeCN	4	2	230	99 (98)[d]

[a] Reaction conditions: copper (4 mg), benzaldehyde (0.12 mmol) and ethylcyanoacetate (0.1 mmol). [b] Cu leaching measured by ICP-OES. [c] Determined *via* HPLC-UV using commercial product **3** as the reference standard. [d] Isolated yield.

Table 3 Scope of the condensation[a]

Entry	Reactant	Reactant	Product	Yield[b] (%)
1	2	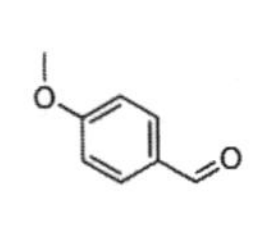	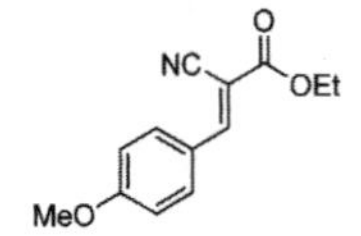	99[c] (98[d])
2	2	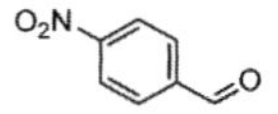	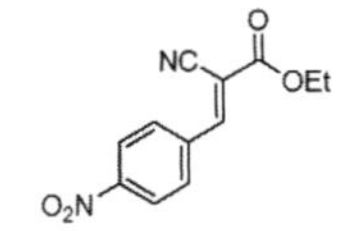	99 (87[d])
3	2	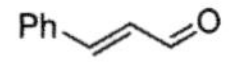	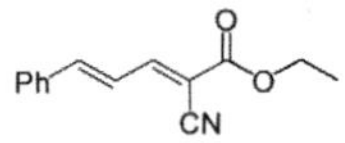	90[c] (86[d])
4	2	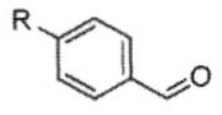 R_1= Cl, R_2= Br, R_3= I	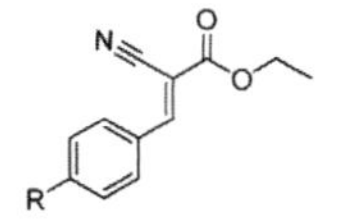	R_1 85 (73[d]), R_2 90 (68[d]), R_3 98 (93[d])
5	1	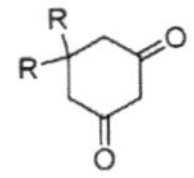 2 eq. R_a= H,H, R_b= Me,Me	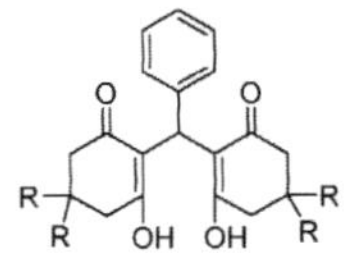	R_a 98[e] (97[d]), R_b 93[e] (89[d])
6	1	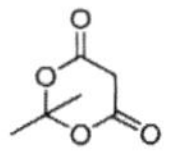	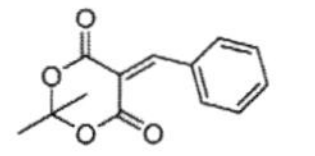	57[e]
7	2	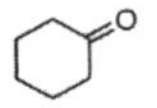	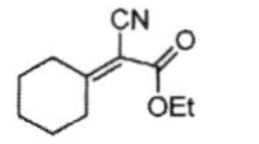	87[e]
8	1	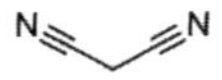	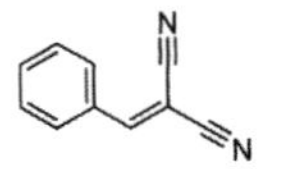	99[f] (99[d])
9	1	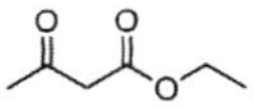	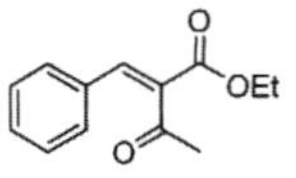	23[g]
10	1	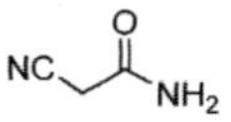	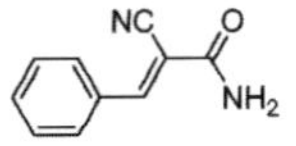	87[c] (85[d])

[a] Reaction conditions: Cu (20 mg), aldehyde (0.21 mmol) and active methylene (0.2 mmol) in EtOH for 6 h at 56 °C. [b] Yield determined *via* HPLC-UV or GC-FID. [c] Stirred for 16 h. [d] Isolated yield, only the *E*-isomer was detected. [e] 6 h in DMF. [f] Stirred for 1 h at RT. [g] Stirred for 15 h in DMSO at 70 °C, *E*/*Z* = 65% : 35%.

and cyanoacetate **2** in ethanol to **3** (entry 8, Table 1) with enhanced yields. Compared to other heterogeneous catalysts for the Knoevenagel condensation, the unmodified, commercially available copper does not need any special treatment, synthesis procedure or sophisticated storage and is the base stock in many laboratories. Notable is the fact that silver, gold, $ZnCl_2$ and brass showed only low activities compared to elemental copper (entries 9–12). To optimize the reaction conditions the catalyst loading was varied, and as expected higher loading resulted in higher yields (up to 99%; entries 1–5, Table 2). It should be noted that an equivalent amount of copper compared to the reagents is not imperative to result in full conversion (entry 6). Optimization tests with different solvents clearly favoured aprotic, polar solvents over non-polar solvents (entries 7–12). These results are in line with the general literature on Knoevenagel condensation, namely with the first step in the reaction mechanism which is the generation of an anion at the α-position of the carbonyl group followed by enolate formation.[7] After optimizing the reaction conditions with the model substrates, a variety of different aldehydes and active methylene compounds were reacted (Table 3). As already summarised by Tietze *et al.*, all condensations using **2** as the substrate exclusively yielded the *E*-isomer.[7] Other active methylene compounds were tested, such as Meldrum's Acid (entry 6) and analogues (yielding the bis-adduct, entry 5), the very reactive malononitrile (99% yield in 1 h, entry 8) and the less reactive ethylacetoacetate (23% yield after 15 h, entry 9). The different reactivities of these compounds are well known and correlate with their ability of stabilizing the corresponding anion.[7] To gain more detailed insight into the reaction mechanism, further experiments concerning the nature of the copper catalyst have been conducted.

A possible first hypothesis was that copper leaching, *i.e.* soluble Cu(I) or Cu(II) species, could be the reason for catalytic activity. Several Cu(I/II) compounds have been tested (entries 1–5, Table 4) to investigate the activity of copper ions in solution. None of these copper species showed substantial activity, instead of a high measured solvated Cu amount (entries 2 and 3). Also, there are reports of metal-alkoxide catalysed Knoevenagel condensations, and activity of *in situ* generated CuOMe may also be possible.[11] However, if the solid catalyst is filtered (Fig. 1), conversion stops immediately. This leads to the conclusions that either there is a very unstable homogeneous reactive species formed on the surface of Cu or leaching is not the reason for catalytic activity. Hence, a second hypothesis can be formulated around Cu surface based catalysis, where the yield is expected to correlate with the specific surface area (SSA) of the catalyst. Cu(D) has dendritic structures between 0.5–1 μm (see Fig. S4, ESI†) and thus a relatively low BET surface area of 0.2 $m^2\ g^{-1}$. While a tenfold larger amount of Cu(D) resulted in a tenfold

Table 4 Copper containing reference substances as catalysts for the Knoevenagel condensation[a]

Entry	Catalyst	Time (h)	SSA (m^2 10^{-4})	Cu (ppm)	Yield[b] (%)
1	Cu(II)$_2$(OH)$_2$CO$_3$	2	—	6	2
2	Cu(II)Cl$_2$	16	—	740	1
3	Cu(I)Cl	16	—	460	3
4	Cu(II)O	20	—	6	4
5	Cu(I)$_2$O	20	—	12	14
6	Cu(D)	2	8	44	32
7[c]	Cu(D)	2	80	114	99
8	Cu(Hol)	2	84	62	65
9	Cu(NPA)	2	168	176	96
10	Cu(NPB)	2	636	197	99

[a] Reaction conditions: catalyst (4 mg), benzaldehyde (0.12 mmol) and ethyl cyanoacetate (0.1 mmol) in EtOH at 56 °C. [b] Yield determined *via* HPLC-UV (MS) using commercial product **3** as the reference standard. [c] 40 mg of Cu(D) used.

Table 5 Effect of surface activation of Cu(0) on the reaction rate[a]

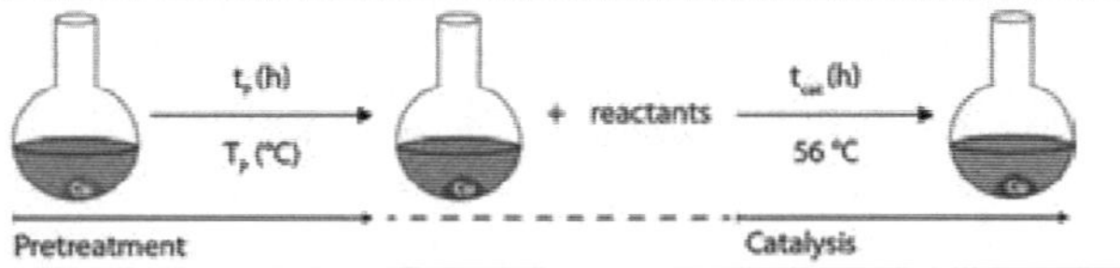

Entry	Pretreatment	t_P (h)	T_P (°C)	t_{cat} (h)	Solvent	Yield[b] (%)
1	—	—	—	2	EtOH	32
2	Reduction w/FA[c]	1	RT	2	EtOH	49
3	Reduction w/H_2	2	60	2	EtOH	52
4	—	—	—	20	Toluene	3
5	Reduction w/H_2	2	110	20	Toluene	53

[a] Reaction conditions: pre-treated catalyst (4 mg), benzaldehyde (0.21 mmol) and ethyl cyanoacetate (0.2 mmol) at 56 °C. [b] Yield determined *via* HPLC-UV (MS) using commercial product **3** as the reference standard. [c] In 2% formic acid.

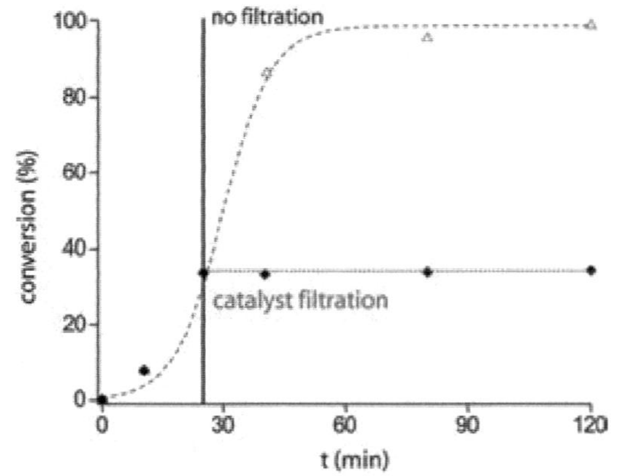

Fig. 1 Knoevenagel condensation of benzaldehyde (0.12 mmol) and ethyl cyanoacetate (0.1 mmol) with copper Cu(D) (4 mg) at 56 °C in DMSO: after 20 minutes half of the reaction volume was extracted and filtered and further subjected to the reaction conditions without a solid catalyst (black diamonds).

larger surface area, leaching was only increased by factor 2.5 (entries 6 and 7), yet the yield substantially changed from 32% to 99%. This is another argument against a mechanism based on solvated Cu species, since the relatively small amount of additional leached species, for example in the case of Copper hollow-spheres (Cu(Hol), entry 8, Table 4), can hardly be responsible for the large increase in yield. Thus, a Cu surface based mechanism can be assumed. However, the nature of the surface state required for catalytic action is still unknown. Thus, surface activation experiments were done: the copper catalyst was pre-treated under different reducing conditions, in order to reduce copper oxide species. This resulted in higher yields, most obviously for toluene, where the pre-treated copper yielded 53% product after 20 h, while the untreated copper did not catalyze the reaction at all (entries 4 and 5, Table 5). This finding leads to the conclusion that a non-oxidized Cu(0) surface reacts with the substrates. Kinetic measurements support the surface catalyzed mechanism too, as a zero-order kinetic model fits well in the beginning of the reaction (see Fig. S1 and S2 in the ESI†). An initial TOF of 0.88 s^{-1} was calculated using optimal conditions (Fig. 1, 41% yield after 30 min) and the conservative assumption that every Cu(0) atom on the surface is an active site. Further mechanistic studies are a subject of ongoing research. Moreover, similar copper(0) species with different BET surface areas (see Table S1 in the ESI†), such as copper hollow spheres Cu(hol),[12] copper nanoparticles A Cu(NPA) and copper nanoparticles B Cu(NPA) were tested (entries 8–10). A correlation between the surface area and conversion could be confirmed. Also, an experiment was conducted to prove the reusability of the catalyst. The standard setup in DMSO afforded constant high yields after 1 h for 5 consecutive cycles (Fig. S1, ESI†).

To illustrate the practical synthetic utility of our method, a 100 gram scale experiment was performed (Scheme 1). Benzaldehyde **1** (81.2 mL, 0.8 mol) was reacted with ethyl cyanoacetate **2** (84.8 mL, 0.8 mol) and 16.00 ± 0.001 g of Cu(D) in EtOH at 56 °C for 16 h. After filtration through aluminum oxide and recrystallization, 91% (146 g) pure isolated yield was obtained and almost all of the solid catalyst (15.94 ± 0.001 g) could be recovered. This satisfying result highlights the simplicity of both catalyst and work-up and hence is of interest for industrial scale Knoevenagel reactions. It should be noted that this kind of upscale experiment is difficult to perform with soluble base catalysts as large amounts of solvent and neutralization agent have to be used.

In summary, we have developed a simple and mild method for the Knoevenagel condensation using commercially available semi-noble copper. In this catalysis, tedious separation procedures are not necessary and no traces of base remain. Further experiments revealed that a Cu(0) surface is necessary to catalyze the condensation and, surprisingly, other noble metals such as gold and silver were not active. Moreover, a substantial scale up (>100 g) experiment demonstrated simple applicability.

Financal support was provided by ETH Zurich and the Swiss Science Foundation (No. 200021-150179).

Notes and references

1 (*a*) E. Nakamura and S. Mori, *Angew. Chem., Int. Ed.*, 2000, **39**, 3750; (*b*) J. S. Johnson and D. A. Evans, *Acc. Chem. Res.*, 2000, **33**, 325; (*c*) D. A. Evans, D. M. Barnes, J. S. Johnson, R. Lectka, P. von Matt, S. J. Miller, J. A. Murry, R. D. Norcross, E. A. Shaughnessy and

K. R. Campos, *J. Am. Chem. Soc.*, 1999, **121**, 7582; (*d*) H. J. Cristau, P. P. Cellier, J. F. Spindler and M. Taillefer, *Chem. – Eur. J.*, 2004, **10**, 5607; (*e*) A. Alexakis and C. Benhaim, *Eur. J. Org. Chem.*, 2002, 3221; (*f*) O. Daugulis, H.-Q. Do and D. Shabashov, *Acc. Chem. Res.*, 2009, **42**, 1074; (*g*) I. P. Beletskaya and A. V. Cheprakov, *Coord. Chem. Rev.*, 2004, **248**, 2337.

2 P. E. Fanta, *Chem. Rev.*, 1946, **38**, 139.

3 (*a*) D. M. T. Chan, K. L. Monaco, R. P. Wang and M. P. Winters, *Tetrahedron Lett.*, 1998, **39**, 2933; (*b*) D. A. Evans, J. L. Katz and T. R. West, *Tetrahedron Lett.*, 1998, **39**, 2937; (*c*) P. E. Fanta, *Synthesis*, 1974, 9; (*d*) H. H. Hodgson, *Chem. Rev.*, 1947, **40**, 251; (*e*) P. Y. S. Lam, C. G. Clark, S. Saubern, J. Adams, M. P. Winters, D. M. T. Chan and A. Combs, *Tetrahedron Lett.*, 1998, **39**, 2941.

4 (*a*) H. C. Kolb, M. G. Finn and K. B. Sharpless, *Angew. Chem., Int. Ed.*, 2001, **40**, 2004; (*b*) C. W. Tornøe, C. Christensen and M. Meldal, *J. Org. Chem.*, 2002, **67**, 3057.

5 (*a*) K. C. Nicolaou, P. G. Bulger and D. Sarlah, *Angew. Chem., Int. Ed.*, 2005, **44**, 4442; (*b*) K. M. Bratlie, H. Lee, K. Komvopoulos, P. Yang and G. A. Somorjai, *Nano Lett.*, 2007, **7**, 3097; (*c*) G. C. Bond and D. T. Thompson, *Catal. Rev.*, 1999, **41**, 319.

6 (*a*) L. R. Madivada, R. R. Anumala, G. Gilla, S. Alla, K. Charagondla, M. Kagga, A. Bhattacharya and R. Bandichhor, *Org. Process Res. Dev.*, 2009, **13**, 1190; (*b*) C. A. Martinez, S. Hu, Y. Dumond, J. Tao, P. Kelleher and L. Tully, *Org. Process Res. Dev.*, 2008, **12**, 392; (*c*) S. D. Walker, C. J. Borths, E. DiVirgilio, L. Huang, P. Liu, H. Morrison, K. Sugi, M. Tanaka, J. C. S. Woo and M. M. Faul, *Org. Process Res. Dev.*, 2011, **15**, 570; (*d*) E. Knoevenagel, *Ber. Dtsch. Chem. Ges.*, 1898, **31**, 2596.

7 L. F. Tietze and U. Beifuss, in *Comprehensive Organic Synthesis*, ed. B. M. T. Fleming, Pergamon, Oxford, 1991, pp. 341–394.

8 (*a*) A. Corma, V. Fornes, R. M. Martín-Aranda, H. Garcia and J. Primo, *Appl. Catal.*, 1990, **59**, 237; (*b*) A. Corma and R. M. Martín-Aranda, *J. Catal.*, 1991, **130**, 130; (*c*) B. C. Ranu and R. Jana, *Eur. J. Org. Chem.*, 2006, 3767; (*d*) C. Su, Z.-C. Chen and Q.-G. Zheng, *Synthesis*, 2003, 0555; (*e*) E. M. Schneider, R. A. Raso, C. J. Hofer, M. Zeltner, R. D. Stettler, S. C. Hess, R. N. Grass and W. J. Stark, *J. Org. Chem.*, 2014, **79**, 10908; (*f*) N. Mase and T. Horibe, *Org. Lett.*, 2013, **15**, 1854; (*g*) Y. A. Sonawane, S. B. Phadtare, B. N. Borse, A. R. Jagtap and G. S. Shankarling, *Org. Lett.*, 2010, **12**, 1456.

9 R. N. Grass, E. K. Athanassiou and W. J. Stark, *Angew. Chem., Int. Ed.*, 2007, **46**, 4909.

10 F. M. Koehler, N. A. Luechinger, D. Ziegler, E. K. Athanassiou, R. N. Grass, A. Rossi, C. Hierold, A. Stemmer and W. J. Stark, *Angew. Chem., Int. Ed.*, 2009, **48**, 224.

11 (*a*) V. S. R. Rajasekhar Pullabhotla, A. Rahman and S. B. Jonnalagadda, *Catal. Commun.*, 2009, **10**, 365; (*b*) S. Komiya, T. Sone, Y. Usui, M. Hirano and A. Fukuoka, *Gold Bull.*, 1996, **29**, 131.

12 (*a*) N. Kränzlin, S. Ellenbroek, D. Durán-Martín and M. Niederberger, *Angew. Chem., Int. Ed.*, 2012, **51**, 4743; (*b*) N. Kränzlin and M. Niederberger, *Adv. Mater.*, 2013, **25**, 5599.

This journal is © The Royal Society of Chemistry 2015

455851 Reaction Route Analysis of Mo Based Shift Catalyst

Abstract:

We evaluated a reaction route of sulfur tolerant shift catalyst which is expected to apply to IGCC (Integrated coal Gasification Combined Cycle) with CO_2 capture. In the conventional many papers, the elementary reaction as the origin of the MoS_2 to be constructed in a) MoS_2 + H_2O MoSO + H_2S, b) MoSO + CO MoS + CO_2, and c) MoS + H_2O MoSO + H_2O has been proposed. In this study, we have clarified that there is a elementary reaction d) MoS+H_2S MoS_2+H_2 instead of c), and H_2S contributes to reaction progress.

Copyright © American Institute of Chemical Engineers. All rights reserved.

456199 Decrosslinking and Decomposition of Silane-Crosslinked Polyethylene in Supercritical Methanol and Water

Abstract:

Decrosslinking and Decomposition of Silane-Crosslinked Polyethylene in Supercritical Methanol and Water

Giyoung Hong, Jebin Ryu, Hee Suk Woo, Won-Su Son, Kuan Wwan Park and Youn-Woo Lee

In decades, the necessity of plastic recycling has risen due to increasing concerns about environmental pollution and oil depletion. The chemical recycling which recovers monomers or starting material from plastics by depolymerization or decomposition is a promising strategy because the recycled monomers obtained can be used for any application regardless of the property of the original polymer before recycling. However, to utilize this advantage of chemical recycling, the recovery of monomers having good quality with high yield is essential. Many researchers have reported that supercritical fluid technology will be the promising methods for recycling of the plastics because of its characteristic properties. For example, Water and alcohol are representative polar solvents at ambient conditions, however, their polarity decreases dramatically under supercritical conditions. so that supercritical water and alcohol can be applied as good solvent to treat nonpolar polymer. In this study, the decrosslinking reaction of silane-crosslinked polyethylene (S-XLPE) in supercritical fluids was investigated. Various experiments have been performed under conditions of different temperature, pressure, reaction time, kind of solvents, weight ratio of solvent to polymer in batch reactor system. The change of gel content and molecular weight along with reaction temperature and reaction time was analyzed and the kinetic parameters were calculated from the experimental data. The complete decrosslinking of S-XLPE was achieved at several conditions, and the reaction time for the decrosslinking was shortened as the reaction proceed. The decrosslinking rate of S-XLPE was fitted well with the first order reaction model. The molecular weight of de-crosslinked polyethylene (DXPE) decreased as reaction temperature and reaction time increased when the reaction temperature exceeded 360°C. The decrease of molecular weight of DXPE was also fitted well with the first order reaction model when it was assumed that the decreasing rate was linearly dependent on the molecular weight of DXPE.

Copyright © American Institute of Chemical Engineers. All rights reserved.

456236 Regeneration of Fatty Acid-Adsorbed γ-Alumina Using Supercritical Methanol

Abstract:

Vegetable oil is both a major source of nutrition and an important fuel source for production of biodiesel. Triglycerides (TG), diglycerides (DG), monoglycerides (MG), free fatty acids (FFA), and other impurities are components of most vegetable oil. Deacidification, which refers to removal of FFA in vegetable oil, is an important refining process for reducing the high level of FFA that may promote rancidity of oils. Common deacidification method is neutralization with alkali, generally sodium hydroxide solution. However, neutralization via the alkaline method is not suitable for vegetable oils that contain a high concentration of FFA because saponification of vegetable oil due to alkaline treatment results in high loss of oils during the washing step. It is reported that the refining loss reaches three times the FFA content during the deacidification of soybean and cottonseed oil. Adsorption of FFA using activated alumina could be an alternative method of the alkaline neutralization. A chromatographic process using activated alumina at room temperature without physical contact between the oils and alkaline reagent leading to decreased loss of oils has been proposed. However, regeneration of spent activated alumina is energy intensive process because chemisorption of carboxylic acids on alumina can occur via surface esterification with hydrated alumina surfaces. In this study, we introduce a novel method for regeneration of the fatty acid-adsorbed γ-alumina using supercritical methanol. The adsorbed fatty acid converted to fatty acid methyl ester (FAME) through transesterification reaction. This method is based on a difference in the affinity of a fatty acid and the corresponding fatty acid methyl ester (FAME) for γ-alumina. Palmitic acid was selected as a model fatty acid. Batch-type reactors were used to investigate the effect of the operating parameters (temperature, methanol to γ-alumina weight ratio, and reaction time) on supercritical methanol regeneration. Reaction product was analyzed by gas chromatography. Almost all of the adsorbed palmitic acid was desorbed at temperatures above 300 , or when the weight ratio of methanol to γ-alumina was higher than 75:1, and/or if the reaction time was longer than fifteen minutes. The crystal structure and BET surface area of γ-alumina was unchanged after supercritical methanol regeneration. The developed technique requires

relatively lower operating temperature, is eco-friendly, and generates fuel, although the process is more complex than the thermal regeneration method.

Copyright © American Institute of Chemical Engineers. All rights reserved.

456294 Equilibrium Shifting of Esterification with Water-Permselective Zeolite Membrane

Abstract:

Ethyl acetate (EtOAc) is one of the important chemical products and used as solvent for ink, synthetic resin, adhesive agent and so on. EtOAc is usually synthesized by the following esterification between ethanol (EtOH) and acetic acid (AcOH) in liquid phase.

AcOH + EtOH = EtOAc + H_2O

It is well known that the esterification is equilibrium-limited reaction. So far, shifting the equilibrium point has been attempted by incorporating membrane separation of water. There are two schemes for water removal: one is the vapor permeation (VP) method to remove the water vapor in the vapor phase at equilibrium with the liquid reacting mixture; the other is the pervaporation (PV) method to remove the water directly from the liquid reacting mixture. This work focused on the vapor phase esterification in flow-type membrane reactor, where water vapor is selectively separated with vapor permeation (VP) using a CHA type zeolite membrane. Advantages in this vapor system, carried out at higher temperature, should be: (1) reaction rate will become larger than that in liquid phase operated at lower temperature, and (2) a higher rate of water vapor permeation (permeability) can be obtained. As an overall, a higher productivity of EtOAc with increasing reaction and permeation rates can be expected.

Copyright © American Institute of Chemical Engineers. All rights reserved.

456296 Pd/CuO Bifunctional Catalytic Membrane for Direct Synthesis of Phenol from Benzene

Abstract:

Synthesis of phenol, an important intermediate of the organic chemicalã€€product, is presently carried out in a multistep process, where a large amount of energy is consumed for the separation and purification. Therefore inexpensive and environmentally benign one-step synthesis process is very much expected. As a new synthetic route of benzene to phenol, our group has examined a Pd membrane reactor being able to realize the direct hydroxylation. In this system, the active oxygen species are formed by the reaction between the dissociated hydrogen permeated through Pd membrane and the adsorbed oxygen, participating to the direct conversion of benzene to phenol. Aim of this study is to improve the reaction efficiency much more by introducing another idea.

Pd film is formed on a porous Î±-Al_2O_3 tube. The hydroxylation is considered to occur near the palladium film on the alumina support side. So, if an effective hydroxylation catalyst is loaded on the Î±-Al_2O_3 support tube, it will be expected to increase the reaction efficiency. In this study, a co-catalytic effect by adding copper oxide, which is reported to be useful, is examined.

Copyright © American Institute of Chemical Engineers. All rights reserved.

456850 Kinetics and Spectroscopic Studies of Methane Partial Oxidation on Ni and Rh Catalysts

Abstract:

Kinetics and Spectroscopic studies of Methane Partial Oxidation on Ni and Rh Catalysts

<u>Cun Wen</u>, Jason Hattrick-Simpers, Jochen Lauterbach,*

University of South Carolina, Columbia, SC, 29208, USA

**Corresponding author:* lauteraj@cec.sc.edu

Recent exploration of shale gas has led to shockwaves in both energy markets and academic research fields. On the basis of same amount of energy content, shale gas has a much lower price (about 70% lower) compared with that of crude oil, even with todayfs low crude-oil price at $40 per barrel. The price difference between shale gas and crude oil has revived interests to convert methane, as the major component in shale gas, to gasoline and diesel. Methane partial oxidation (POM), in particular, is an preferred processed due to its optimum H_2 to CO ratio of 2 for downstream Fischer-Tropsch synthesis and mild heat release. However, the most active catalyst for POM is Rh, which is a nobel metal catalyst with a much higher price tag compared with the next best candidate in-line, Nickle. Here, fundamental studies that compares the reaction kinetics and reaction mechanism between Ni and Rh will be presented, and the possibility of design an Ni catalyst possessing comparable activity of Rh will be discussed.

Series of Rh and Ni catalysts using *f*Á-Al_2O_3 as supports were synthesized by following literature.[1, 2] The structures of two catalyst were characterized with ICP, XRD, BET, TEM, to make sure they have comparable metal loadings, particle size, surface area. Their catalytic activity for POM were compared under the same reaction conditions. Consistently with literature,[1, 3] the Rh catalysts showed activity toward POM (T_{50}= 502 ‹C) at a temperature about 100 ‹C lower than that on Ni catalysts (T_{50}= 605 ‹C). The catalytic activity of Rh and Ni catalysts were widely reported to be limited by methane dissociation as the rate-determining step.[4, 5] Therefore, methane-temeprature programmed reduction experiments were design to compare how easily methane can be activated and dissociated by Rh and Ni catalysts. In contrary to the catalytic activity tests, the Rh and Ni catalysts possessed comparable starting temperature for CH_4 dissociation at around 200‹C, as shown in Figure 1. Furthremore, Ni catalysts had no activity toward POM until 450 ‹C, at which temperature CH_4 could readily dissociate on the Ni catalysts accroding to CH_4-TPR. More detailed kinetics and spectroscopic results indicate that Ni catalysts can be partially oxided under the POM condition with CH_4 to O_2 ratio of 0.5, and loss the activity toward POM. More importantly, methodologies to increase the resistence of Ni to oxidation will be discussed in this presentation.

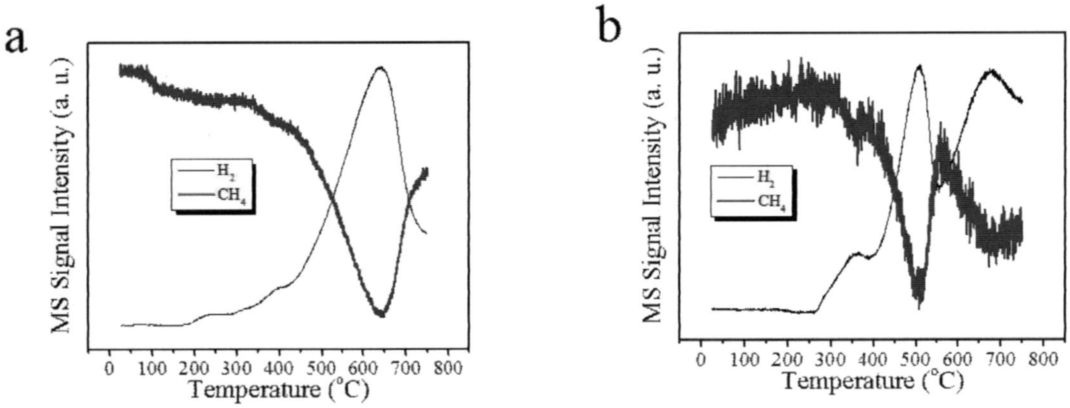

Figure 1. Methane-temperature programmed reduction on a) Rh/Al_2O_3 and b) Ni/Al_2O_3 catalysts.

References:

1. Li, J.-M., et al., *Effect of Rh loading on the performance of Rh/Al2O3 for methane partial oxidation to synthesis gas.* Catalysis Today, 2008. **131**(1–4): p. 179-187.

2. Jin, R., et al., *Mechanism for catalytic partial oxidation of methane to syngas over a Ni/Al2O3 catalyst.* Applied Catalysis A: General, 2000. **201**(1): p. 71-80.

3. Dissanayake, D., et al., *Partial oxidation of methane to carbon monoxide and hydrogen over a Ni/Al2O3 catalyst.* Journal of Catalysis, 1991. **132**(1): p. 117-127.

4. Wei, J. and E. Iglesia, *Isotopic and kinetic assessment of the mechanism of reactions of CH4 with CO2 or H2O to form synthesis gas and carbon on nickel catalysts.* Journal of Catalysis, 2004. **224**(2): p. 370-383.

5. Wei, J. and E. Iglesia, *Structural requirements and reaction pathways in methane activation and chemical conversion catalyzed by rhodium.* Journal of Catalysis, 2004. **225**(1): p. 116-127.

Copyright © American Institute of Chemical Engineers. All rights reserved.

456966 Developing Catalysts for the Partial Electrochemical Oxidation of Methane

Abstract:

Developing Catalysts for the Partial Electrochemical Oxidation of Methane

Michael J. Boyd, Chris Hahn, Thomas F. Jaramillo

Department of Chemical Engineering, Stanford University

Partial oxidation of methane to methanol and other oxygenates is an appealing alternative to methane steam reforming and Fischer-Tropsch reactions. Traditional heterogeneous catalyst systems require high temperatures to overcome the activation barriers for methane oxidation, leading to complete oxidation of methane to carbon dioxide. Electrocatalysis, however, offers precise control of catalyst-free surface energies via applied potential, and is a promising approach to high conversion and selectivity for methane partial oxidation at low temperatures. Our work investigates the activity of Pt and several other transition metal catalyst materials for low temperature electrochemical methane oxidation. The effects of mass transport are assessed via the use of multiple reactor designs, including rotating disk electrode, gas diffusion electrode, and membrane electrode assembly. Several methods of detecting products, including Nuclear Magnetic Resonance, Gas Chromatography, and Differential Electrochemical Mass Spectrometry, are employed to identify interesting active materials. Catalysts are characterized ex-situ via Scanning Electron Microscopy, X-ray Photoelectron Spectroscopy, and X-ray Diffraction. Up till now we have observed methane adsorption on noble metal surfaces at low potentials, followed by oxidation when sweeping to higher potentials.

Copyright © American Institute of Chemical Engineers. All rights reserved.

456308 The Characteristics of Diesel Steam Reforming at Pressurized Environment for the Operation of Solid Oxide Fuel Cell

Abstract:

Hydrogen energy has been emerged as one of the new regeneration energy with high potential. Also, diesel with high hydrogen energy density has been attracted for extracting hydrogen sources at fuel cell fields. There are three predominant reforming methods for this process; steam reforming (SR), partial oxidation (POX) and auto thermal reforming (ATR). Among them, SR is regard as the most appropriate method at pressurized environment because it doesnâ€™t need to require compressors that consumes much power. In other words, it only needs pumps to apply pressure to the reformer. With regard to overall system efficiency, SR is significantly beneficial compared to POX and ATR. Furthermore, SR method yields the most amount of methane (CH_4) in reforming methods. This CH_4 produced from the SR method can be used as fuel source in SOFC because SOFC is operated at high temperature.

In particular, CH_4 in reformate gas can be used to control temperature in SOFC by internal reforming. Endothermic CH_4 internal reforming can keep the heat balance with the exothermic electrical reaction generated from SOFC. Consequently, the moderate CH_4 amounts are necessary in SOFC. The amount of CH_4 production can be increased in accordance with the pressure increase in reforming reactor based on Le Chatelierâ€™s principle.

Â In this paper, SR of commercial diesel at pressurized environment for the operation of SOFC was investigated through experiment on various conditions. The characteristics of pressurized SR were identified as the CH_4 production increase and catalytic reaction improvement. The amount of CH_4 was augmented with pressure increase in reactor. Also, catalytic activity and stability was improved with kinetic increase as pressurizing reactor.

Copyright © American Institute of Chemical Engineers. All rights reserved.

457230 Combined Process for Hydrogen Production from Coke Oven Gas (COG)

Abstract:

Steel industry has been clearly classified as a mass CO_2 emission sector because approximately 2 ton CO_2 per 1 ton steel is produced. Almost carbon dioxide in the steel industry are emitted from pig iron making process, because C (Carbon) is employed for a reducing agent of iron ore ($2Fe_2O_3$+3C à 4Fe + $3CO_2$) and intensive energy should be provided to make molten iron.

A blast furnace is the typical facility for producing pig iron and moving bed reactor operated at high temperature. Sintered iron ore and coke (carbon resource) are fed to the top of blast furnace for producing pig iron and coke is used for energy source and reducing agent. The reducing reaction of iron ore is divided into direct reduction and indirection reduction. In case of direct reduction, coke and iron ore are directly converted into molten pig iron with solid-solid state reaction at high temperature. Therefore, direct reduction between iron ore and coke is usually occurred at the bottom part of the blast furnace. Alternatively, carbon monoxide (CO) is basically used for reducing iron ore into other iron oxide form during the indirect reduction. As a result of using coke as energy source and reducing agent, mass carbon dioxide should be produced in the steel industry.

Due to operating ETS (Emission Trading Scheme) in South Korea, steel making company has been allocated challengeable CO_2 emission reductions. To response to challengeable CO_2 emission reductions, energy saving and hydrogen reduction of iron ore have an attention as key technologies for reducing greenhouse in steel industry. Especially, CO_2 emissions will be reduced by substituting carbon with hydrogen as the reducing agent, because the product for hydrogen reduction of iron ore is H_2O. For adopting hydrogen reduction of iron ore in a steel industry, large amount of hydrogen should be provided to hydrogen reduction process.

COG (Coke Oven Gas) which is one of by-product gases is mainly consisted of 25% CH4, 55% H2, and 7% CO in steel mill. Because more than 50% hydrogen is included in COG, COG has been considered as an attractive hydrogen resource. A large amount of hydrogen can commerically be produced with H2-PSA, CH4(in COG)-H2O reforming, water gas shift (WGS) process. In this study, hydrogen production technologies using COG will be introduced.

Copyright © American Institute of Chemical Engineers. All rights reserved.

457415 Glycerol Hydrogenolysis with Reduced Co_2AlO_4 Nanoparticles Produced By Flame Spray Pyrolysis

Abstract:

Biodiesel is currently one of the most important biofuels (Martino et al., 2008). However, about 10-wt% of glycerol is formed as a by-product during the transesterification process of vegetable oil to biodiesel (Zhou et al., 2008). Therefore the catalytic conversion of glycerol to added value chemicals has been intensively studied.

The catalytic-alumina particles are typically prepared in a solution followed by thermal treatments in air (Dumond et al., 2007). However, the thermal treatments are likely to induce migration of cobalt into alumina forming an aluminate spinel that cannot be reduced below 800 ¼C. Co_3O_4 and γ-Al_2O_3 also have isotype crystal structures which enables the migration of ions from cobalt oxide into the underlying support. The optimization of dispersion and degree of reduction is therefore demanding.

In the current study Co_2AlO_4 nanoparticles were synthesized using flame spray pyrolysis (FSP). In order to activate Co particles the produced powder was reduced using a temperature programmed reduction (TPR) at 850 ¼C for 1h. The catalytic activity of the powder was then tested for glycerol hydrogenolysis using a quartz fixed bed down flow reactor at 573 K for 6h. The properties of the powder were characterized before and after the reduction using electron microscopy and X-ray diffraction. The surface characterization was carried out using X-ray photoelectron spectroscopy.

According to BET analysis the total surface area of the as-synthesized Co_2AlO_4 powder was around 174 m^2/g with a mean pore diameter of 10 nm. Figure 1A show a TEM images of the produced particles. The particles were spherical with the particle size between 50 nm and 200 nm. The particle surface consists of Co, Al and O. The atomic ratio of Co/Al on the surface is 0.27 which is lower than ratio of nominal $CoAl_2O_4$ (0.5).

A TEM image of the reduced powder is shown in Figure 1B. The appearance of Co nanoparticles with sizes between 3 and 8 nm can be observed, as well as the formation of shell-like structure. The reduction also caused a significant decrease in Co/Al atomic ratio down to 0.11. This indicates the reduction of exposed Co atoms in the aluminate to metallic Co.

Based on the TPR profile the reduction of $CoAl_2O_4$ is a two stage process. At temperatures between 200 and 250 ¼C Co_3O_4 is first reduced to CoO which is then further reduced to metallic Co. The second stage of the reduction takes place at temperatures between 500 and 800 ¼C. At this temperature range the oxidized cobalt species (Co^{2+} and Co^{3+}) which are strongly interacting with the support are reduced to metallic Co (Arnodly and Moulijin, 1985).

The reduced Co_2AlO_4 consisting of Co and Al_2O_3 was observed to promote dehydration and dehydrogenation of glycerol. The main products were hydroxyacetone, lactic acid, and lactide. However, the glycerol conversion dropped from 80 % to around 35 % after 4 h of reaction, indicating deactivation of the catalyst. The deactivation is most likely due to the deposition of carbon species on the surface of the catalyst powder during the dehydration reactions.

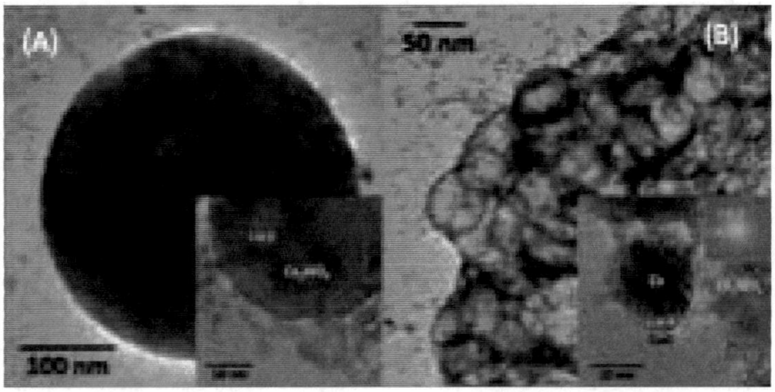

Figure 1. TEM images of the Co_2AlO_4 particles (A) as-synthesized and (B) after the reduction.

This work was supported by the strategic funding of the University of Eastern Finland (Namber).

P. Arnoldy and J.A. Moulijn (1985) *J. Catal.* **93**, 38.

F. Dumond et al. (2007) *J. Phys. Chem. C* 111, 4780.

D.S. Martino et al. (2008) *Energy Fuels* **22**, 207-217.

C.H. Zhou et al. (2008) *Chem. Soc. Rev.* **37**, 527-549.

Copyright © American Institute of Chemical Engineers. All rights reserved.

458156 Catalytic Selective Oxidation in Biomass Conversion

Abstract:

In the global attempt to reduce carbon footprint, the Chemical and Petrochemical Industry faces the problem to replace the currently used fossil feedstock with renewable resources, reduce energy consumption and to intensify and integrate the processes to be more carbon efficient. In all three issues, catalysis will be the key to a successful transformation. Knowledge-based development and implementation of catalytic technology will help to process the novel feedstock, reduce the energy required for maintaining the desired process and improve the carbon efficiency of the targeted synthesis routes. In this seminar, two examples will be discussed to illustrate our efforts in the last few years in biomass transformation.

The first example is a classic heterogeneous catalysis approach in which an integrated experimental and computational investigation reveals that the surface lattice oxygen of copper oxide activates the formyl Câ€"H bond in glucose and incorporates itself into the glucose molecule to oxidize it to gluconic acid. The reduced CuO catalyst regains its structure, morphology and activity upon re-oxidation. The activity of lattice oxygen is shown to be superior to that of the chemisorbed oxygen on the metal surface and the hydrogen abstraction ability of the catalyst is correlated with the adsorption energy. Based on the present investigation, it is suggested that surface lattice oxygen is critical for the oxidation of glucose to gluconic acid, without further breaking down the glucose molecule into smaller fragments, due to Câ€"C cleavage. Using CuO as the catalyst, excellent yield of gluconic acid is also obtained for the direct oxidation of cellobiose and polymeric cellulose, as biomass substrates.

The selective oxidation can be effectively catalyzed using noble metal nanoparticles. Choosing an appropriate support with well-defined structure and suitable surface chemistry is a feasible and effective approach to make metal nanoparticles with specific size, shape, structure, and to avoid agglomeration leading to catalytic deactivation. With the intention of successfully synthesizing more efficient heterogeneous catalysts for the selective oxidation, it is extremely important to elucidate the effect of metal nanoparticle configuration, the novel nano-structured support on the catalytic activity for selective oxidation with particular highlights on the interactions between both metal-metal and metal-support. The second example shows the Au-Pd bimetallic catalyst and its application in selective oxidation of HMF to FDCA. The unique structure of selected support is found to improve the stabilization of the metals in the preparation as well as the reaction process. Furthermore, surface chemistry (metal-support interaction) and synergetic effect in the case of metal alloy catalysts (metal-metal interaction) play a crucial role in controlling the catalytic performance of as-prepared catalysts. Various characterizations will be carefully conducted to look into the insights of these catalystsâ€™ physicochemical properties.

Copyright © American Institute of Chemical Engineers. All rights reserved.

458187 Highly Dispersed Nickel within Mesochannels of SBA-15 for CO Methanation with Enhanced Activity and Excellent Thermostability

Abstract:

In this work, several nickel-based SBA-15 catalysts were prepared by surfactant-assisted impregnation method and applied in CO methanation reaction. The characterization results showed that Ni species were uniformly distributed in the ordered mesoporous channels of SBA-15 by adding proper surfactants. The morphologies of the unroasted catalysts and spent catalysts after 100h stability test were also investigated. The surfactant can hinder the migration of metal ions during drying and inhibit the aggregation of NiO during calcination, which ultimately improve the nickel dispersion. Consequently, the Ni/SBA-15 catalysts prepared with surfactant-assisted impregnation method exhibited superior activity and excellent thermostability owing to the moderate interaction between nickel and support, high nickel surface area and the confinement of support. The 10%Ni/S15(3) catalyst achieved the best activity with 100% CO conversion and about 99.9% CH_4 yield at 350â„ƒ, 0.3Mpa and 15000mL/g/h.

Copyright © American Institute of Chemical Engineers. All rights reserved.

458467 Fischer Tropsch Synthesis: A Study of the Selectivity to Light Hydrocarbons

Abstract:

The effect of co-feeding nitrogen in Fischer–Tropsch synthesis (FTS) in a fixed bed reactor over a 10%Co/TiO_2 catalyst was investigated. The partial pressure of CO and H_2 in the feed to the reactor was kept constant, while the partial pressure of N_2 was switched between a low and high value. We plotted the data in the standard formats typically used, but also in two other ways: Yao plot is a plot of P_{n+1}/O_{n+1} versus P_n/O_n where for us n is a carbon number from n=2 to 5; Lu plot is a plot of the relationship between normalized mole fractions of O_n, O_{n+1} and P_n. in ternary plot.

A comparison of the results obtained indicates that there was almost no effect on the overall reaction rate, but that there was a significant, positive effect on the product selectivity. A lower selectivity to the undesired light hydrocarbons was observed when co-feeding N_2. From Yao plot, we find that the data lies on two straight lines, one for n=2 and the other for n>2. Ternary plots show that the reactor operation tends to a stable point. Such results cannot be explained by classic kinetics alone but reaction equilibrium among some species, vapour-liquid equilibrium (VLE), boiling, or a combination of these factors may be contributing to the observed behaviour in FT. The laboratory data indicates that the selectivity to the (C_5–C_{19}) fraction is particularly enhanced by co-feeding N_2 in FTS.

Copyright © American Institute of Chemical Engineers. All rights reserved.

458476 Arrays of Coupled Cells Performing Chemical Computing with and without Parallelism Using Discrete Turing Patterns

Abstract:

We focus on modeling of dynamic behavior enabling chemical computing with the use of glycolytic oscillatory reaction in arrays of coupled cells with various topology, where the kinetic and transport parameters are set so that the system displays discrete reaction-diffusion patterns of Turing type.

We analyze two, three and four coupled cells in a linear and in a cyclic array. Each cell is modeled as a CSTR, coupling is assumed via membranes. Glycolysis is modeled using core model of glycolysis by Moran and Golbeter (1984), where ATP is transformed to ADP through phosphofructokinase and negative feedback is provided via lower parts of the glycolytic reaction chain whereby ADP is transformed back to ATP. We assume that rate coefficients of positive and negative feedback steps are modified via presence of other metabolites, pH and temperature, and thus may be elevated up to 9-fold the values at usual laboratory conditions. Transport coefficients are assumed equal for both ATP and ADP and this value is taken as a variable parameter.

Analysis of stability and bifurcations of stationary states provides parameter ranges corresponding to regions of coexistence of multiple discrete Turing patterns with oscillations, which is in turn used to examine transitions between discrete Turing patterns and oscillations using carefully targeted perturbations by ATP.

We propose a chemical computing technique capable of parallelism with a central knockout perturbation system, which replaces our previously published local knockout perturbation system. These techniques use transitions among discrete Turing patterns to achieve advanced logic functions and their plasticity during a chemical computing process.

In particular, we propose a chemical computing technique operating with a single data thread based on two coupled cells working as an XOR gate and using transitions among discrete Turing patterns and oscillations. This type of gate can be used to build a central processing unit.

Copyright © American Institute of Chemical Engineers. All rights reserved.

459161 Double Bond Isomerization of 1-Butene Using Supported $AlBr_3$ Catalysts

Abstract:

Isomerization of olefins is an important process in the chemical industry that is known to be catalyzed by acids and bases [1], [2]. When supported on slightly acidic supports like alumina or silica, strong Lewis acids like aluminum halides form both Lewis and Brønsted acid sites. In particular, reports have shown that supported $AlCl_3$ approaches what is known as superacid strength [3, 4], which even can isomerize alkanes like n-butane [5]. $AlBr_3$ is another strong Lewis acid that, in principle, can be supported on the same supports like silica and alumina, producing solid acid catalysts.

Here, we present the synthesis of supported $AlBr_3$ catalysts using various supports like silica, alumina and mesoporous silicas (MCM-41, SBA-15), and its potential use for double bond isomerization of 1-butene. Although the complete isomerization was expected, results show that only double bond isomerization was observed even at 100 °C. This we believe could have been due to the formation of weak Brønsted acid sites instead of the expected, strong Brønsted acid sites. At 100 °C, pyridine DRIFTS qualitatively shows that there are far more Brønsted acid sites than Lewis acid sites. Loading of the active material ($AlBr_3$), type of support, and strength of acid sites affected the conversion of 1-butene.

1. Hubert, A.J. and H. Reimlinger, *The Isomerization of Olefins Part II. Thermal and Catalytic Isomerization of Olefins using Acids, Metals, Metal Complexes, or Boron Compounds as Catalysts.* Synthesis, 1970. **1970**(08): p. 405-430.

2. Hubert, A.J. and H. Reimlinger, *The Isomerization of Olefins Part I. Base-Catalysed Isomerization of Olefins.*

Synthesis, 1969. **1969**(03): p. 97-112.

3. Drago, R.S., S.C. Petrosius, and C.W. Chronister, *Characterization and improvements in the synthesis of the novel solid superacid AlCl2(SG)n.* Inorganic Chemistry, 1994. **33**(2): p. 367-372.

4. Drago, R.S., S.C. Petrosius, and P.B. Kaufman, *Alkylation, isomerization and cracking activity of a novel solid acid, AlCl2(SG)n.* Journal of Molecular Catalysis, 1994. **89**(3): p. 317-327.

5. Fuentes, G.A., J.V. Boegel, and B.C. Gates, *n-butane isomerization catalyzed by supported aluminum chloride.* Journal of Catalysis, 1982. **78**(2): p. 436-444.

Copyright © American Institute of Chemical Engineers. All rights reserved.

459444 Kinetic Model on Conversion of D Fructose to 5 Hydroxymethylfurfural.

Abstract:

The use of fossil resources has led to environment issues related to CO_2emissions. As a result, a sustainable bio-economy requires the efficient utilization of all the ingredient of biomass. One of the sources from biomass is sugar. Many useful reaction product are available from sugar under existing of acid.

As well known, D-fructose is transformed into 5-hydroximethylfurfural (HMF) by dehydration and then, HMF is changed to levulinic acid (LA) and/or formic acid (FA) by hydration. The target product of HMF is a starting material for the synthesis of industrially important products. For example, HMF is a starting materials such as 2, 5-furandicarboxylic acid, 2,5-furandicarboxylic acid, etc., and new polymer materials.

For many years, method of HMF has been developed, but the procedure is not established yet, because the reactions mechanism are very complicated and hence it is difficult to obtain the HMF selectively. Furthermore, the kinetic study of HMF formation seems to be performed not in detail. The main purpose of the study is to establish high-yield method for the synthesis of HMF based on kinetic consideration. The reactions were performed experimentally under conditions of various hydrochloric acid concentrations at room temperature. The concentrations history was determined by NMR technique. Using try and error method, the reaction rate constants were determined by comparing concentration change experimentally obtained with theoretical curves evaluated from several kinetic models.<span">It should be noted from the analysis of the constants that some particular reaction rate was affected by the acidic catalyst concentration. By controlling the specific reactions, production rate of HMF seems to be controlled.

The detail discussions will be presented at the conference.

Reference

[1] Boy A. Fachri, Ria M. Abdilla, Henk H. van de Bovenkamp, Carolus B. Rasrendra,,and Hero J. Heeres."Experimental and Kinetic Modeling Studies on the Sulfuric AcidCatalyzed Conversion of D Fructose to 5 Hydroxymethylfurfural and Levulinic Acid in Water." ACS Sustainable Chem. Eng 3 (2015): 3024–3034.

Copyright © American Institute of Chemical Engineers. All rights reserved.

459709 Oxidative Coupling of Methane over Zeolite Supported Na-Mn-W

Abstract:

Oxidative Coupling of Methane over Zeolite Supported Na-Mn-W

Naseem S. Hayek and Oz M. Gazit*

Chemical Engineering, Israel Institute of Technology-Technion, Haifa, Israel *ozg@technion.ac.il*

The proven reserves of natural gas are rapidly increasing.[1] Currently, it is a highly underutilized hydrocarbon source for chemicals and liquid fuels synthesis, except for steam reforming.[2] One possible route is the direct conversion of methane, the main component of natural gas, to Ethane (C_2H_6) and Ethylene (C_2H_4) via oxidative coupling (Eq. 1).

Oxidative coupling of methane (OCM) has received much attention since the fundamental work of Hinsen et al.[3] and Keller et al.[4] This is because the prospect of direct conversion of methane to C_2 molecules is highly attractive, being a vital building block in the industry with an expected increase demand. Moreover, OCM is an exothermic reaction and therefore not limited by thermodynamics.

Since 1981, hundreds of materials have been tested as catalysts.[5,6] However, until now only a few catalysts have shown the potential for being economically viable for incorporation in a large scale process. The main challenge for the catalyst is the need to be highly selective for C_2 products, overcoming total oxidation of CH_4, which is by far the thermodynamically favored reaction route under these conditions (Eq. 2).

(1)

$$CH_4 + \frac{1}{2}O_2 \rightarrow C_2H_6 \text{ or } C_2H_4 + H_2O$$

$$\Delta H^0_{298} = -140\frac{kJ}{mol}$$

$$CH_4 + 2O_2 \rightarrow CO_2 + 2H_2O \qquad (2)$$

$$\Delta H^0_{298} = -890\frac{kJ}{mol}$$

Mn-Na_2WO_4/SiO_2 catalyst has been found to have remarkable stability at the required reaction conditions.[7] Moreover, this catalyst obtained performance of ~33% CH_4 conversion at 80% C_2 selectivity giving a C_2 yield of ~25%.[8] However, the performance needs to be improved in order to achieve above 30% yield, the indicated threshold for practical applications.[9]

Many variations in the composition of the metals based on this catalyst platform have been tested in an effort to understand structure-function relationship.[10-14] What is known suggests a strong synergetic effect for all three components (Na, Mn, W) on silica, where the absence of either one of them leads to inferior catalytic performance.[8] For example, sodium induces the transition of amorphous silica to α-cristobalite. This is considered to be crucial for obtaining above mentioned catalytic performance, despite the dramatic 2 orders of magnitude loss in surface area. This phase transition is thought to improve the dispersion of the active site, however, the dispersion of which of the components or both remains an open question.

In this study we investigate the fundamental aspects of active site dispersion for the Mn/Na/W based catalyst, while avoiding the phase transition of silica. This is done using the post-synthetic modification of zeolites (i.e. dealumination) to form open T sites, in which the active metals can be atomically dispersed regardless of phase transition.[15-18] Furthermore, the use of zeolites supports provides high surface area, a crystalline structure and high thermal stability. As control, we use amorphous silica-alumina to parallel the various Si/Al ratios used as zeolites.

Preliminary results indicate that contrary to native zeolite, dealuminated zeolite undergoes phase transition to α-cristoballite, upon impregnation with Na_2WO_4 and calcination at 850°C, analogous to silica supported catalyst. This is accompanied by a dramatic loss in surface area from ~680 m^2/g to ~2.5 m^2/g. Alumina is found to preserve its surface area and crystalline structure after the same impregnation and calcination progress. Amorphous silica-alumina is found to keeps its amorphous structure at low Si/Al ratios, while it transforms to α-cristoballite at high Si/Al ratios. Manganese atoms have been post-synthetically dispersed in the open T sites following the dealumination process. The Mn-incorporated zeolite was characterized by FTIR, XRD, XPS, UV-vis, Chemisorption and Physisorption. Dispersion of tungsten and catalytic evaluation of the different catalysts in a tubular fixed-bed reactor are in progress.

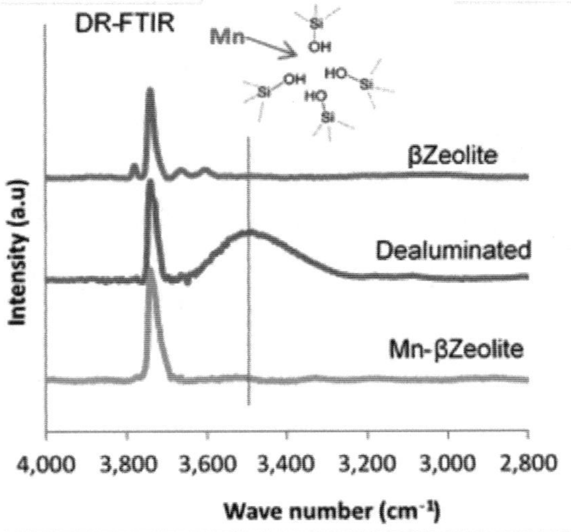

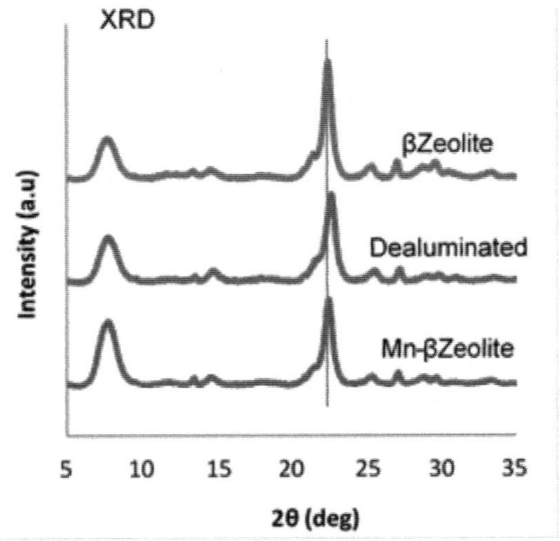

References

1 PB. in *BP Statistical Review of World Energy* (2011).

2 Lunsford, J. H. Catalytic conversion of methane to more useful chemicals and fuels: a challenge for the 21st century. *Catalysis Today* **63**, 165-174 (2000).

3 Hinsen, W. & Baerns, M. Oxidative Coupling of Methane to C2-Hydrocarbons in the Presence of Different Catalysts. *Chemiker-Zeitung* **107**, 223-226 (1983).

4 KellerG, E. & Bhasin, M. Synthesis of ethylene via oxidative coupling of methane. *JCatal* **73**, 9-19 (1982).

5 Maitra, A. M. Critical performance evaluation of catalysts and mechanistic implications for oxidative coupling of methane. *Applied Catalysis A: General* **104**, 11-59 (1993).

6 Zavyalova, U., Holena, M., Schlögl, R. & Baerns, M. Statistical Analysis of Past Catalytic Data on Oxidative Methane Coupling for New Insights into the Composition of High-Performance Catalysts. *ChemCatChem* **3**, 1935-1947 (2011).

7 Liu, H. *et al.* Scale up and stability test for oxidative coupling of methane over Na 2 WO 4-Mn/SiO 2 catalyst in a 200 ml fixed-bed reactor. *Journal of Natural Gas Chemistry* **17**, 59-63 (2008).

8 Palermo, A., Vazquez, J. P. H., Lee, A. F., Tikhov, M. S. & Lambert, R. M. Critical influence of the amorphous silica-to-cristobalite phase transition on the performance of Mn/Na 2 WO 4/SiO 2 catalysts for the oxidative coupling of methane. *Journal of catalysis* **177**, 259-266 (1998).

9 San Su, Y., Ying, J. Y. & Green, W. H. Upper bound on the yield for oxidative coupling of methane. *Journal of Catalysis* **218**, 321-333 (2003).

10 Palermo, A., Vazquez, J. P. H. & Lambert, R. M. New efficient catalysts for the oxidative coupling of methane. *Catalysis letters* **68**, 191-196 (2000).

11 Malekzadeh, A. *et al.* Correlation of electrical properties and performance of OCM MO x/Na 2 WO 4/SiO 2 catalysts. *Catalysis Communications* **2**, 241-247 (2001).

12 Ji, S.-f. *et al.* The relationship between the structure and the performance of Na-W-Mn/SiO 2 catalysts for the oxidative coupling of methane. *Applied Catalysis A: General* **225**, 271-284 (2002).

13 Mahmoodi, S., Ehsani, M. & Ghoreishi, S. Effect of promoter in the oxidative coupling of methane over synthesized Mn/SiO 2 nanocatalysts via incipient wetness impregnation. *Journal of Industrial and Engineering Chemistry* **16**, 923-928 (2010).

14 Yildiz, M. *et al.* Support material variation for the Mn x O y-Na 2 WO 4/SiO 2 catalyst. *Catalysis Today* **228**, 5-14 (2014).

15 Wolf, P., Hammond, C., Conrad, S. & Hermans, I. Post-synthetic preparation of Sn-, Ti-and Zr-beta: a facile route to water tolerant, highly active Lewis acidic zeolites. *Dalton Transactions* **43**, 4514-4519 (2014).

16 Tang, B. *et al.* A procedure for the preparation of Ti-Beta zeolites for catalytic epoxidation with hydrogen peroxide. *Green Chemistry* **16**, 2281-2291 (2014).

17 Dzwigaj, S. & Che, M. Incorporation of Co (II) in dealuminated BEA zeolite at lattice tetrahedral sites evidenced by XRD, FTIR, diffuse reflectance UV-vis, EPR, and TPR. *The Journal of Physical Chemistry B* **110**, 12490-12493 (2006).

18 Dzwigaj, S. *et al.* Incorporation of vanadium species in a dealuminated β zeolite. *Chem. Commun.*, 87-88 (1998).

460076 The Effect of the Reaction Environment on the Surface Acidity of t-ZrO_2 (101) Investigated By DFT

Abstract:

Zirconium oxide is an important ceramic material with a variety of commercial applications. This metal oxide has been successfully used as catalyst or catalyst support due to its thermal stability and acid-basic properties. In particular, the tetragonal phase of ZrO_2 has been repeatedly identified as the most catalytically active phase.[1] However, the assignment of specific catalytic performance is difficult without complete understanding of chemical properties of the catalysts at a molecular level.

In this study, the acidity of the most stable (101) facet of tetragonal zirconia (t-ZrO_2) was investigated using density functional theory (DFT) by means of ammonia adsorption. In order to cover a broader range of experimental conditions, different surface models were carefully selected to account for reductive or humid environments, i.e., stoichiometric terminations, oxygen vacancy defects, and hydroxylated surfaces were employed to characterize the acidity of the (101) facet of t-ZrO_2. Our results reveal that the acidity of t-ZrO_2 (101) is mainly due to the electron deficient metal atoms (Lewis acid sites) exhibiting an adsorption energy of ammonia (E_{ads}) up to 120 kJ/mol on a stoichiometric, vacancy-free surface. An oxygen vacancy site leads to a stronger NH_3-Zr interaction causing the adsorbate to dissociate and release up to 190 kJ/mol upon dissociative adsorption. In contrast, a humid environment leads to the formation of surface hydroxyl groups subsequent to the dissociation of water molecules on acid-base pairs; however, this only creates weak Brønsted acid sites with E_{ads} < 40 kJ/mol. Ammonia adsorption microcalorimetry and DRIFTS-pyridine desorption experiments are in very good agreement with our DFT calculations.[2] This study provides detailed information on the acidity of ZrO_2, which is a catalytically important property affecting conversion and selectivity of chemical reactions. Furthermore, the use of different model surfaces accounting for diverse

environmental conditions is particularly relevant, since it provides insights on tuning the surface acidity by adjusting reaction conditions accordingly.

REFERENCES:

[1] K. Samson, M. Sliwa, R.P. Socha, K. Gora-Marek, D. Mucha, D. Rutkowska-Zbik, J.-F. Paul, M. Ruggiero-Mikolajczyk, R. Grabowski, J. Sloczynski, Influence of ZrO_2 Structure and Copper Electronic State on Activity of Cu/ZrO_2 Catalysts in Methanol Synthesis from CO2, ACS Catal. 4 (2014) 3730â€"3741.

[2] J.M. Arce-Ramos, L.C. Grabow, B.E. Handy, M.G. CÃ¡rdenas-Galindo, Nature of Acid Sites in Silica-Supported Zirconium Oxide: A Combined Experimental and Periodic DFT Study, J. Phys. Chem. C. 119 (2015) 15150â€"15159. doi:10.1021/acs.jpcc.5b02394.

Copyright © American Institute of Chemical Engineers. All rights reserved.

459667 Promoter Effects of Residual K and Mn in Cobalt Supported Carbon Spheres Catalysts for Fischer Tropsch Synthesis

Abstract:

Carbon spheres (CSs) constitute a set of flexible and valuable properties when used as heterogeneous catalyst supports due to the inert nature of the Carbon spheres. These support materials may also be used as model support materials for the study of promoter effects in small amounts on the metal precursor. The surface of CSs must often contains N, O or H atoms in functional groups which may play pivotal roles in catalyst preparation by altering the acid-base and hydrophilic character of the carbon material hence creating metal anchoring sites. The effect of functionalising agents ($KMnO_4$ and HNO_3) on the surface properties of carbon spheres (CSs) produced from acetylene gas by a chemical vapour deposition method, was examined. Cobalt was loaded onto the differently functionalised CSs using the deposition precipitation (DPU) method and the Co/CS catalysts were tested for Fischer-Tropsch (FT) activity. FT studies, revealed differences in activity and selectivity for the HNO_3 and $KMnO_4$ functionalised catalysts. Investigation of the role of Mn (0.05%) and K (0.05%) on Co/CS FT catalysts functionalised with HNO_3 revealed that these differences were due to the presence of residual MnO_4^- ions and K^+ ions remaining after the surface functionalisation reactions.

Copyright © American Institute of Chemical Engineers. All rights reserved.

460024 Enhanced Electro-Kinetics of CO_2 Generation for Ethanol Oxidation Using Catalysts with a Partially Oxidized Pt and Rh Core and a SnO_2 Shell

Abstract:

Direct ethanol fuel cell (DEFC) is a promising technology for generating electricity via the electro-oxidation of liquid ethanol. Its implementation requires the development of anode catalysts capable of producing CO_2 and yielding 12-electron transfer through breaking C-C bond of ethanol. To evaluate the capability to break C-C bond over catalysts, here we designed an electrochemical cell equipped with a CO_2 microelectrode, through which CO_2 partial pressure and current density could be obtained simultaneously. Then CO_2 generation rate and selectivity as functions of potential were calculated in 0.5M H_2SO_4/0.5 M ethanol solution. In this work, the Pt/Rh/Sn catalysts with different Pt and Rh ratios were synthesized using surfactant free approach. Reduced Pt/Rh/Sn samples were prepared by reducing the as-made Pt/Rh/Sn catalysts in H_2/Ar mixture. Both as-made and reduced samples were characterized by transmission electron microscopy (TEM), scanning transmission electron microscope-electron energy loss spectroscopy (STEM-EELS), X-ray diffraction (XRD), and X-ray absorption spectroscopy (XAS). We demonstrated that the as-made sample was the tri-phase $PtRhO_x$-SnO_2 catalysts with a partially oxidized Pt and Rh core and a SnO_2 shell, and the reduced sample was the bi-phase PtRh-SnO_2 catalysts with a metallic PtRh alloy core and a SnO_2 shell. The comprehensive kinetics studies showed that $PtRhO_x$-SnO_2 catalysts coincided with a 2.5-fold increase in the CO_2 generation rate towards ethanol oxidation reaction at 0.35V, compared with PtRh-SnO_2 and commercial Pt. $PtRhO_x$-SnO_2 also exhibited lower onset potential of CO_2 generation and higher CO_2 selectivity than PtRh-SnO_2. These *in situ* studies provided insight on the design of a new genre of electro-catalysts with a partially oxidized noble metal.

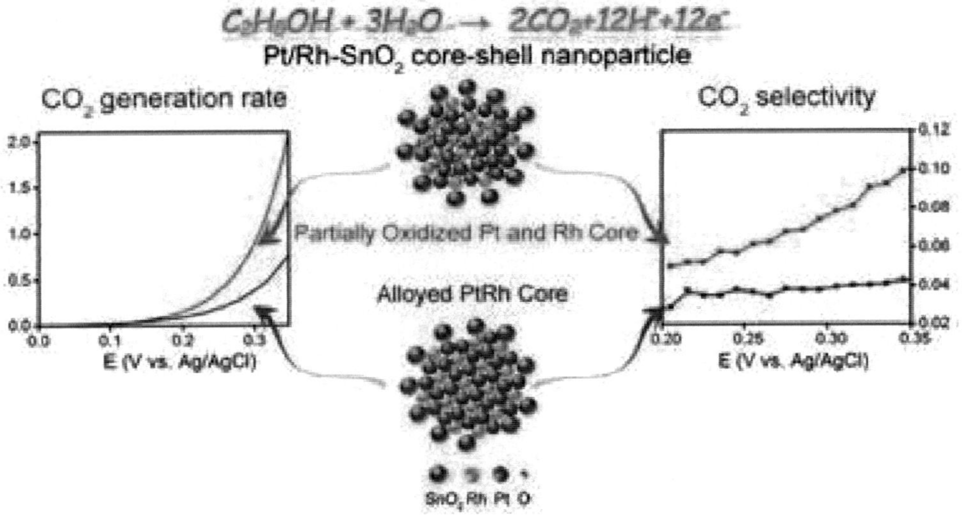

Copyright © American Institute of Chemical Engineers. All rights reserved.

460032 Development of Novel Catalysts and Reactor Configurations for Methane Dry Reforming

Abstract:

The conversion of CO_2 into fuels and useful chemicals has been intensively pursued for renewable and green energy production. At the same time, CO_2 conversion into valuable products is the most efficient way to reduce CO_2 fingerprint. However, CO_2 is a highly stable molecule almost chemically inert, thus making difficult its conversion under normal conditions. Among the viable alternatives, dry reforming of methane, DRM (CH_4+ $CO_2 \leftrightarrow 2H_2 + 2CO$, $\Delta H°298 = 247.3$ kJ/mol) is an efficient route to convert CO_2 into syngas, which can be used to produce liquid fuels and platform chemicals via Fischer–Tropsch (FT) synthesis [1].

Unfortunately, DRM is inevitably accompanied by catalysts deactivation due to carbon deposition and metal sintering making necessary the development of highly active and robust heterogeneous catalysts to overcome the deactivation issues. Together with efficient catalysts, the overall performance can be enhanced by using non-conventional reactor configurations. In this work, we have designed new catalysts for methane dry reforming and we have analysed the optimum reaction conditions and conceptual design of reactor configurations aiming to maximise the catalysts' stability and the syngas production while protecting the catalyst from coke deposition.

Copyright © American Institute of Chemical Engineers. All rights reserved.

460663 Modification of Molybdenum Phosphide Catalyst Surface with Fischer-Tropsch and Alkali Metals for the Production of Alcohols from Syngas

Abstract:

Ethanol today is used as a high value fuel blend, over 95% of the gasoline in the United States contains at least 10% ethanol according to the Alternative Fuels Data Center. Currently there are two predominant methods of producing ethanol, fermentation of sugars from corn and sugar cane and through hydration of petroleum-based ethylene. Both methods are expensive and consume large amounts of energy but there is no viable alternative at the moment. A possible solution to this problem could be using transition metals doped with promoters as catalyst to hydrogenate syngas and convert it into higher alcohols such as ethanol. Syngas can be readily obtained through the gasification of biomass, coal and natural gas. I am studying a Molybdenum phosphide system and how the modification of its surface with Fischer-Tropsch metals and alkali metals will influence the activity and selectivity of the catalyst towards the production of alcohols from syngas. Various catalyst synthesis methods are being investigated as different methods produce different metal phases on the surface of the catalyst. Thus far Molybdenum phosphide has shown promising results, tuning the surface with promoters can lead to a stable and highly active catalyst that can become commercially viable.

Copyright © American Institute of Chemical Engineers. All rights reserved.

460118 Formation of Rhodium Nanoclusters Fabricated from $Rh(CO)_2(acac)$ on Graphene/Cu(111)

Abstract:

The adsorption of an organometallic compound ($Rh(CO)_2(acac)$) and formation of Rh nanoclusters on graphene/Cu(111) surface were investigated with ultrahigh vacuum scanning tunneling microscopy (UHV-STM). Graphene layers were successfully prepared on Cu(111) by thermal dissociation of ethylene (C_2H_4). Due to weak interaction between graphene monolayer and Cu(111) substrate, several graphene moiré patterns with different periodicities (e.g. 2.42 and 3.86 nm) and different rotational alignments were observed. Subsequently, Rh nanoclusters were prepared on the surface by chemical vapor deposition of $Rh(CO)_2(acac)$. Findings of this study confirmed that the graphene moiré pattern is the preferential nucleation site and Rh nanoparticles were regularly trapped at the "fcc" hollow sites.

Copyright © American Institute of Chemical Engineers. All rights reserved.

461671 Methanol Synthesis from CO_2 in Slurry Bubble Column Reactors: A Modelling/ Experimental Approach

Abstract:

<u>**Methanol Synthesis from CO2 in Slurry Bubble Column Reactors: A Modelling/ Experimental Approach**</u>

Methanol is a common feedstock in the production of many industrial products and is the main reactant in the production of dimethyl ether (DME), which itself can be transformed into olefins and other commodity chemicals. DME is also being considered as a clean fuel for LPG based engines. Methanol is usually produced in fixed bed catalytic reactors (ICI, Mitsubishi, Lurgi). The reaction is highly exothermic, leading to hot spots in the reactor and catalyst sintering in the fixed beds. This work is part of a wider study investigating the production of methanol from CO_2 in bubble column reactors (employing colloidal catalysts), where heat may be better distributed throughout the reactor. To this effect two slurry bubble column reactors have been designed and commissioned (1m and 2m tall with a volume of 1L and 2L, respectively) to investigate CO_2 hydrogenation in such systems. The focus of this work is to develop a calibrated model that accounts for the residence time distribution of gas in the reactor and allows mass transfer coefficients as well as Henry's constants to be determined.

The reactor system (Figure 1) is operated in semi-batch mode, whereby reactant gases (CO_2 and hydrogen) and product vapour (methanol and water) interact with a colloidal catalyst suspended in an inert solvent (squalane). A gas circulation loop recycles permanent (reactant) gases whereas product is condensed and collected. The slurry bubble column reactors can be operated up to 300°C and 100bar.

Figure 2 shows the exit tracer signal of an Argon pulse from the 1L reactor system. It was found that the residence time distribution is a strong function of the carrier flow rate. We postulate that recirculation currents within the solvent

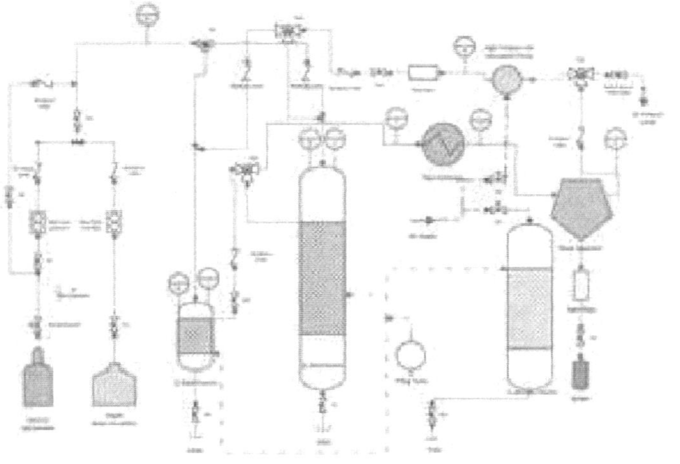

Figure SEQ Figure * ARABIC 1: Reactor System

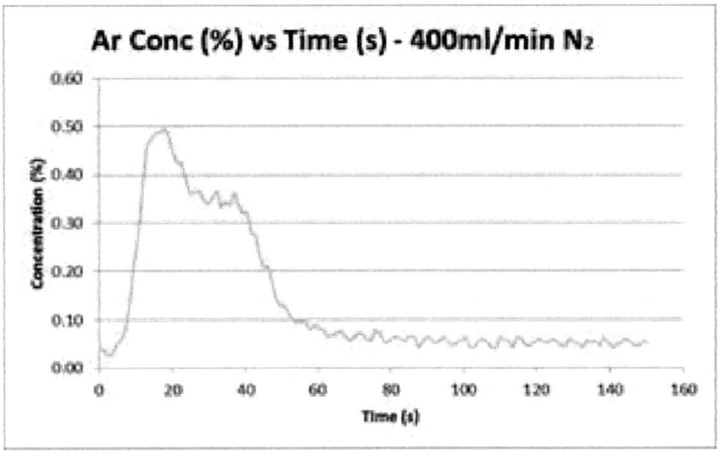

Figure SEQ Figure * ARABIC 2: Exit Tracer Signal of Argon Pulse In response to a 30s Input Pulse (max. concentration = 2.5%)

Copyright © American Institute of Chemical Engineers. All rights reserved.

461717 *Contributions of Heterogeneous Catalysis to Renewable Energy: an Electrochemical Perspective*

Abstract:

Nowadays, research in sustainable energy sources has received considerable attention. Residual biomass is considered a potential source of neutral carbon that can be transformed into biofuels, mainly for transportation and chemicals production, through thermochemical processes for the generation of chemical precursors (ketones, acids, alcohols, furan compounds) which are characterized by high oxygen content [1]. One of the technological issues of lignocellulosic biomass is its structural configuration who has hydrophilic character and high oxygen content, that causes a decrease in the heat capacity and generates corrosion, being necessary to remove oxygen and increase the H/C ratio [2]. An alternative to these problems is use heterogeneous catalysts which are more selective, stable and environmentally friendly, in terms of biomass conversion. In example, HMF and furfural are very versatile chemical precursors of liquid fuels through a series of processes including dehydration, condensation and hydrogenation reactions. The aldol compounds are transformed into a mixture of linear hydrocarbons C_9 - C_{15}, using bifunctional metal catalysts [3]. Bioethanol as fuel in direct fuel cells, which is produced from biomass as a waste product from biorefineries is also a good example of transforming lignocellulosic biomass into chemical energy or electrical energy [4]. The breaking of the C-C is the rate determining step in the mechanism of electrooxidation, being the palladium metals the most active for the oxidation of alcohols under basic metal together with, Sn and Ru. These are only few examples.

In this work the high contributions from the catalysis (in this case Pd, Sn and Ru) for the electrooxidation of bioethanol to obtain electricity is well discussed and supported on experimental data including the synthesis, characterization and the fuel cell evaluation. It is shown how through catalysis it is possible to produce energy at high quality using hydrated ethanol that is cheaper in the market.

References

[1] K. S. Triantafyllidis, A. A. Lappas, M. Stcker. The role of catalysis for the sustainable production of bio-fuels and bio-chemicals. First edition 2013 Elsevier.

[2] P. Mäki-Arvela, T. Salmi, B. Holmbom, S. Willför, D.Y. Murzin, Synthesis of sugars by hydrolysis of hemicelluloses. A review, Chem. Rev. 111 (2011

[3] Sudipta De, Basudeb Saha, Rafael Luque. Hydrodoxigenation processes: Advanced on catalityc transformation of biomass-deriveted plattform chemicals into hydrocarbon fuels. Bioresource Technology. 178 (2015) 108–118

[4]]. R. M. Modibedi, T. Masombuka, and M. K. Mathe, "Carbon supported Pd–Sn and Pd–Ru–Sn nanocatalysts for ethanol electro-oxidation in alkaline medium," International Journal of Hydrogen Energy, vol. 36, no. 8, pp. 4664–4672, Apr. 2011

Copyright © American Institute of Chemical Engineers. All rights reserved.

461492 Kinetic Monte Carlo Simulation ofÂ Propylene Epoxidation with Au/TiO_2/SiO_2 Catalysts

Abstract:

Propylene oxide is currently produced using three different types of commercial processes: (1) chlorohydrin process; (2) propylene oxide â€“ styrene monomer (PO-SM) process; and (3) BASF-Dow hydrogen peroxide (HPPO) process. The first two processes pose significant environmental challenges, due to the production of either chlorinated or peroxycarboxylic acid waste. The BASF-Dow process uses hydrogen peroxide as the oxidant and TS-1 (titanium silicalite-1) as the catalyst. This process allows for a selectivity as high as 95%.Â However, on a molar basis, propylene oxide and hydrogen peroxide have similar market values, and the production of H_2O_2 is associated with environmental risks. Therefore, direct oxidation of propylene by oxygen has received considerable attention. Using supported gold-based nano-catalyst to produce propylene oxide directly from propylene using molecular oxygen as the oxidant provides an alternative, clean, and potentially more efficient route (compared to the current methods). For some time, it has been recognized that Au supported by titanium silicalite (TS-1) zeolite has the ability to catalyze propylene epoxidation. When propylene, oxygen, and hydrogen are co-fed, hydrogen peroxide can be readily formed on the surface of gold, and in turn, selectively oxidizes propylene molecules that are adsorbed on TiO_2 to propylene oxide. Isolated Ti active sites are preferred for high selectivity towards propylene oxide, since propylene oxide molecules that adsorb on adjacent Ti sites lead to catalyst deactivation and the formation of unwanted byproducts. Although the selectivity is very high (>90%), even the best catalysts found to date still suffer from multiple challenges, including low propylene conversion (<10%), poor stability, and inefficient usage of H_2 (<50%). Therefore, significant improvement with regard to these issues is necessary.

We are currently exploring the use of Au-based bimetallic catalysts on oxide supports for direct propylene epoxidation (with H_2 and O_2) in order to achieve higher conversion, while maintaining selectivity. In order to do this, we are

pursuing a combined computational and experimental investigation to clearly understand the structure/performance relationships of the catalyst material. The will allow us to identify the kinetically important steps in the propylene oxide formation mechanism, and consequently, the key step(s) involved in controlling product selectivity, stability, and overall activity.

Based on our on-going experimental work, a comprehensive atomistic model has been developed to quantify the underlying mechanisms which dictate the observed performance characteristics. The model is based on the kinetic Monte Carlo (KMC) simulation technique, which is well-grounded in the energetic and kinetic parameters extracted from DFT calculations. However, the kinetic rate parameters governing the steps in the reaction mechanism differ by several orders-of-magnitude, leading to stiffness in the traditional stochastic KMC approach. Thus, a multi-scale KMC technique is implemented, which is based on a quasi-steady state assumption of the fast system events. This allows rapid system propagation through time, and we are able to reach experimental time scales (while atomistic resolution is still preserved). Thus, we are able to make direction connections between the catalyst details (metal nanoparticle size, catalyst loading, etc.) and operating variables (temperature, partial pressure of gasses, etc.) and the resulting turnover frequency (TOF) of the catalyst for propylene epoxidation.

Copyright © American Institute of Chemical Engineers. All rights reserved.

461825 Comparison of Metal-Support Interaction on Pd/α-Fe_2O_3 Nanorod and Pd/α-$Fe_{2-X}Mn_xO_3$ nanorod Catalysts

Abstract:

Comparison of metal-support interaction on Pd/α-Fe_2O_3 nanorod and Pd/α-$Fe_{2-x}Mn_xO_3$ nanorod catalysts

Chao Wang, Emily Freeman, Jochen Lauterbach*

Department of Chemical Engineering, University of South Carolina, Columbia, SC, 29208, U.S.A.

*lauteraj@cec.sc.edu

Iron oxide supported Pd catalysts have been studied as model systems for the effect of strong metal–support interaction (SMSI) in CO oxidation [1-2]. Such interaction can lead to electron transfer from Pd to Fe_2O_3, oxygen transfer from Fe_2O_3 to reduced Pd during reaction, or formation of encapsulated Pd structures. Doping a secondary metal into a metal oxide support can greatly improve CO oxidation activity, which has been reported for Pd-Ce based system [3], while such effects are rarely investigated in Pd-Fe systems. The metal-support interaction of Pd and doped iron oxide is not yet understood. The surface structure and exposed crystal planes of metal oxide support, on the other hand, can also affect catalytic activity [4]. To gain a better understanding on the SMSI of Pd-Fe system, the interface of Pd and Fe based supports need to be well controlled and are expected to be identical in the catalysts.

In this study, we synthesized Pd supported on α-Fe_2O_3 nanorods and Mn doped α-Fe_2O_3 nanorods with

well controlled morphology and surface structure. CO oxidation was used as a probe reaction to evaluate structure-activity relationships. Pd/α-Fe_2O_3 nanorod shows almost 100% CO conversion at 150°C, but zero conversion at 75°C. With Mn doping into α-Fe_2O_3, Pd/α-$Fe_{2-x}Mn_xO_3$ nanorod shows much higher activity at lower temperature and achieves almost 100% CO conversion at 75°C. Detailed catalysts surface structure will be studied by high resolution transmission electron microscopy (HRTEM) to evaluate exposed facets of Fe and Fe-Mn supports, Pd particle size and structure. Electronic structure and oxidation state of Pd are characterized by X-ray photoelectron spectroscopy (XPS). The low temperature redox property and reaction mechanism are studied by temperature-programmed reduction (TPR), CO-O_2 pulse reaction and in-situ diffuse reflectance IR Fourier transform spectroscopy (DRIFTs) to give a more comprehensive understanding on metal-support interaction.

References

1. L. Liu, F. Zhou, L. Wang, X. Qi, F. Shi, Y. Deng, J. Catal. 274 (2010) 1-10.

2. R. Naumann d'Alnoncourt, M. Friedrich, E. Kunkes, D. Rosenthal, F. Girgsdies, B. Zhang, L. Shao, M. Schuster, M. Behrens, R. Schlögl, J. Catal. 317 (2014) 220–228.

3. C. Wang, E. Sasmaz, C.Wen, J. Lauterbach, Catal. Today 258 (2015) 481–486.

4. C. Wen, D. Dunbar, X. Zhang, J. Lauterbach, J. Hattrick-Simpersm, Chem. Commun. 50 (2014) 4575-4578.

Copyright © American Institute of Chemical Engineers. All rights reserved.

462071 Titanium-Based, Hierarchical, Flow-through Catalyst for Water Electrolysis Fabricated Via a Combination of Additive Manufacturing and Anodization

Abstract:

This work proposes a metal-based catalyst that can potentially overcome the hurdles towards industrial-scale hydrogen production including: (1) high cost of precious metals, (2) simplicity and scalability of fabrication of catalysts and electrodes, and (3) enabling large-scale, continuous hydrogen production.

A titanium-based catalyst was fabricated via combining 3D printing and in-situ anodization in order to grow highly-ordered TiO_2 nanotubes (NTs) onto a microporous 3D Ti-metal base as current collector. The performance of the reported 3D Ti/TiO_2 electrocatalyst was tested for hydrogen-evolution reaction (HER) via water electrolysis.

The proposed electroatalyst features the following advantages: (1) the fabrication method for the 3D Ti/TiO_2 composite is simple and scalable; (2) the 3D Ti/TiO_2 composite features a flow-through architecture that not only enables more efficient utilization of the surface area of the catalyst but that could enable large-scale, continuous-flow hydrogen-production reactor technologies; and (3) minimization of electrical resistance between active surface area (TiO_2 NTs) and current collector.

Copyright © American Institute of Chemical Engineers. All rights reserved.

462290 Co-Oxidation of CO and Hydrocarbons on Pd/Ceria-Zirconia/Al2O3 Three-Way Catalysts

Abstract:

Co-oxidation of CO and Hydrocarbons on Pd/Ceria-Zirconia/Al2O3 Three-way Catalysts

Wendy Lang[1], Michael P. Harold[1*], Yisun Cheng[2], Carolyn Hubbard[2], Manish Sharma[2], and Paul Laing[2]

[1]*Department of Chemical and Biomolecular Engineering, University of Houston, Houston, TX*

[2]*Research and Innovation Center, Ford Motor Company, Dearborn, MI*

*mharold@uh.edu

Three-way catalyst (TWC) light-off performance is critical in cost effectively meeting future emission standards. Improvements in catalyst light-off (LO) should include catalyst modifications that minimize inhibition effects of exhaust species such as CO and hydrocarbons (HCs), lowering precious group metal (PGM) loading, or using a lower cost PGM such as Pd. Alumina (Al_2O_3) and zirconia (ZrO_2) supports provide high surface area and stability. Furthermore, ceria (CeO_2) supports for PGMs provide the key benefit of oxygen storage, compensating for deviations from stoichiometric TWC operating conditions by supplying oxygen for CO and HC oxidation under rich transient conditions. Studies have demonstrated that inclusion of ceria in PGM catalysts enhances CO and HC conversion and minimizes self-inhibition; this improved activity is mainly attributed to ceria containing more efficient oxygen activation sites at the PGM-support interface relative to other support materials such as Al_2O_3 [1-4].

Monolith catalysts having washcoats with 1 wt% Pd, 3 g/in^3 loading, 400 cpsi and formulations Pd/Al_2O_3, Pd/CZO, and Pd/CZO/Al_2O_3 (17 wt% CZO, balance Al_2O_3) were synthesized. Catalyst performance studies were conducted to elucidate the role of Pd and support (CZO, Al_2O_3) compositions and loadings on inhibition during CO and C_3H_6 LO under near-stoichiometric conditions. Transient and steady-state bench scale reactor studies with simulated exhaust gas mixtures were conducted to extract oxidation kinetics, compare LO behavior, and develop a predictive model for understanding and optimizing catalyst performance.

Catalyst activities were compared via temperature-programmed oxidation experiments. Figure 1 compares CO LO curves using feed concentrations of 0.5% CO and 1% CO ($\lambda = 1.01$ in both cases). Using all three catalysts, LO temperatures increase with feed CO concentration, showing that CO oxidation is self-inhibiting. With increasing CZO content, LO temperatures decrease, demonstrating the promotional effects of using ceria.

Figure 2 shows C_3H_6 LO using feed concentrations of 250 ppm C_3H_6 and 500 ppm C_3H_6 ($\lambda = 1.01$). Like, CO, C_3H_6 is self-inhibiting. The shoulder feature in the LO curves using Pd/Al_2O_3 and Pd/CZO/Al_2O_3 is speculated to be due to a change in the oxidation state of Pd and/or the accumulation of inhibiting carbonaceous surface species. LO begins at lower temperatures and the shoulder becomes less prominent with increasing CZO content and is absent with Pd/CZO.

Comparing the individual oxidations of 1% CO and 500 ppm C_3H_6 in Figures 1 and 2 to LO behavior during their co-oxidation (1% CO + 500 ppm C_3H_6) in Figure 3, LO temperatures for both species are higher in the mixture, i.e. CO and C_3H_6 are mutually-inhibiting. During co-oxidation, CO lights off before C_3H_6 and the C_3H_6 shoulder does not appear, suggesting that CO inhibits C_3H_6 adsorption and the accumulation of carbonaceous species on the catalyst. Once CO lights off and temperatures increase, C_3H_6 oxidation proceeds. Significantly lower LO temperatures are achieved using Pd/CZO, and higher conversions of CO and C_3H_6 are reached at lower temperatures using catalysts containing CZO.

Differential kinetics studies were conducted to quantify reaction orders and activation energies. The reaction order with respect to CO was found to be approximately −1 using Pd/Al_2O_3 and $Pd/CZO/Al_2O_3$ and 0 to slightly negative using Pd/CZO, consistent with literature values [5-7]. The activation energies for CO oxidation was found to be 125-149 kJ/mol using Pd/Al_2O_3 and $Pd/CZO/Al_2O_3$, and 29-44 kJ/mol using Pd/CZO. Using the Pd/Al_2O_3 catalyst, the reaction order with respect to C_3H_6 was found to be approximately −1 and activation energy 77 kJ/mol.

Ongoing work involves the use of a mechanistic-based kinetic model incorporated into a low-dimensional monolith model [8-9]. The global reactor models are used with CO and C_3H_6 oxidation on Pd with steps involving ceria. The 1+1 dimensional model formulation comprises coupled species balances with transverse-average equations having external and internal mass transfer coefficients. As shown in Figure 4, the model verifies the experimental LO of CO on Pd/Al_2O_3. For example, it validates additional experimentally-observed oxidation behavior such as the switch in order of ignition between individual and mixture oxidations on Pd/Al_2O_3 and $Pd/CZO/Al_2O_3$ as well as the promotional effect of ceria.

Steady-state and transient CO and C_3H_6 oxidation experiments using fresh Pd/Al_2O_3, Pd/CZO, and $Pd/CZO/Al_2O_3$ monolith catalysts were conducted to compare LO behavior and kinetic parameters (i.e. reaction orders and activation energies). CO and C_3H_6 individual and mixture LO behavior was improved with increasing catalyst ceria content. A reactor model confirms experimentally-observed oxidation behavior and predicts catalyst performance. Finally, experiments involving other hydrocarbons including acetylene and toluene are being conducted to investigate inhibiting during LO, extract kinetic parameters for oxidation reactions, and provide a basis for validating modeling simulations.

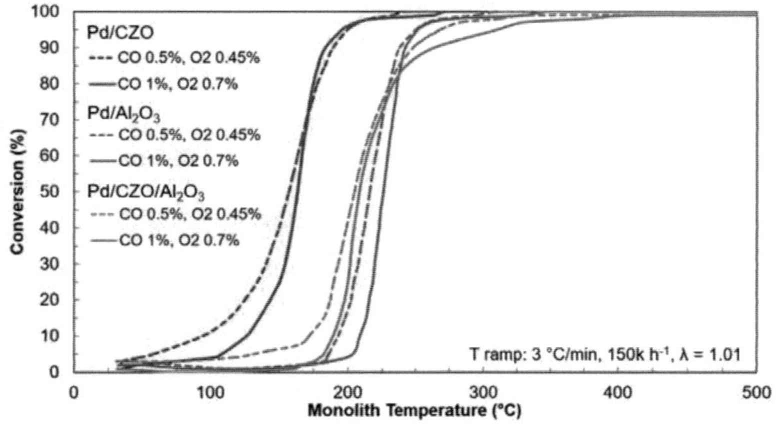

Figure 1: CO conversion (%) versus monolith temperature using fresh Pd/CZO, Pd/Al_2O_3, and $Pd/CZO/Al_2O_3$ catalysts for feed concentrations of 0.5% CO (λ = 1.01) and 1% CO (λ = 1.01).

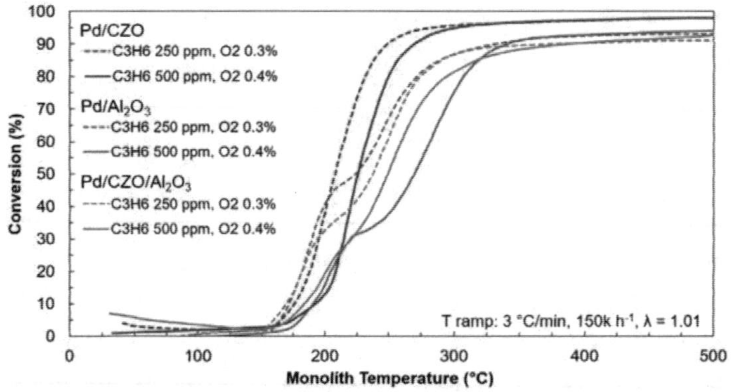

Figure 2: C_3H_6 conversion (%) versus monolith temperature using fresh Pd/CZO, Pd/Al_2O_3, and $Pd/CZO/Al_2O_3$ catalysts for feed concentrations of 250 ppm C_3H_6 (λ = 1.01) and 500 ppm C_3H_6 (λ = 1.01).

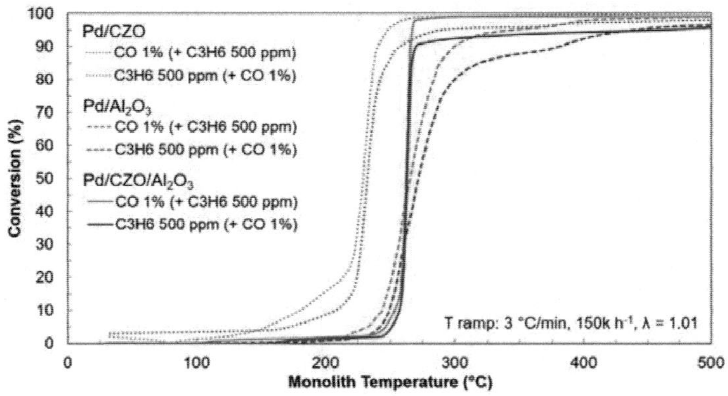

Figure 3: CO, C_3H_6 conversion (%) versus monolith temperature using fresh Pd/CZO, Pd/Al_2O_3, and Pd/CZO/Al_2O_3 catalysts for feed concentration of 500 ppm C_3H_6, 1% CO (λ = 1.01).

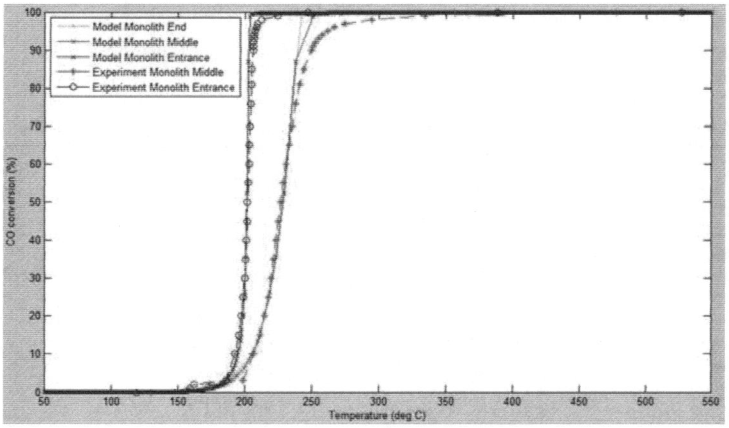

Figure 4: CO conversion (%) versus temperature using fresh Pd/Al_2O_3 catalysts for experimental and simulated feed concentration of 1% CO (λ = 1.01).

References

[1] Fernandez-Garcia, M.; Martinez-Arias, A.; Iglesias-Juez, A.; Hungria, A.B.; Anderson, J.A.; Conesa, J.C.; and Soria, J. *App. Cat. B* 2001, **31**, 39-50.

[2] Martinez-Arias, A., Fernandez-Garcia, M.; Iglesias-Juez, A.; Hungria, A.B.; Anderson, J.A.; Conesa, J.C.; and Soria, J.. *App. Cat. B* 2001, **31**, 51-60.

[3] Sharma, S.; Hegde, M.S.; Das, R.N.; and Pandey, M. *App. Cat. A* 2008, **337**, 130-137.

[4] Harmsen, J.M.A.; Hoebink, H.B.J.; and Shouten, J.C. *Chem. Eng. Sci.* 2001, **56**, 2019-2035.

[5] Yao, Y-.F. Y. *J. Catal.*, 1984, **87**, 152-162.

[6] Cant, N.W.; Hicks, P.C.; and Lennon, B.S. *J. Catal.* 1978, **54**, 372-383.

[7] Rainer, D.R.; Koranne, M.; Vesecky, M.; and Goodman, D.W. *J. Phys. Chem. B* 1997, **101**, 10769-10774.

[8] Joshi, S.; Harold, M.P.; Balakotaiah, V. *AIChE J.,* 2009, **55,** 1771-1783.

[9] Raj, R.; Harold, M.P.; Balakotaiah, V. *Chem. Eng. J.*, 2015, **281**, 322-333.

Copyright © American Institute of Chemical Engineers. All rights reserved.

462316 Understanding Gas Evolution Dynamics on Planar Photoelectrodes

Abstract:

Gas bubbles evolved at the surface of an electrode can result in significant loss in efficiency due to increasing surface coverage of the active material by gas bubbles. Therefore, it becomes crucial to comprehensively understand gas evolution dynamics in these systems so that operating conditions may be chosen and electrodes engineered to minimize the corresponding losses in efficiencies. Scanning photocurrent microscopy (SPCM) can be used to determine the change in photocurrent response of an area of interest and provides a powerful tool for quantifying photocurrent losses associated with individual bubbles. In this study, we employ SPCM as a tool for deconvoluting the optical influence of bubbles from their influence on ohmic- and kinetic losses of a photoelectrode surface. As a basis for this work, Si-based photocathodes are studied, and SPCM is used to systematically investigate the influence of varying bubble size on the performance of these photoelectrodes under varied conditions. This study provides much needed insight into the dynamics of gas bubbles on photoelectrode surfaces and sets the stage for quantitatively modeling bubble-related losses for various photoelectrode geometries and operating conditions.

Copyright © American Institute of Chemical Engineers. All rights reserved.

462164 Multiple H_2O and NH_3 Adsorbed on Cu-SSZ-13: XANES and XES Study from First Principles

Abstract:

Nitrogen oxides (NOx) are one of the main air pollutants present in the exhaust from diesel engines, which are popular for vehicle transportation due to their efficiency and durability [1]. However, NOx emission control is a challenge in these "lean-burn" engines [2]. As is well known, the selective catalytic reduction (SCR) of NOx with NH3 is a reaction between the NO, NO2, and O2 oxidants and the NH3 reductant to form N2 and H2O. Copper-exchanged small-pore micro-crystalline materials with the chabazite structure (Cu/CHA), such as Cu-SSZ-13, display excellent catalytic activity and hydrothermal stability in SCR of NOx, as has been shown in a number of recent studies [3, 4]. Although Cu-SSZ-13 have been commercialized as diesel after-treatment catalysts, the fundamental chemical and physical properties need to be characterized in order to aid in the design of new and better catalysts [5].

X-ray absorption spectroscopy (XAS) is a versatile tool to determine the oxidation state and the local structure of Cu in Cu/CHA. Cu K-edge X-ray absorption near edge spectra (XANES) has been widely used to study the properties of Cu-SSZ-13 because of its high activity and selectivity in the NH3 SCR of NOx. It is demonstrated that XANES is extremely sensitive to the local structure of Cu ions [6, 7]. However, the exact relationship between XANES features and local structure is still unclear. In this work, we examine how the geometry of a Cu+ ion with a linear configuration and Cu2+ ion with a square planar configuration in Cu-SSZ-13 correlate with the XANES features. When water and ammonia binds to a Cu+ ion with a linear configuration, a strong intensity peak around 8983 eV in Cu K-edge XANES appears. When two ammonia molecules bind at a Cu+ ion, it is also found that the intensity of the peak around 8983 eV decreases as one decreases the N-Cu-N bond angle from 180 to 100 degree (see Figure 1). By analyzing the corresponding PDOS, it is found that the splitting of 4p state correlates with the N-Cu-N bond angle and the corresponding decrease in the K-edge intensity. Concerning the case of Cu2+ ions with a square planar configuration bounded to four ammonia molecules, it is found that there are two feature peaks around 8986 and 8993 eV in their Cu K-edge XANES. By studying this simple model, we find that changing the N-Cu-N angle between one of the bonded NH3 and the Cu-centered ion from 180 to 100 degree results in a decrease in the intensity of both peaks around 8986 and 8993 eV. It is also found that the broken of square planar configuration results in the delocalization of 4p state.

An important limitation of XANES is that it is difficult to distinguish the nitrogen and oxygen atoms in the proximity of the metal center. X-ray emission spectroscopy (XES) could overcome this limitation. It is reported that the Kbeta" peak in XES of a Cu-SSZ-13 sample presents a blue shift with the formation of Cu-N bond [8]. This can be understood by the fact that the Kbeta" peak originates from the ligand 2s to a metal 1s crossover transition and the nitrogen 2s level lies higher in energy than the one for oxygen. By comparing the XES results between H2O and NH3 adsorbed on Cu-SSZ-13, it is found that the Kbeta" peak of Cu-O bond appears at lower energy than that of Cu-N bond. In addition, the coordination number of Cu ions significantly affects on the energy difference between the position of the Kbeta" peak. By performing a density of states (DOS) analysis, these results can be understood at the electronic level. Overall, our careful elucidation of the spectroscopic properties of Cu ions in Cu-SSZ-13 via theoretical techniques can assist us in interpreting the XANES experimental results so as to gain a deeper understanding of the SCR of NOx with NH3 throughout its catalytic cycle.

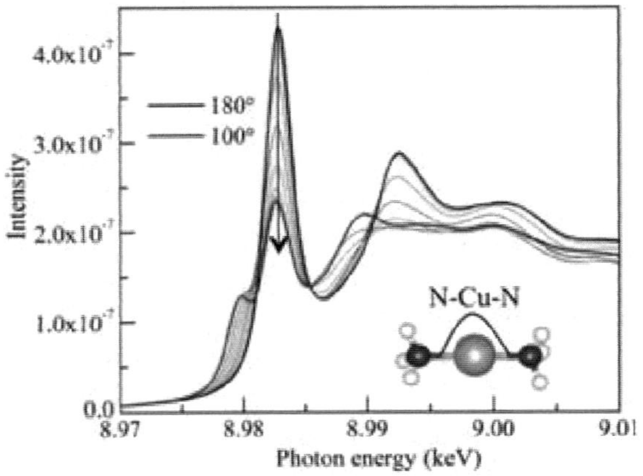

Figure 1. Variation of Cu K-edge XANES for 2 NH3 adsorbed on Cu+-SSZ-13 with changing the angle of N-Cu-N (as shown insert in the right bottom) from 180 degree (black line) to 100 degree (blue line) with decrement of 10 degree.

References

[1] U. Deka, I. Lezcano-Gonzalez, B.M. Weckhuysen, A.M. Beale, Local Environment and Nature of Cu Active Sites in Zeolite-Based Catalysts for the Selective Catalytic Reduction of NOx, ACS Catal., 3 (2013) 413-427.

[2] S.i. Matsumoto, Catalytic Reduction of Nitrogen Oxides in Automotive Exhaust Containing Excess Oxygen by NOx Storage-Reduction Catalyst, CATTECH, 4 (2000) 102-109.

[3] J.H. Kwak, R.G. Tonkyn, D.H. Kim, J. Szanyi, C.H. Peden, Excellent activity and selectivity of Cu-SSZ-13 in the selective catalytic reduction of NOx with NH3, J. Catal., (2010) 1-4.

[4] L. Ma, Y. Cheng, G. Cavataio, R.W. McCabe, L. Fu, J. Li, Characterization of commercial Cu-SSZ-13 and Cu-SAPO-34 catalysts with hydrothermal treatment for NH3-SCR of NOx in diesel exhaust, Chem. Engin. J., 225 (2013) 323-330.

[5] F. Gao, J. Kwak, J. Szanyi, C.F. Peden, Current Understanding of Cu-Exchanged Chabazite Molecular Sieves for Use as Commercial Diesel Engine DeNOx Catalysts, Top. Catal., 56 (2013) 1441-1459.

[6] R. Zhang, J. Szanyi, F. Gao, J.-S. McEwen, The interaction of reactants, intermediates and products with Cu ions in Cu-SSZ-13 NH3 SCR catalysts: An energetic and ab initio X-ray absorption modeling study, Catal. Sci. Technol., 2016, DOI: 10.1039/C5CY02252E.

[7] R. Zhang, K. Helling and J.-S. McEwen, Ab initio X-ray absorption modeling of Cu-SAPO-34: Characterization of Cu exchange sites under different conditions, Catal. Today, 267 (2016) 28-40.

[8] F. Giordanino, E. Borfecchia, K.A. Lomachenko, A. Lazzarini, G. Agostini, E. Gallo, A.V. Soldatov, P. Beato, S. Bordiga and C. Lamberti, Interaction of NH3 with Cu-SSZ-13 Catalyst: A Complementary FTIR, XANES, and XES Study, J. Phys. Chem. Lett., 5 (2014) 1552-1559

Copyright © American Institute of Chemical Engineers. All rights reserved.

462570 Catalyst Morphology Engineering: Towards a Better Understanding of the Effects of Surface Structure and Mass Transport in Copper Electrodes for the Electrochemical CO_2 Reduction Reaction

Abstract:

The modern global energy economy and chemical industry are both heavily dependent on fossil resources, but with rising concerns over the depletion of these resources and their negative impacts on the environment, there has been a strong push for the development of sustainable, renewable alternatives. One potential solution is the electrochemical reduction of carbon dioxide (CO_2) to fuels and chemicals. However, the CO_2 reduction reaction (CO_2RR) is notoriously difficult to catalyze. To date, only copper has shown a propensity to produce high value products such as ethylene and ethanol, but copper still requires large overpotentials (excess energy input) to reduce CO_2 and generates a variety of products requiring separation.[1,2]

Recent literature reports have shown that modifying catalyst morphology can be a path towards improved efficiency and selectivity for the CO_2 reduction reaction.[3,4] However, the origin of these improvements is not well understood since nanostructured morphologies can alter both the crystal surface structure and the local mass transport. While single crystal studies have clearly shown that the CO_2RR activity and selectivity are facetâ€"dependent[1], the influence of transport phenomena has only been characterized to a limited degree.[5] It is thought though that in mass transportâ €"limited systems, accumulation of interfacial OH^- generated during the CO_2RR can lead to a higher local pH, a phenomenon which has been attributed to increased selectivity towards products such as ethylene and ethanol.[4,6] Due to the random, disordered nature of nanostructured catalysts that have been previously reported, it has been difficult to

decouple these two potential impacts of catalyst morphology. In this work, catalyst morphology engineering is used in an attempt to isolate and explore the effects of mass transport/pH on the CO_2reduction reaction. There are two main methods being used: nanostructuring the catalyst itself and depositing an inert, porous layer onto a planar catalyst. The first method will leverage the unique processability of silicon to create highly ordered vertical nanowire arrays with tunable control over the nanowire length, diameter, and pitch, which will then be coated with copper, allowing us to examine the impacts of the nanowire dimensions and surface area. The second method will utilize mesoporous oxides and membranes to create a diffusion barrier layer on top of copper, perturbing the local mass transport to the catalyst surface.

References:

[1] Y. Hori, *Modern Aspects of Electrochemistry*. **2008**, 89

[2] K. Kuhl et al., *Energy Environ. Sci.* **2012**, 5, 7050

[3] C.W. Li et al., *J. Am. Chem. Soc.* **2012**, 134, 7231

[4] D. Ren et al., *ACS Catalysis*. **2015**, 5, 2814

[5] M. Singh et al., *Phys. Chem. Chem. Phys.* **2015**, 17, 18924

[6] R. Kas et al., *ChemElectroChem*. **2015**, 2, 354

Copyright © American Institute of Chemical Engineers. All rights reserved.

462671 A Parametric Study of Biodiesel Production Under Ultrasounds

Abstract:

Biodiesel, a vegetable oil-derived fuel, can be used as a partial or complete substitute to diesel oil. The main argument for its usage in internal combustion engines is its net CO_2 balance which is considerably reduced compared to diesel fuel of fossil origin. A systematic study of ultrasound continuous biodiesel production using canola oil was conducted in the presence of methanol and sodium methoxide as catalyst. Effects of various reaction parameters such as residence time, catalyst concentration, reaction temperature, ultrasounds amplitude and power, methanol/oil molar ratio were analyzed. Fatty acid methyl esters were produced rapidly by using ultrasound assisted transesterification. In typical conditions (35°C) conversion to FAME higher than 80% could be reached at residence time as low as 20s. The parametric study allowed to establish that the effect of ultrasound wave on transesterification reaction rate is localized in a very small volume surrounding the sonotrode tip. This unprecedented conclusion has significant consequences for the design of the large scale continuous flow biodiesel production reactor.

Copyright © American Institute of Chemical Engineers. All rights reserved.

462943 Data Auditing: Analysis Techniques for the Compatibility of Experimental Data for Molecular-Level Kinetic Models

Abstract:

Data Auditing: Analysis techniques for the compatibility of experimental data for molecular-level kinetic models

Scott R. Horton*, Triveni Billa, Pratyush Agarwal, Juan Lucio, and Michael T. Klein

Department of Chemical & Biomolecular Engineering, University of Delaware

Mayuresh Sahasrabudhe and Chandra Saravanan (Saru)

Reliance Industries

In modeling industrial reactors, vast quantities of data are often available for years of reactor operation. These data allow for the evaluation and modeling of reactor performance through catalyst deactivation and feedstock variation. The sheer quantity of data also gives the motivation for automated methods of analysis of the data and their utility in detailed kinetic models. In this work, automated methods are developed and applied to five years of industrial data on hydrocarbon processing to select data sets for model evaluation and parameter optimization.

A software tool has been developed to perform an â€˜auditâ€™ of the data. First, the software searches for incompatibilities and outliers within the entire set of data. This allows for the identification of erroneous data entry or equipment failure. For models where solution time is an issue, the user might want to tune parameters on a sub-set of the entire data sets. To this end, a second feature of the app is the ability to find representative data sets within the full set of data. To select these data sets, the algorithm analyzes the measurable inputs to find data sets which best span the range of possible experimental values. This allows the user to tune kinetic parameters to the full range of possible data with a limited number of data sets.

The final set of features in this app analyze the compatibility of the experimental data with the reaction network in the molecular-level kinetic model. For instance, this app, without kinetic parameter tuning, will inform the user if the output measurements are possible given the input measurements and reaction network. In the case where the analysis returns red flags, the user can then analyze the structure of the reaction network and the validity of the experimental measurements. Because this analysis is performed before tuning, incompatibilities of data and reaction network can be discovered and remedied in moments rather than days-weeks.

Copyright © American Institute of Chemical Engineers. All rights reserved.

463009 Mn@Ptn (n = 1, 2, and 3) Core-Shell Nanoparticles: Synthesis, Characterization and Electrocatalysis

Abstract:

We report the synthesis of Mn@Ptn (n = 1, 2, 3) core-shell nanoparticles (NPs) and their electrocatalytic properties. The syntheses are achieved by one-pot sonochemical reactions between Pt(acac)2 and Mn(acac)2 in ethylene glycol. The shell-thickness of the NPs can be controlled from 1 monolayer (ML) to 2 ML through variation of the composition of reacting solution. X-ray diffractometry, transmission electron microscope (TEM), and energy dispersive X-ray spectroscopy (EDS), and inductively couple plasma-absorption emission spectroscopy are used for the analyses of the crystal structure, particle size, morphology, and composition, respectively. The formation of core-shell NPs is confirmed by EDS line profiles coupled with scanning TEM and the electronic states of Pt are analyzed by X-ray photoelectron spectroscopy. The NPs are investigated as electrocatalysts in oxygen reduction reaction (ORR). The linear sweep voltammograms of Mn@Ptn catalysts show unprecedentedly high activity in ORR with the mass activity 14-16 times larger than that of pure Pt NP. Within the Mn@Ptn series, the activity increases as the Pt shell thickness increases. This trend and theoretical calculations on the binding energies of H or O atom on Pt/Mn model structures suggest that the ORR on Mn@Ptn electrocatalysts follows the associative mechanism. The durability test data, measured by ORR activity after electrochemical cycling, show that Mn@Ptn samples have improved retention of kinetics from that of Pt. Especially, the Mn@Pt3 sample with 2 ML thick Pt-shells shows remarkably high stability compared with the other two Mn@Ptn samples as well as pure Pt catalyst

Copyright © American Institute of Chemical Engineers. All rights reserved.

463460 Modeling and Simulation of the Transient Reaction Behavior in Reactive Sorption Enhanced Reforming of Methane with Nano CaO-Based Sorbents for Hydrogen Production

Abstract:

Modeling and Simulation of the Transient Reaction Behavior in Reactive Sorption Enhanced Reforming of Methane with Nano CaO-based Sorbents for Hydrogen Production

Haoliang Ping, Sufang Wu*

College of Chemical and Biological Engineering, Zhejiang University, Hangzhou 310027, Zhejiang, P. R. China

* Corresponding author E-mail: wsf@zju.edu.cn

Abstract

Reactive sorption enhanced reforming (ReSER) process employing the nano CaO-based sorbents for in situ CO_2 sorption offered a promising high purity hydrogen production technique with a single-reactor.

However, the coupling of steam methane reforming reaction, water shift reaction and CO_2 sorption reaction in one reactor makes it much more difficult to simulate the concentrations of components and temperature distribution along the axial of the reactor exactly. [1-2] The key issue is the synergy effect between steam methane reforming (SMR) reaction rate and CaO carbonation rate in the reactor. In order to investigate the above mentioned synergy effect, a two-dimensional axisymmetric pseudo-homogeneous model was set up to describe the transient performance of the

reactions within a fixed bed. The mathematic model was numerically solved by COMSOL software with the finite element methods (FEM) to obtain transient methane conversion, concentration of each outlet gas compositions and temperature profile change with time regards the variation of the rate constant of CaO-based sorbents carbonation (k_{carb}), CO_2 sorption rate and SMR reactions rate.

The simulation results demonstrated that a methane conversion of 96% and a purity of 98% of hydrogen yield (on dry basis) could be obtained by ReSER process, while the methane conversion was 76% and the hydrogen purity was only 74% via steam methane reforming (SMR) process without sorption enhancement under the same reaction conditions. A comparison between the predicted and the experimental data was made and the average relative deviation was 3.86%, which means the model accurately depicted the ReSER process and the simulation results were reliable.

Moreover, the simulation results demonstrated that the sorption rate of the CaO-based sorbents played a decisive role in the overall efficiency of ReSER process because an enhanced CO_2 sorption rate would promote the rate of SMR reactions and then resulted in a higher methane conversion as well as the produced hydrogen purity. The methane conversion with nano-CaO sorbent (with k_{carb}=0.35 s^{-1}) was 5% higher than the one with limestone (with k_{carb} = 0.5 s^{-1}) and the yield hydrogen purity was enhanced by 3%. Therefore, sorbents with faster CO_2 sorption rate resulted in an enhanced CH_4 conversion and a higher hydrogen purity. An optimum reaction temperature of 600 ¡ãC could assure a hydrogen purity of 98%. Alternatively, the purity of produced hydrogen declines form 96% to 89% with the increase of the reaction pressure from 1bar to 5bar due to the suppressed CH_4 conversion.

Keywords: reactive sorption enhanced reforming, hydrogen, CO_2, CaO-based sorbent, simulation

References:

[1] J. Xu and G. F. Froment, AIChE Journal, 1989, 35, 97-103.

[2] D. K. Lee, I. H. Baek and W. L. Yoon, Chemical Engineering Science, 2004, 59, 931-942.

Copyright © American Institute of Chemical Engineers. All rights reserved.

463590 Transition Metal Modified Zirconium Phosphates for the Oxygen Evolution Reaction

Abstract:

Transition Metal Modified Zirconium Phosphates for the Oxygen Evolution Reaction

Joel Sanchez+, Ieva Narkeviciute+, Mario Ramos++, Jorge Colón++, Thomas F. Jaramillo+

+*Department of Chemical Engineering, Stanford University, Stanford, California, 94305*

++*Department of Chemistry, University of Puerto Rico, Río Piedras, Puerto Rico 00931*

A sustainable energy supply in the future will depend on innovative breakthroughs regarding the design of efficient, reliable, and inexpensive systems that allow for the conversion and storage of renewable energy sources. Water electrolysis provides a suitable method to generate hydrogen and oxygen gas with the scalable potential to store intermittent renewable energy needed to power a sustainable energy grid. Converging towards this reality, the efficiency of electrolyzers must be optimized through the development of highly active and stable catalysts in order to reduce the kinetic overpotenial losses associated with driving the oxygen evolution reaction (OER).[1] Currently, due to acid-aggressive and strong corrosive conditions in which OER takes place, the only active and stable catalyst are precious metals (IrO_2 and RuO_2) and the implementation of acid stable non-precious metal catalyst is unknown.

Recent theory has shown that confining catalyst materials within nanoscale dimensions can lead to improvements for the OER by introducing reaction channels that break scaling relations through increased selectivity. Therefore, in this study, we focus on leveraging the versatility of nanomaterials by taking advantage of the layered structure of Zirconium Phosphate (ZrP).[2] ZrP is an acidic and inorganic cation exchange material which is stable at low pH and high temperatures with the ability to intercalate molecular species with dimensions up to 24 Å.[3,4] In base, some of the most active and stable catalysts consist of Fe, Ni, Co, their oxides, or mixtures thereof; in this study we investigate ZrP modified with a range of transition metal oxides known to catalyze the OER.[5] The modified ZrP materials are characterized through X-ray powder diffraction, X-ray photoelectron spectroscopy, and scanning electron microscopy coupled with energy dispersive X-ray spectroscopy and then further tested for OER activity.

References:

1. 1. Burke, M. S., et al. *Chemistry of Materials*, **2005**, *27*(22), 7549â€"7558.
2. 2. Doyle, A., et al. *ChemCatChem*, **2015**, 7, 738-742.
3. 3. Santiago, M.E.B, et al. *Langmuir*, **2007**, 23, 7810-7817.
4. 4. Santiago, M.E.B, et al. *Langmuir*, **2012**, 28, 4447-4452.
5. 5. McCrory, C.C.L, et al. *Journal of the American Chemical Society*, **2015**, 137, 4347-4357.

Copyright © American Institute of Chemical Engineers. All rights reserved.

463595 Methanol Oxidation on Pristine and Doped MoO_3 (010): A DFT and Microkinetic Analysis

Abstract:

Methanol oxidation is employed as a probe reaction to rigorously evaluate catalytic properties of molybdenum trioxide (MoO_3), a reducible oxide that exhibits a fascinating interplay of catalytic chemistry and structural transformations. As a first step towards this goal, the catalytic cycle is studied on the (010) basal planes, which exist during the early stages of methanol oxidation. The reaction mechanism is investigated with electronic structure calculations using the BEEF-vdW and HSE06 functionals along with mean field microkinetic modeling. Considered pathways include vacancy formation and oxidation, monomolecular dehydrogenation steps on reduced and non-reduced surfaces, bimolecular reactions between dehydrogenated intermediates, and precursor steps that form hydrogen molybdenum bronze phase (H_yMoO_{3-x}). Methanol dissociation begins with C-H or O-H scission, leading to two distinct pathways, with the O-H route found to be kinetically and thermodynamically more favorable. Dehydrogenation of CH_2O* to CHO* is kinetically slower as compared with desorption (by a factor of $1.4x10^5$), leading to complete selectivity towards CH_2O. The most favorable pathway has three kinetically significant elementary steps: (1) C â€" H scission of CH_3O*, (2) recombination of dissociated OH* to form H_2O*, and (3) dissociative adsorption of $O_{2Â}$ which oxidizes the surface. (1) and (2) promote methanol oxidation, while (3), oxidation of the reduced surface is rate inhibiting. Reaction orders with respect to gas phase methanol and oxygen are 1.0 and -0.5 respectively.

The surface state of a reducible oxide can have a direct impact on Mars van Krevelen-type catalysis. Thus, the microkinetic analysis is applied towards understanding the relationship between catalysis and surface reducibility by varying the oxygen partial pressure. As a consequence of too few vacancies limiting CH_2O and vacancy formation at atmospheric pressure, and too low O_2* coverage limiting surface oxidation in Ultra High Vacuum (UHV), the reaction

rate follows a volcano dependency across a range of oxygen pressure, with the optimum rate located where surface oxidation is neither limiting nor inhibiting the overall rate. Thus, at the volcano maximum, the reaction order with respect to gas phase oxygen is close to zero. Finally, a Constable â€" Cremer relationship (compensation effect) between the apparent prefactor and apparent activation energy is observed as the mechanism changes around the top of the volcano.

A simplified analytical form of the rate expression, which is valid for a wide range of oxygen pressure, is proposed in this contribution. The kinetic and thermodynamic constants in this expression can be tuned through geometric (strain, coordination environment) and electronic modifications (substitutional doping) of the (010) surface of MoOÂ suggesting strategies to enhance catalytic activity of the basal planes of MoO_3. The DFT and microkinetic modeling approach demonstrated here is generally suitable for analysis of catalytic mechanisms on reducible oxide surfaces.

Copyright © American Institute of Chemical Engineers. All rights reserved.

463604 Simultaneous Reaction Identification and Parameter Estimation

Abstract:

Understanding the chemical mechanism through which a reaction occurs is crucial to the design, optimization, and control of a reactive system. Elucidating a possible reaction path, and identifying the associated kinetic rate parameters, routinely relies on the expertise of an individual to identify possible reaction pathways and intermediate species. A curve-fitting procedure is then used to identify kinetic rate parameters, and quantify uncertainty in their estimation [1]. Consideration of alternative reactions or mechanisms is then performed in an exhaustive fashion.

We present an optimization-based approach that utilizes a mixed-integer non-linear programming (MINLP) formulation to identify a set of reactions that describe the chemistry observed in a steady-state continuous stirred-tank reactor, and estimate the associated kinetic rate parameters. This approach utilizes a superset of chemical reactions, which can contain different mechanistic pathways for the same reaction. A curve-fitting procedure, similar to one described previously for subset selection in multiple linear regression [2], is then used to minimize an objective that balances the complexity of the model with its fit in order to identify a probable subset of reactions. The utility of this methodology is demonstrated on a number of case studies.

References

[1] Warren E. Stewart and Michael Caracotsios. 2008. Computer-Aided Modeling of Reactive Systems. Wiley-Interscience, New York, NY, USA.

[2] Cozad, A., N. V. Sahinidis, and D. C. Miller, Automatic learning of algebraic models for optimization, AIChE Journal, **60**, 2211-2227, 2014.

Copyright © American Institute of Chemical Engineers. All rights reserved.

463649 Design and Synthesis of Porous Carbon-Based Catalysts for the Electrochemical Reduction of O_2 Via Either 2 e^- or 4 e^- Route

Abstract:

To improve the cost-effectiveness, efficiency and scalability, high performance catalysts made from low cost, earth-abundant materials need to be developed. Carbon is a unique element in its ability to provide a vast range of bonding environments for heteroatoms. Carbon-based catalysts also show wide tunability, such as degree of graphitization, ultra-high surface area, adjustable pore volume and pore size distribution. [1] This leads to a tremendous flexibility for synthesizing materials with catalytic active sites of many possible structural and electronic configurations that are necessary for different electrochemical reactions. The ability to precisely control carbon structure and composition is crucial in achieving high performance carbon catalysts for any desirable application.

In this work, we explored new synthetic methods that provide high versatility in tuning carbon structure and chemical composition. In this approach, a rigid conjugated polymer network is used as the carbon precursor, which prevents pore collapsing during carbonization while allowing incorporation of various heteroatoms, such as N, B and metal ions. This facile synthetic approach allows us to synthesize the target promising structures with optimized surface motifs to catalyze reactions with targeted efficiency, stability, and selectivity, predicted by theory. In particular, the use of these porous carbon materials as electrocatalysts for oxygen reduction reaction (ORR) is explored. O_2 can be either reduced to H_2O via a 4 e- route with potential application in fuel cells or to H_2O_2, which is one of the most important industrial chemicals in the world, via a 2 e- route. [2-3] Density functional theory (DFT) modeling shows that that defective carbons with various heteroatom doping show different catalytic activity towards ORR. With the guidance of theory, we are able to use our approach to control carbon structure and composition to

synthesize promising carbon catalysts. With various techniques, such as X-ray absorption spectroscopy, gas sorption, high resolution TEM, SEM and XPS, we are able to systematically characterize our carbon materials and this allows us to achieve clear understanding of the structure property relationship and design carbon materials with high electrocatalytic performance.

[1] L. Dai, Y. Xue, L. Qu, H.-J. Choi and J.-B. Baek, *Chemical reviews*, 2015, **115**, 4823-4892.

[2] K. N. Wood, R. O'Hayre and S. Pylypenko, *Energy & Environmental Science*, 2014, **7**, 1212-1249.

[3] J. M. Campos Martin, G. Blanco Brieva and J. L. Fierro, *Angewandte Chemie International Edition*, 2006, **45**, 6962-6984.

Copyright © American Institute of Chemical Engineers. All rights reserved.

463845 Numerical Investigation on Dispersion Phenomena in Packed Bed Reactors with Low Tube-to-Particle Diameter Ratio

Abstract:

Numerical investigation on dispersion phenomena in packed bed reactors with low tube-to-particle diameter ratio

Nico Jurtz[1], Gregor Wehinger[1], Thomas Eppinger[2], Matthias Kraume[1]

[1]Chair Chemical and Process Engineering, TU Berlin, Berlin, Germany;

[2]CD-adapco, Nuremberg, Germany;

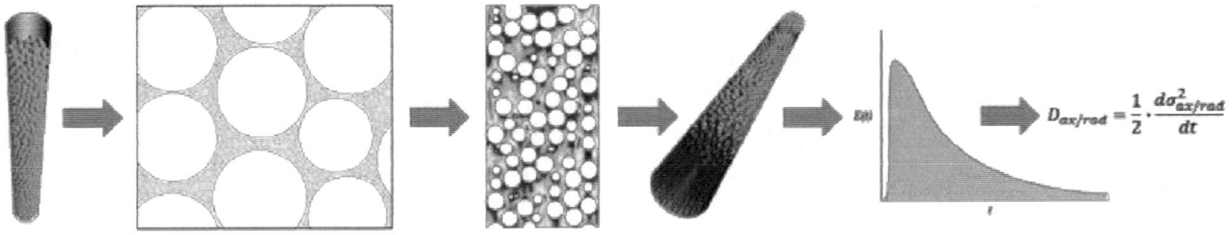

Abstract

Packed bed reactors (PBR) are widely used in the chemical and process industry. They play a key role in the heterogeneous catalysis or in novel process concepts like the chemical looping. A quantity to characterize the mass transfer in PBR's is the dispersion coefficient. But correlation to determine the dispersion coefficient fail for reactors with a low tube-to-particle diameter ratio due to local wall effects that play an important role (Delgado, 2006).

The flow field of that special reactor configuration can be determined by the use of spatially resolved CFD-simulations using the DEM-CFD coupled approach from (Eppinger, 2011). Based on the calculated flow field a numerical residence time analysis is conducted using the Lagrangian parcel approach. Compared to other passive scalar based approaches this method is beneficial in terms of accuracy and calculation time. The dispersion coefficient can subsequently be calculated based on the simulated residence time distribution function (RTD).

Validation studies show that the presented method allows an accurate and efficient prediction of the RTD. Further simulations show the effect of diameter ratio, Reynolds number and particle shape on the RTD and the dispersion coefficient. Effects like channeling, stagnant zones and back mixing will be detected by analyzing the shape of the RTD and localized by evaluating the calculated flow field.

References

Delgado, J., 2006. A critical review of dispersion in packed beds. Heat Mass Transfer 42, 279-310

Eppinger, T., Seidler, K., Kraume, M., 2011. DEM-CFD simulations of fixed bed reac-tors with small tube to particle diameter ratios. Chem. Eng. J. 166, 324-331.

Copyright © American Institute of Chemical Engineers. All rights reserved.

463785 Modeling Syngas Production for Miniaturized Gas-to-Liquids Application: Extending Operation Window By Process Improvement from Different Scales

Abstract:

The scale-down of gas-to-liquids (GTL) has practical significance in developing renewable energy systems and mitigating green-house gas (GHG) emissions. Examples include on-site conversion of associated gas to liquid fuels to prevent flaring at remote onshore/offshore oil fields, and utilization of biogas/biomass or other domestic waste to produce valuable chemicals in distributed plants. Economy of such small-scale applications can benefit from their use of cheap, "waste" feedstock, but is very likely restricted by the device investment. This dictates the reactor operation at high space velocity with respect to reactor volume, resulting in a non-equilibrium operating regime, which is distinct from large-scale processes.

Herein, we focus on miniaturized GTL using modular microchannel reactors for associated gas utilization. Both two major stages of the process, syngas production and the following Fischer-Tropsch synthesis, remain challenging during scale-down due to the intrinsic complexity of the multi-scale transport and heterogeneous reactions. The problems involved can be characterized as intrinsic chemistry/physics on the micro-scale, internal mass transfer on the meso-scale and external heat and mass transport and pressure drop control on the macro-scale [1]. We modeled catalytic combustion assisted syngas production in an integrated microchannel reactor. The aim is to illustrate the combined effects of process improvement from an integral perspective of the entire process, as well as to highlight co-optimization of the reactor and catalyst of such systems.

To account for the effects of multiscale improvement measures, we developed a hybrid modeling scheme that decoupled the governing equations at different scales in order to save computational time in parametric study. The macro-scale phenomena were treated in conventional CFD framework and coupled to the lower scales. On the meso-scale, the catalyst washcoat was modelled as isotropic porous media. The effective diffusivity was calculated by different models specific to the pore size distribution. We derived a meso-scale effectiveness factor submodel to describe internal mass transfer limitation under typical reaction conditions. On the micro-scale, detailed surface chemistry for methane reforming over Ni [2] and catalytic combustion over Pt [3] were used, which can be related to the metal dispersion of catalysts.

Numerical simulations on the full, heat-coupled processes in a modular microchannel reactor disclosed the reactor operation maps by varying fuel flowrate for different reactant GHSVs. It is found that the volumetric CO yield increases with GHSV before fuel breakthrough (incomplete conversion), which almost never happens because of the ultra-fast intrinsic kinetics. In contrast, GHG emission rate slowly increases with GHSV. For a specific GHSV, the GHG emission rate has a minimum ranging from 70% to 80% when the fuel flowrate is adjusted to some medium value. These observations suggest that extending operable GHSV and enhancing GHG emission control should be the primary optimization targets for process improvement from an economic point of view. At low GHSV, the reactor has a wide operation window with respect to material stability and methane conversion. With increasing GHSV, however, the operation window narrows down quickly, mainly due to hotspot formation, which prohibits further possible increase of GHSV. Multiscale measures prove efficient to enhance the thermal uniformity, for instance, by increasing/reducing the activity of reforming/combustion catalyst, using a thicker, more heat conductive wall, and shortening the reactor length. Still at high GHSV, increasing fuel flowrate shows minor effect on raising methane conversion, indicating the dominant role of internal mass transfer. Introducing a small amount of transport pores on the scale of 1 μm into the catalyst coating can significantly improve intra-coating transport and further increase volumetric CO yield in the operable GHSV range. The results here indicate that the degree of freedom in process improvement can be more than envisioned through purposefully tailoring catalyst and reactor characteristics from micro-, meso- to macro-scale.

References

[1] R. Guttel, and T. Turek, "Improvement of Fischer-Tropsch Synthesis through Structuring on Different Scales", Energy Technol., 4, pp.44-54, 2016.

[2] L. Maier, B. Schädel, K. Herrera Delgado, S. Tischer and O. Deutschmann, "Steam reforming of methane over nickel: development of a multi-step surface reaction mechanism", Top. Catal., 54, pp.845-858, 2011.

[3] O. Deutschmann, R. Schmidt, F. Behrendt and J. Warnat, "Numerical modeling of catalytic ignition", Proceedings of the Combustion Institute, pp.1747-1754, 1996.

Copyright © American Institute of Chemical Engineers. All rights reserved.

464142 Effect of ZSM-11 Crystal Size on Methanol-to-Hydrocarbon Reactions

Abstract:

ZSM-11 (MEL) is an emerging catalyst for a wide variety of applications. The MEL framework is very similar to ZSM-5 (MFI), which is one of the most widely used zeolite catalysts in industry. Recent studies have shown that ZSM-11 has better catalytic performance in several different reactions compared to ZSM-5 [1-3], possibly due to improved intra-crystalline diffusion within the less-tortuous straight channels of the MEL topology. However, there are relatively few systematic studies of ZSM-11 synthesis, specifically with respect to tailored physicochemical properties such as size, shape, and Si/Al ratio (SAR). Here, we will present a systematic study of ZSM-11 catalyst preparation and testing for C1 applications. Through the judicious selection of synthesis parameters, we successfully developed a facile protocol to generate a library of ZSM-11 catalysts with tunable size and SAR. More precise modification in ZSM-11 crystal morphology (e.g. aspect ratio) can be accomplished through the use of zeolite growth modifiers (ZGMs), which are molecules or macromolecules that modulate the anisotropic growth rate of crystals with concomitant alterations in bulk habit [4, 5]. These methods for tailoring zeolite crystallization will be discussed along with catalytic studies using methanol-to-hydrocarbon (MTH) reactions. Collectively, our findings are part of a broader initiative to establish quantitative structure-performance relationships as a platform for optimizing zeolite catalysts.

References

[1] Kustova, M. Y.; Rasmussen, S. B.; Kustov, A. L.; Christensen, C. H., *Applied Catalysis B-Environmental* **2006**, *67* (1-2), 60-67.

[2] Zhang, L.; Liu, H.; Li, X.; Xie, S.; Wang, Y.; Xin, W.; Liu, S.; Xu, L., *Fuel Processing Technology* **2010**, *91* (5), 449-455.

[3] Bleken, F.; Skistad, W.; Barbera, K.; Kustova, M.; Bordiga, S.; Beato, P.; Lillerud, K. P.; Svelle, S.; Olsbye, U., *Physical Chemistry Chemical Physics* **2011**, *13* (7), 2539-2549.

[4] Lupulescu, A. I.; Rimer, J. D., *Angewandte Chemie International Edition* **2012**, *51* (14), 3345-3349.

[5] Lupulescu, A. I.; Kumar, M.; Rimer, J. D., *Journal of the American Chemical Society* **2013**, *135* (17), 6608-6617.

Copyright © American Institute of Chemical Engineers. All rights reserved.

463929 Effect of Dehydration on Dmc Synthesis Via Oxidative Carbonylation over Ceria Based Catalysts

Abstract:

In this study, ceria- based catalysts were prepared for DMC (dimethyl carbonate) synthesis and reaction conditions were evaluated for finding the optimal reaction route. The catalytic reactions were studied under various experimental conditions at a continuous reactor (fixed bed) and a batch reaction (autoclave). In order to find optimal catalysts for DMC synthesis, co-catalysts (or dehydration agent), calcination temperature and Cu(II) impregnation amount were evaluated. The oxidative carbonylation using methanol, carbon monoxide and oxygen and the direct synthesis using methanol and carbon dioxide were introduced for producing DMC. Following the law of Le Chatelier, the dehydration reaction was applied for enhancing the reactivity (methanol conversion) as removing water during the reaction. 2-cyanopyridine, as a chemical dehydration agent, was used. In the case of the oxidative carbonylation, methanol conversion rate increased from 15.1% to 38.7% and the DMC selectivity increased from 88.8% to 98.8%. In the case of the direct synthesis, methanol conversion rate increased from 1.0% to 77.8% and the DMC selectivity increased from 41.2% to 100.0%. The optical, physical, and chemical characteristics of synthesized catalysts were analyzed using BET, XRD, SEM, EDX, and TGA before and after reaction and breakthrough tests.

Copyright © American Institute of Chemical Engineers. All rights reserved.

464456 Neural Network Predictions of Oxygen Interactions on a Dynamic Pd Surface

Abstract:

Implementation of artificial neural networks (NNs) are becoming increasingly common for surface chemistry applications. Such models are trained to higher-level /ab-initio/ calculations and are capable of achieving arbitrary levels of accuracy. The most common applications thus far have been specialized for either bulk or surface structures of one or two chemical components. NNs designed for the robust prediction of adsorbate interactions with dynamic surface constructions still remain challenging to create. Furthermore, since the training methods for NNs are so flexible, they are also easily expanded upon as demonstrated in previous work.

The purpose of this work is two fold:

1. To determine methodologies for producing a neural network capable of accurately reproducing properties of oxygen interactions with a dynamic Pd fcc(111) surface.
2. To establish a database of high quality PdO DFT calculations to use as a basis for future work, such as the inclusion of a third chemical species.

Copyright © American Institute of Chemical Engineers. All rights reserved.

464355 Novel Catalytic Etherification Reaction of Glycerol to Shortâ €“Chain Polyglycerol

Abstract:

Zeolite Y sodium, a support base material for the catalyst, has been pre-conditioned for pore cleansing and opening in a carbon dioxide medium, both subcritical and supercritical. Target ion impregnation with a basic compound has been subsequently applied for catalytic activity increase. The synthesized catalyst has been evaluated in terms of its catalytic activity and efficiency of the etherification of glycerol to polyglycerol whose desired products are diglycerol and triglycerol. The chemical and physical properties of the catalyst, both before and after the reaction, have been examined using X-ray diffraction (XRD), scanning electron microscope (SEM) and accelerated surface and porosity analyzer (ASAP). The selectivity towards the desired products, the glycerol conversion, and the yield of the reaction have been determined, using gas chromatography-mass spectrometry (GC/MS). The current study could significantly impact the way glycerol, an over-abundant byproduct of biodiesel industry, is converted to short-chain polyglycerol, a value-added product.

Copyright © American Institute of Chemical Engineers. All rights reserved.

464450 Selective Electrocatalytic Hydrogenation of Furfural on Metal Electrodes

Abstract:

Electrocatalytic hydrogenation of biomass-derived compounds is an emerging route for biofuel and chemical production [1â€"3]. Furfural is a dehydration product of xylose that can serve as a building block for many biorenewable fuels and chemicals including valuable reduction products furfuryl alcohol and 2-methylfuran. Furfuryl alcohol has served as a precursor for various synthetic fibers, rubbers, resins and other chemicals, while 2-methylfuran is a good candidate to replace popular biofuels such as ethanol because of its higher heating value and higher octane number. Reduction of furfural is mainly obtained through heterogeneous vapor phase catalytic reactions. However, such routes typically require precious metal catalysts, high pressure external hydrogen gas and relatively high temperatures which increase operation costs. The electrocatalytic process is an alternative route, which can be driven by renewable electricity sources, does not require external hydrogen, and can be performed at ambient conditions [1â €"2].

In this talk, the electrocatalytic hydrogenation of furfural in acidic aqueous solutions on metal foil electrodes will be presented. Among all the metals tested, copper was unique toward 2-methylfuran production with high selectivity and electron efficiency. Other metals generated mainly furfuryl alcohol or products of dimerization reactions, or suffered from low electron efficiency resulting from the competing hydrogen evolution reaction. With an emphasis on copper, the effects of applied electrode potential, electrolyte composition and pH, and other reaction conditions on the efficiency and selectivity of furfural conversion were studied. For example, 2-methylfuran selectivity increased with decreasing pH and increasing electrode potentials. Dimerization products were majorly formed at conditions of higher pH and greater initial furfural concentrations. Cyclic voltammetry and bulk electrolysis studies elucidated the reaction pathways for formation of 2-methylfuran and furfuryl alcohol on copper electrodes. This work opens up opportunities to use earth-abundant materials and mild reaction conditions for efficient conversion of biorenewable chemicals.

1. Nilges, P. and U. Schroder, *Electrochemistry for biofuel generation: production of furans by electrocatalytic hydrogenation of furfurals.* Energy & Environmental Science, 2013. **6**, 2925.

2. Xin, L., et al., *Electricity storage in biofuels: selective electrocatalytic reduction of levulinic acid to valeric acid or gamma-valerolactone.* ChemSusChem, 2013. **6**, 674.

3. Green, S.K., et al., *The electrocatalytic hydrogenation of furanic compounds in a continuous electrocatalytic membrane reactor.* Green Chemistry, 2013. **15**, 1869.

Copyright © American Institute of Chemical Engineers. All rights reserved.

464873 The Selective Production of Octene By the Metathesis of 1-Pentene

Abstract:

Linear α-olefins are key components for the production of LLDPE or lubricant and the demand for hexene-1 and octene-1 increased enormously in recent years. Initially linear α-olefins were produced from paraffinic wax via thermal cracking, which gives a large propotion of alkenes-1 as the primary product, togather with alkenes. However, the efficiency of linear α-olefins synthesis is severely limited by thermodynamics. To overcome this limitation several processes were developed in the last decade to synthesize hexene-1 and octene-1 selectively. In recent, alkene metathesis is one of the most attractive process for the production of linear α-olefins. Alkene metathesis reaction is defined as the catalytic exchange of alkylidene groups in alkenes. For instance, two molecules of pentene-1 can react to one molecule of octene and one molecule of ethene. Olefin metathesis was first commercialized in petroleum reformation for the synthesis of higher olefins from the products alpha-olefins from the Shell higher olefin process (SHOP) under high pressure and high temperatures. Modern catalysts can be used for a variety of specialized organic compounds and monomers. Modern applications include the synthesis of pharmaceutical drugs, macrocyclic crownophanes the manufacturing of high-strength materials, the production of propylene, the preparation of cancer-targeting nanoparticles, and the conversion of renewable plant-based feedstocks into hair and skin care products. Because of the relative simplicity of olefin metathesis, it often creates fewer undesired by product and hazardous wastes than alternative organic reactions. This reaction can be catalysed by homogeneous or by hetrogeneous catalysts but in industrial applications heterogeneous catalsis is usually preferred. However, hrtrogeneous metathesis catalysts loose their catalytic activity during the reaction.

For the efficient production of octene by the metathesis of pentene-1, numerous catalyst system employed in this reaction includes Mo-based catalyst impregnated on γ-Al_2O_3, Mo-based monolith, MoO_3/Al_2O_3 catalyst promoted by SiO_2 activated by hv, and CO-precipitated $MoO_3/TiO_2/ZrO_2$. In recent, rhenium-based catalysts had attracted much attention brcause of high activity and tolerance of impurities. It is well known

that noble metal catalysts are shown high catalytic performance, but the high market price of noble metals renders their industrial application quite questionable. Therefore, it is more practical from an industrial standpoint to develop an improved non-novel metal-based catalyst.

One of the most problem in the metathesis reaction is deactivation of catalysts. The deactivation time is in the large od hours till several weeks depending on catalyst system and reaction conditions. The poisoning of active centre due to trace impurities in the feed components is one main cause od catalyst deactivation. Poisons of catalyst are mainly polar substances, e.g. water or oxygenates. Additionlly oxygen, organic bases and diene can lead to catalyst deactivation.

In this work, octene synthesis by the metathesis of pentene-1 was examined over a series of Re_2O_7/Al_2O_3 catalysts ($Re_2O_7/ZnO/Al_2O_3$, $Re_2O_7/ZrO_2/Al_2O_3$, $Re_2O_7/Al_2O_3/B_2O_3$, $Re_2O_7/V_2O5/Al_2O_3$) prepared by co-precipitation method at various pH conditions in a flow type fixed bed reactor to evaluate the catalyst performance. The effect of promoters on the performance of Re_2O_7/Al_2O_3 catalysts were investigated in detail, with an aim of enhancing catalytic activity and improving stability of the catalyst. N_2O chemisorption experiments were conducted to investigate the copper surface area on the Re_2O_7/Al_2O_3 catalysts. Temperature programmed reduction (TPR) measurements were carried out to investigate copper dispersion and reducibility.

Copyright © American Institute of Chemical Engineers. All rights reserved.

465066 Connected Nanoparticle Catalysts Having a Porous, Hollow Capsule Structure As Carbon-Free Electrocatalysts for Polymer Electrolyte Fuel Cells

Abstract:

We developed connected carbon-free platinum-iron (PtFe) nanoparticle catalysts with porous hollow capsule structure (Fig. 1) as oxygen-reduction-reaction (ORR) electrocatalysts for polymer electrolyte fuel cells (PEFCs).[1] This catalysts consist of a beaded network by connected PtFe nanoparticles with a crystallite size of 6~7 nm and a chemically-ordered (face centered tetragonal) superlattice structure. The beaded network is electrically conductive; thus, carbon-supports can be removed from catalyst layers in PEFCs. We have demonstrated that an MEA prepared using a carbon-free cathode of connected PtFe-nanoparticle catalysts was highly durable against start-up/shut-down operations because of the elimination of carbon-corrosion problems. Moreover, an ORR specific activity of a connected PtFe catalyst is about 9 times higher than that of a commercial Pt-nanoparticle catalyst supported on carbon black (Pt/C). Recent results will be explained.

[1] T. Tamaki, H. Kuroki, S. Ogura, T. Fuchigami, Y. Kitamoto, and T. Yamaguchi, *Energy Environ. Sci.*, 8, 3545-3549 (2015)

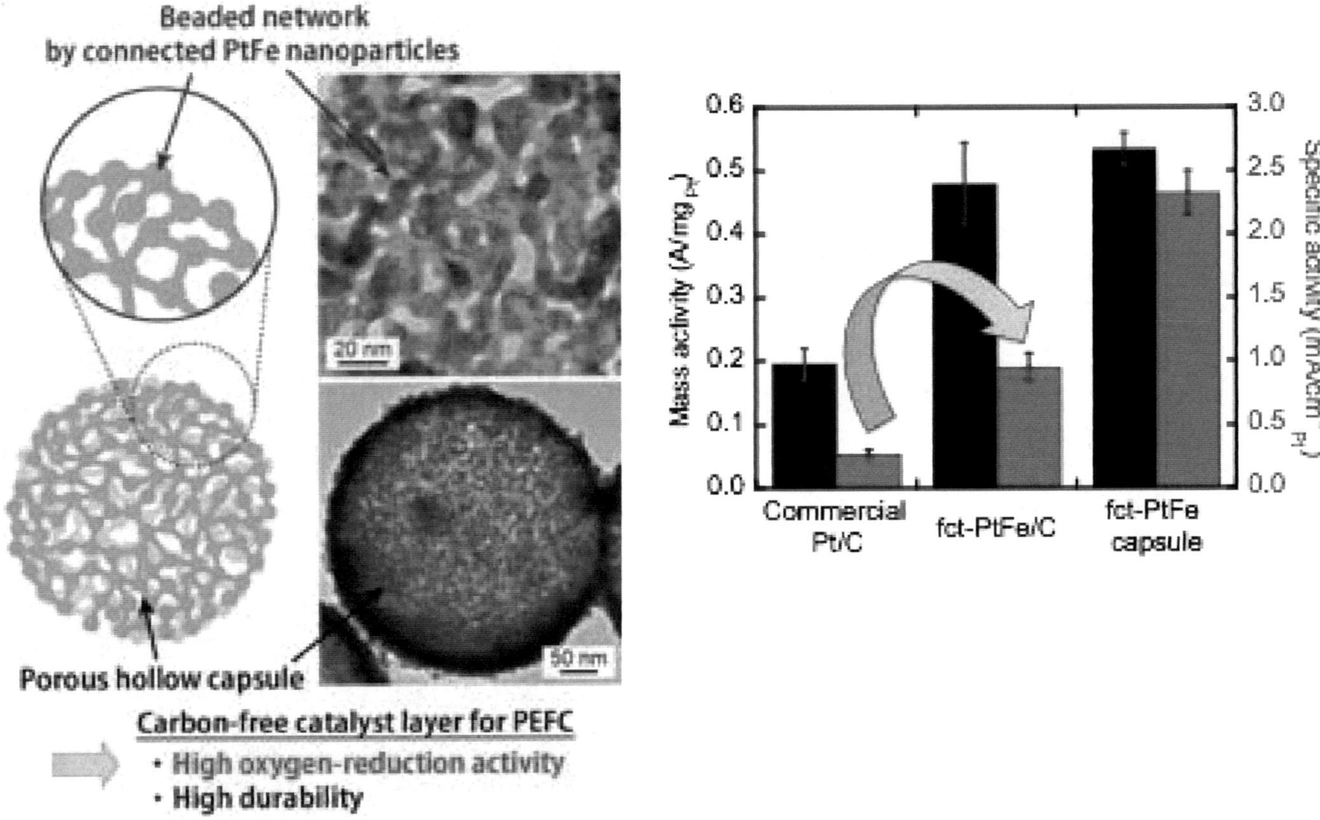

Fig. 1. Schematic illustration and TEM images of a carbon-free connected PtFe-nanoparticle catalyst with porous hollow capsule structure and its ORR activity compared with catalysts on carbon.

Copyright © American Institute of Chemical Engineers. All rights reserved.

464635 Synthesis of a Highly Active Hierarchically Ordered Micro-Meso Porous Titanosilicate *Via* Statistically Optimized Synthesis

Abstract:

An efficient method to optimize the synthesis of titanosilicates is discussed, along with strategies to unveil critical structure-function relationships. Titanosilicates are industrially relevant, versatile heterogeneous catalysts used in alkene epoxidation, ammoximation and aromatic diol production.[1] Therefore, synthesis of new types of titanosilicates, targeting specialized applications, is a highly active field of research. However, their synthesis is challenging, due to requirement of expensive, specialized conditions and extreme sensitivity to changes in the synthesis environment. This, in turn, results in low catalytic performance.[2]

Therefore, a systematic, statistical approach to probe optimal synthesis conditions is most desirable. The Doehlert matrix (DM) statistical model is introduced as an effective tool for optimizing the synthesis of titanosilicates. The DM is a second-order statistical model, widely used in analytical sciences, to quickly determine optimal experimental conditions when many parameters play a role simultaneously.[3] However, it has not yet gained prominence in the field of catalysis. Often, one condition is changed at a time, missing out on synergistic effects.

The DM approach was used to investigate the optimal synthesis conditions of a novel, hierarchically ordered, micro-mesoporous titanosilicate material (MMTS) with the aim of achieving superior catalytic performance. Catalytic characterization of MMTS was conducted *via* epoxidation of cyclohexene with *tert*-butylhydroperoxide (TBHP). The results reveal that the catalytic performance of MMTS is significantly enhanced by the optimization experiments, with unprecedented catalytic activity, measured as turn over frequency (TOF) in mols of cyclohexene converted per mol Ti

and per unit of time. The relationship between catalytic performance and properties of the material, such as the amount of framework Ti, and the nature of the pore network, is thoroughly discussed. The enhanced catalytic performance, together with critical structure-function insights, demonstrates the potential of using a DM approach in catalysis.

1. Perera, A. S.; Coppens, M.-O., Titano-silicates: highlights on development, evolution and application in oxidative catalysis. In *Catalysis*, Spivey, J. J., Ed. RSC Publishing: 2016; Vol. 28, pp 119-143

2. Corma, A., From microporous to mesoporous molecular sieve materials and their use in catalysis. *Chem. Rev.* **1997,** 97, 2373-2419.

3. Doehlert, D. H., Uniform shell designs. *Applied Statistics* **1970,** 19, 231â€“239.

Copyright © American Institute of Chemical Engineers. All rights reserved.

464677 Effects of Cations on Activity and Selectivity in Carbon Dioxide Electroreduction

Abstract:

Efficient heterogeneous electrocatalytic reduction of carbon dioxide to fuels and useful chemicals with renewably-produced electricity would both reduce the atmospheric CO_2 concentration and provide an alternative to fossil fuels. While electrolyte cation effects have been studied extensively with polycrystalline transition metal catalysts, cation effects have not been explored on well-defined, epitaxially-grown electrocatalyst surfaces. We will determine the effects of a variety of cations â€“ including alkali metals, alkali earth metals, and organic bases â€“ on the activity and product selectivity of epitaxially-grown transition metal electrocatalysts for the CO_2 reduction reaction. The samples are tested using an electrochemical flow cell previously designed by our group, and we will use *in situ* vibrational spectroscopy to detect intermediates on the working electrode surface. In light of our preliminary results, we anticipate that larger metal cations on stepped surfaces will exhibit higher current densities, more selectivity for oxygenated products, and higher C-C coupling activity.

Copyright © American Institute of Chemical Engineers. All rights reserved.

465263 Study of Reactivity of CO+NO Reaction on Supported Pd Catalyst

Abstract:

Palladium (Pd) is the primary component in most automotive three-way catalysts (TWCs). Under slightly lean conditions, CO in the exhaust emissions can be oxidized to CO_2. The light-off of CO oxidation by O_2 follows an S-shaped curve. Under stoichiometric conditions, where there is a lack of molecular O_2 in the exhaust emissions, the conversion of CO to CO_2 can be accomplished by the CO+NO reaction, where NO is reduced by CO to N_2, to achieve the simultaneous removal of CO and NO. However, we found that when CO is oxidized by NO, instead of O_2, the CO light-off curve has a concave shape. This phenomenon has been further investigated by comparing Al_2O_3 and Ce-Zr-O supported Pd catalysts with focus on the effects of different catalyst pre-treatment methods, catalyst aging, reactant feed (CO/NO) ratio, as well as the operating conditions such as heating and cooling on the reactivity of CO+NO reaction. DRIFTs experiments were carried out to study the CO+NO conversion to CO_2+N_2O, instead of the target products CO_2+N_2.

Copyright © American Institute of Chemical Engineers. All rights reserved.

STUDY OF OXIDATION REACTIONS IN PRELIMINARY ZONE OF STABILIZATION PROCESS OF CARBON FIBER PRODUCTION LINE

KHASHAYAR BADII, MINOO NAEBE, GELAYOL GOLKARNARENJI, HAMID KHAYYAM

Institute for Frontier Materials, Carbon Nexus , Deakin University, Victoria 3216, Australia

16AIChE Annual Meeting, San Francisco, CA

DEAKIN UNIVERSITY AUSTRALIA Worldly

ABSTRACT:

Carbon fiber (CF), with low weight and high mechanical properties, has found a wide range of applications during the last three decades. Therefore, the optimization of its process can decrease its expenses and increase its applications. Although, stabilization, carbonization, and graphitization processes are controlling the quality of final CF product, stabilization process, as the most time and energy consuming step, is the main key to commercialize CF with more than 50% energy consumption. In order to optimize stabilization process, understanding the chemical circumstances and the mechanism of chemical reactions is very important due to the availability of sufficient tools to design required chemical units. Whilst, cyclization, dehydrogenation, oxidation and crosslinking are the possible chemical mechanisms in stabilization production line, the role of oxidation reactions, particularly in the first zone of process is unknown due to the nature of precursors.

In the present study, the set of oxidation reactions in the first zone of the process has been investigated for conversion of polyacrylonitrile (PAN) precursor to oxidized PA N fiber (OPF) and its effects on required parameters for mathematical modeling of process has been studied.

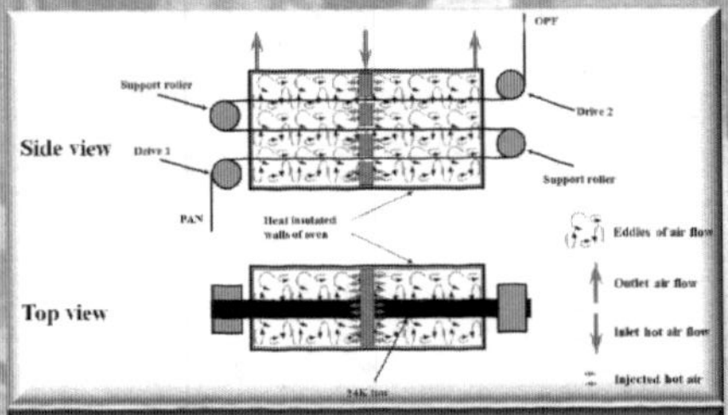

Side and top schemas of single tow stabilization oven

Bundle model

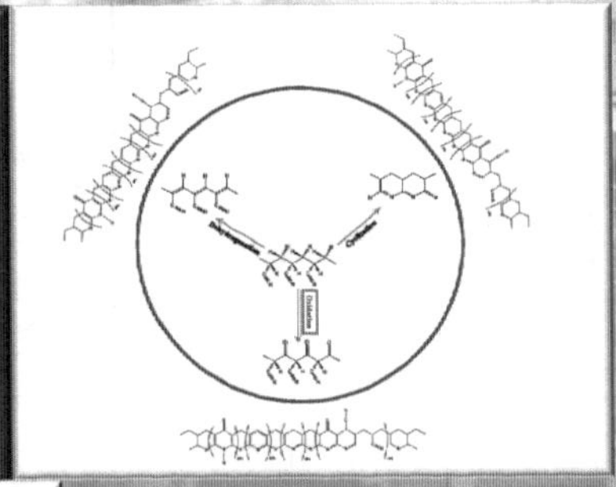

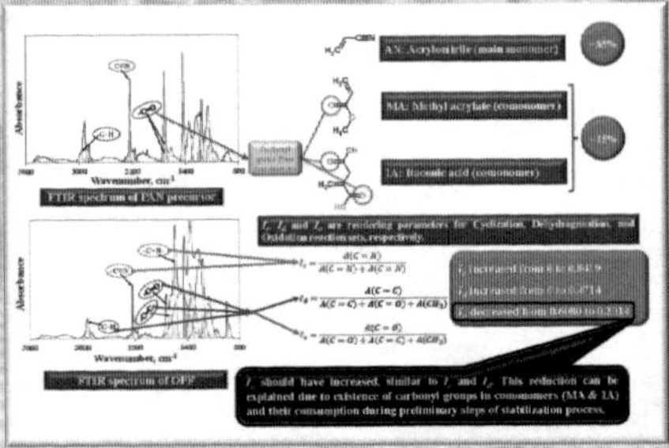

Modified kinetics model (Modified bundle model):

1. A set of four reaction mechanisms has been developed.
2. Adding a -C=O group consumption reaction set
3. Although, -C=O groups in comonomers (MA and IA) are different from OPF, they assumed similar in modified kinetics model.

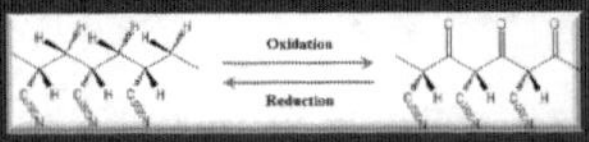

MODEL DEVELOPMENT (3D):

$$\left(v_x\frac{\partial C_i}{\partial x}+v_y\frac{\partial C_i}{\partial y}+v_z\frac{\partial C_i}{\partial z}\right)=-\left[\frac{\partial J_{x_i}}{\partial x}+\frac{\partial J_{y_i}}{\partial y}+\frac{\partial J_{z_i}}{\partial z}\right]+r_i$$

v_x, v_y, and v_z are velocities (m/s), C_i is concentration of functional group i and O_2 (kgmol/s).

Fick's first law: $J_{x_i}=-D_e\frac{\partial C_i}{\partial x}$ & $r_i=\sum_{j=1}^{n}k_iC_jC_{O_2}$ is reaction rate

D_e is effective diffusion coefficient (m^2/s) & ρ is density, kg/m^3

$$\begin{cases}v_y=v_z=0\\J_{x_i}=J_{y_i}=J_{z_i}=0\\\rho=cte\end{cases}$$ Functional groups (fiber, solid)

$$\begin{cases}v_x=v_y=v_z=0\\\rho=cte\\D_e=cte\end{cases}$$ Free O_2 (gas)

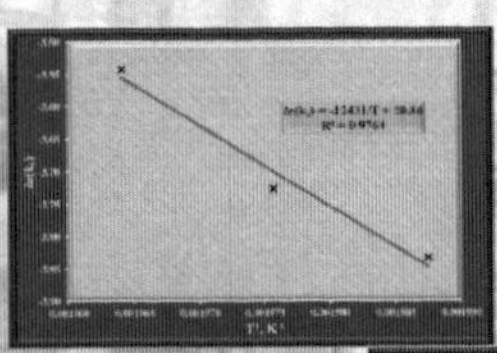

Prediction of cyclization reaction constant (k_1) parameters

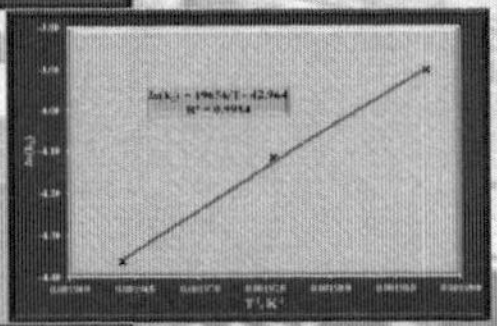

Prediction of dehydrogenation reaction constant (k_2) parameters

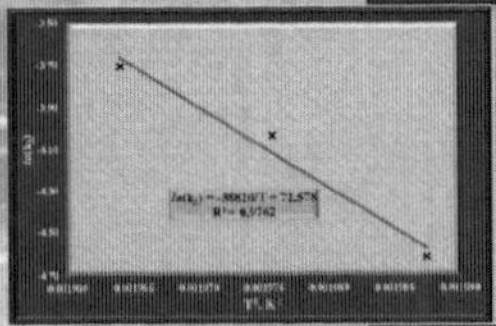

Prediction of oxidation reaction constant (k_3) parameters

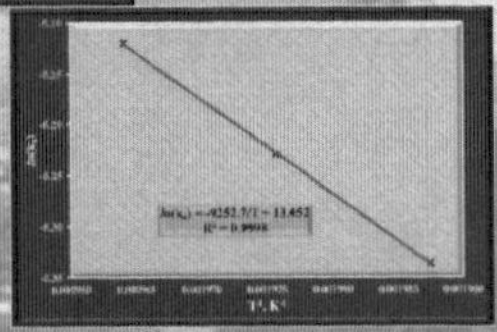

Prediction of reduction reaction constant (k_4) parameters

	From four sets model		From DSC[#]	
Parameters	k_0, $m^3kmol^{-1}s^{-1}$	E_a, MJ/kmol	k_0, $m^3kmol^{-1}s^{-1}$	E_a, MJ/kmol
k_1, $m^3kmol^{-1}s^{-1}$	$1.147\times10^{+09}$	+103.351	$4.718\times10^{+11}$	159.060
k_2, $m^3kmol^{-1}s^{-1}$	2.193×10^{-19}	-163.420	$4.718\times10^{+11}$	159.060
k_3, $m^3kmol^{-1}s^{-1}$	$3.313\times10^{+31}$	+322.749	$9.122\times10^{+15}$	190.060
k_4, $m^3kmol^{-1}s^{-1}$	$4.660\times10^{+05}$	+76.927	-	-

[#]DSC: Differential Scanning Calorimetry.

CONCLUSION:

The results present:

1. The rate of reaction increases with temperature for cyclization, oxidation and reduction reaction sets and their developments are faster in higher temperatures.
2. Temperature has reverse effect on dehydrogenation reaction rate and it decreases at higher temperatures.
3. The reaction parameters are different from DSC results due to the nature of stabilisation process as a diffusion-limited process. This has not been considered in DSC analysis methods.

CARBON nexus

465100 Understanding of Reactivity of Asphaltene in H-Donor Solvents

Abstract:

Asphaltene is a representative refractory compound in heavy oils, adopting poly-aromatic hydrocarbon cores connected with many alkyl chains and heteroatoms and metals, which often causes coke formation and precipitation upon the reaction severity. Considering the heavy oil trend in refinery feedstock, understanding of reactivity of asphaltene in the course of various hydrocracking condition is of great importance in achieving effective conversion into lighter fraction. In this work, the hydro-conversion of asphaltene at different reaction severity of temperature and pressure was investigated. The reaction progress of asphaltene was measured by carbon-NMR and C-7 solvent extraction method. In particular, the reactivity of asphaltene was controlled in the presence of H-donor solvents. It was demonstrated that the asphaltene was converted by 85% into lighter fractions like middle distillates and gas oils without coke formation upon the hydrocracking at 370°C and 100 bar with dispersed catalyst of MoS_2. Further improvement in hydroconversion of asphaltene was achieved in the presence of H-donor solvent.

Copyright © American Institute of Chemical Engineers. All rights reserved.

465544 Ordered Mesoporous Metal Oxide Supports Grown Around Nanocatalysts for High Temperature Stability

Abstract:

Ordered Mesoporous Metal Oxide Supports Grown Around Nanocatalysts for High Temperature Stability

Victor Miller,[1] Joshua J. Willis,[1] Matteo Cargnello[1,2]

[1]Department of Chemical Engineering, Stanford University, Stanford, CA 94035, United States

[2]SUNCAT Center for Interface Science and Catalysis, Stanford University, Stanford, CA 94305

Designing robust, active catalysts for clean chemical transformations is key to a wealthier, greener future. However, an industrially useful catalyst must meet a number of criteria, including high stability and active site control. Preserving catalytic activity in high-temperature regimes with rationally designed supports enables the exploitation of previously unstable yet otherwise promising catalytic materials. This is all the more critical as catalysts for desirable reactions are often designed from expensive materials. This study investigates such supports with high-surface area monodisperse nanoparticles.

We prepare these catalyst-support systems by growing mesoporous metal oxides around monodisperse nanoparticles, followed by calcination in air. The supports are chosen to be thermally and chemically stable, while allowing for diffusion of products and reactants in the pores to reach the embedded catalytic surfaces. Current species are monodisperse Palladium nanoparticles; embedded in the current oxides of interest, mesoporous alumina and silica. The ideal design will prevent catalyst sintering or agglomeration, likely through immobilization, and enable mass transfer. Further, successful embedding may allow for future 3D manipulation of catalytic active sites, yielding unparalleled control. This may also enhance selectivity, the primary challenge in reactions involving methane. Successful embedding with growth involve dissolving micelle-forming surfactants in solution, partitioning catalytic particles into the micelles, and using the micelles as templates for the growth of the metal oxide support. We investigate the stability and activity of nanoparticles embedded into mesoporous supports. We observe the effect of manipulating the pore and nanocrystal size, the activity at different temperatures, and following different heat treatments and pretreatments. Such a library of activity related to pore sizes matched to nanoparticle sizes could be used to design stable, high surface area catalysts.

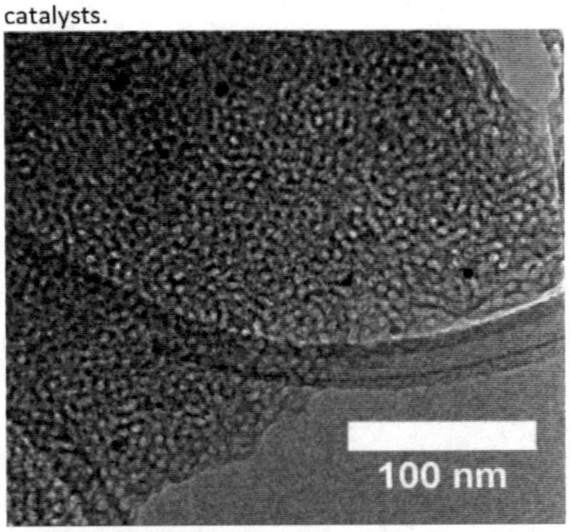

Fig. 1: TEM image of a particle of alumina grown around Pd nanoparticles (referred to as Pd@Alumina). Particles average 7.1nm wide.

- Abstract Text

Copyright © American Institute of Chemical Engineers. All rights reserved.

465481 Ionic Compounds for Carbon Dioxide Electroreduction

Abstract:

The electrochemical reduction of carbon dioxide could serve as a carbon neutral source of fuels and chemicals in a sustainable energy economy. For carbon dioxide reduction intermediates which bind to the surface through similar interactions, adsorption energies of the intermediates tend to scale[1]. These scaling relations dictate a large overpotential for electrochemical carbon dioxide reduction. Breaking these scaling relations would require selectively stabilizing intermediates. This motivates the exploration of materials beyond metals and metal alloys. We have chosen to first explore metal phosphide and molybdenum sulfide materials. We synthesize the materials using a general methodology of physical vapor deposition of thin metal films, followed by vapor assisted conversion in a tube furnace[2,3]. The samples are tested using an electrochemical flow cell previously designed by our group[4]. We characterized the materials before and after electrochemical testing by X-ray photoelectron spectroscopy (XPS) to confirm their identity and observe changes in composition induced during testing. We will report our findings on the activity of these materials for electrochemical carbon dioxide reduction.

1. Peterson, A. A., & Norskov, J. K. (2012). Activity Descriptors for CO 2 Electroreduction to Methane on Transition-Metal Catalysts. *J. Phys, Chem. Lett.*, *3*, 251â€“258.
2. Hellstern, T. R., Benck, J. D., Kibsgaard, J., Hahn, C., & Jaramillo, T. F. (2016). Engineering Cobalt Phosphide (CoP) Thin Film Catalysts for Enhanced Hydrogen Evolution Activity on Silicon Photocathodes. *Adv. Energy Mater.*, *6*(4), 1â€“8.
3. Benck, J. D., Lee, S. C., Fong, K. D., Kibsgaard, J., Sinclair, R., & Jaramillo, T. F. (2014). Designing active and stable silicon photocathodes for solar hydrogen production using molybdenum sulfide nanomaterials. *Adv. Energy Mater.*, *4*(18), 1â€“8.
4. Kuhl, K. P., Cave, E. R., Abram, D. N., & Jaramillo, T. F. (2012). New insights into the electrochemical reduction of carbon dioxide on metallic copper surfaces. *Energy Environ. Sci.*, *5*, 7050.

Copyright © American Institute of Chemical Engineers. All rights reserved.

465598 Catalytic Upgrading of Olefin ResiduesÂ withÂ Odd Number of Carbon Atoms

Abstract:

Linear alpha olefins (LAOs) are important chemical compounds with a variety of applications such as surfactants, detergents, and co-monomers for production of polyethylene. LAOs are commonly produced by two methods: ethylene oligomerization and Fischer-Tropsch synthesis followed by purification process. LAOs derived from Fischer-Tropsch (F-T) process, unlike oligomerization process, are composed of olefin mixtures of both even and odd carbon numbers. Because there is little demand for these odd numbered alpha-olefins in the market, more technical efforts to be converted into higher value-added compounds are required. In this work, the Rh catalysts supported on the various metal oxides were prepared and applied to the catalytic upgrading reaction of 1-heptene. The reactions were carried out in 100 cc autoclave reactor under different reaction conditions (temperature, pressure, and feed ratio). The effect of supported Rh catalysts on the catalytic performance and reaction pathway in the 1-heptene upgrading reaction was investigated. To elucidate the catalytic performance over the prepared Rh catalysts, various characterization method such as NH_3/CO_2-TPD (temperature-programmed desorption), XRD (X-ray diffractometer), and TPR (temperature-programmed reduction) measurements were conducted.

Copyright © American Institute of Chemical Engineers. All rights reserved.

465554 Co-Processing of Heavy Residue Oil and Biomass Using Supercritical Organic Solvents

Abstract:

The use of biomass as energy resources can increase energy security and reduce the need to use fossil fuels because it is abundant in most countries. Furthermore, utilization of biomass has received significant attention as a renewable energy source. Co-processing of biomass or biofuels with petroleum feedstock using conventional oil refinery processes seems to be an attractive alternative for the fuel production because the infrastructure investment can be saved by utilizing present facilities. In this study, co-processing of wood biomass (microcrystalline cellulose or oil palm empty fruit bunch fibers (EFB)) and vacuum residue (VR) was carried out by using sub- and supercritical hydrocarbon solvents. The effects of Fe_3O_4 catalyst and hydrogen on the co-processing performance were evaluated in terms of residue conversion, coke formation and product distribution. In addition, the results were compared with the results of only VR upgrading. The co-processing of VR/EFB could achieve a residue conversion of 86.4 wt. % with only 0.8 wt. % of coke formation at a specific operating condition and solvent. In the presence of H_2, the Fe_3O_4 catalyst contributed toward suppressing coke formation. The liquid products were completely soluble in the applied solvents. It implied that the products consisted of soluble compounds in low polar solvents.

Copyright © American Institute of Chemical Engineers. All rights reserved.

465745 Development of Pt/Al_2O_3 Catalytic System for the Production of Furanic Compounds from Biomass Derived Itaconic Acid

Abstract:

The needs for the innovative production of platform materials are increased due to the limitation of fuels and the effort to retain the sustainable hydrocarbon sources for chemical products. Biomass-derived feedstocks could be converted to chemical products by the hydrogenation and dehydration processes and be used as platform chemicals [1]. Thus, biomass could be candidate for renewable feedstocks and fuels. Among the various platform chemicals, furanic compounds are widely used as organic solvents and biofuel precursors. Methyl-tetrahydrofuran (MTHF) is a commercially used solvent that is produced from renewable resources. Also, methyl-γ-butyrolactone (MGBL) is used as solvent and reactant for processes. In these reasons, the conversion from biomass to furanic compounds is challenging tasks for both scientific and industrial aspects. The furanic compounds production from biomass-derived itaconic acid (IA) was reported by Geilen *et al.*by using Ru based homogeneous catalyst [2]. Ru catalysts, ligands and additives were used for conversion of IA into furanic compounds. IA was converted into furanic compounds by successive hydrogenation and dehydration processes. However the application of heterogeneous catalysts has not been reported yet for the IA conversion into the furanic compounds. The heterogeneous catalytic process is usually easier to handle than homogeneous catalytic process and leads to a lower risk for the environment.

Thus, we introduced the heterogeneous metal catalysts for producing methyl-γ-butyrolactone (MGBL) and 3-methyltetrahydrofuran (3-MTHF) as target materials. Ruthenium, palladium, platinum and iridium metal loaded catalysts were synthesized by wetness impregnation method with γ-Al_2O_3 as support. The activity of the catalyst was measured in batch reactor using 1 g of the catalyst. The IA conversion was carried out in the liquid phase: 0.05 M of IA, 200 ml of 1,4-dioxane solvent and prepared catalyst were placed into the reactor. The reaction pressure was 140 bar $H_2$and the temperature was 260°C and the mixture was stirred

with 1000 rpm for 18 hours. In the hydrogenation and dehydration of IA to MGBL and 3-MTHF using synthesized catalyst, methylsuccinc acid (MSA) and methylsuccinic anhydride (MSAN) were produced as intermediate chemicals.

Among the synthesized metal catalysts, Pt catalyst (3 wt.% Pt/γ-Al_2O_3) achieved the highest activity on the conversion of IA into 3-MTHF. In order to increase the hydrogenation and dehydration of carbonyl groups (C=O bond) in MSAN and MGBL, we further applied different kinds of Pt metal precursors to the Pt/γ-Al_2O_3 catalysts. It is well known that precursors of metal influence the properties of catalyst, which could promote the catalytic activity [3]. Thus, we introduced two kinds of Pt precursors, $Pt(NH_3)_4(NO_3)_2$ and H_2PtCl_6 on γ-Al_2O_3. In the case of *chlorided* catalyst (H_2PtCl_6), the chlorine remained after both calcination and reduction pretreatments. Also in this case, the lower 3-MTHF selectivities were exhibited than the case of *chloride free* catalyst ($Pt(NH_3)_4(NO_3)_2$) in all reduction temperature. These differences resulted from the differences of acidity and the amounts of H_2adsorbed on catalyst surface [4].

As a result, the highest 3-MTHF selectivity (44.1%) was obtained in the Pt catalyst prepared using $Pt(NH_3)_4(NO_3)_2$ precursor reduced at 400°C. The remain chlorine decreased the catalytic activity, and the catalyst based on $Pt(NH_3)_4(NO_3)_2$ showed the highest activity on the hydrogenation and dehydration of carbonyl C=O bond in the MGBL. The additional characterization of catalyst will be performed and discussed. In the production of the chemicals from biomass-derived materials, these catalysts would be used for the active and energy efficient way for selective chemical production.

References

1. J.N. Chheda et al., *Catal. Today*, **123** (2007) 59
2. Frank M. A. Geilen et al., *Angew. Chem.*, **122** (2010) 5642
3. M. Consonni et al., *J. Catal.*, **188** (1999) 165
4. M. Paulis et al., *J. Catal.*, **199** (2001) 30

Copyright © American Institute of Chemical Engineers. All rights reserved.

465817 Synthesis of a Bi-Functional Catalyst and Application in One-Pot Conversion of Fructose into Oxygenated C_9 and C_{15} Hydrocarbons

Abstract:

Long-chain saturated hydrocarbons (C_9 and C_{15}) have high energy density, excellent cold flow properties and similar boiling-point range as fuels for diesel engines and jets. These alkanes can be one of the alternatives to the conventional and limited fossil fuels resources. These fuel-grade hydrocarbons can be produced from renewable biomass feedstocks via step-wise hydrolysis, dehydration, aldol condensation and hydrogenation reactions. Various homo- and heterogeneous catalysts have been developed for these reactions separately but are limited with issues of downstream waste generation and poor recycling performance. To address these issues, an attractive approach is to directly produce the liquid alkanes from the biomass feedstocks (C_5 and C_6carbohydrates) through a combination of the above-mentioned reactions. Thus, a lucrative strategy for facile one-pot conversion of carbohydrates catalyzed by a bi-functional catalyst was developed by eliminating the purification and isolation of furanic intermediate.

A bi-functional acid-base catalyst with silica-coated magnetic core was prepared through a series of functionalization steps. Amorphous silica layer was initially coated on crystalline iron oxide magnetic nanoparticles (Fe_3O_4 NPs) via solvothermal method. Two catalytic sites were then introduced on $SiO_2@Fe_3O_4$ NPs via sequential silanization of two types of tri-ethoxy silanes. For the acidic active sites, the $SiO_2@Fe_3O_4$ NPs was silanized with a thiol (-SH) terminated silane wherein its subsequent oxidation generated the sulfonic (SO_3H) acid catalytic sites (SO_3H-$SiO_2@Fe_3O_4$). An epoxide terminated silane was then immobilized on SO_3H-$SiO_2@Fe_3O_4$ on which triazabicyclodecene (TBD) was attached via epoxide ring cleavage to generate the base catalytic sites, completing the bi-functional catalyst TBD-SO_3H-$SiO_2@Fe_3O_4$.

The prepared bi-functional catalyst and its precursors were characterized by XRD, HRTEM, EDAX, TGA, FT-IR and VSM analyses. Meanwhile, the catalytic performance of TBD-SO_3H-$SiO_2@Fe_3O_4$ was rigorously tested using fructose as substrate and its reusability for long-term use was also investigated. Results reveal that the developed catalyst can facilitate a cascade reaction involving dehydration of fructose and aldol condensation of furanic intermediate, 5-hydroxymethylfurfural (5-HMF) in one-pot. In particular, the prepared heterogeneous catalyst provided a synergistic catalytic activity for two-step chemical conversions of fructose into oxygenated precursors of C_9 and C_{15}alkanes. In this manner, tedious separation and purification steps of intermediate 5-HMF was avoided.

This work was supported by the National Research Foundation of Korea (NRF) grant funded by the Ministry of Education (No. 2009-0093816) and by the Ministry of Science, ICT & Future Planning (No. 2015R1C1A2A01054605).

Copyright © American Institute of Chemical Engineers. All rights reserved.

465685 Nano-Sized Tungsten Carbide Catalyst for Upgrading Extra-Heavy Oil

Abstract:

Nowadays, extra-heavy oil has been received attention as alternative oil source as crude oil is being depleted. Extra-heavy oil is composed largely of resin and asphaltene which are macromolecules unable to use as fuel or chemical resources. These materials should be cracked into light oils to utilizing extra-heavy oil. The cracking process involves pyrolysis or hydrocracking. In these processes, some radicals can be produced and activate the formation of massive molecules like coke. Formation of coke reduces available carbon which means quantity of carbon atoms that compose liquid products. And also coke induces deactivation of catalysts and blocks pores of supports so it decreases catalytic activity. In this case, the provision of protons by hydrogenation process could suppress formation of coke. So it is obvious that hydrogenation process could perform important role in upgrading progress of extra-heavy oil.

Platinum, one of the most important element in catalysis, has outstanding performance in catalytic reaction involving hydrogen. For this reason, platinum had been used for hydrogenation process to saturate unsaturated hydrocarbons in spite of its high price. But, because of poisoning by sulfur, platinum could not be used with reactants such as petroleum which include sulfur. In 20th century, Levy and Boudart reported tungsten carbide, composed of tungsten and carbon atoms, could provide similar catalytic activity with platinum[1] because of its electron structure. But unlike platinum, tungsten carbide has sulfur-resistance when it is exposed to sulfur in high temperature[2]. In this respect, tungsten carbide has been considered as an alternative material of platinum for catalytic processes accompany with sulfuric reactants. In this case, tungsten carbides are used as hydrogenation catalysts and dispersed catalysts for the extra-heavy oil upgrading reaction.

The size of the catalyst particles should be small to use as dispersed catalysts, but obtaining small size tungsten carbide is difficult because tungsten carbide is formed in very high temperature[2]. In this temperature, sintering of tungsten particles becomes a serious problem. In traditional method, a ball-milling process is necessary to obtain small size tungsten carbide particles. But the ball-milling method is energy and time consuming process and its products contaminate milling machine[3]. Recently, Sean T. Hunt reported the method which could restrict sintering while the carburization process. In this process, silica is used as impediment between the tungsten particles[4]. As a result, nano-sized tungsten carbide could be produced in this method without any contamination. In this study, the nano-sized tungsten carbide catalyst, produced via tungsten oxide-silica core-shell based on the Hunt's method, was applied to the extra-heavy oil upgrading process.

The reactivity of catalysts was measured in a 100-mL batch-type autoclave reactor. 30g of feedstock was charged into a reactor and the catalyst was added to the reactor. Vacuum residue was used as a feedstock, provided by the Energy Process Laboratory, SK Innovation. After charging the reactants, the reactor was purged three times with 10 bar of H_2 gas to remove air. Then the reactor was pressurized to 70 bar with hydrogen at 27 , followed by heating to 400 at a ramp rate of 12.5 /min. Reaction was carried out at 400 for 4 hours.

For the bulk catalyst, tungsten carbide (2 μm, Aldrich, ≥99%) was used. In case of nano-sized tungsten carbide, tungsten isopropoxide in isopropanol (5 wt. %, Alfa) and tetraethyl orthosilicate (reagent grade, 98%, Aldrich) were used as tungsten and silicon precursor. These precursors were injected into reverse microemulsion and aged over 16.5 hours. Aged sample was separated by centrifuge at 10000 rpm for 10 min and washed once with acetone. After separation, the sample was dried over 12 hours at 50 and calcined for 1 hour at 450 . Finally, calcined sample was carburized with 21% CH_4/H_2 gas at 835 and SiO_2 shell was removed in ammonium bifluoride solution.

From XRD and TEM data, nano-sized tungsten carbide formed in SiO_2 shell was observed. The thickness of SiO_2 shell was 23 nm and diameter of nano-sized tungsten carbide was 5 nm which was 1/400 size of bulk tungsten carbide. In this study, the reaction were performed without catalyst, with bulk tungsten carbide and with nano-sized tungsten carbide to compare catalytic activities. The result was analyzed by using simulated distillation (SIMDIS) and SARA (Saturate-Aromatic-Resin-Asphaltene) analysis. Tungsten carbide catalyst showed high performance for upgrading extra-heavy oil to produce liquid products. In the reaction with tungsten carbide, decrease of pressure, which means hydrogen uptake, was observed. As expected, increment of liquid fraction and decrease of solid (coke) fraction was indicated when tungsten carbide catalyst was applied due to lower energy barrier for H_2 dissociative adsorption.

[1] Levy and Boudart, *Science*, **181** (1973) 547-549

[2] Zhe Cheng et al., *J. Electrochem. Soc.*, **153** (2006) A1302-A1309

[3] Zaoxue Yan et al., *Sci. Rep.*, **3** (2013), 1646

[4] Sean T. Hunt et al., *Angew. Chem. Int. Ed.*, **53** (2014) 5131-5136

Copyright © American Institute of Chemical Engineers. All rights reserved.

466035 Three-Dimensional Fluid Dynamic Analysis of a Newly Designed 100 cm^2 Single Cell with Internal Flow Channels for Molten Carbonate Fuel Cells

Abstract:

Molten carbonate fuel cells (MCFCs) were developed to produce clean and efficient power conversion. MCFCs are one of the promising sources of renewable energy for the high-efficiency cogeneration of electricity and heat with minimal environmental impact [1]. MCFCs are operated at a high temperature of approximately 650 °C. Because of the high temperature, precious metal catalysts are not required during the operation. MCFCs employ an electrolyte of molten carbonate salt mixtures. Carbonate ions act as an electric charge carrier.

The current collector lies between electrodes and cell frames. The current collector forms gas flow channels. The matrix and electrolyte separate anode gases (H_2, CO_2 and H_2O) and cathode gases (Air and CO_2). In the anode side, H_2 reacts with Carbonate ions from electrolytes to produce water (H_2O), carbon dioxide (CO_2). Current generated by electro chemical reaction of the MCFC flows through the contact area between the current collector and the electrodes. The geometry of the current collector and gas flow channel affects the flow characteristics which have a large effect on the performance of fuel cells. Also, the arrangements of the shielded slot plate influence gas flow characteristics [2].

In this work, a newly designed 100 cm^2 cell frames with internal flow channels for molten carbonate fuel cells (MCFCs) was investigated using three-dimensional fluid dynamic analysis. Also, the effect of the current collector structure on the performance of molten carbonate fuel cells (MCFCs) was investigated. From the simulation, pressure drop, flow field, and gas mole fractions inside the cell frame were studied.

In the operation of the single cell, a Ni - 5wt% Al anode, a lithiated NiO cathode, and γ-$LiAlO_2$ matrices were used. For the electrolyte, Li_2CO_3 and K_2CO_3 (62:38) were used. The single cell was operated at 620°C. The gas utilizations for the anode side (H_2) and the cathode side (CO_2, O_2) were fixed to 0.4 at 150mA/cm^2.

Cell frames with inner flow channels were utilized. The cell frame with internal flow channels has advantages in controlling the temperature of the cell frame and components. In addition, the gas uniformity of the gas flow channel increases by separating the gas flow line in the cell frame. The gas flow line outside the cell frame acts as a pre-heater. The size of the cell frame was 130 mm (length), 130 mm (width), and 30 mm (height). The active area of the single cell was 100cm^2. In order to make close contacts between current collector sand electrodes, a sealing pressure of 0.2 MPa was applied to the cell frame using a hydraulic pressure.

To investigate the effect of the current collector on the performance of MCFCs. Different types of current collectors were employed in experiments and simulation. In this work, three types of current collectors such as a sheet with sheared protrusions (shielded slot plate) and perforated sheets were employed. In case 1 and case 2, the shielded slot plates were used as a current collector. The difference between case 1 and case 2 is the stacking direction of the shielded slot plate. In case 3, the perforated sheet was used as a current collector.

In the simulation, the commercial CFD code (COMSOL multi-physics, COMSOL Inc.) was employed in order to calculate the steady state such as the energy balance, species balance, momentum and continuity equations. In the simulation, the polarization model by Yuh and Selman [ref] was adopted. In this polarization model, total polarizations were the sum of anode polarization resistance (R_a), cathode polarization resistance (R_c) and ohmic resistance (R_{ohm}). Each polarization resistance was correlated using linear multiple regression in various gas conditions [3].

In previous studies, the gas flow channels were considered to be a porous media [4]. The flow characteristics in a porous media were modeled with Darcy's law. In the simulation model with Darcy's law, detailed flow characteristics of flow channels were not considered. In addition, the heat transfer between the current collector and gases was not properly considered in the simulation model. Instead of using porous media approach, the flow channels were three-dimensionally modeled.

Three types of heat transfer models such as natural convection between the heat chamber and the cell frame, forced convection between the gas and the gas flow channel and conduction between the cell frame and the MCFC components were considered in the simulation. Thermal properties of the mixture gases of the anode and cathode gas were calculated with respect to the temperature, gas compositions, and electro-chemical reactions.

For the verification of the simulation model, experimental results were compared with the simulation results. The performance of the cell can be predicted precisely. The distributions of the gas concentration, current density, and polarization components were investigated using the simulation results. The cell frame with internal flow channels showed improved thermal characteristics. Cell frames with internal flow channels were operated without any hot-spots. From the simulation results and experimental results, it was found that the current collector structure, which makes large gas open area to electrodes, results in effective gas supply to electrodes, and as a result, the performance of the fuel cells increases.

References

[1] R. O'Hyare, S. W. Cha, W. Collela, F. B. Prinz. Fuel cells - Fundamentals. New York: John Wiley & Sons; 2006

[2] Li X, Sabir I. Review of bipolar plates in PEM fuel cells: Flow-field designs. International Journal of Hydrogen Energy. 2005;30:359-71.

[3] Yuh C, Selman J. The Polarization of Molten Carbonate Fuel Cell Electrodes I. Analysis of Steady State

Polarization Data. Journal of the Electrochemical Society. 1991;138:3642-8.

[4] Kim H, Bae J, Choi D. An analysis for a molten carbonate fuel cell of complex geometry using three-dimensional transport equations with electrochemical reactions. International Journal of Hydrogen Energy. 2013;38:4782-91.

Copyright © American Institute of Chemical Engineers. All rights reserved.

466497 Fischer-Tropsch Synthesis: Simulation of the Diffusion of H_2, CO, and Hydrocarbons in Liquid Phase Water and Hydrocarbons

Abstract:

Fischer-Tropsch Synthesis: Simulation of the diffusion of H_2, CO, and hydrocarbons in liquid phase water and hydrocarbons

Xiaojun Lu, Diane Hildebrandt, and David Glasser

Material and Process Synthesis; College of Science, Engineering, and Technology, University of South Africa

Abstract

As the major products of the Fishcer-Tropsch synthesis, H_2O and liquid hydrocarbons have been discovered by researchers to have different effects to both the reaction rate and product selectivity. The effects of hydrocarbons on the performance of the FT reaction are: slowing down the overall reaction rate, increasing the selectivity to CH_4 and favouring the selectivity to the alkanes; while H_2O improves the overall reaction rate and increases the selectivity to alkenes. Although the phenomena have been clearly shown, the mechanism for the different roles of hydrocarbons and H_2O on the reaction rate and product selectivity are not completely understood. Some researchers have attributed such effects to the catalyst properties or to the mass transportation. A recent publication by Hibbitts and Iglesia *etc.* [1] explained the effect of H_2O on the performance of Fischer-Tropsch reaction from the catalytic perspective at the reaction mechanism level.

Most of the experimental results suggest that these effects should not be attributed to the change of the catalyst properties as the changes are fast and reversible. Before the catalysts cause a change in kinetics, the concentration of the reactants or the products close to the surface of the catalyst need to change first. Therefore the mass transfer plays a very important role in the change of the overall reaction rate and product selectivity. The experimental and simulation data for the diffusion of the reactants (H_2 and CO) in the liquid hydrocarbons cannot be used to interpret the experimental results observed for the effects from liquid hydrocarbons and H_2O. The diffusion coefficient of H_2 is always larger than that of CO, in both the liquid hydrocarbons and the H_2O. To further complicate matters the data for the diffusion of H_2 and CO, in H_2O, at high temperature and pressure is not available in the literature. Due to nonideality, water formed by reaction is not entirely in the gas phase while the proportion of water in the liquid phase depends on the operating conditions and the composition of liquid phase hydrocarbons. Our experimental work, [2] conducted in a CSTR with a supported cobalt catalyst for FT reaction, suggests that a considerable proportion of water is in the liquid phase and that temperature has a big effect in this proportion. This work will simulate the diffusion of both H_2 and CO in liquid H_2O and liquid hydrocarbons under the experimental conditions. It will also investigate the diffusion of hydrocarbons out of the liquid layer coated on the catalyst. We expect this study will give us a clear and fundamental understanding of the different effects of hydrocarbons and H_2O on both reaction rate and product selectivity in the Fischer-Tropsch reaction.

Keywords: Fischer-Tropsch Synthesis, H_2 diffusion, CO diffusion, diffusion of hydrocarbons

Reference:

1. Hibbitts, D. D.; Loveless, B. T.; Neurock, M.; Iglesia, E. Mechanistic Role of Water on the Rate and Selectivity of Fischer¨CTropsch Synthesis on Ruthenium Catalysts. *Angew. Chem.* **2013**, *125*, 12499-12504.

2. Lu, X.; Zhu, X.; Hildebrandt, D.; Glasser, D.; Liu, X. A New Way to Look at Fischer-Tropsch Synthesis Using Flushing Experiment. *Ind. Eng. Chem. Res.* **2011**, *50*, 4359¨C4365.

Copyright © American Institute of Chemical Engineers. All rights reserved.

466588 Resolved Particle Fixed-Bed CFD with Microkinetics for Methanol Oxidation on Silver

Abstract:

Resolved Particle Fixed-bed CFD with Microkinetics for Methanol Oxidation on Silver

Behnam Partopour, Anthony G. Dixon

Department of Chemical Engineering, Worcester Polytechnic Institute,

Worcester, MA, USA, 01609

Formaldehyde is one of the highest volume chemicals with annual production of 46.4 million tons as of 2012. One of the major production processes for formaldehyde is methanol oxidation on silver. The reaction occurs at high temperature while carbon dioxide is produced as a dominant by-product. A microkinetics model was proposed and validated by Andreasen et al.[1], to describe the reaction mechanism. The reaction parameters were highly affected by the temperature, species concentrations, and reactor configuration. Therefore, it is desirable to study the mechanism under realistic operating conditions. Recently, we have developed two approaches to couple the microkinetics model with computational fluid dynamics. Unlike the general approaches that couple CFD with microkinetics by setting the reaction as boundary condition for the fluid phase[2], here we integrate the kinetics and transport by defining scalars inside the solid particles. The first method is using the reduced microkinetics with general rate expressions which are generated using the conventional reaction engineering assumptions. The second approach is to couple the detailed kinetics without any assumption about the elementary steps using splines regression with look-up tables.[3] In this approach the microkinetics model is solved under the steady-state condition initially. Then the reaction rates are

mapped into quadratic splines which are used for reaction rate evaluation during the CFD simulations. The simulations are carried out for a randomly packed bed of several hundreds of particles. It is observed that the reduced microkinetic model deviates from the original full microkinetic model under some extreme operating conditions. Moreover, diffusion limitations, axial and radial heat and transport profiles and their effects on the selectivity are studied. It is shown that the combustion is favorable at lower temperatures. Therefore, formaldehyde mass fraction increases along the tube while inside of the upstream particles (closer to the inlet) formaldehyde concentration decreases. It is also observed that when the oxygen mass fraction approaches zero inside the particles the overall reaction rate becomes zero. However, in the regions where methanol concentration becomes very low but oxygen and formaldehyde are present, combustion speeds up. These observations can be used to guide operation of the reactor to improve selectivity.

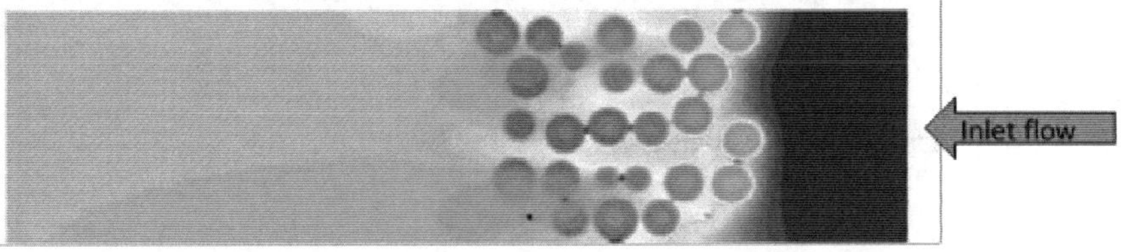

Contours of formaldehyde mass fraction on a mid-plane through the fixed bed

References

1. Andreasen A., Lynggaard H, Stegelmann C, Stoltze P. A microkinetic model of the methanol oxidation over silver. *Surf Sci*. 2003;544(1):5-23.

2. Wehinger GD, Eppinger T, Kraume M. Detailed numerical simulations of catalytic fixed-bed reactors: Heterogeneous dry reforming of methane. *Chem Eng Sci*. 2015;122:197-209.

3. Partopour B, Dixon AG. Computationally efficient incorporation of microkinetics into resolved-particle CFD simulations of fixed-bed reactors. *Comput Chem Eng*. 2016;88:126-134.

Copyright © American Institute of Chemical Engineers. All rights reserved.

466636 Experimental and Theoretical Analysis of Carbon Dioxide Absorption in Aqueous Sodium Hydroxide in the Millichannels

Abstract:

Various types of two phase flow patterns (Sobieszuk, P., Aubin, J., & Pohorecki, R. (2012) in gas-liquid systems have been studied by various researchers (Sobieszuk, P., Aubin, J., & Pohorecki, R. (2012). One of the applications of gas liquid flows is in gas absorption. Carbon dioxide absorption is an important step in today's context since the gas plays an important role in global warming. The results of kinetics reported in the literature have many contradictions in the quantitative kinetics. This is due to the inherent coupling of the physical and chemical phenomena involved in the absorption process in the presence of solutes in the aqueous phase. The physico-chemical effects have been lumped together in a parameter in earlier approaches. This is the primary cause for the discrepancies in the reported measurements where the intrinsic kinetics was not delineated.

In this work, an absorption study of CO_2in an aqueous solution of NaOH is carried out in slug flow in a milli channel. The well defined flow behavior in the micro channel helps to separate out the effects of hydrodynamic mass transfer resistance and the kinetics of absorption. This would help identify the rate determining step to help us better design absorption systems. Slug flow is usually found to be present at low flow rates of gas. (Abadie, T., Aubin, J., Xuereb, C., & Legendre, D. (2012)).Accordingly, in this work also, slug flow is observed at low flow rates and the absorption study is carried out in this flow regime.

Experiments are carried out in a T-junction circular cross-section milli-channels of 2 mm inner diameter using the 99.9 v/v% pure CO_2 and the aqueous solution. The flow rates of gas and liquid are maintained with the help of a mass flow controller (MFC) and a Harvard syringe pump respectively. Liquid phase composition is carried out by Warder's method of titration. This helps us estimate the concentration of

various species in the liquid and consequently the rate of gas absorption.

A model is developed for predicting the slug length variation along the channel. This depicts the rate of absorption of CO_2 in the sodium hydroxide solution. The theoretical model and the experimental data are compared for various operating conditions and they are found to be in close agreement.

Keywords: milli-channels, slug flow, circular channels, two phase flow.

REFERENCES:

1. Sobieszuk, P., J. Aubin, and R. Pohorecki (2012). Hydrodynamics and mass transfer in gas-liquid flows in microreactors. *Chemical Engineering & Technology, 35(8), 1346–1358.*
2. Abadie, T., Aubin, J., Legendre, D., & Xuereb, C. (2012). Hydrodynamics of gas-liquid Taylor flow in rectangular microchannels *Microfluidics and Nanofluidics, 12, 355-369.*

Copyright © American Institute of Chemical Engineers. All rights reserved.

Comparison of the performance on the 60 wt% Fe_2O_3/Al_2O_3 oxygen carrier under the various reducing atmosphere for chemical looping combustion

16AIChE Annual Meeting, San Francisco, CA

Ching Ying Huang and **Yau-Pin Chyou***

Institute of Nuclear Energy Research, Longtan, TAIWAN (ROC)

Introduction

- At present, there is general assent on the need of reducing the emissions of the greenhouse gas, especially CO_2, in order to mitigate climate change.
- Chemical-Looping Combustion (CLC) has been identified as a promising technology to carry out CO_2 capture at low cost.
- Chemical looping combustion (CLC) technology **is a low-cost, high energy efficiency technique** to capture and separate CO_2.
- The chemical looping combustion process consists of two interconnected reactors, i.e. fuel reactor (FR) and air reactor (AR), which is shown in Fig. 1.
  - By oxygen carrier, the mechanisms of CLC reaction are
    (1) **FR:** $C_nH_{2m} + (2n+m)\ Me_xO_y \rightarrow n\ CO_2 + m\ H_2O + (2n+m)\ Me_xO_{y-1}$, and
    (2) **AR:** $2\ Me_xO_{y-1} + O_2 \rightarrow 2\ Me_xO_y$

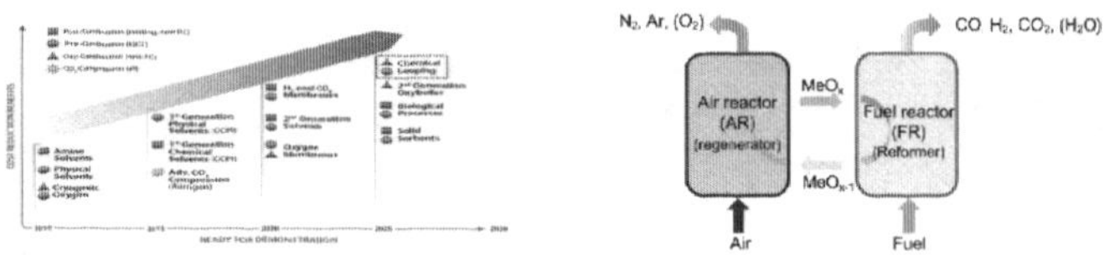

Fig. 1 CLC process.

- Oxygen ratio (R_o) of oxygen carrier describes the maximum oxygen of the oxygen carrier could be transferred during between the air and the reducing (fuel) reaction.

$$R_o(\text{oxygen ratio}) = \frac{m_{ox} - m_{red}}{m_{ox}}$$

m_{ox}: mass of metal-oxide; m_{red}: mass of reduced metal-oxide

Experimental

- A set of chemical looping combustion test system (Figure 2), which contains **gas simulation unit, LabVIEW software control system and thermogravimetric analyzer** (TGA), has been established.
  - to simulate different fuel gas composition.
  - to observe the weight change of oxygen carriers, for assessing the oxygen transport capacity.
  - LabVIEW automatically switch reduction / oxidation gas to perform multiple redox cycle test.
- The oxygen carrier was composed of **iron-oxide** and supported on alumina for increasing the mechanical strength of the oxygen carrier by deposition-precipitation method.

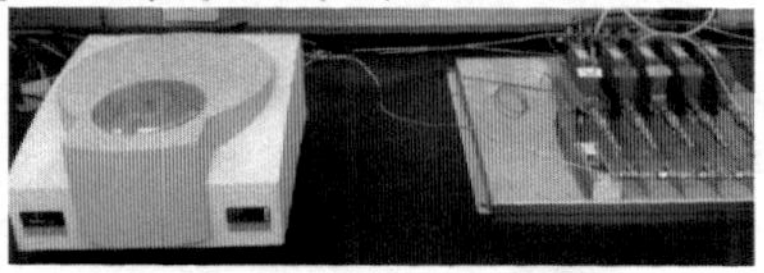

Figure 2. chemical looping combustion test system.

Results and Discussion

- *Material's Physical Properties*
  - The 60wt%Fe_2O_3/Al_2O_3 oxygen carrier was calcined at 950ºC for 6 hours, which surface area was around 35.77 m²/g.
  - The iron oxide recovery of oxygen carrier has been more than 95% by ICP analysis. The differences between experimental and theoretical are caused from repeated washing filter step by preparing oxygen carrier.

Table 1. Metal loading and surface area of the oxygne carriers

Oxygen carrier	MeO_x loading (%)	Sintering temperature (ºC)	BET surface area (m²/g)
60wt%Fe_2O_3/Al_2O_3	57.51	950	35.77

- *Chemical Properties*
  - The oxygen transport capacity and chemical stability of the 60wt%Fe_2O/Al_2O_3 oxygen carrier was evaluated by a thermo-gravimetric analyzer (TGA) under different atmospheres.
    - Comparisons of the oxygen transport capacities of the 60wt%Fe_2O/Al_2O_3 oxygen carrier under different atmospheres reveal that the oxygen transport capacity in hydrogen was the best than the counterpart with the syngas or CH_4, while the oxygen transport capacities were 15.4%, 14.8% and 14.3%, respectively(Fig. 4).
    - It is reported in literature that Fe_2O_3 to Fe_3O_4 is the only reduction in the iron oxide system that is able to convert methane completely to carbon dioxide and water at high temperatures. So, in multiple-cycle tests of the Fe_2O_3/Al_2O_3 oxygen carrier, the **first stage** was selected to do test.
    - After the 20th redox cycling tests, it is found that the stability was very good under syngas and hydrogen atmosphere. The oxygen transport capacity did not decay during redox cycle test. However, the stability of the 60wt%Fe_2O/Al_2O_3 oxygen carrier decreased with the number of cycles under methane atmosphere. (Fig. 5(a)~(c)).

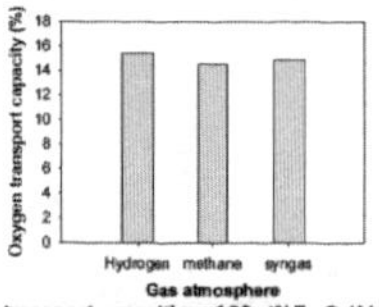

Fig. 4. The oxygen transport capacities of 60wt%Fe_2O_3/Al_2O_3 oxygen carrier under various atmosphere

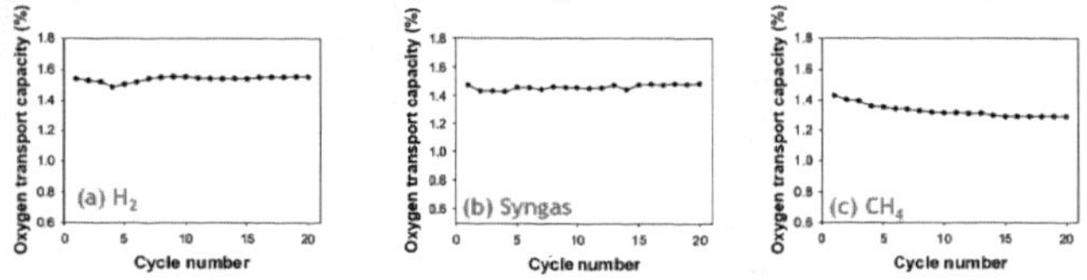

Fig. 5 The oxygen transport capacities of the 60wt%Fe_2O_3/Al_2O_3 for 20 redox cycles conducted with alternately H_2/synas/CH_4 or air.

Conclusion

The 60wt%Fe_2O_3/Al_2O_3 oxygen carrier was successfully prepared by deposition-precipitation method. The oxygen transport capacity of oxygen carrier under hydrogen atmosphere was the best, which was 15.4% by TGA analyzer testing. However, in practical application, the Fe-based oxygen carrier was only able to convert methane completely to carbon dioxide and water in the first reduction stage. The 60wt%Fe_2O_3/Al_2O_3 oxygen carrier could be reduced and oxidized by altering CH_4, H_2, syngas/air to achieve multi-redox cycles. From the 20th redox cycling tests, it was found that the stabilities of the 60wt%Fe_2O_3/Al_2O_3 oxygen carrier under hydrogen and syngas atmospheres were very good. The oxygen transport capacities were no significant degradation. Consider the **environmental friendliness, low cost and high stability**, the 60 wt%Fe_2O_3/Al_2O_3 exhibits a promising potential oxygen carrier for chemical looping combustion.

Acknowledgements

The authors are grateful to the financial support from governmental R&D budget allocated to the Institute of Nuclear Energy Research (INER), Atomic Energy Council (AEC), and the Bureau of Energy (BoE), Ministry of Economic Affairs (MoEA) of Taiwan, ROC.

466822 Dimerization of Bicyclo[2.2.1]Hepta-2,5-Diene over Zeolite Catalysts

Abstract:

Norbornadiene (NBD, bicyclo[2.2.1]hepta-2,5-diene) dimer can be used as a high-energy densty fuel. Dimerization of NBD needs to be promoted by a catalyst. Recently, the homogeneous catalysts have been used in the NBDD manufacturing process. If a homogeneous catalyst is used, the cost of a catalyst is high and the separation process is quite complicated. If a homogeneous catalyst is replaced with a heterogeneous catalyst in the NBDD manufacturing process, recovery and reuse of a catalyst are possible and therefore the cost of a catalyst can be innovatively reduced[6]. The aim of the study is to verify availability of zeolite and mesoporous materia as heterogeneous catalysts for NBD dimerization. In this work, we conducted a research for the dimerization of NBD over a solid catalyst. Zeolite beta and the mesoporous materials produced from zeolite beta ($MMZ_{H\beta}$) were used for the dimerization of NBD. A batch reactor(160 ml) was used for mechanical agitation. Zeolite, NBD and toluene as a solvent were put into the batch reactor and the reactor was purged with nitrogen. The reaction was conducted at 150 – 300 ºC. Cooling water was supplied through cooling coils in the reactor in order to keep constant temperature. The products were collected every three hours and analyzed by GC. Zeolite used as a catalyst was purchased from Zeolyst Co. Hβ zeolite was decomposed in the alkali solution and a mplating agent was added to the solution. Then, a mesoporous material was produced by hydrothermal synthesis. The catalyst which had both micropore and mesopore was named $MMZ_{H\beta}$. Though the amount and the strength of acid sites over $MMZ_{H\beta}$ decreased compared with zeolite beta, the $MMZ_{H\beta}$ catalyst shows high reaction activity and selectivity in a dimerization of NBD, which can be attributed to the large pore size of $MMZ_{H\beta}$ which is favorable for the diffusion of reactant, intermediate, and product.

Copyright © American Institute of Chemical Engineers. All rights reserved.

467004 Benchmarks for Adsorption on Transition Metal Oxide Surfaces: A Comparison of DFT to Experimental Data

Abstract:

Density functional theory (DFT) is commonly used to investigate adsorption and reaction processes on surfaces. While it is widely used, the applicability of DFT to highly correlated transition metal oxide systems with unpaired d-electrons is problematic, especially when using plane wave basis sets. A primary issue is the lack of quality experimental data for adsorbate and reaction systems on well-defined (single crystal) transition metal oxide surfaces to provide benchmarks for the applicability of current DFT methods and the future development of new methods. We have established experimental benchmarks using temperature programmed desorption (TPD) for the heat of adsorption of carbon monoxide on MnO(100) for comparison to DFT simulations on terraces and step defects, using Perdew-Burke-Ernzerhof (PBE) functionals along with DFT+U, DFT-D2, and DFT-D3 packages. Ongoing experimental work with MnO(100) will also provide experimental heat of adsorption data for CO_2 and H_2O on this high spin $3d^5$ transition metal oxide surfaces.

Copyright © American Institute of Chemical Engineers. All rights reserved.

467010 Gold-Supported Cerium-Doped NiO_X Catalysts for Water Oxidation

Abstract:

The development of high-performance catalysts for the oxygen evolution reaction (OER) is critical for converting renewable electricity to fuels and chemicals cost-efficiently.[1] Herein we report the significant enhancement in the OER activity of electrodeposited NiO_X films resulting from the dual effects of using a gold metal support and a cerium dopant.[2] This $NiCeO_X$-Au catalyst achieves high OER activity in an alkaline environment, and is among the most active OER electrocatalysts reported to date. Based on theoretical modelling coupled with experimental observations, we attribute the activity to a combination of geometric, electronic and support effects, where highly active under-coordinated sites at the oxide-support interface are modified by the local chemical binding environment and by doping the host Ni oxide with Ce. The $NiCeO_X$-Au catalyst is further demonstrated in a device context by coupling it with a nickel-molybdenum hydrogen evolution catalyst in a water electrolyser setup, which delivers 50 mA consistently at 1.5 V over 24 hours of continuous operation. This work deepens the understanding of complex multi-component catalytic systems by showing how electronic, geometric and support effects can be combined to enhance catalytic activity.

References

[1] Yang, Z., Zhang, J., Kintner-Meyer, M.C.W., Lu, X., Choi, D., Lemmon, J.P. & Liu, J. Electrochemical energy storage for green grid. *Chem. Rev.* **111**, 3577-3613 (2011).

[2] Ng, J.W.D., Garcia-Melchor, M., Bajdich, M., Charthranont, P., Kirk, C., Vojvodic, A., Jaramillo, T.F. Gold-Supported cerium-doped NiO_X catalysts for water oxidation. *Nat. Energy*. **1,** 16053 (2016).

Copyright © American Institute of Chemical Engineers. All rights reserved.

467030 Scale up Study of Fischer Tropsch Synthesis Reactor with Reduced CO_2 emissions Using Cu-K Promoted Silica Supported Fe Catalyst

Abstract:

Fischer Tropsch Synthesis (FTS) is a technology to produce liquid fuels from syngas (mixture of H_2 & CO obtained from coal, petcoke and biomass) as an alternate to crude oil. However, the quantity of emitted CO_2 from the FTS is one of the serious disadvantages of the process. Choice of operating parameters of reactor is one of the methods for reduction of CO_2 emissions from FTS. Parametric studies were conducted in a once through fixed bed reactor with mixture of H_2 and CO in a temperature range of 215-245°C, the pressure range of 5-25 bar, the space velocity range of 1000-4000 cc/g cat/h and a H_2/CO feed ratio of 0.7/1–3/1 (mol/mol) to investigate the influence of the above process parameters on the performance of in-house developed Cu-K promoted silica supported Fe based catalyst in terms of its hydrocarbon & CO_2 formation rate, selectivity of olefins to paraffins, liquid hydrocarbon product distribution and properties.

The reaction rate, one of the most important parameters, plays a significant role in scale up of FTS reactor and industrial design. It was observed that FTS reaction rate (hydrocarbon formation rate) increased with increase in temperature, pressure and GHSV. However, FTS reaction rate passed through optimum with increase in H_2/CO feed ratio from 0.7/1–3/1 (mol/mol). CO_2 formation rate (Water Gas Shift activity rate) increased with increase in temperature and GHSV, while decreased with increase in pressure, H_2/CO molar ratio. With increase of temperature and space velocity, selectivity of olefins to paraffins increased while the reverse was observed with increase in pressure and H_2/CO molar ratio. Further, liquid hydrocarbon product distribution, chain growth probability and properties were investigated on the basis of liquid products collected after completion of 1000 h run at the above operating domain. The experimental results show that the chain growth probability and liquid product properties depend on operating conditions as well as composition of catalyst. From the above analysis, the optimum operating conditions for higher hydrocarbon formation and lower CO_2 formation rate of developed Cu-K promoted and silica supported Fe based catalyst are identified for scale up to industrial reactor. This is done in order to explore the prospect of obtaining higher amount of synthetic crude fuel & lighter olefins products that can be integrated into existing refineries & petrochemical complex, in line with the main aims and objectives of the project.

Copyright © American Institute of Chemical Engineers. All rights reserved.

467265 A Cost-Free Catalyst for CO_x-Free Hydrogen Production: Utilization of Red Mud for Ammonia Decomposition

Abstract:

Here we utilize red mud (RM), a by-product of the Bayer process for aluminum production (produced up to 2.5 ton for each ton of aluminum produced), as an ammonia decomposition catalyst to present an almost cost-free catalyst as a green alternative to the high-cost Ru-based catalysts. Upon acid digestion at different conditions followed by calcinations at different temperatures modified red mud (MRM) samples with different chemical composition and physical properties were obtained. Data measured for ammonia decomposition on these MRM samples show high and stable hydrogen production rates. For instance, an MRM sample prepared by acid digestion of RM at 220 °C in HCl followed by reduction at 700 °C showed stable hydrogen production rates of 72 and 196 mmol H_2 min^{-1} g_{cat}^{-1} for more than 72 h at 700 °C measured at the corresponding space velocities of 72 000 and 240 000 cm^3 NH_3 h^{-1} g_{cat}^{-1}, respectively. These results are the highest ever reported for a non-noble metal catalyst for stable hydrogen production from ammonia. Detailed characterization of the samples revealed that Fe_3N species formed by the conversion of metallic iron during the induction period are the active species for the reaction. Data further indicate that there is a direct correlation between the size of Fe_3N nanoparticles and the hydrogen production rate. These results offer a clean solution for the utilization of one of the most hazardous and highly-produced industrial wastes for generating CO_x-free clean energy.

This work is supported by the Scientific and Technological Research Council of Turkey (TUBITAK), 1003-Primary Subjects R&D Funding Program (Project number: 213M028). A.U. acknowledges the BAGEP Award of the Science Academy of Turkey. We thank Koç University Surface Science and Technology Center (KUYTAM) for help with sample characterization.

Copyright © American Institute of Chemical Engineers. All rights reserved.

467284 Crystallization of One-Dimensional Zeolites: Elucidating Nonclassical Mechanisms of Growth and the Role of Structure Direction

Abstract:

Crystallization of One-dimensional Zeolites: Elucidating Nonclassical Mechanisms of Growth and the Role of Structure Direction

Rui Li and Jeffrey D. Rimer

Department of Chemical and Biomolecular Engineering, University of Houston, 4726 Calhoun Road, Houston, TX, 77204, jrimer@central.uh.edu

Zeolites possess well-defined pore systems that give rise to shape selectivity for chemical reactions, thus providing an essential advantage for their use as catalysts in commercial applications. The performance of zeolites (i.e., activity and lifetime) is closely aligned with the selection of internal pore geometry and dimensions, which can influence diffusion, adsorption, and reaction pathways. Design of optimized zeolite catalysts *a priori* is challenging owing to the complexity of synthesis that is derived in part from the unknown role of amorphous precursors, organic structure-directing agents (OSDAs), and the pathways of crystallization. Previous studies have demonstrated that one-dimensional (1D) zeolites, such as zeolite L (LTL type), grow via nonclassical mechanisms involving the assembly of so-called *worm-like particles* (WLPs), which are bulk amorphous precursors that serve as growth units during crystallization.[1] Evidence of crystallization by particle attachment (or CPA) is mounting for a range of materials that include biominerals[2], metal oxides,[3] and zeolites.[4] In this relatively new area of research, there are many knowledge gaps between the design of synthesis conditions and the resulting physicochemical properties of the materials.[5] One interesting phenomenon regarding nonclassical crystallization of zeolites is the existence of WLPs, which are common to many different framework types.[1] Here, we will discuss how two widely-used 1D zeolites,[6,7] LTL and TON, crystallize from similar growth solutions that comprised of amorphous WLP precursors. Despite the similarity of their precursor solutions, we will show how these two structures involve distinct growth pathways that can be manipulated by the judicious selection of synthesis parameters (e.g., temperature, alkaline content, OSDA, etc.). We will present our work on the design of OSDAs for zeolite TON crystallization. Moreover, we will discuss mechanisms of LTL and TON crystallization and demonstrate how facile, economically-viable methods can be used to produce zeolites with improved properties for a wide range of industrial applications.

References

(1) Kumar, M.; Li, R.et al., *Chemistry of Materials,* **2016**, *28*, 1714-1727.

(2) Politi, Y.; Arad, T.et al., *Science,* **2004**, *306*, 1161-1164.

(3) Penn, R. L.; Banfield, J. F. *Science,* **1998**, *281*, 969-971.

(4) Lupulescu, A. I.; Rimer, J. D. *Science,* **2014**, *344*, 729-732.

(5) De Yoreo, J. J.; Gilbert, P. U. P. A.et al., *Science,* **2015**, *349*.

(6) Nacamuli, G. J.,US6143166 A, 2000.

(7) Perego, C.; Carati, A. *Zeolites and zeolite-like materials in industrial catalysis*; In *Zeolites: From Model Materials to Industrial Catalysts*; Čejka, J., Peréz-Pariente, J., Eds.; Transworld Research Network: 2008, p 357-389.

Copyright © American Institute of Chemical Engineers. All rights reserved.

467318 Enhancing the Value of Detailed Kinetic Models through the Development of Interrogative Software Applications

Abstract:

Chemical kinetics models are required to face day-to-day issues in an industrial setting due to stringent product quality, process optimization and environmental constraints. Models help in tracking plant operations and give research direction for improvement of the process. As the complexity of the model increases, it becomes difficult to understand the complete set of detailed implications it conveys. Moreover, a process engineer often requires a rapid calculation of model results in terms of plant measurable input and output variables. These issues can be especially challenging in the case of detailed kinetics models, where the once-through solution time can be of order tens of seconds. This motivated the present work which is focused on the development of user-friendly software applications to interrogate detailed kinetic models.

A user-friendly I/O (Input/Output) converter application permits execution of molecular-level composition and kinetic models in a manner which focuses on measurable inputs and outputs. This app was used to efficiently study the impact of different feed streams, operating conditions, hydrogen partial pressure and catalyst flow rate on performance of a catalytic reactor. Such I/O converter apps are process-specific as the app is dependent on the process I/O. A second application, the KME Reactor Flowsheet tool, was developed to build kinetic models based on data from a pilot or commercial reactor which can split feed streams and perform reactor bypasses. It also includes different features like splitters, heaters and reactor blocks and also can handle multiple kinetic models.

The functionality of these applications is tested to study the behavior of a catalytic reforming unit. The inputs to the model are distillation, overall and detailed PIONA, density, and reactor conditions. The outputs from the model are light-gas yields, detailed PIONA, coke, and temperature drops across each reactor bed. The I/O converter and flowsheeting applications were used to study the impact of feed variation, partial pressure of hydrogen and catalyst

circulation rate on yield, composition and coke concentration in the product.

Copyright © American Institute of Chemical Engineers. All rights reserved.

467341 Tailor-Designed Oxidation Catalysts for Low Temperature Combustion Engines

Abstract:

Stricter fuel economy standards have led to the automotive industry moving forward with higher efficiency lean combustion modes. These lean combustion modes need to meet environmental policies placed on the vehicle emissions before they are able to be commercially implemented. Much lower exhaust temperatures and higher CO and hydrocarbon concentrations lead to a significant challenge in reducing the emissions of these lean combustion modes, as current catalytic exhaust systems are not able to give adequate activity. This study looks at the performance of various oxidation catalysts under these fuel conditions. Catalysts containing Pt and Pd and various ratios thereof supported on Al_2O_3 washcoated onto a cordierite monolith were tested under simulated exhaust conditions. An understanding of the different oxidation properties for different reactant species of each metal can be seen in these results, which can be used to guide us in designing superior oxidation catalysts in order to meet the environmental regulations.

Copyright © American Institute of Chemical Engineers. All rights reserved.

467550 Reducing Greenhouse Gasses Using Low Temperature AEM Electrochemical Devices

Abstract:

Climate change is potentially the largest area of concern in the scientific community today. The global carbon dioxide emissions are 35 giga tons per year and climbing, and account for the largest contribution to the greenhouse gas problem.[1] In addition, global anthropogenic methane emissions total 350 million tons annually and have 25 times the greenhouse gas effect of carbon dioxide.[2] This results in wasteful flaring of the energy rich methane produced as a byproduct of the fossil fuel industry. This research has utilized a two pronged approach to the greenhouse gas problem applying electrochemical techniques utilizing anion exchange membranes (AEM) to decrease greenhouse gasses release into the atmosphere. Methane emission is decreased by converting it into methanol, the worldâ€™s 5th largest commodity by volume. Methanol can also be used to synthesize numerous products, and has be touted as an important energy carrier of the future. It has a high energy density and is a liquid at atmospheric conditions, which make it ideal for stable transportation and storage that is compatible with existing petroleum infrastructure, which is quite the opposite of methane gas.[3] Methane activation and conversion is typically accomplish through the energy intensive syngas pathway, along with many intermediate reaction steps that result in an inelegant and energetically circuitous process. Allowing for control of the catalyst surface free energy, electrochemical methods have the potential to reduce the thermal and overall energy barrier to convert methane to oxygenates, and more specifically methanol or formaldehyde production may be reduced to a single step.[4] In this research methane activation is shown on a nickel oxide anode catalyst in AEM devices utilizing both hydroxide and carbonate ions and the oxygen donor.

Similar AEM electrochemical devices can be used for carbon dioxide separation, specifically with (bi)carbonate as the mobile ion.[5, 6] In a typical carbonate-based device, oxygen is reduced with carbon dioxide at the cathode, generally through an indirect pathway first to hydroxide, then (bi)carbonate. This pathway results in a mixed anion composition throughout the AEM where three very different anions can carry the charge across the membrane. Bicarbonate and carbonate (especially bicarbonate) have lower conductivities than hydroxide, but this cannot alone explain the 2 to 3 order of magnitude performance loss from a carbonate-free cell. In a carbon dioxide pump, the oxygen evolution reaction at the anode results in evolved and separated carbon dioxide. The cell potential and resulting half reactions influence the mixed transport and the amount of carbonate, bicarbonate, and hydroxide anions transported. In the carbon dioxide pump a relationship is observed between the evolution of carbon dioxide at the anode and the flow of electrons in the cell. This relationship between current and evolved CO_2 is indicative of the anion ratio and the exchange mechanisms occurring at the anode the cathode. In this study, a peak current density of 100 mA/cm^2 was reached in CO_2 pumping mode. This performance was achieved with radiation grafted anion exchange membranes and ionomers.[7]It was observed that performance was very sensitive to AEM humidification; that there is a delicate balance between maintaining a high water content in the membrane to facilitate anion transport and preventing catalyst and GDL flooding. The cell anion balance was examined in these high performing cells. The order of magnitude increase in the carbonate/bicarbonate transport observed in this work significantly increases the viability of membrane based carbon dioxide pumps and fuel conversion devices. However, our results also point to advances that are needed in both catalysts and membranes to increase the faradaic efficiency in high current density carbonate devices.

1. IPCC, *Climate Change 2014: Impacts, Adaptation, and Vulnerability. Part A: Global and Sectoral Aspects. Contribution of Working Group II to the Fifth Assessment Report of the Intergovernmental Panel on Climate Change Field, C.B., V.R. Barros, D.J. Dokken, K.J. Mach, M.D. Mastrandrea, T.E. Bilir, M. Chatterjee, K.L. Ebi, Y.O. Estrada, R.C. Genova, B. Girma, E.S. Kissel, A.N. Levy, S. MacCracken, P.R. Mastrandrea, and L.L. White (eds.)]*, p. 1132, Cambridge University Press, Cambridge, United Kingdom and New York, NY, USA (2014).

2. S. Menon, K. L. Denman, G. Brasseur, A. Chidthaisong, P. Ciais, P. M. Cox, R. E. Dickinson, D. Hauglustaine, C. Heinze, E. Holland, D. Jacob, U. Lohmann, S. Ramachandran, Leite da Silva Dias, Pedro, S. C. Wofsy and X. Zhang, in *Climate Change 2007 â€" The Physical Science Basis: Contribution of Working Group I to the Fourth Assessment Report of the IPCC*, S. Solomon, Editor, p. 498, Cambridge University Press, 2007 (2007).

3. G. A. Olah, A. Goeppert and S. K. Surya-Prakash, *Beyond oil and gas: the methanol economy*, Wiley-VCH (2009).

4. N. Spinner and W. E. Mustain, *J. Electrochem. Soc.*, 159, 12 (2012).

5. W. A. Rigdon, T. J. Omasta, C. Lewis, M. A. Hickner, J. R. Varcoe, J. N. Renner, K. E. Ayers and W. E. Mustain, *Journal of Electrochemical Energy Conversion and Storage.*, February SI (2017). DOI: 10.1115/1.4033411.

6. J. Landon and J. R. Kitchin, *J.Electrochem.Soc.*, 157, 8 (2010).

7. S. D. Poynton, R. C. Slade, T. J. Omasta, W. E. Mustain, R. Escudero-Cid, P. Ocón and J. R. Varcoe, *Journal of Materials Chemistry A.*, 2, 14 (2014).

Copyright © American Institute of Chemical Engineers. All rights reserved.

467676 Hydrogen Production By Steam Reforming of Bio-Oil Derived Acetic Acid over Ni Based Zr-SBA-15 Type Mesoporous Catalysts

Abstract:

Acetic acid (AcOH) is a renewable, non-toxic and a safe bio-based hydrogen carrier. It is one of the main components of bio-oil, which can be produced by pyrolysis of biomass. Acetic acid can also be produced from biomass through fermentation. Hence, it is considered as an attractive renewable resource for the production of synthesis gas and/or hydrogen, through a catalytic steam reforming process [1]. As a result of steam reforming process of AcOH, a gas mixture which is mainly composed of H_2, CO and CO_2, is produced. Product distribution and the ratio of CO/CO_2 in the produced syngas is expected to be strongly related to the catalyst composition and reaction conditions. This gas mixture can be used as a fuel in solid oxide fuel cells or to obtain high purity hydrogen for chemical applications. Syngas thus produced can also be used as a resource for the production of valuable chemicals and fuels through Fischer-Tropsch or dimethyl ether synthesis routes. In such cases, a gas composition with equimolar quantities of H_2 and CO is generally preferred. Production of syngas from acetic acid mainly involves steam reforming (R.1) and water gas shift reactions (R.2). Occurrence of thermal decomposition, methanation, decarboxylation and coke formation reactions would also contribute to the product distributions and catalyst deactivation.

$CH_3COOH + 2H_2O \leftrightarrow 4H_2 + CO_2$ (R.1)

$CO + H_2O \leftrightarrow H_2 + CO_2$ (R.2)

Synthesis of active, stable and selective catalysts for steam reforming of acetic acid is a hot research area for the development of new renewable energy processes. Nickel is one of the most active catalytic metals for reforming reactions. Type of support material is expected to have strong influence on the performance of Ni based catalysts in steam reforming of AcOH. Invention of silicate structured mesoporous materials, like MCM-41 and SBA-15, opened new avenues in catalysis research. These materials, with ordered pore structures, were reported to be less susceptible to deactivation due to coke formation than the conventional microporous catalyst supports. They also cause less resistance to diffusion of reactants to the active sites. However, some of the silicate structured mesoporous materials were known to have low hydrothermal stability in the presence of water vapor. Zirconia, with its high thermal and chemical stability attracted the attention of catalysis researchers in recent decades. Strong interaction of Ni clusters with zirconia and quite high water dissociation capacity of zirconia were reported as attractive properties of this material in steam reforming reactions. More recently, zirconia incorporated SBA-15 was reported as a highly stable catalyst support for steam reforming of ethanol [2]. In the present study, zirconia incorporated SBA-15 type mesoporous materials with different Zr/Si ratios were synthesized following a one-pot hydrothermal route, characterized and used as the catalyst support materials for the synthesis of new Ni based catalysts to be used in steam reforming of AcOH. Some of these Ni based catalysts were modified by addition of tungsten, during the impregnation step. Catalytic performances of these materials were tested in steam reforming of AcOH at 750 ^oC. Synthesized Zr-SBA-15 type support materials were shown to have ordered mesopore structures with high surface area values. For instance, the material containing 25% zirconia in SBA-15 (25Zr-SBA-15) had a surface area of 608 m^2/g, while some decrease of surface area was observed as a result of Ni and W impregnation. For instance, the catalysts containing 5% Ni (5Ni@25Zr-SBA-15) and 5% Nİ+10% W (5Ni-10W@25Zr-SBA-15) had surface area values of 561 m^2/g and 151 m^2/g, respectively. Catalytic activity tests proved that the performance of the Zr-SBA-15 supported Ni based catalysts were highly stable and they also showed very high activity in steam reforming of acetic acid, giving complete conversion at temperatures over 700 ^oC, at space time of 0.072 $s.g.ml^{-1}$. Product distributions were shown to be strongly influenced by the composition of the catalyst. In the case of 5Ni@25Zr-SBA-15, syngas produced at 750 ^oC contained about 54 % H_2, 22% CO, 20% CO_2 and 4% CH_4. These results indicated that thermal decomposition of AcOH to methane and CO_2 was minimized over this catalyst. Results were considered to be highly promising from the point of view of production of hydrogen rich syngas. It was most interesting to observe that modification of this catalyst by the addition of tungsten caused significant changes in the product distribution. For instance, syngas produced over 5Ni-10W@25-SBA-15 at the same reaction conditions, contained equimolar quantities of H_2 and CO (about 47.5 % each) with very small amounts of CO_2 and CH_4 (about 3% and 2%, respectively). Production of a syngas with such a composition was considered to be highly attractive from the point of view of a resource gas for DME and Fischer-Tropsch synthesis. Activity test results obtained with different catalyst compositions and at different reaction conditions, as well as the characterization results of the synthesized catalytic materials, will be reported in this presentation.

Acknowledgements: TUBITAK Project (214M578) and TUBA.

References:

[1] Goicoechea S., Kraleva E., Sokolov S., Schneider M., Pohl M.M., Kockmann N., Ehrich H., Appl. Catal. A: Gen. 514 (2016) 182.

[2] Arslan A., Gunduz S., Dogu T., Int. J. Hydrogen Energy, 39 (2014) 18264.

Copyright © American Institute of Chemical Engineers. All rights reserved.

467732 Effect of Fast Cycling on the Formation of HC Intermediates and NOx Conversion Using H_2, Propene or Propane As Reductants

Abstract:

Effect of Fast Cycling on the Formation of HC Intermediates and NOx Conversion Using H_2, Propene or Propane as Reductants

Allen Wei-Lun Ting, MengMeng Li, Michael P. Harold and Vemuri Balakotaiah

Department of Chemical & Biomolecular Engineering, University of Houston, Houston, TX 77204, USA

Abstract
A combined experimental and modeling study of fast cycling NOx storage and reduction for emissions control of lean burn gasoline and diesel vehicles is conducted to provide mechanistic insight. Experiments reveal notable NOx conversion enhancement for cycle times less than 10 seconds (fast cycling) compared to long cycle times of a minute (Fig. 1, Perng et al. Catalysis Today 231 (2014) 125–134 [HMP1] [TA2]).

The mechanism of the fast cycling has been analyzed through both experiments and simulations with H_2, propene, or propane as the reductant and ceria-free and ceria-containing NOx storage catalyst ($Pt/BaO/Al_2O_3$). For H_2, more frequent storage and regeneration under fast cycling increases the storage capacity available proportionally when storage and reaction is capacity limited, with negligible effect when kinetically limited. The mechanism explains the increase in NOx conversion over a wide temperature range and corresponding decrease in the NH_3 selectivity. The cycle-averaged catalyst temperature under fast cycling increases because of the combined effect of dispersion effect between lean and rich feed and the enhancement of O_2 storage ability by ceria.

The "Di-Air" system proposed by Toyota researchers shows an overall increase in NOx conversion when using propene as the reductant under fast cycling[HMP3] [TA4] (Bisaiji et al., SAE 2011-01-2089). Modeling results show that fast cycling enhances the formation of a thermally-stable HC intermediates from propene and NO. This results in a substantial increase the conversion at high temperature (Fig. 2). However, using propane as the reductant shows a negative effect. Modeling results predict this inverse behavior by treating the dehydrogenation reaction from propane to propene to be the rate limiting step to reduce NOx.

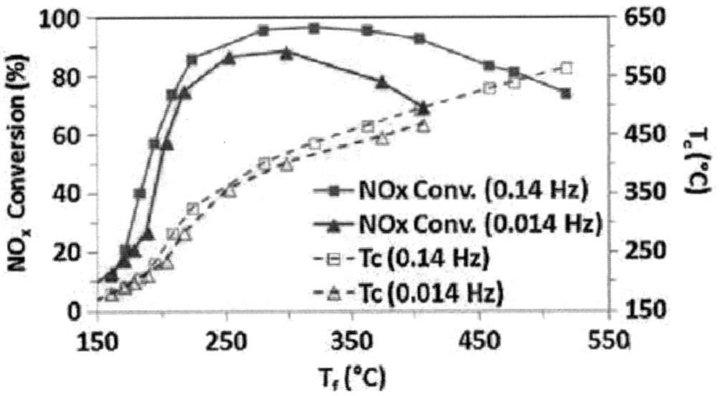

Fig. 1 Cycle averaged NOx conversion and catalyst temperature vs. feed temperature for two different frequencies. Feed: 700 ppm NO 5% O_2, rich phase contain 1.75% C_3H_6.

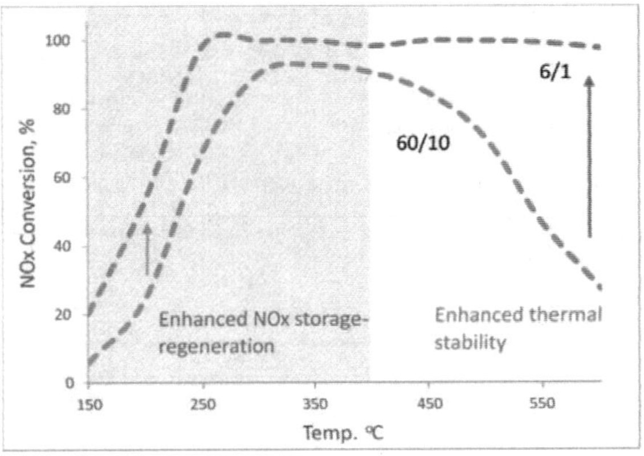

Fig. 2 Simulated isothermal cycle averaged NOx conversion vs. catalyst temperature for two different frequencies. Feed: 700 ppm NO 5% O_2, rich phase contain 1.75% C_3H_6.

[HMP1]Provide reference

[TA2]

[HMP3]Provide reference

[TA4]

Copyright © American Institute of Chemical Engineers. All rights reserved.

467735 Calcium Oxide As in-Situ Catalyst for Steam Gasification of Bio-Char

Abstract:

With increasing demand for energy and the environmental impact from fossil fuel use, energy recovery from waste biomass such as forestry residue and agricultural waste has attracted increasing attentions. Thermal treatment is the most commonly used conversion method to upgrade energy quality from biomass, among which pyrolysis is regarded as an economical and promising process to reach high efficiency and low emission. By pyrolysis, biomass can be converted into bio-oil, bio-char and gaseous fraction. As a by-product of biomass pyrolysis, bio-char can be burned for energy recovery, prepared as activated carbon, or soil amendment. Recently, gasification of bio-char has gained increasing interest. With steam introduced as gasification agent, bio-char exhibits effective ability to produce syngas with a high hydrogen content, which is regarded as a clean, low carbon future source of energy.

Calcium oxide (CaO) has been proven to be an effective and economical in-situ catalyst in biomass gasification, in order to further improve hydrogen production from syngas. The loaded CaO to the gasifier can take effect via several pathways. Firstly, CaO can act as sorbent for CO_2 absorption to form $CaCO_3$. Following the decrease in CO_2 partial pressure, the water gas shift reaction is enhanced towards hydrogen production. Secondly, since the CaO carbonation reaction is exothermic, the releasing heat could also provide necessary energy to facilitate the endothermic gasification process. Besides, tar formation is a main issue during biomass gasification due to the fact that it may decrease the system efficiency and even block and damage downstream gasifier. Another potential advantage of the CaO additives is that it could catalytically contribute to tar cracking and accelerate the reaction between tar and H_2O, therefore significantly reduce the tar content of the product gas from a gasifier. The tar cracking could result in gas yield increase.

Accordingly, in this study, steam gasification of bio-char in a fluidized bed reactor has been investigated, using CaO as in-bed catalyst. The purpose of the study is to evaluate the catalytic gasification performance of bio-char for hydrogen-rich gas production and tar removal. The influence of CaO/wood mass ratio, steam flowrate and reaction temperature is studied to search for an optimal working condition. Results reveal that increasing the ratio of CaO/wood is effective to enhance H_2 concentration and yield in the syngas. Simultaneously the tar yield is reduced remarkably due to the enhanced reforming reactions. Increasing the steam flowrate is found to be beneficial to shift the water gas shift reactions and improve the reactivity of CaO towards H_2 production. An excessive steam injection leads to an opposite effect, and a maximum H_2 output is achieved at steam flowrate of 160 g/h. Temperature also plays a great role on the yield and product composition. A higher temperature enhances the H_2 production and tar reduction. In conclusion, the results indicate that there is a strong potential for producing hydrogen-rich gas from bio-char by catalytic steam gasification with inexpensive and abundant CaO as catalyst. The data obtained can be served as leverage for future development on biomass to energy field.

Copyright © American Institute of Chemical Engineers. All rights reserved.

467848 Use of Stability Diagrams to Predict Catalyst Speciation during Fischer Tropsch Reduction and Reaction

Abstract:

An understanding of the speciation of iron catalyst precursor during reduction is an important piece of knowledge in Fischer-Tropsch (FT) reaction. The thermodynamics of the precursor speciation is governed by the reducing gases used, and this tends to influence the activity, selectivity and the catalyst life span. Economic benefits can potentially be obtained if the same syngas used for the normal FT reaction can also be used to activate the catalyst at the normal FT reaction temperature.

During catalyst activation or FT synthesis, the interaction of reducing agents or syngas with Fe-based catalyst results in the formation of several gaseous components such as CO, H_2, CO_2, H_2O, and CH_4. The partial pressure of each gaseous component determines the predominant state of the catalyst. Catalyst speciation happens due to different partial pressures of the gaseous components, and as a result, the stable iron phases formed during FT synthesis are those that are in equilibrium with the gas composition. Although the ratio of P_{H2}/P_{H2O} and P_{CO}/P_{CO2} or the partial pressure of water and carbon dioxide does not have any appreciable effect on the deactivation by oxidation on the FT reaction rate in Cobalt catalysts, in the iron based catalyst deactivation by oxidation is predominant. Based on thermodynamic calculations of the FT system, the maximum allowable oxygen partial pressure during catalyst activation is 10^{-45} bar. The oxygen partial pressure in therefore dependent on the PH_2/PH_2O and PCO_2/P_{CO} ratios, and the equilibrium constants. Plots of O_2 partial pressure against PH_2/PH_2O , PCO_2/P_{CO} ratios have been used to predict the stability of the different species.

Copyright © American Institute of Chemical Engineers. All rights reserved.

467973 Modeling the Sensitivity of Design Parameters for Photo-Electrochemical Cells Containing a Semiconductor Absorber and Metal Photocatalysts

Abstract:

The magnitudes of energetic losses in photo-electrochemical cells are highly dependent on the nature of the semiconductor, electrocatalyst and the integration of these two building blocks. For example, one can reduce overpotential losses with the addition of an electrocatalyst, while at the same time diminishing the optical efficiency of photon to useful electron conversion. The interconnected nature of energetic losses in photo-electrochemical devices creates challenges in the design of systems that reduce losses simultaneously.

In this contribution, we present a model that integrates optical and kinetic properties for photo-electrochemical cells. We have applied this model to analyze the sensitivity of various experimental parameters on system performance. Provided with these insights we will discuss the most significant parameters, and suggest where research efforts can make the largest impact in improving efficiencies in photo-electrochemical cells.

Copyright © American Institute of Chemical Engineers. All rights reserved.

468034 Enhanced Carbon Sequestration with Cascade Enzymtic Reactions

Abstract:

Carbon capture capacity is an important factor of the feasibility of carbon capture and sequestration process. Almost complete conversion would be achieved by chemical carbon capture while most biocatalytic reactions reach much lower degree of conversion as limited by reaction equilibrium. In order to achieve high conversion rate in biocatalytic reactions, cascade enzyme reaction designed for carbon capture in a homogeneous solution can further the conversion of substrate, thus improving the overall carbon capture capacity. By integrating the reactions of two key enzymes of RTCA cycle, isocitrate dehydrogenase and aconitase, we demonstrate that intensified carbon capture can be realized through such cascade enzymatcic reactions. With cascade reaction, 0.5 mol carbon could be captured at pH 6 for each mole of substrate applied, compared with only 0.2 mol carbon with single isocitrate dehydrogenase. Overal it manifests tat carbon capture capacity of biocatalytic reactions cab be intensified with cascade reactions.

Copyright © American Institute of Chemical Engineers. All rights reserved.

468068 Global Optimization of Initial Conditions of RAFT Polymerization Process

Abstract:

This works presents a survey on global optimization methods applied to the operation of a polymerization batch reactor. Usually operational variables as temperature and pressure are used to optimize reaction time and conversion using fixed initial concentrations and parameters usually inherited from the lab work. In this effort, the initial conditions are optimized as a first step toward the optimization of the process. To achieve this goal the method starts with a comprehensive solution for the RAFT (Reversible Addition-Fragmentation chain Transfer) polymerization in a batch reactor using the moment method model. The complex model provides concentrations Initiator Monomer, dead and slept polymer and polydispersity and average molecular weight and lenght among others. The model is then used to perform optimization of the initial operating conditions using Differential Evolution (DE), Simulated Annealing (SA) algorithm, Particle Swarm (PS) and the Barter Algorithm (BA). With this information it is intended to implement a methodology for process optimization to enhance its use in industry. The RAFT process is known to provide a chemical control of the most important parameters, however if it is intended to use in the industry it is necessary to propose control systems and optimal initial conditions and operation to ensure product quality and therefore its sustainability and profitability.

Copyright © American Institute of Chemical Engineers. All rights reserved.

468115 The Role of Co-Catalysts in the Metathesis of Butenes to Propylene

Abstract:

The metathesis of ethylene and 2-butenes over a WO_3/SiO_2 catalyst is an important industrial reaction for the production of on-purpose propylene to supplement traditional supply routes such as steam cracking and fluid catalytic cracking (FCC). Over the years, various metal oxides have been proposed as catalysts with the most common being oxides of tungsten, rhenium and molybdenum supported on silica, alumina or Si-Al. In addition to the primary catalysts, co-catalysts comprising of various compounds including metal oxides, hydroxides and nitrates such as CaO, MgO, Y_2O_3, $Ca(OH)_2$ and $NaNO_3$ have been shown to enhance metathesis activity.

The primary objective of this research study is to develop an industrially pertinent co-catalyst for the valorization of C4 raffinate streams to propylene via cross- and self-metathesis as opposed to conventional ethylene metathesis. In this study, we examined the influence of adding various co-catalysts to study their effect on the rate of propylene formation. Various preparation techniques were employed to synthesize the co-catalysts. Catalytic performance was correlated to surface area and pore volume (BET), basicity measurements (CO_2-TPD), chemical and electronic states measurements (XPS), particle morphology (SEM) and crystallinity (XRD).

The obtained results prove that appropriately synthesized co-catalysts with controlled microstructures when added to a primary metathesis catalyst, augment the metathesis performance. A novel reaction mechanism explaining the contribution of activators to the formation of the metal-carbene species will be presented. This research effort provides insight into the design of fine-tuned and catalytically active metathesis co-catalyst systems.

© Copyright 2016, Saudi Aramco. All rights reserved.

Copyright © American Institute of Chemical Engineers. All rights reserved.

468205 Elucidation of the Mechanisms for Direct Synthesis on Pd, Aupd, and Agpt Clusters

Abstract:

Hydrogen peroxide (H_2O_2), an environmentally benign oxidant, is currently produced by the auto-oxidation of anthraquinones (AO), yet direct synthesis (DS) is a greener alternative with the potential to make H_2O_2 using less energy (i.e., less concentration and separation steps) than AO. However, DS currently suffers from low H_2O_2 selectivities (< 60% H_2O_2

of acids and halides.[1] In order to achieve DS that can compete financially, higher selectivities (> 90% H_2O_2) must be obtained, especially in the absence of toxic or caustic additives. We aim to reach this goal by developing a fundamental understanding of the DS mechanism on Pd and AuPd in order to elucidate the manner in which AuPd catalysts give higher H_2O_2 selectivities than do Pd catalysts.[2] Additional work examined tunable octahedron AgPt clusters to demonstrate the effects of changing the coordination of Pt surface atoms. This information will guide the rational design of alternative catalysts, made from more abundant materials that meet or exceed selectivities achieved on AuPd.

Steady-state H_2O_2 and H_2O formation rates were measured using a fixed bed, plug flow reactor as functions of H_2 pressure (25-400 kPa), O_2 pressure (25-400 kPa), and temperature (273-337 K) on silica-supported, 6-11 nm diameter Pd and AuPd clusters with varying Au to Pd ratio. These data are inconsistent with Langmuirian mechanisms for H_2O_2 formation and instead show that H_2O_2 formation likely occurs by kinetically relevant proton-electron transfer to chemisorbed hydroperoxy (OOH**) on Pd and AuPd. This hypothesis is consistent with results from pure Pd[3] and Au_xPd clusters (where x denotes the bulk molar ratio of Au to Pd), which show that H_2O_2 will readily form by DS only in protic solutions, regardless of the

Au to Pd ratio. Rate data suggest that H_2O forms by kinetically relevant O-O bond cleavage in OOH** surface intermediates. Using Au_1Pd catalysts increases rates of H_2O_2 formation over pure Pd, while subsequent addition of Au (up to $Au_{12}Pd$) decreases rates of both H_2O_2 and H_2O formation (with H_2O rates decreasing more than H_2O_2 rates). In addition, clusters with high Au to Pd ratios (e.g., $Au_{12}Pd$) show low rates of H_2O_2 hydrogenation, as measured directly by feeding H_2O_2 to the system in the presence of H_2. These changes are due, in part, to greater activation enthalpies for H_2O_2 ($\Delta H^{\ddagger}_{H2O2}$) and H_2O ($\Delta H^{\ddagger}_{H2O}$) formation on Au_1Pd than on pure Pd clusters. Further addition of Au, however, does not continue to increase $\Delta H^{\ddagger}_{H2O}$ values. This observation suggests that the five-fold increase in H_2O_2 selectivity on $Au_{12}Pd$ over pure Pd must also result from a decrease in the number of Pd ensembles that act as sites for O-O bond rupture during primary and secondary reaction pathways. Work by our group has shown that undercoordinated Pd sites tend to form H_2O over H_2O_2.[3] We intend to further probe this phenomenon by tuning AgPt octahedral clusters to have Pt preferentially migrate to either coordinated or undercoordinated surface sites. The improved understanding of the mechanisms behind DS, the factors that make Au_xPd such an ideal catalyst, and the importance of the position of active metal within bimetallic clusters will help guide the rational design of selective catalysts for DS.

(1) Samanta, C. *Appl. Catal., A* **2008**, *350*, 133.

(2) Landon, P. et al. *Chem. Chem. Phys* **2003**, *5*, 1917.

(3) Wilson, N. M.; Flaherty, D. W. *J. Am. Chem. Soc.* **2016**, *138*, 574.

Copyright © American Institute of Chemical Engineers. All rights reserved.

Development of a Computationally Efficient Model of a Packed-Bed Membrane Reactor for the Oxidative Coupling of Methane

d|b|t|a

Process Dynamics and Operations Group

Alberto Penteado[1,*], Mi Jin Kim[1], Hamid Reza Godini[1], Erik Esche[1], Eva Speelmanns[1] and Jens-Uwe Repke[1]

1 Process Dynamics and Operations Group, Technical University Berlin, Sekr. KWT 9, Str.17. Juni 135, Berlin, 10623, Germany
* albertopenteado@mailbox.tu-berlin.de

Introduction

Oxidative Coupling of Methane – OCM

- Attractive path for producing ethylene from natural gas
- Limited CH_4 conversion and C_2 selectivity
- Increased separation costs
- Exothermic reactions → Hot spot formation
- Investigated reactors within UniCat:
  - Packed Bed Reactor (PBR)
  - Fluidized Bed Reactor (FBR)
  - **Packed Bed Membrane Reactor (PBMR)**
  - Fluidized Bed Membrane Reactor (FBMR)

PBMR: O_2 dosing; Lower $CH_4 : O_2$; Avoid hot spots; Higher S_{C2} and Y_{C2}; Higher complexity; Less industrial experience

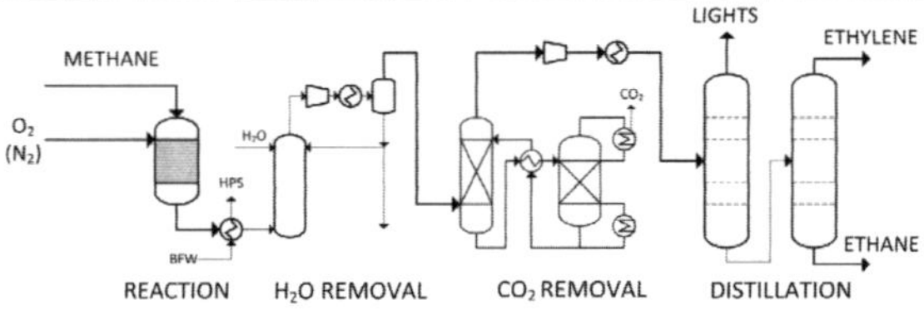

Fig. 1 – Process Flow Diagram for the OCM Process

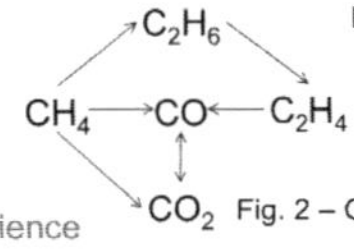

Fig. 2 – OCM 10 Reactions Network based on Stansch 1997[1]

Packed Bed Membrane Reactor (PBMR)

PBMR Modeling

- Kinetic correlations by Stansch 1997[1] for La_2O_3/CaO catalyst
- Knudsen's diffusion through porous ceramic membrane

1-D

- No radial gradients
- Overprediction of reactor's performance
- Easy implementation
- Fast Computation
- Wide usability

2-D

- Radial gradients
- Accurate prediction of reactor's performance
- Complex implementation
- Slower computation
- Limited usability e.g. flowsheet simulation or optimization

Question:
Can the 1-D model be tuned for improved results?

11 Tuning parameters: η_r Reactions' rate tuning factor; f Diffusion tuning factor
Fine tuning: Individual fitting to every condition
Broad tuning: Single fitting to all conditions

Implementation in Matlab

- Simulations performed using ode23s
- Parameter tuning using lsqnonlin and Levenberg-Marquardt algorithm. Objective is the difference between model and experimental X_{CH4}, S_{C2} and Y_{C2} from Godini 2014[2]

Fig. 3 – Schematic Representation of the PBMR with simulation conditions and main model equations

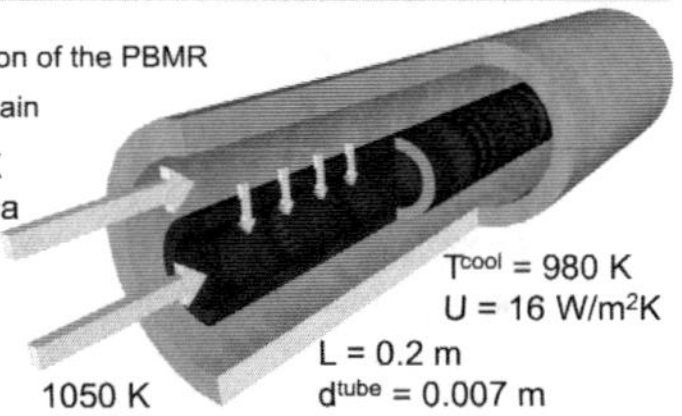

$CH_4 : O_2$ — 1050 K, 102 kPa

2 : 1	3 : 1
0.25 O_2 0.75 N_2	0.17 O_2 0.83 N_2
0.5 CH_4 0.5 N_2	0.5 CH_4 0.5 N_2

1050 K, 102 kPa

T^{cool} = 980 K
U = 16 W/m²K
L = 0.2 m
d^{tube} = 0.007 m
d^{shell} = 0.017 m

$$\frac{dF_i^{tube}}{dz} = A_\emptyset^{tube} cr_i - A_S^{tube} f_i^{diff} \qquad \frac{dF_i^{shell}}{dz} = A_S^{tube} \cdot f_i^{diff}$$

$$cr_i = \varphi_{cat}\rho_{cat} \sum_{\substack{r=1 \\ r\neq 7}}^{r=10} (\nu_{i,r} \cdot \boldsymbol{\eta}_r r_r) + (1-\varphi_{cat})\nu_{i,r=7} \cdot \boldsymbol{\eta}_{r=7} r_{r=7}$$

$$f_i^{diff} = \frac{\boldsymbol{f} \cdot D_i}{\delta_{memb}} \left(\frac{p_i^{tube}}{RT^{tube}} - \frac{p_i^{shell}}{RT^{shell}} \right) \qquad D_i = \frac{2}{3}\frac{\epsilon}{\tau} r_P \left(\frac{8RT^{tube}}{\pi \cdot M_i} \right)$$

Fig.4 – Experimental set-up for the PBMR used by Godini 2014[2]

Simulation Results

Results	1-D No Tuning	1-D Fine Tuning	1-D Broad Tuning	2-D[2*] No Tuning
Average Deviation from Experiments[2*]	15.9%	0.1%	7.7%	14.1%
Average CPU time** (s)		0.87		728

* Experimental and 2-D model results from Godini 2014[2] for $CH_4:O_2$ ratios of 2:1 and 3:1
** CPU: Intel Xeon® E5-2620 0 @ 2.00 GHz and 16 GB of RAM

- Fine tuning allows for the precise description of experimental data, but requires multiple parameter sets
- Tuning does not allow for insights on the actual phenomena in the reactor, thus it is not adequate for detailed design

Conclusions and Outlook

- ✓ 1-D Model effectively tuned to fit experimental data
- ✓ Fast computation time allows for its wide usability
- ➢ Implement the model into flowsheet simulations e.g. Chemcad
- ➢ Extend experimental points to allow for a wider operation range
- ➢ Development of a detailed 2-D model, including kinetics for the $Mn\text{-}Na_2WO_4/SiO_2$ catalyst, for detailed design and scale-up

Literature

1. Z. Stansch, L. Mleczko, M. Baerns, 1997, Comprehensive kinetics of oxidative coupling of methane over the La2O3/CaO catalyst. Ind. Eng. Chem. Res. , 36, 2569 – 2579
2. H.R. Godini, S. Xiao, M. Kim, N. Holst, S. Jaso, O. Görke, J. Steinbach, G. Wozny, 2014, Experimental and model-based analysis of membrane reactor performance for methane oxidative coupling: Effect of radial heat and mass transfer. J. Ind. Eng. Chem., 20, 1993-2002

CAPES

SCIENCE WITHOUT BORDERS

unicat

Technische Universität Berlin

www.dbta.tu-berlin.de
info@dbta.tu-berlin.de

468586 Experimental and Kinetic Modeling of Lean-Rich Switching Study over a Modified Twc System

Abstract:

Experimental and kinetic modeling of Lean-Rich switching study over a modified TWC system

Mengmeng Li, Sam Malamis, Michael P. Harold*, William Epling**

Department of Chemical and Biomolecular Engineering, University of Houston, Houston, TX 77204

* *mharold@uh.edu,* ** *wsepling@central.uh.edu*

The most recent CAFE standard for automobile requires an average fuel economy of 54.5 miles per gallon by 2025 [1]. At the same time, more stringent emission rules enacted by the EPA for 2017-18 will require an 80% reduction in non-methane organic gases (NMOG) plus NOx from current Tier 2 Bin 5 levels [2]. Gasoline vehicles that operate under both lean-burn and stoichiometric combustion hold promise since lean-burn combustion is more fuel efficient than conventional stoichiometric combustion. However, elimination of NOx ($NO+NO_2$) by the traditional three-way catalyst (TWC) is not possible under lean conditions. Advanced NOx reduction strategies must be applied to exploit the more efficient lean-burn engine combustion such as the three-way catalyst with a NOx storage function (TWNSC). The TWNSC stores NO_x storage during lean operation and reduces the trapped NOx during stoichiometric operation. To date, there have been only a few studies that report TWNSC. In this study a predictive TWNSC model is developed to predict TWNSC performance and in so doing provide operational insight and optimization. A model of a catalytic monolith channel was developed that incorporates a TWNSC kinetic model into a low-dimensional model formulation [3Ð5]. The convection-diffusion-reaction equations are averaged in the transverse direction which replaces the transverse gradient term with an algebraic term that contains an internal mass transfer coefficient. Overall heat and mass transfer coefficients are used to represent the heat and mass transfer in the transverse direction between the bulk of fluid and solid phase. The TWNSC kinetics consists of a combination of elementary steps and global reactions. Following our previous experimental study of NOx reduction by propylene over a multifunctional NOx trap catalyst [6], the hybrid model includes the kinetics of the oxidation of CO, H_2, NO and hydrocarbons, of reduction of NO/NO_2 by CO, H_2 and representative hydrocarbons, and of storage of NOx and O_2. Fig. 1 shows the reaction network including 10 main paths that described the chemistry in qualitative terms.

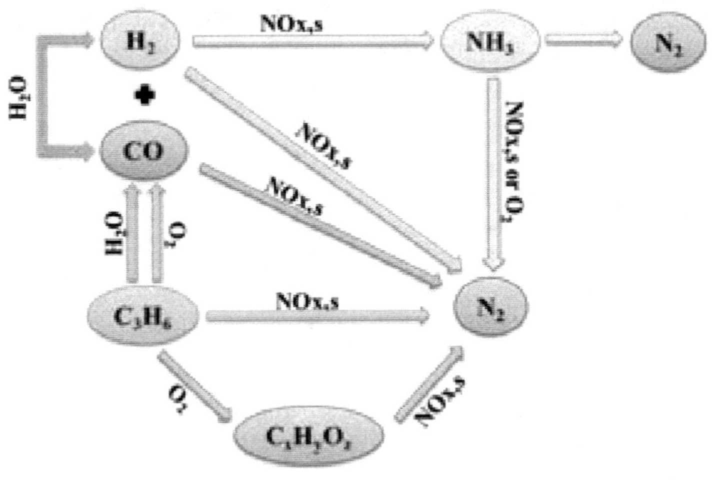

Figure 1. Simplified reaction network of NOx reduction using propylene as reductant[6]

A step-wise approach was used to develop kinetics for the oxidations of CO, H_2 and propylene. First, global kinetics are fitted tuned to transient light-off experiments following Raj et al. [5]. For example, the Langmuir-Hinshelwood expression is given by

$$R_{co} = \frac{k_1 X_{CO} X_{O_2}}{(1 + K_{CO} X_{CO})^2} \quad (1)$$

Fig. 2 shows the fitted light-off curve for CO oxidation and the comparison between experimental and model data. The model shows a reasonable fit and captures the dynamical feature of CO TPO (temperature programmed oxidation) experiment.

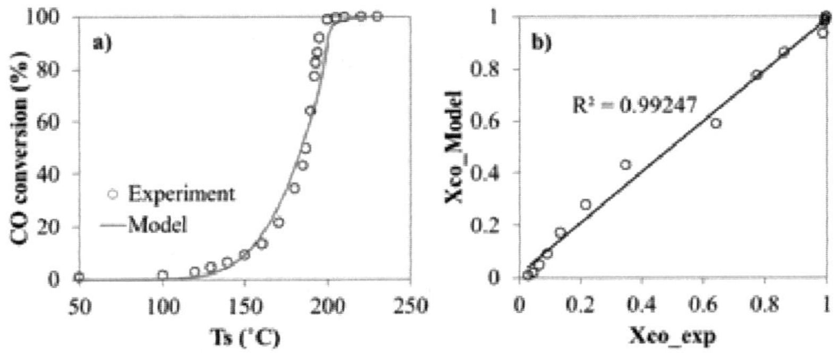

Figure 2. (a) Fitted light-off curve of CO oxidation (b) comparison between experimental and model-predicted conversions. [Conditions: 0.5% CO, 0.5% O; temperature ramp: 3ûC/min]

The rate expressions for the co-oxidation of CO, H_2, and C_3H_6 include coupling effects. These are combined with a dual-site NOx storage model and incorporated into species balances.

We will show how the model is effective in capturing the main trends in the lean-rich switching data. This study advances the understanding of TWNSC catalyst and provides guidance for optimizing catalyst formulation and operation strategies.

Reference

[1] http://www.nhtsa.gov/About+NHTSA/Press+Releases/2012/Obama+Administration+Finalizes+Historic+54.5+mpg+Fuel+Efficiency+Standards.

[2] T. Johnson, SAE Int. J. Engines. 6 (2014) 699Ð715.

[3] S.Y. Joshi, M.P. Harold, V. Balakotaiah, AIChE J. 55 (2009) 1771Ð1783.

[4] D. Bhatia, R.D. Clayton, M.P. Harold, V. Balakotaiah, Catal. Today. 147 (2009) 250Ð256.

[5] R. Raj, M.P. Harold, V. Balakotaiah, Chem. Eng. J. 281 (2015) 322Ð333.

[6] M. Li, V.G. Easterling, M.P. Harold, Catal. Today. 267 (2016) 177Ð191.

Copyright © American Institute of Chemical Engineers. All rights reserved.

468592 Aqueous-Phase Hydrodechlorination of Trichloroethene over Pd-Supported on Swellable Organically-Modified Silica (SOMS)

Abstract:

Aqueous-phase Hydrodechlorination of Trichloroethene over Pd-supported on Swellable Organically-modified Silica (SOMS)

Gokhan Celik[1], Hyuntae Sohn[1], Seval Gunduz[1], Saurabh A. Ailawar[1], Paul Edmiston[2], Umit. S. Ozkan[1*]

[1]The Ohio State University, Columbus, Ohio 43210 (United States)

[2]The College of Wooster, Wooster, Ohio, 44691 (United States)

*ozkan.1@osu.edu

The widespread contamination of water by chlorinated ethenes such as trichloroethene (TCE) and perchloroethene (PCE) is a growing environmental concern [1]. The high level of toxicity and the carcinogenic effects of TCE pose a serious threat to human health and the environment. According to a study conducted by Environmental Protection Agency, TCE was detected in 91 out of 945 drinking-water and groundwater samples [2]. Hydrodechlorination (HDC) is an efficient way of removing chlorinated compounds from water. It is an elimination-based remediation technique in which chlorinated compounds react with hydrogen and are catalytically converted to Cl-free hydrocarbons and hydrogen chloride. There has been on-going research on HDC of chlorinated compounds where promising catalytic activities have been obtained with the palladium-based state-of-the-art catalysts. However, catalyst deactivation due to anionic groundwater constituents such as sulfur species, nitrates, carbonates, and chlorides is a recurring problem [3,

4]. For industrial organizations to adapt catalytic HDC as their remediation technique, deactivation issues due to the anionic species need to be addressed.

Herein, we report the use of a new class of materials, namely swellable organically-modified silica (SOMS) as a catalyst support for HDC reactions. SOMS is extremely hydrophobic and has a high affinity for absorbing organics. While absorption of organics is in progress, SOMS swells in such a way that its volume physically expands to more than 3-6 times of its original size [5-7]. Active metals incorporated within the SOMS matrix in its expanded state are protected from dissolved anions present in water. Hydrophobic nature of these materials will also create a repulsive force against water and its constituents. This will result in an additional level of protection of active sites. SOMS can have the potential to address the deactivation issues involved in catalytic HDC.

In this study, Pd-incorporated SOMS (Pd/SOMS) and Pd/Al_2O_3, for comparison purposes, have been tested for HDC of TCE. Reaction experiments and characterization studies including N_2 physisorption, solid-state NMR, laser RAMAN and infrared spectroscopy, extended X-ray absorption fine structure (EXAFS), and inductively coupled plasma optical emission spectrometer (ICP-OES) have been performed to understand the poison-resistant characteristics of Pd/SOMS.

References

[1] R.E. Doherty, A History of the Production and Use of Carbon Tetrachloride, Tetrachloroethylene, Trichloroethylene and 1,1,1-Trichloroethane in the United States: Part 1--Historical Background; Carbon Tetrachloride and Tetrachloroethylene, Environmental Forensics, 1 (2000) 69-81.

[2] Toxilogical Profile for Trichloroethylene, in, Public Health Service,Agency for Toxic Substances and Disease Registry, U.S. Department of Health and Human Services, 2014.

[3] N. Munakata, M. Reinhard, Palladium-catalyzed aqueous hydrodehalogenation in column reactors: Modeling of deactivation kinetics with sulfide and comparison of regenerants, Applied Catalysis B: Environmental, 75 (2007) 1-10.

[4] G.V. Lowry, M. Reinhard, Pd-Catalyzed TCE Dechlorination in Water: Effect of [H2](aq) H2-Utilizing Competitive Solutes on the TCE Dechlorination Rate and Product Distribution, Environmental Science & Technology, 35 (2001) 696-702.

[5] C.M. Burkett, P.L. Edmiston, Highly swellable solâ€"gels prepared by chemical modification of silanol groups prior to drying, Journal of Non-Crystalline Solids, 351 (2005) 3174-3178.

[6] C.M. Burkett, L.A. Underwood, R.S. Volzer, J.A. Baughman, P.L. Edmiston, Organicâ€"Inorganic Hybrid Materials that Rapidly Swell in Non-Polar Liquids: Nanoscale Morphology and Swelling Mechanism, Chemistry of Materials, 20 (2008) 1312-1321.

[7] P.L. Edmiston, L.A. Underwood, Absorption of dissolved organic species from water using organically modified silica that swells, Separation and Purification Technology, 66 (2009) 532-540.

Copyright © American Institute of Chemical Engineers. All rights reserved.

468666 Understanding the Mechanisms and Key Parameters That Influence the Performance of Composite Semiconductor/Electrocatalyst Photoelectrodes

Abstract:

The integration of metal nanoparticle electrocatalysts with semiconductor photoelectrodes is ubiquitous in the field of photoelectrochemistry. For semiconductor photoelectrodes with long charge carrier diffusion lengths and good optical absorption, these metal nanoparticles hinder optical efficiency. Innovative semiconductor-electrocatalyst integration techniques can improve these efficiencies, but often introduce unforeseen enhancement/loss mechanisms.

In this contribution, we identify an integration technique to simultaneously improve the kinetic and optical efficiencies of planar silicon/platinum nanoparticle photoelectrodes. We also discuss our theoretical work on identifying key parameters in this system to explain overall device efficiency.

Copyright © American Institute of Chemical Engineers. All rights reserved.

468737 Photochemical Reactions on Plasmonic Metal Nanoparticles: Underlying Mechanism of Charge Extraction from Nanoparticles to Adsorbates

Abstract:

Plasmonic metal nanoparticles represent a new class of photocatalysts that harvest the energy of visible photons through excitation of localized surface plasmon resonance (LSPR). The relaxation of these excited surface plasmons can lead to the transfer of energetic charge carriers to adsorbates which in turn promote chemical transformations on the surface of the metal nanoparticles.[1] Multiple mechanisms have been proposed for the energetic charge carrier transfer process. The conventional model of photoexcitation in metals suggests the formation of energetic charge carriers in the metal, rapid thermalization, and subsequent transfer of â€œhotâ€ charge carriers to adsorbates. More recent studies point instead towards the direct transfer of charge from the photo-excited metal to adsorbates avoiding the formation and thermalization of charge carriers in the metal.[2]

In this presentation, we study the flow of charge for a photo-excited plasmonic metal-adsorbate system in which the presence of the adsorbate opens a new channel for the direct, resonant flow of charge from the metal to the adsorbate. This direct transfer of charge results in the selective heating of the adsorbate molecule without exciting charge carriers in the metal thereby avoiding the heating and loss of energy to the metal nanoparticle. This suggests that the transfer of energy to adsorbates in these photo-excited metal-adsorbate systems can be very efficient. We outline a conceptual model that explains these results and close by discussing the consequences of this energy flow in plasmonic metal-adsorbate systems for applications in photocatalysis.

(1) Linic, S.; Aslam, U.; Boerigter, C.; Morabito, M. Photochemical Transformations on Plasmonic Metal Nanoparticles. *Nat. Mater.* **2015**, *14* (6), 567â€“576.

(2) Boerigter, C.; Campana, R.; Morabito, M.; Linic, S. Evidence and Implications of Direct Charge Excitation as the Dominant Mechanism in Plasmon-Mediated Photocatalysis. *Nat. Commun.* **2016**, *7*, 10545.

Copyright © American Institute of Chemical Engineers. All rights reserved.

468743 Tailoring the Physicochemical Properties of ZSM-5 with Zeolite Growth Modifiers

Abstract:

Tailoring the Physicochemical Properties of ZSM-5 with Zeolite Growth Modifiers

Wei Qin and Jeffrey D. Rimer

Chemical and Biomolecular Engineering, University of Houston, 4726 Calhoun Rd., Houston, Texas 77204 USA

Zeolite ZSM-5 is a common heterogeneous catalyst in refinery and petrochemical processes, owing to its excellent shape-selectivity, strong Brønsted acidity, and high thermal/hydrothermal stability. It has an MFI structure with 3-dimensional pores consisting of interconnecting straight and sinusoidal channels. Molecular diffusion is much faster in the straight channels oriented along the [010] direction. To this end, it is desirable to selectively control the [010] thickness of ZSM-5 crystals.

Here, we will present a novel approach to tailor ZSM-5 morphology that incorporates zeolite growth modifiers (ZGMs). The use of ZGMs has proven to be an efficient method to tailor the morphology of silicalite-1, which is the siliceous isostructural analogue of ZSM-5.[1] For our study of ZSM-5 synthesis in the presence of ZGMs, we assessed a library of modifiers previously screened for slicalite-1 to determine the influence of aluminum incorporation within the MFI framework. It was observed that subtle changes in the synthesis conditions can alter ZGM binding specificity to different crystallographic faces of ZSM-5. Using atomic force microscopy (AFM), we show that chemical force microscopy can be used to efficiently probe modifier-crystal interactions in order to elucidate ZGM molecular recognition for different zeolite surfaces.[2]

We also investigated the effects of ZGMs on the initial stages of amorphous precursor self-assembly and evolution (pre-nucleation) as well as crystal growth using a combination of two techniques: small angle X-ray scattering (SAXS) [3] and dynamic light scattering (DLS). We observed that polycations have a unique effect of promoting precursor aggregation during the induction period.

Through the judicious selection of ZGMs under different synthesis conditions, we were able to synthesize ZSM-5 crystals with various sizes and morphologies. We will present details of these syntheses as well as catalytic tests of ZSM-5 catalysts with a model reaction to gain a better understanding of the relationship between ZSM-5 morphology and catalyst performance. Collectively, this work provides quantitative guidelines for designing synthesis protocols to tailor ZSM-5 crystallization and improve catalyst properties for a wide range of applications.

(1) Lupulescu, A. I.; Rimer, J. D. *Angew. Chemie - Int. Ed.* **2012**, *51* (14), 3345–3349.

(2) Lupulescu, A. I.; Rimer, J. D. *Science.* **2014**, *729.*

(3) Fedekyo, J.; Rimer, J. D.; Lobo, R. et al. *J. Physical Chemistry C.* 2004, 108.

Copyright © American Institute of Chemical Engineers. All rights reserved.

468825 Hotspot Mitigation in an Annular Microchannel Reactor (AMR) for Hydrogen and/or Syn Gas Production

Abstract:

The potential economic impact of a portable, modular and efficient technology for converting logistical fuels (e.g., natural gas, methanol or ammonia) into hydrogen on-site or in-the-field processing of natural gas to liquid fuels via syngas production is significant. Heat-exchanger microreactors provide novel process intensification routes through breakthrough heat transfer rates per reaction volume, enhancing the capacity and thermal efficiency of heat-transfer limited reactions including methane steam reforming or ammonia decomposition. In both applications, supplying heat from a separate combustion reaction (e.g., pairing methane steam reforming with methane combustion) promises autothermal operation for maximum portability.

This study focuses on a way to mitigate local hotspot formation when coupling methane steam reforming with catalytic methane combustion in a heat-exchanger microreactor. Previous studies have shown that the annular microchannel reactor (AMR), originally developed by Power + Energy, can be powered by catalytic combustion of methane achieving up to 67% overall thermal efficiency. However, hotspot formation was found to be an issue for catalyst stability. The addition of reforming catalyst to the inlet region of the AMR acts as a reaction heat sink providing hotspot control and adding further reforming capacity. Different catalyst thicknesses ranging from 30-60 microns were used and catalyst length ranged from 10-30 mm. The additional catalyst results in a temperature drop of over 100°C at the inner wall of the annulus, as well as lowering peak temperature in the reactor by over 70°C. Despite the temperature drop, the reactor performance is maintained.

Copyright © American Institute of Chemical Engineers. All rights reserved.

468832 Catalytic Activity and Sulfur Dependence of Co/Mo Nanoparticle Catalysts in WGS Reaction

Abstract:

The global interest in water-gas shift (WGS) reaction to produce pure hydrogen from fossil resources increased significantly in recent years. However, conventional WGS catalysts are easily deactivated by sulfur compounds which are ubiquitous in syngas. On the other hand, MoS_2 is a sulfur-dependent phase that shows WGS and HDS (Hydrodesulfurization) activity. Relatively low WGS activity of pure MoS_2 can be improved by adding Co as a structural promoter. In addition, promoted nanoparticles possessing different crystal sizes are expected to show differences not only in the WGS activity, but also sulfur dependence.

Several different size of Co/Mo catalysts were synthesized in our laboratory and characterized by scanning electron microscopy (SEM), transmission electron microscopy (TEM), X-ray diffraction (XRD), thermo gravimetric analysis (TGA), physic- and chemisorption, and temperature programmed reduction (TPR) in H_2. The XPS analysis was performed to determine the oxidation states and chemical composition in the surface region. The Raman spectrum was employed to characterize the nature of bulk phases (ordered and disordered) present in fresh and used catalysts. The CO chemisorption was used to detemine active metal surface areas, while the TPR analysis probed whether the addition of Co facilitated the reduction of active metals at lower temperatures. The TGA analysis was employed to study the interactions between Co, Mo and support as a function of particle size and composition ratios for the Co/Mo nanoparticles. The XRD was employed to estimate the size and phase composition of Co/Mo nanoparticles. The size and shape of supported Co/Mo nanoparticles was further probed by TEM. The WGS catalytic activity of these catalysts was evaluated in the WGS reaction at 398-823 K under atmospheric pressure in a fixed-bed micro reactor employing the feed of 10% CO and 20% H_2O (balance He). The Co/Mo catalysts were pre-activated in H_2S abundant atmosphere (1 vol. % H_2S balanced by H_2) for 8 hours. The improvements in the WGS catalytic activity was

correlated to the size, shape and the composition ratios of the Co/Mo nanoparticles.

Copyright © American Institute of Chemical Engineers. All rights reserved.

468871 Optimized $CeZrO_2$-Modification of the Metal-Foam-Coated Pd-Rh/Al_2O_3 Catalyst for CH_4 Bi-Reforming Reaction

Abstract:

Optimized $CeZrO_2$-modification of the metal-foam-coated Pd-Rh/Al_2O_3 catalyst for CH_4 bi-reforming reaction

Kiseok Kim[1*], Partho Sarothi Roy[2], Jinwoo Song[3,4], Ji Man Kim[3]

[1*]*School of Chemical Engineering, Yeungnam University, Gyeongsan, 712-749, South Korea*

[2]*CE-CERT, University of California Riverside, CA 92507, USA*

[3]*Department of Chemistry, Sungkyunkwan University, Suwon, 440-746, South Korea*

[4]*R & D Center, Heesung Catalysts Corp., 91 Somanggongwon-ro, Siheung, 429-848, South Korea*

[*]*Corresponding Author; Fax: +82-53-810-4631 E-mail:kimks@yu.ac.kr*

Reforming of biogas by steam to syn gas is important for viable conversion of renewable energy resources because biogas consists primarily of the most simple and inexpensive hydrocarbon (CH_4) and the most stable carbon oxide (CO_2), both of which are major greenhouse gases. Steam-reforming of biogas is considered to be more beneficial than CO_2 (dry)-reforming of biogas, for steam is a reforming reagent that can alleviate catalyst fouling caused by carbon deposition which tends to be aggravated in dry reforming environment. To optimize the reaction condition and catalyst for steam-reforming of biogas will be more systematic if it is performed in the context of CH_4 bi-reforming, i. e., reforming of CH_4 in the presence of two reagents, carbon dioxide and steam. Some of our previous works on CH_4 bi-reforming have been performed for developing the process operating condition over the nano-scale, active and stable Pd-Rh catalyst supported on CeZrO2-modified Al_2O_3that is coated on a metal foam substrate designed to fit SOFC fuel processor application [1, 2].

In the present study, $CeZrO_2$-modification of Al_2O_3 support was optimized with respect to catalyst activity and stability as well as syn gas product yield and selectivity for CH_4 bi-reforming based on the reference operating condition obtained from Aspen Plus process simulation and previous experiments. $CeZrO_2/Al_2O_3$ weight ratio of the Pd-Rh catalyst support was varied as 0/100, 15/85, 25/75, 35/65, 50/50 and 75/25, and the catalyst performances were evaluated and compared in terms of CH_4 and CO_2 conversions, H_2/CO ratio of product syn gas, transient or accumulated coke deposition, and process thermal efficiency. CH_4 bi-reforming reaction runs were performed in a heat exchanger platform type reactor in the modes of transient activity screening and 200 h on-stream stability test. The fresh and used catalysts were characterized by BET surface area and porosity measurements, noble metal dispersion, and SEM as well as TEM analysis. Optimum relative abundance of $CeZrO_2$ within the alumina support was addressed for enhancing CH_4 bi-reforming performance of the Pd-Rh/Al_2O_3 catalyst.

References

1. P. S. Roy, M. S. Kang and K. Kim, Effects of Pd-Rh composition and $CeZrO_2$-modification of Al_2O_3 on performance of metal-foam-coated Pd-Rh/Al_2O_3 catalyst for steam reforming of model biogas, *Catal. Lett.* 144, 2021(**2014**).
2. P. S. Roy, A. S. K. Raju and K. Kim, Influence of S/C ratio and temperature on steam reforming of model biogas over a metal-foam-coated Pd-Rh/($CeZrO_2$-Al_2O_3)catalyst, *Fuel* 139, 314(**2015**).

Copyright © American Institute of Chemical Engineers. All rights reserved.

468885 Liquid-Phase Stability of ZSM-5: Role of De-Alumination and De-Silication

Abstract:

Zeolites are microporous aluminosilicate structures commonly used for catalysis in environmental, fuel upgrading and biomass industries. Recent advances to improve process intensity and energy efficiency motivate interest in zeolite catalysis in the presence of liquid water or water-rich liquid reaction mixtures. However, many zeolites are prone to framework degradation under aqueous phase conditions. We focus our attention on ZSM-5, a commonly used zeolite for hydrothermal catalytic cracking of fatty acids and model fuels that has proven to be more robust than other zeolites such as HY and Beta. Literature studies to date have found ZSM-5 to be stable for at least 6 hours at 200°C; however many chemical reactions of interest require temperatures >250°C. In our study, we have found that ZSM-5 breaks down in the liquid phase at temperatures greater than 250 °C, allowing us to investigate the degradation mechanism and determine methods for stabilizing ZSM-5. We exposed ZSM-5 with a silica/alumina ratio of 38:1 to liquid water for three hours at temperatures ranging from 200Â°C to 450Â°C. Post-run, the catalyst was characterized with a battery of complementary techniques including XRD, IR, TEM, XPS and 27Al MQMS NMR. ZSM-5 degradation occurs due to a combination of acid site loss at intermediate temperatures followed by surface degradation at higher temperatures. Specifically, we observed improvements in the ZSM-5 backbone stability at temperatures greater than waterâ€™s critical point as the familiar properties of liquid water (i.e., high dielectric and auto-ionization) shift dramatically near the critical point. Therefore, we recommend that studying ZSM-5 hydrothermal stability both above and below waterâ€™s critical temperature might provide deep insight into the role of water in ZSM-5 degradation.

Copyright © American Institute of Chemical Engineers. All rights reserved.

468953 Process Intensification of Ethanol Dehydration Under Liquid Phase Conditions

Abstract:

Replacement of petroleum-derived chemicals with renewable alternatives has potential to improve the sustainability of the chemical industries, mitigate global CO_2 emissions, and provide a bridge to an economy featuring cost-effective biofuels. Because most of the needed infrastructure is already in place, chemical upgrading of inexpensive fermentation products is especially attractive. Specifically, acid-catalyzed dehydration of ethanol to produce ethylene is a commercial process which has been demonstrated using a variety of acidic catalysts. One common upgrading catalyst is ZSM-5 zeolite, a three dimensional silica-alumina microporous framework made of internal channels with Brønsted acid sites. ZSM-5 catalyzed ethanol dehydration is highly selective for the desired ethylene product at lower operating temperatures relative to Syndol®, the catalyst currently used commercially for ethanol dehydration. In this work, we have examined continuous ethanol dehydration in the liquid phase, exploring a region of thermodynamic phase space that has received scant attention compared to the vapor phase.

Operating under liquid phase conditions has many potential benefits, including manipulation of product distribution, reduced separations burden, increased process intensity, decreased coke formation rates, and superior heat management for the highly endothermic reaction. As seen in Figure 1, the favored is butanol, a potential drop-in biofuel which is more than twice as valuable as ethylene according to current commodity prices. On the other hand, ethanol conversion does not suffer under liquid phase conditions compared to the vapor phase. Because the density of liquid reaction mixture is more than 100-times greater than the vapor phase at reaction temperatures, the result is an improvement in process intensity – that is moles of ethylene produced by gram of catalyst – or nearly 2 orders of magnitude. In terms of coke formation, operation under liquid phase conditions shows a clear reduction in catalyst coking. Finally, engineering calculations show superior heat transfer to the scaled-up liquid-phase reactor compared to the vapor-phase reactor. Based on this analysis, we conclude that liquid-phase ethanol upgrading has commercial promise and that liquid-phase reactions in general deserve increased attention.

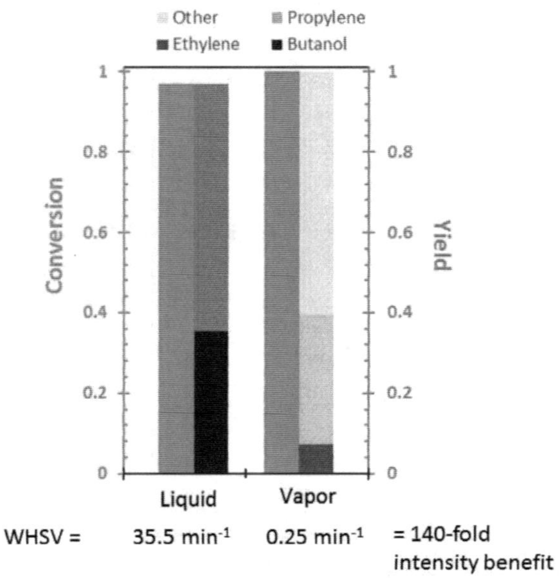

Figure 1: Product selectivity from ethanol dehydration operated under liquid phase conditions at 390°C

Copyright © American Institute of Chemical Engineers. All rights reserved.

468980 Development of Fischer-Tropsch Synthesis in Microreactor

Abstract:

Fischer-Tropsch Synthesis (FTS) is the process that converts synthesis gas (syngas, a mixture of carbon monoxide and hydrogen) into a wide range of long chain hydrocarbons and oxygenates in the presence of catalysts such as iron or cobalt. The products of the reaction include jet fuel, naphtha and diesel. The goal of this study is to design and construct a microchannel reactor that demonstrates a low selectivity to methane and high yield of naphtha and diesel range hydrocarbons. The benefit of the implementation of FTS in a microreactor include: better flow residence time uniformity, temperature control, and catalyst utilization. Better flow uniformity will achieve more uniformly selective FTS reactor products, the temperature control will narrow the product distribution and catalyst utilization will improve the yield of hydrocarbon products.

The investigation of FTS in a microreactor will examine the effects of operation conditions such as pressure, temperatures, residence time and flow ratio on the performance of the reactor. In this contribution, we will present important experimental features of the FTS reaction in a microreactor as well as summarize some recent insights on the FTS microreactor product distribution.

Copyright © American Institute of Chemical Engineers. All rights reserved.

469134 Au@M_xO_y Core-Shell Nanoparticles As Catalysts for the Oxygen Evolution Reaction

Abstract:

One of the most imperative challenges facing society today is the development of a sustainable energy economy. While renewable energy sources such as solar and wind can provide more than enough energy to meet current and projected global demand, the intermittency of these sources is problematic. A promising strategy to mitigate this intermittency issue is to convert and store energy in the form of chemical bonds. Electrochemical water splitting is one such promising method in which renewable energy is converted into a storable, high purity, carbon-free hydrogen fuel. Unfortunately, the anodic oxygen evolution reaction (OER) in electrolyzer devices suffers from slow kinetics. In alkaline conditions, first row transition metals have shown high activity and stability for the OER, however, high overpotentials are still required.[1] Therefore, before electrolyzers can become industrially feasible solutions, high activity, device-ready catalysts for the OER must first be developed.

Recently, Au has been found to have beneficial effects on the activity of transition metal catalysts for the OER. Firstly, Au can alter catalyst electronic structure and allow access to higher activity metal oxide phases as was observed for the Au/Mn-oxide system.[2] Additionally, it has been postulated that Au can reduce overpotential requirements for Mn and Co oxides by participating directly in the OER mechanism.[3] In a recent report, Au was also found to enhance the activity of NiCe based catalysts.[4] Lastly, as a high conductivity support, Au can increase the conductivity of insulating metal oxide materials whose high intrinsic activity is thwarted by poor electronic properties.[5] Capitalizing on these beneficial effects, the Au-core metal oxide-shell nanoparticle structure has the potential to achieve enhanced activity in a device-ready form. This nanostructure also provides an increased number of active sites and greater material utilization than thin films and can achieve higher metal loadings without hitting conductivity limitations.

Herein we present the synthesis, characterization and electrochemical performance of Au-core metal oxide-shell

(Au@M_xO_y) nanoparticles as OER electrocatalysts. Single metal and alloy oxide nanoparticles with and without Au cores were synthesized with high uniformity via wet chemical methods. Particle morphology and composition were analyzed using transmission electron microscopy (TEM), scanning TEM energy dispersive spectroscopy (STEM-EDS), scanning electron microscopy (SEM), and x-ray photoelectron spectroscopy (XPS). Electrochemical activity and stability were evaluated in a three electrode rotating disk arrangement in Fe-free conditions for non-Fe based catalysts.[6] The effect of the Au core is evaluated and catalyst performances are compared.

---

[1] C.C.L. McCrory, et al., *J. Am. Chem. Soc.* 13, 4347â€“4357 (2015).

[2] Y. Gorlin, et al., *J. Am. Chem. Soc*. 13, 4920-4926 (2014).

[3] R. Frydendal, et al., *ChemCatChem* 7, 149-154 (2015).

[4] J. W. D. Ng, et al., Nature Energy 1, (2016).

[5] M. S. Burke, et al., *Chemistry of Materials* 27, 7549-7558 (2015).

[6] L. Trotochaud, et al., *J. Am. Chem. Soc.* 136, 6744-6753 (2014).

Copyright © American Institute of Chemical Engineers. All rights reserved.

469267 Electrochemical CO_2 Reduction in an Oxygen-Ion Conducting Solid Oxide Electrolyzer Cell (SOEC)

Abstract:

It is generally accepted that there is a strong correlation between global CO_2 emissions and energy consumption [1]. Due to its increasing levels, conversion of CO_2 into valuable products has become an important objective in recent years. High temperature electrolysis of CO_2 in the presence of H_2O is considered a promising route for syngas ($CO+H_2$) production, which could be used for Fischer-Tropsch synthesis.

Solid oxide electrolyzer cells (SOECs) used in high temperature CO_2 electrolysis are basically solid oxide fuel cells (SOFCs) operated in reverse [2]. In the present study, an oxygen-ion conducting electrolyzer type reactor was used for the electrolysis of carbon dioxide into carbon monoxide and oxygen using external electricity. The SOEC reactor used in this study consists of a button cell, i.e., an electrode-electrolyte assembly, and an alumina tube. Yttria-stabilized zirconia (YSZ) disc which has a thickness of 125 μm, La-doped strontium titanate (LST) type perovskite and $(La_{0.8}Sr_{0.2})_{0.95}MnO_{3-ï}$· (LSM) mixed with YSZ were used as electrolyte, cathode and anode, respectively. The activity and stability of synthesized perovskite was compared with Ni-YSZ type commercially available electrode material. In electrocatalytic CO_2 reduction experiments, pure carbon dioxide was flowed to the cathode side and data was collected at a current range of 0-5 mA at 600°C.

The La-doped strontium titanate type perovskite material used in this study was synthesized via modified Pechini method [3]. The characteristics of the synthesized material were investigated using several techniques such as X-ray diffraction (XRD), X-ray photoelectron spectroscopy (XPS), diffuse reflectance infrared spectroscopy (DRIFTS), CO_2-TPO (temperature-programmed oxidation with CO_2), and electronic conductivity measurements.

Temperature-programmed oxidation (TPO) was performed on post-reaction button cells (both on LST and Ni-YSZ cells) in order to compare the coke deposition on cathode during CO_2 reduction.

The aim of this study is to show that high-temperature electrocatalytic CO_2 reduction in a SOEC type reactor is a promising, energy efficient alternative for syngas production. The results obtained so far indicated that La-doped strontium titanate perovskite materials can be considered as promising cathodes for CO_2 electro-reduction and using perovskites as cathode material can significantly enhance the durability of the system by lowering the coke deposition.

References:

[1] International Energy Outlook 2013, Report # DOE/EIA-0484, 2013; release date: July 25, 2013.

[2] F. Bidrawn, G. Kim, G. Corre, J.T.S. Irvine, J.M. Vohs, R.J. Gorte, Electrochemical and Solid-State Letters 11 (2008) B167-B170.

[3] X. Lu, T.S. Pine, D.R. Mumm, J. Brouwer, Solid State Ionics 178 (2007) 1195.

Copyright © American Institute of Chemical Engineers. All rights reserved.

469342 The Mechanism of CO_2 Hydrogenation over Pd/Al_2O_3: *A Steady State Isotopic Transient Kinetic Analysis (SSITKA)/Operando-FTIR Spectroscopy Investigation*

Abstract:

The mechanism of CO_2 hydrogenation over Pd/Al_2O_3: *a steady state isotopic transient kinetic analysis (SSITKA)/operando-FTIR spectroscopy investigation*

Xiang Wang, János Szanyi*

Institute for Integrated Catalysis, Pacific Northwest National Laboratory, Richland, WA 99352, USA

**Corresponding author: janos.szanyi@pnnl.gov*

1. Introduction

Understanding the critical steps involved in the heterogeneous catalytic CO_2 reduction has attracted a lot of attention recently. In order to fully understand the mechanism of this reaction the determination of both the rate-determining steps and reaction intermediates are vital. Steady-State Isotopic Transient Kinetic Analysis (SSITKA) is one of the most powerful techniques used to investigate the elementary steps under steady-state reaction conditions. This technique provides valuable information on mean resident lifetime of surface intermediates, surface concentrations of adsorbed reactant species and an upper bound of the turnover frequency.[1-3] Coupling SSITKA with *operando*-FTIR spectroscopy allows us to discriminate between active and spectator species present on the catalytic surface under steady state reaction conditions. Most of the reaction cells applied in these types of studies possess a large volume leading to long gas hold-up times, which makes it impossible to do kinetic study on the surface species. In this work, we used a transmission IR cell which has an internal reactor volume of 0.48 cm^3, providing a short effective response time. The fast time grants us the ability to obtain kinetic parameters of surface species and relate a particular surface intermediate to a given reaction product by comparison with kinetic parameters from SSITKA.

On Pd/Al_2O_3 catalyst, our previous results showed that adsorbed CO was hydrogenated to methane while H-assisted formate decomposition can yield gaseous CO.[4] Those results were obtained from transient switches between hydrogen and reaction gases rather than under steady state reaction conditions. Therefore, it is difficult to understand the roles of adsorbed CO and formate on the surface of Pd/Al_2O_3 under steady state reaction conditions. In the present work, for the first time, *operando* SSITKA experiments coupled with transmission FTIR, mass spectrometry (MS) and gas chromatography (GC) were performed to probe both the chemical nature and kinetics of reactive intermediates over a 5% $Pd\text{-}Al_2O_3$ catalyst and provide a clear mechanistic picture of the CO_2 hydrogenation reaction by revealing the rate-determining steps for $CH_4$and CO production under steady state conditions.

2. Experimental

The 5 wt% Pd/Al_2O_3 catalyst was prepared on a commercial $\gamma\text{-}Al_2O_3$ powder (Condea, BET surface area=200 m^2/g) by the incipient wetness method using $Pd(NH_3)_4(NO_3)_2$ as the precursor. The reactor was loaded with 40 mg catalyst, of which 13 mg were pressed onto a tungsten mesh centered in the optical path of a transmission FTIR spectrometer. Steady-state CO_2 hydrogenation was carried out in the temperature range of 533-573 K. The total gas flow was kept at 10 ml/min, and the gas compositions were switched between $H_2/CO_2/Ar$ and $H_2/^{13}CO_2$ (4 ml/min H_2, 1 ml/min CO_2 or $^{13}CO_2$and 1 ml/min Ar with dilution of He). The effluent of the cell was analyzed by a MS that was calibrated by a GC.

3. Results and discussion

Figure 1 shows normalized real-time signals for the decay and increase of methane (a) and carbon-monoxide (b) in the effluent at 533 K reaction temperature after the feed gas was switched at 0 s from $CO_2/H_2/Ar$ mixture to $^{13}CO_2/H_2$ mixture. The fast decay of the Ar gas concentration and the feed gas indicates that the gas phase hold-up of the system is negligible. With increasing temperature, the decay of CH_4 and CO get faster.(not shown) By integration under the decay curves , the mean surface-residence times $_{CH4}$ and $_{CO}$), the abundance of adsorbed surface intermediates leading to CH_4 and CO products $_{CH4}$ and $_{CO}$) at 533-573 K were obtained and summarized in Table 1. At low temperature, CO_2 methanation is slower than the reverse water-gas shift reaction, but became faster as the temperature was increased over 563 K. The total amount of intermediates converted to CH_4 and CO is 3.50.2 μmol at 533-573 K. The similar apparent activation energies obtained for the hydrogenation of adsorbed CO and for the formation of CH_4 indicates that the hydrogenation of CO is the rate-determining step during the CO_2methanation reaction. Moreover, the similar apparent activation energies estimated for the consumption of adsorbed formates (FTIR) and for the formation of CO (MS), indicates that the H-assisted decomposition of formates is the rate determining step in the reverse water/gas shift reaction.

Figure 1. Normalized response of (a) CH_4 and $^{13}CH_4$ products and (b) CO and ^{13}CO products as functions of time.

Table 1. Mean surface-residence time $_{CH4}$ and $_{CO}$, abundance $_{CH4}$ and $_{CO}$ of adsorbed surface intermediates leading to $CH_4$and CO products.

4. Conclusions

The amount of active sites for both CH_4 and CO products is around 3.50.2 μmol. The CO_2 methanation and reverse water shift reactions share the same formate intermediate. The H-assisted decomposition of adsorbed formate is the rate-determining step for the CO production, whereas the hydrogenation of adsorbed CO is the rate-determining step for the formation of CH_4. Once the adsorbed CO formed on Pd, it will convert to CH_4. At low temperatures (563 K), formate decomposition to CO (P-CO) is faster than the hydrogenation of adsorbed CO to CH_4 (P-CH_4). However, P-CH_4 becomes faster than P-CO as the temperature is increased above 563 K. The faster P-CH_4 leads to the conversion of more adsorbed CO (produced in the decomposition of formates) to form CH_4 rather than desorb as CO. In other words, more of the total amount of intermediates (3.50.2 μmol) will form CH_4rather than CO with increasing reaction temperature.

References

1. B.C. McClaine, R.J. Davis, J. Catal., 211 (2002) 379-386.

2. J. Gao, X.H. Mo, J.G. Goodwin, J. Catal., 275 (2010) 211-217.

3. C. Ledesma, J. Yang, D. Chen, A. Holmen, ACS Catal., 4 (2014) 4527-4547.

4. *X. Wang, H. Shi, JH Kwak, J. Szanyi, ACS Catal.*, 5 (2015), 6337–6349

Copyright © American Institute of Chemical Engineers. All rights reserved.

469485 Machine-Learning-Augmented Chemisorption Model for Metal Oxides

Abstract:

Perovskites with the ABO_3 type structure have attracted enormous research efforts in recent years due to their high activity for catalyzing oxygen evolution in (photo)-electrochemical systems. The thermodynamically and kinetically unfavorable oxygen evolution reaction (OER) is considered as a bottleneck for many energy conversion processes such as metal-air batteries and photocatalytic water splitting cells[1]. We aim to design cost-effective metal oxides for catalyzing the sluggish OER. It is, however, very time-consuming and costly to search for highly optimized metal oxides by high-throughput experiments and/or quantum-chemical calculations due to the vast materials space of composition and geometry.

In this poster, we extend our previous study[2] on metal catalysis to metal oxides by developing a machine-learning-augmented chemisorption model that captures adsorption properties of active sites on metal oxides. To design effective features that the machine-learning algorithms can use to 'learn' and predict properties of metal oxides, we applied a feature engineering process on the catalyst database. We explore electronic based features such as the B-site metal e_g band filling, d-band upper edge, oxidation state, d-orbital splitting, ligand strength, and easily accessible structural based features such as the coordination number, together with the intrinsic properties of metal atoms. Using those features, the machine-learning model optimized with available *ab initio*adsorption energies on perovskite (001) surfaces can capture adsorption energies of *O, *OH, and *OOH intermediates with the root mean squared errors (RMSE) smaller than the DFT-GGA calculation error of 0.2 eV. Compared with the traditional high-throughput computational and experimental trial-and-error approach, the machine-learning chemisorption models exhibit great potential in accelerating the discovery of high-performance oxide materials for electrocatalysis.

1. Suntivich, J., May, K. J., Gasteiger, H. A., Goodenough, J. B. & Shao-Horn, Y. A Perovskite Oxide Optimized for Oxygen Evolution Catalysis from Molecular Orbital Principles. *Science* **334,**1383–1385 (2011).

2. Ma, X., Li, Z., Achenie, L. E. K. & Xin, H. Machine-Learning-Augmented Chemisorption Model for CO2 Electroreduction Catalyst Screening. *J. Phys. Chem. Lett.* **6,** 3528–3533 (2015).

Copyright © American Institute of Chemical Engineers. All rights reserved.

469595 Cobalt-Copper Alloy Nanoparticle Catalysts for Higher Alcohol Synthesis from Syngas

Abstract:

Higher alcohols, those with two or more carbons, are of great interest for their applications as both liquid fuels and synthetic feedstocks. While synthesis gas (a mixture of CO and H2) has long been used to synthesize long-chain hydrocarbon fuels via the Fischer-Tropsch process, an analogous process for the direct production of higher alcohols from syngas still faces many challenges. Theoretical studies and prior work in the field have shown that cobalt-cobalt alloys are among the most selective catalysts for the higher alcohol synthesis reaction [1-3], but further improvements are necessary. In this work, the authors synthesize cobalt-copper nanoparticles via a controlled liquid-phase synthesis to give greater control of particle size and composition and evaluate higher alcohol synthesis performance of the catalysts.

Co2.5Cu alloy nanoparticles were synthesized by a polyol synthesis in which metal acetates are reduced in refluxing diethylene glycol with a polyvinylpyrrolidone capping agent to control particle size. The nanoparticle catalysts were then supported on a variety of metal oxides and tested in a packed bed reactor under Fischer-Tropsch conditions (250° C, 40 bar, 2:1 H2:CO) to evaluate their activity and selectivity. The best performing sample, supported on alumina, had a 11.3% carbon selectivity towards higher alcohols, primarily ethanol. Nanoparticle composition and structure, both before and after reaction, were confirmed ex situ using a combination of transmission electron microscopy (TEM), energy dispersive spectroscopy (EDS), x-ray photoelectron spectroscopy (XPS), and x-ray diffraction (XRD). A consistent alloy phase and composition was observed prior to reaction, suggesting the synthesis successfully reduced the metal precursors to the cobalt-copper alloy phase. After reaction, new phases were observed to have formed via metal segregation and cobalt carbide formation. The alumina-supported sample, which had the highest performance, also showed minimal formation of these new phases, indicating a possible correlation between the observed performance and the extent of alloy degradation. Future investigations are focused on improving the stability of the desired alloy phase under reaction conditions.

[1] A.J. Medford, A.C. Lausche, F. Abild-Pedersen, B. Temel, N.C. Schjødt, J.K. Nørskov, F. Studt, Top. Catal. 57 (2013) 135-142.
[2] P. Courty, D. Durand, E. Freund, A. Sugier, J. Mol. Catal. 17 (1982) 241-254.
[3] V. Subramani, S.K. Gangwal, Energ. Fuel. 22 (2008) 814-839.

Copyright © American Institute of Chemical Engineers. All rights reserved.

469749 A Novel Method for Synthesis of Pt/Momc Catalyst for PEM Fuel Cell Application

Abstract:

Commercialization of proton exchange membrane fuel cells (PEMFC) has become an important challenge since platinum (Pt), which is being used as the primary catalyst, is highly expensive. The ordered mesoporous carbons (OMCs), due to their very high surface area compared to conventional carbon back, are considered to be excellent substitute to improve the dispersion and efficient utilization of the Pt in the electro-catalysts. However, the OMCs prepared using reported methods suffer from lower electrical conductivity due to their amorphous structure making them unusable as Pt support in fuel cell. The aim of this research is to synthesize modified ordered mesoporous carbon (mOMC) supports using M/SBA-15 (M: Ni, CO, Fe, W) as hard template prepared via impregnation method. The mOMCs with high surface area, improved electrical conductivity, and durability are obtained by optimizing the parameters employed in the synthesis processes of mOMC. The synthesized mOMC is characterized by BET, TEM, XRD, Raman spectroscopy and cyclic voltammetry. The catalysts will be tested for performance and polarization on 5 cm^2 membrane electrode assembly (MEAs) under controlled experimental conditions using well equipped Fuel Cell Testing Station (Model 850, Scribner Associates Inc.). Some of these results will be presented in this paper.

Copyright © American Institute of Chemical Engineers. All rights reserved.

469964 Lattice Strain Engineering of Tin Oxides for Energy-Efficient CO_2 Electroreduction

Abstract:

With dwindling fossil fuels and rising concern about their impacts on climate, electrochemical reduction of CO_2 has attracted researchers' attention as it has the potential to utilize the abundant greenhouse gas in the Earth's atmosphere and store intermittent energy from solar panels and wind turbines in chemical bonds[1,2]. A commercially viable catalyst for CO_2 electroreduction should meet the cost-effective requirement while possessing high efficiency and selectivity. Experimental studies showed that tin oxides (SnOx) have the ability to selectively convert CO_2 to formate ($HCOO^-$), although the reaction mechanism is not clear with respect to the nature of active sites and the process occurs at prohibitively high overpotentials (~1 V) [3].

The goal of this study is to understand the reaction mechanism of CO_2 electroreduction on SnOx and leverage that knowledge to design metal/SnOx composite catalysts with superior activity. In this poster, we will discuss the reaction mechanism of CO_2 reduction on SnOx obtained from first-principles density functional theory calculations, focusing on the competing pathways leading to the production of CO, H_2, and $HCOO^-$. We found that the outcome of CO_2 electroreduction on SnOx is highly dependent on its surface structure, which can explain a wide range of faradaic efficiencies for formate production on SnOx reported in different studies. By engineering the lattice strain of SnOx, we illustrated that the electrocatalytic reactivity can be further optimized for formate production.

1. Hori, Y. in *Modern Aspects of Electrochemistry* (eds. Vayenas, C. G., White, R. E. & Gamboa-Aldeco, M. E.) 89–189 (Springer New York, 2008).

2. Ma, X., Li, Z., Achenie, L. E. K. & Xin, H. Machine-Learning-Augmented Chemisorption Model for CO2 Electroreduction Catalyst Screening. *J. Phys. Chem. Lett.* **6,** 3528–3533 (2015).

3. Chen, Y. & Kanan, M. W. Tin Oxide Dependence of the CO2 Reduction Efficiency on Tin Electrodes and Enhanced Activity for Tin/Tin Oxide Thin-Film Catalysts. *J. Am. Chem. Soc.* **134,** 1986–1989 (2012).

Copyright © American Institute of Chemical Engineers. All rights reserved.

469995 Furfuryl Alcohol Dehydration over Metal Oxide Catalysts

Abstract:

ABSTRACT:

After U.S. Department of Energy reported top 12 template chemicals (C_3~C_6 open and close ring structures) that can be produced from lignocellusic biomass in 2004, it has been extensively investigated the utilization and production of these building blocks and intermediate chemicals. One of the attractive chemical product during biomass conversion, especially hemicellulose decomposition, is furfuryl alcohol (FA, $C_5H_6O_2$) which can be produced via selective hydrogenation of furfural. It has the capacity to be converted into FA oligomers (dimer and trimer) through the dehydration reaction which could possibly form biofuel via hydrodeoxygenation reaction. Previously we reported homogenous acid catalyzed FA conversion reaction for oligomer formation using different spectroscopic techniques such as Raman and IR spectrometer. However, heterogeneous catalysts should be considered to replace homogeneous catalysts to avoid catalyst separation and regeneration issues. Despite the extensive experimental and theoretical studies on metal oxide catalyzed alcohol dehydration reactions, fundamental understating of the FA dehydration reaction over metal oxides is still unclear.

In this study, the effect of different metal oxide catalysts (Al_2O_3, TiO_2, ZrO_2, Nb_2O_5, and SiO_2), surface areas, and crystalline structure on the FA dehydration catalytic activity and productsâ€™ selectivity has been extensively investigated. Combining spectroscopic and analytical techniques including X-ray diffraction, Raman spectroscopy, FTIR spectroscopy, BET, and Gas Chromatography (GC and GC/MS) have been applied under ex-situ and in-situ conditions. Furthermore, Operando Diffuse Reflectance Infrared Fourier Transform Spectroscopy (DRIFT) have been used to identify surface intermediates and to understand FA dehydration mechanisms over the metal oxides.

KEYWORDS: In-situ, Operando, Spectroscopy, Metal Oxides, Furfuryl Alcohol dehydratio***n***

ACKNOWLEDGMENT:

We gratefully acknowledge the financial support for this study from the Department of Materials Science & Engineering and the Program in Chemical and Molecular Engineering at Stony Brook University through startup research funding, as well as the funding from National Science Foundation (NSF-CBET-1546647).

Copyright © American Institute of Chemical Engineers. All rights reserved.

470184 Low-Cost, Potassium-Based, Catalytic Biomass Cookstove for Improved Indoor Air Quality

Abstract:

An estimated 2.5 billion people regularly cook with biomass fuels such as wood or charcoal. Emissions from biomass cookstoves contribute to global climate change, indoor air quality issues, and related health effects. Exposure to high indoor air pollutant levels from cooking with biomass fuels is responsible for an estimated 1.6 million deaths annually and about 3% of the global burden of disease. Recently developed forced-air and â€œrocketâ€ stoves offer improvements but are unable to consistently meet World Health Organization (WHO) guidelines for indoor air quality.

Similar to the evolution of emissions controls for automobiles, advanced biomass cookstoves have progressed to the point where inclusion of an oxidation catalyst is the logical next step. However, the widely used noble-metal catalysts are prohibitively expensive. Recent work has identified potassium as an inexpensive soot oxidation catalyst, but potassium is limited by its stability. In this work, additional elements were used to stabilize potassium and promote its catalytic activity. Thermogravimetric analysis (TGA) was used to determine both catalytic activity and catalyst stability. The most promising catalysts were then supported on cordierite monoliths and tested in a cookstove operated under common cooking conditions. Emissions monitoring equipment was used to demonstrate real world emissions (CO and soot) reduction by nearly 90% compared to traditional open fire stoves.

Copyright © American Institute of Chemical Engineers. All rights reserved.

470219 Effect of Poisoning on Low Temperature Methane Dry Reforming Pt-Ni-Mg/Ceria-Zirconia Catalysts

Abstract:

Silica poisoning of low temperature methane reforming catalyst comprised of platinum (0.16wt. %) and/or nickel (1.4 wt. %) and magnesium (1.0 wt. %) immobilized onto a ceria-zirconia support was studied. This was done to determine the effect of siloxane containing landfill gas which ultimately can decompose to silica over the reforming catalyst. The catalyst was poisoned using three different silica amounts corresponding to 1 week, 1 month and 6months concentrations found in landfill gas. XRD patterns demonstrated that silica was present in the catalyst as a depression in the lower 2theta range appeared. Temperature-programmed reduction showed that addition of silica adversely affected the reduction temperature. At the highest amount, the reduction temperature increased from 242°C to 315°C. Dry reforming of the poisoned catalyst also showed a decrease in catalytic activity where the lowest CH_4 and CO_2 conversion (X_{10}) temperatures were achieved at 586.8°C and 563.8°C respectively for the poisoned versus at 454°C and 437°C respectively for the un-poisoned platinum-containing catalyst.

Copyright © American Institute of Chemical Engineers. All rights reserved.

470307 Understanding Direct NOx Decomposition over Pdo/SiO_2: Influence of Reaction Temperature and Pretreatment

Abstract:

More stringent emission requirements highly demand advanced NOx abatement methods to purify NOx from emission exhaust. While the automotive industry continues to improve existing 3-way catalyst systems as well as developing new technologies for emission control under lean conditions, direct NO_x decomposition (dNOd) provides an ultimate solution as NO_x is converted into environmental friendly N_2 and O_2 without the need of a secondary reducing agent. However, the discovery of an effective catalyst to drive the kinetically sluggish dNOd reaction remains the biggest hurdle for scientific research. Several types of catalysts such as perovskites,[1] rare earth oxides,[2] ceria based oxides[3] were reported for direct NOx decomposition. However, these catalysts are typically only active at temperatures higher than the emission exhaust. For instance, although platinum group catalysts (PGMs) exhibit decent activity at temperatures higher than 500 °C, at lower temperatures they suffer from severe deactivation issue.[4,5] A mechanistic study about dNOd on PGM is clearly beneficial to provide fundamental insights for the design of novel catalysts. For this purpose we used PdO supported on silica as a model system to investigate dNOd in a broad temperature region from 100 to 800 °C, focusing on the effect of structural and surface properties of PdO/SiO_2 on the catalytic performance. Using a series of characterization techniques such as XRD and XPS, the dNOd mechanism was revealed to be highly dependent on reaction temperatures, suggesting different strategies should be applied to improve the performance at different temperature regions. In addition, the activity was critically affected by the pretreatment of the catalyst before the reaction. Our results provide the first-hand evidence about a reaction with both scientific significance and technical importance. The insights from our work towards the design of novel dNOd catalysts will also be discussed.

References

1. J. Zhu, A. Thomas, *Appl. Catal. B,* 92, 2009, 225

2. H. Masaki, et al, *J. Alloys and Comp.* 451, 2008, 406

3. N. Drenchev, et al. *Appl. Catal. B,*138â€“ 139, 2013, 362

4. S. Naito, et al. *Catal. Today* ,73, 2002, 355

5. J.D. Consul, et al. *J. Mol. Catal. A,* 246, 2006, 33

Copyright © American Institute of Chemical Engineers. All rights reserved.

470555 Metal Distribution after Drying of Supported Catalysts with High Metal Loadings

Abstract:

Supported catalysts are essential components in a variety of industrial processes, ranging from catalytic converters to production of new drugs. They are generally required because of their high surface area and high mechanical and thermal stabilities. The performance of a catalytic process is intimately related to the catalyst design - uniform, egg-yolk, egg-shell and egg-white metal profiles. Although catalyst preparation and catalytic processing have been investigated for many years, many aspects of catalyst manufacturing are still not fully understood and in industry the design of catalysts is usually based on trial and error, which is expensive and time-consuming, and does not offer assurances on the final results.

It is generally believed that the metal profile is controlled by the conditions that are applied during impregnation, however experiments have shown that drying can also significantly impact the metal distribution within the support. Therefore, to achieve a desired metal profile we need to understand both impregnation and drying. Controlling the impregnation and drying conditions can enhance catalyst performance, and minimize the production of useless batches that have to be disposed, or recycled.

In this work we have investigated the metal distribution during the high concentration impregnation and subsequent drying of supported catalysts. Previous work found that high concentration impregnation (3-4 molar) and drying of Ni/Alumina catalysts yield a uniform metal distribution. At intermediate concentrations (~1 molar), the Ni\Alumina catalysts yielded egg-shell profiles. Theoretical models to simulate the impregnation and drying processes of Ni/Alumina had been previously developed. By not taking into account the crystallization process at higher concentrations, the drying simulation was able to obtain uniform metal distributions. Essentially, the simulation assumes that the metal remains molten throughout the drying process and that solidification during cooling does not affect the metal distribution post-drying. Good comparisons between the simulation and experimental results were obtained for Ni/Alumina. However, experimental results also yielded uniform metal distributions when drying was performed below the melting point of the metal salt when crystallization effects during drying should have been significant.

Mechanisms for the uniform metal distributions were investigated, including metal salt melting point depression due to the microporous support structure and pore blockage. However, we propose deliquescence as novel mechanism for the uniform metal distributions. We hypothesize that deliquescence, or the tendency of a crystalline solid to become a liquid, occurs at all drying temperatures due to the high relative humidity within the pore structures, allowing the metal salt to remain in a liquid state as if it was molten. Metal salts with lower solubility do not deliquesce as readily and were found to not yield uniform metal profiles at higher concentrations.

Our experiments have allowed us to better understand the fundamental mechanisms that occur during drying, and to develop a strategy that can generate desired metal profiles. Although the results presented are based on a particular metal/support system, they serve to provide physical insight into the fundamentals of the drying processes.

Copyright © American Institute of Chemical Engineers. All rights reserved.

470613 Glycerol Steam Reforming for Hydrogen Production over Co_xNi_y-MCM-41 and Co_xNi_y-SBA-15 Catalysts

Abstract:

Recent technological advances in fuel cells technology demand search for efficient methods to produce hydrogen, especially from bio-based resources. One encouraging approach is to expand steam reforming to produce H_2 from glycerol due to its low cost, availability and high hydrogen/carbon ratio. However, limited studies have been reported in the literature investigating the different catalysts for steam reforming of glycerol (GSR). This study focusses on the development of Co and Ni based mono- and bimetallic catalysts with high surface area MCM-41 and SBA-15 as the support materials. The prepared catalysts are characterized using TGA-DSC, N_2 adsorption-desorption, XRD, TEM, ICP-OES, FTIR and H_2-TPR techniques. Catalysts possess high surface area in the range of 540-750 m^2/g and pore sizes of 4.8 -5.9 nm, depending on the type of metal and its loading, and support. The catalysts are tested for their activity, selectivity and long-term stability for GSR in a tubular fixed bed reactor at different temperatures (450 â€“ 700 Â°C) and glycerol/water molar ratios (1/9 â€“ 1/18). The highlights of these studies will be presented.

Copyright © American Institute of Chemical Engineers. All rights reserved.

470614 Development of a Graphene Supported Iron-Oxide Nanocatalyst for Fischer Tropsch Synthesis

Abstract:

In order to provide for increasing global energy demand, alternative sources of energy need to be developed. Fischer Tropsch Synthesis is one such technology that can produce a wide variety of hydrocarbon products from gasified biomass feedstock. However, Fischer Tropsch Synthesis faces a several significant hurdles before it can be considered a reliable alternative source of liquid fuels. Mainly, the selectivity of Fischer Tropsch Synthesis to a narrow range of liquid hydrocarbons is difficult to achieve on a large scale, thus leading to expensive cracking and separation steps to purify the final product. As a result, catalysts with unique structures and chemical properties have been studied extensively to develop a catalyst with higher selectivity or conversion to a narrower range of hydrocarbon products. Due to the unique properties of graphene, including high surface area (>2600m^2/g theoretical), high thermal conductivity (2000-4000W/m-K), and high electrical conductivity, it is hypothesized that a graphene support will facilitate heat transfer and electron mobility along the catalyst surface, thus improving the catalyst effectiveness in overall yield. As a result, graphene supported iron-oxide nanocatalyst was developed and tested to determine the graphene supportâ€™s effect on liquid fuel production through Fischer Tropsch Synthesis.

Copyright © American Institute of Chemical Engineers. All rights reserved.

466696 Microwave-Intensified Synthesis of Biodiesel Catalyzed with Graphene Oxide

Abstract:

Biodiesel (BDF), the fatty acid methyl ester (FAME) derived by reaction of alcohol with carbon-neutral plant or vegetable oils, has attracted increased attention recently as an alternative to petroleum diesel. Conventional approach to BDF production employs homogeneous alkali catalysts such as NaOH and KOH, requiring complex and costly purification process. This method also generates enormous amount of wastewater, and its treatment adds up to already high production cost. In this study, the use of functional carbon-based catalyst in combination with microwave irradiation was investigated as an alternative to the conventional method. Oleic acid and rapeseed oil were used as model compounds for free fatty acids and triglycerides, respectively. The graphene oxide catalysts were prepared by the commonly used Hummersâ€™ method, then dried and finely ground. In a typical run, pre-determined amounts of oleic acid or rapeseed oil, methanol and GO were placed in a 100-mL three-necked round flask. The flask was placed inside the microwave apparatus, then microwave power of 200~600 W was irradiated continuously for 1 to 5 min under total reflux. The obtained products were analyzed of its composition by GC-FID. The effects of parameters such as MeOH-to-feedstock ratio, microwave power, irradiation time were then investigated. The maximum conversion of oleic acid to FAME of about 99 % was obtained. The results were then compared to other intensification techniques such as the use of ultrasonic irradiation.

Copyright © American Institute of Chemical Engineers. All rights reserved.

470779 Syngas Production Via Chemical Looping

Abstract:

Syngas Production via Chemical Looping

Amey More, Saurabh Bhavsar, Charlie Hansen and Götz Veser

Department of Chemical Engineering, University of Pittsburgh, Pittsburgh, PA

Abstract:

'Chemical Looping Combustion' is a clean combustion technology, which enables fossil fuel combustion with inherent CO_2 capture based on the cyclic oxidation and reduction of an oxygen carrier. While most efforts in chemical looping (CL) are focused on combustion, we previously demonstrated the application of the "CL principle"—the periodic oxidation and reduction of a metal oxide to couple two independent redox reactions—to the activation of CO_2 via reduction to CO [1]. In the present contribution, we investigate and compare CO_2 activation via CL in two different operating modes: In the first scheme, CO_2 reduction is coupled with CH_4 oxidation by using mixtures of Fe and Ni (as alloys or simple physical mixtures) to produce CO and syngas product streams [2]. In the second operating scheme, monometallic Ni carriers are utilized to catalytically crack CH_4, producing pure H_2 streams. The solid carbon deposits are then burnt off with CO_2, overall producing separate CO and H_2 product streams.

Supported metal oxide carriers were synthesized, and a combination of thermo-gravimetric analysis (TGA), X-ray diffraction (XRD), and electron microscopy (TEM) was used to identify structure-reactivity correlations for systematic carrier design. Gas phase conversions and selectivities were determined in periodic fixed bed reactor operation. We

find that Fe-Ni alloys can indeed show good activity towards both methane activation and CO_2 reduction, and that the weak oxidant CO_2 allows controlled oxidation of Fe–Ni alloys, which enables selective oxidation of CH_4 to syngas. Remarkably, a simple physical mixture of Fe and Ni far exceeds the reactivity of an equivalent alloy carrier, which can be traced back to a synergistic gas-phase coupling between the two carrier fractions. For the second proposed scheme, we find a strong dependence on the support material, which affects particle size and distribution of the active metal species. Overall, our investigations demonstrate the potential of CO_2 as a "soft" oxidant which enables selective oxidation reactions via novel, intensified chemical looping processes.

[1] M. Najera, et al, *Chem. Eng. Res. Des.* 89 (2011) 1533; Bhavsar, et al, *Chem. Eng. Technol.* 35 (2012) 1281; S. Bhavsar and G. Veser, *Energy Fuels* 27 (2013) 2073; A. More, S. Bhavsar, and G. Veser. *Energy Technology* (2016) AOP. ADDIN EN.REFLIST

Copyright © American Institute of Chemical Engineers. All rights reserved.

470787 Controlling Carbon Affinity with Transition Metal Ceramics and Intermetallic Alloys to Modify the Degree of Olefin and Aromatic Activation

Abstract:

Investigating how to control the carbon affinity and the degrees of activation of olefins and aromatics is of vital interest to the production of fuels and chemicals from petroleum and biomass feedstocks. Moderate activation of the olefin/aromatic is needed to inhibit unselective reactions and the breakdown of the valuable reactant. Transition metals (TM) mixed with p-block elements â€" TM ceramics (TM + B, C, N, O, S, P, Se, etc.) and intermetallics (TM + post TM elements) â€" exhibit a wide range of surface chemistry towards carbon that can be systematically tuned to control selectivity and catalytic activity in the activation of unsaturated hydrocarbons (HCs). The presentation will focus on ceramics and intermetallics and the general trends we encountered in the activation of the olefin/aromatic. Through a combined experimental/theoretical approach, we investigated the effect of p-block element selection and elemental ratio on surface chemistry, the tunability of carbon affinity and the degrees of molecule activation, and the connection of the surface chemistry to the electronic structures and the changes of work function.

Copyright © American Institute of Chemical Engineers. All rights reserved.

471265 Tuning Catalytic Activity for Meerwein-Ponndorf-Verley (MPV) Reaction through Designing Catalytic Sites

Abstract:

Catalytic materials are essential to upgrading biomass into usable commodity chemicals and fuels, which requires increasing the hydrogen content and decreasing the oxygen content. An important reaction for catalytic upgrading is the Meerwein-Ponndorf-Verley reaction that sacrifices an inexpensive alcohol (e.g., isopropanol)Â to convert a ketone into an alcohol. This important reaction has recently been investigated widely with a number of different insights being revealed.[1â€"4] Through using these insights, we have designed a novel catalytic material. In this poster presentation, we will present the synthesis, characterization, and catalytic activity of our novel catalytic material. The material has been characterized using a battery of standard methods, including advanced spectroscopy methods. Importantly, we find that the catalytic activity for hindered substrates is increased. Based on these insights, we suggest new formulations of the catalyst that could achieve higher catalytic activity.

References

[1] H.Y. Luo, D.F. Consoli, W.R. Gunther, Y. Román-Leshkov, J. Catal. 320 (2014) 198.

[2] A. Corma, M.E. Domine, L. Nemeth, S. Valencia, J. Am. Chem. Soc. 124 (2002) 3194.

[3] A. Corma, M.E. Domine, S. Valencia, J. Catal. 215 (2003) 294.

[4] M. Boronat, A. Corma, M. Renz, J. Phys. Chem. B 110 (2006) 21168.

Copyright © American Institute of Chemical Engineers. All rights reserved.

471079 Synthesis and Characterization of Au Nanoparticles for Ethanol Oxidation: Effect of Acidity and Support Structure

Abstract:

Introduction

Gold catalysts have recently turned out to be the most interesting catalysts in alcohol oxidation with environmentally friendly oxidizers such as molecular oxygen. Among different supports for gold, zeolites are promising due to their high thermal stability, presence of large surface area, and micropores that may hinder the sintering of Au. Furthermore, gold catalysts can be used for catalytic oxidation of hydroxyl and carbonyl groups in molecules from biomass.

However, tuning the gold nanoparticles size is one of the challenging steps to maximize the reaction rate [1]. In order to overcome this challenge, efforts have been made to prepare catalysts with gold particle sizes smaller than 10 nm using zeolites (H-Y-12, H-Y-80, H-Beta-25, H-Beta-150 and H-Beta-300) together with Al_2O_3 and SiO_2via deposition-precipitation method. This method has been the most frequently employed method for Au modification and involves the use of tetrachloroauric (III) acid as a precursor of Au. In addition, some alumina, carbon, titania and silica catalysts modified with gold were also tested as commercial catalysts. In this work, was studied. the effect of support structure and acidity, pH and aging time of catalyst preparation step, on gold cluster size

Furthermore, the catalysts were used in gas phase ethanol oxidation utilizing fixed bed tubular reactor. Different reagents like primary linear (1-butanol, 1-octanol) [2] [3] or aromatic (benzylic alcohol) [4] [5] alcohols and polyols (glycerol, glycol, glucose) [6] [7] [8] have been under research. Acetaldehyde and ethyl acetate were the goal products which are needed as intermediates and end products in pharmaceutical and food industry. Activity and stability of these catalysts depends on different parameters including the structure and acidity of the support, as well as on the specific interaction between the gold and the support. The effect of different parameters including Au particle size, type of precursor and reaction temperature on catalytic activity and selectivity were studied.

Experimental

The catalysts were prepared using deposition-precipitation method by changing the pH between 6.5 and 10.5 and aging time from 3h to one day to achieve catalysts with 1 to 3 wt% Au loading and smaller than 10 nm of Au clusters. Chlorine ions from gold precursor ($HAuCl_4$) responsible for sintering of gold particles were washed away followed by drying of the catalysts at 100°C and calcining at 300°C for 3h. The catalysts were characterized thoroughly by TEM, SEM/EDXA, ICP, nitrogen-physisorption, FTIR, XRD, and XPS.

In order to use the catalysts for the reaction, they were sieved to have the particle sizes between 63 to 125 μm. The catalysts were diluted with sand to avoid pressure drop. The reaction was run at 125 to 250°C. Oxidation reaction was performed by using He gas as a carrier. Organic and inorganic products were detected by utilizing an online GC.

Results and discussions

Au mean sizes and size distributions were mainly affected by the acidity and structure of the catalyst supports. For instance, zeolite gold modified catalysts exhibited bigger gold clusters (6-9 nm) and broader size distributions (2-18 nm). The narrowest size distributions and smallest Au particles were found on Au/Al_2O_3 which showed 1.5-4 nm range with mean size of 2.1 nm. It should be noted that a slightly wider Au size range and mean compared to Au/Al_2O_3 characterized Au/SiO_2. The influence of acidity on Au particle size was investigated by comparing gold nano sizes on Au/H-Y-12 and Au/H-Y-80 obeying the same preparation conditions. The results suggested that smaller Au particles were formed on Au/H-Y-80 as a less acidic catalyst than Au/H-Y-12. Besides, H-Beta-25 which is more acidic than H-Y-12, formed very large Au particles (>100μm) which are almost non active phase; behaviour of the catalyst in terms of activity and selectivities was the same as proton form of the support without gold. Thus, it was confirmed that high acidic zeolite supports could not be used for formation of Au clusters with less than 10 nm.

Another parameter which affected the Au size was pH of the solution. Au formed on H-Y-80 at pH of 10.5 was about 1.3 nm smaller than Au formed on H-Y-80 at pH of 9. Aging time of the solution also influenced the Au particle size; an increase from 1 to 3 h reduced the Au average size from 9.3 to 6.3 nm.

Selectivities of the catalysts showed that strong acidity of the support is not even needed for this reaction due to the formation of mainly diethyl ether as a side product.

Figure below shows the ethanol conversion as a function of temperature (125-250°C). A comparison of the effect of gold size on activity exhibited that reductions in size from 4 to 2 nm on alumina can double the activity of the catalyst.

Fig. Conversions of gold catalysts at different temperatures

The importance of acidity in ethanol oxidation reaction was investigated in this work by preparing and testing gold catalysts using Y zeolites (H-Y-12 and H-Y-80). TOF of Au/H-Y-80 (80=SiO_2/Al_2O_3molar ratio) was noticeably higher than Au/H-Y-12. Thus, it was concluded that Y zeolite with high acidities are not suitable for this reaction due to showing lower activities.

Besides, among the tested catalysts (Fig), deactivation studies were performed for Au/SiO_2which indicated promising results in terms of activity and selectivity. This catalyst was tested at 250°C for fifteen days. Characterizations were performed thoroughly by TEM, EDXA, ICP, nitrogen physisorption and XPS for the spent catalyst. The activity and selectivity results illustrated no deactivation. No sintering (TEM) and change in gold oxidation state (XPS) was observed but a slight reduction in surface area was attributed to the formation of carbon residues on the support mesopores.

References

[1] Simakova et al. J.Phys.Chem. C 115 (2011) 1036-1043

[2] Enache et al. Science 311 (2006), 362-365

[3] Li et al. Appl. Catal. A 433-434 (2012) 146-151

[4] Cao et al. Chem. Eng. J. (2011) 734-743

[5] Mertens et al. J. Mol. Catal. A: Chem. 313 (2009) 14-21

[6] Prati et al. Top. Catal. 44 (2007) 319-324

[7] Comotti et al. J. Mol. Cat. A: Chem. 251 (2006) 89-92

[8] Bianchi et al. Top. Catal. 13 (2000) 231-236

Copyright © American Institute of Chemical Engineers. All rights reserved.

471310 Electrochemical Control of the Nitrogen Cycle: Ammonia Synthesis from Nitrates

Abstract:

The electrochemical conversion of nitrates or nitrogen to ammonia and other useful products, offers an appealing alternative approach to synthesis via the energy intensive Haber-Bosch process. Nitrates are considered a waste material and pollutant from fertilizer run-off which may be able to be converted into various useful products such as ammonia, hydrazine, nitrous oxide, nitric oxide, nitrogen dioxide, or other useful nitrogenous materials. Our work involves developing electrocatalytic control and establishing trends which are important to the fundamental understanding of nitrogen based electrochemistry. Preliminary results show the development of screened catalysts at various applied potentials, leading to significant differences in current efficiency toward specific products. We aim to combine our work with theoretical models to continue to improve the activity, selectivity, and understanding of these electrochemical processes.

Copyright © American Institute of Chemical Engineers. All rights reserved.

471531 Theoretical Study on Oxidative Coupling of Methane Using MgO

Abstract:

The recent surge in natural gas production, fueled by the exploitation of shale gas reserves, is providing an abundant supply of methane. Tapping into this supply by converting methane into higher value chemicals can significantly impact the chemical industry by adding new supplies of raw materials. However, methane activation remains a major challenge. It is difficult to break the strong and localized C-H bond. In the event the bond is broken, it leads to products more reactive than methane, making it challenging to terminate at desired products without going to complete combustion products. Oxidative coupling of methane offers the potential of selectively converting methane to ethane/ethylene.

The work of Ito and Lunsofrd [1] ignited the interest in this field when they showed that Li-doped MgO can produce ethane/ethylene from methane. The recent surge of natural gas production renewed the interest in this field. Much of the work has focused on the impact of Li-doping on MgO activity. Recently, it was suggested that the effect of Li-doping on MgO is mainly exposing under-coordinated sites that are more reactive. Schwach et al. [2] showed that MgO in its pure form can have appreciable activity, and that MgO morphology can greatly impact its reactivity. Despite much experimental and theoretical work on the subject, there are still many open questions. In this work, we attempt to answer some of these questions. We utilize Density Functional Theory (DFT) to study the impact of surface geometry. We also investigate dopant effect on MgO activity. BEEF-vdw exchange correlation functional is used. Results suggest that surface geometry can have a pronounced effect on methane activation. Developed scaling relations show that hydrogen binding can be used a descriptor to understand the activation process.

[1] Ito, T. Lunsford, J. Synthesis of ethylene and ethane by partial oxidation of methane over lithium-doped magnesium oxide *Nature* 314 (1985) 721 - 722

[2] Schwach, P. Frandsen, W. Willinger, M. Schlögl, R. Trunschke, R. Structure sensitivity of the oxidative activation of methane over MgO. *Journal of Catalysis* 329 (2015) 560â€"573

Copyright © American Institute of Chemical Engineers. All rights reserved.

471585 A Highly Active and Stable $IrO_x/SrIrO_3$ Catalyst for the Oxygen Evolution Reaction

Abstract:

Electrochemical water oxidation plays a key role in processes such as water splitting and CO_2 reduction which are highly studied for conversion of renewable energy sources into fuels and chemicals. However, water oxidation catalysts suffer from slow kinetics, and previously reported catalysts are limited to >330 mV overpotential to reach 10 mA/cm^2_{oxide} of intrinsic activity when normalized to actual catalyst surface area.[1, 2] Furthermore, there exists a significant need to develop water oxidation catalysts that operate in acidic electrolyte as there are only a few current candidates.

We present investigation of a novel water oxidation catalyst, $IrO_x/SrIrO_3$, which has demonstrated remarkable activity and stability in acidic electrolyte.[3] Thin, crystalline films of $SrIrO_3$ are deposited using pulsed laser deposition and the active catalyst is formed via Sr leaching from surface layers. This leaching process leaves behind IrO_x layers which significantly outperform rutile IrO_2 and RuO_2, the only other OER catalysts to have reasonable stability and activity in acidic electrolyte. XPS and inductively coupled plasma results indicate leaching of Sr from the material surface during initial electrochemical testing in acidic electrolyte, which likely contribute to material restructuring at the surface and result in the activity increase observed over the first 2 hours of electrochemical testing. Additionally, the low surface roughness of these films allows activity normalization to the catalyst surface area (measured using atomic force microscopy) and reveals only 270 â€“ 290 mV overpotential to reach 10 $mA/cm^2_{oxide\_AFM}$ during extended testing. Lastly, density functional theory calculations reveal possible surface structures that could be formed during Sr leaching, and suggest the formation of highly active surface layers with IrO_3 or anatase IrO_2 motifs.

1. C. C. L. McCrory, S. Jung, I. M. Ferrer, S. M. Chatman, J. C. Peters and T. F. Jaramillo, *JACS*, 2015, **137**, 4347-4357.

2. M. S. Burke, L. J. Enman, A. S. Batchellor, S. Zou and S. W. Boettcher, *Chemistry of Materials*, 2015, **27**, 7549-7558.

3. L. C. Seitz, C. F. Dickens, K. Nishio, Y. Hikita, J. Montoya, A. Doyle, C. Kirk, A. Vojvodic, H. Y. Hwang, J. K. Norskov and T. F. Jaramillo, *In review.*, 2016.

Copyright © American Institute of Chemical Engineers. All rights reserved.

471381 Engineering Interfaces for the Activation and Stabilization of Photovoltaic-Grade Thin Film Light Absorbers for Photoelectrochemical Hydrogen Production

Abstract:

The next key challenge for the proliferation of renewable energy technologies is scalable storage of the energy captured from these often intermittent and diffuse energy sources. One promising option for readily transportable energy storage is the production of a chemical fuel from an abundant feedstock. Toward this end, our work focuses on the development of stable photoelectrochemical (PEC) materials for the light-induced catalytic decomposition of water to hydrogen gas. Such systems combine both energy capture and energy storage into one device, but currently suffer from either low efficiencies or poor stability under illumination. In collaboration with Dr. Nicolas Gaillard (U of Hawaii), photo-active electrocatalytic arrays that optimally match the solar spectrum and can sustain operation over technologically meaningful time scales are being developed. In the Jaramillo group, our focus has been on understanding the complex interfaces between the light-absorber, protective film, catalyst and electrolyte in the PEC device.

The chalcopyrite class of materials has shown promise for photovoltaic (PV) applications given these materialsâ€™ utility as potentially scalable and bandgap-tunable thin film light absorbers. Accordingly, Dr. Gaillardâ€™s group has developed thin films of $CuGa(S,Se)_2$ that approach the ideal bandgap for the front absorber in a tandem dual-absorber array for PEC; as a p-type semiconductor, these films function well as photocathodes for carrying out the hydrogen evolution reaction (HER). However, the material is neither intrinsically stable in aqueous electrolyte nor intrinsically active as an HER catalyst. Accordingly, attempts have been made to activate and stabilize the photocathodes using thin films of oxides (such as TiO_2) and sulfides (like MoS_2) that have proven valuable as either protection layers, electrocatalysts, or both. Mechanisms for degradation have been probed via post-mortem electron microscopy and

elemental analysis. Furthermore, work has been conducted to investigate the incorporation of n-type emitter layers (such as CdS and ZnOS) in order to maximize the photovoltage produced by the photoelectrode array.

Copyright © American Institute of Chemical Engineers. All rights reserved.

471412 Engineering Ultra-Small Nanochannel Electrocatalyst Towards the Improvement in Oxygen Evolution and Oxygen Reduction Reaction

Abstract:

The oxygen evolution (OER) and oxygen reduction reaction (ORR) offer a promising method of scalable energy storage and grid-scale power generation that could facilitate widespread adaptation of solar and wind power. However, the main challenge of these reactions is the sluggish kinetics, which caused by the scaling relation between adsorption energies of two reaction intermediates: *OH and *OOH species.[1] A theoretical study has proposed a scheme to circumvent these scaling relationships by defining a nanoscopic catalyst channel that capitalizes on different interactions between the intermediates and the catalyst owing to confinement.[2] However, such ultra-small nanochannel electrocatalyst has not been demonstrated experimentally.

Here in, we present a top-down nanofabrication scheme of nanoscopic electrocatalytic channels with tunable opening sizes from 10 nm to as small as 1 nm. The process utilizes standard nanofabrication techniques such as photolithography, atomic layer deposition, and ion-beam milling, and can be transferred to various catalyst systems. The performance of this proof-of-concept nanochannel electrocatalysts are investigated for the oxygen evolution and oxygen reduction reactions.

(1) Man, I. C.; Su, H.-Y.; Calle-Vallejo, F.; Hansen, H. A.; Martínez, J. I.; Inoglu, N. G.; Kitchin, J.; Jaramillo, T. F.; Nørskov, J. K.; Rossmeisl, J. *ChemCatChem* **2011,** *3* (7), 1159-1165.

(2) Doyle, A. D.; Montoya, J. H.; Vojvodic, A. *ChemCatChem* **2015,** *7* (5), 738-742.

Copyright © American Institute of Chemical Engineers. All rights reserved.

471895 An Integrated 'One-Box' Process for Hydrogen Production

Abstract:

An Integrated 'One-Box' Process for Hydrogen Production

Ashkan Garshasbi[1], Doug Parsley[2], Richard J. Ciora[2], Jr., Paul KT Liu[2], and Theodore T. Tsotsis[1]

[1]The Mork Family Department of Chemical Engineering and Materials Science, University of Southern California, 925 Bloomwalk, Los Angeles, CA 90089-1211, USA.

[2]Media and Process Technology Inc, 1155 William Pitt Way, Pittsburgh, PA 15238-1360.

High-temperature gas separations (GS) with inorganic membranes have attracted attention recently. In particular, the use of such membranes in membrane reactors (MR) has the potential to enhance process intensification and to increase energy savings and/or product yield. Though the potential benefits of high-temperature GS and/or MR processes are substantial, commercialization still remains elusive. A major technical barrier is the lack of robust inorganic membranes and full-scale modules which are suitable for use at the high-temperature and high-pressure conditions required. This team recently proposed and tested in the laboratory a technology, termed the "one-box" process, that employs carbon molecular sieve membranes (CMSM) in an MR for the water gas shift reaction of coal- and/or biomass-derived syngas. The process combines syngas contaminant removal and CO conversion into a single unit to produce high-purity hydrogen for power generation and/or chemical use.

Towards technology commercialization, full-scale multi-tubular CMSM modules have been constructed and characterized comprehensively and their effectiveness has been demonstrated. Our successful preparation of the full-scale multi-tubular inorganic membrane module overcomes some of the technical barriers which hamper the commercial implementation of inorganic membranes. Further multiple field-tests of the full-scale CMSM modules were conducted at the US National Carbon Capture Center (NCCC) coal gasification facility under conditions suitable for the "one-box" process, and this presentation will detail the results of these tests. During continuous use in raw coal-derived and/or biomass-derived syngas, the CMSM successfully rejected tar-like species present in the syngas without any evidence of fouling. Our current efforts focus on the lab-scale testing of reactive separations using these membranes in a broader range of pressure and temperatures akin to the real process, and on the field-testing of the "one-box" process by incorporating the catalyst used in the prior lab-scale study into the full-scale CMSM modules developed.

Copyright © American Institute of Chemical Engineers. All rights reserved.

471904 Novel Catalysts for Photocatalytic Conversion of CO_2/H_2O and CO_2/CH_4 Systems to Syngas

Abstract:

Oral Presentation

Novel Catalysts for the Photocatalytic Conversion of CO_2/H_2O and CO_2/CH_4 Systems to Syngas

Karishma Piler, Paul Bernazzani, Cristian Bahrim, and Tracy J Benson.

In accordance with EPA, Carbon dioxide accounts for ~ 81 % of all U.S. greenhouse gas emissions in 2014. According to NASA global climate change, for the first time in the recorded history CO_2 levels have surpassed 400 ppm in 2013. While many researchers are searching for capture and storage methods for reducing CO_2 emissions, identifying methods to effectively convert CO_2 to saleable products, CO_2 can then become a valuable feed stock for sustainable energy production. This research will present the use of pairing of Titania nanotubes with single walled carbon nanotubes in photocatalytic conversion of carbon dioxide in water/CO_2 and methane/CO_2 systems into useful fuels and chemicals. The reaction was carried out in a continuous flow annular heterogeneous reactor in the presence of UV and visible light.

Titanium dioxide as a photocatalyst has been studied and implemented for a long period of time, chiefly in the production of hydrogen and other chemicals through water splitting, decomposition of bacteria and organic waste, and carbon dioxide conversion. Titanium dioxide has been doped with various metals and non-metals to engineer the band gap for effective light harvesting and to increase the time of electron-hole pair recombination so as to bring in a large increase in the efficiency of energy production through photocatalysis.

In this research, titania nanotubes have been used because of their high specific surface area for interaction and optimum pore size for diffusion of active species. Also, because of their nanotube architecture, the recombination effect is greatly reduced. The reasons for implementing carbon nanotubes is because of their properties such as large surface area, one dimensional electron transfer, and large electron storage capacity easily accepting photo-excited electrons. They also act as photosensitizers for TiO_2. They can adsorb twice the CO_2 adsorbed compared to activated carbon, another significant asset. The reason behind pairing of titanium dioxide nanotubes and carbon nanotubes was to observe and implement the synergistic effects which can result from combination of their unique properties in photocatalytic conversion of CO_2 to useful products.

The pairing of titanium dioxide nanoparticles and metallic carbon nanotubes along with providing several surface irregularities which act as trap states for electrons, it also leads to a distinctive quantum effect called the schottky barrier, a space charge separation region at the interface of TiO_2-CNT that increases the recombination time of electron-hole pairs. This results in TiO_2 acquiring a strong positive charge and metallic single walled carbon nanotube (SWCNT) a strong negative charge. The Ti-O-C bond results in the shift from UV to Visible wavelength region.

In this presentation, we will present the characterizations of the photo-catalyst, including FTIR, Raman, TEM, and XRD, as well as the reaction results, namely the conversion of CO_2. Reactions were conducted at 100 â€" 120 Â°C, and a GC-TCD was used to analyze the conversion of CO_2 and the formation of CO, H_2, and low molecular weight hydrocarbon gases.

Copyright © American Institute of Chemical Engineers. All rights reserved.

471956 Computational Modelling of the Fischer-Tropsch Synthesis within a Microreactor

Abstract:

The Fischer-Tropsch Synthesis (FTS) is a set of chemical reactions that involves the catalytic hydrogenation of carbon monoxide to various liquid hydrocarbons. It has received considerable interest of late, primarily driven by the recent natural gas boom in the United States and Russia. Natural gas, which is chiefly methane, can be steam reformed to synthesis gas (a mix of CO and H_2) before being fed to a FTS reactor for conversion to more valuable liquid hydrocarbons such as diesel and naphtha. The purpose of this study is to construct a theoretical model that successfully incorporates both microkinetics and diffusion to simulate the FTS reactions within a microreactor.

This mathematical model will use COMSOL Multiphysics to simulate the FTS reactions on the surface of a cobalt-based catalysis. The aim of this computational model will be to use COMSOL to extract reaction kinetics from measured product distributions in the process as implemented in a local microstructured synthesis reactor. From this data, the rate determining steps will be determined to compare with our own estimates from quantum mechanical computational work done at the DFT level. In this contribution, 2-D models of the microreactor will be presented along with steady-state calculations of the final products from reaction.

Copyright © American Institute of Chemical Engineers. All rights reserved.

472076 Hydroxyl Radical Scavenging Activity of Silver Nanoparticles: Enhancing Photocatalytic Reduction and Decreasing Photocatalytic Oxidation Rate

Abstract:

Metallic nanoparticles assisted photocatalysis has been emerging to improve the efficiency of many important processes like water splitting, reduction of CO_2 with H_2O to form hydrocarbon fuels, and degradation of organic molecules.[1] The efficiency of photocatalytic materials can be influenced in the presence of plasmonic metallic nanoparticles due to both radiative energy and electron transfer among them. In this study, the catalytic efficiency of TiO_2 in the presence of silver nanoparticles is evaluated using one reduction and one oxidation reaction. We observed that the presence of Ag nanoparticles (~50nm) with TiO_2(~30nm) significantly enhance the important photocatalytic reduction like hydrogenation of CO_2 derived bicarbonate to produce value-added chemicals such as formate in comparison to same quantity of TiO_2 alone under the solar light (solar simulator, 1.5 AM). A high yield of formate (production rate 160 micromol/gm TiO_2.hr.), was achieved after reducing sodium bicarbonate in glycerol–water solution using Silver-TiO_2 nanostrctures. However, the same photocatalytic system combining silver and TiO_2 nanostructures adversely affect the degradation/ oxidation rate of Rhodamine B one of the important industrial pollutants. Further experiments with terephthalic acid (a probe for hydroxyl radical) reveal that presence of silver nanoparticles decrease TiO_2 induced hydroxyl radical, which is required for photocatalytic oxidation of Rhodamine B. On the other hand, scavenging hydroxyl radical may have favorable effect on formate production as strongly oxidizing hydroxyl radicals may oxidize formate to produce •$CO2^-$, hence decreasing overall yield.[2] To understand the mechanism better, we studied the Ag nanoparticle/TiO2-catalyzed photooxidation of Rhodamine B under two different excitation wavelengths, one corresponding to the bandgap of TiO_2 (365 nm) and other corresponding to the plasmonic peak of Ag nanoparticles (450

nm). This enables us to separate processes associated with the plasmon resonance from those associated with electronic transitions in the TiO_2 itself. This study indicates that plasmonic effect may not be important for the scavenging of hydroxyl radical by silver nanoparticles. From this study, we conclude that adding metallic nanoparticles with TiO_2 does not always have unequivocal benefits. In general, understanding of charge transfer among metallic nanoparticles and photocatalysts under different geometric configuration can provide useful mechanistic information that may be used in the improvement of photocatalytic systems for the reduction of CO_2 also stored in the form of carbonates.

1. Hou, W.; Cronin, S. B., A Review of Surface Plasmon Resonance-Enhanced Photocatalysis. *Advanced Functional Materials* **2013,** *23* (13), 1612-1619.

2. Molinari, A.; Samiolo, L.; Amadelli, R., EPR spin trapping evidence of radical intermediates in the photo-reduction of bicarbonate/CO2 in TiO2 aqueous suspensions. *Photochemical & Photobiological Sciences* **2015,** *14* (5), 1039-1046.

Copyright © American Institute of Chemical Engineers. All rights reserved.

472082 Sputter Deposition of Silver-Copper Nanoparticles for the Oxygen Reduction Reaction

Abstract:

The co-sputtering of bimetallic nanoparticles opens up exciting new directions in the field of heterogeneous catalysis. In addition to precise control over nanoparticle size, composition, and loading, this vapor phase deposition technique could allow for the â€œfreezing-inâ€ of phases which are not thermodynamically stable in the bulk and which may have interesting properties beyond those of their parent materials. These nanoparticulate alloy phases are of interest to researchers in heterogenous catalysis, as they allow for more precisely tuned adsorbate binding energies than are offered by bulk phase materials. Alloying copper into silver nanoparticles, for example, has been predicted to enhance silverâ€™s activity for the electrochemical oxygen reduction reaction (ORR). Here we demonstrate the ability to deposit size selected silver-copper nanoparticles from 5 to 20 nm across a full range of silver-copper compositions, and we show an enhancement in catalytic activity of the bimetallic nanoparticles for the ORR in base compared to pure silver.

Copyright © American Institute of Chemical Engineers. All rights reserved.

472187 Rare Earth Element Doped Titania (TiO_2) for Heterogeneous Photocatalytic Applications

Abstract:

Titanium dioxide or titania (TiO_2) is a well-studied photocatalyst. The major drawback of titania is its bandgap. Having a very high band gap (~3.0 eV for rutile; ~3.2 for anatase), titania is only responsive towards UV light. There is interest in enhancing the light absorbing properties of titania by doping various metals through upconversion phenomenon. We hereby studied the effect of doping erbium (Er) and ytterbium (Yb) ions on the photocatalytic effects of titania. For our purpose, we used Rose Bengal as a model pollutant and phenol as a real pollutant. Efforts have been spent in understanding the phenomena of upconversion, adsorption, recombination kinetics etc. and its implications towards the photocatalytic degradation rates of pollutants. We prepared pure anatase samples along with those with erbium doping concentrations of 2% and ytterbium doping concentrations of 0%, 10%, 15% and 20% (all in elemental basis) in a hydrothermal process. These samples were characterized for crystal structure, surface properties, composition, particle size, band gap and band edge positions. Band gaps and band edges were measured using diffuse reflectance mode in UV-visible spectroscopy and by Electrochemical Impedance Spectroscopy (EIS) respectively. Computational calculations using density functional theory (DFT) have been conducted to probe the crystal structure of the doped material as well as their band structure. Vienna ab-initio Simulation Package (VASP) has been used for the purpose. The aqueous phase photocatalytic reduction rate constants were quantified for both the organic liquids under different light sources. The 2% Er doped TiO_2 sample performed ~ 3 times better than pure TiO_2 under simulated solar light. To further elucidate the effect of different light sources, LEDs with different wavelengths (405 nm, 530 nm, 660 nm and 940 nm) were used. We could obtain the photo reduction of phenol and Rose Bengal under 405 nm light. Plausible explanations are made for these photocatalytic reduction results based on suppressed charge carrier kinetics and a close correlation with the bandgaps and band edges (both from experiments and computations).

Copyright © American Institute of Chemical Engineers. All rights reserved.

472253 Experimental and Theoretical Studies of CO_2 Adsorption on Hydrotalcite

Abstract:

This work investigates experimentally the adsorption kinetics of CO_2 on hydrotalcite, and puts forward a simple two-step adsorption reaction mechanism that forms the modeling foundations of the process used to collect the experimental data. The experimental data collection process consists of pure CO_2 gas being continuously fed to and removed from a semi-batch reactor in which a given amount of adsorbent material is enclosed within a Thermogravimetric Analysis (TGA) apparatus. The obtained experimental data consist of adsorbent weight gain measurements as function of time. A nonlinear, multi-parameter model describing the time evolution of the adsorption process is developed. At its core is a novel two-step mechanism, that involves CO_2 adsorption onto multiple types of active sites. One to one and onto maps between the adsorption/desorption kinetic constants and parameters describing the time evolution curves are developed, which in turn allow the resulting nonlinear parameter regression problem to be solved for its global optimum. Uncertainty intervals containing the resulting kinetic constant values are also developed, corresponding to the variability of the experimental data.

Copyright © American Institute of Chemical Engineers. All rights reserved.

472441 First-Principles Study for Controlling the Cation Segregation on the Surface of Perovskite Oxide Catalysts

Abstract:

Controlling the surface chemistry of perovskite oxide is a key challenge for enhancing the oxygen reduction reaction (ORR) kinetics of solid oxide fuel cell (SOFC) cathodes. However, segregation of A-site cation often generates secondary phase on the perovskite oxide surface, thereby deteriorating the reactivity and stability in ORR. Although the issue for suppressing the cation segregation has been raised by many researchers, the underlying mechanism is still not clear.

In this study, a key driving force of A-site cation segregation was elucidated for the two classes of perovskite-type materials using density functional theory (DFT) calculations; alkaline earth metal (Ba, Sr, and Ca)-doped $LaBO_3$ (B = $Cr_{0.50}Mn_{0.50}$, Mn, Fe, $Co_{0.25}Fe_{0.75}$, Co, and Ni) and $SrTi_{1-x}Fe_xO_3$. In both cases, elastic energy of A-site cation is a dominant factor governing the segregation phenomena, which depends on the size of constituent ions in the perovskite oxide. Based on these results, we suggest simple ways how to suppress the cation segregation by reducing the elastic energy. We believe that our results will be useful to rationally design the SOFC cathodes.

Copyright © American Institute of Chemical Engineers. All rights reserved.

472457 Low Temperature Hydrogen Production from Kinetically Limited NH_3 Decomposition in Palladium Membrane Reactor

Abstract:

It is desired that decomposition of ammonia as a promising chemical hydrogen carrier is carried out at lower temperatures, whereby making possible a highly efficient hydrogen recovery from ammonia. Although the equilibrium conversion is almost 100% in the range of 623–723K, the decomposition remains in a low conversion, so that insufficient activity of present catalysts should be improved. This study attempts to prove that this kinetically limited decomposition using a ruthenium supported catalyst can also be enhanced in the palladium membrane reactor by selective hydrogen separation.

The Langmuir-Hinshelwood types of rate equations, seven reaction models, for ammonia decomposition using a Ru/SiO_2 or Ru/Al_2O_3 catalyst were derived and compared with experimental results. As the result, the combinative desorption of nitrogen atom was found to be the rate determining step in the range of 623–723 K. One dimensional model for a palladium membrane reactor made clear that the kinetic enhancement differed considerably according to the reaction model (rate expression)

In the ammonia decomposition employing a membrane reactor with a thinner palladium composite membrane tube, a large increase in conversion and hydrogen recovery compared with the conventional packed reactor could be obtained.

Copyright © American Institute of Chemical Engineers. All rights reserved.

472498 Fast Prediction of CO Oxidation Activity on Rare-Earth Metal Doped CeO_2

Abstract:

Fast Prediction of CO Oxidation Activity on Rare-Earth Metal Doped CeO_2

-

Kyeounghak Kim[1], JeongDo Yoo[2], Siwon Lee[2], WooChul Jung[2], and Jeong Woo Han[1,*]

[1]*Department of Chemical Engineering, University of Seoul, Seoul 130-743, Republic of Korea*

[2]Department of Materials Science and Engineering, Korea Advanced Institute of Technology, Daejeon 305-701, Republic of Korea

*E-mail: jwhan@uos.ac.kr

Owing to high oxygen storage capacity and high stability over the wide range of temperatures, CeO_2 has been widely applied to support materials for the electrode of solid oxide fuel cells or diverse oxidation catalysts. While there have been several studies for understanding the catalytic activity according to the rare-earth (RE) metal dopants on CeO_2 nanoparticles, it is hard to find the detailed reaction mechanism and a simple descriptor which enables to rapidly estimate the activity. In this study, our combinatorial study of experiments and theories shows the catalytic activities of CO oxidation are in order of CeO_2 > $Ce_{0.8}Pr_{0.2}O_2$ (PDC) > $Ce_{0.8}Nd_{0.2}O_2$ (NDC) > $Ce_{0.8}Sm_{0.2}O_2$ (SDC). Our

density functional theory calculations demonstrated that CO oxidation reaction on RE-doped CeO_2(111) is well described via the Mars-Van Krevelen (MvK) mechanism (Fig. 1). We found that the ionic radius of dopants can be used as a simple descriptor to predict the whole reaction activity (Fig. 2). Based on our results, we suggest that La might be a promising dopant to enhance the catalytic activity of CO oxidation. Our results will provide useful insight to unravel and predict the CO oxidation activity of ceria based catalysts.

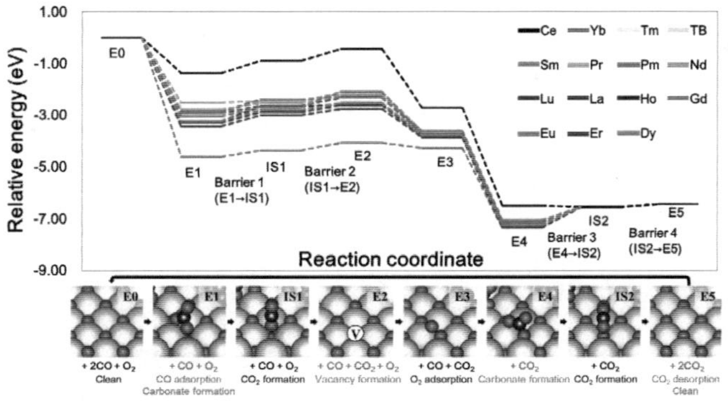

Figure 1 Relative energy diagram of CO oxidation on RE-doped CeO_2(111).

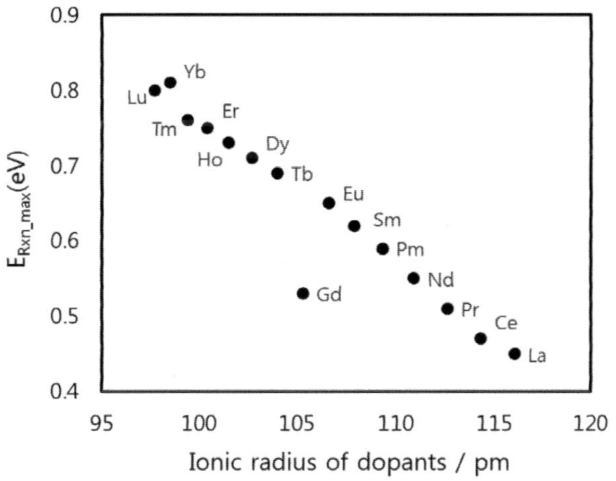

Figure 2 Linear relationship between maximum reaction energy for CO oxidation and ionic radius of dopants on RE-doped CeO_2(111).

Copyright © American Institute of Chemical Engineers. All rights reserved.

472554 Diagnostics of Field-Aged Three-Way Catalyst (TWC) on Stoichiometric Natural Gas Engines

Abstract:

Natural gas is the most attractive alternate fuel for gasoline and diesel due to its competitive price resulting from its abundant availability and lower GHG emissions resulting from its combustion. In stoichiometric natural gas engines, three-way catalysts (TWCs) are widely employed to convert the exhaust pollutants CO, HC (including CH_4) and NO_x to CO_2, N_2 and H_2O. The TWCs contain Platinum Group Metals (PGM) as the active centers for various reactions and ceria-zirconia ($CeZrO_x$) material provides oxygen storage capacity (OSC), which acts as a buffer for maintaining stoichiometric conditions under broad air to fuel ratios. The activity of PGM and OSC components of TWC degrade upon prolonged real life operation. The degradation could arise due to various reasons such as high temperature hydrothermal exposure of catalysts for prolonged periods and masking/poisoning of the catalytic sites by, for example, chemical contaminates originating from lube oil consumption and/or coolant leakage. Isolating the impact of individual degradation mechanisms on TWC performance is imperative for developing improved catalysts, controls and performance recovery methodologies.

From laboratory hydrothermal aging studies, a linear relationship between NO_x conversions and OSC was shown. Real-world aged TWC catalysts have shown significant decrease in surface area and OSC and can be attributed to hydrothermal aging. In addition, SEM and EDX analysis confirmed the presence of P, Na, K and S containing species concentrated on the inlet portion of the real-world aged TWC and their concentrations rapidly decreased along the axial length of the catalyst. Lower NO_xconversion was observed on the sample having the above-mentioned chemical deposits, e.g. inlet portion, while the performance of the rest of the catalyst unit was minimally impacted. An acid washing methodology was developed to remove the chemical contaminates from the real-world TWC which restored its performance.

In this contribution we describe the i) laboratory evaluation methodologies, ii) real-world aged catalyst characterization, iii) aging mechanisms and iv) quantification and correlation between NO_xconversion and OSC all of which are important to develop TWC and controls.

Copyright © American Institute of Chemical Engineers. All rights reserved.

The beneficial effect of fructose on the catalytic oxidation of glucose to gluconic acid

*Fabiana S. Lima ; João Guilherme R. Poço **

Centro Universitário da Fundação Educacional Inaciana "Padre Sabóia de Medeiros" - FEI,
Av. Humberto de Alencar Castelo Branco, 3.972 - Bairro Assunção - São Bernardo do Campo - SP, Brasil.
* jgrpoco@fei.edu.br

16AIChE Annual Meeting, San Francisco, CA

centro universitário FEI

Introduction

The saccharides do not have their limited use only as a sweetener. They can be applied as raw material in the production of organic acids, like gluconic acid, which is produced by the oxidation of glucose. The gluconic acid and gluconate salts are considered safe by the Food and Drug Administration (FDA) having wide applicability not only in the food industry, but also in chemical, pharmaceutical and textile industries [1,2].

The catalytic oxidation is usually developed using pure oxygen or air as the oxidizing agent, and usually uses noble metals (Pt, Pd, Au) supported catalysts or bimetallic supported catalysts, in which have Bi, Tl, Sn and Co are used as promoters [3]. The selectivity to gluconic acid decreases in reactions conducted at pH ≥ 9.5 due to the formation of by-products [4], such as glycolic, acetic and formic acids [5]. Additionally, fructose is formed by the isomerization of glucose [6].

This study presents a new methodology for obtaining gluconic acid from glucose and fructose mixtures subjected to catalytic air oxidation using Pd-Pt-Bi/C commercial catalyst. The data obtained indicate that the presence of fructose positively influenced the yield of glucose to gluconic acid which increases from 79 % to 90 %.

Experimental

a) Catalytic oxidation

Materials

Commercial catalyst (Evonik™)
4 % Pd + 1 % Pt + 5 % Bi on activated charcoal.

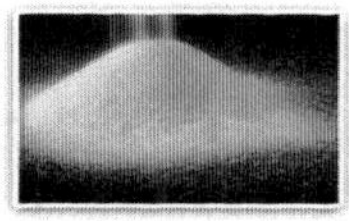

Materials

D-glucose (Merck™)
D-fructose (Sigma-Aldrich™)

- The reactions were performed in a 500 mL glass reactor equipped with a mechanical stirrer (2030 rpm).
- The pH was monitored using a pH meter (in-situ) and manually controlled by the addition of 2 M NaOH by means of a burette connected to a peristaltic pump and then to the reactor.
- During the reactions, air was pumped to the reactor and distributed in the solution through a diffuser.
- The temperature was kept constant using a thermostatic bath coupled to the reactor jacket and the pressure was atmospheric (0.921 ± 0.015 atm).
- Oxygen concentration was evaluated using Quimis Q758P

Experimental conditions:

Glucose solutions	Mixture of glucose and fructose solutions
pH 8.5, 9.5 and 10.5	pH 8.5 and 9.5
55 and 60 °C	55 and 60 °C

b) Product characterization

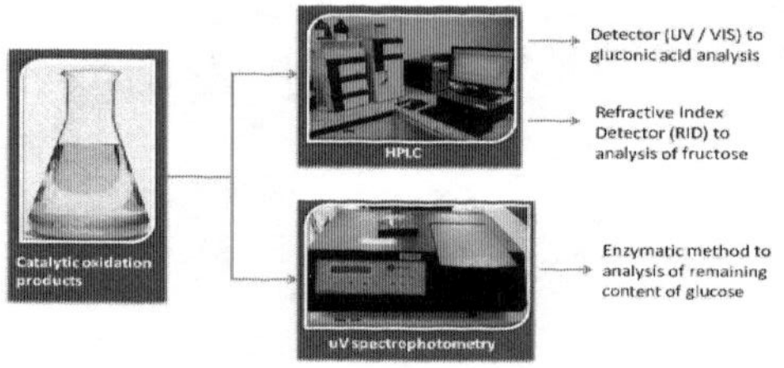

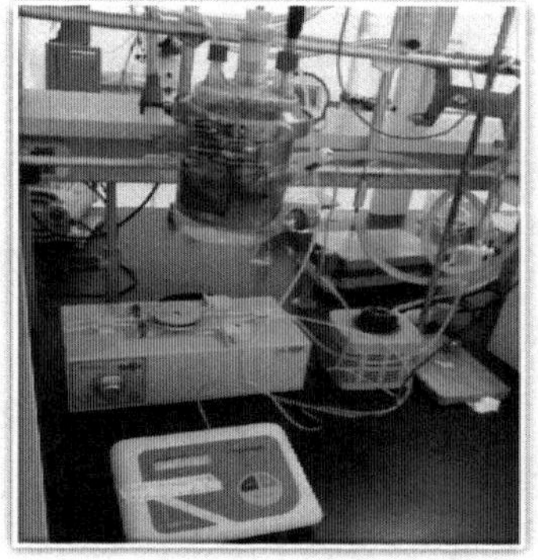

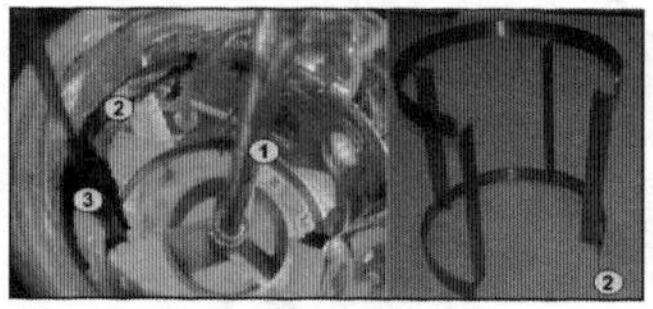

View of the Process and internals of the reactor

Acknowledgements

Centro Universitário da Fundação Educacional Inaciana "Padre Sabóia de Medeiros" - FEI

Results

Table 1 - Results of the catalytic oxidation

Initial Concentrations ($g.L^{-1}$)		Conditions		Final Concentrations ($g.L^{-1}$)			X_{Gr}	Y_{GA}	$S_{GA/H+}$
Glucose	Fructose	pH	T (°C)	Gluconic Acid	Glucose	Fructose	(%)	(%)	(%)
120	-	8,5	55	32.30	21.02	-	76.65	38.65	39.54
120	-	8,5	60	47.75	20.50	-	77.29	56.52	58.83
120	-	9.5	55	66.58	21.08	-	76.79	78.86	83.98
120	-	9,5	60	63.84	14.79	-	83.57	70.06	78.13
120	-	10.5	55	57.37	11.12	30.69	87.83	59.04	75.31
120	-	10.5	60	47.11	10.68	27,47	88,16	43,15	57,65
60	60	8.5	55	24.80	16.47	66.07	67.23	60.59	61.23
120	120	8.5	55	48.40	52.61	122.66	46.15	88.74	94.48
60	60	9.5	55	43.10	13.03	59.24	73.76	97.18	93.31
120	120	9.5	55	59.30	42.86	122.67	55.00	93.55	97.92

Note: T (temperature); XGR (conversion of glucose), YGA (yield of gluconic acid), SGA/H+ gluconic acid (selectivity with respect to other acids); CPd-Bi-Pt / C: 1.2 g / L; Reaction time 70 minutes.
Maximum absolute uncertainties: Concentrations of Gluconic Acid 1.56 g/L; Glucose 1.11 g/L and Fructose: 3.23 g/L., XGR: 2.31 %; YGA 3.64 %; SAG / H +: 2.92 %.

Figure 1 - Results of conversions of glucose to gluconic acid

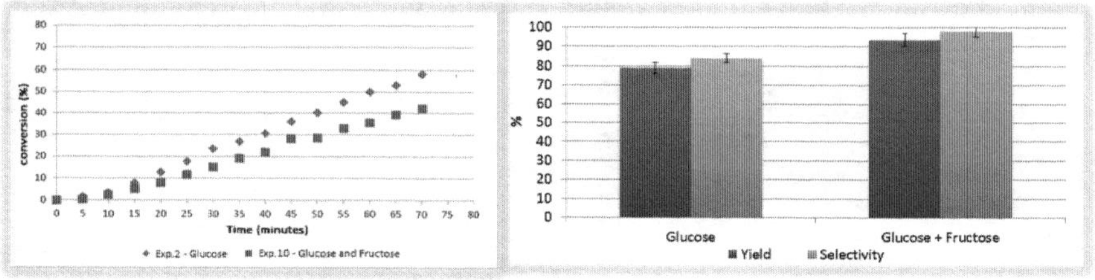

a) Stoichiometric conversions of glucose over time acids b) Yield and selectivity of glucose to gluconic acid

Note: Glucose 120 g/L and fructose 120 g/L; pH 9.5, 55 °C; absolute maximum uncertainty Stoichiometric conversions (XAGE) 1.68 %

Conclusion

Based on the results, it is evident that the optimal conditions for the catalytic oxidation were established at pH 9.5 and 55 °C, which show the highest concentration of gluconic acid, yield and selectivity.

The results obtained indicate that the presence of fructose positively affected the reaction, since a significant increase in yield and selectivity was observed.

The results still present the possibility of using inverted sucrose as raw material, which is much more cost saving than glucose, increasing revenues of gluconic acid production.

References

[1] Wong, C; Wonh, K.; Chen, X. Applied microbiology and biotechnology 78.6 (2008): 927-938.
[2] Delidovich, I.V. et al. Catalysis letters 140.1-2 (2010): 14-21.
[3] Liu, J.; Cui, Z. Journal of Membrane Science 302.1 (2007): 180-187.
[4] Önal, Yücel et al. Journal of Catalysis 223.1 (2004): 122-133.
[5] Onda, A. et al. Applied catalysis A: general 343.1 (2008): 49-54.
[6] Biella, S. Prati, L; Rossi, M. Journal of catalysis 206.2 (2002): 242-247.

472654 Modeling of Microreactors for Chemoselective Capture of Trace Volatile Organic Compounds

Abstract:

We report a microreactor approach for chemoselective capture of trace carbonyl compounds in air and exhaled breath. The Microreactors were fabricated on silicon wafer by microelectromechanical systems (MEMS) technology. There were thousands of micropillars in the microreactor to distribute gas flowing through the microreactor. The micropillar surfaces were functionalized with a quaternary ammonium aminooxy salt for chemoselective capture of trace carbonyl compounds by means of oximation reactions. 500 mL to 1L gaseous samples with known amounts of carbonyl compounds in Tedlar bags were used for characterization of capture efficiencies. Carbonyl compounds captured by the microreactors were analyzed by Fourier transform-ion cyclotron resonance (FT-ICR) mass spectrometry (MS). Carbonyl compounds including aldehydes and ketones are ubiquitous in air, exhaled breath, aerosols of electronic cigarette and conventional cigarette smoke. Some carbonyl compounds in exhaled breath are related to disease metabolic processes. Quantitative analysis of trace carbonyl compounds is critical for identification of disease metabolic markers in exhaled breath. One challenge of the microreactor approach was that the capture efficiencies of carbonyl compounds were significantly affected by the gas flow rate. Higher capture efficiencies of carbonyl compounds (95%) require lower sample flow rate, thus longer sample process time. Longer sample process time results in degradation of carbonyl compounds in the sampling bags. This work investigated the effects of the gaps between the micropillars and microreactor length on capture efficiencies of carbonyl compounds at different flow rates. A reaction engineering model was developed to understand the effects of the microreactor length and the gap between the micropillars on the capture efficiencies of carbonyl compounds at different sample flow rate. Breath samples were used to validate the capture efficiencies of the microreactors with different length at various sample flow rates.

Copyright © American Institute of Chemical Engineers. All rights reserved.

472875 Trace Levels of Copper in Carbon Materials Show Significant Electrochemical CO_2 Reduction Activity

Abstract:

Carbon materials are frequently used as supports for electrocatalysts because they are conductive and have high surface area. However, recent studies have shown that these materials can contain significant levels of metallic impurities that can dramatically alter their electrochemical properties. Here, the electrocatalytic activity of pure graphite (PG), graphene oxide (GO), and carbon nanotubes (CNT) dispersed on glassy carbon (GC) are investigated for the electrochemical CO_2 reduction reaction (CO2RR) in aqueous solution. It was observed that PG, GO, and CNT dispersed on GC all exhibit significant electrochemical activity that can be ascribed to impurities of Ni, Fe, Mn, and Cu. The level of Cu in GO can be particularly high and is the cause for the appearance of methane in the products produced over this material when it is used for the CO2RR. Washing these supports in ultrapure nitric acid is effective in removing the metal impurities and results in a reduction in the electrochemical activity of these forms of carbon. In particular, for GO, nearly all of the catalytically relevant metals can be removed. Electrochemical deposition of Cu on GO and PG supported on GC, and on GC itself, increased both the electrochemical activity of these materials and the production of methane via CO2RR. Particularly high rates of methane formation per unit of Cu mass were obtained for Cu electrodeposited on GO and PG supported on GC. We suggest that this high activity may be due to the preferential deposition of Cu onto defects present in the graphene sheets comprising these materials.

Copyright © American Institute of Chemical Engineers. All rights reserved.

472987 Engineering the Electronic Structure of Mxenes Using Density Functional Theory: Strain and Support Effect

Abstract:

Engineering the electronic structure of MXenes using Density Functional Theory: strain and support effect

MXenes (M = Transition Metal, X = C or N) are two-dimensional sheets, consist of M_(x) and X_(x-1).[1] They are one of the top emerging materials in recent years. Previous studies demonstrate that the electronic structure of MXenes can be well engineered by controlling the surface termination. In this, we used density function theory(DFT) to investigate how the electronic structure can be engineered by **strain** and **support**. In this poster, we will present several examples including 1) band structure engineering 2) catalytic activity tuning. Ultimately, we will have a general physical picture describing the tunability of MXenes by strain and support, providing a guideline for designing novel MXene-based catalysts.

(1) M. Naguib, V.D. Mochalin, M.W. Barsoum and Y. Gogotsi, *Adv. Mater.* **26**, 992 (2014).

Copyright © American Institute of Chemical Engineers. All rights reserved.

473030 Effect of Carbon Addition in the Slurry Phase Hydrocracking of Vacuum Residue

Abstract:

The slurry phase hydrocracking is widely known suitable for heavy oil upgrading like vacuum residue. However, the dispersed catalysts often suffer from deactivation upon reaction severity. This study focuses on maintaining the stability and activity of dispersed catalyst in the course of slurry phase hydrocracking of vacuum residue. Various reaction conditions including reaction temperature, H_2 pressure, and time were applied to investigate the factors influencing the structure and activity of oil-dispersed MoS_2 catalyst. In particular, the addition effect of carbon materials in the hydrocracking of vacuum residue was examined. Characterizations of catalyst stability were made by TEM, EXAFS, and ICP-AES. Repetitive reaction tests for VR hydrocracking revealed that reaction severity of temperature above 420°C and H_2 pressure less than 80 bar led to the growth of MoS_2 slab, resulting in deactivation. In contrast, the addition of carbon materials considerably enhanced the catalytic stability and activity, with maintain the MS2 slab size less than 10 nm.

Copyright © American Institute of Chemical Engineers. All rights reserved.

473097 Dispersed Sulfide Catalysts for Hydrocracking of Vacuum Residue

Abstract:

<span" roman"="Roman"" new="New">Recent decrease in light petroleum crude has spurred research in heavy-oil upgrading technologies. Among the heavy oils like petroleum residue and bitumen, vacuum residue (VR) occupying a half of heavier crude oil, has been a target to be upgraded into middle distillates. Hydrocracking is one of the most promising processes with high conversion rate of heavy oil to distillates. In particular, unsupported oil-dispersible nano-catalyst in VR HCK, in which enhanced VR HCK was for oil-dispersible catalyst system over the supported catalyst system, due to the better accessibility toward active centers of the catalyst. This study presents the hydrocracking and hydrogenation activity of bimetallic sulfide on the slurry phase hydrocracking of VR. The catalysts were synthesized by the ligand stabilization method with trioctylphosphine oxide (TOPO) used as coordinating agent. The reaction was conducted in an autoclave reactor at 10 MPa $H_2$and 623~693K for 1.5~24hr. The feedstock was prepared by blending a VR (30g) and synthesized catalysts or recycled one (360ppm metal basis). The characterization was conducted with X-ray absorption fine structure (XAFS) method. The EXAFS spectra were recorded at beamline 8C and 10C of the Pohang accelerator laboratory.

Copyright © American Institute of Chemical Engineers. All rights reserved.

Oxidative-Extractive Desulfurization of Model Diesel Oil By CTA-type Dawson type Emulsion Catalysts

F. Banisharif[a], M.R. Dehghani[a, 1], M. Parvari[a], J.M. Campos-Martin[b]
[a] *Chemical engineering department, Iran university of science and technology, Narmak, Tehran, Iran*
[b] *Grupo de Energía y Química Sostenibles (EQS) Instituto de Catálisis y Petroleoquímica, CSIC. Marie Curie, 2 Cantoblanco, 28049 Madrid, Spain*

ABSTRACT

For the first time, in this study, the CTA-type vanadium substituted Dawson-type emulsion catalysts, $[CTA]_{6+x}P_2W_{18-x}V_xO_{62}$ (x = 0, 1, 3, 5), were successfully synthesized. They were characterized by means of ICP, FT-IR, UV spectroscopy, XRD and XPS. $[CTA]_{11}P_2W_{13}V_5O_{62}$, chosen as a best catalyst among all prepared samples, was used to extractive- catalytic oxidative desulfurization (ECODS) of dibenzothiophene (DBT) from two kinds of model diesel oil with H_2O_2 (50 wt-%) as an oxidant and three kinds of solvents (deep eutectic solvent (DES), ionic liquid (IL), ethylene glycol (EG)). The Taguchi robust was used to study the effect of parameters (solvent type, volume ratio of solvent to oil, temperature and contact time) on the ECODS. The results showed that The ECODS removed 95% of DBT from model diesel oil containing only *iso*-octane under optimized conditions, IL with the volume ratio of solvent to oil 1:4, 50 °C and contact time 60 minutes for model oil containing only *iso*-octane as a solvent. The existence of toluene (20 v-%) reduced the efficiency of system and removal of DBT to ~94 % under determined optimized conditions.

Keywords: Oxidative-extractive desulfurization, Vanadium substituted Dawson-type emulsion catalyst, Model diesel oil

1. INTRODUCTION

Sulfur containing compounds in fuels can be converted to SOx during combustion and are responsible for the formation of some environmental pollution.[1,2] To date, hydrodesulfurization (HDS) have been applied in refinery plant to remove S-containing compounds from fuels. However, removal of aromatic S-containing compounds such as dibenzothiophene (DBT), which are the most refractory S-containing compounds present in fuel, are one of challenges in the HDS.[1,2] The formal HDS requires high temperature, high pressure, high active catalysts. So, the operating cost of the hydrogenation plants goes up. Under such circumstances, one of the promising alternative to HDS is oxidative desulfurization (ODS) process. The operating conditions are mild (atmospheric pressure and temperature lower than 100 °C) and no hydrogen is necessary for ODS whereby S-containing compounds are oxidized to sulfones and sulfoxides which are eliminated in the subsequent extraction process.[3-5] It is improved when oxidation and extraction are in one-pot system (extractive-catalytic oxidative desulfurization (ECODS)). It has been proven by previous works that ECODS shows a higher efficiency compared to the ODS.[4-7] The choice

[1] **Corresponding author:** Tel.: +98 2177240496; Fax: +98 2173227772. E-mail address: m_dehghani@iust.ac.ir (M.R. Dehghani).

of high efficient solvents is important in ECODS. In comparison to conventional toxic solvents, ionic liquids (ILs) exhibit many desirable properties and show great potential in various research areas as well as the high S-extractive ability.[8-11] Among outlooks for ILs desulfurization, there are some disadvantage, including the production of corrosive component, non-stability to moisture and air, etc. Deep eutectic solvents (DESs) as a new type of solvent have attracted great attention as well as ILs. They not only possess the advantages of ILs, but compared with ionic liquids, they also share many excellent properties and simple synthetic process without using other organic solvents.[6-11] The literature review on ECODS demonstrated that some polyoxometalates (POMs) are efficient desulfurization catalysts.[12-25] The Dawson-type POMs have attracted few attentions in comparison with other kinds of POMs in ECODS.[15-35] Previous works showed that the introduction of vanadium in POMs framework improved the desulfurization process.[7] In formal biphasic ECODS system, desulfurization occurs in the bulk of the polar phase, and the disadvantage of these systems is phase-transfer limitation.[20-25] Therefore, the use of surfactants to form emulsions has attracted wide attention in recent years.[20-35] A new approach to phase transfer agent is surfactant based catalysts, which combine together a POM anion and a quaternary ammonium cation.[20-25]

In this work, to search for efficient ECODS systems, we investigated the ECODS composed of various types of solvents (IL ([BMIM]-methyl SO_4), ethylene glycol (EG), DES (ChCl/EG with mole ratio ChCl:EG = 1:3)) and vanadium-substituted Dawson-type emulsion catalyst $[CTA]_{6+x}P_2W_{18-x}V_xO_{62}$ (x = 0, 1, 3, 5) for the first time by means of Taguchi experimental design.

2. EXPERIMENTAL

2.1. Chemical and Reagent

Dibenzothiophene and H_2O_2 (aqueous solution, 50 wt. %) were purchased from Sigma-Aldrich. All of the other reagents and solvents used in this work are available commercially and were used as received.

2.2. Preparation of Emulsion Catalyst

$H_{6+x}P_2W_{18-x}V_xO_{62}$ (x= 1, 3, and 5) heteropoly acids were prepared according to the similar method reported in the literature.[11] Emulsion catalysts $[cetrimonium]_{6+x}P_2W_{18-x}V_xO_{62}$ (x= 1, 3, and 5) were synthesized according to the literature.[12]

2.3. Preparation of Deep Eutectic Solvent

The DES used in this work, ChCl/ ethylene glycol (EG) with mole ratio of ChCl to EG, 1:3 was prepared according to the similar method reported in the literature.[13]

2.4. Characterization of Catalyst

Transmission FT-IR spectra of emulsion catalysts were measure as 3% KBr pellets on Shimadzu-8400S FT-IR spectrometer. The chemical analysis of each sample was determined by Inductively Coupled Plasma Spectroscopy ICP (ICPS S7000, Shimadzu) and ECS 4010 CHNS-O elemental Analyzer, Costech Analytical Technologies, Inc., Italy. The Shimadzu UV-2101 PC spectrometer equipped with a diffuse-reflectance was used to record UV-vis DRS patterns at 190-800 nm in a step-scan mode with a step of 2 nm. A Philips PW-1800 diffractometer using Cu Kα radiation (λ= 0.15406 nm) at 40 kV and 30 mA was used to determine the XRD patterns over an angular range of $5^o<2\theta<60^o$ with a step size (2θ) of 0.04^o and a count time of 2 seconds per step. XPS spectrum was obtained by using VG Escalab 200 R with Al/Mg X-ray source.

2.4. Desulfurization

CODS and ECODS were performed on a model fuel (approximately 500 ppmw S in iso-Octane). In a CODS run, the solution was heated to 60 °C. The emulsion catalysts and the oxidant were added to the above model fuel simultaneously at a magnetic stirring speed of 1000 r/min. In a typical run of ECODS (introduced by Taguchi experimental design table), a water bath was heated to a desired temperature. The mixture, containing model fuel and solvent was stirred vigorously using a stir bar at reaction temperature for 3 min to reach the desired temperature. Then the emulsion catalyst was added to the mixture follow by 50 wt. % H_2O_2 (in desired O/S mole ratio). In each process, the sample was taken out and cooled down to stop the reaction. The emulsion catalyst in the sample was separated by centrifugation.

The performance of reaction in each part was analyzed by a gas chromatography (Agilent, 7890A) coupled with a FID detector using a capillary column (HP-5, 30m×0.32mm×0.25μm).

3. RESULTS AND DISCUSSION

3.1. Characterization of Emulsion Catalysis

The elemental analysis confirmed that the element composition of emulsion catalyst was approximately the same as the calculated values. (Found: C, 24.37; H, 4.51; N, 1.48; P, 0.96; W, 49.26; V, 0.79. Calculated for $[CTA]_7P_2W_{17}VO_{62}$: C, 25.66; H, 4.73; N, 1.58; P, 1.00 ; W, 50.28; V, 0.82. Found: C, 29.89; H, 5.53; N, 1.82; P, 0.92; W, 41.77; V, 2.27. Calculated for $[CTA]_9P_2W_{15}V_3O_{62}$: C, 31.46; H, 5.79; N, 1.92; P, 0.95; W, 42.31; V, 2.35. Found: C, 34.91; H, 6,46; N, 2.12; P, 0.88; W, 33.94; V, 3.62. Calculated for $[CTA]_{11}P_2W_{13}V_5O_{62}$: C, 36.75; H, 6.77; N, 2.26; P, 0.91; W, 35.05; V, 3.74.).

The FT-IR spectrum of the prepared catalyst has the characteristic peaks in a range of 700-1100 cm^{-1} (1090 cm^{-1} related to P=O, 912 and 960 cm^{-1} M=O, 912 and 780 cm^{-1} related to M-O-M (M=W, V), 1468 and 2850 and 2918 cm^{-1} related to vibration of CTA cation) which indicate that the compounds possess Dawson-type POM structure. [32-33]

The UV spectrum of all catalyst presents bands at 198 nm, 214 nm (charge transfer from O to P), 278 and 378 nm (the incorporation of vanadium into the Dawson-type anion) attributed to the Dawson-type structure and in agreement with the results reported previously.[32-33]

The XRD patterns of the Dawson-type catalyst shows that the main positions (2θ) of the XRD peaks which are typical features of the Dawson-type structure, are appeared in the region of 16-22°, 25-30° and 35-42° in XRD spectra of synthesized emulsion catalysts.[17]

XPS analysis shows that the binding energy of $V2p_{3/2}$ (~517eV) and $W4f_{7/2}$ (~35eV) are attributed to V^{5+} and W^{6+}. The position of peaks of $V2p_{3/2}$ and $W4f_{7/2}$ of vanadium substituted Dawson-type heteropoly acid and potassium salts are the same. So, It can be said that tungstate and vanadium are in the higher level of oxidation level in $[P_2W_{18-x}V_xO_{64}]^{(6+x)-}$ and introduction of vanadium in the frame of heteropoly anion has no effect on oxidation level of tungstate.[15,32-33]

3.2. CODS Operation

The catalytic activity of $[CTA]_{6+x}P_2W_{18-x}V_xO_{62}$ (x=1, 3, 5) was in the order of V5 (80.22% removal of DBT)> V3(70.16% removal of DBT)> V1(60.42% removal of DBT). In fact, the substitution of W^{6+} with V^{5+} results in creation of more strongly reactive lattice oxygen associated to the W-O-V species. Meanwhile, V^{n+} can be readily reduced to $V^{(n-1)+}$ and improves the catalytic activity of catalyst.[19,20]

3.3. ECODS of Model Fuel by the Best Performance Catalyst

The emulsion catalyst along with H_2O_2 (50 wt-%) was used for extractive catalytic oxidative desulfurization (ECODS) of two kinds of model fuel oil (*iso*-octane and *iso*-octane containing 20 v% toluene with 500 ppmw total S, DBT was selected as a S-containing component). Three kinds of solvent including deep eutectic solvent (DES, ChCl: EG with mole ratio 3 (EG) to 1(ChCl)), ionic liquid (IL, 1-butyl-3-methyl imidazolium methyl sulfate (BMIM-sulfate)) and ethylene glycol (EG) were utilized under mild conditions for extraction process. Using Taguchi experimental design, effective parameters such as solvent type, volume ratio of solvent to oil, temperature and contact time on the ECODS were studied. All chosen parameters were studied at three levels by means of L_9 orthogonal array (**Table 1**).[29] The concentration of catalyst in all entries of **Table 1** was 7.5 g/L. The efficiency of system containing IL was much better than other systems. The results showed that the EG ECODS system removed DBT better than DES systems.

Table 1. L_9 orthogonal array experiment designed used for optimization of ECODS

Entry	Type of Solvent	Volume Ratio Solvent/Oil	T (°C)	Time (min)	DBT Removal %	
					N	Y
1	DES	1:4	40	20	85.16	78.09
2	DES	1:6	50	40	85.51	80.56
3	DES	1:8	60	60	84.10	79.15
4	EG	1:4	50	60	91.17	87.63
5	EG	1:6	60	20	88.69	86.57
6	EG	1:8	40	40	84.81	83.39
7	IL	1:4	60	40	94.70	94.70
8	IL	1:6	50	60	92.58	91.87
9	IL	1:8	40	20	90.81	90.45

N: Model oil: *iso*-Octane with 500 ppmw S
Y: Model oil: *iso*-Octane with 500 ppmw S and 20 V-% Toluene
Catalyst concentration: 7.5 g/L

ANOVA analysis (**Table 2**) was used to find out the optimized conditions. The result showed that after type of solvent, the volume ratio of solvent to oil was more important than other parameter. Our results showed that for the case of paraffinic fuel model (*iso*-octane) using IL at solvent/oil volume ratio of 1:4 gives the best performance at 50 °C after 60 minutes contact time, while in the case of paraffinic-aromatic model fuel (80 % *iso*-octane, 20 % toluene, volume basis) higher temperature (60 °C) was required to meet the same condition. The predicted optimized conditions was not similar to any entry of **Table 1** for both model diesel oils. So, the validation test was performed. The results of validation tests approved that the predicted optimized removal of DBT were approximately reputational. The nearly 95 % of DBT was removed under optimized conditions from model diesel oil containing only *iso*-octane. The existence of toluene reduced the efficiency of system and removal of DBT to ~94 % under determined optimized conditions.

Table 2. The results of ANOVA for optimization the ECODS of model oil

Model Oil (*iso*-Octane, 500 ppmw S)					
Factor	DOF (f)	Sum of Square (S)	Variance (V)	Pure Sum (S)	Percent P (%)
Type of Solvent	2	91.323	45.661	91.323	75.761
Volume ratio solvent to oil	2	21.76	10.88	21.76	18.052
Temperature	2	5.417	2.708	5.417	4.494
Time	2	2.033	1.016	2.033	1.687
Other/Error	0	-	-	-	-
Total	8	120.541	-	-	100.00%
Model Oil (80 v% *iso*-Octane, 20 v% Toluene, 500 ppmw S)					
Type of Solvent	2	256.378	128.189	256.378	92.036
Volume ratio solvent to oil	2	10.373	5.186	10.373	3.724
Temperature	2	9.021	4.51	9.021	3.238
Time	2	2.787	1.393	2.787	1
Other/Error	0	-	-	-	-
Total	8	278.561	-	-	100 %

4. CONCLUSION

For the first time, the CTA-type vanadium substituted Dawson-type emulsion catalysts, were successfully prepared to use in extractive-oxidative desulfurization (ECODS) of model diesel oil. The results showed that increase the number of vanadium in POM frame improves the efficiency of CODS. So, $[CTA]_{11}P_2W_{13}V_5O_{62}$, chosen as a best performance emulsion catalyst. In second step, the best Dawson-type catalyst was used to ECODS of dibenzothiophene (DBT) from two kinds of model diesel oil. The Taguchi robust was used to study the effect of parameters on the ECODS. The ECODS optimized conditions was IL with the volume ratio of solvent to oil 1:4, 50 °C and contact time 60 minutes for model oil containing only *iso*-octane as a solvent. The existence of 1-ring aromatic compound, toluene, with volume ratio 20 v% was changed ECODS optimized conditions to IL with volume ratio solvent to oil 1:4, 60 °C and contact time 60 minutes. The nearly 95 % of DBT was removed under optimized conditions from model diesel oil containing only *iso*-octane. The existence of toluene reduced the efficiency of system and removal of DBT to ~94 % under determined optimized conditions.

REFERENCES

(1) Zongxuan, J.; Hongying, L.; Yogna, L.; Can, L. *Chinese J. Catal.* **2011**, *32*, 707-715.

(2) Baeza, P.; Aguila, G.; Vargas, G.; Ojeda, J.; Araya, P. *Appl. Catal. B Environ.* **2012**, *111-112*, 133-140.

(3) Wang, L.; Cai, H.; Li, S.; Mominou, N. *Fuel* **2013**, *105*, 752-756.

(4) Campos-Martin, J.M.; Capel-Sanchez, M.C.; Perez-Presas, P.; Fierro, J.L.G. *J. Chem. Technol. Biotechnol.* **2010**, *85*, 879-890.

(5) Ma, C.; Chen, D.; Liu, F.; Sun, X.; Xiao, F.; Dai, B. *RSC Adv.* **2015**, *5*, 96945-96952.

(6) Campos-Martin, J.M.; Capel-Sanchez, M.C.; Fierro, J.L.G. *Green Chem.* **2004**, *6*, 557-562.

(7) Capel-Sanchez, M.C.; Campos-Martin, J.M.; Fierro, J.L.G. *Energy Environ. Sci.* **2010**, *3*, 328-333.

(8) Stanger, K.J.; Angelici, R.J.; *Energy Fuel.* **2006**, *20*, 1757-1760.

(9) Capel-Sanchez, M.C.; Perez-Presas, P.; Campos-Martin, J.M.; Fierro, J.L.G. *Catal. Today* **2010**, *157*, 390-396.

(10) Shojaei, A.F.; Rezvani, M.A.; Loghmani, M.H. *Fuel Process. Technol.* **2014**, *118*, 1-6.

(11) Rezvani, M.A.; Oveisi, M.; Nia Asli, M.A. *J. Mol. Catal. A Chem.* **2015**, *410*, 121-132.

(12) Trakarnpruk, W.; Rujiraworawut, K. *Fuel Process. Technol.* **2009**, *90*, 411-414.

(13) Rezvani, M.A.; Zonoz, F.M. *J. Ind. Eng. Chem.* **2015**, *22*, 83-91.

(14) Mirhoseini, H.; Taghdiri, M. *Fuel* **2016**, *167*, 60-67.

(15) Ammam, M.; Fransaer, J. *J. Solid State Chem.* **2011**, *184*, 818-824.

(16) Misono, M. *Surf. Sci. Catal.* **2013**, *176*, 97-154.

(17) Wang, C.; Bu, X.; Ma, J.; Liu, C.; Chou, K.; Wang, X.; Li, Q. *Catal. Today* **2016**, *274*, 82-87.

(18) Park, D.R.; Kim, H.; Jung, J.C.; Lee, S.H.; Song, I.K. *Catal. Commun.* **2008**, *9*, 293-298.

(19) Zou, C.; Zhao, P.; Ge, J.; Qin, Y.; Luo, P. *Fuel* **2013**, *104*, 635-640.

(20) Omwoma, S.; Gore, C.T.; Ji, Y.; Hu, C.; Song, Y-F. *Coord. Chem. Rev.* **2015**, *286*, 17-29.

(21) Ueda, T.; Nishimoto, Y.; Saito, R.; Ohnishi, M.; Nambu, J.-I. *Inorganics* **2015**, *3*, 355-369.

(22) Yuhao, S.; Jingfu, L.; Enbo, W. *Inorg. Chim. Acta* **1986**, *117*, 23-26.

(23) Yu, F.; Wang, R. *Chem. Lett.* **2014**, *43*, 834-836.

(24) Lü, H.; Zhang, Y.; Jiang, Z.; Li, C. *Green Chem.* **2010**, *12*, 1954-1958.

(25) Lü, H.; Ren, W.; Liao, W.; Chen, W.; Li, Y.; Suo, Z. *Appl. Catal. B Environ.* **2013**, *138-139*, 79-83.

(26) Lü, H.; Gao, J.; Jiang, Z.; Jing, F.; Yang, Y.; Wang, G.; Li, C. *J. Catal.* **2006**, *239*, 369-375.

(27) Huang, W.; Zhu, W.; Li, H.; Shi, H.; Zhu, G.; Liu, H.; Chen, G. *Ind. Eng. Chem. Res.* **2010**, *49*, 8998-9003.

(28) Zhu, W.; Huang, W.; Li, H.; Zhang, M.; Jiang, W.; Chen, G.; Han, C. *Fuel Process. Technol.* **2011**, *92*, 1842-1848.

(29) Taguchi, G. *Systems of Experimental Design*; UNIPUB/Kraus International Publications: New York, 1987.

(30) Wang, E.B.; Gao, L.H.; Liu, J.F.; Liu, Z.X.; Yan, D.H. *Acta. Chem. Sinica.* **1986**, *46*, 757-762.

(31) Ueda, T.; Komatsu, M.; Hojo, M. *Inorg. Chem. Act.* **2003**, *344*, 77-84.

(32) Pope, M.T. *Heteropoly and Isopoly Oxometalates*; Springer: Berlin, 1983; pp 128-141.

(33) Dablemont, C.; Hamaker, C.G.; Thouvenot, R.; Sojka, Z.; Che, M.; Maatta, E.A.; Proust, A. *Chem. Eur. J.* **2006**, *12*, 9150-9160.

(34) Hill, C.L.; Gall, R.D. *J. Mol. Catal. A Chem.* **1996**, *114*, 103-111.

(35) Ribeiro, S.; Barbosa, A.D.S.; Gomes, A.C.; Pillinger, M.; Gonçalves, I.S.; Cunha-Silva, L.; Balula, S.S. *Fuel Process. Technol.* **2013**, *116*, 350-357.

473466 Conversion of Carbon Dioxide to Syngas Using Transition Metal Dichalcogenide Catalysts

Abstract:

Conversion of Carbon Dioxide to Syngas Using Transition Metal Dichalcogenide Catalysts

Mohammad Asadi, Pedram Abbasi, Amin Salehi-Khojin

Department of Mechanical and Industrial Engineering, University of Illinois at Chicago, Chicago, IL, 60607, USA.

Abstract:

Carbon dioxide (CO_2) concentration in the atmosphere has drastically increased in the past 50 years and is expected to reach the maximum level in the next few decades (2060) due to fossil fuels consumption using existing infrastructure[1]. It is recognized that the CO_2 level can be controlled by sequestrating[2,3] or converting it into useful energy-rich products[4,5]. The conversion approach is more desirable since it not only offers an efficient path for CO_2 remediation but also provides a sustainable energy source. Among various remediation techniques, electrochemical reduction of carbon dioxide (CO_2) offer convenient ways to recycle CO_2 into fuels[4,5]. However, it has been advanced far more slowly in the last two decades, mainly due to insufficient performance of existing catalysts[6]. Fundamentally, the electronic structure of commonly

used catalysts is not well-suited for the electrochemical CO_2 reduction reaction[7]. The high density of active d-orbital electrons near the Fermi level together with a low work function are essential to catalyze this process[8]. However, the majority of existing catalysts are far below the ideal limits of these properties.

In a search of cost and energy efficient catalysts, here, we report results for transition metal dichalcogenide (TMDC) nanoflakes (NFs) that exhibit extraordinary catalytic activity for CO_2 reduction in the ionic liquid (IL) 1-ethyl-3-methylimidazolium tetrafluoroborate (EMIM-BF_4). Our results indicate that the tungsten diselenide (WSe_2) is the most active catalyst for CO_2 reduction and exhibits more than two orders of magnitude better conversion performance compared to commonly used catalysts in the literature. Atomic level study of WSe_2 NFs before and after electrochemical experiments using scanning transmission electron microscopy (STEM) indicate that; (i) mainly tungsten (W) atoms on the edge sites are responsible for this remarkable activity and (ii) the edge sites remain stable during the reaction. Our electrochemical impedance spectroscopy (EIS) and work function measurements also reveal that the high electrocatalytic activities of TMDC NFs mainly rely upon the fast electron transfer to the intermediate complex resulting in the low overpotential and high current density for conversion of CO_2 to the syngas.

References:

(1) Davis, S. J.; Caldeira, K.; Matthews, H. D. Future CO2 Emissions and Climate Change from Existing Energy Infrastructure. *Science.* **2010**, *329*, 1330–1333.

(2) Haszeldine, R. S. Carbon Capture and Storage: How Green Can Black Be *Science.* **2009**, *325*, 1647–1652.

(3) Lastoskie, C. Caging Carbon Dioxide. *Science.* **2010**, *330*, 595–596.

(4) Medina-Ramos, J.; Dimeglio, J. L.; Rosenthal, J. Efficient Reduction of CO_2 to CO with High Current Density Using in Situ or Ex Situ Prepared Bi-Based Materials. *J. Am. Chem. Soc.* **2014**, *136*, 8361–8367.

(5) Rosen, B. a; Salehi-Khojin, A.; Thorson, M. R.; Zhu, W.; Whipple, D. T.; Kenis, P. J. a; Masel, R. I. Ionic Liquid-Mediated Selective Conversion of CO to CO at Low Overpotentials. *Science.* **2011**, *334*, 643–644.

(6) Gao, D.; Zhou, H.; Wang, J.; Miao, S.; Yang, F.; Wang, G.; Wang, J.; Bao, X. Size-Dependent Electrocatalytic Reduction of CO_2 over Pd Nanoparticles. *J. Am. Chem. Soc.* **2015**, 150306114146006.

(7) Li, Y.; Chan, S. H.; Sun, Q. Heterogeneous Catalytic Conversion of CO_2 : A Comprehensive Theoretical Review. *Nanoscale.* **2015**, 8663–8683.

(8) Asadi, M.; Kumar, B.; Behranginia, A.; Rosen, B. a; Baskin, A.; Repnin, N.; Pisasale, D.; Phillips, P.; Zhu, W.; Haasch, R.; *et al.* Robust Carbon Dioxide Reduction on Molybdenum Disulphide Edges. *Nat. Com.* **2014**, *5*, 4470.

Copyright © American Institute of Chemical Engineers. All rights reserved.

474007 On the Stability of Cu-Chabazite SCR Catalysts

Abstract:

Zeolites exchanged with different transition metal cations, primarily Cu or Fe, are being used as catalyst for removing NOx from the effluent of internal combustion engines operating in an excess of oxygen, i.e. Diesel or lean-burn gasoline engines. This is accomplished via Selective Catalytic Reduction (SCR) with NH_3. The last generation of catalysts involves the use of zeolites with the Chabazite structure, such as SSZ-13 or SAPO-34. These materials have high NOx reduction activity, good selectivity towards N_2 and high hydrothermal stability. The last parameter is usually determined via accelerated testing protocols that show that the chabazite structure is robust at high temperatures.

Although much work has centered upon the stability of the zeolite framework, less work has been reported upon the stability and structural changes undergone by the metal centers responsible for the activity of zeolite catalysts. We are interested in understanding what are these structures and how they evolve as the catalysts age. To do so, we perform isothermal reaction experiments under realistic SCR conditions and analyze the transients of operation as a function of water presence. We follow the metal species in-situ using UV-Vis spectroscopy, and ex-situ using X-ray diffraction, MAS-NMR, H_2-TPR, and EPR in the case of Cu.

In this work we report studies done under isothermal SCR aimed at understanding how the structure and siting of Cu cations in different zeolites influences NH_3-SCR operation, and to determine their stability in the absence of framework modifications. We present results using chabazites, both natural and synthetic (SSZ-13), and compare them with ZSM-5 and Mordenite. We also present deactivation models for the different zeolites that correlate changes in activity with those in cation-siting experienced as a function of time-on-stream.

Copyright © American Institute of Chemical Engineers. All rights reserved.

474055 Visible Light Photoactivity of Ion Implanted Titania Nanotubes

Abstract:

Titania nanotubes with spiral wall structure were prepared by hydrothermal synthesis. The prepared nanotubes are active for photocatalytic oxidation of organic molecules in water. Ion implantation of nonmetallic elements was used to modify the electronic bandgap structure of the nanotubes to obtain visible light active titania nanotube photocatalysts. The nature and chemistry of the implanted ions, the ion fluence and post-implantation treatment were shown to affect the bandgap structure and visible light photocatalytic activity of the ion-implanted titania nanotubes. UV-Vis, Raman and FTIR spectroscopy studies were done to examine the chemistry and optical properties of the prepared photocatalysts. Activity studies were done using 2,4-dichlorophenol (DCP) photocatalytic oxidation under visible light irradiation.

Copyright © American Institute of Chemical Engineers. All rights reserved.

474064 Enhancement of Photochemical Activity of One Dimensional TiO_2 By Targeted Decoration of Transition Metals

Abstract:

The application of one-dimensional TiO_2 nanotubes is largely dependent on the photo-generation of electron-hole pairs (excitons) and the longevity of the excitons to conduct through the material. In this current research, the TiO_2 nanotubes are functionalized with different transition metals to extend the activity in visible range of the solar spectra. Two different approaches are followed to successfully decorate the nanotubes with the metal clusters: one pot alkaline hydrothermal approach is used to locate the metal clusters on the surface of the nanotube and metal complex solution is used to deposit within the lumen of the nanotubes. The specific location of the metals triggered specific optical and chemical properties as seen in UV-Vis, Raman and FTIR spectroscopy. The new material shows excellent performance for visible light photocatalytic oxidation of refractory 2, 4 dichlorophenol (DCP) pollutant.

Copyright © American Institute of Chemical Engineers. All rights reserved.

474073 New Materials Technology for Sustainable Removal of Moisture from Air

Abstract:

A new material is developed which is able to efficiently remove and harvest water from air with little or no energy requirement has important local and global implications. Such technology is expected to significantly lower the energy cost of air cooling, reduce greenhouse gas emission and save money. The new material was made by confinement of low water activity materials (LWA) within superhydrophilic matrix that can allow sustained, removal of moisture and harvesting of clean water from air. The low water activity creates a chemical potential gradient for water vapor near the surface of the materials, while the superhydrophilic matrix promotes micro-condensation and wicks excess water before penetrating into the LWA material. Thus, sustained water harvesting can be attained at roughly 0.5kg/kg-1day at 60% relative humidity at 25°C, without the need for refrigeration cooling or heat regeneration. This provides a potential solution to energy demand in space cooling and conditioning.

Copyright © American Institute of Chemical Engineers. All rights reserved.

472749 Towards Industry-Scale Measurements of Reactor Gas-Liquid Mass-Transfer Phenomena

Abstract:

In many gas–liquid contactors or reactors, mass transfer from the gas phase to the liquid phase is a major player in the reactor's overall production and/or economics. This talk presents the case for research to intensify mass transfer in bioreactors that upgrade natural gas. The mass transfer rate can be enhanced by increasing interfacial area *a*, by increasing reactor pressure, and/ or by intensifying the liquid-side mass transfer coefficient k_L. In the first part of this work the goal is to achieve a volumetric mass transfer coefficient ($k_L a$) above 2,000 1/hr with better energy efficiency than current industrial options. To increase mass transfer rate to $k_L a$ > 2000 1/hr, microbubbles (< 200 micron) were generated in a 100L bubble column reactor while maintaining low power consumption. The volumetric surface area is inversely proportional to the bubble diameter. Therefore, microbubbles offer larger gas–liquid interfacial areas and longer residence times compared to conventional larger bubbles. Gas-liquid-transfer coefficient ($k_L a$) and bubble dispersion hydrodynamics were measured under a spectrum superficial liquid and gas velocities. Non-chemical experimental methods for measuring reactor volume-averaged mass transfer coefficient are limited to small values of $k_L a$. In order to measure $k_L a$ accurately in the range >2000 1/hr, a chemical reactant is necessary to buffer the rise of dissolved gas. Further, control of the reaction rate is necessary in order to ensure the Hatta number and other governing parameters of the process are appropriate (Gaddis 1999, Sharma 1969), which will be explained in this talk. Here we present new data and methods of the [Na_2SO_3, $CoSO_4$, O_2] system for $k_L a$ measurement. Sulfite measurements of $k_L a$ are not uncommon in industry and literature, but are hindered by sensitivity to impurity of the catalyst, which necessitates costly and prohibits experiments in metal reactors. In the second part of this work the goal is to improve the sulfite measurement of $k_L a$ via a novel electrowinning method that allows in-situ cleaning (CIB) of the test solution in the reactor. This new method allows for use of less expensive reagents and possibly for use of

stainless steel reactor materials, both of which greatly facilitate the [Na_2SO_3, $CoSO_4$, O_2] system in larger industrial reactors. We present data from successful tests of this electrowinning system, and new kinetics data to constrain the reaction order with respect to the gas *m*, order with respect to the liquid *n*, and order with respect to cobalt *l*. It was observed that electrowinning reduces background impurities of the chemical system, causing a decrease in absorption rate by a factor of O(10) and allowing use of less expensive sulfite. The electrowining method was scaled up from bench-top (0.5L) to the larger reactor (100L). The electrowinner design and results are presented.

References:

Gaddis, E.S. 1999 Mass transfer in gas–liquid contactors. Chemical Engineering and Processing. 38, 503–510.

Sharma, M.M, Dankwerts, P.V. 1969 Chemical methods of measuring interfacial area and mass transfer coefficients in two-fluid systems. British Chemical Engineering. 15, 4, 522-528.

Copyright © American Institute of Chemical Engineers. All rights reserved.

Paper Number: 476691

Kinetic Study for Hydrogenation of Toluene to Methylcyclohexane over Titanuim Substituted SBA-15 Supported Pt-Pd Catalysts

Shyamal Roy, Dilip Kumar Mondal, Chandona Mondal*

Department of Chemical Engineering, Jadavpur University, Kolkata-700032, West Bengal, India

Abstract: SBA-15, Ti-SBA-15 materials were synthesized and employed as supports for Pt-Pd catalyst. Supports and catalysts were thoroughly characterized by using inductively coupled plasma atomic emission spectroscopy (ICP-AES), X-ray diffraction (XRD), energy dispersive X-ray analysis (EDX), temperature-programmed desorption (TPD), Brunauer-Emmett-Teller (BET) surface area, and CO chemisorptions techniques. The detail kinetic study of hydrogenation of toluene over a (0.15 wt %) Pt-(0.15 wt %) Pd / SBA-15 and (0.15 wt %) Pt- (0.15 wt %) Pd / Ti-SBA-15 was performed in a continuous upflow stainless steel catalytic fixed bed reactor at varied hydrogen to toluene mole ratio, weight hourly space velocity (WHSV), hydrogen partial pressure and temperature. It was observed that toluene conversion was increased on increasing hydrogen to toluene mole ratio and H_2 partial pressure, and decreased with increasing WHSV. The conversion is dependent on temperature and shows a well-defined maximum. It was found that incorporation of Ti into the SBA-15 structure increased the toluene conversion and reducibility of the catalysts. It was found that (0.15 wt %) Pt- (0.15 wt %) Pd / Ti-SBA-15 catalysts showed better performance in hydrogenation of toluene than the (0.15 wt %) Pt- (0.15 wt %) Pd / SBA-15 catalyst.

Keywords: SBA-15, Ti-SBA-15, Pt-Pd/ Ti-SBA-15, activation energy.

1. Introduction

The increased attention paid to catalytic hydrogenation in the oil refining industry is due in part to legislation regarding the maximum contents of sulfur, aromatic compounds, and alkenes in traffic fuels.[1,2] Aromatics in diesel increase the particle emissions in gases[3] and they have the further effect of lowering the fuel quality.[2] Alkenes in gasoline are undesired in larger amounts because of higher NOx and CO formation and the higher reactivity of evaporative emissions in comparison to their saturated counterparts.[4] Tightening legislation in regards to vehicle exhaust emissions has forced the industry to improve motors and catalytic converters and also the quality of fuels. Lead free and low-sulfur fuels have been introduced and CO, SO_x, and NO_x emissions have been significantly reduced. The major concern remaining is particle emissions.[5] Particles are reported[6] to correspond to aromatic compounds, and the reduction of these is the next target. The faster growth in the demand for diesel fuel relative to gasoline, together with today's higher quality requirements and lower grade raw materials, creates a need for improved dearomatization processes.[7] Better knowledge of reactions and rigorous reactor and kinetic models is essential to new process development and optimization. Gas-phase dearomatization has been widely studied, both on metal sulfides (Co/Ni and Mo/W) and on group VIII metals (Pt, Pd, Ni, and Ru).[8] Publications on liquid phase reaction kinetics, in turn, are few[7,9], even though industrial dearomatization takes place in the liquid phase in three-phase reactors.

The drawbacks of the conventional hydrotreating catalysts include severe operating conditions such as high temperature, high pressure, low space velocity, and hydrogen to feed mole ratios to achieve acceptable aromatic reduction. The hydrodearomatization reaction is reversible and at normal hydrotreating conditions, complete conversion is not possible because of equilibrium limitations. Furthermore, conventional hydrotreating catalysts comprising sulfided mixed oxides (Ni-Mo, Ni-W, Co-Mo) can only accomplish moderate levels of aromatic saturation under typical hydrotreating conditions in a single-stage operation.[10,11] Using these systems, increasing operation severity (temperature and pressure) do not result in deep levels of aromatics saturation because of thermodynamic limitations. Efficient approaches towards performance improvement of the catalysts used in such processes include:

Email: shyamal.roy@rutgers.edu

changing the active phase (carbide, nitride, phosphide, etc.), using a different catalyst preparation method (co-precipitation, chemical vapor deposition, etc.) or applying new types of catalyst supports (CNTs, mesoporous silicas such as Santa Barbara Amorphous, SBA-15). These ordered mesoporous materials are considered ideal hosts for nanoparticles due to their large surface area and well defined pore structure. Most importantly, their pore sizes can be tuned within the nanometer range and the pore surface can be functionalised with various metal oxides and organic groups. Among these materials, SBA-15 silica has been extensively studied due to their 2D structure with interconnected pores.[12] SBA-15 is a mesoporous silica molecular sieve with a high surface area, tunable uniform hexagonal channels ranging from 5 to 30 nm and thick framework walls (3–6 nm), which is characteristic for it being thermally and hydro-thermally robust.[13,14] Such attractive textural properties of SBA-15 makes it a potential alternative to the commonly used γ-Al_2O_3 support due to the fact that its pore size can be manipulated via controlling the synthesis parameters, while maintaining a relatively high surface area. Noble-metal based catalysts are preferred for aromatic saturation since they can work at lower temperatures, thus avoiding the thermodynamic constraints encountered with the sulfided oxides. However, noble-metal hydrogenation catalysts are sensitive and get poisoned by small amounts of sulfur and nitrogen containing hetero organic compounds normally present in the feed.[15] It is reported that high intrinsic activity and sulfur tolerance may be enhanced by deposition of noble metal on acidic supports. Recently, Song[16] proposed a design concept for unique acidic zeolites as supports for noble metals that utilizes shape-selective exclusion, hydrogen spillover, and two types of sulfur resistance. Such zeolite supports can be used to prepare bimodal distributions of noble metal particles. Pt-Pd bimetallic supported on large pore β-zeolites have been claimed[17] to display dearomatization/cetane improvement. Lee and Rhee[18] recently reported a simultaneous decrease in the metal dispersion and hydrogenation activity of this catalyst, induced by platinum sulfide formation followed by coke deposition. The large pore zeolites preferably allow fast diffusion and reaction of bulky aromatic compounds. It is known that noble metals deposited on acidic supports show higher turnover frequencies (TOF) compared to those supported on nonacidic supports. This has been related either to the polarization of the metal particles by nearby cations or to a metal-support interaction, which might induce a partial electron transfer between the metal and the oxide ions[19] or between the metal and the zeolite protons.[20,21] In the latter case, the formation of the more reactive metal-proton adducts [Mn-Hz]z+ has been proposed.[20,21] The positive effect of the acidity on the noble metal sulfur resistance was explained by the changes in the metal electron deficiency induced by metal/support interactions.[22-24] This effect leads to the formation of electron-deficient metal sites, which, in turn, lowers the strength of the sulfur-metal bond.[21,22] However, high acidity of zeolite is also known to favor excessive cracking and coke formation. In order to avoid it without suppressing the sulfur tolerance of catalysts, less acidic supports such as silica-alumina or non acidic supports has been recently investigated. For example, Reinhoudt et al.[25] suggested the superior performance of silica-alumina supported Pt-Pd catalyst for second-stage deep hydrodesulfurization process. In contrast, Fujikawa et al.[26] reported that bimetallic Pt-Pd/silica-alumina does not improve sulfur tolerance as compared to the monometallic Pt or Pd catalysts although it significantly enhances the intrinsic activity for hydrodearomatization. Miller et al.[27] investigated the potential for developing nonacidic sulfur tolerant catalyst. Sidhpuria et al.[28] reported rhodium supported on Hβ-zeolite catalyst for toluene hydrogenation in the literature. Recently, Boahene et al.[29] investigated hydrotreating of coker light gas oil using Ni/HPMo catalysts.

In the present study noble metals Pt, Pd supported on SBA-15 and Ti-SBA-15 have been explored as a hydrodearomatization catalyst. Toluene was used as a model compound to simulate the aromatics in diesel fuels, because the hydrogenation of toluene is known to be more difficult than that of benzene and naphthalene. This could be related to a decrease in the resonance energy per aromatic ring as well as to differences in the π-electron cloud density in the aromatic ring as a result of the inductive effect of the methyl group.[29] Hydrogenation activity of toluene over Pt-Pd / SBA-15 and Pt-Pd / Ti-SBA-15 catalysts was studied in a fixed bed reactor at varied hydrogen to toluene mole ratio, temperature, pressure and toluene feed WHSV. A modification of SBA-15 silica was made by frame modification with Ti+4 ions. Ti-containing SBA-15 sample was prepared using Ti-isopropaxide as Ti precursor, tri-block copolymer, poly (ethylene glycol)-poly (propylene glycol)-poly (ethylene glycol) (PEG-PPG-PEG) (Aldrich), as the surfactant and tetraehoxysilane (TEOS) as the silica source. To our knowledge, no published work has been published using Pt-Pd / Ti-SBA-15 hydrotreating catalyst and Ti-SBA-15 was synthesized using Ti- isopropaxide. Thus, the aim of this work was to study the catalyst activity using Ti modified SBA-15 supported Pt, Pd catalysts and compare the catalyst activities using Pt-Pd / SBA-15 catalysts. For comparison, Pt-Pd / SBA-15 catalyst was prepared by excess solution impregnation method.

2. Experimental

2.1. Synthesis of SBA-15 and Ti-SBA-15

Preparation of Ti-SBA-15 supports used the following chemicals: tetraethylorthosilicate (TEOS, 98%, Aldrich), tri-block copolymer, poly (ethylene glycol)-poly (propylene glycol)-poly (ethylene glycol) (PEG-PPG-PEG) (Aldrich), isopropanol (Aldrich, >99.5%), titanium isopropaxide (98+%, Acros), tetraamine platinum nitrate (99.9%, Aldrich) and tetraamine palladium (II) nitrate (10 wt% solution in water, Aldrich). Titanium was incorporated into the mesoporous SBA-15 material by direct hydrothermal synthesis routes.

The preparation of mesoporous silica SBA-15 was carried out according to the method of Zhao et al.[13] In the synthesis procedure, 16 g amphiphilic tri-block copolymer, poly (ethylene glycol)-poly (propylene glycol)-poly (ethylene glycol) was first dissolved in 120 ml of water and 480 ml of 2(N)HCl (37 wt %, Merck) solution at 40° C for 2 h under continuous stirring. After addition of 36.4 ml of TEOS the reaction mixture was stirred first at 40°C for 24 h and then at 80°C for 48 h in a Teflon-lined steel autoclave. The resulting solid was filtered, washed with deionized water, dried at 100°C for overnight and calcined in air at 550° C for 6 h at a heating rate of 2°C/min to decompose the tri-block copolymer.

Ti-SBA-15 material with Si/Ti molar ratio 20 was synthesized hydrothermally.[30] TEOS and titanium isopropaxide were been used as silicon and titanium sources, respectively. The tri-block copolymer (PEG-PPG-PEG) was used as the structure-directing agent, whereas concentrated 2(N)HCl was used to adjust the pH. The synthesis was carried out similarly as for pure SBA-15, but with the Ti source added after the introduction of TEOS. The titanium isopropaxide was pre-dissolved in a solution of 2-propanol (99.7 %, Merck). The resulting samples were washed with deionized water and dried overnight at 100° C. Once dry, the resulted white powder was calcined at 550° C for 6 h.

2.2. Synthesis of Pt-Pd/SBA-15 and Pt-Pd/Ti-SBA-15

Pt (0.15 wt %) and Pd (0.15 wt %) impregnated on SBA-15 and Ti-SBA-15 supports were prepared by excess solution impregnation (ESI) method. In a typical procedure, the calculated quantity of tetra-amine platinum nitrate and tetra-amine palladium (II) nitrate were dissolved in 100 ml of double distilled water, which was stirred for 30 min at pH ca.9.5, followed by the slow addition of support to the solution. The solution was kept for 4 h at room temperature under stirring conditions. The solvent was evaporated on a rotary vacuum evaporator and then the catalysts were calcined at 673 K for 4 h in an oxygen flow of 2 ml/min.

2.3. Catalyst activity measurements

Hydrogenation of toluene was performed in a continuous-up flow stainless steel catalytic fixed bed reactor (Metal: 316 SS, Diameter: 1.5 cm, Length: 50 cm) with the ability to operate up to 150 bar pressure and 923 K temperature, equipped with high pressure piston pump (Knauer, Germany). The schematic diagram is shown in Figure 1. The catalyst extrudates (Diameter: 1.5 mm, Length: 3 mm) were kept at the centre of the reactor throughout the kinetic study.

The bulk density of the catalyst was 0.785 g / cc and exact length of the catalytic bed was 15 cm when the amount of catalyst is 10 g. The remaining space of the catalyst bed at the top and the bottom of reactor were filled with glass beads (2-3 mm diameter) to avoid the entrance and exit effects. Catalyst was dried at 403 K for 48 h and reduced in situ by hydrogen at 553 K for 6 h. The reactor was brought to the desired temperature before the reactant was introduced. Toluene hydrogenation was carried out at different temperature (423-498 K), hydrogen pressure (5-35 bar), hydrogen to toluene mole ratio (2-12) and feed WHSV (1-6 h-1). Liquid sample was withdrawn from the sample-collecting vessel and analyzed by gas chromatograph (Agilent Technologies 6890 N, USA) equipped with a flame ionization detector (FID) and DB-Petro, 122-10A6 column (100 m length, 0.25 diameter and coated with 0.50 □m thick stationary phase of 5 % diphenyl and 95 % dimethyl siloxane) at appropriate times. The initial GC column temperature was increased from 313 to 473 K at the rate of 2 K / min. Helium gas (3.5 mL / min) was used as a carrier gas.

The temperature of injection port and FID were kept constant at 473 K during product analysis. The retention times for different compounds were determined by injecting pure compounds under identical GC conditions. Activity is described in terms hydrogenation of toluene using a specific reaction rate according to Equation (1):

$$r_i = \frac{X_i F}{W_{cat}} \quad (1)$$

where ri is the specific rate (mol / g. h), Xi the conversion of reactant i (i = toluene), F is the molar flow rate of the reactant i (mol / h), and W is the catalyst weight (g).

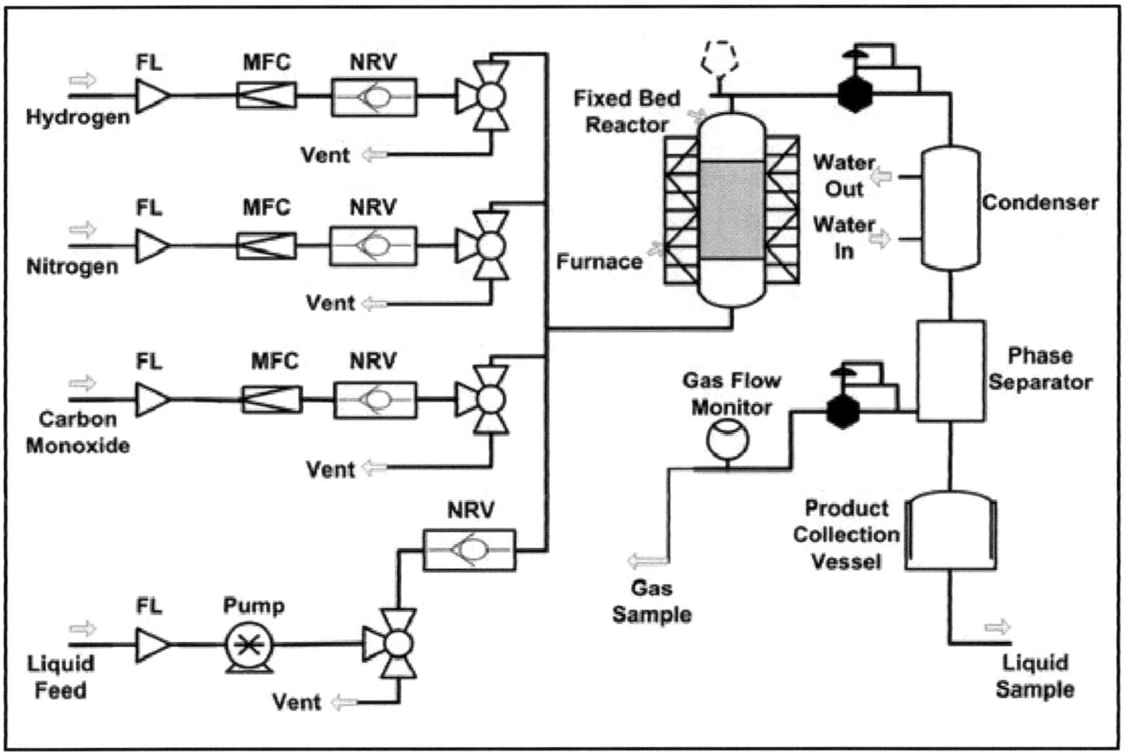

Figure 1. Schematic diagram of the catalytic fixed bed reactor.

Toluene conversion at the end of the 30 min of operation was taken as a steady state conversion. It is also essential for kinetic studies that the catalytic reactor operates in the differential mode. The reactor exhibited no temperature or pressure gradients across the catalyst bed and there are no diffusional limitations for the transport of reactant and product molecules, in this condition.31

Hydrogenation of toluene was observed to occur readily within studied temperature range and the selectivity for methylcyclohexane was found to be between 30 and 100%. No other products like, 1, 2- and 1, 3-dimethyl cyclopentane and ethylcyclopentane were also observed due to isomerization and cracking reactions. Experimental runs were repeated 5 times and the results compared with those of the previous runs in order to ascertain the reproducibility of the data obtained and the error was found to be ± 2%.

2.4. Characterization

2.4.1. Inductively Coupled Plasma Atomic Emission Spectroscopy

The chemical compositions of the catalysts (Pt, Pd) were determined by inductively coupled plasma atomic emission spectroscopy (ICP-AES), using an IRIS Intrepid (Thermo Elemental) instrument.

Table 1. Chemical Compositions,[a] Textural Properties[b] of SBA-15 and Ti-SBA-15 Supported Pt–Pd Catalysts, Average Metal Particle Size,[c] Metal Dispersion,[d] Acidity[e] of Reduced Pt–Pd/SBA-15 and Pt–Pd/Ti-SBA-15 Catalysts, Rate, and Activation Energy of the Hydrogenation of Toluene

Catalyst/ support	BET (m^2/g)	Total pore vol (cm^3/g)	Pore dia (nm)	Metal dispersion (%)	Metal particle size (nm)	Acidity using NH_3 (μmol/g)	Pt (wt %)	Pd (wt %)	TiO_2 (wt %)	Rate [gmol/(g_{cat}-min)]	Activation energy (kJ/mol)
SBA-15	747	1.488	5–8.6			155.75					
Ti-SBA-15	630	0.88	5–8.7			325.55	-	-	1.85		
Pt–Pd/SBA-15_{cal}	560	0.845	5–8.3	88	1.3	340.58	0.145	0.14		1.6×10^{-4}	45.55
Pt-Pd/Ti-SBA-15_{cal}	590	0.856	5.5–8.65	96	1.4	355.75	0.147	0.145	1.83	1.92×10^{-4}	40.25
Pt–Pd/SBA-15_{spent}	440	0.742	5–7.9		1.7						
Pt–Pd/Ti-SBA-15_{spent}	575	0.822	5–8.58		1.5						

[a]Chemical compositions of the catalysts (Pt, Pd) were determined by inductively coupled plasma atomic emission spectroscopy (ICP-AES)

[b]BET specific area, pore volume, and average pore diameter as measured by N_2 adsorption–desorption isotherms at 77 K.

[c]Average metal particle size of reduced at 573 K samples as measured by CO chemisorptions.

[d]Metal dispersion using CO chemisorptions.

[e]Acidity of reduced samples and bare carriers as measured by TPD of adsorbed NH_3.

The solid samples were first digested (in a mixture of HCl and H2O2) in a microwave oven for 2 h, and then aliquots of solution were diluted to 50 ml using deionized water. The elemental compositions of calcined Pt-Pd/SBA-15 and Pt-Pd / Ti-SBA-15 catalysts were measured by ICP-AES and are given in Table 1. It indicates that the compositions of most of the catalysts closely match the targeted values.

2.4.2. Measurement of N_2 adsorption-desorption isotherms

The BET surface area, pore volume and pore size distribution of the samples were measured with an Autosorb-1C instrument using low temperature N_2 adsorption- desorption isotherms. Before measuring, the sample was degassed in vacuum at 200°C. The surface area was computed from these isotherms using the multipoint Brunauer-Emmett-Teller (BET) method based on the adsorption data in the partial pressure P/P0 range from 0.01 to 0.2. The value of 0.1620 nm^2 was taken for the cross-section of the physically adsorbed N_2 molecule. The mesopore volume was determined from the N_2 adsorbed at a P/P_0 = 0.4. The total pore volume was calculated from the amount of nitrogen adsorbed at P/P0 = 0.95, assuming that adsorption on the external surface was negligible compared with adsorption in pores. The pore diameter and pore volume were determined using the Barret-Joyner-Halenda (BJH) method. In all cases, correlation coefficients above 0.999 were obtained.
The specific surface area (SBET), average pore volume and average pore diameter of supports, calcined and spent (after reaction) catalysts are listed in Table 1. The SBA-15 support shows the highest specific area value in comparison with the other samples (747 m^2 /g). The specific area follows the order: SBA-15 > Ti-SBA-15 > Pt-Pd/Ti-SBA-15cal > Pt-Pd/Ti- SBA-15spent > Pt-Pd/SBA-15cal > Pt-Pd/SBAspent. The Ti-SBA-15 shows the higher average pore diameter value (5-8.7 nm). The average pore diameter follows the order: Ti-SBA-15 > Pt-Pd/Ti-SBA-15cal > SBA-15 > Pt-Pd/Ti-SBA-15spent > Pt-Pd/SBA-15cal> Pt-Pd/SBA-15spent. Finally, the average pore volume follows the order: SBA-15 > Ti-SBA-15 > Pt-Pd/Ti-SBA-15cal > Pt-Pd/SBA-15cal > Pt-Pd/Ti-SBA-15spent > Pt-Pd/SBA-15spent.
The specific SBET values of the supports decrease upon incorporation of the Pt and Pd components (Table 1). The highest specific area value is observed in the calcined Pt-Pd / Ti-SBA-15 sample in comparison with calcined Pt-Pd / SBA-15 catalyst. The average pore diameter in the calcined Pt-Pd / Ti-SBA-15, Pt-Pd / SBA-15 samples decrease slightly with the deposition of the active components ca. 0.05- 0.3 nm, respectively. By contrast, the Pt-Pd / Ti-SBA-15 sample almost preserves the pore diameter in comparison with the homologous Ti-SBA-15 support. Once again, the lower pore volume is observed in the calcined Pt-Pd / SBA-15cal and spent Pt-Pd / SBA-15spent samples. Besides, it seems that the type of mesostructured material has an important role in the structural parameters of the supported catalysts.
The Pt-Pd/SBA-15 catalyst shows the higher decrease in specific area of the sample might indicate that, under the conditions of hydrogenation employed in this study, larger carbon, Pt and Pd particles are formed, which block the pores. This is an agreement with the pore diameter value. Pore diameter decreases ca. 0.7 nm in Pt-Pd/SBA-15spent sample, whereas in the Pt-Pd/Ti-SBA-15spent sample almost it remains the same. Comparing the structural parameters of the calcined catalysts with the spent catalyst in the hydrogenation of toluene, we can note that the spent Pt-Pd / Ti-SBA-15 catalysts largely preserve the structural parameters of the freshly calcined state. Contrary to this sample, one can observe an important decrease in the structural parameters in the spent Pt-Pd/SBA-15 catalyst. The origin of this behavior is not very clear, but it proves, that the Pt-Pd / SBA-15 catalyst undergo major changes in it's structural parameters during the reaction and this should be reflected in their catalytic performance.

2.4.3. CO Chemisorptions

Volumetric CO chemisorptions isotherms at 298 K were recorded separately in order to estimate metal dispersion using an Autosorb-1C instrument. For CO chemisorptions catalyst was pretreated and reduced by H_2 as for the catalytic experiments and then out gassed at 9×10^{-6} Torr. After cooling the sample to 303 K, CO was admitted at pressures below 20 Torr. Since in the region of low pressures, the CO isotherm can be described by the Langmuir equation, only one isotherm was recorded. The extrapolation of its linear part to zero pressure gives the amount of CO adsorbed in monolayer. Metal particle size was estimated assuming CO: Pt (Pd) = 1:1 surface stoichiometry. In the bimetallic catalysts, platinum was assumed to behave like palladium and metal dispersion was calculated from the adsorption branch of isotherm by employing the strong / difference extrapolation formalism showed maximum number of active catalyst sites about 96 % for Pt (0.15 wt %)-Pd (0.15 wt %) / Ti-SBA-15 catalyst (Table 1).

2.4.4. X-ray Diffraction Analysis

The broad-angle X-ray diffraction (XRD) patterns of the samples were measured using a Bruker D8 Advance Powder diffractometer with a Gemonochromator producing a monochromatic Cu Kα radiation. The scanning was made from 5 to 80° with a scan rate of 2°/min. In all cases, the generator was operated at 40 kV and 30 mA. To avoid the problem of illuminated areas, all the samples were measured using the same sample holder. The wide-angle X-ray diffraction patterns of the SBA-15, Pt-Pd/SBA-15, and Pt-Pd/Ti-SBA-15 catalyst samples exhibited a broad line between the 2θ values of 20-30°, which was attributed to the siliceous amorphous SiO_2 phase.[32] No reflections belonging to TiO_2 were observed, indicating the homogeneous dispersion of TiO_2 in the framework of mesoporous silica material.[32]

2.4.5. Energy Dispersive X-ray Analysis

Quantitative compositional analysis was carried out with an energy-dispersive X-ray analysis (EDX) system attached to the electron microscope, which was operated at 25 kV. Determination of the chemical composition was based on the average analytical data of individual particles and the result of EDX analyses of SBA-15, Ti- SBA-15 support, Pt-Pd/SBA-15, and Pt-Pd/Ti-SBA-15 catalysts in the form of energy peaks at different applied voltages showed the presence of silica (Si), platinum (Pt), palladium (Pd), titanium (Ti), and oxygen in the catalyst and no nitrogen was found. This indicates that during calcinations all the nitrate decomposed completely to their respective state.

2.4.6. Temperature-Programmed Desorption

Temperature programmed desorption (TPD) of ammonia over the catalysts was carried out in a quartz reactor (i.d. = 4.5 mm) packed with about 0.1 g of catalyst from 323 to 883 K, at a linear heating rate of 20 K/min in a flow of moisture free helium (40 mL/ min). The adsorbate desorbed in the TPD was measured quantitatively by a TCD detector and TPD results were given in Table 1. Before carrying out the TPD, the catalyst was pretreated in situ at 883 K for 1 h in a flow of helium. The ammonia was chemisorbed on the catalyst at 373 K by saturating the catalyst with the adsorbate at lower temperature (323-333 K) and then desorbing the physically adsorbed adsorbate in a flow of pure helium (40 mL/min) at 373 K for 1 h.

3. Result and Discussions

3.1. Effect of hydrogen to toluene mole ratio on conversion

The effect of the hydrogen to toluene mole ratio in the range of 2-12 was studied at 498 K (Figure 2). Toluene conversion increased up to a hydrogen/toluene mole ratio of 10 and then remained constant. Toluene conversion is increased due to the increase in the concentration of hydrogen on the support surface in the vicinity of the metal particles. Hence, the metal-support interface region could contribute to the overall activity of the toluene hydrodearomatisation if the concentration of adsorption sites for toluene is high enough and the hydrogen transport rate is sufficiently rapid.[28]

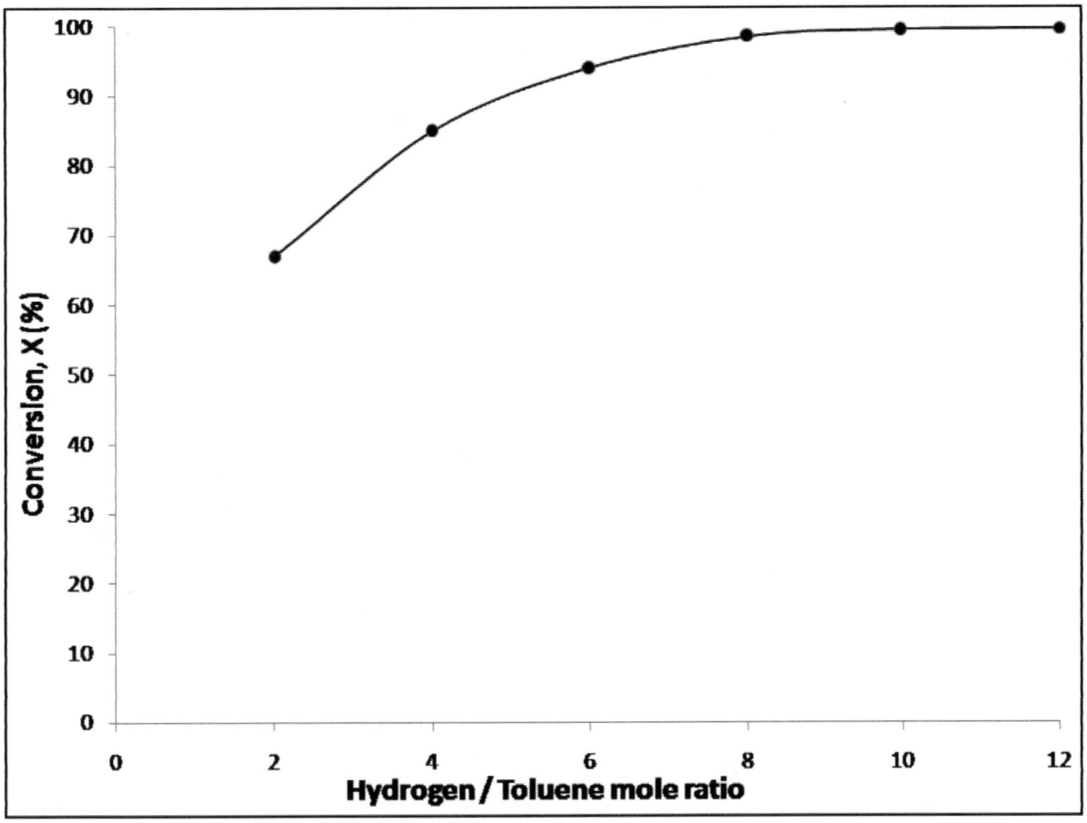

Figure 2. Effect of hydrogen to toluene mole ratio on conversion at T:498 K, P:15 bar, WHSV: 2 h-1, Catalyst: Pt (0.15 wt %)–Pd (0.15 wt %)/Ti-SB-15.

3.2. Effect of feed weight hourly space velocity on toluene conversion

Figure 3 depicts the effect of feed WHSV on the conversion of toluene at 498 K temperature, 15 bar pressure and hydrogen to toluene mole ratio 10. Toluene conversion was found to decrease with an increase in feed WHSV of toluene that can be explained by two different ways. With increase in liquid feed rate, wetted fraction of the catalyst as well as the gas-liquid and liquid-solid mass transfer coefficients also increase.[33] At lower liquid feed rate, catalyst particles are partially wetted, and under these conditions, conversion increases due to direct transfer of the gas-phase reactant to the catalyst surface, already wetted internally due to the capillary forces. Therefore, with an increase in liquid feed rate an increase in the wetted fraction is expected to retard the conversion, while an increase in the external mass transfer coefficients will enhance the conversion. Another possible reason for observed decrease in conversion is due to the shorter residence time of reactants under high feed WHSV. However, the former explanation is applicable to the large catalyst bed but in this study low amount of catalyst around 10 g (smaller catalyst bed) was used, hence the latter explanation is more appropriate for this study.[28]

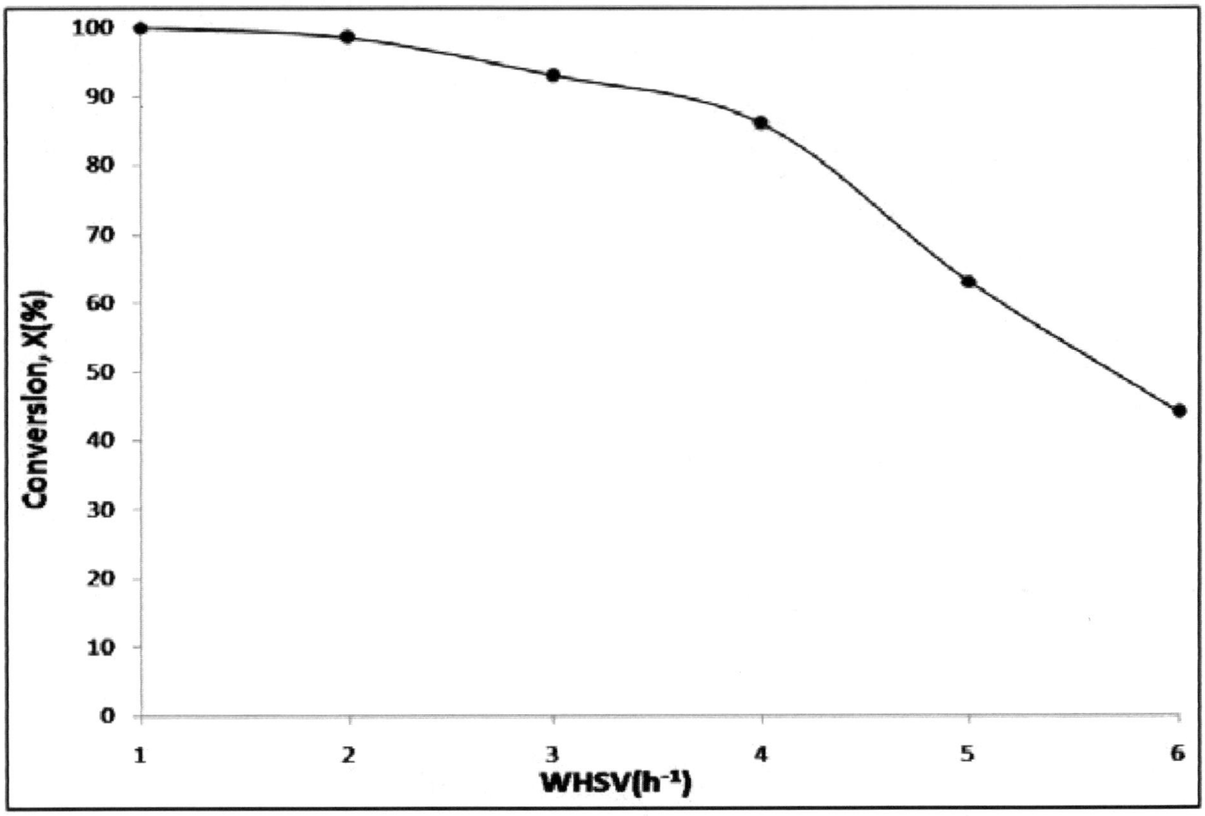

Figure 3. Toluene conversion vs WHSV plot in a fixed bed reactor at T:498 K, P:15 bar.

3.3. Effect of hydrogen pressure on toluene conversion

Figure 4 shows the plot of toluene conversion at 498 K as a function of pressure using hydrogen as a carrier gas. From the plot, it can be observed that toluene conversion increased with pressure for up to 15 bar beyond which it remains fairly constant up to 35 bar. The initial increase in toluene conversion between 5 and 15 bar can be attributed to the fact that within this pressure range increasing the pressure enhanced the adsorption of toluene on the active sites, leading to increased toluene conversion. Further increase in pressure beyond 15 bar only resulted in a slight increase in conversion probably because the adsorption of toluene had already reached saturation with the occupation of most of the active sites. Hydrogen is adsorbed along with toluene and takes part in the reaction. It should be noted that, because of its smaller size, the diffusion of hydrogen is faster than that of toluene. Thus, there are competing reactions where hydrogen occupies some of the active sites and dissociates into atomic hydrogen which participates in the reactions. This competing adsorption, diffusion, and activation of hydrogen molecule on the active sites may lower toluene conversion. This phenomenon can also be explained on the basis of Bronsted and Lewis acidity. The presence of hydrogen may lead to a decrease in Bronsted acidity and thus decrease in the concentration of intermediates benzylic carbocations, leading to lower toluene conversion.[28,34]

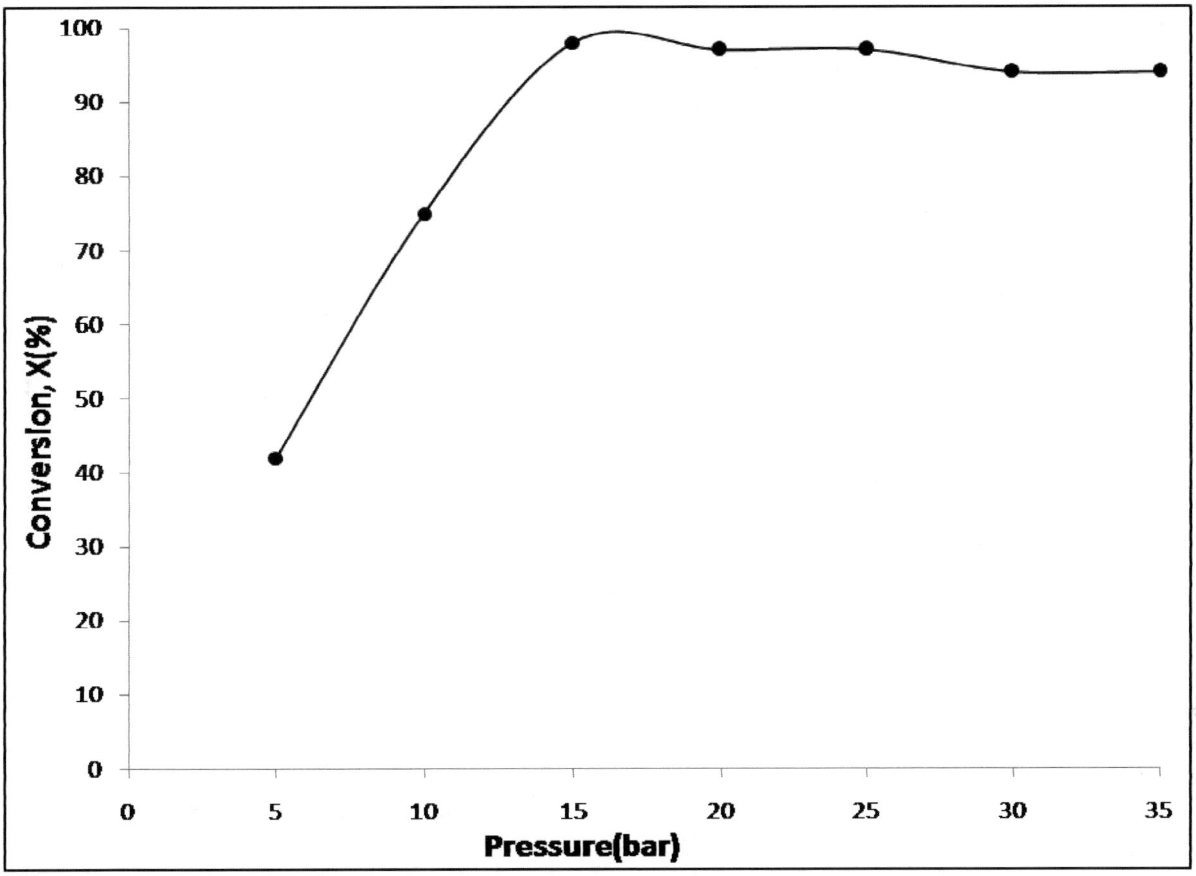

Figure 4. Toluene conversion vs pressure plot in a fixed bed reactor at WHSV: 2 h−1, T: 498 K.

3.4. Effect of temperature on toluene conversion

Low mechanical strength and collapsing of the structure at elevated temperature in the presence of water are the main drawbacks of mesoporous materials limiting their use as catalysts. In order to investigate the stability of the synthesized materials under real reaction conditions, the Pt (0.15 wt %)-Pd (0.15 wt %) / SBA-15 and Pt (0.15 wt %)-Pd (0.15 wt %) / Ti-SBA-15 were tested in toluene hydrogenation (Figure 5). Only methylcyclohexane was found as the reaction product. The toluene conversion of 100 % was achieved over the Pt (0.15 wt %)-Pd (0.15 wt %) / Ti-SBA-15 catalyst at 498 K, pressure 15 bar and WHSV=2 h^{-1}. The catalytic activity was stable during the long-term operation (5 h). The measurements performed after the reaction revealed that the specific surface area of the spent Pt (0.15 wt %)-Pd (0.15 wt %) / Ti-SBA-15 catalyst dropped from 590 m^2 /g to 575 m^2/g (Table 1), and the average metal particle size only slightly increased from 1.4 to 1.5 nm (Table 1) but in case of spent Pt (0.15 wt %)-Pd (0.15 wt %) / SBA-15 catalyst the specific surface area dropped significantly from 560 m^2 /g to 440 m^2 /g and average metal particle size increased from 1.3 to 1.7 nm (Table 1). The Ti-SBA-15 support and Pt, Pd nanoparticles confined within the mesopores under the mentioned reaction conditions were seen to be quite resistant against structure collapsing and metal sintering.

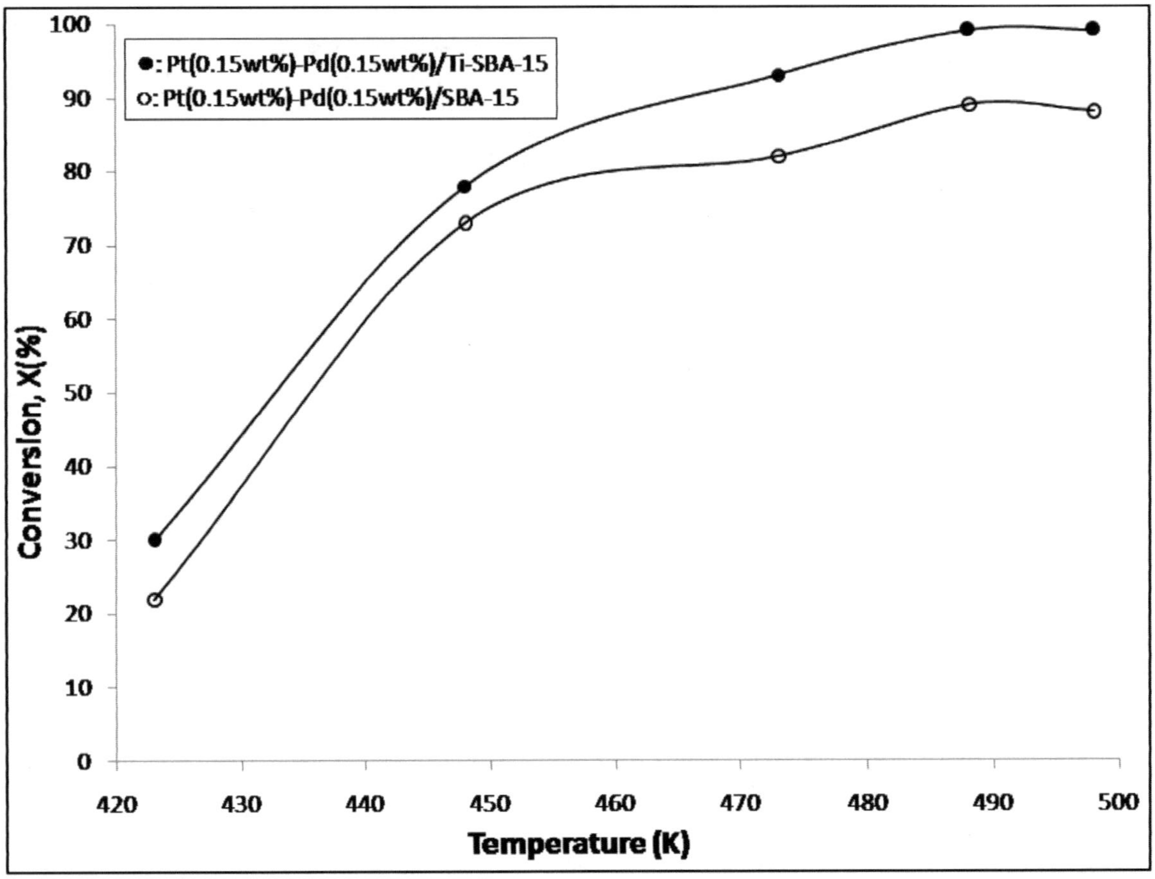

Figure 5. Toluene conversion vs temperature plot in a fixed bed reactor at WHSV: 2 h−1, P:15 bar..

The reaction rate constants were calculated from the experimental data and the results are plotted as ln (k) vs 1/T as shown in Figure 6 for Pt (0.15 wt %)-Pd (0.15 wt %) / SBA-15 and Pt (0.15 wt %)-Pd (0.15 wt %) / Ti-SBA-15 catalysts separately. The activation energy (E_2) of the reaction using Pt (0.15 wt %)-Pd (0.15 wt %) / SBA-15 catalyst was found to be 45.55 kJ/mol while the activation energy (E_1) using the Pt (0.15 wt %)-Pd (0.15 wt %) / Ti-SBA-15 catalyst was found to be 40.25 kJ/mol. The value of activation energy for Pt (0.15 wt %)-Pd (0.15 wt %) / Ti-SBA-15 catalyst is lower than the activation energy of Pt (0.15 wt %)-Pd (0.15 wt %) / SBA-15 catalyst indicating higher activity of Pt (0.15 wt %)-Pd (0.15 wt %) / Ti-SBA-15 catalyst. The activation energy obtained in the present study is compared with the activation energy reported in the literature for hydrogenation of toluene. Navarro et al.[35] reported that activation energy was found to be 129.2 kJ/mol for hydrogenation of toluene over Pt-Pd / Silica catalyst at 523 K. Coughlar and Keane[36] reported that the activation energy was found to be 78.2 kJ/mol for hydrogenation of toluene over nickel loaded Y-zeolite. It can be noticed that value of activation energies using Pt (0.15 wt %)-Pd (0.15 wt %) / Ti-SBA-15 catalyst for hydrogenation of toluene in the present study are roughly one fourth to half of that reported in literature. This is due to the fact that in the present study hydrogenation was carried out using a catalyst which facilitates a chemical reaction proceeds by a different pathway with a lower energy barrier.

Activity is described in terms hydrogenation of toluene using a specific reaction rate at 498 K, 15 bar pressure and 5 h on-stream according to Eq. (1) are given in Table 1. It is observed from Table 1 that reaction rate for toluene hydrogenation is higher using Pt (0.15 wt %)-Pd (0.15 wt %) / Ti-SBA-15 catalyst than that of Pt (0.15 wt %)-Pd (0.15 wt %) / SBA-15 catalyst. The hexagonal nature of the mesoporous materials and their thermal and hydrothermal stability are increased when transition metals are anchored into the framework of mesoporous silicate. Substitution of foreign ion (Al^{+3}, Ti^{+4}, Zr^{+4}) into the silicate framework is an efficient route in order to enhance acidity and stability of the mesoporous silica.

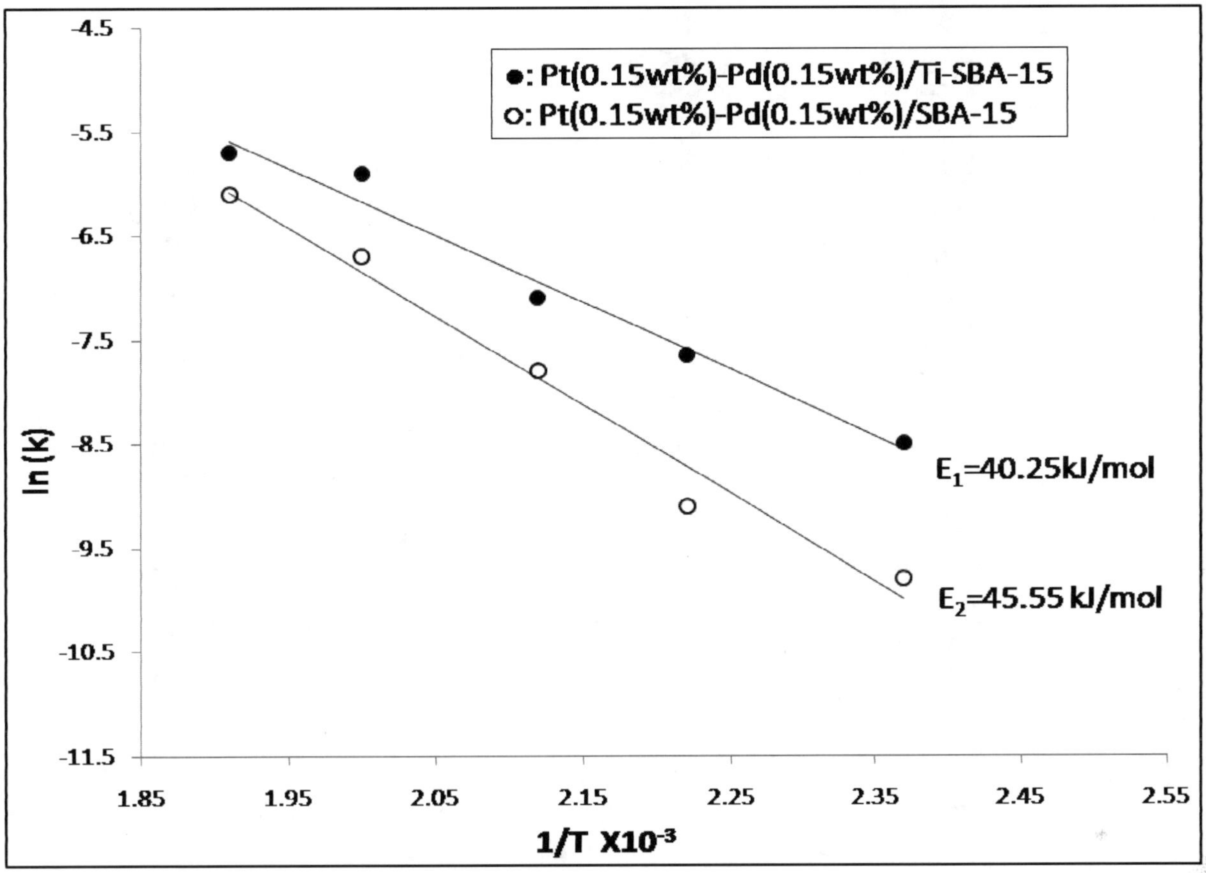

Figure 6. Arrhenius plot of ln(k) as a function of 1/T for Pt(0.15wt %)−Pd(0.15wt%)/SBA-15 and Pt(0.15wt%)−Pd(0.15wt%)/Ti-SBA-15 catalysts.

It can be further noticed from Table 1 that addition of Ti on SBA-15 support increases the metal dispersion and acidity. Some authors have reported that Al, Ti, or Zr-containing SBA-15 provided a better dispersion of metal species as compared to the pure SBA-15 and alumina-supported catalysts.[37,38]

3.5. Performance of SBA-15 supported catalysts

Figure 7 shows that the highest catalytic performance is observed using Pt (0.15 wt %)-Pd (0.15 wt %) / Ti-SBA-15 catalyst. The results of this study indicate that the addition of titanium to the SBA-15 support increased the catalytic performance of the samples in the reaction of the hydrogenation of toluene. The presence of the vanadium has a higher effect on hydrothermal stability and dispersion of the Pt and Pd metals in comparison with the Pt (0.15 wt %)-Pd (0.15 wt %) / SBA-15 catalyst (Table 1). Pt (0.3 wt %) / Ti-SBA-15 and Pd (0.3 wt %) / Ti-SBA-15 catalysts show a low conversion. It is widely accepted that the optimum catalytic performance of the system largely depends on the synergetic interactions (the effect of two metals taken together is greater than the sum of their separate effect at the same doses) established between Pt and Pd.[39] A kinetic model of hydrogenation of toluene may be proposed in the future.

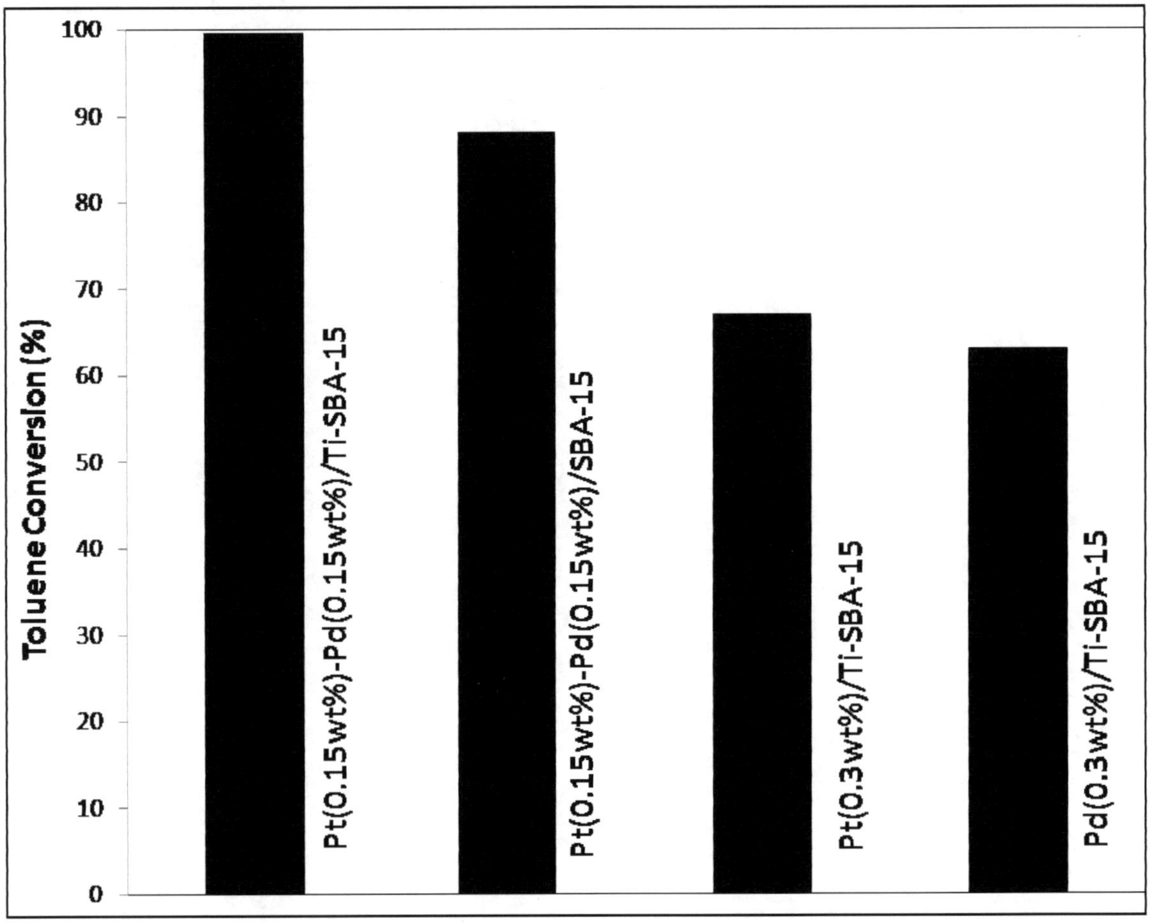

Figure 7. Conversion of toluene to methylcyclohexane over Pt (0.15 wt%)-Pd (0.15 wt%)/Ti-SBA-15, Pt (0.15 wt%)–Pd (0.15 wt%)/SBA-15, Pt (0.3 wt %)/Ti-SBA-15 and Pd (0.3 wt%)/Ti-SBA-15 catalysts at P:15 bar, T:498 K.

4. Conclusions

The following conclusions can be drawn from this study.

1. Under similar reaction conditions, toluene conversion was found to be higher when Pt-Pd / Ti-SBA-15 instead of Pt-Pd / SBA-15 catalyst were used for hydrogenation reaction. In all cases, toluene conversion measured at 498 K was found to be increased with pressure over 5-15 bar, after which it remained practically constant up to 35 bar.
2. The type of mesostructured material has an important effect on the dispersion of both Pt and Pd species. The highest dispersion is observed for the catalysts prepared on the Ti-SBA-15 material, and their sample had the higher catalytic performance in the hydrogenation of toluene under conditions employed in this work. The higher catalytic activity in this sample was related to the dispersion of both Pt and Pd species and the acidity in this sample.
3. The catalyst Pt (0.15 wt%)-Pd (0.15 wt%)/Ti-SBA-15 demonstrated stable activity and no metal agglomeration during toluene hydrogenation at 498 K for 5 h.
4. Activation energy was found to be 40.25 kJ/mol using Pt (0.15 wt%)-Pd (0.15 wt%)/Ti-SBA-15 catalyst indicates that the catalyst has higher activity.

The effect of feed WHSV, hydrogen pressure, hydrogen to toluene mole ratio and temperature on hydrogenation of toluene into methylcyclohexane was studied. The high surface area (630–747 m^2/g) and pore diameters in the range of mesopores make them suitable carriers for better and uniform distribution of metal species when the Pt and Pd is used as precursor. Activation energy markedly increases in the case of Pt-Pd/SBA-15 catalyst. Our approach of catalytic techniques furthers the knowledge required for the rational design of dedicated catalysts.

Acknowledgments

The work reported in this article was financially supported by a research grant of Chemical Engineering Department, Jadavpur University, Kolkata, West Bengal, India. We also acknowledge Prof. Asoke Kumar Biswas (Former Head and Professor, Chemical Engineering Department, Indian Institute of Technology, Kharagpur, India) for useful discussion.

References

[1] L. Allen, A. Suchanek, Hart's Fuel Technol. Manage. 8 (1998) 14-16.
[2] M. Hunter, A. Gentry, S. Lee, L. Groeneveld, C. Olivier, D. Pappal, The NPRA 2000 Annual Meeting, San Antonio, TX, Mar 26-28, 2000; Paper AM-00-18.
[3] D. Karonis, E. Lois, S. Stournas, F. Zannikos, Energy Fuels 12 (1998) 230-238.
[4] S.-X Wang, M. Shao, Y.-H. Zhang, J. Environ. Sci. 13 (2001) 265-271.
[5] M. Walsh, Platinum Met. Rev. 44 (2000) 22-30.
[6] D. Karonis, E. Lois, S. Stournas, F. Zannikos, Energy Fuels 12 (1998) 230-238.
[7] B. Cooper, B. Donnis, Appl. Catal., A 137 (1996) 203-223.
[8] S. Toppinen, T. K. Rantakyl, T. Salmi, J. Aittamaa, Ind. Eng. Chem. Res. 35 (1996) 1824-1833.
[9] U. K. Singh, M. Vannice, AIChE J. 45 (1999) 1059-1071.
[10] A. Stanislaus, B. H. Cooper, Catal. Rev. Sci. Eng. 36 (1994) 75-123.
[11] B. H. Cooper, A. Stanislaus, P. N. Hannerup, Diesel ACS National Meeting, San Francisco, April 1992.
[12] R. Ryoo, C. H. Ko, M. Kruk, V. Antochshuk, M. Jaroniec, J. Phys. Chem. B 104 (2000) 11465-11471.
[13] D. Zhao, J. Feng, Q. Huo, N. Melosh, B.F. Chmelka, G.D. Stucky, J. Am. Chem. Soc. 120 (1998) 6024-6036.
[14] Y. Cao, J. C. Hu, P. Yang, W. L. Dai, K. N. Fan, Chem. Commun. 7 (2003) 908-909.
[15] J. Barbier, E. Lamy-Pitara, P. Marecot, J. P. Boitiaux, J. Cosyns, F. Verna, Adv. Catal. 37 (1990) 279-318.
[16] C. Song, X. Ma, Appl. Catal. B: Environ. 41 (2003) 207-238.
[17] S. G. Kukes, F. T. Clark, D. Hopkins, WIPO Patent, WO 94/19429, 1994.
[18] J. K. Lee, H. K. Rhee, J. Catal. 177 (1998) 208-216.
[19] A. De Mallmann, D. J. Barthomeuf, J. Chim. Phys. Phys. -Chim. Biol. 87 (1990) 535-538.
[20] S. T. Homeyer, W. M. H. Schatler, J. Catal. 117 (1989) 91-101.
[21] W. M. H. Schatler, A.Yu. Stakheev, Catal. Today 12 (1992) 283-295.
[22] M.V. Landau, V. Ya Kruglikov, N.V. Goncharova, O. D. Konoval'chikov, G. D. Chukin, B. V. Smirnov, V. I. Malevich, Kinet. Catal. 17 (1976) 1281.
[23] P. Marecot, J. R. Mahoungou, J. Barbier, Appl. Catal., A: Gen. 101 (1993) 143-149.
[24] J. T. Miller, D. C. Koningsberger, J. Catal. 162 (1996) 209-219.
[25] H. R. Reinhoudt, R. Troost, A. D. van Langenveld, J. A. R. van Veen, S. T. Sie, J. A. Moulijn, Stud. Surf. Sci. Catal. 127 (1999) 251-258.
[26] T. Fujikawa, K. Idei, T. Ebihara, H. Mizuguchi, K. Usui, Appl. Catal., A: Gen. 192 (2000) 253-261.
[27] J. T. Miller, B. L. Mojet, D. E. Ramaker, D. C. Koningsberger, Catal. Today 62 (2000) 101-114.
[28] B. K. Sidhpuria, A. P. Parikh, P. Bahadur, V. R. Jasra, Ind. Eng. Chem. Res. 47 (2008) 4034- 4042.
[29] P.E. Boahene, K. Soni, A.K. Dalai, J. Adjaye, Appl. Catal., B: Environ. 101 (2011) 294-305.
[30] V. Parvulescu, C. Anastasescu, C. Constantin, B.L. Su, Catal. Today 78 (2003) 477-485.

[31] B. Mitra, J.P. Chakraborty, D. Kunzru, AIChE J. 57 (2011) 3480-3495.
[32] T.A. Zepeda, B. Pawelec, J.L.G. Fierro, T.J. Halachev, J. Catal. 242 (2006) 254–269.
[33] R.V. Chaudhari, R. Jaganathan, S.P. Mathew, C. Julcour, H. Delmas, AIChE J. 48 (2002) 110-125.
[34] N. S. Gnep, M. Guisnet, Appl. Catal. 1 (1981) 329-340.
[35] R.M. Navarro, B. Pawelee, J.M. Trejo, R.M. Mariscale, J.L.G. Fierro, J. Catal. 189 (2000) 184-194.
[36] B. Coughlan, A.M. Keane, Catal. Lett. 5 (1990) 101-112.
[37] D.P. Sawant, A. Vinu, N.E. Jacob, F. Lefebvre, S.B. Halligudi, J. Catal. 235 (2005) 341-352.
[38] O.Y. Gutierrez, G.A. Fuentes, C. Salcedo, T. Klimova, Catal. Today 116 (2006) 485-497.
[39] M. J. Tiernan, O. E. Finlayson, Appl. Catal., B: Environ. 19 (1998) 23-35.

473328 Aqueous Phase Catalytic Conversion of Methane to Methanol on Bimetallic Catalysts Under Mild Conditions

Abstract:

The direct conversion of methane to methanol has long been considered the holy grail of catalysis due to the inertness of the C-H bond in methane and the ease of over-oxidation of the product. Methanol has the highest energy density among all the C1 oxygenates and it is also an important chemical feedstock in large demand. The current industrial processes to produce methanol from methane involves multiple reaction steps by first converting methane to syngas followed by the methanol synthesis reaction, and such process is only economical at large scale. A direct single step conversion circumvents the energy demanding steps, which makes it highly attractive. Despite extensive research, a fully catalytic system using molecular oxygen as oxidant with heterogeneous catalysts at low temperatures has not yet been reported. Liquid- phase conversion of methane to methanol is either conducted using H_2O_2 as oxidant[1] or using strong acid as the solvent[2] which limits their further applications. In the gas phase, although several zeolite- based catalysts have been shown to be active for this reaction[3], the process requires a separate hydrolysis step to react and extract methanol from the active center, thus it is a non-catalytic process. The separate extraction step is not needed in the case of a liquid- phase system. Obviously, an aqueous phase catalytic system using O_2 is highly desirable.

Herein we report that iridium- based bimetallic catalysts in zeolites can directly convert methane to methanol with high selectivity in aqueous phase using O_2 as the oxidant and CO as a promoter. The monometallic Ir/ZSM-5 catalyst shows a relatively low activity, whereas the addition of a second metal significantly promotes both the activity and selectivity of the catalyst, leading to a high methanol yield. The promoter itself has a negligible activity and there is clearly a synergistic effect between iridium and the second metal. This study opens a new route for the development of methane- to- methanol catalysts using bimetallic systems.

References:

1. C. Hammond, M. M. Forde, M. H. Ab Rahim, et al. *Angew. Chem. Int. Ed.* 51, (2012) 5129

2. R.A. Periana, O. Mironov, D. Taube, et al. *Science* 301, (2003) 814

3. E. M. Alayon, M. Nachtegaal, M. Ranocchiari, et al. *Chem. Comm.* 48, (2012) 404

Copyright © American Institute of Chemical Engineers. All rights reserved.

468503 Gas-Phase Catalytic Oxidation of Methane into Methanol over Copper-Exchanged Zeolites with Molecular Oxygen

Abstract:

The direct catalytic conversion of methane to liquid oxygenated compounds, such as methanol or dimethyl ether, using molecular oxygen at low temperature is a grand challenge in C-H activation that has never been met.Â We report the first demonstration of direct, catalytic oxidation of methane into methanol with molecular oxygen over copper-exchanged zeolites at low reaction temperatures. Reaction kinetics studies show sustained catalytic activity and high selectivity over copper-exchanged ZSM-5 under mild conditions.Â Transient and steady state measurements with isotopically labeled molecules confirm catalytic turnover. Catalytic activity is observed for a variety of commercially available zeolite topologies. The catalytic rates and apparent activation energies are significantly affected by the zeolite topology, with caged-based zeolites (e.g., Cu-SSZ-13) showing the highest rates.Â Reactivity and characterization data demonstrate that the mono-(Î¼-oxo) dicupric site in copper exchanged ZSM-5 active for stoichiometric methane oxidation is not the catalytic site.

Copyright © American Institute of Chemical Engineers. All rights reserved.

470111 Bi-Reforming of Methane at Low Temperatures over Pt or Pd (Ni-Mg/ceria-zirconia) Catalysts

Abstract:

Bi-reforming of methane was studied over platinum (0.16 wt. %) or palladium (0.13wt. %) and/or nickel (1.4 wt. %) and magnesium (1.0 wt. %) immobilized onto a ceria-zirconia support. XRD patterns demonstrated that the support was a cubic fluorite structure and Ni and Mg were immobilized onto it. Temperature-programmed reduction showed that addition of Pt or Pd substantially decreased the reduction temperature with Pd doped catalysts having lower reduction temperatures than Pt doped catalysts. CO_2 desorption studies indicated that the metals had an impact on the basicity of the catalysts with Pd catalysts having more basic sites as more carbon dioxide was adsorbed. Bi-reforming studies showed that addition of Pt or Pd improved the H_2:CO ratio compared to dry reforming alone. In addition both Pd and Pt catalysts had comparable reaction results (dry reforming), however Pt catalysts had marginally higher TOFâ€™s (2.69-4.74 s^{-1}) compared to Pd catalysts (2.40-4.58 s^{-1}). Ongoing studies have shown that the addition of steam has improved the H2:CO ratio at low temperatures.

Copyright © American Institute of Chemical Engineers. All rights reserved.

469620 Understanding the Role of Fe in Bi-Metallic Ni-Fe Dry Reforming Catalysts: A Combined *in-Situ* XAS-XRD Study

Abstract:

One major drawback of Ni-based dry reforming (CH_4 + CO_2 â†' 2CO + $2H_2$) catalysts is coke formation.[1,2] In order to overcome this limitation, bi-metallic Ni-Fe catalysts have been developed, synthesized via a hydrotalcite-based precursor. Here, Ni and Fe are embedded in a thermally stable $Mg_xAl_yO_z$ matrix. The formation of alloyed Ni-Fe nanoparticles was confirmed via TEM-EDX, XRD and XAS measurements. The catalytic performance of monometallic Ni and Fe as well as the Ni-Fe bimetallic catalysts was evaluated at 650 Â°C. It was found that the bi-metallic catalysts, in particular, Ni_4Fe_1 (2.6 h^{-1} of TOF_{CH4}) have a high activity and stability, whereas fast deactivation and a low catalytic activity was observed for monometallic Ni (1.8 h^{-1} of TOF_{CH4}) and Fe (0.4 h^{-1} of TOF_{CH4}) catalysts, respectively. *In-situ* XAS and XRD analysis in combination with Raman, TEM, TGA and TPO allowed us to elucidate the underlying reason for the high activity and stability of the bimetallic Ni-Fe catalyst. The formation of FeO was observed in bimetallic Ni-Fe catalysts via *in-situ* XAS measurement, whereas metallic Ni^0 was maintained in both Ni and Ni-Fe catalysts. Furthermore, using combined in-situ XAS and XRD experiments, we could demonstrate that FeO removes the carbon deposited via a redox mechanism (C + FeO â†' CO + Fe).

References

1. V.C.H. Kroll, H.M. Swaan, S. Lacombe, C. Mirodatos, *J. Catal.*,1996, 2, 387-398
2. E. Ruckenstein and Y.H. Hu, *J. Catal.*,1996, 2, 230-238
3. S. A. Theofanidis, V. V. Galvita, H. Poelman, G. B. Marin, *ACS Catal.*, 2015, 5, 3028-3039.

Copyright © American Institute of Chemical Engineers. All rights reserved.

471764 Selective Deposition of Ni on Mo_2c That Has Been Supported on Al_2O_3: Enhanced Stability for Dry Reforming of Methane

Abstract:

In recent years, there has been much greater incentive to reduce greenhouse gas (GHG) emissions, CH_4 and CO_2 being the most potent offenders [1]. Biogas, which contains both GHG's, is a largely under-utilized and passively emitted resource that may be used for energy production; therefore utilizing biogas represents not only an environmental benefit, but also a significant economic opportunity. One way to utilize biogas and provide a distributable source of energy is to catalytically convert it into liquid fuels. This may be achieved by first (dry) reforming CH_4 with CO_2 to form CO and H_2

production reaction such as Fischer-Tropsch or dimethyl ether (DME) synthesis. Although there are a few cases in which dry reforming of biogas has been scaled to demonstration-sized facilities [2], it has not attained full scale production. One of the main barriers is catalyst deactivation due to coking [3]. In this presentation, we address this issue by both choice of support and preparation method of Ni-based catalysts.

Oxide-supported Ni catalysts have been modified by others using Mo_2C in appropriate amounts to improve stability and activity [4, 5]. The enhanced stability has been attributed to a redox cycle where Mo_2C is oxidized by CO_2 which is then followed by recarburization from the coke deposited on Ni to re-form Mo_2C [4]. However, no efforts have been made to ensure intimate contact between Ni and Mo_2C when both are supported on a metal oxide surface. We demonstrate that Ni may be selectively deposited on Mo_2C aggregates that have been previously supported on Al_2O_3; the enhanced interaction between Ni and Mo_2C improve stability and provide more efficient utilization of the catalyst. Electroless deposition (ED) is used to

selectively deposit Ni on Mo_2C by activation of a reducing agent on the Mo_2C surface to reduce a Ni^{2+} precursor in solution. Strong electrostatic adsorption (SEA) of the Ni^{2+} precursor on the Al_2O_3 surface is prevented by adjusting the pH of the solution to maintain the Al_2O_3 surface in a positive state. Thus, it is critical to avoid SEA of Ni on Al_2O_3 and to deposit Ni only on Mo_2C.

The resulting catalysts have been evaluated for dry reforming of methane; activities and stabilities are compared with catalysts prepared by conventional co-impregnation. XRD, TPR, and CO chemisorption have been used to determine bulk and surface compositions. The selectivity of Ni deposition on Mo_2C has been confirmed by STEM images.

References

[1] Inventory of U.S. Greenhouse Gas Emissions and Sinks: 1990 – 2014, in: E.P. Agency (Ed.), 2016.

[2] A. Corradini, J. McCormick, Process and system for converting biogas to liquid fuels, in, Google Patents, 2012.

[3] C. Papadopoulou, H. Matralis, X. Verykios, Utilization of biogas as a renewable carbon source: dry reforming of methane, in: Catalysis for Alternative Energy Generation, Springer, 2012, pp. 57-127.

[4] S. Zhang, C. Shi, B. Chen, Y. Zhang, Y. Zhu, J. Qiu, C. Au, Catalytic role of β-Mo 2 C in DRM catalysts that contain Ni and Mo, Catalysis Today, 258 (2015) 676-683.

[5] W. Li, Z. Zhao, P. Ren, G. Wang, Effect of molybdenum carbide concentration on the Ni/ZrO 2 catalysts for steam-CO 2 bi-reforming of methane, RSC Advances, 5 (2015) 100865-100872.

Copyright © American Institute of Chemical Engineers. All rights reserved.

473851 Optimization of Metal Dopants in Pyrochlore Catalysts for the Dry Reforming of Methane: Combining Computational and Experimental Methods

Abstract:

The advent of advanced drilling technologies has significantly increased methane production rates world-wide, and further increases are possible if cost effective technologies can be developed to remove or make use of the CO_2 species present in many untapped natural gas deposits. Therefore, the dry reforming of methane (DRM) has received special attention as it provides a route for the production of fuels and chemicals via syngas, by employing methane and CO_2 as feedstocks. The trial and error optimization of catalysts for DRM has thus far yielded a modicum of promising catalysts. For example, Rhodium-substituted lanthanum zirconate pyrochlores (LRhZ) have been shown to exhibit high catalytic activity and long-term thermal stability for DRM. To further develop lower cost DRM pyrochlore catalysts with even greater catalyst lifetimes, an in silico optimization study was undertaken that built upon our previous density functional theory and microkinetic modeling studies of Rh-doped pyrochlore catalysts. Computational analysis of one of the rate determining steps (the CHO dehydrogenation reaction, which lies on the favored CH_4 dehydrogenation pathway) suggested Pd as an effective co-dopant to reduce the activation barrier for this step. Thus, a bimettalic Rh-Pd co-doped lanthanum zirconate pyrochlore (Rh-Pd-LZ) was synthesized, characterized, and tested for DRM activity. The Rh-Pd-LZ catalyst exhibited high activity for DRM and H_2 to CO product ratios close to unity at both moderate and high temperatures, which evidences the fostering of DRM and inhibition of the undesired reverse water gas shift reaction (RWGS, CO_2 + H_2 â‡Œ CO + H_2O).

Copyright © American Institute of Chemical Engineers. All rights reserved.

472924 The Investigation of the Catalytic Partial Oxidation of Methane in a Shock Tube

Abstract:

The Investigation of the Catalytic Partial Oxidation of Methane in a Shock Tube

Robyn E. Smith[1], Marco J. Castaldi[1], Kenneth Brezinsky2

[1]Department of Chemical Engineering, The City College of New York, City University of New York, New York, NY 10031

[2]Department of Mechanical and Industrial Engineering, The University of Illinois at Chicago, Chicago, IL 60607

Introduction

The societal and environmental drive to become an energy and resource conscious world has lead to significant research concentrating on uncovering new methods of producing and utilizing energy. This can be done successfully by using catalytic processes to improve current and generate new methods. The production of synthesis gas from methane is one example, where the existing technology utilizes the steam reforming of methane to generate syn-gas, however this method requires continuous energy input. Conversely, the catalytic partial oxidation of methane, which also produces syn-gas, generates energy as a product of reaction, therefore is a more energy conscious method.

However, the chemical kinetic mechanism of the catalytic partial oxidation of methane has long been debated in scientific literature. This is a seemingly simple reaction proceeding through either partial oxidation followed by complete oxidation[1–5] or complete oxidation followed by reforming steps[6–12]. This dispute is due to the complexity of the catalytic cycle, which becomes further convoluted when comparing the reaction performance data over various catalyst conformations. This study focuses on the effects that catalyst conformation has on the performance of the catalytic partial oxidation of methane using a unique shock tube apparatus to generate short reaction times in stagnant reaction conditions.

Experimental Results

In these experiments, a mixture of dilute methane and oxygen were reacted in a shock tube at temperatures and pressures ranging from 1030-1200 K and 2.3-2.8 bar. The reaction times varied between 2.3 and 6.4 milliseconds. Experiments over a platinum on γ-alumina catalyst in a conventional configuration was compared to those using a fixed catalyst. For direct comparison, the performance of each catalyst was defined by the conversion of methane to products per catalyst weight.

Product gas analysis unveiled that the performance of the conventional catalyst was two orders of magnitude higher than that of the fixed catalyst. The results also showed that methane conversion corresponded to the production of carbon dioxide without the production of carbon monoxide for all temperatures, pressures, reaction times and both catalyst conformations. This second finding suggests that the mechanism for the catalytic partial oxidation of methane over both catalyst configurations goes through complete oxidation followed by reforming steps.

References

1. Song, X., Williams, W. R., Schmidt, L. D. & Aris, R. H--H Reactions. 1129–1137 (1990).

2. Bharadwaj, S. S. & Schmidt, L. D. Catalytic partial oxidation of natural gas to syngas. *Fuel Process. Technol.* **42,** 109–127 (1995).

3. Hofstad, K. H., Rokstad, O. A. & Holmen, A. Partial oxidation of methane over platinum metal guaze. **36,** 25–30 (1996).

4. Horn, R. & Schlogl, R. Methane Activation by Heterogeneous Catalysis. *Catal. Letters* 23–39 (2014). doi:10.1007/s10562-014-1417-z

5. Fathi, M., Monnet, F., Schuurman, Y., Holmen, a. & Mirodatos, C. Reactive Oxygen Species on Platinum Gauzes during Partial Oxidation of Methane into Synthesis Gas. *J. Catal.* **190,** 439–445 (2000).

6. Tong, G. C. M., Flynn, J. & Leclerc, C. A. A dual catalyst bed for the autothermal partial oxidation of methane to synthesis gas. *Catal. Letters* **102,** 131–137 (2005).

7. Tsang, S. C., Claridge, J. B. & Green, M. L. H. Recent advances in the conversion of methane to synthesis gas. *Catal. Today* **23,** 3–15 (1995).

8. Mattos, L. V., Rodino, E., Resasco, D. E., Passos, F. B. & Noronha, F. B. Partial oxidation and CO2 reforming of methane on Pt/Al2O3, Pt/ZrO2, and Pt/Ce-ZrO2 catalysts. *Fuel Process. Technol.* **83,** 147–161 (2003).

9. Mhadeshwar, a. B., Aghalayam, P., Papavassiliou, V. & Vlachos, D. G. Surface reaction mechanism development for platinum-catalyzed oxidation of methane. *Proc. Combust. Inst.* **29,** 997–1004 (2002).

10. Aghalayam, P. *et al.* A C1 mechanism for methane oxidation on platinum. *J. Catal.* **213,** 23–38 (2003).

11. Bui, P.-A., Vlachos, D. G. & Westmoreland, P. R. Catalytic ignition of methane/oxygen mixtures over platinum surfaces: comparison of detailed simulations and experiments. *Surf. Sci.* **385,** L1029–L1034 (1997).

12. Degroote, A. & Froment, G. Simulation of the catalytic partial oxidation of methane to synthesis gas. *Appl. Catal. A Gen.* **138,** 245–264 (1996).

Copyright © American Institute of Chemical Engineers. All rights reserved.

470916 Single-Atom Active Sites for Natural Gas to Liquid Fuel Conversion

Abstract:

In this work, we will discuss the study of large macrocyclic molecules, transition metal phthalocyanines, as the single-atom metal active sites for the methane catalytic oxidation. The hypothesis of this work is that single-atom metal active sites (analogous to metal-oxo species of monooxygenase enzyme) can be created on a porous sheet of transition metal phthalocyanine (TMPc) deposited on 2D graphene substrate. We modeled this catalytic reaction process using ab initio density functional theory (DFT) to develop an understanding of functionalization, reaction energetics, and reaction mechanism at the atomic level. Our results indicate that this material will have significant advantages over conventional catalysts for the selective oxidation of natural gas to methanol conversion process at mild reaction condition. Our results also indicate that chemical doping of substrate can play an important role on catalytic activity of TMPc surface by donating/backdonating necessary electrons from/to the substrate and create additional activity on the active metal sites during methane oxidation reaction. This active sites will provide not only a low energy barrier for C-H bond dissociation but also a high selectivity towards the desired methanol product. This information will be helpful to develop a novel catalyst that will have significant advantages over conventional catalysts for the selective oxidation of natural gas to methanol conversion process at mild condition.

Acknowledgement:

DFT calculation work was also supported from NSF-XSEDE resources under allocation number [TG-DMR140131].

Copyright © American Institute of Chemical Engineers. All rights reserved.

451022 Strategies for the Encapsulation and Stabilization of Monodisperse Au Clusters within Zeolites

Abstract:

The encapsulation of nanoparticles within microporous solids can improve the inherent instability of highly dispersed nanoclusters by sequestering them within voids that prevent their coalescence with other clusters, while also exploiting spatial constraints to restrict the size to which they can grow [1,2]. Confinement within such voids also precludes access by certain reactants or poisons to active sites present within intracrystalline voids, retains undesired products until they convert to smaller species that can egress by diffusion, or stabilizes specific transition states; in all cases, these enabling strategies are dictated by the size of the voids or channels in a specific microporous framework [2]. The encapsulation of such clusters within small-pore and medium-pore zeolites typically cannot be achieved through post-synthetic exchange, because solvated metal precursors are too large to enter the apertures of these zeolites [3].

This study describes successful strategies and guiding principles for the synthesis of small (1-2 nm) and monodisperse Au clusters protected against coalescence and poisoning by their uniform dispersion throughout the void space of LTA (0.42 nm apertures) and MFI (0.55 nm apertures) zeolites. The synthesis protocols involve hydrothermal zeolite crystallization around cationic Au^{3+} precursors. These precursors are stabilized by mercaptosilane ligands, which bind with the metal cations via their terminal sulfur atoms and protect against premature reduction or precipitation during the nucleation and growth of the zeolite frameworks. The silane groups in these ligands simultaneously promote the formation of siloxane bridges with nucleating silicate oligomers, thus enforcing the uniform dispersion of precursors through each crystal and the ultimate size homogeneity of the encapsulated clusters after thermal treatments required for ligand removal. UV-visible spectra show that Au^{3+} forms Au^0 clusters in O_2 or H_2 over a narrow temperature range that sets the dynamics of nucleation and growth and thus the mean size and uniformity of the clusters. H_2 treatments at the mildest temperatures within this range lead to small clusters uniform in size (1.0-2.0 nm) that are evident in electron micrographs, and to clean and accessible metal surfaces as demonstrated by the infrared spectra of chemisorbed CO. Their unprecedented size and monodispersity are preserved after oxidative treatments (>773 K) that significantly sinter Au clusters on mesoporous supports. Oxidative dehydrogenation rates of small (ethanol) and large (isobutanol) alkanols and the poisoning of unprotected clusters by organosulfur titrants show that >90% of the Au surfaces reside within intracrystalline LTA and MFI voids. This study illustrates how confinement favors small, uniquely stable, and monodisperse clusters, even for Au, a metal prone to cluster growth at conditions often required for its catalytic use.

X-Ray diffractograms of the synthesized zeolites confirmed the formation of the intended frameworks (LTA, MFI) and the absence of large (>10 nm) metal clusters. Elemental analysis showed that the incorporation of the ligand-stabilized metal precursors into the solids recovered was essentially complete and gave metal contents consistent with those predicted from synthesis reagents. The intensity of the surface plasmon resonance band in the UV-visible spectra of these zeolites was monitored during heating in O_2 or H_2 to track the formation of Au^0 clusters from the Au^{3+} precursors occluded within the intracrystalline voids. The intensity of this band serves as a diagnostic of Au^0 clusters larger than 2 nm, the diameter at which they incipiently exhibit plasmon resonance [4]. The H_2 treatments led to a sharp increase in the intensity of the plasmon resonance band between 540-623 K, reflecting the reduction of Au^{3+} by H_2 and the subsequent nucleation of Au^0 clusters throughout the zeolites. Transmission electron micrographs (TEM) showed that H_2 reduction treatments at 573 K form small Au clusters (LTA: 1.3 nm, MFI: 2.0 nm) narrowly distributed in size and embedded uniformly throughout zeolite crystals. These zeolite-encapsulated clusters retained their sizes even after subsequent treatments up to 773-823 K in O_2 or H_2. Such stability reflects the constraints imposed on cluster growth by kinetic hurdles brought forth by the large number and small size of intervening zeolite apertures that separate nucleation points during the decomposition processes; we also surmise that a limiting size is imposed by thermodynamic barriers originating from the energy required to disrupt the zeolite framework to accommodate clusters larger than a few cages or channel intersections in LTA or MFI, respectively. Au clusters dispersed on mesoporous supports such as TiO_2 or SiO_2, in contrast, form larger clusters (>10 nm) that are heterogeneously distributed in size after similar or even milder thermal treatments [5,6]. The zeolite encapsulation technique described here thus allows the systematic study of Au clusters at reaction conditions that strongly favor the sintering of Au particles into larger and much less reactive agglomerates.

Infrared bands of CO chemisorbed on Au clusters in the zeolite samples were examined to assess the cleanliness and accessibility of their metal surfaces. This infrared technique was used as an alternative to chemisorptive titrations, which are challenging to implement with Au clusters because of high dissociation activation barriers for common titrants, such as H_2 or O_2 [7]. The integrated intensities of Au-CO infrared bands on Au-MFI and Au-LTA samples, normalized by the number of surface Au atoms (as estimated by TEM), were compared to those measured on Au/SiO_2 samples prepared with methods known to fully remove contaminating reagents and form clean metal surfaces [8]. Normalized integrated Au-CO band intensities on Au-LTA and Au-MFI samples prepared using mercaptosilane ligands were essentially identical to those on Au/SiO_2 at all CO pressures, consistent with Au surfaces that are fully accessible and free of synthetic debris.

The selective placement of the Au clusters within the LTA zeolite (0.42 nm apertures) was confirmed by measuring oxidative dehydrogenation (ODH) turnover rates of ethanol (0.40 nm kinetic diameter, [3]) and isobutanol (0.55 nm kinetic diameter, [3]). The windows in LTA allow ethanol diffusion into the void space but exclude isobutanol; thus, comparison of ODH turnover rates for these small and large alkanols on unrestricted surfaces (e.g. Au/SiO_2) and on Au-LTA can be used to confirm that the exposed Au atoms reside predominantly within the LTA void space. Ethanol ODH rates on Au-LTA samples were also measured in the presence of thiophene, an organosulfur moiety that irreversibly titrates Au metal surfaces [3]. Thiophene (0.46 nm, [9]) is too large to enter the apertures of LTA; consequently, the extent to which thiophene decreases ethanol ODH rates provides a rigorous measure of the fraction of the Au surfaces accessible to thiophene and thus unprotected by intracrystalline voids. MFI channels (0.55 nm apertures) cannot exclude isobutanol or thiophene from intracrystalline voids. Consequently, encapsulation in MFI was instead probed by determining the fraction of Au cluster surfaces that are protected by titration with an organosulfur molecule (dibenzothiophene; DBT, 0.9 nm, [9]) that cannot enter MFI crystals.

The ratio of the ethanol to isobutanol ODH turnover rate on Au-LTA (133) was much larger than that on Au/SiO_2 (1.5), showing that the LTA zeolite framework indeed precludes access of isobutanol to encapsulated Au surfaces. Ethanol ODH rates on Au clusters supported by mesoporous SiO_2 were almost fully suppressed by exposure to thiophene or DBT, consistent with unrestricted access of the sulfur compounds to these unconstrained clusters. LTA-encapsulated Au, in contrast, retained >90% of its initial ethanol ODH rate following exposure to thiophene. Analogous experiments that exposed Au clusters within MFI to DBT also showed that >95% of the initial ODH activity was preserved in the presence of the sulfur titrant. This evident size selectivity confirms that the Au clusters reside predominantly within the LTA or MFI zeolite voids.

This work provides a general synthetic procedure to encapsulate Au clusters within the voids of small-pore and medium-pore zeolites. The shape selectivity shown is yet another demonstration of the reactant selection properties that have made zeolites such ubiquitously useful catalysts, while also demonstrating how confinement can lead to highly dispersed and uniquely stable Au clusters that are uniform in size, making such clusters useful at temperatures previously inaccessible in their synthesis and catalytic use.

The authors acknowledge the Chevron Energy Technology Company for financial support.

[1] Weisz, P.; Frilette, V.; Maatman, R.; Mower, E. *J. Catal.* **1962**, *1*, 307.

[2] Wu, J. C. S.; Goodwin, J. G.; Davis, M. *J. Catal.* **1990**, *125*, 488.

[3] Wu, Z.; Goel, S.; Choi, M.; Iglesia, E. *J. Catal.* **2014**, *311*, 458.

[4] Peng, S.; McMahon, J. M.; Schatz, G. C.; Gray, S. K.; Sun, Y. *Proc. Natl. Acad. Sci. U.S.A.* **2010**, *107*, 14530.

[5] Otto, T.; Zones, S. I.; Iglesia, E. *J. Catal.*, in press.

[6] Veith, G. M.; Lupini, A. R.; Pennycook, S. J.; Mullins, D. R.; Schwartz, V.; Bridges, C. A.; Dudney, N. J. *J. Catal.* **2009**, *262*, 92.

[7] Carabineiro, A. A. C.; Nieuwenhuys, B. E.; *Gold Bull.* **2009**, *42*, 288.

[8] Zhu, H.; Ma, Z.; Clark, J. C.; Pan, Z.; Overbury, S. H.; Dai, S. *Appl. Catal., A* **2007**, *326*, 89.

[9] Van de Voorde, B.; Hezinova, M.; Lannoeye, J.; Vandekerkhove, A.; Marszalek, B.; Gil, B.; Beurroies, I.; Nachtiqall, P.; De Vos, D. *Phys. Chem. Chem. Phys.* **2015**, *17*, 10759.

Copyright © American Institute of Chemical Engineers. All rights reserved.

465672 Crystallization of One-Dimensional Zeolites: Elucidating Mechanisms of Growth and the Role of Structure Direction

Abstract:

Crystallization of One-dimensional Zeolites: Elucidating Mechanisms of Growth and the Role of Structure Direction

Rui Li and Jeffrey D. Rimer

Department of Chemical and Biomolecular Engineering, University of Houston, 4726 Calhoun Road, Houston, TX, 77204, jrimer@central.uh.edu

Zeolites possess well-defined pore systems that give rise to shape selectivity for chemical reactions, thus providing an essential advantage for their use as catalysts in commercial applications. The performance of zeolites (i.e., activity and lifetime) is closely aligned with the selection of internal pore geometry and dimensions, which can influence diffusion, adsorption, and reaction pathways. Design of optimized zeolite catalysts *a priori* is challenging owing to the complexity of synthesis that is derived in part from the unknown role of amorphous precursors, organic structure-directing agents (OSDAs), and the pathways of crystallization. Previous studies have demonstrated that one-dimensional (1D) zeolites, such as zeolite L (LTL type), grow via nonclassical mechanisms involving the assembly of so-called *worm-like particles* (WLPs), which are bulk amorphous precursors that serve as growth units during crystallization.[1] Evidence of crystallization by particle attachment (or CPA) is mounting for a range of materials that include biominerals[2], metal oxides,[3] and zeolites.[4] In this relatively new area of research, there are many knowledge gaps between the design of synthesis conditions and the resulting physicochemical properties of the materials.[5] One interesting phenomenon regarding nonclassical crystallization of zeolites is the existence of WLPs, which are common to many different framework types.[1] Here, we will discuss how two widely-used 1D zeolites,[6,7] LTL and TON, crystallize from similar growth solutions that comprised of amorphous WLP precursors. Despite the similarity of their precursor solutions, we will show how these two structures involve distinct growth pathways that can be manipulated by the judicious selection of synthesis parameters (e.g., temperature, alkaline content, OSDA, etc.). We will present our work on the design of OSDAs for zeolite TON crystallization. Moreover, we will discuss mechanisms of LTL and TON crystallization and demonstrate how facile, economically-viable methods can be used to produce zeolites with improved properties for a wide range of industrial applications.

References

(1) Kumar, M.; Li, R.et al., *Chemistry of Materials,* **2016**, *28*, 1714-1727.

(2) Politi, Y.; Arad, T.et al., *Science,* **2004**, *306*, 1161-1164.

(3) Penn, R. L.; Banfield, J. F. *Science,* **1998**, *281*, 969-971.

(4) Lupulescu, A. I.; Rimer, J. D. *Science,* **2014**, *344*, 729-732.

(5) De Yoreo, J. J.; Gilbert, P. U. P. A.et al., *Science,* **2015**, *349*.

(6) Nacamuli, G. J.,US6143166 A, 2000.

(7) Perego, C.; Carati, A. *Zeolites and zeolite-like materials in industrial catalysis*; In *Zeolites: From Model Materials to Industrial Catalysts*; Čejka, J., Per¨¦z-Pariente, J., Eds.; Transworld Research Network: 2008, p 357-389.

Copyright © American Institute of Chemical Engineers. All rights reserved.

471700 A Method to Screen Nanoporous Catalysts for Transition-State Selectivity

Abstract:

3-dimensional nanoporous materials such as zeolites and metal-organic frameworks are used as catalysts for some of the most important industrial chemical reactions. The unique ability of these materials to selectively promote certain transformations originates from the subtle (mis)match between structures of the catalyst frameworks and those of the reacting species, and is usually classified into reactant, product, and transition-state selectivity. While the first two types have been successfully studied or even predicted at a large scale using Monte Carlo and molecular dynamics methods, analyzing transition-state selectivity requires much more computationally expensive quantum chemical techniques, and is limited to examination of a few sites on a case-by-case basis. Drawing on our findings from investigating zeolite-catalyzed reactions involved in catalytic cracking, we introduce a method based on the configurational-bias Monte Carlo algorithms that allows us to sample transition-state structures over the entire pore space of a material, and show that it is able to yield energies in good correlation with those obtained by density-functional theory calculations. The low-energy configurations identified in the sampling procedure can optionally be refined automatically for better accuracy via higher-level calculations. This method is easily applicable to long-chain molecules with large numbers of conformational degrees of freedom, for which current quantum chemical techniques can be very challenging to use. In addition, it also allows us to screen known zeolite structures for the wide array of different reactions involved in energy conversion processes.

Copyright © American Institute of Chemical Engineers. All rights reserved.

455924 Synthetic Control of Framework Aluminum Distribution in Zeolites and Consequences for Acid and Redox Catalysis

Abstract:

Synthetic Control of Framework Aluminum Distribution in Zeolites and Consequences for Acid and Redox Catalysis

John R. Di Iorio[1], Young Gul Hur[2], Rajamani Gounder[1]*

[1]School of Chemical Engineering, Purdue University, 480 Stadium Mall Drive, West Lafayette, IN 47907

[2] Department of Chemical and Biological Engineering, Korea University, 145, Anam-ro, Seongbuk-gu, Seoul 136-701, South Korea

*rgounder@purdue.edu

Zeolites contain different arrangements of framework Al atoms ($Al-O(-Si-O)_x-Al$) between paired (x = 1, 2) and isolated ($x \geq 3$) configurations, defined functionally by their ability to exchange cationic species of different valence (e.g., $M^{n+} = Cu^{2+}$, $[CuOH]^+$) and with different stoichiometry (M^{n+}/Al) quantifiable using methods to selectively titrate residual protons with NH_3 [1]. CHA zeolites synthesized in hydroxide media (Si/Al = 15-30) using only N,N,N-trimethyl-1-adamantylammonium cations ($TMAda^+$) as the organic structure directing agent (SDA) contain only isolated Al atoms [2], which exchange monovalent (e.g. $[CuOH]^+$, Na^+, NH_4^+) but not divalent (e.g., Co^{2+}, Cu^{2+}) cations. Addition of inorganic Na^+ cations to zeolite synthesis solutions as a cooperative SDA, with other synthesis parameters held constant (e.g., Si/Al, pH, total cationic charge: $[Na^+ + TMAda^+]/Al$), crystallized SSZ-13 zeolites of fixed Si/Al ratio but with numbers of paired Al sites that increased linearly with the Na^+ incorporated in the crystalline product. Equimolar amounts of Na^+ and $TMAda^+$ in the synthesis solution, but with varying Si/Al ratios (5-25), crystallized SSZ-13 zeolites with a paired Al density consistent with a random Al distribution subject to Lowenstein's rule [3]. Manipulating the ratio of high charge density inorganic (Na^+) and low charge density organic ($TMAda^+$) cations in crystallization media, which determines the solution cationic charge density, systematically influences the distribution of framework Al atoms between paired and isolated sites, which determines the anionic charge density of the zeolite framework. We have used mixtures of organic and inorganic SDAs in zeolite crystallization media and charge density mismatch concepts to systematically control the distribution of framework Al atoms in CHA and other zeolite frameworks. The catalytic consequences of paired and isolated Al arrangements in zeolites, which influence Brønsted acid strength and extraframework cation speciation, are investigated using the acid-catalyzed dehydration of methanol to dimethyl ether [4] and Cu-catalyzed NOx selective catalytic reduction with NH_3 [5]. These synthetic techniques provide routes to systematically control the atomic arrangement of Al heteroatoms and introduce structural and catalytic diversity into zeolites of a given topology at fixed elemental composition.

[1] Di Iorio, J. R., Bates, S. A., Verma, A. A., Delgass, W. N., Ribeiro, F. H., Miller, J. T., Gounder, R., *Topics in Catalysis*, 2015, 58, 424-434

[2] Di Iorio, J. R., Gounder, R., *Chemistry of Materials*, 2016, 28, 2236−2247

[3] Bates, S. A., Verma, A. A., Paolucci, C., Parekh, A. A., Anggara, T., Yezerets, A., Schneider, W. F., Miller, J.T., Delgass, W. N., Ribeiro, F. H., *Journal of Catalysis*, 2014, 312, 87-97

[4] Jones, A. J., Carr, R. T., Zones, S. I., Iglesia, E., *Journal of Catalysis*, 2014, 312, 58-68

[5] Paolucci, C., Parekh, A. A., Khurana, I., Di Iorio, J. R., Li, H., Albarracin-Caballero, J. D., Shih, A., Anggara, T., Delgass, W. N., Miller, J.T., Ribeiro, F. H., Gounder, R., Schneider, W. F., *Journal of the American Chemical Society*, 2016, DOI: 10.1021/jacs.6b02651

Copyright © American Institute of Chemical Engineers. All rights reserved.

465979 Design and Synthesis of IFR-Type Zeolites with Controlled Locations of Framework Aluminum with the Aid of Theoretical Calculation

Abstract:

The location of Al in zeolites, a class of crystalline microporous aluminosilicates widely utilized in several applications, is one of the key factors determining their physicochemical properties [1, 2]. Rationally designed synthesis of zeolites with controlled Al locations is one of the ultimate goals in the zeolite syntheses, but is very challenging due to the limitation of analytical techniques and unclarified crystallization mechanisms. In this work, synthesis of zeolite with controlled Al locations has been attempted by selecting an IFR topology as a model zeolite. The computational investigation suggested that its organic structure-directing agents (OSDAs) are likely to be fitted inside the zeolite and slightly change the relative stability of Al sites. IFR-type zeolites synthesized with different OSDAs under an identical condition were characterized by ^{27}Al NMR with the aid of DFT calculations. The results showed that Al locations of IFR-type zeolites can be tuned in accordance with energies derived from zeolite-OSDA complexes. This combined experimental and computational approach would lead to rational synthesis of zeolites with controlled Al locations.

[1] Davis, M. E. Chem. Mater. 2014, 26 (1), 239â€“245.

[2] K. Muraoka, W. Chaikittisilp, T. Okubo, J. Am. Chem. Soc., doi: 10.1021/jacs.6b01341.

Copyright © American Institute of Chemical Engineers. All rights reserved.

451570 A Mixed-Solvent Approach to Delaminate Layered Zeolitic Precursors

Abstract:

Two dimensional zeolitic materials built in layered precursors serve as building blocks to synthesize new catalysts, which are equipped with molecular-sized crystalline micropores that enable molecular sieving effects as well as large mesopores that allow fast mass transfer of reactants. Because these zeolitic nanosheets are held together tightly by strong hydrogen bonds, which are mediated by structure-directing agents (SDAs) present in the interlayer region, the first step generally used to obtain zeolitic nanosheets is to weaken strong interactions by intercalation of surfactants or removal of SDAs by acid treatment. Our strategy is to use a mixture of readily available solvents where one type of solvent extracts SDAs from the interlayer while the other interacts with the hydroxyl groups on the surface of zeolitic nanosheets and prevent their excessive condensation after removal of SDAs. The resultant material allows further processing such as facile delamination, swelling under mild conditions, and intercalation of other guest molecules. Here, to demonstrate our approach, we report results from treatment of as-made ITQ-1, a layered precursor of pure-silica MWW-type molecular sieve, with *N,N*-dimethylformamide (DMF) and ethylene glycol (EG). As-made ITQ-1 was synthesized with dual SDAs of *N,N,N*-trimethyl-1-adamantammonium ($TMAda^+$) and hexamethyleneimine (HMI) according to the literature method [1] and subjected to the treatment of a DMF-EG mixture with various compositions at 423 K. Data from elemental analysis, ^{13}C MAS NMR spectroscopy, thermogravimetric analysis, and powder X-ray diffraction indicate that treatment of as-made ITQ-1 with a DMF-EG mixture preferentially removes $TMAda^+$ from the interlayer region although the SDA is charged and thus its removal requires another cation to maintain charge neutrality. Subsequent shear mixing of the solvent-treated ITQ-1 in a high-speed mixer yielded a material, which exhibits a 3-fold higher external surface area and 2.5-fold larger mesopore volume than ITQ-1 after calcination as characterized by N_2adsorption experiments. In contrast, shear mixing of as-made ITQ-1 without the solvent treatment barely increased the external surface area, indicating that the solvent treatment renders the layered precursor labile to shear forces.

[1] Camblor, M. A., Corma, A., Díaz-Cabañas, M. J., Baerlocher, C., Synthesis and structural characterization of MWW type zeolite ITQ-1, the pure silica analog of MCM-22 and SSZ-25. *J. Phys. Chem. B 102*, 44-51 (1998).

Copyright © American Institute of Chemical Engineers. All rights reserved.

465741 Engineering Highly Active Brookite Titania Nanorods for Sustainable Hydrogen Production

Abstract:

Titanium dioxide (TiO_2, titania) nanomaterials are the most studied photocatalysts because of their abundancy, non-toxicity, stability and activity. Anatase and rutile polymorphs have been deeply investigated because they are the energetically favored forms at the nanoscale and in the bulk, respectively. Brookite, instead, has been rarely studied, despite theoretical and experimental data support its higher activity in some photocatalyzed transformations. By taking advantage of methods to produce 1-D phase-pure brookite nanorods, we show not only that brookite is a much more active photocatalyst than anatase, but also that its activity is length-dependent. In particular, we attribute the high activity to both the 1-D nature of the nanostructure, which favors electron-hole separation, and to the defective structure of the rods, which are subject to strain that causes favorable changes in electron-hole recombination processes. By tuning length and strain we are able to prepare samples that show among the highest rates of hydrogen production from biomass-derived compounds under solar simulated irradiation reported to date.

Copyright © American Institute of Chemical Engineers. All rights reserved.

460638 Hydrogen Production Via Photocatalytic Water Splitting over Mesoporous Pt/TiO_2

Abstract:

Mesoporous TiO_2 semiconductors were synthesized by sol-gel technique, using a triblock-copolymer-templating, named Pluronic F-127. In order to minimize the electron-hole pair recombination, Pt was added as an electron trap in different percentages (*i.e.* 1.00 wt. %, 2.50 wt. %) during the sol-gel synthesis of the TiO_2 semiconductors. This was followed by calcination at 500 °C and 550 °C. Semiconductors' activity was evaluated in a bench-scale Photo-CREC-Water II (PCWII) photoreactor. In this sense, water splitting was carried out under near-UV light for hydrogen molecule production in the PCWII Reactor using ethanol at 2 % v/v as a renewable organic scavenger (electron donor). Ethanol acts as a good hole scavenger. The PCWII consisted of a 2.65 L internally irradiated slurry annular reactor with a 15 W T8 Black-Light-Blue (F15T8/BLB) lamp. The optimal photocatalyst loading was 0.15 g L^{-1} in acid medium (pH = 4.0). The volume of the aqueous suspension was 6.0 L. Thereby the present study shows that mesoporous TiO_2 with 2.50 wt. % of Pt exhibits the best activity for hydrogen production under the same photoreaction conditions. This semiconductor material showed: a) An anatase phase with XRD, b) A reduced band gap at 2.34 eV was calculated, employing UV-Vis spectroscopy and a Tauc plot, c) 10-30 nm TiO_2 particles with SEM-EDX, d) A 140 m^2/g BET surface area as well as 10 nm and 20 nm bimodal pore size distributions. Furthermore, the absorbed photons on the photocatalysts have been studied by employing macroscopic balances. These macroscopic energy balances considered for incident, scattered and transmitted near-UV photons. In this respect, evaluated Quantum Yields (QY) displayed a promising 0.27 (27%) when utilizing mesoporous 2.50 wt. % Pt/TiO_2-550°C.

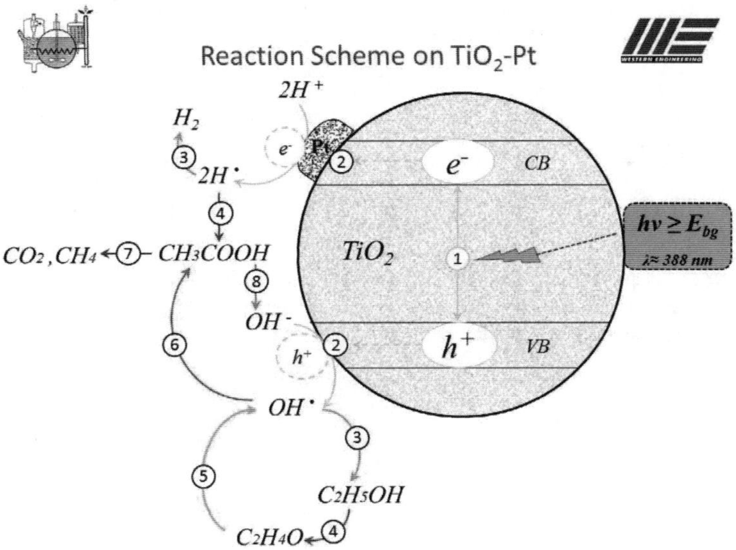

Copyright © American Institute of Chemical Engineers. All rights reserved.

467466 How Ceria Redox Materials Split H_2o and CO_2 with Concentrated Solar Energy into H_2, CO, and O_2

Abstract:

Heterogeneous catalyst design has been revolutionized in the past several years by the discovery of adsorption-energy scaling relations on transition-metal catalysts. These and correlated activation energies of chemical reactions are understood to limit the catalytic activity of a surface for a number of interesting reactions, such as production of H_2 and CO via splitting of H_2O and CO_2. Metal oxide redox materials exchange oxygen with a gas phase and store it in the solid state. This allows circumventing scaling relations that limit catalytic reactions, as a chemical reaction may be split into reduction and oxidation steps that are conducted at different process conditions. This work employs electronic structure calculations to investigate the reaction mechanism of CO_2 and H_2O splitting with ceria. Conceptually, non-stoichiometric ceria is reduced at elevated temperature with concentrated solar energy, liberating O_2. In a second step, the lattice oxygen is replenished via oxidation with H_2O and CO_2, yielding renewable H_2 and CO (syngas). Syngas is the precursor for synthetic liquid hydrocarbon fuel, such as diesel, gasoline and kerosene. The presented computations outline the effect of reduction and oxidation temperatures, partial pressure of O_2, and metal doping at the ceria surface on the free energy landscape of the two studied chemical reactions. Limits of isothermal processing as well redox trade-offs when controlling the reduction potential of the oxide surface are discussed.

Copyright © American Institute of Chemical Engineers. All rights reserved.

472043 Methane Conversion to Syngas Using Dopant Modified Metal Oxide Composites in Chemical Looping Reforming

Abstract:

Metal oxide composites are used as enabling materials for operation in many energy conversion systems including shale gas and coal gas reforming using chemical looping systems. When metal oxide composites are involved in redox reactions, ionic diffusion takes place which is accompanied by the creation and diffusion of defects that decide such their morphology and chemical properties. Rare earth element dopants can affect the redox reactions by providing alternative reaction sites, thus change the ionic diffusivity and reaction reactivity of metal oxides. In this study, we examine at nanoscale to delineate the characteristics of undoped and doped iron oxide that undergo methane reforming reaction. 0-2% lanthanum doped iron oxides are fabricated by modified wet chemical synthesis. XRD and SEM/EDS characterization implies that the dopant changes lattice parameters of metal oxides whose surface energy is changed. The atomistic thermodynamics methods and ab initio calculations are carried out to investigate the interaction between methane molecules and undoped/doped iron oxide. It is found that dissociation of C-H bond in the CH-La species is the key step for CH_4 partial oxidation reaction, and low-fold O atoms neighboring dopants in metal oxide composites could more readily dissociate from external surface to create oxygen vacancies. Moreover, the presence of rare earth element dopants can significantly decrease the barrier of CH_4 disociation, hence the reaction between CH_4 and oxygen carrier is promoted, which is in good agreement with our experimental results. The fundamental insight into the mechanism of cyclic redox reactions of La dopant modified metal oxide composites at nanoscale revealed in this study can be applied to other rare earth element dopant modified metal oxide systems and to guide future efforts to improve oxygen carriers for methane conversion.

Copyright © American Institute of Chemical Engineers. All rights reserved.

471655 A New Thin Layered Structural Coating on a Metal Substrate for Enhanced Hydrogen Production from Steam Methane Reforming

Abstract:

The recent increase in the availability of hydrocarbons, especially methane, from shale gas exploration has led to renewed interest in converting or upgrading methane to higher value chemicals. Steam reforming is still the most developed and widely practiced industrial process for converting methane to hydrogen/syngas and its derivatives. It has no oxygen requirement and the most favorable H2/CO ratio for methanol and ammonia production of all available technologies. However, opportunities still remain to lower the energy requirement of steam reforming, including development of more efficient catalyst systems in terms of higher activity and effectiveness factors. Recently, Alloy Surfaces Co. Inc. developed an intrinsically bound thin-layered skeletal catalyst adhered on a metal substrate. These structures having relatively uniform coating thickness allow the physical manipulation into a highly active complex monolith structure that greatly enhances the catalyst effectiveness and interfacial heat and mass transport without applying an intermediate washcoat. A collaboration among City University of New York, City College, Alloy Surfaces, Co. Inc., Innoveering, and Université catholique de Louvain has been established to develop novel catalyst systems to more efficiently convert methane. Recently the team studied and analyzed this catalyst developed for hydrogen generation via steam methane reforming. Results have demonstrated that this catalyst is intrinsically more active than the conventional Ni catalyst and shows an increased selectivity toward CO2 and H2 and higher WGS rates. Methane-to-steam ratios ranging from 2.0 to 6.0 were tested over a temperature range between 400°C to 900°C at pressures of 1.0, 1.5, 3.0 bar with GHSV from 50,000 to 350,000 h-1. Mass balances were closed to within 2% and methane conversions were maintained below 20% of the potential equilibrium conversions possible at the particular test conditions. Results demonstrate the catalyst is capable of producing H2/CO ratios up to 425 from a reactant steam to carbon ratio of 3 at 515oC with no measureable change in activity over the time tested. The WGS reaction is about an order of magnitude faster than on the conventional nickel catalyst with nearly a factor of 2 enhancement of the CH4 + 2 H2O = CO2 + 4 H2 reaction. This presentation will show the results obtained over a wide range of conditions. A

kinetic model with rate parameters was derived from the experiments with a specially designed microreactor and used to quantify the enhance selectivity toward hydrogen and carbon dioxide under typical commercial operating conditions. Finally some details will be given on how the enhanced performance is achieved.

Copyright © American Institute of Chemical Engineers. All rights reserved.

460805 Promotion Mechanisms of Iron Oxide-Based High Temperature-Water Gas Shift (HT-WGS) Catalysts By Chromium and Copper

Abstract:

The Cr and Cu promotion mechanisms of high temperature water-gas shift (HT-WGS) iron oxide catalysts, synthesized by co-precipitation, were investigated as a function of reaction conditions. XRD and *in situ* Raman characterization showed that the initial calcined catalysts consisted of the Fe_2O_3 (hematite) bulk phase and transformed to the Fe_3O_4 (magnetite) phase during the HT-WGS reactions. *In situ* NAP-XPS and HS-LEIS surface analysis revealed that Cr was surface enriched as Cr^{+6} for the initial catalyst and reduced to Cr^{+3} during the HT-WGS reactions, with the Cr^{+3} dissolving into the bulk iron oxide lattice forming a solid solution. *In situ* NAP-XPS, XANES and HS-LEIS characterization indicated that Cu was initially present as Cu^{+2} cations dissolved in the Fe_2O_3 bulk lattice and reduced to metallic Cu^0 nanoparticles (~3nm) on the external surface of the iron oxide support during the HT-WGS reactions. *In situ* HS-LEIS surface analysis also suggests that ~1/3 of the surface of the Cu nanoparticles was covered by a FeO_x overlayer. The CO-TPR probe demonstrated that Cr does not chemically promote the iron oxide catalyst and that only Cu is a chemical promoter for the iron oxide HT-WGS catalysts. The Cu promoter introduces new catalytic active sites that enhance the reaction rates of the WGS reactions.

Copyright © American Institute of Chemical Engineers. All rights reserved.

472372 Catalytic Ammonia Decomposition to Hydrogen and Nitrogen on Supported NanosizedÂ and Single-Site CobaltÂ Catalysts: Effects of Atomicity, Ligand Environment and Support

Abstract:

Hydrogen is one of the most widely used chemicals in the world. Petroleum refining and ammonia production are the major fields with the largest demand for hydrogen whereas automotive fuel is an emerging sector with the hydrogen demand growing incredibly fast. The total expected value of the whole hydrogen market is 152 billion USD by 2021. The main disadvantage of using pure hydrogen is the difficulty associated with its storage and transport.

Ammonia, on the other hand, is a hydrogen rich fuel (17.65 wt% hydrogen) that can be liquefied, stored and transported easily. As a liquid it has a volumetric hydrogen density about 45% higher than liquid hydrogen. To use this hydrogen, ammonia has to be decomposed over a catalyst to produce the desired clean fuel - hydrogen (H_2) along with nitrogen (N_2) a non-toxic, non-greenhouse gas that has no contribution to carbon emissions.

The most active catalysts used nowadays for ammonia decomposition are carbon nanotube-supported ruthenium nanoparticles. They are expensive and not always straightforward to produce. Therefore, finding new catalysts for ammonia decomposition that contain Earth-abundant metals utilized with atomic efficiency could be a viable alternative to expensive ruthenium catalysts. Moreover, gaining fundamental knowledge about metal-catalyzed ammonia decomposition, starting from the single-atom presents an important challenge that would allow us to establish more precise structure-catalytic property relationships, as well as create a pathway to produce more active/stable catalysts.

Herein, we demonstrate for the first time that the single-site silica-supported cobalt catalyst is an effective and stable catalyst for ammonia decomposition to nitrogen and hydrogen with activity starting at 480 °C and high TOF at elevated temperatures. The catalyst retains its single-site nature up until 630 °C, as evidenced by XPS, Raman and Diffuse-Reflectance UV-VIS spectroscopic measurements. Reduction of the catalyst takes place above 630 °C (with accompanying color change from blue to black) producing a reduced supported cobalt or cobalt nitride Co_xN_y phase, which is even more active in catalytic NH_3 decomposition, with activity above 390 °C. We discuss the proposed mechanism of NH_3 decomposition by single-site Co(II)/SiO_2 catalyst: even though the catalyst itself is very robust and it cannot be reduced or oxidized below 630 °C, it has unique activity towards nitrogen-containing molecules like ammonia and pyridine; those molecules chemisorb on Co(II) with the concomitant change of distorted tetrahedral environment of CoO4 unit to square pyramidal and octahedral as evidenced by Diffuse Reflectance UV-VIS. The catalytic activity of Co(II)/silica catalyst is assessed considering the previous report by A. Hock et al which indicates Co(II)/SiO_2 is an active and selective catalyst for propane dehydrogenation at temperatures close to 600 °C. Based on the similarity of N-H and C-H bond strengths, the ability of Co(II)/SiO2 to break strong N-H bonds therefore signifies a general trend of this catalyst to break strong X-H bonds. However, the single-site cobalt catalyst is less active than extended metal or metal nitride surface, exemplifying the important fundamental and mechanistic difference between the regimes of operation for the single-site and nanosized supported Co catalysts for NH_3 decomposition under given conditions.

Moreover, we show that the ligand environment around cobalt center can be modified on a single-atom scale: from CoO_4 to CoN_4 by incorporating single Co(II) atoms into mesoporous carbon nitride C_3N_4. This change in the ligand environment is not favourable for ammonia decomposition, with activity starting at much higher temperature compared to analogous Co(II)/SiO_2 catalyst.

Additionally, we demonstrate that the same dehydrogenation principle can be applied to ethylamine $C_2H_5NH_2$, with the Co(II)/SiO_2 catalyst capable of selectively dehydrogenating the mixed RCH_2-NH_2 bond of ethylamine to produce acetonitrile CH_3CN.

To get a more general and fundamental understanding of the catalytic behavior of mononuclear Co(II) species embedded into silica, we tested it in two bench mark reactions for which nanosized supported cobalt and cobalt oxide particles are predominantly used: CO and CO_2 hydrogenation. CO hydrogenation activity starts at temperatures above 380-400 °C with the main product being methane (~90%) and minor C_2 products at 2 MPa CO/H_2 pressure; CO_2 hydrogenation of CO_2/H_2 mixture at 1 bar leads to reverse water-gas shift reaction and formation of CO as the main CO_2 reduction product, detectable at temperatures above 300 °C.

In conclusion, Co(II)/SiO_2 was shown to be catalytically active for ammonia decomposition as well as carbon monoxide/carbon dioxide hydrogenation. The direct comparison of two catalysts with atomically dispersed and nanodispersed supported metal atoms/particles once again exemplifies the fundamental difference between catalysis by single metal-atoms (cationic species) and traditional nanosized supported catalysts.

Copyright © American Institute of Chemical Engineers. All rights reserved.

471230 Selective Activation of Primary C-H Bonds of Propane on Pdo(101)

Abstract:

Achieving selective C-H bond cleavage is critical for developing catalytic processes that transform small alkanes to value-added products. In this talk, I will discuss our recent investigations that clarify the molecular-level origin for an exceptionally strong preference for propane to dissociate on the crystalline PdO(101) surface via primary C-H bond cleavage. From experiments using different propane isotopologues, we find that more than 90% of the adsorbed propane molecules that dissociate on PdO(101) during temperature programmed reaction spectroscopy react via cleavage of primary C-H bonds. Using reflection absorption infrared spectroscopy (RAIRS) and density functional theory (DFT) calculations, we show that the adsorbed propane Ïƒ-complexes preferentially adopt geometries on PdO(101) in which only primary C-H bonds datively interact with the surface Pd atoms at low propane coverages and are thus activated under typical catalytic reaction conditions. We show that a propane molecule achieves maximum stability on PdO(101) by adopting a bidentate geometry in which a H-Pd dative bond forms at each CH_3 group. We also find that adsorbed propane molecules undergo a configurational change with increasing coverage, and begin to preferentially coordinate to the surface through the CH_2 group at sufficiently high coverage. Our results demonstrate that structural registry between the molecule and surface can strongly influence the selectivity of a metal oxide surface in activating alkane C-H bonds, and may thus provide guidance in designing catalysts capable of promoting selective alkane conversions.

Copyright © American Institute of Chemical Engineers. All rights reserved.

458416 Methane Activation on Subsurface Boron Doped Copper: First Principles Investigation of the Structure, Activity and Electronic Properties of the Catalyst

Abstract:

Methane activation on subsurface boron doped copper: First principles investigation of the Structure, Activity and Electronic properties of the catalyst

Quang Thang Trinh,[1,2] Samir H. Mushrif [1,*]

[1]School of Chemical and Biomedical Engineering, Nanyang Technological University, 62 Nanyang Drive, Singapore 637459, Singapore.

[2]Cambridge Centre for Advanced Research in Energy Efficiency in Singapore (CARES), Campus for Research Excellence and Technological Enterprise (CREATE), Singapore 138602, Singapore.

[*]Email: SHMushrif@ntu.edu.sg

Abstract: Methane, the primary component of commercially exploitable resources like natural gas, shale gas and gas hydrates, with great potential to be converted to a variety of value added chemicals, still remains hugely underutilized.[1,2] Although being a commercial catalyst in the synthesis of methanol from syngas, or in the fabrication of graphene by chemical vapor deposition, high activation barriers on Copper (Cu) for C-H bond activation limits its application in methane utilization.[2,3] However in this study, using first principles calculations, we predict that Copper

catalyst, when doped with a monolayer of subsurface Boron (Cu-B), can efficiently activate the C-H bond of Methane and can facilitate the C-C coupling reaction. Boron binds strongest in the subsurface octahedral site of Cu and the thermodynamic driving force for B diffusion from on-surface to the sub-surface position in Cu is stronger than that for Ni. Since, Ni doped with subsurface B was experimentally synthesized, [4,5] we suggest that Cu doped with subsurface B (Cu-B) is also synthesizable. The presence of the monolayer subsurface B creates corrugated step-like structure on Cu surface and significantly brings down the methane C-H activation barrier from 170 kJ/mol on Cu(111) to only 74 kJ/mol on Cu-B. One of the major issues in methane activation on other transition metals is the complete decomposition of methane to carbon; but it is not the case for Cu-B. The dissociation of surface CH_2 on Cu-B is inhibited due to the high activation barrier of 161 kJ/mol. Moreover, C-C coupling between CH_2 fragments is feasible, with the activation barrier of only 53 kJ/mol; suggesting that Methane activation on Cu-B catalyst might have higher selectivity towards olefins. Evaluation of C-H activation on Cu(110) surface which has similar step-like structure as Cu-B shows that the geometrical/corrugation effect of Cu-B surface is not solely responsible for its high activity. Further analysis on Bader charge and Density of states reveal that there is a charge transfer from B to Cu, which synergistically promotes the C-H activation together with the geometrical/corrugation effect and makes the Cu-B as active for C-H activation as other expensive transition metals (Rh and Ni).[6]

References:

1. Lunsford, J. H. Catal. Today **2000**, 63, 165.
2. Tang, P.; Zhu, Q.; Wu, Z.; Ma, D. *Energy Environ. Sci.* **2014**, *7*, 2580.
3. Varghese, J. J.; Trinh, Q. T.; Mushrif, S. H. *Catal. Sci. & Technol.* **2016**. Advance article. **DOI**: 10.1039/C5CY01784J.
4. Xu, J.; Chen, L.; Tan, K. F.; Borgna, A.; Saeys, M. *J. Catal.* **2009**, *261*, 158.
5. Trinh, Q. T.; Nguyen, A. V.; Huynh, D. C.; Pham, T. H.; Mushrif, S. H. *Catal. Sci. & Technol.* **2016**, in press, **DOI:** 10.1039/C6CY00358C.
6. Zhao, Z.-J.; Chiu, C.-C.; Gong, J. *Chem. Sci.* **2015**, *6*, 4403.

Copyright © American Institute of Chemical Engineers. All rights reserved.

454388 Combined DFT and Kinetic Monte Carlo Studies of Formic Acid Decomposition on Au_{18} Nanoclusters

Abstract:

Formic acid (FA) is a promising non-toxic, organic hydrogen (H_2) storage material (4.4 wt. % H_2). Being an abundant byproduct when producing levulinic acid from biomass, it has been envisioned as an *in situ* H_2 source for converting levulinic acid to γ-valerolactone, which is a valuable precursor for production of various fuels and chemicals.[2] Additionally, it may also be utilized in fuel cells, e.g. in the direct formic acid fuel cell.[3] To unlock its potential however, we need to find ways to quickly and selectively release the stored hydrogen from FA. Using Au/SiC catalysts, Mavrikakis and co-workers showed experimentally that FA decomposition activity increases with decreasing Au particle size.[4] Based on detailed Density Functional Theory (DFT) calculations, they concluded that extended Au(111)/(211)/(100) surfaces were not the active sites for the reaction, and that sub-nanometer Au clusters with highly undercoordinated sites were likely to be responsible for FA decomposition activity.[4] Based on these findings, we calculated potential energy surfaces for FA decomposition on Au_n (n=3–25) clusters using DFT. It was found that Au_{18} had a potential energy profile that could best reproduce the experimental results when placed in a microkinetic model (MKM), and thus likely contains the active site for the reaction.

In this work, we further probe the activity of Au_{18} clusters by performing comprehensive Kinetic Monte Carlo (KMC) simulations with our DFT results. KMC simulations outrank mean-field models in terms of fidelity, by being able to capture spatial inhomogeneities in the site distribution of Au_{18}, as well as correlations due to adsorbate ordering. We indeed noticed strong adsorbate-adsorbate interactions for this system. To factor in their effects on the reaction energetics, we employ a cluster expansion, parameterizing the interaction energies of adsorbates in terms of point and multi-body interactions.[5]

Our simulations reveal that the TOF of H_2 on Au_{18} is much closer to the experimental TOF compared to extended Au(211)/(100) surfaces,[4] and is thus a likely model of the active site for FA decomposition. A triangular ensemble of three Au atoms, each with a coordination number of five (CN=5 sites), is found to be the active site. Our work will offer valuable insights into the molecular pathways of FA decomposition, and serve to guide future efforts to optimize the shape and size of Au catalysts for the FA decomposition reaction.

References

1. Sipilä, K., Kuoppala, E., Fagernäs, L. & Oasmaa, A. Characterization of biomass-based flash pyrolysis oils. *Biomass and Bioenergy* **14,** 103–113 (1998).

2. Alonso, D. M., Bond, J. Q. & Dumesic, J. A. Catalytic conversion of biomass to biofuels. *Green Chem.* **12,** 1493–1513 (2010).

3. Alonso, D. M., Wettstein, S. G. & Dumesic, J. A. Bimetallic catalysts for upgrading of biomass to fuels and chemicals. *Chem. Soc. Rev.* **41,** 8075–8098 (2012).

4. Singh, S. *et al.* Formic acid decomposition on Au catalysts: DFT, microkinetic modeling, and reaction kinetics experiments. *AIChE J.* **60,** 1303–1319 (2014).

5. Schmidt, D. J., Chen, W., Wolverton, C. & Schneider, W. F. Performance of cluster expansions of coverage-dependent adsorption of atomic oxygen on Pt(111). *J. Chem. Theory Comput.* **8,** 264–273 (2012).

Copyright © American Institute of Chemical Engineers. All rights reserved.

467699 Mechanistic Study of C-O Â Bond Rupture on Ruthenium Phosphide Surface: Transient and Steady State Measurements

Abstract:

Transition metal phosphide (TMP) catalysts are selective and active towards C-O bond rupture during oxygenate hydrodeoxygenation (HDO), however, the manner in which C-O bond rupture mechanisms and intrinsic barriers differ between transition metals and TMP are not well understood. Here, we present results from the decomposition of formic acid (e.g., DCOOH, a model oxygenate) on pristine and P-modified Ru(0001) surfaces using a combination of surface science methods. DCOOH adsorbs on Ru(0001) and forms a formate intermediate, which decomposes either by dehydration, initiated by C-O bond rupture followed by C-H bond rupture to form CO, or by dehydrogenation through a direct C-H bond rupture to form CO_2. The P_x-Ru(0001) surfaces (x is the coverage of P-atoms in monolayer equivalents) are prepared by dosing PH_3 gas onto Ru(0001) followed by a 1200 K flash anneal. The composition and structure of the resulting surfaces are characterized by Auger electron spectroscopy and low-energy electron diffraction. P_x-Ru(0001) surfaces are chemically characterized using CO and NH_3 temperature programmed desorption (TPD). TPD results demonstrate that the addition of P-atoms to Ru(0001) decreases the binding energy of CO by 12 Â± 2 kJ mol^{-1} at Î¸ < 0.33 ML and that of NH_3 by 11 Â± 2 kJ mol^{-1} at Î¸ < 0.1 ML compared with Ru(0001). These TPD data suggest that P-atoms introduce an electronic effect which decreases the extent of electron exchange between Ru(0001) surface atoms and adsorbates, which is consistent with density function theory (DFT) calculations for Ni_2P that show net charge transfer from Ni atoms to P-atoms.[1] Temperature programmed reaction (TPR) and reactive molecular beam scattering (RMBS) of DCOOH were used to determine how barriers and selectivities for C-O bond rupture differ under transient and steady-state conditions on Ru(0001) and P_x-Ru(0001) surfaces. TPR spectra of DCOOH show that formate intermediates decompose at higher temperatures on $P_{0.43}$-Ru(0001) and have activation barriers of 14 ± 2 kJ mol^{-1} greater than on Ru(0001). These differences suggest that the electronic effect of P-atom decreases the tendency for electrons to back donate from Ru atoms to anti-bonding orbitals of the formate intermediate, and thus, increases activation energies. P-atom addition may, however, also enhance the stability of formate intermediates by geometric effects. TPR spectra of DCOOH show that the ratio of CO to CO_2 produced decreases as P to Ru increases, which suggests that C-O bond rupture is affected by the addition of P-atoms more than C-H bond rupture. RMBS of DCOOH on Ru(0001) and $P_{0.43}$-Ru(0001) surfaces support this finding and show that activation energies of C-O bond rupture and C-H bond rupture (measured from 500 - 700 K) are greater on $P_{0.43}$-Ru(0001) by 28 ± 5 kJ mol^{-1} and 11 ± 7 kJ mol^{-1}, respectively, than on Ru(0001). Steady-state CO to CO_2ratios, measured as a function of temperature, are consistent with those determined by TPR of DCOOH, and together, these data demonstrate that the addition of P-atoms to Ru(0001) favors dehydration pathway at temperature above ~ 475 K compared to Ru(0001). In summary, P-atom addition to Ru(0001) increases the activation energy of C-O bond rupture and alters the metal surface-adsorbate interaction via electronic and/or geometric effects that stabilize formate intermediates. These findings may provide useful information for the rational design of TMP catalysts by enhancing C-O bond rupture, which will increase bio-mass conversion efficiency to platform chemicals.

(1) Liu, P.; Rodriguez, J. A.; Asakura, T.; Gomes, J.; Nakamura, K. *J. Phys. Chem. B* **2005**, *109*, 4575.

Copyright © American Institute of Chemical Engineers. All rights reserved.

471953 Active Site Determination for CO Oxidation on Al_2O_3 Supported Pt Nanoparticle Catalysts By *in-Situ* Quantitative FTIR Measurements

Abstract:

CO Oxidation on noble metal surfaces has been widely studied since the 1920â€™s and is of fundamental importance to the field of surface science and heterogeneous catalysis in addition to many important industrial applications. Detailed kinetics studies suggest that at low temperatures (<250Â°C), the reaction proceeds via Langmuir-Hinshelwood kinetics, where surface is nearly covered with CO and the reaction rate is limited by the availability of vacant Pt sites for O_2 dissociation. Interestingly, CO oxidation on supported Pt nanoparticles under CO-poisoned conditions has been shown to be a structure insensitive reaction, where turnover frequencies do not change significantly with variations in nanoparticle size and as a result of surface structure. However, theoretical analyses of the proposed CO oxidation mechanism on Pt surfaces have predicted the reaction should exhibit structure sensitivity, where lower index facets of Pt are predicted to be more active.

Recently it has been shown through *in-situ* Transmission electron microscopy (TEM) and X-ray absorption spectroscopy (XAS) that the presence of CO can induce significant Pt nanoparticle restructuring. However, the potential role of restructuring in explaining the anomalous structure insensitivity for CO oxidation on Pt, has not been identified, due to the lack of a quantitative correlation between the concentrations of various types of active sites measured under reaction conditions to the areal reaction rate. In this work, we utilize *in-situ* quantitative DRIFTS measurements to measure the concentration of CO bound to under-coordinated (UC) and well-coordinated (WC) Pt sites for various Pt nanoparticle catalysts with sizes ranging from 2-20 nm on oxide supports at room temperature and under typical reaction conditions. This allows for a correlated analysis relating the turnover frequency of CO oxidation over Pt catalysts and the concentration of WC and UC Pt sites. The results demonstrate that CO oxidation on Pt nanoparticle surfaces is structure sensitive, where WC Pt atoms are the active site, but that this effect is masked by

adsorbate-induced restructuring of Pt nanoparticle surfaces under CO oxidation conditions. We also investigate the kinetics of the temperature-dependent and transient evolution of CO-induced catalyst reconstruction to gain insights into the activation of catalyst restructuring. Our results provide a complete description of the observed trends in the structure sensitivity of CO oxidation on Pt nanoparticles of various sizes, in agreement with theoretical predictions.

Copyright © American Institute of Chemical Engineers. All rights reserved.

467678 Competing Forces in Chiral Catalytic Surface Chemistry: Enantiospecificity Versus Enantiomer Disproportionation

Abstract:

Equilibrium adsorption of gas phase mixtures of D- and L-alanine (Ala), serine (Ser), phenylalanine (Phe) and aspartic acid (Asp) onto the naturally chiral Cu{3,1,17}$^{R\&S}$ and achiral Cu(111) and Cu(100) surfaces has been studied by both experiment and DFT-based modeling. Isotopically labelled *L-amino acids ($HO_2{}^{13}CCH(NH_2)$-R) and unlabeled D-amino acids allow mass spectrometric enantiodifferentiation of the adsorbed species during temperature programmed decomposition, following equilibrium adsorption. Measurements of the relative equilibrium coverages of D- and *L-amino acids on the Cu{3,1,17}$^{R\&S}$ surfaces, $q_{D/R}/q_{*L/R} = q_{*L/S}/q_{D/S}$, at gas phase partial pressure ratios of P_{*L}/P_D = 0.5, 1, and 2 indicate that the D-Ala and *L-Ala conglomerate phases are more energetically stable than a D*L-Ala racemate phase, but that their adsorption energies are not measurably enantiospecific, DDE_{DL} » 0. In contrast, the adsorption of D-Asp and *L-Asp on the Cu{3,1,17}$^{R\&S}$ surfaces exhibits measurable enantiospecificity, DDE_{DL} » 3 kJ/mole. The surprising observation is that even on achiral surfaces such as Cu{111}, exposure to non-racemic mixtures of D-Asp and *L-Asp in the gas phase, P_{*L}/P_D = 0.5 or 2, can result in dramatic enhancement of enantiomeric excess on the surface, q_D/q_{*L} = 0.06 and 16, respectively.

In general, we show that the enantiospecific adsorption of chiral molecules on chiral surfaces is dictated by two competing forces the enantiospecificity of adsorption energetics and the propensity of enantiomers to disproportionate into homochiral (conglomerate) or heterochiral (racemate) clusters. These phenomena have been studied by measuring the surface enantiomeric excess, ee_s, of mixtures of chiral amino acids adsorbed on Cu surfaces and in equilibrium with gas phase mixtures of varying enantiomeric excess, ee_g. Ala adsorption on Cu{3,1,17}$^{R\&S}$ surfaces is non-enantiospecific, $ee_s = ee_g$, because Ala enantiomers do not interact with either the surface or one another enantiospecifically. Asp adsorbs enantiospecifically on the Cu{3,1,17}$^{R\&S}$ surfaces; ee_s ' ee_g, even during exposure to a racemic mixture in the gas phase, $ee_g = 0$. Exposure of the achiral Cu{111} surface to non-racemic Asp, ee_g ' 0, results in local amplification of enantiomeric excess, $|ee_s| > |ee_g|$, as a result of homochiral disproportionation. Finally, in spite of the fact that the Cu{653}$^{R\&S}$ surface are chiral, the adsorption of Asp mixtures is dominated by homochiral disproportionation of adsorbed enantiomers rather than enantiospecific adsorbate-surface interactions, $|ee_s| > |ee_g|$. All of these types of behavior are captured by a Langmuir-like adsorption isotherm that also describes competition between enantiospecific adsorption and both homochiral (conglomerate) and heterochiral (racemate) clustering of adsorbed molecules (Figure).

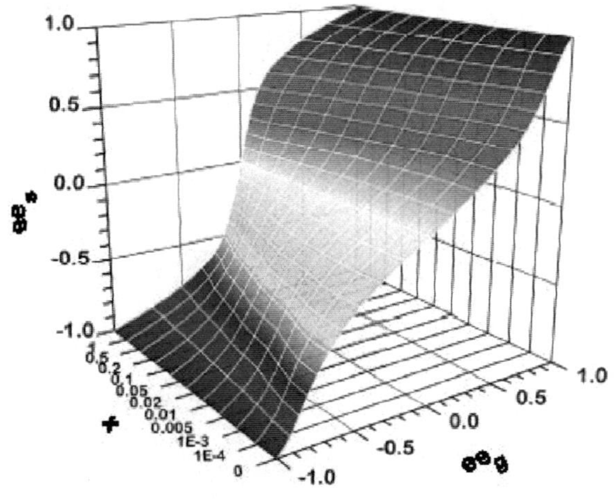

Figure. The surface enantiomeric excess, ee_s, versus gas phase enantiomeric excess, ee_g, for adsorption of enantiomer mixture under conditions in which the dominant surface interactions vary from disproportionation into conglomerate clusters, x = 0, to disproportionation into racemate clusters, x = 1.

Copyright © American Institute of Chemical Engineers. All rights reserved.

453354 On the Reaction Pathways and Intermediates of Selective Ringopening of Furanics By Iridium

Abstract:

As the chemical economy gradually shifts from fossil fuels to renewable resources, it becomes important to develop technologies to upgrade renewable platform molecules to final products of industrial relevance. It has been shown the renewable heterocyclic molecules 5-hydroxymethylfurfural and furfural, can be produced in high yields by depolymerizing and dehydrating lignocellulosic biomass. These heterocyclics can undergo reduction of their side groups with a subsequent hydrogenolysis to produce liquid fuel additives and value added products for the chemical industry. Selective hydrogenolysis of these heterocyclic molecules at the C–O bond in the aromatic ring has been shown to be a viable route towards the production of long-chain hydrocarbons, alcohols, and polyols. The selective hydrogenolysis of furfural can result in the production of 1,5-pentanediol (1,5-PeD), a valuable precursor for the production of polyesters and polyurethanes, with current research showing iridium to have a high selectivity for this process.

While the hydrogenolsysis of cyclic compounds to long chain alcohols and polyols has been known for decades, the factors that control the selectivity remain either uknown or debated. It has long been thought the hydrogenolysis of aromatic cyclic compounds proceeds after saturation of the aromatic ring; however recent studies have shown that full ring saturation is not required. Futhermore, any aromatic ring side groups are capable of undergoing reduction and oxidation, and the impact of such reactions on the selectivity are unknown. In the current study, we combine density functional theory and high resolution electron energy loss spectroscopy (HREELS) to probe the hydrogenolysis of saturated and unsaturated heterocylic molecules. The reduction of the aromatic ring side group in conjunction with the selective hydrogenation of the aromatic ring shifts the selectivity towards either terminal or secondary diols, with the selectivity being determined via the rearrangement of the electronic density of the aromatic π-system. Furthermore, there is a competition between the kinetics and thermodynamics governing the hydrogenolysis, indicating a strong temperature and time effect on the selectivity. Comparison of DFT and HREELS spectra allow for several key intermediates to be identified and characterized on the Ir(111). These insights may allow for the development of more efficient catalysts and processes that can render renewable chemicals commercial.

Copyright © American Institute of Chemical Engineers. All rights reserved.

470935 Controllable Solution Combustion Synthesis of Nanoscale α/β-Bi_2O_3 and Its Catalytic Application

Abstract:

Among the four different polymorphs of bismuth trioxide, α- and β- Bi_2O_3 show unique physical and chemical properties . In the present work, Bi_2O_3 nanoparticles of different phases (α and β) were selectively synthesized via simple, rapid and energy efficient solution combustion synthesis (SCS). Bismuth nitrate was used as the oxidizer, while tartaric acid, glycine, EDTA and sucrose were employed as fuels. By changing types of fuels, ratios of fuel/oxidizer and heating temperature, specific phases (α and β) of Bi_2O_3 were controlled and particle sizes (5-50 nm) of these materials were tailored. The formation mechanism was proposed by thermodynamic calculations. Stable α-Bi_2O_3 forms when the combustion temperature is higher than 350 °C while metastable β-Bi_2O_3 forms at temperature lower than 300 °C. Using the SCS products as catalytic supports, series of Pt/SCS-Bi_2O_3 catalysts were prepared and tested for glycerol oxidation. These results were compared to our prior studies of Pt-Bi catalysts placed on activated carbon and other supports.

Copyright © American Institute of Chemical Engineers. All rights reserved.

472321 A Combined Theoretical and Experimental Investigation on the Degradation of Organophosphorus Chemical Warfare Agents on ZnO_{1-X}

Abstract:

Metal oxide nanoparticles (MONP) have long been known to act as reactive sorbents for the degradation of chemical warfare agents (CWA). However, contrary to nature, which is capable of producing enzymes that can detoxify the same CWA at ambient conditions, MONP are stoichiometric and require elevated temperatures in order to remove the products from the surface. Herein vacancy induced MONP are examined for the degradation of organophosphorus agents by both experimental and theoretical means. Oxygen vacancies are introduced into MONP of zinc oxide (ZnO_{1-X}) and characterized by Raman spectroscopy, x-ray photospectroscopy (XPS), UV-vis spectroscopy, transmission electron microscopy (TEM) and x-ray diffraction (XRD). The reactivity of ZnO_{1-X} is evaluated using diisopropyl fluorophosphate (DFP) with solid state NMR, indicating that the vacancy induce structures exhibit a 25% increase in reactivity compared to the parent MONP. In parallel, a detailed mechanistic study revealing the path of degradation is investigated utilizing plane-wave density functional theory (DFT). Comparative studies are performed evaluating the stability of an oxygen vacancy on either the 002 or 100 surface planes of zinc oxide, as well as in the bulk and surface, revealing the latter to be the most stable. Binding energies were calculated using phosphine as a surrogate molecule to represent the surface reaction between ZnO_{1-X} and organophosphorus CWA's. The results obtained here not only provide a mechanistic insight towards CWA degradation, but will aid in the future development of more efficient catalyst against CWA's.

Copyright © American Institute of Chemical Engineers. All rights reserved.

JOC The Journal of Organic Chemistry

Article

pubs.acs.org/joc

Magnetic Superbasic Proton Sponges Are Readily Removed and Permit Direct Product Isolation

Elia M. Schneider, Renzo A. Raso, Corinne J. Hofer, Martin Zeltner, Robert D. Stettler, Samuel C. Hess, Robert N. Grass, and Wendelin J. Stark*

Institute for Chemical and Bioengineering, Department of Chemistry and Applied Biosciences, ETH Zürich, Wolfgang-Pauli Strasse 10, CH-8093 Zurich, Switzerland

Ⓢ *Supporting Information*

ABSTRACT: Workup in organic synthesis can be very time-consuming, particularly when using reagents with both a solubility similar to that of the desired products and a tendency not to crystallize. In this respect, reactions involving organic bases would strongly benefit from a tremendously simplified separation process. Therefore, we synthesized a derivative of the superbasic proton sponge 1,8-bis(dimethylamino)naphthalene (DMAN) and covalently linked it to the strongest currently available nanomagnets based on carbon-coated cobalt metal nanoparticles. The immobilized magnetic superbase reagent was tested in Knoevenagel- and Claisen−Schmidt-type condensations and showed conversions of up to 99%. High yields of up to 97% isolated product could be obtained by simple recrystallization without using column chromatography. Recycling the catalyst was simple and fast with an insignificant decrease in catalytic activity.

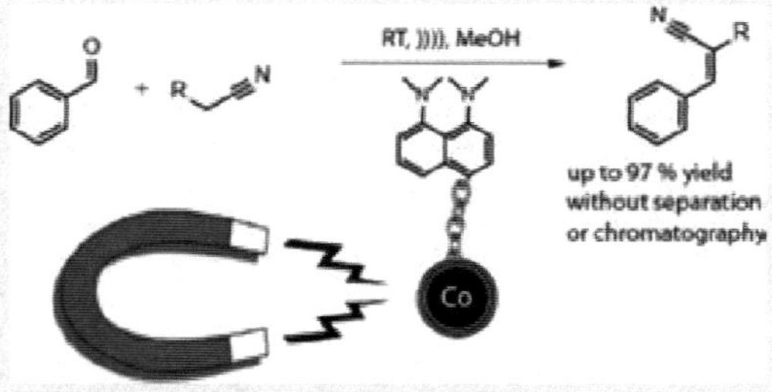

■ INTRODUCTION

Organosuperbases, such as the classic proton sponge 1,8-bis(dimethylamino)naphthalene (DMAN),[4] Verkade's base (proazaphosphatrane), and 1,8-bis(tetramethylguanidino)-naphthalene (TMGN) have become very important reagents in organic chemistry in past years.[7] Their exceptional basicity is associated with high levels of kinetic activity in proton-exchange reactions. These properties are often manifested by low nucleophilicity, making this kind of compound interesting for a wide scope of reactions,[8−11] and even heterogeneous cases are known.[1,2] However, the workup of such reactions on the laboratory level can be difficult, time-consuming, and expensive. Similarly, in industry, a broad variety of condensation reactions are applied. At present, free bases, such as sodium hydroxide, potassium hydroxide, and various organic bases, are the most commonly used, and they often lead to corrosion, significant waste production, and difficult workup. Thus, a chemically stable magnetic base with the attributes necessary for rapid quantitative separation would be very useful for industrial applications, resulting in significant solvent savings, a reduced amount of time spent, a decreased need for expensive equipment, and reusability of the separation reagent. Moreover, a ready-to-separate reagent is interesting for its potential use in the synthesis of high-quality products, where impurity carryover defines product performance. Magnetic nanoparticles have fascinated scientists for several decades and have been used in a plethora of applications,[14] such as for drug delivery,[15] in cancer treatments,[16] and as contrast agents for magnetic resonance imaging.[17] In chemical synthesis, magnetic nanoparticles have recently gained attention in the field of catalysis[18] because they combine a large surface area with the potential for a simple separation process. However, many magnetic reagents and surface linkers for reagent attachment are unstable in acidic-, basic-, or organic-solvent-containing reaction media. The exceptional chemical stability of carbon-coated metal nanoparticles[19] is based on a crystalline, graphene-like carbon surface that effectively prevents core oxidation. The all-carbon surface further allows covalent functionalization of the particle surface by using commercially available aryl diazonium salts.[20] At present, a number of promising recyclable, stable, metal-based magnetic catalysts have been proposed.[21−31] Most recently, alkene hydrogenation palladium catalysts[32,33] (so-called "catch-and-release" systems[34]) have been developed. In this work, we present an organic superbase, coupled to magnetic nanoparticles, with stability amenable for use under such challenging reaction conditions. We also show that such an easy-to-separate reagent is useful in a number of condensation reactions and simplifies workup and product isolation.

■ RESULTS AND DISCUSSION

Preparation and Characterization of the Magnetic Base. Magnetic, nanosized DMAN was prepared from chemically modified magnetic nanoparticles, a linker, and a derivative of DMAN. For covalent attachment of the base to the magnetic particle (**1**), the bifunctional linker hexamethylene diisocyanate (**4**) was first reacted with the primary amine of 4′-

Received: August 18, 2014

benzylamine-derivatized, graphene-coated metal nanoparticles (C/Co@amine) (3) to yield C/Co@hexamethyleneisocyanate (5). This intermediate contains a stable *N*-alkylurea bond and an active isocyanate moiety. Subsequently, C/Co@hexamethyleneisocyanate (5) was reacted in situ with 4-amino-1,8-bis(dimethylamino)naphthalene (DMAN-NH_2, 6), which was prepared according to the literature.[35] This afforded a magnetic, covalently bound DMAN (C/Co@DMAN (7)) in proper yield with a base capacity of 0.11 ± 0.01 mmol g^{-1}. For an illustration of the synthesis, see Scheme 1. The chemical identity of the herein synthesized reagent C/Co@DMAN (7) was proven by diffuse reflectance infrared Fourier transform spectroscopy (DRIFTS) and elemental microanalysis (C, H, N; Figure S1 and Table S1). The fairly strong urea-stretching vibration at 1690 cm^{-1} clearly confirms urea formation in the case of C/Co@hexamethyleneisocyanate (5). The successful attachment of DMAN to the isocyanate linker of C/Co@hexamethyleneisocyanate (5) was investigated with IR spectroscopy and elemental microanalysis (Table S1). A substantial increase in nitrogen content ($\Delta N = 0.82\%$) as well as an increase in carbon content ($\Delta C = 3.43\%$) is in line with successful functionalization. Vibrating sample magnetometry (VSM) measurements (at room temperature) showed an overall magnetization of 139.6 emu g^{-1}, which is in good agreement with the expected values, namely, a little lower than the saturation magnetization of nonfunctionalized C/Co (lower content of organics or carbon; 158 emu g^{-1}).[36] It is notable that the saturation magnetization of this material is still much higher than even the strongest conventional magnetite–silica nanoparticles (30–50 emu g^{-1}).[37]

The high saturation magnetization is of relevance because it permits rapid and easy separation.[38] To confirm the chemical stability and robustness of C/Co@DMAN (7), we recorded transmission electron microscopy (TEM) images of nanocarrier system 7 before and after reaction runs. Particles remained spherical and uniformly sized, in the diameter range of 20 to 60 nm (Figure S2). The reactivity of the herein proposed magnetic base C/Co@DMAN (7) was compared to reactions using similar catalysts from the literature. Another reference material, polystyrene-based resin-supported DMAN (**S2**), was synthesized (Scheme S2), fully characterized (Figure S3 and Table S2), and used for comparison.

Reliable recycling and separation of a reagent requires a robust confirmation of covalent attachment. Therefore, we used several test reactions (Scheme S1) to prove the covalent nature of DMAN's attachment to the nanomagnets. As a first control experiment, C/Co@hexamethyleneisocyanate (5), which bears one free, active isocyanate moiety, was reacted with pure (not aminated) DMAN. The lack of a free amine group in DMAN does not allow covalent attachment. Using the same synthesis and purification/washing procedure as for the herein proposed reagents, product analysis (Table S1) revealed the absence of both nitrogen incorporation and physisorption effects. As a separate control experiment, C/Co@hexamethylene-isocyanate (5) was first quenched with water, leading to a primary amine (i.e., the linker was deliberately rendered nonfunctional). Nonfunctional **5** was then exposed to functionalized, amine-bearing DMAN derivative **6**; product analysis after applying the same purification/washing procedure as in the preparation of the functional magnetic base again confirmed the absence of reactivity (Table S1). These two experiments show that both the linker and the amination of the DMAN are necessary for the covalent binding of a DMAN moiety to the nanomagnet.

Scheme 1. Synthesis of Immobilized Magnetic Superbase 7[a]

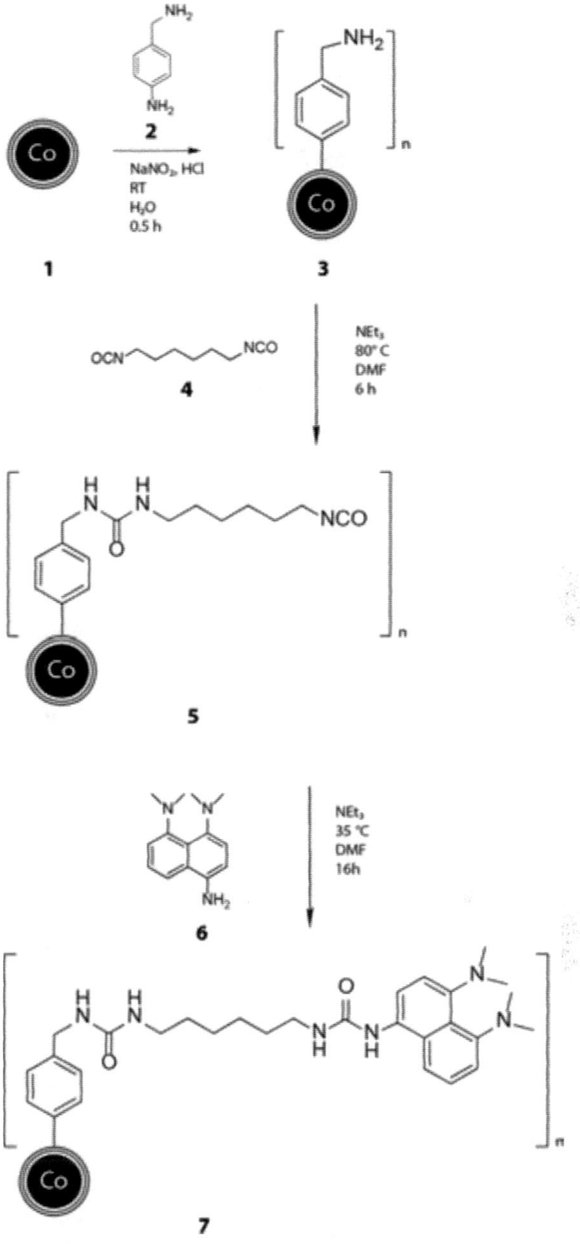

[a]C/Co (1) was coupled to 4-aminobenzylamine (2) via diazonium chemistry to yield C/Co@amine (3), which was further reacted with hexamethylendiisocyanate (4), resulting in C/Co@hexamethyleneisocyanate (5), bearing an active isocyanate moiety. C/Co@hexamethyleneisocyanate (5) was coupled with 4-amino-1,8-bis(dimethylamino)naphthalene (6) to yield C/Co@DMAN (7).

The experiments further show the absence of base physisorption, a potential experimental error in such reagent-anchoring studies.

Knoevenagel and Claisen–Schmidt Condensation. The literature states that DMAN indeed has catalytic activity in Knoevenagel condensation.[39] After identity confirmation, the catalytic activity of C/Co@DMAN (7) for the Knoevenagel

Table 1. Knoevenagel Condensation of Benzaldehyde and Malononitrile under Different Reactions Conditions

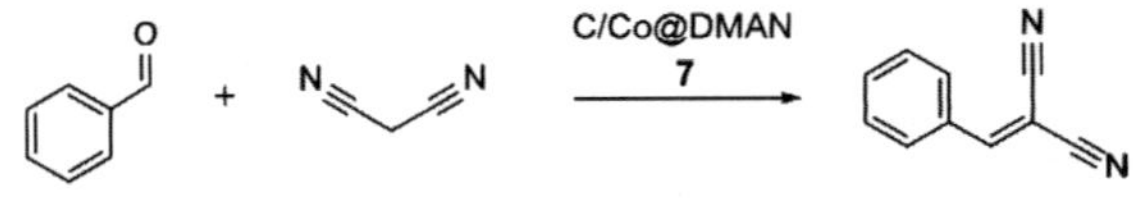

8 9 10

entry	catalyst	catalytic amount	solvent	time (h)	conversion (%)[a]	isolated yield (%)[b]	purity (%)
1	7	2 mmol %	toluene	7.5	63	61	99
2	7	2 mmol %	H_2O	7.5	99 (2)[c]	97	94
3	7	2 mmol %	MeOH	7	93 (85)	89	95
4	7	2 mmol %	MeOH[d]	4.5	98 (95)	91	96
5	7	0.2 mmol %	MeOH[d]	6	92 (90)	91	92
6	DMAN[e]	2 mmol %	MeOH	6	99	63[f]	96
7	S2	2 mmol %	toluene	6	99	97	98
8			MeOH	5.5	30	g	
9	C/Co	100 mg	MeOH	7.5	90	77	86
10	$CoCl_2$	5 mg	MeOH	6	33	g	

[a]Conversion was determined by HPLC with the reference product as the standard (benzaldehyde/malononitrile = 1:1). [b]Isolated yield after recrystallization. [c]Numbers in parentheses represent the percent conversion without washing the particles once with 3 mL of toluene. [d]Reaction carried out using an ultrasound bath. [e]Experiment was carried out using 2 mmol % of the free base. [f]Isolated yield after column chromatography (DCM/MeOH = 20:1). [g]No direct product isolation possible without column chromatography.

Table 2. Knoevenagel Condensation with Various Substrates Catalyzed by C/Co@DMAN (7)

entry[a]	basic substrate	aldehyde / ketone	product	temperature (°C)	conversion (%)[b]	isolated yield (%)
1	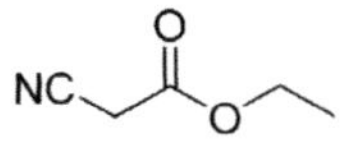	8	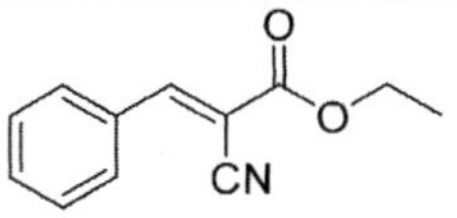	40	96[c]	95 [99][d]
2	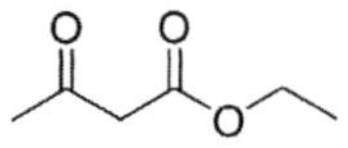	"	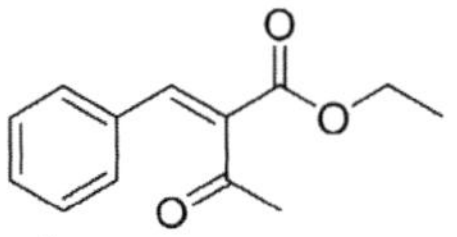	40	7	-[e]
3	9	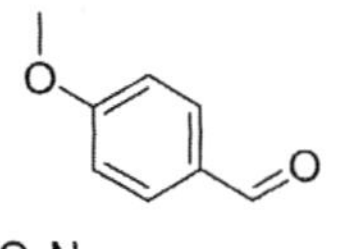	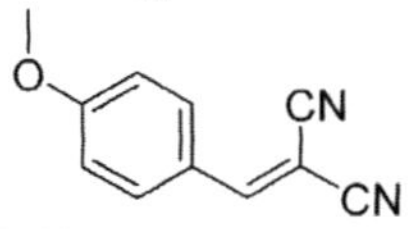	40	99	96 [91][d]
4	"	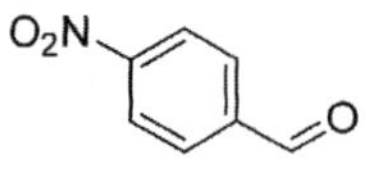	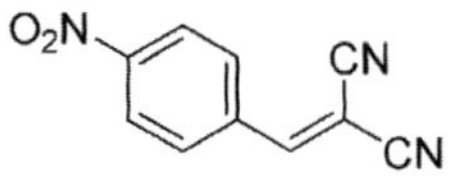	40	83	73[f] [96][d]
5	"	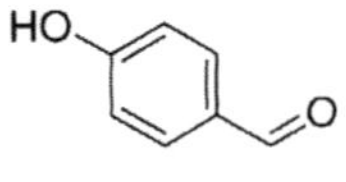	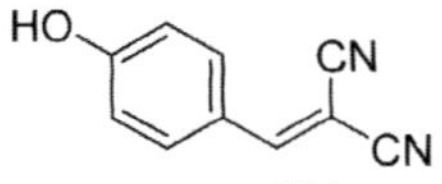	40	8	-[e]
6	"	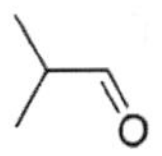	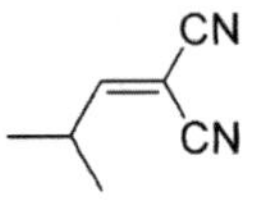	RT	81	-[e]
7	"	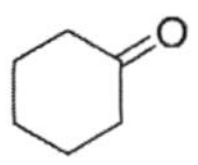	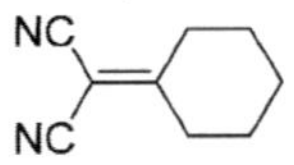	40	37	-[e]

[a]Reaction conditions: catalyst loading, 2 mmol %; basic substrate, 0.4 mmol; aldehyde or ketone, 0.4 mmol; MeOH, 5 mL; 16 h. [b]GC conversion. [c]After three consecutive cycles, conversion was still 95%. [d]Purity detected by NMR is given in brackets []. [e]No direct product isolation was possible without column chromatography. [f]Isolated yield after recrystallization (DCM/hexane).

reaction of benzaldehyde (8) with malononitrile (9) was evaluated in different solvents at room temperature (Table 1, entries 1−3). If toluene was used as a solvent, then only 61% yield could be obtained. However, in water, quantitative conversion was achieved after 7.5 h. Still, the catalyst had to be washed once with toluene to release the product from the carbene surface. Therefore, methanol was tested as a solvent with promising results (Table 1, entries 3−5). The use of ultrasound during catalysis is known to ensure good reagent dispersion and consistently leads to excellent yields at shorter contact times. Indeed, when the reaction was conducted in an ultrasonication bath, a higher yield (91%) was obtained in a shorter time span (Table 1, entry 4). If the catalyst loading was decreased 10-fold, then similar yields were detected but a longer reaction time was necessary (Table 1, entry 5). Alternatively, running the same reaction using free DMAN where product workup and purification were not considered, quantitative conversion was obtained after 6 h (Table 1, entry 6). However, column chromatography had to be used to separate the catalyst residues from the product, which resulted in a significant drop in isolated yield. Magnetic base C/Co@DMAN (7) was separated within seconds, and recrystallization of the crude product (leading to the pure product) was achieved within minutes. Using polystyrene-immobilized proton sponge **S2** gave good yields as well, but filtration of the polymer was tedious and not very solvent-efficient. It is remarkable that even in the absence of any catalysts for the chosen reaction a low conversion (30%) was detected after 5.5 h (Table 1, entry 8). More surprisingly, the chosen reaction showed enhanced conversion (90%) in the presence of nonmodified (i.e., naked carbon surface) nanoparticles (Table 1, entry 9). To study further this unexpected finding, we included additional experiments. One probable hypothesis is based on leached cobalt ions functioning as a source of activity (hypothesis 1). Also plausible is that the carbon surface itself acts as an active catalyst (hypothesis 2). To exclude the first hypothesis, the reaction was carried out with a deliberate addition of cobalt(II) chloride ($CoCl_2$; Table 1, entry 10), the species that is supposedly leaching from the particles.[40] The addition of cobalt salts did not significantly affect the conversion (Table 1, entry 10; conversion = 33%); therefore, we do not assign significant activity of ionic cobalt in this kind of reaction. With regard to the second hypothesis, however, it is known that carbon-based surfaces (e.g., graphene, carbon nanotubes (CNT), carbon nanorods (CNR), etc.) indeed have been reported to support catalytic reactions (so-called carbocatalysis).[41,42] Thus, our finding is in line with similar carbon-surface activity and is subject to further extended studies by our group.

In the next part of the investigation, we tested different substrates (Table 2, entries 1−7). Ethyl cyanoacetate (Table 2, entry 1) reacted smoothly with benzaldehyde, and the product (ethyl α-cyanocinnamate) was detected in high yield and purity (96 and 99%, respectively) after 16 h at 40 °C; even after three consecutive cycles, the conversion was similarly high (95%). The less acidic and thus more demanding substrate ethyl acetoacetate (Table 2, entry 2) was not very active, and only low yields could be obtained. Also, different aldehydes and ketones were reacted with malononitrile. 4-Methoxy benzaldehyde performed very well (Table 2, entry 2), whereas the electron-poor nitro substrate was slightly less active (Table 2, entry 3). Surprisingly, 4-methoxy benzaldehyde reacted quickly, whereas 4-hydroxy benzaldehyde showed a slow conversion rate (Table 2, entry 6). An explanation may be the formation of hydrogen bonds with the solvent; one can imagine a shell of solvent molecules shielding the substrate from the catalyst. Nonaromatic substrates such as isobutyraldehyde and cyclohexanone were tested as well (Table 2, entries 6 and 7), and isobutyraldehyde performed better because of steric reasons (i.e., hindrance of nucleophilic attack on ketones).

Kinetic plots (Figures 1 and 2) confirmed the different reaction rates of C/Co@DMAN (7) when different solvents and mixing conditions (stirring vs sonication) were used. As expected, the reaction rate increased when the superbase was mixed more thoroughly with the reactants (i.e., a higher degree of mixing). To put the performance of C/Co@DMAN (7) in an appropriate context and to provide a fair comparison, we measured reaction kinetics for free DMAN (no mass-transfer limitation expected) and polystyrene-bound superbase **S2** (under optimal conditions in toluene) (Figure 2). C/Co@DMAN (7) initially catalyzes the chosen reaction at least as well as free DMAN, which leads to comparable or even slightly higher efficiency after 1 h of reaction time. The conversions converge with time, leading to high final conversion for all catalysts. The immobilized magnetic base C/Co@DMAN (7) can therefore be considered tp be directly competitive with existing or alternative catalytic systems. The minor changes in activity (i.e., rapid first part) may be a result of the altered electronic structure (note that DMAN was derivatized at position 4 to anchor the nanoparticle).

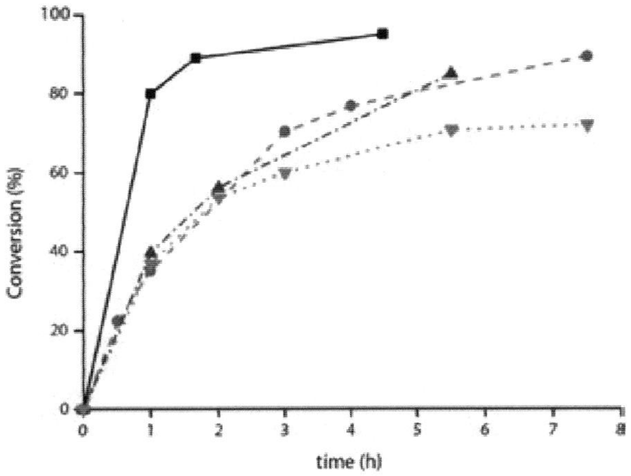

Figure 1. Kinetic plots for the Knoevenagel condensation of benzaldehyde (1.5 equiv) with malononitrile (1 equiv). All reactions used 2.5 mmol % 7 unless otherwise specified. Solvents and conditions for each reaction include ■, methanol and ultrasonication; ●, toluene and stirring; ▲, methanol and stirring; and ▼, C/Co instead of 7, methanol and stirring.

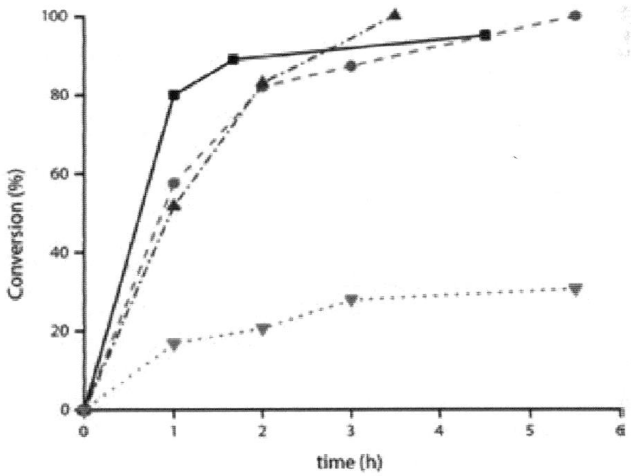

Figure 2. Kinetic plots for reactions with different catalysts (2.5 mmol %) under optimal conditions for each: ■, 7 in methanol, using ultrasonication (taken from Figure 1); ●, free DMAN in methanol; ▲, polymer-bound DMAN (**S2**) in toluene; ▼, without catalyst in methanol.

To determine the relevance of and to illustrate a broad applicability of magnetic organic bases such as C/Co@DMAN (7), another type of condensation reaction (Claisen–Schmidt) between acetophenone and benzaldehyde was tested (Table 3).

Table 3. Claisen–Schmidt Condensation of Acetophenone and Benzaldehyde

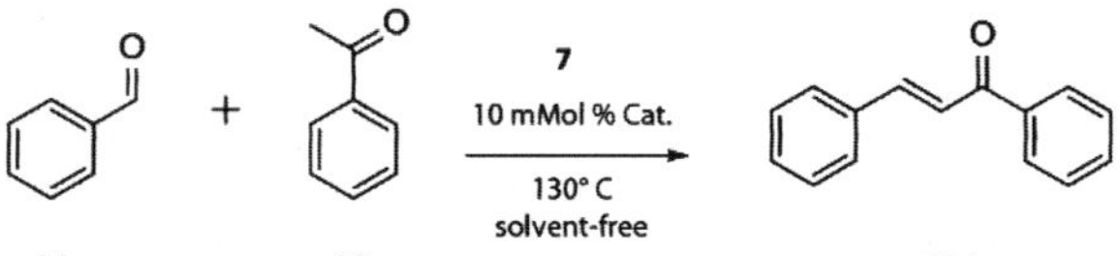

11 12 13

entry	catalyst	run	conversion (%)[a]	isolated yield (%)[b]	time (h)	Co (ppm)[c]
1	7	1	66	62 [97.6]	16	19
2	7	2	78	*d*	20	16
3	7	3	71	*d*	18	5
4	7	4	68	*d*	18	44
5	7	5	76	75 [98.6]	20	64
6	DMAN	1	<1	*e*	20	
7	Co/C	1	58	*e*	21	42
8		1	<1	*e*	21	
9	**S2**	1	<1	*e*	20	

[a]Conversion was determined by HPLC using the reference product as the standard. [b]Isolated yield after solvent evaporation; purity as detected by NMR is given in brackets []. [c]ICP-OES detection limit for Co = 0.2 ppb. [d]No product isolation has been carried out. [e]No direct product isolation was possible without column chromatography.

This well-known reaction yields *trans*-chalcone, an important precursor for flavonoids that are widely used in medicine as anti-inflammatory,[43] antiviral,[44] and anticancer agents.[45] Immobilized magnetic base C/Co@DMAN (7) catalyzed the reaction with a yield of up to 75% and high levels of purity. Additionally, the same catalyst was reused in five iterative runs with similar yields. The harsh temperature (130 °C) leads to the conclusion that C/Co@DMAN (7) indeed remains stable using covalent-linker systems (Table 3, entries 1–5). Surprisingly, free DMAN does not catalyze the above condensation reaction (Table 3, entry 6; yield = 1%). This behavior is explainable considering the pK_a values. Because acetophenone has a higher pK_a value (pK_a = 25)[46] than does DMAN (pK_a = 12.1),[2] it can be concluded that the basicity of DMAN is not the key factor governing this catalysis. Furthermore, this result is in line with earlier work in the literature. Corma et al. experimentally investigated a similar case and reported that heterogeneous base catalysts work considerably better than free bases.[47]

Comparison to Known Solid-Supported Catalytic Base Reagents. In the recent literature, a number of excellent examples describe immobilized base catalysts.[1,2,6,13,47,48] Magnetic supports have been realized using modified magnetite and cobalt ferrite with good yields (Table 4, entries 6–7). In comparison to metal-based C/Co@DMAN (7), the oxidic supports exhibit a much lower saturation magnetization (133 emu/g for C/Co@DMAN (7), ca. 70 emu/g for magnetite[49]). This is a direct result of the inherently lower level of magnetization of most oxides. This value is important because it correlates directly with separation efficiency/speed and has recently enabled magnetic metal nanoparticles to purify water on a ton per hour scale.[50] In a typical laboratory setting, the improved magnetization facilitates workup, as demonstrated

Table 4. Comparison to Different Solid Support Base Reagents

entry	catalyst/base	saturation magnetization (emu g^{-1})	reaction	solvent	desired product	loading (mmol %)	time (h)	conversion (%)	conversion after three cycles (%)
1	7	139	Knoevenagel condensation	benzaldehyde	α-ethyl-2-cyanocinnamate	2	16	96	95
2	PS-MCM-41-DMAN[1]		Knoevenagel condensation	EtOH	α-ethyl-2-cyanocinnamate	0.5	7	99	85
3	DMAN/SiO_2-0.5[2]		Knoevenagel condensation	EtOH	α-ethyl-2-cyanocinnamate	1	6	90	85
4	TMGN/SiO_2-0.5[2]		Knoevenagel condensation	EtOH	α-ethyl-2-cyanocinnamate	1	6	100	97
5	7	139	Knoevenagel condensation	MeOH	benzylidene malononitrile	2	4.5	98	96
6	$CoFe_2O_4$@Si–N1-propylethane-1,2-diamine[3]	30[5]	Knoevenagel condensation	benzene	benzylidene malononitrile	2.5	0.33	100	100
7	Fe_2O_3@HAP@SiImidazoliumsalt[6]	55[12]	Knoevenagel condensation	H_2O	benzylidene malononitrile		1	98	97
8	7	139	Claisen–Schmidt condensation	acetophenone	*trans*-chalcone	10	20	78	71
9	DMAN/SiO_2-0.5[2]		Claisen–Schmidt condensation	acetophenone	*trans*-chalcone	1	24	95	
10	TMGN/SiO_2-0.5[2]		Claisen–Schmidt condensation	acetophenone	*trans*-chalcone	1	14	98	
11	Fe_3O_4-DABCO[13]	<60	Morita–Baylis–Hillman	MeOH	2-(hydroxy(4-nitrophenyl)methyl)acrylonitrile	20	5	88	88

using commercially available silica–iron oxide particles and the herein described metal reagent (Figure 3). The high level of magnetization of the carbon-coated cobalt allows clean, quantitative separation of the organic base within seconds. Established nonmagnetic immobilized catalyst supports are frequently based on zeolites and silica beads. The active site/moiety is predominantly attached by a siloxane-type linker.[51,52] These catalytic systems usually show excellent performance and reusability (Table 4, entries 2–4, 9–10). In the case of harsh reaction conditions (e.g., strong base), the weakest of these otherwise well-performing materials are the siloxane-type moieties. It is well known that extremely acidic or basic environments (i.e., pH <4 or >9) hydrolyze silane linkers. In contrast, the herein presented dialkyl urea moiety exhibits a high stability over a broad range of pH.

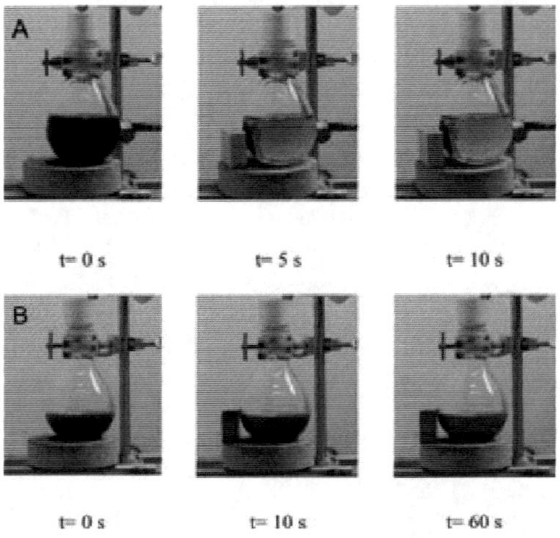

Figure 3. Recovery of organic base bound to magnetic nanoparticles using a low-cost external magnet. Model system in H_2O; same particle concentration as used in all experiments (0.7 mg mL^{-1}). A, metal-based reagents derived from Co/C; B, silica-coated Fe_3O_4. Note the different times above and below.

To confirm this statement experimentally, we directly compared siloxane linkers to the herein used C/Co@DMAN (7) in a number of representative acidic and basic environments. After 24 h at room temperature and pH 4, more than 20% of the siloxane linkers were cleaved, but less than 5% of the urea-type linkers were cleaved (Figure 4). At pH 9.5, the dialkyl urea linkers afforded lower stability (about 10% cleaved), but it is notable that under the same conditions >75% of the siloxane linkers were cleaved. Furthermore, the stability of these linkers was tested at elevated temperatures in different solvents (toluene and water). The dialkyl urea linkers performance under these conditions was superior to that of their silane-based pendants (Figure 4). The intrinsic stability of these particles in acidic or basic media has already been investigated[19] and has also been proven to be superior to that of silica-coated magnetite.

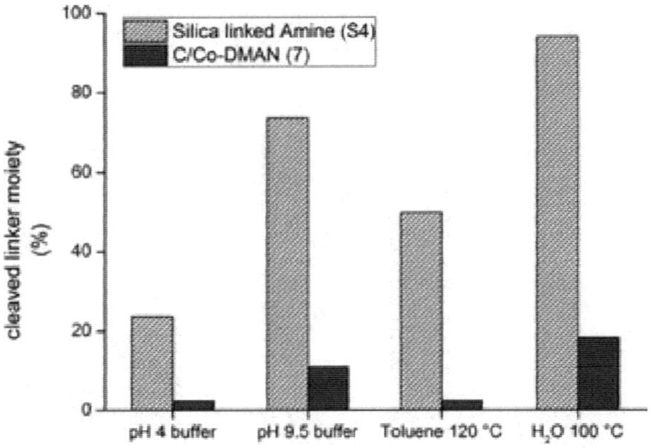

Figure 4. Immobilization linker stability tests under harsh reaction conditions overnight. The amount of cleaved linker was determined by analyzing the nitrogen content of the materials before and after exposure using quantitative elemental microanalysis. pH 4 buffer, 0.1 M potassium hydrogen phthalate with HCl; pH 9.5 buffer, 0.1 M carbonate-bicarbonate.

■ CONCLUSIONS

We have described the preparation of a covalently bound, chemically stable, magnetic nanoparticle-immobilized organic superbase and its extensive characterization using quantitative elemental microanalysis, infrared spectroscopy, vibrating sample magnetometry, and transmission electron microscopy. A number of control reactions were conducted to prove that the reagent was indeed chemically linked to the nanoparticles, and these confirmed the importance of linker type and reliability under relevant, harsh reaction conditions. We further demonstrated the experimental advantages of magnetic bases in a number of Knoevenagel condensations, reporting quantitative conversion and rapid product isolation in minutes. The demanding Claisen–Schmidt condensation afforded conversions of around 70% at 130 °C. The base catalyst was separated from the reaction mixtures in less than 1 min and could be recycled through at least five iterative runs. A quantitative comparison between siloxane linkers used in many immobilization studies and the herein described carbon shell-based chemistry revealed the important role of linker stability when designing reliable reagents for rough, everyday laboratory use.

■ EXPERIMENTAL SECTION

General Experimental Details. Carbon-coated cobalt nanoparticles[20] were suspended by the use of an ultrasonic bath and subjected to various reactions. After a reaction or a pretreatment step, nanoparticles were recovered from the reaction mixture with the aid of a conventional magnet (neodymium-based magnet, side length = 12 mm). Silica-coated magnetite nanoparticles (350 nm) were purchased from a chemical supplier. Commercially available reagents were used as received. All air-sensitive reactions were carried out under an argon atmosphere. The nanoparticles were analyzed by FT-IR spectroscopy (5% in KBr) and elemental microanalysis. HPLC measurements were performed (particle size, 5 μm; column size, 2.1 × 150 mm^2) to determine the conversion of the catalytic reactions. GC–MS measurements were performed (capillary, 30 m × 250 × 0.5 μm^2) with split injection (250 °C; ratio = 50:1; injection volume, 1 μL) and a temperature program (80 °C for 2 min; increase at 20 °C min^{-1} until 250 °C). The cobalt concentration present in solution was measured by inductively coupled plasma–atomic emission spectroscopy (ICP-OES).

Synthesis of C/Co@amine (3). Carbon-coated cobalt nanoparticles (1) were purchased from a chemical supplier. The particles were functionalized according to reported procedures (i.e., 3 g of **1** was suspended in 20 mL of H_2O, and 4-aminobenzylamine (**2**) (0.7 mL, 5.3 mmol) were added). Then sodium nitrite (0.8 g, 11.5 mmol) was added to the slurry. The vessel was put in the sonication bath, and 2 mL of concentrated HCl was slowly added dropwise. Upon completion of the reaction (30 min), the nanoparticles were recovered from the reaction mixture with a magnet and washed with toluene (2 × 10 mL), EtOH (1 × 10 mL), and acetone (2 × 10 mL) and dried for 24 h at 50 °C in vacuo. FT-IR: 1603, 1503, 1015, 831 cm^{-1}.

Synthesis of C/Co@hexamethyleneisocyanate (5). C/Co@ amine (**3**, 2 g) was degassed three times in a Schlenk flask. Fifty milliliters of dry DMF, hexamethylendiisocyanate (HDMI) (6 mL, 37 mmol, 97 equiv, 98%), and triethylamine (NEt_3, 0.01 mL, 99%) were added under an argon atmosphere. Then the solution was dispersed in a sonication bath for 5 min. The dispersion was heated to 70 °C and stirred for 6 h. Upon completion of the reaction, the nanobeads were recovered from the reaction mixture with a magnet and washed with anhydrous DMF (2×). FT-IR: 2927, 2857, 1689, 1553, 1018 cm^{-1}.

Synthesis of C/Co@DMAN (7). Freshly synthesized C/Co@ hexamethyleneisocyanate (**5**, 2 g) was used directly in a 250 mL Schlenk flask under an argon atmosphere. Forty milliliters of dry DMF and triethylamine (NEt_3, 0.01 mL, 99%) were added. The solution was dispersed in a sonication bath for 5 min. 4-Amino-1,8-bis-(dimethylamino)naphthalene (DMAN-NH_2 (**6**), 1 g, 4.3 mmol) was degassed three times in a Schlenk flask, dissolved in 10 mL of anhydrous DMF, and added dropwise to the C/Co-HMDI solution. The dispersion was heated to 35 °C and stirred overnight (16 h). Upon completion of the reaction, the nanoparticles were recovered from the reaction mixture with a magnet, washed with DMF (2 × 10 mL), EtOH (1 × 10 mL), acetone (2 × 10 mL), and dried for 24 h at 50 °C in vacuo. FT-IR: 2931, 2857, 2783, 1676, 1532, 1249 cm^{-1}.

Synthesis of Polystyrene-Supported DMAN (S2). Chloromethyl polystyrene (1 g, 2% DVB, 100–200 mesh, 0.9–1.5 mmol/g,) was used in a 50 mL Schlenk flask. Twenty milliliters of anhydrous DMF and triethylamine (NEt_3, 0.01 mL, 99%) were added under an argon atmosphere. DMAN-NH_2 (**6**) (0.6 g, 2.58 mmol) was degassed three times in a Schlenk flask, dissolved in 5 mL of anhydrous DMF, and added dropwise to the polystyrene slurry. The dispersion was heated to 40 °C and stirred overnight (20 h). Upon completion of the reaction, the functionalized polymer was filtered and intensively washed with acetone, 0.1 M NaOH, H_2O, EtOH, and acetone and dried for 24 h at 50 °C in vacuo. FT-IR: 3030, 2927, 2775, 1944, 1874, 1726, 1602 cm^{-1}.

Synthesis of Silica-Amine (S4). Silica gel 230–400 mesh (1 g) was placed in a 50 mL Schlenk flask and degassed three times. Twenty milliliters of anhydrous toluene was added, and the slurry was heated to 60 °C. *N*1-(2-Aminoethyl)-*N*2-(3-(trimethoxysilyl)propyl)ethane-1,2-diamine (0.3 mL, 11.6 mmol) was added dropwise to the solution, which then was stirred for 24 h. The resulting solid was filtered with 50 mL of toluene, 100 mL of DCM, and 20 mL of Et_2O and dried for 24 h at 50 °C in vacuo. FT-IR: 3289, 2940, 2828, 1863, 1474, 1109 cm^{-1}.

General Procedure for Knoevenagel Condensation Reactions. A mixture of solvent (5 mL) and C/Co@DMAN (**7**, 0.1 g, 0.01 mmol) was sonicated in an ultrasonication bath for 5 min in a Schlenk flask and then stirred at room temperature. Benzaldehyde (0.5 mmol) and malononitrile (0.5 mmol) were added, and the progress of the reaction was monitored by HPLC (MS to ensure product identity, UV at 280 nm for quantification) or GC-FID using a commercially available product as a reference.

*Benzylidenemalononitrile (**10**).* Isolated product was obtained by recrystallization in hot EtOH and hexane, followed by filtration. ^1H NMR ($CDCl_3$, 200 MHz, 25 °C): δ 7.86–7.82 (m, 2H), 7.71 (s, 1H), 7.57–7.43 (m, 3H).

α-Cyanocinnamic Acid Ethyl Ester. Pure isolated product was obtained by evaporation at 40 °C. ^1H NMR ($CDCl_3$, 200 MHz, 25 °C): δ 8.19 (s, 1H), 7.9 (m, 2H), 7.47–7.43 (m, 3H), 4.37–4.27 (q, 2H), 1.37–1.30 (t, 3H).

2-(4-Methoxybenzylidene)malononitrile. Pure isolated product was obtained by evaporation at 40 °C. ^1H NMR ($CDCl_3$, 200 MHz, 25 °C): δ 7.86–7.75 (m, 2H), 7.58 (s, 1H), 6.96–6.92 (m, 2H), 3.84 (s, 3H).

2-(4-Nitrobenzylidene)malononitrile. Isolated product was obtained by evaporation of the solvent, followed by recrystallization in DCM/hexane. ^1H NMR ($CDCl_3$, 200 MHz, 25 °C): δ 8.73 (s, 1H), 8.45–8.42 (m, 2H), 6.96–6.92 (m, 2H).

General Procedure for Claisen–Schmidt Condensations. A solvent-free mixture of acetophenone (2 mL) and C/Co@DMAN (**7**, 0.1 g, 0.01 mmol) was sonicated in an ultrasonication bath for 5 min and then stirred at 130 °C under a nitrogen atmosphere. Benzaldehyde (0.1 mmol) was added, and the progress of the reaction was monitored by HPLC (MS to ensure product identity, UV at 280 nm for quantification) using a commercially available product as a reference.

*1,3-Diphenyl-2-propen-1-one (**13**).* Isolated product was obtained by evaporation in high vacuum at 65 °C. ^1H NMR ($CDCl_3$, 200 MHz, 25 °C): δ 7.93–7.97 (m, 2H), 7.18–7.71 (m, 10H).

General Washing/Drying Procedure for Catalyst Recovery. Used catalyst was washed by sonication (5 min) with the reaction solvent of the catalytic reaction and acetone (2×) and dried for 4 h at 50 °C in vacuo.

General Procedure for Linker Stability Tests. A sample (30 mg) was placed in a 10 mL round-bottomed flask, and solvent (10 mL) was added. The solution was dispersed via ultrasonication bath and then shaken overnight. The solid sample was washed by either sonication (magnetic samples) or filtration (silica samples) with EtOH, H_2O, and acetone and dried for 4 h at 50 °C in vacuo.

■ ASSOCIATED CONTENT

Supporting Information

Spectroscopic data, elemental analysis data, IR spectra, TEM and SEM micrographs of the magnetic and polymeric compounds, scheme depicting the test reactions carried out to verify covalent linkage, and linker stability tests with a duration of 1 week. This material is available free of charge via the Internet at http://pubs.acs.org.

■ AUTHOR INFORMATION

Corresponding Author

*E-mail: wstark@ethz.ch. Web: www.fml.ethz.ch. Fax: +41 44 633 10 83. Tel: +41 44 632 09 80.

Notes

The authors declare the following competing financial interest(s): W.J.S. and R.N.G. declare financial interests, because they are shareholders of TurboBeads LLC, a company active in magnetic nanoparticles.

■ ACKNOWLEDGMENTS

We thank Andreas Dutly and Nikita Kobert for GC measurements, ICB/ETH Zurich, the EU-ITN network Mag(net)icFun (PITN-GA-2012-290248), and the Swiss National Science Foundation (no. 200021-150179) for financial support.

■ REFERENCES

(1) Corma, A.; Iborra, S.; Rodriguez, I.; Sanchez, F. *J. Catal.* **2002**, *211*, 208.

(2) Gianotti, E.; Diaz, U.; Velty, A.; Corma, A. *Eur. J. Inorg. Chem.* **2012**, 5175.

(3) Phan, N. T. S.; Jones, C. W. *J. Mol. Catal. A: Chem.* **2006**, *253*, 123.

(4) Alder, R. W.; Bryce, M. R.; Goode, N. C.; Miller, N.; Owen, J. *J. Chem. Soc., Perkin Trans. 1* **1981**, 2840.

(5) Rondinone, A. J.; Samia, A. C. S.; Zhang, Z. J. *J. Phys. Chem. B* **1999**, *103*, 6876.

(6) Zhang, Y.; Zhao, Y.; Xia, C. *J. Mol. Catal. A: Chem.* **2009**, *306*, 107.

(7) Ishikawa, T. *Superbases for Organic Synthesis*; John Wiley & Sons: Chichester, U.K., 2009; p 1.

(8) Gwaltney, S. L.; Sakata, S. T.; Shea, K. J. *J. Org. Chem.* **1996**, *61*, 7438.

(9) Ireland, R. E.; Liu, L. B.; Roper, T. D.; Gleason, J. L. *Tetrahedron* **1997**, *53*, 13257.

(10) Nielsen, M.; Jacobsen, C. B.; Jørgensen, K. A. *Angew. Chem., Int. Ed.* **2011**, *50*, 3211.
(11) Tur, F.; Saá, J. M. *Org. Lett.* **2007**, *9*, 5079.
(12) Mori, K.; Kanai, S.; Hara, T.; Mizugaki, T.; Ebitani, K.; Jitsukawa, K.; Kaneda, K. *Chem. Mater.* **2007**, *19*, 1249.
(13) Luo, S.; Zheng, X.; Xu, H.; Mi, X.; Zhang, L.; Cheng, J.-P. *Adv. Synth. Catal.* **2007**, *349*, 2431.
(14) Gao, J. H.; Gu, H. W.; Xu, B. *Acc. Chem. Res.* **2009**, *42*, 1097.
(15) Xu, Z. P.; Zeng, Q. H.; Lu, G. Q.; Yu, A. B. *Chem. Eng. Sci.* **2006**, *61*, 1027.
(16) Pankhurst, Q. A.; Connolly, J.; Jones, S. K.; Dobson, J. *J. Phys. D: Appl. Phys.* **2003**, *36*, R167.
(17) Huh, Y.-M.; Jun, Y.-w.; Song, H.-T.; Kim, S.; Choi, J.-s.; Lee, J.-H.; Yoon, S.; Kim, K.-S.; Shin, J.-S.; Suh, J.-S.; Cheon, J. *J. Am. Chem. Soc.* **2005**, *127*, 12387.
(18) Schätz, A.; Reiser, O.; Stark, W. J. *Chem.—Eur. J.* **2010**, *16*, 8950.
(19) Schaetz, A.; Zeltner, M.; Michl, T. D.; Rossier, M.; Fuhrer, R.; Stark, W. J. *Chem.—Eur. J.* **2011**, *17*, 10565.
(20) Grass, R. N.; Athanassiou, E. K.; Stark, W. J. *Angew. Chem., Int. Ed.* **2007**, *46*, 4909.
(21) Baruwati, B.; Guin, D.; Manorama, S. V. *Org. Lett.* **2007**, *9*, 5377.
(22) Gleeson, O.; Tekoriute, R.; Gun'ko, Y. K.; Connon, S. J. *Chem.—Eur. J.* **2009**, *15*, 5669.
(23) Keller, M.; Collière, V.; Reiser, O.; Caminade, A.-M.; Majoral, J.-P.; Ouali, A. *Angew. Chem., Int. Ed.* **2013**, *52*, 3626.
(24) Polshettiwar, V.; Baruwati, B.; Varma, R. S. *Chem. Commun.* **2009**, 1837.
(25) Riente, P.; Yadav, J.; Pericas, M. A. *Org. Lett.* **2012**, *14*, 3668.
(26) Schätz, A.; Long, T. R.; Grass, R. N.; Stark, W. J.; Hanson, P. R.; Reiser, O. *Adv. Funct. Mater.* **2010**, *20*, 4323.
(27) Stevens, P. D.; Fan, J. D.; Gardimalla, H. M. R.; Yen, M.; Gao, Y. *Org. Lett.* **2005**, *7*, 2085.
(28) Stevens, P. D.; Li, G. F.; Fan, J. D.; Yen, M.; Gao, Y. *Chem. Commun.* **2005**, 4435.
(29) Zeltner, M.; Schaetz, A.; Hefti, M. L.; Stark, W. J. *J. Mater. Chem.* **2011**, *21*, 2991.
(30) Cantillo, D.; Moghaddam, M. M.; Kappe, C. O. *J. Org. Chem.* **2013**, *78*, 4530.
(31) Zheng, Y.; Stevens, P. D.; Gao, Y. *J. Org. Chem.* **2005**, *71*, 537.
(32) Kainz, Q. M.; Linhardt, R.; Grass, R. N.; Vile, G.; Perez-Ramirez, J.; Stark, W. J.; Reiser, O. *Adv. Funct. Mater.* **2014**, *24*, 2020.
(33) Linhardt, R.; Kainz, Q. M.; Grass, R. N.; Stark, W. J.; Reiser, O. *RSC Adv.* **2014**, *4*, 8541.
(34) Wittmann, S.; Schätz, A.; Grass, R. N.; Stark, W. J.; Reiser, O. *Angew. Chem., Int. Ed.* **2010**, *49*, 1867.
(35) Ozeryanskii, V. A.; Pozharskii, A. F. *Russ. Chem. Bull.* **1997**, *46*, 1437.
(36) Zeltner, M.; Grass, R. N.; Schaetz, A.; Bubenhofer, S. B.; Luechinger, N. A.; Stark, W. J. *J. Mater. Chem.* **2012**, *22*, 12064.
(37) Lu, A. H.; Salabas, E. L.; Schuth, F. *Angew. Chem., Int. Ed.* **2007**, *46*, 1222.
(38) Schumacher, C. M.; Herrmann, I. K.; Bubenhofer, S. B.; Gschwind, S.; Hirt, A.-M.; Beck-Schimmer, B.; Guenther, D.; Stark, W. J. *Adv. Funct. Mater.* **2013**, *23*, 4888.
(39) Rodriguez, I.; Sastre, G.; Corma, A.; Iborra, S. *J. Catal.* **1999**, *183*, 14.
(40) Schumacher, C. M.; Grass, R. N.; Rossier, M.; Athanassiou, E. K.; Stark, W. J. *Langmuir* **2012**, *28*, 4565.
(41) Machado, B. F.; Serp, P. *Catal. Sci. Technol.* **2012**, *2*, 54.
(42) Schaetz, A.; Zeltner, M.; Stark, W. J. *ACS Catal.* **2012**, *2*, 1267.
(43) Gomes, A.; Couto, D.; Alves, A.; Dias, I.; Freitas, M.; Porto, G.; Duarte, J. A.; Fernandes, E. *BioFactors* **2012**, *38*, 378.
(44) Cushnie, T. P. T.; Lamb, A. J. *Int. J. Antimicrob. Agents* **2005**, *26*, 343.
(45) Woo, H. D.; Kim, J. *PLoS One* **2013**, *8*, e75604.
(46) Novak, M.; Loudon, G. M. *J. Org. Chem.* **1977**, *42*, 2494.
(47) Climent, M. J.; Corma, A.; Iborra, S.; Primo, J. *J. Catal.* **1995**, *151*, 60.
(48) Karaoğlu, E.; Baykal, A.; Şenel, M.; Sözeri, H.; Toprak, M. S. *Mater. Res. Bull.* **2012**, *47*, 2480.
(49) Goya, G. F.; Berquó, T. S.; Fonseca, F. C.; Morales, M. P. *J. Appl. Phys.* **2003**, *94*, 3520.
(50) Rossier, M.; Schreier, M.; Krebs, U.; Aeschlimann, B.; Fuhrer, R.; Zeltner, M.; Grass, R. N.; Günther, D.; Stark, W. J. *Sep. Purif. Technol.* **2012**, *96*, 68.
(51) Corma, A.; Garcia, H. *Adv. Synth. Catal.* **2006**, *348*, 1391.
(52) Han, Y.-J.; Stucky, G. D.; Butler, A. *J. Am. Chem. Soc.* **1999**, *121*, 9897.

461122 Selective Hydrogenolysis of Furfuryl Alcohol to 2-Methylfuran over Reduced Co-Fe-Al Mixed Metal Oxides

Abstract:

Taylor Sulmonetti, Bo Hu, Pradeep Agrawal, Christopher W. Jones

Selective Hydrogenolysis of Furfuryl Alcohol to 2-Methylfuran Over Reduced Co-Fe-Al Mixed Metal Oxides

Furfural and furfuryl alcohol have shown promise as major biomass platform chemicals, since they can be converted into a variety of value added products. Through side chain hydrogenolysis, furfuryl alcohol can be converted into 2-methylfuran, a potential fuel additive. Few monometallic catalysts show selective conversion of furfuryl alcohol towards 2-methylfuran; therefore, bifunctional and bimetallic catalysts have been widely pursued for this purpose in recent years. In this investigation, low cost, earth-abundent metal/metal oxide composites were utilized for the creation of 2-methylfuran through vapor phase conversion of furfuryl alcohol. An array of Co-Fe-Al catalysts were synthesized through the calcination of layered double hydroxides, and subsequently reduced in H2 to produce materials that contained both metallic and oxide domains. Vapor phase flow reactions were conducted after reduction to evaluate the reactivity of the reduced mixed metal oxides with varying Co/Fe ratios. The introduction of Fe into the mixed metal oxide matrix facilitated further reduction of the Co species to produce more catalytically active sites. Additionally, with the incorporation of Fe, the selectivity of 2-methylfuran reached about 85%, which is comparable or superior to prior reports using other catalytic compositions. Further characterization utilizing an array of spectroscopic experiments, including XAS and XPS, indicated the emergence of both metallic Co and Fe phases upon reduction. Lastly, the morphology was investigated through STEM to support the hypothesis that the catalysts were composed of well-dispersed mixed metal oxides containing metallic domains.

Copyright © American Institute of Chemical Engineers. All rights reserved.

471191 Supported, Homogeneously Alloyed Bimetallic Nanoparticles By Electrostatic Adsorption

Abstract:

Bimetallic catalysts are often important for chemical reactions due to their enhanced performance over monometallic catalysts [1]. Incorporating a secondary metal into a system can benefit the catalytic system through ensemble, bifunctional, and electronic effects. The most prevalent method for synthesizing supported catalysts is dry impregnation (DI), also known as incipient wetness, in which a solution containing a specific amount of metal precursor is added to just fill the pore volume of the support [2]. While the procedure is simple and the metal loading is precisely known, limitations of DI preparations often include larger particle sizes and less intimate metal-support interactions. Moreover, bimetal catalysts produced through DI often exhibit a poor degree of alloying between the metals.

Our goal was to demonstrate a simple, effective method to synthesize highly dispersed, well alloyed bimetallic nanoparticles using strong electrostatic adsorption (SEA). SEA is achieved by charging the support by controlling the pH of the solution. A metal precursor that is oppositely charged can then be adsorbed onto the surface of the charged support. Using a simultaneous adsorption of either two cationic or anionic precursors allows excellent bimetallic interactions. The combinations of multiple bimetallic systems (Cu, Co, Ni, Pd, and Pt) prepared by SEA were compared to equivalent DI catalysts. Particle sizes and the degree of alloying were determined through electron micrographs, elemental mapping of nanoparticles, and powder XRD.

Results from XRD show the bimetallic catalysts produced by SEA were highly dispersed (<2 nm), while the DI catalysts exhibited large XRD peaks (>5nm). Initial elemental maps and STEM images indicate the SEA catalysts had a higher degree of alloying compared to the DI catalyst, and particle agglomeration was avoided for the SEA catalyst. A complete comparison of the bimetal systems prepared by DI and SEA will be discussed.

References

[1] Sankar, M., et al. (2012). "Designing bimetallic catalysts for a green and sustainable future." Chemical Society Reviews **41**(24): 8099-8139.

[2] Cho, H.-R. and J. R. Regalbuto (2015). "The rational synthesis of Pt-Pd bimetallic catalysts by electrostatic adsorption." Catalysis today **246**: 143-153.

Copyright © American Institute of Chemical Engineers. All rights reserved.

460867 Chemoselective Transfer Hydrogenation of Nitroarenes Catalyzed By Highly Dispersed, Supported Nickel Nanoparticles

Abstract:

Aromatic amines are useful intermediates in the preparation of dyes, pharmaceuticals, and agricultural chemicals, which can be obtained by the reduction of aromatic nitro compounds. A variety of procedures involving metal catalysts and other reducing agents are available for this purpose; however, the selective reduction of a nitro group in the presence of other reducible functional groups in a molecule is a challenging task. In addition, reduction of aromatic nitro compounds often stops at an intermediate stage, yielding hydroxylamines, hydrazines, azoarenes, or azoxyarenes as byproducts. In this study, a highly active nickel (Ni) nanoparticles catalyst supported on porous silica gel particles was prepared by atomic layer deposition. Chemoselective reduction of nitroarenes was studied using the prepared Ni/SiO_2 as the catalyst and hydrazine hydrate as a hydrogen donor. Different kinds of nitroarenes were converted to the corresponding anilines with high yields. The high activity of the catalysts could be a result of the highly dispersed Ni nanoparticles.

Copyright © American Institute of Chemical Engineers. All rights reserved.

Controllable Synthesis of Highly Dispersed Cu/SiO_2 Catalysts in Supercritical CO_2 and their Application in the Hydrogenation of Dimethyl Oxalate to Ethylene Glycol

*Qin-Qin Xu, Guo-Yue Qiao and Jian-Zhong Yin**

State Key Laboratory of Fine Chemicals, School of Chemical Machinery, Dalian University of Technology, Dalian 116024, China

* Corresponding Author: Jian-Zhong Yin, jzyin@dlut.edu.cn ; Tel. and Fax. +86-411-84986274

Abstract: Highly dispersed Cu nanostructures were synthesized within mesoporous silica in supercritical CO_2 using inorganic Cu precursors. Co-solvent was found to have significant influence on the metal loading and the nanostructure morphologies of the nanocomposites. Although the inorganic precursor has large solubility in ethanol, 1,3 propylene glycol, n-butyl alcohol and methanol, only big nanoparticles outside the nanochannels were obtained when the first three co-solvents were used, while zero metal loading was obtained when methanol was used as co-solvent. However, when ethylene glycol or the mixture of ethylene glycol and water were used as co-solvents, large amounts of short nanotubes were found inside the nanochannels with large metal loading. It indicates that the role of the co-solvent may have many aspects, such as enhancement of the precursor dissolution, change of the interaction between the precursor and the substrates and so on.

Key Words: supercritical carbon dioxide; catalyst; mesoporous silica; co-solvent; ethylene glycol

1. Introduction

Ethylene glycol (EG) is widely used as antifreeze and precursors in the production of polyesters. One of the most promising approach to EG is coal-based production technology via dimethyl oxalate (DMO) hydrogenation. Preparation of highly active catalysts plays a key role in the synthesis of EG through the coal chemical industry. Silica supported copper catalyst was found to be the most efficient one in this reaction[1-3]. There are many methods to prepare these supported metal nanocomposites, such as impregnation[4], sol-gel processing[5], chemical vapor impregnation[6] and so on. There are some disadvantages for the traditional methods to prepare these copper-based catalysts, for example, the active components are usually poorly dispersed for the impregnation method and the metal loading is quite low for the ion exchange method. Ammonia-evaporation method is characterized by highly dispersed active components with high metal loading, however, ammonia is very toxic. Herein, it is very important to develop some new and effective methods to prepare these catalysts.

Supercritical fluid deposition (SCFD) is an effective and environmentally benign technique to synthesize supported nanocomposites. This method involves the dissolution of metal precursor in supercritical fluids (SCFs), the diffusion and impregnation of SCF solution into the pores of the substrates and adsorption of metal precursor onto the walls of the channels, followed by thermal or chemical reduction in or after depressurization process. The low surface tension of SCF will favor the impregnation of precursors into the nano-scale channels of the substrates. Furthermore, the ordered pores of the substrate won't be collapsed because no organic solvents are left after depressurization, and the post-treatment process is quite simply compared to other preparing methods.

Supercritical carbon dioxide ($scCO_2$) is the most widely used SCF due to its abundance, low cost, non-flammability, non-toxicity and in particular the mild critical pressure and temperature ($T_C = 31.1°C$, $P_C = 7.38$ MPa). SCFD method was first proposed by Watkins based on chemical vapor deposition (CVD) technique in 1995[7]. After that, it has attracted growing attentions by many researchers. They used this method to deposit thin metal films (Cu, Au, Pt, Pd, Co, Ni etc.) onto a wide range of substrates such as silicon wafer, metal, ceramic, polymer and so on[6,8-12]or incorporate metallic nanoparticles (Pt, Pd, Au etc.) on different inorganic substrates such as carbon aerogel, carbon black, Al_2O_3, silica aerogel[13-16], graphene[17,18] etc.

However, organometallic compounds were usually used as precursors because they have relatively large solubilities in $scCO_2$. In order to overcome the limitation of using the expensive and toxic organometallic compounds, inorganic salt was used as precursor with the addition of appropriate co-solvent by some groups. Liu and coworkers used metal nitrates dissolved in $scCO_2$/ethanol system to deposit metal and metal oxides on carbon nanotubes[16,19,20] . Hu et al. prepared 4A-zeolite supported silver nanoparticles with the assistance of $scCO_2$ and hydrogen reduction using $AgNO_3$ as precursor[21]. Yin and his co-workers have made much effort towards the cheap SCFD process based on inorganic precursors and appropriate co-solvents in recent years [22-26].

In this study, $Cu(NO_3)_2$ was successfully deposited into mesoporous silica SBA-15 using $scCO_2$ as the solvent and different co-solvents. Single co-solvent including methanol, ethanol, ethylene glycol, 1,3 propylene glycol and n-butyl alcohol was investigated to find out the most proper one. After that, considering the competition adsorption agent, a small amount of water or dilute acid was added into the most proper co-solvent and formed the dual co-solvent to prepare the different samples. TEM, N_2 adsorption-desorption isotherm and ICP analysis were performed to characterize the metal loading and the microstructure of these samples.

2. Experimental

2.1 Materials

Mesoporous SBA-15 was prepared according to the published literatures[27]. The synthesized SBA-15 has a BET surface area of 807 m^2/g, total pore volume of 0.87 m^3/g and average pore diameter of 4.3 nm. $Cu(NO_3)_2 \cdot 3H_2O$ was purchased from TianJin DaMao chemicals Co. Ltd.. Carbon dioxide (>99%) was obtained from DaLian GuangMing gas Co. Ltd.. Ethanol, methanol, ethylene glycol, 1,3 propylene glycol and n-butyl alcohol of analytical grade were supplied by TianJin FuYu fine chemicals Co. Ltd..

2.2 Deposition process and characterizations

The schematic of the experiment and the details of the deposition process were described elsewhere [23,24,28]. Briefly, a certain amount of precursor and co-solvent were placed at the bottom of the reactor. The substrate was held in a stainless steel basket which was fixed in the upper part of the reactor. Then the reactor was connected to the lines, preheated to the experimental temperature and charged into CO_2 using a piston pump until the pressure of the system got up to the required pressure. Keeping the pressure and temperature for a required time, and then the reactor was depressurized slowly. After that, $Cu(NO_3)_2$@SBA-15 was obtained and subjected to calcinations (CuO@SBA-15) and H_2 reduction (Cu@SBA-15). The products were characterized by transmission electron microscopy (TEM), N_2 adsorption-desorption isotherms and Inductively coupled plasma (ICP).

3. Results and Discussion

3.1 Synthesis of CuO@SBA-15 using different co-solvents

SCFD method involves the dissolution of precursor in $scCO_2$, the diffusion and impregnation of the supercritical solution into the pores of the substrates and adsorption of the precursor onto the walls of the channels, followed by depressurization. Therefore, the dissolution of precursors in $scCO_2$ is essential to the successful preparation in this method. However, inorganic salts are not soluble in $scCO_2$. As we know that, $scCO_2$ has a strong ability to dissolve organic solutes and these organic solutes may act as co-solvent to enhance the ability of $scCO_2$ to dissolve the inorganic salts.

In this study, five kinds of organic solvents were tried to use as co-solvents respectively. Five samples were prepared using 200 mg SBA-15, 200 mg $Cu(NO_3)_2 \cdot 3H_2O$, 1 mL co-solvent at deposition temperature of 50°C, pressure of 18 MPa and deposition time of 3h. The metal loading were 14.4%, 8.3%, 0%, 16.0%, and 7.6% for the samples prepared using ethanol, ethylene glycol, methanol, 1,3 propylene glycol and n-butyl alcohol as co-solvents respectively. However, it was interesting that there were only big nanoparticles aggregated outside of the nanochannels for the three samples when ethanol, 1,3 propylene glycol and n-butyl alcohol were used as co-solvent (Figure 1a, 1c and 1d), while there were some nanoparticles and short nanorods inside the nanochannels with ethylene glycol as co-solvent (Figure 1b).

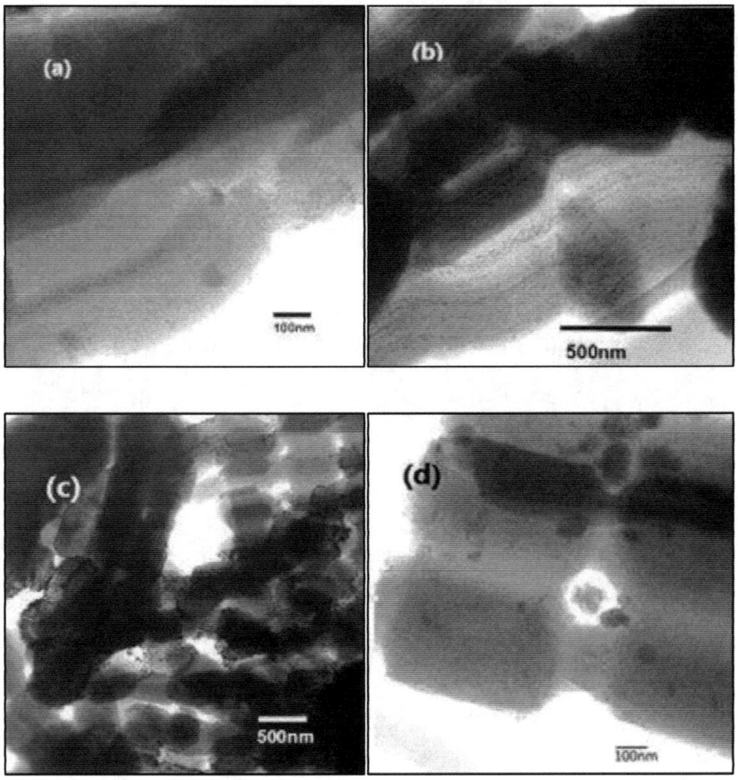

Figure 1 TEM images of samples prepared by different co-solvents (a) ethanol; (b) ethylene glycol; (c) 1,3 propylene glycol; (d) n-butyl alcohol

As stated above, the dissolution of inorganic salts in $scCO_2$ is essential to the deposition process and all the five co-solvents could play the role of the solubility enhancement. However, nanostructures could be confined within the nanochannels only when the ethylene glycol was used as co-solvent. It seems that the deposition process was not only dependent on the solubility of the precursor in the system, but also greatly influenced by the interaction between the co-solvent and the substrates. This phenomenon was also observed when silver nanostructures supported nanocomposites were prepared in $scCO_2$ and co-solvent system[25].

3.2 Synthesis of CuO@SBA-15 using dual co-solvent

Yin and his co-workers[24] found that compared to the case of only ethanol as the co-solvent, the employment of dual co-solvent of ethylene glycol and ethanol led to the significant reduction of deposition pressure and time when Ag@SBA-15 was prepared in $scCO_2$. In their subsequent work[25], they found that well dispersed and continuous nanowires were anchored into the nanochannels of SBA-15 in only 5 minutes at 8.5 MPa with a metal loading of 5.8 wt%. Furthermore, the metal loading increased with the increase of deposition time while the morphology was not changed. In summary, co-solvent played an important role in the adsorption of the precursor on the substrates. Besides the solubility enhancement, the dielectric constant, surface tension, viscosity of the solution and the interaction among the precursor, co-solvent and substrate were different when different co-solvents were used, which would affect the adsorption and dispersion of the precursor on the substrate.

Water is usually thought as a nonideal co-solvent, because it has a small solubility in $scCO_2$ due to its strong polarity. However, the solubility of $Cu(NO_3)_2$ in water is quite large. Therefore, the mixture of ethylene glycol and water was applied as co-solvent. Water was used to enhance the dissolution of precursor in co-solvent and ethylene glycol was added here to enhance the dissolution of the co-solvent in $scCO_2$. Furthermore, Yin et al.[29] enlightened by the traditional impregnation method and considered the addition of a small amount of dilute acid as the competition adsorption agent. Herein, we investigated the feasibility of preparing Cu/SBA-15 using the dual co-solvent of ethylene glycol and water as well as ethylene glycol and dilute acid.

When the sample was prepared at the temperature of 50°C, pressure of 18MPa, deposition time of 3h using the mixture of 1mL ethylene glycol and 0.5mL water as co-solvent, more uniform and dispersed nanoparticles were found inside the nanochannels of SBA-15, and the metal loading was increased to 12.7% (8.3% for only ethylene glycol co-solvent). It seems that addition of water will favor the deposition of precursors onto the substrates. When 1 mL ethylene glycol and 0.5 mL dilute HNO_3 was used as co-solvent, the metal loading was 13.8%. As can be seen in the TEM images (Figure 2), the nanochannels of the substrates were more complete filling after addition of water or dilute acid as co-solvent. What's more, the nanostructures were more uniform dispersed using the dual co-solvents.

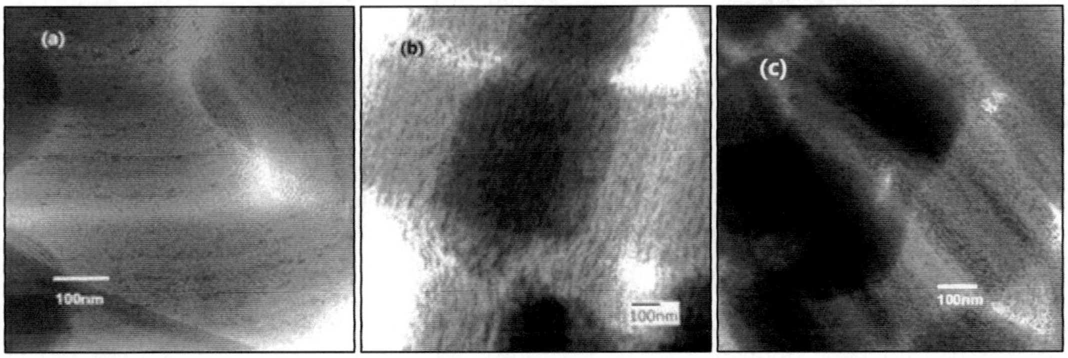

Figure 2 TEM images of samples prepared by different co-solvents: (a) ethylene glycol; (b) ethylene glycol and water; (c) ethylene glycol and dilute acid

The N_2 adsorption isotherms and the pore size distribution of SBA-15 and CuO@SBA-15 with different co-solvents shows typical steps in the 0.5-0.8 relative pressure(p/p_0) range which is characteristic of a highly ordered mesoporous porous lattice. The BET surface area, pore volume and the most probable pore size derived from the BJH adsorption branch were shown in Table 1.

Table 1 The N_2 adsorption-desorption analysis results of the samples

Sample	Co-solvent	BET surface area (cm^2/g)	Pore volume (cm^3/g)	Average pore size (nm)
SBA-15	-	807	0.87	4.3
CuO@ SBA-15	ethylene glycol	473	0.61	5.2
CuO@ SBA-15	ethylene glycol +HNO_3	519	0.62	4.8
CuO@ SBA-15	ethylene glycol +water	547	0.64	4.7

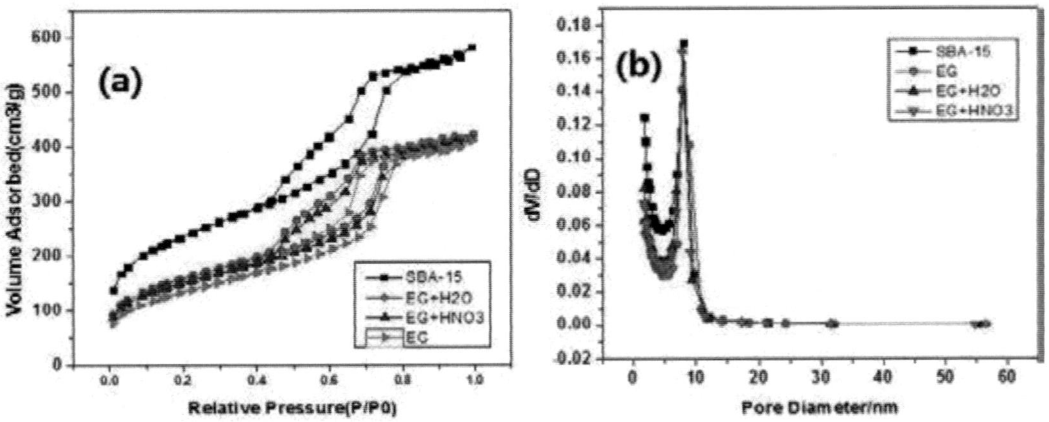

Figure 3 The N_2 adsorption isotherm(a) and pore size distribution curve(b) of the nanocomposites prepared using different co-solvents

3.3 The hydrogenation of dimethyl oxalate to ethylene glycol

This part of the work is just underway and the results will be reported later on.

4. Conclusions

In summary, SCFD is an effective method to deposit inorganic precursor into the nano-scale channels of mesoporous silica using appropriate co-solvent. It was found that co-solvent had a significant influence on the metal loading and the morphology of the nanostructure. Ethylene glycol or the mixture of ethylene glycol and water, the mixture of ethylene glycol and dilute acid are proved to be effective co-solvent to deposit $Cu(NO_3)_2$ into the nanochannels of SBA-15, while ethanol, methanol, 1,3 propylene glycol and n-butyl alcohol were not beneficial to the deposition. The underlying reason for different co-solvents, the operating parameters influencing the deposition results and the hydrogenation of dimethyl dxalate to ethylene glycol are underway.

Acknowledgments

The work was financially supported by the National Natural Science Foundation of China (21376045, 21506027), Chinese Postdoctoral Science Foundation (2015M571307) and the Fundamental Research Funds for the Central Universities.

References

1. Gong J, Yue H, Zhao Y, et al. Synthesis of ethanol via syngas on Cu/SiO2 catalysts with balanced Cu0-Cu+ sites. *Journal of the American Chemical Society.* Aug 29 2012;134(34):13922-13925.
2. Ma X, Chi H, Yue H, et al. Hydrogenation of dimethyl oxalate to ethylene glycol over mesoporous Cu-MCM-41 catalysts. *AIChE Journal.* 2013;59(7):2530-2539.
3. Yue H, Zhao Y, Zhao L, et al. Hydrogenation of dimethyl oxalate to ethylene glycol on a Cu/SiO2/cordierite monolithic catalyst: Enhanced internal mass transfer and stability. *AIChE Journal.* 2012;58(9):2798-2809.
4. Huang MH, Choudrey A, Yang PD. Ag nanowire formation within mesoporous silica. *Chem. Commun.* 2000(12):1063-1064.
5. Bekyarova E, Kaneko K. Structure and physical properties of tailor-made Ce,Zr-doped carbon aerogels. *Adv. Mater.* Nov 2000;12(21):1625-1628.
6. Miller JM, Dunn B, Tran TD, Pekala RW. Deposition of ruthenium nanoparticles on carbon aerogels for high energy density supercapacitor electrodes. *J. Electrochem. Soc.* Dec 1997;144(12):L309-L311.
7. Watkins JJ, McCarthy TJ. Polymer/Metal Nanocomposite Synthesis in Supercritical CO2. *Chem. Mater.* Nov 1995;7(11):1991-1994.
8. Blackburn JM, Long DP, Cabanas A, Watkins JJ. Deposition of conformal copper and nickel films from supercritical carbon dioxide. *Science.* Oct 2001;294(5540):141-145.
9. Cabanas A, Long DP, Watkins JJ. Deposition of gold films and nanostructures from supercritical carbon dioxide. *Chem. Mater.* May 2004;16(10):2028-2033.
10. Morere J, Tenorio MJ, Torralvo MJ, Pando C, Renuncio JAR, Cabanas A. Deposition of Pd into mesoporous silica SBA-15 using supercritical carbon dioxide. *J. Supercrit. Fluids.* Mar 2011;56(2):213-222.
11. Karanikas CF, Watkins JJ. Kinetics of the ruthenium thin film deposition from supercritical carbon dioxide by the hydrogen reduction of Ru(tmhd)(2)cod. *Microelectron. Eng.* Apr 2010;87(4):566-572.
12. Haruki M, Li SK, Qian G, Watkins JJ. Reactive deposition of cobalt using bis(2,2,6,6-tetramethyl-3,5-heptanedionato) cobalt(II) from supercritical carbon dioxide. *Journal of Supercritical Fluids.* Jan 2016;107:189-195.
13. Saquing CD, Cheng TT, Aindow M, Erkey C. Preparation of platinum/carbon aerogel nanocomposites using a supercritical deposition method. *J. Phys. Chem. B.* Jun 2004;108(23):7716-7722.

14. Zhang Y, Kang DF, Saquing C, Aindow M, Erkey C. Supported platinum nanoparticles by supercritical deposition. *Ind. Eng. Chem. Res.* May 2005;44(11):4161-4164.

15. Erkey C. Preparation of metallic supported nanoparticles and films using supercritical fluid deposition. *J. Supercrit. Fluids.* Jan 2009;47(3):517-522.

16. Sun ZY, Liu ZM, Han BX, Miao SD, Miao ZJ, An GM. Decoration carbon nanotubes with Pd and Ru nanocrystals via an inorganic reaction route in supercritical carbon dioxide-methanol solution. *J. Colloid Interface Sci.* Dec 2006;304(2):323-328.

17. Meng Y, Su FH, Chen YZ. Synthesis of nano-Cu/graphene oxide composites by supercritical CO2-assisted deposition as a novel material for reducing friction and wear. *Chemical Engineering Journal.* Dec 2015;281:11-19.

18. Zhao J, Yu H, Liu ZS, Ji M, Zhang LQ, Sun GW. Supercritical Deposition Route of Preparing Pt/Graphene Composites and Their Catalytic Performance toward Methanol Electrooxidation. *Journal of Physical Chemistry C.* Jan 2014;118(2):1182-1190.

19. Liu ZM, Han BX. Synthesis of Carbon-Nanotube Composites Using Supercritical Fluids and Their Potential Applications. *Adv. Mater.* Feb 2009;21(7):825-829.

20. Sun ZY, Liu ZM, Han BX, An GM. Supercritical carbon dioxide-assisted deposition of tin oxide on carbon nanotubes. *Mater. Lett.* Sep 2007;61(23-24):4565-4568.

21. Hu XS, Bai J, Hong HL, Li CP. Supercritical carbon dioxide anchored highly dispersed silver nanoparticles on 4A-zeolite and selective oxidation of styrene performance. *Crystengcomm.* 2016;18(14):2469-2476.

22. Yin J-Z, Xu Q-Q, Wang A-Q. Controlled Growth of Copper Nanoparticles and Nanorods in the Channels of Sba-15 by Supercritical Fluid Deposition. *Chemical Engineering Communications.* 2010;197(4):627-632.

23. Xu Q-q, Wang Y-q, Wang A-q, Yin J-z. Systematical study of depositing nanoparticles and nanowires in mesoporous silica using supercritical carbon dioxide and co-solvents: morphology control, thermodynamics and kinetics of adsorption. *Nanotechnology.* 2012;23(28):285602.

24. Xu Q-Q, Zhang C-J, Zhang X-Z, Yin J-Z, Liu Y. Controlled synthesis of Ag nanowires and nanoparticles in mesoporous silica using supercritical carbon dioxide and co-solvent. *The Journal of Supercritical Fluids.* 2012;62:184-189.

25. Xu QQ, Ma YL, Gang X, Yin JZ, Wang AQ, Gao JJ. Comprehensive study of the role of ethylene glycol when preparing Ag@SBA-15 in supercritical CO2. *Journal of Supercritical Fluids.* Aug 2014;92:100-106.

26. Xu QQ, Xu G, Yin JZ, Wang AQ, Ma YL, Gao JJ. Preparation of Superhighly Dispersed Co3O4@SBA-15 with Different Morphologies in Supercritical CO2 with the Assistance of Dilute Acids. *Industrial & Engineering Chemistry Research.* Jun 2014;53(25):10366-10371.

27. Zhao D. Triblock Copolymer Syntheses of Mesoporous Silica with Periodic 50 to 300 Angstrom Pores. *Science.* 1998;279(5350):548-552.

28. Xu Q-Q, Ma Y-L, Gang X, Yin J-Z, Wang A-Q, Gao J-J. Comprehensive study of the role of ethylene glycol when preparing Ag@SBA-15 in supercritical CO2. *The Journal of Supercritical Fluids.* 2014;92:100-106.

29. Xu Q-Q, Xu G, Yin J-Z, Wang A-Q, Ma Y-L, Gao J-J. Preparation of Superhighly Dispersed Co3O4@SBA-15 with Different Morphologies in Supercritical CO2with the Assistance of Dilute Acids. *Industrial & Engineering Chemistry Research.* 2014;53(25):10366-10371.

469138 Single-Site Iridium Complexes on MgO Powder:Â Spectroscopic, Microscopic, and Catalytic Evidence of Dependence of Properties on MgO Surface Sites

Abstract:

Supports for Metal Catalysts Are Often High-Area Metal Oxides That Present Highly Nonuniform Surface Sites for Bonding of Metals. in Attempts to Understand How Specific Support Surface Sites Bond with Metal Cations, We Investigated Single-Site Iridium on Powder MgO Treated to Control the Support Surface Sites, Including Faces and Defect Sites. the Catalysts Were Synthesized from the Precursor $Ir(C_2H_4)_2$(acetylacetonate), and the MgO Treatment Conditions Were Selected to Generate Various Surface Sites for Bonding of the Iridium. X-Ray Absorption Spectra Characterizing the Supported Iridium Complexes Identify Distinct Irâ€“Support Surface Site Contributions on the Various MgO Supports and for Iridium Loadings in the Range of 0.1 to 1.0 Wt%. Probing of the Supported Iridium Species with CO (to form iridium

Copyright © American Institute of Chemical Engineers. All rights reserved.

473351 Design of Palladium-Gold Catalysts for Liquid Phase Selective Alkyne Hydrogenation with Single Atom Alloy Strategy

Abstract:

Gold (Au) nanoparticles (NPs) catalysts have been found to be very selective in catalyzing the partial hydrogenation of alkynes, however, are known for low activity in hydrogen activation. We applied single atom alloy (SAA) strategy in design of gold catalysts for selective hydrogenation of alkynes to alkenes, which is a very important process for producing polymer grade alkenes. In Pd-Au SAA, atomically dispersed palladium (Pd) atoms are added into the surface of gold NPs that activate the gold for hydrogenation and maintain the high selectivity to alkenes. Previous studies have demonstrated the bi-functional hydrogenation mechanism on Pt-Cu SAAs that Pt atoms dissociate H_2 and Cu surface catalyzes the hydrogenation. Here we proposed that Pd atoms can improve the H_2 activation activity of gold NPs but do not affect the highly selective hydrogenation chemistry on gold.

We have investigated the selective hydrogenation performance of Pd-Au SAAs. The Pd-Au SAAs enhance activity in liquid phase partial hydrogenation of 1-hexyne and phenylacetylene. The Pd-Au SAA NPs were synthesized by sequential reduction in ethylene glycol with PVP as stabilizing ligand. ATR-IR of adsorbed CO indicates there is no bridge CO on Pd-Au SAA NPs, which proves the formation of SAA. The batch reactor hydrogenation tests show more than 80% selectivity to partial hydrogenation products at full conversion and at least 5-fold increase in activity by doping Au NPs with 0.4% Pd atoms. The preparation, reactivity, and characterization of the alloy nanoparticles with ATR-IR, electron microscopy and X-ray absorption techniques will be presented.

Acknowledgments: This work was supported as part of the Integrated Mesoscale Architectures for Sustainable Catalysis, an Energy Frontier Research Center funded by the U.S. Department of Energy, Office of Science, Basic Energy Sciences, under award #DESC0012573.

Copyright © American Institute of Chemical Engineers. All rights reserved.

473391 Selective Dehydrogenation of Ethanol to Acetaldehyde and Hydrogen on a Highly Diluted Nicu Alloy

Abstract:

The development of catalysts with a high dispersion of an active site for the production of targeted products selectively is a timely subject. Various highly diluted alloys or single atom catalysts have been reported to be highly active for a variety of catalytic reactions. For example atomically dispersed Au species supported on $ZnZrO_x$ oxides can catalyze the dry ethanol dehydrogenation reaction.[1] PtCu single atom alloy catalysts exhibit high activity for several selective hydrogenation and dehydrogenation reactions.[2,3] Moreover, nanoporous Au materials modified with highly dispersed Ag species exhibit unique catalytic performance in the oxidative coupling of alcohols to esters.[4]

Here we report on the selective dehydrogenation of ethanol to acetaldehyde and hydrogen, an important industrial reaction to produce acetaldehyde which is an important commodity chemical, and also valuable hydrogen, using the single atom alloy strategy. Ni-containing Cu nanoparticles (NiCu NPs) and nanoporous Cu (np-NiCu) materials were found very active and selective for this reaction. Although monometallic Cu catalysts are also active, deactivation of Cu was observed. Adding a small amount of Ni to Cu forming NiCu highly diluted alloys dramatically increases the catalytic activity of copper. The kinetic studies show the apparent activation energy decreases from ~70 kJ/mol for Cu to ~45 kJ/mol for NiCu highly diluted alloys. Furthermore adding a small amount of Ni to Cu also significantly improves the stability of these catalysts, likely due to the stabilization of Cu against sintering. Our characterization data show that Ni is highly dispersed or even atomically dispersed on the Cu surface. Such highly dispersed Ni or isolated Ni atoms play a key role in the dehydrogenation of ethanol. This work opens a new route for the use of highly diluted alloys for alcohol dehydrogenation reactions. Furthermore the comparison of the catalytic performance of highly diluted alloy NPs with nanoporous materials is used to guide the design of the novel mesoporous catalyst architectures for selective dehydrogenation or oxidation reactions.

References

1. C. Wang, G. Garbarino, L. F. Allard, F. Wilson, G. Busca, M. Flytzani-Stephanopoulos, *ACS Catal.*, 2016, 6, 210â€"218.

2. F. Lucci, J. Liu, M. Marcinkowski, M. Yang, M. Flytzani-Stephanopoulos, C. E. H. Sykes, *Nat. Commun.*6, 8550 (2015).

3. J. Shan, F. R. Lucci, J. Liu, M. El-Soda, M. D. Marcinkowski, L. F. Allard, C. E. H. Sykes, M. Flytzani-Stephanopoulos, *Surface Science*(2016), DOE: 10.1016/j.susc.2016.02.010.

4.Â Branko Zugic, Stavros Karakalos, Kara J. Stowers, Monika M. Biener, Juergen Biener, Robert J. Madix, and Cynthia M. Friend, *ACS Catal.*, 2016, 6, 1833â€"1839.

Copyright © American Institute of Chemical Engineers. All rights reserved.

464993 Heterobimetallic Complexes As Controlled Precursors for Atomically Dispersed, Promoted Methane Oxidation Catalysts

Abstract:

Heterobimetallic Complexes as Controlled Precursors for Atomically Dispersed, Promoted Methane Oxidation Catalysts

Joseph Maalouf, Jay Schwalbe, Joshua J. Willis, Eric A. Stach[2], Matteo Cargnello

[1]Department of Chemical Engineering and SUNCAT Center for Interface Science and Catalysis,Stanford University, Stanford, CA 94305, USA

[2]Center for Functional Nanomaterials, Brookhaven National Laboratory, Upton, NY 11973, USA

Thanks to improvements in fracking technology, methane has become a cheap and abundant commodity that is gaining a prominent position in the energy scenario. However, there is a crucial need for improved methane combustion catalysts, as low temperature methane oxidation can reduce emissions from natural gas leaks, and reduce levels of toxic gases released from flaring, such as NO_x , SO_x and CO. In this study, a method for producing a highly active catalyst that combines both promotion and atomic dispersion is presented.

Promoted, atomically disperse catalysts are extremely difficult to produce, especially with traditional impregnation methods. These techniques rely on a probabilistic adsorption of a compound onto a support, with no real control over distribution and composition, but promotion often requires two metals to be in extremely close vicinity to one another. The issue is only exacerbated when attempting to achieve atomic dispersion, as this often requires extremely low weight loadings, meaning that statistical occurrences of the metal and promoter depositing onto the support with a high interface are extremely low.

Here, we show that by using a pre-formed, atomically precise Palladium Cerium heterobimetallic precursor, it is possible to achieve atomic dispersion and keep the active phase (Pd) and the promotor (Ce) in close proximity. To fully exploit this method, support engineering was also performed. Alumina was functionalized with mixtures of triethoxy(octyl)silane (TEOOS) and (3-mercaptopropyl) trimethoxysilane (MPTMS) in a TEOOS to MPTMS ratio much greater than one, in order to produce atomic scale islands of MPTMS, which bind to the Pd in the complex, facilitating atomic dispersion and hindering clumping via diffusion across the surface. Indirect evidence of atomic dispersion was supported by HAADF-STEM characterization. These samples were then calcined to remove any organic ligands and tested for activity, showing rates that were at least 5 times higher than those obtained with catalysts prepared via impregnation of the isolated Pd and Ce precursors.

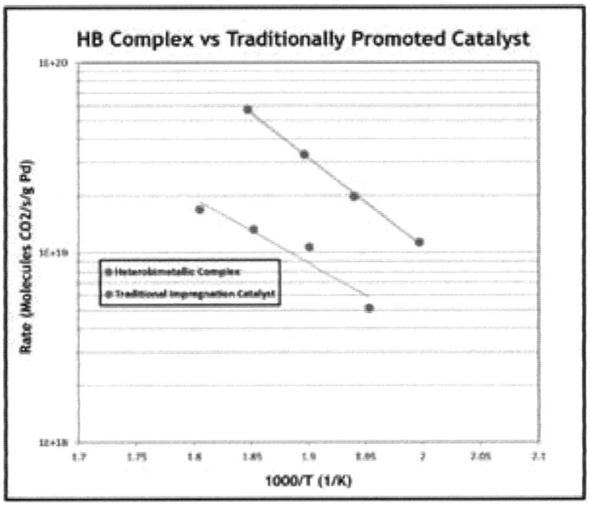

(5) Reaction Conditions: Total flow of 66.7 mL/min with 0.5% CH_4 and 2% O_2 with a GHSV of 200,000 h^{-1}. Weight Loading (Pd) : HB Complex 0.1w%, Co-Imp 1w%

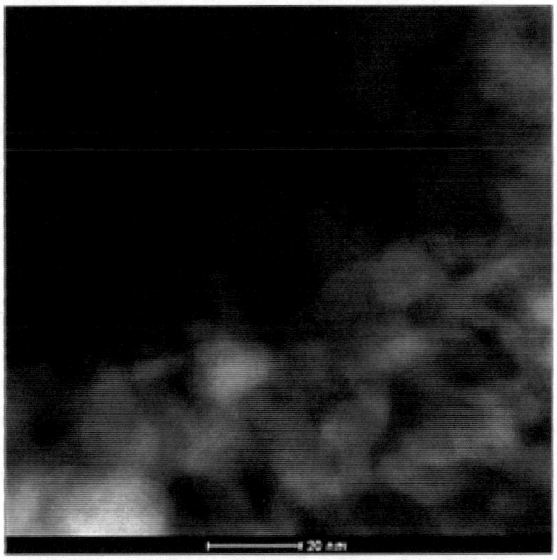

(3) HAADF-STEM image of PdCe Complex on functionalized Alumina (0.1 w%). Calcined at 300°C for 5 hours.

Copyright © American Institute of Chemical Engineers. All rights reserved.

468276 Site-Specific IR Spectroscopy for Analyzing Pt Structure Sensitivity from Single Atoms to Fully Coordinated Surfaces

Abstract:

Significant efforts have been focused on understanding the structure sensitivity of Pt catalysts from the limit of well-coordinated Pt atoms in terraces on large particles, down to single Pt atoms coordinated directly to oxide supports for the CO oxidation and water gas shift reactions. Often X-ray based spectroscopy and transmission electron microscopy provide ensemble measurements of structural features of supported Pt catalysts that are used to correlate the reactivity of various active sites on different supports. Recently, it has been demonstrated that *site-specific* CO probe molecule FTIR spectroscopy is a powerful approach to identify the reactivity of various Pt active site geometries on supported Pt catalysts, overcoming shortcomings of characterization techniques that rely on ensemble measurements. However, the use of FTIR to study the reactivity of various Pt active sites requires unequivocal assignment of spectral features, which can vary as a function of support composition. Here we will present results of rigorous FTIR-based analysis of molecular adsorption, temperature programmed studies, and oxidation reactions for Pt catalysts on various oxide supports prepared through conventional dry impregnation and strong electrostatic adsorption (SEA) wet impregnation techniques with weight loadings ranging from ultra-low to more typical (0.05 - 2%). Through these approaches we are able to directly differentiate the reactivity of various Pt active sites, ranging across Pt single atoms, well-coordinated Pt metal sites, and Pt oxide clusters while simultaneously exploring the role of the support in modifying the reactivity of these active sites.

Copyright © American Institute of Chemical Engineers. All rights reserved.

473773 Atomically Dispersed Pt Catalysts on Polyphosphomolybdic Acid: Structural Features and Catalytic Properties

Abstract:

Single-atom catalyst (SAC), referring to a class of catalysts where bare metal atoms are atomically dispersed or anchored on the support, has recently emerged as a new frontier in heterogeneous catalysis.[1] A key challenge encountered, however, is to stabilize isolated metals on a support without compromising catalytic activity.

In coordination chemistry, catalytically active metal complexes in zero or low valent state often adopt four-coordinate square-planar or tetrahedron geometry. By applying this principle, we have developed stable, high loading (close to 1 wt%), Pt_1 single-atom catalyst on phosphomolybdic acid modified active carbon, by anchoring Pt on the four-fold hollow sites on PMA. Each Pt atom is stabilized by four oxygen atoms in a distorted square-planar geometry, with Pt slightly protruding from the oxygen planar surface. Pt is positively charged, easily absorbs hydrogen, and exhibits excellent performance in the hydrogenation of nitrobenzene and cyclohexanone. The system described here is likely to be extended to a number of stable SACs with superior catalytic activities.[2]

1. a) Flytzani-Stephanopoulos, M.; Gates, B.C. *Annu. Rev. Chem. Biomol. Eng.* **2012**, *3*, 545-574; b) Thomas, J.M. *Nature* **2015**, *525*, 325-326.

2. Zhang, B.; Asakura, H.; Zhang, J.; Zhang, J.; De, S.; Yan, N. *Angew. Chem. Int. Ed.,* **2016**, DOI: 10.1002/anie.201602801R1.

Copyright © American Institute of Chemical Engineers. All rights reserved.

472560 Understanding Site Isolation of Pd and Ni in Zn-Based Bulk Intermetallics during the Selective Hydrogenation of Alkynes

Abstract:

The catalytic semi-hydrogenation of acetylene to produce ethylene is a common method for the removal of trace acetylene (~1%) in ethylene feed streams destined for ethylene polymerization. An effective catalyst for this reaction converts all of the acetylene to ethylene without further conversion of ethylene to ethane such that there is a net increase in the amount of ethylene. Pd-Ag alloys, and more recently, intermetallic Pd-Ga compounds, demonstrate high selectivity towards ethylene and long-term stability. Improved selectivity is a result of isolation of active Pd hydrogenation sites which reduces over-hydrogenation to form ethane, oligomerization products, and coke formation on the catalyst surface. Replacing Pd-based catalysts with base metal Ni-based catalysts would be highly beneficial in terms of cost and environmental impact. Bulk intermetallic catalysts contain little structural and compositional variance, a property that is not easily attainable with supported catalysts. We report on the catalytic selectivity of unsupported bulk intermetallic Ni-Zn and Pd-Zn catalysts for acetylene semi-hydrogenation. We demonstrate the addition of Ni to Zn improves selectivity to ethylene due to a reduction in acetylene oligomerization products rather than ethane over-hydrogenation. The most selective catalysts had the lowest Ni content with a g-brass bulk structure. Structural investigation by neutron diffraction demonstrated the presence of well-defined Ni (or Pd) trimers in the g-brass structure with the number of these trimers increasing with Ni (or Pd) content. The Ni alloys are active for H_2-D_2 exchange, but all surfaces (i.e, different Ni content) are indistinguishable with respect to their catalytic behavior. In the case of Pd-Zn alloys with the g-brass structure, there are apparent differences in the H_2-D_2 rates and ethylene hydrogenation as the Pd content (and therefore number of Pd trimers) increases. While the Ni and Pd trimers are stable in the bulk structure, the differences in the surface stability of these trimers in the Zn matrix and the corresponding percentage of the trimer-containing surface present in a Wulff reconstruction explain the disparate catalytic results in the seemingly related systems. We demonstrate bulk intermetallics are useful systems to screen compositional and structural (from site-isolated to well-defined 2D clusters) variance on catalytic behavior. Transforming the bulk intermetallic structure into its nanoscale analog still remains a critical challenge for catalyst synthesis.

Copyright © American Institute of Chemical Engineers. All rights reserved.

463555 Mechanism of Oxygen Activation during Selective Catalytic Reduction of NO with NH_3 in Cu-SSZ-13 Zeolites

Abstract:

Cu-SSZ-13 zeolites are used commercially in diesel engine exhaust after-treatment for abatement of NOx pollutants via selective catalytic reduction (SCR) with NH_3. Yet, the mechanistic details of the redox reactions they catalyze, and specifically of the Cu(I)->Cu(II) oxidation half-reaction, are not well understood. A detailed understanding of the SCR reaction mechanism and Cu active site requirements would provide insight into their catalytic performance and specific guidance on how to synthesize materials with improved low temperature (<473 K) reactivity.

The standard SCR reaction uses O_2 as the oxidant ($4NH_3 + 4NO + O_2 \rightarrow 6H_2O + 4N_2$) and involves a Cu(I)/Cu(II) redox cycle, with Cu(II) reduction mediated by NO and NH_3 [1], and Cu(I) oxidation mediated by NO and O_2. In contrast, the fast SCR reaction ($4NH_3 + 2NO + 2NO_2 \rightarrow 6H_2O + 4N_2$) uses NO_2 as the oxidant. Here, we show through a coordinated effort involving theory and experiment, that the fast and standard SCR reactions are not linked via common reactive intermediates but instead proceed with different active site requirements and by different elementary reaction steps.

Low temperature (437 K) standard SCR reaction kinetics depend strongly on the spatial density of Cu ions, varied by changing the Cu/Al and Si/Al ratio of Cu-SSZ-13 zeolites. Facilitated by NH_3 solvation [2], mobile Cu(I) complexes can dimerize with other Cu(I) complexes within diffusion distances to activate O_2, as demonstrated through X-ray absorption spectroscopy and density functional theory calculations. Monte Carlo simulations are used to define average Cu-Cu distances, ab initio metadynamics simulations to quantify Cu mobility, and climbing-image nudged elastic band calculations to calculate Cu diffusion energies between chabazite cages and activation of O_2 by Cu complexes. These theoretical results are validated by kinetic and spectroscopic experiments.

In contrast with O_2-assisted oxidation reactions, NO_2 oxidizes single Cu(I) complexes with similar kinetics among samples of varying Cu spatial density.Â These findings demonstrate that low temperature standard SCR is strongly dependent on Cu spatial density and requires NH_3Â solvation to mobilize Cu(I) sites in order to activate O_2, while fast SCR uses NO_2Â to oxidize single Cu sites and complete the catalytic redox cycle at one Cu site for both reduction and oxidation half-reactions.

(1) Paolucci, C.; Verma, A. A.; Bates, S. A.; Kispersky, V. F.; Miller, J. T.; Gounder, R.; Delgass, W. N.; Ribeiro, F. H.; Schneider, W. F.Â *Angew. Chem., Int. Ed.* 2014, 53, 11828â"11833.

(2) C. Paolucci, A. A. Parekh, I. Khurana, J. R. Di Iorio, H. Li, J. D. Albarracin Caballero, A. Shih, T. Anggara, W. N. Delgass, J. T. Miller, F. H. Ribeiro, R. Gounder and W. F. Schneider *Journal of the American Chemical Society*, (**2016**) DOI: 10.1021/jacs.6b02651

Copyright © American Institute of Chemical Engineers. All rights reserved.

468514 Light Hydrocarbon Reactions on Solid Bronsted Acids: Confinement and Acid Strength Effects on Ion-Pair Transition States and Their Precursors

Abstract:

Light hydrocarbons are converted on solid Brønsted acids via oligomerization, isomerization, β-scission, hydride transfer and cyclization reactions; these reactions are mediated by ion-pair transition states (TS). The relative stability of these ion-pairs, consisting of organic cations derived from reactants and inorganic anions formed by deprotonation of the inorganic catalyst, relative to their relevant precursors determine the rate of each reaction and their prevalence over the others. The anion contributes to TS stability by (i) van der Waals interactions conferred by the geometry of its surroundings (confinement) and (ii) its willingness to accept the negative charge (acid strength) and to interact electrostatically with the organic cation. The cation, in contrast, determines TS stability through its ability to accept the proton and then interact with the anion. These confinement and electrostatic effects can be probed using reactions mediated by TS that differ in size, shape or charge to various extents from each other and from their relevant kinetic precursors; density function theory (DFT) methods that account for the pertinent dispersive interactions further provide detailed information that is inaccessible from experiments alone, such as how the cation and anion conform to each other.

Here, we report turnover rates (per proton) for C_2-C_4 alkene oligomerization reactions and for the incorporation of C_4-C_5 alkanes via hydride transfer on acids of different acid strength and confining environments (TON, MFI, BEA, MOR, FAU zeolites; amorphous silica-alumina (SiAl), Keggin polyoxometalate clusters (POM) clusters on SiO_2). C-C bond formation and hydride transfer rates show that kinetically-relevant steps involve reactions of an alkane or alkane to alkene-derived alkoxides present as saturation coverages, consistent with *in-situ* infrared spectra and DFT estimates of activation free energy barriers and of the stability of alkoxide intermediates. These kinetic data, obtained over a very broad range of reactant pressures, allow a systematic comparison between theory and experiment and also accurate estimates of alkoxide adsorption constants. In doing so, we provide a quantitative assessment of the effects of surface curvature and of alkene size and structure on alkoxide stabilities, which indicates bulkier alkoxides become sterically hindered within smaller, concave environments (TON) as compared to larger pore environments (MOR, FAU, SiAl, POM).

The rate constants for oligomerization and hydride transfer increase exponentially as the deprotonation energy (DPE) of the solid acid decreases and the stability of its conjugate anion concomitantly increases. These effects reflect DFT-derived transition states that differ in charge from their alkoxide precursors. Rates constants also generally increase with increasing TS size for both oligomerization and hydride transfer on each zeolite framework, because of the combined effects of the greater stability of the larger TS carbenium ions and their more effective contact with the void walls. DFT treatments show that zeolite frameworks distort locally, so as to enhance van der Waals contacts at the expense of a slight distortion of the framework lattice, which ultimately becomes too costly as the TS reach the size of the confining voids. For example, the lattice of TON (0.57 nm channel) locally moves closer to the TS for smaller TS (ethene dimerization; 0.46 nm diameter) and moves away for larger TS (isobutene dimerization; 0.58 nm diameter); the distortions become too costly, however, when the TS is much larger than the void (isobutane-C_6 hydride transfer; 0.81 nm diameter), consistent with a lower rate constant for TON than for larger pore environments. These energy compromises are also critical in alkoxide formation. DFT-calculated alkoxide energies indicate bulkier alkoxides distort the framework more in smaller, concave environments (TON) than larger pore ones (MOR, HPW).

This array of transition states and their precursors—formed from a range of reactants and catalysts—exploits the diversity in size, shape and charge in solid acid upgrading of alkenes/alkanes and provides unprecedented clarity of the complex relationship between organic moieties and inorganic catalysts when combined with state-of-the-art theoretical methods. By developing descriptors that relate reactivity and selectivity to transition state properties (intrinsic stability of the organic cation, electrostatic interaction with the anion (acid strength) and stabilization via confinement provided by the flexible inorganic framework), we aim to extrapolate these relationships to other transition states and acid catalysts than those studied here.

Financial support from BP through the ICC Program and National Science Foundation Graduate Research Fellowship Programs and computational resources from National Science Foundation's Extreme Science and Engineering Discovery Environment (XSEDE; ACI- 1053575) are gratefully acknowledged.

Copyright © American Institute of Chemical Engineers. All rights reserved.

470142 Development of a Practical Catalyst for Terminal Olefin Epoxidation Using Organic Hydroperoxide As Oxidant

Abstract:

More than 30 % of worldwide propylene oxide production relies on an organic hydroperoxide as oxidant and grafted titanium centers on a siliceous support as catalyst. The industrial catalyst support is an amorphous and poorly defined silica network. Here, we investigate repercussions of a crystalline silaceous framework as support on the catalyst-site activity, selectivity, and long-term stability. Our data demonstrates that a crystalline microporous framework as support for grafted Ti centers can function at a higher per-gram-activity and selectivity when compared with a conventional amorphous catalyst, at similar loadings of isolated Ti sites, and also highlights the importance of localized framework structure on catalyst function.

Copyright © American Institute of Chemical Engineers. All rights reserved.

468649 Mesoporous Nb-Silicates As Ethylene Epoxidation Catalysts: Novel Strategies to PassivateÂ Bronsted Acidity

Abstract:

Ethylene oxide (EO) is one of the widely used chemical intermediates with current global production capacity of 20.5 million metric tons/yr. The present Ag-based catalytic process has a large carbon footprint due to CO_2 formation as byproduct. The search for alternate EO technology that suppresses such CO_2 formation continues to receive attention. Recently, we reported that Nb-TUD-1 catalysts are highly active and selective (â‰¥99%) toward EO formation with no CO_2 formation as byproduct.[1] However, H_2O_2 (oxidant) decomposition and significant Nb leaching are major drawbacks that must be overcome for this catalytic process to be practically viable. Computational studies suggest that passivation of Brønsted acid sites can reduce H_2O_2 decomposition and metal leaching.

This report focuses on developing improved strategies to neutralize Brønsted acid sites including base treatment and covalent capping of catalysts. The effects of such improved catalysts on catalyst performance (epoxidation activity, catalyst recyclability, H_2O_2 utilization and resistance to metal leaching) were systematically investigated. Different capping agents were investigated with Nb-TUD-1 catalysts, aimed at virtually eliminating the Brønsted (B) acidity of the catalyst while retaining only the Lewis (L) acidity. Capping of Nb-OH and Si-OH groups in the catalyst using two organic bases (tetraethylammonium hydroxide and di-isopropylethyl amine) significantly enhanced H_2O_2 utilization efficiency (70% and 90%, respectively), but the resistance to metal leaching was only moderately improved. In order to increase hydrolytic stability, four different capping groups, viz., methyl (-Me), trimethylsilyl (-$SiMe_3$), t-butyl (-CMe_3) and benzyl (-CH_2Ph), with moderate to high hydrolytic stability were investigated as capping groups. The capped catalyst using benzyl group shows the best performance in terms of reducing H_2O_2 decomposition and metal

leaching. It displays significantly reduced H_2O_2 decomposition (0.0-0.26 %) after a 5 h run. It also shows very high EO selectivity (~99 %) and a dramatic reduction in Nb leaching (< 3% after 4 cycles compared to >75% after one cyce with the non-benzylated catalyst). The benzylated catalyst was recycled 4 times with minimum loss of catalytic activity. These results clearly suggest that passivation of Brønsted acid sites either by base treatment or by capping with suitable organic groups increases EO selectivity, H_2O_2 utilization efficiency and resistance to metal leaching compared to the untreated parent catalyst. The benzylated Nb-TUD-1 catalyst in particular shows immense potential to be a viable epoxidation catalyst.

References:

1. Yan, W.; Ramanathan, A.; Patel, P. D.; Maiti, S. K.; Laird, B. B.; Thampson, W. H.; Subramaniam, B. J. Catal. 336 (2016) 75-84.

Copyright © American Institute of Chemical Engineers. All rights reserved.

472099 Investigation of the Acid Catalyzed Formaldehyde-Olefin Condensation Reaction

Abstract:

New carbon-carbon bonds can be formed via the "ene" reaction, a pericyclic process between an alkene having an allylic hydrogen atom (an "ene") and an electron-deficient multiple bond (an enophile) to form two sigma-bonds with transposition of the pi-bond [1] (see Scheme 1).

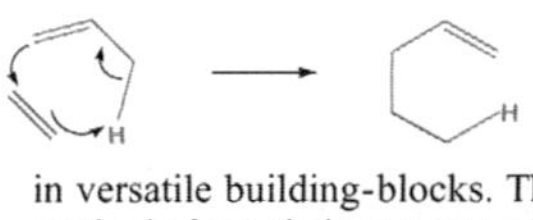

The Prins reaction is an example of the "ene" reaction referring to condensation of formaldehyde with olefins [2,3]. Functionalization of simple olefins in this way results in versatile building-blocks. The commercial availability of lower olefins, especially those formed by on-purpose methods from their corresponding alkanes (C3 and C4), coupled with the versatility of formaldehyde as a one-carbon electrophile make this reaction potentially important. Prins condensation can form unsaturated alcohols, diols, alkyl dioxanes, pyran skeleton compounds and dienes. The reaction is an acid catalyzed addition of aldehydes to olefins and is traditionally catalyzed by homogeneous mineral acids such as sulphuric acid or homogeneous Lewis acids like $SnCl_4$, BF_3 and $ZnCl_2$ [4, 5].

We have investigated solid Lewis acid catalysis for the reaction. Acidic zeolite Beta catalysts have been synthesized and evaluated in the condensation of formaldehyde with propylene. Specifically, framework Sn-, Zr-beta, and Zn-Beta —the last prepared by ion exchange— were prepared and characterized. The catalytic activity was tested by autoclaving formaldehyde and propylene using an appropriate solvent at a temperature range of 120-180^{0}C. We have found that Zn-Beta is the most active catalyst for this reaction. Different reaction pathways are followed (see Scheme 2) producing valuable products. The product spectra can be divided into two categories; compounds with four carbon atoms and compounds having five carbon atoms. C-4 products include the unsaturated alcohol, 3-buten-1-ol, which

dehydrates to form 1,3-butadiene. Reaction of excess formaldehyde with propylene forms the alkyl dioxane, 1,3-Dioxane, 4-methyl- (C-5 product). This compound can potentially hydrolyzed to afford 1,3-butanediol and CH_2O (a non-favored pathway under our experimental conditions). Prins cyclization [6] (reaction between 3-buten-1-ol and formaldehyde) leading to the synthesis of tetrahydropyran-ol, which subsequently dehydrates to the corresponding dihydropyran is the dominant reaction route. Based on these data, it seems necessary to exert an accurate control of the experimental conditions and the catalyst used in order to selectively drive the reaction in a specific direction. On-going research is directed towards optimization of reaction conditions to minimize important undesired side reaction pathways as well as evaluation of different pore size zeolites to induce shape selectivity effects.

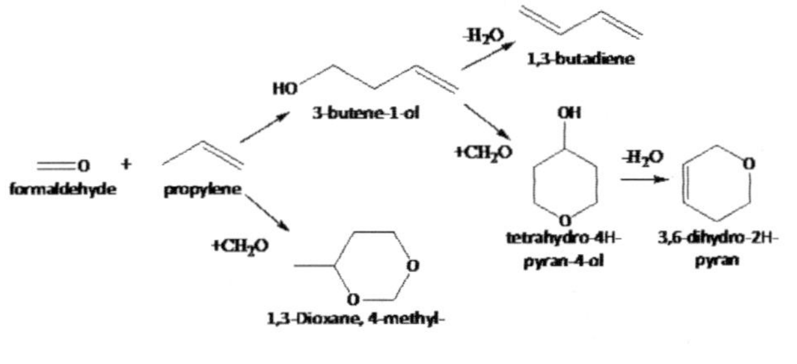

References

[1] M. Yamanaka and K. Mikami, Helvetica Chimica Acta, **85** (2002) 4264

[2] P. C. Bloys van Treslong Prins, Chem. Weekkbl. 1919, 1510

[3] P. C. Bloys van Treslong Prins, Chem. Weekkbl. 1919, 1072

[4] D. R. Adams and S. P. Bhatanagar, Synthesis **10** (1977) 661

[5] W. Fitzky, U.S. Patent 2,325,760, 1943

[6] C. Olier, M. Kaafarani, S. Gastaldi, M. P. Bertrand, Tetrahedron **66** (2010) 413

Copyright © American Institute of Chemical Engineers. All rights reserved.

468998 Experimental and Computational Interrogation of Fast SCR Mechanism and Active Sites on H-Form SSZ-13

Abstract:

The performance of cation-exchanged zeolites for the selective catalytic reduction (SCR) of NOx is sensitive to the identity of the cations as well as the composition of gaseous feed. While the redox function of a metal-exchanged zeolites (most notably Cu-SSZ-13) is necessary for so-called "standard" SCR (4 NO + 4 NH3 + O2 → 2 N2 + 6 H2O) activity, proton-exchanged zeolites, like H-SSZ-13, are comparable in performance to their metal-exchanged cousins for "fast" SCR (2 NO + 2 NO2 + 4 NH3 → 2 N2 + 6 H2O) [1]. Further, residual exchanged protons (Brønsted acid sites) are present in all metal-exchanged zeolites, and these acid sites may have a role even in standard SCR [2]. For these reasons, it is important to understand the nature and the reactivity of exchanged proton under SCR conditions.

Here we report the results of an integrated density functional theory (DFT) and kinetics analysis of H-SSZ-13 under fast SCR conditions. Rates, rate orders, and apparent activation energies collected under differential conditions reveal distinct kinetic regimes suggestive of different intermediates and/or coverages. First-principles thermodynamics models based on ab initio dynamics distinguish correspondingly different NH3 coverage regimes, and in particular identify NH3 physisorbed to NH4+ as a key low-temperature intermediate. Further, we use non-equilibrium dynamics methods to determine activation energies for rate-limiting steps in the catalysis. Based on these results, we propose a kinetic model that explains the available experimental data, including in particular the changes across temperature regimes. The results highlight the importance of considering realistic reaction conditions in both measuring and modeling catalytic function.

Reference:

[1] Beale, A.M., Gao, Feng., Lezcano-Gonzalez, I., Peden, C.H.F., Szanyi, J., Chem. Soc. Rev., 2015,44,

7371-7405

[2] Paolucci, C., Verma, A.A., Bates, S.A., Kispersky, V.F., Miller, J.T., Grouder, R., Delgass, W.N., Ribeiro, F.H., and Schneider , W.F., Angew. Chem. Int. Ed., 2014, 53, 44

Copyright © American Institute of Chemical Engineers. All rights reserved.

466681 Quantification of External and Pore Mouth Acid Sites in Unit-Cell Thick Pillared MFI and Pillared MWW Zeolites

Abstract:

The quantification of acid sites on the external surface and in pore mouth of pillared MFI (PMFI) and pillared MWW (PMWW or MCM-36) zeolites was studied. A combined techniques of organic base titration, atomic layer deposition (ALD) of silicon (ALD-Si) and alkylation of benzyl alcohol in mesitylene were carried out to identify the acid site locations, their quantities and catalytic behaviors. The organic base titrations with triphenylphosphine (TPP, ? nm, only access to external acid sites) and 2,6-di-tert-butylpyridine (DTBP, ? nm, access to both external and pore mouth acid sites), respectively, in the course of methanol dehydration reactions indicated that PMFI contains 10% external and 26% pore mouth active sites, while PMWW consists of 22% and 45%, respectively. The ALD-Si treatment of PMFI and PMWW zeolites led to loss of external surface acid sites, which is consistent with that determined from TPP titrations. The catalysis tests in the absence and presence of TPP and DTBP, respectively, comparatively indicated the catalytic behaviors of external and pore mouth acid sites in PMFI and PMWW zeolites. The alkylation and etherification reaction rates as a function of cumulative DTBP addition suggested Brønsted acid sites with different strengths on the external surface or in pore mouth environments of the pillared zeolites, which influenced the external etherification reaction, but not as significantly as the alkylation reaction. The PMWW and PMFI zeolites are 2-dimensional (2D) catalytically active materials with single- or near single-unit-cell thick 2D microporous layers which expose higher number of external surface and pore mouth acid sites than their microporous counterparts. The present study exemplified for the first time to differentiate and quantify surface acid sites (external and pore mouth) in the 2D unit-cell thick zeolites.

Copyright © American Institute of Chemical Engineers. All rights reserved.

460664 Green Application of Hydrogenation Catalyst Yn-1 on C_4 Cut

Abstract:

Green Application of Hydrogenation Catalyst YN-1 on C_4 cut

zhou Du

Yanshan Branch of Beijing Research Institute of Chemical Industry£¬Sinopec, Beijing, China£¬102500

Abstract:

YN-1 Ni-based hydrogenation catalyst adopts TiO_2-Al_2O_3 as its carrier in the shape of trilobe in cross section. It is well known that a significant synergistic interaction exists between Ni and titanium oxide, which makes the catalyst be endowed with comparatively high activity. More importantly, the catalyst can be operated at the lowest tolerable temperature due to the synergistic effect; thereby the catalyst acquires higher selectivity and minimum side reaction. Ni, as the active species, possesses superior resistance to sulfur, especially to mercaptan. The larger pores in the catalyst significantly favor diminishing the resistance in the mass transfer process during the hydrogenation reaction. Its physical property is illustrated in Table 1.

Table 1 Physical property of YN-1 hydrogenation catalyst

Item	Property
Color	Green
Shape	Trilobe Extrudate
Size	4-7 mm
Bulk density	0.6-0.8 g/ml
Crushing strength:	>18 N/mm
Active component	Nickel

The YN-1 Ni-based catalyst is suitable for selective hydrogenation of diverse distillation cuts (C_4 ¨C C_{10}) of pyrolysis gasoline containing high content of impurities such as sulfur, arsenic and nitrogen. In this paper, its green application in hydrotreating C_4 cut will be discussed.

YN-1 catalyst has been successfully commercialized in 40Kton/a Butadiene Tail gas Hydrogenation unit in Maoming division of Sinopec, China in July, 2015. This is the first industrial application which turns the C_4 tail gas from Butadiene Extraction unit into ethylene cracking feedstock in China. The tail gas full of C_4 olefins£¨like butadiene£© and alkynes(like ethyl acetylene and vinyl acetylene) once was sent to torch to be lighted as waste gas. With the development of our new C_4 hydrotreating technology, it¡¯s now mixed with toluene and sent into hydrogenation reactor to undergo hydrogenation reaction. After that, the olefins in it are fully saturated and are then introduced to ethylene cracking unit after toluene separation. The schematic of C_4 cut hydrogenation technology is shown in Figure1. The hydrogenation product which is rich in butane produces 13 wt% more ethylene than ordinary cracking feed like naphtha, which in turn promote the ethylene production and protect environment.

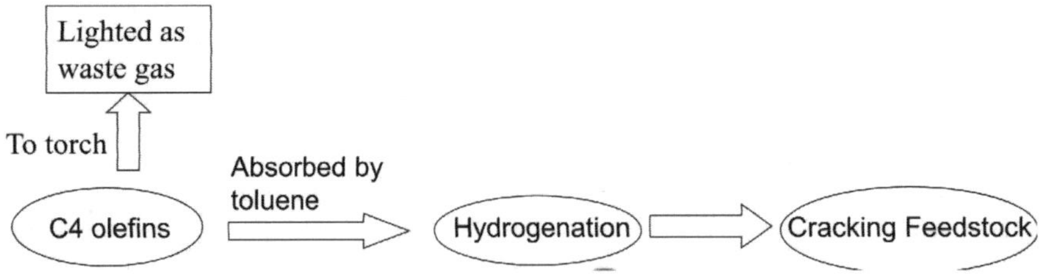

Figure1 The schematic of C_4 cut hydrogenation technology

After successful start up under room temperature in the plant, the hydrogenation product¡¯s olefin content (including alkynes) is controlled bellow 0.3wt%, while that of feed is about 40wt%, indicating that YN-1 catalyst exerts excellent activity under low temperature on C_4 cut saturation.

Key words:

green application; hydrogenation; tail gas; ethylene production ; C_4 cut

Copyright © American Institute of Chemical Engineers. All rights reserved.

450870 "Molybdenum Carbide As a Potential Catalyst for Olefins Hydrogenation"

Abstract:

There is a vital necessity in the suppression of olefins from upgraded heavy oil fractions. These deleterious molecules can form gum deposits on equipment and polymerize in pipelines over time during oil transportation, resulting in the fouling of heat exchangers and possible formation of coke. Most of the catalysts used so far in olefins hydrogenation, are expensive noble metals, such as Pt, Pd and Ru. In addition, these catalysts are vulnerable to sulfur poisoning, especially when used with heavy oil feedstocks. Novel inexpensive molybdenum carbide (Mo_2C) catalysts with nanocrystalline size domains have been synthesized in our laboratory at lower temperatures than previously reported.[1,2] Our cost-effective synthetic approach now allows us to obtain good control over physical properties and structure of the carbide along with catalyst scale-up. The prepared material showed high activity towards the selective hydrogenation of aromatics at temperatures below 200 ^{o} C, producing methyl-cyclohexane, when toluene was used as a model molecule in a bench scale hydrogenation setup.[2,3]

More suprisingly, our preliminary results using the above mentioned catalyst showed that it is possible to hydrogenate aromatics at very low pressures (< 100 psi) while still maintaining an excellent selectivity and catalyst stability. Activity of the proposed catalyst towards aromatics at low pressures and temperatures has directed our attention that it could be possible to hydrogenate olefins at even milder reaction conditions, as olefins hydrogenation is generally thermodynamically favored at lower temperatures and pressures compared to aromatics. Thus, heterogeneous catalytic hydrogenation of 1-octene as an olefinic model molecule was tested on bulk cubic Mo_2C. The selective conversion of 1-octene to n-octane occurred in a timely manner. In addition, the cubic Mo_2C active phase transformed into the hexagonal Mo_2C phase during both toluene and 1-octene hydrogenation reactions, as confirmed by XRD

analysis; however, without affecting the activity and selectivity towards the desired hydrogeneted products. Another designed carbide catalyst, which is the cubic Mo_2C impregnated onto commercially available alumina spheres, has also been tested for 1-octene hydrogenation. Decreasing the carbide active phase amount by its impregnation on alumina did not show a decrease in catalyst activity or selectivity when compared to the bulk cubic Mo_2C. This shows that less amounts of carbide active phase can be used when this carbide is supported on alumina towards olefins hydrogenation.

References

[1] K. Leary, J. Michaels, A. Stacy, J. Catal. 101 (1986) 301–313.

[2] G. Vitale, M. Frauwallner, C. Scott, P. Pereira-Almao, J. Appl. Catal. A: Gen. 408 (2011) 178-186.

[3] G. Vitale, H. Guzman, M. Frauwallner, C. Scott, P. Pereira-Almao, Catal. Today. 250 (2015) 123-133.

Copyright © American Institute of Chemical Engineers. All rights reserved.

469821 Olefin Oligomerization on Carbon Supported Cobalt Catalyst: Identification of the Reaction Pathway

Abstract:

In this presentation we will discuss the reaction pathway involved in oligomerization of liquid 1-butene to linear octenes (in 70-85% selectivity) using carbon supported cobalt. The liquid products were characterized by 2 dimensional GCMS which allowed us to better understand the complicated reaction pathway occurring with this catalyst. The linear octene products decreased in selectivity according to 3-octene > *trans*-2-octene> *cis*-2-octene > 4-octene. Methyl-heptenes including *trans/cis*-5-methyl-2-heptene > *trans/cis*-5-methyl-3-heptene > *trans*-3-methyl-2-heptene (at the lowest conversion) were the other major products. The product distribution suggests the reaction pathway involves a head- to -head coupling of two 1-butene molecules to form internal linear octenes. Head- to -tail coupling of two 1-butene molecules or a coupling between 1-butene and 2-butene forms the observed methyl-heptenes. The rate of head- to -head coupling is higher than the rate of head- to -tail or the rate of 1-butene to 2-butene coupling as indicated by the higher selectivity of linear octenes. We will then compare the reaction pathway of this catalyst for 1-butene conversion to the conversion of other olefins to obtain a better mechanistic understanding of this catalyst system.

Copyright © American Institute of Chemical Engineers. All rights reserved.

454187 Oxidative Dehydrogenation of Propane to Propylene over $VO_x/CaO\text{-}\gamma Al_2O_3$ Catalysts

Abstract:

Propylene is one of the important feed stocks for the chemical industry to produce different valuable products. There are three major commercial processes available for propylene production, including steam cracking, catalytic cracking (FCC) and catalytic dehydrogenation. Steam cracking process consumes a large amount of energy, which accounts for 70 % of the overall production cost. The coke formation during the cracking of heavy hydrocarbon molecules is another drawback of the catalytic cracking process. This causes severe process operational problems, especially fouling, which requires frequent plant shut-downs for cleaning. The catalytic dehydrogenation, also suffers from the problem of coke formation and high energy requirement as a result of the endothermic nature of the reaction.

In this regard the oxidative dehydrogenation (ODH) of propane to propylene is more attractive due to its low operational cost and minimal environmental impacts. The abundant availability of propane in natural gas and refinery off gases can make the propylene production by ODH propane even sustainable as compared to the conventional processes. The most important advantage is the exothermic nature of the reaction, which requires no additional energy to accelerate the reaction. In this study the ODH of propane to propylene is investigated with a new vanadium catalyst supported on $CaO\text{-}\gamma Al_2O_3$ under gas phase oxygen free atmosphere. The catalysts are synthesized with different $CaO/\gamma Al_2O_3$ ratios keeping vanadium loading at 10 percent. The prepared catalysts are characterized using various physicochemical techniques. Raman spectroscopy reveals that the catalyst have monovanadate and polyvanadate surface species of VO_x with minute crystal particles of V_2O_5. FTIR and XRD analysis confirm the presence of V_2O_5, CaO and Al_2O_3 in the catalyst. The catalysts shows stable reduction and re-oxidation behavior in repeated TPR and TPO cycles, respectively. NH_3-TPD shows that catalyst acidity decreases with increasing the CaO content. The

NH_3-TPD kinetics analysis reveals that the activation energy of desorption increases with higher CaO, indicating higher active site-support interaction. The ODH of propane experiments are conducted in a fluidized CREC Riser Simulator under gas phase oxygen free conditions. Among the studied catalyst, $VOx/CaO\text{-}Al_2O_3$ (1:1) displays highest propane conversion (65 %) and propylene selectivity (85%) and the low CO_x due to its excellent oxygen carrying capacity, balanced acidity and moderate active site-support interactions.

Copyright © American Institute of Chemical Engineers. All rights reserved.

455192 On-Purpose Propylene Production from Light Hydrocarbons with MFI-Zeolite/Metal-Oxide Composite Catalysts

Abstract:

Propylene has become increasingly significant as a raw material in the petrochemical industry because of its increasing demand. The global consumption of propylene has continued to expand at growth rate of 4-5 wt% annually and conventional thermal-cracking of naphtha feedstock can no longer meet the demand due to its low propylene selectivity (Weight ratio of propylene to ethylene: ca. 0.5). In addition, thermal steam-cracking processes are unfavorable in terms of energy consumption. From these technological backgrounds, it is necessary to establish any alternative technology for producing propylene efficiently from widely available feedstocks such as light naphtha. Though catalytic cracking of light hydrocarbons over zeolites (e.g., ZSM-5) in fixed-bed operation have been actively investigated as a promising choice for on-purpose propylene production, this method has not been commercialized because stable catalysts, being applicable to fixed-bed reactor, are still undeveloped. In the present study, unique composite catalysts [1,2], consisting of Fe-Ga-Al-MFI zeolites and metal oxides, have been developed to demonstrate efficient propylene production from light hydrocarbons at moderate temperatures (< 600°C) in fixed-bed mode. Both characteristics of these catalysts and their excellent performance are summarized below.

1. MFI-type zeolites containing iron, gallium and aluminum at optimized ratios (Fe-Ga-Al-MFI), giving high propylene selectivity, were synthesized as main component. The Fe-Ga-Al-MFI zeolites combined with metal-oxide binder (e.g., SiO_2, Al_2O_3) were employed as composite catalysts for practical use.

2. The composite catalysts exhibited the following excellent performance in *n*-hexane cracking: (A) High propylene productivity at lower temperatures than 600°C (Overall propylene yield: 20-30 wt%, Weight ratio of propylene to ethylene: 1.5-2.0); (B) High stability in fixed-bed operation (Lifetime of Fe-Ga-Al-MFI/SiO_2 catalyst: ca. 1,000 h).

3. Energy consumption in reaction unit in *n*-hexane cracking without dilution over the Fe-Ga-Al-MFI/SiO_2 catalyst at 565°C was estimated to be reduced by ca. 60% compared to conventional thermal-cracking, because no thermal energy for heating diluent such as steam, i.e., apparent heat, was needed as well as its low temperature.

On these experimental basis, it was suggested that catalytic cracking of light hydrocarbons in fixed-bed mode is feasible as an efficient method for on-purpose propylene production by using the present zeolite-based composites.

References

[1] S. Hodoshima, A. Motomiya, S. Wakamatsu, R. Kanai, F. Yagi, *Res. Chem. Intermed.* **2015**, 41(12), 9615-9626.

[2] S. Hodoshima, A. Motomiya, S. Wakamatsu, R. Kanai, F. Yagi, Micropor. Mesopor. Mater., in press.

Copyright © American Institute of Chemical Engineers. All rights reserved.

461220 Pt-Promoted Supported Gallium Catalyst for Dow Fluidized Catalytic Dehydrogenation Process

Abstract:

On-purpose propylene production from propane dehydrogenation (PDH) provides a solution to resolving the shortage in propylene supply resulted from the recent feedstock shift to crack more light gases in steam cracking. Although there are existing propane dehydrogenation technologies available, improvement spaces are present in propylene selectivity and the associated energy efficiency. Here we introduce a new PDH technology, Dow PDH, a circulating fluid bed technology based process capable to achieve 45% propane conversion at approximately 93 mol% selectivity to propylene. The Dow process utilizes a Pt promoted supported Ga catalyst with a proprietary reactor and regenerator design. In this presentation, we will focus on catalyst performance and stability. The function of Pt and Ga in PDH chemistry will be discussed with supporting evidence from various characterization techniques and model catalyst investigation.

Copyright © American Institute of Chemical Engineers. All rights reserved.

461850 Dehydroaromatization of Ethane over Zinc Containing Zeolites

Abstract:

Production of light alkanes such as methane or ethane has increased significantly due to the use of hydraulic fracturing for the extraction of shale gas. As a result, the cost of natural gas has plummeted creating an opportunity converting light alkanes to more valuable products such as olefins and aromatics. Metals such as Zn(II) or Ga(III) exchanged into aluminosilicate zeolites have showed promising properties for the aromatization of light alkanes [1-4].

Zn-exchanged zeolites (MFI, BEA) were investigated for the aromatization of ethane. Different amounts of zinc (0-10 %wt), incorporated by ion exchange or incipient wetness impregnation, have been exchanged into ZSM-5 and beta zeolites with different SiO_2/Al_2O_3 (23 to 50) and investigated using a plug-flow microreactor at 500 °C and atmospheric pressure. The measurements indicate that lower ratios of SiO_2/Al_2O_3 and higher zinc content lead to higher ethane conversion. Product selectivity is dependent of zinc content and zeolite composition and structure.

The dehydroaromatization of ethane requires a bifunctional catalyst, in which the initial ethane dehydrogenation to ethylene is catalyzed by the metal sites (H^+/ZSM5 does not activate ethane at similar reaction conditions) and the oligomerizationd and aromatization of olefins is catalyzed by acid sites. However, extra-framework zinc increases the selectivity towards aromatic products. Based on the amounts of zinc, there appear to be different types of zinc sites, which have different activities towards aromatization: low zinc contents show higher stability, lower conversion with dominant selectivity to ethylene while samples with high zinc contents show lower stabilities, higher conversion with high selectivity (~ 50%) towards aromatics. The role of different zinc sites such as isolated Zn ions at the ion exchange positions or Zn_xO_y for the aromatization and methods to stabilize those sites will be further investigated.

References:

1. Biscardi, J.A. and E. Iglesia, Reaction Pathways and Rate-Determining Steps in Reactions of Alkanes on H-ZSM5 and Zn/H-ZSM5 Catalysts. Journal of Catalysis, 1999. 182(1): p. 117-128.

2. Biscardi, J.A., G.D. Meitzner, and E. Iglesia, Structure and Density of Active Zn Species in Zn/H-ZSM5 Propane Aromatization Catalysts. Journal of Catalysis, 1998. 179(1): p. 192-202.

3. Almutairi, S.M.T., et al., Structure and Reactivity of Zn-Modified ZSM-5 Zeolites: The Importance of Clustered Cationic Zn Complexes. ACS Catalysis, 2012. 2(1): p. 71-83.

4. Schulz, P. and M. Baerns, Aromatization of Ethane over Gallium-promoted H-ZSM-5 Catalysts. Applied Catalysis, 1991. 78(1): p. 15-29.

Copyright © American Institute of Chemical Engineers. All rights reserved.

466221 Direct Synthesis of Propene from Ethene Feedstock: Investigation of Catalytic Concepts

Abstract:

The production capacities of polypropylene and propylene oxide, cumene and acrylic acid have grown over the past decades leading to an increased demand in propene. The most common source of propene is currently the cracking of naphtha, which is not sufficient to meet the present demand [1]. A promising alternative is the direct conversion of ethene to propene (ETP-reaction). The suggested reaction mechanism for this overall reaction consists of three steps [2]. The first step is the dimerization of ethene to 1-butene. Secondly 1-butene is isomerized. In the final third step the produced 2-butene reacts with unconverted ethene via a metathesis reaction and gives propene [2].

Ni loaded aluminized MCM-41 is a promising catalyst for the described reaction scheme, because it is capable of supporting all three steps. However, this catalyst represents a complex integrated system.

In previous works different preparation methods of the catalyst have been developed [2, 3]. In this presentation a decoupled approach is followed. For this, a two-catalyst/two-reactor concept is studied, in which the metathesis step is separated from the dimerization and isomerization (Fig. 1). As a consequence the system gains degrees of freedom with respect to residence time, temperature and specific feed composition.

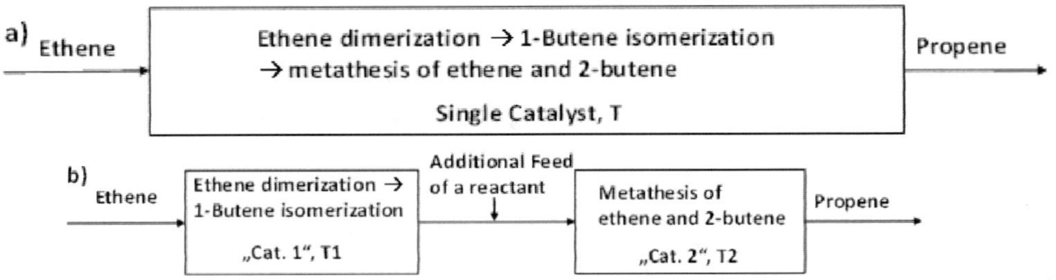

Fig. 1: Scheme of (a) an integrated catalytic system performing simultaneously all reactions using a dedicated single catalyst and (b) a decoupled two-catalyst/two-reactor concept using two sequentially applied catalysts.

The support MCM-41 and its aluminated counterpart with different Si/Al ratios were synthesized using two different procedures, reported in references [3, 4]. The incorporation of Ni and the most common metathesis metals Re, Mo and W onto the supports was realized using the template ion exchange (TIE) method and incipient wetness impregnation (IWI) method [3]. The catalysts were characterized by XRD and N_2-physisorption.

The testing of the catalyst activity was carried out in a fixed bed reactor. For the experimental investigation of the ETP-reaction a fraction of ethene in N_2 was fed into the reactor. Then, investigations of metathesis catalysts were conducted with an equimolar mixture of ethene and 2-butene (*trans*-butene) at atmospheric pressure. The temperature for the catalytic testing was modified in the interval from 50 to 475 °C by increments of 25 °C. The feed and reaction products were analyzed by online gas chromatography and mass spectrometry.

Objective of a carried out kinetic investigation is the identification of conditions for optimal conversion of the reactants and yield of propene. Additionally it is of interest to suppress the production of byproducts, such as pentene and hexene. While working with hydrocarbons, furthermore, the formation of coke and thereby the blocking of active sites should be minimized in order to prevent deactivation processes.

The Ni catalysts ("Cat. 1") for the ETP-reaction revealed high activities with high specific conversions of the reactants and a selectivity of approximately 25 % for propene only at temperatures higher than 400 °C. Severe deactivation of the Ni/(Al)MCM-41 was observed, with time on stream. Though it has been shown that this catalyst can be regenerated [3] in a separate oxidative step.

In the approach of decoupled reactions Re has shown interesting activities in the metathesis step at lower temperatures ("Cat. 2"). Especially the aluminized MCM-41 with incorporated Re gave an equivalent selectivity as the Ni-catalyst at 200 °C (Fig. 2).

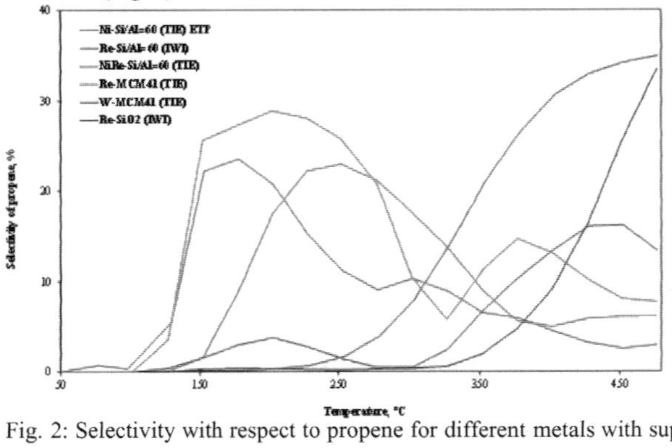

Fig. 2: Selectivity with respect to propene for different metals with supports as a function of temperature, x_{C2H4}=2.5% and x_{C4H8}=2.5%; W/F=2580 kg·s/m^3.

Main goal of the current work is the derivation of a kinetic model in order to optimize the reactor systems.

Acknowledgement

We would like to thank the DAAD with its PROALMEX program and Proyecto apoyado por el Fondo Institucional de CONACYT (FOINS) for their financial support.

[1] D. Liu, W. C. Choi, N. Y. Kang, Y. J. Lee, H. S. Park, C.-H. Shin, Y.-K. Park, Catal. Today, 226 (2014) 52-66.

[2] M. Iwamoto, Molecules, 16 (2011) 7844-7863.

[3] L. Alvarado Perea, T. Wolff, P. Veit, L. Hilfert, F.T. Edelmann, C. Hamel, A. Seidel-Morgenstern, J. Catal. 305 (2013) 154-168.

[4] T. Lehmann, T. Wolff, C. Hamel, P. Veit, B. Garke, A. Seidel-Morgenstern, Microporous Mesoporous Mater. 151 (2012) 113-125.

Copyright © American Institute of Chemical Engineers. All rights reserved.

459971 Olefin Metathesis By Supported WO_x/SiO_2 Catalysts

Abstract:

The kinetics of propylene metathesis by supported WO_x/SiO_2 catalysts were studied using *in situ* XAS (XANES/EXAFS), temperature programmed surface reaction (TPSR) and steady-state experiments. The *in situ* XAS experiments with well-defined catalysts provide information about the reducibility of different WO_x phases (surface dioxo and mono-oxo WO_x species and WO_3 crystalline nanoparticles). Various TPSR experiments and steady-state partial pressure experiments allow for derivation of kinetic equations and determination of important kinetic parameters. Although WO_3 crystals are more reducible upon contact with olefins, and thus reduce desorption activation energy (E_d), they do not contribute to metathesis reactivity and increase apparent activation energy (E_{app}). The steady-state propylene partial pressure experiments show the first-order dependence of reactivity on propylene partial pressure. This is also confirmed by C_3= or C_2=/C_4 TPSR experiments which provide the same peak temperature (T_p) value. Ethylene (C_2=) cannot create sites unlike propylene (C_3=) or 2-butene (C_4=).

Copyright © American Institute of Chemical Engineers. All rights reserved.

468866 Spectroscopic Investigation of Pd Structures Formed over CeO_2 and MnO_x-CeO_2 Supports

Abstract:

In order to meet fuel economy and U.S. EPA Tier 3 emission standards, efficient catalysts that can operate at lower exhaust temperatures and higher CO and hydrocarbon levels are required. Addition of metals to ceria support increases its catalytic activity significantly, in which Pd/CeO_2 catalysts are shown to oxidize CO even at ambient and lower temperatures. Promotion of CeO_2 support with cations can facilitate the activation of gas phase oxygen and increase the reducibility of the catalyst. Previously we have shown that Pd-doped MnO_x-CeO_2 solid-solution catalysts achieved 100% CO conversion at 50°C, which was 60% better than the Pd/CeO_2 tested at the same conditions. High activity of Pd/MnO_x-CeO_2 was found to be related with the ionic Pd^{2+} species, and high oxygen storage capacity and mobility of the MnO_x-CeO_2 solid-solution.

One of the existing challenges for Pd-deposited CeO_2-based systems is the active sites responsible for CO oxidation at low temperatures. In this work, in-situ x-ray absorption fine structure (XAFS) experiments were performed over Ce and Mn solid solution catalysts to determine the change in the local structure of Pd during CO oxidation. Both Pd^{2+} ionic and PdO structural models were applied to fit the extended x-ray absorption fine structure (EXAFS). Additional characterization experiments through x-ray photoelectron spectroscopy (XPS), raman spectroscopy and in situ diffuse reflectance infrared Fourier transform spectroscopy (DRIFTS) were conducted to explain Pd formation during CO oxidation. Our results indicated that both Pd^{2+} ionic structures and highly dispersed PdO nanoclusters were forming on CeO_2 and MnO_x-CeO_2 supports. The close interaction of Pd^{2+} ionic structure and PdO phase possibly resulted in high activity for CO oxidation.

Copyright © American Institute of Chemical Engineers. All rights reserved.

472771 Tailoring the Catalytic Activity of Perovskites forÂ Oxygen Reduction and Evolution

Abstract:

Perovskite materials have drawn broad scientific attention in the past decades in fields such as solid-state fuel cell, photo-electrolysis, photovoltaic and so on. In this talk, we used density functional theory (DFT) to investigate how the surface catalytic activity can be finely tuned by engineering the structure and composition of perovskites. The talk is consisted of two major parts:

1) **Mechanistic study of oxygen incorporation on $La_xSr_{1-x}FeO_{3-\P}$ (001)**: we explicitly examined oxygen incorporation reaction on $La_xSr_{1-x}FeO_{3-\P}$ (001) with different surface termination and oxygen-stoichiometry. Detailed about comparison between theory and experiment will be also discussed in the talk.

2) **Surface activity tailoring for oxygen evolution reaction (OER)**: computational screening was carried out to understand the correlation between structure and function of perovskite for oxygen evolution reaction. Prediction of promising structures will be suggested to experiments for test.

Our study provides a general guideline for designing efficient perovskites catalysts with much less loading of noble metal and improved stability.

Copyright © American Institute of Chemical Engineers. All rights reserved.

461329 Origin of the Support Effect for Methanol Oxidation for Supported Vanadium Oxide Catalysts

Abstract:

The selective oxidation of methanol was investigated over monolayer supported vanadium oxide catalysts as a function of the specific oxide support (CeO_2, ZrO_2, TiO_2, Al_2O_3, and SiO_2). The preferential formation of HCHO reflects the redox nature of the catalytic surface vanadia sites. The methanol oxidation reaction rate was found to vary by a factor of 10^3 as the support was varied ($CeO_2 > ZrO_2 > TiO_2 > Al_2O_3 > SiO_2$). The combination of CH_3OH-Temperature Programmed Surface Reaction (TPSR) spectroscopy and steady-state methanol oxidation studies allowed for quantitative determination of the adsorption thermodynamic (K_{ads}, Î"H_{ads}, and A_{ads}) and surface reaction kinetic (k_{rds} and E_{act}) parameters to better understand the origin of the support effect. For the non-silica catalysts, the TOF was found to be primarily dependent on the CH_3OH equilibrium adsorption constant (K_{ads}) involving O-H bond breaking. For the supported VO_x/SiO_2 catalyst, the TOF was affected by an increase in E_a for C-H bond scission of the surface methoxy as well as a decrease in K_{ads}. The TOF, k_{rds} and K_{ads} methanol oxidation parameters generally decreased with increasing electronegativity of the oxide support cation. These findings demonstrate that, with the exception of the SiO_2 support, that it is the adsorption step and not the surface reaction step that is responsible for the orders of magnitude in the reaction rate variation of methanol oxidation reaction rate by the different oxide support ligands.

Copyright © American Institute of Chemical Engineers. All rights reserved.

459802 Oxygen Transmission in Solid Phase——MnO_2 promoted Phase-Pure M1 Movnbte Oxide for Ethane Oxidative Dehydrogenation

Abstract:

MoVNbTe oxide is among the most promising catalysts for oxidative dehydrogenation of ethane (ODHE) process. Vanadium ions come in two valence states in the catalyst system: V^{5+} and V^{4+}. The V^{5+} ions in M1 phase are considered as the paraffin activating sites. Therefore, increasing the V^{5+} concentration is the most effective way to improve the catalytic performance.

High temperature oxygen treatment is the traditional way to oxidize the elements in metal oxide. But considering the complex structure of the MoVNbTe catalyst, high temperature oxygen treatment would destroy the catalyst crystal structure, while contacting oxygen molecule under 400 °C has limited effectiveness.

To break the limit of gas-solid oxygen transmission, we have used oxygen plasma at room temperature to insert oxygen into the catalyst surface through a gas-solid reaction by raising the activity of oxygen donor [1]. In another work, we have introduced CeO_2 to the catalyst system to form a nanocomposite with the M1 catalyst, thus oxidizing the vanadium on the M1 catalyst surface through a self-redox solid-state reaction [2].

In this work, to break the limit of gas-solid oxygen transmission, manganese oxide (MnOx) is introduced into the catalyst system. Mn is an element whose valence state is variable, and is easy to react with oxygen molecule. By mixing MnOx xerogel with M1 catalyst and calcinating the mixture at 400 °C in air for 6 h, the oxygen molecule in air is inserted into MnOx. Meanwhile, the oxygen transmission happens from MnO_x to M1 phase, increasing the abundance of V^{5+} on the catalyst surface by a self-redox solid-state reaction. The SEM and TEM characterization show the composite of MnO_2 on M1 catalyst surface. The increase of V^{5+} concentration is proved by XPS characterization. Corresponding to the valence state increment, the catalyst promoted by MnO_2 gives an improved catalytic performance, e.g., making a 10% increase of ethane conversion in ODHE process at 400 °C. As a comparison, without the introduction of MnOx, calcinating the the M1 catalyst alone in air can only give a 3% increase of ethane conversion under the same condition. In this work, manganese oxide plays a role of 'oxygen bridge', allowing the oxygen molecule in gas phase to be transmitted to the catalyst phase more easily through a gas-solid-solid path during calcination.

It is anticipated that this insight into oxygen transmission in phase-pure M1 MoVNbTe oxide catalyst can give more inspiration in redox catalysts.

1. X. Chen, B. Chu, Q. Yang, H. An, Y. Cheng, Valence variation of phase-pure M1 MoVNbTe oxide by plasma treatment for improved catalytic performance in oxidative dehydrogenation of ethane. *RSC Adv.* 2015; 5: 91295-91301

2. B. Chu, H. An, T.A. Nijhuis, J. C. Schouten, Y. Cheng, A self-redox pure-phase M1 MoVNbTeOx/CeO2 nanocomposite as a highly active catalyst for oxidative dehydrogenation of ethane. *J Catal.* 2015; 329: 471–478.

Copyright © American Institute of Chemical Engineers. All rights reserved.

461855 On the Mechanistic Aspects of Mg_6MnO_8-Based Redox Catalysts for Oxidative Dehydrogenation of Ethane Via a Chemical Looping Scheme

Abstract:

Redox catalysts, which perform selective or deep oxidation reactions utilizing lattice oxygen, have become increasingly attractive due to their potential to facilitate high energy conversion efficiency with low CO_2 and NO_x emissions. Oxidative dehydrogenation of ethane (ODH) is one very promising application of chemical looping. In chemical looping-ODH, ethane is partially oxidized by a redox catalyst to form water and ethylene. The regeneration of lattice oxygen lost to water formation provides the heat necessary to drive the endothermic dehydrogenation reaction. Although such an approach possesses numerous advantages relative to conventional cracking, achieving industrially relevant conversion in ODH while suppressing deep oxidation is not trivial. Additionally, thermal cracking of ethane to ethylene and hydrogen can be significant at temperatures where chemical looping normally operates. It is therefore important to ensure facile supply of oxygen from the lattice without permitting significant oligomerization or deep oxidation of ethylene. As such, detailed mechanistic understanding of redox catalyst in ODH is highly desirable to support rational catalyst design and optimization.

Previously, we have identified that Mg_6MnO_8, a mixed oxide with a cation deficient rocksalt structure, is an excellent model redox catalyst for ODH. It is capable of supplying lattice oxygen at rates comparable to the rate of hydrogen formation via thermal cracking of ethane. Unpromoted Mg_6MnO_8 is highly efficient for deep oxidation of ethane, producing primarily of $CO_2$and water. When promoted with alkali salts, however, it can produce ethylene with exceptional selectivity.

The current work investigates the mechanistic aspects of (promoted) Mg_6MnO_8 redox catalysts. Extensive

characterizations including TPR, TPD, XPS, *in-situ* XRD, *in-situ* DRIFTS, Raman Spectroscopy, and pulse reactions are performed. The data indicate that alkali salt promoters induce changes in the bulk and near surface properties of the Mg/MnO system and suppress activation and deep oxidation of ethane and ethylene. Meanwhile, the promoted redox catalysts allow facile combustion of hydrogen, leading to high ethylene yield.

Copyright © American Institute of Chemical Engineers. All rights reserved.

456090 Experimental and Theoretical Investigations of Propene Ammoxidation to Acrylonitrile over Bismuth Molybdate

Abstract:

Propene ammoxidation over $Bi_2Mo_3O_{12}$ was investigated to elucidate product (acrylonitrile, acetonitrile, HCN, acrolein, N_2, etc.) formation pathways. Propene consumption rate is first order in propene and zero order in ammonia and oxygen partial pressures, has an activation energy (E_a = 22 kcal/mol) comparable to that for propene oxidation, suggesting that the rate-limiting step for both reactions is the same. Two N-containing species are relevant at ammoxidation conditions: adsorbed NH_3 on surface Bi^{3+} ions that reacts with vinyl alkoxide species to form products with C-N bonds, and metastable Mo-NH_2 groups that are responsible for NH_3 oxidation to N_2. The proposed reaction mechanism and model of the reaction kinetics captures the experimental trends in product distribution as a function of partial pressures and temperature. DFT calculations of the elementary steps comprising the reaction pathway support the proposed mechanism, as well as revealing what processes control product selectivity.

Copyright © American Institute of Chemical Engineers. All rights reserved.

468447 Dioxygen Activation Pathways in Selective Oxidations Catalyzed By Metal Oxides

Abstract:

The oxidative dehydrogenation (ODH) of hydrocarbons and oxygenates is an essential route to convert undervalued feedstocks into marketable chemicals [1,2]. ODH reactions on reducible metal oxides occur via Mars-van Krevelen redox cycles at lattice O-atoms (O*) with H-abstractions from the organic reactants as the kinetically-relevant step, followed by rapid re-oxidation of the reduced centers by O_2 [3]. Compared with the relatively well established reduction part of the cycle, the re-oxidation steps have received much less attention in spite of their consequences for selectivity in ODH reactions [4]. The activated dioxygen intermediates [4,5] formed via O_2 activations can mediate O-insertion steps for species that act as intermediates in undesired combustion reactions [4,6]. Here, we investigate plausible O-insertion routes in ODH reactions by examining propene epoxidation during methanol-O_2 reactions on Mo-based Keggin polyoxometalate (POM) clusters.

Propene epoxidation (PO) products were detected from O_2-C_3H_6 reactants only in the presence of CH_3OH on Mo-based Keggin clusters, consistent with the involvement of O_2-derived species generated on reduced centers that form as CH_3OH ODH intermediates. Two possible activated dioxygen intermediates, peroxo (OO*) and hydroperoxo (HOO*) species, may form via O_2 activation at reduced centers consisting of O-vacancies (*) and surface hydroxyls (OH*), respectively. DFT indicates that these two pathways differ in their kinetic response to O_2 concentrations, because OO* formation is limited by the formation of the O-vacancy formation step, which does not depend on O_2 pressure; in contrast, HOO* formation is mediated by H-transfer from hydroxyls to O_2 as the kinetically-relevant step, making HOO* concentrations proportional to O_2 pressure. Measured PO formation rates from CH_3OH-O_2-C_3H_6 reactants (453 k, 0.1-4 kPa CH_3OH, 1-60 kPa O_2, 1-35 kPa C_3H_6) were proportional to O_2 pressures at low pressures (10 < kPa), suggesting that HOO* act as the main O_2-derived epoxidation intermediates. DFT calculations also indicate the low activation barrier ($\Delta E^{\ddagger}$ = 32 kJ/mol) for C_3H_6 reactions with HOO* sites. These HOO* intermediates can also react with OH* to form OO* and H_2O, but such steps occur with much higher barriers ($\Delta E^{\ddagger}$ = 81 kJ/mol) than for OOH* reactions with C_3H_6. In contrast, OO* species, formed via either O_2 reactions with vacancies or from OOH*, preferably activate C-H bonds in CH_3OH ($\Delta E^{\ddagger}$ = 31 kJ/mol) over O-insertion reactions with C_3H_6 ($\Delta E^{\ddagger}$ = 39 kJ/mol). The DFT-derived barriers for CH_3OH C-H activation on OO* ($\Delta E^{\ddagger}$ = 31 kJ/mol) are much lower than for the respective reaction with O* species ($\Delta E^{\ddagger}$ = 69 kJ/mol) for a give O-atom in a POM cluster, suggesting that such steps occur rapidly after O_2 activation to consume the OO* species and complete a catalytic ODH turnover. Alternate steps involving O-atom migration from OO* to reduced centers require multiple steps, each with significant larger barriers ($\Delta E^{\ddagger}$ = 104 kJ/mol) than for direct C-H activation in CH_3OH by OO* .

The first-order dependence of PO formation rates from CH_3OH-O_2-C_3H_6 mixtures on C_3H_6 and O_2 pressures at low pressures, approached a zero-order kinetic regime on both C_3H_6 and O_2; rates also increased with increased CH_3OH pressures along with the concomitant increase in ODH rates. These trends are consistent with the proposed reaction steps for PO occurring on the HOO* sites that are generated from hydroxyl centers by O_2, where the formation rate is in steady-state with consumption rates of OOH* via PO or decomposition to OO* and H_2O. These data, and their mechanistic interpretation, indicate that the ratio of O-insertion to C-H activation rates for a given C_3H_6/CH_3OH reactant ratio would increase as the HOO*/OO* increases, because of their relative reactivities in these two reactions. Indeed, the presence of H_2O increased measured PO/ODH ratios, because O-vacancies (*) react with H_2O to form OH*, thus causing a concomitant increase in HOO*/OO* ratios during O_2 activation on oxide surfaces. These mechanistic inferences, derived from the scavenging of O_2-derived species by propene, taken together with theoretical treatments, illustrate their complexity and selectivity consequences, while providing guidance for catalysts and materials that can enhance or inhibit the relative contributions from pathways mediated by HOO* and OO* intermediates. They also provide fundamental insights into how isolated vacancies, formed as intermediates in ODH reactions, react with the O_2 co-reactants to complete ODH catalytic turnovers.

The authors acknowledge the financial support of the U.S. Department of Energy, Office of Science, Office of Basic Energy Sciences (DE-AC05-76RL0-1830) and computational resources from EMSL at Pacific Northwest National Laboratory (PNNL) and from XSEDE supported by the National Science Foundation.

References

[1] T. Punniyamurthy, S. Velusamy, J. Iqbal, Chem. Rev. 105 (2005) 2329.

[2] R.K. Grasselli, Top. Catal. 21 (2002) 79.

[3] P. Deshlahra, R.T. Carr, S.H. Chai, E. Iglesia, ACS Catal. 5 (2015) 666.

[4] X. Rozanska, E. Kondratenko, J. Sauer, J. Catal. 256 (2008) 84.

[5] a. Goodrow, A.T. Bell, J. Phys. Chem. C 111 (2007) 14753.

[6] Y. V Geletii, C.L. Hill, R.H. Atalla, I.A. Weinstock, J. Am. Chem. Soc. 128 (2006) 17033.

Copyright © American Institute of Chemical Engineers. All rights reserved.

464561 Tuning the Dehydration Activity of Alcohols Using Phosphonic Acid Self-Assembled Monolayers

Abstract:

One of the biggest problems in heterogeneous catalysis is that of poor selectivity. Often dopants are utilized to tune the properties of various catalysts to improve selectivity or activity. Over the last decade self-assembled monolayers (SAMs) have been developed as a means to control the selectivity of reactions on supported metal catalysts. We sought to utilize a similar approach for metal oxide catalysts. We developed a simple and rapid approach to functionalize anatase titania and other oxides with phosphonic acid SAMs. These SAMs are exceptionally robust, resisting thermal degradation up to 400°C. We investigated the selectivity of these catalysts for reactions of alcohols on titania. When 1-propanol is reacted over titania three main pathways are possible: dehydration, dehydrogenation, and condensation. At low temperatures (~250°C), dehydrogenation is favored on the native catalyst, closely followed by condensation. However, upon functionalization of titania with phosphonic acids with various alkyl tails (methyl, octadecyl, benzyl, etc), dehydration becomes the dominant pathway. More importantly, depending on the functionality of the alkyl tail acid (e.g. 4-aminobenzylphosphonic vs 3-fluorobenzylphosphonic acid), the activity of dehydration can be tuned over a wide range, allowing far higher dehydration activities than on the uncoated catalysts. Using reactor and surface studies, we will demonstrate in this presentation how the selectivity and activity of metal oxide catalysts (i.e. anatase titania) can be tuned for certain constituents.

Copyright © American Institute of Chemical Engineers. All rights reserved.

470921 Single-Facet-Dominated Titania and Zirconia As Catalysts or Supports for Bio-Alcohol Dehydration and Oxidative Dehydrogenation

Abstract:

Early transition metal oxides, used as catalysts or catalyst supports, play a significant role in acid/base and redox reactions, which are among the most common reaction types in upgrading biomass-derived molecules and intermediates to fuels/chemicals [1]. A key piece of knowledge for designing high-performance metal oxide catalysts is a molecular-level understanding of structure-activity correlations for desired reactions. However, the chemical and structural complexity of metal oxides makes it challenging to unambiguously describe these materials at an atomic-level, even more so to identify and tailor active surface ensembles. Simplified, well-defined model catalysts are highly desirable in this regard. Recent studies using VO_x/CeO_2 catalysts for methanol oxidative dehydrogenation (ODH) reveal that the catalytic performance is highly dependent on the exposed facet of the CeO_2 support. This strong dependence is attributed to oxygen defect density difference for different facets, which impart great influences on redox properties of the VO_x active centers and activation energies for methoxyl group decomposition, the rate limiting step for this particular reaction [2]. However, studies on the roles for different oxide facets on catalytic systems are still at a very early stage.

Research described in this work investigates the facet-dependent properties of two widely used metal oxides, TiO_2 and ZrO_2. Single-facet-dominated anatase titania were synthesized using hydrothermal methods. These TiO_2 nanocrystals exhibited well-defined shapes of truncated tetragonal bipyramid, rod, and nano-sheet, with the exposed dominated facets of (101), (010), and (001), respectively. Using a similar hydrothermal method, monoclinic ZrO_2 nanocrystals were also prepared in shapes of spheres and rods. In contrast to TiO_2, these ZrO_2 nanocrystals possessed more complex facets. The spheres were composed of (111),Â (1Ì…11), and (110) facets, while the rod appeared to be dominated by (011) and (001) facets on the longitudinal side and small portions of other facets on the ends. The synthesized TiO_2 and ZrO_2 model catalysts were tested for the dehydration of isopropanol, bio-diols, and triols to probe their acid activities. They were also employed as supports of VO_x for the methanol ODH to evaluate the redox activities of the catalysts. A series of microscopic and spectroscopic techniques, such as scanning/transmission electron microscopy (SEM/TEM), X-ray diffraction (XRD), Raman and solid state nuclear magnetic resonance (NMR) spectroscopies were utilized to identify the active domains and correlate the structural properties of different facets to the catalytic performance. The key findings from the characterization studies will also be discussed in this talk.

Reference

[1] Chen, K.; Bell, A. T.; Iglesia, E. *Journal of Catalysis* **2002**, *209*, 35.

[2] Li, Y.; Wei, Z.; Gao, F.; Kovarik, L.; Peden, C. H. F.; Wang, Y. *Journal of Catalysis* **2014**, *315*, 15.

Copyright © American Institute of Chemical Engineers. All rights reserved.

468347 Mechanistic Insights into Carbon-Oxygen Bond-Breaking on Mo_2c Catalysts: Ethanol Dehydration

Abstract:

Molybdenum carbide catalysts have been shown to be active catalysts for the deoxgenation of biomass-derived oxygenates, an important reaction for the production of fuels and chemicals via ex situ catalytic fast pyrolysis (CFP).[1-3] Under typical ex situ CFP conditions, the Mo_2C catalyst is likely highly covered in oxygen, and possesses metallic-like sites, oxygen vacancy sites and acidic sites, which may each play a different role in controlling the observed activity and selectivity of deoxygenation reactions. To elucidate the role of these various sites, we apply first principles electronic structure calculations to investigate the reaction mechanism of ethanol dehydration on Mo_2C. Temperature programmed reaction experiments have shown that ethanol dehydration occurs at low temperatures (150 °C) on Mo_2C to produce primarily ethylene and water. In this work, we elucidate the reaction pathways and active sites for C-O bond-breaking in ethanol dehydration as well as the relationship between the energetics of the competing pathways, the Mo_2C surface termination, and the oxygen coverage under reaction conditions.

1. Schaidle, J. A..; Blackburn, J.; Farberow, C. A.; Nash, C.; Steirer, K. X.; Clark, J.; Robichaud, D. J.; and Ruddy, D. A.; *ACS Catal.*, **2016**, *6*, 1181.

2. Bej, S. K. and Thompson, L. T.; *Appl. Catal. A.* **2004**, *264*, 141.

3. Sullivan, M. M.; Chen, J.-C.; Bhan, A. *Catal. Sci. Tech.* **2016**, 6, 602.

Copyright © American Institute of Chemical Engineers. All rights reserved.

472583 Elucidating Particle Size Effects on the Hydrodeoxygenation (HDO) of Phenolics By Micro-Pulse Experiments and DFT Calculations

Abstract:

The catalytic hydrodeoxygenation (HDO) of phenolic compounds is very important for the upgrading of bio-oil as transportation fuels. This study investigated mechanistic aspects of HDO of different phenolic compounds such as m-cresol and anisole over different metal catalysts (Pt, Rh, Ru and Fe) using a micro-pulse reaction system, in which the catalyst is flushed with hydrogen all the time and a small amount of feed is passed through the catalyst bed in each pulse. This reactor system thus allows monitoring the deactivation of different sites with great precision and study the activity of very active, but quickly deactivating sites. Here, the effects of the metal particle size on the HDO of phenolic compounds were studied. The TOF for both the ring hydrogenation and the HDO of m-cresol and anisole were calculated over the metal catalysts with different particle size. Density functional theory calculations were also carried out to explore the reaction pathways at both the terrace sites and the steps sites to understand the particle size effects on the HDO reactions at the molecular level. Different mechanisms including the direct deoxygenation (DDO), and the tautomerization-hydrogenation-dehydration paths were examined and analyzed. It was found that the plausible pathway for the HDO of phenolic compounds may be different at the terrace and step sites which leads to the different catalytic activity and product distribution at the metal catalysts with different particle size.

Copyright © American Institute of Chemical Engineers. All rights reserved.

461983 Adsorption of Hetero-Bifunctional Urea on Ge(100)-2x1 Surface

Abstract:

Organic functionalization of Group IV semiconductor surfaces is of interest due to the need for new pathways in surface chemistry modification. This is important in applications such as molecular layer deposition, molecular electronics, and biosensors where the ability to control and tune surface properties requires a deep understanding of the interactions between molecules and solid surfaces. In this study, adsorption of the heterobifunctional urea molecule on the Ge(100)-2x1 surface was investigated. Both the amine and carbonyl group of the urea molecule are known to react with the Ge surface. The aim of this study is to determine if any preferential reaction pathways exist and to understand the driving forces toward the final products.

Density functional theory (DFT) calculations suggest that NH_2 dissociation is the most thermodynamically favorable pathway for the single reaction. The reaction can occur through two possible precursor states: a nitrogen dative bond with the surface or an oxygen dative bond that can further undergo an enolization reaction. Interestingly, the oxygen dative bond is 7 kcal/mol more stable than the nitrogen bond, suggesting a preference for the latter reaction pathway. Furthermore, calculations show that the dual reactions provide less stabilization gain and a higher kinetic cost than single reactions, as the dual NH_2 dissociation has an activation barrier greater than 30 kcal/mol and only provides 5 kcal/mol of additional exothermicity, indicating that the urea molecule will likely react only through a single functional group. X-ray photoelectron spectroscopy (XPS) and multiple internal reflection Fourier transform infrared (MIR-FTIR) spectroscopy were used to determine the final reaction products. Results suggest that urea adsorbs on Ge(100) forming a mix of surface products. One of the products can be identified by the downshift in binding energy of the N(1s) XP peak and the Ge-H stretching mode in IR as NH_2-dissociated urea on Ge. This assignment is also consistent with the presence of a carbonyl group in the IR and XPS spectra, which is expected to remain unchanged in this surface configuration. Another reaction product exists that is evident by second O(1s) and C(1s) XPS peaks downshifted from that of the parent urea molecule, suggesting a loss of the carbonyl group by a reaction with a more electropositive atom. Moreover, coverage results support our DFT findings by suggesting that each urea molecule will occupy a Ge dimer by reacting through a single functional group per molecule.

Copyright © American Institute of Chemical Engineers. All rights reserved.

462687 Suppressing Pd Catalyst Deactivation in Alcohol Oxidation Using Thiolate Self-Assembled Monolayer Modification

Abstract:

The selective oxidation of alcohols is of great importance in the production of fine chemicals and pharmaceuticals. Heterogeneous Pt-group catalysts have been widely investigated for this process because of the ability to activate molecular oxygen and alcohols. However, these catalysts have often been reported to deactivate rapidly. One of the frequently reported factors responsible for deactivation is the strongly adsorbed species on the catalyst surface, which can be generated by decarbonylation and dimerization of the aldehyde product.

Recent studies have shown that thiolate self-assembled monolayers (SAMs) can be applied as Pt-group catalyst modifiers for hydrogenation reactions. One of the advantages of these modifiers is that their surface coverage can be tuned in a straightforward way: the coverage mainly depends on the bulkiness of the tail group. Though thiolate SAMs may block active sites to poison the catalysts, they have also been shown to limit deposition of carbonaceous species on the surface to increase the catalytic activity. However, this group of modifiers have not yet been applied for Pt-group catalysts in oxidation reactions. It is therefore desirable to use alcohol oxidation reactions as a probe to investigate the thiolate SAM modifiers for deactivation control.

Here, we show that the deactivation of a supported palladium catalyst can be successfully limited by thiolate SAMs modification in the liquid phase oxidation of trans-2-hexen-1-ol. While the native Pd catalyst suffered from rapid deactivation, the oxidation rate was improved on the modified catalysts, particularly at high conversion, without a significant change in selectivity to the desirable aldehyde product. This reactivity and selectivity on the modified catalyst was maintained during two additional recycle reactions. Kinetic studies combined with infrared spectroscopy indicated that the aldehyde product generated surface poisons through decarbonylation and dimerization, which can be suppressed by thiolate SAMs. The effect was controllable by varying the thiolate surface coverage through use of either 1-adamantanethiol (AT) or 1-octadecanethiol (C18) SAMs. This result suggests that thiolate SAMs can be a promising tool for enhancing the catalytic activity in low-temperature alcohol oxidation reactions.

Copyright © American Institute of Chemical Engineers. All rights reserved.

463941 Facile Formation of Methyl Acrylates By Cross Coupling of Unsaturated Alcohols-Aldehydes with Methanol on Au(110)

Abstract:

A range of reactions is promoted by gold catalysts, and their application to catalysis has received significant attention in the past years. Gold-based catalysts have been shown to be highly selective for the oxidative coupling of alcohols and aldehydes to organic esters and various selective oxidation and hydrogenation reactions of organic molecules. It has been reported by our group that the competitive binding of reactive intermediates governs the selectivity in the coupling of dissimilar hydrocarbons on goldâ€"both in model studies and over nanoporous Au (npAu) under catalytic conditions

In the present study we use temperature programmed reaction spectroscopy (TPRS) and scanning tunneling microscopy (STM), to study the esterification of methanol due to cross-coupling reactions with small unsaturated oxygenates on a model single crystal surface - Au(110). We identify reaction intermediates and their surface stability on the Au single crystal and compare the ester optimum selectivity with studies under flow conditions on npAu.

In the first part of this research we demonstrate the catalytic production of methyl acrylate and methyl methacrylate over O/Au(110) interface, via the coupling of unsaturated alcohols (allyl alcohol and methallyl alcohol) with methanol under ultra high vacuum (UHV) conditions. We show that the alcohols can be used to generate aldehydes in-situ with high selectivity, without oxidation of the unsaturated C=C bond, from which coupling with methanol to form acrylates follows. The product selectivity can be tuned by changing the relative concentrations of the unsaturated alcohol and methanol; the high relative concentrations of methanol which were found to be necessary to maximize acrylate selectivity, indicate that the C=C bond and methyl group stabilize the allyloxy intermediate on the surface (compared to saturated alcohols).

In the second part of our research we study the oxygen-assisted esterification of acrolein and methacrolein with methanol on Au to form their respective acrylates. Following the same experimental procedure with the alcohol cross-coupling part, TPRS was used to establish the mechanism and conditions for the facile esterification of acrolein and methacrolein. Selectivities of ~95% were achieved on oxygen pre-covered Au (110) with methanol conversions of nearly 100 % under appropriate conditions. Under flow conditions on npAu, methyl acrylate and methyl methacrylate were formed with optimum selectivities ~20%. The origin of this low activity is shown to be due to poisoning of active sites by adsorbed acrylates.

Pursuant to this poisoning effect, STM studies were performed on the acrylate and methacrylate on Au(110) at room temperature, formed by the reaction of acrolein and methacrolein with oxygen. In both cases ordered structures of the adsorbed intermediates were observed, indicating strong intermolecular interactions. The reactive intermediates are present on the surface at room temperature, and lead to the destruction of the 1x2 reconstruction of Au(110), confirming that they are strongly bound with Au atoms. The formation of these dense structures almost certainly accounts for the higher stability of these intermediates on gold, therefore, contributing to its blocking of the surface sites.

It is shown that model studies of the methanol cross-coupling reactions on single-crystal gold surfaces, provided information about the oxidative cross-coupling of allyl and methallyl alcohols with methanol which can be achieved under mild conditions without competing attack of the unsaturated C=C bond. Optimum cross-coupling selectivities are achieved when the surface concentrations of methoxy and allyloxy/methallyloxy intermediates are relatively equal, requiring methanol mole fractions above 0.99 due to the stabilizing effects of the C=C bond (on allyloxy) and an additional methyl group (on methallyloxy).

The esterification of unsaturated aldehydes with methanol can also be achieved selectively at low temperature by oxygen-assisted coupling on Au(110). However. Lower conversion and selectivity was observed in steady flow under catalytic conditions with nanoporous Au catalysts. The lower activity observed for npAu (*vs* Au(110)), is due to the buildup of adsorbed acrylates. The model studies of the reaction mechanism indicate that facile esterification of acrolein and methacrolein with methanol can be achieved with ~95% selectivity provided the oxidation of the aldehyde can be suppressed.

The critical aspect of designing efficient heterogeneous catalytic reactions by determining the factors responsible for the relative stability of intermediates on the surface of the catalyst is addressed. It was found that the presence of highly stable species, which do not lead to productive reactions, inhibits desired reactions and leads to the suppression of catalytic activity. The catalytic processes proposed in this research can provide an alternative, green route for the production of acrylates and methacrylates and can further extend the principles governing coupling activity/selectivity on gold surfaces to unsaturated organic molecules.

Copyright © American Institute of Chemical Engineers. All rights reserved.

467100 Reactivity and Selectivity Descriptors for the Activation of C-H Bonds in Hydrocarbons and Oxygenates on Metal Oxides

Abstract:

Reactivity and Selectivity Descriptors for the Activation of C-H Bonds in Hydrocarbons and Oxygenates on Metal Oxides

Prashant Deshlahra and Enrique Iglesia*

Department of Chemical Engineering, University of California, Berkeley, CA 94720

*iglesia@berkeley.edu

C-H bond activations at lattice O-atoms on oxides mediate some of the most important chemical transformations of small organic molecules [1-2]. The relations between molecular and catalyst properties and C-H activation energies are discerned in this study for the diverse C-H bonds prevalent in C_1-C_4 hydrocarbons and oxygenates using lattice O-atoms of polyoxometalate (POM) clusters with a broad range of H-atom abstraction properties. These activation energies determine, in turn, attainable selectivities and yields of desired oxidation products, which differ from reactants in their C-H bond strength. Brønsted-Evans-Polanyi (BEP) linear scaling relations [3-4] predict that C-H activation energies depend solely and linearly on the C-H bond dissociation energies (BDE) in molecules and on the H-atom addition energies (HAE) of the lattice oxygen abstractors. These relations omit critical interactions between organic radicals and surface OH groups that form at transition states that mediate the H-atom transfer, which depend on both molecular and catalyst properties (Fig. 1a); they also neglect deviations from linear relations caused by the lateness of transition states. Thus, HAE and BDE values, properties that are specific to a catalyst and a molecule in isolation, represent incomplete descriptors of reactivity and selectivity in oxidation catalysis (Fig. 1b). These effects are included here through crossing potential formalisms that account for the lateness of transition states in estimates of activation energies from HAE and BDE and by estimates of molecule-dependent, but catalyst-independent, parameters that account for diradical interactions that differ markedly for allylic and non-allylic C-H bonds (Fig. 1c). The systematic ensemble-averaging of activation energies for all C-H bonds in a given molecule show how strong abstractors and high temperatures decrease an otherwise ubiquitous preference for activating the weakest C-H bonds in molecules, thus allowing higher yields of products with C-H bonds weaker than in reactants than predicted from linear scaling relations based on molecule and abstractor properties [5]. Such conclusions contradict the prevailing guidance to improve such yields by softer oxidants and lower temperatures, a self-contradictory strategy, given the lower reactivity of such weaker H-abstractors. The diradical-type interactions, not previously considered as essential reactivity descriptors in catalytic oxidations, may expand the narrow yield limits imposed by linear free energy relations by guiding the design of solids with surfaces that preferentially destabilize allylic radicals relative to those formed from saturated reactants at C-H activation transition states.

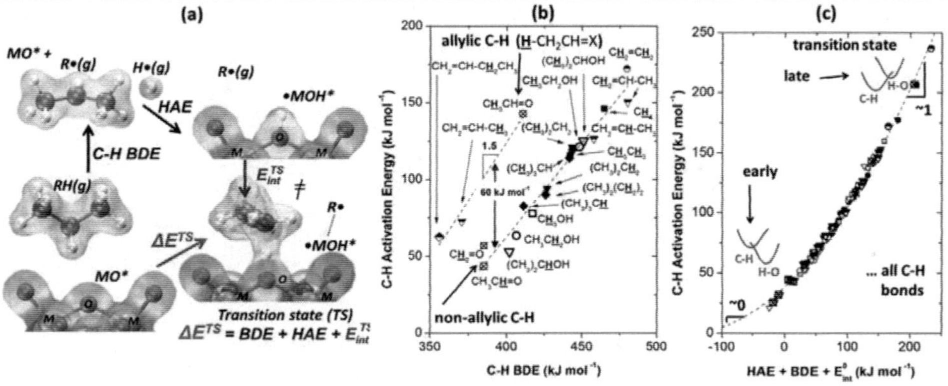

Figure 1. (a) A thermochemical cycle description of the C-H activation transition-state energy relative to a gaseous organic reactant and a surface metal oxide site (MO*) as a sum of, (i) the energy required to separate H atom from the C atom in the reactant to form a radical species (C-H BDE), (ii) the energy of adding the H atom to the O atom of MO* (HAE), (iii) and interaction energy between the radical (R•) and hydroxylated metal oxide (•MOH*) at the transition state (E_{int}^{TS}). **(b)** DFT-derived C-**H** bond activation energies in alkanes (closed symbols), alkanols (open symbols), alkenes (half-filled symbols) and alkanals (crossed symbols) at a specific O-atom location in $H_3PMo_{12}O_{40}$ POM cluster as a function of their C-**H** bond dissociation energy (BDE). **(c)** DFT-derived C-H bond activation energies as a function of the sum of the C-H BDE, of reactants, the HAE of different abstractors and the product state interaction energy (E_{int}^{0}). Dashed curves represent regressed values of the activation energies to the functional form described by C-H O-H crossing-potential models for C-H activation.

Acknowledgments

This work was supported by the U.S. Department of Energy, Office of Science, Office of Basic Energy Sciences, under contract number DE-AC05-76RL0-1830. Computational facilities were provided by the Environmental Molecular Science Laboratory (EMSL) at Pacific Northwest National Laboratory (PNNL), a DOE Office of Science User Facility, under proposal number 48772.

References

[1] Mamedov, E. A.; Corberán, V. C. *Appl. Catal. A: General* **1995**, *127*, 1-40.

[2] Labinger, J. A.; Bercaw J. E. Nature 2002, 417, 507-514.

[3] Bronsted, J. N. *Chem. Rev.* **1928**, *5*, 231-338.

[4] Evans, M. G.; Polanyi, M. *Trans. Faraday Soc.* **1938**, *34*, 11-24.

[5] Zboray, M.; Bell, A. T.; Iglesia, E. *J. Phys. Chem. C* **2009**, *113* 12380-12386.

Copyright © American Institute of Chemical Engineers. All rights reserved.

467134 Selective Oxidation of n-Butane to 1-Butanol over Transition Metal Catalysts Encapsulated By Metal-Organic Frameworks

Abstract:

Density functional theory (DFT) is used to model reaction pathways of the oxidation of n-butane to 1-butanol over several different metal alloy and pure metal catalysts under steric constraints, and the implications on designing MOF-encapsulated catalysts are discussed. MOFs are porous crystalline solids comprised of metal-based nodes connected by organic "linker" molecules. Recently, they have been successfully grown around metal nanoparticles, creating porous networks at the MOF/catalyst interface, which have been shown to promote the regioselective hydrogenation of *trans*-1,3-hexadiene to 3-hexene. In this work, we explore the use of sterically constrained catalysts for selective alkane oxidation. We are specifically interested in n-butane oxidation to 1-butanol, which is important in the pharmaceuticals, energy, and specialty chemicals industries. It is a useful reaction for studying regioselectivity since 2-butanol is thermodynamically more stable than 1-butanol. In this work, we investigate n-butane oxidation to 1-butanol, 2-butanol, 1-butanal, and 1-butene oxide on the (111) facets of Ag_3Pd, Cu_3Pd, PdZn, Cu, and Pd. We use a "surrogate pore" comprised of helium atoms to model the MOF pores. We compute the energies of reaction intermediates and transition states involved in 90 total reactions using DFT, and we develop scaling relationships linking these energies to more simple "descriptors" of catalytic activity. These relationships are used in coordination with microkinetic modeling. We identify the best descriptors for this mechanism along with the dominant reaction pathways and the catalytic rates as functions of the descriptor values. The descriptor values that optimize activity and selectivity toward 1-butanol are identified, and predictions about which materials will be optimal for this reaction along with predictions about their stabilities are discussed.

Copyright © American Institute of Chemical Engineers. All rights reserved.

467091 Highly Selective Fischer-Tropsch Synthesis for Production of JP-8 Jet Fuel from Coal and Coal/Biomass Mixtures

Abstract:

An indirect liquefaction process employing advanced process intensification approaches is being developed to reduce costs and accelerate commercialization of a coal to liquid (CTL) process for converting coal and coal-biomass mixtures to JP-8 jet fuel. In this Air Force and DOE/NETL sponsored cooperative agreement, Southern Research is leading a team consisting of IntraMicron, Inc., Southwest Research Institute, and Nexant, Inc., along with Chevron as selective Fischer-Tropsch (FT) catalyst supplier and the National Carbon Capture Center (NCCC) as the host site. The process reduces the number of parasitic syngas cleaning steps and eliminates costly upgrading of wax produced in conventional indirect liquefaction processes, resulting in improved economics and lower lifecycle greenhouse gas (GHG) emissions for CTL-based jet fuel production. These improvements directly reduce capital costs of indirect CTL processes, resulting in significantly improved economics and cost competitiveness of CTL when compared to petroleum refining. Specific objectives for FT experiments are to demonstrate a selective and active second generation cobalt-zeolite hybrid Fischer-Tropsch (FT) catalyst from Chevron that predominantly produces liquid range C5-C20 hydrocarbons with high productivity (5X conventional FT catalysts) and jet fuel selectivity while eliminating the production and associated handling of solid C21+ wax. In fall of 2015, Southern operated a bench-scale FT reactor with a 2â€ diameter heat exchange reactor tube using IntraMicron technology in a nominal 5 gallons per day skid-mounted fixed-bed installed at NCCC. In those experiments, both productivity and selectivity goals were achieved during a 300 hour experiment. In fall 2016, another experiment is planned with the reactor diameter increased to 4â€. This work will compare the results from the two long-term tests of the hybrid FT jet-selective catalyst using the skid-mounted fixed-bed FT reactor system at NCCC.

Copyright © American Institute of Chemical Engineers. All rights reserved.

468121 Composition of Bio-Crude Oils from Hydrothermal Liquefaction of Algae in Batch Reactor

Abstract:

Due to the limitations of food, farmland, and weather, novel biofuels produced from hydrothermal liquefaction of biomass has attracted much attention. Algae-derived oils from hydrothermal liquefaction are promising because of high algae growth rates, strong CO_2-mitigation potential, and avoidance of drying requirements. In hydrothermal liquefaction, water at subcritical states (270-350°C and 80-173 bar) catalyzes depolymerization of lipids, proteins and carbohydrates in algae, leading to relatively high yields of bio-crude oil that can be valorized to transportation fuels. In this study, we investigated the influences of operating conditions (temperatures of 310-350 °C, residence times of 5-60 min, and solid algae contents of 5-10 wt.%) on the composition of bio-crude oil from hydrothermal liquefaction of *Nannochloropsis salina* and *Galdieria sulphuraria* using a variety of analytical methods. Organic molecules with molecular masses of ~150-700 Da were observed by Fourier transform ion cyclotron resonance mass spectroscopy (FT-ICR/MS) for comprehensive qualitative compositional description that employed three complimentary ionization methods. Free fatty acids in the HTL oil were quantitated by gas chromatography mass spectroscopy (GC/MS). Higher heating values and CHNS elemental content were also measured by bomb calorimeter and elemental analysis to obtain information about energy recovery relative to previous studies. Bio-crude oil characterization results will be used to optimize hydrothermal liquefaction operating conditions in a 1.8 L batch reactor and to select initial conditions for a pilot scale, continuous flow reactor.

Copyright © American Institute of Chemical Engineers. All rights reserved.

469356 De-Alloyed Platinum-Bismuth Nanoparticles As Highly Active Electrocatalysts for Dimethyl Ether Oxidation and Oxygen Reduction

Abstract:

Dimethyl ether (DME) is a clean-burning alternative to diesel fuel that meets strict emissions standards in combustion engines. DME is currently derived from natural gas but can also be synthesized from biomass on-site and is thus a sustainable and highly distributed energy source. DME has properties similar to propane and can utilize established fueling infrastructure and handling procedures. In addition to its use in internal combustion engines, the convenience of DME generation and transport can be exploited to overcome the high-pressure infrastructure requirements of hydrogen fuel in more efficient electrochemical fuel cells. Fuel cell-driven vehicle efficiencies are among the highest of all vehicle technologies.

Relative to other direct liquid fuels envisioned for use in electrochemical cells, DME is: (1) more energy-dense than hydrogen, (2) non-toxic compared to methanol, (3) a lower dipole moment than methan which limits fuel crossover, (4) the simplest ether which can be more efficiently oxidized than ethanol through electrocatalytic C-O bond cleavage, and (5) unlike ethanol, does not interfere with world food production. A key stumbling block to the commercialization of DME fuel cells has always been identification of a suitable electrocatalyst. The most active electrocatalyst presently available for DME oxidation is highly dispersed commercial precious metal Pt/Ru nanoparticles supported on carbon. A mass activity no greater than 1.3 A/g_{Pt} is observed at peak potential and is more than an order of magnitude less than that obtained from hydrogen.

We here discuss the synthesis of a novel DME oxidation electrocatalyst that has *more than an order of magnitude* greater activity per unit mass of Pt than state-of-the art. The electrocatalyst is composed of de-alloyed $PtBi_2$ nanoparticles (NPs). The NP pre-cursor material formed without the use of toxic organic solvents, difficult-to-

synthesize organic precursors, or dangerous hydrazine reducing agent typically needed to produce bismuth metal. Synthesis involves the use of stannous chloride as both reducing and stabilizing agent in aqueous suspension and is unprecedented in that Sn-Bi ligand formation was unknown prior to our work. Electrochemical and chemical de-alloying of these nanoparticles is shown to alter the surface atomic coordination relative to pure Pt and produce unusually high electrochemical areas and activities. DME electrooxidation activities greater than 30 A/g_{Pt} are observed in sulfuric acid at peak potential while the activity of the corresponding cathodic oxygen reduction reaction (ORR) increases to over 15 A/g_{Pt} at 0.9V vs. RHE in perchloric acid. ORR activity is comparable to that obtained with recent dealloyed Pt-Cu catalysts but in a one-pot synthesis approach that avoids the need for a high pressure hydrogen reduction step during synthesis. Despite the significant increases in mass-specific activity observed relative to pure Pt, the area-specific activity of the dealloyed electrocatalyst is lower. This observation is discussed in terms of the corresponding changes in surface atom coordination and electrolyte effects.

Copyright © American Institute of Chemical Engineers. All rights reserved.

Modification of Cu-Based Catalyst System for Co-production of Methanol and Dimethylether

Lujie Ye and Sunggyu Lee

Sustainable Energy and Advanced Materials Laboratory
Department of Chemical and Biomolecular Engineering
Ohio University
Athens, OH 45701

EXTENDED ABSTRACT

Dimethylether (DME) is the simplest ether, which has outstanding physical and chemical properties in many areas including oxygenated fuels and industrial chemicals. Since DME only contains C, H, and O elements, no NO_X and SO_X emission is generated during the combustion of DME. The high oxygen content of DME also alleviates the emission of CO as well, which is a common product of incomplete combustion. It is also reported that DME has a higher cetane number (CN) than the conventional petrodiesel. Therefore, DME is regarded as a promising substitute for petrodiesel as well as a cetane enhancer. Besides the advantages of using DME as fuel, the recently developed processes of producing lower olefins like ethylene and propylene [1], gasoline range boiling hydrocarbons [2], industrial chemicals such as oxygenates [3], methyl acetate [4], and ethylene glycol intermediate like dimethoxyethane (DMET) [5] make a strong case for DME to become a promising building block chemical for the synthesis chemistry.

Co-production of methanol and DME under a dual catalyst system of $Cu/ZnO/Al_2O_3$ and γ-Al_2O_3 has been developed for years. However, deactivation of the methanol synthesis catalyst is nontrivial, thus rendering a practical applicability problem of the system. This problem is at least partially caused by the growth of the copper crystallite size of the catalyst during the synthesis reaction. It is reported that the crystallite size of Cu could increase from an average of 37 angstroms to 109 angstroms after thermal aging under CO-free or CO_2-free syngas only for short periods of time-on-stream [6]. The aged catalyst only provides 78% of the original reactivity [7]. Besides the accelerated growth of the catalyst's copper crystallite size, deactivation of the catalyst can also be triggered by accumulation (i.e., build-up) of water in the catalyst pores, which could promote the mechanical breakdown of the catalyst while reducing the overall reaction conversion. Therefore, understanding of the detailed catalytic mechanism and developing a more robust catalyst are the key challenges for improving the efficiency of the DME production. In this study, supercritical fluid treatment is performed as a pre-treatment step for the unreduced catalyst of $CuO/ZnO/Al_2O_3$ to eventually improve the catalytic reactivity of the reduced (activated) catalyst.

The supercritical treatment is carried out in a stainless steel autoclave-based reactor system. Chemically pure CO_2 is compressed by a booster pump and stored in a CO_2 storage vessel. A certain metered amount of CO_2 will be fed into the reactor from the bottom (Figure 1). Prior to the supercritical treatment, the catalyst loaded reactor will be purged and rinsed by gaseous CO_2 three times to remove the residual moisture and air. Then, the system will be pressurized to a desired supercritical condition gradually, according to the estimated P-V-T values using the Peng-Robinson equation of state. Different temperatures, pressures, and treatment times are compared to find the most effective and beneficial combinations for this treatment step.

After the supercritical fluid treatment, a portion of ZnO is transformed to $ZnCO_3$, which is reported to be beneficial to the methanol synthesis [8] [9]. The cumulative volume of pores for the treated catalyst increases about 20 percent on an average, while the internal surface area of the treated catalyst increases about 40 percent. At the same time, the crystallite size of copper is found to be decreased by about 30 percent after the supercritical CO_2 treatment.

Acknowledgment

This work was partially supported by a research grant from the Korea Institute of Science and Technology (KIST). The authors are grateful for the support.

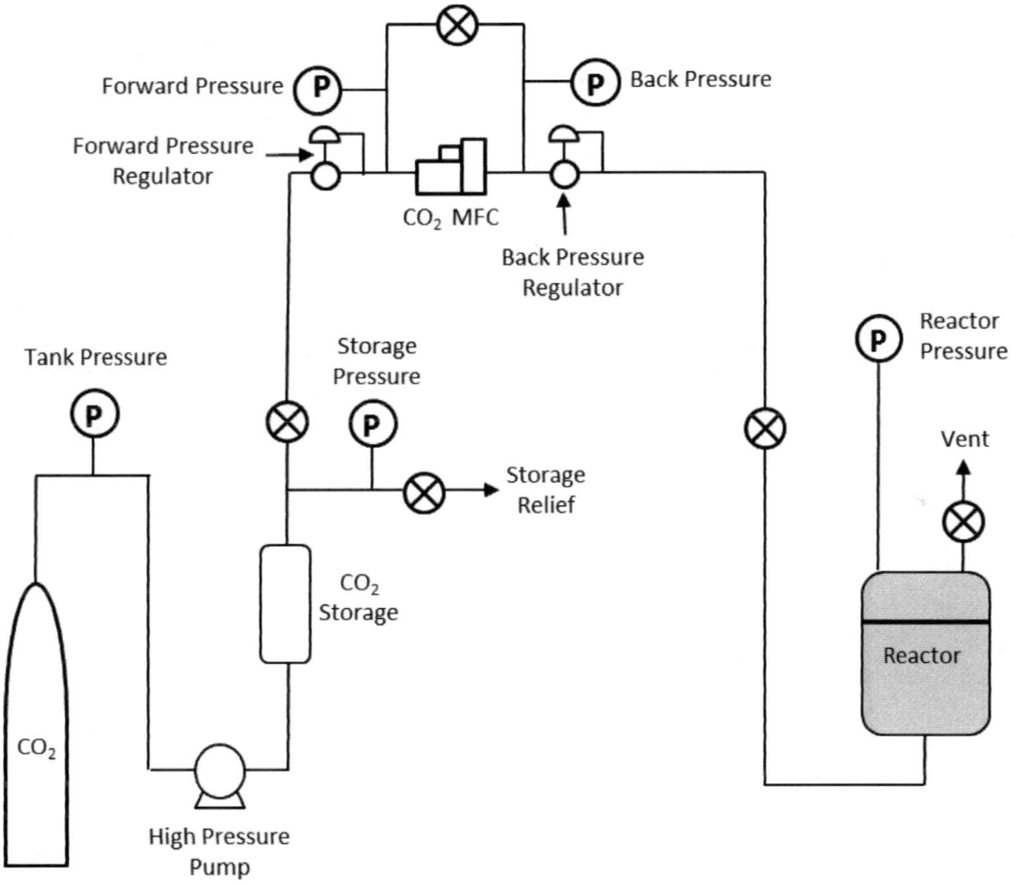

Figure 1. A process flow diagram of CO_2 supercritical treatment system (Courtesy of Chad Able, SEAM Lab, Ohio University)

References

[1] A. S. Al-Dughaither and D. L. Hugo, "Neat dimethyl ether conversion to olefins (DTO) over HZSM-5: Effect of SiO2/Al2O3 on porosity, surface chemistry, and reactivity," *Fuel,* vol. 138, pp. 52-64, 2014.

[2] P. Haro, F. Trippe, R. Stahl and E. Henrich, "Bio-syngas to gasoline and olefins via DME–a comprehensive techno-economic assessment," *Applied Energy,* vol. 108, pp. 54-65, 2013.

[3] G. Liu, B. Yan and B. Chen, "Technical review on jet fuel production," *Renewable and Sustainable Energy Reviews,* vol. 25, pp. 59-70, 2013.

[4] P. Cheung, A. Bhan, G. J. Sunley and E. Iglesia, "Selective carbonylation of dimethyl ether to methyl acetate catalyzed by acidic zeolites," *Angewandte Chemie International,* vol. 45, no. 10, pp. 1617-1620, 2006.

[5] A. Sardesai, T. Tartamella and S. Lee, "Synthesis of hydrocarbons from dimethyl ether: selectivtties towards light hydrocarbons," *Fuel Science & Technology International,* vol. 14, no. 5, pp. 703-712, 1996.

[6] S. Lee, Methanol Synthesis Technology, Boca Raton, FL: CRC Press, 1990.

[7] B. G. Lee and S. Lee, "Regeneration of liquid phase methanol synthesis catalyst," *Fuel Science and Technology Int'l,* vol. 5, no. 9, pp. 587-612, 1991.

[8] V. Parameswaran, S. Lee and S. Wender, "Roles of carbon dioxide in methanol synthesis," *Fuel Science and Technology International,* vol. 7, no. 8, pp. 1021-1057, 1989.

[9] V. Parameswaran, S. Lee and S. Wender, "The role of water in methanol synthesis," *Fuel Science and Technology International,* vol. 7, no. 7, pp. 899-918, 1989.

465206 Design of a Sabatier Reactor for CO_2 Conversion into Synthetic Methane

Abstract:

Design of a Sabatier reactor for CO_2 conversion into synthetic methane

Duo Sun and David Simakov

Department of Chemical Engineering, University of Waterloo, Waterloo, ON N2L 3G1, Canada

Converting CO_2-reach streams (e.g., biogas, fermentation off-gas, flue gas) into synthetic natural gas (SNG) via Sabatier reaction (CO_2 hydrogenation into CH_4) is an attractive route for CO_2 utilization [1-3]. There is a potential to decrease CO_2 emissions significantly and, at the same time, to reduce the consumption of fossil natural gas. To make the process sustainable, H_2 which is required for this reaction should have zero or very low carbon footprint. This requirement is achievable if H_2 is produced via water electrolysis using renewable electricity (hydro, wind, solar) or surplus (e.g., nuclear) electricity. Electrolysis can be done in a highly efficient way using polymer electrolyte membrane (PEM) electrolyzes [4]. However, a number of technological issues remain to be resolved, including catalyst deactivation and thermal management (Sabatier reaction is highly exothermic) [5]. The study presented herein focuses on the design of a highly efficient and compact Sabatier reactor for CO_2 hydrogenation into synthetic natural gas.

The general concept is a heat-exchanger type packed bed with active cooling for efficient heat removal. Compressed air, steam, or molten salt can be used as a coolant; molten salts are excellent heat transfer fluids due to their high heat capacity and thermal conductivity [6-9]. To identify the optimal configuration, we performed numerical simulations, using a transient pseudo-homogeneous model [10]. We searched for optimal operating conditions to maximize methane yield and to prevent at the same time reactor overheating and catalyst deactivation. We have identified several configurations that allow highly efficient CO_2 hydrogenation, providing CO_2 conversions and CH4 yields higher than 90%. A preliminary techno-economic evaluation demonstrates the potential of this approach for commercialization.

References

[1] Rönsch, S. et al. Review on methanation – From fundamentals to current projects. *Fuel* 2016, 166, 276-296.

[2] Schaaf, T. et al. Methanation of CO_2 – storage of renewable energy in a gas distribution system. *Energy Sustain. Soc.* 2014, 4, 29.

[3] Götz, M et al. Renewable Power-to-Gas: A technological and economic review. *Renew. Energ.* 2016, 85, 1371-1390.

[4] Gahleitner, G. Hydrogen from renewable electricity: An international review of power-to-gas pilot plants for stationary applications. *Int. J. Hydrogen Energ.* 2013, 38, 2039-2061.

[5] Kao, Y.-L.; Lee, P.-H.; Tseng, Y.-T.; Chien, I.-L.; Ward, J.-D. Design, control and comparison of fixed-bed methanation reactor systems for the production of substitute natural gas. *J. Taiwan Inst. Chem. Eng.* 2014, 45, 2346-2357.

[6] Simakov, D. S. A.; Wright, M. M.; Ahmed, A.; Mokheimer, E. M. A.; Román-Leshkov, Y. Solar thermal reforming of natural gas: a review on chemistry, catalysis and system design. *Catal. Sci. Technol.* 2015, 5, 1991-2016.

[7] Said, S. A. M.; Waseeuddin, M.; Simakov, D. S. A. A review on solar reforming systems. *Renew. Sust. Energ. Rev.* 2016, 59, 149-159.

[8] Said, S. A. M.; Simakov, D. S. A.; Mokheimer, E. M. A.; Habib, M. A.; Ahmed, S.; Waseeuddin, M.; Román-Leshkov, Y. Computational fluid dynamics study of hydrogen generation by low temperature methane reforming in a membrane reactor. *Int. J. Hydrogen Energ.* 2015, 40, 3158-3169.

[9] Said, S. A. M.; Simakov, D. S. A.; Waseeuddin, M.; Román-Leshkov, Y. Solar molten salt heated membrane reformer for natural gas upgrading and hydrogen generation: A CFD model. *Sol. Energy* 2016, 124, 163-176.

[10] Simakov, D. S. A.; Sheintuch, M. Design of a thermally balanced membrane reformer for hydrogen production. *AICHE J.* 2008, 54, 2735-2750.

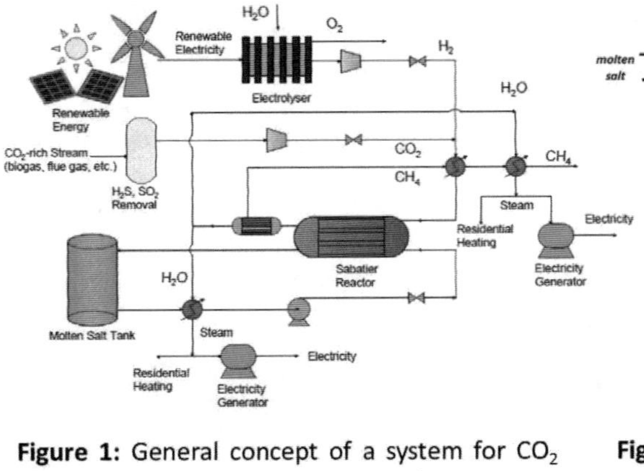

Figure 1: General concept of a system for CO_2 hydrogenation into synthetic natural gas.

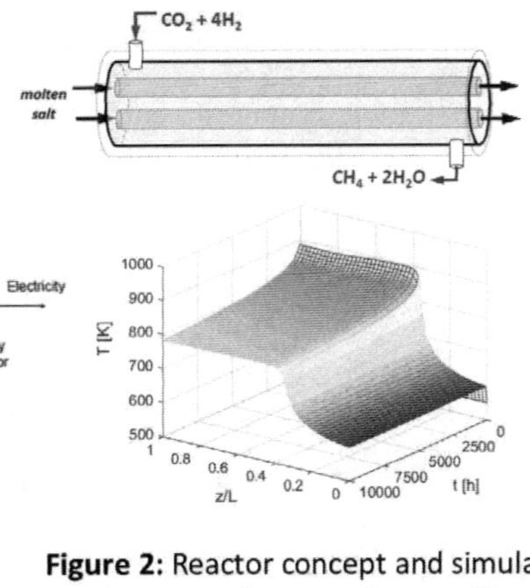

Figure 2: Reactor concept and simulated spatiotemporal temperature profile.

Copyright © American Institute of Chemical Engineers. All rights reserved.

462795 Steam Reforming of Dimethyl Ether over a Novel Plate-Type γ-Al_2O_3/Al Monolith Supported Cu-Based Bi-Functional Catalyst

Abstract:

Nowadays, the increasing global demand of energy accelerates the depletion of fossil fuels, thus boosting the extensive exploration of new energy alternatives. Hydrogen is considered as a clean energy and steam reforming of dimethyl ether (DME) is regarded as a promising method to produce hydrogen for fuel cell applications. Unlike many other fuels (e.g. methanol, methane and bio-ethanol), DME is non-toxic and provides high H/C ratio and high energy density. Moreover, the well-developed infrastructure of LPG can readily be used for the distribution of DME due to their similar physical properties. Generally, steam reforming of dimethyl ether (DME SR) (Eq. (1)) can be expressed as a combination of DME hydrolysis (Eq. (2)) to methanol over a solid acid catalyst and steam reforming of methanol (MSR, Eq. (3)) over a metallic catalyst. Therefore, a bi-functional catalyst consisting of acid sites and MSR sites is required for overall DME SR process.

DME steam reforming: $CH_3OCH_3 + 3H_2O \leftrightarrow 6H_2 + 2CO_2 \quad H_r = +135kJ/mol$ (1)

DME hydrolysis: $CH_3OCH_3 + H_2O \leftrightarrow 2CH_3OH \quad H_r = +37kJ/mol$ (2)

MeOH steam reforming: $CH_3OH + H_2O \leftrightarrow 3H_2 + CO_2 \quad H_r = +49kJ/mol$ (3)

In our previous works,[1-3] a plate-type γ-Al_2O_3/Al monolith was prepared through anodization technology. The structured monolith was proven to have an excellent activity in DME hydrolysis. Moreover, the anodic alumina supported Cu composite catalyst exhibited a satisfied performance in DME SR system.

In this study, the self-reduction behavior of fresh Cu/γ-Al_2O_3/Al catalyst during DME SR was further studied and the self-activation mechanism was proposed based on XRD and XPS analysis. However, the Cu/γ-Al_2O_3/Al catalyst showed a poor thermal stability at 350 °C due to the sintering of copper. As an approach, nickel was added to the Cu-based catalyst, and then the effect of nickel loading and chemical state on the activity of Cu/Ni/γ-Al_2O_3/Al catalyst was investigated. It was found that the proper amount of nickel could improve the dispersion of copper species, and thus enhancing the catalytic activity. For catalysts with lower nickel loadings, the thermal stability of Cu-based catalyst could not be improved. While, for catalysts with higher nickel loadings, part of nickel species would aggregate and block the active sites of copper. Moreover, a small amount of metallic Ni appeared, which resulted in the occurrence of side-reaction of CO methanation. Finally, the durability evaluation of optimized Cu/Ni/γ-Al_2O_3/Al was carried out under critical conditions (400 °C) and the results showed that it has an excellent stability during a 120 h test, which demonstrated that the novel plate-type γ-Al_2O_3/Al monolith supported Cu and Ni composite catalyst was an excellent catalyst for DME SR. Furthermore, it would be very promising for the application of micro-channel reformer for the domestic fuel cell system as for its outstanding shape flexibility and catalytic performance.

References

[1] F.Y. Fan, Q. Zhang, J.J. Xu *et al.*, Catal. Today, 216 (2013) 194-9.

[2] Q. Zhang, F.Y. Fan, G.M. Xu *et al.*, Int. J. Hydrogen Energy, 38 (2013) 10305-14.

[3] Q. Zhang, J.J. Xu, F.Y. Fan *et al.*, Fuel Process. Technol., 119 (2014) 52-9.

Copyright © American Institute of Chemical Engineers. All rights reserved.

462167 Low Temperature CO Oxidation with a Single Site Pt Catalyst Supported on the "29" Cu_xo/Cu(111) Surface: Elucidating the Chemical Nature of the Pt Active Sites

Abstract:

The ability to create highly dispersed, single site noble metal catalysts is of significant interest as such catalysts would be highly atom efficient and have the potential to significantly reduce the cost of catalysts for key reactions, such as low temperature CO oxidation and water-gas shift. While the presence and catalytic activity of various noble metal single site catalysts has been reported,[1-5] there is still disagreement over whether atomically dispersed metal atoms are indeed catalytically active or if the observed catalytic activity of single site catalysts is due to metal nanoparticles.[6, 7] Such disagreements create a crucial need for the development of well-defined single site supported catalysts with an accompanying experimentally accurate theoretical model in order to correctly determine the chemical nature of the catalytically active sites. To this end, we have studied low temperature CO oxidation on Pt single site catalysts supported on the "29" Cu_xO/Cu(111) surface. By first thoroughly characterizing the "29" Cu_xO/Cu(111) surface, we were able to determine the chemical nature of the Pt sites active for CO oxidation.

In order to have a truly accurate model how and where the single Pt atoms bind to the "29" Cu_xO/Cu(111) surface, we must have an accurate model of the "29" Cu_xO/Cu(111) structure itself. This was accomplished by characterizing both the clean "29" Cu_xO/Cu(111) surface[8] and the adsorption of CO on the "29" Cu_xO/Cu(111). First, for the characterization of the clean "29" Cu_xO/Cu(111) surface, the comparison of experimental and theoretical and STM images showed that the Cu_xO layer was formed from 6 fused hexagonal rings, each with 6 O and 6 Cu atoms. Furthermore, we determined that there are 5 O adatoms present on the Cu(111) surface in the center of the Cu_xO rings. By developing a phase diagram for various Cu_xO species under the conditions used to create the "29" Cu_xO/Cu(111) surface, we showed this model for the "29" Cu_xO/Cu(111) surface was the most stable structure under the experimental conditions. Second, the "29" Cu_xO/Cu(111) surface structure was further verified through the adsorption of CO on said surface. Depending on the amount of CO exposed to the "29" Cu_xO/Cu(111) surface, 6 different CO ordered structures were observed experimentally to form. Through an extensive comparison with theoretical STM images, the nature of the ordered structures was elucidated. These comparisons confirm the structure of the "29" Cu_xO/Cu(111) surface.

With the exact structure known for the "29" Cu_xO/Cu(111) surface, we can then accurately characterize and model the binding of the catalytically active Pt sites to the oxide surface. Experimental STM images show that the Pt exists as single atoms on the "29" Cu_xO/Cu(111) surface. An electronic analysis of the Pt atoms on the "29" Cu_xO/Cu(111) surface, coupled with experimental x-ray photoelectron spectroscopy (XPS), show that these Pt atoms exist in the metallic Pt state. Furthermore, depending on the temperature used when Pt is added to the "29" Cu_xO/Cu(111) surface, experimental STM images show that there are two distinct Pt sites: an active CO oxidation site with Pt dosed below room temperature and an inactive CO oxidation site with Pt dosed at room temperature. By modeling the adsorption of CO on the single Pt sites on the "29" Cu_xO/Cu(111) surface, we determined that the binding strength of CO on the Pt single atoms is inversely proportional to the binding strength of Pt on the "29" Cu_xO/Cu(111) surface (i.e. CO is more weakly bound on Pt sites where the oxide binds the Pt atoms strongly). These results suggest that the catalytically active Pt single sites might be meta-stable, diffusing to more stable, yet inactive, sites as the temperature is raised. Overall, our careful elucidation of the chemical nature of the Pt single atom sites on the "29" Cu_xO/Cu(111) surface via both theoretical and experimental techniques can assist in the design and optimization of atom efficient Pt catalysts.

References:

(1) Yang, M.; Li, S.; Wang, Y.; Herron, J.A.; Xu, Y.; Allard, L.F.; Lee, S.; Huang, J.; Mavrikakis, M.; Flytzani-Stephanopoulos, M. *Science* 2014, *346*, 1498.

(2) Kwak, J.H.; Hu, J.; Mei, D.; Yi, C.-W.; Kim, D.H.; Peden, C.H.F.; Allard, L.F.; Szanyi, J. *Science* 2009, *325*, 1670.

(3) Zhou, X.; Yang, W.; Chen, Q.; Geng, Z.; Shao, X.; Li, J.; Wang, Y.; Dai, D.; Chen, W.; Xu, G.; Yang, X.; Wu, K. *The Journal of Physical Chemistry C* 2016, *120*, 1709.

(4) Dvorak, F.; Farnesi Camellone, M.; Tovt, A.; Tran, N.D.; Negreiros, F.R.; Vorokhta, M.; Skala, T.; Matolinova, I.; Myslivecek, J.; Matolin, V.; Fabris, S. *Nature Communications* 2016, *7*, 10801.

(5) Fu, Q.; Saltsburg, H.; Flytzani-Stephanopoulos, M. *Science* 2003, *301*, 935.

(6) Ding, K.; Gulec, A.; Johnson, A.M.; Schweitzer, N.M.; Stucky, G.D.; Marks, L.D.; Stair, P.C. *Science* 2015, *350*, 189.

(7) Stephens, I.E.L.; Elias, J.S.; Shao-Horn, Y. *Science* 2015, *350*, 164.

(8) Therrien, A.J.; Zhang, R.; Lucci, F.R.; Marcinkowski, M.D.; Hensley, A.J.R.; McEwen, J.-S.; Sykes, E.C.H. *The Journal of Physical Chemistry C*2016, 10.1021/acs.jpcc.6b01284.

Copyright © American Institute of Chemical Engineers. All rights reserved.

457933 Identifying the Active Site for the Water-Gas Shift Reaction over TiO_2 Supported Pt Catalysts: Single Pt Atoms Versus Pt Clusters

Abstract:

Atomically dispersed noble metal catalysts on support surfaces offer maximum utilization of expensive noble metals that have potential applications in a variety of industrial chemical reactions. Fabrication of stable single atom catalysts has, especially in the case of Pt, always been a challenging task because of their high mobility that can lead to sintering under realistic reaction conditions. A few recent studies reported successful synthesis of single Pt atom catalysts on various supports which also showed excellent activity for CO oxidation and water-gas shift (WGS) at low temperatures. In spite of these advances, the exact nature and function of the active sites in these highly active catalysts is still a matter of debate and disagreements have been reported regarding the active site structures being single metal atoms or small metal nanoparticles. In this work, using a combination of DFT and microkinetic modeling techniques, we investigated the WGS activity of Pt catalysts supported on TiO_2 (110) considering different active sites at a Pt cluster/TiO_2 interface and a positively charged single Pt atom. Our goal is to determine under which conditions a specific active site is responsible for the majority of the observed experimental behavior.

We find that positively charged single Pt atoms, if stabilized on a TiO_2 surface, can be as active or even more active as nanoclusters for the WGS at low temperatures (< 573 K). Corner interface Pt sites become most active at temperatures above 573 K. These results agree with the experimental report by Flytzani-Stephanopoulos group (*J. Am. Chem. Soc.* **2015**, *137*, 3470) that positively charged single Pt atoms on TiO_2 are active for the WGS reaction. In a contradictory report, Stair and coworkers (*Science* **2015**, *350*, 189) claimed that the active site is the Pt nanoparticle and not the single Pt atoms based on the fact that they observed the disappearance of only the Pt nanoparticle related CO peak/spectrum and no change in the Pt single atom related CO peak during WGS reaction at a temperature of up to 300 °C. In our single atom model, one CO remains on the Pt atom throughout the catalytic cycle which can explain the non-disappearance of the Pt single atom related CO peak observed by the Stair group. In fact, this strongly adsorbed CO reduces the adsorption strength of the second CO molecule on Pt which reacts easily with the neighboring oxygen atom to form CO_2. In other words, the support, one CO and one H atom act as ligands on the positively charged Pt atom. Overall, we find that the redox pathway operates on different types of active sites at the Pt/TiO_2 (110) interface and oxygen vacancies on the TiO_2 surface play a significant role in WGS activity.

Copyright © American Institute of Chemical Engineers. All rights reserved.

460768 Trends in Single Transition Metal Atom/Oxide Support Interaction from Density Functional Theory

Abstract:

Supported single metal atoms offer intriguing properties as heterogeneous catalysts. Typically supported on oxide surfaces, single-atom catalysts (SACs) may offer enhanced reactivity and selectivity while allowing for perfect dispersion of the expensive active metal component. The activity and stability of these single atoms is impacted by the strength and character of their interaction with the underlying oxide support. Stronger interfacial binding between the atom and oxide support would help promote high dispersion during preparation and reduce the rate of sintering. Herein, we apply Density Functional Theory to evaluate the strength of binding between late transition metals and a series of oxide single crystal surfaces. We observe correlations between the strength of interfacial binding and the metalâ€™s oxidation enthalpy and the supportâ€™s reducibility. We also observe metal-metal interactions between some single metal atoms and supports. We develop an empirical correlation to predict metal-support interactions strength based on the individual metal oxidation enthalpy, oxide vacancy formation energy, and metal-metal bonding parameters.

Copyright © American Institute of Chemical Engineers. All rights reserved.

469030 CO Oxidation Kinetics on Supported Ir Single Atoms

Abstract:

$MgAl_2O_4$ supported Ir single atoms were synthesized and characterized by multiple *in situ/operando* techniques. We show that Ir single atoms are stabilized by strong covalent bonds with the support as evidenced by x-ray absorption spectroscopy (XAS). The O and Al and the hydroxyl groups from the support act as ligands affecting the Ir electronic properties as shown by XAS and *in situ* diffuse reflectance infrared Fourier transform spectroscopy (DRIFTS) of adsorbed CO. DRIFTS of adsorbed CO shows the formation of a remarkably stable $Ir(CO)_3$. The $Ir(CO)_3$ structure is supported by chemisorption measurements and DFT calculations. The mechanistic consequences of the $Ir(CO)_3$ on CO oxidation kinetics will be presented and the reaction mechanism will be compared with that on Ir nanoparticles.

Copyright © American Institute of Chemical Engineers. All rights reserved.

472090 Copper Atoms Stabilized on Reduced Titanium Oxides Catalyze Carbon Monoxide Oxidation

Abstract:

Copper catalysts haven been extensively studied as robust and efficient heterogeneous catalysts. We demonstrated that copper atoms stabilized on reduced titanium oxides are highly active and stable in CO oxidation. TiO_2 supports were pretreated with hydrogen under different temperatures. Copper was introduced through solid state reaction. A strong interaction between copper (II) and titanium (III) was identified through H_2-TPR and CO-TPR. The existence of copper (I) was attributed to the enhanced reaction rate and improved stability.

Copyright © American Institute of Chemical Engineers. All rights reserved.

462577 the Role of Au in Au-Catalyzed, Solvent-Free, Selective Epoxidation of Cyclooctene with Molecular Oxygen

Abstract:

Au catalyzes solvent-free epoxidation of cyclooctene with molecular oxygen with a high selectivity (>85%) for cyclooctene oxide. The catalytic effect is found to be independent of the Au precursor tested, which include Au/SiO_2, AuCl, and $AuCl_3$. There is an induction period before the reaction proceeds with a significant rate, and the catalytic significant species is solubilized Au that is formed under reaction conditions. The solubilized Au exhibits strong fluorescence in the UV-vis region, and is present in concentrations around 10-100 ng/mL. The fluorescence peak maxima are in the range around 400 nm, and shift depending on the concentrations of Au, reaction products, and other intentionally added chelating agents. The solubilized Au is present as atomic clusters mostly less than 10 atoms in size. These clusters are in a dynamic equilibrium with solubilized Au atoms, and the size distribution depends on the concentration of various species in solution. The solubilized Au is essential in initiating the reaction. After initiation, the reaction is dominated by the chain reaction of homogeneous hydrocarbon autooxidation, resulting in an apparent turnover frequency based on solubilized Au as high as ~3000 s^{-1}.

Copyright © American Institute of Chemical Engineers. All rights reserved.

472575 Structure, Reactivity and Selective Tuning of Well-Defined Rh Complexes in HY Zeolites

Abstract:

Structure, Reactivity and Selective Tuning of Well-defined Rh Complexes in HY Zeolites

K. Khivantsev,[a] A. D. Vityuk,[a] H. A. Aleksandrov,[b] G. N. Vayssilov,[b] O. S. Alexeev,[a,*] M. D. Amiridis[a]

[a]*Department of Chemical Engineering, University of South Carolina, Columbia, 29208, USA*

[b]*Faculty of Chemistry and Pharmacy, University of Sofia, Sofia, BG-1126, Bulgaria*

**Corresponding author: alexeev@cec.sc.edu*

1. Introduction

Rhodium is one of the most important metals in homogeneous and heterogeneous catalysis, with Rh carbonyl complexes catalyzing a number of reactions of industrial importance primarily in the liquid phase, including hydroformylation, hydrogenation, hydrosilylation, dehydrogenation, and carbonylation.[1] The search for new heterogeneous Rh catalysts that resemble the properties of their homogeneous analogs has constituted an active area of research for years. Such catalytic materials could offer not only easy separation of products from a catalyst but also the opportunity more efficiently perform the same organic reactions in flow rather than in batch reactors. Metal oxides frequently used as catalyst supports have nonuniform surfaces, and therefore, a variety of binding sites with different structural and electronic properties. As a result, metal complexes grafted on such supports yield catalytic sites that are nonuniform in structure and composition and exhibit different electronic properties, impacting substantially their catalytic performance. In this respect, the use of zeolites as supports is more promising, since these crystalline materials offer highly ordered arrays of binding sites for metal complexes and, therefore, allow for the preparation of catalytic materials with well-defined and nearly uniform structures.

Our goals were to prepare catalytic materials incorporating highly uniform and molecular in nature $Rh(CO)_2$ complexes anchored to the zeolite framework, to precisely examine the structure and chemical properties of such complexes, and to determine how the reactivity of the ligands in these complexes can be used for the surface mediated synthesis of other Rh species. FTIR, XPS, STEM, EXAFS, mass spectrometry, isotope labeling, and DFT calculations were used to monitor ligand exchange reactions and to understand the structure and composition of the species formed at the molecular level. New results presented herein demonstrate selective pathways for the conversion of HY zeolite-supported $Rh(CO)_2$ into well-defined and structurally uniform $Rh(CO)(C_2H_4)$, $Rh(NO)_2$, and $Rh(CO)(H)_x$ complexes and disclose the stability, chemical reactivity, and catalytic performance of these complexes for the hydrogenation of C_2H_4.

2. Experimental

1% $Rh(CO)_2$/HY samples were prepared by slurrying a $Rh(CO)_2(C_5H_7O_2)$ precursor with dealuminated HY zeolites (Si/Al=2.7-30 atomic) in dry n-pentane under N_2 at 25°C for 16 h. Ligand exchange reactions taking place during exposure to NO, C_2H_4, and H_2 were followed by in-situ FTIR at 25°C and atmospheric pressure. Isotopically labeled ^{13}CO and D_2 were used to confirm assignments of FTIR bands and to identify the nature of the complexes formed. An Inficon Transpector 2 residual gas analyzer equipped with a MDC variable leak valve was used for the analysis of products formed during ligand exchange reactions. XAS spectra were collected at beamline 4-1 at SSRL. XPS measurements were conducted using a Kratos AXIS Ultra DLD system. STEM images were recorded using a JEOL-2100F instrument equipped with a CEOS aberration corrector on the illuminating optics and operating at 200 keV. Kinetic measurements were performed in a quartz tubular micro reactor with various feed compositions, while products were analyzed with an on-line GC (HP 7890 A) system. Periodic DFT calculations were performed with the PW91 exchange-correlation functional using a Vienna *ab initio* simulation package (VASP).

3. Results and discussion

We found that grafting of $Rh(CO)_2$(acac) on dealuminated HY zeolites leads to the formation of two types of $Rh(CO)_2$ species with characteristic ν_{CO} bands at 2117/2053 and 2113/2048 cm^{-1} that have similar structural properties (evidenced by EXAFS). However, their thermal stabilities are different, and the fraction of each species formed strongly depends on the Si/Al ratio, as zeolites with a higher Al content favor the formation of the latter species in larger amounts. Our FTIR results strongly suggest that $Rh(CO)_2$ species with the ν_{CO} bands at 2113/2048 cm^{-1} cannot be linked to unreacted and partially reacted $Rh(CO)_2$(acac) complexes or to the formation of $Rh(CO)_2(H_2O)_x$ species. DFT calculations further indicate that binding sites of different nature in dealuminated faujasites are responsible for their formation.

Carbonyl ligands in both types of zeolite-grafted $Rh(CO)_2$ complexes are capable of reacting with gas phase C_2H_4 and NO to form $Rh(CO)(C_2H_4)$ and $Rh(NO)_2$ species, respectively. Nevertheless, the conversion rate is substantially higher for $Rh(CO)_2$ complexes with the ν_{CO} bands at 2117/2053 cm^{-1}, suggesting that electronic properties of Rh sites control the reactivity of carbonyl ligands. Our results further show that the exposure of $Rh(CO)(C_2H_4)$ species to H_2 selectively yields well-defined $Rh(CO)(H)_x$ complexes. The $Rh(CO)(H)_x$ species thus formed are stable, site-isolated, mononuclear, bound to oxygen atoms of the zeolite framework (evidenced by EXAFS), and are characterized by a set of well-defined ν_{CO} and ν_{RhH} bands in their FTIR spectra. DFT results further confirm that experimentally observed ν_{CO} and ν_{RhH} bands correspond to Rh carbonyl hydride complexes incorporating dissociated and molecular H_2 and having different locations in the zeolite framework. In fact, HY zeolite-supported $Rh(CO)(H)_x$ species described herein are the first known heterogeneous analogs of molecular rhodium carbonyl hydride complexes, which are believed to be key intermediates for the hydroformylation and hydrogenation of alkenes in solution.

The hydride ligands in grafted $Rh(CO)(H)_x$ can be displaced by CO or N_2 to form $Rh(CO)_2$ and $Rh(CO)(N_2)$ complexes, respectively, although the displacement of hydrides with N_2 is slow and non-selective. In contrast, C_2H_4 does not displace the hydride ligands in $Rh(CO)(H)_x$, but reacts with them to form C_2H_6 and unstable Rh(CO) intermediates. The latter species readily react with C_2H_4 from the gas phase to form the more stable $Rh(CO)(C_2H_4)$ complexes. Our results further suggest that $Rh(CO)(H)_x$ species lose hydride ligands at approximately 130°C, with unstable Rh(CO) species formed during this process. Subsequent redistribution of CO between Rh(CO) species leads to the formation of more stable $Rh(CO)_2$ complexes and Rh sites free of CO ligands. Furthermore, the HY zeolite-grafted $Rh(CO)_2$, $Rh(CO)(C_2H_4)$, $Rh(CO)(H)_x$ species were found to be catalytically active for the hydrogenation and dimerization of C_2H_4 at 25°C. Since both reactions occur simultaneously, experiments with the selective blockage of Rh coordination sites with bulky ligands and support sites with $(CH_3)_3SiCl$ allowed us to clarify the role and function of the Rh and support sites during catalysis and tune the selectivity of ethylene transformations. Finally, we demonstrated that NO ligands attached to Rh sites exhibit a different type of surface chemistry. Regardless of this fact, however, $Rh(NO)_2$/HY was also found to be active in the hydrogenation of C_2H_4, which is a rare example of alkene hydrogenation catalysis by grafted rhodium nitrosyl complexes.

4. Conclusions

We have shown that the use of organometallic precursors and highly crystalline supports (zeolites) provides an excellent opportunity for the surface mediated synthesis of a family of grafted $Rh(CO)_2$, $Rh(NO)_2$, $Rh(CO)(C_2H_4)$, $Rh(CO)(H)_x$ complexes, all of which are well-defined, site-isolated, molecular in nature, and catalytically active. We have established that coordination environment of Rh sites controls the catalytic activity of these complexes in the conversion of alkenes. This is an example of fine tuning catalytic properties of supported metal sites by altering the ligand environment on a single atom scale. Thus, a new pathway is opened to examine the role of such complexes and the resulting structure-reactivity relationships for a wide spectrum of industrially relevant catalytic applications.

References

1. 1. R. Franke, D. Selent, A. Börner, Chem. Rev. 112 (2012) 5675-5732.

Copyright © American Institute of Chemical Engineers. All rights reserved.

462655 Structure-Property Relationships of Palladium Catalyzed Methane Complete Combustion Using Uniform Nanoparticles

Abstract:

Structure-Property Relationships of Palladium Catalyzed Methane Complete Combustion using Uniform nanoparticles

Joshua J. Willis, Emmett Goodman, Matteo Cargnello

[1]Department of Chemical Engineering and SUNCAT Center for Interface Science and Catalysis, Stanford University, Stanford, CA 94305, USA

There is a critical need for reducing emissions of methane, the second most prevalent greenhouse gas. Homogeneous combustion of methane, as in flaring, results in the release of toxic gases such as NO_x, SO_x and CO, and current methods of catalytic emission control are still ineffective at low temperatures (<300°C) and in the presence of steam. Improving methane combustion catalysts would affect several technologies that have to deal with methane emissions. While many materials have been investigated for their methane combustion activity, palladium is widely accepted as one of the best metal catalysts. Unfortunately, it is unclear which phase of palladium (Pd metal, Pd oxide or a combination of both) is most active for this reaction; furthermore, there is a need to utilize Pd to the best possible extent given its precious nature. To gain a fundamental understanding of this system and provide insights into the active sites to further improve the activity and stability of Pd-based catalysts, control over the size, shape and composition of the metal nanocrystals is essential.

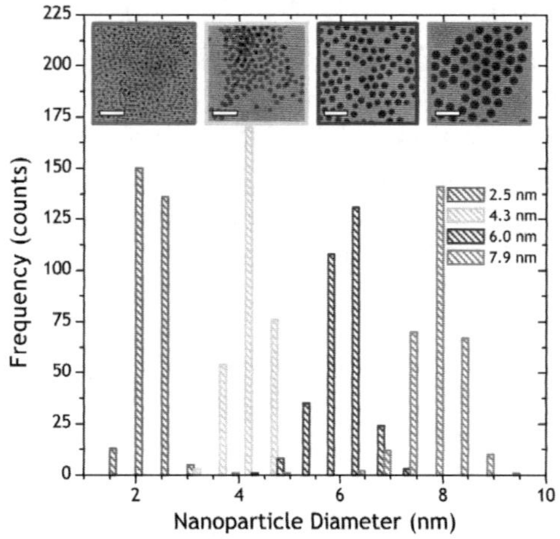

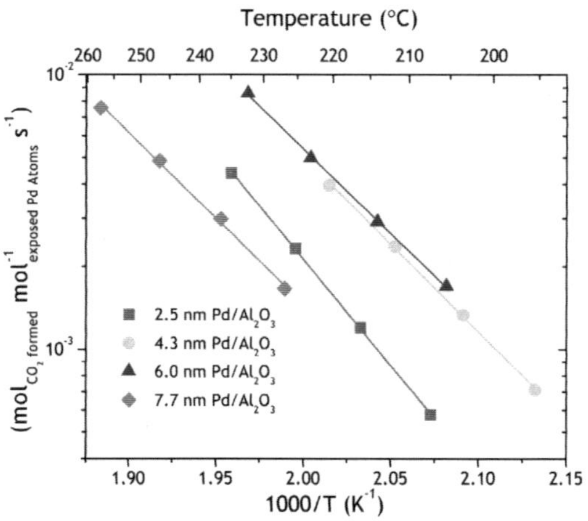

In this study, highly monodisperse palladium nanocrystals, prepared via high-temperature colloidal synthesis, are used to investigate the structure-activity relationships of palladium-catalyzed methane complete combustion in the low size regime (2-8nm). By deposition of the nanocrystals onto high-surface area supports (Al_2O_3, CeO_2, MgO and SiO_2; all with similar surface areas) and activation using a fast thermal annealing process, highly monodisperse, stable catalysts are obtained. These materials are used to draw structure-activity relationships under realistic methane complete combustion conditions. Kinetic characterization demonstrates that the activity is dependent on the oxidation state of the Pd phase, which is also affected by the size of the Pd nanocrystals. Furthermore, the effect of steam is systematically investigated as a function of size and support. These new observations are made possible through the use of tightly controlled Pd nanocrystals. Our work provides clear elements for improving the activity of Pd-based combustion catalysts and in general a framework for understanding structure-activity relationships using highly uniform catalysts under realistic reaction conditions.

Copyright © American Institute of Chemical Engineers. All rights reserved.

466079 Characterizing Active Sites on Chirally-Modified Supported Noble Metal Catalysts for Asymmetric Hydrogenation through Selective Poisoning

Abstract:

The chemical and structural surface heterogeneity of supported noble metal catalysts impedes a detailed description of the active sites. The catalytic performance of a supported noble metal catalyst in terms of the activity and the selectivity towards a specific surface reaction is affected by the adsorption of substrate and product molecules at the catalytically active surface sites. Heterogeneous hydrogenations in the liquid phase exhibit a complex adsorption behavior and the molecular structure of the substrate, the surface properties of the catalyst and the nature of the solvent primarily control the surface processes at the liquid-solid interface. A true *in situ* approach to study active sites on a supported noble metal catalyst consists in selectively poisoning of the noble metal surface using small amounts of strongly adsorbing substances such as thiols.[1,2] In this work, we made an attempt to study active sites in heterogeneous asymmetric hydrogenations on chirally-modified supported Pd catalyst. We examined the effect of active site poisoning on the asymmetric hydrogenation of 4-methoxy-6-methyl-2-pyrones on cinchonine-modified Pd/TiO_2using different organosulfur compounds.

In heterogeneous asymmetric hydrogenations, enantioselectivity is only induced at chirally-modified sites, while remaining unmodified surface sites lead to racemic product.[3]The highly specific interplay between co-adsorbed chiral modifier and prochiral substrate further increases the complexity of such poisoning experiments, and various organosulfur compounds with different adsorption behavior are needed to shed light on the complex surface processes. However, such studies have the potential to provide valuable insight into the distribution and intrinsic catalytic performance of racemic and chirally modified sites. Furthermore, they might pave the way to optimization strategies for asymmetric heterogeneous hydrogenations.

The effect of a specific thiol on the catalytic performance was investigated by adding the thiol and the chiral modifier cinchonine simultaneously to the reaction slurry, prior to the substrate. Using *operando*attenuated total reflection infrared spectroscopy, that is combining catalytic performance and spectroscopic measurement of the same catalyst sample, we were able to correlate the adsorption geometry and number of strongly bound organosulfur compounds at the active sites to the measured activity and enantioselectivity of the catalytic system. Our current results indicate that thiol adsorption has a strong impact on the catalytic activity, while leaving enantioselectivity unaffected. In contrast, the heterocycle thiophene disrupts the catalytic activity and the enantioselectivity. In general, the molecular structure of the organosulfur compound was found to play a crucial role for the poisoning behavior.

(1) Panthi, B.; Mukhopadhyay, A.; Tibbitts, L.; Saavedra, J.; Pursell, C. J.; Rioux, R. M.; Chandler, B. D. *ACS Catalysis* **2015**, *5*, 2232.

(2) Marshall, S. T.; O'Brien, M.; Oetter, B.; Corpuz, A.; Richards, R. M.; Schwartz, D. K.; Medlin, J. W. *Nature Materials* **2010**, *9*, 853.

(3) Mallat, T.; Orglmeister, E.; Baiker, A. *Chem. Rev.* **2007**, *107*, 4863.

Copyright © American Institute of Chemical Engineers. All rights reserved.

471040 Carbon Decoration in Supported Palladium Catalysts: Discrepancy in STEM-Chemisorption Particle Sizes

Abstract:

Introduction

Metal-support interactions have been known to stabilize metal nanoparticles in harsh reaction environments or cause hindrances in the chemisorption surface area during chemical reactions [1, 2]. Although it is necessary for the active sites to remain exposed without any poisoning to allow the maximum adsorption of reactants in heterogeneous reactions, it is also vital to maintain the structural stability in chemical reactions at elevated temperatures. Carbon-supported palladium catalysts find a wide variety of applications in the fine chemicals industry, as electrodes in fuel cells and in a large number of hydrogen-mediated aqueous phase catalytic reactions. While it has been frequently seen in the literature that active sites in the metal are blocked by carbonaceous deposits [3], there is also ample evidence that carbon coatings are frequently used to coat oxide supports and metal nanoparticles for enhancing their hydrothermal stability in aqueous phase hydrogenation reactions at elevated temperatures [4] due to their inherent hydrophobicity and inertness.

During the preparation of palladium catalysts on carbon black in our laboratory, it has been found that well dispersed catalysts are formed as confirmed with Scanning Transmission Electron Microscopy (STEM) and X-Ray Diffraction (XRD) but the chemisorption of hydrogen is substantially suppressed on the catalysts leading to a disagreement between particle sizes obtained by STEM and chemisorption [5]. This work focusses on understanding the STEM-chemisorption discrepancy based on the hypothesis that during the preparation of carbon supported palladium nanoparticles, the metal surface becomes coated from the surface/sub-surface/interstitial carbon that migrates from the support to the metal nanoparticles or with a partial surface blockage by residual chloride from the chloride precursor used. This suppresses the hydrogen chemisorption and leads to larger particle sizes as compared to transmission electron microscopy TEM/X-ray diffraction (XRD). The discrepancy is also believed to be dependant on the type of carbon used, the functional groups on the surface, the precursor chosen and pre- treatment conditions of the support.

Materials and Methods

Strong Electrostatic Adsorption (SEA) was used to deposit Pd-nanoparticles onto two sets of different carbons: VXC72 (carbon black) and Darco G60 (activated carbon) that were oxidized and heat-treated to different temperatures. The pre-treated catalysts were then dried at 120°C and reduced in hydrogen at 180°. The catalysts were characterized for particle sizes using STEM, XRD and chemisorption. Systematic Temperature Programmed Oxidation (TPO) studies were conducted to burn off the surface carbon and pre and post TPO chemisorption sizes were compared. Chloride and non-chloride precursors were also compared to determine the effect of chloride ions on the discrepancy.

Results and Discussion

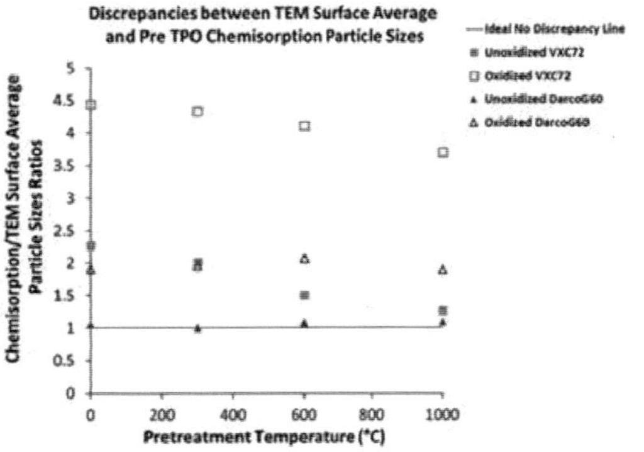

Figure 1

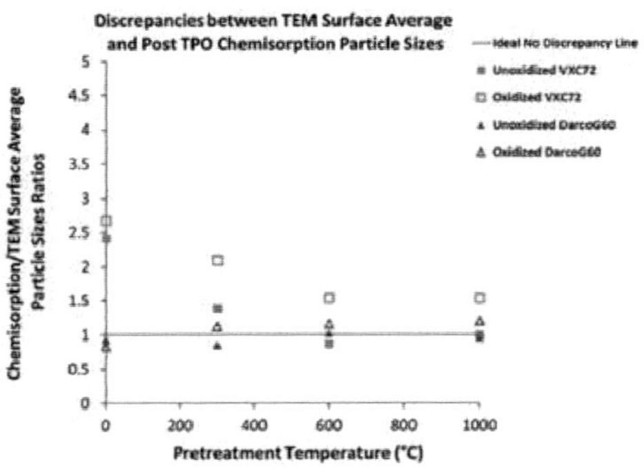

Figure 2

To understand the discrepancy better, the ratio of chemisorption and STEM surface average sizes were plotted against the pretreatment temperatures. The discrepancy for both DarcoG60 and VXC72 is the greatest when they are oxidized and decreases with the increase in pretreatment temperature which confirms that there is a relation between the surface oxygen groups and carbon decoration. It is also seen that the unoxidized Darco series did not show any discrepancy. Incidentally, this was the only series of samples that had very large Pd particles (>10 nm) with wide standard deviations (Fig 1). After the TPO burn off experiment, the discrepancy is reduced by 50-60% for the highest discrepancy (oxidized carbon samples) (Fig 2). For the unoxidized Darco sample, there was no change in the ratio since the discrepancy did not exist in the first place. However, it is seen that burning off the surface carbon does not recover the chemisorption surface completely. This led to the comparison of results using chloride and non-chloride precursors which showed that chemisorption gives a 30-50% reduction in the discrepancy for the nitrate precursor as compared to chloride precursor. It may, thus, be concluded that carbon decoration coupled with chloride poisoning is responsible for the discrepancy in the STEM-chemisorption particle sizes.

References

[1] D. Nerhing, H. Dreyer. Chem.Tech. 1960, 12, 343.

[2] G.M. Schwab. Adv. Catalysis, 1978, 27, 1-22.

[3] P. Forzatti , L. Lietti, Catalysis Today, 1999, 52,165-181.

[4] H. N. Pham, A. E. Anderson, R. L. Johnson, T. J. Schwartz,B. J. O'Neill, P. Duan, K. Schmidt-Rohr, J. A. Dumesic and A. K. Datye, ACS Catalysis, DOI: 10.1021/acscatal.5b00329 ACS Catal. 2015, 5, 4546−4555.

[5] J.M.M.Tengco, Y.K.Lugo-Jose, J.R.Monnier, J.R.Regalbuto, Catalysis Today, 2015, 246, 9.

Copyright © American Institute of Chemical Engineers. All rights reserved.

465628 100-Fold Rate Enhancement Caused By Dioxygen Ligands on a Supported, Single-Site Tetrairidium Cluster Catalyst for Ethylene Hydrogenation

Abstract:

The bonding of oxygen to transition-metal centers is of central importance from both a fundamental and technological perspective, in biological and synthetic catalysts. Yet the complexity of these systems usually prevents direct structure-function elucidation. Working with a silica-supported tetrairidium cluster as a simplified model catalyst, we investigated the relationship between oxygen ligands bonded at metal centers of the basal plane of the cluster and the reactivity of a metal atom located at the productive apical site for ethylene hydrogenation. The Ir_4 cluster was sterically stabilized with three bulky *tert*-butyl-calix[4]arene phosphine ligands, which are bound at the basal Ir sites, and as reported previously, it exhibited a low activity for ethylene hydrogenation at 323 K and 1 bar (5% C_2H_4, 16% H_2, balance He). However, after 24 h of O_2 treatment, a two order of magnitude increase in activity was observed. The oxygen ligands increased the activity of the apical Ir site for both ethylene hydrogenation and H-D exchange in the reaction of H_2 with D_2 in this system. In contrast, creation of open sites at the apical Ir atom with a mild oxidant, trimethylamine N-oxide, did not increase the catalytic activity substantially. Thus, we infer that the oxygen bonded to basal plane Ir atoms of the Ir_4 tetrahedron served as a ligand that enhanced the catalytic activity of the neighboring productive apical site for ethylene hydrogenation. Reaction orders for ethylene hydrogenation were found to be -0.24 and 0.60 for ethylene and H_2, respectively. The near zero and half rate orders for ethylene and H_2 are characteristic of a mechanism whereby H_2 is able to bond noncompetitively to a surface site where ethylene cannot. IR spectra determined the following characteristics of the terminal carbonyl ligands during a sequence of ethylene hydrogenation catalysis, oxygen treatment, ethylene hydrogenation catalysis, and recarbonylation treatment. A blue shift of approximately 20 cm^{-1} observed during the first and third steps indicates the formation of a hydride species, evident by an Ir-H band at 2110 cm^{-1}. The terminal carbonyl band returned to its original position in the second and fourth steps, accompanied by the loss of the Ir-H band. Distinct ethyl bands formed during the second ethylene hydrogenation step, after the removal of apical carbonyl ligands. The decarbonylation occurring in the first three steps corresponds to a loss of two terminal and bridging carbonyl ligands from the cluster, and the subsequent recarbonylation was only partial, consistent with bonding of oxygen to the cluster, which was confirmed by EXAFS and Raman spectra, the latter showing the presence of iridium peroxo species with bands at 541 and 713 cm^{-1}. EXAFS and IR spectra both gave evidence of isosbestic points during the oxidation, showing that it was a stoichiometric reaction. The stability of the supported Ir_4 clusters after first and second catalysis was confirmed with aberration-corrected STEM imaging. DFT calculations support the irreversible bonding of dioxygen ligands to the basal plane Ir atoms of the cluster and their electronic effects leading to a significant increase in the rate of ethylene hydrogenation as a result of decreasing the energy barrier and increasing the thermodynamic driving force for ethylâ€™s reductive elimination to form ethane. This step is inferred to be rate limiting, as supported by IR evidence of ethyl and hydride ligands present concomitantly during the second catalysis step.

Copyright © American Institute of Chemical Engineers. All rights reserved.

456438 Structure–Function Relationships for Graphene-Supported Pt Nanoclusters

Abstract:

Platinum-based catalysts play an important role in energy conversion technologies, particularly, in hydrogen-based or methanol-based proton exchange membrane fuel cells (PAFC). Graphene-supported Pt nanoclusters were recently found to be promising catalysts due to their enhanced catalytic activity and tolerance to CO poisoning, as well as their long-term stability toward sintering. However, structure–function relationships that govern the improved electrocatalytic activity in these materials are still not well understood. Here, we employ a combination of empirical potential simulations and density functional theory (DFT) calculations to investigate the structure–function relationships of small Pt_n (n=2-80) clusters on model carbon (graphene) supports. A Tersoff-Brenner style empirical potential, was employed within a Genetic Algorithm to investigate the global-minimum structures of Pt clusters in the size range of N=2-80 on pristine as well as defective graphene supports. Point defects in graphene strongly anchor Pt clusters and also appreciably affect geometries of small clusters, which we characterize via various structural metrics such as the gyration radius, average bond length, and average coordination number. Through selected *ab initio* studies, we find a consistent trend for charge transfer from Pt clusters to defective graphene supports resulting in the lowering of the cluster *d*-band center. This lowering of the cluster *d*-band center has been shown previously to result in weaker CO adsorption as well as reduced barriers for CO oxidation. A key finding from the structural analysis is that the fraction of potentially active surface sites in supported clusters is maximized for stable Pt clusters in the size range of 20-30 atoms, which provides a useful design criterion for optimal utilization of the precious metal.

Copyright © American Institute of Chemical Engineers. All rights reserved.

465371 Role of the Support and Reaction Conditions on the Vapor-Phase Deoxygenation of *M*-Cresol over Pt/C and Pt/TiO_2 Catalysts

Abstract:

1. Introduction

Biomass deconstruction using fast pyrolysis offers a promising route for the production of renewable bio-oil. However, the high oxygen content of bio-oil contributes to a number of undesirable characteristics, and bio-oil must be upgraded before it is suitable for use as a drop-in transportation fuel or blendstock.[1] *Ex-situ* catalytic fast pyrolysis (CFP) provides a route for bio-oil upgrading in which pyrolysis vapors are catalytically deoxygenated in the presence of hydrogen prior to condensation.[2-4] Within our group, we are exploring a dual fixed-bed *ex-situ* CFP reactor system because it offers greater catalyst flexibility over a fluidized bed system, thus providing greater control over chemistry and product composition.[3] In the first fixed-bed reactor, we are specifically targeting the deoxygenation of recalcitrant phenolic compounds. However, a clear understanding of the mechanistic details for the deoxygenation of these compounds has yet to be achieved, and questions remain about the role of the support (especially for reducible metal oxides) and effect of reaction conditions. To probe these questions, the vapor-phase deoxygenation of *m*-cresol, a model compound representative of the lignin-derived components contained within bio-oil, was investigated over Pt/C and Pt/TiO_2 catalysts using a combination of experimental and computational techniques.[5]

2. Experimental

Carbon- and TiO_2-supported Pt were prepared from $Pt(NH_3)_4(NO_3)_2$ via standard incipient-wetness impregnation methods. Each catalyst was characterized using powder X-ray diffraction, NH_3 temperature programmed desorption, inductively coupled plasma optical emission spectroscopy, transmission election microscopy, N_2 physisorption, and CO pulse chemisorption. The performance of each catalyst was evaluated in a packed bed reactor under conditions representative of hydrotreating (HT; 250 °C, 2.0 MPa) and *ex-situ* CFP (CFP; 350 °C, 0.5 MPa). In both cases, a 8:1 molar ratio of molecular hydrogen to *m*-cresol was maintained for the duration of the reaction period. Quantitative analysis was carried out using gas chromatographs equipped with flame ionization and thermal conductivity detectors, which had been calibrated with standards of known concentrations. In all cases, the mass balance closure was > 90%. Calculations were performed using the Vienna Ab initio Simulation Package (VASP) 5.3. The periodic density functional calculations used the projector augmented wave (PAW) potentials and an energy cutoff of 400 eV for the plane wave basis set for all calculations. The Perdew-Burke-Ernzerhof (PBE) generalized gradient corrected functional was used for all periodic calculations with Monkhorst-Pack 5×5×1 *k*-point sampling for metals and 3×3×1 *k*-point sampling for metal oxides.

3. Results and discussion

Toluene was the dominant product for both catalysts under *ex-situ* CFP conditions, with higher toluene selectivity observed over Pt/TiO2 as compared to Pt/C. 3-methylcyclohexanone, 3-methylcyclohexanol, and 3-methylcyclohexane were also observed over each catalyst. Under HT conditions, the production of 3-methylcyclohexanone and 3-methylcyclohexanol was highly favorable over both catalysts. Regardless of condition, Pt/TiO2 exhibited a TOF (metal site basis) that was more than double that of Pt/C.

The energetics of ring-hydrogenation (HDO), direct deoxygenation (DDO), and tautomerization (TAU) mechanisms were calculated over hydrogen-covered Pt(111) and oxygen vacancies on the surface of TiO_2(101). The results show that ring-hydrogenation to 3-methylcyclohexanone and 3-methylcyclohexanol is the most energetically favorable pathway over Pt(111). Over TiO_2(101), tautomerization and direct deoxygenation to toluene were identified as possible additional routes, although with slightly higher barriers. These calculations are consistent with the experimental results in which Pt/TiO_2 was shown to be more active on a metal site basis and exhibited comparatively high selectivity to toluene at 350 °C relative to Pt/C. Based on these findings, it is likely that the reactivity of Pt/TiO_2 and Pt/C is driven by the active phase at 250 °C while contributions from the TiO_2 support enhance deoxygenation at 350 °C.

4. Conclusions

The activity and selectivity observed during the deoxygenation of *m*-cresol was shown to be dependent on the choice of support and reaction conditions. Pt/TiO_2 was more active on a Pt site basis and exhibited higher selectivity to toluene than Pt/C. Computational modeling of *m*-cresol over Pt(111) and TiO_2(101) surfaces was used to develop a reaction mechanism that describes deoxygenation pathways over the active phase and identifies additional pathways that deoxygenate *m*-cresol to toluene over oxygen vacancies on TiO_2. These results suggest that synergistic effects between hydrogenation catalysts and reducible metal oxide supports provide access to additional deoxygenation pathways that are not accessible to the active phase alone. The insight gained from this study aids in the identification of fundamental catalyst requirements for deoxygenation reactions and informs catalyst development for the production of advanced hydrocarbon fuels from the pyrolysis of biomass.

References

1. M. Saidi, F. Samimi, D. Karimipourfard, T. Nimmanwudipong, B. Gates, M. Rahimpour, *Energy Environ. Sci.* 7 (2014) 103-129.
2. D. A. Ruddy, J. A. Schaidle, J. R. Ferrell, J. Wang, L. Moens, J. E. Hensley, *Green Chem.* 16 (2014) 454-490.
3. A. Dutta, J. A. Schaidle, D. Humbird, F. G. Baddour, A. Sahir, *Top. Catal.* 59 (2016) 2-18.
4. M. B. Griffin, F. G. Baddour, S. Habas, D. A. Ruddy, J. A. Schaidle, *Top. Catal.* 59 (2016) 124-137.
5. M. B. Griffin, G. Ferguson, D. A. Ruddy, M. Biddy, G. T. Beckham, J. A. Schaidle, *ACS Catal.* 6 (2016) 2715-2727.

Copyright © American Institute of Chemical Engineers. All rights reserved.

466660 Understanding Solvent Effects over Supported Metal Catalysts: Ketone Hydrogenation over Ru/SiO_2

Abstract:

Ketone hydrogenation over supported Ru catalysts is frequently reported to possess strong solvent effects, where different solvents are found to enhance the hydrogenation rate to varying extents. While there is no consensus as to the order in which different functionalities will enhance the rate through solvation, water is consistently reported as the solvent that provides the greatest promotional effect. Several qualitative explanations have been proposed for the observed solvent effects, where the majority propose that the kinetics of the hydrogenation are rendered more favorable in the presence of a solvent. To evaluate this claim quantitatively the kinetics of ketone hydrogenation in the absence of solvents must first be established, which can be better studied in the absence of any solvent.

To this end, the kinetics of the gas phase hydrogenation of C_3-C_5 ketones over Ru/SiO_2 catalysts was examined over a range of partial pressures and temperatures ranging from 303-476 K. A single thermodynamically consistent microkinetic model is developed to fit the observed experimental trends over all conditions and chain length, where a step-wise addition of hydrogen Horuiti-Polanyi mechanism is adopted. The reaction is found to proceed through a reactive alkoxide intermediate on the surface, where the addition of hydrogen to the alkoxide forming the alcohol on the surface is found to be the rate determining step. The developed model is then expanded to include the presence of multiple solvent functionalities on the surface including water, alcohols, alkanes and ethers. Multiple side reactions due to the presence of the solvents were observed, which lead to erroneous solvent effect observations if their presence is not accounted for. With the exception of water, no substantial promotional effect on the rate in the presence of any solvent was observed. Solvents are found to inhibit the rate through competitive adsorption on the surface; active sites are occupied which would otherwise facilitate ketone hydrogenation. Water is found to provide a promotional effect only at relatively high water/ketone activity ratios, which is thought to be due to the water itself participating in the reaction as a hydrogen donor. The consequence of ignoring thermodynamic non-idealities and catalyst stability are also examined in the context of observing an apparent solvent effect.

Copyright © American Institute of Chemical Engineers. All rights reserved.

468380 Hammett Study: Combined Experimental-Computational Approach to Understanding Electronic Metal-Support Interaction

Abstract:

Metal-support interactions may include electron transfer between the two components that alters catalytic activity; however, probing this phenomenon is challenging. Using a combined experimental-computational approach, we investigate how the underlying support affects the electronics of the active site of a supported Au catalyst. We employ a *Hammett study* using benzyl alcohol oxidation to show that reaction rate is dependent on the identity of benzyl alcohol ring substituents (–OCH3, -CH3, -Cl, and –CF3). Further, the changes in substrate reactivity are well correlated with the sigma$^+$ Hammett parameter of the substituted alcohols. Hammett studies on Au/Al_2O_3, Au/SiO_2, Au/TiO_2 and Au/ZnO show that the slope of this correlation, ρ, changes significantly with the underlying support, ranging from -0.36 for Au/ZnO to -0.87 for Au/Al_2O_3. The slope ρ can be interpreted as a semi-quantitative measure of the degree of charge build-up in the transition state, and the degree of influence of the support on the electronics of the Au nanoparticle. We use Density Functional Theory (DFT) to further investigate these relationships. We calculate the oxidation reaction energetics (to benzaldehyde) of para-substituted (CF_3 and OCH_3) benzyl alcohols over unsupported Au clusters with varying charges. The ring substituents have less effect on the rate of oxidation (indicative of the low ρ value) on electron rich clusters, consistent with a rate determining step of a hydride transfer to the Au nanoparticle. We will also show support properties that correlate with the slope ρ, and methods to quantify these electronic interactions using our experimental/computational data.

Copyright © American Institute of Chemical Engineers. All rights reserved.

480928 Quantification of Bronsted and Lewis Acid Sites on Zeolites in the Presence of Solvents

Abstract:

Many biomass derived feedstocks are highly oxygenated, meaning reactions cannot be conducted in gas phase without risking significant decomposition. However, the liquid phase incurs additional interactions and thermodynamic considerations which are not well understood. Fundamental understanding of solvent effects, including how solvent choice affects catalyst structure, catalyst acidity, and product distributions is still lacking. To provide experimental insight, attenuated total reflection (ATR) Fourier transform infrared spectroscopy (FTIR) was used to study the in-situ interactions between solvent, substrate, and catalyst. Specifically, pyridine was used as a probe molecule to characterize the relative ease of protonation by proton exchanged zeolites in solution. This has been used to understand how solvent choice affects substrate protonation, zeolite acidity, and activity in Brønsted acid catalyzed reactions. However, the FTIR technique is limited by its inability to quantify adsorbed and bulk species without integrated molar extinction coefficients (IMEC), or the amount of signal per mole of adsorbed or bulk species. A method has been established for vapor phase IMECs, but liquid phase IMECs are still missing. We develop a simple method to obtain IMECs of pyridine adsorbed on zeolites in the liquid phase in an ATR cell. Densities of Bronsted and Lewis acid sites on H/Y, H/Beta, and H/ZSM-5 of varying Si/Al ratios in solvents water, acetonitrile, and ethanol have been quantified with this method.

Copyright © American Institute of Chemical Engineers. All rights reserved.

478866 Synthesis of Itaconic Acid Ester Analogues via Self-aldol Condensation of Ethyl Pyruvate Catalyzed by Hafnium BEA Zeolites

Abstract:

Lewis acidic zeolites are used to synthesize unsaturated dicarboxylic acid esters via aldol condensation of keto esters. Hafnium-containing BEA (Hf-BEA) zeolites catalyze the condensation of ethyl pyruvate into diethyl 2-methyl-4-oxopent-2-enedioate and diethyl 2-methylene-4-oxopentanedioate (an itaconic acid ester analog) with a selectivity of ca. 80% at ca. 60% conversion in a packed-bed reactor. The catalyst is stable for 132 h on stream, reaching a turnover number of 5110 molEP molHf-1. Analysis of the dynamic behavior of Hf-BEA under flow conditions and studies with Na-exchanged zeolites suggest that Hf(IV) open sites possess dual functionality for Lewis and Brønsted acid catalysis.

Copyright © American Institute of Chemical Engineers. All rights reserved.

469702 Probing Reactions at the Solid-Liquid Interface By *Operando* Solid-State NMR Spectroscopy

Abstract:

Reactions of biomass-derived carbohydrates and phenolics in porous catalysts take place at solid-liquid interfaces where the local composition of the fluid phase can be very different from that of the bulk solution. Many such reactions take place in semi-aqueous solvent mixtures whose micro-heterogeneity is enhanced inside the catalyst pores. Solid-state NMR spectroscopy can probe molecular dynamics at solid-liquid interfaces, however, the demanding reaction conditions used in hydrothermal biomass transformations are particularly challenging for NMR methods. Customized chemically-resistant high pressure rotors were adapted to withstand elevated temperatures and pressures. Magic-angle-spinning spectra were recorded to monitor the populations of adsorbed molecules and measure the kinetics of their transformations. Examples involving glucose-fructose isomerization in basic zeolites and hydrogenolysis of aryl ethers will be presented, in which selective isotope labeling (^{13}C, ^{2}H) was used to enhance sensitivity and provide mechanistic information.

Copyright © American Institute of Chemical Engineers. All rights reserved.

468939 Direct Observation of Active Sites of Ru Catalyst in Hydrodeoxygenation of Furfural

Abstract:

Surface and subsurface oxygen species have been recognized as species of great significance in understanding surface reaction mechanisms in noble metals. Recent theoretical investigations in our group has shown the importance of surface oxygen vacancies in the furfuryl alcohol hydrodeoxygenation (HDO) reaction over ruthenium oxide structures.[1] In this work, we present direct evidence of the coexistence of metal and oxide (strained Ru_xO_y) on Ru catalyst by atomic resolution high angle annular dark field (HAADF) scanning transmission electron microscopy (STEM). This structure is further confirmed by X-ray absorption spectroscopic analyses of fresh, used and *in situ* reduced Ru/C catalyst. The results show that the fresh Ru/C catalyst is predominately $RuO_2$with small quantities of metallic Ru, indicative of particles with metal-core and oxide shell structure.

In order to directly correlate the strained Ru_xO_y structure to the reaction of furfural HDO we performed in *operando* X-ray absorption spectroscopy at beamline X18B, NSLS, BNL. All the experiments were performed with a custom-made XAS cell, which was charged with 1wt% furfural-isopropanol and pressurized with 400 psi He, heated up to defined temperatures (140 °C, 160 °C and 180 °C). Based on both near edge structure (XANES) and extended X-ray absorption fine structure (EXAFS) analyses of the data, we conclude that the confirmed strained Ru_xO_y structure can be directly correlated to the 2-methylfuran (MF) production from furfural via catalytic transfer hydrogenation. However, once the surface is dominated by metallic Ru, MF formation is suppressed.

[1]. Mironenko and Vlachos, *J. Am. Chem. Soc*, **2016**, Accepted

Copyright © American Institute of Chemical Engineers. All rights reserved.

464426 Single Center Reactivity of Cu(II) Sites in Copper Substituted CHA Zeolite Followed By *in-Situ* Electron Paramagnetic Resonance Spectroscopy

Abstract:

Copper exchanged zeolites are active for the selective catalytic reduction of nitrogen oxides with ammonia.[1] Recently small-pore zeolites and zeotypes with the CHA framework topology has attracted much attention and has been commercialized for use in the abatement of NO_x in diesel exhaust due to favorable properties such as high stability towards hydrothermal ageing and high resistance towards the presence of hydrocarbons. At the same time the simple CHA zeolite structure greatly simplifies the spectroscopic characterization of the materials and allows investigation of one type of site at the time.

Electron paramagnetic resonance spectroscopy (EPR) is very sensitive towards paramagnetic Cu^{2+} and offers the possibility for both quantification and speciation of Cu^{2+} sites with unrivaled sensitivity.[2] Recent results obtained on a Cu-CHA zeolite material with Si/Al = 14 and Cu/Al = 0.46 shows that the elucidation of the different copper species is possible using EPR by exposing the Cu-CHA to different gas mixtures.[2,3]

The time-resolved development of the different Cu^{2+} sites was followed and quantified by EPR in order to follow the oxidation and reduction of the individual sites in the steps of the SCR cycle.[3] In this work Cu-CHA with low concentrations of copper (Cu/Al < 0.1) was exposed to different selective catalytic reduction (SCR) relevant gas mixtures under *in-situ* conditions at high space velocities (400,000 h^{-1}) to resolve the reactivity of each specific copper site. It is shown that during exposure to SCR relevant gas mixtures the different extra-framework Cu species present in Cu-CHA react differently and several examples of this will be presented.

It is shown that the rate of oxidation of Cu^{+} to Cu^{2+} is highly dependent of the coordination environment of Cu, and that there is a pronounced difference in the reactivity of the two possible Cu sites in the 6mr of Cu-CHA that can be distinguished based on their distinct EPR signal. Thus the Cu^{2+} species belonging to both EPR active site 1 and site 2 can be assigned and quantified on the basis of their EPR spectrum. The difference in the 6mr Cu sites are suggested to be due to a difference in the position of the two aluminium in the ring, that can be separated by either one or two Si T-sites.

The sample is reduced in NH_3/NO to Cu^{+}, flushed with He, and then exposed *in-situ* to either NO/O_2/He or O_2/He at 200°C. For one type of Cu sites in the 6mr, Cu^{+} cannot be oxidized to EPR observable Cu^{2+} in a flow of O_2/He. Additionally, the rate of oxidation in NO/O_2/He depends on the concentration of NO. The other type of 6mr Cu^{2+} site, however, is oxidized fast in both flows. A difference in the reactivity of the two sites can also be observed during reduction in NO/NH_3 and during reduction in H_2.

A discussion of the implications for the activity in the SCR reaction for each of these sites in these materials will be presented and the results will be correlated with the recent suggestion for a consistent reaction cycle for the NH_3-SCR reaction.[4]

[1] U. Deka et. al. *ACS Catal.* **2013**, 3, 413.

[2] A. Godiksen et. al *J. Phys. Chem. C.* **2014**, 118, 23126.

[3] A. Godiksen et. al. *Top. Catal.* **2016**, Accepted

[4] T. V. W. Janssens et. al. *ACS Catal.* **2015**, 5, 2832.

Copyright © American Institute of Chemical Engineers. All rights reserved.

472977 In-Situ Characterization of Intermetallic Ni_5Ga_3 Catalyst for CO_2 Reduction to Methanol

Abstract:

Recently, a NiGa intermetallic catalyst for the reduction of CO_2 to methanol gathered great attention because it showed better activity and selectivity than CuZnAl-based conventional catalysts[1,2]. DFT calculation and catalytic activity tests pointed out how the intermetallic nature of this catalyst plays a major role for activity and selectivity to methanol. However, a more detailed in-situ characterization that confirms the theoretical results is still missing. In fact, it is well know that nanoparticles morphology and oxidation state are strongly related to the gas environment, temperature and pressure and nanoparticles may undergo dynamic changes under catalytic conditions[3].

We recently performed operando HERFD-XAS tests on Ni_5Ga_3/SiO_2. The tests were conducted under the same conditions employed in a labscale tubular reactor (CO_2/H_2=1; P=1-10bar; T= 200-300 °C; GHSV=6000 h^{-1}) and comparable catalytic performances were revealed. The right intermetallic phase was obtained by ex-situ reduction at 700 °C, as confirmed by XRD, and the samples were pre-reduced at 400 °C in-situ before the analysis. After in-situ reduction at 400 °C the analysis of the XANES region showed that while Ni was completely reduced, Ga was only partially reduced (as also evidenced by ETEM analysis;). Remarkably, we can conclude that upon reduction at 400 °C the Ni is in a reduced metallic state and when the sample is exposed to the CO_2/H_2gas mixture both Ni and Ga show a minimal change in oxidation state. These results obtained through bulk sensitive technique, such as HERFD-XAS, for bimetallic nanoparticles of ~8 nm may indicate a surface sites oxidation change during catalysis. Moreover, surface segregation effects cannot be ruled out. Promising preliminary APXPS data on the reducibility of the catalyst clearly showed the presence of oxidized Ga upon exposure to air, while no signals related to Ni was detected, confirming that

the surface is covered by a layer of Ga oxide.

References

1) F. Studt et al. Nature Chemistry 6, 320-324 (2014).

2) I. Sharafutdinov et al. J. Catal. 320, 77-88 (2014).

3) C Holse, J. Phys. Chem. C 119, 2804-2812 (2015).

Copyright © American Institute of Chemical Engineers. All rights reserved.

453294 Growth of Encapsulating Carbon on Supported Pt Nanoparticles Studied By in Situ TEM

Abstract:

Supported platinum nanoparticles are used extensively to promote the transformation of alkanes to a variety of products via reactions such as dehydrogenation, dehydroaromatization, and isomerization. These processes are often accompanied by formation of carbonaceous deposits that lead to blockage of the active sites and ultimately to catalyst deactivation [1]. Because coke formation on Pt is an important problem, considerable effort has been devoted to characterizing coke formed under different conditions and developing an understanding of the factors that affect coke formation on Pt. Transmission electron microscopy (TEM) has proven to be a particularly informative technique for this purpose because it enables visualization of the relationship of the coke with respect to the Pt nanoparticles with atomic resolution and can provide insights into the processes involved in the nucleation and growth of carbon. The development of techniques for obtaining TEM images of catalysts in the presence of a low pressure of hydrocarbon gas (1-10 mbar) and at elevated temperatures has opened the way for making *in situ* observations of carbon growth as a function of time [2].

In contrast to studies of carbon growth on 3d transition metals, only a few *in situ* studies have been reported for Pt [3]. Encapsulation of 3d metal nanoparticles by multiple carbon layers has been proposed to occur via precipitation of carbon dissolved in the bulk of the nanoparticle at lower temperatures. However, in view of the low solubility of carbon in Pt, it remains an open question pertaining to how multiple carbon layers form on top of each other and what the role of the Pt surface is during carbon encapsulation. To address this issue, the onset of carbon growth and the initial encapsulation of Pt nanoparticles by multiple layers of carbonaceous matter are examined by means of *in situ* TEM [4]. Such *in situ* HRTEM experiments were done to investigate the formation of carbon layers on MgO-supported Pt nanoparticles upon exposure of the catalyst to ethylene at elevated temperature. Despite the impressive capability of *in situ* TEM to monitor carbon formation on working catalysts with atomic scale resolution, beam-induced processes remain challenging. An imaging strategy was carefully established in order to suppress beam-assisted carbon growth and to minimize beam-induced sputtering of grown carbon layers. Hereby, *in situ* TEM reveals via time-resolved image series that multiple partial layers of carbon formed without the Pt particle being completely encapsulated by the first carbon layer. Additionally, particles became rounder as a function of reaction time, suggesting that coking caused significant restructuring on Pt nanoparticles. A closer look at the surface of the nanoparticles revealed that, in some cases, steps formed spontaneously in the surface and assisted in nucleation and growth of carbon layers.

[1] J. Barbier, G. Corro, Y. Zhang, Appl. Catal. 13 **(1985)** 245-255.

[2] S. Helveg, C. Lopez-Cartes, J. Sehested, P.L. Hansen, B.S. Clausen, J.R. Rostrup-Nielsen, F. Abild-Pederson, J.K. Nørskov, Nature 427 **(2004)** 426-429.

[3] Z. Peng, F. Somodi, S. Helveg, C. Kisielowski, P. Specht, A.T. Bell, J. Catal. 286 **(2012)** 22-29.

[4] J. Wu, S. Helveg, S. Ullmann, Z. Peng, A.T. Bell, J. Catal. 338 **(2016)** 295-304.

Copyright © American Institute of Chemical Engineers. All rights reserved.

467022 Enhancing Oxygen Reduction Activity on Pt Monolayer Electrocatalysts through Selective Tuning of Ligand and Lattice Effects

Abstract:

Platinum exhibits superior activity toward the oxygen reduction reaction (ORR) compared with other elements, making it an attractive cathode material for hydrogen fuel cell applications. The main problems with pure platinum electro-catalysts are large activation losses, limited stability, and high material cost. These deficiencies have greatly inhibited the development of efficient, reliable, and inexpensive proton exchange membrane fuel cells (PEM FCs). Previous experimental and computational studies have shown that it is possible to improve the performance of platinum cathodes by tuning the chemical reactivity of catalytic sites through geometric and electronic effects induced by alloying. This has led to the development of a number of Pt-based alloy electrocatalysts for this reaction.

Since ORR rates are strongly tied to the near-surface nanostructure, having precise atomistic control over the local site environment enables selective control of chemical ligand and lattice strain effects present in alloys. In this work, we describe the design of a class of Pt monolayer alloy catalysts with a core-shell structure, which enables independent control over these effects that influence ORR activity. Guided by quantum chemical calculations, we prepared

Au_x

changing the composition within the particle core. Extensive characterization and electrochemical testing of these materials was able to determine that alloys with increased Cu content experience a compressive strain that weakens OH binding energy. By tuning the strength of OH adsorption energy, it is possible to identify alloys with superior catalytic activity (up to four-fold increase in rate) compared to pure Pt/C when normalized on a site basis. Additionally, the observed ORR rates across this alloy series were consistent with predicted activity trends forecasted by our computational model.

Copyright © American Institute of Chemical Engineers. All rights reserved.

468856 Engineering the Active Site for Oxygen Electroreduction in Fuel Cells

Abstract:

Polymer electrolyte membrane fuel cells (PEMFCs) are very attractive as potentially zero-emission sources of power. However, the high platinum loadings required to compensate for the slow kinetics of the oxygen reduction reaction (ORR) impede the widespread uptake of PEMFCs. In order to reduce the Pt loading, researchers have intensively studied alloys of Pt with late transition metals such as Ni and Co. However, these compounds typically degrade under fuel cell conditions, due to dealloying. In contrast, alloys of Pt and lanthanides present exceptionally negative alloying energy [1-2], which should increase their resistance to degradation.

Polycrystalline Pt_5Gd shows a 5-fold increase in ORR activity [2], relative to Pt. The active phase consists of a compressed Pt overlayer formed by acid leaching. Mass-selected Pt_xGd nanoparticles present a significant ORR activity enhancement as compared to pure Pt nanoparticles, Pt_xGd 8 nm showing 3.6 A (mg Pt)$^{-1}$ mass activity [3,4]. The activity of Pt_xGd nanoparticles correlates strongly with compressive strain [3]. In order to study the trends in ORR activity for different levels of compression, we studied eight novel polycrystalline Pt-lanthanide and Pt-alkaline earth electrocatalysts [5]. Pt_5Tb is the most active polycrystalline Pt-based catalyst ever reported. All the materials present a 3 to 6-fold activity enhancement over Pt. Notably, the experimental ORR activity (kinetic current density measured at 0.9 V vs. the reversible hydrogen electrode, j_k (0.9 V vs. RHE)) versus the bulk lattice parameter determined by X-ray diffraction follows a â€œvolcanoâ€ relation. We use the lanthanide contraction to control strain effects and tune the electrocatalytic activity, stability and reactivity.

References

[1] I.E.L. Stephens, A.S. Bondarenko, U. GrÃ¸nbjerg, J. Rossmeisl, I. Chorkendorff, *Energy Environ. Sci.* **2012**, *5*, 6744.

[2] M. Escudero-Escribano, et al., *J. Am. Chem. Soc.* **2012**, *130*, 16476.

[3] A. Velazquez-Palenzuela, et al. *J. Catal.* **2015**, *328*, 297.

[4] C.M. Pedersen, M. Escudero-Escribano, A. Velazquez-Palenzuela, L.H. Christensen, I. Chorkendorff, I.E.L. Stephens, *Electrochim. Acta* **2015**, *179*, 647.

[5] M. Escudero-Escribano, et al., *Science* **2016**, *352*, 73.

Copyright © American Institute of Chemical Engineers. All rights reserved.

461298 Engineering Core-Shell Nanoparticles of Noble Metals on Transition Metal Carbides As Next-Generation Catalysts in Energy Applications

Abstract:

Noble metal (NM) catalysts critically enable many existing and emerging technologies, such as catalytic converters, reforming, and fuel cells. However, their scarcity and high cost necessitate the development of catalytic systems with significantly reduced NM loadings, increased activity, and improved durability. Here we present a new class of catalysts based on the transition metal carbide nanoparticles coated with atomically-thin noble metal monolayers. The core-shell structure of the catalysts was attained through a high-temperature self-assembly method by carburizing mixtures of noble metal salts and transition metal oxides encapsulated in removable silica templates. While the materials synthesis was guided by computational quantum chemistry, advanced spectroscopic techniques were applied to corroborate the exact architecture of resulting nanoformulations. The developed synthesis method allowed for precise control of the final nanoparticle architecture, including particle size, monolayer coverage, and heterometallic composition.

Exceptional catalytic performance of the core-shell catalysts was demonstrated in a number of thermo- and electrocatalytic reactions. Carbon-supported tungsten carbide nanoparticles coated with Pt or bimetallic PtRu monolayers exhibited enhanced resistance to sintering and CO poisoning, achieving an order of magnitude increase in specific activity over commercial catalysts for methanol electrooxidation after 10,000 cycles. Moreover, sub-monolayer Pt coverages were able to activate tungsten carbide nanoparticles for hydrogen evolution at low overpotentials and impart hydrogen oxidation activity. The developed catalysts also demonstrated remarkable sinter resistance, and the core-shell structure remained stable at high temperatures under various atmospheres. Collectively, these core-shell catalysts present a novel platform to drastically enhance utilization of noble metals while substantially enhancing their reactivity and stability.

Copyright © American Institute of Chemical Engineers. All rights reserved.

460816 Optimal Pattern of Bimetallic Electrocatalysts for Efficient Conversion of Water and Carbon Dioxide to Hydrocarbons and Oxygenates

Abstract:

The electrochemical conversion of water and CO_2 to renewable fuels offers a sustainable route to store intermittent solar energy and recycle CO_2. Of the various electrocatalysts investigated to date, Ag and Au are known to be highly selective to CO and H_2, and Cu is the metal exhibiting the highest selectivity to hydrocarbons and oxygenated compounds. However, the onset potentials for the production of hydrocarbons and oxygenates over Cu is ~ -1 V vs RHE. The higher overpotential for CO_2 reduction over Cu is partly due to the higher activation barrier for the transfer of the first two electrons to produce CO. Therefore, the overpotential to produce hydrocarbons from CO is lower as compared to the reduction of CO_2. Catalysts such as Au and Ag can reduce CO_2 to CO at lower overpotential ~0.4 V vs RHE, which can be directly fed to Cu for the efficient conversion of CO to hydrocarbons. Such tandem operation requires optimal spatial arrangement of bimetallics- Ag and Cu. There are several factors that need to be considered to find the optimal pattern of such bimetallic catalysts. As the fractional area of Ag increases, the CO formation and thereby its utilization by Cu increases. However, the increase in the area of Ag also increases the net production of H_2, which in turn decreases the net production of hydrocarbons. In the case of higher Ag to Cu area ratio, the utilization of CO by Cu decreases as the fractional area of Cu decreases. This talk will discuss the design principles and the optimal configurations of bimetallic electrocatalysts for the efficient conversion of CO_2 to hydrocarbons and oxygenates.

Copyright © American Institute of Chemical Engineers. All rights reserved.

456834 Supported Cobalt Catalysts with Preferentially Exposed Crystal Facets for CO_2 Hydrogenation

Abstract:

Supported Cobalt Catalysts with Preferentially Exposed Crystal Facets for CO_2 Hydrogenation.

Cun Wen, Juan Jimenez, Maryorie Santos, Jason Hattrick-Simpers, Jochen Lauterbach,*

University of South Carolina, Columbia, SC, 29208, USA

**Corresponding author: lauteraj@cec.sc.edu*

Global climate change has been widely acknowledged to be a result of rising CO_2 concentration in the atmosphere.[1-3] However, curbing CO_2 emission is hindered by the high thermodynamic stability of CO_2, which leads to low reactivity in most reactions.[4] Our recent studies have demonstrated that by tuning the surface crystal facets of Cobalt catalysts, catalysts with {110} facets have superior activity toward CO_2 hydrogenation compared with Co with {111} and {001} facets. More importantly, our previous studies have shown that the Co catalysts with {110} facets have an unprecedented self-healing functionality during a hydrogenation reaction.[5] Similar to other faceted catalysts reported in literature, the highly active and stable Cobalt catalysts are not supported and lack the capability to meet industrial requirements for high heat and mass transfer, facilitated reactor packing, mechanic strength, and so on. Here, a novel methodology leading to supporting the faceted Co catalysts without sabotaging the referentially exposed surface crystal facet will be discussed in this presentation.

In this study, Co catalysts for CO_2 hydrogenation are based on series of cobalt oxide catalysts that we have recently published for CO hydrogenation.[5] The Co nanorods, for instance, have predominant {110} and {001} facets, while the Co nanoparticles have {111} and {001} facets. The Co nanorods start to show activity toward CO_2 hydrogenation (130‹C) at 95 ‹C lower than that (225 ‹C) for Co nanoparticles. For practical applications, the faceted Co catalysts needs to be supported to meet industrial specifications. As a proof or principle, Al_2O_3 supported Co_3O_4 nanorods are synthesized and tested for CO_2 hydrogenation. The Al_2O_3 supported Co_3O_4 nanorods have comparable catalytic activity and product selectivity to that of pure Co_3O_4 nanorods, as shown in Fig. 1. This result indicates that after supporting the faceted Co_3O_4 nanorods with metal oxide, the superior catalytic performance can be preserved. More detailed structure characterizations, including XRD, STEM-HAADF, BET, and ICP, indicate that the preferential crystal facets exposure is preserved after supported on Al_2O_3.

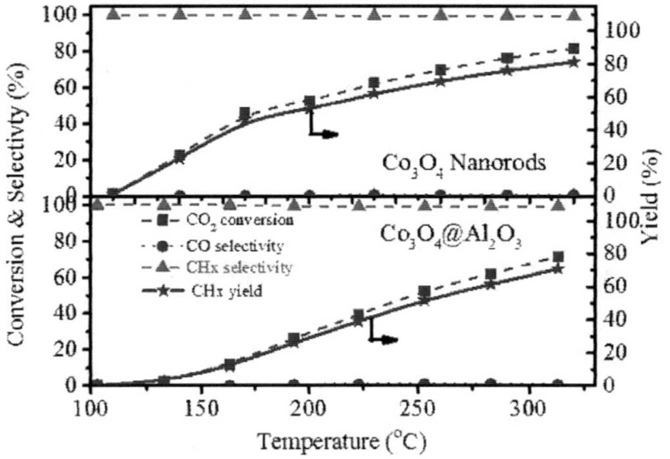

Figure 1. Comparison between the catalytic performance of a) Co nanorods and b) Co nanorods@Al_2O_3 in CO_2 hydrogenation.

References

1. T.F. Stocker, A. Schmittner, Nature, 388 (1997) 862-865.

2. M. Meinshausen, N. Meinshausen, W. Hare, et al., Nature, 458 (2009) 1158-1162.

3. E. Dlugokencky, P. Tans, in, NOAA/ESRL.

4. H. Arakawa, M. Aresta, J.N. Armor, et al., Chem Rev, 101 (2001) 953-996.

5. C. Wen, D. Dunbar, X. Zhang, et al., Chem Commun, 50 (2014) 4575-4578.

Copyright © American Institute of Chemical Engineers. All rights reserved.

467893 Tunable Thermodynamic Activity of $La_xSr_{1-x}Mn_yAl_{1-y}O_{3-\delta}$ ($0 \leq x \leq 1$, $0 \leq y \leq 1$) Perovskites for Solar-Driven Splitting of CO_2 and H_2O

Abstract:

To guide the development of advanced metal oxides for solar-driven splitting of CO_2 and H_2O into CO, H_2 (syngas), and O_2, we investigate the equilibrium thermodynamics of a family of perovskites, i.e. $La_xSr_{1-x}Mn_yAl_{1-y}O_{3-\delta}$ ($La_{0.6}Sr_{0.4}Mn_{0.4}Al_{0.6}O_{3-\delta}$, $La_{0.6}Sr_{0.4}Mn_{0.8}Al_{0.2}O_{3-\delta}$, $La_{0.4}Sr_{0.6}Mn_{0.6}Al_{0.4}O_{3-\delta}$, $La_{0.4}Sr_{0.6}Mn_{0.8}Al_{0.2}O_{3-\delta}$, $La_{0.2}Sr_{0.8}Mn_{0.8}Al_{0.2}O_{3-\delta}$) and Ca-doped $La_{0.6}Ca_{0.4}Mn_{0.8}Al_{0.2}O_{3-\delta}$ for comparison. Oxygen nonstoichiometry measurements in the temperature range 1573 to 1773 K and oxygen partial pressure range 0.206 to 180.015 mbar O_2 showed tunable reduction extents increasing with increasing Sr content. Maximum oxygen nonstoichiometry of 0.32 is established by $La_{0.2}Sr_{0.8}Mn_{0.8}Al_{0.2}O_{3-\delta}$ at 1773 K and 2.369 mbar O_2. The partial molar enthalpy, entropy and Gibbs free energy were extracted from the experimental data using defect models based on the two simultaneous redox couples Mn^{4+}/Mn^{3+} and Mn^{3+}/Mn^{2+} which describe the measured oxygen nonstoichiometry of all perovskites investigated in this work. While most perovskite compositions showed decreasing partial molar enthalpy with increasing oxygen nonstoichiometry, this desirable feature is most pronounced for $La_{0.6}Sr_{0.4}Mn_{0.4}Al_{0.6}O_{3-\delta}$. Compared to state-of-the-art ceria, partial molar enthalpy and entropy values were found to be lower for all investigated perovskites, resulting in thermodynamically more favorable reduction but less favorable oxidation with CO_2, implying energy penalties due to larger temperature swings or large amounts of excess CO_2 or H_2O required. Generally, we observe changing the ratios of A-cations and B-cations influences redox thermodynamics significantly, while changing the ratio of B-cations alone has an only a minor effect on the material's thermodynamics. Using electronic structure theory, we conclude with a practical methodology to

estimate the thermodynamic activity of perovskites in CO_2 and H_2O splitting.

Copyright © American Institute of Chemical Engineers. All rights reserved.

468673 Material Prediction for Carbon Dioxide Conversion to Carbon Monoxide Using Reverse Water Gas Shift Chemical Looping (RWGS-CL) Process

Abstract:

Carbon dioxide capture and utilization is of prime interest nowadays. The world's major energy provider – the fossil fuel is not only the source of environmental concerns, but also, its decreasing reserves and fluctuating crude prices make it almost inevitable that we turn to reliable and sustainable renewable energy. Reduction of ambient carbon dioxide and converting it to useful hydrocarbon fuels thereby, presents a novel approach to harvesting solar energy. Amongst the different ways of capturing carbon dioxide and further converting it to useful fuels, the solar thermochemical process pose to be the most economic option.

In general, the solar thermochemical CO_2 reduction is a two-step cycle, where CO_2 is converted to CO over metal oxides, while utilizing the solar energy to provide for the high temperature requirements of the process. Over the years, perovskite oxides have proven worthy for this reaction. In this process, a perovskite oxide of the form (ABO_3) is first reduced to $ABO_{3-\delta}$ at high temperatures, while in the second step, it is oxidized back to its stoichiometric form while the CO_2 is converted to CO. This CO can then be further hydrogenated to hydrocarbon fuels. A modified approach of this process is the use of renewable hydrogen during the reduction of the perovskite oxides. This process is named as reverse water gas chemical looping (RWGS CL) and has the ability to create oxygen vacancies at much lower temperatures (~500 ^{o}C). The success of these CO_2 conversion processes lies greatly on the perovskite oxide's oxygen vacancy formation capacities. The oxygen vacancy formation characteristic is greatly dominated by the materials composition. Perovskite oxides have the ability to accommodate one or more elements on the 'A' and 'B' sites, thereby forming structures like $A1_{(1-x)}A2_xBO_3$, $AB1_{(1-y)}B2_yO_3$ and $A1_{(1-x)}A2_xB1_{(1-y)}B2_yO_3$

along with ABO_3. It is this oxygen vacancy forming ability of these materials that determines their CO_2 conversion capabilities.

We hence, undertook a screening study to predict the appropriate materials for this purpose. We chose four different elements like lanthanum, strontium, barium and calcium for the 'A' site while the 3d transition metals along with aluminum and gallium for the 'B' sites. Using bulk oxygen vacancy formation energy as the descriptor for this process, we were able to predict the materials that can convert CO_2 to CO using RWGS-CL process. For this purpose, we performed density functional theory based calculations using Vienna an-initio Simulation Package (VASP). We restricted ourselves to only cubic structures, as it had been shown that oxygen vacancy formation trends across different compositions were invariant of crystal structures. For the experimental validation, we performed temperature programmed experiments to observe the formation of oxygen vacancies and carbon monoxide in the two steps of RWGS-CL process. Intrinsic material property driven models to predict oxygen vacancy formation and the CO_2 conversion capability are also being developed.

Copyright © American Institute of Chemical Engineers. All rights reserved.

472236 The Influence of Water on Metal/Oxide Perimeter Sites for the Selective Conversion of Phenolic and Furanic Species over Ru/TiO_2

Abstract:

Catalytic upgrading of biomass relies on the selective conversion of C-O bonds contained in phenolic and furanic species. Metals such as Pd or Ru supported on reducible oxides such as TiO_2 have shown significant promise for these selective transformations, with enhancement in both activity and selectivity reported[1-3]. Reasons for the enhanced activity proposed in the literature include promoter effects of the metal to create defects on the support, selective cleavage at the perimeter of the metal particle, and strong metal support interactions via decoration of the metal particle. Because several potential active sites are created under reaction conditions for this family of catalysts, the true nature of the active site required for the desirable transformation is often speculated without concrete proof. Here we use a wide range of catalysts with varying metal particle size and TiO_2 phase to clearly identify the active sites for each transformation via a combination of characterization and reaction kinetics. Metal particle sizes are estimated both with microscopy as well as structure insensitive reactions to probe the presence of a TiO_2 overlayer on metal particles, and the resulting influence on TOF values. We demonstrate that the pathway from guaiacol to toluene follows a complex route, with the metal serving as a promoter to create Lewis acid sites on the support necessary for transalkylation, while the metal/oxide perimeter sites are responsible for phenol C-O bond cleavage [1-2]. DFT calculations confirm the experimental evidence, and a mechanism is proposed. We use catalysts synthesized with carefully controlled distances between reduced metals particles and oxide clusters separated by a controlled distance across a conductive bridge to quantify the role of the promoter effect vs. the direct metal/oxide interface on furfural conversion. Finally, we discuss the positive interaction of these perimeter sites with water, enhancing the rate of cresol conversion and catalyzing the ring rearrangement of furfural.

[1]Boonyasuwat, S., Omotoso, T., Resasco, D.E. and Crossley, S.P., 2013. Conversion of guaiacol over supported Ru

catalysts. *Catalysis letters*, *143*(8), pp.783-791.

[2]Omotoso, T., Boonyasuwat, S. and Crossley, S.P., 2014. Understanding the role of TiO_2 crystal structure on the enhanced activity and stability of Ru/TiO 2 catalysts for the conversion of lignin-derived oxygenates. *Green Chemistry*,*16*(2), pp.645-652.

[3] Resasco, D.E. and Crossley, S.P., 2015. Implementation of concepts derived from model compound studies in the separation and conversion of bio-oil to fuel. *Catalysis Today*, *257*, pp.185-199.

Copyright © American Institute of Chemical Engineers. All rights reserved.

462526 Olefin Oligomerization over Cobalt Catalysts: Effect of Catalyst Support

Abstract:

The decreasing price and increasing supply of light olefins are driving renewed interest in heterogeneous, oligomerization catalysts for the upgrading of light olefins to oligomer products for use in the production of polyethylene, plasticizers and surfactants. While commercially-viable, homogeneous catalysts are currently employed, the ability to operate with heterogeneous catalysts has the potential to decrease both operational costs and the environmental impact of the industry. Common problems associated with heterogeneous, oligomerization catalysts, compared to homogeneous complexes, include lower activity, fast deactivation, and poor selectivity towards linear products. In this presentation we will discuss the linear oligomerization of 1-butene over cobalt-based catalysts. Liquid-phase oligomerization of 1-butene was carried out at low temperatures in a continuous flow reactor (<100°C). Complex reaction product mixtures were accurately characterized by two-dimensional gas chromatography (GCxGC-MS). A selectivity of >70% linear oligomers was achieved at conversion as high as 50%. We will describe the effects that support materials have on catalyst activity and selectivity. The dynamic nature of the active site was characterized by XRD, XAS, and in-situ Raman spectroscopy.

Copyright © American Institute of Chemical Engineers. All rights reserved.

456473 Interfacial Perimeter Sites of Au-TiO_x in Isopropanol Decomposition

Abstract:

Au-TiO_x interfacial perimeter sites was found to be very active for dehydrogenation to acetone by comparing Au/SiO_2, TiO_x/SiO_2 and TiO_x/[Au/SiO_2] in isopropanol decomposition. Although metal-support perimeter sites have been regarded active in different catalytic reactions,[1] dehydrogenation of isopropanol at these sites has only been addressed recently.[2, 3] Two TiO_x coverage, corresponding to 0.5 and 3 monolayer equivalent of TiO_2 and loadings of Ti 0.8-1.1 and 4.1-5.3 wt.%, respectively, were prepared on SiO_2 or Au/SiO_2. TiO_x/SiO_2, which possessed only Lewis acid sites as determined by FTIR spectroscopy and pyridine adsorption, catalyzed only dehydration of isopropanol to propene. The ratio of propene formation rate on 3MonoTiO_2/SiO_2 to that on Ti/SiO_2 was 8. Unlike TiO_x/SiO_2, acetone as well as propene was produced over Au/SiO_2 and TiO_x/[Au/SiO_2] with a high acetone selectivity of 50-85%. However, Au/SiO_2 was much less active than TiO_x/[Au/SiO_2]. The rates of propene and acetone formation on TiO_x/[Au/SiO_2] were 3-16 and 8-9 times, respectively, higher than on Au/SiO_2. In addition, relative to TiO_x/SiO_2, the propene formation rates on TiO_x/[Au/SiO_2] was 35-57% lower. These results suggested that the active sites for dehydrogenation of isopropanol were at the Au-TiO_x interfacial perimeters. The decreases in propene formation rates may be contributed to the creation of Au-TiO_x interfacial perimeters at the expense of TiO_x free from Au. The kinetic results of reaction order and activation energies will also be presented.

1. Y. Y. Wu, N. A. Mashayekhi and H. H. Kung, *Catal. Sci. Technol.*, 2013, 3, 2881-2891.

2. Z. Martinez-Ramirez, J. A. Gonzalez-Calderon, A. Almendarez-Camarillo and J. C. Fierro-Gonzalez, *Surf. Sci.*,

2012, 606, 1167-1172.

3. M. C. Holz, K. Kahler, K. Tolle, A. C. van Veen and M. Muhler, *Phys. Status Solidi B*, 2013, 250, 1094-1106.

Copyright © American Institute of Chemical Engineers. All rights reserved.

471937 Theoretical Investigation of Supported Utra-Thin Bilayer, Double Bilayer and Trilayer Metal-Oxides: Structure, Stabilty and Catalytic Activity

Abstract:

In the last decade, a number of experiments have shown that ultra-thin layers of transition metal oxides (TMOs) can be stabilized when interfaced with precious metal supports such as Au(111) and Pt(111) or Ir(100). Moreover, gold supported Co/Ni/Mn-based catalysts have been experimentally proven to exhibit higher oxygen evolution reaction (OER) activities than other metal supported oxide catalysts. Recetly, we have explained the synergistic effect of contact with gold support for Ce-doped NiOx catalyst (10.1038/nenergy.2016.53). Furtermore, we have linked experimentally and theoretically a high H2O reactivity of ultra-thin cobalt oxide nanoislands supported on Au(111) to high edge reactivity of the bilayer systems. The calculated trends in the thermodynamic stability of ultrathin supported oxide films reveals that single metal adsortion energy is a reasonable descriptor for the stability of the overlayers, but larger errors appear for films with large lattice mismatch (10.1021/acs.jpcc.6b01404). I will also present results for the the comparative analysis of cobalt oxide nanoisland stability on three related noble metal surfaces: Au(111), Pt(111) and Ag(111). Fundamental understanding in terms of surface polarity, lattice mismatch and charge trasfer properties will be discussed as well as applications to other catalystics process that OER such as hydrogen evolution reaction (HER) and CO2 reduction reaction (CO2RR).

Copyright © American Institute of Chemical Engineers. All rights reserved.

460607 Controlling Selectivity of Metal Catalysts By Using Self-Assembled Monolayers on the Oxide Support

Abstract:

Application of phosphonate-based self-assembled monolayers (SAMs) for chemoselective hydrogenation and hydrodeoxygenation has been investigated. Here, our approach was to control the near-surface environment of active metals via covalent attachment of organic ligands to the support. The SAMs were formed by immersing catalysts such as Pt/Al_2O_3 and Pd/Al_2O_3 into a phosphonate solution followed by annealing and washing. Control experiments indicated that SAMs were selectively formed on the support rather than the metal. Uncoated and SAM-coated catalysts were used in several reactions including the gas-phase hydrodeoxygenation of benzyl alcohol and furfuryl alcohol. Depending on the structure of the organic ligand tethered to the support, a significant improvement in selectivity towards desired products, toluene and methyl furan, respectively, could be observed. Importantly, the catalytic activity was not compromised by the SAMs, and in fact was increased for certain coatings. Diffuse infrared Fourier transform spectroscopy conducted after CO adsorption showed no apparent change in the site distribution after coating with the SAMs. These results indicate that the performance of supported metal catalysts can be significantly altered via attachment or organic ligands to the support, complementing prior work in which SAMs have been directly attached to the metal.

Copyright © American Institute of Chemical Engineers. All rights reserved.

468901 In-Situ Grown Metal Nanocatalysts from Oxide Host Lattices

Abstract:

Catalysis plays a central role in energy conversion and chemical production. High activity and robustness under operational condition is required for practical utilization of catalysts. Especially for high temperature applications such as solid oxide fuel cells and methane steam reforming, catalyst stability is an important issue. This talk presents three different catalyst systems for efficient energy conversion and chemical production. First, I will discuss metal nanoparticles formed by partial decomposition of oxide. This phenomenon, metal exsolution, has been known for some time and is used in formulations of automotive emission control catalysts. The ability to re-dissolve and exsolve the metal via redox cycling has a merit of nanoparticle regeneration. However, the mechanism by which the transition metal is exsolved from the oxide host is still poorly understood. My recent contribution looks into the mechanism for the evolution of the unique metal-oxide dual phase structure. I have used well-defined model systems and detailed structural analysis to characterize the nucleation and exsolution of Ni particles. I will show how interesting properties of resistance against coarsening and carbon fiber formation are linked to the particle-in-a-pit morphology. Second catalyst system is for membrane reactor application. Syngas production via methane steam reforming is an important industrial process by which hydrogen is produced in mass scale. However, it requires high energy input due to endothermic nature of the reaction. Using oxygen transport membrane, one can combine air separation and methane partial oxidation to reduce energy cost. I will show how catalyst can increase the oxygen flux by reducing the surface resistance, and expand the applicability of exsolution concept to proton conducting membranes.

Copyright © American Institute of Chemical Engineers. All rights reserved.

467916 Designing Catalysts for Enhanced Pairwise Selectivity of Parahydrogen Addition to Unsaturated Hydrocarbons

Abstract:

Parahydrogen induced polarization (PHIP) nuclear magnetic resonance (NMR) was used to investigate the effects of metal-support interactions on the pairwise addition of hydrogen to propene. In pairwise addition, two hydrogen atoms originating from the same hydrogen molecule are added across the propene double-bond. The PHIP NMR signal enhancement, which is proportional to the fraction of the total conversion occurring by pairwise addition, was measured for titania-supported iridium (Ir/TiO_2) catalysts. These enhancements were compared to those obtained for Ir/Al_2O_3 and Ir/SiO_2. While the PHIP NMR signal enhancement is important for magnetic resonance imaging, it is limited in hydrogenation reactions over heterogeneous catalysts where hydrogen atoms easily diffuse across the metal particle on the catalyst surface upon dissociative chemisorption.

To keep the particle size of Ir on the support small and to minimize hydrogen diffusion and loss of pairwise selectivity, the Ir loading was kept low (0.5 % by weight). However, this was not sufficient to yield significant pairwise addition of hydrogen to propene. Therefore, the Ir/TiO_2 catalyst was reduced at 500 °C, instead of the typical 200 °C reduction temperature, to induce strong metal support interactions. After this treatment a â˘¼20-fold increase in the pairwise selectivity was observed, and this effect could be completely reversed by oxidation followed by re-reduction at 200 Â °C. X-ray photoelectron spectroscopy (XPS) data reveal partial reduction of the TiO_2 support, and Scanning Transmission Electron Microscopy (STEM) data reveal flattening of Ir particles after high-temperature reduction. Non-interacting supports, such as Al_2O_3 and SiO_2, did not show this behavior. It was also shown that chloride ions on the catalyst surface could increase the pair-wise selectivity for the Ir/TiO_2 catalyst, possibly by blocking hydrogen diffusion on the catalyst surface. Other methods of designing catalysts to optimize the pairwise hydrogen selectivity

over supported metal catalysts are underway.

Copyright © American Institute of Chemical Engineers. All rights reserved.

468060 TiO2 Promoted Gold-Based Catalysts for Propylene Epoxidation

Abstract:

Propylene oxide (PO) is a chemical intermediate of great value and in high demand. It is used to produce useful polymeric materials. A direct catalytic epoxidation of propylene with hydrogen and oxygen to PO represents an environmentally friendly process and can be fulfilled over gold-titanium based catalysts. Despite past studies, the gold-based catalysts suffer from low conversion and poor stability. Several factors govern the activity and stability the catalysts. Isolation of the titanium was found to significantly reduce the cracking of propylene to ethanal and carbon dioxide.[1-2] However, these sites could gradually deactivate due to the change of titanium oxidation state and coordination, leading to low selectivity and coke formation. In addition, gold nanoparticles appear to sinter at high temperature, resulting in low activity.

In this work, the geometric and electronic effects of gold-titanium based catalysts are studied in detail using a number of *in situ* characterization techniques, including *in situ* X-ray absorption spectroscopy and *in situ* Raman spectroscopy. The loading and location of titanium are precisely controlled by means of atomic layer deposition (ALD). Strategy has been developed to enhance the stability of the gold-based catalysts. The gold-based catalyst regenerated has been studied by applying different post-treatment of the spent catalysts.

References

[1] S. T. Oyama, *Mechanisms in Homogeneous and Heterogeneous Epoxidation Catalysis*. Elsevier, 2011.

[2] E. E. Stangland, B. Taylor, R. P. Andres, and W. N. Delgass, â€œDirect Vapor Phase Propylene Epoxidation over Deposition-Precipitation Gold-Titania Catalysts in the Presence of H2/O2: Effects of Support, Neutralizing Agent, and Pretreatment,â€ *J. Phys. Chem. B*, vol. 109, no. 6, pp. 2321â€“2330, Feb. 2005.

Copyright © American Institute of Chemical Engineers. All rights reserved.

471786 Adsorbate-Mediated Strong Metal-Support Interactions in Oxide Supported Rh Catalysts

Abstract:

Optimization of supported metal catalysts predominantly focuses on engineering the metal site, where physical insights based on extensive theoretical and experimental contributions have enabled rational design of active sites. While it is well known that supports can influence the catalytic properties of metals, physically transparent insights into how metal-support interactions can be exploited to optimize metal active site properties are lacking. We utilize *in-situ* FTIR/XAS/EELS spectroscopy and *in-situ* HRTEM microscopy to identify and characterize a support-effect in oxide-supported heterogeneous Rh catalysts, where strongly bound adsorbates (HCO_x) on reducible oxide supports (TiO_2 and Nb_2O_5) induce oxygen-vacancy formation in the support, causing HCO_x functionalized support encapsulation of Rh nanoparticles. The encapsulation layer is permeable to reactants, stable under reaction conditions, and strongly influences the catalytic properties of Rh, enabling rational and dynamic tuning of CO_2 reduction selectivity.

Copyright © American Institute of Chemical Engineers. All rights reserved.

460568 Understanding Active Sites in Ethanol Synthesis on PtCo Catalysts Prepared Using Atomic Layer Deposition

Abstract:

Production of ethanol and other C_{2+} oxygenates from syngas has brought continued interest over the past decades due to their importance as alternative transportation fuels and industrial feedstocks. Currently, no catalyst meets the selectivity and activity requirements for industrial-scale production of ethanol. In this work, we synthesize transition metal catalysts using atomic layer deposition (ALD) to study this reaction. Using ALD, we can introduce metalorganic precursors in the gas phase and deposit catalytic materials on surfaces layer-by-layer. Through controlled deposition of metal layers, we create model alloy catalysts to gain insights into the mechanism of syngas conversions to ethanol and other oxygenates. In particular, platinum cobalt alloys have been reported as a potential catalyst candidate. We create PtCo catalysts using ALD and compare their catalytic performance and structure to catalysts prepared using traditional methodologies. The addition of Pt using ALD to Co catalysts results in increased selectivity towards methanol and higher alcohols and decreased selectivity towards longer chain hydrocarbon products. Using *in situ* diffuse reflectance infrared Fourier transform spectroscopy, we demonstrate that the interplay of bridging and linear CO species is important in carrying out oxygenate synthesis on PtCo catalysts.

Copyright © American Institute of Chemical Engineers. All rights reserved.

452585 The Role of the Support for Pt Catalysts during the Water-Gas Shift Reaction

Abstract:

The Role of the Support for Pt Catalysts during the Water-Gas Shift Reaction

Viktor J. Cybulskis, Yanran Cui, Mayank Shekhar, Juan J. Lov—n Quintana, W. Nicholas Delgass, Fabio H. Ribeiro *

School of Chemical Engineering, Purdue University, West Lafayette, IN 47907

Because of its relative simplicity, the water-gas shift (WGS) reaction is an ideal system for studying catalytic chemistry on supported noble metal surfaces at the molecular level. There is a general consensus that the most active WGS catalysts are bi-functional in nature with contributions from both the dispersed metal and as well as the support in activating CO and H_2O, respectively [1, 2]. On supported Pt catalysts, the WGS turnover frequency (TOF) per surface metal has been shown to be independent of the Pt particle size [3, 4], but can vary by a factor of 100 depending upon the reducibility of the support as shown in Figure 1. For this study, we combined *operando* infrared spectroscopy (IR) and transient kinetics with $CO/^{13}CO$ and H_2O/D_2O isotope switches to examine the carbon and hydrogen reactive pools on Pt/Al_2O_3, Pt/ZrO_2, Pt/CeO_2, and $Na\text{-}Pt/Al_2O_3$ while correlating catalyst performance with surface features related to CO and H_2O activation.

The presence of a hydrogen/deuterium (H/D) kinetic isotope effect (KIE) along with time-resolved IR spectra of transient surface species, reveals that O-H bond breaking is involved in the rate-determining step on alkali-free Pt catalysts. The addition of Na to these Pt-containing materials leads to the formation of new H_2O activation sites that

modify the rate-determining O-H dissociation step such that nearly all of the CO adsorbed on the Pt surface is able to participate in the reaction. The implications of these findings on the elementary kinetic steps for WGS, the number of true catalytic sites, and the variation in WGS TOF on supported Pt catalysts will be discussed.

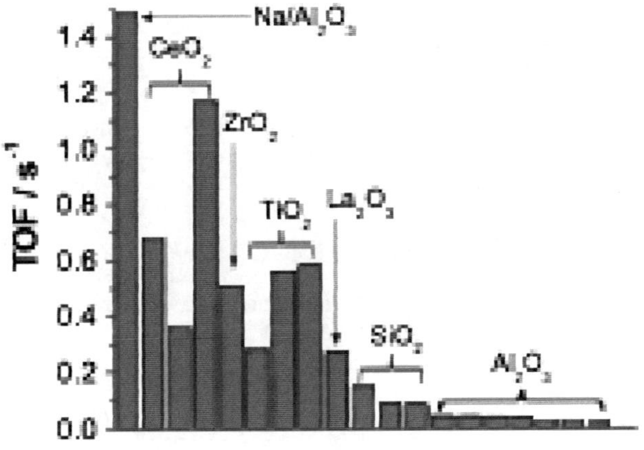

Figure 1. Comparison of WGS TOF per surface Pt at 300 ¡C (22% H_2O, 7% CO, 37% H_2, 9% CO_2) for various supports.

REFERENCES

[1] C. M. Kalamaras, P. Panagiotopoulou, D. I. Kondarides, and A. M. Efstathiou, J. Catal. 264 (2009) 117.

[2] G. Jacobs, P. M. Patterson, U. M. Graham, D. E. Sparks, and B. H. Davis, Appl. Catal. A: Gen. 269 (2004) 63.

[3] P. Panagiotopoulou, and D. I. Kondarides, Catal. Today 112 (2006) 49.

[4] M. Shekhar, J. Wang, W.-S. Lee, W. D. Williams, S.M. Kim, E. A. Stach, J. T. Miller, W. N. Delgass, and F. H. Ribeiro, J. Am. Chem. Soc. 134 (2012) 4700.

Copyright © American Institute of Chemical Engineers. All rights reserved.

463739 *Operando* Attenuated Total Reflection-Infrared Spectroscopic Detection of Catalytic Solid-Liquid Interfaces: Design Aspects of Spectroscopic Cells

Abstract:

In heterogeneous catalysis, optimal conditions for the catalytic reaction and for a spectroscopic investigation may deviate strongly. However, coupling catalytic and spectroscopic measurements of the same sample can be advantageous (relevance of the spectroscopic data, signal correlation, etc.) and finding experimental conditions to achieve complementary measurements is an essential part of *operando* spectroscopic characterization. Furthermore, the application of *operando* spectroscopy requires the design of spectroscopic reactor cells, which mimic the conditions of a conventional reactor providing similar catalytic performance and allow surface-sensitive detection of the catalyst in its working state.[1]

Currently, the most suitable spectroscopic technique for *in situ* spectroscopic study of catalytic solid-liquid interfaces is attenuated total reflection-infrared (ATR-IR) spectroscopy, which allows for the detection of adsorbates on technical powder catalysts even in the presence of strongly absorbing solvent.[2] In the context of *operando* spectroscopic characterization of a catalytic solid-liquid interface, the application of supported polycrystalline catalysts and their deposition as a thin layer is crucial for achieving realistic catalytic performance.[3] In this presentation, recent results on *operando*spectroscopic characterization of heterogeneous hydrogenation catalysis in liquid-phase are presented, and design aspects for the construction of ATR-IR spectroscopic cells suitable for the detection of catalytic solid-liquid interfaces under working conditions are discussed.

For *operando* monitoring of the catalytic solid-liquid interface during semi-batch hydrogenation a fixed-bed flow-through reactor was integrated into a recycle loop.[4] The catalytic performance and spectroscopic detection of various spectroscopic reactor cells with different inlet and reactor geometries were assessed using a fast reaction, the asymmetric hydrogenation of the C=C bond in phenylcinnamic acid on CD-modified Pd/TiO_2. In a slurry reactor operated at atmospheric hydrogen pressure under vigorous stirring a turnover-frequency (TOF) of 1200 $mol_{product}$ mol_{Pd}^{-1} $h^{-1}$and an enantiomeric excess (ee) of 45% can be achieved in isopropanol.

In general, commercially available ATR-IR spectroscopic cells have a hole-drilled inlet and a reactor geometry with rounded-off edges. Employing such a flow-through reactor cell very sluggish mass exchange was observed at the catalyst bed leading to a severely mass transport limited hydrogenation (TOF < 20 h^{-1}) and a deteriorated enantioselectivity (ee on the level of 25%). A custom-made spectroscopic cell with a slit-inlet and a rectangular reactor shape has significantly better fluid flow dynamics, which is reflect in the higher catalytic performance with a TOF of 450 $h^{-1}$and an ee of 45%. The spectroscopic measurement of the deposited catalyst sample is also very sensitive to the mass transport and different, catalytically more relevant surface processes were captured using the new spectroscopic cell.

In addition to the inlet geometry, the influence of other design parameters like the height of the reactor cell, the properties of the supported catalyst, the liquid flow rate and the quantity of deposited catalyst was investigated. The mass transfer in the porous catalyst layer depending on the reactor geometry was also investigated using mathematical modelling, and computational and experimental results are compared.

References

[1] N. E. Tsakoumis, A. P. E. York, D. Chen, M. Ronning, *Catal. Sci. Tech.* **2015**, *5*, 4859-4883.

[2] F. Zaera, *Chem. Soc. Rev.* **2014**, *43*, 7624-7663.

[3] L. Rodríguez-García, K. Hungerbühler, A. Baiker, F. Meemken, *Catal. Today.*, **2016**, *in press*.

[4] F. Meemken, P. Müller, K. Hungerbühler, A. Baiker, *Rev. Sci. Instrum.* **2014**, *85*, 17.

Copyright © American Institute of Chemical Engineers. All rights reserved.

472855 In Situ and Operando Drifts Characterization of Catalytic Sites on Metal-Doped KIT-6 during Alcohol Conversion Reactions

Abstract:

Large pore KIT-6 materials doped with transition metals (e.g., Zr, Nb, W) have been shown to be active and selective catalysts in various reactions including acid catalyzed reactions involving alcohol moieties. In this presentation, we will explore metal doped KIT-6 catalyst active sites during reaction with model compounds such as ethanol and isopropanol by means of in situ temperature-programmed surface reaction (TPSR) Fourier transform infrared spectroscopy (TPSR-FTIR) and operando FTIR-GC supported by density functional theory calculations. We will show, for example, that on the Zr-doped KIT-6, the catalyst surface is predominantly composed of strong Zr Lewis sites which favor the selective formation of primarily olefin (e.g., ethylene, propylene) and ether (e.g., diethyl ether, diisopropyl ether) from the corresponding alcohol. Although the catalysts are quite stable, the in situ TPSR FTIR and operando FTIR-GC results suggest that only a fraction of the catalyst Lewis acid sites is involved during alcohol conversion whereas the majority of catalytic sites are likely precursors to coke which originates from alcohol reaction intermediates rather than from the subsequent reaction of products. This study offers insights on metal doped KIT-6 catalyst active sites of importance for the conversion of reactants containing alcohol moieties such as those present in bioderived feedstocks such as glycols, sugars, and lignin.

This work is supported by the National Science Foundation, grant No OIA-1539105

Copyright © American Institute of Chemical Engineers. All rights reserved.

464756 *Operando* spectroscopy during Ethylene Polymerization By Supported CrO_x/SiO_2: Role of Promoters

Abstract:

Supported catalysts for the ethylene polymerization reaction have been studied since the discovery of the CrO_x/SiO_2 catalyst system in the 1950s. The addition of small amounts of promoter oxides, such as TiO_x, ZrO_x, or AlO_x, has been seen to have such effects as increasing the catalytic activity or altering polymer properties. For example, the inclusion of a few percent TiO_x has been observed to broaden the molecular weight index. There have been only a few studies reported over the years, so very little is understood about the fundamental structural and mechanistic details. The current studies extend previous work done on the un-promoted catalyst and are focused on elucidating the role of the promoter oxides used in ethylene polymerization.

To address the molecular details of ethylene polymerization by supported and promoted CrO_x/SiO_2 catalysts, modern *in situ* molecular spectroscopic techniques were applied before and during reaction conditions. The initial oxidized supported CrO_x/SiO_2 catalyst was found to only contain isolated chromia sites (high UV-vis edge energy value) and two distinct surface chromia sites (dioxo $(O=)_2CrO_2$ and mono-oxo $O=CrO_4$). The ratio of dioxo:mono-oxo chromia species was quantified by C_2H_4-TPSR and found to differ depending on the promoter oxide. *In situ* Raman spectroscopy during ethylene polymerization revealed that the surface dioxo CrO_4 site reduces and becomes activated more easily than the surface mono-oxo CrO_5 site. The C_2H_4-TPSR study revealed a higher reactivity of the dioxo surface CrO_4 species and the mono-oxo surface CrO_5 species. The corresponding UV-vis experiments verified the reduction of Cr^{+6} to Cr^{+3} for both the dioxo and mono-oxo surface chromia sites during ethylene polymerization. The condensed polyethylene product formed on the catalysts during the polymerization reaction was monitored with *in situ*

IR spectroscopy and key surface reaction intermediates were detected. The catalysts promoted with TiO_x and ZrO_x exhibited the best polyethylene production, and they were about equal, while the production of polyethylene by the catalyst promoted with AlO_x was even less than the un-promoted catalyst. This poster will address the role of the promoters for the ethylene polymerization reaction.

Copyright © American Institute of Chemical Engineers. All rights reserved.

471706 Evidence for CO Insertion in Alkane and Alkenes Production through Fischer-Tropsch Reaction

Abstract:

Despite extensive studies on the mechanism of the catalytic hydrogenation of CO according to the Fischer (FT) Tropsch synthesis, a detailed understanding of the reaction mechanism responsible for the formation of long chain hydrocarbons is still missing. The so-called C-C coupling and CO-insertion mechanisms remain the two main proposed mechanisms for FT. The distinguishing feature of these two mechanisms is whether cleavage of the C-O bond in the CO molecule occurs before (leading to C-C coupling) or after incorporation of the C_1 monomer (CO-insertion) into the growing hydrocarbon chain. We provide here experimental evidence for the CO-insertion mechanism via Chemical Transient Kinetics (CTK) studies, developed by our group for operando-type investigations at atmospheric pressures and above.

Chemical transients, triggered by abrupt changes in the molecular fluxes of the reactants, provide relaxation-type information on the response behavior of catalysts relative to these changes. A careful calibration of the method and reactor system allows counting atom amounts of species adsorbing on the surface from the point in time of switching the fluxes till the occurrence of a steady state (build-up phase). Similar information can be gleaned from the reverse procedure by returning from the steady state of the reaction to non-catalytic conditions (back-transients). Furthermore, from the occurrence of delay times in the formation of reactants and products during the build-up, key features of the mechanistic steps can be established. Previous studies with Co/MgO as well as pure Co and Ni allowed the H, C, and O coverages during the build-up to be evaluated for the CO hydrogenation under atmospheric pressure conditions. This way it was shown that the monolayer limit was surpassed even before reaching the steady state, with the amounts of O atoms always exceeding those of C atoms. Thus, the catalytic surface did not provide metallic sites anymore under steady state conditions.

Furthermore, no hydrocarbon chain lengthening was ever observed without gaseous CO being built-up in the reactor. Most interestingly, the "Anderson Schulz-Flory" chain lengthening probability α calculated from the build-up species clearly showed proportionality to the CO pressure which is in agreement with a CO insertion mechanism. On the contrary, a non-monotonous behavior as a function of accumulating carbon (or CH_x) was observed as would be expected if CH_x insertion or C-C coupling were occurring. With respect to elucidating the CO hydrogenation mechanism toward different products, the target bond of monomer insertion is as important as insertion of the monomer itself. In our previous studies, no water was being produced despite the build-up of considerable oxygen and hydrogen amounts on the surface. Accordingly, it was concluded that hydroxyl (OH) groups are most probably being formed on the surface and CO is being inserted into the O-H or O-R bond of surface hydroxyl or alkoxyl.

Despite the considerable number of arguments favoring a CO insertion mechanism in our model studies, the debate of C-C coupling versus CO insertion is still ongoing and far from being solved. In the present CTK study we provide further evidence for the occurrence of a CO insertion mechanism by choosing bimetallic "CoMn" and ternary "CoMnTi" catalysts. Both these catalysts were prepared according to the oxalate route, as developed in our laboratory, in the absence of a generic support material.

Switching molecular fluxes from H_2 to $CO+H_2$ at atmospheric pressure conditions caused reactants and products to be produced with characteristic delay times as compared to the Ar noble gas (as reference). The first species to appear in the gas phase with a delay time of less than 5 s was methane. Interestingly, besides paraffins also olefins were produced in our studies. These species were always observed time-correlated with gaseous CO. For a "$Co_1Mn_1Ti_{0.1}$" catalyst (atomic ratios 1:1:0.1) containing Co metal and Mn (as Mn_5O_8 according to TEM results) in equal amounts and a non-stoichiometric TiO_x phase with much smaller abundance, the delay time for the appearance of gaseous CO could be as large as 40 s. Moreover, alkanes and alkenes products occurred in sequence of their carbon number. The large delay times encountered in our study have most probably to be associated with the presence of extended oxidic phases, Mn_5O_8 and TiO_x. Strong Metal Support Interaction (SMSI) effects may be invoked to explain the finding.

In back-transients, the CO outlet flux dropped exponentially with the same time constant as Ar. As the Ar decay is characteristic of the reactor volume emptying, we conclude that CO was irreversibly chemisorbed on the catalyst surface during steady state reaction conditions. We also note that once the CO inlet flow was stopped to initiate the back-transients, C_{2+} alkanes and alkenes decreased quickly with similar time constants (yet much longer than those for CO and Ar). This is in agreement with CO being the insertion monomer since no chain lengthening happened without gaseous CO. Finally, plotting the chain lengthening probability vs. the CO partial pressure resulted in a linear behavior. Quite differently, a non-monotonous behavior was observed when the chain lengthening probability was plotted vs. the titrated carbon amounts.

It is also remarkable that the formation of H_2O and CO_2 in our CTK studies with "CoMnTi" were largely delayed in comparison to CO and hydrocarbons. Consequently, while the catalyst took up major amounts of oxygen and hydrogen atoms during the build-up stage, the absence of water production would appear OH groups were being abundantly formed during these initial reaction times. We therefore speculate that surface hydroxyl (or alkoxyl) groups are the target for CO insertion to form formate-type species. Such species could be easily hydrogenated to form paraffins, olefins, or even oxygenates as observed when the CO hydrogenation is being performed at high pressures over "CoMnTi" ternary catalysts.

Keywords

Fischer–Tropsch Synthesis, Chemical Transient Kinetics, CO Insertion

Copyright © American Institute of Chemical Engineers. All rights reserved.

469448 Scaled Degree of Rate Control: Identifying Elementary Steps That Control Differences in Performance of Transition Metal Catalysts

Abstract:

Detailed mechanistic knowledge of catalytic reaction networks and identification of key parameters that control their kinetics, such as rate limiting steps, are of central importance to the fast discovery of optimum catalyst material. Microkinetic models based on Density Functional Theory (DFT) calculated kinetic and thermodynamic parameters have been used to investigate the mechanism of complex reaction systems, predict kinetic behavior of different catalytic material and identify kinetically relevant parameters. With the fast evolution of quantum chemical computational methods, it has become apparent that microkinetic analysis of heterogeneous catalytic reactions is an indispensable part of catalyst design process.

A commonly used method in the determination of a rate limiting or selectivity controlling step in a complex kinetic network is the "Degree of Rate Control" (DoRC) method, a sensitivity analysis approach first proposed by Campbell.[1] The DoRC simplifies complex reactions to identify a few steps that are crucial to performance, thereby facilitating the identification of novel catalysts. However, due to the assumption of parametric independence, the DoRC analysis is not rigorous for heterogeneous catalytic reactions where surface species and transition state energies are related through parametric correlations, such as linear scaling (LS) and Brønsted-Evans-Polanyi (BEP) relations.

Here we propose a computational approach for the identification of elementary steps that are responsible for the differences of catalytic behavior of transition metal catalysts, by implicitly incorporating inherent parametric correlations in DoRC calculations. We introduce the concept of "scaled degree of rate control", (S-DoRC) which is closely related to Nørskov and Bligaard's "degree of catalytic control".[2] Rather than identifying all steps, which could influence the rate or selectivity of a catalytic reaction as in the DoRC analysis, the S-DoRC only identifies the relevant rate or selectivity controlling steps that are tunable within the confines of LS and BEP relations. We derive a formulation for the S-DoRC, which is simple to implement as a modification of DoRC, given the expansive and previously published parametric correlations.

1. Campbell, C. T. Finding the Rate-Determining Step in a Mechanism. *J. Catal.* **204,** 520–524 (2001).

2. Nørskov, J. K., Bligaard, T. & Kleis, J. Rate control and reaction engineering. *Science* **324,** 1655–6 (2009).

Copyright © American Institute of Chemical Engineers. All rights reserved.

472089 Effective Coordination Based Descriptors for Rational Design of Metal Nanocatalysts

Abstract:

Effective Coordination based Descriptors for Rational Design of Metal Nanocatalysts

Xianfeng Ma, Siwen Wang and Hongliang Xin*

Department of Chemical Engineering, Virginia Tech,

Blacksburg, VA, 24061

hxin@vt.edu

Session: Catalysis and Reaction Engineering Division ¨C Rational Catalyst Design

One major challenge for designing metal nanocatalysts is to know a priori the appropriate surface structure for a given catalytic reaction. Understanding the effects of particle morphology and composition on the interaction of surface atoms with adsorbates is of pivotal importance for identifying optimal active sites and engineering nanoparticles with maximized fraction of such sites. Many efforts have been made aiming to predict the binding energy of an adsorbate at metal surfaces using the electronic and/or geometric factors of the adsorption site (termed descriptor).[2,3] While these models have been successfully used as reactivity descriptors for either extended-metal surfaces or simple metal nanoparticles, their extension to complex particle systems with varying broken-bond strains and/or metal ligands remains elusive due to a formidable computational cost or the lack of an explicit consideration of interatomic interactions.

In this work, we present the effective coordination number (CN_e) as an interaction-aware reactivity descriptor for metal nanocatalysts[4]. The CN_e, quantified by interatomic coupling matrix elements between the site of interest and its all neighbors within a certain cutoff radius, provides a robust description of CO, O_2, and O adsorption energies on Au metal nanoparticles of varying sizes and shapes attributed to its explicit consideration of broken-bond strains (see Fig. 1 for CO as an example). Importantly, the CN_e has a solid physiochemical basis via a direct connection to the moment characteristics (e.g., center) of occupied density of states projected onto valence orbitals of the adsorption site. Furthermore, the CN_e shows promise as a general descriptor for predicting adsorption properties of core-shell and alloyed structures of Au nanoparticles with d^{10} metal ligands (e.g., Cu and Ag). The approach can be readily extended to understand and predict reactivity trends of large and more complex metal catalysts with defects, impurities, transition-metal additions, supports, etc, and thus provides a general basis for rational design of metal nanocatalysts.

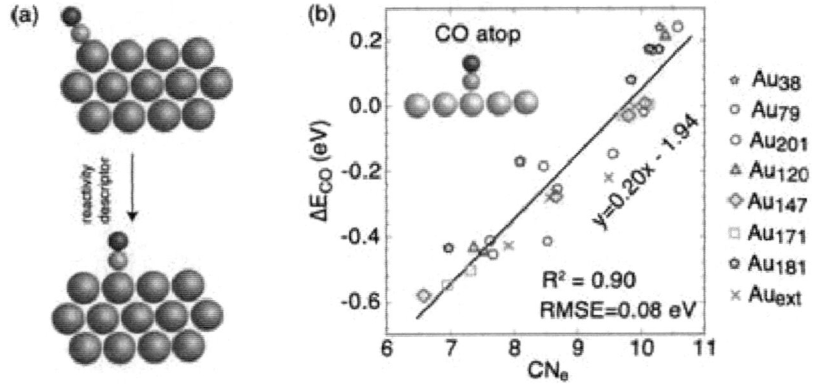

FIG. 1. Adsorption energies of *CO atop described by CN_e on various surface sites of Au nanoparticles and extended surfaces. The statistics of linear regression (solid line) are also given.

References:

1. Calle-Vallejo, F. *et al.* Finding optimal surface sites on heterogeneous catalysts by counting nearest neighbors. *Science* **350,** 185¨C189 (2015).

2. Hammer, B. & Nørskov, J. K. Electronic factors determining the reactivity of metal surfaces. *Surf. Sci.* **343,** 211¨C220 (1995).

3. Calle-Vallejo, F. *et al*. Fast Prediction of Adsorption Properties for Platinum Nanocatalysts with Generalized Coordination Numbers. *Angew. Chem. Int. Ed.* **53,** 8316¨C8319 (2014).

4. Ma, X., Wang, S. & Xin, H. Effective Coordination Number as a Reactivity Descriptor for Coinage Metal Nanocatalysts. *Submitted* (2016).

Copyright © American Institute of Chemical Engineers. All rights reserved.

472719 Tuning Surface Carbon Affinity and Olefin Activation in Intermetallic Alloys for the Semi-Hydrogenation of Acetylene to Ethylene

Abstract:

Understanding and developing inexpensive materials that exhibit appropriately tunable surface chemical reactivity towards carbon and unsaturated hydrocarbons is critical in many olefin production and functionalization reactions. Intermetallic alloys -- transition metals mixed with larger p-block elements -- present a special class of materials that exhibit highly tunable surface chemistry as well as improved stability. This study focuses on understanding how the size of the p-block element affects orbital overlap, the strength of bulk bonding, and the residual surface reactivity towards carbon-carbon double bonds. A combined experimental and theoretical investigation of NiB, NiAl, NiGa, NiIn, and NiTl materials in the semi-hydrogenation of acetylene to ethylene has been performed. Clear correlations between orbital overlap, surface reactivity towards ethylene, and catalytic performance were discovered. Further study of the NiAl system with respect to element ratio in the solid revealed that additional tunability of these materials is readily achieved.

Copyright © American Institute of Chemical Engineers. All rights reserved.

463606 Scaling Relationships at Bifunctional Metal/Oxide Interfaces: A Case Study of Doped Au/MgO

Abstract:

Transition metal nanoparticles dispersed on metal oxides possess interesting catalytic properties through bifunctional catalytic sites located at the interface between the metal and oxide. These sites enhance catalytic activity in comparison with individual metal or oxide sites, for several reactions such as the Water Gas Shift reaction (WGS) [1] and methanol synthesis [2]. Although experimental and theoretical investigations have observed bifunctionality at the metal/oxide interface, strategies enabling computational design of these catalysts have not been developed. Linear Scaling Relationships (LSRs), which relate adsorption energies of reaction intermediates to atomic species known as descriptors, are an important part of the descriptor-based catalyst design strategy. They have been reported for transition metals [3], metal oxides [4] and zeolites [5] in prior literature. In this talk, we show that LSRs can be developed at a metal/oxide interface using the Au/MgO (100) system as a representative model for a metal/oxide catalyst. This choice is motivated by the small strain between Au and MgO (100), thereby preventing additional complications arising from nanoparticle relaxation. Electronic properties of the interface are modified through cationic (Mg substituted with Na, Cu, Zn, Fe, Al, Ga, In, Sn, Zr, Ti, Mo, W, V, Cr) and anionic doping (O substituted with C, N, S, F and oxygen vacancies). Based on the dopants valence with respect to the host atom, electrons/holes are released to the interface, modifying adsorption strengths of reaction intermediates. Adsorption energies of a series of molecular fragments (COOH*, O_2*, OH*, OOH*, CH_3O, CH_2O*, CHO*, CH_2OH*, CHOH*, NH
involved in catalytic reactions at metal-oxide interfaces, are shown to scale linearly with the adsorption strength of O*. Thus, the oxygen binding energy is proposed as a descriptor these chemistries. All dopants, irrespective of their location in the oxide matrix, and their formal oxidation state, fall on the same LSR for each adsorbate. Slopes of the LSRs are greater than that established via adsorbate bond counting arguments, which are valid for metals [3], metal oxides [4] and zeolites [5]. Moreover, the slope is dependent on the coordination of the bifunctional adsorbate fragments that are bound to both MgO and Au. Closed shell species (CH_2O*, O_2*) which do not follow LSRs on transition metals, are shown to do so at the interface. Finally, we discuss the physical origin of the LSRs in terms of electronic and electrostatic properties of these systems.

1. Shekhar, M.; Wang, J; Lee, W. S.; Williams, W. D.; Kim, S. M.; Stach, E. A.; Miller, J. T.; Delgass, W.N.; Ribeiro, F.H.; *J. Am. Chem. Soc.* **2012**, *134*, 4700.

2. Graciani, J.; Mudiyanselage, K; Xu, F.; Baber, A.; Evans, J.; Senanayake, S.D.; Stacchiola, D.J.; Liu, P.; Hrbek, J.; Sanz, J.F.; Rodriguez, J.A.; *Science* **2014**, *345*, 546.

3. Pedersen, F. A.; Greeley, J; Studt, F.; Rossmeisl, J.; Munter, T. R.; Moses, P.G.; Skúlason, E.; Bligaard, T.; Nørskov, J.K.; *Phys. Rev. Lett.* **2007**, *99*, 016105.

4. Fernández, E. M.; Moses, P.G.; Totfelund, A.; Hansen, H.A.; Martínez, J.I.; Pedersen, F.A.; Kleis, J.; Hinnemann, B.; Rossmeisl, J.; Bligaard, T.; Nørskov, J.K.; *Angew. Chem. Int. Ed.* **2008**, *47*, 4683.

5. Brogaard, R.Y.; Wang, C-M.; Studt, F.; *ACS Catal.* **2014**, *4*, 4504.

Copyright © American Institute of Chemical Engineers. All rights reserved.

465565 Rational Design of Nanostructured Metallic Surfaces Via Mathematical Optimization

Abstract:

Recent advances in nanotechnology have enabled the fabrication of materials with precision on the micro- or even nano-scale. These precisely fabricated materials have properties that are wildly different from their bulk counterparts, enhancing useful functionalities in a number of application contexts. In particular, the rational design of nanostructured metallic surfaces has received attention due to the potential for improving the catalysis of many important reactions. In the case of the oxygen evolution reaction in fuel cells, for example, several studies have indicated that slight changes in the structure of a metallic surface can lead to a dramatic improvement in the achievable rate of reaction [1].

Progress in our understanding of metallic surfaces for heterogeneous catalysis has brought us to the point where novel materials can be proposed by design *in silico* [2]. However, due to the combinatorial nature of how atoms can arrange themselves on crystalline lattices, the best structures are unintuitive and likely impossible to identify without a rigorous decision-making approach. Fortunately, this material design problem can be formalized and addressed as a mathematical optimization problem such as those traditionally cast and solved in many process systems engineering contexts. In essence, one seeks to identify a material design (i.e., a specific instantiation of the given class of materials of interest) that maximizes some metric of performance.

The proposed design approach capitalizes on the fact that the rate for the limiting step(s) in the overall conversion process can ultimately be correlated to the microstructure of a surface site, typically in the form of a volcano plot [2,3]. Importantly, the activation energy of a chemical reaction can be predicted from the adsorption energies of transition states [4]. Therefore, we can translate local site characteristics to adsorption energies (e.g., as provided by density functional theory computations [5,6]) and to activation energies. Using these demonstrated relationships, we can predict reaction rates of metal surface sites and, thus, be in position to rigorously design periodic surface patterns that maximize the total reactivity on a per area basis.

In this work, we show how correlations that link performance (e.g., catalytic activity) to site descriptors, such as commonly available volcano plots, can be used to determine the optimal (i.e., of highest performance) periodic surfaces of transition metal crystals via a mathematical optimization model. As an illustration, we will use the average neighbor coordination number, chosen both due to its recent and emerging appearance in the literature [6,7] as well as due to its interest from the mathematical modeling viewpoint. We should, however, emphasize that our model can be readily adapted to handle a variety of other site descriptors as well.

Using our optimization model, we conducted extensive computational studies involving an array of crystallographic lattices and structure-function relationships. More specifically, we considered three face-centered cubic (FCC) lattices oriented such that the design surface is aligned with the {1,0,0}, {1,1,0} and {1,1,1} crystallographic planes, respectively, one body-centered cubic (BCC) lattice aligned with the {1,0,0} plane, and one hexagonal close-packed (HCP) lattice aligned with the {0,0,0,1} plane. These design lattices were combined with a set of volcano plots spanning a range of ideal site descriptors in order to identify optimal surface patterns relating to a gamut of catalytic systems.

Our proposed approach validates certain surface patterns that were previously known to be optimal but also reveals a number of non-intuitive designs. We also demonstrate that the optimal surface patterns vary significantly with the target application, which strengthens the need to keep developing those fabrication capabilities that can yield higher levels of structural control.

[1] Stamenkovic V., Mun B. S., Mayrhofer K. J. J., Ross P. N., Markovic N. M., Rossmeisl J., Greeley J., Nørskov J. K., "Changing the activity of electrocatalysts for oxygen reduction by tuning the surface electronic structure," *Angewandte Chemie - International Edition*, 45(18):2897–2901, 2006.

[2] Nørskov J. K., Bligaard T., Rossmeisl J., Christensen C. H., "Towards the computational design of solid catalysts," *Nature Chemistry*, 1(1):37–46, 2009.

[3] Hammer B., Norskov J. K., "Theoretical Surface Science and Catalysis — Calculations and Concepts," *Advances in Catalysis*, 45:71–129, 2000.

[4] Nørskov J. K., Bligaard T., Logadottir A., Bahn S., Hansen L. B., Bollinger M., Bengaard H., Hammer B., Sljivancanin Z., Mavrikakis M., Xu Y., Dahl S., Jacobsen C. J. H., "Universality in Heterogeneous Catalysis," *Journal of Catalysis*, 209(2):275–278, 2002.

[5] Taylor M. G., Austin N., Gounaris C. E., Mpourmpakis G., "Catalyst Design Based on Morphology- and Environment-Dependent Adsorption on Metal Nanoparticles," *ACS Catalysis*, 5(11):6296–6301, 2015.

[6] Calle-Vallejo F., Loffreda D., Koper M. T. M., Sautet P., "Introducing structural sensitivity into adsorption-energy scaling relations by means of coordination numbers," *Nature Chemistry*, 7(5):403–410, 2015.

[7] Calle-Vallejo F., Martínez J. I., García-Lastra J. M., Sautet P., Loffreda D., "Fast prediction of adsorption properties for platinum nanocatalysts with generalized coordination numbers," *Angewandte Chemie - International Edition*, 53(32):8316–8319, 2014.

Copyright © American Institute of Chemical Engineers. All rights reserved.

459191 Screening of Active Site Motifs for C-H Bond Activation in Nanoporous Catalysts

Abstract:

Given the availability of new natural gas resources, direct transformation of methane to higher value commodity chemicals or liquids fuels is likely to revolutionize the energy and chemicals industry. However, the high economic barriers associated with methane activation have hindered the development of such technologies. Many materials including zeolites, metal organic frameworks, graphenes, metal surfaces, oxides, porous organic polymers and nanoparticles have been shown to be feasible for C-H bond activation, but a unified understanding of the activation process is not available. Although the material properties, reactants and proposed active sites for the above systems are very different, we show that a unifying feature is the radical-like geometry of the transition state (TS) during reaction. In this work, we discuss a universal model for predicting C-H activation barriers, where the TS energy scales linearly with the hydrogen affinity energy. Furthermore, by combining this scaling approach with a thermodynamic analysis for the number of active sites, we illustrate a catalyst screening strategy using a simple descriptor for various classes of materials. This analysis identifies a subset of candidate zeolites and porous organic polymers that are likely to be active and selective for partial methane oxidation. Although we focus on the methane oxidation reaction in this work, our active site motif approach is likely to be transferable to other reactions and will be of a broader interest to nanoporous and surface catalysis community.

Copyright © American Institute of Chemical Engineers. All rights reserved.

460276 Design of Functionalized Metal Organic Frameworks for CO_2 Hydrogenation:the Effects of MOF Topology and Functional Group

Abstract:

CO_2 capture coupled with chemical recycling of CO_2 to renewable fuels and valuable chemicals is a promising approach to reduce carbon emissions. An ideal process would combine both capture and chemical reduction of CO_2, taking advantage of process intensification. Metal organic frameworks (MOFs), a class of porous crystalline materials, have great potential for CO_2 capture. We have combined the adsorption selectivity of MOFs with functional groups having chemical activity toward CO_2 hydrogenation to produce a single material with the potential to both capture and convert CO_2. We have covalently bound catalytic functional groups based on Lewis pairs (LPs), which are organic molecules possessing both Lewis acid and base sites, to the linkers of a MOF. Our initial work utilized a highly stable and CO_2-selective Zr based MOF known as UiO-66 to create a heterogeneous catalyst, UiO-66-P-BF_2.[1,2] This new catalyst retains the catalytic activity of LPs and also avoids the limitations typical for homogeneous catalysts, such as difficulty with catalyst recycling and product separation. The energy favorable pathway for CO_2 hydrogenation to produce formic acid (FA) over UiO-66-P-BF_2 has two steps as revealed by density functional theory (DFT) calculation: (i) dissociation of H_2 to hydridic and protic hydrogens and (ii) CO_2 reacting with hydridic and protic hydrogens to produce FA.[1,2]

However, one drawback of UiO-66-P-BF_2 is that CO_2 binds much stronger with Lewis acid-base sites than H_2, which could poison the catalytically active sites for H_2 dissociation. Additionally, a high density of LPs in the pore could lead to mutual quenching, which would deactivate the catalyst. In order to overcome this difficulty, we design a new catalyst we call UiO-67-NBF_2. Our DFT results suggest it has high selectivity for H_2 dissociation and activity for CO_2 hydrogenation to produce methanol. In addition, we have found that the framework of the MOF plays an important role in tuning the adsorption of H_2 and CO_2 at the functional groups. We have also examined a different MOF (MIL-140B) functionalized with NBF_2 in order to study the effect of MOF topology on CO_2 hydrogenation. Furthermore, we have screened hundreds of LPs searching for promising candidates for CO_2 hydrogenation based on their ability of selective dissociation of H_2, rather than adsorption of CO_2.

1. Ye, J.; Johnson, J. K. Design of Lewis Pair-Functionalized Metal Organic Frameworks for CO_2 Hydrogenation. *ACS Catal.* **2015**, 2921-2928.

2. Ye, J.; Johnson, J. K. Screening Lewis Pair Moieties for Catalytic Hydrogenation of CO_2 in Functionalized UiO-66. *ACS Catal.* **2015,** *5* (10), 6219-6229.

Copyright © American Institute of Chemical Engineers. All rights reserved.

473867 The Nature of Heterogeneous Palladium Catalyzed Cross-Coupling Reactions on Novel Carbon Support Systems

Abstract:

Palladium-catalyzed carbon-carbon bond forming reactions have emerged as one of the most important chemical transformations in the preparation of complex organic molecules. The Suzuki-Miyaura cross-coupling reaction has gained particular prominence in pharmaceutical applications due to its mild reaction conditions and broad application across a wide range of functionalized substituents. These reactions are most often carried out under homogeneous conditions that can lead to product contamination, extensive purifications and loss of expensive metals and ligands. Pdn clusters supported on graphene (Pd/G) present an attractive alternative to homogeneous catalysis. These materials offer excellent catalytic activity with extremely high turnover frequencies, minimal leaching and remarkable recyclability opening a new paradigm in cross coupling reactions. Here we demonstrate that unlike previously reported examples of solid-supported metal catalysts, Pd/G operates as a heterogeneous catalyst where the enhanced binding to defect sites precludes leaching of the metal nanoparticles. In addition, the conductive nature of the support allows charge exchange that lowers the reaction barriers offering a highly effective catalyst with exceptional recyclability.

Copyright © American Institute of Chemical Engineers. All rights reserved.

481047 Electrostatic Adsorption of Pt onto Carbon Nanotubes and Fibers

Abstract:

Strong Electrostatic Adsorption (SEA) has been demonstrated as a simple, scientific method to prepare well dispersed Pt nanoparticles over typical forms of carbon: activated, black, and graphitic carbons [1]. Many varieties of specialty carbons have been invented in the last few decades including multi-walled nanotubes, nanofibers, graphene nanoplatelets, etc. In this work we explore whether SEA can be applied to these specialty carbons for the synthesis of Pt nanoparticles.

Over a number of oxidized and unoxidized multiwalled nanotubes and nanofibers the point of zero charge (PZC) was measured and the uptake of anionic Pt complexes (Pt hexachloride, $[PtCl_6]^{2-}$, and cationic Pt complexes (platinum tetraammine, $[Pt(NH_3)_4]^{2+}$) as functions of final pH were surveyed. Pt nanoparticles on the various supports were synthesized at the optimal pH and were characterized by XRD and STEM.

The specialty carbons displayed volcano-shaped uptake curves typical of electrostatic adsorption for both Pt anions at low pH and Pt cations at high pH. However, the regimes of uptake often did not correspond to the measured PZC. It was seen that the PZC of many of the carbons could be changed with washing, and so was likely affected by residual impurities of the manufacturing process. This renders the measured PZC of these specialty carbons unreliable for predicting anion and cation uptake. On the other hand, the anion and cation uptake curves provide an â€œeffectiveâ€ PZC and do indicate the optimal pH for the synthesis of ultrasmall nanoparticle synthesis.

Reference

[1] Hao X, Barnes S, Regalbuto J.R, Journal of Catalysis, 2011, 279, 48.

Copyright © American Institute of Chemical Engineers. All rights reserved.

459145 Impact of Mixed Metal Oxides Supports on the Durability and Low-Temperature Performance of Pd-Based Diesel Oxidation Catalysts

Abstract:

Future diesel oxidation catalysts (DOC) will need to perform effectively at increasingly low exhaust temperatures; this so-called â€œ150 Â°C challengeâ€ (i.e., achieve over 90% conversion below 150 °C) arises from continued improvements in diesel engine efficiency. Recently, there has been progress made in designing catalytic materials with enhanced low-temperature oxidation performance (e.g., Au and Ag nanoparticles combined with Cu, Mn, or Fe). However, significant technical barriers exist for implementing these novel materials into practice. For instance, these novel materials tend to lose performance rather quickly under harsh reaction environments typical to automotive emissions control due to hydrothermal aging and poisoning. Alternatively, increasing precious metal loading (e.g., Pt) can improve the low-temperature performance of current commercial catalysts, but that approach is cost-prohibitive. We have recently reported the impact of ZrO_2 supports on CO and C_3H_6 oxidation, sulfur tolerance, and hydrothermal stability [1]. Strong interaction between Pd and ZrO_2 manifested a greater thermal stability shown by good oxidation performance even after aging at 800 and 900°C for 16 h. On the other hand, Pd/SiO_2 suffered significant performance loss due to Pd particle coarsening. A higher and more complete coverage of SiO_2 by ZrO_2 could have led to a higher dispersion and stability of Pd. This work confirms the potential of developing Pd-based oxidation catalysts with enhanced durability and low-temperature activity using ZrO_2-SiO_2 and $CeZrAl_2O_3$ mixed oxide supports. In this work, the accessible surface area of ZrO_2 was enhanced by depositing a layer of ZrO_2 on SiO_2 spheres, leading to the formation of a core@shell structure, with SiO_2 located in the core and ZrO_2 on the shell. The SiO_2@ZrO_2 core-shell structure was verified using TEM, EDX and the surface area of the material was 220 m^2/g, much higher than the surface area of bulk ZrO2 (~58 m^2/g). Most importantly, the SiO_2@ZrO_2 support was able to maintain its surface area

even after thermal treatments in air at 700 and 800°C. Furthermore, Pd/$CeZrAl_2O_3$ catalysts were synthesized by impregnating CeZr aqueous nanoparticles onto Al_2O_3 via an incipient wetness method followed by a pH controlled impregnation of a $Pd(NH_3)_4(NO_3)_2$ precursor leading to Pd nanoparticles in a close proximity to CeO_x. Catalytic performance of both Pd/SiO_2@ZrO_2 and Pd/$CeZrAl_2O_3$ catalysts, with Pd loadings varying from 1 to 2 and 4 wt.%, was measured under the following conditions: temperature increased from 100 to 500 °C at 2 °C/min, total flow rate 333 mL/min, 12% O_2, 6% H_2O, 6% CO_2, 400 ppm H_2, 2000 ppm CO, 100 ppm NO, 1667 ppm C_2H_4, 1000 ppm C_3H_6 and 333 ppm C_3H_8 in Ar balance. The performance of Pd catalysts was evaluated in fresh and hydrothermally aged states. Overall, Pd/$CeZrAl_2O_3$ catalyst showed considerably higher CO and hydrocarbon oxidation performance than Pd/SiO_2@ZrO_2 in all states studied. The 2 wt.% Pd/$CeZrAl_2O_3$ has shown to have the best low temperature performance and hydrothermal durability. On the other hand, 4 wt.% Pd/$CeZrAl_2O_3$ suffered significant performance loss due to Pd particle coarsening. This work confirms the potential of developing Pd-based oxidation catalysts with enhanced durability and low-temperature activity using ZrO_2-SiO_2 and $CeZrAl_2O_3$ mixed oxide supports.

References

1. Kim, M.-Y.; Kyriakidou, E.A.; Choi, J.-S.; Toops, T.J.; Binder, A.J.; Thomas, C.; Parks II, J.E.; Schwartz, V.; Chen, J.; Hensley, D.K. Appl. Catal. B 187 (2016) 181.

Copyright © American Institute of Chemical Engineers. All rights reserved.

474320 Tuning Acidity of Mesoporous Silicates with Earth-Abundant Metals and Beneficial Effects during Catalytic Dehydration of Linear Alcohols

Abstract:

Catalytic dehydration of biomass-derived alcohols is capable of providing new olefin sources for the chemical industry. Recently, Hensley *et al.* reported production of an ethanol-rich mixed alcohol stream from bio-syngas as an alternative to conventional biological pathways.[1] A suitable dehydration catalyst with tunable acidity and pore size is desirable to produce olefins from such bio-alcohol streams. We synthesized metal-incorporated mesoporous silicates Zr-KIT-6 which exhibited mainly Lewis acidity and were fairly active in catalyzing ethanol dehydrations. However, comparative studies showed that while Zr-KIT-6 materials are highly selective for olefin production, they are much less active than catalysts that exhibit predominantly Brønsted acidity, such as H-ZSM-5.[2] Therefore we explored the addition of alternative metal Nb as well as co-addition of Zr and Nb into KIT-6 framework in an effort to enhance/tune the acidity and thereby the activity of the catalysts. While Nb-KIT-6 exhibits good activity at 350 °C, the selectivity for ethylene is fairly low compared to Zr-KIT-6[2], γ-alumina[3] or HZSM-5[2]. Kinetic study on Nb-KIT-6 showed higher activation energy for ethylene production of (169.3 ± 6.7) kJ/mol, than the values of (79 ± 1) kJ/mol on Zr-KIT-6 and 95.6 kJ/mol on γ-alumina. These results along with those obtained with bimetallic mesoporous silicates, some of which exhibit remarkable synergy in intensifying the olefin yield, will be presented.

1. Hensley JE, Lovestead TM, Christensen E, Dutta A, Bruno TJ, McCormick R. Compositional Analysis and Advanced Distillation Curve for Mixed Alcohols Produced via Syngas on a K-CoMoSx Catalyst. *Energy & Fuels.* June 20, 2013 2013;27:3246-3260.

2. Nash CP, Ramanathan A, Ruddy DA, et al. Mixed alcohol dehydration over Brønsted and Lewis acidic catalysts. *Applied Catalysis A: General.* January 25, 2016 2016;510:110-124.

3. DeWilde JF, Chiang H, Hickman DA, Ho CR, Bhan A. Kinetics and Mechanism of Ethanol Dehydration on γ-Al2O3: The Critical Role of Dimer Inhibition. *ACS Catalysis.* April 5, 2013 2013;3:798-807.

Copyright © American Institute of Chemical Engineers. All rights reserved.

467181 Designed Synthesis of Co_3O_4 Total Oxidation Catalysts

Abstract:

In recent years, nanocrystalline cobalt oxides have received increased attention due to their unprecedented activity as low temperature CO oxidation catalysts and their potential for use in various hydrocarbon oxidation reactions. [1] However, despite high levels of interest in these materials and a growing body of work devoted to synthesizing size and shape controlled cobalt oxide nanostructures, observations of catalytic performance and reported structure-activity relationships are often varied. This is likely due to the multitude of Co_3O_4 catalytic properties, including lattice oxygen mobility, Co^{3+} reducibility and Co^{3+} surface enrichment related to the size, morphology, and faceting of the nanostructures. Additionally, the existence of mesoporous, polycrystalline, and defected Co_3O_4 nanostructures and the very diverse set of synthesis methodologies employed further confuses the elucidation of important structural properties and synthesis conditions. [2]â€"[4] Because the intricate relationships between synthesis, structural properties, and performance in various systems are not exhaustively understood, our ability to utilize Co_3O_4 oxidation catalysts under various reaction schemes and operating conditions is limited by how quickly and precisely we can explore the relevant parameter space.

In our work, a statistical design and analysis platform was used to develop cobalt oxide based oxidation catalysts prepared via one pot metal salt reduction, a widely used and versatile synthesis technique. An emphasis was placed upon understanding the effects of synthesis conditions, such as heating regimen and Co^{2+} concentration on the metal salt reduction mechanism as well as the resultant nanomaterial properties, such as size, crystal structure, and crystal faceting. This was accomplished by carrying out XRD, TEM, and FTIR studies on synthesis intermediates and products. Additionally, high-throughput experimentation was employed to study and optimize the performance of Co_3O_4 oxidation catalysts over a wide range of synthesis and reaction conditions using a 16-channel fixed bed reactor equipped with a parallel infrared imaging system. Specifically, Co_3O_4 nanomaterials of varying properties were evaluated for their performance as CO oxidation and hydrocarbon oxidation catalysts. Figure-of-merits including light-off temperature, activation energy, and stability, which were measured and mapped back to the aforementioned catalyst properties and synthesis conditions. Statistical analysis methods were used to elucidate significant property-activity relationships as well as the design rules relevant in the synthesis of active catalysts.

References

[1] X. Xie, W. Shen, Nanoscale, vol. 1, no. 1 (2009) pp. 50.

[2] L. Hu, Q. Peng, L. Yadong, J. Am. Chem. Soc. 130 (2008) pp 16136â€"16137.

[3] C. Liu, Q. Liu, L. Bai, A. Dong, G. Liu, S. Wen, J. Mol. Catal. A Chem. 370 (2013) pp 1â€"6.

[4] X. Xie, Y. Li, Z.-Q. Liu, M. Haruta, W. Shen, Nature, vol. 458, no. 7239 (2009) pp 746-749.

Copyright © American Institute of Chemical Engineers. All rights reserved.

461206 Ring-Opening of Furfuryl Alcohol Towards 1,5-Pentanediol over Reduced Cu-Co-Al Mixed Metal Oxides

Abstract:

Taylor Sulmonetti, Bo Hu, Pradeep Agrawal, Christopher W. Jones

Ring-Opening of Furfuryl Alcohol towards 1,5-Pentanediol Over Reduced Cu-Co-Al Mixed Metal Oxides

1,5-pentanediol, a valuable specialty chemical and precursor to polyesters, is typically produced through petroleum resources. In consideration of the development of alternate, renewable feedstocks, recent interest has arisen in the ring-opening of biomass-derived furanic compounds towards suck alkane diols. Furanic compounds such as furfuryl alcohol are key biomass platform chemicals that have shown promise is reducing societyâ€™s dependence on fossil resources. Recent literature has emerged on the ring-opening of furans utilizing bifunctionalÂ meta/metal oxide catalysts. However, these catalysts typically incorporate precious metals, with Ir-ReO_x and Rh-ReO_x as typical examples, being less attractive than lower cost earth-abundent metal catalysts. In this work, we sought to investigate the ring-opening of furfuryl alcohol towards 1,5-pentanediol in liquid phase batch conditions under H_2 pressure. Cu-Co-Al mixed metal oxides were synthesized through the calcination of layered double hydroxides, and reduced in H_2 to create bifunctional metal/metal oxide materials prior to reaction. Kinetic studies of multiple Cu-Co-Al catalysts with varying Cu/Co ratios and varying reduction temperatures were conducted, and the results suggested an optimum Cu/Co ratio and reduction temperature for promoting ring-opening pathways. Various characterization methods such as temperature programmed reduction supported the hypothesis that the incorporation of Cu facilitates the reduction of Co. Other techniques, including XRD and STEM, further supported the idea that well-dispersed mixed metal oxides were formed. *In situ* XAS was carried out to probe the electronic environment of the metal atoms upon reduction in H_2, which suggested the emergence of both metallic Cu and Co phases imbedded into an oxide matrix. Ultimately, yields of about 48% for

1,5-pentanediol were observed, which suggests that the use of non-precious mixed metal oxides for furanic compound ring-opening is a promising direction warranting further exploration.

Copyright © American Institute of Chemical Engineers. All rights reserved.

460091 Mechanistic Studies on the Direct Synthesis of H_2O_2 on Pd and Aupd Clusters

Abstract:

Hydrogen peroxide (H_2O_2), an environmentally benign oxidant, is currently produced by the auto-oxidation of anthraquinones (AO), yet direct synthesis (DS) is a greener alternative with the potential to make H_2O_2 using less energy than AO. However, DS currently suffers from low H_2O_2 selectivities (< 60% H_2O_2 (identified as an effective catalyst for DS) in the absence of acids and halides.[1] In order to achieve economically viable DS, higher selectivities (> 90% H_2O_2) must be obtained, especially in the absence of toxic or caustic additives. We aim to reach this goal by developing a fundamental understanding of the DS mechanism on Pd and AuPd in order to elucidate the manner in which AuPd catalysts give higher H_2O_2 selectivities than do Pd catalysts.[2] This information will guide the rational design of alternative catalysts, made from more abundant materials, that meet or exceed selectivities achieved on AuPd.

Steady-state H_2O_2 and H_2O formation rates were measured using a fixed bed, plug flow reactor as functions of H_2 pressure (25-400 kPa), O_2 pressure (25-400 kPa), and temperature (273-337 K) on silica-supported, 6-11 nm diameter Pd and AuPd clusters with varying Au to Pd ratio. These data are inconsistent with Langmuirian mechanisms for H_2O_2 formation and instead show that H_2O_2 formation likely occurs by kinetically relevant proton-electron transfer to chemisorbed hydroperoxy (OOH**) on Pd and AuPd. This hypothesis is consistent with results from pure Pd[3] and Au_xPd clusters (where x denotes the bulk molar ratio of Au to Pd), which show that H_2O_2 will readily form by DS only in protic solutions, regardless of the Au to Pd ratio. Rate data suggest that H_2O forms by kinetically relevant O-O bond cleavage in OOH** surface intermediates. Changing the catalysts from pure Pd to Au_1Pd increases rates of H_2O_2 formation,

while subsequent addition of Au (up to $Au_{12}Pd$) decreases rates of both H_2O_2 and H_2O formation (with H_2O rates decreasing more than H_2O_2 rates). In addition, clusters with high Au to Pd ratios (e.g., $Au_{12}Pd$) show low rates of H_2O_2 hydrogenation, as measured directly by feeding H_2O_2 to the system in the presence of H_2. These changes are due, in part, to greater activation enthalpies for H_2O_2 ($\Delta H^{\ddagger}_{H2O2}$) and H_2O ($\Delta H^{\ddagger}_{H2O}$) formation on Au_1Pd than on pure Pd clusters. Further addition of Au, however, does not continue to increase $\Delta H^{\ddagger}_{H2O}$ values. This observation suggests that the five-fold increase in H_2O_2 selectivity on $Au_{12}Pd$ over pure Pd must also result from a decrease in the number of Pd ensembles that act as sites for O-O bond rupture during primary and secondary reaction pathways. The improved understanding of the mechanisms behind DS and the factors that make Au_xPd such an ideal catalyst will help guide the rational design of selective catalysts for DS by our group and others.

Copyright © American Institute of Chemical Engineers. All rights reserved.

453326 Influence of Sn and Bi on Pt-Catalyzed Aqueous Phase Oxidation of 1,6-Hexanediol

Abstract:

In this study, a series of carbon-supported PtM (M = Sn, Bi) bimetallic catalysts with various M/Pt ratios was prepared and evaluated in the oxidation of HDO by dioxygen in the aqueous phase. Extensive characterization of the catalysts before and after reaction as well as detailed kinetic measurements (including a kinetic isotope effect) were performed to enable correlation of catalyst performance to structure and composition.

To prepare the PtSn/C catalysts, the Pt and Sn precursors were reduced simultaneously by $NaBH_4$ in a slurry of activated carbon. The catalysts were then treated in H_2 at 673 K to induce the formation of PtSn alloy. The PtBi/C catalysts were prepared by selectively reducing $Bi(NO_3)_3$ in an aqueous solution onto supported Pt nanoparticles by H_2. The HDO oxidation reaction was performed under reaction conditions that eliminated artifacts from mass transfer limitations.

The catalysts collected after synthesis as well as those collected after pretreatment and reaction were extensively characterized by N_2-physisorption, H_2-chemisorption, X-ray diffraction, scanning electron microscopy, (scanning) transmission electron microscopy, energy-dispersive X-ray spectroscopy, X-ray photoelectron spectroscopy, inductively coupled plasma – atomic emission spectroscopy and X-ray absorption spectroscopy.

Based on the extensive characterization of the fresh and used catalysts, we propose a model for the structural evolution of PtSn and PtBi catalysts under pretreatment and reaction conditions. A PtSn/C catalyst prepared by $NaBH_4$ reduction was composed of SnO_x species that were well dispersed on the carbon support and the Pt nanoparticles, with negligible formation of Pt-Sn alloy. After the H_2 treatment at 673 K, Pt-Sn alloy particles were observed on the sample, which could be explained by the enhanced interaction between adjacent Pt and Sn species in H_2under high temperature. The oxidation of HDO in liquid water over the alloyed particles facilitated a phase segregation of Pt and Sn. A significant structural evolution of PtBi/C was also observed. During catalyst preparation, Bi was selectively deposited on Pt particles in the outer layer of the carbon support. However, under conditions of pretreatment and reaction, substantial migration of Bi species on the carbon support was observed.

The impact of adding Sn to Pt on the catalysis of HDO oxidation depended on the structure of the catalyst. The results suggest that the Pt-SnO_x interface modestly promotes the oxidation activity whereas alloying Pt with Sn inhibits activity under 1 MPa O_2. The impact of adding Bi to Pt depended on the dioxygen pressure used during HDO oxidation. Under high O_2 pressure, similar activity was observed on Pt/C and PtBi/C, whereas Bi significantly promoted the oxidation activity of Pt under low O_2 pressure. Indeed, the reaction order with respect to O_2 was zero over PtBi/C compared to 0.75 over Pt/C in the range of 0.02 – 0.2 MPa. Rate measurements in D_2O solvent revealed a moderate isotope effect (TOF_H/TOF_D = 1.4) on HDO oxidation over Pt/C under 0.02 MPa O_2 whereas a negligible isotope effect was observed on PtBi/C, indicating that the promotional effect of Bi was likely related to the formation of surface hydroxyl groups from reaction of O_2 and H_2O.

In summary, this work presents evidence for severe restructuring of PtSn and PtBi catalysts during pretreatment and alcohol oxidation in liquid water and demonstrates that the impact of added promoters on Pt depends on catalyst structure and reaction conditions.

Copyright © American Institute of Chemical Engineers. All rights reserved.

466526 CeO_2 Supported Metal Oxide Catalysts for NO_x Removal

Abstract:

ABSTRACT:
It has been found that lean burn and diesel engines have much greater efficiency in terms of fuel consumption and power compared to their gasoline engine counterparts. However, due to the high air to fuel ratio of the engine, greater amounts of NOx is produced therefore these more efficient engines are struggling to meet emission regulations set by the Environmental Protection Agency (EPA). Studies have shown that ammonia-based selective catalytic reduction (NH3-SCR) with the use of zeolite catalyst have shown efficient NOx conversion to nitrogen (N2) and water. Due to the high efficiency of lean burn and diesel engine, there is a demand to identify a catalyst that has both high durability for the temperature ranges that a diesel engine operates in and relatively cheap to manufacture. Currently, conversion of NOx requires the use of expensive precious metal loaded catalysts, ergo the need to develop a cheaper solution is in high demand in order to meet the regulation. Various metal-exchanged zeolites-base metal catalysts such as Cu- and Fe-exchanged ZSM-5 (MFI) have been investigated for SCR catalytic activity and stated that the effective temperature ranges for Cu- and Fe-ZSM-5 are 250– 450°C and 350– 650°C, respectively. But De-NOx efficiency at low temperature is still challenging to meet the emission regulation. Mn doped ceria showed improved activity at lower temperature than zeolite based catalysts.
Ceria (CeO2¬) has been studied since it has an oxygen reservoir capacity, which stores and release oxygen via the redox shift between Ce4+ and Ce3+ under oxidizing and reducing conditions, respectively. Metal-doped CeO2 enhanced its oxygen mobility and provide activated oxygen to the reactant. In this study, a series of CeO2-supported transition metal (Mn, Co and Cu) catalysts were investigated for NH3-SCR and NO oxidation. Catalysts were prepared by using the incipient wetness impregnation method (IM) at different surface species loadings (0.5, 1, 2, 10 & 30 wt.%). Operando experiment condition was applied to understand (1) the relationship between catalyst structure and activity/selectivity, (2) surface reaction, and (3) reaction mechanisms. Several spectroscopic techniques, such as XRD, Raman spectrometer and FTIR, were also applied under ex-situ/in-situ conditions.

ACKNOWLEDGMENT
We gratefully acknowledge the financial support for this study from SBU/BNL 2015 SEED grant and the Department of Materials Science & Engineering at Stony Brook University (SBU) through start-up research funding as well as Brookhaven National Lab and Advance Energy Centre at SBU for laboratory facilities.

Copyright © American Institute of Chemical Engineers. All rights reserved.

460048 Mechanistic Insights on Catalytic Ethanol Steam Reforming

Abstract:

Conversion of bio-derived oils and feedstock with high concentrations of oxygen-containing hydrocarbons into chemicals is a highly desirable objective for environmental and economic sustainability. The production and use of chemicals synthesized from oxygenated feedstocks primarily derived from bio-based sources is projected to grow from $3.6 billion in 2011 to $12.2 billion by 2021. However, no efficient technologies exist for such a conversion because current chemistries rely on catalysts that have been developed for processing natural gas and petroleum fractions with low oxygenate concentrations. To advance progress in chemical synthesis from bio-derived feeds the fundamentals of adsorption and reactivity of oxygenates over metal catalysts, such as Rh, must be better understood. This will enable development of new technologies for oxygenates with promising catalysts formulations.

Ethanol steam reforming (ESR) has been widely explored as a test case in order to identify appropriate catalyst materials for oxygenate reforming as well as to elucidate the reaction scheme governing the process. A primarily classic approach to reaction model development has previously been employed, focusing on measured quantities of the products over a range of test conditions combined with a proposed set of equilibrium reactions to match observed experimental data. Through work of this type, it has been shown that there are myriad competing reactions contributing to the overall mechanistic understanding of ESR. However, in a recent review on catalytic ESR by Hou *et al.*, the authors highlight that there is no agreed upon reaction pathway for the overall ethanol steam reforming process. Here we seek to provide new insights on the reaction pathways involved in steam reforming over a Rh-based catalyst.

The low-temperature ESR reaction mechanism over a supported Rh/Pt catalyst was investigated. An unprecedented level of understanding with respect to the dominant reaction pathways, the contribution of each metal to the product distribution, and the role of the support was achieved. Both the recombination of C-species on the surface of the catalyst as well as preservation of the C-C bond within ethanol are responsible for C_2 product formation. The onset of ethylene, a common byproduct observed after catalyst deactivation, does not occur until incomplete ethanol conversion is observed. In addition, we quantitatively show that 57% of observed ethylene is formed directly through ethanol dehydration. Finally we provide clear evidence that oxygen in the silica-zirconia support likely constitutes 10% of the CO formed during reaction.

Copyright © American Institute of Chemical Engineers. All rights reserved.

470893 Effect of Reduced Adsorption Strength with Bimetallic Platinum Overlayer Catalysts

Abstract:

The growing demand for energy resources has stimulated extensive research on converting biomass into transportation fuels and valuable chemicals to alleviate the excessive consumption of fossil fuels. Platinum-based catalysts have shown excellent reactivity and stability for various biomass conversion reactions. However, literature studies have suggested that the strong adsorption of H_2 or CO reforming products would block platinum surface active sites and reduce the catalyst reactivity. One possible option to solve this problem is bimetallic overlayer catalysts. Bimetallic overlayer catalysts are widely studied recently due to their unique adsorption properties. For example, a Pt overlayer on top of Ni or Co was been demonstrated computationally and experimentally to display a reduced d-band center energy and thus weakened adsorption strength for H_2 and CO [1]. In this respect, overlayer catalysts with reduced binding strength of H_2 and CO are expected to show increased catalytic reactivity compared to pure Pt in reactions where strong H_2 or CO product binding are proposed to inhibit the reactivity. Moreover, Pt overlayer catalysts suggest possibility of taking advantages of catalytic properties of Pt metal, while increasing its reactivity through modifying its adsorption strength as desired.

In this research, 5 bimetallic platinum overlayer catalysts, Ni@Pt/SiO_2-Al_2O_3, Co@Pt/SiO_2-Al_2O_3, Ir@Pt/SiO_2-Al_2O_3, Ni@Pt/SiO_2, and Cu@Pt/SiO_2 (core@shell) have been synthesized. These combinations were chosen based on previous computational studies predicting the reduced adsorption strength of platinum overlayer over these bulk metals [3]. The goal of this work is to examine the catalytic reactivity of these platinum overlayer catalysts in two biomass reactions: aqueous phase glycerol hydrodeoxygenation and furfural hydrogenation. Both reactions were conducted with an excessive amount of hydrogen present, suggesting that the Pt site blockage by strong H_2 adsorption is likely to affect Pt reactivity. Therefore, Pt overlayer catalysts with reduced adsorption strength are expected to show higher catalytic reactivity in both reactions compared to pure Pt.

Monometallic catalysts and non-structured alloys were prepared by incipient wetness impregnation. The directed deposition synthesis technique was used to synthesize all overlayer catalysts by selectively depositing the overlayer (platinum) metal on the base metal without it depositing onto the support surface [2]. H_2 chemisorption and ethylene hydrogenation were used to studied the adsorption strength of all synthesized overlayer catalysts compared to pure Pt. Hydrogen isotherms at various temperatures were measured and hydrogen heats of adsorption were calculated using the Clausius-Clapeyron equation. All overlayer catalysts showed a lower hydrogen heat of adsorption compared to pure Pt. Ethylene hydrogenation reactivity, which is a strong function of H_2 surface coverage and H_2 adsorption strength, was also tested as a descriptor reaction. All platinum overlayer catalysts showed reduced turnover frequencies compared to pure platinum, implying lower hydrogen surface coverage and weaker hydrogen adsorption. To conclude the H_2 chemisorption and ethylene hydrogenation results, all overlayer catalysts showed decreased hydrogen adsorption strength, consistent with previous computation predictions [3].

Aqueous phase hydrodeoxygenation of glycerol using Ni@Pt, Co@Pt, and Ir@Pt supported on silica-alumina showed enhanced TOF in glycerol conversion and hydrocarbon production rate compared to their monometallic counterparts and non-structured bimetallic alloys. This improvement in TOF could be correlated with chemisorption and ethylene hydrogenation results, as more surface active sites are exposed by reducing the adsorption strength of hydrogen or carbon monoxide. In terms of selectivity, Ni@Pt and Co@Pt showed higher selectivity toward C-C bond cleavage, while Ir@Pt showed higher selectivity toward C-O bond cleavage. In furfural hydrogenation, two silica-supported overlayer catalysts, Cu@Pt and Ni@Pt, showed higher TOF in furfural conversion compared to pure Pt and corresponding parent metals. With Cu@Pt, both high selectivity toward furfuryl alcohol and high reactivity in furfural conversion were achieved at the same time, indicating the advantage of overlayer catalysts in furfural hydrogenation reactions.

In conclusion, several bimetallic overlayer catalysts have been synthesized using the directed deposition technique. Through the characterization of hydrogen chemisorption and ethylene hydrogenation, it is demonstrated that the hydrogen adsorption on the surface was significantly reduced, which is consistent with computational predictions and previous research observations. Biomass reaction results showed that the reactivity has been enhanced with overlayer catalysts, likely due to the weakened hydrogen or carbon monoxide adsorption and exposing more surface active sites. This demonstrated the viability of preparing overlayer catalysts aiming at efficient and effective biomass conversion.

[1] J.G. Chen, C.A. Menning, M.B. Zellner, Surface Science Reports 63 (2008) 201-254.

[2] M.D. Skoglund, C.L. Jackson, K.J. McKim, H.J. Olsen, S. Sabirzyanov, J.H. Holles, Applied Catalysis A: General 467 (2013) 355-362.

[3] E. Christoffersen, P. Liu, A. Ruban, H.L. Skriver, J.K. Norskov, Journal of Catalysis 199 (2001) 123-131.

Copyright © American Institute of Chemical Engineers. All rights reserved.

452591 Tuning Nanoparticle Alloys to Enhance C-H Bond Activation for the Catalytic Dehydrogenation of Ethane

Abstract:

Tuning Nanoparticle Alloys to Enhance C-H Bond Activation for the Catalytic Dehydrogenation of Ethane

Viktor J. Cybulskis[1], James R. Gallagher[2], Han-Ting Tseng[1], Brandon Bukowski[1], Zhenwei Wu[1], Evan Wegener[1], A. Jeremy Kropf[2], Bruce Ravel[3], Jeffrey Greeley[1], Fabio H. Ribeiro[1], Jeffrey T. Miller[1]

[1] Chemical Engineering, Purdue University, West Lafayette, IN 47907, USA

[2] Chemical Sciences and Engineering, Argonne National Laboratory, Argonne, IL 60439, USA

[3] Materials Measurement Laboratory, National Institute of Standards and Technology, Gaithersburg, MD 20899, USA

Supported Pt catalysts are widely used in a number of industrial hydrocarbon processes, including hydrogenation, isomerization, naphtha reforming, and dehydrogenation reactions, due to their affinity for paraffinic C-H bonds [1]. While structure insensitive reactions, such as alkane dehydrogenation, can occur on isolated metal sites, larger nanoparticle ensembles are known to catalyze side reactions, such as cracking and hydrogenolysis [2]. Recent work by Childers et al. [3] has demonstrated that Zn addition to SiO_2-supported Pt catalysts enhances propylene selectivity during propane dehydrogenation by forming an intermetallic alloy that effectively isolates Pd surface sites via a geometric effect.

In the present study, we demonstrate how the addition of Zn to Pt/SiO_2 leads to near 100% ethylene selectivity during

ethane dehydrogenation (EDH) at 600 ¡C by forming a high-symmetry Pt_1Zn_1 alloy with isolated Pt sites that effectively eliminate concomitant C-C bond cleavages. Kinetic measurements and *in situ* resonant inelastic X-ray scattering (RIXS) experiments indicate that Zn also modifies the electronic structure of Pt and increases the EDH turnover frequency (TOF) per exposed surface Pt by a factor of ten compared to monometallic Pt/SiO_2. The molecular level insight obtained from this study provides a model that suggests control of the geometric structure of the active sites affects alkene selectivity, while control of the promoter affects the adsorbate binding strength and TOF.

References:

[1] J. J. H. B. Sattler, J. Ruiz-Martinez, E. Santillan-Jimenez, and B. M. Weckhuysen, Chem. Rev. 114 (2014) 10613.

[2] M. Boudart, G. Djega-Mariadassou, Kinetics of Heterogeneous Catalytic Reactions, Princeton, 1984, pp. 155-193.

[3] D.J. Childers, N.M. Schweitzer, S.M.K. Shahari, R.M. Rioux, J.T. Miller, R.J. Meyer, J. Catal. 318 (2014) 75.

Copyright © American Institute of Chemical Engineers. All rights reserved.

472592 First Principles Light Alkane Dehydrogenation on Pt:Main Group Alloys and the Effect of Hydrogen Spectators

Abstract:

The effects of alloying platinum with transition metals and main group elements on the kinetics and thermodynamics of dehydrogenation and coke formation pathways during light alkane dehydrogenation have been studied using density functional theory. Due to the recent shale boom, light alkane prices have dropped significantly; ethane is now the same price as methane and is flared in some areas. Light alkane dehydrogenation to olefins can add significant value to and reduce greenhouse gas emissions from hydrocarbon processes that generate alkanes by converting low value commodity fuels to high-value chemical and polymer precursors.

Supported Pt catalysts are known to be active for light alkane dehydrogenation, but the high temperatures required by these endothermic reactions leads to significant coke formation and deactivation. Numerous Pt alloys have been experimentally shown to decrease coke formation and deactivation, including with Sn, Ga, In, Cu, Au, and Re. In addition, low amounts of hydrogen feed have also been experimentally shown to increase selectivity of ethylene over coke. Using periodic density functional theory we have studied the dehydrogenation for an extensive number of alloys and under hydrogen coadsorption. We show that hydrogen spectators decrease the ethylene binding energy by competing with adsorption locations, thus making it easier to desorb. Hydrogen spectators also increase the ethylene dehydrogenation activation energy barrier making it harder to continue dehydrogenating to carbonaceous species providing support to hydrogen feed increasing selectivity. In addition we have calculated potential energy surfaces on numerous alloys to better understand the reduction in surface carbon formation during ethane dehydrogenation. At low concentrations, the main group elements are the only species that decrease the binding energy of carbon, and presumably carbonaceous species in general. These elements affect Pt purely though electronic effects at low alloy ratios while at higher concentrations geometric effects also play an increasingly important role. The electronic effects of Bi in Pt_3Bi/Pt suggest higher ethylene selectivity than either Pt_3Sn/Pt or Pt_3In/Pt, but at higher alloy concentrations, the geometric effects of In in PtIn/Pt suggest an higher selectivity than PtBi/Pt.

Copyright © American Institute of Chemical Engineers. All rights reserved.

468055 Effect of Boron Promotion on the Performance of Platinum Catalysts for Propane Dehydrogenation

Abstract:

The activity, stability and selectivity of supported Pt catalysts in light alkane dehydrogenation reactions are often optimized by the addition of promotors such as Sn, Zn, Ga, In and Ge. Such promotors for example enhance the catalyst lifetime by reducing carbon deposition (coking) and/or particle sintering,[1] often at the expense of catalyst activity.

Recently, our group has reported that the addition of small amounts boron enhances the stability of Ni[2] and Co[3] catalysts without affecting their activity and selectivity. Density Functional Theory (DFT) calculations show that boron binds strongly at sites where deleterious carbon deposition nucleates. By titrating these sites with boron, carbon deposition and growth becomes inhibited. Calculations furthermore suggest that this effect might be rather general.[4] Indeed, the effect of boron promotion on the selectivity of Pd hydrogenation catalysts was recently reported by Tsang et al.[5] and by Studt et al.[6] Also in these studies, it is suggested that boron binds strongly at subsurface sites and hence modifies the surface electronic properties of the Pd catalyst.

Based on these insights, we evaluated the effect of boron promotion for Pt-based catalysts in propane dehydrogenation reactions. DFT calculations show that, although boron binds weaker to Pt than to Ni and Co, boron binds sufficiently strong at Pt subsurface octahedral sites to be stable against removal by hydrogen or carbon under dehydrogenation conditions (600ËšC and 1 atm) with Î"G_{rxn} of 157 and 178 kJ/mol, respectively. Boron-promoted Pt/SiO_2 and Pt/Al_2O_3 catalysts were prepared with 3 wt. % Pt and with boron loadings ranging from 0.0 and 1.5 wt.%. To evaluate the effect of boron, propane dehydrogenation reactions were performed at 600 ËšC and atmospheric pressure with C_3H_8/H_2/He (1/1/3). After reaction, the amount of deposited carbon was determined by oxidation.

For both series of catalysts, Pt/B-SiO_2 and Pt/B-Al_2O_3, the same trend was observed: the presence of small amounts of boron indeed improves the catalyst stability and reduces the amount of coke deposition. For example, coke formation is reduced threefold by the addition of 1 wt% boron to Pt/Al_2O_3. To elucidate the effect of boron promotion, several characterization techniques were used.

1. Siddiqi G, Sun P, Galvita V, Bell AT. Catalyst performance of novel Pt/Mg(Ga)(Al)O catalysts for alkane dehydrogenation. *Journal of Catalysis.* 9/9/ 2010;274(2):200-206.

2. Xu J, Chen L, Tan KF, Borgna A, Saeys M. Effect of boron on the stability of Ni catalysts during steam methane reforming. *Journal of Catalysis.* 1/25/ 2009;261(2):158-165.

3. Tan KF, Chang J, Borgna A, Saeys M. Effect of boron promotion on the stability of cobalt Fischerâ€"Tropsch catalysts. *Journal of Catalysis.* 5/16/ 2011;280(1):50-59.

4. Nandula A, Trinh QT, Saeys M, Alexandrova AN. Origin of Extraordinary Stability of Square-Planar Carbon Atoms in Surface Carbides of Cobalt and Nickel. *Angewandte Chemie International Edition.* 2015;54(18):5312-5316.

5. Chan CWA, Mahadi AH, Li MM-J, et al. Interstitial modification of palladium nanoparticles with boron atoms as a green catalyst for selective hydrogenation. *Nat Commun.* 12/19/online 2014;5:5787-5795.

6. Yoo JS, Zhao Z-J, NÃ¸rskov JK, Studt F. Effect of Boron Modifications of Palladium Catalysts for the Production of Hydrogen from Formic Acid. *ACS Catalysis.* 2015/11/06 2015;5(11):6579-6586.

Copyright © American Institute of Chemical Engineers. All rights reserved.

450717 Physical Properties Controlling Heterogeneous Catalyst Stability in Supercritical Water

Abstract:

Catalysis in supercritical water (SCW) offers a potentially "greener" approach than many alternatives for carrying out reactions such as gasification of wet biomass, biofuel production, desulfurization of heavy crude oil, and destruction of environmental pollutants. Near the critical point of H_2O (374°C, 22.1 MPa), its ion product and dielectric constant are highly tunable with temperature and density. These properties can strongly influence reaction rates and selectivities [1]. By coupling the tunable solvent properties of SCW with a heterogeneous catalyst, we aim to expand the design space for engineering novel reaction pathways for energy production and waste destruction. A key challenge for applying heterogeneous catalysts in SCW is the limited understanding of catalyst stability across conditions that correspond with the significant changes in the solvent properties [2]. While thermodynamic models can predict process conditions where metal oxidation and dissolution would be problematic [3-4], experiments are needed to validate the models and provide rates for the instability mechanisms.

We examined the stabilities of noble and transition metals, metal oxide catalyst supports, and transition metal carbides and nitrides. The catalysts were tested in both batch and flow conditions and the SCW density was varied from 0-0.5g/mL to observe the influence of the solvent properties on stability. The composition and structure of the fresh catalyst samples were characterized using x-ray diffraction, N_2 physisorption, and electron microscopy techniques and then characterized again after exposure to the SCW environment. The aqueous and gaseous compositions exiting the flow reactor were analyzed using inductively coupled plasma atomic emission spectroscopy and gas chromatography, respectively. Overall, the experimentally observed hydrothermal stabilities agreed with trends predicted in the thermodynamic models. In addition, we found that the oxidative or reductive strength of the reaction substrate will influence the dissolution rate and maximum solubility of the catalyst. These results provide design criteria for creating high-stability catalysts for hydrothermal applications.

References

1. N. Akiya, P.E. Savage, Chem. Rev. 102 (2002) 2725-2750.

2. T.M. Yeh, J.G. Dickinson, A. Franck, S. Linic, L.T. Thompson, P.E. Savage, Technol. Biotechnol. 88 (2013) 12-24.

3. E.L. Shock and H.C. Helgeson, Geochimica et Cosmochimica Acta 52 (1988) 2009-2036.

4. K. Sue, T. Adschiri, K. Arai, Ind. Eng. Chem. Res. 41 (2002) 3298-3306.

Copyright © American Institute of Chemical Engineers. All rights reserved.

462665 Systematic Investigation of Earth-Abundant Transition Metal Promoters for Pd-Catalyzed Methane Complete Combustion

Abstract:

Systematic Investigation of Earth-abundant Transition Metal Promoters for Pd-Catalyzed Methane Complete Combustion

Joshua J. Willis, Ian Andrew Naccarella, Helen Yan, Joseph Maalouf, Kester Wade, Matteo Cargnello

Department of Chemical Engineering and SUNCAT Center for Interface Science and Catalysis, Stanford University, Stanford, CA 94305, USA

There is a critical need for reducing emissions of methane, the second most prevalent greenhouse gas. Palladium is widely accepted the best metal catalyst. Still rates are unsatisfactory below 400°C where new technologies for methane utilization (e.g. low-temperature combustion engines) are currently operating. Also, steam dramatically deactivates Pd-based catalysts through the formation of less active hydroxide phases. Thus, there is a critical need to utilize Pd to the best possible extent and improve its activity and stability. Earth-abundant transition metals (Mn, Fe, Co, Ni and Cu) as promoters are a promising route to optimizing Pd utilization and destabilizing the formation of Pd hydroxide.

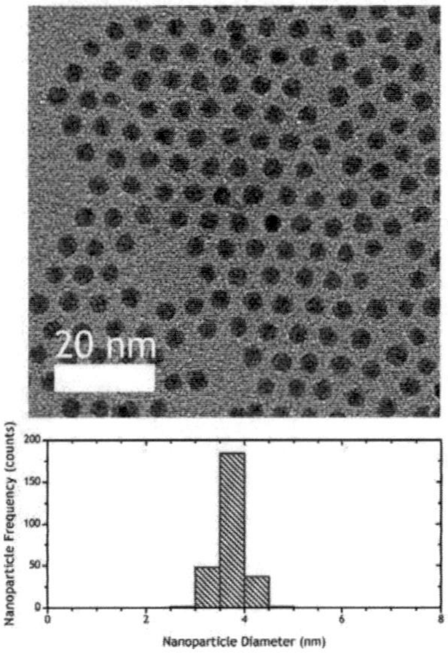

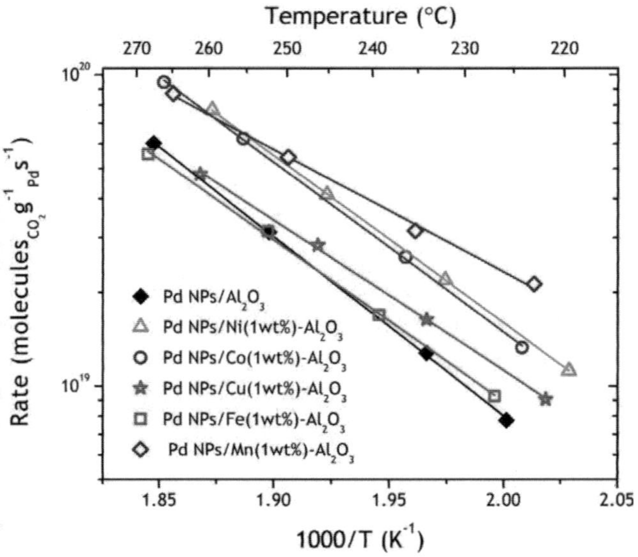

In order to perform systematic studies of promoting effects in Pd-based catalysts, here we present the results of using highly monodisperse palladium nanocrystals as the active phase. In this way, any variation in catalytic activity can be directly attributed to the promoter structure and properties. By deposition of the nanocrystals onto promoted γ-alumina (M-Al_2O_3) and activation using a fast thermal annealing process, a library of Pd/M-Al_2O_3 is obtained. By controlling the Pd nanocrystal structure and size, promoter-activity relationships can be drawn. Kinetic characterization demonstrates a unique promoting effect for Pd-Mn systems for methane complete combustion, with decrease activation energy and a three-fold increase in rates compared to the corresponding Pd/M-Al_2O_3. Experiments in the presence of steam also show pronounced effects of the promoters on the activity of the Pd-only catalyst. Our work provides clear elements for improving the activity of Pd-based combustion catalysts and in general a framework for understanding promoter-activity relationships using highly uniform catalysts under realistic reaction conditions.

Copyright © American Institute of Chemical Engineers. All rights reserved.

468785 Unraveling the Nature of Boundary Sites of Metal-on-Oxide Catalysts

Abstract:

The rise of density functional theory (DFT) over the past few decades has enabled great progress in rational catalyst design, especially for extended surfaces of transition metals [1]. However, insights from such metal-only models do not always translate to real catalysts, which usually consist of metal nanoparticles dispersed on high surface area supporting phases. In this contribution, we examine how the inclusion of a support phase can lead to deviations in reactivity compared to a metal-only model. Specifically, we investigate the adsorption behavior of simple adsorbates at the boundary of several late transition metals supported on MgO(100). We model the system as quasi one dimensional metal nanowires on the oxide, similar to the model proposed by *Molina et al* [2], shown in Fig. 1. Our results show that the role of the oxide is highly variable and strongly depends on the choice of metal and adsorbate. For some metal-adsorbate combinations, the presence of the support results in enhanced adsorption at the perimeter sites compared to the unsupported metal nanowire. In other cases, adsorption is destabilized when the metal is supported by the MgO. We have traced the origins of these stabilization/destabilization effects to a combination of multiple structural and electronic perturbations, including steric interactions, a shift in the metal d-band center, and redox energy gained by charge transfer facilitated by the oxide. Moreover, we show that in many cases, while the oxide has a destabilizing effect on the adsorption of an atomic adsorbate, *A,* it has a stabilizing effect on the adsorption of its hydrogenated form, AH_x. We use these results to discuss possible implications for scaling relationships developed on metal-only models.

[1] J. K. Norskov, Abild-Pedersen, F., Studt, F., & Bligaard, T. (2011). *PNAS*, 108(3), 937-943.

[2] Molina, L., & Hammer, B. (2003). *Phys. Rev. Lett.,* 90(20), 206102.

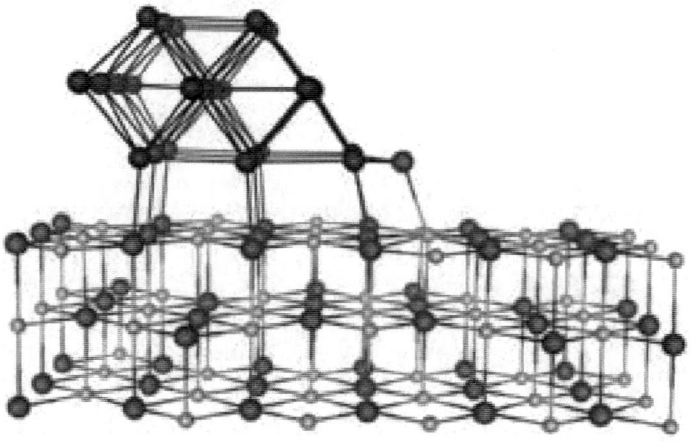

Figure 1. Structural model of a Pd nanowire supported on MgO(100). An oxygen atom adsorbed at the metal-oxide boundary is also shown.

Copyright © American Institute of Chemical Engineers. All rights reserved.

456710 Identification of Active Sites for NaY Catalyzed Methyl Lactate Dehydration

Abstract:

The dwindling reserve of crude oil and surge in natural gas production is rapidly changing the mix of the carbon source pool for the production of fuels and chemical feedstocks, and in turn creating shortages of several key commodity chemicals, e.g., propylene and butadiene. The shortage of certain commodity chemicals, such as propylene, drives up their prices, which in turn raises the cost of the downstream chemicals, such as acrylic acid. In this regard, lignocellulosic biomass derived feedstocks, e.g., lactic acid and its esters, can potentially bridge the gap. Currently, the commercial fermentation process using biomass-derived sugars can achieve a lactic acid (or its esters) yield of up to 90%. The absence of efficient and selective catalyst for lactic acid dehydration is the main missing link in the production of renewable acrylic acid. The primary roadblock for the rational design of catalysts for lactic acid dehydration is the lack of the mechanistic understanding of the nature of active sites and mechanistic steps leading to the selective removal of the a-hydroxyl group by dehydration. Through kinetic and in-situ spectroscopic investigations, we identify the dehydration reaction proceeds through dissociative adsorption, acid-mediated dehydration, and associative desorption steps. These mechanistic insights will guide the design of selective catalysts for this reaction.

Copyright © American Institute of Chemical Engineers. All rights reserved.

460600 Using Non-Covalent Interactions to Control Availability of Different Catalyst Sites

Abstract:

Furfuryl oxygenates are among the main oxygenates produced from biomass deconstruction, and it is desirable to upgrade components such as furfural and furfuryl alcohol to value-added fuels and chemicals. Furfuryl alcohol can undergo hydrodeoxygenation (HDO) to methyl furan, dehydrogenation to furfural, and decarbonylation (DC) to furan, among other reactions. It has been proposed that DC and HDO are carried out on different types of Pd sites. Thus, controlling the relative amount of terrace and edge sites is a potential method to manipulate catalyst performance. Recently, thiolate self-assembled monolayers (SAMs) on catalysts have been considered as selectivity modifiers. In this study, experimental and theoretical techniques were combined to study the effect of thiolate chain length on Pd surface active site distribution and furfuryl alcohol HDO/DC selectivity.

Hexanethiol (C6), decanethiol (C10), tetradecanethiol (C14) and octadecanthiol (C18)-coated Pd/Al_2O_3 were used to investigate trends in selectivity with chain length. Addition of the C6 thiol modifiers was found to slightly improve methyl furan selectivity from 5% to 15%. As the tail length was increased from C6 to C18, the selectivity to methyl furan improved dramatically because the active sites for DC were selectively blocked by the SAMs. The nature of these sites was investigated by infrared spectroscopy during CO adsorption. We found that the longer thiolate tails strongly decreased the availability of contiguous active sites was decreased much more dramatically than step and edge sites. DFT calculations indicated that short thiols prefer to occupy Pd edge sites because of a stronger covalent bond between low-coordination Pd atoms and the thiol group. However, through the stronger van der Waals interaction between thiolate chains on terrace sites, longer thiol molecules were found to preferentially bind on (111) surfaces, selectively poisoning those sites. DFT models of the surface reaction chemistry revealed that reaction at steps between thiolate-covered terraces could serve as active sites for HDO. These results show that the self-organizing properties of alkanethiols can be used to selectively poison specific sites on supported metal catalysts, leading to improved reaction selectivity.

Copyright © American Institute of Chemical Engineers. All rights reserved.

466485 Cu/Chabazite Catalyst to Meet the Chemical and Mechanical Challenges for NO_x Removal

Abstract:

ABSTRACT:

Global attention has been drawn towards the utilization of lean burn and diesel engines as oppose to gasoline, due to the fact that they are more efficient in terms of fuel consumption and power. These more efficient engines, however, are still challenging to meet the emission regulation of harmful gas emission called nitrogen oxide (NO_x) due to excessive air-to-fuel ratio. The use of an ammonia-based selective catalytic reduction, NH_3-SCR, through the application of zeolite catalysts has been proven efficacious in terms of converting the harmful gases into nitrogen (N_2) and water (H_2O). Ultimately, we come to the task of finding catalysts that would be durable and cheap for the use of a diesel engine. Metal-exchanged molecular sieves with a Chabazite (CHA) structure has been extensively studied because of their superior NOx reduction activity, selectivity and hydrothermal stability (100 to 800°C). Nevertheless, understanding of the structure-reactivity relationships such as the chemical nature and location of the catalytically active sites within the CHA framework, reaction intermediates, and reaction pathways of the SCR process are still not well understood.

In this study, the effect of Cu loading (0.5~6 wt%) and reaction temperatures on the catalytic activity (NH_3-SCR and NO oxidation) and productsâ€™ selectivity were examined over a series of Cu exchanged CHA catalysts. A series of Cu exchanged CHA catalysts were synthesized by using the incipient wetness impregnation method and several spectroscopic techniques were applied to understand the relationship between catalysts structure and activity. In addition to the commercial zeolite catalysts, further studies of the catalyst will be done by preparation of Cu-

exchanged two-dimensional zeolite model systems then examining the interaction of relevant molecule with the Cu-site. In this work, we will carry out experiments by using infrared reflection absorption spectroscopy (IRRAS) and ambient pressure XPS at the CSX-2 beamline of NSLS-II at Brookhaven National Laboratory.

ACKNOWLEDGMENT

We gratefully acknowledge the financial support for this study from SBU/BNL SEED grant and the Department of Materials Science & Engineering at Stony Brook University (SBU) through start-up research funding as well as Brookhaven National Lab and Advance Energy Centre at SBU for laboratory facilities.

Copyright © American Institute of Chemical Engineers. All rights reserved.

471030 Effect of the Preparation Method on the Porosity of Amorphous Silica-Alumina to be Used in Hydrocracking Reactions

Abstract:

EFFECT OF THE PREPARATION METHOD ON THE POROSITY OF AMORPHOUS SILICA-ALUMINA TO BE USED IN HYDROCRACKING REACTIONS

Angélica L. Coconubo-Díaz[1A], Alexander Guzmán-Monsalve[2B], Luz M. Ballesteros-Rueda[1C]

1: Centro de Investigaciones en Catálisis, Escuela de Ingeniería Química, Carrera 27 Calle 9 Universidad Industrial de Santander, Bucaramanga, Santander, Colombia

2: Km 7 Vía Piedecuesta, Instituto Colombiano del Petróleo, Piedecuesta, Santander, Colombia

e-mail: angelica.coconubo@gmail.com [A], alexander.guzman@ecopetrol.com.co [B], luzmabal@uis.edu.co [C]

ABSTRACT

Catalyst for heavy oil and vacuum gas oil hydrocracking (HCK) require high pore diameter to treat the bulky molecules present in such kind of feedstocks, in order to allow the access of large molecules into the active sites of the catalyst, avoiding diffusional limitations [1, 2]. Amorphous AluminoSilicates (ASAs) are widely used as catalyst support in the heavy oil refining industry because of its physico-chemical properties (acidity and large pore size) [1, 3, 4]. On the other hand, several authors have stated that the synthesis methods of ASAs impacts the textural properties such as surface area and pore size [4]. Particulary, in the sol-gel method, the control of the hydrolysis process and condensation is very important to obtain the desired gel [3, 5].

In the present study, bifunctional catalysts consisting of nickel-molybdenum (NiMo) supported on amorphous silica-alumina were prepared using two different sol-gel methods. It is well known that the characteristics of the obtained oxides depend on the sol-gel parameters such as precursor molecules, concentration, solvent type, temperature, amount of water and pH [6, 7]. The sol-gel methods used were: sol-gel using PolyEthylene Glycol (PEG) as an organic template [8], and sol-gel with gel skeletal reinforcement [9]. In the first method the amount of PEG was varied taking into account the amount of silica precursor used which, in this study, was TetraEthyl OrthoSilicate (TEOS); on the other hand, in the second method was varied the amount of TEOS used in the reinforcement solution (RS). In addition, in both methods the calcination were carried out at two different values of temperature, in order to compare its effect in the textural properties of the supports. The supports (ASAs) and the NiMo catalysts were characterized by means of N_2-physisorption, X-ray diffraction, SEM-spectroscopy, and NH_3-adsorption; in order to determine their chemical and physical properties.

Amorphous silica-alumina prepared by the first method, showed both type of porosities: micro and mesopores, and it was found that surface area and mesoporosity area increased while the amount of PEG added increased as well. On the other hand, ASAs prepared by the second method showed only mesoporosity, meaning that RS helps avoiding the formation of micropores. Additionally, surface area, pore volumen and pore size, increased while increasing the TEOS amount in the RS. Furthermore, it was found, that the calcination temperature affects the porosity of the material using both methods, since the surface area decreased when the highest calcination temperature was used likely ocurring by sinterization of the material.

Keywords: *Hydrocracking, Amorphous Aluminosilicates, Ni-Mo Catalyst, Sol-Gel method, Polyethylene Glycol, Gel Skeletal Reinforcement, Vacuum Gas Oil.*

REFERENCES

[1] LEYVA, C., *et al*, "Activity and Surface Properties of NiMo/SiO_2-Al_2O_3 Catalysts for Hydroprocessing of Heavy Oils," *Appl. Catal. A Gen.*, vol. 425–426, pp. 1–12, 2012

[2] PASHKOVA, V. O., SARV, P., and DEREWIŃSKI, M., "Composite Porous Materials Containing Zeolitic Domains Prepared by Controlled Partial Recrystallization of Amorphous Aluminosilicates," *Stud. Surf. Sci. Catal.*, vol. 170, pp. 289–296, 2007.

[3] ISHIHARA, A., *et al*, "Catalytic Properties of Amorphous Silica-Alumina Prepared using Malic Acid as a Matrix in Catalytic Cracking of n-dodecane," *Appl. Catal. A Gen.*, vol. 388, pp. 68–76, 2010.

[4] CAILLOT, M., *et al*, "Synthesis of Amorphous Aluminosilicates by Grafting: Tuning the Building and final Structure of the Deposit by Selecting the Appropriate Synthesis Conditions," *Microporous Mesoporous Mater.*, vol. 185, pp. 179–189, 2014.

[5] OKAMOTO, Y., *et al*, "A study on the Preparation of Supported Metal Oxide Catalysts using JRC-reference Catalysts. I. Preparation of a molybdena–alumina catalyst. Part 4. Preparation parameters and impact index," *Appl. Catal. A Gen.*, vol. 170, pp. 359–379, 1998.

[6] ALI, M.A., TATSUMI, T., and MASUDA, T., "Development of Heavy Oil Hydrocracking Catalysts using Amorphous Silica-Alumina and Zeolites as Catalyst Supports," *Appl. Catal. A Gen.*, vol. 233, pp. 77–90, 2002.

[7] PÉREZ-MARTÍNEZ, D. J., GAIGNEAUX, E. M., and GIRALDO, S. A., "Improving the Selectivity to HDS in the HDT of Synthetic FCC Naphtha using Sodium Doped Amorphous Aluminosilicates as Support of CoMo Catalysts," Appl. Catal. A Gen., vol. 421–422, pp. 48–57, 2012.

[8] ISHIHARA, A., *et al*, "Preparation of Amorphous Silica-Alumina using Polyethylene Glycol and its role for Matrix in Catalytic Cracking of n-dodecane," *Appl. Catal. A Gen.*, vol. 478, pp. 58–65, 2014.

[9] ISHIHARA, A., HASHIMOTO, T., and NASU, H., "Large Mesopore Generation in an Amorphous Silica-Alumina by Controlling the Pore Size with the Gel Skeletal Reinforcement and its Application to Catalytic Cracking," *Catalysts*, vol. 2, pp. 368–385, 2012.

Copyright © American Institute of Chemical Engineers. All rights reserved.

472494 Insight into Supported Metal Catalyst Stability By Quantifying Thermodynamic Interactions at the Solid-Liquid Interface

Abstract:

Industrial applications of supported late transition metal catalysts demand economic and scalable synthesis of these catalysts and current synthetic methods lack precision in terms of size, shape and compositional control. Moreover, supported metal catalysts suffer from poor stability, manifested in the form of sintering (i.e., particle growth) during reaction. The proper selection of the oxide support is of great importance to ensure high dispersion, activity and selectivity of the nanoparticles. The ability of these supports to enhance the dispersion of the active metal on their surface and control their morphology and sintering kinetics is fundamentally related to the nature and strength of the metalâ€"metal oxide interaction at the time of adsorption.Â In this work, we have utilized isothermal titration calorimetry (ITC), a technique capable of quantifying the thermodynamic description (D*G*, D*H*, D*S*, *n* (stoichiometry)) of transition metal association with a support material in a single experiment. After providing a brief introduction to ITC and methods of catalyst synthesis, we will discuss our results to quantify the electrostatic interactions between solvated transition metal ions and charged amphoteric metal oxide surface. Within this interaction-type, we have studied both refractory and reducible metal oxides. With a reducible metal oxide, ceria, we demonstrate a potentially new mechanism of adsorption, which may describe the successful stabilization of noble metals enabling maintenance of small sized nanoparticles compared to other oxide supports. In addition to ITC, bulk uptake studies have aided in quantifying the amount of metal precursor adsorbed on the support surface and equilibrium isotherms describe the uptake behavior and may provide insight for predicting long term stability of the nanoparticles. In the second half of the talk, we discuss the adsorption of transition metal oxide and hydroxide nanoparticles in the galleries of of Nb-based perovskites. ITC was used to quantitatively rank the strength of adsorption between the metal nanoparticle and their propensity to sinter, as assessed by in-situ, high-temperature transmission electron microscopy. In both examples, we will emphasize this initial interaction at the solid-liquid interface is important and conveys a history effect to the catalyst that is evident during post-processing (drying, calcination and reduction). The estimated thermodynamic parameters are expected to quantify the type of bonding at the interface, shed light on the binding mechanism and the growth and sintering kinetics of supported catalysts.

Copyright © American Institute of Chemical Engineers. All rights reserved.

469846 Simulation and Experimental Studies of Dry Catalyst Impregnation for Improved Content Uniformity and Scale up

Abstract:

Dry catalyst impregnation of active metals onto a porous catalyst support is an important step in the preparation of heterogeneous catalyst. In a typical dry impregnation process, metal solutions are sprayed over a particulate bed in a mixing vessel until the pore volume is reached. The inter-particle variability of the impregnated liquids inside the particles and metal content may significantly affect the activity and selectivity of the resulting catalyst. Current scale-up practices lead to poor fluid distribution and inhomogeneity in metal content. The aim of this work is to understand the dynamic behavior of the particles under the spray nozzle, which is essential for desired content uniformity, and to develop a scale-up model for the dry impregnation process.

In this work, Discrete Element Method (DEM) simulations coupled with a novel algorithm allowing the transfer of metal solution to and between particles were used in combination with geometrically equivalent experiments to model dry catalyst impregnation. Metal adsorption is considered in the simulation for accurate metal/liquid transfer. Once a droplet of metal solution, containing the metal ions, comes in contact with a support particle, the droplet penetrates into the pores. The deposition of the metal ions takes place during the equilibration of the liquid phase inside the pores and solid surface of the pores. The adsorption mechanism can be modeled by Langmuir isotherm, which depicts a relationship between the extent of adsorption of the surface (or the fraction of the adsorption sites occupied) and the molar concentration of the solution. Solving Langmuir model coupled with the mass balance, the equilibrated absorbed metal and the concentration of metal in the liquid phase can be calculated. Simulation results are compared with experiments with identical setup, where nickel nitrate hexahydrate was dissolved in aqueous solution and used as the metal precursor. Experiments were conducted at a rotation speed of 9 rpm and a spray rate of 1.7 L/hr. Good agreement was observed in the metal content in particles along the axis of rotation.

Two dimensionless numbers are used to characterize the scale up of the system: Froude number (Fr) and Spray Flux number (Î¨). The Froude number is defined as the ratio of inertial force to gravitational force. The calculation of Fr gives the correlation between rotation speed (Ï‰) and vessel size (R). Spray Flux number measures the density of spray droplet in the spray zone. The location and area of spray zone can significantly affect the overall content uniformity. Results in the cylindrical vessel show that a homogeneous liquid distribution with optimal particle mixing is obtained by minimizing the spray rate and Î¨ for an optimal Fr number. It is found that keeping Fr number at the optimal value and lower spray rates increase the content uniformity of the resulting particles. The ability to control dry impregnations and establish effective models for large batches can significantly reduce the amount of time required per batch while simultaneously achieving a more homogeneous distribution of metal ions.

Copyright © American Institute of Chemical Engineers. All rights reserved.

470807 Reversible Ionic Surfactants for the Preparation of Monodisperse, Supported Nanoparticles

Abstract:

The preparation of monodisperse supported nanoparticles remains a non-trivial task. Organic surfactants are usually required to passivate the surface of nanoparticles to limit their growth to control the size of the particles. Calcination or annealing is typically required following the deposition of the particles onto a catalyst support to remove the surfactant. This calcination/annealing typically causes growth of the nanoparticle, destroying the monodispersity created during the synthesis. This undesired growth can additionally lead to a decline in the catalytic activity of the nanoparticles.

We present the use of Reversible Ionic Surfactants (RevIS) for the preparation of supported, monodisperse nanoparticles. RevIS are switchable molecules which display drastically different properties when an external physical stimulus is applied. The RevIS can be “turned-on” to template nanoparticles to aid in the synthesis and then “turned-off” to release a base nanoparticle for deposition onto a catalyst support. The RevIS contains an amine functionality which is capable of reacting with CO_2 to form the Ionic Surfact. The templating synthesis via a hypothesized reverse-micelle process allows for the creation of monodisperse nanoparticles controllable by varying the size of the RevIS or the continuous phase. The template is broken by "reversing" the RevIS by simply heating the solution to drive off the CO_2 in the presence of a catalyst support. The deposition of a bare nanoparticle allows for the formation of highly active supported nanoparticles that do not require calcination thus maintaining the monodispersity of the nanoparticles. Supported gold nanoparticles prepared using RevIS are more active than their traditionally prepared counter-parts while requiring less processing thus maintaining the size of the nanoparticle catalysts throughout the preparation.

Copyright © American Institute of Chemical Engineers. All rights reserved.

472347 Beneficial Influence of Surface Capping Agents on the Stability and Activity of Metal/Oxide Catalysts

Abstract:

Capping agents have long been used as chemical additives to control the size, morphology, and composition of nanoparticle catalysts during their synthesis. Strong interactions between surface atoms and capping agents can stabilize structures with greater abundance of particular facets and surface sites, which drastically improve reaction rates or selectivity. These catalysts often require some mild treatment or conditioning (extensive rinsing with solvent, UV-Ozone, or low temperature calcination) to free more active sites by removing surface capping agents from the surface without inducing dramatic surface reconstructions.

Alternatively, capping agents can also improve catalytic activity and selectivity by reducing surface coverage of reactive intermediates and preventing undesirable bimolecular reactions, which can lead to coking. Previously, our group has deposited different capping agents (thiols, silanes, phosphonates) to control the orientation of chemical species on active phase of different metal and metal oxide catalysts to drastically enhance the selectivity compared to uncoated catalysts. All these beneficial effects are contingent on the stability of capping agents, which can degrade over time.

In this work, we discuss recent results that have shown the deposition of capping agents on the inactive phase can also have a beneficial impact on the stability of metal nanoparticles deposited on oxide supports. Improved stability results from the modification of physical properties of the catalyst following the addition of different tail groups on capping agents. Coated Pd/Al_2O_3 catalysts were showed to be more stable in highly acidic solutions, under hydrothermal conditions, and at elevated temperatures under reaction conditions.

Copyright © American Institute of Chemical Engineers. All rights reserved.

AUTHOR INDEX

Abbasi, Pedram 73, 754
Abdala, Paula 778
Abdelrahman, Omar A. 891
Abdulhadi, Abdullah 108
Abiev, Rufat 248
Abild-Pedersen, Frank 523, 726, 910
Abolhasani, Milad 506
Abram, David N. 184
Abrokwah, Richard 716
Adjiman, Claire 439
Agarwal, Manish 454
Agarwal, Pratyush 635, 676
Ager, Joel W. 743, 903
Agrawal, Pradeep K. 816, 934
Aguirre-Soto, Alan 47
Ahmadi, Masoudeh 63
Ahmed, Sheikh F 240
Ahmed, Usama 175
Aho, Atte 722
Ailawar, Saurabh 219, 692
Akter, Nusnin 937, 948
Al Dahhan, Muthanna 87
Al Hinai, Abdullah 177
Alalouni, Mohammed R. 686
Alanazi, Yousef 503, 701, 732
Albarracin, Jonatan 298, 546
Alba-Rubio, Ana 364
Al-Bazzaz, Hamza 164
Alcorn, Brett 712
Al-Dahhan, Muthanna 157, 164
Aleksandrov, Hristiyan A. 881
Alexander, Anne-Marie C. 214
Alexander, Vineet 164
Alexeev, Oleg S. 881
Ali, Shaikh A. 686
Alimoradi, Sirwan 106
AlJama, Hassan 523, 726
Almalki, Fahad 490, 798
Almkhelfe, Haider 369, 561
Alonso, David Martin 398
Al-Otaibi, Bader 108
Alphones, Arokiaraj 243
Al-Rifai, Noor 422
AlSalihi, Sara 716
Alshafei, Faisal H. 686
Alvarez, Dra Mayra Garcia 606
Alvarez, Pedro 165
Alvarez, Pedro J. 127
Amama, Placidus B. 369, 561
Amiridis, Michael D. 881
Ammal, Salai C. 379, 876
Amundsen, Ted J. 712
An, Hang 854
An, Hongmei 138, 740
An, Wei 497
Anderst, Emily 400
Angelopoulos, Anastasios 869
Anggara, Trunojoyo 209, 631
Annesini, Maria Cristina 149
Ansah, Emmanuel 407
Ansari, Manizheh 759
Apaja, Vesa 287
Aponte, Yira 213
Aranda, Donato 708
Arbag, Huseyin 679
Arca, Serhat 100
Arce-Ramos, Juan Manuel 616
Ardo, Shane 569
Arellano-Garcia, Harvey 619
Argyle, Morris D. 578
Aristizábal, Valentina 623
Arnadottir, Liney 701, 732
Aroh, Kosi 482
Arturo, Steven G. 430
Artyushkova, Kateryna 71
Asadi, Mohammad 17, 73, 754
Asano, Shusaku 508
Asgari, Nazli 240
Asiaee, Alireza 227
Askar, Khalid 779
Aslam, Umar 694
Asthagiri, Aravind 545, 799
Atanassov, Plamen 71
Athariboroujeny, Motahare 921
Auerbach, Scott M. 888
Austin, Natalie 405
AuYeung, Nick 503
Avanesian, Talin 922
Ayandiran, Afees 845
Aymonier, Cyril 505
Azzam, Sara 316
Baah, David 365
Bababrik, Mohammed Reda 88
Back, Seoin 461
Badii, Khashayar 654
Bae, Joongmyeon 604
Bae, Minseok 604
Bagusetty, Abhishek 544
Bahrim, Cristian 731
Bai, Peng 788
Baiker, Alfons 885, 918
Bajdich, Michal 40, 336, 672, 910
Bakare, Idris 845
Baker, Jonathan G. 338
Balakotaiah, Vemuri 91, 143, 680
Balbuena, Perla B. 556, 565
Baldauf-Sommerbauer, Georg 452
Balsara, Nitash 344
Baltrusaitis, Jonas 322, 458
Banerjee, Arghya 392, 404

AUTHOR INDEX

Banerjee, Ritubarna 886, 930
Banerjee, Sanjoy 759
Banholzer, William F. 363
Banisharif, Farhad 747
Bao, Zhenan 72, 641
Barim, S. Bengisu 357
Barnett, Kevin J. 122, 363
Barrio, Laura 301
Barton, David J. 847
Basdogan, Yasemin 463
Basu, Suddhasatwa 454
Bates, Jason 494
Bauer, Tobias 247, 248
Baxter, Jason B. 792
Beatriz, Fidalgo 420
Becker, Henning 301
Beckham, Gregg T. 331, 889
Behr, Arno 170
Behravesh, Erfan 722
Beletskiy, Evgeny 880
Belhseine, Yasmeen 124
Bell, Alexis T. 119, 320, 344, 348, 349, 360, 429, 553, 743, 856, 899, 903
Beltran, Saul Perez 565
Benjamin, Kenneth M. 227
Benson, Shawn 106
Benson, Tracy J. 731
Bent, Stacey F. 338, 862, 916
Berkson, Zachariah 787
Bernasko, Dorothy 622
Bernazzani, Paul 731
Berson, R. Eric 332
Beuvier, Thomas 505
Bhattacharyya, Debasis 673
Bhavsar, Saurabh 719
Bhethanabotla, Venkat R. 735, 906
Bhethanabotla, Vignesh 735
Bhoorasingh, Pierre L. 428
Biancardi, Alessandro 549, 919
Biddinger, Elizabeth J. 234, 487
Biddy, Mary 889
Biener, Juergen 830
Biener, Monika 830
Biernacki, Joseph 94
Biernacki, Joseph J. 419
Billa, Triveni 635, 676
Binder, Andrew J. 931
Binous, Housam 845
Bishop, Cody 715
Bligaard, Thomas 378
Boateng, Akwasi A. 192
Bobba, Pallavi 215
Bodenschatz, Cameron 500
Boerigter, Calvin 694
Boes, Jacob R. 383, 646
Bollas, George M. 576
Bond, Jesse Q. 325, 395, 403, 891
Boningari, Thirupathi 488
Boolchand, Punit 869
Borghard, William G. 953
Borgna, Armando 121
Born, Dirk 98
Borne, Isaiah 107
Boscoboinik, Jorge A. 937, 948
Bossard, Peter 696
Bowers, Clifford R. 913
Boyd, Michael 603
Brandstädter, Willi 147
Bravo-Suarez, Juan J. 919
Brentzel, Zachary 122, 363
Brewer, Catherine E. 225, 868
Brezinsky, Kenneth 781
Briggs, Nicholas M. 907
Brinson, Bruce E. 416
Briquet, Ludovic 96
Broadbelt, Linda J. 430, 495
Brown, Avery 75
Brown, Robert 331
Browning, Barbara 165
Bruce, David A. 780
Brückner, Angelika 792
Bruijnincx, Pieter 120, 471
Brunelli, Nicholas 540, 721
Bryant, Kristin 954
Buesser, Beat 517
Bui, Hao 564
Bukhovko, Maxim P. 686
Bukowski, Brandon 494, 940
Bukur, Dragomir B. 557
Bullard, Lisa G. 114
Burgos, Juan C. 556
Burt, Samuel P. 363
Busygin, Igor 887
Butcher, Holly 696
Cai, Jianghuai 107
Cakiryilmaz, Nurbanu 679
Calvignac, Brice 505
Camaioni, Donald M. 125
Campbell, Charles T. 125, 400
Campos-Martin, Jose M. 747
Candelaria, Stephanie 70, 374
Candia, Sandra Varinia Bernales 221
Caniaz, Ramazan Oguz 100
Cao, Chenxi 643
Cao, Enhong 422
Cao, Fei 231
Cao, Guoqiang 879
Cao, Xiaoyan 315
Caratzoulas, Stavros 199

AUTHOR INDEX

Cárdenas-Galindo, María-Guadalupe 616
Cardona, Carlos A. 623
Cardona-Martinez, Nelson 370, 396
Cargnello, Matteo 453, 511, 512, 535, 656, 792, 831, 884, 944
Caricato, Marco 549, 919
Carneiro, Juliana S. A. 335
Carpenter, Andrew 712
Carr, Shane F. 385
Carrero, Carlos 844, 908
Carter, Emily A. 372
Cashman, Derek 94
Caspi, Daniel D. 480
Castaldi, Marco 796
Castaldi, Marco J. 781, 938
Castro, Angel 685
Cattaneo, Stefano 422
Cave, Etosha R. 184
Celik, Fuat E. 288, 807, 941
Celik, Gokhan 214, 219, 692
Cerbelli, Stefano 149
Chada, Joseph P. 844, 908
Chadderdon, David 7, 286, 289, 648
Chadderdon, Xiaotong 286, 648
Chaikittisilp, Watcharop 790
Chakrabarti, Anisha 920
Chakrabarti, Debanjan 441
Chakthranont, Pongkarn 633, 672, 702, 729
Chamas, Ali 895
Chan, Karen 460, 464, 465, 547
Chan, Maria K. Y. 384
Chan, Xiaojun 220, 711
Chandler, Bert D. 892
Chang, Chun-Chih 541
Chang, Seong-Cheol 663
Chang, SiWei A. 802
Changi, Shujauddin M. 479
Chatterton, Lindsey 366
Chatzidimitriou, Anargyros 325
Chaudhari, Raghunath V. 215, 222, 327, 469
Chavez, Steven 200, 683, 693
Che, Fanglin 514, 533
Chen, Benjamin W J 801
Chen, Binghui 34
Chen, Daniel 243
Chen, Elaine 92
Chen, Han 671
Chen, Hao 715
Chen, Jingguang G. 201, 805
Chen, Leanne D. 29
Chen, Li 416
Chen, Shucheng 72, 641
Chen, Wan-Ting 105
Chen, Xiaowen 333
Chen, Xiaoyin 653
Chen, Xin 854
Chen, Yizheng 742
Chen, Yuan 859
Chen, Zhihua 72
Chen, Zhu 373
Chen, Zihua 641
Cheng, Feng 225, 868
Cheng, Xun 486
Cheng, Yan 520, 522
Cheng, Yi 520, 522, 643, 854
Cheng, Yisun 203, 627
Cheng, Yongqiang 491
Cheng, Zhuo 795
Chi, Yong 681
Chimenton, Ricardo José 606
Chizallet, Céline 96
Chmelka, Bradley F. 787
Cho, Hong Je 541
Choi, Eun-Kyung 583
Choi, Jae-Soon 300
Choi, Juhee 799
Choi, Ki-Hyouk 108
Choi, Yongsuck 192
Choksi, Tej S. 380, 639, 925
Choong, Catherine 121
Chorkendorff, Ib 901
Choudhury, Nujhat 238
Choudhury, Pabitra 783
Chowdhury, Sanchari 733
Christopher, Phillip 468, 803, 832, 915, 922
Chu, Bozhao 854
Chubukov, Boris 427
Chun, Dong Hyun 649, 658
Chun, Hee-Joon 287
Chung, Wook-Jin 661
Chyou, Yau-Pin 669
Ciora, Richard J. 730, 736
Clark, Ezra 553
Clark, Ezra L. 743, 903
Clary, Jacob M. 567
Clayborne, Andre 287
Clemens, Bruce M. 734
Cline, Brandon 867
Co, Anne 285, 703
Coan, Patrick 911
Coley, Connor 506
Collinge, Greg 579, 921
Colón, Jorge L. 638
Conrad, Sabrina 350
Contreras-Mora, Jose 123, 930
Copéret, Christophe 778
Coppens, Marc-Olivier 513, 651
Coppernoll, Aaron 588
Cordon, Michael J. 539

AUTHOR INDEX

Costa, Karina K. 158
Cournoyer, Justin T. 879
Cowen, Kenneth 301
Cox, David F. 671
Crossley, Steven 89, 233, 431, 907
Crow, Allison M. 862
Crowley, Stephen 938
Cui, Kai 330
Cui, Yanran 917
Cui, Zheng 868
Currier, Neal 138, 297, 740
Cutlip, Michael B. 111
Cybulskis, Viktor 917, 940
Dade, William 716
Daemen, Luke L. 491
Dagnall, Katelyn A. 288
Dai, Ning 131
Dalai, Eswar Prasad 673
Damodara, Vijaya 243
Damouny, Christine Warwar 531
Dana, Alon Grinberg 517
Danby, Andrew 469
Dang, Dan 643
Dang, Thanh 400
Das, Anirban 335
Das, Biswapriya 673
Das, Satyen Kumar 673
Dasgupta, Anish 834
Dashliborun, Amir Motamed 244
Dasireddy, Venkata 442
Datta, Ravindra 75
Datye, Abhaya 217, 315, 936
Datye, Abhaya K. 229
Dauenhauer, Paul J. 390, 393, 541
Davis, Andrew W. 796
Davis, Kirsten 331
Davis, Mark E. 538
Davis, Robert J. 216, 314, 323, 936
Daw, C. Stuart 415
Daza, Yolanda 906
De Jong, Ed 471
De Jong, Johannes 524
De Lasa, Hugo 213, 793
De Vasconcelos, Bruna Rego 528
De Wilde, Juray 796
Dean, Anthony M. 519
Dean, James 463
Debefve, Louise 553, 828, 887
deGrazia, Janet 110, 117
Dehghani, Mohammad Reza 747
Deka, Dhruba Jyoti 285, 703
Dekui, Shen 420
Delgass, W. Nicholas 298, 546, 835, 917, 945
Demir, Bengi 398
DeRita, Leo 832
Deshlahra, Prashant 857, 865
Deshmane, Vishwanath 709, 716
Deshmukh, Soumitra 301
Deshpande, Nitish 540, 721
Deskins, N. Aaron 75
Devocht, Brigitte R. 432
Dharanipragada, Aditya N.V.R. 524
Di Iorio, John R. 298, 546, 789, 835
Diamanti, Aikaterini 439
Diao, Weijian 197
Díaz, Angélica Liliana Coconubo 949
Diaz, Rosa 218
Dickens, Colin 462, 466, 547, 727
Dinh, Kimberly 776
Diulus, John 799
Dixit, Jagdev Kumar 673
Dixon, Anthony G. 92, 340, 667
Dixon, David A. 828, 887
Doan-Nguyen, Vicky 792
Dogu, Doruk 285, 703
Dogu, Gulsen 679
Dogu, Timur 679
Dominguez, Bernardo Castro 340
Dong, Jun 681
Doren, Douglas J. 839
Dorfi, Anna 630
Doyle, Andrew 336, 462, 466, 727
Drochner, Alfons 98
Du, Zhou 842
Duan, Pu 229, 315
Dubois, Jean-Luc 190
Ducousso, Marion 681
Dumesic, James 231, 315, 362, 364, 394, 398
Dumesic, James A. 122, 363
Dunwell, Marco 185
Duong, Nhung 861
Durgadevi, A. 668
Duyar, Melis S. 484, 620
Eastman, Stephen 281
Edake, Mahesh 190
Edgar, James H. 93
Edmiston, Paul 219, 692
Egle, Tobias 830
Elkasabi, Yaseen 192
Ellis, Lucas 858
Elly, Michael 111
Elnabawy, Ahmed 510
Elsayed, Nada 713, 777
Elwell, Anthony 713
Embry, Matthew C. 479
Entesari, Nazanin 341
Epling, William 290, 677
Epling, William S. 292, 689
Eppinger, Thomas 570, 642
Eränen, Kari 722

AUTHOR INDEX

Erdem, Ezgi ... 357
Erkey, Can ... 100, 357
Ertler, Daniel ... 887
Esche, Erik ... 688
Escudero-Escribano, Maria ... 33, 702, 901
Esposito, Daniel V. ... 376, 630
Eswaranandam, Ees ... 347
Evans, Scott ... 243
Ewing, Christopher S. ... 544
Ezbiri, Miriam ... 905
Falcone, Derek D. ... 216, 936
Falconer, John ... 117
Falconer, John L. ... 567
Fan, Feiyue ... 874
Fan, Jonathan A. ... 795
Fan, Liang-Shih ... 795
Fan, Wei ... 541
Fang, Yuxin ... 486
Fanzhong, Ma ... 659
Farberow, Carrie A. ... 860
Farouk, Tanvir I ... 240
Fatemi, Naghmeh ... 432
Fazlollahi, Farhad ... 132
Feaster, Jeremy T. ... 184
Felischak, Matthias ... 145, 849
Feng, Lian-Fang ... 592
Ferdousi, Shammi Akter ... 756, 757, 758
Ferguson, Glen Allen ... 889
Fester, Jakob ... 910
Fisher, Galen B. ... 295
Flaherty, David W. ... 321, 687, 802, 935
Flake, John C. ... 486
Fleischman, Samuel D. ... 708
Fletcher, Thomas ... 132
Flytzani-Stephanopoulos, Maria ... 775, 829, 830
Fogler, H. Scott ... 109
Fornasiero, Paolo ... 792
Fornero, Esteban L. ... 942
Fortner, Edward ... 243
Foston, Marcus ... 895
Fredrickson, Kurt ... 552, 744
Freeman, Emily ... 625
Freund, Hannsjörg ... 95, 144, 148
Friend, Cynthia M. ... 406, 830, 864
Fry, Andrew ... 239
Ftouni, Jamal ... 120
Fu, Xiao-an ... 742
Fuentes, Gustavo A. ... 755
Fuller, Anshuman ... 285
Fulton, John L. ... 125
Funkenbusch, LiLu ... 4
Furuya, Eiji ... 612
Fushimi, Rebecca ... 566
Gagliardi, Laura ... 221
Gaillard, Nicolas ... 728
Galindo, Amparo ... 439
Gallagher, James ... 218, 940
Gallo, Alessandro ... 898
Galloway, Benjamin ... 212
Gallup, Jennifer ... 202
Galvita, Vladimir ... 524, 942
Gamliel, David P. ... 576
Gan, Zhuoran ... 485
Ganase, Zara ... 439
Gangwal, Santosh ... 867
Ganzer, Gunnar ... 148
Gao, Feng ... 209, 291, 631, 840, 859
Gao, Hanyu ... 430
García-Melchor, Max ... 672, 910
Garrabos, Yves ... 505
Garrick, Taylor R. ... 197
Garshasbi, Ashkan ... 730
Gaspard, Pierre ... 498
Gates, Bruce C. ... 221, 553, 828, 887
Gauthier, Joseph ... 462
Gavriilidis, Asterios ... 422
Gazit, Oz M. ... 531
Geantet, Christophe ... 165
Gellman, A.J. ... 426
Gellman, Andrew J. ... 804
Genuino, Homer ... 39, 120, 471
Gerceker, Duygu ... 394
Getman, Rachel ... 500, 866
Ghale, Kushal ... 400
Ghoniem, Ahmed F. ... 90
Ghoroi, Chinmay ... 562
Gibaud, Alain ... 505
Gibbons, Brenna ... 734
Giles, Stephen A. ... 199
Gill, Rajinder ... 796
Giordano, Nicolás A. Grosso ... 542, 837
Glasser, Benjamin ... 715
Glasser, David ... 609, 666, 682
Gleaves, John T. ... 566
Godiksen, Anita ... 897
Godini, Hamid ... 688
Goldsmith, Bryan R ... 5, 78
Goldsmith, C. Franklin ... 241, 354, 518
Golkarnarenji, Gelayol ... 654
Göltl, Florian ... 350, 536, 546
Gómez, Alvaro ... 623
Gómez, Sergio A. ... 755
Gomez-Ballesteros, Jose L. ... 556
Gong, Jian ... 297, 740
Goodman, Emmett ... 535, 884
Goodpaster, Jason ... 349, 429
Gordon, Thomas ... 792
Gorimbo, Joshua ... 682
Gorte, Raymond J. ... 493, 541
Goulas, Konstantinos ... 230, 493, 896

AUTHOR INDEX

Gould, Nicholas 543, 893
Gounaris, Chrysanthos E. 926
Gounder, Rajamani 298, 494, 539, 546, 789, 835
Govedarica, Zora 366
Goyal, Amit 366
Grabow, Lars C. 616
Grady, Brian P. 369
Grambow, Colin 440
Grant, Eliana 439
Grass, Robert N. 593, 808
Gray, Jake T 533
Gray, Michel 470
Gray, Stephen 384
Greeley, Jeffrey P. 287, 380, 494, 639, 925, 940, 945
Green, William H. 107, 440, 517, 521
Greenlee, Lauren F. 70, 374
Gregor, D. 642
Griffin, Michael B. 889
Grilc, Miha 472
Grillo, Fabio 564
Groehn, Arto J. 427
Groenenboom, Mitchell 353, 383
Grolman, Eric 425
Grossardt, Jason 369
Gu, Tian 143
Gu, Tonghan 502
Gu, Xiang-Kui 335
Gu, Xue-Ping 592
Guangsheng, Luo 591
Gudiyella, Soumya 107
Guerra, Patricia 223
Guliants, Vadim 697
Gumidyala, Abhishek 233, 431
Gümüslü, Gamze 221, 426
Gunasooriya, G. T. Kasun Kalhara 392
Gunduz, Seval 219, 285, 692, 703
Gunugunuri, Krishna Reddy 714
Guo, Mengqing 795
Guo, Mond 470
Guo, Zhe 74
Gupton, Frank 929
Gutiérrez, Oliver 125
Ha, Su 514, 533
Haase, Stefan 247, 248
Hable, Robert 106
Hack, John H. 216
Hagelin-Weaver, Helena 913
Hahn, Christopher 184, 633, 652, 657
Haider, M. Ali 454
Hajari, Amit 281
Haller, Gary L. 137
Ham, Hyung Chul 663
Hamel, Christof 424, 849
Hamidipour, Mohsen 244
Hammouti, Abdelkader 96
Han, Chonghun 175, 176
Han, Jeong Woo 207, 737, 739
Han, Jeongsik 670
Han, Jonghee 663
Han, Kehang 521
Han, Wenfeng 308
Handy, Brent E. 616
Hanna, Brian S. 686
Hanselman, Christopher L. 926
Hansen, Charles 719
Hanspal, Sabra 323
Hao, Pengxiao 863
Harden, Tiffany 487
Hare, Bryan J. 906
Harker, Audrey 553
Harold, Michael 627
Harold, Michael P 128, 203, 292, 680, 689
Harpool, Scott 503
Harris, James W. 218, 539
Hartman, Ryan L. 159, 477
Hasan, Mohammad Mehedi 756, 757, 758
Hashimoto, Akito 228
Hatsukade, Toru 184
Hatton, T. Alan 502
Hattrick-Simpers, Jason 456, 602, 904, 933
Haudek, Kevin 115
Hayek, Naseem S. 531, 613
Hazlett, Melanie 677
He, Jiayue 231, 363
He, Peng 537
He, Ping 90
He, Xiang 485
He, Yang 720
Head-Gordon, Martin 349, 429
Headrick, Sierra 783
Heagy, Michael D. 733
Hecker, William C 578
Hellgardt, Klaus 104, 423, 622
Hellstern, Thomas R. 657, 728
Helveg, Stig 899
Hendon, Christopher H. 364, 902
Henkelman, Graeme 509
Hensley, Alyssa 382, 400, 498, 875
Hermans, Ive 350, 363, 536, 844, 908
Hernández-Salgado, Gabriela 755
Hernández-Terán, M. Eugenia 755
Hernley, Paul 200, 683, 693
Héroguel, Florent 563
Herrera, Carolina 868
Herrera, Valeria 907
Herrmann, Stanley 139
Heyden, Andreas 379, 403, 876
Heydt, Thomas 98
Hiebel, Fanny 864
Higgins, Christopher 332

AUTHOR INDEX

Higgins, Drew 42
Hii, Mimi 423
Hikita, Yasuyuki 466, 727
Hildebrandt, Diane 609, 666, 682
Hillmyer, Marc A. 390
Hirose, Takuya 345
Hischier, Illias 427
Hoard, John 295
Hocke, Elisabeth 455
Hodoshima, Shinya 846
Hofer, Corinne 808
Hoffman, Adam 828
Hoffman, Graham 868
Hohn, Keith L. 369, 561
Holles, Joseph 939
Hong, Giyoung 598, 599
Hong, Yongchun 382
Honkala, Karoliina 287
Hook, Alec 941
Horton, Scott R. 418, 635, 676
Hossain, Mohammad Mozahar 845
Hou, Zhen 676
Hoyt, David W. 895
Hu, Bo 816, 934
Hu, Chuntian 477
Huang, Benjamin 314, 936
Huang, Ching-Ying 669
Huang, Kefeng 122, 231, 363
Huang, Qilong 554
Huang, Qunxing 681
Hubbard, Carolyn 203, 627
Huber, George W. 122, 231, 363, 844, 908
Hungerbuehler, Konrad 885, 918
Hunt, Sean T. 364, 902
Huo, Chao 308
Huo, Jiajie 229
Hur, Young Gul 662, 789
Huš, Matej 442
Hutchings, Graham J 422
Huynh, Dang Chinh 450
Hwang, Harold Y. 466, 727
Hwang, Sang Youp 208
Hwang, Taeho 78
Ibrahim, Gasim 954
Iglesia, Enrique 139, 311, 784, 836, 857, 865
Iisa, Kristiina 194
Ilias, Shamsuddin 709
Illner, Markus 76
Imtiaz, Qasim 778
Inturi, Siva Nagi Reddy 488
Ioannidis, Efthymios 381
Ipek, Bahar 536
Isaksen, Oliver 897
Ishikawa, Tatsuya 612
Itoh, Naotsugu 345, 600, 601, 738
Izadpanah, Arash 104
Jaén, Juan J. Delgado 792
Jamal, Adeel 440
Jan, Oliver 368
Janet, Jon Paul 381
Jang, Seung Soon 501
Janik, Michael 198, 947
Janik, Michael J. 877, 892
Janmanchi, Krishna 588
Jansen, Tim 165
Jantasee, Sasiradee 346
Janvelyan, Nare 830
Jaramillo, Thomas F. 72, 184, 466, 484, 620, 633, 638, 641, 652, 657, 672, 702, 708, 725, 727, 728, 729, 734, 898
Jarboe, Laura 331
Jarvis, Jacqueline 868
Jeffrey, Miller 218
Jehng, Jih-Mirn 458
Jenness, Glen 805
Jenness, Glen R. 11, 118, 230
Jensen, Klavs F. 482, 483, 504, 506
Jentoft, Friederike C. 367
Jeon, Jong Ki 670
Jeong, Byung-Hun 670
Jeong, Gwangsik 662
Jeong, Heondo 649, 658
Jeong, Kwan-Hyoung 670
Jeong, Soon Kwan 583
Ji, Jingjing 624
Jia, Hongfei 714
Jiajun, Zhang 420
Jiang, Chengjun 818
Jiang, Jimeng 224
Jimenez, Juan 456, 904
Jin, Bo 310
Jin, Xin 215, 327
Jocz, Jennifer 943
Johnson, J. Karl 928
Johnson, Karl 544
Johnson, Martin D. 479
Johnson, Matthew 243
Johnson, Robert 315
Johnson, Robert L. 229, 326
Jokiniemi, Jorma 606
Jones, Christopher W. 501, 816, 934
Jongerius, Anna L. 184
Jörke, Andreas 424
Joseph, Babu 713, 777
Joseph, Kristeen Esther 393
Josephson, Alexander J 237
Joshi, Saurabh Y. 138
Jovanovic, Goran 503, 701, 732
Jung, Sungyup 234
Jung, WooChul 737, 739

AUTHOR INDEX

Jung, Yousung 461
Jurtz, Nico 570, 642
Jystad, Amy 549, 919
Kageyama, Naoki 432
Kaiser, Markus....95
Kalaga, Dinesh V....759
Kalamaras, Christos 677
Kaldor, Andy 696
Kale, Matthew 41, 803
Kaliaguine, Serge 634
Kamasamudram, Krishna....138, 740
Kamaz, Mohanad....347
Kampfe, Sara....301
Kang, Joo 847
Kang, Sung Bong....294
Kang, Sungbong 677
Kanitkar, Swarom 611
Kapil, Nidhi 513
Karagoz, Secgin....343
Karaiskakis, Alexandros 487
Karakalos, Stavros....864
Karamad, Mohammadreza....465
Karanjkar, Pranav U....231, 363
Karhunen, Tommi 606
Karim, Ayman M....878, 896
Karroum, Hafsa 571
Katikaneni, Sai P....604
Kato, Takafumi 738
Katz, Alexander 542, 553, 837, 887
Kawakita, Takashi 612
Kawi, Sibudjing....121
Kaxiras, Efthimios 352, 406
Keels, Jayson 779
Keith, John A....353, 383, 463
Kelleher, Patrick 137
Key, Haley 389
Khan, Saif A. 478, 502
Khayyam, Hamid....654
Khinast, Johannes G. 715
Khivantsev, Konstantin V. 798, 881
Khokhar, Munir D....686
Khorasgani, Elmira Soghrati....121
Khosravian, Homa 621
Khoury, Christine 531
Khurana, Ishant....298, 546, 835
Kibsgaard, Jakob....484, 620, 633, 657, 708, 725
Kida, Tetsuya 228, 718
Kiely, Christopher 120
Kienberger, Marlene....346
Kikuchi, Shuhei....345, 600
Kilos, Beata A....847
Kim, Chan Hun....662
Kim, Changsoo....176
Kim, Do Heui....220
Kim, Dongyeon....604
Kim, Doo-Wook....659
Kim, Gookhee....605
Kim, Hak Joo....583
Kim, Hyunuk....645
Kim, Ji Man....698
Kim, Jong Jin....670
Kim, Ki Chul....501
Kim, Ki-Duk....745, 746
Kim, Kiseok....698
Kim, Kyeongsu....176
Kim, Kyeounghak....739
Kim, Mi Jin....688
Kim, Minkyu....545
Kim, Sung Min....778
Kim, Sung-Ho....745, 746
Kim, Taeho Roy....641
Kim, Taejin....220, 711, 853, 937, 948
Kim, YongWoo....220
Kimmel, Yannick C....201
Kinaci, Alper....384
Kirk, Charlotte....466, 547, 672, 727
Kistler, J. D....887
Kitchin, John R....383, 646
Kizilel, Riza....357
Klein, Michael T. 418, 635, 676
Klyushin, Alexander....216
Knop-Gericke, Axel....216, 797
Koberstein, Jeffrey T....376
Koel, Bruce....373
Koel, Bruce E....371
Kohara, Kento....601
Kommoss, Bjoern....455
Kondratyuk, Petro....426
Konstantinov, Ivan....430
Koo, Kee Young....649, 658
Koob, Brittany....867
Koretsky, Milo D....116
Koriakin, Anton....659
Kostetskyy, Pavlo....496
Kovarik, Libor....124, 859, 898
Kramer, Nathaniel....285
Kraume, Matthias....570, 642
Krause, Jonas....590
Krishna, Siddarth H....231, 363
Kronawitter, Coleman....371, 373
Kropf, A. Jeremy....940
Krumm, Christoph....393
Kruse, Norbert....573, 579, 921
Kuboyama, Toshiharu....508
Kuhl, Kendra P....184
Kuhn, John....735, 777
Kuhn, John N....713, 906
Kuhn, Simon....169
Kuila, Debasish....716
Kulik, Heather J....381

AUTHOR INDEX

Kulkarni, Ambarish R. 523, 726, 927
Kumar, Anand 21, 636
Kumar, Ashok 138
Kumar, Gaurav 892, 947
Kumar, Narendra 722
Kumar, Nitin 611
Kumar, Prashant 896
Kumar, Rajeev 468
Kumari, Neetu 454
Kuncharam, Bhanu Vardhan Reddy 92
Kung, Harold H. 142, 880, 909
Kung, Mayfair 880
Kunsa, Tyler 712
Kuo, Chih-Ting 105
Kuroki, Hidenori 650
Kurtoglu, Samira Fatma 674
Kurtz, Richard L. 491
Kwon, Hyunguk 737
Kwon, Kyung 365
Kwon, Stephanie 857
Kwon, Youngkook 743
Kyriakidou, Eleni A. 300, 931
Labrador, Natalie 376
Lacroix, Maxime 165
Lähde, Anna 606
Lai, Lawrence 107
Laing, Paul 203, 627
Lambrecht, Daniel S. 544
Lance, Michael J. 931
Landers, Alan 657
Landers, John 807
Landon, Mark 245
Lang, Wendy 203, 627
Lange, Ruediger 247, 248, 282
Larachi, Faical 244
Latimer, Allegra A. 523, 726, 927
Lauritsen, Jeppe 910
Laursen, Siris 377, 720, 924
Lauterbach, Jochen 206, 456, 575, 602, 625, 851, 904, 933
Lawagon, Chosel P. 661
Le Brun, Niccolo 104
Le Monnier, Benjamin 563
Le, Thuy T. 489, 644
Lecoutre, Carole 505
Le-Doux, Travis 225, 868
Lee, Chang-Ha 183, 659
Lee, Chang-Whan 663
Lee, Choul-Ho 670
Lee, JaeHa 220
Lee, Jae-Hwan 660
Lee, Joo-Hyeong 108
Lee, Joo-Youp 133
Lee, Kwan-Young 583, 660, 662
Lee, Siwon 739
Lee, Sunggyu 647
Lee, Wonyoung 737
Lee, Woo-Sung 183
Lee, Yong-Kul 655, 745, 746
Lee, Youn-Woo 598, 599
Lei, Yu 624, 914
Leonard, Brian 196
Leonard, Kevin C. 375
Leonarduzzi, Daniele 301
Lercher, Johannes A. 125, 492
Leung, Andrew 423
Lewis, Connor A. 678
Lewis, Jennifer 894
Li, Can 135
Li, Dongmei (Katie) 196, 566
Li, Fanxing 855
Li, Guo-Ling 551
Li, Hang 830
Li, Hui 209, 209, 351, 546, 631, 631, 835
Li, Jian-Feng 34
Li, Junhui 297, 740
Li, Lin 306
Li, Meifang 134
Li, Mengmeng 292, 680, 689
Li, Mengwei 775
Li, Mingxiao 742
Li, Qi 742
Li, Rui 675
Li, Sha 801
Li, Shiwen 942
LI, Sichi 840
Li, Tianyang 520, 522
Li, Wei 134
Li, Wenzhen 286, 289, 648
Li, Xianchang 243
Li, Xiang 626
Li, Xinxin 376
Li, Xu 561
Li, Yanxin 193
Li, Yawei 334
Li, Yongdan 330
Li, Zheng 284, 707
Li, Zhongtang 341
Liang, Xinhua 818
Liauw, Marcel A. 391
Licht, Rachel B. 856
Licht, William 310
Lien, Chih-Heng 947
Liesche, Georg S. 97
Likozar, Blaž 442, 472
Lim, Chern-Hooi 459
Lima, Fabiana Dos Santos 741
Lin, Hongfei 126
Lin, Hongkun 506
Lin, John 652

AUTHOR INDEX

Lin, Pin Ann 556
Lindsay, Mark J. 578
Ling, Chen 714
Linic, Suljo....200, 683, 693, 694, 900
Lira, Carl T. 115
Liu, Bin 93, 286
Liu, Dongxia 841
Liu, Dupeng 222
Liu, Huazhang....308
Liu, Jilei 829, 830
Liu, Jingyue 880
Liu, Jiuxu....226
Liu, Paul K.T. 730
Liu, Qiuli....192, 235, 817
Liu, Rui 92
Liu, Song....93
Liu, Xiaoyang 497
Liu, Xinying....682
Liu, Xue 715
Liu, Yukun....376
Liu, Zhouyang 133
Llorca, Jordi 606
Lo, Cynthia S. 385
Loane, Letícia Pezzi Martins....173
Lobaccaro, Peter 743
Lobo, Raul F. 195, 393, 536, 839, 848
Lonergan, William W. 309
López-Curiel, Julio C. 755
Lou, Helen 243
Louie, Ying Lin 119
Love, Alyssa....350
Lowd, Jack....736
Lu, Li....120
Lu, Xiaojun....666, 682
Lu, Yubing....878
Lu, Zheng....624, 914
Lucero, Andrew 867
Lucio-Vega, Juan 418, 635, 676
Lugo, Michael....796
Lui, Alison 107
Lum, Yanwei 743, 903
Lundgaard, Keld T 378
Luo, Guangsheng 507, 587
Luo, Jinyong....138
Luo, Lin 847
Luss, Dan 128
Luterbacher, Jeremy S. 324, 397, 563
Lux, Susanne 452
Lwin, Soe 850
Lyu, Yimeng 653
Ma, Rui 340
Ma, Xianfeng....878, 923
Ma, Yi Hua 340
Maag, Alex....699, 700
Maalouf, Joseph 831, 944
Mackus, A.J.M. 338
Madix, Robert J. 406, 864
Mae, Kazuhiro 508
Magrini, Kim 194
Mahboub, Mohammad Jaber Darabi 190
Maiti, Debtanu....735, 906
Maiti, Sanat Chandra 562
Maiti, Swarup 838
Majumdar, Paulami 925
Maki, Taisuke 508
Malamis, Sam 292, 689
Malefyt, Amanda P. 113
Malek, Andre....847
Mallick, Kwonit....868
Mamun, Osman 403
Mandic, Milos 557
Manoli, Kyriakos....79
Manousiouthakis, Vasilios 343, 736
Mao, Liaoyuan....193
Maravelias, Christos T. 122, 231, 363
Mardilovich, Ivan 340
Margossian, Tigran 778
Marin, Guy B. 432, 524
Marre, Samuel 505
Marrero-Vazquez, Michelle M. 370, 396
Martin, Christopher B. 243
Martinez-Hernandez, Angel 755
Masiello, Caroline A. 127
Matanovic, Ivana....71
Matin, Md. Abdul....636
Matsubu, John C. 915
Maurin, Jérémy 165
Mavrikakis, Manos....394, 401, 510, 585, 801
Mazal, Tobias....230
Mazur, Zachary 105
McCabe, Kevin 867
McCrea, Sean....913
McCrum, Ian T. 198
McDanel, Katherine 117
McEnaney, Joshua 725
McEwen, Jean-Sabin....209, 382, 400, 498, 514, 533, 579, 631, 875, 921
McKay, Ian 792
McKee, William C. 339
Medina, Francesc 606
Medlin, J. Will 117, 858, 863, 911, 947, 955
Medlin, Michelle....117
Meemken, Fabian 885, 918
Mehar, Vikram....799
Mehdad, Ali 848
Mehrabadi, Bahareh Alsadat Tavakoli 197
Mehta, Prateek 945
Mei, Donghai....492
Meintel, Kathryn 481
Mellmer, Max A. 398

AUTHOR INDEX

Mendes, Laiza....708
Merchant, Shamel S....517
Merrill, John....115
Merz, Juliane....590
Meyer, Katja E. Binkley....285, 703
Miao, Yu....503
Michalsky, Ronald....794, 905
Michel, Carine....536
Mikkola, Jyri-Pekka....722
Milina, Maria....364, 902
Miller, Jeffrey....835, 940
Miller, Jeffrey T....298
Miller, Juanita....225
Miller, Victor V....656
Minette, Florent....796
Mingle, Kathleen....933
Minh, Doan Pham....528
Mironenko, Alexander V....118, 230, 493
Mitani, Makoto....729
Mittal, Neha....661
Miyake, Garret....459
Modesto-Lopez, Luis balam....606
Modroukas, Dean....796
Mohammad, Abdul Salam....94
Moher, Eric D....479
Mondal, Chandona....760
Mondal, Dilip Kumar....760
Moniri, Saman....900
Monnier, J.R....123
Monnier, John....490
Monnier, John R....197, 779
Monsalve, Alexander Guzmán....949
Montemore, Matthew M....10, 352, 406
Montgomery, Stephen....717
Montini, Tiziano....792
Montoya, Joseph H....402, 460, 466, 727
Moon, Jong-Ho....645
More, Amey....719
Mori, Milton....158
Moschetta, Eric G....501
Moskowitz, Benjamin....853
Mossin, Susanne....897
Motagamwala, Ali Hussain....398
Mote, Dhananjay R....388
Moteki, Takahiko....321
Moyo, Mahluli....617
Mpourmpakis, Giannis....405, 496
Mueller, Christoph....778
Mukai, Shin R....586, 791
Mukarakate, Calvin....194
Mukhopadhyay, Ahana....952
Muleja, Adolph....609
Mullins, Michael....4
Mungma, Nuttakul....346
Muraoka, Koki....790
Murmura, Maria Anna....149
Murphy, Brian....946
Murray, Christopher B....493, 792
Murugesan, Vijayakumar....470
Murzin, Dmitry Yu....722
Musgrave, Charles B....459, 567
Mushrif, Samir H....404, 450, 800
Mustain, William E....678
Muzika, František....610
Naccarella, Ian....944
Naebe, Minoo....654
Nakajima, Hikaru....479
Nakayama, Ryutaro....508
Nakhla, George....79
Nam, In-Sik....294
Nam, Sung Bang....294
Nan, Yao....373
Nan, Yue....226
Nan, Zhipeng....869
Nandiwale, Kakasaheb....469
Naoki, Takagiwa....601
Narayanan, Badri....384
Narkeviciute, Ieva....638
Narsimhan, Karthik....776
Nash, Connor....860
Neal, Luke....14, 855
Neimark, Alexander V....807
Neurock, Matthew....319, 788
Ng, Desmond....672
Nguyen, Anh Vu....450
Ni, Mingjiang....681
Nigra, Michael M....513, 651
Nikacevic, Nikola....557
Nikolic, Daliborka....145
Nikolla, Eranda....335
Nimlos, Mark....194
Nirmalakhandan, Nagamany....868
Nishio, Kazunori....466, 727
Nisola, Grace M....661
Nitopi, Stephanie....633
Noh, Kyung-Jong....207
Nordlund, Dennis....641, 898
Nørskov, Jens....72, 465, 512, 551
Norskov, Jens K....378, 462, 523, 547, 641, 726, 927
Nørskov, Jens K....402, 460, 466, 727
Northrup, Scott....94
Notestein, Justin M....136
Nwabara, Uzoma....653
Nzihou, Ange....528, 681
O'Connor, Nolan....877
Ofosu, Thomas....919
Ogasawara, Hirohito....898
Ogino, Isao....586, 791
Oh, Hyun-Taek....183
Oh, Jiwoo....604

AUTHOR INDEX

Oh, Se H. 294
Oh, Tae-Sik 912
Okolie, Chukwuemeka 124
Okrut, Alexander 542, 837, 887
Oktar, Nuray 679
Okubo, Tatsuya 790
Olsson, Louise 299
Omasta, Travis J. 678
Omotoso, Taiwo 907
Onal, Isik 548
Orazov, Marat 538
Ortiz-Arroyo, Arturo 685
Otto, Trenton 784
Ouyang, Xiaoying 837
Oyama, S. Ted 432
Oyarzún, Bernardo 99
Ozkan, Umit S. 214, 219, 285, 692, 703
Padak, Bihter 212, 238, 240
Padhye, Richa 66
Pagan-Torres, Yomaira J. 370, 396
Paleg, Sarah W. 186
Palermo, Andrew 553, 887
Palm, David W. 728
Palumbo, Aaron W. 427
Pan, Hanqing 733
Pan, Li 799
Pan, Yung-Tin 687
Panthani, Matthew G. 289
Paolucci, Christopher 60, 209, 298, 351, 546, 631, 835
Parekh, Atish A. 209, 298, 835
Park, Dae Sung 393
Park, Hai Woong 605
Park, Han-Byul 655
Park, Ki Tae 583
Park, Kuan Hwan 598
Park, Young Cheol 645
Parks, James E. 300, 931
Parsley, Doug 730
Partopour, Behnam 667
Parulkar, Aamena 540, 721
Parvari, Matin 747
Patience, Gregory 190
Patriarca, Evan 544
Patterson, Brent 295
Paulsen, Alex D. 712
Pavlišic, Andraž 442
Paz, Neil 868
Peck, Torin 714
Peden, Charles H. F. 291, 859
Peng, Guowen 394
Peng, Po-Yu 128
Peng, Zhenmeng 208, 899
Pennington, Ashley M. 288
Penteado, Alberto 688
Perea, Leo Alvarado 849
Pereira-Almao, Pedro 584, 843
Perera, Ayomi S. 651
Pérez, Fernando 685
Perras, Frédéric 540, 721
Peters, Baron 78
Petkovska, Menka 145
Pfennig, Toni 326
Pham, Hien 315, 936
Pham, Hien N. 229
Pham, Thanh Huyen 450
Phillips, Eric M. 480
Piler, Karishma 731
Ping, Haoliang 637
Pinho, Bruno 159
Pitault, Isabelle 165
Plessow, Phillip 910
Poco, Joao G. R. 741
Poelman, Hilde 524
Pogrzeba, Tobias 76, 421
Pohar, Andrej 442
Pohl, Marga-Martina 792
Polo-Garzon, Felipe 780
Pommerenck, Justin 503, 701, 732
Porosoff, Marc D. 6, 457
Potdar, Aditi 169
Poudyal, Samiksha 377
Pradhan, Sivaram 416
Prasad, Subramanian 787
Prasad, Vinay 441
Priebe, Jacqueline 792
Pritchard, Jay 301
Pruski, Marek 540, 721
Punetha, Ashok K. 108
Pushpavanam, S. 668
Qarzouh, Muneef 108
Qi, Long 895
Qian, Linping 880
Qian, Wanyi 105
Qian, Xianghong 236, 347
Qiao, Guo-Yue 819
Qin, Lang 795
Qin, Wei 695
Qiu, Kaipei 651
Quang, Dang Viet 177
Quinn, Joseph 200, 683, 693
Quinn, Thalia 783
Quintana, Juan Lovón 917
Quitain, Armando 228, 718
Radich, James G. 717
Ragauskas, Arthur J. 328
Raghavan, Ashwin 90
Rahmati, Mahmood 132, 578
Ramakumar, S. S. V. 673
Ramanathan, Anand 469, 838, 932

AUTHOR INDEX

Ramasamy, Karthikeyan K. 470
Ramasubramaniam, Ashwin 888
Ramirez, Emilio 415
Ramirez, Maria 919
Ramirez-Reina, Tomás 619
Ramos, Adela E. 906
Ramos, Mario 638
Ranade, Vivek V. 388
Rao, Vivek M. 245
Rasmussen, Søren B. 897
Ratnakar, Ram R. 91
Ravel, Bruce 940
Rawal, Saurin 339
Ray, Ajay K. 79
Raybaud, Pascal 96
Razzak, Shaikh 845
Rea, Anita 439
Reddick, Ian 503
Regalbuto, John R. 192, 197, 235, 490, 779, 817, 886, 930
Regauskas, Art 329
Rehmet, Christophe 522
Reihani, Amin 295
Reitzmann, Andreas 147
Ren, Yang 218
Rentz, Nikki S. 70, 374
Repke, Jens-Uwe 688
Resasco, Daniel 88, 861
Resasco, Daniel E. 89
Resende, Fernando 368
Reynolds, Ronald A 928
Rezaei, Fateme 281
Rhee, Chang-Houn 605
Ribeiro, Fabio 209, 218, 940
Ribeiro, Fabio H. 298, 539, 546, 835, 917
Rieg, Carolin 400
Rigdon, William A. 678
Rimer, Jeffrey D. 489, 644, 675, 695
Rioux, Robert M. 834, 952
Riscoe, Andrew 453
Rivera-Goyco, Christian G. 370, 396
Roberts, Christopher 568
Roberts, Christopher B. 582
Robota, Heinz J. 301
Rocha, Tulio 797
Rodriguez-Garcia, Laura 885, 918
Roe, David 568, 582
Rogers, Jessica 844, 908
Roine, Jorma 722
Roling, Luke T. 510, 585
Rolland, Matthieu 96
Román-Leshkov, Yuriy 364, 776, 894, 902
Romero, Carolina 623
Rorrer, Julie 320
Rosenfeld, Devon C. 844, 908
Rover, Marjorie 331
Rowe, Scott C. 427
Rownaghi, Ali A. 281
Roy, Partho S 698
Roy, Shyamal 760
Rozmyslowicz, Bartosz 324
Ruan, Hao 333
Ruddy, Daniel A. 889
Rueda, Luz Marina Ballesteros 949
Runnebaum, Ron C. 887
Russ, David C. 332
Russell, Daniel 735
Ryan, Matthew 459
Ryou, Youngseok 220
Ryu, Jebin 598, 599
Sabyrov, Kairat 31
Saeys, Mark 392, 942
Safdari, Mohammad-Saeed 132, 578
Saha, Shibely 196, 566
Sahasrabudhe, Mayuresh 635, 676
Sahinidis, Nick 640
Salavati-Fard, Taha 839
Salehi-Khojin, Amin 73, 754
Salmi, Tapio 246, 722
Salonen, Jarno 722
Salvachua, Davinia 331
Samad, Jadid 366
Sanchez, Joel 638
Sanchez-Forero, Diana 158
Sandberg, Robert B. 460
Sandoval, Tania E. 862
Sangodkar, Rahul P. 787
Sankar, Meenakshisundaram 422
Sankaranarayanan, Subramanian K.R.S. 384
Santos, Haroldo 880
Santos, Maryorie C. 904
Saraçi, Erisa 195
Saravanan, Chandra 635, 676
Saravanan, Karthikeyan 463
Sarazen, Michele L. 20, 836
Sasaki, Mitsuru 228, 718
Sasaki, Takashi 597
Sasmaz, Erdem 38, 206, 575, 851, 933
Sathitsuksanoh, Noppadon 332
Sato, Takafumi 345, 600, 601, 738
Saunders, Steven R. 954
Sautet, Phillippe 96, 283, 350, 399, 536
Savage, Phillip E. 101, 224, 943
Sayahpour, Baharak 73
Schaidle, Joshua 186, 889
Schaidle, Joshua A. 860
Schaub, Tanner 868
Schembecker, Gerhard 590
Scheppe, Karalyn 105
Schideman, Lance 105

AUTHOR INDEX

Schlogl, Robert 216, 797
Schmidt, Marcel 76, 421
Schmidt-Rohr, Klaus 229, 315
Schneider, Elia 593, 808
Schneider, William 298, 631, 840
Schneider, William F. 209, 351, 546, 835, 945
Schneiderman, Deborah 390
Schomäcker, Reinhard 76, 421
Schouten, Jaap C. 173
Schreiber, Igor 610
Schreiberova, Lenka 610
Schubert, Markus 244, 247
Schultheis, Stephan 455
Schwalbe, Jay 792, 831
Schwank, Johannes W. 653
Schwartz, Daniel 858
Schwartz, Daniel K. 863
Schwartz, Nicholas R. 712
Schwartz, Thomas J. 362
Schweitzer, Jean-Marc 96
Scott, Susannah L. 78, 895
Scuseria, Gustavo E. 416
Sebakhy, Khaled O. 584, 843
Seebauer, Edmund G. 554
Seidel-Morgenstern, Andreas 145, 424, 849
Seitz, Linsey C. 35, 466, 727
Semenov, Sergey N. 59
Sen, Fatih 384
Sen, Irem 426
Senftle, Thomas P. 26, 372, 877
Senkan, Selim 451
Senthamaraikkannan, Gouthami 441
Seritan, Stefan 78
Serov, Alexey 71
Serrano, Benito 793
Servia, Alberto 160
Seshadri, Vikram 417
Seuffert, Wolfgang 147
Shacham, Mordechai 111
Shackelford, Joseph 783
Shafei, Emad 108
Shahane, Saurabh 483
Shaikh, Sohel K. 686
Shan, Junjun 54, 775, 829, 830
Shan, Nannan 286
Shang, Zeyu 818
Shanks, Brent H. 229, 326, 361, 362
Sharada, Shaama Mallikarjun 378
Shareghi, Adam 503
Shariff, Humayun 87, 157
Sharkey, Bryan E. 367
Sharma, Manish 203, 627
Sharma, Renu 556
Sharma, Virender K. 79
Sharrock, Patrick 528
Shaughnessy, Kevin H. 477
Shchukarev, Andrey 722
Sheehan, James 101
Shekhar, Mayank 917
Shekhar, Shashank 480
Shen, Chun 587
Shen, Yangyang 953
Shen, Yufeng 489, 644
Shi, Daming 191
Shi, Hongbo 888
Shi, Honghong 327
Shi, Xinjian 551
Shih, Arthur 298
Shinde, Kiran 634
Shirobe, Taichi 791
Shuai, Li 25
Siahrostami, Samira 46, 72, 551, 641
Siebenhofer, Matthaeus 346, 452
Sievers, Carsten 124
Sikes, Hadley D. 47
Silva Jr, João L. 158
Simakov, David 146, 873
Simonetti, Dante 316
Sinclair, Robert 641
Singh, Aayush R. 402
Singh, Joseph 916
Singh, Meenesh 429, 903
Singh, Nirala 51, 125
Sinha, Nishant 454
Sinha-Majumdar, Sreshtha 214
Slakman, Belinda L. 428
Smirniotis, Panagiotis 488
Smith, Joseph D. 245
Smith, Robyn 781
Smith, Ryan 331
Smolin, Sergey 792
Snider, Jonathan 708, 898
Snyder, Joshua 202, 334
Sohn, Hyuntae 219, 692
Sokaras, Dimosthenis 641, 898
Solovyov, Andrew 542, 887
Somorjai, Gabor A. 516
Son, Won-Su 598, 599
Song, Hua 537
Song, James 291
Song, Jinwoo 698
Song, Kyung-Ho 583
Song, Yang 125
Song, Yuanjun 720, 924
Sooknoi, Tawan 431
Sosa, Fabricio Guayaquil 793
Spanjers, Charles S. 390
Speelmanns, Eva 688
Spinner, Neil S. 678
Spivey, James J. 491, 611

AUTHOR INDEX

Sprunger, Phillip T. 491
Sriram, Vishnu 133
Stach, Eric A. 831
Stagg-Williams, Susan M. 106
Stamatakis, Michail 99, 496, 801
Stangland, Eric E. 401
Starace, Anne 194
Stark, Wendelin J. 593, 808
Stavitski, Eli 124
Stears, Brien 847
Steele, Matthew 115
Steinfeld, Aldo 794, 905
Steiniger, Alexzander 733
Steinmann, Stephan 283
Stenfeldt, Marie 299
Stephens, Ifan E.L. 901
Stevens, James 74
Strah, Neja 442
Strickler, Alaina 702
Studt, Felix 400, 523, 641, 927
Sturm, Belinda S.M. 106
Su, Yaxin 242
Subramaniam, Bala 215, 222, 327, 469, 838, 932
Suidan, Makram 488
Sulmonetti, Taylor 816, 934
Sultana, Khondker 709
Sumigawa, Yoshifumi 718
Sun, Duo 146, 873
Sun, Junming 68
Sundmacher, Kai 97
Suzuki, Tomoko 597
Suzuki, Yukei 586
Svenningsen, Glen 468
Svetlov, Stanislav 248
Sykes, E. Charles H. 875
Szanyi, Janos 209, 291, 631, 704, 840
Taboada-Serrano, Patricia 626
Tackett, Brian M. 201
Tadepalli, Sunitha 74
Taifan, William 322
Takacs, Michael 905
Takahashi, Misaki 738
Takaoka, Masaki 597
Tamaki, Takanori 650
Tampy, Geatesh 74
Tan, Qiaohua 861
Tan, Shuai 196, 566
Tang, Lei 573
Tang, Maureen H. 202
Tang, Michael 465
Tang, Yadan 138
Tang, Yuanjun 681
Tao, Miao 608
Taranto, Osvaldir P. 158
Tassone, Christopher J. 511
Tate, Jenae 561
Tavlarides, Lawrence L. 226
Tayakout, Mélaz 165
Taylor, Sara Jo 389
Teerikoski, Sakari 246
Teixeira, Andrew 483, 504
Teixeira, Ivo 496
Teng, Xiaowei 618
Tengco, John Meynard M. 197
Terejanu, Gabriel 379
Thakkar, Harshul 281
Thapa, Prem 327
Theiler, David 905
Theofanidis, Stavros Alexandros 524
Theopold, Klaus 195
Theriot, Jordan 459
Therrien, Andrew 875
Thiyagarajan, Shanmugam 471
Thomas, Cyril 137
Thomassen, Leen C.J. 169
Thompson, Levi T. 186, 943
Thybaut, Joris W. 432
Tierney, Eric 518
Timko, Michael T. 75, 107, 223
Ting, Allen Wei-Lun 680
To, Anh T. 553
To, John 641
Toch, Kenneth 432
Todic, Branislav 557
Tomassone, M. Silvina 953
Tompsett, Geoffrey 107, 223
Toops, Todd J. 300, 931
Toste, Dean 320
Traverso, Andrew 701, 732
Trinh, Quang Thang 450, 800
Trogadas, Panagiotis 651
Tsai, Charlie 402
Tsang, Shik Chi Edman 496
Tsapatsis, Michael 896
Tseng, Han-Ting 218, 940
Tsotsis, Theodore 341, 343, 730, 736
Tsouris, Costas 626
Tucker, Melvin 333
Turner, C. Heath 624, 914
Turney, Damon 759
Ulissi, Zachary 465
Ullmann, Sven 899
Unocic, Raymond 936
Urban-Lurain, Mark 115
Utsunomiya, Ryuji 508
Uzun, Alper 674
Vahdat, Nader 365
Vahdatzaman, Maral 647
Vaid, Radhe K. 479
Valla, Julia A. 576

AUTHOR INDEX

Valle, Eduardo 484, 620
Van Beek, Wouter 778
Van Cleve, Timothy 8, 900, 955
Van Der Schaaf, J. 173
Van Der Waal, Jan 471
Van Duin, Adri 93
Van Es, Dan 471
Van Haveren, Jacco 471
Van Norman, Staci A. 567
Van Ommen, J. R. 564
Vann, Tyler 907
Vara, Madeline 510
Varma, Arvind 13, 189, 387, 806
Vasiliadou, Efterpi 839
Vasudevan, Kiran 202
Vayssilov, Georgi N. 881
Vazquez-Ramos, Leida M. 370
Vega-Vila, Juan C. 539
Vengsarkar, Pranav S. 568
Venkatesh, Archana 422
Vennestrøm, Peter N. R. 897
Verma, Anuj A. 480
Verstraete, Jan J. 96
Veser, Götz 544, 719
Vidonish, Julia E. 127
Villano, Stephanie 519
Virk, Preetinder S. 438
Vitale, Gerardo 584, 843
Vityuk, Artem D. 881
Vlachos, Dion 393, 839
Vlachos, Dionisios 493
Vlachos, Dionisios G. 118, 199, 230, 805, 896
Vogel, Herbert 98, 455
Vohs, John 191
Vojvodic, Aleksandra 336, 402, 466, 552, 672, 727, 744, 852, 910
Vömel, Lukas T. 391
Von Jouanne, Annette 503
Vorholt, Andreas J. 170
Voss, Jenny 573
Voss, Johannes 378
Votsmeier, Martin 98
Vu, Anh 236
Wachs, Anthony 96
Wachs, Israel 797, 853
Wachs, Israel E. 458, 548, 850, 920
Wade, Kester 944
Wagner, John 113
Walker, Eric 64, 379
Walter, Eric D. 291, 895
Wan, Weiming 805
Wander, Brook 316
Wang, Aiyong 291
Wang, Alex 344
Wang, Bin 88, 233, 431, 555
Wang, Chao 206, 625, 851
Wang, Di 297, 740
Wang, Haihui 342
Wang, Hongliang 333
Wang, Huamin 859
Wang, Jia-Jun 592
Wang, Jiamin 386
Wang, Kai 507, 591
Wang, Kun 519
Wang, Lan 195
Wang, Lijun 407
Wang, Lucun 566
Wang, Meng 492
Wang, Peijian 507, 591
Wang, Ping 684
Wang, Richard Baochang 99
Wang, Shuai 311
Wang, Siwen 284, 710, 923
Wang, Wei-Ning 485
Wang, Xiang 704
Wang, Xiaohong 806
Wang, Xue 510
Wang, Yilin 291
Wang, Yong 382, 498, 859
Wang, Yujun 587
Wang, Yuran 50, 894
Wang, Zhen 880
Wang, Zi 491
Warmeling, Helge 170
Washton, Nancy M. 895
Watts, Michael 347
Weaver, Jason F. 799
Weber, Adam 429
Weber, Ariane 421
Weckhuysen, Bert 120, 471
Weeranoppanant, Nopphon 483
Wegener, Evan 218, 940
Wehinger 642
Wei, Yanying 342
Weidner, John W. 197
Weimer, Alan W. 427, 567
Weiss-Hortala, Elsa 681
Wellendorff, Jess 378
Wen, Cun 1, 206, 456, 602, 904, 933
Wen, Zhiyou 331
Wendt, Jost O.L 239
Wentworth, Travis 299
Werner, Sebastian 147
Werner-Zwanziger, Ulrike 787
West, Alan 630
West, Richard H. 354, 428, 518
Westmoreland, Phillip R. 389, 417
Wette, Melissa 734
White, Colin 106
Wickramasinghe, Ranil 236, 347

AUTHOR INDEX

Wiebenga, Michelle....294
Wiggins, Gavin....415
Wilburn, Monique....290
Wilhite, Benjamin....696
Willauer, Heather D....6, 457
Williams, Christopher....123
Willis, Joshua....453, 511, 512, 656, 831, 884, 944
Wilson, Neil M....687, 935
Wilson, Zachary....640
Wiser, Artur....98
Wolf, Patrick....350
Wolff, Tanya....849
Wong, Andrew....817
Wong, Breanna....498
Wong, Hsi-Wu....77
Wong, Michael S....416
Wong, Sze Wing....479
Wong, Wai Kuan....478
Wongkaew, Akkarat....197
Woo, Hee Suk....598, 599
Worku, Dereje....709
Wrasman, Cody....512
Wu, Gang....337
Wu, Ivy....289
Wu, Jason....899
Wu, Liheng....511, 535
Wu, Su-Fang....637
Wu, Xiao-Yu....22
Wu, Yi Y. (Chloe)....909
Wu, Yiqing....841
Wu, Zhenwei....218, 940
Wu, Zili....458, 914
Wyman, Charles....468
Xia, Shunxiang....684
Xia, Younan....510
Xiang, Yizhi....573
Xiao, Yang....13, 189, 387, 806
Xie, Jiahan....936
Xie, Jingyi....561
Xie, Mingquan....144
Xie, Pei....591
Xie, Tianjun....500
Xin, Hongliang....284, 386, 707, 710, 878, 900, 923
Xin, Yan....913
Xin, Zhong....608
Xiong, Haifeng....315
Xu, Bingjun....185, 543, 893, 946
Xu, Chao-Zhong....592
Xu, Jiayi....369
Xu, Lang....401
Xu, Mingyuan....795
Xu, Qin-Qin....819
Xu, Rui....568, 582, 717
Xu, Ye....339, 400, 486, 550
Xu, Zhuoran....844, 908
Yablonsky, Gregory S....566
Yakobov, Roman....759
Yalcin, Ozgen....548
Yamafuku, Saya....228
Yamaguchi, Takeo....650
Yamamoto, Tetsuya....479
Yan, Helen....944
Yan, Ning....232, 833
Yan, Wenjuan....327
Yan, Yushan....185, 199
Yang, Bin....329, 333
Yang, Dong....221
Yang, Guangxing....618
Yang, Haishen....459
Yang, Hong....687
Yang, Nuoya....916
Yang, Rachel A....288
Yang, Xiaofang....373
Yang, Xiaokun....126
Yang, Yanhui....607
Yao, Hongbao....587
Yao, Yali....609, 682
Yap, Swee Kun....478
Ye, Jianchao....830
Ye, Jingyun....928
Ye, Lujie....870
Yeh, Alexander....542
Yeh, Benjamin....928
Yeh, Yao-wen....373
Yelvington, Paul E....712
Yeung, King Lun....756, 757, 758
Yezerets, Aleksey....138
Yi, Nan....879
Yin, Jian-Zhong....819
Yin, Kehua....936
Yokochi, Alex....701
Yokochi, Alexandre....503, 732
Yokozawa, Tohru....479
Yoo, JeongDo....739
Yoo, Jong Suk....512, 726
Yoon, Sung-Pil....663
Youn, Min Hye....583, 649, 658
Young, Zachary....323
Yu, Jingye....541
Yu, Jun....221
Yu, Kun....131
Yu, Miao....798
Yu, Xiaoxiao....426
Yu, Ying....495
Yuan, Joshua....329
Yun, SeongUk....697
Yun, Yongju....804
Yung, Matthew M....194
Yurchekfrodl, Eric....208
Yusuf, Seif....855

AUTHOR INDEX

Zahid, Umer 175
Zaker, Azadeh 223
Zeltner, Martin 593, 808
Zemichael, Fessehaye 622
Zeng, Chun 327
Zeng, Zhenhua 32
Zhan, Zhonghua 239
Zhang, Changlin 208
Zhang, Chen 939
Zhang, Conrad 193
Zhang, Hua 34
Zhang, Jing 911
Zhang, Jisong 504, 507
Zhang, Kechun 390
Zhang, Liang 36, 509, 744, 852
Zhang, Lu 859
Zhang, Nian 643
Zhang, Nuowei 34
Zhang, Peng 105
Zhang, Qi 874
Zhang, Qian 201
Zhang, Renqin 48, 209, 631, 875
Zhang, Shengjie 828, 887
Zhang, Wen 347
Zhang, Xiankuan 307
Zhang, Ximing 467
Zhang, Yuanhui 105
Zhang, Yunya 197
Zhang, Z.Conrad 416
Zhao, Chuanlin 550
Zhao, Dongting 844, 908
Zhao, Evan W. 913
Zhao, Jingkai 134
Zhao, Jin-mo 416
Zhao, Meng 327
Zhao, Peng 373
Zhao, Qing 381
Zhao, Zheng 88, 89
Zheng, Feng 799
Zheng, Haibin 913
Zheng, Jinbao 34
Zheng, Quanxing 369
Zheng, Weiqing 19, 896
Zheng, Xiaolin 551
Zheng, Yang 291, 840
Zhou, Xiao-Dong 174
Zhu, Chunxiang 576
Zhu, Hongda 932
Zhu, Jiazhou 866
Zhu, Minghui 797
Zhu, Xiaoxiang 310
Zhu, Zibin 874
Zhuang, Libin 342
Zivanic, Ljiljana 557
Zohour, Bahman 451
Zohra, Mohammad Abu 177
Zolghadr, Ali 419
Zones, Stacey I. 542, 784, 837
Zougheib, Nicole 807
Zuo, Wilbur Y. 77
Zygourakis, Kyriacos 127

AIChE
120 Wall Street, FL 23
New York, NY 10005-4020

ISBN 978-1-5108-3429-3

9 781510 834293

CURRAN ASSOCIATES INC.
proceedings.com